UNITED STATES
USMIL
MILITARY AIRCRAFT SERIALS

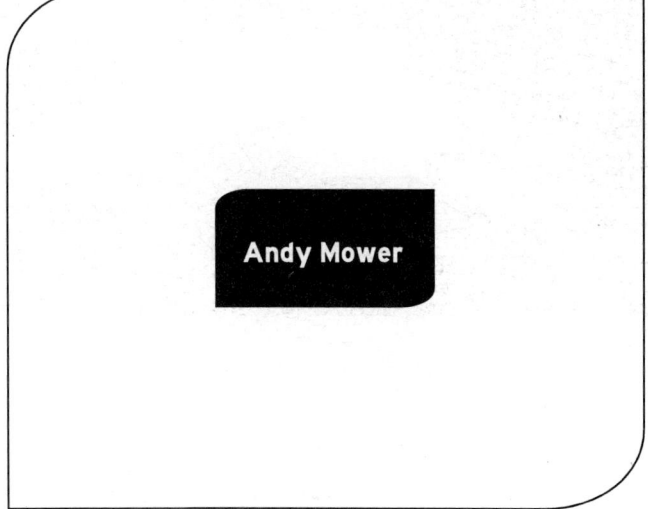

Andy Mower

An AVIATION ASSOCIATES publication

UNITED STATES MILITARY AIRCRAFT SERIALS

first edition	(as US Military Aircraft to Europe)	May 1979
second edition	(as US Military Aircraft to Europe)	May 1981
third edition	(as US Military Aircraft to Europe)	June 1982
fourth edition	(as United States Military Aircraft Serials)	May 1983
reprinted	(as United States Military Aircraft Serials)	May 1984
fifth edition	(as United States Military Aircraft Serials)	April 1985
sixth edition		December 1993

ISBN 0 947755 13 6

published by

AVIATION ASSOCIATES

P.O. Box 3201 London E7 8RJ

CONTENTS

PHOTOGRAPHS

CREDITS: Most of the above photographs were taken by Gordon P. O'Kane who also acquired, begged or borrowed others. The remainder were from the Author's collection.

INTRODUCTION

This sixth edition, now formally titled USMIL to acknowledge common usage, introduces several changes which it is hoped will meet with readers' approval. The first is an entirely new section entitled 'PRESERVED AIRCRAFT'. This attempts to list, town by town, state by state, all known remaining non-operational aircraft of U.S. military origin whether or not they are not listed in the main text. This includes both genuinely 'preserved' museum exhibits, gate guards and dumped and derelict aircraft. The list was compiled from a variety of sources and personal sightings, some rather ancient, so any revised sightings or reports for inclusion in this section in future editions will be most gratefully received.

Known replicas and civil aircraft in spurious military colours are NOT included.

The second area of change coincides with the renaming of the MILITARY AIRCRAFT STORAGE AND DISPOSITION CENTER (MASDC), located at Davis Monthan Air Force Base at Tucson, Arizona as the AIRCRAFT MAINTENANCE AND REGENERATION CENTER (AMARC). In this edition, aircraft known to have been resident during October 1987, or to have arrived since then, are shown in the text as 'AMARC' in the unit column and the park code (an internally allocated serial) is listed, where known, under the remarks column. Where aircraft were LAST KNOWN at this location, but were not confirmed during October 1987 (i.e. broken up, re-issued or sold prior to that date), then the term 'MASDC' is shown in the unit column. Any comments from readers concerning the current status of such aircraft will also be most welcome for inclusion in later editions.

Finally, when USMIL first appeared, the practice of the USAF five figure serial application was commonplace. However things change, and in recognition of the varying styles of presentation by the USAF now in use, it has been decided to quote the full FISCAL and year serial allocation for each aircraft, REGARDLESS of the presentation style actually used.

Andy Mower   November 1993

TYPE DESIGNATIONS

1962 TRI-SERVICE SYSTEM

Since the introduction of the Tri-Service aircraft designation system in 1962, all military aircraft in the United States of America have shared a common system. It was based on the previous Air Force\Army system which was itself originally established in 1924 and then modified in 1948. The uniform Tri-Service system was established under Department of Defence Directive Number 4505.6 and was dated July 6th, 1962. The US Air Force was made responsible for establishing procedures and maintaining and assigning all designations. The application of this system is achieved by co-ordination with the Departments of the Army and Navy. Additionally the Air Force was made responsible for maintaining current lists of 'popular' names for production service aircraft.

The 1962 system basically employs a single letter to define the prime role of the aircraft (the PRIMARY MISSION SYMBOL), followed by a sequential MODEL NUMBER within that role, and is suffixed by a single SERIES LETTER to indicate major modifications within the model range. This arrangement holds good for all fixed wing aircraft, helicopters, missiles, probes, drones, rockets and airships.

PRIMARY MISSION SYMBOL

This is (usually) a single letter to indicate the prime use of the vehicle:-

A	Attack	R	Rocket
B	Bombardment	S	Anti-Submarine
C	Cargo\Transport	SR	Strategic Reconnaissance
D	Drone		(used only by SR71\B71)
E	Electronics Installation	T	Trainer
F	Fighter	TR	Tactical Reconnaissance
FB	Fighter Bomber		(used on U2 update)
	(used on FB111A only)	U	Utility
H	Helicopter	V	VTOL or VSTOL (used only with
M	Missile		modified mission symbol)
N	Probe	X	Experimental and Research
O	Observation	Z	Airship
P	Patrol		

MODEL NUMBER

The primary mission symbol is followed by a one, two or three figure number to indicate the total of models of any particular type. It should be noted that as primary mission symbols B, C, F and H are deemed to have reached high numbers these sequences have reverted to ONE in a second series. Model numbers are allocated to real and 'paper' aeroplanes, that is design projects and concepts that for one reason or another are never actually built.

SERIES LETTER

Having established primary mission symbol and model number, this is then suffixed by a series letter starting at 'A'. The letters are issued alphabetically (excluding I and O) to each PROPOSED design modification of the type. For this reason not all series letters will be applied to actual aircraft, but may be alloted to a design proposal on the airframe which was eventually not built or modified.

MODIFIED MISSION SYMBOL

Some aircraft not now operating within the primary mission symbol, or requiring clarification (i.e. H and V types) of use may receive a modified mission symbol as a prefix to the primary mission symbol. These prefixes are:-

A	Attack	Q	Drone
C	Cargo\Transport	R	Reconnaissance
D	Drone Director	S	Anti-Submarine
E	Electronics Installation	T	Trainer
H	Search, Rescue and Recovery	U	Utility
K	Tanker	V	Administration\Staff VIP
L	Cold Weather operations	W	Weather
O	Observation		

It should be noted that the symbol K was originally designated as a PRIMARY MISSION SYMBOL (K = Kerosene Tanker), but in practice has never been used as such. Sometimes TWO modified mission symbols are allocated (e.g. VKC135B, but again in practice this would normally be shortened to VC135B).

STATUS PREFIX

The primary mission symbol or modified mission symbol (OR BOTH) may be further pefixed by a status prefix, namely:-

G   Ground Instructional AirframeXExperimental or Research
J   Temporary Special TestYPrototype or pre-production
N   Permanent Special Test

Example: Boeing NKC135A Stratotanker

Boeing	=	manufacturers name
N	=	status prefix (Permanent Special Test)
K	=	modified mission symbol (tanker)
C	=	primary mission symbol (Cargo)
135	=	135th allocation of C symbol
A	=	first series lettter of C135
Stratotanker	=	Air Force name for KC135

CURRENT TRENDS

Whilst commonality has many attributes, it should be remembered that inter-service rivalry has always been strong. Each service will always consider its needs and requirements paramount, both in funding (FISCAL ALLOCATION) as well as in operations. Where more than one service purchases one type of airframe for assorted uses and requirements; application of the 'correct' type designation may prove difficult. The Beech 200 King Air is known as the C12 Huron in Air Force use, but entered service with the US Army as the RU21J Ute, later also being referred to as the U25. It has now been decided to refer to all Beech 200 models as the C12.

Conversely certain allocations of primary mission symbols and model numbers have been extremely apt. Allocation of C9 to the erstwhile civilian MDD DC9 airliner exemplifies this, as does allocation of the designation KC10 to the tanker version of the DC10 airliner. Here of course one has to ignore the fact that the C10 was ORIGINALLY a second series allocation to the ill-fated HANDLEY-PAGE HP137 JETSTREAM, although this airframe has been resurected successfully as the BAe Jetsream 31 and 41. Renumbering of the B types in a second series was fortuitous for the advanced manned bomber project; B1B sounds much more snappy than plain old B72!

An area of confusion is occuring where aircraft are procured second-hand, either as a cost saver where the test installation is more important than the airframe or where aircraft are seized in drugs operations and later appropriated for service use. Frequently they are not 'slotted in' where a similar type has already seen service use, or worse spurious type designators are used. Most ex-drug Queen Airs ARE being noted as U8F Seminoles (but often with false or compromised serials). However, Cessna 310s, Shorts 330s and even Jetstreams (albeit Century conversions) are referred to by their civilian designations. Even more confusing are the UC880, a GD/Convair 880 for special tests, and the unique US Navy Fairchild UC27B, which was assumed to be a correct designation until the US Army failed to apply the designation to two Fokker F27 Friendships acquired for the Golden Knights parachute team, and also when the C27 designation was 'issued' to a brand new type procured for service use, the Alena-Crysler C27A Spartan. The Golden Knights' Friendships have now been allocated a new type designator (C31).

SERIAL SYSTEMS

It is not intended to give a complete history of the application of serials (tail numbers) to United States military aircraft, which in any case may confuse rather than enlighten enthusiasts or general readers not familiar with the various systems; instead an indication of the type of presentation to be found on aircraft listed in this book is given as comprehensively as possible.

There are three basic systems. A joint U.S.A.F./U.S. Army range of serials is based upon the financial year (or in Amreican terminology Fiscal Year) in which the purchase of the aircraft is proposed by the President of the United States (and is subsequently sent to Congress for approval). This system has been in use in its present form since 1921. Second is a series shared by the U.S. Navy and U.S. Marines which commenced at 00001 in 1940 (issued by the Bureau of Aeronautics Department) and is still in use today. During this period serials have been issued consecutively as required, reaching over 161,000 by 1981. Two previous series had existed for Navy aircraft, commencing in 1911 and 1933 respectively. Finally the U.S. Coast Guard issue their own four figure serial to aircraft they operate, despite being a U.S. Navy reserve organisation and coming under the U.S. Navy in times of war. In practice this system has not always been used. Many aircraft transferred from Air Force or Navy use retain their original serials (or a modified or shortened form of the serials). The four figure system of serials originating in 1936. Now that this service generally receives new equipment (despite the continued practice of allocating a Navy serial for procurement) the practice of modifying the original serial has been discontinued.

Air Force/Army

Currently there are two main method of presentation for serials, each unfortunately with several (often unofficial) sub-varieties. Firstly there is the 'five figure tail numbers'.

example: 40612   C141B Starlifter

This indicates that this was the 612th serial issued in the Fiscal Year (FY) in a year ending in'4', in this case 1964. This serial can also be written fully as 64-612. Obviously, reverting to the '40612' presentation, it will be noted that as only the last figure of the year is used (and also note ZEROS are used to 'pad' the number if necessary to make a five figure number) this system will repeat itself every ten years:-

example: 54-612   =   40612   =   Fairchild C123B Provider
                   54-612   =   40612   =   Lockheed C141B Starfighter
                   54-612   =   40612   =   Northrop F5E Tiger II

It is highly unlikely that one type of aircraft will stay in production for such a timespan; and therefore the chance of two aircraft of the same type ten years apart having the same serial can be discounted. However, to combat the possibility of two different types with the same serial, the USAF introduced an O- for obsolete prefix to the serial in 1955 for aircraft as they become ten years old. This system itself became obsolete in 1972.

Reference has already been made to the 'padding' of serial with zeros to make up the 'five figure tail number':-

example: First procurement in 1969   =   69-1   =   90001
                   10th procurement in 1969   =   69-10   =   90010
                   100th procurement in 1969   =   69-100   =   90100
                   1000th procurement in 1969   =   69-1000   =   91000

As will be readily appreciated, problems arise if over 9,999 aircraft are procured in any single Fiscal Year:-

example: 10,000th procurement in 1969   =   69-10000 =   10000

It will be noted that in such circumstances, the last figure if the Fiscal Year does not appear as part of the tail number. Up to 1966 serials were allocated on an 'as ordered' basis. Procurement with the war in Vietnam underway, and the inclusion of missiles and drones, often swelled this total each year beyond 10,000 units; and serial repetition in the ranges above 10,000 became possible and common every year. Where such aircraft in these ranges are listed in this book, the actual fiscal year is quoted in the remarks column. It was decided NOT to use the full procurement serial (viz. 64-612) since, except for a small stencil usually near the nose of the aircraft on the port side (known as the Technical Data Block - TDB), this method of presentation is rarely used on the tail of the aircraft. However, it is recommended that this block of information (giving full aircraft type details) is inspected if the opportunity presents itself. To summarise, be suspicious of five figure tail numbers commencing with the figure '1'; it may be the last figure of a FY61, FY71, or even FY81 serial, or it may be the first figure of a 10,000 plus of ANY fiscal year!

Technical Data Block - (TDB), this method of presentation is rarely used on the tail of the aircraft. However, it is recommended that this block of information (giving full aircraft type details) is inspected if the opportunity presents itself. To summarise, be suspicious of five figure tail numbers commencing with the figure '1'; it may be the last figure of a FY61, FY71, or even FY81 serial, or it may be the first figure of a 10,000 plus of ANY fiscal year!

From FY67 (Fiscal Year 1967) the US Army extricated its serials from within the USAF batches, and commenced its own sequence each year at 15,000 (therefore the full serial without the Fiscal Year is displayed as the tail number), causing some considerable duplication and confusion.

example:
$$68\text{-}15214 \text{ to } 68\text{-}15778 \quad = 15214 \text{ to } 15778 \quad = \text{UH1H Iroquois}$$
$$\text{but } 69\text{-}15214 \text{ to } 69\text{-}15778 = 15214 \text{ to } 15778 = \text{UH1H Iroquois}$$
$$\text{and } 70\text{-}15700 \text{ to } 70\text{-}15778 = 15700 \text{ to } 15778 = \text{UH1H Iroquois}$$

It can be seen that we have a situation in two consecutive years where over five hundred serials were duplicated on the aircraft type!

In FY71 the US Army commenced issuing serials from 20,000, and have progressed consecutively since then without regard to the change in Fiscal Year, reaching approximately 26,000 to date.
This system is now being phased out and serials for Army aircraft are now being allocated within USAF batches.

Since 1967, missiles and drones generally do not appear to have had FY serials in the aircraft ranges allotted to them. The previous inclusion of these had contributed to the yearly total of procurements frequently exceeding 10,000.

The final variation of the 'five figure tail number' presentation was first applied in 1972. but now appears to be falling from favour. This simply involved the total omission of the last figure of the Fiscal Year and its replacement by a zero. This meant that the first serial issued in 1972 (72-1),instead of appearing as 20001 in fact became 00001. This too has caused sone confusion:-

example:
$$\text{fifth allocation of } 1972 = 72\text{-}5 = 00005 = 20005$$
$$\text{fifth allocation of } 1973 = 73\text{-}5 = 00005 = 30005$$
$$\text{fifth allocation of } 1974 = 74\text{-}5 = 00005 = 40005$$
$$\text{fifth allocation of } 1975 = 75\text{-}5 = 00005 = 50005$$

The use of this style of presentation was not universal during this period, as was not necessarily applied to the example above. Where such presentation IS used on an aircraft listed in this book, the revised method of presentation is given and the actual Fiscal Year is quoted in the remarks column. As with serials commencing with the figure '1' great care should be taken where the serial starts with a zero to establish that the aircraft was procured in FY70 or it is in fact a procurement during 1972-1975 with the revised presentation.

The second style of serial presentation also has several variations in display. Application of this style is confined mainly to fighter and transport aircraft, although exceptions do exist. This basically consists of the two letters 'AF' (for Air Force) above the 1st two figures of the Fiscal Year, followed by the 'last three' of the serials:-

example:
AF
$_{77}248$ = 77-248 = 70248
or = 77-1248 = 71248
or even = 77-2248 = 77248 and so on

It should be noted that the last three figures of the serial are equal in height to the AF above 77 lay out. The immediate obvious drawbacks to this method of presentation is that a repetition occurs every one thousand serials within the Fiscal Year. A development of this style of presentation is to give all FOUR figures of the serial, and has appeared on some aircraft, although primarily those based in the United States.

example:
AF
$_{55}2816$ = 55-2816 = 52816

Unfortunately this system is sometimes incorrectly applied, often far too frequently, using the basic 'five figure tail numbers' instead of taking into account the last two of the Fiscal Year:-

example:
AF   AF
$_{68}234$ = 68-1234 = 81234 can be painted as 81234

Enthusiasts will soon become familiar with the batches of serials using a particular style of presentation, and a glance at the five-figure presentation given in this book will enable them quickly to identify individual aircraft, regardless of the mode of serial presentation used.

Navy/Marines

As already mentioned, the US Navy commenced its 'third series' of serial at 00001 in 1940, reaching 99999 in 1945 and continuing into six figures with 100000 at that time. Since then a further 61,000 serials have been issued to the Navy and Marines aircraft (plus overseas MAP and Coast Guard orders). At first these serials may seem unwieldy, especially as relatively few of them are to be seen in Europe, but in comparison with the varying Air Force systems of presentation, they are beautifully simple and uncomplicated. The size of the serial may considerably, but it is usually found on the tail or rear of the fuselage, often beneath the tailplane. If placed here the aircraft designation is often found printed above it.

Coast Guard

All aircraft currently operating (with the exception of the VC11A Gulfstream II) display a four figure serial in the range 1200 to 1600. Serials for Coast aircraft are issued consecutively in this range.

For presentation purposes in this book a change has been made from the 'five figure style' used in all previous editions. To reflect the changing styles and presentations on aircraft in recent years ALL serials are now quoted in full (incluling the Fiscal Year for USAF and Army aircraft) regardless of the actual style on individual aircraft.

ACKNOWLEDGEMENTS

The Author wishes to acknowledge the following publications which he has consulted in the preparation of this volume:-

U.S. Military Designations and Serials since 1909   John M. Andrade
United States Air Force 1992 and 1993   Mach Three Plus
British Aviation Review   British Aviation Research Group

Finally, special thanks to Eddie Leon (my partner), who assisted in several essential ways in preparation of the finished article and putting up with the constant excuses for delays.

AGWM   December 1993

ABBREVIATIONS & ACRONYMS

AAF	Army Air Field
AB	Air Base
ADC	Aerospace Defence Command
ADS	Air Demonstration Squadron (The Thunderbirds)
ADTC	Armament Development and Test Center
AFA	Air Force Academy
AFB	Air Force Base
AFCC	Air Force Communications Command
AFFTC	Air Force Flight Test Center
AFOG	Air Force Orientation Group
AFRES	Air Force Reserve
AFS	Air Force Station
AFSC	Air Force Systems Command
AFSWC	Air Force Special Weapons Center
ALC	Air Logistics Command
AMARC	Aircraft Maintenance and Regeneration Center
ANG	Air National Guard
ANGB	Air National Guard Base
AP	Airport
ARSC	Aircraft Repair and Supply Center (USCG)
ARTS	Aerial Refueled Tanker System (KC135)
ASATD	Anti-Submarine Aircraft Test Directorate (NAWC)
AT	Air Terminal
AVCRAD	Aviation Classification Repair Activity Depot (Army)
AvCo	Aviation Company (Army)
AVN	Aviation Regiment (Army)
BDRT	Battle-damage repair trainer
Bu	Bureau (number)
BuA	Bureau of Aeronautics (USN)
b\u	broken up
CAF	Canadian Armed Forces
CGAS	Coast Guard Air Station
c\n	construction number
CINCLANT	Commander in Chief, Atlantic Fleet
CINCPAC	Commander in Chief, Pacific Fleet
CINCSOUTH	Commander in Chief, Southern Europe
CNETRA	Chief of Naval Education and Training
CNO	Chief of Naval Operations
CONR	Commander, Naval Air Reserve
CV\CVN	US Carrier pennant number prefix (USN)
CVW	Carrier Air Wing (USN)
det	detachment
ECM	Electronics Counter Measures
FAA	Federal Aviation Agency (US civil)
FAA	Fleet Air Arm (Royal Navy)
FDS	Flight Demonstration Squadron (Blue Angels\USN)
FLSW	Fleet Logistics Support Wing (USN)
FMS	Foreign Military Sales
FOL	Forward Operating Location
FORSCOM	Forces Command (US Army)
FRG	Federal Republic of Germany
FTRG	Fleet Tactical Readiness Group (USN)
FY	Fiscal Year
GIA	Ground Instructional Airframe
HQ	Headquarters
IAAFA	Inter-American Air Forces Academy
IAP	International Airport
IDF\AF	Israel Defence Force\Air Force
LTV	Ling-Temco-Vought
MAAG	Military Air Advisory Group
MAC	Military Airlift Command
MAP	Military Aid Program
MAP	Municipal Airport
MASDC	Military Aircraft Storage and Disposition Center
MCAS	Marine Corps Air Station
NAF	Naval Air Facility
NARF	Naval Air Rework Facility
NARU	Naval Air Reserve Unit
NAS	Naval Air Station
NASA	National Aeronautics and Space Administration
NATTC	Naval Air Technical Training Center

NAWC-AD	Naval Air Warfare Center-Aircraft Division
NAWC-WD	Naval Air Warfare Center-Weapons Division
NAWS	Naval Air Weapons Station
NCSC	Naval Coastal Systems Center
NFWS	Naval Fighter Weapons School
NG	National Guard (US Army)
NMIMT	New Mexico Institute of Mining Technology
NRL	Naval Research Laboratory
NS	Naval Station
NSWC	Naval Strike Warfare Center
NTC	National Training Center (US Army)
NTTC	Naval Technical Training Center
NWC	Naval Weapons Center
OTS	Officer Training School (USAF)
PACAF	Pacific Air Force
PMRF	Pacific Missile Research Facility
PMTC	Pacific Missile Test Center
RAAF	Royal Australian Air Force
RAF	Royal Air Force
RNAS	Royal Navy Air Station
RoCAF	Republic of China Air Force
RoKAF	Republic of Korea Air Force
RWATD	Rotary Wing Air Test Directorate (NAWC\USN)
SAC	Strategic Air Command
SATD	Strike Aircraft Test Directorate (NAWC\USN)
SLEP	Service Life Extension Program (USN carriers)
SoC	Struck off Charge
SVAF	South Vietnam Air Force
TAC	Tactical Airlift Command
TPS	Test Pilots School (NAWC\USN)
TRADOC	Training and Doctrine Command (US Army)
TSTI	Texas State Technical Institute
TTC	Technical Training Center
UK	United Kingdom
USAAEFA	US Army Aviation Experimental Flight Activity
USAATC	US Army Aviation Test Center
USAFE	US Air Forces in Europe
USCG	US Coast Guard
USCENTCOMUS	Central Command
USCS	US Customs Service
USEUCOM	US European Command
USFS	US Forestry Service
USLANTCOMUS	Atlantic Command
USPACOM	US Pacific Command
USREDCOM	US Readiness Command
USSOCOM	US Southern Command
USSOCOM	US Special Operations Command
USSPACECOM	US Space Command
USAREUR	US Army Europe
USMC	United States Marine Corps
USN	United States Navy
wfu	withdrawn from use
w.o.	written off

UNITED STATES AIR FORCE

A7 CORSAIR

c\n   A7D: D.1 to D.459.
      A7K: K.1 to K.30.

c\n	type	status	location
67-14582	YA7	AMARC	AE231
67-14583	YA7D	412TW	ED AFFTC
67-14584	YA7D	AMARC	AE230
67-14585	A7D	w.o.	
67-14586	A7D	w.o.	
68-8220	A7D	TTC	Chanute
68-8221	A7D	w.o.	
68-8222	A7D	412TW	ED AFFTC
68-8223	GA7D	TTC	Chanute
68-8224	A7D	w.o.	
68-8225	GA7D	dump	Dallas NAS
68-8226	GA7D	TTC	Chanute
68-8227	A7D	w.o.	
68-8228	GA7D	scrapped	
68-8229	A7D	preserved	D-M AFB
68-8230	GA7D	TTC	Chanute
68-8231	A7D	SoC	
69-6188	A7D	TTC	Lowry
69-6189	A7D	w.o.	
69-6190	GA7D	TTC	Chanute
69-6191	A7D	412TW	ED AFFTC
69-6192	A7D	preserved	USAF Museum
69-6193	GA7D	TTC	Chanute
69-6194	A7D	AMARC	AE018
69-6195	A7D	AMARC	AE008
69-6196	GA7D	dump	Dallas NAS
69-6197	A7D	BDRT	Baltimore AP
69-6198	A7D		
69-6199	A7D	AMARC	AE125
69-6200	A7D		
69-6201	A7D	BDRT	Bradley ANGB
69-6202	A7D	stored	San Juan, PR
69-6203	A7D	AMARC	AE103
69-6204	A7D	w.o.	
69-6205	A7D	w.o.8.3.80	
69-6206	A7D	w.o.11.12.81	
69-6207	A7D	w.o.20.10.87	
69-6208	A7D		
69-6209	A7D	AMARC	AE009
69-6210	A7D	AMARC	AE127
69-6211	A7D	w.o. .7.72	
69-6212	A7D	AMARC	AE162
69-6213	A7D	AMARC	AE039
69-6214	A7D	AMARC	AE169
69-6215	A7D	AMARC	AE122
69-6216	A7D	412TW	ED AFFTC
69-6217	A7D	AMARC	AE227
69-6218	A7D	AMARC	AE112
69-6219	A7D	AMARC	AE116
69-6220	A7D	w.o.	
69-6221	A7D		
69-6222	A7D		
69-6223	A7D	AMARC	AE049
69-6224	A7D	w.o.	
69-6225	A7D	AMARC	AE089
69-6226	A7D	AMARC	AE026
69-6227	A7D	AMARC	AE110
69-6228	A7D	w.o.	
69-6229	A7D	AMARC	AE087
69-6230	A7D		
69-6231	A7D		
69-6232	A7D	AMARC	AE140
69-6233	A7D	AMARC	AE049
69-6234	A7D	preserved	England AFB
69-6235	A7D	BDRT	Rick'backer ANGB
69-6236	A7D	AMARC	AE124
69-6237	A7D		
69-6238	A7D	w.o.	
69-6239	A7D		
69-6240	A7D	w.o.26.10.78	
69-6241	A7D		
69-6242	A7D		
69-6243	A7D	AMARC	AE050
69-6244	A7D	AMARC	AE171
70-0929	A7D	AMARC	AE076
70-0930	A7D	w.o.	
70-0931	A7D	stored	Sioux Falls
70-0932	A7D	AMARC	AE178
70-0933	A7D	AMARC	AE188
70-0934	A7D	AMARC	AE082
70-0935	A7D	AMARC	AE174
70-0936	A7D	AMARC	AE226
70-0937	A7D		
70-0938	A7D	w.o.12.73	
70-0939	A7D	AMARC	AE118
70-0940	A7D	AMARC	AE164
70-0941	A7D	AMARC	AE145
70-0942	A7D	AMARC	AE107
70-0943	A7D	AMARC	AE167
70-0944	A7D	AMARC	AE069
70-0945	A7D	w.o.25.5.73	
70-0946	A7D		
70-0947	A7D	w.o.	
70-0948	A7D		
70-0949	A7D	w.o.17.2.73	
70-0950	A7D	w.o.	
70-0951	A7D	AMARC	AE092
70-0952	A7D	AMARC	AE117
70-0953	A7D	stored	San Juan, PR
70-0954	A7D	w.o.	
70-0955	A7D	AMARC	AE126
70-0956	A7D	SoC	
70-0957	A7D	AMARC	AE146
70-0958	A7D	AMARC	AE135
70-0959	A7D	AMARC	AE238
70-0960	A7D	AMARC	AE081
70-0961	A7D	AMARC	AE013
70-0962	A7D	AMARC	AE075
70-0963	A7D	stored	Sioux City AP, Ia
70-0964	A7D	AMARC	AE086
70-0965	A7D	w.o.27.8.85	
70-0966	A7D		
70-0967	A7D		
70-0968	A7D	w.o.31.1.78	
70-0969	A7D		
70-0970	A7D	preserved	AFM
70-0971	A7D	AMARC	AE067
70-0972	A7D	AMARC	AE077
70-0973	A7D	Pima Museum	
70-0974	A7D	AMARC	AE113
70-0975	A7D	AMARC	AE027
70-0976	A7D	AMARC	AE064
70-0977	A7D	w.o.	
70-0978	A7D	AMARC	AE182
70-0979	A7D	AMARC	AE163
70-0980	A7D	AMARC	AE016
70-0981	A7D	AMARC	AE203
70-0982	A7D		
70-0983	A7D	AMARC	AE242
70-0984	A7D	AMARC	AE150
70-0985	A7D	AMARC	AE137
70-0986	A7D	AMARC	AE105
70-0987	A7D	AMARC	AE065
70-0988	A7D	AMARC	AE202
70-0989	A7D	AMARC	AE218
70-0990	A7D		
70-0991	A7D		
70-0992	A7D	AMARC	AE151

Serial	Type	Status	Location
70-0993	A7D	AMARC	AE068
70-0994	A7D		
70-0995	A7D		
70-0996	A7D		
70-0997	A7D	w.o.	
70-0998	A7D	preserved	McClellan AFB
70-0999	A7D	AMARC	AE134
70-1000	A7D	125FS	OK OK ANG
70-1001	A7D		
70-1002	A7D	preserved	Gt.Pittsburgh IAP
70-1003	A7D	AMARC	AE155
70-1004	A7D	AMARC	AE031
70-1005	A7D	AMARC	AE030
70-1006	A7D	AMARC	AE023
70-1007	A7D	AMARC	AE017
70-1008	A7D		
70-1009	A7D	AMARC	AE111
70-1010	A7D	AMARC	AE025
70-1011	A7D	AMARC	AE106
70-1012	A7D		
70-1013	A7D	AMARC	AE011
70-1014	A7D		
70-1015	A7D	w.o.	
70-1016	A7D	AMARC	AE204
70-1017	A7D	AMARC	AE015
70-1018	A7D	w.o.	
70-1019	A7D	preserved	Myrtle Beach AFB
70-1020	A7D		
70-1021	A7D	AMARC	AE217
70-1022	A7D	AMARC	AE108
70-1023	A7D	AMARC	AE090
70-1024	A7D	w.o.	
70-1025	A7D	AMARC	AE201
70-1026	A7D	AMARC	AE080
70-1027	A7D	AMARC	AE100
70-1028	A7D	preserved	Tulsa IAP
70-1029	A7D	AMARC	AE084
70-1030	A7D	AMARC	AE193
70-1031	A7D	AMARC	AE190
70-1032	A7D	w.o.	
70-1033	A7D	AMARC	AE045
70-1034	A7D	AMARC	AE154
70-1035	A7D	BDRT	McEntire ANGB
70-1036	A7D	AMARC	AE136
70-1037	A7D		
70-1038	A7D	AMARC	AE232
70-1039	YA7	FLTV	Fort Worth
70-1040	A7D		
70-1041	A7D	AMARC	AE029
70-1042	A7D	AMARC	AE138
70-1043	A7D	w.o. .11.82	
70-1044	A7D	AMARC	AE208
70-1045	A7D		
70-1046	A7D	BDRT	Cheyenne AP, Wy
70-1047	A7D	AMARC	AE172
70-1048	A7D	AMARC	AE060
70-1049	A7D	AMARC	AE035
70-1050	A7D	preserved	SiouxFallsAP, SD
70-1051	A7D	w.o.	
70-1052	A7D		
70-1053	A7D		
70-1054	A7D	w.o.30.11.91	
70-1055	A7D		
70-1056	A7D	AMARC	AE133
71-0292	A7D	AMARC	AE012
71-0293	A7D	AMARC	AE185
71-0294	A7D	AMARC	AE194
71-0295	A7D	AMARC	AE207
71-0296	A7D	w.o.19.9.84	
71-0297	A7D	stored	Tulsa IAP
71-0298	A7D	AMARC	AE048
71-0299	A7D	w.o.	
71-0300	A7D	w.o.1.5.78	
71-0301	A7D	AMARC	AE210
71-0302	A7D	w.o.	
71-0303	A7D	AMARC	AE038
71-0304	A7D	AMARC	AE041
71-0305	A7D	w.o.3.5.73	
71-0306	A7D	w.o.	
71-0307	A7D	AMARC	AE098
71-0308	A7D	AMARC	AE191
71-0309	A7D	AMARC	AE206
71-0310	A7D	w.o.24.12.72	
71-0311	A7D	AMARC	AE094
71-0312	A7D	w.o.2.12.72	
71-0313	A7D	w.o.	
71-0314	A7D	AMARC	AE141
71-0315	A7D		
71-0316	A7D	w.o.11.1.73	
71-0317	A7D	AMARC	AE214
71-0318	A7D	AMARC	AE225
71-0319	A7D	w.o.	
71-0320	A7D	w.o.15.1.83	
71-0321	A7D	AMARC	AE187
71-0322	A7D	w.o.	
71-0323	A7D	w.o.30.11.91	
71-0324	A7D		
71-0325	A7D	AMARC	AE078
71-0326	A7D	w.o. . .82	
71-0327	A7D	AMARC	AE097
71-0328	A7D	w.o.23.12.73	
71-0329	A7D		
71-0330	A7D	AMARC	AE006
71-0331	A7D	w.o.25.7.78	
71-0332	A7D	AMARC	AE211
71-0333	A7D		
71-0334	A7D	AMARC	AE183
71-0335	A7D	AMARC	AE034
71-0336	A7D	w.o.	
71-0337	A7D		
71-0338	A7D	AMARC	AE102
71-0339	A7D	preserved	Barksdale AFB
71-0340	A7D	w.o.	
71-0341	A7D	AMARC	AE085
71-0342	A7D	412TW	ED AFFTC
71-0343	A7D	AMARC	AE143
71-0344	YA7F	412TW	ED AFFTC
71-0345	A7D		
71-0346	A7D	w.o.30.8.74	
71-0347	A7D	GIA	Rickenbacker AFB
71-0348	A7D	w.o.7.2.78	
71-0349	A7D	AMARC	AE235
71-0350	A7D	AMARC	AE006
71-0351	A7D	w.o.20.9.76	
71-0352	A7D	AMARC	AE168
71-0353	A7D	AMARC	AE223
71-0354	A7D	AMARC	AE160
71-0355	A7D	w.o. .10.77	
71-0356	A7D	w.o.6.10.81	
71-0357	A7D	w.o.	
71-0358	A7D	AMARC	AE120
71-0359	A7D	AMARC	AE059
71-0360	A7D		
71-0361	A7D	w.o.28.4.83	
71-0362	A7D	AMARC	AE142
71-0363	A7D	w.o.	
71-0364	A7D		
71-0365	A7D	AMARC	AE221
71-0366	A7D		
71-0367	A7D		
71-0368	A7D	w.o.10.7.85	
71-0369	A7D	AMARC	AE073
71-0370	A7D	AMARC	AE234
71-0371	A7D	AMARC	AE044
71-0372	A7D	w.o.5.8.75	
71-0373	A7D	w.o.	
71-0374	A7D	AMARC	AE042
71-0375	A7D	w.o.	
71-0376	A7D	AMARC	AE063
71-0377	A7D	AMARC	AE170
71-0378	A7D	w.o.	
71-0379	A7D	AMARC	AE158
72-0169	A7D	AMARC	AE047
72-0170	A7D	AMARC	AE088
72-0171	A7D	AMARC	AE192
72-0172	A7D	w.o.	
72-0173	A7D		
72-0174	A7D	w.o.	
72-0175	A7D	preserved	Tinker AFB
72-0176	A7D	AMARC	AE219
72-0177	A7D		
72-0178	A7D	preserved	Springfield MAP
72-0179	A7D		
72-0180	A7D	AMARC	AE132
72-0181	A7D	w.o.14.1.80	
72-0182	A7D	AMARC	AE128

Serial	Type	Status	Notes
72-0183	A7D	w.o.	
72-0184	A7D	w.o.22.6.88	
72-0185	A7D	w.o.	
72-0186	A7D	AMARC	AE037
72-0187	A7D	w.o.17.1.81	
72-0188	A7D	AMARC	AE079
72-0189	A7D	destroyed	12.1.81
72-0190	A7D	AMARC	AE129
72-0191	A7D	AMARC	AE115
72-0192	A7D	gate	Richmond IAP
72-0193	A7D	AMARC	AE123
72-0194	A7D	AMARC	AE040
72-0195	A7D	AMARC	AE091
72-0196	A7D	w.o.	
72-0197	A7D		
72-0198	A7D		
72-0199	A7D	AMARC	AE109
72-0200	A7D	AMARC	AE176
72-0201	A7D	AMARC	AE159
72-0202	A7D	AMARC	AE175
72-0203	A7D	w.o.	
72-0204	A7D	w.o.	
72-0205	A7D	AMARC	AE173
72-0206	A7D	AMARC	AE161
72-0207	A7D	w.o.26.10.76	
72-0208	A7D	AMARC	AE028
72-0209	A7D	AMARC	AE046
72-0210	A7D	w.o.	
72-0211	A7D	preserved	Toledo Ex.IAP,Oh
72-0212	A7D		
72-0213	A7D	preserved	Des MoinesIAP,Ia
72-0214	A7D	AMARC	AE212
72-0215	A7D	AMARC	AE179
72-0216	A7D	AMARC	AE215
72-0217	A7D	AMARC	AE147
72-0218	A7D	AMARC	AE199
72-0219	A7D	destroyed 12.1.81	
72-0220	A7D	AMARC	AE114
72-0221	A7D	destroyed 12.1.81	
72-0222	A7D	destroyed 12.1.81	
72-0223	A7D	AMARC	AE061
72-0224	A7D	AMARC	AE074
72-0225	A7D	AMARC	AE197
72-0226	A7D	AMARC	AE043
72-0227	A7D	AMARC	AE186
72-0228	A7D		
72-0229	A7D	AMARC	AE184
72-0230	A7D		
72-0231	A7D		
72-0232	A7D	AMARC	AE014
72-0233	A7D	w.o.9.2.82	
72-0234	A7D		
72-0235	A7D	AMARC	AE166
72-0236	A7D	AMARC	AE104
72-0237	A7D	AMARC	AE229
72-0238	A7D		
72-0239	A7D	AMARC	AE022
72-0240	A7D	stored	Tulsa IAP
72-0241	A7D	AMARC	AE224
72-0242	A7D	AMARC	AE152
72-0243	A7D	AMARC	AE099
72-0244	A7D		
72-0245	A7D	stored	Kirtland AFB
72-0246	A7D	w.o.	
72-0247	A7D		
72-0248	A7D	AMARC	AE177
72-0249	A7D	w.o.6.2.83	
72-0250	A7D	AMARC	AE093
72-0251	A7D	AMARC	AE071
72-0252	A7D	AMARC	AE096
72-0253	A7D	AMARC	AE180
72-0254	A7D		
72-0255	A7D	AMARC	AE148
72-0256	A7D		
72-0257	A7D	w.o.5.8.88	
72-0258	A7D	AMARC	AE101
72-0259	A7D		
72-0260	A7D	AMARC	AE072
72-0261	A7D		
72-0262	A7D		
72-0263	A7D	w.o.1.5.78	
72-0264	A7D	w.o.	
72-0265	A7D	w.o.	
73-0992	A7D		
73-0993	A7D	w.o.8.11.75	
73-0994	A7D	destroyed 12.1.81	
73-0995	A7D	w.o.9.12.77	
73-0996	A7D		
73-0997	A7D	AMARC	AE200
73-0998	A7D	w.o.1.5.74	
73-0999	A7D	stored	Rick'backer ANGB
73-1000	A7D	AMARC	AE144
73-1001	A7D	w.o.12.4.79	
73-1002	A7D		
73-1003	A7D	w.o.	
73-1004	A7D	AMARC	AE181
73-1005	A7D	destroyed 12.1.81	
73-1006	A7D		
73-1007	A7D	AMARC	AE033
73-1008	A7D	AMARC	AE189
73-1009	A7D	stored	McConnell AFB
73-1010	A7D		
73-1011	A7D	AMARC	AE222
73-1012	A7D	w.o.10.9.85	
73-1013	A7D		
73-1014	A7D	AMARC	AE239
73-1015	A7D		
74-1737	A7D		
74-1738	A7D		
74-1739	A7D	preserved	Ellesworth AFB
74-1740	A7D	AMARC	AE153
74-1741	A7D	AMARC	AE024
74-1742	A7D	AMARC	AE057
74-1743	A7D	AMARC	AE066
74-1744	A7D	AMARC	AE157
74-1745	A7D	AMARC	AE240
74-1746	GA7D	TTC	Lowry
74-1747	A7D	AMARC	AE149
74-1748	A7D	destroyed 12.1.81	
74-1749	A7D	AMARC	AE083
74-1750	A7D	AMARC	AE130
74-1751	A7D		
74-1752	A7D		
74-1753	A7D	w.o.	
74-1754	A7D	w.o.	
74-1755	A7D	destroyed 12.1.81	
74-1756	A7D	stored	Tulsa IAP
74-1757	A7D		
74-1758	A7D		
74-1759	A7D	AMARC	AE241
74-1760	A7D		
75-0386	A7D	AMARC	AE205
75-0387	A7D	AMARC	AE198
75-0388	A7D		
75-0389	A7D		
75-0390	A7D	AMARC	AE007
75-0391	A7D	w.o.5.7.88	
75-0392	A7D	AMARC	AE020
75-0393	A7D		
75-0394	A7D	preserved	Tucson ANGB
75-0395	A7D		
75-0396	A7D	w.o.11.1.92	
75-0397	A7D	AMARC	AE209
75-0398	A7D		
75-0399	A7D	AMARC	AE228
75-0400	A7D		
75-0401	A7D	AMARC	AE139
75-0402	A7D	AMARC	AE058
75-0403	A7D		
75-0404	A7D	w.o. .4.77	
75-0405	A7D	AMARC	AE131
75-0406	A7D	stored	Sioux City AP, Ia
75-0407	A7D		
75-0408	A7D	BDRT	Barnes MAP, Ma
75-0409	A7D	AMARC	AE019
79-0460	A7K	AMARC	AE036
79-0461	A7K	AMARC	AE196
79-0462	A7K	AMARC	AE195
79-0463	A7K	AMARC	AE165
79-0464	A7K	AMARC	AE053
79-0465	A7K	AMARC	AE054
79-0466	A7K	AMARC	AE021
79-0467	A7K	AMARC	AE055
79-0468	A7K	AMARC	AE070
79-0469	A7K	AMARC	AE119
79-0470	A7K	AMARC	AE121

79-0471	A7K		
80-0284	A7K	AMARC	AE052
80-0285	A7K	AMARC	AE051
80-0286	A7K	AMARC	AE236
80-0287	A7K	AMARC	AE056
80-0288	A7K	AMARC	AE062
80-0289	A7K		
80-0290	A7K	AMARC	AE213
80-0291	A7K		
80-0292	A7K	w.o.30.5.90	
80-0293	A7K		
80-0294	A7K		
80-0295	A7K	AMARC	AE243
81-0072	A7K	AMARC	AE156
81-0073	A7K		
81-0074	A7K	AMARC	AE220
81-0075	A7K	SoC	
81-0076	A7K	AMARC	AE233
81-0077	A7K	AMARC	AE237

A10 THUNDERBOLT II

71-1369	YA10A		
71-1370	YA10A	USAF Museum	
73-1664	YA10B	preserved	Edwards AFB
73-1665	GYA10A	TTC	Sheppard
73-1666	YA10B	TTC	Sheppard
73-1667	GYA10A	preserved	England AFB
73-1668	A10A	w.o.11.3.82	
73-1669	A10A	w.o.8.8.77	
75-0258	A10A	AMARC	AC020
75-0259	A10A	w.o. 6. 9.59	
75-0260	GA10A	TTC	Sheppard
75-0261	A10A	AMARC	ACOO1
75-0262	A10A	355W	DM
75-0263	A10A	GIA	Shaw AFB
75-0264	A10A	AMARC	AC064
75-0265	A10A	AMARC	AC163
75-0266	A10A	w.o.17.9.79	
75-0267	A10A	AMARC	AC153
75-0268	A10A	355W	DM
75-0269	A10A	AMARC	AC071
75-0270	A10A	GIA	McChord AFB
75-0271	A10A	w.o.27. 7.79	
75-0272	A10A	AMARC	AC073
75-0273	A10A	GIA	Shaw AFB
75-0274	A10A	GIA	McChord AFB
75-0275	A10A	AMARC	AC165
75-0276	A10A	AMARC	AC133
75-0277	A10A	AMARC	AC015
75-0278	A10A	AMARC	AC062
75-0279	A10A	AMARC	AC127
75-0280	A10A	AMARC	AC042
75-0281	A10A	AMARC	AC146
75-0282	A10A	BDRT	DavisM'than AFB
75-0283	A10A	355W	DM
75-0284	A10A	AMARC	AC005
75-0285	A10A	AMARC	AC149
75-0286	GA10A	GIA	DavisM'than AFB
75-0287	A10A	AMARC	AC150
75-0288	A10A	AMARC	AC135
75-0289	A10A	BDRT	Eielson AFB
75-0290	A10A	AMARC	AC129
75-0291	A10A	AMARC	AC036
75-0292	A10A	AMARC	AC161
75-0293	A10A	AMARC	AC077
75-0294	A10A	w.o.3.6.77	
75-0295	A10A	w.o.16.10.78	
75-0296	A10A	AMARC	AC155
75-0297	A10A	AMARC	AC063
75-0298	A10A	AMARC	AC151
75-0299	A10A	AMARC	AC162
75-0300	A10A	AMARC	AC037
75-0301	A10A	GIA	Nellis AFB
75-0302	A10A	w.o.10.7.81	
75-0303	A10A	AMARC	AC139
75-0304	A10A	AMARC	AC125
75-0305	A10A	BDRT	Eielson AFB
75-0306	A10A	AMARC	AC092
75-0307	A10A	AMARC	AC006
75-0308	A10A	GIA	Pope AFB
75-0309	OA10A	AMARC	AC075
76-0512	OA10A	BDRT	B.Creek AP,Mi
76-0513	A10A	w.o.20.11.79	
76-0514	A10A	AMARC	AC157
76-0515	A10A	355W	DM
76-0516	OA10A	BDRT	NAS Willow Grove
76-0517	A10A	w.o.13.12.78	
76-0518	GA10A	TTC	Lowry
76-0519	A10A	BDRT	R-G AFB
76-0520	A10A	46FS	BD AFRes
76-0521	A10A	BDRT	Barksdale AFB
76-0522	A10A	AMARC	AC159
76-0523	OA10A	355W	DM
76-0524	A10A	AMARC	AC091
76-0525	A10A	w.o.15. 8.78	
76-0526	A10A	45FS	IN AFRes
76-0527	A10A	AMARC	AC027
76-0528	A10A	w.o.10.11.80	
76-0529	OA10A	AMARC	AC166
76-0530	GA10A	GIA	R-G AFB
76-0531	A10A	AMARC	AC084
76-0532	OA10A	AMARC	AC076
76-0533	OA10A	AMARC	AC104
76-0534	OA10A	AMARC	AC134
76-0535	A10A	45FS	IN AFRes
76-0536	A10A	AMARC	AC141
76-0537	A10A	355W	DM
76-0538	A10A		
76-0539	A10A	46FS	BD AFRes
76-0540	A10A		
76-0541	A10A	w.o.10. 6.83	
76-0542	A10A	w.o.30. 9.79	
76-0543	OA10A	w.o.19.2.91	
76-0544	A10A	AMARC	AC143
76-0545	A10A	w.o.12. 7.83	
76-0546	A10A	w.o.4.9.86	
76-0547	OA10A	355W	DM
76-0548	A10A	AMARC	AC050
76-0549	OA10A	AMARC	AC044
76-0550	GA10A	GIA	Spangdahlem AB
76-0551	A10A	BDRT	R-G AFB
76-0552	A10A	46FS	BD AFRes
76-0553	GA10A	GIA	Spangdahlem AB
76-0554	A10A	AMARC	AC089
77-0177	OA10A	AMARC	AC118
77-0178	OA10A	355W	DM
77-0179	A10A	w.o.23.6.80	
77-0180	A10A	w.o.8.8.78	
77-0181	A10A	AMARC	AC102
77-0182	A10A	w.o.	
77-0183	OA10A	355W	DM

Serial	Type	Operator	Code
77-0184	A10A	AMARC	AC098
77-0185	OA10A	355W	DM
77-0186	OA10A	AMARC	AC128
77-0187	OA10A	AMARC	AC140
77-0188	A10A	AMARC	AC154
77-0189	A10A	355W	DM
77-0190	OA10A	355W	DM
77-0191	OA10A	AMARC	AC065
77-0192	A10A	AMARC	AC117
77-0193	A10A	AMARC	AC114
77-0194	A10A	AMARC	AC169
77-0195	A10A	AMARC	AC004
77-0196	A10A	104FS	MD MD ANG
77-0197	OA10A	w.o.27.2.91	
77-0198	OA10A	AMARC	AC110
77-0199	GA10A	TTC	Sheppard
77-0200	OA10A	355W	DM
77-0201	A10A	AMARC	AC059
77-0202	A10A	AMARC	AC138
77-0203	A10A	w.o.11.12.87	
77-0204	OA10A	AMARC	AC072
77-0205	A10A		
77-0206	A10A	w.o.22.5.79	
77-0207	OA10A	AMARC	AC100
77-0208	A10A	AMARC	AC081
77-0209	A10A	355W	DM
77-0210	OA10A	AMARC	AC074
77-0211	A10A	AMARC	AC147
77-0212	OA10A	AMARC	AC083
77-0213	OA10A	AMARC	AC087
77-0214	A10A	AMARC	AC095
77-0215	A10A	w.o.12.12.78	
77-0216	OA10A	355W	DM
77-0217	A10A	AMARC	AC121
77-0218	OA10A	355W	DM
77-0219	OA10A	AMARC	AC164
77-0220	A10A	AMARC	AC086
77-0221	A10A	preserved	BaltimoreState AP
77-0222	OA10A	AMARC	AC101
77-0223	OA10A	355W	DM
77-0224	OA10A	AMARC	AC056
77-0225	A10A	AMARC	AC002
77-0226	A10A	AMARC	AC003
77-0227	A10A	AMARC	AC131
77-0228	A10A	45FS	IN AFRes
77-0229	A10A	AMARC	AC116
77-0230	A10A	AMARC	AC115
77-0231	A10A	AMARC	AC109
77-0232	A10A	AMARC	AC105
77-0233	A10A	AMARC	AC051
77-0234	A10A	AMARC	AC103
77-0235	A10A	AMARC	AC010
77-0236	A10A	AMARC	AC009
77-0237	A10A	AMARC	AC047
77-0238	A10A	AMARC	AC028
77-0239	A10A	AMARC	AC099
77-0240	A10A	AMARC	AC093
77-0241	A10A	AMARC	AC049
77-0242	A10A	AMARC	AC048
77-0243	A10A	w.o.8.1.87	
77-0244	A10A	45FS	IN AFRes
77-0245	A10A	AMARC	AC096
77-0246	A10A	AMARC	AC034
77-0247	A10A	AMARC	AC031
77-0248	OA10A	103FS	PA PA ANG
77-0249	A10A	AMARC	AC097
77-0250	A10A	AMARC	AC012
77-0251	A10A	AMARC	AC170
77-0252	A10A	derelict	NAS Willow Grove
77-0253	A10A	w.o.7.7.79	
77-0254	A10A	AMARC	AC029
77-0255	A10A	preserved	NAS New Orleans
77-0256	A10A	AMARC	AC085
77-0257	A10A	AMARC	AC152
77-0258	A10A	w.o.9.1.81	
77-0259	A10A	preserved	Duxford, UK
77-0260	A10A	AMARC	AC132
77-0261	A10A		
77-0262	A10A	AMARC	AC030
77-0263	A10A	AMARC	AC046
77-0264	GA10A	GIA	SpangdahlemAB
77-0265	OA10A	355W	DM
77-0266	A10A	AMARC	AC137
77-0267	A10A	AMARC	AC144
77-0268	A10A	AMARC	AC148
77-0269	A10A	AMARC	AC088
77-0270	OA10A	355W	DM
77-0271	A10A	AMARC	AC094
77-0272	A10A	AMARC	AC082
77-0273	A10A	AMARC	AC145
77-0274	A10A	AMARC	AC130
77-0275	A10A	AMARC	AC142
77-0276	A10A	AMARC	AC011
78-0582	A10A	46FS	BD AFRes
78-0583	A10A	131FS	MAMA ANG
78-0584	A10A	118FS	CT CT ANG
78-0585	A10A	w.o.9.7.82	
78-0586	A10A	118FS	CT CT ANG
78-0587	A10A	AMARC	AC111
78-0588	A10A	w.o.17.11.80	
78-0589	A10A	AMARC	AC018
78-0590	A10A	w.o.17.11.80	
78-0591	OA10A	363FW	SW
78-0592	A10A	AMARC	AC017
78-0593	A10A	23W	FT
78-0594	A10A	AMARC	AC025
78-0595	A10A	AMARC	AC057
78-0596	A10A	23W	FT
78-0597	A10A	23W	FT
78-0598	A10A	23W	FT
78-0599	A10A	23W	FT
78-0600	A10A	23W	FT
78-0601	A10A	w.o.27.8.79	
78-0602	A10A	w.o.26.8.83	
78-0603	A10A	AMARC	AC167
78-0604	A10A		
78-0605	A10A	303FS	KC AFRes
78-0606	A10A	AMARC	AC112
78-0607	A10A	dump	Norton AFB
78-0608	A10A	131FS	MAMA ANG
78-0609	A10A	stored	Battle Creek AP
78-0610	A10A	w.o.6.5.80	
78-0611	A10A	131FS	MAMA ANG
78-0612	A10A	131FS	MAMA ANG
78-0613	A10A	118FS	CT CT ANG
78-0614	A10A	131FS	MAMA ANG
78-0615	A10A	118FS	CT CT ANG
78-0616	A10A	131FS	MAMA ANG
78-0617	A10A	131FS	MAMA ANG
78-0618	A10A	131FS	MAMA ANG
78-0619	A10A	118FS	CT CT ANG
78-0620	A10A		
78-0621	A10A	118FS	CT CT ANG
78-0622	A10A	AMARC	AC024
78-0623	A10A	w.o.9.2.84	
78-0624	A10A	131FS	MAMA ANG
78-0625	A10A	118FS	CT CT ANG
78-0626	A10A	104FS	MDMD ANG
78-0627	A10A	104FS	MDMD ANG
78-0628	A10A	131FS	MAMA ANG
78-0629	A10A	118FS	CT CT ANG
78-0630	A10A	131FS	MAMA ANG
78-0631	A10A	303FS	KC AFRes
78-0632	A10A	131FS	MAMA ANG
78-0633	A10A	118FS	CT CT ANG
78-0634	A10A	104FS	MDMD ANG
78-0635	A10A	118FS	CT CT ANG
78-0636	A10A	104FS	MDMD ANG
78-0637	A10A	104FS	MDMD ANG
78-0638	A10A	118FS	CT CT ANG
78-0639	A10A	118FS	CT CT ANG
78-0640	A10A	131FS	MAMA ANG
78-0641	A10A	118FS	CT CT ANG
78-0642	A10A	131FS	MAMA ANG
78-0643	A10A	118FS	CT CT ANG
78-0644	A10A	131FS	MAMA ANG
78-0645	A10A	w.o.	
78-0646	A10A	118FS	CT CT ANG
78-0647	A10A	131FS	MAMA ANG
78-0648	A10A		
78-0649	A10A	104FS	MDMD ANG
78-0650	A10A	104FS	MDMD ANG
78-0651	A10A	354FS	TC
78-0652	A10A	354FS	TC
78-0653	A10A	104FS	MDMD ANG
78-0654	A10A	AMARC	AC045

Serial	Type	Unit	Code/Base
78-0655	A10A	23W	FT
78-0656	A10A	AMARC	AC124
78-0657	A10A	46FS	BD AFRes
78-0658	A10A	103FS	PA PA ANG
78-0659	A10A	131FS	MA MA ANG
78-0660	A10A	AMARC	AC060
78-0661	A10A	w.o.23.10.80	
78-0662	OA10A	AMARC	AC090
78-0663	A10A		
78-0664	A10A	AMARC	AC052
78-0665	A10A	AMARC	AC016
78-0666	A10A	AMARC	AC120
78-0667	A10A	AMARC	AC119
78-0668	A10A	AMARC	AC023
78-0669	A10A	23W	FT
78-0670	A10A	355W	DM
78-0671	A10A	354FS	TC
78-0672	A10A	AMARC	AC079
78-0673	A10A	355W	DM
78-0674	A10A	23W	FT
78-0675	A10A	AMARC	AC108
78-0676	A10A	23W	FT
78-0677	A10A	AMARC	AC055
78-0678	A10A	AMARC	AC168
78-0679	A10A	23W	FT
78-0680	A10A	AMARC	AC040
78-0681	A10A	USAF Museum	
78-0682	A10A	104FS	MDMD ANG
78-0683	A10A	104FS	MDMD ANG
78-0684	A10A	354FS	TC
78-0685	A10A	354FS	TC
78-0686	OA10A	AMARC	AC080
78-0687	A10A		
78-0688	A10A	23W	FT
78-0689	A10A	104FS	MDMD ANG
78-0690	A10A	354FS	TC
78-0691	A10A	104FS	MDMD ANG
78-0692	A10A	104FS	MDMD ANG
78-0693	A10A	104FS	MDMD ANG
78-0694	A10A	104FS	MDMD ANG
78-0695	A10A	46FS	BD AFRes
78-0696	A10A	46FS	BD AFRes
78-0697	A10A	23W	FT
78-0698	A10A	AMARC	AC054
78-0699	A10A	AMARC	AC078
78-0700	A10A	46FS	BD AFRes
78-0701	A10A	46FS	BD AFRes
78-0702	A10A	104FS	MDMD ANG
78-0703	A10A	104FS	MDMD ANG
78-0704	A10A	104FS	MDMD ANG
78-0705	A10A	104FS	MDMD ANG
78-0706	A10A	354FS	TC
78-0707	A10A	118FS	CT CT ANG
78-0708	OA10A	103FS	PA PA ANG
78-0709	A10A	46FS	BD AFRes
78-0710	A10A	AMARC	AC013
78-0711	A10A		
78-0712	A10A	118FS	CT CT ANG
78-0713	A10A	AMARC	AC107
78-0714	A10A	AMARC	AC068
78-0715	A10A	AMARC	AC123
78-0716	A10A	46FS	BD AFRes
78-0717	A10A	104FS	MDMD ANG
78-0718	A10A	104FS	MDMD ANG
78-0719	A10A	104FS	MDMD ANG
78-0720	A10A	104FS	MDMD ANG
78-0721	A10A	131FS	MAMA ANG
78-0722	A10A	w.o.16.2.91	
78-0723	A10A	w.o.9.2.85	
78-0724	A10A	AMARC	AC033
78-0725	OA10A	363FW	SW
79-0082	A10A	104FS	MDMD ANG
79-0083	A10A	w.o.8.5.81	
79-0084	A10A	118FS	CT CT ANG
79-0085	A10A	46FS	BD AFRes
79-0086	A10A	104FS	MDMD ANG
79-0087	A10A	104FS	MDMD ANG
79-0088	A10A	104FS	MDMD ANG
79-0089	A10A		
79-0090	A10A	303FS	KC AFRes
79-0091	A10A	303FS	KC AFRes
79-0092	A10A	303FS	KC AFRes
79-0093	A10A	303FS	KC AFRes
79-0094	A10A	47FS	BD AFRes
79-0095	A10A	47FS	BD AFRes
79-0096	A10A	AMARC	AC032
79-0097	A10A	preserved	Myrtle Beach AFB
79-0098	A10A	AMARC	AC035
79-0099	A10A	AMARC	AC022
79-0100	A10A	preserved	Barnes MAP, Ma
79-0101	A10A	AMARC	AC053
79-0102	A10A	AMARC	AC106
79-0103	A10A	118FS	CT CT ANG
79-0104	A10A	131FS	MAMA ANG
79-0105	A10A	47FS	BD AFRes
79-0106	A10A	47FS	BD AFRes
79-0107	A10A	303FS	KC AFRes
79-0108	A10A	118FS	CT CT ANG
79-0109	A10A	303FS	KC AFRes
79-0110	A10A	303FS	KC AFRes
79-0111	A10A	303FS	KC AFRes
79-0112	A10A	AMARC	AC014
79-0113	A10A	303FS	KC AFRes
79-0114	A10A	303FS	KC AFRes
79-0115	A10A	AMARC	AC019
79-0116	A10A	preserved	D-M AFB
79-0117	A10A	303FS	KC AFRes
79-0118	A10A	303FS	KC AFRes
79-0119	A10A	303FS	KC AFRes
79-0120	A10A	303FS	KC AFRes
79-0121	A10A	303FS	KC AFRes
79-0122	A10A	303FS	KC AFRes
79-0123	A10A	303FS	KC AFRes
79-0124	A10A	AMARC	AC058
79-0125	A10A		
79-0126	A10A	AMARC	AC069
79-0127	A10A	AMARC	AC026
79-0128	A10A	AMARC	AC061
79-0129	A10A	172FS	BC MI ANG
79-0130	A10A	w.o.16.2.91	
79-0131	A10A	AMARC	AC070
79-0132	A10A	AMARC	AC021
79-0133	A10A	AMARC	AC067
79-0134	A10A	47FS	BD AFRes
79-0135	A10A	23W	FT
79-0136	A10A	47FS	BD AFRes
79-0137	A10A	AMARC	AC066
79-0138	A10A	23W	FT
79-0139	A10A	23W	FT
79-0140	A10A	AMARC	AC041
79-0141	A10A	23W	FT
79-0142	A10A	47FS	BD AFRes
79-0143	A10A	47FS	BD AFRes
79-0144	A10A	47FS	BD AFRes
79-0145	A10A	47FS	BD AFRes
79-0146	A10A	46FS	BD AFRes
79-0147	A10A	47FS	BD AFRes
79-0148	A10A	47FS	BD AFRes
79-0149	A10A	47FS	BD AFRes
79-0150	A10A	47FS	BD AFRes
79-0151	A10A	47FS	BD AFRes
79-0152	A10A	47FS	BD AFRes
79-0153	A10A	47FS	BD AFRes
79-0154	A10A	47FS	BD AFRes
79-0155	A10A	47FS	BD AFRes
79-0156	A10A	w.o.26.9.85	
79-0157	A10A	23W	FT
79-0158	A10A	AMARC	AC126
79-0159	A10A	23W	FT
79-0160	OA10A	AMARC	AC136
79-0161	A10A	w.o.24.2.84	
79-0162	A10A	23W	FT
79-0163	A10A	AMARC	AC122
79-0164	A10A	303FS	KC AFRes
79-0165	A10A	118FS	CT CT ANG
79-0166	A10A	AMARC	AC038
79-0167	A10A	57FW	WA
79-0168	A10A	57FW	WA
79-0169	A10A	57FW	WA
79-0170	A10A	57FW	WA
79-0171	A10A	57FW	WA
79-0172	A10A	57FW	WA
79-0173	A10A	118FS	CT CT ANG
79-0174	A10A	104FS	MDMD ANG
79-0175	A10A	104FS	MDMD ANG
79-0176	A10A	AMARC	AC039

Serial	Type	Unit	Code
79-0177	A10A	ALC	SM Sacramento
79-0178	OA10A	363FW	SW
79-0179	OA10A	363FW	SW
79-0180	A10A	47FS	BD AFRes
79-0181	A10A	23W	FT
79-0182	A10A	AMARC	AC043
79-0183	A10A	354FS	TC
79-0184	OA10A	w.o.12.11.92	
79-0185	A10A	45FS	IN AFRes
79-0186	OA10A	363FW	SW
79-0187	A10A	46FS	BD AFRes
79-0188	A10A	354FS	TC
79-0189	OA10A	363FW	SW
79-0190	A10A	355W	DM
79-0191	A10A	45FS	IN AFRes
79-0192	OA10A	363FW	SW
79-0193	A10A	23W	FT
79-0194	A10A	104FS	MDMD ANG
79-0195	A10A	355W	DM
79-0196	A10A	355W	DM
79-0197	OA10A	363FW	SW
79-0198	A10A	354FS	TC
79-0199	A10A	355W	DM
79-0200	A10A	23W	FT
79-0201	A10A	355W	DM
79-0202	A10A	355W	DM
79-0203	A10A	23W	FT
79-0204	A10A	23W	FT
79-0205	A10A	355W	DM
79-0206	OA10A	363FW	SW
79-0207	OA10A	355W	DM
79-0208	A10A	w.o.29.6.85	
79-0209	A10A	46FS	BD AFRes
79-0210	A10A	355W	DM
79-0211	A10A	45FS	IN AFRes
79-0212	A10A	w.o.22.1.87	
79-0213	A10A	23W	FT
79-0214	A10A	46FS	BD AFRes
79-0215	A10A	354FS	TC
79-0216	A10A	23W	FT
79-0217	A10A	AMARC	AC160
79-0218	A10A	AMARC	AC156
79-0219	OA10A	103FS	PA PA ANG
79-0220	A10A	AMARC	AC007
79-0221	A10A	AMARC	AC158
79-0222	A10A	w.o.28.7.83	
79-0223	A10A	23W	FT
79-0224	A10A	AMARC	AC008
79-0225	A10A		
80-0140	OA10A	363FW	SW
80-0141	A10A	355W	DM
80-0142	A10A	355W	DM
80-0143	A10A	355W	DM
80-0144	OA10A	363FW	SW
80-0145	A10A	355W	DM
80-0146	A10A	103FS	PA PA ANG
80-0147	A10A	355W	DM
80-0148	A10A	w.o.22.3.82	
80-0149	A10A	23W	FT
80-0150	A10A	355W	DM
80-0151	OA10A	354FS	TC
80-0152	OA10A	103FS	PA PA ANG
80-0153	A10A	23W	FT
80-0154	A10A	w.o.25.8.81	
80-0155	A10A	355W	DM
80-0156	A10A	103FS	PA PA ANG
80-0157	OA10A	363FW	SW
80-0158	A10A	45FS	IN AFRes
80-0159	A10A	355W	DM
80-0160	A10A		
80-0161	A10A	23W	FT
80-0162	A10A	355W	DM
80-0163	OA10A	354FS	TC
80-0164	A10A	354FS	TC
80-0165	A10A	354FS	TC
80-0166	A10A	46FS	BD AFRes
80-0167	OA10A	354FS	TC
80-0168	A10A	355W	DM
80-0169	A10A	355W	DM
80-0170	OA10A	363FW	SW
80-0171	A10A	45FS	IN AFRes
80-0172	OA10A	363FW	SW
80-0173	A10A	355W	DM
80-0174	A10A	w.o.7.4.83	
80-0175	A10A	23W	FT
80-0176	OA10A	355W	DM
80-0177	A10A	354FS	TC
80-0178	A10A	46FS	BD AFRes
80-0179	A10A	355W	DM
80-0180	A10A	45FS	IN AFRes
80-0181	A10A	45FS	IN AFRes
80-0182	A10A	w.o.21.9.81	
80-0183	A10A		
80-0184	OA10A	103FS	PA PA ANG
80-0185	A10A	57FW	WA
80-0186	A10A	354FS	TC
80-0187	A10A	46FS	BD AFRes
80-0188	A10A	303FS	KC AFRes
80-0189	A10A	355W	DM
80-0190	A10A	355W	DM
80-0191	A10A	46FS	BD AFRes
80-0192	A10A	45FS	IN AFRes
80-0193	A10A	w.o.12.12.83	
80-0194	OA10A	363FW	SW
80-0195	A10A	57FW	WA
80-0196	OA10A	103FS	PA PA ANG
80-0197	A10A	46FS	BD AFRes
80-0198	A10A	w.o.25.3.88	
80-0199	A10A	23W	FT
80-0200	A10A	57FW	WA
80-0201	A10A	303FS	KC AFRes
80-0202	A10A		
80-0203	A10A	355W	DM
80-0204	A10A	355W	DM
80-0205	A10A	45FS	IN AFRes
80-0206	A10A	355W	DM
80-0207	OA10A	355W	DM
80-0208	OA10A	363FW	SW
80-0209	A10A	23W	FT
80-0210	A10A	355W	DM
80-0211	A10A	57FW	WA
80-0212	OA10A	355W	DM
80-0213	A10A	51W	OS
80-0214	OA10A	103FS	PA PA ANG
80-0215	A10A	45FS	IN AFRes
80-0216	A10A		
80-0217	A10A		
80-0218	OA10A	103FS	PA PA ANG
80-0219	A10A	BDRT	Alconbury, UK
80-0220	OA10A	355W	DM
80-0221	A10A	172FS	BC MI ANG
80-0222	A10A	172FS	BC MI ANG
80-0223	A10A	23W	FT
80-0224	OA10A	354FS	TC
80-0225	A10A	57FW	WA
80-0226	A10A	57FW	WA
80-0227	OA10A	103FS	PA PA ANG
80-0228	OA10A		
80-0229	OA10A	363FW	SW
80-0230	OA10A	103FS	PA PA ANG
80-0231	A10A	w.o.6.2.90	
80-0232	OA10A	103FS	PA PA ANG
80-0233	OA10A	355W	DM
80-0234	A10A	355W	DM
80-0235	A10A	355W	DM
80-0236	A10A	355W	DM
80-0237	OA10A	363FW	SW
80-0238	OA10A	343W	AK
80-0239	A10A	51W	OS
80-0240	OA10A	343W	AK
80-0241	A10A	51W	OS
80-0242	A10A	57FW	WA
80-0243	A10A	51W	OS
80-0244	A10A	51W	OS
80-0245	A10A	51W	OS
80-0246	A10A	354FS	TC
80-0247	A10A	51W	OS
80-0248	A10A	w.o.2.2.91	
80-0249	A10A	51W	OS
80-0250	OA10A	103FS	PA PA ANG
80-0251	A10A	51W	OS
80-0252	A10A	23W	FT
80-0253	A10A	51W	OS
80-0254	OA10A	343W	AK
80-0255	A10A	172FS	BC MI ANG
80-0256	A10A	172FS	BC MI ANG

80-0257	A10A	172FS	BC MI ANG		81-0956	A10A	52FW	SP
80-0258	A10A	172FS	BC MI ANG		81-0957	A10A	w.o.8.12.88	
80-0259	OA10A	343W	AK		81-0958	A10A	57FW	WA
80-0260	A10A	172FS	BC MI ANG		81-0959	A10A	57FW	WA
80-0261	A10A				81-0960	A10A	52FW	SP
80-0262	A10A	172FS	BC MI ANG		81-0961	A10A	52FW	SP
80-0263	A10A	172FS	BC MI ANG		81-0962	A10A	52FW	SP
80-0264	A10A	172FS	BC MI ANG		81-0963	A10A	52FW	SP
80-0265	A10A	172FS	BC MI ANG		81-0964	OA10A	363FW	SW
80-0266	A10A	172FS	BC MI ANG		81-0965	A10A	52FW	SP
80-0267	A10A	172FS	BC MI ANG		81-0966	A10A	52FW	SP
80-0268	A10A	172FS	BC MI ANG		81-0967	OA10A	363FW	SW
80-0269	A10A	172FS	BC MI ANG		81-0968	A10A	w.o.29.8.83	
80-0270	A10A	355W	DM		81-0969	OA10A	343W	AK
80-0271	A10A				81-0970	OA10A	343W	AK
80-0272	A10A				81-0971	A10A	51W	OS
80-0273	OA10A	103FS	PA PA ANG		81-0972	A10A	w.o.29.8.83	
80-0274	A10A				81-0973	A10A	51W	OS
80-0275	OA10A	103FS	PA PA ANG		81-0974	A10A	172FS	BC MI ANG
80-0276	A10A	45FS	IN AFRes		81-0975	A10A	172FS	BC MI ANG
80-0277	OA10A	363FW	SW		81-0976	A10A	52FW	SP
80-0278	OA10A	355W	DM		81-0977	A10A	52FW	SP
80-0279	A10A	45FS	IN AFRes		81-0978	A10A	52FW	SP
80-0280	OA10A	355W	DM		81-0979	OA10A	363FW	SW
80-0281	A10A	355W	DM		81-0980	A10A	52FW	SP
80-0282	A10A	23W	FT		81-0981	OA10A	103FS	PA PA ANG
80-0283	A10A	51W	OS		81-0982	A10A	52FW	SP
81-0939	OA10A	363FW	SW		81-0983	A10A	52FW	SP
81-0940	OA10A	103FS	PA PA ANG		81-0984	A10A	52FW	SP
81-0941	A10A	355W	DM		81-0985	A10A	52FW	SP
81-0942	A10A				81-0986	A10A	w.o.22.12.88	
81-0943	A10A	355W	DM		81-0987	OA10A	363FW	SW
81-0944	A10A				81-0988	A10A	52FW	SP
81-0945	A10A	57FW	WA		81-0989	A10A	ALC	SM Sacramento
81-0946	A10A	57FW	WA		81-0990	OA10A	363FW	SW
81-0947	OA10A	363FW	SW		81-0991	A10A	52FW	SP
81-0948	OA10A	363FW	SW		81-0992	A10A	52FW	SP
81-0949	OA10A	103FS	PA PA ANG		81-0993	A10A	172FS	BC MI ANG
81-0950	A10A	355W	DM		81-0994	A10A	172FS	BC MI ANG
81-0951	A10A	52FW	SP		81-0995	OA10A	343W	AK
81-0952	A10A	52FW	SP		81-0996	A10A	172FS	BC MI ANG
81-0953	OA10A	363FW	SW		81-0997	A10A	172FS	BC MI ANG
81-0954	A10A	52FW	SP		81-0998	A10A	172FS	BC MI ANG
81-0955	OA10A	103FS	PA PA ANG		82-0646	A10A	52FW	SP
					82-0647	OA10A	103FS	PA PA ANG
					82-0648	A10A	355W	DM
					82-0649	A10A	52FW	SP
					82-0650	A10A	52FW	SP
					82-0651	A10A	51W	OS
					82-0652	A10A	51W	OS
					82-0653	OA10A	363FW	SW
					82-0654	A10A	52FW	SP
					82-0655	A10A	52FW	SP
					82-0656	A10A	52FW	SP
					82-0657	OA10A	363FW	SW
					82-0658	A10A	52FW	SP
					82-0659	OA10A	103FS	PA PA ANG
					82-0660	A10A	23W	FT
					82-0661	A10A	23W	FT
					82-0662	OA10A	354FS	TC
					82-0663	A10A	23W	FT
					82-0664	A10A	23W	FT
					82-0665	A10A	355W	DM

A37 DRAGONFLY

c\n	A37A:	conversions from T37B 55-4303, 55-4305, 55-4306, 55-4308 to 55-4316, 55-4318 to 55-4320, 56-3464 to 56-3475, 56-3477 to 56-3479, 56-3481 to 56-3484, 56-3486, 56-3487, 56-3489 to 56-3491 renumbered as 40017 to 40064.
	A37B:	43058, 43059, 43074, 43076, 43079, 43082, 43087, 43088, 43090, 43092, 43095 to 43100, 43102 to 43105, 43109, 43110, 43112 to 43120, 43129, 43162 to 43164, 43167, 43169, 43170, 43193, 43194, 43196, 43200, 43201, 43203, 43204, 43209, 43211 to 43217, 43220, 43223 to 43234, 43238, 43241 to 43243, 43259 to 43262, 43264 to 43271, 43273 to 43275,

43277, 43281, 43283, 43286 to 43288, 43294, 43296, 43297, 43299, 43303 to 43309, 43313 to 43327, 43354, 43369, 43389, 43393 to 43398, 43402 to 43416, 43419 to 43425, 43429 to 43446, 43451 to 43456, 43468, 43469, 43480, 43481, 43484, 43485, 43492 to 43494, 43503 to 43506, 43512 to 43515, 43520 to 43525, 43530 to 43534, 43539.

Serial	Type	Note	Detail
67-14503	A37A	w.o.3.11.67	
67-14504	A37A	scrapyard	Consolidated
67-14505	A37A	scrapyard	Consolidated
67-14506	A37A	w.o.11.3.68	
67-14507	A37A	to Vietnam	
67-14508	A37A	w.o.21.8.67	
67-14509	A37A	scrapyard	Consolidated
67-14510	A37A	scrapyard	Consolidated
67-14511	A37A	SoC	
67-14512	A37A	SoC	
67-14513	A37A	w.o.19.9.68	
67-14514	A37A	scrapyard	Consolidated
67-14515	A37A		SoC
67-14516	A37A	scrapyard	Consolidated
67-14517	A37A	SoC	
67-14518	A37A	w.o.17.4.68	
67-14519	A37A	scrapyard	Consolidated
67-14520	A37A	SoC	
67-14521	A37A	w.o.3.3.68	
67-14522	GA37A	SoC	
67-14523	A37A	w.o.4.12.69	
67-14524	GA37A		
67-14525	GA37A		
67-14526	A37A	w.o.17.12.69	
67-14527	A37A	w.o.23.5.68	
67-14528	A37A	to N128RA	Deer Valley AP,Az
67-14529	A37A	to Vietnam	
67-14530	A37A	scrapyard	Consolidated
67-14531	A37A	scrapyard	Consolidated
67-14532	A37A	SoC	
67-14533	A37A	SoC	
67-14534	A37A	SoC	
67-14535	A37A	SoC	
67-14536	A37A	SoC	
67-14537	A37A	SoC	
67-14538	A37A	SoC	
67-14539	A37A	SoC	
67-14540	A37A	SoC	
67-14541	A37A	scrapyard	Consolidated
67-14776	A37A	to Vietnam	
67-14790	NOA37B	412T	WEDAFFTC
67-14792	A37A		
67-14823	A37B	SoC	
68-7911	OA37B	SoC	
68-7912	OA37B		
68-7927	OA37B		
68-7929	A37B	SoC	
68-7932	OA37B	SoC	
68-7935	A37B	SoC	
68-7940	A37B	SoC	
68-7941	A37B	SoC	
68-7943	A37B	SoC	
68-7945	A37B	SoC	
68-7948	A37B	SoC	
68-7949	A37B	SoC	
68-7950	A37B	SoC	
68-7951	OA37B	SoC	
68-7952	OA37B		
68-7953	A37B	SoC	
68-7955	A37B	SoC	
68-7956	A37B	SoC	
68-7957	A37B	SoC	
68-7958	A37B	SoC	
68-7962	A37B	SoC	
68-7963	A37B	SoC	
68-7965	OA37B	MASDC	
68-7966	OA37B	SoC	
68-7967	OA37B	MASDC	
68-7968	A37B	SoC	
68-7969	A37B	SoC	
68-7970	OA37B	w.o.18.4.85	
68-7971	OA37B		
68-7972	OA37B	SoC	
68-7973	OA37B	SoC	
68-10778	A37B	w.o.27.6.71	
68-10811	A37B	SoC	
68-10812	A37B	SoC	
68-10813	A37B	SoC	
68-10816	OA37B	w.o.	
68-10818	OA37B	SoC	
68-10819	A37B	SoC	
69-6348	A37B	w.o.1.5.72	
69-6349	A37B	w.o.18.6.72	
69-6351	A37B	w.o.26.3.71	
69-6355	A37B	w.o.17.3.72	
69-6356	A37B	w.o.1.2.71	
69-6358	A37B	SoC	
69-6359	A37B	w.o.23.7.72	
69-6364	OA37B	SoC	
69-6366	A37B	SoC	
69-6367	OA37B	SoC	
69-6368	OA37B		
69-6369	OA37B	MASDC	
69-6370	OA37B		
69-6371	OA37B		
69-6372	OA37B		
69-6375	GA37B	IAAFA	
69-6378	A37B		
69-6379	A37B		
69-6380	A37B		
69-6381	A37B	MASDC	
69-6382	A37B		
69-6383	OA37B	SoC	
69-6384	OA37B	SoC	
69-6385	A37B		
69-6386	OA37B		
69-6387	OA37B	SoC	
69-6388	OA37B		
69-6389	OA37B		
69-6393	OA37B	SoC	
69-6396	A37B		
69-6397	OA37B	AMARC	AB023
69-6398	OA37B	SoC	
69-6414	A37B		
69-6415	OA37B	w.o.	
69-6416	A37B		
69-6417	A37B		
69-6419	OA37B	SoC	
69-6420	OA37B	SoC	
69-6421	A37B		
69-6422	OA37B		
69-6423	OA37B	to Colombia	
69-6424	OA37B		
69-6425	OA37B	SoC	
69-6426	OA37B	SoC	
69-6428	OA37B		
69-6429	OA37B		
69-6430	OA37B		
69-6432	OA37B		
69-6436	OA37B	to Colombia	
69-6438	OA37B	AMARC	AB024
69-6441	OA37B	SoC	
69-6442	OA37B	MASDC	
69-6443	OA37B	AMARC	AB010
70-1279	OA37B		
70-1281	OA37B	FMS	
70-1282	OA37B	FMS	
70-1284	A37B		
70-1288	A37B	to Colombia	
70-1289	OA37B	to Colombia	
70-1290	A37B	to Columbia FAC2157	
70-1291	OA37B	FMS	
70-1292	A37B		
70-1293	OA37B	preserved	Batt.Creek AP,Mi
70-1294	OA37B	to El Salvador	
70-1298	OA37B	to Dominican Rep.3706	
70-1299	A37B		
70-1300	OA37B	MASDC	
70-1301	OA37B		
70-1302	OA37B		
70-1303	OA37B		
70-1304	GOA37B	IAAFA	
70-1305	A37B	to Colombia	
70-1306	A37B	FMS	
70-1307	OA37B	FMS	
70-1308	OA37B		

Serial	Type	Unit/Notes	Location
70-1309	A37B		
70-1310	NA37B	412TW	ED AFFTC
70-1311	OA37B	to Dominican Rep.3702	
70-1312	OA37B	MASDC	
71-0816	OA37B	SoC	
71-0818	OA37B		
71-0831	A37B		
71-0851	A37B		
71-0858	A37B		
71-0859	OA37B	AMARC	AB009
71-0860	OA37B	AMARC	AB016
71-0861	A37B		
71-0862	OA37B	SoC	
71-0863	OA37B	AMARC	AB013
71-0867	OA37B		
71-0868	OA37B	SoC	
71-0869	OA37B		
71-0870	OA37B	SoC	
71-0871	OA37B	SoC	
71-0872	A37B		
71-0873	OA37B	AMARC	AB032
71-1409	OA37B	w.o.24.6.86	
71-1410	OA37B		
71-1411	OA37B	SoC	
71-1412	OA37B		
71-1413	OA37B	SoC	
71-1414	OA37B	SoC	
71-1415	OA37B	SoC	
71-1416	OA37B	SoC	
73-1056	A37B		
73-1057	A37B	w.o.16.1.76	
73-1058	A37B	SoC	
73-1059	OA37B		
73-1060	OA37B		
73-1061	OA37B	AMARC	AB008
73-1062	OA37B		
73-1063	OA37B		
73-1064	OA37B		
73-1065	A37B		

Serial	Type	Unit/Notes	Location
73-1066	OA37B	AMARC	AB025
73-1067	OA37B	SoC	
73-1068	A37B	to El Salvador	
73-1069	OA37B		
73-1070	A37B	to El Salvador	
73-1071	OA37B	SoC	
73-1072	OA37B	412TW	ED AFFTC
73-1073	OA37B	AMARC	AB014
73-1074	A37B	to El Salvador	
73-1075	OA37B		
73-1076	A37B	to El Salvador	
73-1077	OA37B	AMARC	AB012
73-1078	OA37B	MASDC	
73-1079	OA37B		
73-1080	OA37B		
73-1081	A37B	SoC	
73-1082	OA37B	MASDC	
73-1083	OA37B		
73-1084	OA37B		
73-1085	A37B	to El Salvador	
73-1086	A37B	SoC	
73-1087	OA37B	MASDC	
73-1088	GOA37B	IAAFA	
73-1089	A37B	to El Salvador	
73-1090	NA37B	412TW	ED AFFTC
73-1091	OA37B		
73-1092	OA37B	MASDC	
73-1093	OA37B		
73-1094	OA37B		
73-1095	OA37B		
73-1096	OA37B	MASDC	
73-1097	OA37B		
73-1098	GOA37B	AMARC	AB026
73-1099	A37B	w.o.24.3.79	
73-1100	OA37B		
73-1101	OA37B	AMARC	AB034
73-1102	OA37B	AMARC	AB033
73-1103	OA37B	MASDC	
73-1104	OA37B		
73-1105	OA37B		
73-1106	A37B	SoC	
73-1107	A37B	SoC	
73-1108	OA37B		
73-1109	OA37B	w.o.25.3.88	
73-1110	OA37B		
73-1111	OA37B	AMARC	AB011
73-1112	OA37B	AMARC	AB020
73-1113	A37B	SoC	
73-1114	A37B	412TW	ED AFFTC
73-1115	A37B	SoC	

B1 LANCER

Serial	Type	Unit/Notes	Location
74-0158	GB1A	GIA	Hanscom AFB
74-0159	B1A	w.o.29.8.84	
74-0160	GB1A	TTC	Lowry
76-0174	B1A	USAF Museum	
82-0001	B1B	GIA	Ellsworth AFB
83-0065	B1B	96W	DY
83-0066	B1B	96W	DY
83-0067	B1B	96W	DY
83-0068	B1B	96W	DY
83-0069	B1B	96W	DY
83-0070	B1B	96W	DY
83-0071	B1B	96W	DY

Serial	Type	Unit/Notes	Location
84-0049	B1B	412TW	ED AFFTC
84-0050	B1B	96W	DY
84-0051	B1B	96W	DY
84-0052	B1B	w.o.28.9.87	
84-0053	B1B	96W	DY
84-0054	B1B	96W	DY
84-0055	B1B	96W	DY
84-0056	B1B	96W	DY
84-0057	B1B	96W	DY
84-0058	B1B	96W	DY
85-0059	B1B	96W	DY
85-0060	B1B	96W	DY

85-0061	B1B	28BW	EL		86-0101	B1B	384BW	OZ
85-0062	B1B	96W	DY		86-0102	B1B	28BW	EL
85-0063	B1B	w.o.8.11.88			86-0103	B1B	96W	DY
85-0064	B1B	28BW	EL		86-0104	B1B	28BW	EL
85-0065	B1B	96W	DY		86-0105	B1B	319BW	GF
85-0066	B1B	96W	DY		86-0106	B1B	w.o.30.11.92	
85-0067	B1B	96W	DY		86-0107	B1B	319BW	GF
85-0068	B1B	412TW	ED AFFTC		86-0108	B1B	319BW	GF
85-0069	B1B	96W	DY		86-0109	B1B	96W	DY
85-0070	B1B	96W	DY		86-0110	B1B	319BW	GF
85-0071	B1B	96W	DY		86-0111	B1B	319BW	GF
85-0072	B1B	96W	DY		86-0112	B1B	319BW	GF
85-0073	B1B	28BW	EL		86-0113	B1B	319BW	GF
85-0074	B1B	96W	DY		86-0114	B1B	319BW	GF
85-0075	B1B	96W	DY		86-0115	B1B	384BW	OZ
85-0076	B1B	w.o.18.11.88			86-0116	B1B	319BW	GF
85-0077	B1B	28BW	EL		86-0117	B1B	319BW	GF
85-0078	B1B	28BW	EL		86-0118	B1B	319BW	GF
85-0079	B1B	28BW	EL		86-0119	B1B	319BW	GF
85-0080	B1B	384BW	OZ		86-0120	B1B	319BW	GF
85-0081	B1B	384BW	OZ		86-0121	B1B	319BW	GF
85-0082	B1B	96W	DY		86-0122	B1B	319BW	GF
85-0083	B1B	28BW	EL		86-0123	B1B	319BW	GF
85-0084	B1B	28BW	EL		86-0124	B1B	384BW	OZ
85-0085	B1B	28BW	EL		86-0125	B1B	384BW	OZ
85-0086	B1B	28BW	EL		86-0126	B1B	384BW	OZ
85-0087	B1B	28BW	EL		86-0127	B1B	384BW	OZ
85-0088	B1B	28BW	EL		86-0128	B1B	384BW	OZ
85-0089	B1B	28BW	EL		86-0129	B1B	384BW	OZ
85-0090	B1B	28BW	EL		86-0130	B1B	384BW	OZ
85-0091	B1B	28BW	EL		86-0131	B1B	384BW	OZ
85-0092	B1B	28BW	EL		86-0132	B1B	96W	DY
86-0093	B1B	28BW	EL		86-0133	B1B	28BW	EL
86-0094	B1B	28BW	EL		86-0134	B1B	384BW	OZ
86-0095	B1B	96W	DY		86-0135	B1B	384BW	OZ
86-0096	B1B	28BW	EL		86-0136	B1B	384BW	OZ
86-0097	B1B	319BW	GF		86-0137	B1B	384BW	OZ
86-0098	B1B	28BW	EL		86-0138	B1B	384BW	OZ
86-0099	B1B	28BW	EL		86-0139	B1B	384BW	OZ
86-0100	B1B	96W	DY		86-0140	B1B	384BW	OZ

B2 'STEALTH BOMBER'

c\n   AV-1 to AV-6.

82-1066	B2A	412TW	ED AFFTC		82-1069	B2A	412TW	ED AFFTC
82-1067	B2A	412TW	ED AFFTC		82-1070	B2A	412TW	ED AFFTC
82-1068	B2A	412TW	ED AFFTC		82-1071	B2A	412TW	ED AFFTC

A total of 132 aircraft have been requested

B52 STRATOFORTRESS

c\n   16248, 16249, 16491 to 16503, 16838 to 16887, 17159 to 17183, 464001 to 464019 (Wichita built), 17184 to 17233, 464020 to 464027, 17263 to 17339, 464028 to 464083, 17408 to 17488, 464084 to 464467.

Serial	Type	Status	Location
49-0230	YB52		
49-0231	YB52		
52-0001	YB52A		
52-0002	YB52A		
52-0003	NB52A	Pima Museum	
52-0004	RB52B	scrapped	
52-0005	RB52B	preserved	Lowry AFB
52-0006	RB52B	scrapped	
52-0007	RB52B	scrapped	
52-0008	NB52B	NASA	Edwards
52-0009	RB52B	scrapped	
52-0010	RB52B	scrapped	
52-0011	RB52B	scrapped	
52-0012	RB52B	scrapped	
52-0013	RB52B	preserved	Kirtland AFB
52-8710	RB52B	scrapped	
52-8711	RB52B	preserved	Offutt AFB
52-8712	RB52B	scrapped	
52-8713	RB52B	scrapped	
52-8714	RB52B	TTC	Chanute
52-8715	RB52B	scrapped	
52-8716	RB52B	scrapped	
53-0366	RB52B	scrapped	
53-0367	RB52B	scrapped	
53-0368	RB52B	scrapped	
53-0369	RB52B	scrapped	
53-0370	RB52B	scrapped	
53-0371	RB52B	scrapped	
53-0372	RB52B	scrapped	
53-0373	B52B	dump	Edwards AFB
53-0374	B52B	scrapped	
53-0375	B52B	scrapped	
53-0376	B52B	scrapped	
53-0377	RB52B	scrapped	
53-0378	RB52B	dump	Edwards AFB
53-0379	RB52B		
53-0380	B52B	scrapped	
53-0381	B52B	scrapped	
53-0382	B52B	scrapped	
53-0383	B52B	scrapped	
53-0384	B52B	scrapped	
53-0385	B52B	scrapped	
53-0386	B52B	scrapped	
53-0387	B52B	scrapped	
53-0388	B52B	scrapped	
53-0389	B52B	scrapped	
53-0390	B52B	scrapped	
53-0391	B52B	scrapped	
53-0392	B52B	scrapped	
53-0393	B52B	scrapped	
53-0394	B52B	Air Force Museum	
53-0395	B52B	scrapped	
53-0396	B52B	scrapped	
53-0397	B52B	scrapped	
53-0398	B52B	scrapped	
53-0399	B52C	scrapped	
53-0400	B52C	AMARC	BC205
53-0401	B52C	AMARC	BC156
53-0402	B52C	AMARC	BC206
53-0403	B52C	AMARC	BC109
53-0404	B52C	AMARC	BC170
53-0405	B52C	AMARC	BC180
53-0406	B52C		
53-0407	B52C	AMARC	BC176
53-0408	B52C	AMARC	BC178
54-2664	B52C	MASDC	
54-2665	B52C	AMARC	BC159
54-2666	B52C		
54-2667	B52C		
54-2668	B52C	AMARC	BC191
54-2669	B52C	AMARC	BC158
54-2670	B52C	MASDC	
54-2671	B52C	AMARC	BC194
54-2672	B52C	AMARC	BC185
54-2673	B52C	AMARC	BC204
54-2674	B52C	AMARC	BC160
54-2675	B52C	AMARC	BC168
54-2676	B52		
54-2677	B52C	AMARC	BC174
54-2678	B52C	AMARC	BC184
54-2679	B52C	AMARC	BC163
54-2680	B52C	MASDC	
54-2681	B52C	AMARC	BC200
54-2682	B52C		
54-2683	B52C	AMARC	BC196
54-2684	B52C	AMARC	BC198
54-2685	B52C	AMARC	BC172
54-2686	B52C	AMARC	BC192
54-2687	B52C	AMARC	BC166
54-2688	B52C	AMARC	BC182
55-0049	B52D	AMARC	BC245
55-0050	B52D	w.o.21.12.72	
55-0051	B52D	AMARC	BC263
55-0052	B52D	MASDC	
55-0053	B52D	AMARC	BC219
55-0054	B52D	AMARC	BC268
55-0055	B52D	MASDC	
55-0056	B52D	w.o.3.1.73	
55-0057	B52D	preserved	Maxwell AFB
55-0058	B52D	w.o.11.12.74	
55-0059	B52D	AMARC	BC278
55-0060	B52D		
55-0061	B52D	w.o.21.12.72	
55-0062	B52D	preserved	K.I. Sawyer AFB
55-0063	B52D	preserved	Carswell AFB
55-0064	B52D	AMARC	BC232
55-0065	B52D		
55-0066	B52D	AMARC	BC300
55-0067	B52D	Pima Museum	
55-0068	B52D	preserved	Lackland AFB
55-0069	B52D	AMARC	BC291
55-0070	B52D	AMARC	BC312
55-0071	B52D	preserved	Mobile, Al
55-0072	B52D	AMARC	BC220
55-0073	B52D	AMARC	BC298
55-0074	B52D	AMARC	BC277
55-0075	B52D	AMARC	BC296
55-0076	B52D	AMARC	BC214
55-0077	B52D	AMARC	BC317
55-0078	B52D	w.o.30.10.81	
55-0079	B52D	AMARC	BC294
55-0080	B52D	AMARC	BC303
55-0081	B52D	AMARC	BC216
55-0082	B52D	AMARC	BC...
55-0083	B52D	preserved	AF Academy
55-0084	B52D	AMARC	BC327
55-0085	B52D	preserved	Robins AFB
55-0086	B52D	AMARC	BC288
55-0087	B52D	AMARC	BC305
55-0088	B52D	AMARC	BC297
55-0089	B52D		
55-0090	B52D	AMARC	BC301
55-0091	B52D	AMARC	BC292
55-0092	B62D	AMARC	BC316
55-0093	B52D		
55-0094	B52D	preserved	McConnell AFB
55-0095	GB52D	TTC	Chanute
55-0096	B52D	AMARC	BC259
55-0097	B52D	w.o.	
55-0098	B52D	MASDC	
55-0099	GB52D	GIA	Andersen AFB
55-0100	B52D	preserved	Andersen AFB
55-0101	B52D	AMARC	BC290
55-0102	B52D		
55-0103	B52D		
55-0104	B52D	AMARC	BC284
55-0105	B52D	AMARC	BC329
55-0106	B52D	AMARC	BC269
55-0107	B52D	AMARC	BC295
55-0108	B52D		
55-0109	B52D	AMARC	BC251
55-0110	B52D	w.o.22.11.72	
55-0111	B52D	AMARC	BC289
55-0112	B52D		
55-0113	B52D	AMARC	BC314
55-0114	B52D		
55-0115	B52D		
55-0116	B52D		
55-0117	B52D	AMARC	BC244
55-0673	B52D	AMARC	BC308
55-0674	B52D	AMARC	BC330
55-0675	B52D	AMARC	BC302
55-0676	B52D	w.o.17.7.69	
55-0677	B52D	preserved	Willow Run AP
55-0678	B52D	MASDC	
55-0679	GB52D	preserved	March AFB

Serial	Type	Status	Location
55-0680	B52D	AMARC	BC234
56-0580	B52D	AMARC	BC299
56-0581	B52D	AMARC	BC218
56-0582	B52D	AMARC	BC235
56-0583	B52D	AMARC	BC223
56-0584	B52D	w.o.26.12.72	
56-0585	B52D	preserved	Edwards AFB
56-0586	B52D	decoy	Guam
56-0587	B52D	AMARC	BC307
56-0588	B52D	AMARC	BC280
56-0589	GB52D	TTC	Sheppard
56-0590	B52D	AMARC	BC261
56-0591	B52D		
56-0592	B52D	AMARC	BC273
56-0593	B52D		
56-0594	B52D	w.o.19.10.78	
56-0595	B52D	w.o.7.7.67	
56-0596	B52D	AMARC	BC285
56-0597	B52D		
56-0598	B52D	AMARC	BC213
56-0599	B52D	w.o.27.12.72	
56-0600	B52D	AMARC	BC310
56-0601	B52D	w.o.8.7.67	
56-0602	B52D	AMARC	BC325
56-0603	B52D	TTC	Lowry
56-0604	B52D	AMARC	BC240
56-0605	B52D		
56-0606	B52D	AMARC	BC281
56-0607	B52D		
56-0608	B52D	w.o.18.12.72	
56-0609	B52D	AMARC	BC267
56-0610	B52D		
56-0611	B52D	AMARC	BC221
56-0612	B52D	preserved	Castle AFB
56-0613	B52D	AMARC	BC228
56-0614	B52D	AMARC	BC324
56-0615	B52D	AMARC	BC246
56-0616	B52D	MASDC	
56-0617	B52D	AMARC	BC326
56-0618	B52D	MASDC	
56-0619	B52D	AMARC	BC222
56-0620	B52D	AMARC	BC207
56-0621	B52D	AMARC	BC286
56-0622	B52D	w.o.20.12.72	
56-0623	B52D	AMARC	BC260
56-0624	B52D	AMARC	BC226
56-0625	B52D	w.o.	
56-0626	B52D	AMARC	BC253
56-0627	B52D	w.o.7.7.67	
56-0628	B52D	destroyed 25.5.84	
56-0629	B52D	preserved	Barksdale AFB
56-0630	B52D	MASDC	
56-0631	B52E	AMARC	BC085
56-0632	B52E	AMARC	BC211
56-0633	B52E		
56-0634	B52E	AMARC	BC116
56-0635	B52E	MASDC	
56-0636	B52E	AMARC	BC275
56-0637	B52E		
56-0638	B52E	MASDC	
56-0639	B52E	AMARC	BC149
56-0640	B52E	AMARC	BC140
56-0641	B52E	MASDC	
56-0642	B52E	MASDC	
56-0643	B52E	MASDC	
56-0644	B52E	AMARC	BC131
56-0645	B52E	AMARC	BC115
56-0646	B52E	AMARC	BC113
56-0647	B52E	MASDC	
56-0648	B52E	MASDC	
56-0649	B52E	scrapped	
56-0650	B52E	AMARC	BC129
56-0651	B52E	AMARC	BC127
56-0652	B52E	MASDC	
56-0653	B52E	AMARC	BC118
56-0654	B52E	MASDC	
56-0655	B52E		
56-0656	B52E	AMARC	BC144
56-0657	B52D	preserved	Ellsworth AFB
56-0658	B52D	AMARC	BC276
56-0659	B52D	preserved	DM-AFB
56-0660	B52D	AMARC	BC320
56-0661	B52D		
56-0662	B52D	destroyed 24.5.84	
56-0663	B52D	AMARC	BC282
56-0664	GB52D	GIA	Andersen AFB
56-0665	B52D	USAF Museum	
56-0666	B52D	AMARC	BC321
56-0667	B52D	AMARC	BC323
56-0668	B52D	AMARC	BC306
56-0669	B52D	w.o.	
56-0670	B52D	AMARC	BC279
56-0671	B52D	AMARC	BC287
56-0672	B52D	AMARC	BC313
56-0673	B52D	AMARC	BC233
56-0674	B52D	w.o.26.12.72	
56-0675	B52D	AMARC	BC247
56-0676	B52D	preserved	Fairchild AFB
56-0677	B52D	w.o.30.7.72	
56-0678	B52D	AMARC	BC252
56-0679	B52D	AMARC	BC293
56-0680	B52D	destroyed	.5.84
56-0681	B52D		
56-0682	B52D	AMARC	BC256
56-0683	B52D	preserved	Whiteman AFB
56-0684	B52D	AMARC	BC315
56-0685	B52D	preserved	Dyess AFB
56-0686	B52D	AMARC	BC318
56-0687	B52D	preserved	Orlando AP, Fl
56-0688	B52D	destroyed 5.84	
56-0689	B52D	preserved	Duxford, UK
56-0690	B52D	AMARC	BC304
56-0691	B52D	AMARC	BC249
56-0692	B52D	preserved	Kelly AFB
56-0693	B52D	AMARC	BC230
56-0694	B52D	AMARC	BC319
56-0695	B52D	preserved	Tinker AFB
56-0696	B52D	preserved	Travis AFB
56-0697	B52D	AMARC	BC328
56-0698	B52D	AMARC	BC311
56-0699	B52E	MASDC	
56-0700	B52E	MASDC	
56-0701	B52E	MASDC	
56-0702	B52E	MASDC	
56-0703	B52E	MASDC	
56-0704	B52E	AMARC	BC145
56-0705	B52E	AMARC	BC114
56-0706	B52E	AMARC	BC123
56-0707	B52E	AMARC	BC152
56-0708	B52E	TTC	Chanute
56-0709	B52E	scrapped	
56-0710	B52E	MASDC	
56-0711	B52E	MASDC	
56-0712	B52E	MASDC	
57-0014	B52E		
57-0015	B52E	AMARC	BC109
57-0016	B52E	MASDC	
57-0017	B52E	AMARC	BC148
57-0018	B52E		
57-0019	B52E		
57-0020	B52E	AMARC	BC134
57-0021	B52E	MASDC	
57-0022	B52E	MASDC	
57-0023	B52E	MASDC	
57-0024	B52E	AMARC	BC112
57-0025	B52E	MASDC	
57-0026	B52E	MASDC	
57-0027	B52E	AMARC	BC132
57-0028	B52E		
57-0029	B52E		
57-0030	B52F		
57-0031	B52F	AMARC	BC162
57-0032	B52F	AMARC	BC264
57-0033	B52F	AMARC	BC239
57-0034	B52F	AMARC	BC227
57-0035	B52F	AMARC	BC271
57-0036	B52F		
57-0037	B52F	AMARC	BC102
57-0038	B52F	preserved	Oklahoma City
57-0039	B52F	AMARC	BC165
57-0040	B52F		
57-0041	B52F		
57-0042	GB52	TTC	Chanute
57-0043	B52F		
57-0044	B52F		
57-0045	B52F	AMARC	BC246

Serial	Type		
57-0046	B52F	AMARC	BC105
57-0047	B52F	w.o.18.6.65	
57-0048	B52F	derelict	Lowry AFB
57-0049	B52F	AMARC	BC059
57-0050	B52F		
57-0051	B52F	AMARC	BC252
57-0052	B52F	AMARC	BC229
57-0053	B52F	AMARC	BC095
57-0054	B52F	AMARC	BC100
57-0055	B52F	AMARC	BC195
57-0056	B52F	AMARC	BC157
57-0057	B52F	AMARC	BC186
57-0058	B52F	AMARC	BC255
57-0059	B52F	AMARC	BC179
57-0060	B52F	AMARC	BC169
57-0061	B52F	AMARC	BC...
57-0062	B52F	AMARC	BC175
57-0063	B52F	AMARC	BC242
57-0064	B52F	AMARC	BC173
57-0065	B52F	AMARC	BC177
57-0066	B52F	AMARC	BC083
57-0067	B52F	AMARC	BC099
57-0068	B52F		
57-0069	B52F	AMARC	BC243
57-0070	B52F		
57-0071	GB52F	TTC	Sheppard
57-0072	B52F	AMARC	BC265
57-0073	B52F		
57-0095	B52F	AMARC	BC110
57-0096	B52F	AMARC	BC135
57-0097	B52F	AMARC	BC133
57-0098	B52F	AMARC	BC130
57-0099	B52F	AMARC	BC136
57-0100	B52F	AMARC	BC121
57-0101	B52F	scrapped	
57-0102	B52F	MASDC	
57-0103	B52F	AMARC	BC119
57-0104	B52F	MASDC	
57-0105	B52F	MASDC	
57-0106	B52F	MASDC	
57-0107	B52F	MASDC	
57-0108	B52F	AMARC	BC143
57-0109	B52F	MASDC	
57-0110	B52F	scrapped	
57-0111	B52F		
57-0112	B52F	AMARC	BC142
57-0113	B52F		
57-0114	B52F		
57-0115	B52F	AMARC	BC125
57-0116	B52F	MASDC	
57-0117	B52F		
57-0118	B52F	AMARC	BC122
57-0119	NB52E	AFSC\AFFTC	
57-0120	B52F	AMARC	BC107
57-0121	B52F	AMARC	BC120
57-0122	B52F	MASDC	
57-0123	B52F	AMARC	BC146
57-0124	B52F	MASDC	
57-0125	B52F	MASDC	
57-0126	B52F	AMARC	BC141
57-0127	B52F	MASDC	
57-0128	B52F	AMARC	BC124
57-0129	B52F	AMARC	BC147
57-0130	B52F	MASDC	
57-0131	B52F	AMARC	BC150
57-0132	B52F	AMARC	BC117
57-0133	B52F	MASDC	
57-0134	B52F		
57-0135	B52F	MASDC	
57-0136	B52F	AMARC	BC128
57-0137	B52F		
57-0138	B52F	MASDC	
57-0139	B52F	MASDC	
57-0140	B52F	MASDC	
57-0141	B52F		
57-0142	B52F	AMARC	BC248
57-0143	B52F	AMARC	BC164
57-0144	B52F		
57-0145	B52F	AMARC	BC250
57-0146	B52F		
57-0147	B52F	AMARC	BC224
57-0148	B52F	MASDC	
57-0149	B52F		
57-0150	B52F	AMARC	BC217
57-0151	B52F	AMARC	BC193
57-0152	B52F	AMARC	BC183
57-0153	B52F	MASDC	
57-0154	B52F	AMARC	BC258
57-0155	B52F	AMARC	BC103
57-0156	B52F		
57-0157	B52F		
57-0158	B52F		
57-0159	B52F	MASDC	
57-0160	B52F	AMARC	BC097
57-0161	B52F	AMARC	BC203
57-0162	B52F	AMARC	BC197
57-0163	B52F	AMARC	BC098
57-0164	B52F		
57-0165	B52F	MASDC	
57-0166	B52F		
57-0167	B52F		
57-0168	B52F	AMARC	BC209
57-0169	B52F	MASDC	
57-0170	B52F	AMARC	BC262
57-0171	B52F		
57-0172	B52F		
57-0173	B52F		
57-0174	B52F	AMARC	BC158
57-0175	B52F	AMARC	BC107
57-0176	B52F	AMARC	BC101
57-0177	B52F	MASDC	
57-0178	B52F	AMARC	BC171
57-0179	B52F	w.o.18.6.65	
57-0180	B52F	AMARC	BC096
57-0181	B52F		
57-0182	B52F	MASDC	
57-0183	B52F	AMARC	BC208
57-6468	B52G	preserved	Offutt AFB
57-6469	GB52G	TTC	Sheppard
57-6470	B52G	AMARC	BC...
57-6471	B52G	AMARC	BC...
57-6472	B52G	AMARC	BC...
57-6473	B52G	AMARC	BC439
57-6474	B52G	AMARC	BC...
57-6475	B52G	AMARC	BC378
57-6476	B52G		
57-6477	B52G	AMARC	BC...
57-6478	B52G	AMARC	BC334
57-6479	B52G	w.o.16.10.84	
57-6480	B52G	AMARC	BC...
57-6481	B52G	w.o.20.12.72	
57-6482	B52G	w.o.16.12.82	
57-6483	B52G	AMARC	BC...
57-6484	B52G	AMARC	BC333
57-6485	B52G	AMARC	BC...
57-6486	B52G	AMARC	BC377
57-6487	B52G	AMARC	BC...
57-6488	B52G	93BW	CA
57-6489	B52G	AMARC	BC343
57-6490	B52G	AMARC	BC433
57-6491	B52G	AMARC	BC...
57-6492	B52G	AMARC	BC437
57-6493	B52G	w.o.3.9.75	
57-6494	B52G	w.o.	
57-6495	B52G	AMARC	BC...
57-6496	B52G	w.o.	
57-6497	B52G	93BW	CA
57-6498	B52G	416BW	GR
57-6499	B52G	AMARC	BC...
57-6500	B52G	AMARC	BC331
57-6501	B52G	AMARC	BC...
57-6502	B52G	AMARC	BC340
57-6503	B52G		
57-6504	B52G	AMARC	BC...
57-6505	B52G	AMARC	BC...
57-6506	B52G	AMARC	BC342
57-6507	B52G	w.o.26.1.83	
57-6508	B52G	AMARC	BC404
57-6509	B52G		
57-6510	B52G	AMARC	BC...
57-6511	B52G	AMARC	BC...
57-6512	B52G	AMARC	BC...
57-6513	B52G	AMARC	BC335
57-6514	B52G	AMARC	BC...
57-6515	B52G	416BW	GR
57-6516	B52G	AMARC	BC387

57-6517	B52G	AMARC	BC347		58-0237	B52G	AMARC	BC382
57-6518	B52G	AMARC	BC...		58-0238	B52G	AMARC	BC379
57-6519	B52G	AMARC	BC...		58-0239	B52G	416BW	GR
57-6520	B52G	366W	MO		58-0240	B52G	93BW	CA
58-0158	B52G	AMARC	BC...		58-0241	B52G	AMARC	BC...
58-0159	B52G	AMARC	BC...		58-0242	B52G	366W	MO
58-0160	B52G	AMARC	BC...		58-0243	B52G	AMARC	BC371
58-0161	B52G	w.o.11.4.84			58-0244	B52G	AMARC	BC...
58-0162	B52G	AMARC	BC...		58-0245	B52G	AMARC	BC...
58-0163	B52G	93BW	CA		58-0246	B52G	w.o.18.12.72	
58-0164	B52G	416BW	GR		58-0247	B52G	AMARC	BC...
58-0165	B52G	416BW	GR		58-0248	B52G	93BW	CA
58-0166	B52G	AMARC	BC438		58-0249	B52G	AMARC	BC382
58-0167	B52G	AMARC	BC...		58-0250	B52G		
58-0168	B52G	AMARC	BC...		58-0251	B52G	AMARC	BC341
58-0169	B52G	w.o.20.12.72			58-0252	B52G	AMARC	BC...
58-0170	B52G	416BW	GR		58-0253	B52G	42BW	LZ
58-0171	B52G	AMARC	BC...		58-0254	B52G	AMARC	BC...
58-0172	B52G	AMARC	BC332		58-0255	B52G	42BW	LZ
58-0173	B52G	AMARC	BC...		58-0256	B52G	w.o.17.1.66	
58-0174	B52G	w.o.8.2.74			58-0257	B52G	42BW	LZ
58-0175	B52G	AMARC	BC...		58-0258	B52G	93BW	CA
58-0176	B52G	AMARC	BC...		59-2564	B52G	AMARC	BC376
58-0177	B52G	AMARC	BC381		59-2565	B52G	93BW	CA
58-0178	B52G	AMARC	BC...		59-2566	B52G	416BW	GR
58-0179	B52G	AMARC	BC436		59-2567	B52G	416BW	GR
58-0180	B52G	w.o.			59-2568	B52G	416BW	GR
58-0181	B52G	AMARC	BC...		59-2569	B52G		
58-0182	B52G	AMARC	BC...		59-2570	B52G	366W	MO
58-0183	B52G	Pima Museum			59-2571	B52G	AMARC	BC...
58-0184	B52G	AMARC	BC385		59-2572	B52G	93BW	CA
58-0185	B52G	preserved	Eglin AFB		59-2573	B52G	42BW	LZ
58-0186	B52G	AMARC	BC...		59-2574	B52G	w.o.	
58-0187	B52G	w.o.			59-2575	B52G	AMARC	BC375
58-0188	B52G	w.o.			59-2576	B52G	w.o.	
58-0189	B52G	AMARC	BC337		59-2577	B52G		
58-0190	B52G	w.o.			59-2578	GB52G	TTC	Sheppard
58-0191	B52G	93BW	CA		59-2579	B52G	AMARC	BC395
58-0192	B52G	42BW	LZ		59-2580	B52G	AMARC	BC...
58-0193	B52G	AMARC	BC435		59-2581	B52G	AMARC	BC...
58-0194	B52G	AMARC	BC...		59-2582	B52G	AMARC	BC382
58-0195	B52G	42BW	LZ		59-2583	B52G	416BW	GR
58-0196	B52G	w.o.			59-2584	B52G	AMARC	BC...
58-0197	B52G	42BW	LZ		59-2585	B52G	42BW	LZ
58-0198	B52G	w.o.			59-2586	B52G	412TW	ED AFFTC
58-0199	B52G	AMARC	BC...		59-2587	B52G	AMARC	BC336
58-0200	GB52G	TTC	Sheppard		59-2588	B52G	93BW	CA
58-0201	B52G	w.o.18.12.72			59-2589	B52G	AMARC	BC...
58-0202	B52G				59-2590	B52G	AMARC	BC...
58-0203	B52G	366W	MO		59-2591	B52G	AMARC	BC...
58-0204	B52G	AMARC	BC384		59-2592	B52G	AMARC	BC338
58-0205	B52G	AMARC	BC...		59-2593	B52G	w.o.3.2.91	
58-0206	B52G	42BW	LZ		59-2594	B52G	AMARC	BC...
58-0207	B52G	AMARC	BC...		59-2595	B52G	93BW	CA
58-0208	B52G				59-2596	B52G	42BW	LZ
58-0209	B52G	w.o.20.8.80			59-2597	B52G	w.o.29.11.82	
58-0210	B52G	93BW	CA		59-2598	B52G	366W	MO
58-0211	B52G	AMARC	BC...		59-2599	B52G	93BW	CA
58-0212	B52G	366W	MO		59-2600	B52G	w.o.	
58-0213	B52G	93BW	CA		59-2601	B52G	416BW	GR
58-0214	B52G	AMARC	BC...		59-2602	B52G		
58-0215	B52G	w.o.			60-0001	B52H	2W	LA
58-0216	B52G	42BW	LZ		60-0002	B52H	410BW	KI
58-0217	B52G	AMARC	BC...		60-0003	B52H	410BW	KI
58-0218	B52G	42BW	LZ		60-0004	B52H	rework	Kelly AFB
58-0219	B52G	w.o.			60-0005	B52H	416BW	GR
58-0220	B52G		BC346		60-0006	B52H	w.o.30.5.74	
58-0221	B52G				60-0007	B52H	2W	LA
58-0222	B52G	AMARC	BC...		60-0008	B52H	410BW	KI
58-0223	B52G	AMARC	BC...		60-0009	B52H	2W	LA
58-0224	B52G	AMARC	BC339		60-0010	B52H	2W	LA
58-0225	B52G	preserved	Griffiss AFB		60-0011	B52H		
58-0226	B52G	42BW	LZ		60-0012	B52H	410BW	KI
58-0227	B52G	AMARC	BC...		60-0013	B52H	92BW	FC
58-0228	B52G	w.o.			60-0014	B52H	2W	LA
58-0229	B52G	AMARC	BC...		60-0015	B52H	92BW	FC
58-0230	B52G	42BW	LZ		60-0016	B52H	2W	LA
58-0231	B52G	416BW	GR		60-0017	B52H	2W	LA
58-0232	B52G	AMARC	BC344		60-0018	B52H	2W	LA
58-0233	B52G	93BW	CA		60-0019	B52H	2W	LA
58-0234	B52G	AMARC	BC...		60-0020	B52H	2W	LA
58-0235	B52G	412TW	ED AFFTC		60-0021	B52H	416BW	GR
58-0236	B52G	AMARC	BC...		60-0022	B52H	92BW	FC

60-0023	B52H	92BW	FC		61-0002	B52H	2W	LA
60-0024	B52H	416BW	GR		61-0003	B52H	92BW	FC
60-0025	B52H	416BW	GR		61-0004	B52H		
60-0026	B52H	410BW	KI		61-0005	B52H	92BW	FC
60-0027	B52H	w.o.			61-0006	B52H	92BW	FC
60-0028	B52H				61-0007	B52H		
60-0029	B52H				61-0008	B52H	92BW	FC
60-0030	B52H	416BW	GR		61-0009	B52H	92BW	FC
60-0031	B52H	2W	LA		61-0010	B52H		
60-0032	B52H	2W	LA		61-0011	B52H		
60-0033	B52H	416BW	GR		61-0012	B52H	410BW	KI
60-0034	B52H	5BW	MT		61-0013	B52H	2W	LA
60-0035	B52H				61-0014	B52H	416BW	GR
60-0036	B52H	410BW	KI		61-0015	B52H	416BW	GR
60-0037	B52H	2W	LA		61-0016	B52H		
60-0038	B52H	410BW	KI		61-0017	B52H	92BW	FC
60-0039	B52H	w.o.1.4.77			61-0018	B52H	92BW	FC
60-0040	B52H	w.o.5.12.88			61-0019	B52H		
60-0041	B52H	2W	LA		61-0020	B52H	416BW	GR
60-0042	B52H	2W	LA		61-0021	B52H		
60-0043	B52H	2W	LA		61-0022	B52H	92BW	FC
60-0044	B52H	92BW	FC		61-0023	B52H	5BW	MT
60-0045	B52H	92BW	FC		61-0024	B52H		
60-0046	B52H	92BW	FC		61-0025	B52H	rework	Kelly AFB
60-0047	B52H	rework	Kelly AFB		61-0026	B52H	92BW	FC
60-0048	B52H	416BW	GR		61-0027	B52H	2W	LA
60-0049	B52H	2W	LA		61-0028	B52H	2W	LA
60-0050	B52H	412TW	ED AFFTC		61-0029	B52H		
60-0051	B52H				61-0030	B52H	w.o.	
60-0052	B52H				61-0031	B52H	5BW	MT
60-0053	B52H				61-0032	B52H	2W	LA
60-0054	B52H	410BW	KI		61-0033	B52H	w.o.14.11.75	
60-0055	B52H				61-0034	B52H	410BW	KI
60-0056	B52H	410BW	KI		61-0035	B52H	92BW	FC
60-0057	B52H	5BW	MT		61-0036	B52H		
60-0058	B52H	92BW	FC		61-0037	B52H	w.o.	
60-0059	B52H				61-0038	B52H	2W	LA
60-0060	B52H	416BW	GR		61-0039	B52H	92BW	FC
60-0061	B52H				61-0040	B52H	92BW	FC
60-0062	B52H	2W	LA					
61-0001	B52H	92BW	FC					

C5 GALAXY

c\n 0001 to 0131.

66-8303	C5A	w.o.17.10.70			68-0225	C5A	439AW	AFRes
66-8304	C5A	439AW	AFRes		68-0226	C5A	137ALS	NY ANG
66-8305	C5A	433AW	AFRes		68-0227	C5A	w.o.27.9.74	
66-8306	C5A	433AW	AFRes		68-0228	C5A	w.o.29.8.90	
66-8307	C5A	433AW	AFRes		69-0001	C5A	60AW	
67-0167	C5A	439AW	AFRes		69-0002	C5A	433AW	AFRes
67-0168	C5A	433AW	AFRes		69-0003	C5A	439AW	AFRes
67-0169	C5A	137ALS	NY ANG		69-0004	C5A	433AW	AFRes
67-0170	C5A	137ALS	NY ANG		69-0005	C5A	439AW	AFRes
67-0171	C5A	433AW	AFRes		69-0006	C5A	433AW	AFRes
67-0172	C5A	scrapped	4.87		69-0007	C5A	433AW	AFRes
67-0173	C5A	137ALS	NY ANG		69-0008	C5A	137ALS	NY ANG
67-0174	C5A	137ALS	NY ANG		69-0009	C5A	137ALS	NY ANG
68-0211	C5A	439AW	AFRes		69-0010	C5A	60AW	
68-0212	C5A	137ALS	NY ANG		69-0011	C5A	439AW	AFRes
68-0213	C5A	433AW	AFRes		69-0012	C5A	137ALS	NY ANG
68-0214	C5A	436AW			69-0013	C5A	439AW	AFRes
68-0215	C5A	439AW	AFRes		69-0014	C5A	60AW	
68-0216	C5A	433AW	AFRes		69-0015	C5A	137ALS	NY ANG
68-0217	C5A	436AW			69-0016	C5A	433AW	AFRes
68-0218	C5A	w.o.4.4.75			69-0017	C5A	439AW	AFRes
68-0219	C5A	439AW	AFRes		69-0018	C5A	60AW	
68-0220	C5A	433AW	AFRes		69-0019	C5A	439AW	AFRes
68-0221	C5A	433AW	AFRes		69-0020	C5A	439AW	AFRes
68-0222	C5A	439AW	AFRes		69-0021	C5A	137ALS	NY ANG
68-0223	C5A	433AW	AFRes		69-0022	C5A	439AW	AFRes
68-0224	C5A	137ALS	NY ANG		69-0023	C5A	60AW	

69-0024	C5A	60AW			85-0008	C5B	60AW	
69-0025	C5A	60AW			85-0009	C5B	436AW	
69-0026	C5A	60AW			85-0010	C5B	60AW	
69-0027	C5A	436AW			86-0011	C5B	436AW	
70-0445	C5A	433AW	AFRes		86-0012	C5B	60AW	
70-0446	C5A	433AW	AFRes		86-0013	C5B	436AW	
70-0447	C5A	439AW	AFRes		86-0014	C5B	97AMW	
70-0448	C5A	439AW	AFRes		86-0015	C5B	97AMW	
70-0449	C5A	60AW			86-0016	C5B	97AMW	
70-0450	C5A	433AW	AFRes		86-0017	C5B	436AW	
70-0451	C5A	60AW			86-0018	C5B	60AW	
70-0452	C5A	436AW			86-0019	C5B	97AMW	
70-0453	C5A	436AW			86-0020	C5B	97AMW	
70-0454	C5A	436AW			86-0021	C5B	60AW	
70-0455	C5A	436AW			86-0022	C5B	97AMW	
70-0456	C5A	436AW			86-0023	C5B	436AW	
70-0457	C5A	60AW			86-0024	C5B	97AMW	
70-0458	C5A	436AW			86-0025	C5B	436AW	
70-0459	C5A	60AW			86-0026	C5B	60AW	
70-0460	C5A	436AW			87-0027	C5B	97AMW	
70-0461	C5A				87-0028	C5B	60AW	
70-0462	C5A	60AW			87-0029	C5B	436AW	
70-0463	C5A	436AW			87-0030	C5B	60AW	
70-0464	C5A	436AW			87-0031	C5B	436AW	
70-0465	C5A	436AW			87-0032	C5B	60AW	
70-0466	C5A	436AW			87-0033	C5B	436AW	
70-0467	C5A	436AW			87-0034	C5B	60AW	
83-1285	C5B	436AW			87-0035	C5B	436AW	
84-0059	C5B	436AW			87-0036	C5B	60AW	
84-0060	C5B	60AW			87-0037	C5B	436AW	
84-0061	C5B	436AW			87-0038	C5B	60AW	
84-0062	C5B	60AW			87-0039	C5B	436AW	
85-0001	C5B	436AW			87-0040	C5B	60AW	
85-0002	C5B	60AW			87-0041	C5B	436AW	
85-0003	C5B	436AW			87-0042	C5B	60AW	
85-0004	C5B	60AW			87-0043	C5B	436AW	
85-0005	C5B	436AW			87-0044	C5B	60AW	
85-0006	C5B	60AW			87-0045	C5B	436AW	
85-0007	C5B	436AW						

C9 NIGHTINGALE

c\n 47241, 47242, 47295 to 47300, 47366, 47367, 47448, 47449, 47467, 47471, 47475, 47495, 47536 to 47538, 47540, 47541, 47668, 47770, 47771.

67-22583	C9A	374AW	YJ		71-0874	C9A	375AW	
67-22584	C9A	375AW			71-0875	C9A	375AW	
67-22585	C9A	375AW			71-0876	C9A	435AW	SACEUR
67-22586	C9A	w.o.16.9.71			71-0877	C9A	375AW	
68-8932	C9A	374AW	YJ		71-0878	C9A	435AW	
68-8933	C9A	374AW	YJ		71-0879	C9A	435AW	
68-8934	C9A	375AW			71-0880	C9A	435AW	
68-8935	C9A	375AW			71-0881	C9A	435AW	
68-10958	C9A	375AW			71-0882	C9A	435AW	
68-10959	C9A	375AW			73-1681	C9C	89AW	
68-10960	C9A	375AW			73-1682	C9C	89AW	
68-10961	C9A	375AW			73-1683	C9C	89AW	

C10 EXTENDER

c\n 48200 to 48251, 48303 to 48310.

79-0433	KC10A	4580G		79-1710	KC10A	4580G
79-0434	KC10A	4580G		79-1711	KC10A	4580G

79-1712	KC10A	4580G	
79-1713	KC10A	4580G	
79-1946	KC10A	22ARW	
79-1947	KC10A	22ARW	
79-1948	KC10A	22ARW	
79-1949	KC10A	22ARW	
79-1950	KC10A	22ARW	
79-1951	KC10A	22ARW	
82-0190	KC10A	w.o.17.9.87	
82-0191	KC10A	22ARW	
82-0192	KC10A	4W	SJ
82-0193	KC10A	22ARW	
83-0075	KC10A	4580G	
83-0076	KC10A	22ARW	
83-0077	KC10A	4W	SJ
83-0078	KC10A	22ARW	
83-0079	KC10A	4580G	
83-0080	KC10A	22ARW	
83-0081	KC10A	4580G	
83-0082	KC10A	4580G	
84-0185	KC10A	22ARW	
84-0186	KC10A	4580G	
84-0187	KC10A	22ARW	
84-0188	KC10A	4580G	
84-0189	KC10A	22ARW	
84-0190	KC10A	4580G	
84-0191	KC10A	22ARW	
84-0192	KC10A	4580G	
85-0027	KC10A	22ARW	
85-0028	KC10A	4580G	
85-0029	KC10A	4W	SJ
85-0030	KC10A	4W	SJ
85-0031	KC10A	4W	SJ
85-0032	KC10A	4580G	
85-0033	KC10A	4580G	
85-0034	KC10A	4580G	
86-0027	KC10A	4580G	
86-0028	KC10A	4W	SJ
86-0029	KC10A	4W	SJ
86-0030	KC10A	4W	SJ
86-0031	KC10A	4W	SJ
86-0032	KC10A	4W	SJ
86-0033	KC10A	4W	SJ
86-0034	KC10A	4W	SJ
86-0035	KC10A	4W	SJ
86-0036	KC10A	4W	SJ
86-0037	KC10A	4W	SJ
86-0038	KC10A	4W	SJ
87-0117	KC10A	22ARW	
87-0118	KC10A	22ARW	
87-0119	KC10A	22ARW	
87-0120	KC10A	22ARW	
87-0121	KC10A	4W	SJ
87-0122	KC10A	4W	SJ
87-0123	KC10A	4W	SJ
87-0124	KC10A	4W	SJ

C12 KING AIR

c\n   BD-1 to BD-30, BD-24 (reserialled), BP-40 to BP-45, BL-73 to BL-112, UC-1 to UC-6.

73-1205	C12C	MAAG	Buenos Aires	76-0158	C12C	MAAG	Djakarta
73-1206	C12C	USMTM	Dhahran	76-0159	C12C	MAAG	Brasilia
73-1207	C12C	USMTM	Dhahran	76-0160	C12C	MAAG	Riyadh
73-1208	C12C	MAAG	Khartoum	76-0161	C12C	MAAG	Bogota
73-1209	C12C	USArmy	SCANG	76-0162	C12C	MAAG	Kinshasa
73-1210	C12C	MAAG	Tegucigalpa	76-0163	C12C	MAAG	Canberra
73-1211	C12A	w.o.31.1.79		76-0164	C12C	USMTM	Dhahran
73-1212	C12C	89AW		76-0165	C12C	MAAG	Tegucigalpa
73-1213	C12C	89AW		76-0166	C12C	MAAG	Islamabad
73-1214	C12C	MAAG	Bangkok	76-0167	C12C	to76-3239	
73-1215	C12C	MAAG	Bangkok	76-0168	C12C	MAAG	Banjul
73-1216	C12C	MAAG	Ankara	76-0169	C12C	USMTM	Dhahran
73-1217	C12C	MAAG	Manila	76-0170	C12C	USMTM	Dhahran
73-1218	C12C	MAAG	Souda	76-0171	C12C	MAAG	Rabat

76-0172	C12C	MAAG	Manila		84-0165	C12F	58ALS	
76-0173	C12C	MAAG	Ankara		84-0166	C12F	58ALS	
76-3239	C12C	89AW			84-0167	C12F	374AW	Det3
83-0494	C12D	MAAG	Monrovia		84-0168	C12F	374AW	Det3
83-0495	C12D	MAAG	Islamabad		84-0169	C12F	18W	ZZ
83-0496	C12D	MAAG	Quito		84-0170	C12F	18W	ZZ
83-0497	C12D	MAAG	Buenos Aires		84-0171	C12F	18W	ZZ
83-0498	C12D	MAAG	Mogadishu		84-0172	C12F	18W	ZZ
83-0499	C12D	MAAG	Mexico City		84-0173	C12F	18W	ZZ
84-0143	C12F	375AW			84-0174	C12F	3W	AK
84-0144	C12F	459ALS	Det3 N5801D		84-0175	C12F	459ALS	Det3 N5803F
84-0145	C12F	375AW			84-0176	C12F	459ALS	Det3 N5819T
84-0146	C12F	375AW			84-0177	C12F	459ALS	
84-0147	C12F	3W	AK		84-0178	C12F	459ALS	
84-0148	C12F	3W	AK		84-0179	C12F	459ALS	
84-0149	C12F	3W	AK		84-0180	C12F	459ALS	
84-0150	C12F	457ALS			84-0181	C12F	457ALS	
84-0151	C12F	457ALS			84-0182	C12F	184FS	FS AR ANG
84-0152	C12F	3W	AK		84-0484	C12F	175FS	SD SD ANG
84-0153	C12F	459ALS	Det3 N58280		84-0485	C12F	165ALS	KY ANG
84-0154	C12F	458ALS	Det2		84-0486	C12F	173RS	NE ANG
84-0155	C12F	458ALS	Det2		84-0487	C12F	118FS	CT CT ANG
84-0156	C12F	458ALS	Det2		84-0488	C12F		
84-0157	C12F	458ALS	Det2		84-0489	C12F	110FS	SL MO ANG
84-0158	C12F	457ALS	ex N58022		86-0078	C12J	184FG	KS ANG
84-0159	C12F	18W	ZZ		86-0079	C12J	124FS	IA IA ANG
84-0160	C12F	18W	ZZ		86-0080	C12J	192RS	NV ANG
84-0161	C12F	58ALS			86-0081	C12J	139ALS	NY ANG
84-0162	C12F	58ALS			86-0082	C12J	116ARS	WA ANG
84-0163	C12F	58ALS			86-0083	C12J	101FS	MA ANG
84-0164	C12F	58ALS						

C17 GLOBEMASTER III

c\n   T-1, P-1 to 10.

87-0025	YC17A	412TW	ED AFFTC		89-1192	C17A	437AW
88-0265	C17A	412TW	ED AFFTC		90-0532	C17A	
88-0266	C17A	412TW	ED AFFTC		90-0533	C17A	
89-1189	C17A	412TW	ED AFFTC		90-0534	C17A	
89-1190	C17A				90-0535	C17A	
89-1191	C17A						

a further 200 aircraft have been requested

C18

c\n   19518, 19382, 19384, 19583, 19391, 19581, 19238, 19380. Civil aircraft identities unknown.

81-0891	EC18B	4950TW	ASD		81-0896	EC18B	4950TW	ASD
81-0892	EC18B	4950TW	ASD		81-0897	C18A	derelict	Greenville, Tx
81-0893	EC18D	4950TW	ASD		81-0898	C18A	4950TW	ASD
81-0894	EC18B	4950TW	ASD		N131EA	TC18E	552ACW	
81-0895	EC18D	4950TW	ASD		N132EA	TC18E	552ACW	

C20 GULFSTREAM III\IV

c\n   III: 382, 383, 389, 456, 458, 465, 470, 468, 475 to 478.
      IV :   1192.

83-0500	C20A	86W		86-0202	C20B	89AW
83-0501	C20A	86W		86-0203	C20B	89AW
83-0502	C20A	86W		86-0204	C20B	89AW
85-0049	C20C	89AW\US Army		86-0205	C20B	89AW
85-0050	C20C	89AW\US Army		86-0206	C20B	89AW
86-0200	C20B	89AW		86-0403	C20C	89AW
86-0201	C20B	89AW		91-0108	C20H	89AW

C21 LEARJET

c\n   35A-509 to 35A-573, 35A-575 to 35A-583, 35A-585, 35A-574, 35A-587, 35A-588, 35A-584, 35A-586,
      35A-624, 35A-625, 35A-628, 35A-629.

84-0063	C21A	375AW		84-0104	C21A	458ALS	Det4
84-0064	C21A	458ALS	Det4	84-0105	C21A	458ALS	Det4
84-0065	C21A	375AW		84-0106	C21A	458ALS	Det4
84-0066	C21A	375AW		84-0107	C21A	458ALS	Det4
84-0067	C21A	375AW		84-0108	C21A	458ALS	Det3
84-0068	C21A	375AW		84-0109	C21A	458ALS	Det3
84-0069	C21A	375AW		84-0110	C21A	458ALS	Det3
84-0070	C21A	375AW		84-0111	C21A	458ALS	Det3
84-0071	C21A	374AW		84-0112	C21A	458ALS	Det3
84-0072	C21A	375AW		84-0113	C21A	457ALS	Det1
84-0073	C21A	457ALS		84-0114	C21A	457ALS	Det1
84-0074	C21A	457ALS		84-0115	C21A	457ALS	Det1
84-0075	C21A	457ALS		84-0116	C21A	457ALS	Det1
84-0076	C21A	457ALS		84-0117	C21A	457ALS	Det1
84-0077	C21A	457ALS		84-0118	C21A	457ALS	Det3
84-0078	C21A	457ALS		84-0119	C21A	457ALS	Det3
84-0079	C21A	457ALS		84-0120	C21A	457ALS	Det3
84-0080	C21A	458ALS	Det2	84-0121	C21A	w.o.14.1.87	
84-0081	C21A	HQ-USEUCOM		84-0122	C21A	457ALS	Det2
84-0082	C21A	HQ-USEUCOM		84-0123	C21A	457ALS	Det2
84-0083	C21A	HQ-USEUCOM		84-0124	C21A	457ALS	Det2
84-0084	C21A	86W		84-0125	C21A	457ALS	Det2
84-0085	C21A	86W		84-0126	C21A	459ALS	
84-0086	C21A	86W		84-0127	C21A	459ALS	
84-0087	C21A	458ALS	Det1	84-0128	C21A	459ALS	
84-0088	C21A	458ALS	Det1	84-0129	C21A	459ALS	
84-0089	C21A	458ALS	Det1	84-0130	C21A	459ALS	Det1
84-0090	C21A	458ALS	Det1	84-0131	C21A	459ALS	Det1
84-0091	C21A	458ALS	Det1	84-0132	C21A	459ALS	Det1
84-0092	C21A	458ALS	Det1	84-0133	C21A	459ALS	Det1
84-0093	C21A	458ALS	Det1	84-0134	C21A	459ALS	Det2
84-0094	C21A	458ALS	Det1	84-0135	C21A	459ALS	Det2
84-0095	C21A	458ALS	Det1	84-0136	C21A	459ALS	Det2
84-0096	C21A	458ALS	Det2	84-0137	C21A	459ALS	Det2
84-0097	C21A	458ALS	Det2	84-0138	C21A	459ALS	Det2
84-0098	C21A	4950TW	AFSC\ASD	84-0139	C21A	310ALS	
84-0099	C21A	458ALS	Det2	84-0140	C21A	457ALS	Det3
84-0100	C21A	458ALS	Det2	84-0141	C21A	457ALS	Det1
84-0101	C21A	374AW		84-0142	C21A	375AW	
84-0102	C21A	374AW		86-0374	C21A	201ALS	HQ-ANG
84-0103	C21A	458ALS	Det4	86-0375	C21A	201ALS	HQ-ANG
				86-0376	C21A	201ALS	HQ-ANG
				86-0377	C21A	201ALS	HQ-ANG

C22

c\n 18811, 18813, 18816, 18817, 21946, 18362.

83-4610	C22B	201ALS	HQ-ANG	83-4616	C22B	201ALS	HQ-ANG
83-4612	C22B	201ALS	HQ-ANG	83-4618	C22C	201ALS	SOUTHCOM
83-4615	C22B	201ALS	HQ-ANG	84-0193	C22A	AMARC	CU001

C23 SHERPA

c\n SH.3100 to SH.3102, SH.3104, SH.3106, SH.3107, SH.3109 to SH.3120.

83-0512	C23A	412TW	ED AFFTC	84-0465	C23A	to USFS
83-0513	C23A	412TW	ED AFFTC	84-0466	C23A	to US Army
84-0458	C23A	412TW	ED AFFTC	84-0467	C23A	to US Army
84-0459	C23A	to USFS		84-0468	C23A	to US Army
84-0460	C23A	to USFS		84-0469	C23A	to USFS
84-0461	C23A	to US Army		84-0470	C23A	to US Army
84-0462	C23A	to USFS		84-0471	JC23A	to US Army
84-0463	C23A	to US Army		84-0472	C23A	to USFS
84-0464	C23A	to US Army		84-0473	C23A	to US Army

C25

c\n 23824, 23825.

82-8000	VC25A	89AW		92-9000	VC25A	89AW

C26 METRO III\IVC

c\n AC-734B, AC-737B, AC-740B, AC-742B, AC-743B, AC-745B, AC-747B, AC-749B, AC-751B, AC-753B, AT-549 plus later deliveries.

86-0450	C26A	111FS	TX ANG	86-0453	C26A	190FS	ID ANG
86-0451	C26A	194FS	CA ANG	86-0454	C26A	169FS	IL IL ANG
86-0452	C26A	153ARS	MS ANG	86-0455	C26A	103FS	PA PA ANG

90-0529	C26B		
90-0530	C26B	178FS	ND ANG
90-0531	C26B	171FS	MI ANG
91-0502	C26B		
91-0503	C26B		
91-0504	C26B	120FS	CO CO ANG
91-0505	C26B		
91-0506	C26B		
91-0507	C26B		
91-0508	C26B	125FS	OK OK ANG
91-0509	C26B		
91-0510	C26B		
91-0511	C26B		
91-0512	C26B		
91-0513	C26B		
91-0514	C26B	134FS	VT ANG
91-0515	C26B	199FS	HI ANG

note: 91-0515 is compromised by the same allocation to a TG3A glider.

C27 SPARTAN

Chrysler\Alenia (Fiat) G222

90-0170	C27A	310ALS		
90-0171	C27A	310ALS		
90-0172	C27A	310ALS		
90-0173	C27A			
90-0174	C27A			
90-0175	C27A			
90-0176	C27A			
90-0177	C27A			
90-0178	C27A			
90-0179	C27A			
91-0103	C27A			
91-0104	C27A	310ALS		
91-0105	C27A	310ALS		
91-0106	C27A			
91-0107	C27A			

C29

c\n   258129, 258131, 258134, 258154, 258156, 258158.

88-0269	C29A	to FAA	N94	88-0272	C29A	to	FAA N97
88-0270	C29A	to FAA	N95	88-0273	C29A	to	FAA N98
88-0271	C29A	to FAA	N96	88-0274	C29A	to	FAA N99

C130 HERCULES

c\n	
C130A:	3001 to 3204.
YC130:	1001, 1002.
C130A:	3217 to 3231.
C130B:	3501 to 3528, 3530 to 3532, 3534 to 3541, 3543 to 3547, 3549 to 3553, 3556 to 3561, 3563, 3568, 3571, 3569, 3576, 3579, 3581, 3586, 3570, 3585, 3584, 3591, 3593, 3596, 3597,3600, 3602 to 3604, 3610, 3611, 3613, 3612, 3614, 3617, 3618, 3620 to 3622, 3624 to 3626, 3628 to 3630, 3633 to 3635, 3637, 3639, 3642, 3643, 3646 to 3649, 3652 to 3656, 3667 to 3669, 3650 (ex USCG).
C130E:	3609, 3651, 3659, 3662, 3663, 3681, 3687, 3688, 3706, 3712 to 3717, 3720.
C130B:	3670 to 3679, 3682, 3683, 3689 to 3692.

C130E:	3729 to 3732, 3735 to 3739, 3743, 3744, 3746 to 3748, 3752 to 3762, 3770 to 3772, 3774 to 3780, 3782 to 3812, 3814 to 3830.
C130B:	3697, 3702, 3707, 3708, 3721, 3722.
C130E:	3813, 3831 to 3848, 3850 to 3857, 3859, 3860, 3872, 3873, 3861 to 3870, 3874 to 3877, 3879 to 3883, 3888, 3889, 3894, 3895, 3884 to 3887, 3890 to 3893, 3903, 3904, 3896 to 3902, 3905 to 3945, 3947 to 4010, 4013, 4014, 4017 to 4019, 4021 to 4025, 4027 to 4035, 4039, 4040, 4043 to 4049, 4056 to 4059, 4062, 4063, 4065, 4068, 4071, 4074, 4077, 4080, 4083, 4086, 4079, 4085, 4087, 4090.
C130H\P:	4036 to 4038, 4055, 4072, 4073, 4081, 4082, 4084, 4088, 4089, 4094, 4097 to 4099, 4102 to 4104, 4106 to 4108, 4110 to 4112, 4116, 4120, 4121, 4123, 4125 to 4127, 4130 to 4133, 4135, 4138 to 4143, 4150 to 4152, 4155 to 4157, 4161 to 4166, 4173 to 4175, 4179, 4183 to 4187, 4255, 4260, 4265.
C130E:	4314 to 4351.
C130N:	4363, 4367, 4368, 4370 to 4372, 4374 to 4382.
C130E:	4340 to 4349, 4351 to 4354, 4356, 4357, 4359, 4360, 4404, 4410, 4413 to 4415, 4417 to 4421, 4423 to 4426, 4428, 4429, 4434, 4435, 4499, 4500, 4502, 4504 to 4506, 4509, 4510, 4517, 4519, 4521, 4527.
C130H:	4513 (ex USCG), 4542 to 4550, 4553 to 4555, 4557, 4559, 4563, 4564, 4568, 4571, 4573, 4574, 4579, 4585, 4592, 4596 to 4598, 4603, 4604, 4611, 4613, 4616, 4617, 4629, 4621, 4623, 4627, 4631, 4640, 4641, 4643, 4645, 4646, 4651, 4654, 4657, 4658, 4663, 4666, 4669, 4670, 4675, 4681, 4682, 4687, 4688, 4693, 4644, 4647, 4655, 4659, 4667, 4671, 4678, 4694, 4699, 4700, 4703, 4705, 4711, 4718, 4722, 4730, 4735, 4815, 4817 to 4823, 4852, 4854 to 4860, 4900, 4902, 4903, 4905, 4906, 4908, 4910, 4943, 4939, 4941, 4942, 4944 to 4946, 4968, 4970, 4971, 4973, 4975, 4977, 4979, 4982, 5008, 5012, 5014, 5018, 5007, 5010, 5013, 5016, 5004, 5038, 5039, 5043, 5044, 5046, 5047, 5049 to 5052.
C130B:	unknown (ex USCG).
C130H:	5041, 5042, 5053, 5054, 5073, 5074, 5077, 5079, 5080, 5083, 5086, 5089, 5071, 5072, 5075, 5076, 5078, 5081, 5082, 5084, 5094, 5097, 5098, 5100, 5102, 5105, 5110, 5113, 5093, 5095, 5096, 5099, 5101, 5103, 5111, 5112, 5026, 5091, 5092, 5115, 5117, 5118, 5122 to 5135, 5162 to 5169, 5173, 5202, 5210, 5154 to 5161, 5236, 5237, 5243, 5244, 5228 to 5233, 5198, 5199, 5201, 5203 to 5205, 5188, 5190, 5192 to 5197, 5216 to 5218, 5220, 5221, 5223, 5265, 5266, 5256, 5257, 5259, 5261, 5262, 5240 to 5242, 5245 to 5251, 5294, 5238, 5238, 5278, 5282 to 5292, 5294, 5293, 5295 to 5297.
C130H:	On order 5310, 5312 to 5315, 5321 to 5338, 5343, 5348 to 5355.

53-3129	AC130A	711SOS	AFRes		55-0020	C130A	w.o. .5.62	
53-3130	C130A	destroyed 20.2.58			55-0021	DC130A	to USN	Bu158228
53-3131	GC130A	derelict	Wichita Falls		55-0022	NC130A	46TW	ET AFMC\MSD
53-3132	C130A	to Mexico			55-0023	C130A	preserved	Dyess AFB
53-3133	NC130A	AMARC	CF021		55-0024	C130A	AMARC	CF062
53-3134	C130A	AMARC	CF010		55-0025	C130A	to Peru	
53-3135	C130A	to Mexico			55-0026	C130A	AMARC	CF086
53-3396	YC130	scrapped 10.60			55-0027	C130A	to Mexico	
53-3397	YC130	scrapped 4.62			55-0028	C130A	to Mexico	
54-1621	GC130A	scrapped 1986			55-0029	AC130A	711SOS	AFRes
54-1622	C130A	MASDC			55-0030	C130A	to Peru	
54-1623	AC130A	711SOS	AFRes		55-0031	C130A	to Mexico	
54-1624	C130A	AMARC	CF009		55-0032	C130A	GIA	Fort Lewis
54-1625	C130A	w.o.21.4.70			55-0033	C130A	AMARC	CF...
54-1626	C130A	USAF Museum			55-0034	C130A	MASDC	
54-1627	AC130A	AMARC	CF015		55-0035	C130A	to Mexico	
54-1628	AC130A	711SOS	AFRes		55-0036	C130A	AMARC	CF073
54-1629	JC130A	w.o.24.5.69			55-0037	GC130A	TTC	Chanute
54-1630	AC130A	711SOS	AFRes		55-0038	C130A	w.o.18.9.65	
54-1631	C130A	to N117TG			55-0039	C130A	w.o.1.7.65	
54-1632	C130A	AMARC	CF017		55-0040	AC130A	AMARC	CF018
54-1633	C130A	to Tchad			55-0041	C130A	AMARC	CF068
54-1634	C130A	AMARC	CF...		55-0042	C130A	w.o.1.7.65	
54-1635	C130A				55-0043	C130A	w.o.18.6.72	
54-1636	C130A	to Bolivia	TAM-64		55-0044	C130A	w.o.28.3.72	
54-1637	C130A	GIA	GoodfellowAFB		55-0045	C130A	to Vietnam	
54-1638	C130A	to Mexico			55-0046	AC130A	711SOS	AFRes
54-1639	C130A				55-0047	C130A	MASDC	
54-1640	C130A	AMARC	CF088		55-0048	C130A	MASDC	
55-0001	C130A	to Vietnam			56-0468	C130A	w.o.9.8.86	
55-0002	C130A	w.o. .4.75			56-0469	AC130A	711SOS	AFRes
55-0003	C130A	to Honduras	FAH557		56-0470	C130A	AMARC	CF071
55-0004	C130A	AMARC	CF085		56-0471	C130A	AMARC	CF087
55-0005	C130A	to Vietnam			56-0472	C130A	w.o.27.5.69	
55-0006	C130A	to Vietnam			56-0473	C130A	to N473TM	
55-0007	C130A	to Bolivia	TAM-65		56-0474	C130A	w.o. .8.63	
55-0008	C130A	to Peru			56-0475	C130A	GIA	Chicago IAP
55-0009	C130A	w.o.15.7.67			56-0476	C130A	to Vietnam	
55-0010	C130A	to Tchad			56-0477	C130A	w.o.22.5.68	
55-0011	AC130A	711SOS	AFRes		56-0478	C130A	to N116TG	
55-0012	C130A	MASDC			56-0479	C130A	to Mexico	
55-0013	C130A	to Vietnam			56-0480	C130A	w.o.16.4.68	
55-0014	AC130A	711SOS	AFRes		56-0481	C130A	AMARC	CF069
55-0015	C130A	to Honduras	FAH558		56-0482	C130A	to Vietnam	
55-0016	C130A	to Vietnam			56-0483	C130A	to Peru	
55-0017	C130A	to Vietnam			56-0484	C130A	to USFS	N137FF
55-0018	C130A	AMARC	CF081		56-0485	C130A	AMARC	CF090
55-0019	C130A	GIA	Hanscomb AFB		56-0486	C130A	AMARC	CF...

Serial	Type	Status	Notes
56-0487	C130A	to USFS	
56-0488	C130A	w.o. .10.62	
56-0489	C130A	to Vietnam	
56-0490	C130A	w.o.21.12.72	
56-0491	DC130A	to USN	Bu158229
56-0492	C130A	w.o. .5.64	
56-0493	C130A	preserved	D-M AFB
56-0494	C130A		
56-0495	C130A	AMARC	CF097
56-0496	C130A	to N8053R	
56-0497	C130A	MASDC	
56-0498	C130A		
56-0499	C130A	w.o.13.12.69	
56-0500	C130A	to N223MA	
56-0501	C130A	w.o.22.1.85	
56-0502	C130A	w.o.8.12.65	
56-0503	C130A	AMARC	CF...
56-0504	C130A	w.o.2.10.80	
56-0505	C130A	to Vietnam	
56-0506	C130A	w.o.3.66	
56-0507	C130A	to N8055R\N45S	
56-0508	C130A	to Mexico	
56-0509	AC130A	711SOS	AFRes
56-0510	C130A	w.o.10.4.70	
56-0511	C130A	to USFS	
56-0512	GC130A	Richards Gebauer AFB	
56-0513	C130A	AMARC	CF066
56-0514	DC130A	to USN as 560514	
56-0515	C130A	w.o.21.12.65	
56-0516	C130A	w.o.10.4.70	
56-0517	GC130A	IAAFA	
56-0518	C130A	preserved	Lt. Rock AFB
56-0519	C130A	to Vietnam	
56-0520	C130A	GIA	Scott AFB
56-0521	C130A	to Vietnam	
56-0522	C130A	711SOS	AFRes
56-0523	C130A	AMARC	CF082
56-0524	C130A		
56-0525	C130A		
56-0526	C130A	w.o. .9.58	
56-0527	DC130A	AMARC	CF030
56-0528	C130A	w.o.2.9.58	
56-0529	C130A	AMARC	CF091
56-0530	C130A	to USFS	N131FF
56-0531	C130A	123FS	OR ANG
56-0532	C130A	to Vietnam	
56-0533	C130A	w.o.24.11.69	
56-0534	C130A	to USFS	N132FF
56-0535	C130A	to USFS	N133FF
56-0536	C130A	to Bolivia	TAM-66
56-0537	C130A	to N537TM	
56-0538	C130A	to USFS	N134FF
56-0539	GC130A	GIA	Lt. Rock AFB
56-0540	C130A	to USFS	N135FF
56-0541	C130A	to USFS	N136FF
56-0542	C130A	to Vietnam	
56-0543	C130A	AMARC	CF080
56-0544	C130A	AMARC	CF084
56-0545	C130A	GIA	Marietta, GA
56-0546	C130A	w.o. .5.62	
56-0547	C130A		
56-0548	C130A	w.o.12.5.68	
56-0549	C130A	w.o.2.3.67	
56-0550	C130A		
56-0551	C130A	to Peru	
57-0453	C130A	AMARC	CF102
57-0454	C130A	w.o.27.7.75	
57-0455	C130A	to Peru	
57-0456	C130A	AMARC	CF...
57-0457	C130A	AMARC	CF083
57-0458	C130A	AMARC	CF074
57-0459	C130A	to N135HP	
57-0460	C130A	preserved	NASM
57-0461	DC130A	545TG	
57-0462	C130A	GIA	McDill AFB
57-0463	C130A		
57-0464	C130A		
57-0465	C130A	to Vietnam	
57-0466	C130A	to N466TM	
57-0467	C130A	w.o.12.10.67	
57-0468	C130A	w.o. .5.59	
57-0469	C130A	711SOS	AFRes
57-0470	C130A	to Peru	
57-0471	GC130A	TTC	Sheppard
57-0472	C130A	MASDC	
57-0473	C130A	to Tchad	
57-0474	C130A	to Bolivia	
57-0475	C130A	w.o.24.4.65	
57-0476	C130A	to Honduras	FAH559
57-0477	C130A	BDRT	Savannah IAP
57-0478	C130A	BDRT	Robins AFB
57-0479	C130A	to USFS	
57-0480	C130A	to Bolivia	
57-0481	C130A	to Bolivia	
57-0482	C130A	to N8026J\N133HP	
57-0483	C130A	GIA	Norfolk
57-0484	C130D	to Peru	383
57-0485	C130D	GIA	Minneapolis
57-0486	GC130D	TTC	Sheppard
57-0487	C130D	to Honduras	FAH556
57-0488	C130D	AMARC	CF016
57-0489	GC130D	TTC	Sheppard
57-0490	GC130D	TTC	Chanute
57-0491	C130D	to Peru	
57-0492	C130D	AMARC	CF046
57-0493	C130D	Pima Museum	
57-0494	C130D	AMARC	CF040
57-0495	C130D	w.o.5.6.72	
57-0496	DC130A	to USN	Bu570496
57-0497	DC130A	to USN	Bu570497
57-0510	C130A		
57-0511	C130A	to N134HP	
57-0512	C130A	to N118TG	
57-0513	C130A	GIA	Pope AFB
57-0514	C130A	preserved	Selfridge ANGB
57-0515	C130A		
57-0516	C130A		
57-0517	C130A	to N9539G	
57-0518	C130A	to N9691N	
57-0519	C130A		
57-0520	C130A	to USFS	N138FF
57-0521	C130A	to Bolivia	
57-0522	C130A	GIA	W-P AFB
57-0523	C130A	to N9539Q	
57-0524	C130A	BDRT	RAF Mildenhall
57-0525	C130B		
57-0526	C130B	545TG	
57-0527	JC130B	to Turkey	
57-0528	C130B	AMARC	CF094
57-0529	C130B	AMARC	CF115
58-0711	C130B	187ALS	WY ANG
58-0712	NC130B	to NASA	N707NA
58-0713	C130B	AMARC	CF0096
58-0714	C130B		
58-0715	C130B	AMARC	CF...
58-0716	C130B	AMARC	CF...
58-0717	JC130B	to Colombia	
58-0718	C130B	w.o.6.10.69	
58-0719	C130B	w.o. .1.65	
58-0720	C130B	AMARC	CF...
58-0721	C130B	w.o.1.2.75	
58-0722	C130B	w.o.16.4.67	
58-0723	C130B	to Greece	
58-0724	C130B	to Singapore	720
58-0725	C130B	AMARC	CF095
58-0726	C130B	to Colombia	
58-0727	GC130B	TTC	Sheppard
58-0728	C130B	156ALS	NC ANG
58-0729	C130B	156ALS	NC ANG
58-0730	C130B	w.o. .1.65	
58-0731	C130B		
58-0732	C130B	w.o.6.2.92	
58-0733	C130B	to Ecuador	
58-0734	C130B	187ALS	WY ANG
58-0735	C130B	to Colombia	
58-0736	C130B	to Turkey	
58-0737	C130B	w.o.9.6.67	
58-0738	C130B		
58-0739	C130B	to Pakistan	58739
58-0740	GC130B	IAAFA	
58-0741	C130B	to Argentina	
58-0742	C130B	to Chile	
58-0743	C130B	w.o.17.2.68	
58-0744	C130B	AMARC	CF...
58-0745	C130B	w.o. .10.61	
58-0746	C130B	AMARC	CF...

Serial	Type	Unit/Fate	Code
58-0747	C130B		
58-0748	C130B	to Indonesia	T1301
58-0749	C130B	to Philippine	
58-0750	C130B	AMARC	CF...
58-0751	C130B	156ALS	NC ANG
58-0752	C103B	to Chile	
58-0753	C130B	156ALS	NC ANG
58-0754	C130B	187ALS	WY ANG
58-0755	C130B	AMARC	CF092
58-0756	C130B	to Singapore	721
58-0757	C130B	to Colombia	
58-0758	C130B	AMARC	CF113
59-1524	C130B		
59-1525	C130B		
59-1526	C130B	to Oman	
59-1527	C130B	to Turkey	
59-1528	C130B	156ALS	NC ANG
59-1529	C130B		
59-1530	C130B	303ALS	AFRes
59-1531	C130B		
59-1532	C130B		
59-1533	C130B	156ALS	NC ANG
59-1534	C130B	w.o. .5.61	
59-1535	C130B		
59-1536	C130B	156ALS	NC ANG
59-1537	C130B	303ALS	AFRes
59-5957	C130B	AMARC	CF132
60-0293	C130B	w.o.17.6.67	
60-0294	C130B		
60-0295	C130B	to Uruguay	
60-0296	C130B	to Greece	
60-0297	C130B	w.o.12.5.68	
60-0298	C130B	w.o.26.4.68	
60-0299	C130B	AMARC	CF122
60-0300	C130B	to Greece	
60-0301	C130B	to Jordan	141
60-0302	C130B	to Jordan	142
60-0303	C130B	to Greece	
60-0304	C130B	to Jordan	140
60-0305	C130B	to Indonesia	T1311
60-0306	C130B	to Indonesia	T1312
60-0307	C130B	w.o.17.2.67	
60-0308	C130B	to Jordan	143
60-0309	C130B	to Indonesia	T1313
60-0310	C130B		
61-0948	C130B	to Greece	
61-0949	C130B	156ALS	NC ANG
61-0950	C130B	156ALS	NC ANG
61-0951	C130B		
61-0952	C130B	AMARC	CF120
61-0953	C130B	w.o.29.3.66	
61-0954	C130B	303ALS	AFRes
61-0955	C130B	w.o. .10.66	
61-0956	C130B	303ALS	AFRes
61-0957	C130B	303ALS	AFRes
61-0958	C130B		
61-0959	C130B		
61-0960	C130B	to Turkey	
61-0961	C130B	to Philippine	
61-0962	C130B	AMARC	CF055
61-0963	C130B	to Turkey	
61-0964	C130B	to Argentina	
61-0965	C130B	w.o.23.6.69	
61-0966	C130B	187ALS	WY ANG
61-0967	C130B	w.o.13.4.68	
61-0968	C130B	303ALS	AFRes
61-0969	C130B	AMARC	CF...
61-0970	C130B	w.o.9.1.66	
61-0971	C130B	to Uruguay	
61-0972	C130B	w.o.6.1.66	
61-2083	HC130B	AMARC	CF041 USCG
61-2358	C130E	115ALS	CA ANG
61-2359	C130E	115ALS	CA ANG
61-2360	WC130E	815ALS	AFRes
61-2361	C130E	109ALS	MN ANG
61-2362	C130E	314AW	
61-2363	C130E	314AW	
61-2364	C130E	731ALS	AFRes
61-2365	WC130E	815ALS	AFRes
61-2366	WC130E	815ALS	AFRes
61-2367	C130E	115ALS	CA ANG
61-2368	C130E	96ALS	AFRes
61-2369	C130E	314AW	
61-2370	C130E	115ALS	CA ANG
61-2371	C130E	314AW	
61-2372	C130E	115ALS	CA ANG
61-2373	C130E	w.o.8.6.88	
61-2634	C130B	to Turkey	
61-2635	C130B	187ALS	WY ANG
61-2636	C130B	156ALS	NC ANG
61-2637	C130B	w.o.29.4.69	
61-2638	C130B	156ALS	NC ANG
61-2639	C130B		
61-2640	C130B	156ALS	NC ANG
61-2641	C130B	w.o.19.3.66	
61-2642	C130B	w.o.21.2.72	
61-2643	C130B	AMARC	CF...
61-2644	C130B	w.o.28.11.68	
61-2645	C130B	to Ecuador	
61-2646	C130B	to Pakistan	61-2646
61-2647	C130B	303ALS	AFRes
61-2648	C130B	to Pakistan	61-2648
61-2649	C130B	w.o.8.10.67	
62-1784	C130E	154ALS	AR ANG
62-1785	C130E	w.o.6.9.68	
62-1786	C130E	109ALS	MN ANG
62-1787	C130E	154ALS	AR ANG
62-1788	C130E	154ALS	AR ANG
62-1789	C130E	731ALS	AFRes
62-1790	C130E	154ALS	AR ANG
62-1791	EC130E	7ACCS	KS
62-1792	C130E	115ALS	CA ANG
62-1793	C130E	115ALS	CA ANG
62-1794	C130E	731ALS	AFRes
62-1795	C130E	154ALS	AR ANG
62-1796	C130E	to Israel	4X-FBE
62-1797	C130E	w.o.3.5.72	
62-1798	C130E	154ALS	AR ANG
62-1799	C130E	115ALS	CA ANG
62-1800	C130E	w.o. .12.69	
62-1801	C130E	115ALS	CA ANG
62-1802	C130E	w.o.31.7.70	
62-1803	C130E	731ALS	AFRes
62-1804	C130E	154ALS	AR ANG
62-1805	C130E	w.o.5.6.72	
62-1806	C130E	96ALS	AFRes
62-1807	C130E	731ALS	AFRes
62-1808	C130E	731ALS	AFRes
62-1809	EC130E	w.o.24.4.80	
62-1810	C130E	731ALS	AFRes
62-1811	C130E	115ALS	CA ANG
62-1812	C130E	109ALS	MN ANG
62-1813	C130E	w.o.19.2.72	
62-1814	C130E	w.o.3.3.68	
62-1815	C130E	w.o.15.7.67	
62-1816	C130E	63ALS	AFRes
62-1817	C130E	109ALS	MN ANG
62-1818	EC130E	7ACCS	KS
62-1819	C130E	435AW	
62-1820	C130E	731ALS	AFRes
62-1821	C130E	314AW	
62-1822	C130E	435AW	
62-1823	C130E	731ALS	AFRes
62-1824	C130E	154ALS	AR ANG
62-1825	EC130E	7ACCS	KS
62-1826	C130E	115ALS	CA ANG
62-1827	C130E	314AW	
62-1828	C130E	435AW	
62-1829	C130E	109ALS	MN ANG
62-1830	C130E	731ALS	AFRes
62-1831	C130E	w.o.2.69	
62-1832	EC130E	7ACCS	KS
62-1833	C130E	115ALS	CA ANG
62-1834	C130E	96ALS	AFRes
62-1835	C130E	96ALS	AFRes
62-1836	EC130E	7ACCS	KS
62-1837	C130E	109ALS	MN ANG
62-1838	C130E	731ALS	AFRes
62-1839	C130E	96ALS	AFRes
62-1840	C130E	w.o.2.10.66	
62-1841	C130E	w.o.20.4.74	
62-1842	C130E	115ALS	CA ANG
62-1843	MC130E	1SOS	
62-1844	C130E	96ALS	AFRes
62-1845	C130E	w.o.15.10.73	
62-1846	C130E	109ALS	MN ANG

Serial	Type	Unit	Notes
62-1847	C130E	96ALS	AFRes
62-1848	C130E	96ALS	AFRes
62-1849	C130E	731ALS	AFRes
62-1850	C130E	731ALS	AFRes
62-1851	C130E	115ALS	CA ANG
62-1852	C130E	96ALS	AFRes
62-1853	C130E	w.o.12.8.72	
62-1854	C130E	w.o.17.5.72	
62-1855	C130E	8SOS	
62-1856	C130E	109ALS	MN ANG
62-1857	EC130E	7ACCS	KS
62-1858	C130E	63ALS	AFRes
62-1859	C130E	167ALS	WV ANG
62-1860	C130E	731ALS	AFRes
62-1861	C130E	w.o.25.6.68	
62-1862	C130E	115ALS	CA ANG
62-1863	EC130E	7ACCS	KS
62-1864	C130E	109ALS	MN ANG
62-1865	C130E	w.o.15.11.67	
62-1866	C130E	731ALS	AFRes
62-3487	C130B		
62-3492	C130B	to Pakistan	62-3492
62-3493	C130B	303ALS	AFRes
62-3494	C130B	to Pakistan	62-3494
62-3495	C130B	to Oman	
62-3496	C130B	to Turkey	
63-7764	C130E	328ALS	AFRes
63-7765	C130E	314AW	
63-7766	C130E	w.o.28.4.78	
63-7767	C130E	314AW	
63-7768	C130E	314AW	
63-7769	C130E		
63-7770	C130E		
63-7771	C130E	314AW	
63-7772	C130E	w.o.12.3.67	
63-7773	EC130E	193SOS	PA PA ANG
63-7774	C130E	to Israel	4X-FBF
63-7775	C130E	w.o.18.4.72	
63-7776	C130E		
63-7777	C130E	167ALS	WV ANG
63-7778	C130E	314AW	
63-7779	C130E	327ALS	AFRes
63-7780	C130E	w.o.2.1.69	
63-7781	C130E	314AW	
63-7782	C130E	143ALS	RI ANG
63-7783	EC130E	193SOS	PA PA ANG
63-7784	C130E	314AW	
63-7785	MC130E	1SOS	
63-7786	C130E	314AW	
63-7787	C130E	w.o.15.4.78	
63-7788	C130E	143ALS	RI ANG
63-7789	C130E	w.o.23.5.69	
63-7790	C130E	314AW	
63-7791	C130E	314AW	
63-7792	C130E	167ALS	WV ANG
63-7793	C130E	314AW	
63-7794	C130E	314AW	
63-7795	C130E	314AW	
63-7796	C130E	314AW	
63-7797	C130E	w.o.25.3.65	
63-7798	C130E	w.o.6.5.72	
63-7799	C130E	314AW	
63-7800	C130E	167ALS	WV ANG
63-7801	C130E	GIA	LT. Rock AFB
63-7802	C130E	w.o. .9.74	
63-7803	C130E	374AW	YJ
63-7804	C130E	23W	FT
63-7805	C130E	815ALS	AFRes
63-7806	C130E	314AW	
63-7807	C130E	317AW	
63-7808	C130E	314AW	
63-7809	C130E	23W	FT
63-7810	C130E	to Israel	4X-FBG
63-7811	C130E	143ALS	RI ANG
63-7812	C130E	167ALS	WV ANG
63-7813	C130E	23W	FT
63-7814	C130E	67SOS	
63-7815	EC130E	193SOS	PA PA ANG
63-7816	EC130E	193SOS	PA PA ANG
63-7817	C130E	815ALS	AFRes
63-7818	C130E	167ALS	WV ANG
63-7819	C130E	374AW	YJ
63-7820	C130E	314AW	
63-7821	C130E	317AW	
63-7822	C130E		
63-7823	C130E	327ALS	AFRes
63-7824	C130E	143ALS	RI ANG
63-7825	C130E	135ALS	MD ANG
63-7826	C130E	327ALS	AFRes
63-7827	C130E	w.o.15.11.67	
63-7828	EC130E	193SOS	PA PA ANG
63-7829	C130E	23W	FT
63-7830	C130E	314AW	
63-7831	C130E	314AW	
63-7832	C130E	303ALS	AFRes
63-7833	C130E	327ALS	AFRes
63-7834	C130E	327ALS	AFRes
63-7835	C130E	314AW	
63-7836	C130E	314AW	
63-7837	C130E	374AW	YJ
63-7838	C130E	314AW	
63-7839	C130E	314AW	
63-7840	C130E	143ALS	RI ANG
63-7841	C130E	314AW	
63-7842	C130E	1SOS	
63-7843	C130E	to Israel	4X-FBH
63-7844	C130E	to Israel	4X-FBI
63-7845	C130E	317AW	
63-7846	C130E	317AW	
63-7847	C130E	154ALS	AR ANG
63-7848	C130E	327ALS	AFRes
63-7849	C130E	23W	FT
63-7850	C130E	314AW	
63-7851	C130E	167ALS	WV ANG
63-7852	C130E	815ALS	AFRes
63-7853	C130E		
63-7854	C130E	314AW	
63-7855	C130E	to Israel	4X-FBK
63-7856	C130E	303ALS	AFRes
63-7857	C130E	314AW	
63-7858	C130E	167ALS	WV ANG
63-7859	C130E	143ALS	RI ANG
63-7860	C130E	314AW	
63-7861	C130E	314AW	
63-7862	C130E	to Israel	4X-FBL
63-7863	C130E	328ALS	AFRes
63-7864	C130E	314AW	
63-7865	C130E	374AW	YJ
63-7866	C130E	314AW	
63-7867	C130E	327ALS	AFRes
63-7868	C130E	143ALS	RI ANG
63-7869	EC130E	193SOS	PA PA ANG
63-7870	C130E	to Israel	4X-FBM
63-7871	C130E	317AW	
63-7872	C130E	167ALS	WV ANG
63-7873	C130E	to Israel	4X-FBN
63-7874	C130E	314AW	
63-7875	C130E	w.o.15.5.68	
63-7876	C130E	314AW	
63-7877	C130E	167ALS	WV ANG
63-7878	C130E	w.o. .9.66	
63-7879	C130E	374AW	YJ
63-7880	C130E	314AW	
63-7881	C130E	w.o.7.10.92	
63-7882	C130E	314AW	
63-7883	C130E	327ALS	AFRes
63-7884	C130E	317AW	
63-7885	C130E	435AW	
63-7886	C130E	w.o.10.66	
63-7887	C130E	314AW	
63-7888	C130E	314AW	
63-7889	C130E	143ALS	RI ANG
63-7890	C130E	317AW	
63-7891	C130E	314AW	
63-7892	C130E	327ALS	AFRes
63-7893	C130E	314AW	
63-7894	C130E	314AW	
63-7895	C130E	135ALS	MD ANG
63-7896	C130E	314AW	
63-7897	C130E	167ALS	WV ANG
63-7898	C130E	8SOS	
63-7899	C130E	317AW	
63-9810	C130E	23W	FT
63-9811	C130E	314AW	
63-9812	C130E	314AW	
63-9813	C130E	314AW	

Serial	Type	Unit	Notes
63-9814	C130E	314AW	
63-9815	C130E	193SOS	PA PA ANG
63-9816	C130E	193SOS	PA PA ANG
63-9817	EC130E	193SOS	PA PA ANG
64-0495	C130E	317AW	
64-0496	C130E	317AW	
64-0497	C130E	374AW	YJ
64-0498	C130E	317AW	
64-0499	C130E	317AW	
64-0500	C130E	Lockheed	D4 AFLC
64-0501	C130E	w.o.28.4.92	
64-0502	C130E	435AW	
64-0503	C130E	374AW	YJ
64-0504	C130E	317AW	
64-0505	C130E	w.o.9.12.72	
64-0506	C130E	to N.....	
64-0507	C130E	to N.....	
64-0508	C130E	w.o.25.4.72	
64-0509	C130E	to Israel	4X-FBO
64-0510	C130E	135ALS	MD ANG
64-0511	C130E	w.o.31.5.66	
64-0512	C130E	154ALS	AR ANG
64-0513	C130E	314AW	
64-0514	C130E	135ALS	MD ANG
64-0515	C130E	135ALS	MD ANG
64-0516	C130E	to Israel	4X-FBP
64-0517	C130E	317AW	
64-0518	C130E	314AW	
64-0519	C130E	314AW	
64-0520	C130E	135ALS	MD ANG
64-0521	C130E	135ALS	MD ANG
64-0522	C130E	w.o.29.2.68	
64-0523	MC130E		
64-0524	C130E	314AW	
64-0525	C130E	317AW	
64-0526	C130E	115ALS	CA ANG
64-0527	C130E	435AW	
64-0528	C130E	to Israel	4X-FBQ
64-0529	C130E	317AW	
64-0530	C130E	314AW	
64-0531	C130E	4950TW	AFMC\ASD
64-0532	C130E	w.o.8.9.78	
64-0533	C130E	314AW	
64-0534	C130E	374AW	YJ
64-0535	C130E	314AW	
64-0536	C130E	w.o.2.10.70	
64-0537	C130E	317AW	
64-0538	C130E	314AW	
64-0539	C130E	317AW	
64-0540	C130E	317AW	
64-0541	C130E	317AW	
64-0542	C130E	317AW	
64-0543	C130E	w.o.12.5.82	
64-0544	C130E	135ALS	MD ANG
64-0545	C130E	w.o.3.69	
64-0546	C130E	to Sweden	84001
64-0547	C130E	w.o.9.12.67	
64-0548	C130E	w.o.15.10.67	
64-0549	C130E	w.o.12.3.85	
64-0550	C130E	435AW	
64-0551	MC130E	8SOS	
64-0552	WC130E	815ALS	AFRes
64-0553	WC130E	815ALS	AFRes
64-0554	WC130E	815ALS	AFRes
64-0555	MC130E		
64-0556	C130E	374AW	YJ
64-0557	C130E	314AW	
64-0558	C130E	w.o.5.12.72	
64-0559	MC130E	8SOS	
64-0560	MC130E	314AW	
64-0561	MC130E		
64-0562	MC130E	8SOS	
64-0563	MC130E	w.o.25.11.67	
64-0564	MC130E	w.o.26.2.81	
64-0565	MC130E	1SOS	
64-0566	MC130H		
64-0567	MC130E	8SOS	
64-0568	MC130E	8SOS	
64-0569	C130E	314AW	
64-0570	C130E	317AW	
64-0571	MC130E	1SOS	
64-0572	MC130E	8SOS	
64-14852	HC130P	542CTW	
64-14853	HC130P	71RQS	
64-14854	HC130P	542CTW	
64-14855	HC130H	304RQS	AFRes
64-14856	HC130P	304RQS	AFRes
64-14857	HC130H	545TG	
64-14858	HC130P	17SOS	
64-14859	EC130H		
64-14860	HC130P	304RQS	AFRes
64-14861	WC130H	815ALS	AFRes
64-14862	EC130H	Lockheed	D4 AFLC
64-14863	HC130P	71RQS	
64-14864	HC130P	71RQS	
64-14865	HC130P	304RQS	AFRes
64-14866	C130H	815ALS	AFRes
64-17680	C130E	314AW	
64-17681	C130E	435AW	
64-18240	C130E	435AW	
65-0962	C130H	7ACCS	KS
65-0963	WC130H	815ALS	AFRes
65-0964	C130H	815ALS	AFRes
65-0665	WC130H	w.o.13.10.74	
65-0966	WC130H	815ALS	AFRes
65-0967	C130H	815ALS	AFRes
65-0968	WC130H	815ALS	AFRes
65-0969	C130H	815ALS	AFRes
65-0970	HC130P	304RQS	AFRes
65-0971	HC130P	542CTW	
65-0972	C130H	815ALS	AFRes
65-0973	HC130P		
65-0974	HC130H	102RQS	NY ANG
65-0975	HC130P	542CTW	
65-0976	C130H	815ALS	AFRes
65-0977	C130H	815ALS	AFRes
65-0978	HC130H	102RQS	NY ANG
65-0979	DC130H	545TG	
65-0980	WC130H	815ALS	AFRes
65-0981	HC130H	129RQS	CA ANG
65-0982	HC130P	71RQS	
65-0983	HC130H	129RQS	CA ANG
65-0984	WC130H	815ALS	AFRes
65-0985	C130H	815ALS	AFRes
65-0986	HC130P	542CTW	
65-0987	HC130P	542CTW	
65-0988	HC130P	102RQS	NY ANG
65-0989	EC130H	355W	DM
65-0990	HC130H	w.o.4.2.69	
65-0991	HC130P	9SOS	
65-0992	HC130P	17SOS	
65-0993	HC130P	9SOS	
65-0994	HC130P	9SOS	
66-0211	HC130P	w.o.2.4.86	
66-0212	HC130P	542CTW	
66-0213	HC130P	542CTW	
66-0214	HC130P	w.o.29.7.68	
66-0215	HC130P	9SOS	
66-0216	HC130P	9SOS	
66-0217	HC130P	9SOS	
66-0218	HC130P	w.o.29.7.68	
66-0219	HC130P	542CTW	
66-0220	HC130P		
66-0221	HC130P	129RQS	CA ANG
66-0222	HC130P	102RQS	NY ANG
66-0223	HC130P	67SOS	
66-0224	HC130P	129RQS	CA ANG
66-0225	HC130P	9SOS	
67-7183	C130H	AMARC	CF119USCG1452
67-7184	C130H	310ALS	HW
67-7185	HC130H	AMARC	CF044USCG1454
68-10934	C130E	317AW	
68-10935	C130E	435AW	
68-10936	C130E	w.o.30.11.78	
68-10937	C130E	317AW	PB
68-10938	C130E	435AW	
68-10939	C130E	317AW	
68-10940	C130E	317AW	
68-10941	C130E		
68-10942	C130E	317AW	
68-10943	C130E	435AW	
68-10944	C130E	w.o.28.2.84	
68-10945	C130E	w.o.1.7.87	
68-10946	C130E	w.o.2.11.84	
68-10947	C130E	435AW	
68-10948	C130E	314AW	

68-10949	C130E	314AW		73-1596	C130H	to Canada	130332
68-10950	C130E	314AW		73-1597	C130H	374AW	YJ
68-10951	C130E	w.o.10.12.78		73-1598	C130H	374AW	YJ
69-5819	HC130N	9SOS		73-1599	C130H	to Canada	130333
69-5820	HC130N	67SOS		74-1658	C130H	3W	AK
69-5821	HC130N	17SOS		74-1659	C130H	3W	AK
69-5822	HC130N	17SOS		74-1660	C130H	374AW	YJ
69-5823	HC130N	301RQS	AFRes	74-1661	C130H	374AW	YJ
69-5824	HC130N	301RQS	AFRes	74-1662	C130H	463AW	
69-5825	HC130N	17SOS		74-1663	C130H	463AW	
69-5826	HC130N	67SOS		74-1664	C130H	374AW	YJ
69-5827	HC130N	67SOS		74-1665	C130H	463AW	
69-5828	HC130N	9SOS		74-1666	C130H	463AW	
69-5829	HC130N	301RQS	AFRes	74-1667	C130H	463AW	
69-5830	HC130N	301RQS	AFRes	74-1668	C130H	3W	AK
69-5831	HC130N	542CTW		74-1669	C130H	463AW	
69-5832	HC130N	9SOS		74-1670	C130H	463AW	
69-5833	HC130N	301RQS	AFRes	74-1671	C130H	463AW	
69-6566	C130E	435AW		74-1672	C130H	w.o.21.9.81	
69-6567	AC130H	w.o.31.1.91		74-1673	C130H	463AW	
69-6568	AC130H	16SOS		74-1674	C130H	463AW	
69-6569	AC130H	16SOS		74-1675	C130H	463AW	
69-6570	AC130H	16SOS		74-1676	C130H	3W	AK
69-6571	C130E	w.o.30.3.72		74-1677	C130H	463AW	
69-6572	AC130H	16SOS		74-1678	C130H	w.o.14.4.82	
69-6573	AC130H	16SOS		74-1679	C130H	463AW	
69-6574	AC130H	16SOS		74-1680	C130H	463AW	
69-6575	AC130H	16SOS		74-1681	C130H		
69-6576	AC130H	16SOS		74-1682	C130H	374AW	YJ
69-6577	AC130H	16SOS		74-1683	C130H	w.o.5.2.81	
69-6578	C130E	w.o.12.11.71		74-1684	C130H	374AW	YJ
69-6579	C130E	314AW		74-1685	C130H	374AW	YJ
69-6580	C130E	317AW		74-1686	YMC130H	preserved	Robins AFB
69-6581	C130E	w.o.14.1.81		74-1687	C130H	463AW	
69-6582	C130E	435AW		74-1688	C130H	463AW	
69-6583	C130E	435AW		74-1689	C130H	463AW	
70-1259	C130E			74-1690	C130H	3W	AK
70-1260	C130E	435AW		74-1691	C130H	3W	AK
70-1261	C130E	317AW		74-1692	C130H	3W	AK
70-1262	C130E			74-1693	C130H	GIA	Pope AFB
70-1263	C130E	317AW		74-2061	C130H	463AW	
70-1264	C130E	435AW		74-2062	C130H	3W	AK
70-1265	C130E	317AW		74-2063	C130H	463AW	
70-1266	C130E	317AW		74-2064	C130H	w.o.14.3.80	
70-1267	C130E	317AW		74-2065	C130H	463AW	
70-1268	C130E	317AW		74-2066	C130H	3W	AK
70-1269	C130E	317AW		74-2067	C130H	463AW	
70-1270	C130E	317AW		74-2068	C130H	w.o.28.6.83	
70-1271	C130E	435AW		74-2069	C130H	463AW	
70-1272	C130E	317AW		74-2070	C130H	3W	AK
70-1273	C130E	317AW		74-2071	C130H	3W	AK
70-1274	C130E	435AW		74-2072	C130H	463AW	
70-1275	C130E	317AW		74-2130	C130H	463AW	
70-1276	C130E	317AW		74-2131	C130H	3W	AK
72-1288	C130E	374AW	YJ	74-2132	C130H	463AW	
72-1289	C130E	374AW	YJ	74-2133	C130H	374AW	YJ
72-1290	C130E	374AW	YJ	74-2134	C130H	463AW	
72-1291	C130E	314AW		78-0806	C130H	185ALS	OK ANG
72-1292	C130E	314AW		78-0807	C130H	185ALS	OK ANG
72-1293	C130E	314AW		78-0808	C130H	185ALS	OK ANG
72-1294	C130E	314AW		78-0809	C130H	185ALS	OK ANG
72-1295	C130E	314AW		78-0810	C130H	185ALS	OK ANG
72-1296	C130E	314AW		78-0811	C130H	185ALS	OK ANG
72-1297	C130E	w.o.28.4.75		78-0812	C130H	185ALS	OK ANG
72-1298	C130E	314AW		78-0813	C130H	185ALS	OK ANG
72-1299	C130E	374AW	YJ	79-0473	C130H	144ALS	AK ANG
72-1302	HC130H	Pemco	ex USCG	79-0474	C130H	160FS	AL ANG
73-1580	EC130H	355W	DM	79-0475	C130H	159FS	FL ANG
73-1581	EC130H	355W	DM	79-0476	C130H	157FS	SC ANG
73-1582	C130H	374AW	YJ	79-0477	C130H	158ALS	GA ANG
73-1583	EC130H	355W	DM	79-0478	C130H	199FS	HI ANG
73-1584	EC130H	355W	DM	79-0479	C130H	185ALS	OK ANG
73-1585	EC130H	355W	DM	79-0480	C130H	122FS	LA ANG
73-1586	EC130H	355W	DM	80-0320	C130H	158ALS	GA ANG
73-1587	EC130H	355W	DM	80-0321	C130H	158ALS	GA ANG
73-1588	EC130H	355W	DM	80-0322	C130H	158ALS	GA ANG
73-1589	C130H	to Canada	130329	80-0323	C130H	158ALS	GA ANG
73-1590	EC130H	355W	DM	80-0324	C130H	158ALS	GA ANG
73-1591	C130H	to Canada	130330	80-0325	C130H	158ALS	GA ANG
73-1592	EC130H	355W	DM	80-0326	C130H	158ALS	GA ANG
73-1593	C130H	to Canada	130331	80-0332	C130H	158ALS	GA ANG
73-1594	EC130H	355W	DM	81-0626	C130H	700ALS	AFRes
73-1595	EC130H	355W	DM	81-0627	C130H	700ALS	AFRes

45

Serial	Type	Unit	Operator	Base
81-0628	C130H	700ALS	AFRes	
81-0629	C130H	700ALS	AFRes	
81-0630	C130H	700ALS	AFRes	
81-0631	C130H	700ALS	AFRes	
82-0054	C130H	144ALS	AK ANG	
82-0055	C130H	144ALS	AK ANG	
82-0056	C130H	144ALS	AK ANG	
82-0057	C130H	144ALS	AK ANG	
82-0058	C130H	144ALS	AK ANG	
82-0059	C130H	144ALS	AK ANG	
82-0060	C130H	144ALS	AK ANG	
82-0061	C130H	144ALS	AK ANG	
83-0486	C130H	139ALS	NY ANG	
83-0487	C130H	139ALS	NY ANG	
83-0488	C130H	139ALS	NY ANG	
83-0489	C130H	139ALS	NY ANG	
83-0490	LC130H	139ALS	NY ANG	
83-0491	LC130H	139ALS	NY ANG	
83-0492	LC130H	139ALS	NY ANG	
83-0493	LC130H	139ALS	NY ANG	
83-1212	MC130H	412TW	ED	AFFTC
84-0204	C130H	700ALS	AFRes	
84-0205	C130H	700ALS	AFRes	
84-0206	C130H	142ALS	DE ANG	
84-0207	C130H	142ALS	DE ANG	
84-0208	C130H	142ALS	DE ANG	
84-0209	C130H	142ALS	DE ANG	
84-0210	C130H	142ALS	DE ANG	
84-0211	C130H	142ALS	DE ANG	
84-0212	C130H	142ALS	DE ANG	
84-0213	C130H	142ALS	DE ANG	
84-0454	HC130B	AMARC	ex USCG	
84-0475	MC130H	412TW	ED	AFFTC
84-0476	MC130H	7SOS		
85-0011	MC130H	8SOS		
85-0012	MC130H	15SOS		
85-0035	C130H	357ALS	AFRes	
85-0036	C130H	357ALS	AFRes	
85-0037	C130H	357ALS	AFRes	
85-0038	C130H	357ALS	AFRes	
85-0039	C130H	357ALS	AFRes	
85-0040	C130H	357ALS	AFRes	
85-0041	C130H	357ALS	AFRes	
85-0042	C130H	357ALS	AFRes	
85-1361	C130H	181ALS	TX ANG	
85-1362	C130H	181ALS	TX ANG	
85-1363	C130H	181ALS	TX ANG	
85-1364	C130H	181ALS	TX ANG	
85-1365	C130H	181ALS	TX ANG	
85-1366	C130H	181ALS	TX ANG	
85-1367	C130H	181ALS	TX ANG	
85-1368	C130H	181ALS	TX ANG	
86-0410	C130H	758ALS	AFRes	
86-0411	C130H	758ALS	AFRes	
86-0412	C130H	758ALS	AFRes	
86-0413	C130H	758ALS	AFRes	
86-0414	C130H	758ALS	AFRes	
86-0415	C130H	758ALS	AFRes	
86-0418	C130H	758ALS	AFRes	
86-0419	C130H	758ALS	AFRes	
86-1391	C130H	180ALS	MO ANG	
86-1392	C130H	180ALS	MO ANG	
86-1393	C130H	180ALS	MO ANG	
86-1394	C130H	180ALS	MO ANG	
86-1395	C130H	180ALS	MO ANG	
86-1396	C130H	180ALS	MO ANG	
86-1397	C130H	180ALS	MO ANG	
86-1398	C130H	180ALS	MO ANG	
86-1699	MC130H	7SOS		
87-0023	MC130H	15SOS		
87-0024	MC130H	15SOS		
87-0125	MC130H	15SOS		
87-0126	MC130H	15SOS		
87-0127	MC130H	412TW	ED	AFFTC
87-0128	AC130U	412TW	ED	AFFTC
87-9281	C130H	64ALS	AFRes	
87-9282	C130H	64ALS	AFRes	
87-9283	C130H	64ALS	AFRes	
87-9284	C130H	64ALS	AFRes	
87-9285	C130H	64ALS	AFRes	
87-9286	C130H	64ALS	AFRes	
87-9287	C130H	64ALS	AFRes	
87-9288	C130H	64ALS	AFRes	
88-0191	MC130H	542CTW		
88-0192	MC130H	542CTW		
88-0193	MC130H	7SOS		
88-0194	MC130H			
88-0195	MC130H	15SOS		
88-0264	MC130H	E-Systems		
88-1301	C130H	130ALS	WV ANG	
88-1302	C130H	130ALS	WV ANG	
88-1303	C130H	130ALS	WV ANG	
88-1304	C130H	130ALS	WV ANG	
88-1305	C130H	130ALS	WV ANG	
88-1306	C130H	130ALS	WV ANG	
88-1307	C130H	130ALS	WV ANG	
88-1308	C130H	130ALS	WV ANG	
88-1803	MC130H	E-Systems		
88-2101	HC130H	210RQS	AK ANG	
88-2102	HC130H	210RQS	AK ANG	
88-4401	C130H	95ALS	AFRes	
88-4402	C130H	95ALS	AFRes	
88-4403	C130H	95ALS	AFRes	
88-4404	C130H	95ALS	AFRes	
88-4405	C130H	95ALS	AFRes	
88-4406	C130H	95ALS	AFRes	
88-4407	C130H	95ALS	AFRes	
88-4408	C130H	95ALS	AFRes	
89-0280	MC130H	E-Systems		
89-0281	MC130H			
89-0282	MC130H			
89-0283	MC130H			
89-0509	AC130U	Lockheed	Palmdale	
89-0510	AC130U	412TW	ED	AFFTC
89-0511	AC130U	412TW	ED	AFFTC
89-0512	AC130U			
89-0513	AC130U	Lockheed	Palmdale	
89-0514	AC130U	Lockheed	Palmdale	
89-1051	C130H	105ALS	TN ANG	
89-1052	C130H	105ALS	TN ANG	
89-1053	C130H	105ALS	TN ANG	
89-1054	C130H	105ALS	TN ANG	
89-1055	C130H	105ALS	TN ANG	
89-1056	C130H	105ALS	TN ANG	
89-1181	C130H	105ALS	TN ANG	
89-1182	C130H	105ALS	TN ANG	
89-1183	C130H	105ALS	TN ANG	
89-1184	C130H	105ALS	TN ANG	
89-1185	C130H	105ALS	TN ANG	
89-1186	C130H	105ALS	TN ANG	
89-1187	C130H	105ALS	TN ANG	
89-1188	C130H	105ALS	TN ANG	
89-9101	C130H	757ALS	AFRes	
89-9102	C130H	757ALS	AFRes	
89-9103	C130H	757ALS	AFRes	
89-9104	C130H	757ALS	AFRes	
89-9105	C130H	757ALS	AFRes	
89-9106	C130H	757ALS	AFRes	
90-0161	MC130H			
90-0162	MC130H			
90-0163	MC130H			
90-0164	AC130U			
90-0165	AC130U			
90-0166	AC130U			
90-0167	AC130U	Lockheed	Chino	
90-1057	C130H	105ALS	TN ANG	
90-1058	C130H	105ALS	TN ANG	
90-1791	C130H	164ALS	OH ANG	
90-1792	C130H	164ALS	OH ANG	
90-1793	C130H	164ALS	OH ANG	
90-1794	C130H	164ALS	OH ANG	
90-1795	C130H	164ALS	OH ANG	
90-1796	C130H	164ALS	OH ANG	
90-1797	C130H	164ALS	OH ANG	
90-1798	C130H	164ALS	OH ANG	
90-2103	HC130H	210RQS	AK ANG	
90-9107	C130H	757ALS	AFRes	
90-9108	C130H	757ALS	AFRes	
91-1231	C130H	165ALS	KY ANG	
91-1232	C130H	165ALS	KY ANG	
91-1233	C130H	165ALS	KY ANG	
91-1234	C130H	165ALS	KY ANG	
91-1235	C130H	165ALS	KY ANG	
91-1236	C130H	165ALS	KY ANG	
91-1237	C130H	165ALS	KY ANG	
91-1238	C130H	165ALS	KY ANG	

91-1239	C130H	165ALS	KY ANG
91-1651	C130H	165ALS	KY ANG
91-1652	C130H	165ALS	KY ANG
91-1653	C130H	165ALS	KY ANG
91-2103	HC130H	210RQS	AK AK ANG

91-9141	C130H	328ALS	AFRes
91-9142	C130H	328ALS	AFRes
91-9143	C130H	328ALS	AFRes
91-9144	C130H	328ALS	AFRes

C135 STRATOTANKER

c\n
18168

17234 to 17262, 17340 to 17407, 17489 to 17585, 17725 to 17875, 17931 to 18011, 18088 to 18153, to 18239, 18292, 18333, 18340 to 18350, 18480 to 18568, 18465 to 18479, 18593 to 18669, 18701 to 18705, 18670 to 18673, 18719 to 18736, 18706, 18768 to 18789.

55-3118	EC135K	8TDCS	
55-3119	NKC135A	AMARC	CA...
55-3120	KC135A	4950TW	AFMC\ASD
55-3121	RC135T	w.o.25.2.85	
55-3122	NKC135A	4950TW	AFMC\ASD
55-3123	NKC135A	USAF	Museum
55-3124	NKC135A	TTC	Sheppard
55-3125	EC135Y	CinC	CENTCOM
55-3126	KC135A	scrapped	
55-3127	NKC135A	AMARC	CA040
55-3128	NKC135A	4950TW	AFMC\ASD
55-3129	EC135P	AMARC	CA008
55-3130	KC135A	preserved	March AFB
55-3131	NKC135A	AMARC	CA053
55-3132	NKC135A	4950TW	AFMC\ASD
55-3133	KC135A	w.o.24.9.68	
55-3134	NKC135A	to US Navy as 553134	
55-3135	NKC135E	4950TW	AFMC\ASD
55-3136	KC135A	96W	DY
55-3137	KC135A	AMARC	CA070
55-3138	KC135A	w.o.2.10.68	
55-3139	KC135A	93ARS	
55-3140	KC135A	w.o.19.4.67	
55-3141	KC135E	116ARS	WA ANG
55-3142	KC135A		
55-3143	KC135E	197ARS	AZ ANG
55-3144	KC135A	w.o.8.8.62	
55-3145	KC135E	314ARS	AFRes
55-3146	KC135E	141ARS	NJ ANG
56-3591	KC135A	906ARS	
56-3592	KC135A	w.o.4.10.89	
56-3593	KC135E	133ARS	NH ANG
56-3594	KC135A	AMARC	CA047
56-3595	KC135A	2W	LA
56-3596	NKC135A	to US Navy as 563596	
56-3597	KC135A	w.o.27.2.63	
56-3598	KC135A	w.o.25.11.58	
56-3599	KC135A	w.o.27.6.58	
56-3600	KC135A	2W	LA
56-3601	KC135A	AMARC	CA072
56-3602	KC135A	w.o.25.3.69	
56-3603	KC135A	AMARC	CA029
56-3604	KC135A	117ARS	KS ANG
56-3605	KC135A	w.o.18.11.60	
56-3606	KC135E	132ARS	ME ANG
56-3607	KC135E	151ARS	TN ANG
56-3608	KC135A	AMARC	CA030
56-3609	KC135E	151ARS	TN ANG
56-3610	KC135A	AMARC	CA055
56-3611	KC135E	146ARS	PA ANG
56-3612	KC135E	146ARS	PA ANG
56-3613	KC135A	w.o.19.1.67	
56-3614	KC135A		
56-3615	KC135A	AMARC	CA032
56-3616	KC135A	96W	DY
56-3617	KC135A	93ARS	
56-3618	KC135A	w.o.9.5.62	
56-3619	KC135A	AMARC	CA051
56-3620	KC135A	906ARS	
56-3621	KC135A		
56-3622	KC135E	132ARS	ME ANG

56-3623	KC135E	336ARS	AFRes
56-3624	KC135A	46ARS	
56-3625	KC135A	96W	DY
56-3626	KC135E	133ARS	NH ANG
56-3627	KC135A	AMARC	CA054
56-3628	KC135A	w.o.3.2.60	
56-3629	KC135A	w.o.19.12.69	
56-3630	KC135E	146ARS	PA ANG
56-3631	KC135E	117ARS	KS ANG
56-3632	KC135A	2W	LA
56-3633	KC135A	AMARC	CA037
56-3634	KC135A	AMARC	CA049
56-3635	KC135A	AMARC	CA038
56-3636	KC135A	AMARC	CA031
56-3637	KC135A	AMARC	CA033
56-3638	KC135E	197ARS	AZ ANG
56-3639	KC135A	96W	DY
56-3640	KC135E	132ARS	ME ANG
56-3641	KC135E	117ARS	KS ANG
56-3642	KC135A	906ARS	
56-3643	KC135A	151ARS	TN ANG
56-3644	KC135A	AMARC	CA028
56-3645	KC135E	314ARS	AFRes
56-3646	KC135A	AMARC	CA034
56-3647	KC135A	AMARC	CA044
56-3648	KC135E		
56-3649	KC135A	AMARC	CA065
56-3650	KC135E	133ARS	NH ANG
56-3651	KC135A	AMARC	CA041
56-3652	KC135A	96W	DY
56-3653	KC135A	AMARC	CA025
56-3654	KC135E	132ARS	ME ANG
56-3655	KC135A	w.o.30.7.68	
56-3656	KC135A		
56-3657	KC135A	w.o.25.1.62	
56-3658	KC135E	117ARS	KS ANG
57-1418	KC135R	153ARS	MS ANG
57-1419	KC135R	4530G	
57-1420	KC135R	AMARC	CA046
57-1421	KC135E	116ARS	WA ANG
57-1422	KC135E	63ARS	AFRes
57-1423	KC135E	147ARS	PA ANG
57-1424	KC135A	w.o.17.5.66	
57-1425	KC135E	151ARS	TN ANG
57-1426	KC135E	108ARS	IL ANG
57-1427	KC135A	145ARS	OH ANG
57-1428	KC135E	133ARS	NH ANG
57-1429	KC135E	117ARS	KS ANG
57-1430	KC135R	43ARW	
57-1431	KC135E	141ARS	NJ ANG
57-1432	KC135R	380ARW	
57-1433	KC135E	197ARS	AZ ANG
57-1434	KC135E	116ARS	WA ANG
57-1435	KC135R	4530G	
57-1436	KC135E	133ARS	NH ANG
57-1437	KC135R		
57-1438	KC135E	72ARS	AFRes
57-1439	KC135R	92W	
57-1440	KC135R	18W	ZZ
57-1441	KC135E	108ARS	IL ANG
57-1442	KC135A	w.o.16.1.65	

57-1443	KC135E	132ARS	ME ANG
57-1444	KC135A	w.o.18.5.66	
57-1445	KC135E	141ARS	NJ ANG
57-1446	KC135A	w.o.22.6.59	
57-1447	KC135E		
57-1448	KC135E	168ARS	AK ANG
57-1449	KC135A	w.o.3.2.60	
57-1450	KC135E	132ARS	ME ANG
57-1451	KC135A	168ARS	AK ANG
57-1452	KC135E	197ARS	AZ ANG
57-1453	KC135R	42ARS	
57-1454	KC135R	42ARS	
57-1455	KC135E	151ARS	TN ANG
57-1456	KC135R	4530G	
57-1457	KC135A	w.o.3.2.60	
57-1458	KC135E	108ARS	IL ANG
57-1459	KC135E	46ARS	
57-1460	KC135E	117ARS	KS ANG
57-1461	KC135R	4570G	
57-1462	KC135E	145ARS	OH ANG
57-1463	KC135E	117ARS	KS ANG
57-1464	KC135E	141ARS	NJ ANG
57-1465	KC135E	168ARS	AK ANG
57-1466	KC135A	w.o.8.3.60	
57-1467	KC135A	AMARC	CA039
57-1468	KC135R	336ARS	AFRes
57-1469	KC135R	42ARS	
57-1470	KC135R	126ARS	WI ANG
57-1471	KC135E	132ARS	ME ANG
57-1472	KC135R	42ARS	
57-1473	KC135R	509ARS	
57-1474	KC135R	305ARW	
57-1475	KC135E	197ARS	AZ ANG
57-1476	KC135A	AMARC	CA035
57-1477	KC135A	AMARC	CA036
57-1478	KC135E	151ARS	TN ANG
57-1479	KC135E	336ARS	AFRes
57-1480	KC135E	108ARS	IL ANG
57-1481	KC135E	w.o.20.9.89	
57-1482	KC135E	117ARS	KS ANG
57-1483	KC135R	18W	ZZ
57-1484	KC135E	197ARS	AZ ANG
57-1485	KC135E	151ARS	TN ANG
57-1486	KC135R	905ARS	
57-1487	KC135R	305ARW	
57-1488	KC135R	4570G	
57-1489	KC135A	w.o.13.3.82	
57-1490	KC135A	AMARC	CA050
57-1491	KC135E	132ARS	ME ANG
57-1492	KC135E	168ARS	AK ANG
57-1493	KC135R	509ARS	
57-1494	KC135E	168ARS	AK ANG
57-1495	KC135E	197ARS	AZ ANG
57-1496	KC135E	197ARS	AZ ANG
57-1497	KC135A	150ARS	NJ ANG
57-1498	KC135A	w.o.21.6.63	
57-1499	KC135R	4530G	
57-1500	KC135A	w.o.5.3.74	
57-1501	KC135E	116ARS	ANG
57-1502	KC135R	42ARS	
57-1503	KC135E	151ARS	TN ANG
57-1504	KC135E	63ARS	AFRes
57-1505	KC135E	132ARS	ME ANG
57-1506	KC135E	509ARS	
57-1507	KC135E	141ARS	NJ ANG
57-1508	KC135R	509ARS	
57-1509	KC135E	147ARS	PA ANG
57-1510	KC135E	191ARS	UT ANG
57-1511	KC135E	314ARS	AFRes
57-1512	KC135E	336ARS	AFRes
57-1513	KC135A	w.o.15.10.59	
57-1514	KC135R	126ARS	WI ANG
57-2589	KC135E	55WOF	HQ SAC
57-2590	KC135A	AMARC	CA042
57-2591	KC135A	AMARC	CA...
57-2592	KC135A	AMARC	CA066
57-2593	KC135R	43ARW	
57-2594	KC135E	108ARS	IL ANG
57-2595	KC135E	147ARS	PA ANG
57-2596	KC135A	906ARS	
57-2597	KC135R	19ARW	
57-2598	KC135E	336ARS	AFRes
57-2599	KC135R	4530G	

57-2600	KC135E	116ARS	ANG
57-2601	KC135E	151ARS	TN ANG
57-2602	KC135A	93ARS	
57-2603	KC135E	336ARS	AFRes
57-2604	KC135E	146ARS	PA ANG
57-2605	KC135R	4570G	
57-2606	KC135E	150ARS	NJ ANG
57-2607	KC135E	147ARS	PA ANG
57-2608	KC135E	147ARS	PA ANG
57-2609	KC135A		
58-0001	KC135R	97AMW	
58-0002	KC135A	w.o.31.3.59	
58-0003	KC135E	108ARS	IL ANG
58-0004	KC135R	153ARS	MS ANG
58-0005	KC135E	117ARS	KS ANG
58-0006	KC135E	191ARS	UT ANG
58-0007	EC135P	w.o.3.1.80	
58-0008	KC135E	141ARS	NJ ANG
58-0009	KC135R	126ARS	WI ANG
58-0010	KC135R	153ARS	MS ANG
58-0011	KC135R	305ARW	ARTS equipped
58-0012	KC135E	191ARS	UT ANG
58-0013	KC135E	63ARS	AFRes
58-0014	KC135E	108ARS	IL ANG
58-0015	KC135R	42ARS	
58-0016	KC135R	4530G	
58-0017	KC135E	146ARS	PA ANG
58-0018	KC135R	305ARW	ARTS equipped
58-0019	EC135P	AMARC	CA009
58-0020	KC135E	116ARS	WA ANG
58-0021	KC135R	126ARS	WI ANG
58-0022	EC135P	AMARC	CA011
58-0023	KC135R	93ARS	
58-0024	KC135E	146ARS	PA ANG
58-0025	KC135A	93ARS	
58-0026	KC135A	w.o.17.1.68	
58-0027	KC135R	905ARS	
58-0028	KC135A	46ARS	
58-0029	KC135A	AMARC	CA056
58-0030	KC135R	509ARS	
58-0031	KC135A	w.o.19.3.82	
58-0032	KC135E	150ARS	NJ ANG
58-0033	KC135A	AMARC	CA026
58-0034	KC135R	380ARW	
58-0035	KC135R	384ARS	
58-0036	KC135A	96W	DY
58-0037	KC135E	147ARS	PA ANG
58-0038	KC135R	4530G	
58-0039	KC135Q	w.o.3.6.71	
58-0040	KC135E	150ARS	NJ ANG
58-0041	KC135E	63ARS	AFRes
58-0042	KC135Q	380ARW	
58-0043	KC135E	191ARS	UT ANG
58-0044	KC135E	141ARS	NJ ANG
58-0045	KC135Q	2W	LA
58-0046	KC135Q	380ARW	
58-0047	KC135Q	380ARW	
58-0048	KC135A	w.o.13.3.72	
58-0049	KC135Q	380ARW	
58-0050	KC135Q	380ARW	
58-0051	KC135R	43ARW	
58-0052	KC135E	336ARS	AFRes
58-0053	KC135E	314ARS	AFRes
58-0054	KC135Q	2W	LA
58-0055	KC135Q	9W	BB
58-0056	KC135R	4530G	
58-0057	KC135E	108ARS	IL ANG
58-0058	KC135E	314ARS	AFRes
58-0059	KC135R	153ARS	MS ANG
58-0060	KC135Q	2W	LA
58-0061	KC135Q	380ARW	
58-0062	KC135Q	96W	DY
58-0063	KC135R	42ARS	
58-0064	KC135E	314ARS	AFRes
58-0065	KC135Q	380ARW	
58-0066	KC135R	19ARW	
58-0067	KC135E	108ARS	IL ANG
58-0068	KC135E	108ARS	IL ANG
58-0069	KC135Q	380ARW	
58-0070	GKC135A	TTC	Sheppard
58-0071	KC135Q	9W	BB
58-0072	KC135Q	9W	BB
58-0073	KC135R	509ARS	

Serial	Type	Unit	Notes
58-0074	KC135Q	9W	BB
58-0075	KC135R	4570G	
58-0076	KC135R	72ARS	AFRes
58-0077	KC135Q	380ARW	
58-0078	KC135E	150ARS	NJ ANG
58-0079	KC135R	906ARS	
58-0080	KC135E	191ARS	UT ANG
58-0081	KC135A	AMARC	CA052
58-0082	KC135E	116ARS	WA ANG
58-0083	KC135R	4530G	
58-0084	KC135Q	9W	BB
58-0085	KC135E	336ARS	AFRes
58-0086	KC135Q	9W	BB
58-0087	KC135E	150ARS	NJ ANG
58-0088	KC135Q	2W	LA
58-0089	KC135Q	9W	BB
58-0090	KC135E	314ARS	AFRes
58-0091	KC135A	906ARS	
58-0092	KC135R	509ARS	
58-0093	KC135R	42ARS	
58-0094	KC135Q	9W	BB
58-0095	KC135Q	2W	LA
58-0096	KC135E	314ARS	AFRes
58-0097	KC135A	AMARC	CA043
58-0098	KC135R	93ARS	
58-0099	KC135Q	9W	BB
58-0100	KC135R		
58-0101	KC135A	w.o.29.4.77	
58-0102	KC135R	305ARW	
58-0103	KC135Q	9W	BB
58-0104	KC135R	509ARS	
58-0105	KC135A	906ARS	
58-0106	KC135A	96W	DY
58-0107	KC135E	191ARS	UT ANG
58-0108	KC135E	314ARS	AFRes
58-0109	KC135R	153ARS	MS ANG
58-0110	KC135A	46ARS	
58-0111	KC135E	141ARS	NJ ANG
58-0112	KC135Q	9W	BB
58-0113	KC135R	384ARS	
58-0114	KC135A	96W	DY
58-0115	KC135E	150ARS	NJ ANG
58-0116	KC135E	197ARS	AZ ANG
58-0117	KC135Q	380ARW	
58-0118	KC135R	42ARS	
58-0119	KC135A	906ARS	
58-0120	KC135R	4570G	
58-0121	KC135R	93ARS	
58-0122	KC135R	4530G	
58-0123	KC135R	384ARS	
58-0124	KC135R	305ARW	ARTS equipped
58-0125	KC135Q	9W	BB
58-0126	KC135R	305ARW	ARTS equipped
58-0127	KC135A	w.o.19.9.79	
58-0128	KC135R	100ARW	
58-0129	KC135Q	9W	BB
58-0130	KC135R	126ARS	WI ANG
59-1443	KC135A	w.o.27.8.85	
59-1444	KC135R	145ARS	OH ANG
59-1445	KC135E	116ARS	WA ANG
59-1446	KC135R	153ARS	MS ANG
59-1447	KC135R	63ARS	AFRes
59-1448	KC135E	133ARS	NH ANG
59-1449	KC135A	46ARS	
59-1450	KC135E	133ARS	NH ANG
59-1451	KC135E	63ARS	AFRes
59-1452	KC135E	116ARS	WA ANG
59-1453	KC135R	145ARS	OH ANG
59-1454	KC135A	AMARC	CA...
59-1455	KC135R	153ARS	MS ANG
59-1456	KC135E	141ARS	NJ ANG
59-1457	KC135E	147ARS	PA ANG
59-1458	KC135R	145ARS	OH ANG
59-1459	KC135R		
59-1460	KC135Q	380ARW	
59-1461	KC135R	97AMW	
59-1462	KC135Q	2W	LA
59-1463	KC135R	4530G	
59-1464	KC135Q	380ARW	
59-1465	KC135A	w.o.17.7.67	
59-1466	KC135R	905ARS	
59-1467	KC135Q	380ARW	
59-1468	KC135Q	9W	BB
59-1469	KC135R	4530G	
59-1470	KC135Q	2W	LA
59-1471	KC135Q	2W	LA
59-1472	KC135R	97AMW	
59-1473	KC135E	191ARS	UT ANG
59-1474	KC135R	96W	DY
59-1475	KC135R	43ARW	
59-1476	KC135R	97AMW	
59-1477	KC135E	63ARS	AFRes
59-1478	KC135E	19ARW	
59-1479	KC135E	146ARS	PA ANG
59-1480	KC135Q	9W	BB
59-1481	KC135	to NASA	N930NA
59-1482	KC135R	384ARS	
59-1483	KC135R	42ARS	
59-1484	KC135E	147ARS	PA ANG
59-1485	KC135E	150ARS	NJ ANG
59-1486	KC135A	906ARS	
59-1487	KC135E	108ARS	IL ANG
59-1488	KC135R	97AMW	
59-1489	KC135E	191ARS	UT ANG
59-1490	KC135Q	2W	LA
59-1491	RC135S	w.o.13.1.69	
59-1492	KC135R	4530G	
59-1493	KC135E	132ARS	ME ANG
59-1494	KC135E	w.o.11.1.90	
59-1495	KC135E	18W	ZZ
59-1496	KC135E	147ARS	PA ANG
59-1497	KC135E	150ARS	NJ ANG
59-1498	KC135E	28ARS	
59-1499	KC135E	133ARS	NH ANG
59-1500	KC135A	380ARW	
59-1501	KC135A	379W	
59-1502	KC135E	42ARS	
59-1503	KC135E	141ARS	NJ ANG
59-1504	KC135Q	2W	LA
59-1505	KC135E	133ARS	NH ANG
59-1506	KC135E	147ARS	PA ANG
59-1507	KC135R	28ARS	
59-1508	KC135R	380ARW	
59-1509	KC135R	133ARS	NH ANG
59-1510	KC135Q	96W	DY
59-1511	KC135Q	19ARW	
59-1512	KC135Q	2W	LA
59-1513	KC135Q	2W	LA
59-1514	KC135E	55WOF	ARTS equipped
59-1515	KC135R	384ARS	
59-1516	KC135R	117ARS	KS ANG
59-1517	KC135R	380ARW	
59-1518	EC135K	8TDCS	
59-1519	KC135E	146ARS	PA ANG
59-1520	KC135Q	9W	BB
59-1521	KC135R	384ARS	
59-1522	KC135R	380ARW	
59-1523	KC135Q	9W	BB
60-0313	KC135R	905ARS	
60-0314	KC135R	42ARS	
60-0315	KC135R	126ARS	WI ANG
60-0316	KC135A	116ARS	WA ANG
60-0317	KC135A	w.o.11.10.88	
60-0318	KC135A	19ARW	
60-0319	KC135R	19ARW	
60-0320	KC135R	4530G	
60-0321	KC135R	28ARS	
60-0322	KC135R	19ARW	
60-0323	KC135R	380ARW	
60-0324	KC135R	905ARS	
60-0325	KC135A	2W	LA
60-0326	KC135A	96W	DY
60-0327	KC135E	191ARS	UT ANG
60-0328	KC135R	4530G	
60-0329	KC135R	19ARW	
60-0330	KC135E	w.o.13.2.87	
60-0331	KC135R	43ARW	
60-0332	KC135A	97AMW	
60-0333	KC135R	305ARW	
60-0334	KC135R	126ARS	WI ANG
60-0335	KC135Q	2W	LA
60-0336	KC135Q	18W	ZZ
60-0337	KC135Q	2W	LA
60-0338	KC135Q	w.o.8.2.80	
60-0339	KC135Q	96W	DY
60-0340	KC135A	w.o.8.7.64	

Serial	Type	Unit/Status	Notes
60-0341	KC135R	145ARS	OH ANG
60-0342	KC135Q	9W	BB
60-0343	KC135Q	380ARW	
60-0344	KC135Q	2W	LA
60-0345	KC135Q	96W	DY
60-0346	KC135Q	2W	LA
60-0347	KC135R	384ARS	
60-0348	KC135R	43ARW	
60-0349	KC135R	46ARS	
60-0350	KC135R	43ARW	
60-0351	KC135R	43ARW	
60-0352	KC135A	w.o.10.9.62	
60-0353	KC135A	28ARS	
60-0354	KC135A	w.o.7.12.75	
60-0355	KC135A	96W	DY
60-0356	KC135R	305ARW	ARTS equipped
60-0357	KC135R	305ARW	ARTS equipped
60-0358	KC135R	43ARW	
60-0359	KC135R	19ARW	
60-0360	KC135R	43ARW	
60-0361	KC135A	w.o.13.3.87	
60-0362	KC135R	305ARW	ARTS equipped
60-0363	KC135R	74ARS	AFRes
60-0364	KC135R	74ARS	AFRes ARTS
60-0365	KC135R	366W	MO
60-0366	KC135R	43ARW	
60-0367	KC135R	145ARS	OH ANG
60-0368	KC135A	w.o.6.2.76	
60-0369	GNC135A	TTC	Chanute
60-0370	NC135A	AMARC	CA005
60-0371	NC135A	4950TW	AFMC\ASD
60-0372	C135E	4950TW	AFMC\ASD
60-0373	C135A	w.o.25.6.65	
60-0374	EC135E	4950TW	AFMC\ASD
60-0375	C135E	4950TW	AFMC\ASD
60-0376	C135E	1SW	OK
60-0377	C135A	4950TW	AFMC\ASD
60-0378	C135A	55W	OF
61-0261	EC135L	AMARC	CA020
61-0262	EC135A	preserved	Ellsworth AFB
61-0263	EC135L	AMARC	CA015
61-0264	KC135R	145ARS	OH ANG
61-0265	KC135A	w.o.4.1.65	
61-0266	KC135R	42ARS	
61-0267	KC135R	384ARS	
61-0268	KC135E	314ARS	AFRes
61-0269	EC135L	preserved	Grissom AFB
61-0270	KC135E	72ARS	AFRes
61-0271	KC135E	63ARS	AFRes
61-0272	KC135R	305ARW	
61-0273	KC135A	w.o.17.1.66	
61-0274	EC135P	AMARC	CA010
61-0275	KC135R	905ARS	
61-0276	KC135R	384ARS	
61-0277	KC135R	4570G	
61-0278	EC135A	AMARC	CA048
61-0279	EC135L	AMARC	CA018
61-0280	KC135E	336ARS	AFRes
61-0281	KC135E	197ARS	AZ ANG
61-0282	GEC135H	TTC	Sheppard
61-0283	EC135L	AMARC	CA016
61-0284	KC135R	4530G	
61-0285	EC135H	AMARC	CA012
61-0286	GEC135H	TTC	Sheppard
61-0287	EC135A	preserved	Offutt AFB
61-0288	KC135A	28ARS	
61-0289	EC135A	AMARC	CA022
61-0290	KC135R	905ARS	
61-0291	EC135H	scrapped	
61-0292	KC135R	384ARS	
61-0293	KC135R	305ARW	ARTSequipped
61-0294	KC135R	42ARS	
61-0295	KC135R	97AMW	
61-0296	KC135A	w.o.26.9.76	
61-0297	EC135A	AMARC	CA021
61-0298	KC135R	126ARS	WI ANG
61-0299	KC135R	905ARS	
61-0300	KC135R	4570G	
61-0301	KC135A	w.o.22.10.68	
61-0302	KC135R	509ARS	
61-0303	KC135E	336ARS	AFRes
61-0304	KC135R	384ARS	
61-0305	KC135R	4530G	
61-0306	KC135R	384ARS	
61-0307	KC135R	19ARW	
61-0308	KC135R	384ARS	
61-0309	KC135R	126ARS	WI ANG
61-0310	KC135R	384ARS	
61-0311	KC135R	4570G	
61-0312	KC135R	28ARS	
61-0313	KC135R	100ARW	
61-0314	KC135R	19ARW	
61-0315	KC135R	28ARS	
61-0316	VC135A	w.o.19.3.85	
61-0317	KC135R	906ARS	
61-0318	KC135R	93ARS	
61-0319	KC135A	w.o.28.8.63	
61-0320	KC135R	509ARS	
61-0321	KC135R	100ARW	
61-0322	KC135A	w.o.28.8.63	
61-0323	KC135R	18W	ZZ
61-0324	KC135R	384ARS	
61-0325	KC135R	906ARS	
61-0326	EC135E	4950TW	AFMC\ASD
61-0327	EC135N	CinC	CENTCOM
61-0328	EC135N	w.o.6.5.81	
61-0329	EC135E	4950TW	AFMC\ASD
61-0330	EC135E	4950TW	AFMC\ASD
61-0331	C135B	w.o.13.6.71	
61-0332	C135B	w.o.11.5.64	
61-2662	RC135S	55W	OF
61-2663	RC135S	55W	OF
61-2664	RC135S	w.o.15.3.81	
61-2665	WC135B	5RS	
61-2666	WC135B	E-Systems	
61-2667	WC135B	55W	OF
61-2668	C135C	89AW	Det1
61-2669	C135C	4950TW	AFMC\ASD
61-2670	WC135B	5RS	
61-2671	C135C	preserved	Tinker AFB
61-2672	WC135B	5RS	
61-2673	WC135B	5RS	
61-2674	WC135B	5RS	
62-3497	KC135A	AMARC	CA...
62-3498	KC135R	4530G	
62-3499	KC135R	43ARW	
62-3500	KC135R	126ARS	WI ANG
62-3501	KC135A	AMARC	CA068
62-3502	KC135R	380ARW	
62-3503	KC135R	509ARS	
62-3504	KC135R	906ARS	
62-3505	KC135R	509ARS	
62-3506	KC135R	19ARW	
62-3507	KC135R	97AMW	
62-3508	KC135R	42ARS	
62-3509	KC135R	93ARS	
62-3510	KC135R	4570G	
62-3511	KC135R	145ARS	OH ANG
62-3512	KC135R	509ARS	
62-3513	KC135R	366W	MO
62-3514	KC135R	305ARW	
62-3515	KC135R	305ARW	
62-3516	KC135R		
62-3517	KC135R	380ARW	
62-3518	KC135R	43ARW	
62-3519	KC135R	905ARS	
62-3520	KC135R	19ARW	
62-3521	KC135R	74ARS	AFRes
62-3522	KC135A	w.o.4.3.77	
62-3523	KC135R	19ARW	
62-3524	KC135R	97AMW	
62-3525	KC135A	906ARS	
62-3526	KC135R	97AMW	
62-3527	KC135E	150ARS	NJ ANG
62-3528	KC135R	Boeing	Wichita
62-3529	KC135A	4530G	
62-3530	KC135R	72ARS	AFRes
62-3531	KC135R	145ARS	OH ANG
62-3532	KC135A	AMARC	CA076
62-3533	KC135R	43ARW	
62-3534	KC135R	19ARW	
62-3535	KC135A	w.o.6.4.70	
62-3536	EC135K	w.o.14.9.77	
62-3537	KC135R	43ARW	
62-3538	KC135R	43ARW	
62-3539	KC135A	96W	DY

Serial	Type	Unit	Code
62-3540	KC135R	28ARS	
62-3541	KC135R	305ARW	
62-3542	KC135R	43ARW	
62-3543	KC135R	19ARW	
62-3544	KC135R	42ARS	
62-3545	KC135R	19ARW	
62-3546	KC135R	43ARW	
62-3547	KC135R	4570G	
62-3548	KC135R	305ARW	
62-3549	KC135R	43ARW	
62-3550	KC135R	19ARW	
62-3551	KC135R	509ARS	
62-3552	KC135R	19ARW	
62-3553	KC135R	905ARS	
62-3554	KC135R	19ARW	
62-3555	KC135A	46ARS	
62-3556	KC135R	97AMW	
62-3557	KC135R	19ARW	
62-3558	KC135R	100ARW	
62-3559	KC135R	509ARS	
62-3560	KC135A	906ARS	
62-3561	KC135R	100ARW	
62-3562	KC135R	Boeing	Wichita
62-3563	KC135A	906ARS	
62-3564	KC135R	28ARS	
62-3565	KC135R	905ARS	
62-3566	KC135A		
62-3567	KC135A	46ARS	
62-3568	KC135R	4570G	
62-3569	KC135R	19ARW	
62-3570	EC135G	AMARC	CA024
62-3571	KC135R	145ARS	OH ANG
62-3572	KC135R	509ARS	
62-3573	KC135R	509ARS	
62-3574	KC135A	AMARC	CA069
62-3575	KC135R	43ARW	
62-3576	KC135R	384ARS	
62-3577	KC135R	100ARW	
62-3578	KC135R	4530G	
62-3579	EC135G	AMARC	CA023
62-3580	KC135R	509ARS	
62-3581	EC135C	2ACCS	OF
62-3582	EC135C	2ACCS	OF
62-3583	EC135C	AMARC	CA019
62-3584	EC135J	w.o.29.5.92	
62-3585	EC135C	2ACCS	OF
62-4125	C135B	55W	OF
62-4126		86W	RS
62-4127	C135B	55W	OF
62-4128	RC135X		
62-4129	TC135S	55W	OF
62-4130	C135B	55W	OF
62-4131	RC135W	55W	OF
62-4132	RC135W	55W	OF
62-4133	TC135S		
62-4134	RC135W	55W	OF
62-4135	RC135W	55W	OF
62-4136	C135B	w.o.23.10.62	
62-4137	RC135E	w.o.5.6.69	
62-4138	RC135W	55W	OF
62-4139	RC135W	55W	OF
63-7976	KC135R	905ARS	
63-7977	KC135R	19ARW	
63-7978	KC135R	305ARW	
63-7979	KC135R	97AMW	
63-7980	KC135R	43ARW	
63-7981	KC135R	97AMW	
63-7982	KC135R	43ARW	
63-7983	KC135A	w.o.17.6.86	
63-7984	KC135R	97AMW	
63-7985	KC135R	305ARW	
63-7986	KC135A	906ARS	
63-7987	KC135A	46ARS	
63-7988	KC135R	4530G	
63-7989	KC135A	w.o.8.3.73	
63-7990	KC135A	w.o.31.1.89	
63-7991	KC135R	28ARS	
63-7992	KC135R	43ARW	
63-7993	KC135R	4570G	
63-7994	EC135G	AMARC	CA045
63-7995	KC135R	19ARW	
63-7996	KC135R	305ARW	
63-7997	KC135R	384ARS	
63-7998	KC135A	96W	DY
63-7999	KC135R	905ARS	
63-8000	KC135A	380ARW	
63-8001	EC135G	AMARC	CA017
63-8002	KC135R	19ARW	
63-8003	KC135R	100ARW	
63-8004	KC135R	366W	MO
63-8005	KC135A	2W	LA
63-8006	KC135R	905ARS	
63-8007	KC135R	19ARW	
63-8008	KC135R	19ARW	
63-8009	KC135A	46ARS	
63-8010	KC135A	46ARS	
63-8011	KC135R	905ARS	
63-8012	KC135R	93ARS	
63-8013	KC135R	4530G	
63-8014	KC135R	380ARW	
63-8015	KC135R	43ARW	
63-8016	KC135A	93ARS	
63-8017	KC135R	380ARW	
63-8018	KC135R	42ARS	
63-8019	KC135A	380ARW	
63-8020	KC135R		
63-8021	KC135R	905ARS	
63-8022	KC135R	42ARS	
63-8023	KC135R	43ARW	
63-8024	KC135R	305ARW	
63-8025	KC135R	905ARS	
63-8026	KC135A	46ARS	
63-8027	KC135R	43ARW	
63-8028	KC135R	305ARW	
63-8029	KC135R	126ARS	WI ANG
63-8030	KC135R	905ARS	
63-8031	KC135R	384ARS	
63-8032	KC135R	384ARS	
63-8033	KC135R		
63-8034	KC135A	96W	DY
63-8035	KC135R	509ARS	
63-8036	KC135R	19ARW	
63-8037	KC135R	93ARS	
63-8038	KC135R	509ARS	
63-8039	KC135R	43ARW	
63-8040	KC135R	28ARS	
63-8041	KC135R	43ARW	
63-8042	KC135A	w.o.3.1.65	
63-8043	KC135A	96W	DY
63-8044	KC135A	96W	DY
63-8045	KC135A	46ARS	
63-8046	EC135C	2ACCS	OF
63-8047	EC135C	2ACCS	OF
63-8048	EC135C	2ACCS	OF
63-8049	EC135C	2ACCS	OF
63-8050	EC135C	2ACCS	OF
63-8051	EC135C		
63-8052	EC135C	2ACCS	OF
63-8053	EC135C	2ACCS	OF
63-8054	EC135C	2ACCS	OF
63-8055	EC135J		
63-8056	EC135J	AMARC	CA013
63-8057	EC135J	AMARC	CA014
63-8058	KC135D	168ARS	AK ANG
63-8059	KC135D	168ARS	AK ANG
63-8060	KC135D	168ARS	AK ANG
63-8061	KC135D	168ARS	AK ANG
63-8871	KC135R	305ARW	
63-8872	KC135R	42ARS	
63-8873	KC135R	380ARW	
63-8874	KC135R	28ARS	
63-8875	KC135R	100ARW	
63-8876	KC135R	4530G	
63-8877	KC135A	906ARS	
63-8878	KC135R	97AMW	
63-8879	KC135A	906ARS	
63-8880	KC135R	97AMW	
63-8881	KC135A	46ARS	
63-8882	KC135A	w.o.26.2.65	
63-8883	KC135R	18W	ZZ
63-8884	KC135R	43ARW	
63-8885	KC135A	46ARS	
63-8886	KC135R	509ARS	
63-8887	KC135A	906ARS	
63-8888	KC135A	93ARS	
63-9792	RC135V	55W	OF

64-14828	KC135R	384ARS	
64-14829	KC135R	509ARS	
64-14830	KC135R	305ARW	
64-14831	KC135R	4530G	
64-14832	KC135R	93ARS	
64-14833	KC135A	100ARW	
64-14834	KC135R	4530G	
64-14835	KC135R	93ARS	
64-14836	KC135A	2W	LA
64-14837	KC135A	96W	DY
64-14838	KC135R	380ARW	
64-14839	KC135R	Boeing	Wichita
64-14840	KC135R	4530G	
64-14841	RC135V	55W	OF
64-14842	RC135V	55W	OF
64-14843	RC135V	55W	OF
64-14844	RC135V	55W	OF
64-14845	RC135V	55W	OF
64-14846	RC135V	55W	OF
64-14847	RC135U	55W	OF
64-14848	RC135V	55W	OF
64-14849	RC135U	55W	OF

C137

c\n   17925 to 17927, 18461, 19417, 20630, 20043, 20297.

58-6970	C137B	89AW	67-19417	EC137D	CinC CENTCOM
58-6971	C137B	89AW	72-7000	C137C	89AW
58-6972	C137B	89AW	85-6973	C137C	89AW
62-6000	C137C	89AW	85-6974	C137C	89AW

C140 JETSTAR

c\n   5010, 5026, 5028, 5030, 5032, 5017, 5022, 5024, 5027, 5031, 5034, 5041 to 5045.

59-5958	C140A	preserved Travis AFB	61-2491	C140B	BDRT	Rhein Main AFB
59-5959	C140A		61-2492	C140B	USAF Museum	
59-5960	C140A		61-2493	C140B	AMARC	CL003
59-5961	C140A	w.o.7.11.62	62-4197	C140B	AMARC	CL007
59-5962	C140A		62-4198	C140B	scrapped 12.91	
61-2488	C140B	preserved Robins AFB	62-4199	C140B	AMARC	CL002
61-2489	C140B	Pima Museum	62-4200	C140B	AMARC	CL005
61-2490	C140B	AMARC CL004	62-4201	C140B	stored	McConnell AFB

C141 STARLIFTER

c\n   6001 to 6109,6111 to 6285.

				63-8086	C141B	62AW	
61-2775	NC141A	4950TW	AFMC\ASD	63-8087	C141B	63AW
61-2776	NC141A	4950TW	AFMC\ASD	63-8088	C141B	60AW
61-2777	NC141A	4950TW	AFMC\ASD	63-8089	C141B	62AW
61-2778	C141B	155ALS	TN ANG	63-8090	C141B	438AW
61-2779	NC141A	4950TW	ARTB	64-0609	C141B	97AMW
63-8075	C141B	60AW		64-0610	C141B	437AW
63-8076	C141B	438AW		64-0611	C141B	437AW
63-8077	C141A	w.o.28.8.73		64-0612	C141B	437AW
63-8078	C141B	97AMW		64-0613	C141B	437AW
63-8079	C141B	437AW		64-0614	C141B	183ALS	MS ANG
63-8080	C141B	438AW		64-0615	C141B	437AW
63-8081	C141B	62AW		64-0616	C141B	438AW
63-8082	C141B	62AW		64-0617	C141B	63AW
63-8083	C141B	438AW		64-0618	C141B	437AW
63-8084	C141B	63AW		64-0619	C141B	437AW
63-8085	C141B	63AW		64-0620	C141B	459MAW	AFRes

64-0621	C141B	438AW	
64-0622	C141B	183ALS	MS ANG
64-0623	C141B	438AW	
64-0624	C141B	w.o.12.7.84	
64-0625	C141B	438AW	
64-0626	C141B	438AW	
64-0627	C141B	438AW	
64-0628	C141B	438AW	
64-0629	C141B	437AW	
64-0630	C141B	437AW	
64-0631	C141B	437AW	
64-0632	C141B	183ALS	MS ANG
64-0633	C141B	438AW	
64-0634	C141B	97AMW	
64-0635	C141B	62AW	
64-0636	C141B	AMARC	CR004
64-0637	C141B	459MAW	AFRes
64-0638	C141B	438AW	
64-0639	C141B	438AW	
64-0640	C141B	183ALS	MS ANG
64-0641	C141A	w.o.20.3.75	
64-0642	C141B	97AMW	
64-0643	C141B	60AW	
64-0644	C141B	437AW	
64-0645	C141B	459MAW	AFRes
64-0646	C141B	437AW	
64-0647	C141A	derelict	Richards-Gebaur AFB
64-0648	C141B	AMARC	CR005
64-0649	C141B	437AW	
64-0650	C141B	438AW	
64-0651	C141B	437AW	
64-0652	C141B	w.o.31.8.82	
64-0653	C141B	63AW	
65-0216	C141B	459MAW	AFRes
65-0217	C141B	437AW	
65-0218	C141B	437AW	
65-0219	C141B	60AW	
65-0220	C141B	437AW	
65-0221	C141B	438AW	
65-0222	C141B	155ALS	TN ANG
65-0223	C141B	438AW	
65-0224	C141B	438AW	
65-0225	C141B	63AW	
65-0226	C141B	459MAW	AFRes
65-0227	C141B	62AW	
65-0228	C141B	97AMW	
65-0229	C141B	62AW	
65-0230	C141B	60AW	
65-0231	C141B		
65-0232	C141B	62AW	
65-0233	C141B	AMARC	CR009
65-0234	C141B	60AW	
65-0235	C141B	62AW	
65-0236	C141B	97AMW	
65-0237	C141B	62AW	
65-0238	C141B	60AW	
65-0239	C141B	60AW	
65-0240	C141B	62AW	
65-0241	C141B	63AW	
65-0242	C141B	60AW	
65-0243	C141B	62AW	
65-0244	C141B	62AW	
65-0245	C141B	60AW	
65-0246	C141B	60AW	
65-0247	C141B	60AW	
65-0248	C141B	62AW	
65-0249	C141B	60AW	
65-0250	C141B	60AW	
65-0251	C141B	60AW	
65-0252	C141B	60AW	
65-0253	C141B	62AW	
65-0254	C141B	60AW	
65-0255	C141B	w.o.30.11.92	
65-0256	C141B	60AW	
65-0257	C141B	60AW	
65-0258	C141B	62AW	
65-0259	C141B	60AW	
65-0260	C141B	60AW	
65-0261	C141B	438AW	
65-0262	C141B	AMARC	CR007
65-0263	C141B	62AW	
65-0264	C141B	62AW	
65-0265	C141B	60AW	
65-0266	C141B	437AW	
65-0267	C141B	437AW	
65-0268	C141B	60AW	
65-0269	C141B	437AW	
65-0270	C141B	437AW	
65-0271	C141B	459MAW	AFRes
65-0272	C141B	437AW	
65-0273	C141B	437AW	
65-0274	C141A	w.o.19.8.74	
65-0275	C141B	437AW	
65-0276	C141B	437AW	
65-0277	C141B	62AW	
65-0278	C141B	97AMW	
65-0279	C141B	437AW	
65-0280	C141B	438AW	
65-0281	C141A	w.o.7.9.66	
65-9397	C141B	62AW	
65-9398	C141B	AMARC	CR002
65-9399	C141B	62AW	
65-9400	C141B	97AMW	
65-9401	C141B	437AW	
65-9402	C141B	437AW	
65-9403	C141B	60AW	
65-9404	C141B	62AW	
65-9405	C141B	438AW	
65-9406	C141B	63AW	
65-9407	C141B	w.o.23.3.67	
65-9408	C141B	437AW	
65-9409	C141B	438AW	
65-9410	C141B	AMARC	CR003
65-9411	C141B	438AW	
65-9412	C141B	356ALS	AFRes
65-9413	C141B	438AW	
65-9414	C141B	63AW	
66-0126	GC141B	TTC	Sheppard
66-0127	C141A	w.o.13.4.67	
66-0128	C141B	63AW	
66-0129	C141B	62AW	
66-0130	C141B	183ALS	MS ANG
66-0131	C141B	437AW	
66-0132	C141B	438AW	
66-0133	C141B	438AW	
66-0134	C141B	63AW	
66-0135	C141B	437AW	
66-0136	C141B	63AW	
66-0137	C141B	63AW	
66-0138	C141B	63AW	
66-0139	C141B	63AW	
66-0140	C141B	438AW	
66-0141	C141B	62AW	
66-0142	C141B	w.o.30.11.92	
66-0143	C141B	AMARC	CR001
66-0144	C141B	437AW	
66-0145	C141B	62AW	
66-0146	C141B	97AMW	
66-0147	C141B	60AW	
66-0148	C141B	60AW	
66-0149	C141B	437AW	
66-0150	C141B	w.o.20.2.89	
66-0151	C141B	60AW	
66-0152	C141B	437AW	
66-0153	C141B	459MAW	AFRes
66-0154	C141B	97AMW	
66-0155	C141B	438AW	
66-0156	C141B	62AW	
66-0157	C141B	155ALS	TN ANG
66-0158	C141B	62AW	
66-0159	C141B	62AW	
66-0160	C141B	437AW	
66-0161	C141B	62AW	
66-0162	C141B	438AW	
66-0163	C141B	437AW	
66-0164	C141B	183ALS	MS ANG
66-0165	C141B	62AW	
66-0166	C141B	438AW	
66-0167	C141B	437AW	
66-0168	C141B	437AW	
66-0169	C141B	438AW	
66-0170	C141B	AMARC	CR006
66-0171	C141B	97AMW	
66-0172	C141B	63AW	
66-0173	C141B	438AW	
66-0174	C141B	459MAW	AFRes

66-0175	C141B	63AW		66-7951	C141B	62AW	
66-0176	GC141B	TTC	Sheppard	66-7952	C141B	63AW	
66-0177	C141B	63AW		66-7953	C141B	438AW	
66-0178	C141B	437AW		66-7954	C141B	356ALS	AFRes
66-0179	C141B	63AW		66-7955	C141B	437AW	
66-0180	C141B	97AMW		66-7956	C141B	437AW	
66-0181	C141B	63AW		66-7957	C141B	63AW	
66-0182	C141B	63AW		66-7958	C141B	62AW	
66-0183	C141B	438AW		66-7959	C141B		
66-0184	C141B	63AW		67-0001	C141B	62AW	
66-0185	C141B	183ALS	MS ANG	67-0002	C141B	437AW	
66-0186	C141B	97AMW		67-0003	C141B	62AW	
66-0187	C141B	437AW		67-0004	C141B	437AW	
66-0188	C141B	AMARC	CR008	67-0005	C141B	62AW	
66-0189	C141B	62AW		67-0006	C141A	w.o.28.8.76	
66-0190	C141B	183ALS	MS ANG	67-0007	C141B	438AW	
66-0191	C141B	60AW		67-0008	C141A	w.o.28.8.76	
66-0192	C141B	62AW		67-0009	C141B	62AW	
66-0193	C141B	62AW		67-0010	C141B	437AW	
66-0194	C141B	437AW		67-0011	C141B	437AW	
66-0195	C141B	438AW		67-0012	C141B	437AW	
66-0196	C141B	437AW		67-0013	C141B	438AW	
66-0197	C141B	62AW		67-0014	C141B	437AW	
66-0198	C141B	63AW		67-0015	C141B	63AW	
66-0199	C141B	438AW		67-0016	C141B	437AW	
66-0200	C141B	63AW		67-0017	C141B	w.o.7.3.82	
66-0201	C141B	63AW		67-0018	C141B	62AW	
66-0202	C141B	437AW		67-0019	C141B	438AW	
66-0203	C141B	97AMW		67-0020	C141B	438AW	
66-0204	C141B	438AW		67-0021	C141B	155ALS	TN ANG
66-0205	C141B	62AW		67-0022	C141B	63AW	
66-0206	C141B	62AW		67-0023	C141B	356ALS	AFRes
66-0207	C141B	438AW		67-0024	C141B	155ALS	TN ANG
66-0208	C141B	63AW		67-0025	C141B	438AW	
66-0209	C141B	437AW		67-0026	C141B	437AW	
66-7944	C141B	60AW		67-0027	C141B	438AW	
66-7945	C141B	437AW		67-0028	C141B	62AW	
66-7946	C141B	63AW		67-0029	C141B	w.o.13.1.87	
66-7947	C141B	437AW		67-0030	C141A	w.o.12.11.80	
66-7948	C141B	438AW		67-0031	C141B	60AW	
66-7949	C141B	63AW		67-0164	C141B	60AW	
66-7950	C141B	438AW		67-0165	C141B	438AW	
				67-0166	C141B	97AMW	

E3 SENTRY

c\n   20518, 20519, 21046, 21185, 21206 to 21209, 21250, 21434 to 21437, 21551 to 21556, 21752 to 21757, 22829 to 22837.

71-1407	E3B	552ACW OK		77-0355	E3B	552ACW OK
71-1408	E3B	552ACW OK		77-0356	E3B	552ACW OK
73-1674	JE3C	Boeing		78-0576	E3B	552ACW OK
73-1675	E3B	552ACW OK		78-0577	E3B	552ACW OK
75-0556	E3B	552ACW OK		78-0578	E3B	552ACW OK
75-0557	E3B	552ACW OK		79-0001	E3B	552ACW OK
75-0558	E3B	552ACW OK		79-0002	E3B	552ACW OK
75-0559	E3B	552ACW OK		79-0003	E3B	552ACW OK
75-0560	E3B	552ACW OK		80-0137	E3C	552ACW OK
76-1604	E3B	552ACW OK		80-0138	E3C	552ACW OK
76-1605	E3B	552ACW OK		80-0139	E3C	552ACW OK
76-1606	E3B	552ACW OK		81-0004	E3C	552ACW OK
76-1607	E3B	552ACW OK		81-0005	E3C	552ACW OK
77-0351	E3B	552ACW OK		82-0006	E3C	552ACW OK
77-0352	E3B	552ACW OK		82-0007	E3C	552ACW OK
77-0353	E3B	552ACW OK		83-0008	E3C	552ACW OK
77-0354	E3B	552ACW OK		83-0009	E3C	552ACW OK

E4

c\n	20682 to 20684, 20949.					
73-1676	E4B	1ACCS		74-0787	E4B	1ACCS
73-1677	E4B	1ACCS		75-0125	E4B	1ACCS

E8\J-STARS

c\n	19626, 19574, 24503.					
86-0416	E8A	Grumman Melbourne, Fl		88-0322	YE8B	AMARC
86-0417	E8A	Grumman Melbourne, Fl				

F4 PHANTOM II

62-12200	YF4E	USAF Museum		63-7445	F4C	w.o.16.10.72
62-12201	GRF4C	TTC Chanute		63-7446	F4C	BDRT Ahlhorn AB, FRG
63-7407	NF4C	412TW ED AFFTC		63-7447	F4C	
63-7408	NF4C	preserved Tyndall AFB		63-7448	F4C	ranges Aberdeen, Md
63-7409	NF4C	AMARC FP646		63-7449	F4C	BDRT RAF U. Heyford
63-7410	F4C	w.o.24.1.81		63-7450	F4C	w.o.1.8.72
63-7411	F4C	preserved Ft. SmithMAP, Ar		63-7451	F4C	w.o.17.8.69
63-7412	F4C			63-7452	F4C	AMARC FP089
63-7413	GF4C	TTC Sheppard		63-7453	F4C	BDRT Soesterburg AB
63-7414	F4C	BDRT RAF Woodbridge		63-7454	F4C	AMARC FP079
63-7415	F4C			63-7455	F4C	BDRT McEntire ANGB
63-7416	F4C	w.o. .5.64		63-7456	F4C	w.o.15.1.68
63-7417	GF4C	TTC Sheppard		63-7457	F4C	w.o.20.1.79
63-7418	F4C	AMARC FP075		63-7458	F4C	w.o.24.1.66
63-7419	F4C	BDRT RAF Alconbury		63-7459	F4C	BDRT Langley AFB
63-7420	F4C	AMARC FP295		63-7460	F4C	AMARC FP179
63-7421	F4C	preserved Hermeskeil, FRG		63-7461	F4C	w.o.20.12.67
63-7422	F4C	BDRT Bergstrom AFB		63-7462	F4C	w.o.15.10.73
63-7423	F4C	preserved Speyer, FRG		63-7463	F4C	
63-7424	F4C	preserved Hill AFB		63-7464	F4C	w.o.28.6.68
63-7425	F4C	w.o.8.6.67		63-7465	F4C	ranges Aberdeen, Md
63-7426	GF4C			63-7466	F4C	w.o.24.5.65
63-7427	F4C	w.o.22.11.65		63-7467	F4C	BDRT Leipheim, FRG
63-7428	F4C	AMARC FP320		63-7468	F4C	AMARC FP321
63-7429	F4C	w.o.11.4.67		63-7469	F4C	w.o.16.1.69
63-7430	F4C	w.o.19.8.76		63-7470	F4C	
63-7431	F4C			63-7471	F4C	BDRT RAF Lakenheath
63-7432	F4C	dump McClellan AFB		63-7472	F4C	w.o.27.2.74
63-7433	F4C	BDRT Kadena AB		63-7473	F4C	MASDC
63-7434	GF4C	TTC Sheppard		63-7474	F4C	BDRT Incirlik AB, Turkey
63-7435	F4C	w.o.24.4.68		63-7475	F4C	AMARC FP365
63-7436	F4C	AMARC FP335		63-7476	F4C	w.o.17.2.71
63-7437	F4C			63-7477	F4C	
63-7438	F4C	w.o.1.12.69		63-7478	F4C	
63-7439	F4C	BDRT Tinker AFB		63-7479	F4C	preserved Kingsley Field
63-7440	F4C			63-7480	F4C	w.o.12.10.66
63-7441	F4C	w.o.28.1.69		63-7481	F4C	w.o.3.3.82
63-7442	F4C	dump McClellan AFB		63-7482	F4C	
63-7443	F4C	BDRT England AFB		63-7483	F4C	w.o.4.3.64
63-7444	F4C	w.o.5.12.69		63-7484	F4C	AMARC FP081

Serial	Type	Status	Location
63-7485	F4C		
63-7486	F4C	w.o.7.10.68	
63-7487	GF4C	preserved	Mobile, Al
63-7488	F4C	w.o.24.7.67	
63-7489	F4C	w.o.27.12.67	
63-7490	F4C	AMARC	FP174
63-7491	GF4C	TTC	Sheppard
63-7492	F4C		
63-7493	F4C	w.o.28.4.69	
63-7494	F4C	w.o.5.9.67	
63-7495	F4C		
63-7496	F4C	w.o.4.7.68	
63-7497	F4C	MASDC	
63-7498	F4C	w.o.8.10.69	
63-7499	F4C	w.o.8.10.67	
63-7500	F4C	MASDC	
63-7501	F4C	MASDC	
63-7502	F4C	w.o.10.3.67	
63-7503	F4C	w.o.30.8.68	
63-7504	F4C	w.o.6.5.65	
63-7505	F4C		
63-7506	F4C	MASDC	
63-7507	GF4C	TTC	Sheppard
63-7508	F4C	BDRT	George AFB
63-7509	F4C	w.o.17.9.68	
63-7510	F4C	MASDC	
63-7511	F4C		
63-7512	F4C	BDRT	Aviano AB, Italy
63-7513	F4C	preserved	Gulfport-BiloxiRAP,
63-7514	F4C	BDRT	Osan AB, S. Korea
63-7515	F4C	preserved	Kelly AFB
63-7516	F4C	AMARC	FP028
63-7517	F4C	AMARC	FP334
63-7518	F4C	w.o.19.10.66	
63-7519	F4C	BDRT	George AFB
63-7520	F4C	w.o.26.7.83	
63-7521	F4C	w.o.6.12.66	
63-7522	F4C	MASDC	
63-7523	F4C	ranges	Aberdeen, Md
63-7524	F4C	w.o.28.12.69	
63-7525	F4C	w.o.27.8.66	
63-7526	F4C	w.o.7.4.67	
63-7527	F4C	w.o.18.12.65	
63-7528	F4C	w.o.4.11.69	
63-7529	F4C	AMARC	FP180
63-7530	F4C	AMARC	FP294
63-7531	F4C	w.o.25.4.66	
63-7532	F4C	preserved	Barksdale AFB
63-7533	F4C	w.o.11.12.66	
63-7534	F4C		
63-7535	F4C	w.o.15.7.67	
63-7536	F4C	BDRT	Norvenich AB
63-7537	F4C		
63-7538	F4C	w.o.15.10.67	
63-7539	F4C	w.o.20.5.66	
63-7540	F4C	preserved	Hickam AFB
63-7541	F4C	MASDC	
63-7542	F4C	AMARC	FP136
63-7543	F4C	AMARC	FP140
63-7544	F4C	w.o.9.12.66	
63-7545	F4C	AMARC	FP785\ex N421FS
63-7546	F4C	w.o.11.7.66	
63-7547	F4C	w.o.5.9.67	
63-7548	F4C	w.o.6.6.67	
63-7549	F4C	MASDC	
63-7550	F4C	BDRT	Robins AFB
63-7551	F4C	preserved	Lowry AFB
63-7552	F4C	BDRT	Bergstrom AFB
63-7553	F4C		
63-7554	F4C	w.o.26.6.66	
63-7555	F4C	BDRT	Shaw AFB
63-7556	F4C		
63-7557	F4C	AMARC	FP204
63-7558	F4C	w.o.15.7.67	
63-7559	F4C	preserved	Dobbins AFB
63-7560	F4C	w.o.8.8.66	
63-7561	F4C	w.o.5.9.66	
63-7562	F4C	AMARC	FP048
63-7563	F4C	w.o.30.8.65	
63-7564	F4C	to N422FS	
63-7565	F4C		
63-7566	F4C	AMARC	FP043
63-7567	F4C	to N402FS	
63-7568	F4C	MASDC	
63-7569	F4C	MASDC	
63-7570	F4C	BDRT	Robins AFB
63-7571	F4C	w.o.1.6.66	
63-7572	F4C	w.o.12.1.73	
63-7573	F4C	w.o.15.7.67	
63-7574	F4C	preserved	Shaw AFB
63-7575	F4C	AMARC	FP322
63-7576	F4C	BDRT	Bitburg AB, FRG
63-7577	F4C	w.o.26.6.67	
63-7578	F4C	BDRT	Myrtle Beach AFB
63-7579	F4C	w.o.16.6.65	
63-7580	F4C	w.o.2.2.68	
63-7581	F4C	AMARC	FP305
63-7582	F4C	AMARC	FP134
63-7583	F4C	preserved	Hermeskeil, FRG
63-7584	F4C	preserved	McChord AFB
63-7585	F4C	AMARC	FP182
63-7586	F4C	w.o.22.2.69	
63-7587	F4C	w.o.6.8.66	
63-7588	GF4C	TTC	Sheppard
63-7589	F4C	MASDC	
63-7590	F4C	w.o.6.1.69	
63-7591	F4C	AMARC	FP027
63-7592	F4C		
63-7593	F4C	w.o.12.8.67	
63-7594	F4C		
63-7595	F4C	AMARC	FP247
63-7596	F4C		
63-7597	F4C	w.o.22.9.64	
63-7598	F4C	AMARC	FP034
63-7599	F4C	w.o.22.9.64	
63-7600	F4C	w.o.7.11.67	
63-7601	F4C	BDRT	Robins AFB
63-7602	F4C	dump	Hill AFB
63-7603	F4C	w.o.31.5.67	
63-7604	F4C	w.o.5.11.69	
63-7605	F4C	AMARC	FP102
63-7606	F4C	w.o.27.9.65	
63-7607	F4C	AMARC	FP733\ex N423FS
63-7608	F4C	w.o.2.12.66	
63-7609	F4C	w.o.24.7.70	
63-7610	F4C	BDRT	RAF Lakenheath
63-7611	F4C	preserved	March AFB
63-7612	F4C	w.o.6.10.67	
63-7613	F4C	w.o.16.9.66	
63-7614	F4C	w.o.15.5.65	
63-7615	F4C	preserved	Kelly AFB
63-7616	F4C	w.o.11.1.67	
63-7617	F4C	AMARC	FP065
63-7618	F4C	BDRT	McClellan AFB
63-7619	F4C	w.o.13.1.70	
63-7620	F4C	MASDC	
63-7621	F4C	w.o.31.12.67	
63-7622	F4C		
63-7623	F4C		
63-7624	F4C	AMARC	FP055
63-7625	F4C	BDRT	Hickam AFB
63-7626	F4C	AMARC	FP087
63-7627	F4C	w.o.16.9.66	
63-7628	F4C	BDRT	Elmendorf AFB
63-7629	F4C	AMARC	FP139
63-7630	F4C		
63-7631	F4C	AMARC	FP035
63-7632	F4C	BDRT	Holloman AFB
63-7633	F4C	AMARC	FP183
63-7634	F4C	w.o.11.8.67	
63-7635	F4C	w.o.3.10.72	
63-7636	F4C	w.o.19.1.68	
63-7637	F4C	AMARC	FP132
63-7638	F4C		
63-7639	F4C	w.o.5.8.67	
63-7640	F4C	w.o.13.9.66	
63-7641	F4C	w.o.7.4.67	
63-7642	F4C	w.o.20.9.66	
63-7643	F4C	w.o.17.9.66	
63-7644	F4C		
63-7645	F4C	w.o.8.10.67	
63-7646	F4C	AMARC	FP141
63-7647	F4C		
63-7648	F4C	w.o.26.9.67	
63-7649	F4C	dump	McClellan AFB
63-7650	F4C	AMARC	FP066

Serial	Type	Status	Location
63-7640	F4C	w.o.13.9.66	
63-7641	F4C	w.o.7.4.67	
63-7642	F4C	w.o.20.9.66	
63-7643	F4C	w.o.17.9.66	
63-7644	F4C		
63-7645	F4C	w.o.8.10.67	
63-7646	F4C	AMARC	FP141
63-7647	F4C		
63-7648	F4C	w.o.26.9.67	
63-7649	F4C	dump	McClellan AFB
63-7650	F4C	AMARC	FP066
63-7651	F4C	w.o.8.4.66	
63-7652	F4C	w.o.9.5.67	
63-7653	F4C	w.o.10.3.67	
63-7654	NF4C	AMARC	FP853
63-7655	F4C	AMARC	FP073
63-7656	F4C	w.o.8.3.67	
63-7657	GF4C	TTC	Sheppard
63-7658	F4C	w.o.30.12.67	
63-7659	F4C	w.o.15.7.67	
63-7660	F4C	w.o.3.5.65	
63-7661	F4C	w.o.27.11.70	
63-7662	F4C	AMARC	FP293
63-7663	F4C	w.o.11.2.68	
63-7664	F4C	w.o.30.5.66	
63-7665	F4C	MASDC	
63-7666	F4C	BDRT	KunsanAB
63-7667	F4C	BDRT	Torrejon AB, Spain
63-7668	F4C	w.o.27.1.68	
63-7669	F4C	w.o.22.5.67	
63-7670	F4C		
63-7671	F4C	w.o.18.2.70	
63-7672	F4C		
63-7673	F4C	w.o.12.8.67	
63-7674	F4C	w.o.27.2.69	
63-7675	F4C	w.o.23.1.74	
63-7676	F4C		
63-7677	F4C	w.o.16.4.66	
63-7678	F4C	w.o.7.11.67	
63-7679	F4C	preserved	Portland IAP, Or
63-7680	F4C	w.o.20.11.67	
63-7681	F4C	w.o.22.3.66	
63-7682	F4C	w.o.28.12.67	
63-7683	F4C	AMARC	FP357
63-7684	F4C	w.o.5.4.68	
63-7685	F4C	MASDC	
63-7686	F4C	MASDC	
63-7687	F4C	w.o.19.9.66	
63-7688	F4C	MASDC	
63-7689	F4C	AMARC	FP842\ex N420FS
63-7690	F4C	w.o.17.7.66	
63-7691	F4C	w.o.21.1.67	
63-7692	F4C	w.o.22.5.67	
63-7693	F4C		
63-7694	F4C	w.o.12.9.66	
63-7695	F4C	w.o.28.7.66	
63-7696	F4C	w.o.5.6.85	
63-7697	F4C	w.o.14.6.66	
63-7698	F4C	w.o.6.7.67	
63-7699	F4C	BDRT	RAF Fairford
63-7700	F4C	w.o.4.9.65	
63-7701	F4C	w.o.16.3.68	
63-7702	GF4C	TTC	Sheppard
63-7703	F4C	w.o.2.6.75	
63-7704	F4C	MASDC	
63-7705	F4C		
63-7706	F4C	w.o.11.6.67	
63-7707	F4C	MASDC	
63-7708	F4C	w.o.2.10.69	
63-7709	F4C	w.o.27.1.67	
63-7710	F4C	w.o.12.5.67	
63-7711	F4C	AMARC	FP061
63-7712	F4C	w.o.23.9.65	
63-7713	F4C	w.o.6.7.68	
63-7740	RF4C	w.o.14.9.68	
63-7741	RF4C	AMARC	FP338
63-7742	RF4C	196RS	CA ANG
63-7743	RF4C	w.o.25.6.75	
63-7744	NRF4C	AMARC	FP871
63-7745	RF4C	preserved	Birmingham IAP, Al
63-7746	RF4C	196RS	CA ANG
63-7747	RF4C	AMARC	FP384
63-7748	RF4C	preserved	Shaw AFB
63-7749	RF4C	w.o.19.2.68	
63-7750	RF4C	AMARC	FP673
63-7751	GRF4C		
63-7752	RF4C	AMARC	FP379
63-7753	RF4C	AMARC	FP261
63-7754	RF4C	AMARC	FP783
63-7755	RF4C	w.o.26.6.68	
63-7756	RF4C	w.o.2.2.83	
63-7757	RF4C	AMARC	FP739
63-7758	RF4C	AMARC	FP297
63-7759	RF4C	AMARC	FP362
63-7760	RF4C	BDRT	Shaw AFB
63-7761	RF4C	BDRT	Bergstrom AFB
63-7762	RF4C	AMARC	FP609
63-7763	RF4C	preserved	Bergstrom AFB
64-0654	F4C	w.o.15.10.66	
64-0655	F4C	AMARC	FP044
64-0656	F4C	w.o.7.8.67	
64-0657	F4C	w.o.15.9.68	
64-0658	GF4C	TTC	Sheppard
64-0659	F4C	ranges	Aberdeen, Md
64-0660	F4C	preserved	Niagara Falls IAP
64-0661	F4C	AMARC	FP030
64-0662	F4C	w.o.13.6.67	
64-0663	F4C	w.o.2.12.66	
64-0664	F4C	preserved	Hill AFB
64-0665	F4C		
64-0666	F4C	ranges	Aberdeen, Md
64-0667	F4C	w.o.9.6.67	
64-0668	F4C	w.o.31.5.67	
64-0669	F4C	w.o.10.11.67	
64-0670	F4C	w.o.19.4.67	
64-0671	F4C	w.o.7.6.66	
64-0672	F4C	AMARC	FP135
64-0673	F4C	preserved	Pima Museum
64-0674	F4C	w.o.9.6.65	
64-0675	F4C	BDRT	Moody AFB
64-0676	F4C	w.o.28.6.69	
64-0677	F4C	AMARC	FP060
64-0678	F4C	w.o.20.12.65	
64-0679	F4C	preserved	Misawa AB
64-0680	F4C	w.o.29.4.65	
64-0681	F4C	w.o.4.9.68	
64-0682	F4C		
64-0683	F4C	USAF Museum	
64-0684	F4C	w.o.10.5.69	
64-0685	F4C	w.o.21.6.65	
64-0686	F4C	w.o.17.2.84	
64-0687	F4C	w.o.8.8.66	
64-0688	F4C	w.o.4.5.66	
64-0689	F4C	w.o.1.5.67	
64-0690	F4C	w.o.1.11.69	
64-0691	GF4C	TTC	Sheppard
64-0692	F4C	w.o.26.8.67	
64-0693	F4C	w.o.31.7.67	
64-0694	F4C	AMARC	FP080
64-0695	F4C	w.o.27.2.74	
64-0696	F4C	w.o.18.11.67	
64-0697	F4C	w.o.26.11.67	
64-0698	F4C	w.o.20.12.66	
64-0699	F4C	preserved	Davis Monthan AFB
64-0700	F4C	w.o.29.9.65	
64-0701	F4C	w.o.27.11.67	
64-0702	F4C	w.o.5.10.66	
64-0703	F4C	w.o.8.8.67	
64-0704	F4C	AMARC	
64-0705	F4C	MASDC	
64-0706	F4C	preserved	McClellan AFB
64-0707	F4C	BDRT	RAF Mildenhall
64-0708	F4C	w.o.21.5.67	
64-0709	F4C	w.o.25.10.65	
64-0710	F4C	w.o.12.10.67	
64-0711	F4C	AMARC	FP358
64-0712	F4C	preserved	Camp Mabry, Tx
64-0713	F4C	AMARC	FP230
64-0714	F4C	w.o.25.5.67	
64-0715	F4C		
64-0716	F4C	w.o.17.9.66	
64-0717	F4C	w.o.17.5.66	
64-0718	F4C	w.o.12.7.67	
64-0719	F4C	w.o.3.8.67	
64-0720	F4C	w.o.28.4.67	
64-0721	F4C	w.o.11.7.73	

Serial	Type	Status	Location/Notes
64-0722	F4C	w.o.26.5.66	
64-0723	F4C	w.o.7.12.65	
64-0724	F4C	BDRT	W-P AFB
64-0725	F4C	AMARC	FP052
64-0726	F4C	AMARC	FP039
64-0727	NF4C	AMARC	FP496
64-0728	F4C	w.o.9.6.66	
64-0729	F4C	w.o.28.11.65	
64-0730	F4C	w.o.29.11.65	
64-0731	F4C	w.o.28.4.69	
64-0732	F4C	w.o.15.3.66	
64-0733	F4C	w.o.31.3.67	
64-0734	F4C	w.o.15.7.67	
64-0735	F4C	w.o.7.7.71	
64-0736	F4C	w.o.29.9.66	
64-0737	F4C	w.o.24.4.72	
64-0738	F4C	w.o.21.12.70	
64-0739	F4C	w.o.25.1.69	
64-0740	F4C	w.o.15.3.66	
64-0741	F4C	preserved	Mojave AP\N403FS
64-0742	F4C	w.o.23.2.66	
64-0743	F4C	w.o.12.11.66	
64-0744	F4C	w.o.2.6.66	
64-0745	GF4C		
64-0746	F4C	w.o.30.3.67	
64-0747	F4C	w.o.	
64-0748	F4C	preserved	Little Rock AFB
64-0749	F4C	BDRT	Homestead AFB
64-0750	F4C	ranges	Aberdeen, Md
64-0751	F4C	preserved	Mountain Home AFB
64-0752	F4C	w.o.5.8.67	
64-0753	F4C	w.o.2.12.66	
64-0754	F4C		
64-0755	F4C	w.o.22.11.66	
64-0756	F4C	w.o.23.4.67	
64-0757	F4C	BDRT	Spangdahlem AB
64-0758	F4C	AMARC	FP024
64-0759	F4C	w.o.18.12.74	
64-0760	F4C	w.o.14.5.66	
64-0761	F4C	AMARC	FP138
64-0762	F4C	w.o.15.7.67	
64-0763	GF4C	AFSC	
64-0764	F4C	w.o.28.2.67	
64-0765	F4C	AMARC	FP137
64-0766	F4C		
64-0767	F4C	w.o.27.1.69	
64-0768	F4C	w.o.7.12.67	
64-0769	F4C	w.o.30.3.67	
64-0770	F4C	preserved	S-J AFB
64-0771	F4C	w.o.3.9.66	
64-0772	GF4C	TTC	Sheppard
64-0773	F4C	w.o.23.1.67	
64-0774	F4C	w.o.12.12.67	
64-0775	F4C	AMARC	FP068
64-0776	F4C		
64-0777	F4C	AMARC	FP363
64-0778	F4C	w.o.15.6.67	
64-0779	F4C	w.o.2.6.66	
64-0780	F4C	AMARC	FP133
64-0781	F4C		
64-0782	F4C	w.o.17.12.67	
64-0783	F4C		
64-0784	F4C	AMARC	FP104
64-0785	F4C	derelict	Hill AFB
64-0786	F4C	w.o.11.6.67	
64-0787	F4C	w.o.16.1.75	
64-0788	F4C	w.o.21.3.67	
64-0789	F4C	AMARC	FP062
64-0790	F4C		
64-0791	F4C	BDRT	Bergstrom AFB
64-0792	F4C	preserved	Hickam AFB
64-0793	F4C	BDRT	Hickam AFB
64-0794	F4C	AMARC	FP069
64-0795	F4C	w.o.4.11.67	
64-0796	F4C	AMARC	FP064
64-0797	F4C	w.o.20.1.68	
64-0798	F4C	w.o.28.8.66	
64-0799	F4C	preserved	USAF Adademy
64-0800	F4C	w.o.14.11.68	
64-0801	F4C	w.o.24.2.67	
64-0802	F4C	AMARC	FP037
64-0803	F4C	w.o.20.2.68	
64-0804	F4C	AMARC	FP051
64-0805	F4C	w.o.8.8.69	
64-0806	F4C	preserved	Nellis AFB
64-0807	F4C	w.o.10.12.67	
64-0808	F4C	w.o.25.2.66	
64-0809	F4C	w.o.21.5.66	
64-0810	F4C	w.o.21.1.67	
64-0811	F4C	AMARC	FP063
64-0812	F4C	AMARC	FP118
64-0813	F4C	to Spain	C.12-13
64-0814	F4C	w.o.27.10.68	
64-0815	F4C	BDRT	Savannah IAP, Ga
64-0816	F4C		
64-0817	F4C	preserved	Eglin AFB
64-0818	F4C		
64-0819	F4C	w.o.13.1.69	
64-0820	F4C	to Spain	C.12-37
64-0821	F4C	w.o.20.10.66	
64-0822	F4C	AMARC	FP025
64-0823	F4C		
64-0824	F4C	w.o.21.8.72	
64-0825	F4C	BDRT	Carswell AFB
64-0826	F4C	w.o.30.1.68	
64-0827	F4C	preserved	Mather AFB
64-0828	F4C	AMARC	FP152
64-0829	F4C	USAF Museum	
64-0830	F4C	w.o.23.3.68	
64-0831	F4C	AMARC	FP181
64-0832	F4C	w.o.10.9.66	
64-0833	F4C	w.o.27.12.66	
64-0834	F4C	w.o.10.11.67	
64-0835	F4C	w.o.9.9.67	
64-0836	F4C	AMARC	FP026
64-0837	F4C	w.o.1.9.68	
64-0838	F4C		
64-0839	F4C	w.o.10.3.67	
64-0840	F4C	AMARC	FP088
64-0841	F4C		
64-0842	F4C	w.o.16.11.69	
64-0843	F4C	w.o.17.4.69	
64-0844	F4C		
64-0845	F4C	w.o.19.1.67	
64-0846	F4C	to Spain	C.12-12
64-0847	F4C		
64-0848	F4C	w.o.26.7.67	
64-0849	F4C	w.o.31.3.67	
64-0850	F4C	to Spain	C.12-14
64-0851	F4C		
64-0852	F4C	AMARC	FP208
64-0853	F4C	to Spain	C.12-17
64-0854	F4C	to Spain	C.12-18
64-0855	F4C	to Spain	C.12-20
64-0856	F4C	to Spain	C.12-32
64-0857	F4C	to Spain	C.12-21
64-0858	F4C	to Spain	C.12-29
64-0859	F4C	to Spain	C.12-31
64-0860	F4C	AMARC	FP029
64-0861	F4C	to Spain	C.12-26
64-0862	F4C	to Spain	C.12-27
64-0863	F4C	w.o.11.70	
64-0864	F4C	to Spain	C.12-36
64-0865	F4C	AMARC	FP070
64-0866	F4C	to Spain	C.12-15
64-0867	F4C	to Spain	C.12-9
64-0868	F4C	to Spain	C.12-16
64-0869	NF4C	AMARC	FP495
64-0870	F4C	to Spain	C.12-35
64-0871	F4C	to Spain	C.12-11
64-0872	F4C	to Spain	C.12-19
64-0873	F4C	w.o.5.2.69	
64-0874	F4C	w.o.5.2.69	
64-0875	F4C		
64-0876	F4C	w.o.27.3.70	
64-0877	F4C	to Spain	C.12-23
64-0878	F4C	to Spain	C.12-30
64-0879	F4C	BDRT	Hahn AB, FRG
64-0880	F4C	to Spain	C.12-28
64-0881	F4C	to Spain	C.12-25
64-0882	F4C	to Spain	C.12-38
64-0883	F4C	w.o.14.7.77	
64-0884	F4C	to Spain	C.12-1
64-0885	F4C	w.o.6.5.66	
64-0886	F4C	to Spain	C.12-6
64-0887	F4C	to Spain	C.12-33

Serial	Type	Disposition	Code/Location
64-0888	F4C	AMARC	FP023
64-0889	GF4C		
64-0890	F4C		
64-0891	F4C		
64-0892	F4C	to Spain	C.12-39
64-0893	F4C	AMARC	FP176
64-0894	F4C	to Spain	C.12-24
64-0895	F4C	to Spain	C.12-10
64-0896	F4C	to Spain	C.12-40
64-0897	F4C	w.o.19.1.71	
64-0898	F4C	w.o.25.3.69	
64-0899	F4C	AMARC	FP177
64-0900	F4C	to Spain	C.12-2
64-0901	F4C	w.o.13.9.68	
64-0902	F4C	BDRT	Bergstrom AFB
64-0903	F4C	to Spain	C.12-3
64-0904	F4C	AMARC	FP040
64-0905	F4C		
64-0906	F4C	to Spain	C.12-5
64-0907	F4C	to Spain	C.12-34
64-0908	F4C	AMARC	FP076
64-0909	F4C	to Spain	C.12-4
64-0910	F4C	BDRT	George AFB
64-0911	F4C		
64-0912	GF4C	TTC	Sheppard
64-0913	F4C	BDRT	Kadena AB
64-0914	F4C	w.o.	
64-0915	F4C		
64-0916	F4C	w.o.18.1.71	
64-0917	F4C	BDRT	Ramstein AB
64-0918	F4C	AMARC	FP119
64-0919	F4C		
64-0920	F4C	to Spain	C.12-7
64-0921	F4C	w.o.13.9.67	
64-0922	F4C	BDRT	Sembach AB
64-0923	F4C	w.o.21.3.87	
64-0924	F4C	to Spain	C.12-22
64-0925	F4C	to Spain	C.12-8
64-0926	F4C	AMARC	FP351
64-0927	F4C	w.o.16.1.68	
64-0928	F4C	AMARC	FP077
64-0929	F4D	46TW	ET AFMC\MSD
64-0930	F4D	AMARC	FP530
64-0931	F4D	to Korea	
64-0932	F4D	w.o.5.1.67	
64-0933	F4D	to Korea	
64-0934	F4D	to Korea	
64-0935	F4D	to Korea	
64-0936	F4D	w.o.18.1.74	
64-0937	F4D	AMARC	FP485
64-0938	F4D	AMARC	FP486
64-0939	F4D	AMARC	FP130
64-0940	F4D	w.o.13.7.69	
64-0941	F4D	to Korea	
64-0942	F4D	AMARC	FP484
64-0943	F4D	to Korea	
64-0944	F4D	to Korea	
64-0945	F4D	AMARC	FP292
64-0946	F4D	to Korea	
64-0947	F4D	to Korea	
64-0948	F4D	to Korea	
64-0949	F4D	AMARC	FP488
64-0950	F4D	to Korea	
64-0951	F4D	to Korea	
64-0952	F4D	to N401AV	
64-0953	F4D	AMARC	FP489
64-0954	F4D	w.o.18.12.71	
64-0955	F4D	to Korea	
64-0956	F4D	AMARC	FP491
64-0957	F4D	to Korea	
64-0958	F4D	to Korea	
64-0959	F4D	AMARC	FP407
64-0960	F4D	w.o.22.3.74	
64-0961	F4D	to Korea	
64-0962	F4D	to Korea	
64-0963	F4D	AMARC	FP444
64-0964	F4D	w.o.15.9.69	
64-0965	F4D	to N424FS	
64-0966	F4D	to Korea	
64-0967	F4D	w.o.9.1.67	
64-0968	F4D	AMARC	FP490
64-0969	F4D	w.o.5.12.69	
64-0970	F4D	AMARC	FP131
64-0971	F4D	w.o.14.10.66	
64-0972	F4D		
64-0973	F4D	AMARC	FP125
64-0974	F4D	w.o.6.5.72	
64-0975	F4D	AMARC	FP217
64-0976	F4D	AMARC	FP458
64-0977	F4D	AMARC	FP487
64-0978	F4D	to Korea	
64-0979	F4D	AMARC	FP442
64-0980	F4D	AMARC	FP544
64-0997	RF4C	196RS	CA ANG
64-0998	RF4C	preserved	Lincoln MAP, NB
64-0999	RF4C	AMARC	FP812
64-1000	RF4C	preserved	Bergstrom AFB
64-1001	RF4C		
64-1002	RF4C	w.o.1.3.76	
64-1003	RF4C	w.o.21.11.80	
64-1004	NRF4C	412TW	ED AFFTC
64-1005	RF4C	192RS	NV ANG
64-1006	RF4C	192RS	NV ANG
64-1007	RF4C	w.o.14.9.79	
64-1008	RF4C	w.o.11.12.70	
64-1009	RF4C		
64-1010	RF4C	196RS	CA ANG
64-1011	RF4C	w.o.28.4.70	
64-1012	RF4C	w.o.2.6.65	
64-1013	RF4C	w.o.29.11.73	
64-1014	RF4C	192RS	NV ANG
64-1015	RF4C	w.o.24.2.70	
64-1016	RF4C	AMARC	FP850
64-1017	RF4C	192RS	NV ANG
64-1018	RF4C	w.o.6.5.70	
64-1019	RF4C	AMARC	FP810
64-1020	RF4C		
64-1021	RF4C	192RS	NV ANG
64-1022	RF4C	192RS	NV ANG
64-1023	RF4C	AMARC	FP818
64-1024	RF4C	AMARC	FP436
64-1025	RF4C	w.o.17.4.75	
64-1026	RF4C		
64-1027	RF4C	w.o.24.5.73	
64-1028	RF4C	AMARC	FP623
64-1029	RF4C	AMARC	FP907
64-1030	RF4C	192RS	NV ANG
64-1031	RF4C	106RS	BH AL ANG
64-1032	RF4C		
64-1033	RF4C	AMARC	FP905
64-1034	RF4C	106RS	BH AL ANG
64-1035	RF4C	AMARC	FP906
64-1036	RF4C	w.o.21.5.85	
64-1037	RF4C	w.o.17.9.67	
64-1038	RF4C	AMARC	FP814
64-1039	RF4C	173RS	NE ANG
64-1040	RF4C	w.o.27.7.66	
64-1041	RF4C	106RS	BH AL ANG
64-1042	RF4C	w.o.27.7.67	
64-1043	RF4C	w.o.18.2.68	
64-1044	RF4C	w.o.8.10.90	
64-1045	RF4C	w.o.25.4.66	
64-1046	RF4C	AMARC	FP796
64-1047	RF4C	106RS	BH AL ANG
64-1048	RF4C	w.o.10.12.68	
64-1049	RF4C	w.o.16.9.76	
64-1050	RF4C	192RS	NV ANG
64-1051	RF4C	106RS	BH AL ANG
64-1052	RF4C	w.o.25.6.75	
64-1053	RF4C	AMARC	FP854
64-1054	RF4C	w.o.20.8.66	
64-1055	RF4C	106RS	BH AL ANG
64-1056	RF4C	w.o.30.3.91	
64-1057	RF4C	106RS	BH AL ANG
64-1058	RF4C	106RS	BH AL ANG
64-1059	RF4C	w.o.9.8.67	
64-1060	RF4C	106RS	BH AL ANG
64-1061	RF4C	preserved	Minneapolis AP, Mn
64-1062	RF4C	AMARC	FP821
64-1063	RF4C	173RS	NE ANG
64-1064	RF4C	w.o.7.6.66	
64-1065	RF4C	106RS	BH AL ANG
64-1066	RF4C	AMARC	FP820
64-1067	RF4C	AMARC	FP729
64-1068	RF4C	192RS	NV ANG
64-1069	RF4C	to Spain	

Serial	Type	Disposition	Notes
64-1070	RF4C	to Spain	
64-1071	RF4C	AMARC	FP772
64-1072	RF4C	w.o.12.12.67	
64-1073	RF4C	w.o.16.5.82	
64-1074	RF4C	AMARC	FP773
64-1075	RF4C	196RS	CA ANG
64-1076	RF4C	w.o.25.5.78	
64-1077	RF4C	106RS	BH AL ANG
64-1078	RF4C	w.o.6.1.67	
64-1079	RF4C	192RS	NV ANG
64-1080	RF4C	AMARC	FP904
64-1081	RF4C	preserved	Louisville AP, Ky
64-1082	RF4C	AMARC	FP732
64-1083	RF4C	to Spain	
64-1084	RF4C		
64-1085	RF4C	w.o.11.12.70	
65-0580	F4D	w.o.24.4.67	
64-0581	F4D	AMARC	FP382
65-0582	F4D	to Korea	
65-0583	F4D	AMARC	FP239
65-0584	F4D	AMARC	FP251
65-0585	GF4D	TTC	Sheppard
65-0586	F4D	AMARC	FP541
65-0587	F4D	w.o.5.12.67	
65-0588	F4D		
65-0589	F4D	to Korea	
65-0590	F4D	AMARC	FP492
65-0591	F4D	to Korea	
65-0592	F4D	to Korea	
65-0593	F4D	w.o.17.9.72	
65-0594	F4D	w.o.14.9.69	
65-0595	F4D	AMARC	FP126
65-0596	F4D	AMARC	FP090
65-0597	F4D	AMARC	FP548
65-0598	F4D	AMARC	FP521
65-0599	F4D	w.o.9.8.72	
65-0600	F4D	w.o.20.5.72	
65-0601	F4D	w.o.3.9.76	
65-0602	F4D	w.o.21.4.72	
65-0603	F4D	w.o.12.10.75	
65-0604	F4D	w.o.2.2.68	
65-0605	F4D	to Korea	
65-0606	F4D	w.o.8.7.66	
65-0607	F4D	w.o.5.5.70	
65-0608	F4D		
65-0609	F4D	AMARC	FP467
65-0610	F4D	to Korea	
65-0611	F4D	AMARC	FP536
65-0612	F4D	w.o.11.7.83	
65-0613	F4D	AMARC	FP123
65-0614	F4D	AMARC	FP083
65-0615	F4D	AMARC	FP084
65-0616	F4D	w.o.18.4.70	
65-0617	F4D	AMARC	FP525
65-0618	F4D	w.o.12.1.70	
65-0619	F4D	w.o.2.8.74	
65-0620	F4D	to Korea	
65-0621	F4D	w.o.5.7.83	
65-0622	F4D	to Korea	
65-0623	F4D	to Korea	
65-0624	F4D	to Korea	
65-0625	F4D	w.o.14.3.72	
65-0626	F4D	preserved	Niagara Falls IAP
65-0627	F4D	w.o.4.10.68	
65-0628	F4D	w.o.14.5.70	
65-0629	F4D	AMARC	FP529
65-0630	F4D	to Korea	
65-0631	F4D	AMARC	FP504
65-0632	F4D	w.o.23.10.72	
65-0633	F4D	w.o.4.5.68	
65-0634	F4D	w.o.24.8.69	
65-0635	F4D	AMARC	FP301
65-0636	F4D	w.o.10.8.70	
65-0637	F4D	w.o.25.2.71	
65-0638	F4D	AMARC	FP252
65-0639	F4D	AMARC	FP409
65-0640	F4D	to Korea	
65-0641	F4D	AMARC	FP396
65-0642	F4D	w.o.26.5.73	
65-0643	F4D	AMARC	FP085
65-0644	F4D	AMARC	FP427
65-0645	F4D	w.o.11.11.69	
65-0646	F4D	AMARC	FP528
65-0647	F4D	AMARC	FP551
65-0648	F4D	AMARC	FP178
65-0649	F4D	w.o.10.4.69	
65-0650	F4D	to Korea	
65-0651	F4D	w.o.14.2.69	
65-0652	F4D	AMARC	FP509
65-0653	F4D	w.o.	
65-0654	F4D	AMARC	FP550
65-0655	F4D	AMARC	FP234
65-0656	F4D	w.o.20.4.67	
65-0657	F4D	w.o.23.3.77	
65-0658	F4D	AMARC	FP120
65-0659	F4D	AMARC	FP577
65-0660	F4D	preserved	Maxwell AFB
65-0661	F4D	AMARC	FP114
65-0662	F4D		
65-0663	F4D	w.o.16.2.73	
65-0664	F4D	w.o.21.4.67	
65-0665	F4D	AMARC	FP307
65-0666	F4D	AMARC	FP414
65-0667	F4D	AMARC	FP523
65-0668	F4D	w.o.18.5.67	
65-0669	F4D	w.o.1.4.67	
65-0670	NF4D	AMARC	FP452
65-0671	F4D	AMARC	FP107
65-0672	F4D	AMARC	FP308
65-0673	F4D	w.o.9.5.72	
65-0674	F4D	w.o.15.7.83	
65-0675	F4D	w.o.2.7.67	
65-0676	F4D	AMARC	FP235
65-0677	F4D	AMARC	FP563
65-0678	F4D	to Korea	
65-0679	F4D	to Korea	
65-0680	F4D	AMARC	FP464
65-0681	F4D	preserved	Charleston AFB
65-0682	F4D	AMARC	FP244
65-0683	F4D	AMARC	FP122
65-0684	F4D	AMARC	FP547
65-0685	F4D	w.o.16.7.80	
65-0686	F4D	w.o.4.8.69	
65-0687	F4D	AMARC	FP545
65-0688	F4D	AMARC	FP524
65-0689	F4D	to Korea	
65-0690	F4D	to Korea	
65-0691	F4D	to Korea	
65-0692	F4D	AMARC	FP253
65-0693	F4D	w.o.25.7.70	
65-0694	F4D	to N403AV	
65-0695	F4D	w.o.17.2.77	
65-0696	F4D	to N402AV	
65-0697	F4D	w.o.28.8.78	
65-0698	F4D	AMARC	FP116
65-0699	F4D	AMARC	FP250
65-0700	F4D	AMARC	FP306
65-0701	F4D	AMARC	FP086
65-0702	F4D	AMARC	FP454
65-0703	F4D	AMARC	FP093
65-0704	F4D	to N404AV	
65-0705	F4D	AMARC	FP461
65-0706	F4D	w.o.20.4.68	
65-0707	F4D	AMARC	FP423
65-0708	F4D	w.o.14.3.72	
65-0709	F4D	w.o.17.11.70	
65-0710	F4D	AMARC	FP463
65-0711	F4D	AMARC	FP219
65-0712	F4D		
65-0713	YF4E	412TW	ED AFFTC
65-0714	F4D	AMARC	FP433
65-0715	F4D	to Korea	
65-0716	F4D	AMARC	FP112
65-0717	F4D	w.o.5.2.79	
65-0718	F4D	AMARC	FP300
65-0719	F4D	AMARC	FP115
65-0720	F4D	AMARC	FP110
65-0721	F4D	AMARC	FP415
65-0722	F4D	w.o.10.3.69	
65-0723	F4D	w.o.5.9.67	
65-0724	F4D	w.o.25.6.77	
65-0725	F4D	w.o.24.1.69	
65-0726	F4D	w.o.23.8.67	
65-0727	F4D	w.o.7.10.67	
65-0728	F4D	w.o.26.3.73	
65-0729	F4D	AMARC	FP108

65-0730	F4D	AMARC	FP095		65-0829	RF4C	w.o.2.12.66	
65-0731	F4D	AMARC	FP124		65-0830	RF4C	AMARC	FP829
65-0732	F4D	to Korea			65-0831	RF4C	AMARC	FP771
65-0733	F4D	w.o.23.6.67			65-0832	RF4C	192RS	NV ANG
65-0734	F4D	AMARC	FP438		65-0833	RF4C	106RS	BH AL ANG
65-0735	GF4D	TTC	Sheppard		65-0834	RF4C	w.o.6.6.67	
65-0736	F4D	AMARC	FP443		65-0835	RF4C	to Spain	
65-0737	F4D	AMARC	FP410		65-0836	RF4C		
65-0738	F4D	AMARC	FP515		65-0837	RF4C	AMARC	FP790
65-0739	F4D	AMARC	FP542		65-0838	RF4C	AMARC	FP940
65-0740	F4D	AMARC	FP231		65-0839	RF4C	w.o.7.8.67	
65-0741	F4D	AMARC	FP094		65-0840	RF4C	AMARC	FP819
65-0742	F4D	AMARC	FP508		65-0841	RF4C	to Spain	
65-0743	F4D	AMARC	FP117		65-0842	RF4C	w.o.19.2.68	
65-0744	F4D	AMARC	FP175		65-0843	RF4C	AMARC	FP902
65-0745	F4D	w.o.4.1.73			65-0844	RF4C	w.o.24.11.67	
65-0746	F4D	AMARC	FP106		65-0845	RF4C	AMARC	FP700
65-0747	F4D	AMARC	FP191		65-0846	RF4C	w.o.28.10.68	
65-0748	F4D	AMARC	FP518		65-0847	RF4C	w.o.22.7.71	
65-0749	F4D	AMARC	FP426		65-0848	RF4C	w.o.2.8.67	
65-0750	F4D	w.o.25.8.67			65-0849	RF4C	AMARC	FP770
65-0751	F4D	w.o.26.10.69			65-0850	NRF4C	AMARC	FP872
65-0752	F4D	AMARC	FP101		65-0851	RF4C	to Spain	
65-0753	F4D	AMARC	FP574		65-0852	RF4C	196RS	CA ANG
65-0754	F4D	AMARC	FP071		65-0853	RF4C	173RS	NE ANG
65-0755	F4D	to Korea			65-0854	RF4C	AMARC	FP9..
65-0756	F4D	AMARC	FP313		65-0855	RF4C	w.o.16.10.67	
65-0757	F4D	w.o.7.10.74			65-0856	RF4C	w.o.11.3.72	
65-0758	F4D	AMARC	FP425		65-0857	RF4C	w.o.9.9.67	
65-0759	F4D	AMARC	FP419		65-0858	RF4C	w.o.1.9.68	
65-0760	F4D	AMARC	FP151		65-0859	RF4C	173RS	NE ANG
65-0761	F4D	w.o.26.5.70			65-0860	RF4C	w.o.12.11.70	
65-0762	F4D	to Korea			65-0861	RF4C	w.o.22.6.67	
65-0763	F4D	to N426FS			65-0862	RF4C	w.o.26.9.66	
65-0764	F4D	w.o.29.6.76			65-0863	RF4C	w.o.10.4.70	
65-0765	F4D	AMARC	FP460		65-0864	RF4C	192RS	NV ANG
65-0766	F4D	w.o.2.1.68			65-0865	RF4C	w.o.5.1.68	
65-0767	F4D	AMARC	FP543		65-0866	RF4C	AMARC	FP795
65-0768	F4D	AMARC	FP113		65-0867	RF4C	106RS	BH AL ANG
65-0769	F4D	AMARC	FP260		65-0868	RF4C	AMARC	FP831
65-0770	F4D	w.o.14.4.82			65-0869	RF4C		
65-0771	F4D	w.o.16.4.72			65-0870	RF4C	192RS	NV ANG
65-0772	F4D	AMARC	FP078		65-0871	RF4C	w.o.6.7.66	
65-0773	F4D	AMARC	FP109		65-0872	RF4C	w.o.30.4.67	
65-0774	F4D	AMARC	FP285		65-0873	RF4C	to Spain	
65-0775	F4D	AMARC	FP522		65-0874	RF4C	173RS	NE ANG
65-0776	F4D	w.o.13.8.74			65-0875	RF4C	192RS	NV ANG
65-0777	F4D				65-0876	RF4C	192RS	NV ANG
65-0778	F4D	to Korea			65-0877	RF4C	w.o.12.3.67	
65-0779	F4D	AMARC	FP531		65-0878	RF4C	AMARC	FP804
65-0780	F4D	w.o.12.12.79			65-0879	RF4C	190FS	ID ANG
65-0781	F4D	AMARC	FP568		65-0880	RF4C	w.o.19.11.67	
65-0782	F4D	AMARC	FP471		65-0881	RF4C	AMARC	FP863
65-0783	F4D	AMARC	FP243		65-0882	RF4C	w.o.13.8.67	
65-0784	F4D	w.o.10.5.72			65-0883	RF4C	w.o.16.1.67	
65-0785	F4D	w.o.18.5.78			65-0884	RF4C	106RS	BH AL ANG
65-0786	F4D	to Korea			65-0885	RF4C	w.o.7.10.68	
65-0787	F4D	w.o.7.2.67			65-0886	RF4C	192RS	NV ANG
65-0788	F4D	AMARC	FP258		65-0887	RF4C	w.o.21.7.67	
65-0789	F4D	AMARC	FP538		65-0888	RF4C	w.o.17.1.67	
65-0790	F4D	AMARC	FP099		65-0889	RF4C	w.o.2.10.69	
65-0791	F4D	w.o.29.10.80			65-0890	RF4C	192RS	NV ANG
65-0792	F4D	AMARC	FP570		65-0891	RF4C	w.o.15.12.68	
65-0793	F4D	AMARC	FP468		65-0892	RF4C	192RS	NV ANG
65-0794	F4D	AMARC	FP533		65-0893	RF4C	AMARC	FP815
65-0795	F4D	to Korea			65-0894	RF4C	w.o.17.9.67	
65-0796	F4D	preserved	Dyess, Tx		65-0895	RF4C	w.o.22.7.69	
65-0797	F4D	to Korea			65-0896	RF4C	AMARC	FP625
65-0798	F4D	AMARC	FP129		65-0897	RF4C	192RS	NV ANG
65-0799	F4D	w.o.18.12.71			65-0898	RF4C	190FS	ID ANG
65-0800	F4D	w.o.7.7.72			65-0899	RF4C	w.o.19.11.67	
65-0801	F4D	MASDC			65-0900	RF4C	106RS	BH AL ANG
65-0818	RF4C	w.o.16.1.67			65-0901	RF4C	196RS	CA ANG
65-0819	RF4C	w.o.6.12.66			65-0902	RF4C	173RS	NE ANG
65-0820	RF4C	w.o.16.12.66			65-0903	RF4C	173RS	NE ANG
65-0821	RF4C	w.o.11.12.70			65-0904	RF4C	190FS	ID ANG
65-0822	RF4C	to Spain			65-0905	RF4C	ALC	Ogden
65-0823	RF4C	190FS	ID ANG		65-0906	RF4C	w.o.22.12.68	
65-0824	RF4C	173RS	NE ANG		65-0907	RF4C	173RS	NE ANG
65-0825	RF4C	w.o.24.11.72			65-0908	RF4C	190FS	ID ANG
65-0826	RF4C	AMARC	FP730		65-0909	RF4C	w.o.4.4.68	
65-0827	RF4C	w.o.19.5.70			65-0910	RF4C	190FS	ID ANG
65-0828	RF4C	173RS	NE ANG		65-0911	RF4C	192RS	NV ANG

Serial	Type	Status/Location	Code
65-0912	RF4C	AMARC	FP778
65-0913	RF4C	w.o.8.1.68	
65-0914	RF4C	w.o.5.5.70	
65-0915	RF4C	w.o.18.9.68	
65-0916	RF4C	w.o.16.11.69	
65-0917	RF4C	173RS	NE ANG
65-0918	RF4C	196RS	CA ANG
65-0919	RF4C	w.o.1.2.74	
65-0920	RF4C	AMARC	FP803
65-0921	RF4C	w.o.18.8.69	
65-0922	RF4C	196RS	CA ANG
65-0923	RF4C	173RS	NE ANG
65-0924	RF4C	AMARC	FP689
65-0925	RF4C	AMARC	FP794
65-0926	RF4C	w.o.2.9.76	
65-0927	RF4C	AMARC	FP637
65-0928	RF4C	173RS	NE ANG
65-0929	RF4C		
65-0930	RF4C	w.o.26.7.68	
65-0931	RF4C	173RS	NE ANG
65-0932	RF4C	173RS	NE ANG
65-0933	RF4C	w.o.6.12.67	
65-0934	RF4C	w.o.20.11.77	
65-0935	RF4C	AMARC	FP941
65-0936	RF4C	to Spain	CR.12-41
65-0937	RF4C	to Spain	CR.12-42
65-0938	RF4C	to Spain	CR.12-43
65-0939	RF4C	AMARC	FP9..
65-0940	RF4C		
65-0941	NRF4C	AMARC	FP746
65-0942	RF4C	196RS	CA ANG
65-0943	RF4C	to Spain	CR.12-44
65-0944	RF4C	FSI	Mojave AP
65-0945	RF4C	106RS	BH AL ANG
66-0226	F4D	AMARC	FP526
66-0227	F4D	w.o.7.3.79	
66-0228	F4D	AMARC	FP111
66-0229	F4D	AMARC	FP092
66-0230	F4D	w.o.11.5.72	
66-0231	F4D	w.o.7.4.69	
66-0232	F4D	w.o.6.6.72	
66-0233	F4D	w.o.5.4.69	
66-0234	F4D	AMARC	FP314
66-0235	F4D	w.o.16.1.75	
66-0236	F4D	w.o.9.6.67	
66-0237	F4D	w.o.19.12.71	
66-0238	F4D	w.o.23.8.67	
66-0239	F4D	to Korea	
66-0240	F4D	preserved	Burlington IAP, Vt
66-0241	F4D	w.o.18.12.71	
66-0242	F4D	AMARC	FP513
66-0243	F4D	AMARC	FP286
66-0244	F4D	AMARC	FP236
66-0245	F4D	w.o.25.5.70	
66-0246	F4D	w.o.23.5.68	
66-0247	F4D	w.o.23.8.67	
66-0248	F4D	w.o.17.7.67	
66-0249	F4D	w.o.22.3.85	
66-0250	F4D	w.o.8.11.67	
66-0251	F4D	w.o.2.6.80	
66-0252	F4D		
66-0253	F4D	w.o.19.7.72	
66-0254	F4D	AMARC	FP527
66-0255	F4D	w.o.3.2.67	
66-0256	F4D	w.o.22.11.75	
66-0257	F4D	w.o.10.12.70	
66-0258	F4D	w.o.12.10.68	
66-0259	F4D		
66-0260	F4D	w.o.23.8.67	
66-0261	F4D	AMARC	FP411
66-0262	F4D		
66-0263	F4D	w.o.3.2.71	
66-0264	F4D	w.o.24.10.68	
66-0265	F4D	w.o.22.7.72	
66-0266	F4D	AMARC	FP278
66-0267	F4D	preserved	Homestead AFB
66-0268	F4D		
66-0269	F4D		
66-0270	F4D	AMARC	FP494
66-0271	F4D		
66-0272	F4D	AMARC	FP212
66-0273	F4D	preserved	Homestead, Fl
66-0274	F4D	to Korea	
66-0275	F4D	w.o.28.5.70	
66-0276	F4D	AMARC	FP210
66-0277	F4D	AMARC	FP417
66-0278	F4D	w.o.11.4.79	
66-0279	F4D	AMARC	FP532
66-0280	F4D	preserved	Ellington ANGB
66-0281	F4D	w.o.11.8.72	
66-0282	F4D	w.o.7.7.78	
66-0283	F4D	AMARC	FP480
66-0284	NF4E	AMARC	FP825
66-0285	F4E	w.o.22.6.70	
66-0286	NF4E	412TW	ED AFFTC
66-0287	GF4E	TTC	Lowry
66-0288	F4E	w.o.17.1.71	
66-0289	NF4E	AMARC	FP627
66-0290	F4E	w.o.1.3.73	
66-0291	NF4E	AMARC	FP626
66-0292	F4E	to Turkey	
66-0293	F4E	to Turkey	
66-0294	NF4E	AMARC	FP747
66-0295	F4E		
66-0296	F4E	w.o.4.12.68	
66-0297	F4E	to Turkey	
66-0298	F4E		
66-0299	F4E	w.o.12.5.72	
66-0300	F4E	to Turkey	
66-0301	F4E	to Turkey	
66-0302	F4E	preserved	McDill AFB
66-0303	F4E	to Turkey	
66-0304	F4E	to Turkey	
66-0305	F4E	to Turkey	
66-0306	F4E	AMARC	FP390
66-0307	F4E	to Turkey	
66-0308	F4E	BDRT	Spangdhalem AB
66-0309	F4E	to Turkey	
66-0310	F4E	w.o.8.9.72	
66-0311	F4E	w.o.2.2.68	
66-0312	F4E	to Turkey	
66-0313	F4E	to Israel	
66-0314	F4E	to Turkey	
66-0315	NF4E	AMARC	FP744
66-0316	F4E	w.o.7.9.71	
66-0317	F4E	to Turkey	
66-0318	F4E	to Turkey	
66-0319	NF4E	AMARC	FP748
66-0320	F4E	to Turkey	
66-0321	F4E	w.o.4.6.72	
66-0322	F4E	w.o.9.8.72	
66-0323	F4E	to Turkey	
66-0324	F4E	w.o.15.4.72	
66-0325	F4E	w.o.11.5.72	
66-0326	F4E	w.o.20.3.68	
66-0327	F4E	to Israel	
66-0328	F4E	to Turkey	
66-0329	NF4E	AMARC	FP826
66-0330	F4E	AMARC	FP387
66-0331	F4E	w.o.23.1.73	
66-0332	F4E	w.o.30.9.71	
66-0333	F4E	to Turkey	
66-0334	F4E	to Turkey	
66-0335	F4E	w.o.22.8.69	
66-0336	F4E	to Turkey	
66-0337	F4E	to Egypt	
66-0338	F4E	AMARC	FP385
66-0339	F4E	to Turkey	
66-0340	F4E	to Egypt	
66-0341	F4E	to Egypt	7823
66-0342	F4E	AMARC	FP359
66-0343	F4E	to Egypt	
66-0344	F4E	to Turkey	
66-0345	F4E	to Turkey	
66-0346	F4E	to Turkey	
66-0347	F4E	w.o.22.5.70	
66-0348	F4E	w.o.21.9.79	
66-0349	F4E	to Egypt	
66-0350	F4E	AMARC	FP386
66-0351	F4E	to Turkey	
66-0352	F4E	to Israel	
66-0353	F4E	to Egypt	
66-0354	F4E	to Turkey	
66-0355	F4E	to Turkey	
66-0356	F4E		
66-0357	F4E	AMARC	FP392

Serial	Type	Status	Location
66-0358	F4E	to Egypt	
66-0359	F4E	to Turkey	
66-0360	F4E	to Egypt	
66-0361	F4E	to Turkey	
66-0362	F4E	to Egypt	
66-0363	F4E	w.o.24.10.69	
66-0364	F4E	to Egypt	
66-0365	F4E	w.o.16.7.69	
66-0366	F4E	to Egypt	
66-0367	F4E	w.o.31.7.69	
66-0368	GF4E	TTC	Sheppard
66-0369	F4E	w.o.24.7.69	
66-0370	F4E	to Turkey	
66-0371	F4E	w.o.12.2.72	
66-0372	F4E	AMARC	FP346
66-0373	F4E	to Turkey	
66-0374	F4E	to Turkey	
66-0375	F4E	to Egypt 7804	
66-0376	F4E	AMARC	FP315
66-0377	NF4E	AMARC	FP658
66-0378	F4E	AMARC	FP309
66-0379	F4E	to Turkey	
66-0380	F4E	w.o.3.1.71	
66-0381	F4E	w.o.15.6.68	
66-0382	F4E		
66-0383	RF4C	w.o.10.1.67	
66-0384	RF4C	AMARC	FP824
66-0385	RF4C	w.o.4.6.70	
66-0386	RF4C	w.o.19.10.67	
66-0387	RF4C	w.o.5.9.67	
66-0388	RF4C	w.o.5.6.69	
66-0389	RF4C	AMARC	FP736
66-0390	RF4C	w.o.30.4.68	
66-0391	RF4C	w.o.18.2.68	
66-0392	RF4C	w.o.10.7.71	
66-0393	RF4C	AMARC	FP780
66-0394	RF4C	w.o.19.11.67	
66-0395	RF4C	AMARC	FP502
66-0396	RF4C	w.o.21.12.69	
66-0397	RF4C	GIA	Nellis AFB
66-0398	RF4C	w.o.29.4.68	
66-0399	RF4C	190FS	ID ANG
66-0400	RF4C	106RS	BH AL ANG
66-0401	RF4C	196RS	CA ANG
66-0402	RF4C	196RS	CA ANG
66-0403	RF4C	w.o.2.10.67	
66-0404	RF4C	w.o.30.7.68	
66-0405	RF4C	w.o.17.4.69	
66-0406	RF4C	w.o.16.10.67	
66-0407	RF4C	196RS	CA ANG
66-0408	RF4C		
66-0409	RF4C	w.o.16.4.70	
66-0410	RF4C	AMARC	FP368
66-0411	RF4C	w.o.26.4.67	
66-0412	RF4C	w.o.3.6.68	
66-0413	RF4C		
66-0414	RF4C		
66-0415	RF4C	106RS	BH AL ANG
66-0416	RF4C	w.o.2.11.87	
66-0417	RF4C	106RS	BH AL ANG
66-0418	RF4C	173RS	NE ANG
66-0419	RF4C	AMARC	FP340
66-0420	RF4C	w.o.13.11.70	
66-0421	RF4C	AMARC	FP767
66-0422	RF4C	AMARC	FP903
66-0423	RF4C	196RS	CA ANG
66-0424	RF4C	w.o.25.8.77	
66-0425	RF4C	AMARC	FP901
66-0426	RF4C	w.o.22.8.73	
66-0427	RF4C	AMARC	FP750
66-0428	RF4C	173RS	NE ANG
66-0429	RF4C		
66-0430	RF4C	AMARC	FP900
66-0431	RF4C	w.o.27.2.68	
66-0432	RF4C	w.o.22.11.68	
66-0433	RF4C	AMARC	FP789
66-0434	RF4C	w.o.	
66-0435	RF4C	196RS	CA ANG
66-0436	RF4C	w.o.30.7.70	
66-0437	RF4C	w.o.8.2.68	
66-0438	RF4C		
66-0439	RF4C	w.o.1.11.71	
66-0440	RF4C		
66-0441	RF4C	w.o.26.11.69	
66-0442	RF4C	w.o.20.5.68	
66-0443	RF4C	w.o.5.2.68	
66-0444	RF4C	AMARC	FP811
66-0445	RF4C	w.o.24.11.68	
66-0446	RF4C	AMARC	FP843
66-0447	RF4C	w.o.16.6.68	
66-0448	RF4C	w.o.22.7.80	
66-0449	RF4C	173RS	NE ANG
66-0450	RF4C	w.o.1.8.68	
66-0451	RF4C	w.o.15.2.74	
66-0452	RF4C	AMARC	FP816
66-0453	RF4C	196RS	CA ANG
66-0454	RF4C	w.o.19.2.80	
66-0455	RF4C	BDRT	Bergstrom AFB
66-0456	RF4C	AMARC	FP740
66-0457	RF4C	w.o.31.10.68	
66-0458	RF4C	w.o.26.7.68	
66-0459	RF4C	w.o.4.1.84	
66-0460	RF4C	w.o.1.10.76	
66-0461	RF4C	AMARC	FP674
66-0462	RF4C	w.o.28.3.69	
66-0463	RF4C	AMARC	FP737
66-0464	RF4C	46TW	ET AFMC\MSD
66-0465	RF4C		
66-0466	RF4C	w.o.24.8.68	
66-0467	RF4C	BDRT	Moody AFB
66-0468	RF4C	w.o.16.11.69	
66-0469	RF4C	ALC	Ogden
66-0470	RF4C	AMARC	FP328
66-0471	RF4C	w.o.24.11.72	
66-0472	RF4C		
66-0473	RF4C	AMARC	FP749
66-0474	RF4C	AMARC	FP728
66-0475	RF4C	AMARC	FP734
66-0476	RF4C	AMARC	FP615
66-0477	RF4C	w.o.23.5.86	
66-0478	RF4C	AMARC	FP774
66-7455	F4D	preserved	Aberdeen, Md
66-7456	F4D	AMARC	FP552
66-7457	F4D	AMARC	FP479
66-7458	F4D	AMARC	FP096
66-7459	F4D	AMARC	FP456
66-7460	F4D	AMARC	FP412
66-7461	F4D	AMARC	FP082
66-7462	F4D	w.o.22.5.72	
66-7463	F4D	preserved	USAF Academy
66-7464	F4D	AMARC	FP457
66-7465	F4D	w.o.29.6.78	
66-7466	F4D	AMARC	FP391
66-7467	F4D	AMARC	FP299
66-7468	F4D		
66-7469	F4D	AMARC	FP555
66-7470	F4D	AMARC	FP289
66-7471	F4D	AMARC	FP478
66-7472	F4D	AMARC	FP537
66-7473	F4D	w.o.11.2.71	
66-7474	F4D	w.o.29.4.69	
66-7475	F4D	AMARC	FP339
66-7476	F4D	w.o.	
66-7477	F4D	AMARC	FP121
66-7478	F4D	AMARC	FP459
66-7479	F4D	to Korea	
66-7480	F4D	w.o.24.8.76	
66-7481	F4D	w.o.11.4.67	
66-7482	F4D	w.o.25.8.72	
66-7483	NF4D	to N430FS	
66-7484	F4D	AMARC	FP462
66-7485	F4D	AMARC	FP569
66-7486	F4D	AMARC	FP440
66-7487	F4D	AMARC	FP516
66-7488	F4D	AMARC	FP262
66-7489	F4D	AMARC	FP535
66-7490	F4D	AMARC	FP534
66-7491	F4D	AMARC	FP503
66-7492	F4D	w.o.1.10.69	
66-7493	F4D	w.o.12.6.69	
66-7494	F4D	w.o.23.4.72	
66-7495	F4D	w.o.16.1.68	
66-7496	F4D	AMARC	FP546
66-7497	F4D	AMARC	FP562
66-7498	F4D	AMARC	FP406
66-7499	F4D	w.o.3.12.68	

Serial	Type	Disposition	Code
66-7500	GF4D	TTC	Sheppard
66-7501	F4D	w.o.21.11.72	
66-7502	F4D	AMARC	FP350
66-7503	F4D	w.o.	
66-7504	GF4D	GIA	
66-7505	F4D	to N427FS	
66-7506	F4D	AMARC	FP267
66-7507	F4D	to Korea	
66-7508	F4D	w.o.15.3.68	
66-7509	F4D	AMARC	FP422
66-7510	F4D	w.o.16.6.69	
66-7511	F4D	AMARC	FP187
66-7512	F4D	AMARC	FP394
66-7513	F4D	w.o.2.10.67	
66-7514	F4D		
66-7515	F4D	AMARC	FP211
66-7516	F4D	w.o.9.9.67	
66-7517	F4D	w.o.23.8.67	
66-7518	GF4D	TTC	Sheppard
66-7519	F4D	AMARC	FP366
66-7520	F4D	AMARC	FP148
66-7521	F4D	w.o.28.12.67	
66-7522	F4D	w.o.21.5.71	
66-7523	F4D	w.o.26.11.68	
66-7524	F4D	AMARC	FP416
66-7525	F4D	AMARC	FP097
66-7526	F4D	w.o.18.2.70	
66-7527	F4D	AMARC	FP245
66-7528	F4D	w.o.27.2.68	
66-7529	F4D	AMARC	FP259
66-7530	F4D	w.o.11.9.69	
66-7531	F4D	AMARC	FP312
66-7532	F4D	w.o.28.8.68	
66-7533	F4D	w.o.19.9.67	
66-7534	F4D	w.o.25.9.67	
66-7535	F4D	w.o.11.4.68	
66-7536	F4D	AMARC	FP256
66-7537	F4D	AMARC	FP198
66-7538	F4D	AMARC	FP207
66-7539	F4D	AMARC	FP344
66-7540	F4D	w.o.28.6.68	
66-7541	F4D	w.o.25.4.68	
66-7542	F4D	w.o.13.1.81	
66-7543	F4D	w.o.20.12.69	
66-7544	F4D	AMARC	FP317
66-7545	F4D	AMARC	FP348
66-7546	F4D	w.o.1.10.69	
66-7547	F4D	AMARC	FP404
66-7548	F4D	AMARC	FP163
66-7549	F4D	AMARC	FP373
66-7550	F4D		
66-7551	F4D	AMARC	FP287
66-7552	F4D	AMARC	FP249
66-7553	F4D	AMARC	FP470
66-7554	F4D		
66-7555	F4D	to Korea	
66-7556	F4D	AMARC	FP311
66-7557	F4D	w.o.31.7.67	
66-7558	F4D	AMARC	FP273
66-7559	F4D	w.o.4.3.81	
66-7560	F4D	AMARC	FP155
66-7561	F4D	AMARC	FP224
66-7562	F4D	w.o.6.12.69	
66-7563	F4D	AMARC	FP274
66-7564	F4D	w.o.3.10.67	
66-7565	F4D	w.o.18.9.68	
66-7566	F4D	AMARC	FP408
66-7567	F4D	w.o.17.11.69	
66-7568	F4D	w.o.19.8.69	
66-7569	F4D	w.o.26.5.68	
66-7570	F4D	AMARC	FP167
66-7571	F4D	w.o.19.7.68	
66-7572	F4D	w.o.25.10.74	
66-7573	F4D	w.o.31.12.71	
66-7574	F4D	w.o.14.6.69	
66-7575	F4D	AMARC	FP220
66-7576	F4D	w.o.30.7.72	
66-7577	F4D	to Korea	
66-7578	F4D	AMARC	FP161
66-7579	F4D	AMARC	FP263
66-7580	F4D	w.o.15.2.76	
66-7581	F4D	w.o.18.1.68	
66-7582	F4D	AMARC	FP265
66-7583	F4D	AMARC	FP290
66-7584	F4D	w.o.12.10.70	
66-7585	F4D	w.o.17.11.78	
66-7586	F4D	w.o.15.2.68	
66-7587	F4D	AMARC	FP303
66-7588	F4D	AMARC	FP430
66-7589	F4D	AMARC	FP291
66-7590	F4D	w.o.18.1.70	
66-7591	F4D	AMARC	FP156
66-7592	F4D	w.o.15.3.68	
66-7593	F4D	AMARC	FP221
66-7594	F4D	w.o.28.8.79	
66-7595	F4D	AMARC	FP472
66-7596	F4D	AMARC	FP264
66-7597	F4D	w.o.30.7.72	
66-7598	F4D	w.o.17.4.70	
66-7599	F4D	AMARC	FP...
66-7600	F4D	w.o.22.4.68	
66-7601	F4D	w.o.16.2.72	
66-7602	F4D	w.o.10.4.70	
66-7603	F4D	w.o.15.7.69	
66-7604	F4D	AMARC	FP353
66-7605	F4D	AMARC	FP374
66-7606	F4D	w.o.10.8.73	
66-7607	F4D	AMARC	FP398
66-7608	F4D	to Korea	
66-7609	F4D	AMARC	FP246
66-7610	F4D	AMARC	FP143
66-7611	F4D	w.o.24.10.72	
66-7612	F4D	w.o.18.5.72	
66-7613	F4D	w.o.3.5.70	
66-7614	F4D	AMARC	FP400
66-7615	F4D	AMARC	FP209
66-7616	F4D	w.o.25.4.71	
66-7617	F4D	w.o.28.8.79	
66-7618	F4D		
66-7619	F4D	AMARC	FP142
66-7620	F4D	w.o.4.3.81	
66-7621	F4D	w.o.26.5.72	
66-7622	F4D	w.o.13.4.72	
66-7623	F4D	AMARC	FP401
66-7624	F4D	w.o.19.12.70	
66-7625	F4D	AMARC	FP327
66-7626	F4D		
66-7627	F4D	w.o.7.7.72	
66-7628	F4D	w.o.3.1.71	
66-7629	F4D	AMARC	FP154
66-7630	F4D	w.o.17.5.68	
66-7631	F4D	w.o.16.12.67	
66-7632	F4D	w.o.20.1.70	
66-7633	F4D	AMARC	FP127
66-7634	F4D	AMARC	FP375
66-7635	F4D	AMARC	FP481
66-7636	F4D	w.o.24.6.72	
66-7637	F4D	w.o.6.3.69	
66-7638	F4D	w.o.25.5.76	
66-7639	F4D	w.o.	
66-7640	F4D	AMARC	FP413
66-7641	F4D	AMARC	FP218
66-7642	F4D	AMARC	FP255
66-7643	F4D	w.o.12.5.70	
66-7644	F4D	AMARC	FP160
66-7645	F4D	AMARC	FP144
66-7646	F4D	w.o.28.11.68	
66-7647	F4D	AMARC	FP549
66-7648	F4D	AMARC	FP323
66-7649	F4D	AMARC	FP343
66-7650	F4D	AMARC	FP169
66-7651	F4D	w.o.14.3.72	
66-7652	F4D	AMARC	FP275
66-7653	F4D	w.o.5.8.70	
66-7654	F4D	w.o.7.9.67	
66-7655	F4D	w.o.22.3.71	
66-7656	F4D	AMARC	FP159
66-7657	F4D	w.o.24.4.78	
66-7658	F4D	AMARC	FP431
66-7659	F4D	AMARC	FP196
66-7660	F4D	AMARC	FP356
66-7661	F4D		
66-7662	F4D	AMARC	FP477
66-7663	F4D	AMARC	FP361
66-7664	F4D	AMARC	FP418
66-7665	F4D	w.o.9.8.79	

66-7666	F4D	w.o.1.7.85	
66-7667	F4D	AMARC	FP399
66-7668	F4D	AMARC	FP266
66-7669	F4D	AMARC	FP257
66-7670	F4D	w.o.14.2.68	
66-7671	F4D	w.o.19.7.68	
66-7672	F4D	w.o.6.1.74	
66-7673	F4D	to Korea	
66-7674	F4D	AMARC	FP145
66-7675	F4D	w.o.	
66-7676	F4D	AMARC	FP149
66-7677	F4D	AMARC	FP238
66-7678	F4D	w.o.28.11.72	
66-7679	F4D	AMARC	FP281
66-7680	F4D	w.o.5.7.72	
66-7681	F4D	AMARC	FP162
66-7682	F4D	w.o.25.7.68	
66-7683	F4D	AMARC	FP372
66-7684	F4D	AMARC	FP254
66-7685	F4D	AMARC	FP336
66-7686	F4D	w.o.8.9.69	
66-7687	F4D	AMARC	FP225
66-7688	F4D	AMARC	FP432
66-7689	F4D	AMARC	FPO91
66-7690	F4D	to Korea	
66-7691	F4D	w.o.6.3.69	
66-7692	F4D	AMARC	FP279
66-7693	F4D		
66-7694	F4D	AMARC	FP474
66-7695	F4D	w.o.6.4.70	
66-7696	F4D	AMARC	FP352
66-7697	F4D	w.o.12.7.69	
66-7698	F4D	AMARC	FP271
66-7699	F4D	AMARC	FP347
66-7700	F4D		
66-7701	F4D	AMARC	FP146
66-7702	F4D	AMARC	FP405
66-7703	F4D	w.o.25.7.68	
66-7704	F4D	AMARC	FP153
66-7705	F4D	AMARC	FP345
66-7706	F4D	AMARC	FP326
66-7707	F4D	w.o.10.7.72	
66-7708	F4D	AMARC	FP424
66-7709	F4D	to Korea	
66-7710	F4D	AMARC	FP237
66-7711	F4D	AMARC	FP420
66-7712	F4D	AMARC	FP402
66-7713	F4D	w.o.25.6.74	
66-7714	F4D	AMARC	FP105
66-7715	F4D	to Korea	
66-7716	NF4D	AMARC	FP647
66-7717	F4D	w.o.22.4.71	
66-7718	F4D	AMARC	FP331
66-7719	F4D	w.o.12.3.68	
66-7720	F4D	AMARC	FP103
66-7721	F4D	AMARC	FP455
66-7722	F4D	AMARC	FP421
66-7723	F4D	AMARC	FP241
66-7724	F4D	w.o.	
66-7725	F4D	AMARC	FP324
66-7726	F4D	AMARC	FP367
66-7727	F4D	w.o.12.12.67	
66-7728	F4D	AMARC	FP157
66-7729	F4D	AMARC	FP215
66-7730	F4D	AMARC	FP370
66-7731	F4D	AMARC	FP269
66-7732	F4D	to Korea	
66-7733	F4D	AMARC	FP325
66-7734	F4D	w.o.25.6.81	
66-7735	F4D	AMARC	FP100
66-7736	F4D	w.o.17.12.69	
66-7737	F4D	to Korea	
66-7738	F4D	AMARC	FP283
66-7739	F4D	AMARC	FP318
66-7740	F4D	w.o.21.10.66	
66-7741	F4D	AMARC	FP270
66-7742	F4D	AMARC	FP197
66-7743	F4D	w.o.1.11.77	
66-7744	F4D	w.o.19.12.67	
66-7745	F4D	preserved	Montgomery AP, Al
66-7746	F4D		
66-7747	F4D	to Korea	
66-7748	F4D	w.o.4.11.69	
66-7749	F4D	AMARC	FP242
66-7750	F4D	to Korea	
66-7751	F4D	AMARC	FP193
66-7752	F4D	w.o.26.11.71	
66-7753	F4D	to Korea	
66-7754	F4D	AMARC	FP098
66-7755	F4D	AMARC	FP349
66-7756	F4D	w.o.6.7.68	
66-7757	F4D	w.o.17.12.67	
66-7758	F4D	to Korea	
66-7759	F4D	AMARC	FP473\N428FS
66-7760	F4D	AMARC	FP199
66-7761	F4D	w.o.24.7.70	
66-7762	F4D	to Korea	
66-7763	F4D	w.o.3.12.69	
66-7764	F4D	w.o.9.4.72	
66-7765	F4D	AMARC	FP355
66-7766	F4D	AMARC	FP147
66-7767	F4D	AMARC	FP288
66-7768	F4D	AMARC	FP476
66-7769	F4D	w.o.11.2.68	
66-7770	F4D	w.o.30.7.72	
66-7771	F4D	AMARC	FP201
66-7772	F4D	AMARC	FP268
66-7773	F4D	AMARC	FP337
66-7774	F4D	w.o.17.12.67	
66-8685	F4D	w.o.9.4.69	
66-8686	F4D	w.o.29.6.71	
66-8687	F4D	w.o.4.5.68	
66-8688	F4D	w.o.1.9.68	
66-8689	F4D	w.o.24.9.83	
66-8690	F4D	w.o.28.1.69	
66-8691	F4D	w.o.25.2.71	
66-8692	F4D	w.o.20.9.68	
66-8693	F4D	AMARC	FP158
66-8694	F4D	w.o.25.8.68	
66-8695	F4D	w.o.20.5.68	
66-8696	F4D	w.o.19.3.70	
66-8697	F4D	w.o.20.10.69	
66-8698	F4D	AMARC	FP448
66-8699	F4D	AMARC	FP608
66-8700	F4D	AMARC	FP226
66-8701	F4D	to Korea	
66-8702	F4D	w.o.22.4.70	
66-8703	F4D	w.o.1.2.72	
66-8704	F4D	w.o.10.1.68	
66-8705	F4D	AMARC	FP276
66-8706	F4D	w.o.	
66-8707	F4D	w.o.2.4.70	
66-8708	F4D	w.o.28.10.72	
66-8709	F4D	AMARC	FP164
66-8710	F4D	AMARC	FP222
66-8711	F4D		
66-8712	F4D	w.o.10.9.71	
66-8713	F4D	w.o.29.1.76	
66-8714	F4D	preserved	Carswell AFB
66-8715	F4D	AMARC	FP277
66-8716	F4D	w.o.24.7.68	
66-8717	F4D	w.o.22.2.69	
66-8718	F4D	w.o.11.11.69	
66-8719	F4D	AMARC	FP185
66-8720	F4D	w.o.18.1.68	
66-8721	F4D	w.o.2.5.70	
66-8722	F4D	AMARC	FP229
66-8723	F4D	AMARC	FP354
66-8724	F4D	w.o.24.6.68	
66-8725	F4D	w.o.23.2.68	
66-8726	F4D	w.o.23.5.72	
66-8727	F4D	AMARC	FP192
66-8728	F4D	AMARC	FP272
66-8729	F4D	w.o.9.1.68	
66-8730	F4D	w.o.29.1.76	
66-8731	F4D	w.o.2.9.71	
66-8732	F4D	AMARC	FP223
66-8733	F4D	AMARC	FP213
66-8734	F4D	to Korea	
66-8735	F4D	AMARC	FP200
66-8736	F4D	w.o.26.4.68	
66-8737	F4D	to Korea	
66-8738	F4D	w.o.5.10.72	
66-8739	F4D	AMARC	FP202
66-8740	F4D		
66-8741	F4D	w.o.13.2.72	

Serial	Type	Status	Notes		Serial	Type	Status	Notes
66-8742	F4D	w.o.26.9.72			66-8825	F4D	AMARC	FP332
66-8743	F4D	w.o.13.1.79			67-0208	F4E	to Turkey	
66-8744	F4D	w.o.13.12.71			67-0209	F4E	w.o.13.9.68	
66-8745	F4D	AMARC	FP214		67-0210	F4E	to Turkey	
66-8746	F4D	w.o.10.6.68			67-0211	F4E	to Egypt	
66-8747	F4D	w.o.26.6.73			67-0212	F4E	to Egypt	
66-8748	F4D				67-0213	F4E	to Egypt	
66-8749	F4D	w.o.7.1.73			67-0214	F4E	w.o.25.4.70	
66-8750	F4D	w.o.16.2.71			67-0215	F4E	to Turkey	
66-8751	F4D	w.o.6.10.68			67-0216	F4E	to Turkey	
66-8752	F4D	w.o.11.9.68			67-0217	F4E	to Turkey	
66-8753	F4D	AMARC	FP282		67-0218	F4E	AMARC	FP388
66-8754	F4D	w.o.17.8.68			67-0219	F4E	w.o.12.11.69	
66-8755	F4D				67-0220	F4E	to Egypt	
66-8756	F4D	to Korea			67-0221	F4E	to Turkey	
66-8757	F4D	w.o.29.4.68			67-0222	F4E	to Turkey	
66-8758	F4D	to Korea			67-0223	F4E	AMARC	FP439
66-8759	F4D	to Korea			67-0224	F4E		
66-8760	F4D	w.o.26.2.68			67-0225	F4E	w.o.5.12.73	
66-8761	F4D	w.o.23.1.79			67-0226	F4E	to Turkey	
66-8762	F4D	AMARC	FP168		67-0227	F4E	to Turkey	
66-8763	F4D	w.o.6.1.69			67-0228	F4E		
66-8764	F4D	w.o.30.3.69			67-0229	F4E	AMARC	FP389
66-8765	F4D	to Korea			67-0230	F4E	to Turkey	
66-8766	F4D	w.o.13.4.69			67-0231	F4E	to Egypt	
66-8767	F4D	w.o.22.3.68			67-0232	F4E	to Turkey	
66-8768	F4D	preserved	Bergstrom AFB		67-0233	F4E	to Turkey	
66-8769	F4D	w.o.21.9.72			67-0234	F4E	w.o.27.12.72	
66-8770	F4D	w.o.8.8.70			67-0235	F4E		
66-8771	F4D	w.o.30.3.69			67-0236	F4E	to Egypt	
66-8772	F4D	w.o.17.7.72			67-0237	F4E		
66-8773	F4D	w.o.17.1.69			67-0238	F4E	to Egypt	7811
66-8774	F4D	w.o.27.2.71			67-0239	F4E	to Egypt	
66-8775	F4D	w.o.23.4.70			67-0240	F4E	w.o.10.10.78	
66-8776	F4D	w.o.7.12.78			67-0241	F4E	AMARC	FP360
66-8777	F4D	w.o.4.2.71			67-0242	F4E	to Turkey	
66-8778	F4D	w.o.23.4.68			67-0243	F4E	w.o.27.6.72	
66-8779	F4D	w.o.20.9.77			67-0244	F4E	w.o.30.8.76	
66-8780	F4D	w.o.18.3.68			67-0245	F4E	AMARC	FP446
66-8781	F4D	w.o.6.7.77			67-0246	F4E	AMARC	FP726
66-8782	F4D	AMARC	FP190		67-0247	F4E	w.o.20.1.72	
66-8783	F4D	AMARC	FP475		67-0248	F4E	to Turkey	
66-8784	F4D	w.o.20.1.70			67-0249	GF4E	TTC	Sheppard
66-8785	F4D	w.o.16.9.72			67-0250	F4E	w.o.19.3.82	
66-8786	F4D	AMARC	FP186		67-0251	F4E	to Turkey	
66-8787	F4D	w.o.16.2.71			67-0252	F4E	w.o.23.10.70	
66-8788	F4D	AMARC	FP165		67-0253	F4E	w.o.19.12.69	
66-8789	F4D	AMARC	FP128		67-0254	F4E	w.o.10.10.72	
66-8790	F4D	w.o.20.9.72			67-0255	F4E	w.o.13.9.76	
66-8791	F4D	w.o.3.9.69			67-0256	F4E	AMARC	FP304
66-8792	F4D	w.o.23.3.72			67-0257	F4E	w.o.17.8.70	
66-8793	F4D	w.o.15.11.76			67-0258	F4E	to Turkey	
66-8794	F4D	AMARC	FP150		67-0259	F4E	to Turkey	
66-8795	F4D	w.o.22.9.73			67-0260	F4E	BDRT	Spangdhalem AB
66-8796	F4D	w.o.16.4.69			67-0261	F4E	w.o.9.8.69	
66-8797	F4D				67-0262	F4E	to Turkey	
66-8798	F4D				67-0263	F4E	AMARC	FP376
66-8799	F4D	w.o.12.5.72			67-0264	F4E	to Egypt	7809
66-8800	F4D				67-0265	F4E	to Turkey	
66-8801	F4D	w.o.20.3.68			67-0266	F4E		
66-8802	F4D	AMARC	FP228		67-0267	F4E	w.o.20.1.71	
66-8803	F4D	w.o.22.5.87			67-0268	F4E	to Turkey	
66-8804	F4D	AMARC	FP280		67-0269	F4E	to Turkey	
66-8805	F4D	AMARC	FP166		67-0270	F4E	preserved	McGuire AFB
66-8806	F4D	to Korea			67-0271	F4E		
66-8807	F4D	w.o.20.11.68			67-0272	F4E	to Turkey	
66-8808	F4D	AMARC	FP371		67-0273	F4E	to Turkey	
66-8809	F4D	w.o.29.3.69			67-0274	F4E	to Turkey	
66-8810	F4D	to Korea			67-0275	F4E	AMARC	FP310
66-8811	F4D	w.o.20.6.68			67-0276	F4E	w.o.14.7.81	
66-8812	F4D				67-0277	F4E	w.o.1.7.72	
66-8813	F4D	AMARC	FP232		67-0278	F4E	to Egypt	7835
66-8814	F4D	w.o.1.3.69			67-0279	F4E	w.o.30.6.70	
66-8815	F4D	AMARC	FP184		67-0280	F4E	to Turkey	
66-8816	F4D	AMARC	FP216		67-0281	F4E	w.o.23.4.73	
66-8817	F4D	AMARC	FP240		67-0282	F4E	w.o.9.3.70	
66-8818	F4D	w.o.26.12.71			67-0283	F4E		
66-8819	F4D	AMARC	FP203		67-0284	F4E	w.o.9.1.74	
66-8820	F4D	w.o.15.1.70			67-0285	F4E	to Turkey	
66-8821	F4D	AMARC	FP403		67-0286	F4E	w.o.26.1.69	
66-8822	F4D	w.o.2.8.68			67-0287	F4E	w.o.28.12.74	
66-8823	F4D	w.o.21.8.82			67-0288	F4E	AMARC	FP369
66-8824	F4D	AMARC	FP227		67-0289	F4E	to Egypt	

Serial	Type	Disposition	Code/Location
67-0290	F4E	to Turkey	
67-0291	F4E	w.o.3.8.69	
67-0292	F4E	w.o.27.12.72	
67-0293	F4E	w.o.7.5.70	
67-0294	F4E	w.o.14.1.69	
67-0295	F4E	w.o.8.8.71	
67-0296	F4E	w.o.5.7.72	
67-0297	F4E	w.o.18.6.70	
67-0298	F4E	to Turkey	
67-0299	F4E	AMARC	FP727
67-0300	F4E	w.o.5.12.69	
67-0301	F4E	to Turkey	
67-0302	F4E	to Turkey	
67-0303	F4E	w.o.8.6.72	
67-0304	F4E	to Turkey	
67-0305	F4E	to Egypt	
67-0306	F4E	w.o.6.1.76	
67-0307	F4E	to Egypt	7816
67-0308	F4E	w.o.9.9.85	
67-0309	F4E	to Egypt	
67-0310	F4E	w.o.10.9.71	
67-0311	F4E	AMARC	FP296
67-0312	F4E	w.o.17.8.72	
67-0313	F4E	to Egypt	
67-0314	F4E	w.o.22.11.69	
67-0315	F4E	w.o.5.9.86	
67-0316	F4E	to Turkey	
67-0317	F4E	to Egypt	
67-0318	F4E	to Egypt	
67-0319	F4E	to Turkey	
67-0320	F4E	AMARC	FP435
67-0321	F4E	to Turkey	
67-0322	F4E	to Egypt	
67-0323	F4E	w.o.1.8.70	
67-0324	F4E	AMARC	FP316
67-0325	F4E	w.o.4.8.69	
67-0326	F4E	to Israel	
67-0327	F4E	preserved	Luke AFB
67-0328	F4E	AMARC	FP319
67-0329	F4E	w.o.1.12.70	
67-0330	F4E	w.o.27.6.73	
67-0331	F4E	to Turkey	
67-0332	F4E	to Egypt	7824
67-0333	F4E	AMARC	FP333
67-0334	F4E	to Turkey	
67-0335	F4E	w.o.6.5.71	
67-0336	F4E	to Turkey	
67-0337	F4E	AMARC	FP380
67-0338	F4E	to Turkey	
67-0339	F4E	w.o.5.7.72	
67-0340	F4E	to Israel	
67-0341	F4E	to Egypt	
67-0342	F4E	to Turkey	
67-0343	F4E	AMARC	FP769
67-0344	F4E	to Turkey	
67-0345	F4E	to Greece	
67-0346	F4E	to Israel	
67-0347	F4E		
67-0348	F4E		
67-0349	F4E	AMARC	FP639
67-0350	F4E	to Greece	
67-0351	F4E		
67-0352	F4E	w.o.9.4.69	
67-0353	F4E	to Turkey	
67-0354	F4E	to Turkey	
67-0355	F4E	to Egypt	
67-0356	F4E	AMARC	FP654
67-0357	F4E	w.o.7.1.70	
67-0358	F4E	w.o.21.5.72	
67-0359	F4E	w.o.2.9.71	
67-0360	F4E	to Turkey	
67-0361	F4E	w.o.9.1.86	
67-0362	F4E	to Israel	
67-0363	F4E	w.o.8.5.72	
67-0364	F4E	AMARC	FP622
67-0365	F4E	w.o.13.6.72	
67-0366	F4E	to Egypt	
67-0367	F4E	w.o.21.12.72	
67-0368	F4E	to Israel	
67-0369	F4E		
67-0370	F4E	AMARC	FP743
67-0371	F4E	to Egypt	
67-0372	F4E		
67-0373	F4E	to Egypt	7808
67-0374	F4E	w.o.16.6.73	
67-0375	F4E		
67-0376	F4E	to Turkey	
67-0377	F4E	to Greece	
67-0378	F4E	w.o.5.12.71	
67-0379	F4E		
67-0380	F4E	w.o.18.6.69	
67-0381	F4E	to Greece	
67-0382	F4E	to Turkey	
67-0383	F4E	to Turkey	
67-0384	F4E	49FW	HO
67-0385	F4E	w.o.22.9.72	
67-0386	F4E	w.o.10.5.72	
67-0387	F4E	to Turkey	
67-0388	F4E	to Egypt	
67-0389	F4E	preserved	Ft. Wayne MAP
67-0390	F4E	AMARC	FP642
67-0391	F4E	to Turkey	
67-0392	F4E		
67-0393	F4E	preserved	Hampton, Va
67-0394	F4E		
67-0395	F4E	to Turkey	
67-0396	F4E		
67-0397	F4E	w.o.9.4.70	
67-0398	F4E	to Turkey	
67-0428	RF4C	AMARC	FP699
67-0429	RF4C	196RS	CA ANG
67-0430	RF4C	w.o.10.12.80	
67-0431	RF4C	w.o.11.4.88	
67-0432	RF4C		
67-0433	RF4C	AMARC	FP813
67-0434	RF4C	196RS	CA ANG
67-0435	RF4C		
67-0436	RF4C	w.o.16.5.79	
67-0437	RF4C	w.o.1.8.72	
67-0438	RF4C	AMARC	FP754
67-0439	RF4C	w.o.26.3.80	
67-0440	RF4C	w.o.21.1.87	
67-0441	RF4C	AMARC	FP766
67-0442	RF4C	AMARC	FP753
67-0443	RF4C	AMARC	FP775
67-0444	RF4C	AMARC	FP830
67-0445	RF4C	w.o.7.9.76	
67-0446	RF4C	w.o.12.4.69	
67-0447	RF4C	w.o.30.1.69	
67-0448	RF4C	AMARC	FP686
67-0449	RF4C	196RS	CA ANG
67-0450	RF4C	AMARC	FP724
67-0451	RF4C		
67-0452	RF4C	46TW	ET AFMC\MSD
67-0453	RF4C	AMARC	FP784
67-0454	RF4C	AMARC	FP676
67-0455	RF4C	AMARC	FP341
67-0456	RF4C	192RS	NV ANG
67-0457	RF4C		
67-0458	RF4C	AMARC	FP672
67-0459	RF4C	AMARC	FP298
67-0460	RF4C	w.o.14.1.70	
67-0461	RF4C		
67-0462	RF4C	AMARC	FP698
67-0463	RF4C	190FS	ID ANG
67-0464	RF4C	AMARC	FP377
67-0465	RF4C	AMARC	FP383
67-0466	RF4C	190FS	ID ANG
67-0467	RF4C	AMARC	FP759
67-0468	RF4C	w.o.30.4.81	
67-0469	RF4C	AMARC	FP782
68-0303	F4E	AMARC	FP660
68-0304	F4E	ALC	Ogden
68-0305	F4E	49FW	HO
68-0306	F4E	w.o.28.4.72	
68-0307	F4E	to Turkey	
68-0308	F4E	AMARC	FP621
68-0309	F4E		
68-0310	F4E		
68-0311	F4E		
68-0312	F4E		
68-0313	F4E	to Turkey	
68-0314	F4E	w.o.27.6.72	
68-0315	F4E	w.o.24.6.72	
68-0316	F4E	w.o.30.9.71	
68-0317	F4E	AMARC	FP501

Serial	Type	Fate	Notes		Serial	Type	Fate	Notes
68-0318	F4E	to Greece			68-0405	F4E	to Greece	
68-0319	F4E	to Turkey			68-0406	F4E		
68-0320	F4E	AMARC	FP539		68-0407	F4E		
68-0321	F4E	49FW	HO		68-0408	F4E	to Greece	
68-0322	F4E				68-0409	F4E	to Turkey	
68-0323	F4E				68-0410	F4E	AMARC	FP696
68-0324	F4E	AMARC	FP540		68-0411	F4E	49FW	HO
68-0325	F4E				68-0412	F4E	to Greece	
68-0326	F4E				68-0413	F4E		
68-0327	F4E	w.o.21.1.81			68-0418	F4E	49FW	HO
68-0328	F4E	w.o.15.2.86			68-0419	F4E	w.o.18.2.71	
68-0329	F4E				68-0420	F4E		
68-0330	F4E				68-0421	F4E		
68-0331	F4E	to Israel			68-0422	F4E	w.o.14.11.69	
68-0332	F4E	w.o.14.1.70			68-0423	F4E	AMARC	FP510
68-0333	F4E	to Israel			68-0424	F4E	to Greece	
68-0334	F4E	w.o.19.2.80			68-0425	F4E	w.o.7.1.74	
68-0335	F4E	w.o.2.9.72			68-0426	F4E	to Greece	
68-0336	F4E				68-0427	F4E		
68-0337	F4E	49FW	HO		68-0428	F4E		
68-0338	F4E	preserved	St. Louis IAP		68-0429	F4E		
68-0339	F4E				68-0430	F4E	w.o.25.10.84	
68-0340	F4E	AMARC	FP578		68-0431	F4E		
68-0341	F4E				68-0432	F4E	to Greece	
68-0342	F4E				68-0433	F4E	w.o.10.6.71	
68-0343	F4E	AMARC	FP511		68-0438	F4E	to Greece	
68-0344	F4E				68-0439	F4E		
68-0345	F4E	FSI	Mojave AP		68-0440	F4E	to Greece	
68-0346	F4E	to Turkey			68-0441	F4E		
68-0347	F4E				68-0442	F4E	to Greece	
68-0348	F4E				68-0443	F4E	49FW	HO
68-0349	F4E				68-0444	F4E	to Greece	
68-0350	F4E	to Turkey			68-0445	F4E	to Greece	
68-0351	F4E				68-0446	F4E		
68-0352	F4E	w.o.15.8.72			68-0447	F4E	AMARC	FP648
68-0353	F4E				68-0448	F4E	to Turkey	
68-0354	F4E	to Turkey			68-0449	F4E	FSI	Mojave AP
68-0355	F4E				68-0450	F4E	AMARC	FP653
68-0356	F4E	w.o.9.2.70			68-0451	F4E	w.o.17.2.78	
68-0357	F4E	AMARC	FP768		68-0452	F4E	AMARC	FP554
68-0358	F4E	to Korea			68-0453	F4E		
68-0359	F4E	w.o.8.5.72			68-0458	F4E		
68-0360	F4E				68-0459	F4E	to Korea	
68-0361	F4E	to Greece			68-0460	F4E	AMARC	FP499
68-0362	F4E	w.o.6.5.83			68-0461	F4E		
68-0363	F4E	to Greece			68-0462	F4E	AMARC	FP661
68-0364	F4E	w.o.1.3.77			68-0463	F4E	AMARC	FP466
68-0365	F4E				68-0464	F4E	AMARC	FP582
68-0366	GF4E	TTC	Sheppard		68-0465	F4E	preserved	St. Louis IAP
68-0367	F4E				68-0466	F4E		
68-0368	F4E	w.o.23.2.81			68-0467	F4E		
68-0369	F4E	w.o.10.12.82			68-0468	F4E		
68-0370	F4E	to Korea			68-0473	F4E		
68-0371	F4E	AMARC	FP505		68-0474	F4E	to Turkey	
68-0372	F4E	49FW	HO		68-0475	F4E	w.o.13.11.80	
68-0373	F4E	w.o.20.10.86			68-0476	F4E	MASDC	
68-0374	F4E				68-0477	F4E	w.o.22.8.72	
68-0375	F4E	AMARC	FP556		68-0478	F4E	49FW	HO
68-0376	F4E				68-0479	F4E	w.o.1.9.74	
68-0377	F4E				68-0480	F4E	to Greece	
68-0378	F4E				68-0481	F4E	to Greece	
68-0379	F4E	AMARC	FP559		68-0482	F4E		
68-0380	F4E	to Israel			68-0483	F4E		
68-0381	F4E	to Greece			68-0488	F4E	w.o.1.5.74	
68-0382	F4E	BDRT	March AFB		68-0489	F4E	w.o.30.9.74	
68-0383	F4E				68-0490	F4E	49FW	HO
68-0384	F4E				68-0491	F4E	w.o.12.11.80	
68-0385	F4E	AMARC	FP697		68-0492	F4E	w.o.27.10.71	
68-0386	F4E				68-0493	F4E		
68-0387	F4E				68-0494	F4E		
68-0388	F4E	AMARC	FP557		68-0495	F4E	49FW	HO
68-0389	F4E	AMARC	FP656		68-0496	F4E	to Greece	
68-0390	F4E				68-0497	F4E		
68-0391	F4E	AMARC	FP643		68-0498	F4E	to Turkey	
68-0392	F4E				68-0503	F4E	49FW	HO
68-0393	F4E	to Greece			68-0504	F4E		
68-0394	F4E	to Greece			68-0505	F4E	w.o.1.5.80	
68-0395	F4E	w.o.16.8.80			68-0506	F4E	to Greece	
68-0400	F4E	to Turkey			68-0507	F4E	AMARC	FP500
68-0401	F4E	to Korea			68-0508	F4E		
68-0402	F4E	to Greece			68-0509	F4E	AMARC	FP514
68-0403	F4E	to Turkey			68-0510	F4E	w.o.11.8.70	
68-0404	F4E				68-0511	F4E	AMARC	FP657

Serial	Type		
68-0512	F4E		
68-0513	F4E		
68-0514	F4E		
68-0515	F4E	to Greece	
68-0516	F4E	49FW	HO
68-0517	F4E	to Greece	
68-0518	F4E	w.o.2.9.77	
68-0526	F4E		
68-0527	F4E	to Korea	
68-0528	F4E		
68-0529	F4E	49FW	HO
68-0530	F4E		
68-0531	F4E	49FW	HO
68-0532	F4E		
68-0533	F4E		
68-0534	F4E		
68-0535	F4E	w.o.17.8.84	
68-0536	F4E	AMARC	FP553
68-0537	F4E	w.o.24.1.73	
68-0538	F4E	49FW	HO
68-0548	RF4C	AMARC	FP890
68-0549	RF4C		
68-0550	RF4C		
68-0551	RF4C	AMARC	FP884
68-0552	RF4C	AMARC	FP876
68-0553	RF4C	AMARC	FP885
68-0554	RF4C	BDRT	Ramstein AB
68-0555	RF4C	AMARC	FP897
68-0556	RF4C	w.o.8.6.83	
68-0557	RF4C	AMARC	FP777
68-0558	RF4C	w.o.30.1.75	
68-0559	RF4C	w.o.21.9.71	
68-0560	RF4C		
68-0561	RF4C	AMARC	FP887
68-0562	RF4C	AMARC	FP655
68-0563	RF4C	w.o.18.2.88	
68-0564	RF4C	AMARC	FP881
68-0565	RF4C	AMARC	FP888
68-0566	RF4C	w.o.28.3.79	
68-0567	RF4C	AMARC	FP792
68-0568	RF4C	AMARC	FP735
68-0569	RF4C	AMARC	FP690
68-0570	RF4C		
68-0571	RF4C	AMARC	FP755
68-0572	RF4C	AMARC	FP877
68-0573	RF4C	w.o.20.1.72	
68-0574	RF4C	AMARC	FP677
68-0575	RF4C	106RS	BH AL ANG
68-0576	RF4C	AMARC	FP668
68-0577	RF4C		
68-0578	RF4C	AMARC	FP595
68-0579	RF4C		
68-0580	RF4C	AMARC	FP781
68-0581	RF4C	AMARC	FP756
68-0582	RF4C	AMARC	FP751
68-0583	RF4C	AMARC	FP793
68-0584	RF4C	AMARC	FP758
68-0585	RF4C	AMARC	FP624
68-0586	RF4C	w.o.5.3.74	
68-0587	RF4C	preserved	Hermeskeil, FRG
68-0588	RF4C	w.o.16.5.77	
68-0589	RF4C	AMARC	FP664
68-0590	RF4C	preserved	Bruxelles, Belgium
68-0591	RF4C	w.o.21.1.87	
68-0592	RF4C	AMARC	FP895
68-0593	RF4C	AMARC	FP858
68-0594	RF4C	preserved	Boise AT, Id
68-0595	RF4C	AMARC	FP170
68-0596	RF4C	AMARC	FP171
68-0597	RF4C	w.o.11.12.72	
68-0598	RF4C	w.o.20.4.72	
68-0599	RF4C	AMARC	FP205
68-0600	RF4C	AMARC	FP188
68-0601	RF4C	w.o.21.12.70	
68-0602	RF4C	AMARC	FP172
68-0603	RF4C	BDRT	Bergstrom AFB
68-0604	RF4C	w.o.13.8.72	
68-0605	RF4C	AMARC	FP194
68-0606	RF4C	AMARC	FP189
68-0607	RF4C	AMARC	FP173
68-0608	RF4C	AMARC	FP206
68-0609	RF4C	AMARC	FP195
68-0610	RF4C	w.o.8.10.70	
68-0611	RF4C	w.o.17.4.81	
69-0236	F4G	AMARC	FP606
69-0237	F4G	AMARC	FP731
69-0238	F4G	190FS	ID ANG
69-0239	F4G	AMARC	FP607
69-0240	F4G	w.o.16.6.82	
69-0241	F4G		
69-0242	F4G	AMARC	FP765
69-0243	F4G	190FS	ID ANG
69-0244	F4G	AMARC	FP823
69-0245	F4G	AMARC	FP762
69-0246	F4G	AMARC	FP847
69-0247	F4G	57W	WA
69-0248	F4G	57W	WA
69-0249	F4G		
69-0250	F4G	AMARC	FP798
69-0251	F4G	AMARC	FP832
69-0252	F4G	w.o.8.11.82	
69-0253	F4G		
69-0254	F4G	AMARC	FP852
69-0255	F4G	AMARC	FP764
69-0256	F4E	w.o.27.4.70	
69-0257	F4G	w.o.1.6.81	
69-0258	F4G		
69-0259	F4G		
69-0260	F4G		
69-0261	F4G	190FS	ID ANG
69-0262	F4E	w.o.11.10.73	
69-0263	F4G		
69-0264	F4G	190FS	ID ANG
69-0265	F4G		
69-0266	F4E	w.o.15.9.75	
69-0267	F4G	AMARC	FP860
69-0268	F4E	w.o.17.8.79	
69-0269	F4G	AMARC	FP738
69-0270	F4G	AMARC	FP682
69-0271	F4G	w.o.8.1.81	
69-0272	F4G	190FS	ID ANG
69-0273	F4G		
69-0274	F4G	AMARC	FP763
69-0275	F4G	190FS	ID ANG
69-0276	F4E	w.o.12.10.72	
69-0277	F4G	190FS	ID ANG
69-0278	F4G	57W	WA
69-0279	F4G	AMARC	FP848
69-0280	F4G	w.o.23.5.84	
69-0281	F4G	AMARC	FP865
69-0282	F4E	w.o.21.6.72	
69-0283	F4G	w.o.12.1.81	
69-0284	F4G	AMARC	FP864
69-0285	F4G	57W	WA
69-0286	F4G		
69-0287	F4E	w.o.5.10.72	
69-0288	F4E	w.o.11.9.72	
69-0289	F4E	w.o.4.7.72	
69-0290	F4E	AMARC	FP600
69-0291	F4E		
69-0292	F4G	AMARC	FP874
69-0293	F4G		
69-0294	F4E	to Israel	
69-0295	F4E	to Israel	
69-0296	F4E	to Israel	
69-0297	F4G		
69-0298	F4G		
69-0299	F4E	to Israel	
69-0300	F4E	to Israel	
69-0301	F4E	to Israel	
69-0302	F4E	w.o.	
69-0303	F4G	190FS	ID ANG
69-0304	F4G	AMARC	FP856
69-0305	F4G	190FS	ID ANG
69-0306	F4G	190FS	ID ANG
69-0307	F4G	190FS	ID ANG
69-0349	RF4C	w.o.30.3.76	
69-0350	RF4C	BDRT	Boise ANGB
69-0351	RF4C	w.o.22.9.72	
69-0352	RF4C	AMARC	FP757
69-0353	RF4C	w.o.5.12.78	
69-0354	RF4C	w.o.15.5.75	
69-0355	RF4C	w.o.19.8.72	
69-0356	RF4C	AMARC	FP702
69-0357	RF4C	AMARC	FP896
69-0358	RF4C	AMARC	FP364

Serial	Type	Location	Notes		Serial	Type	Location	Notes
69-0359	RF4C	AMARC	FP741		69-7258	F4G	AMARC	FP861
69-0360	RF4C				69-7259	F4G	w.o.18.3.80	
69-0361	RF4C	AMARC	FP329		69-7260	F4G	AMARC	FP866
69-0362	RF4C	AMARC	FP636		69-7261	QF4G	FSI	Mojave AP
69-0363	RF4C	AMARC	FP752		69-7262	F4G	AMARC	FP817
69-0364	RF4C	w.o.6.11.85			69-7263	F4G	190FS	ID ANG
69-0365	RF4C				69-7264	F4E	w.o.25.9.72	
69-0366	RF4C				69-7265	F4E	w.o.2.10.75	
69-0367	RF4C				69-7266	F4E	w.o.12.9.72	
69-0368	RF4C	AMARC	FP342		69-7267	F4G		
69-0369	RF4C	AMARC	FP776		69-7268	F4G	AMARC	FP788
69-0370	RF4C	AMARC	FP886		69-7269	F4E	w.o.3.7.79	
69-0371	RF4C				69-7270	F4G		
69-0372	RF4C	BDRT	Langley AFB		69-7271	F4E	w.o.27.6.72	
69-0373	RF4C	w.o.18.8.83			69-7272	F4G	190FS	ID ANG
69-0374	RF4C	AMARC	FP330		69-7273	F4E	w.o.27.10.72	
69-0375	RF4C	w.o.21.7.81			69-7286	F4G		
69-0376	RF4C	AMARC	FP805		69-7287	F4G	AMARC	FP857
69-0377	RF4C	w.o.18.6.83			69-7288	F4G	AMARC	FP859
69-0378	RF4C	AMARC	FP878		69-7289	F4G	AMARC	FP867
69-0379	RF4C	AMARC	FP675		69-7290	F4G	190FS	ID ANG
69-0380	RF4C				69-7291	F4E	57W	WA
69-0381	RF4C	w.o.22.9.87			69-7292	F4E	w.o.21.9.79	
69-0382	RF4C	AMARC	FP889		69-7293	F4G	AMARC	FP797
69-0383	RF4C	FSI	Mojave AP		69-7294	F4G	190FS	ID ANG
69-0384	RF4C	AMARC	FP806		69-7295	F4G	57W	WA
69-7201	F4G	AMARC	FP828		69-7296	F4E	w.o.27.6.72	
69-7202	F4G	190FS	ID ANG		69-7297	F4G	190FS	ID ANG
69-7203	F4E	w.o.16.6.71			69-7298	F4G	190FS	ID ANG
69-7204	F4G	190FS	ID ANG		69-7299	F4E	w.o.1.6.72	
69-7205	F4G	w.o.29.3.80			69-7300	F4G	AMARC	FP851
69-7206	F4G	GIA	Nellis AFB		69-7301	QF4G	FSI	Mojave AP
69-7207	F4G	190FS	ID ANG		69-7302	F4G		
69-7208	F4G	190FS	ID ANG		69-7303	F4G	AMARC	FP839
69-7209	F4G	AMARC	FP849		69-7546	F4G	190FS	ID ANG
69-7210	F4G	57W	WA		69-7547	F4E	to Israel	
69-7211	F4G				69-7548	F4E	w.o.6.10.72	
69-7212	F4G				69-7549	F4E	to Israel	
69-7213	F4G	w.o.7.11.79			69-7550	F4G		
69-7214	F4G	AMARC	FP862		69-7551	F4E	190FS	ID ANG
69-7215	F4G	w.o.6.12.76			69-7552	F4E	w.o.7.3.72	
69-7216	F4G	AMARC	FP807		69-7553	F4E	to Israel	
69-7217	F4G	57W	WA		69-7554	F4E	to Israel	
69-7218	F4G	AMARC	FP598		69-7555	F4E	w.o.5.4.82	
69-7219	F4G	w.o.21.11.86			69-7556	F4G		
69-7220	F4G	190FS	ID ANG		69-7557	F4G	190FS	ID ANG
69-7221	F4E	w.o.10.5.72			69-7558	F4G		
69-7222	F4E	w.o.8.12.70			69-7559	F4E	w.o.17.12.71	
69-7223	F4G	w.o.1.6.84			69-7560	F4G	190FS	ID ANG
69-7224	F4E	to Israel			69-7561	F4G	190FS	ID ANG
69-7225	F4E	to Israel			69-7562	F4E	w.o.23.11.71	
69-7226	F4E	to Israel			69-7563	F4E	w.o.8.7.82	
69-7227	F4E	to Israel			69-7564	F4E	w.o.20.4.72	
69-7228	F4G	57W	WA		69-7565	F4E	w.o.9.9.72	
69-7229	F4E	to Israel			69-7566	F4E	AMARC	FP787
69-7230	F4E	w.o.10.5.72			69-7567	F4E	to Israel	
69-7231	F4G	AMARC	FP838		69-7568	F4E	to Israel	
69-7232	F4G	57W	WA		69-7569	F4E	to Israel	
69-7233	F4G	AMARC	FP833		69-7570	F4E	to Israel	
69-7234	F4G	AMARC	FP761		69-7571	F4G	w.o.19.1.91	
69-7235	F4G	85TES	OT Det.6		69-7572	F4G	190FS	ID ANG
69-7236	F4G	BDRT	Spangdahlem AB		69-7573	F4E	w.o.6.10.72	
69-7237	F4E	to Israel			69-7574	F4G	190FS	ID ANG
69-7238	F4E	to Israel			69-7575	F4E	to Israel	
69-7239	F4E	to Israel			69-7576	F4E	to Israel	
69-7240	F4E	to Israel			69-7577	F4E	to Israel	
69-7241	F4E	to Israel			69-7578	F4E	to Israel	
69-7242	F4E	to Israel			69-7579	F4G		
69-7243	F4E	to Israel			69-7580	F4G	190FS	ID ANG
69-7244	F4E	to Israel			69-7581	F4G	AMARC	FP827
69-7245	F4E	to Israel			69-7582	F4G	AMARC	FP786
69-7246	F4E	to Israel			69-7583	F4G	AMARC	FP599
69-7247	F4E	to Israel			69-7584	F4G	w.o.17.12.80	
69-7248	F4E	to Israel			69-7585	F4E	to Turkey	
69-7249	F4E	to Israel			69-7586	F4G	w.o.1.6.82	
69-7250	F4E	to Israel			69-7587	F4G	57W	WA
69-7251	F4G	GIA	Nellis AFB		69-7588	F4G		
69-7252	F4E	AMARC	FP601		69-7589	F4E	w.o.30.9.87	
69-7253	F4G	w.o.6.7.81			71-0237	F4E	AMARC	FP644
69-7254	F4G	AMARC	FP855		71-0238	F4E	AMARC	FP635
69-7255	F4E	to Israel			71-0239	F4E	AMARC	FP669
69-7256	F4G	190FS	ID ANG		71-0240	F4E	AMARC	FP645
69-7257	F4G	AMARC	FP844		71-0241	F4E	w.o.28.12.74	

Serial	Type	Disposition	Code
71-0242	F4E		
71-0243	F4E	AMARC	FP595
71-0244	F4E	w.o.7.1.76	
71-0245	F4E		
71-0247	F4E	AMARC	FP469
71-0248	RF4C	AMARC	FP891
71-0249	RF4C	stored	Ogden ALC
71-0250	RF4C	w.o.16.4.81	
71-0251	RF4C	AMARC	FP880
71-0252	RF4C	AMARC	FP799
71-0253	RF4C	w.o.10.5.79	
71-0254	RF4C	AMARC	FP800
71-0255	RF4C	AMARC	FP808
71-0256	RF4C	w.o.7.6.73	
71-0257	RF4C	w.o.5.3.74	
71-0258	RF4C		
71-0259	RF4C	AMARC	FP893
71-1070	F4E	w.o.25.4.77	
71-1071	F4E	to Israel	
71-1072	F4E	ranges	Eglin, Fl
71-1073	F4E	AMARC	FP649
71-1074	F4E	to Israel	
71-1075	F4E	AMARC	FP670
71-1076	F4E	AMARC	FP617
71-1077	F4E	w.o.	
71-1078	F4E	to Israel	
71-1079	F4E	AMARC	FP591
71-1080	F4E	to Israel	
71-1081	F4E	AMARC	FP628
71-1082	F4E	to Israel	
71-1083	F4E	AMARC	FP691
71-1084	F4E	AMARC	FP681
71-1085	F4E	AMARC	FP687
71-1086	F4E	AMARC	FP630
71-1087	F4E	AMARC	FP684
71-1088	F4E	AMARC	FP671
71-1089	F4E	AMARC	FP616
71-1090	F4E	to Israel	
71-1091	F4E	w.o.7.1.76	
71-1092	F4E	AMARC	FP588
71-1093	F4E	to Israel	
71-1391	F4E		
71-1392	F4E		
71-1393	F4E	to Israel	
71-1394	F4E	to Israel	
71-1395	F4E	to Israel	
71-1396	F4E	to Israel	
71-1397	F4E	AMARC	FP605
71-1398	F4E	to Israel	
71-1399	F4E	to Israel	
71-1400	F4E	to Israel	
71-1401	F4E	to Israel	
71-1402	F4E	to Israel	
72-0121	F4E		
72-0122	F4E	AMARC	FP517
72-0123	F4E	to Israel	
72-0124	F4E	AMARC	FP507
72-0125	F4E	w.o.13.4.78	
72-0126	F4E	BDRT	Misawa AB, Japan
72-0127	F4E	to Israel	
72-0128	F4E	AMARC	FP596
72-0129	F4E	to Israel	
72-0130	F4E	to Israel	
72-0131	F4E	to Israel	
72-0132	F4E	to Israel	
72-0133	F4E	to Israel	
72-0134	F4E	w.o.3.12.77	
72-0135	F4E	AMARC	FP720
72-0136	F4E	AMARC	FP620
72-0137	F4E	to Israel	
72-0138	F4E	to Israel	
72-0139	F4E	AMARC	FP558
72-0140	F4E	AMARC	FP564
72-0141	F4E	AMARC	FP705
72-0142	F4E	AMARC	FP685
72-0143	F4E	AMARC	FP587
72-0144	F4E	AMARC	FP898
72-0145	RF4C	AMARC	FP879
72-0146	RF4C	w.o.24.7.86	
72-0147	RF4C	AMARC	FP892
72-0148	RF4C	w.o.30.1.78	
72-0149	RF4C	AMARC	FP883
72-0150	RF4C	AMARC	FP894
72-0151	RF4C		
72-0152	RF4C	AMARC	FP809
72-0153	RF4C	AMARC	FP822
72-0154	RF4C	AMARC	FP802
72-0155	RF4C	AMARC	FP801
72-0156	RF4C	AMARC	FP882
72-0157	F4E	to Israel	
72-0158	F4E	to Israel	
72-0159	F4E	AMARC	FP666
72-0160	F4E	w.o.21.5.87	
72-0161	F4E	AMARC	FP590
72-0162	F4E	AMARC	FP584
72-0163	F4E	to Israel	
72-0164	F4E	to Israel	
72-0165	F4E	AMARC	FP631
72-0166	F4E	AMARC	FP592
72-0167	F4E	AMARC	FP579
72-0168	F4E	AMARC	FP566
72-1118	F4F	412TW	ED AFFTC
72-1407	F4E		
72-1476	F4E	w.o.3.3.81	
72-1477	F4E	AMARC	FP718
72-1478	F4E	AMARC	FP580
72-1479	F4E	AMARC	FP695
72-1480	F4E	to Israel	
72-1481	F4E	to Israel	
72-1482	F4E	AMARC	FP703
72-1483	F4E	AMARC	FP638
72-1484	F4E	AMARC	FP576
72-1485	F4E	AMARC	FP708
72-1486	F4E	w.o.15.4.81	
72-1487	F4E	to Israel	
72-1488	F4E	to Israel	
72-1489	F4E	AMARC	FP450
72-1490	F4E	AMARC	FP710
72-1491	F4E	to Israel	
72-1492	F4E	to Israel	
72-1493	F4E	AMARC	FP589
72-1494	F4E	AMARC	FP561
72-1495	F4E	to Israel	
72-1496	F4E	to Israel	
72-1497	F4E	to Israel	
72-1498	F4E	to Israel	
72-1499	F4E	to Israel	
73-1157	F4E	to Israel	
73-1158	F4E	to Israel	
73-1159	F4E	to Israel	
73-1160	F4E	AMARC	FP667
73-1161	F4E	to Israel	
73-1162	F4E	to Israel	
73-1163	F4E		
73-1164	F4E	AMARC	FP722
73-1165	F4E	AMARC	FP602
73-1166	F4E	AMARC	FP719
73-1167	F4E	AMARC	FP519
73-1168	F4E	AMARC	FP899
73-1169	F4E	to Israel	
73-1170	F4E	to Israel	
73-1171	F4E	AMARC	FP715
73-1172	F4E	AMARC	FP603
73-1173	F4E	AMARC	FP840
73-1174	F4E		
73-1175	F4E	BDRT	Luke AFB
73-1176	F4E	AMARC	FP870
73-1177	F4E	w.o.2.4.79	
73-1178	F4E	to Israel	
73-1179	F4E	to Israel	
73-1180	F4E	w.o.23.3.82	
73-1181	F4E	AMARC	FP560
73-1182	F4E	AMARC	FP632
73-1183	F4E	AMARC	FP873
73-1184	F4E	AMARC	FP713
73-1185	F4E	AMARC	FP597
73-1186	F4E	AMARC	FP393
73-1187	F4E	AMARC	FP434
73-1188	F4E	AMARC	FP692
73-1189	F4E	AMARC	FP378
73-1190	F4E	to Israel	
73-1191	F4E	to Israel	
73-1192	F4E	w.o.13.4.78	
73-1193	F4E	AMARC	FP565
73-1194	F4E	w.o.	

73-1195	F4E	AMARC	FP453	74-1056	F4E	w.o.26.12.85	
73-1196	F4E	AMARC	FP441	74-1057	F4E	AMARC	FP428
73-1197	F4E	AMARC	FP629	74-1058	F4E	w.o.27.4.77	
73-1198	F4E	AMARC	FP633	74-1059	F4E	AMARC	FP451
73-1199	F4E	AMARC	FP634	74-1060	F4E	AMARC	FP834
73-1200	F4E	w.o.3.9.80		74-1061	F4E	AMARC	FP567
73-1201	F4E	to Israel		74-1620	F4E	AMARC	FP683
73-1202	F4E	to Israel		74-1621	F4E	AMARC	FP618
73-1203	F4E	FSI Mojave AP		74-1622	F4E	AMARC	FP575
73-1204	F4E	FSI Mojave AP		74-1623	F4E	AMARC	FP611
74-0643	F4E	AMARC	FP429	74-1624	F4E	AMARC	FP693
74-0644	F4E	AMARC	FP701	74-1625	F4E	AMARC	FP397
74-0645	F4E	AMARC	FP447	74-1626	F4E	AMARC	FP512
74-0646	F4E	AMARC	FO520	74-1627	F4E	AMARC	FP652
74-0647	F4E	AMARC	FP445	74-1628	F4E	AMARC	FP572
74-0648	F4E	AMARC	FP498	74-1629	F4E	AMARC	FP583
74-0649	F4E	BDRT	Semr.-Johnson AFB	74-1630	F4E	AMARC	FP665
74-0650	F4E	AMARC	FP704	74-1631	F4E	AMARC	FP651
74-0651	F4E	w.o.1.12.78		74-1632	F4E	w.o.18.8.82	
74-0652	F4E	AMARC	FP716	74-1633	F4E	w.o.8.10.87	
74-0653	F4E	AMARC	FP613	74-1634	F4E	AMARC	FP662
74-0654	F4E	AMARC	FP612	74-1635	F4E	AMARC	FP706
74-0655	F4E	AMARC	FP483	74-1636	F4E	AMARC	FP581
74-0656	F4E	AMARC	FP875	74-1637	F4E	AMARC	FP506
74-0657	F4E	AMARC	FP694	74-1638	F4E	AMARC	FP573
74-0658	F4E	preserved	Moody AFB	74-1639	F4E		
74-0659	F4E	AMARC	FP663	74-1640	F4E	AMARC	FP497
74-0660	F4E	w.o.23.3.81		74-1641	F4E	GIA	Ramstein AB
74-0661	F4E	w.o.		74-1642	F4E	AMARC	FP465
74-0662	F4E	AMARC	FP680	74-1643	F4E	AMARC	FP714
74-0663	F4E	AMARC	FP841	74-1644	F4E	AMARC	FP619
74-0664	F4E	AMARC	FP837	74-1645	F4E	AMARC	FP586
74-0665	F4E	AMARC	FP709	74-1646	F4E	w.o.12.5.81	
74-0666	F4E	AMARC	FP723	74-1647	F4E	w.o.22.3.84	
74-1038	F4E	AMARC	FP678	74-1648	F4E	AMARC	FP614
74-1039	F4E	AMARC	FP593	74-1649	F4E	AMARC	FP725
74-1040	F4E	AMARC	FP721	74-1650	F4E	AMARC	FP650
74-1041	F4E	AMARC	FP437	74-1651	F4E	AMARC	FP679
74-1042	F4E	AMARC	FP493	74-1652	F4E	AMARC	FP707
74-1043	F4E	AMARC	FP641	74-1653	F4E	AMARC	FP717
74-1044	F4E	AMARC	FP869	75-0628	F4E	49FW	HO
74-1045	F4E	AMARC	FP836	75-0629	F4E	49FW	HO
74-1046	F4E	w.o.2.3.79		75-0630	F4E	49FW	HO
74-1047	F4E	AMARC	FP604	75-0631	F4E	49FW	HO
74-1048	F4E	AMARC	FP868	75-0632	F4E	49FW	HO
74-1049	F4E	AMARC	FP760	75-0633	F4E	49FW	HO
74-1050	F4E	AMARC	FP845	75-0634	F4E	w.o.21.11.83	
74-1051	F4E	w.o.29.8.77		75-0635	F4E	49FW	HO
74-1052	F4E	AMARC	FP846	75-0636	F4E	49FW	HO
74-1053	F4E	AMARC	FP395	75-0637	F4E	w.o.21.11.83	
74-1054	F4E	w.o.24.3.80		84-0494	F4N	BDRT	RAF Alconbury
74-1055	F4E	AMARC	FP835				

F5 FREEDOM FIGHTER/TIGER II

c\n	F5A:	N.6001 to N.6009, N.6061 to N.6064, N.6066, N6086, N6088.
	F5B:	N.8003 to N.8008, N.8077, N.8085, N.8090 to N8092, N.8100, N.8121, N.8122, N.8126.
	F5E:	R.1001 to R.1028, R.1036, R.1046, R.1047, R.1051, R.1060, R.1062, R.1063, R.1066,
R.1080,		R.1081, R.1087 to R.1089, R.1095, R.1123, R.1163 to R.1177, R.1186 to R.1212, R.1215
to		R.1218, R.1230, R.1231, R.1243 to R.1245, R.1252 to R.1254, R.1260 to R.1262, R.1266,
R.1269		to R.1271, R.1273, R.1275, R.1307 to R.1310, R.1316, R.1317.
	F5F:	W.1001, W.1002, IS.1001 to IS.1006, IH.1019, IH.1009.

59-4987	YF5A	preserved Seattle, Wa		63-8425	F5A	w.o.12.12.65
59-4988	YF5A			63-8426	F5C	w.o.8.8.66
59-4989	YF5A	USAF Museum		63-8427	F5A	w.o.30.6.65
63-8367	NF5A	rebuilt as F5B 65-13071		63-8429	F5C	w.o.7.12.66
63-8368	NF5A	rebuilt as F5B 65-13072		63-8440	GF5B	preserved Kelly AFB
63-8369	NF5A	rebuilt as F5B 65-13073		63-8441	GF5B	
63-8370	NF5A	rebuilt as F5B 65-13074		63-8442	F5B	w.o.
63-8371	F5A	to Thailand		63-8443	GF5B	TTC Chanute
63-8372	F5A	MASDC		63-8444	F5B	to Greece 63-8444
63-8424	F5A	w.o.10.9.66		63-8445	F5B	w.o.

64-13316	F5A	w.o.22.8.66	74-1516	F5E	w.o.22.9.83	
64-13319	F5A	w.o.3.7.66	74-1517	F5E	w.o.21.1.81	
70-1408	F5A	to Turkey 70-1408	74-1518	F5E	w.o.	
71-1033	F5B	to Turkey 71-1033	74-1519	F5E	to USN	Bu741519
71-1417	F5E	to Brazil	74-1528	F5E	to USN	Bu741528
71-1418	F5E	to Brazil	74-1529	F5E	to USN	Bu741529
71-1419	F5E	w.o.9.2.82	74-1530	F5E	to USN	Bu741530
71-1420	F5E	to Brazil	74-1531	F5E	to USN	Bu741531
71-1421	F5E		74-1532	F5E	to Tunisia	
72-0439	F5B	w.o.27.7.82	74-1533	F5E	to USN	Bu162307
72-0440	F5B	to Turkey 72-0440	74-1534	F5E	w.o.19.10.87	
72-0441	GF5B	IAAFA Homestead AFB	74-1535	F5E	to Tunisia	
72-0449	F5B	to Turkey 72-0449	74-1536	F5E	to USN	Bu741536
72-1386	F5E	to Brazil	74-1537	F5E	to USN	Bu741537
72-1387	F5E	to USN Bu721387	74-1538	F5E	to USN	Bu741538
72-1388	F5E	to Morocco	74-1539	F5E	to USN	Bu741539
72-1389	F5E		74-1540	F5E	to USN	Bu741540
72-1390	F5E		74-1541	F5E	to USN	Bu741541
72-1391	F5E	to Brazil	74-1542	F5E	w.o.15.1.86	
72-1392	F5E	to Brazil	74-1543	F5E	to Morocco	
72-1393	F5E	to Brazil	74-1544	F5E	to USN	Bu741544
72-1394	F5E	w.o.29.4.74	74-1545	F5E	to USN	Bu741545
72-1395	F5E	w.o.15.10.86	74-1546	F5E	w.o.	
72-1396	F5E		74-1547	F5E	to USN	Bu741547
72-1397	F5E	w.o.	74-1548	F5E	w.o.17.3.82	
72-1398	F5E	to Brazil	74-1549	F5E	to Morocco	
72-1399	F5E	w.o.	74-1550	F5E	w.o.25.1.82	
72-1400	F5E	to Brazil	74-1551	F5E	to Morocco	
72-1401	F5E	to Brazil	74-1552	F5E	w.o.22.8.78	
72-1402	F5E	to Brazil	74-1553	F5E	to Morocco	
72-1403	F5E	to Brazil	74-1554	F5E	to USN	Bu741554
72-1404	F5E		74-1555	F5E	w.o.29.12.76	
72-1405	F5E		74-1556	F5E	to USN	Bu741556
72-1406	F5E		74-1557	F5E		
73-0846	F5E		74-1558	F5E	to USN	Bu741558
73-0847	F5E		74-1559	F5E	to Tunisia	
73-0855	F5E	to USN Bu730855	74-1560	F5E	to Morocco	
73-0865	F5E	to USN Bu730865	74-1561	F5E		
73-0866	F5E	w.o.6.2.86	74-1562	F5E		
73-0870	F5E	w.o.	74-1563	F5E	to USN	Bu741563
73-0879	F5E	to USN Bu730879	74-1564	F5E	to USN	Bu741564
73-0881	F5E		74-1565	F5E	w.o.20.10.86	
73-0882	F5E		74-1566	F5E	to Tunisia	
73-0885	F5E	to USN Bu730885	74-1567	F5E		
73-0889	F5F	to NASA N550NA	74-1568	F5E	to USN	Bu741568
73-0891	F5F	w.o.22.12.86	74-1569	F5E	to Morocco	
73-0896	F5E	w.o.	74-1570	F5E	to USN	Bu741570
73-0897	F5E	w.o.11.7.83	74-1571	F5E		
73-0899	F5E	w.o.27.7.82	74-1572	F5E	to USN	Bu741572
73-1602	F5B		74-1573	F5E		
73-1603	F5B		74-1574	F5E	to Brazil	
73-1607	F5B	w.o.13.2.86	74-1575	F5E		
73-1635	F5E	to USN Bu731635	75-0612	F5E	to Morocco	
73-1636	F5E		75-0613	F5E	to Morocco	
73-1640	F5E	ALC San Antonio	75-0614	F5E		
74-1484	F5E	w.o.	75-0615	F5E		
74-1505	F5E	w.o.5.5.80	75-0616	F5E	to Thailand	
74-1506	F5E	w.o.14.4.79	75-0617	F5E	to Morocco	
74-1507	F5E	w.o.14.4.79	82-0089	F5F	SoC	
74-1508	F5E		82-0090	F5F	SoC	
74-1509	F5E	w.o.6.10.83	82-0091	F5F	SoC	
74-1510	F5E		83-0072	F5F	w.o.1.6.84	
74-1511	F5E		83-0073	F5F	to Brazil	4806
74-1512	F5E		83-0074	F5F	to Brazil	4807
74-1513	F5E	to Brazil	84-0456	F5F	to USN	Bu840456
74-1514	F5E	to Brazil	84-0457	F5F	to Brazil	4809
74-1515	F5E	to Brazil				

F15 EAGLE

c\n	F15E: E.1 to E.200.				
71-0280	YF15A		71-0282	YF15A	
71-0281	YF15A	preserved Langley AFB	71-0283	YF15A	MDD St. Louis

Serial	Type	Unit	Location		Serial	Type	Unit	Location
71-0284	GF15A	TTC	Sheppard		74-0118	F15A	AMARC	FH034
71-0285	YF15A	MDD	St. Louis		74-0119	GF15A	TTC	Sheppard
71-0286	GF15A	TTC	Chanute		74-0120	F15A	w.o.	
71-0287	YF15A	NASA			74-0121	F15A	128FS	GAGA ANG
71-0288	YF15A				74-0122	F15A	to Israel	
71-0289	YF15A				74-0123	F15A	AMARC	FH053
71-0290	YF15B	412TW	ED AFFTC		74-0124	F15A	preserved	Eglin AFB
71-0291	YF15E	MDD	St. Louis		74-0125	F15A	to Israel	
72-0113	F15A				74-0126	F15A		
72-0114	F15A	to Israel			74-0127	F15A	AMARC	FH032
72-0115	GF15A	TTC	Lowry		74-0128	F15A	AMARC	FH012
72-0116	F15A	to Israel			74-0129	F15A	w.o.28.2.77	
72-0117	F15A	to Israel			74-0130	F15A	325FW	TY
72-0118	F15A	to Israel			74-0131	F15A	BDRT	Lakenheath
72-0119	F15A	USAF Museum			74-0132	F15A	325FW	TY
72-0120	F15A	to Israel			74-0133	F15A	325FW	TY
73-0085	F15A	ALC	RG Warner-Robins		74-0134	F15A	AMARC	FH031
73-0086	F15A				74-0135	F15A	AMARC	FH035
73-0087	F15A	to Israel			74-0136	F15A	w.o.29.12.78	
73-0088	F15A	w.o.14.10.75			74-0137	F15B		
73-0089	F15A	128FS	GAGA ANG		74-0138	F15B	AMARC	FH036
73-0090	F15A	AMARC	FH...		74-0139	F15B	w.o.	
73-0091	F15A	BDRT	Soesterberg AB		74-0140	F15B	199FS	HI ANG
73-0092	F15A	AMARC	FH...		74-0141	F15B	NASA	Dryden
73-0093	F15A	to Israel			74-0142	GF15B	TTC	Sheppard
73-0094	F15A	to Israel			75-0018	F15A	w.o.1.9.79	
73-0095	F15A	BDRT	Spangdahlem AB		75-0019	F15A	128FS	GAGA ANG
73-0096	F15A	AMARC	FH006		75-0020	F15A	AMARC	FH044
73-0097	F15A	w.o.8.2.78			75-0021	F15A	AMARC	FH030
73-0098	F15A				75-0022	F15A	AMARC	FH045
73-0099	F15A				75-0023	F15A	w.o.10.3.80	
73-0100	F15A				75-0024	F15A	128FS	GAGA ANG
73-0101	F15A	to Israel			75-0025	F15A	128FS	GAGA ANG
73-0102	F15A	to Israel			75-0026	F15A		
73-0103	F15A	AMARC	FH007		75-0027	F15A		
73-0104	F15A	to Israel			75-0028	F15A		
73-0105	F15A	to Israel			75-0029	F15A	128FS	GAGA ANG
73-0106	F15A	w.o.15.12.81			75-0030	F15A	128FS	GAGA ANG
73-0107	F15A	to Israel			75-0031	F15A	AMARC	FH055
73-0108	F15B	preserved	Luke AFB		75-0032	F15A	AMARC	FH035
73-0109	F15B				75-0033	F15A	BDRT	Eglin AFB
73-0110	F15B	to Israel			75-0034	F15A	AMARC	FH046
73-0111	F15B				75-0035	F15A		
73-0112	F15B				75-0036	F15A	AMARC	FH049
73-0113	F15B				75-0037	F15A	325FW	TY
73-0114	F15B	85TES	OT		75-0038	F15A	AMARC	FH038
74-0081	F15A	BDRT	Elmendorf AFB		75-0039	F15A	128FS	GAGA ANG
74-0082	F15A	BDRT	Elmendorf AFB		75-0040	F15A	128FS	GAGA ANG
74-0083	F15A				75-0041	F15A		
74-0084	F15A	BDRT	Elmendorf AFB		75-0042	F15A	AMARC	FH039
74-0085	F15A	to Israel			75-0043	F15A	128FS	GAGA ANG
74-0086	F15A	AMARC	FH016		75-0044	F15A	BDRT	Eglin AFB
74-0087	F15A	w.o.24.6.85			75-0045	F15A	BDRT	Eglin AFB
74-0088	F15A				75-0046	F15A	AMARC	FH021
74-0089	F15A	to Israel			75-0047	F15A	128FS	GAGA ANG
74-0090	F15A	w.o.5.11.85			75-0048	F15A	BDRT	Langley AFB
74-0091	F15A				75-0049	F15A	128FS	GAGA ANG
74-0092	F15A	AMARC	FH017		75-0050	F15A	AMARC	FH022
74-0093	F15A	128FS	GA GA ANG		75-0051	F15A	w.o.3.11.81	
74-0094	F15A	w.o.9.9.85			75-0052	F15A	128FS	GAGA ANG
74-0095	F15A	preserved	Tyndall AFB		75-0053	F15A		
74-0096	F15A	AMARC	FH027		75-0054	F15A	128FS	GAGA ANG
74-0097	F15A	128FS	GA GA ANG		75-0055	F15A	BDRT	Langley AFB
74-0098	F15A	to Israel			75-0056	F15A	325FW	TY
74-0099	F15A	AMARC	FH019		75-0057	F15A	128FS	GAGA ANG
74-0100	F15A	AMARC	FH028		75-0058	F15A	128FS	GAGA ANG
74-0101	F15A				75-0059	F15A	w.o.17.4.78	
74-0102	F15A	325FW	TY		75-0060	F15A	AMARC	FH037
74-0103	F15A	AMARC	FH047		75-0061	F15A	128FS	GAGA ANG
74-0104	F15A	AMARC	FH033		75-0062	F15A	AMARC	FH052
74-0105	F15A	AMARC	FH020		75-0063	F15A	w.o.19.12.78	
74-0106	F15A				75-0064	F15A	w.o.28.12.78	
74-0107	F15A	128FS	GA GA ANG		75-0065	F15A	128FS	GAGA ANG
74-0108	F15A				75-0066	F15A	101FS	MA ANG
74-0109	F15A	to Israel			75-0067	F15A	128FS	GAGA ANG
74-0110	F15A	AMARC	FH029		75-0068	F15A	128FS	GAGA ANG
74-0111	F15A	AMARC	FH005		75-0069	F15A	128FS	GAGA ANG
74-0112	F15A	BDRT	Langley AFB		75-0070	F15A	w.o.4.3.80	
74-0113	F15A	199FS	HI ANG		75-0071	F15A	w.o.15.1.92	
74-0114	F15A	325FW	TY		75-0072	F15A	128FS	GAGA ANG
74-0115	F15A	AMARC	FH023		75-0073	F15A	AMARC	FH054
74-0116	F15A	128FS	GAGA ANG		75-0074	F15A	AMARC	FH048
74-0117	F15A	BDRT	Langley AFB		75-0075	F15A	128FS	GAGA ANG

Serial	Type	Unit	Base/Location
75-0076	F15A	w.o.24.11.87	
75-0077	F15A	AMARC	FH018
75-0078	F15A	128FS	GAGA ANG
75-0079	F15A	128FS	GAGA ANG
75-0080	F15B	46TW	ET AFMC\MSD
75-0081	F15B	199FS	HI ANG
75-0082	F15B	325FW	TY
75-0083	F15B	325FW	TY
75-0084	F15B	46TW	ET AFMC\MSD
75-0085	F15B	w.o.6.12.77	
75-0086	F15B	128FS	GAGA ANG
75-0087	F15B	w.o.17.8.84	
75-0088	F15B	128FS	GAGA ANG
75-0089	F15B	AMARC	FH025
76-0008	GF15A	TTC	Sheppard
76-0009	F15A	325FW	TY
76-0010	F15A	101FS	MA ANG
76-0011	F15A	101FS	MA ANG
76-0012	F15A	101FS	MA ANG
76-0013	F15A	w.o.25.7.80	
76-0014	F15A	101FS	MA ANG
76-0015	F15A	101FS	MA ANG
76-0016	F15A	101FS	MA ANG
76-0017	F15A	101FS	MA ANG
76-0018	F15A	101FS	MA ANG
76-0019	F15A	101FS	MA ANG
76-0020	F15A		
76-0021	F15A	101FS	MA ANG
76-0022	F15A		
76-0023	F15A	w.o.15.1.86	
76-0024	F15A	101FS	MA ANG
76-0025	F15A	101FS	MA ANG
76-0026	F15A	325FW	TY
76-0027	F15A	101FS	MA ANG
76-0028	F15A		
76-0029	F15A	101FS	MA ANG
76-0030	F15A	110FS	SL MO ANG
76-0031	F15A	110FS	SL MO ANG
76-0032	F15A	199FS	HI ANG
76-0033	F15A	110FS	SL MO ANG
76-0034	F15A	101FS	MA ANG
76-0035	F15A	w.o.3.6.79	
76-0036	F15A		
76-0037	F15A	325FW	TY
76-0038	F15A	AMARC	FH...
76-0039	F15A	101FS	MA ANG
76-0040	F15A	101FS	MA ANG
76-0041	F15A	110FS	SL MO ANG
76-0042	F15A	325FW	TY
76-0043	F15A	110FS	SL MO ANG
76-0044	F15A	325FW	TY
76-0045	F15A	110FS	SL MO ANG
76-0046	F15A		
76-0047	F15A	w.o.16.5.78	
76-0048	F15A	101FS	MA ANG
76-0049	F15A	101FS	MA ANG
76-0050	F15A	123FS	OR ANG
76-0051	F15A	AMARC	FH...
76-0052	F15A	199FS	HI ANG
76-0053	F15A	w.o.6.7.78	
76-0054	GF15A	TTC	Sheppard
76-0055	F15A		
76-0056	F15A	325FW	TY
76-0057	F15A		
76-0058	F15A	101FS	MA ANG
76-0059	F15A	123FS	OR ANG
76-0060	F15A		
76-0061	F15A	110FS	SL MO ANG
76-0062	F15A	199FS	HI ANG
76-0063	F15A	199FS	HI ANG
76-0064	F15A	199FS	HI ANG
76-0065	F15A	w.o.13.2.81	
76-0066	F15A	123FS	OR ANG
76-0067	GF15A	TTC	Sheppard
76-0068	F15A	199FS	HI ANG
76-0069	F15A	w.o.15.3.90	
76-0070	F15A	123FS	OR ANG
76-0071	F15A	123FS	OR ANG
76-0072	F15A	110FS	SL MO ANG
76-0073	F15A	123FS	OR ANG
76-0074	F15A		
76-0075	F15A	123FS	OR ANG
76-0076	F15A	AMARC	FH014
76-0077	F15A	123FS	OR ANG
76-0078	F15A	110FS	SL MO ANG
76-0079	GF15A	TTC	Sheppard
76-0080	F15A	123FS	OR ANG
76-0081	F15A	w.o.4.2.83	
76-0082	F15A	w.o.6.3.80	
76-0083	GF15A	TTC	Sheppard
76-0084	F15A	199FS	HI ANG
76-0085	F15A	w.o.13.9.79	
76-0086	F15A	412TW	ED AFFTC
76-0087	F15A	110FS	SL MO ANG
76-0088	F15A	BDRT	St Louis IAP, Mo
76-0089	F15A	AMARC	FH015
76-0090	F15A	110FS	SL MO ANG
76-0091	F15A	123FS	OR ANG
76-0092	F15A	123FS	OR ANG
76-0093	F15A	123FS	OR ANG
76-0094	F15A	110FS	SL MO ANG
76-0095	F15A	123FS	OR ANG
76-0096	F15A	AMARC	FH...
76-0097	F15A		
76-0098	F15A	123FS	OR ANG
76-0099	F15A	123FS	OR ANG
76-0100	F15A	AMARC	FH...
76-0101	F15A	46TW	ET AFMC\MSD
76-0102	F15A	123FS	OR ANG
76-0103	F15A	199FS	HI ANG
76-0104	F15A	AMARC	FH...
76-0105	F15A	123FS	OR ANG
76-0106	F15A	123FS	OR ANG
76-0107	F15A	199FS	HI ANG
76-0108	F15A	123FS	OR ANG
76-0109	F15A	123FS	OR ANG
76-0110	GF15A	TTC	Sheppard
76-0111	F15A	123FS	OR ANG
76-0112	F15A	199FS	HI ANG
76-0113	F15A	199FS	HI ANG
76-0114	F15A	199FS	HI ANG
76-0115	F15A	110FS	SL MO ANG
76-0116	F15A	ALC	RGWarner-Robins
76-0117	F15A	199FS	HI ANG
76-0118	F15A	199FS	HI ANG
76-0119	F15A	199FS	HI ANG
76-0120	F15A	199FS	HI ANG
76-0124	F15B	101FS	MA ANG
76-0125	F15B	101FS	MA ANG
76-0126	F15B	110FS	SL MO ANG
76-0127	F15B	32FS	CR
76-0128	F15B	101FS	MA ANG
76-0129	F15B	110FS	SL MO ANG
76-0130	F15B		
76-0131	F15B		
76-0132	F15B	412TW	ED AFFTC
76-0133	F15B	325FW	TY
76-0134	F15B	412TW	ED AFFTC
76-0135	F15B	325FW	TY
76-0136	F15B	325FW	TY
76-0137	F15B	325FW	TY
76-0138	F15B	w.o.1.5.89	
76-0139	F15B	123FS	OR ANG
76-0140	F15B	412TW	ED AFFTC
76-0141	F15B	123FS	OR ANG
76-0142	F15B	123FS	OR ANG
77-0061	F15A	122FS	LA ANG
77-0062	F15A	122FS	LA ANG
77-0063	F15A	122FS	LA ANG
77-0064	F15A	46TW	ET AFMC\MSD
77-0065	F15A	85TES	OT
77-0066	F15A	122FS	LA ANG
77-0067	F15A	122FS	LA ANG
77-0068	F15A	ALC	RGWarner-Robins
77-0069	F15A	122FS	LA ANG
77-0070	F15A	199FS	HI ANG
77-0071	F15A	122FS	LA ANG
77-0072	F15A	w.o.31.10.79	
77-0073	F15A		
77-0074	F15A	199FS	HI ANG
77-0075	F15A	122FS	LA ANG
77-0076	F15A	w.o.13.3.79	
77-0077	F15A	199FS	HI ANG
77-0078	F15A	199FS	HI ANG
77-0079	F15A	199FS	HI ANG
77-0080	F15A	46TW	ET AFMC\MSD

Serial	Type	Unit	Base	
77-0081	F15A	122FS		LA ANG
77-0082	F15A	325FW	TY	
77-0083	F15A	199FS		HI ANG
77-0084	F15A	412TW	ED	AFFTC
77-0085	F15A	325FW	TY	
77-0086	F15A	199FS		HI ANG
77-0087	F15A	32FS	CR	
77-0088	F15A	122FS		LA ANG
77-0089	F15A	32FS	CR	
77-0090	F15A	325FW	TY	
77-0091	F15A	122FS		LA ANG
77-0092	F15A	325FW	TW	
77-0093	F15A	122FS		LA ANG
77-0094	F15A	w.o.9.5.83		
77-0095	F15A			
77-0096	F15A	32FS	CR	
77-0097	F15A	32FS	CR	
77-0098	F15A	32FS	CR	
77-0099	F15A	32FS	CR	
77-0100	F15A	32FS	CR	
77-0101	F15A			
77-0102	F15A	32FS	CR	
77-0103	F15A	32FS	CR	
77-0104	F15A	32FS	CR	
77-0105	F15A	32FS	CR	
77-0106	F15A	32FS	CR	
77-0107	F15A	w.o.6.2.79		
77-0108	F15A	122FS		LA ANG
77-0109	F15A	32FS	CR	
77-0110	F15A	32FS	CR	
77-0111	F15A	32FS	CR	
77-0112	F15A	112FS		LA ANG
77-0113	F15A	32FS	CR	
77-0114	F15A	122FS		LA ANG
77-0115	F15A	32FS	CR	
77-0116	F15A	122FS		MO ANG
77-0117	F15A	122FS		LA ANG
77-0118	F15A	122FS		LA ANG
77-0119	F15A	122FS		LA ANG
77-0120	F15A	122FS		LA ANG
77-0121	F15A	122FS		LA ANG
77-0122	F15A	122FS		LA ANG
77-0123	F15A	122FS		LA ANG
77-0124	F15A	32FS	CR	
77-0125	F15A	325FW	TY	
77-0126	F15A	122FS		LA ANG
77-0127	F15A	325FW	TY	
77-0128	F15A	110FS	SL	MO ANG
77-0129	F15A	325FW	TY	
77-0130	F15A	325FW	TY	
77-0131	F15A	110FS	SL	MO ANG
77-0132	F15A	32FTS	CR	
77-0133	F15A	325FW	TY	
77-0134	F15A	122FS		LA ANG
77-0135	F15A	325FW	TY	
77-0136	F15A	325FW	TY	
77-0137	F15A	128FS		GA GA ANG
77-0138	F15A	325FW	TY	
77-0139	F15A	412TW	ED	AFFTC
77-0140	F15A	110FS	SL	MO ANG
77-0141	F15A	110FS	SL	MO ANG
77-0142	F15A	325FW	TY	
77-0143	F15A	110FS	SL	MO ANG
77-0144	F15A	325FW	TY	
77-0145	F15A	110FS	SL	MO ANG
77-0146	F15A	325FW	TY	
77-0147	F15A	325FW	TY	
77-0148	F15A	122FS	LA ANG	
77-0149	F15A	325FW	TY	
77-0150	F15A	325FW	TY	
77-0151	F15A	325FW	TY	
77-0152	F15A	325FW	TY	
77-0153	F15A			
77-0154	F15B	325FW	TY	
77-0155	F15B	325FW	TY	
77-0156	F15B	325FW	TY	
77-0157	F15B	325FW	TY	
77-0158	F15B	325FW	TY	
77-0159	F15B	122FS		LA ANG
77-0160	F15B	325FW	TY	
77-0161	F15B	46TW	ET	AFMC\MSD
77-0162	F15B	325FW	TY	
77-0163	F15B	32FS	CR	
77-0164	F15B	w.o.21.1.81		
77-0165	F15B	122FS		LA ANG
77-0166	F15B	412TW	ED	AFFTC
77-0167	F15B	w.o.25.6.79		
77-0168	F15B	325FW	TY	
78-0468	F15C	46TW	ET	AFSC\MSD
78-0469	F15C	18W	ZZ	
78-0470	F15C	18W	ZZ	
78-0471	F15C	325FW	TY	
78-0472	F15C			
78-0473	F15C	325FW	TY	
78-0474	F15C	18W	ZZ	
78-0475	F15C	325FW	TY	
78-0476	F15C	18W	ZZ	
78-0477	F15C	18W	ZZ	
78-0478	F15C	18W	ZZ	
78-0479	F15C	18W	ZZ	
78-0480	F15C	18W	ZZ	
78-0481	F15C	w.o.28.12.82		
78-0482	F15C	18W	ZZ	
78-0483	F15C	18W	ZZ	
78-0484	F15C	18W	ZZ	
78-0485	F15C	18W	ZZ	
78-0486	F15C	18W	ZZ	
78-0487	F15C	18W	ZZ	
78-0488	F15C	325FW	TY	
78-0489	F15C	18W	ZZ	
78-0490	F15C	3W	AK	
78-0491	F15C	18W	ZZ	
78-0492	F15C	18W	ZZ	
78-0493	F15C	18W	ZZ	
78-0494	F15C	18W	ZZ	
78-0495	F15C	18W	ZZ	
78-0496	F15C	18W	ZZ	
78-0497	F15C	18W	ZZ	
78-0498	F15C	18W	ZZ	
78-0499	F15C	18W	ZZ	
78-0500	F15C	18W	ZZ	
78-0501	F15C	18W	ZZ	
78-0502	F15C	18W	ZZ	
78-0503	F15C	18W	ZZ	
78-0504	F15C	18W	ZZ	
78-0505	F15C	325FW	TY	
78-0506	F15C	18W	ZZ	
78-0507	F15C	18W	ZZ	
78-0508	F15C	18W	ZZ	
78-0509	F15C	18W	ZZ	
78-0510	F15C	18W	ZZ	
78-0511	F15C	18W	ZZ	
78-0512	F15C	18W	ZZ	
78-0513	F15C	3W	AK	
78-0514	F15C	325FW	TY	
78-0515	F15C	18W	ZZ	
78-0516	F15C	18W	ZZ	
78-0517	F15C	18W	ZZ	
78-0518	F15C	18W	ZZ	
78-0519	F15C	18W	ZZ	
78-0520	F15C	18W	ZZ	
78-0521	F15C	18W	ZZ	
78-0522	F15C	18W	ZZ	
78-0523	F15C	325FW	TY	
78-0524	F15C	w.o.6.4.82		
78-0525	F15C	18W	ZZ	
78-0526	F15C	18W	ZZ	
78-0527	F15C	18W	ZZ	
78-0528	F15C	18W	ZZ	
78-0529	F15C	18W	ZZ	
78-0530	F15C	18W	ZZ	
78-0531	F15C	18W	ZZ	
78-0532	F15C	18W	ZZ	
78-0533	F15C	3W	AK	
78-0534	F15C	18W	ZZ	
78-0535	F15C	325FW	TY	
78-0536	F15C	18W	ZZ	
78-0537	F15C	18W	ZZ	
78-0538	F15C	18W	ZZ	
78-0539	F15C	18W	ZZ	
78-0540	F15C	w.o.28.12.82		
78-0541	F15C	18W	ZZ	
78-0542	F15C	85TES	OT	
78-0543	F15C	18W	ZZ	
78-0544	F15C	18W	ZZ	
78-0545	F15C	18W	ZZ	

Serial	Type	Unit	Code		Serial	Type	Unit	Code
78-0546	F15C	3W	AK		79-0068	F15C	36FW	BT
78-0547	F15C	18W	ZZ		79-0069	F15C		
78-0548	F15C	18W	ZZ		79-0070	F15C	3W	AK
78-0549	F15C	3W	AK		79-0071	F15C	w.o.1.6.83	
78-0550	F15C	3W	AK		79-0072	F15C	36FW	BT
78-0561	F15D	3W	AK		79-0073	F15C	36FW	BT
78-0562	F15D	3W	AK		79-0074	F15C	3W	AK
78-0563	F15D	18W	ZZ		79-0075	F15C	3W	AK
78-0564	F15D	3W	AK		79-0076	F15C	36FW	BT
78-0565	F15D	325FW	TY		79-0077	F15C	36FW	BT
78-0566	F15D	18W	ZZ		79-0078	F15C	36FW	BT
78-0567	F15D	18W	ZZ		79-0079	F15C	3W	AK
78-0568	F15D	325FW	TY		79-0080	F15C	3W	AK
78-0569	F15D				79-0081	F15C	3W	AK
78-0570	F15D	325FW	TY		80-0002	F15C	3W	AK
78-0571	F15D				80-0003	F15C	36FW	BT
78-0572	F15D	325FW	TY		80-0004	F15C	36FW	BT
78-0573	F15D	18W	ZZ		80-0005	F15C	36FW	BT
78-0574	F15D	325FW	TY		80-0006	F15C	3W	AK
79-0004	F15D	to Saudi Arabia 4223			80-0007	F15C	w.o.12.9.81	
79-0005	F15D	to Saudi Arabia 4221			80-0008	F15C	w.o.1.6.83	
79-0006	F15D	to Saudi Arabia 4222			80-0009	F15C	3W	AK
79-0007	F15D	3W	AK		80-0010	F15C	36FW	BT
79-0008	F15D	3W	AK		80-0011	F15C	36FW	BT
79-0009	F15D	3W	AK		80-0012	F15C	36FW	BT
79-0010	F15D	to Saudi Arabia 4224			80-0013	F15C	325FW	TY
79-0011	F15D	36FW	BT		80-0014	F15C	325FW	TY
79-0012	F15D	36FW	BT		80-0015	F15C	36FW	BT
79-0013	F15D	412TW	ED AFFTC		80-0016	F15C	325FW	TY
79-0014	F15D	325FW	TY		80-0017	F15C	w.o.8.11.88	
79-0015	F15C	to Saudi Arabia 4201			80-0018	F15C	3W	AK
79-0016	F15C	325FW	TY		80-0019	F15C		
79-0017	F15C	to Saudi Arabia 4206			80-0020	F15C	3W	AK
79-0018	F15C	to Saudi Arabia 4214			80-0021	F15C		
79-0019	F15C	to Saudi Arabia 4215			80-0022	F15C	36FW	BT
79-0020	F15C	3W	AK		80-0023	F15C	w.o.15.4.92	
79-0021	F15C				80-0024	F15C	3W	AK
79-0022	F15C	36FW	BT		80-0025	F15C	w.o.22.12.82	
79-0023	F15C	to Saudi Arabia 4202			80-0026	F15C	36FW	BT
79-0024	F15C	to Saudi Arabia 4207			80-0027	F15C	57FW	WA
79-0025	F15C	36FW	BT		80-0028	F15C	36FW	BT
79-0026	F15C				80-0029	F15C		
79-0027	F15C	325FW	TY		80-0030	F15C	57FW	WA
79-0028	F15C	to Saudi Arabia 4216			80-0031	F15C	36FW	BT
79-0029	F15C	325FW	TY		80-0032	F15C	w.o.7.1.86	
79-0030	F15C	325FW	TY		80-0033	F15C	57FW	WA
79-0031	F15C	to Saudi Arabia 4203			80-0034	F15C	57FW	WA
79-0032	F15C	to Saudi Arabia 4217			80-0035	F15C	57FS	IS
79-0033	F15C	to Saudi Arabia 4208			80-0036	F15C	w.o.4.1.83	
79-0034	F15C	325FW	TY		80-0037	F15C	w.o.2.1.86	
79-0035	F15C	36FW	BT		80-0038	F15C	57FS	IS
79-0036	F15C	36FW	BT		80-0039	F15C	57FS	IS
79-0037	F15C	36FW	BT		80-0040	F15C	57FS	IS
79-0038	F15C	to Saudi Arabia 4218			80-0041	F15C	57FS	IS
79-0039	F15C	to Saudi Arabia 4212			80-0042	F15C	57FS	IS
79-0040	F15C	w.o.23.6.81			80-0043	F15C	57FW	WA
79-0041	F15C	3W	AK		80-0044	F15C	57FW	WA
79-0042	F15C	3W	AK		80-0045	F15C	57FW	WA
79-0043	F15C	to Saudi Arabia 4213			80-0046	F15C	57FS	IS
79-0044	F15C	w.o.10.4.84			80-0047	F15C	57FS	IS
79-0045	F15C	to Saudi Arabia 4204			80-0048	F15C	57FS	IS
79-0046	F15C	36FW	BT		80-0049	F15C	57FW	WA
79-0047	F15C	3W	AK		80-0050	F15C	57FS	IS
79-0048	F15C	3W	AK		80-0051	F15C	3W	AK
79-0049	F15C	3W	AK		80-0052	F15C	57FS	IS
79-0050	F15C	3W	AK		80-0053	F15C	3W	AK
79-0051	F15C	to Saudi Arabia 4209			80-0054	F15D	3W	AK
79-0052	F15C	to Saudi Arabia 4219			80-0055	F15D	325FW	TY
79-0053	F15C	3W	AK		80-0056	F15D	57FS	IS
79-0054	F15C	3W	AK		80-0057	F15D	57FS	IS
79-0055	F15C	to Saudi Arabia 4210			80-0058	F15D	3W	AK
79-0056	F15C	3W	AK		80-0059	F15D		
79-0057	F15C	36FW	BT		80-0060	F15D	325FW	TY
79-0058	F15C	36FW	BT		80-0061	F15D		
79-0059	F15C	3W	AK		81-0020	F15C	3W	AK
79-0060	F15C	to Saudi Arabia 4211			81-0021	F15C	3W	AK
79-0061	F15C	w.o.7.1.86			81-0022	F15C	1FW	FF
79-0062	F15C	to Saudi Arabia 4220			81-0023	F15C	1FW	FF
79-0063	F15C	to Saudi Arabia 4205			81-0024	F15C	1FW	FF
79-0064	F15C	36FW	BT		81-0025	F15C	1FW	FF
79-0065	F15C	3W	AK		81-0026	F15C	1FW	FF
79-0066	F15C	3W	AK		81-0027	F15C	325FW	TY
79-0067	F15C	w.o.24.10.90			81-0028	F15C	325FW	TY

Serial	Type	Unit	Base
81-0029	F15C		
81-0030	F15C	1FW	FF
81-0031	F15C	1FW	FF
81-0032	F15C	1FW	FF
81-0033	F15C	1FW	FF
81-0034	F15C	1FW	FF
81-0035	F15C	1FW	FF
81-0036	F15C	1FW	FF
81-0037	F15C	1FW	FF
81-0038	F15C	1FW	FF
81-0039	F15C	1FW	FF
81-0040	F15C	1FW	FF
81-0041	F15C	1FW	FF
81-0042	F15C	1FW	FF
81-0043	F15C	3W	AK
81-0044	F15C	1FW	FF
81-0045	F15C	325FW	TY
81-0046	F15C	325FW	TY
81-0047	F15C		
81-0048	F15C	325FW	TY
81-0049	F15C	w.o.25.4.90	
81-0050	F15C	1FW	FF
81-0051	F15C	1FW	FF
81-0052	F15C	57FW	WA
81-0053	F15C	57FW	WA
81-0054	F15C	3W	AK
81-0055	F15C	57FW	WA
81-0056	F15C		
81-0061	F15D	3W	AK
81-0062	F15D	57FW	WA
81-0063	F15D	57FW	WA
81-0064	F15D	325FW	TY
81-0065	F15D	325FW	TY
82-0008	F15C	1FW	FF
82-0009	F15C	1FW	FF
82-0010	F15C	1FW	FF
82-0011	F15C	1FW	FF
82-0012	F15C	1FW	FF
82-0013	F15C	1FW	FF
82-0014	F15C	1FW	FF
82-0015	F15C	1FW	FF
82-0016	F15C	1FW	FF
82-0017	F15C	1FW	FF
82-0018	F15C	1FW	FF
82-0019	F15C	1FW	FF
82-0020	F15C	w.o.	
82-0021	F15C	1FW	FF
82-0022	F15C	1FW	FF
82-0023	F15C	1FW	FF
82-0024	F15C	1FW	FF
82-0025	F15C	3W	AK
82-0026	F15C	325FW	TY
82-0027	F15C	3W	AK
82-0028	F15C	3W	AK
82-0029	F15C	3W	AK
82-0030	F15C	325FW	TY
82-0031	F15C	3W	AK
82-0032	F15C	3W	AK
82-0033	F15C	325FW	TY
82-0034	F15C	325FW	TY
82-0035	F15C	325FW	TY
82-0036	F15C	1FW	FF
82-0037	F15C	1FW	FF
82-0038	F15C	1FW	FF
82-0044	F15D	325FW	TY
82-0045	F15D	325FW	TY
82-0046	F15D	57FW	WA
82-0047	F15D	1FW	FF
82-0048	F15D	325FW	TY
83-0010	F15C	1FW	FF
83-0011	F15C	1FW	FF
83-0012	F15C	1FW	FF
83-0013	F15C	1FW	FF
83-0014	F15C	1FW	FF
83-0015	F15C	1FW	FF
83-0016	F15C	1FW	FF
83-0017	F15C	1FW	FF
83-0018	F15C	1FW	FF
83-0019	F15C	1FW	FF
83-0020	F15C	325FW	TY
83-0021	F15C	w.o.30.11.92	
83-0022	F15C	325FW	TY
83-0023	F15C	1FW	FF
83-0024	F15C	1FW	FF
83-0025	F15C	1FW	FF
83-0026	F15C	1FW	FF
83-0027	F15C	1FW	FF
83-0028	F15C	1FW	FF
83-0029	F15C	1FW	FF
83-0030	F15C	1FW	FF
83-0031	F15C	1FW	FF
83-0032	F15C	1FW	FF
83-0033	F15C	1FW	FF
83-0034	F15C	1FW	FF
83-0035	F15C	1FW	FF
83-0036	F15C	1FW	FF
83-0037	F15C	1FW	FF
83-0038	F15C	1FW	FF
83-0039	F15C	1FW	FF
83-0040	F15C	1FW	FF
83-0041	F15C	1FW	FF
83-0042	F15C	1FW	FF
83-0043	F15C	1FW	FF
83-0046	F15D	1FW	FF
83-0047	F15D	1FW	FF
83-0048	F15D	1FW	FF
83-0049	F15D	1FW	FF
83-0050	F15D	1FW	FF
84-0001	F15C	36FW	BT
84-0002	F15C	36FW	BT
84-0003	F15C	36FW	BT
84-0004	F15C	36FW	BT
84-0005	F15C	36FW	BT
84-0006	F15C	36FW	BT
84-0007	F15C	36FW	BT
84-0008	F15C	36FW	BT
84-0009	F15C	36FW	BT
84-0010	F15C	36FW	BT
84-0011	F15C	85TES	OT
84-0012	F15C	57FW	WA
84-0013	F15C	36FW	BT
84-0014	F15C	36FW	BT
84-0015	F15C	36FW	BT
84-0016	F15C	36FW	BT
84-0017	F15C	ALC	Sacramento
84-0018	F15C	85TES	OT
84-0019	F15C	36FW	BT
84-0020	F15C	36FW	BT
84-0021	F15C	36FW	BT
84-0022	F15C	36FW	BT
84-0023	F15C	36FW	BT
84-0024	F15C	36FW	BT
84-0025	F15C	36FW	BT
84-0026	F15C	36FW	BT
84-0027	F15C	36FW	BT
84-0028	F15C	57FW	WA
84-0029	F15C	57FW	WA
84-0030	F15C	33FW	EG
84-0031	F15C	57FW	WA
84-0042	F15D	w.o. .85	
84-0043	F15D	36FW	BT
84-0044	F15D	36FW	BT
84-0045	F15D	46TW	ET AFMC\MSD
84-0046	F15D	412TW	ED AFFTC
85-0093	F15C	33FW	EG
85-0094	F15C	33FW	EG
85-0095	F15C	33FW	EG
85-0096	F15C	33FW	EG
85-0097	F15C	33FW	EG
85-0098	F15C	33FW	EG
85-0099	F15C	33FW	EG
85-0100	F15C	33FW	EG
85-0101	F15C	33FW	EG
85-0102	F15C	33FW	EG
85-0103	F15C	33FW	EG
85-0104	F15C	33FW	EG
85-0105	F15C	33FW	EG
85-0106	F15C	33FW	EG
85-0107	F15C	33FW	EG
85-0108	F15C	33FW	EG
85-0109	F15C	33FW	EG
85-0110	F15C	33FW	EG
85-0111	F15C	33FW	EG
85-0112	F15C	33FW	EG
85-0113	F15C	33FW	EG
85-0114	F15C	33FW	EG

85-0115	F15C	33FW	EG
85-0116	F15C	33FW	EG
85-0117	F15C	33FW	EG
85-0118	F15C	33FW	EG
85-0119	F15C	33FW	EG
85-0120	F15C	33FW	EG
85-0121	F15C	33FW	EG
85-0122	F15C	33FW	EG
85-0123	F15C	33FW	EG
85-0124	F15C	33FW	EG
85-0125	F15C	33FW	EG
85-0126	F15C	85TES	OT
85-0127	F15C	57FW	WA
85-0128	F15C	57FW	WA
85-0129	F15D	33FW	EG
85-0130	F15D	33FW	EG
85-0131	F15D	33FW	EG
85-0132	F15D	33FW	EG
85-0133	F15D	33FW	EG
85-0134	F15D	33FW	EG
86-0143	F15C	366W	MO
86-0144	F15C	33FW	EG
86-0145	F15C	33FW	EG
86-0146	F15C	33FW	EG
86-0147	F15C	33FW	EG
86-0148	F15C	366W	MO
86-0149	F15C	33FW	EG
86-0150	F15C	366W	MO
86-0151	F15C	366W	MO
86-0152	F15C	366W	MO
86-0153	F15C	w.o.28.12.89	
86-0154	F15C	33FW	EG
86-0155	F15C	366W	MO
86-0156	F15C	33FW	EG
86-0157	F15C	366W	MO
86-0158	F15C	33FW	EG
86-0159	F15C	33FW	EG
86-0160	F15C	33FW	EG
86-0161	F15C	366W	MO
86-0162	F15C	366W	MO
86-0163	F15C	33FW	EG
86-0164	F15C	33FW	EG
86-0165	F15C	33FW	EG
86-0166	F15C	33FW	EG
86-0167	F15C	33FW	EG
86-0168	F15C	33FW	EG
86-0169	F15C	33FW	EG
86-0170	F15C	33FW	EG
86-0171	F15C	33FW	EG
86-0172	F15C	33FW	EG
86-0173	F15C	33FW	EG
86-0174	F15C	33FW	EG
86-0175	F15C	33FW	EG
86-0176	F15C	33FW	EG
86-0177	F15C	57FW	WA
86-0178	F15C	33FW	EG
86-0179	F15C	57FW	WA
86-0180	F15C	33FW	EG
86-0181	F15D	366W	MO
86-0182	F15D	33FW	EG
86-0183	F15E	412TW	ED AFFTC
86-0184	F15E	412TW	ED AFFTC
86-0185	F15E	46TW	ET AFMC\MSD
86-0186	F15E	58FW	LF
86-0187	F15E	58FW	LF
86-0188	F15E	46TW	ET AFMC\MSD
86-0189	F15E	57FW	WA
86-0190	F15E	57FW	WA
87-0169	F15E	58FW	LF
87-0170	F15E	58FW	LF
87-0171	F15E	58FW	LF
87-0172	F15E	58FW	LF
87-0173	F15E	58FW	LF
87-0174	F15E	58FW	LF
87-0175	F15E	58FW	LF
87-0176	F15E	58FW	LF
87-0177	F15E	58FW	LF
87-0178	F15E	58FW	LF
87-0179	F15E	58FW	LF
87-0180	F15E	412TW	ED AFFTC
87-0181	F15E	4W	SJ
87-0182	F15E	4W	SJ
87-0183	F15E	4W	SJ
87-0184	F15E	58FW	LF
87-0185	F15E	58FW	LF
87-0186	F15E	58FW	LF
87-0187	F15E	58FW	LF
87-0188	F15E	58FW	LF
87-0189	F15E	58FW	LF
87-0190	F15E	58FW	LF
87-0191	F15E	58FW	LF
87-0192	F15E	58FW	LF
87-0193	F15E	58FW	LF
87-0194	F15E	58FW	LF
87-0195	F15E	4W	SJ
87-0196	F15E	4W	SJ
87-0197	F15E	58FW	LF
87-0198	F15E	366W	MO
87-0199	F15E	4W	SJ
87-0200	F15E	4W	SJ
87-0201	F15E	4W	SJ
87-0202	F15E	4W	SJ
87-0203	F15E	w.o.1.10.90	
87-0204	F15E		
87-0205	F15E	58FW	LF
87-0206	F15E	58FW	LF
87-0207	F15E		
87-0208	F15E		
87-0209	F15E		
87-0210	F15E		
88-1667	F15E		
88-1668	F15E	4W	SJ
88-1669	F15E	4W	SJ
88-1670	F15E	4W	SJ
88-1671	F15E	4W	SJ
88-1672	F15E	4W	SJ
88-1673	F15E	4W	SJ
88-1674	F15E	4W	SJ
88-1675	F15E	4W	SJ
88-1676	F15E	4W	SJ
88-1677	F15E	57FW	WA
88-1678	F15E	57FW	WA
88-1679	F15E	4W	SJ
88-1680	F15E	58FW	LF
88-1681	F15E	57FW	WA
88-1682	F15E	4W	SJ
88-1683	F15E	4W	SJ
88-1684	F15E	58FW	LF
88-1685	F15E	58FW	LF
88-1686	F15E	4W	SJ
88-1687	F15E	4W	SJ
88-1688	F15E	4W	SJ
88-1689	F15E	w.o.18.1.91	
88-1690	F15E	4W	SJ
88-1691	F15E	4W	SJ
88-1692	F15E	w.o.19.1.91	
88-1693	F15E	4W	SJ
88-1694	F15E	4W	SJ
88-1695	F15E	4W	SJ
88-1696	F15E	4W	SJ
88-1697	F15E	4W	SJ
88-1698	F15E	4W	SJ
88-1699	F15E	4W	SJ
88-1700	F15E	4W	SJ
88-1701	F15E	4W	SJ
88-1702	F15E	4W	SJ
88-1703	F15E	4W	SJ
88-1704	F15E	4W	SJ
88-1705	F15E	366W	MO
88-1706	F15E	4W	SJ
88-1707	F15E	4W	SJ
88-1708	F15E	4W	SJ
89-0471	F15E	4W	SJ
89-0472	F15E	4W	SJ
89-0473	F15E	4W	SJ
89-0474	F15E	4W	SJ
89-0475	F15E	57FW	WA
89-0476	F15E	4W	SJ
89-0477	F15E	58FW	LF
89-0478	F15E	4W	SJ
89-0479	F15E	w.o.10.8.92	
89-0480	F15E	4W	SJ
89-0481	F15E	57FW	WA
89-0482	F15E	4W	SJ
89-0483	F15E	4W	SJ
89-0484	F15E	4W	SJ

89-0485	F15E	4W	SJ		90-0252	F15E	3W	AK
89-0486	F15E	4W	SJ		90-0253	F15E	3W	AK
89-0487	F15E	4W	SJ		90-0254	F15E	3W	AK
89-0488	F15E	4W	SJ		90-0255	F15E	48FW	LN
89-0489	F15E	4W	SJ		90-0256	F15E	48FW	LN
89-0490	F15E	4W	SJ		90-0257	F15E	48FW	LN
89-0491	F15E	4W	SJ		90-0258	F15E	48FW	LN
89-0492	F15E	4W	SJ		90-0259	F15E	48FW	LN
89-0493	F15E	4W	SJ		90-0260	F15E	48FW	LN
89-0494	F15E	4W	SJ		90-0261	F15E	48FW	LN
89-0495	F15E	4W	SJ		90-0262	F15E	48FW	LN
89-0496	F15E	4W	SJ		91-0300	F15E	48FW	LN
89-0497	F15E	4W	SJ		91-0301	F15E	48FW	LN
89-0498	F15E	4W	SJ		91-0302	F15E	48FW	LN
89-0499	F15E	4W	SJ		91-0303	F15E	48FW	LN
89-0500	F15E	4W	SJ		91-0304	F15E	48FW	LN
89-0501	F15E	4W	SJ		91-0305	F15E	48FW	LN
89-0502	F15E	4W	SJ		91-0306	F15E	57FW	WA
89-0503	F15E	4W	SJ		91-0307	F15E	48FW	LN
89-0504	F15E	4W	SJ		91-0308	F15E	48FW	LN
89-0505	F15E	4W	SJ		91-0309	F15E	48FW	LN
89-0506	F15E	366W	MO		91-0310	F15E	48FW	LN
90-0227	F15E	57FW	WA		91-0311	F15E	48FW	LN
90-0228	F15E	57FW	WA		91-0312	F15E	48FW	LN
90-0229	F15E	4W	SJ		91-0313	F15E	48FW	LN
90-0230	F15E	4W	SJ		91-0314	F15E	48FW	LN
90-0231	F15E	4W	SJ		91-0315	F15E	48FW	LN
90-0232	F15E	4W	SJ		91-0316	F15E	48FW	LN
90-0233	F15E	3W	AK		91-0317	F15E	48FW	LN
90-0234	F15E	3W	AK		91-0318	F15E	48FW	LN
90-0235	F15E	3W	AK		91-0319	F15E	48FW	LN
90-0236	F15E	3W	AK		91-0320	F15E	48FW	LN
90-0237	F15E	3W	AK		91-0321	F15E	48FW	LN
90-0238	F15E	3W	AK		91-0322	F15E	48FW	LN
90-0239	F15E	3W	AK		91-0323	F15E	48FW	LN
90-0240	F15E	3W	AK		91-0324	F15E	48FW	LN
90-0241	F15E	3W	AK		91-0325	F15E	48FW	LN
90-0242	F15E	3W	AK		91-0326	F15E	48FW	LN
90-0243	F15E	3W	AK		91-0327	F15E	48FW	LN
90-0244	F15E	3W	AK		91-0328	F15E	48FW	LN
90-0245	F15E	3W	AK		91-0329	F15E	48FW	LN
90-0246	F15E	3W	AK		91-0330	F15E	48FW	LN
90-0247	F15E	3W	AK		91-0331	F15E	48FW	LN
90-0248	F15E	48FW	LN		91-0332	F15E	48FW	LN
90-0249	F15E	3W	AK		91-0333	F15E	48FW	LN
90-0250	F15E	3W	AK		91-0334	F15E	48FW	LN
90-0251	F15E	48FW	LN		91-0335	F15E	48FW	LN

F16 FIGHTING FALCON

c\n
F16A: 61-1 to 61-670.
F16A: KLu 6D-8, 6D-10, 6D-11, 6D-18, 6D-31, 6D-34, 6D-36, 6D-46
F16B: 62-1 to 62-123.
F16B: KLu 6E-5, 6E-10, 6E-11.
F16C: 5C-1 to 5C-610, 1C-1 to 1C-13, 5C-611 to 5C-625, 1C-14 to 1C-384, CC-1 to CC-189.
F16D: 5D-1 to 5D-51, 1V-1, 5D-52 to 5D-84, 1D-1 to 1D-6, 5D-85 to 5D-87, 1D-7 to 1D-78, CD-1 to 40.

72-1567	YF16	preserved Hampton, Va		78-0008	F16A	162FG	AZ ANG
72-1568	YF16	SoC 3.80		78-0009	F16A	w.o.24.1.91	
75-0745	F16A	USAF Museum		78-0010	F16A	121FS	DCDC ANG
75-0746	F16A			78-0011	F16A	107FS	MI MI ANG
75-0747	F16XL	to NASA 848		78-0012	F16A	121FS	DCDC ANG
75-0748	F16A	preserved AF Academy		78-0013	GF16A	TTC	Lowry
75-0749	F16XL	to NASA 849		78-0014	F16A	107FS	MI MI ANG
75-0750	F16AFT	I412TW ED AFFTC		78-0015	F16A	162FG	AZ ANG
75-0751	YF16B			78-0016	F16A	BDRT	Hahn AB
75-0752	F16B	Gen. Dyn.GD Fort Worth		78-0017	F16A		
78-0001	F16A	preserved Langley AFB		78-0018	F16A	107FS	MI MI ANG
78-0002	F16A	w.o.29.11.88		78-0019	F16A	466FS	HI AFRes
78-0003	F16A	107FS MI MI ANG		78-0020	F16A	107FS	MI MI ANG
78-0004	F16A	w.o.11.2.85		78-0021	F16A	466FS	HI AFRes
78-0005	F16A	162FG AZ ANG		78-0022	F16A	466FS	HI AFRes
78-0006	F16A	w.o.1.10.79		78-0023	F16A	w.o.26.3.80	
78-0007	F16A	412TW ED AFFTC		78-0024	F16A	466FS	HI AFRes

Serial	Type	Unit		
78-0025	F16A	466FS	HI	AFRes
78-0026	F16A	162FG		AZ ANG
78-0027	F16A	466FS	HI	AFRes
78-0038	F16A	466FS	HI	AFRes
78-0039	F16A	466FS	HI	AFRes
78-0040	F16A	466FS	HI	AFRes
78-0041	F16A	466FS	HI	AFRes
78-0042	F16A	160FS	AL	AL ANG
78-0043	F16A	466FS	HI	AFRes
78-0044	F16A	121FS	DC	DC ANG
78-0045	F16A	466FS	HI	AFRes
78-0046	F16A	w.o.5.8.81		
78-0047	F16A	466FS	HI	AFRes
78-0048	F16A	w.o.15.1.82		
78-0049	F16A	107FS	MI	MI ANG
78-0050	F16A	466FS	HI	AFRes
78-0051	F16A	466FS	HI	AFRes
78-0052	F16A	466FS	HI	AFRes
78-0053	F16A	466FS	HI	AFRes
78-0054	F16A	46TW	ET	AFMC\MSD
78-0055	F16A	w.o.12.2.86		
78-0056	F16A	162FG		AZ ANG
78-0057	F16A	466FS	HI	AFRes
78-0058	F16A	107FS	MI	MI ANG
78-0059	F16A	107FS	MI	MI ANG
78-0060	F16A	466FS	HI	AFRes
78-0061	F16A	160FS	AL	AL ANG
78-0062	F16A	121FS	DC	DC ANG
78-0063	F16A	162FG		AZ ANG
78-0064	F16A	412TW	ED	AFFTC
78-0065	F16A	466FS	HI	AFRes
78-0066	F16A	466FS	HI	AFRes
78-0067	F16A	w.o.11.5.82		
78-0068	F16A	466FS	HI	AFRes
78-0069	F16A	162FG		AZ ANG
78-0070	F16A	466FS	HI	AFRes
78-0071	F16A	w.o.25.6.80		
78-0072	F16A	w.o.19.6.84		
78-0073	F16A	466FS	HI	AFRes
78-0074	F16A	466FS	HI	AFRes
78-0075	F16A	466FS	HI	AFRes
78-0076	F16A	466FS	HI	AFRes
78-0077	F16B	157FS	SC	SC ANG
78-0078	F16B	w.o.9.8.79		
78-0079	F16B	107FS	MI	MI ANG
78-0080	F16B	412TW	ED	AFFTC
78-0081	F16B	412TW	ED	AFFTC
78-0082	F16B	162FG		AZ ANG
78-0083	F16B	138FS		NY ANG
78-0084	F16B	107FS	MI	MI ANG
78-0085	F16B	412TW	ED	AFFTC
78-0086	F16B	162FG		AZ ANG
78-0087	F16B	138FS		NY ANG
78-0088	F16B	412TW	ED	AFFTC
78-0089	F16B	412TW	ED	AFFTC
78-0090	F16B	46TW	ET	AFMC\MSD
78-0091	F16B	160FS	AL	AL ANG
78-0092	F16B	w.o.23.7.80		
78-0093	F16B			
78-0094	F16B	162FG		AZ ANG
78-0095	F16B	162FG		AZ ANG
78-0096	F16B	412TW	ED	AFFTC
78-0097	F16B	46TW	ET	AFMC\MSD
78-0098	F16B	412TW	ED	AFFTC
78-0099	F16B	412TW	ED	AFFTC
78-0100	F16B	412TW	ED	AFFTC
78-0101	F16B	46TW	ET	AFMC\MSD
78-0102	F16B	160FS	AL	AL ANG
78-0103	F16B	178FS		ND ANG
78-0104	F16B	46TW	ET	AFMC\MSD
78-0105	F16B	w.o.27.3.81		
78-0106	F16B	162FG		AZ ANG
78-0107	F16B	121FS	DC	DC ANG
78-0108	F16B	121FS	DC	DC ANG
78-0109	F16B	162FG		AZ ANG
78-0110	F16B	w.o.29.10.80		
78-0111	F16B	162FG		AZ ANG
78-0112	F16B	w.o.23.3.82		
78-0113	F16B	w.o.25.7.83		
78-0114	F16B	162FG		AZ ANG
78-0115	F16B	162FG		AZ ANG
78-0219	F16A	162FG		KLu J-219
78-0221	F16A	162FG		KLu J-221
78-0222	F16A	162FG		KLu J-222
78-0229	F16A	162FG		KLu J-229
78-0242	F16A	162FG		KLu J-242
78-0245	F16A	162FG		KLu J-245
78-0247	F16A	162FG		KLu J-247
78-0257	F16A	162FG		KLu J-257
78-0263	F16B	162FG		KLu J-263
78-0268	F16B	162FG		KLu J-268
78-0269	F16B	162FG		KLu J-269
79-0288	F16A	157FS	SC	SC ANG
79-0289	F16A	157FS	SC	SC ANG
79-0290	F16A	157FS	SC	SC ANG
79-0291	F16A	157FS	SC	SC ANG
79-0292	F16A	157FS	SC	SC ANG
79-0293	F16A	157FS	SC	SC ANG
79-0294	F16A	157FS	SC	SC ANG
79-0295	F16A	157FS	SC	SC ANG
79-0296	F16A	157FS	SC	SC ANG
79-0297	F16A	157FS	SC	SC ANG
79-0298	F16A	w.o.8.11.82		
79-0299	F16A	157FS	SC	SC ANG
79-0300	F16A	465FS	SH	AFRes
79-0301	F16A	w.o.20.5.82		
79-0302	F16A	157FS	SC	SC ANG
79-0303	F16A	162FG		AZ ANG
79-0304	F16A	157FS	SC	SC ANG
79-0305	F16A	157FS	SC	SC ANG
79-0306	F16A	157FS	SC	SC ANG
79-0307	F16A	107FS	MI	MI ANG
79-0308	F16A	157FS	SC	SC ANG
79-0309	F16A	89FS		DOAFRes
79-0310	F16A	465FS	SH	AFRes
79-0311	F16A	412TW	ED	AFFTC
79-0312	F16A	157FS	SC	SC ANG
79-0313	F16A	w.o.29.6.81		
79-0314	F16A	157FS	SC	SC ANG
79-0315	F16A	w.o.10.4.84		
79-0316	F16A	w.o.10.4.81		
79-0317	F16A	157FS	SC	SC ANG
79-0318	F16A	w.o.27.1.82		
79-0319	F16A	157FS	SC	SC ANG
79-0320	F16A	157FS	SC	SC ANG
79-0321	F16A	157FS	SC	SC ANG
79-0322	F16A	157FS	SC	SC ANG
79-0323	F16A	w.o.7.2.85		
79-0324	F16A	162FG		AZ ANG
79-0325	F16A	157FS	SC	SC ANG
79-0326	F16A	121FS		DCDC ANG
79-0327	F16A	162FG		AZ ANG
79-0328	F16A	107FS	MI	MI ANG
79-0329	F16A	465FS	SH	AFRes
79-0330	F16A	121FS		DCDC ANG
79-0331	F16A	AMARC		FG016
79-0332	F16A	121FS		DCDC ANG
79-0333	F16A	107FS	MI	MI ANG
79-0334	F16A	121FS		DCDC ANG
79-0335	F16A	AMARC		FG013
79-0336	F16A	AMARC		FG...
79-0337	F16A	138FS		NY ANG
79-0338	F16A	w.o.9.9.88		
79-0339	F16A	121FS		DCDC ANG
79-0340	F16A	AMARC		FG012
79-0341	F16A	465FS	SH	AFRes
79-0342	F16A	107FS	MI	MI ANG
79-0343	F16A	w.o.27.12.82		
79-0344	F16A	121FS		DCDC ANG
79-0345	F16A	162FG		AZ ANG
79-0346	F16A	121FS		DCDC ANG
79-0347	F16A	107FS	MI	MI ANG
79-0348	F16A	466FS	HI	AFRes
79-0349	F16A	AMARC		FG...
79-0350	F16A	w.o.18.11.83		
79-0351	F16A	138FS		NYNY ANG
79-0352	F16A	AMARC		FG...
79-0353	F16A	160FS	AL	AL ANG
79-0354	F16A	89FS		DOAFRes
79-0355	F16A	160FS	AL	AL ANG
79-0356	F16A	121FS		DCDC ANG
79-0357	F16A	162FG		AZ ANG
79-0358	F16A	162FG		AZ ANG
79-0359	F16A	465FS	SH	AFRes
79-0360	F16A	107FS	MI	MI ANG
79-0361	F16A	107FS	MI	MI ANG

Serial	Type	Unit	Notes
79-0362	F16A	162FG	AZ ANG
79-0363	F16A	AMARC	FG...
79-0364	F16A	162FG	AZ ANG
79-0365	F16A	89FS	DOAFRes
79-0366	F16A	138FS	NY ANG
79-0367	F16A		
79-0368	F16A	138FS	NY ANG
79-0369	F16A	107FS	MI MI ANG
79-0370	F16A	121FS	DCDC ANG
79-0371	F16A	121FS	DCDC ANG
79-0372	F16A	w.o.1.11.85	
79-0373	F16A	138FS	NY ANG
79-0374	F16A	w.o.20.5.82	
79-0375	F16A	162FG	AZ ANG
79-0376	F16A	465FS	SH AFRes
79-0377	F16A	89FS	DOAFRes
79-0378	F16A		
79-0379	F16A	107FS	MI MI ANG
79-0380	F16A	89FS	DOAFRes
79-0381	F16A	465FS	SH AFRes
79-0382	F16A	138FS	NY ANG
79-0383	F16A	465FS	SH AFRes
79-0384	F16A	465FS	SH AFRes
79-0385	F16A		
79-0386	F16A	w.o.19.1.83	
79-0387	F16A	107FS	MI MI ANG
79-0388	F16A	138FS	NY ANG
79-0389	F16A	162FG	AZ ANG
79-0390	F16A	w.o.4.5.82	
79-0391	F16A	138FS	NY ANG
79-0392	F16A	w.o.9.6.82	
79-0393	F16A	465FS	SH AFRes
79-0394	F16A	138FS	NY ANG
79-0395	F16A	AMARC	FG018
79-0396	F16A	AMARC	FG...
79-0397	F16A	w.o.22.3.88	
79-0398	F16A		
79-0399	F16A	138FS	NY ANG
79-0400	F16A		
79-0401	F16A	138FS	NY ANG
79-0402	F16A	ALC	Ogden
79-0403	F16A	138FS	NY ANG
79-0404	F16A	138FS	NY ANG
79-0405	F16A	AMARC	FG...
79-0406	F16A	138FS	NY ANG
79-0407	F16A	89FS	DOAFRes
79-0408	GF16A	TTC	Lowry
79-0409	F16A	412TW	ED
79-0410	F16B	157FS	SC SC ANG
79-0411	F16B	162FG	AZ ANG
79-0412	F16B	162FG	AZ ANG
79-0413	F16B	46TW	ET AFMC\MSD
79-0414	F16B	162FG	AZ ANG
79-0415	F16B	465FS	SH AFRes
79-0416	F16B	w.o.16.5.85	
79-0417	F16B	162FG	AZ ANG
79-0418	F16B	466FS	HI AFRes
79-0419	F16B	w.o.14.11.91	
79-0420	F16B	138FS	NY ANG
79-0421	F16B	162FG	AZ ANG
79-0422	F16B	162FG	AZ ANG
79-0423	F16B	89FS	DOAFRes
79-0424	F16B	89FS	DOAFRes
79-0425	F16B	162FG	AZ ANG
79-0426	F16B	465FS	SH AFRes
79-0427	F16B	162FG	AZ ANG
79-0428	F16B	162FG	AZ ANG
79-0429	F16B	162FG	AZ ANG
79-0430	F16B	121FS	DC DC ANG
79-0431	F16B	162FG	AZ ANG
79-0432	F16B	107FS	MI MI ANG
80-0474	F16A	89FS	DOAFRes
80-0475	F16A	465FS	SH AFRes
80-0476	F16A	465FS	SH AFRes
80-0477	F16A	w.o.25.9.84	
80-0478	F16A	w.o.10.2.83	
80-0479	F16A	465FS	SH AFRes
80-0480	F16A	89FS	DO AFRes
80-0481	F16A	121FS	DC DC ANG
80-0482	F16A	465FS	SH AFRes
80-0483	F16A	465FS	SH AFRes
80-0484	F16A	w.o.27.11.91	
80-0485	F16A	107FS	MI MI ANG
80-0486	F16A	121FS	DC DC ANG
80-0487	F16A	465FS	SH AFRes
80-0488	F16A	138FS	NY ANG
80-0489	F16A	465FS	SH AFRes
80-0490	F16A	w.o.6.7.82	
80-0491	F16A	121FS	DC DC ANG
80-0492	F16A	162FG	AZ ANG
80-0493	F16A	138FS	NY ANG
80-0494	F16A	107FS	MI MI ANG
80-0495	F16A	107FS	MI MI ANG
80-0496	F16A	AMARC	FG...
80-0497	F16A	465FS	SH AFRes
80-0498	F16A	465FS	SH AFRes
80-0499	F16A	162FG	AZ ANG
80-0500	F16A	121FS	DC DC ANG
80-0501	F16A	107FS	MI MI ANG
80-0502	F16A	162FG	AZ ANG
80-0503	F16A	107FS	MI MI ANG
80-0504	F16A	138FS	NY ANG
80-0505	F16A	107FS	MI MI ANG
80-0506	F16A	465FS	SH AFRes
80-0507	F16A	465FS	SH AFRes
80-0508	F16A	465FS	SH AFRes
80-0509	F16A	465FS	SH AFRes
80-0510	F16A	AMARC	FG015
80-0511	F16A	121FS	DC DC ANG
80-0512	F16A	AMARC	FG...
80-0513	F16A	89FS	DOAFRes
80-0514	F16A	162FG	AZ ANG
80-0515	F16A	89FS	DOAFRes
80-0516	F16A	121FS	DC DC ANG
80-0517	F16A	89FS	DOAFRes
80-0518	F16A	89FS	DOAFRes
80-0519	F16A	89FS	DOAFRes
80-0520	F16A	121FS	DC DC ANG
80-0521	F16A	160FS	AL AL ANG
80-0522	F16A	121FS	DC DC ANG
80-0523	F16A	465FS	SH AFRes
80-0524	F16A	162FG	AZ ANG
80-0525	F16A	160FS	AL AL ANG
80-0526	F16A	121FS	DCDC ANG
80-0527	F16A	162FG	AZ ANG
80-0528	F16A	162FG	AZ ANG
80-0529	F16A	AMARC	FG...
80-0530	F16A	AMARC	FG...
80-0531	F16A	160FS	AL AL ANG
80-0532	F16A	157FS	SC SC ANG
80-0533	F16A	89FS	DOAFRes
80-0534	F16A	121FS	DCDC ANG
80-0535	F16A	89FS	DOAFRes
80-0536	F16A	w.o.24.1.91	
80-0537	F16A	121FS	DCDC ANG
80-0538	F16A	89FS	DOAFRes
80-0539	F16A	89FS	DOAFRes
80-0540	F16A	89FS	DOAFRes
80-0541	F16A	119FS	NJ ANG
80-0542	F16A	171FS	MI ANG
80-0543	F16A	194FS	CA ANG
80-0544	F16A	194FS	CA ANG
80-0545	F16A	119FS	NJ ANG
80-0546	F16A	194FS	CA ANG
80-0547	F16A	136FS	NY ANG
80-0548	F16A	169FS	IL IL ANG
80-0549	F16A	136FS	NY ANG
80-0550	F16A	412TW	ED AFFTC
80-0551	F16A	85TES	OT
80-0552	F16A	136FS	NY ANG
80-0553	F16A	171FS	MI ANG
80-0554	F16A	179FS	MN ANG
80-0555	F16A	119FS	NJ ANG
80-0556	F16A	169FS	IL IL ANG
80-0557	F16A	412TW	ED AFFTC
80-0558	F16A	171FS	MI ANG
80-0559	F16A	111FS	TX ANG
80-0560	F16A	169FS	IL IL ANG
80-0561	F16A	169FS	IL IL ANG
80-0562	F16A	186FS	MT ANG
80-0563	F16A	186FS	MT ANG
80-0564	F16A	w.o.1.12.82	
80-0565	F16A	186FS	MT ANG
80-0566	F16A	179FS	MN ANG
80-0567	F16A	171FS	MI ANG
80-0568	F16A	171FS	MI ANG

Serial	Type	Unit		
80-0569	F16A	136FS		NY ANG
80-0570	F16A	171FS		MI ANG
80-0571	F16A	171FS		MI ANG
80-0572	F16A	194FS		CA ANG
80-0573	F16A	46TW	ET	AFMC\MSD
80-0574	F16A	w.o.20.5.88		
80-0575	F16A	111FS		TX ANG
80-0576	F16A	119FS		NJ ANG
80-0577	F16A	198FS	PR	PR ANG
80-0578	F16A	179FS		MN ANG
80-0579	F16A	171FS		MI ANG
80-0580	F16A	171FS		MI ANG
80-0581	F16A	114FS		OR ANG
80-0582	F16A	194FS		CA ANG
80-0583	F16A	194FS		CA ANG
80-0584	F16A	412TW	ED	AFFTC
80-0585	F16A	119FS		NJ ANG
80-0586	F16A	w.o.22.10.85		
80-0587	F16A	169FS	IL	IL ANG
80-0588	F16A	119FS		NJ ANG
80-0589	F16A	ANG\AFRes Test Center		
80-0590	F16A	119FS		NJ ANG
80-0591	F16A	169FS	IL	IL ANG
80-0592	F16A	194FS		CA ANG
80-0593	F16A	179FS		MN ANG
80-0594	F16A	136FS		NY ANG
80-0595	F16A	w.o.25.1.84		
80-0596	F16A	136FS		NY ANG
80-0597	F16A			
80-0598	F16A	171FS		MI ANG
80-0599	F16A			
80-0600	F16A	w.o.12.1.83		
80-0601	F16A	111FS		TX ANG
80-0602	F16A	198FS	PR	PR ANG
80-0603	F16A	111FS		TX ANG
80-0604	F16A	159FS	FL	FL ANG
80-0605	F16A	186FS		MT ANG
80-0606	F16A			
80-0607	F16A	194FS		CA ANG
80-0608	F16A	159FS	FL	FL ANG
80-0609	F16A	46TW	ET	AFMC\MSD
80-0610	F16A	186FS		MT ANG
80-0611	F16A	194FS		CA ANG
80-0612	F16A	111FS		TX ANG
80-0613	F16A	171FS		MI ANG
80-0614	F16A	194FS		CA ANG
80-0615	F16A	186FS		MT ANG
80-0616	F16A	194FS		CA ANG
80-0617	F16A	w.o.20.1.83		
80-0618	F16A	119FS		NJ ANG
80-0619	F16A	119FS		NJ ANG
80-0620	F16A	194FS		CA ANG
80-0621	F16A	194FS		CA ANG
80-0622	F16A	159FS	FL	FL ANG
80-0623	F16A			
80-0624	F16A	138FS		NY ANG
80-0625	F16B	AMARC	FG...	
80-0626	F16B	107FS	MI	MI ANG
80-0627	F16B	w.o.11.7.83		
80-0628	F16B	162FG		AZ ANG
80-0629	F16B	159FS	FL	FL ANG
80-0630	F16B	162FG		AZ ANG
80-0631	F16B	162FG		AZ ANG
80-0632	F16B	162FG		AZ ANG
80-0633	F16B	412TW	ED	AFFTC
80-0634	F16B	412TW	ED	AFFTC
80-0635	F16B	412TW	ED	AFFTC
80-0636	F16B	186FS		MT ANG
80-0637	F16B	114FS		OR ANG
80-0638	F16B	425FS	LF	
81-0663	F16A	425FS	LF	
81-0664	F16A	w.o.10.5.83		
81-0665	F16A	178FS		ND ANG
81-0666	F16A	186FS		MT ANG
81-0667	F16A	425FS	LF	
81-0668	F16A	111FS		TX ANG
81-0669	F16A	194FS		CA ANG
81-0670	F16A	425FS	LF	
81-0671	F16A	w.o.14.9.89		
81-0672	F16A	119FS		NJ ANG
81-0673	F16A	159FS	FL	FL ANG
81-0674	F16A	186FS		MT ANG
81-0675	F16A	159FS	FL	FL ANG
81-0676	F16A	425FS	LF	
81-0677	F16A	425FS	LF	
81-0678	F16A	425FS	LF	
81-0679	F16A	425FS	LF	
81-0680	F16A	186FS		MT ANG
81-0681	F16A	159FS	FL	FL ANG
81-0682	F16A	159FS	FL	FL ANG
81-0683	F16A	425FS	LF	
81-0684	F16A	179FS		MN ANG
81-0685	F16A	111FS		TX ANG
81-0686	F16A	171FS		MI ANG
81-0687	F16A	425FS	LF	
81-0688	F16A	412TW	ED	AFFTC
81-0689	F16A	194FS		CA ANG
81-0690	F16A	171FS		MI ANG
81-0691	F16A	114FS		OR ANG
81-0692	F16A	w.o.16.11.82		
81-0693	F16A	ANG\AFRes Test Center		
81-0694	F16A	198FS	PR	PR ANG
81-0695	F16A	159FS	FL	FL ANG
81-0696	F16A	159FS	FL	FL ANG
81-0697	F16A	186FS		MT ANG
81-0698	F16A	134FS		VT ANG
81-0699	F16A	178FS		ND ANG
81-0700	F16A	169FS	IL	IL ANG
81-0701	F16A	171FS		MI ANG
81-0702	F16A	136FS		NY ANG
81-0703	F16A	111FS		TX ANG
81-0704	F16A	w.o.3.3.92		
81-0705	F16A	119FS		NJ ANG
81-0706	F16A	171FS		MI ANG
81-0707	F16A	119FS		NJ ANG
81-0708	F16A	171FS		MI ANG
81-0709	F16A	171FS		MI ANG
81-0710	F16A	119FS		NJ ANG
81-0711	F16A	119FS		NJ ANG
81-0712	F16A	159FS	FL	FL ANG
81-0713	F16A	159FS	FL	FL ANG
81-0714	F16A	171FS		MI ANG
81-0715	F16A	119FS		NJ ANG
81-0716	F16A	136FS		NY ANG
81-0717	F16A	w.o.26.1.91		
81-0718	F16A	136FS		NY ANG
81-0719	F16A	171FS		MI ANG
81-0720	F16A	136FS		NY ANG
81-0721	F16A	159FS	FL	FL ANG
81-0722	F16A	186FS		MT ANG
81-0723	F16A	186FS		MT ANG
81-0724	F16A	w.o.15.12.82		
81-0725	F16A	136FS		NY ANG
81-0726	F16A	136FS		NY ANG
81-0727	F16A	136FS		NY ANG
81-0728	F16A	111FS		TX ANG
81-0729	F16A	136FS		NY ANG
81-0730	F16A	w.o.27.1.84		
81-0731	F16A	159FS	FL	FL ANG
81-0732	F16A	159FS	FL	FL ANG
81-0733	F16A	169FS	IL	IL ANG
81-0734	F16A	134FS		VT ANG
81-0735	F16A	136FS		NY ANG
81-0736	F16A	136FS		NY ANG
81-0737	F16A	119FS		NJ ANG
81-0738	F16A	119FS		NJ ANG
81-0739	F16A	169FS	IL	IL ANG
81-0740	F16A	134FS		VT ANG
81-0741	F16A	169FS	IL	IL ANG
81-0742	F16A	134FS		VT ANG
81-0743	F16A	134FS		VT ANG
81-0744	F16A	134FS		VT ANG
81-0745	F16A	w.o.1.5.84		
81-0746	F16A	134FS		VT ANG
81-0747	F16A			
81-0748	F16A	134FS		VT ANG
81-0749	F16A	134FS		VT ANG
81-0750	F16A	w.o.8.8.85		
81-0751	F16A	186FS		MT ANG
81-0752	F16A	134FS		VT ANG
81-0753	F16A	134FS		VT ANG
81-0754	F16A	194FS		CA ANG
81-0755	F16A	159FS	FL	FL ANG
81-0756	F16A	111FS		TX ANG
81-0757	F16A	159FS	FL	FL ANG
81-0758	F16A	119FS		NJ ANG

Serial	Type	Unit		
81-0759	F16A	159FS	FL	FL ANG
81-0760	F16A	194FS		CA ANG
81-0761	F16A	46TW	ET	AFMC\MSD
81-0762	F16A	169FS	IL	IL ANG
81-0763	F16A	171FS		MI ANG
81-0764	F16A	159FS	FL	FL ANG
81-0765	F16A	136FS		NY ANG
81-0766	F16A	w.o.11.3.88		
81-0767	F16A	134FS		VT ANG
81-0768	F16A	134FS		VT ANG
81-0769	F16A	134FS		VT ANG
81-0770	F16A	114FS		OR ANG
81-0771	F16A	114FS		OR ANG
81-0772	F16A	ANG\AFRes Test Center		
81-0773	F16A	178FS		ND ANG
81-0774	F16A	198FS	PR	PR ANG
81-0775	F16A	111FS		TX ANG
81-0776	F16A	136FS		NY ANG
81-0777	F16A	179FS		MN ANG
81-0778	F16A	179FS		MN ANG
81-0779	F16A	169FS	IL	IL ANG
81-0780	F16A	179FS		MN ANG
81-0781	F16A	178FS		ND ANG
81-0782	F16A	178FS		ND ANG
81-0783	F16A	179FS		MN ANG
81-0784	F16A	186FS		MT ANG
81-0785	F16A	179FS		MN ANG
81-0786	F16A	114FS		OR ANG
81-0787	F16A	179FS		MN ANG
81-0788	F16A	93FS		FM AFRes
81-0789	F16A	178FS		ND ANG
81-0790	F16A	170FS	SI	IL ANG
81-0791	F16A	178FS		ND ANG
81-0792	F16A	170FS	SI	IL ANG
81-0793	F16A	179FS		MN ANG
81-0794	F16A	93FS		FM AFRes
81-0795	F16A	179FS		MN ANG
81-0796	F16A	93FS		FM AFRes
81-0797	F16A	111FS		TX ANG
81-0798	F16A	w.o.25.5.90		
81-0799	F16A	179FS		MN ANG
81-0800	F16A	93FS		FM AFRes
81-0801	F16A	114FS		OR ANG
81-0802	F16A	93FS		FM AFRes
81-0803	F16A	179FS		MN ANG
81-0804	F16A	93FS		FM AFRes
81-0805	F16A	179FS		MN ANG
81-0806	F16A	93FS		FM AFRes
81-0807	F16A	179FS		MN ANG
81-0808	F16A	w.o.8.2.85		
81-0809	F16A	114FS		OR ANG
81-0810	F16A	93FS		FM AFRes
81-0811	F16A	114FS		OR ANG
81-0812	F16B	114FS		OR ANG
81-0813	F16B	412TW	ED	AFFTC
81-0814	F16B			
81-0815	F16B	425FS	LF	
81-0816	F16B	412TW	ED	AFFTC
81-0817	F16B	412TW	ED	AFFTC
81-0818	F16B	159FS	FL	FL ANG
81-0819	F16B	186FS		MT ANG
81-0820	F16B	111FS		TX ANG
81-0821	F16B	93FS		FM AFRes
81-0822	F16B	93FS		FM AFRes
82-0900	F16A	93FS		FM AFRes
82-0901	F16A	179FS		MN ANG
82-0902	F16A	184FS	FS	AR ANG
82-0903	F16A	178FS		ND ANG
82-0904	F16A	93FS		FM AFRes
82-0905	F16A	178FS		ND ANG
82-0906	F16A	93FS		FM AFRes
82-0907	F16A	178FS		ND ANG
82-0908	F16A	93FS		FM AFRes
82-0909	F16A			
82-0910	F16A	114FS		OR ANG
82-0911	F16A	184FS	FS	AR ANG
82-0912	F16A			
82-0913	F16A	114FS		OR ANG
82-0914	F16A	93FS		FM AFRes
82-0915	F16A	119FS		NJ ANG
82-0916	F16A	169FS	IL	IL ANG
82-0917	F16A	169FS	IL	IL ANG
82-0918	F16A	704FS		TX AFRes
82-0919	F16A	178FS		ND ANG
82-0920	F16A	184FS	FS	AR ANG
82-0921	F16A	170FS	SI	IL ANG
82-0922	F16A	93FS		FM AFRes
82-0923	F16A	169FS	IL	IL ANG
82-0924	F16A	184FS	FS	AR ANG
82-0925	F16A	w.o.10.11.83		
82-0926	F16A	178FS		ND ANG
82-0927	F16A	184FS	FS	AR ANG
82-0928	F16A	184FS	FS	AR ANG
82-0929	F16A	178FS		ND ANG
82-0930	F16A	111FS		TX ANG
82-0931	F16A	182FS		SATX ANG
82-0932	F16A	171FS		MI ANG
82-0933	F16A	93FS		FM AFRes
82-0934	F16A	169FS	IL	IL ANG
82-0935	F16A	179FS		MN ANG
82-0936	F16A	93FS		FM AFRes
82-0937	F16A			
82-0938	F16A	93FS		FM AFRes
82-0939	F16A			
82-0940	F16A	w.o.15.11.85		
82-0941	F16A	93FS		FM AFRes
82-0942	F16A	170FS	SI	IL ANG
82-0943	F16A	170FS	SI	IL ANG
82-0944	F16A	170FS	SI	IL ANG
82-0945	F16A	171FS		MI ANG
82-0946	F16A	184FS	FS	AR ANG
82-0947	F16A	169FS	IL	IL ANG
82-0948	F16A	704FS		TX AFRes
82-0949	F16A	184FS	FS	AR ANG
82-0950	F16A	178FS		ND ANG
82-0951	F16A	111FS		TX ANG
82-0952	F16A	170FS	SI	IL ANG
82-0953	F16A	198FS	PR	PR ANG
82-0954	F16A			
82-0955	F16A	114FS		OR ANG
82-0956	F16A	178FS		ND ANG
82-0957	F16A	184FS	FS	AR ANG
82-0958	F16A	169FS	IL	IL ANG
82-0959	F16A	w.o.21.11.84		
82-0960	F16A	194FS		CA ANG
82-0961	F16A	111FS		TX ANG
82-0962	F16A	182FS	SA	TX ANG
82-0963	F16A	194FS		CA ANG
82-0964	F16A			
82-0965	F16A	w.o.23.5.89		
82-0966	F16A	169FS	IL	IL ANG
82-0967	F16A	178FS		ND ANG
82-0968	F16A	170FS	SI	IL ANG
82-0969	F16A	111FS		TX ANG
82-0970	F16A	184FS	FS	AR ANG
82-0971	F16A	w.o.25.6.84		
82-0972	F16A	186FS		MT ANG
82-0973	F16A	198FS	PR	PR ANG
82-0974	F16A	134FS		VT ANG
82-0975	F16A	170FS	SI	IL ANG
82-0976	F16A	to NASA	N516NA	
82-0977	F16A	170FS	SI	IL ANG
82-0978	F16A	134FS		VT ANG
82-0979	F16A	186FS		MT ANG
82-0980	F16A	184FS	FS	AR ANG
82-0981	F16A	119FS		NJ ANG
82-0982	F16A	704FS		TX AFRes
82-0983	F16A	178FS		ND ANG
82-0984	F16A	111FS		TX ANG
82-0985	F16A	ALC		Ogden
82-0986	F16A	170FS	SI	IL ANG
82-0987	F16A	111FS		TX ANG
82-0988	F16A	170FS	SI	IL ANG
82-0989	F16A	111FS		TX ANG
82-0990	F16A	134FS		VT ANG
82-0991	F16A	182FS	SA	TX ANG
82-0992	F16A	178FS		ND ANG
82-0993	F16A	93FS		FM AFRes
82-0994	F16A	w.o.13.9.88		
82-0995	F16A	198FS	PR	PR ANG
82-0996	F16A	182FS	SA	TX ANG
82-0997	F16A	111FS		TX ANG
82-0998	F16A	w.o.9.10.86		
82-0999	F16A	704FS		TX AFRes
82-1000	F16A	111FS		TX ANG
82-1001	F16A	111FS		TX ANG

Serial	Type	Unit	Code	Owner
82-1002	F16A	170FS	SI	IL ANG
82-1003	F16A	186FS		MT ANG
82-1004	F16A	170FS	SI	IL ANG
82-1005	F16A	111FS		TX ANG
82-1006	F16A	111FS		TX ANG
82-1007	F16A	704FS		TX AFRes
82-1008	F16A	119FS		NJ ANG
82-1009	F16A	184FS	FS	AR ANG
82-1010	F16A	134FS		VT ANG
82-1011	F16A	704FS		TX AFRes
82-1012	F16A	178FS		ND ANG
82-1013	F16A	184FS	FS	AR ANG
82-1014	F16A	114FS		OR ANG
82-1015	F16A	w.o.13.1.88		
82-1016	F16A	136FS		NY ANG
82-1017	F16A	170FS	SI	IL ANG
82-1018	F16A	184FS	FS	AR ANG
82-1019	F16A	186FS		MT ANG
82-1020	F16A	184FS	FS	AR ANG
82-1021	F16A	186FS		MT ANG
82-1022	F16A	704FS		TX AFRes
82-1023	F16A	186FS		MT ANG
82-1024	F16A	170FS	SI	IL ANG
82-1025	F16A	184FS	FS	AR ANG
82-1026	F16B	171FS		MI ANG
82-1027	F16B	114FS		OR ANG
82-1028	F16B	136FS		NY ANG
82-1029	F16B	w.o.15.11.85		
82-1030	F16B	114FS		OR ANG
82-1031	F16B	114FS		OR ANG
82-1032	F16B	134FS		VT ANG
82-1033	F16B	ANG\AFRes Test Center		
82-1034	F16B	114FS		OR ANG
82-1035	F16B	159FS	FL	FL ANG
82-1036	F16B	178FS		ND ANG
82-1037	F16B	46TW	ET	AFMC\MSD
82-1038	F16B			
82-1039	F16B	114FS		OR ANG
82-1040	F16B	w.o.19.12.91		
82-1041	F16B	179FS		MN ANG
82-1042	F16B	169FS	IL	IL ANG
82-1043	F16B	w.o.		
82-1044	F16B	134FS		VT ANG
82-1045	F16B	w.o.2.5.84		
82-1046	F16B	114FS		OR ANG
82-1047	F16B	412TW	ED	AFFTC
82-1048	F16B	194FS		CA ANG
82-1049	F16B	119FS		NJ ANG
83-1066	F16A	182FS		SATX ANG
83-1067	F16A	w.o.17.12.87		
83-1068	F16A	170FS	SI	IL ANG
83-1069	F16A	704FS		TX AFRes
83-1070	F16A	704FS		TX AFRes
83-1071	F16A	704FS		TX AFRes
83-1072	F16A	170FS	SI	IL ANG
83-1073	F16A	170FS	SI	IL ANG
83-1074	F16A	704FS		TX AFRes
83-1075	F16A	170FS	SI	IL ANG
83-1076	F16A	704FS		TX AFRes
83-1077	F16A	170FS	SI	IL ANG
83-1078	F16A	704FS		TX AFRes
83-1079	F16A	184FS	FS	AR ANG
83-1080	F16A	704FS		TX AFRes
83-1081	F16A	704FS		TX AFRes
83-1082	F16A	w.o.17.11.89		
83-1083	F16A	170FS	SI	IL ANG
83-1084	F16A	170FS	SI	IL ANG
83-1085	F16A	170FS	SI	IL ANG
83-1086	F16A	w.o.27.2.86		
83-1087	F16A	184FS	FS	AR ANG
83-1088	F16A	704FS		TX AFRes
83-1089	F16A			
83-1090	F16A	704FS		TX AFRes
83-1091	F16A	ANG\AFRes Test Center		
83-1092	F16A	170FS	SI	IL ANG
83-1093	F16A	182FS		SA TX ANG
83-1094	F16A	182FS		SA TX ANG
83-1095	F16A	184FS	FS	AR ANG
83-1096	F16A	704FS		TX AFRes
83-1097	F16A	182FS		SA TX ANG
83-1098	F16A	182FS		SA TX ANG
83-1099	F16A	184FS	FS	AR ANG
83-1100	F16A	182FS		SA TX ANG
83-1101	F16A	704FS		TX AFRes
83-1102	F16A	w.o.19.2.93		
83-1103	F16A	182FS		SA TX ANG
83-1104	F16A	182FS		SA TX ANG
83-1105	F16A	182FS		SA TX ANG
83-1106	F16A	182FS		SA TX ANG
83-1107	F16A	170FS	SI	IL ANG
83-1108	F16A	182FS		SA TX ANG
83-1109	F16A	182FS		SA TX ANG
83-1110	F16A	182FS		SA TX ANG
83-1111	F16A	170FS	SI	IL ANG
83-1112	F16A	170FS	SI	IL ANG
83-1113	F16A	182FS		SA TX ANG
83-1114	F16A	182FS		SA TX ANG
83-1115	F16A	derelict Misawa AB		
83-1116	F16A	w.o.4.1.89		
83-1117	F16A	w.o.27.4.85		
83-1118	F16C	412TW	ED	AFFTC
83-1119	F16C	412TW	ED	AFFTC
83-1120	F16C	412TW	ED	AFFTC
83-1121	F16C	184FG		KS ANG
83-1122	F16C	163FS	FW	IN ANG
83-1123	F16C	46TW	ET	AFMC\MSD
83-1124	GF16C	TTC		Sheppard
83-1125	GF16C	TTC		Sheppard
83-1126	F16C	366W		MO
83-1127	GF16C	TTC		Lowry
83-1128	F16C	366W		MO
83-1129	F16C	366W		MO
83-1130	F16C	163FS	FW	IN ANG
83-1131	F16C	366W		MO
83-1132	F16C	366W		MO
83-1133	F16C	184FG		KS ANG
83-1134	F16C	58FW	LF	
83-1135	F16C	184FG		KS ANG
83-1136	F16C	184FG		KS ANG
83-1137	F16C	184FG		KS ANG
83-1138	F16C	184FG		KS ANG
83-1139	F16C	58FW	LF	
83-1140	F16C	58FW	LF	
83-1141	F16C	457FS	TF	AFRes
83-1142	F16C	184FG		KS ANG
83-1143	F16C	412TW	ED	AFFTC
83-1144	F16C	184FG		KS ANG
83-1145	F16C	112FS	OH	OH ANG
83-1146	F16C	58FW	LF	
83-1147	F16C	457FS	TF	AFRes
83-1148	F16C	184FG		KS ANG
83-1149	F16C			
83-1150	F16C	184FG		KS ANG
83-1151	F16C	w.o.3.9.90		
83-1152	F16C	58FW	LF	
83-1153	F16C	366W		MO
83-1154	F16C	366W		MO
83-1155	F16C	457FS	TF	AFRes
83-1156	F16C	184FG		KS ANG
83-1157	F16C	457FS	TF	AFRes
83-1158	F16C	112FS	OH	OH ANG
83-1159	F16C	163FS		FWIN ANG
83-1160	F16C	184FG		KS ANG
83-1161	F16C	112FS	OH	OH ANG
83-1162	F16C	457FS	TF	AFRes
83-1163	F16C	46TW	ET	AFMC\MSD
83-1164	F16C	112FS	OH	OH ANG
83-1165	F16C			
83-1166	F16B	184FS	FS	AR ANG
83-1167	F16B	704FS		TX AFRes
83-1168	F16B	170FS	SI	IL ANG
83-1169	F16B	85TES	OT	
83-1170	F16B	170FS	SI	IL ANG
83-1171	F16B	704FS		TX AFRes
83-1172	F16B	412TW	ED	AFFTC
83-1173	F16B	182FS		SATX ANG
83-1174	F16D	184FG		KS ANG
83-1175	F16D	112FS	OH	OH ANG
83-1176	F16D	412TW	ED	AFFTC
83-1177	F16D	184FG		KS ANG
83-1178	F16D	58FW	LF	
83-1179	F16D	58FW	LF	
83-1180	F16D	184FG		KS ANG
83-1181	F16D	184FG		KS ANG
83-1182	F16D	184FG		KS ANG
83-1183	F16D	184FG		KS ANG

83-1184	F16D	184FG	KS ANG
83-1185	F16D	184FG	KS ANG
84-1212	F16C	184FG	KS ANG
84-1213	F16C	366W	MO
84-1214	F16C	457FS	TF AFRes
84-1215	F16C	457FS	TF AFRes
84-1216	F16C	457FS	TF AFRes
84-1217	F16C	366W	MO
84-1218	F16C	w.o.17.2.91	
84-1219	F16C	112FS	OH OH ANG
84-1220	F16C	184FG	KS ANG
84-1221	F16C	w.o.2.8.88	
84-1222	F16C	112FS	OH OH ANG
84-1223	F16C	366W	MO
84-1224	F16C		
84-1225	F16C	457FS	TF AFRes
84-1226	F16C	457FS	TF AFRes
84-1227	F16C	457FS	TF AFRes
84-1228	F16C	w.o.5.1.89	
84-1229	F16C	184FG	KS ANG
84-1230	F16C	366W	MO
84-1231	F16C	457FS	TF AFRes
84-1232	F16C	w.o.25.7.88	
84-1233	F16C	w.o.12.11.86	
84-1234	F16C	184FG	KS ANG
84-1235	F16C	457FS	TF AFRes
84-1236	F16C	163FS	FW IN ANG
84-1237	F16C	184FG	KS ANG
84-1238	F16C	163FS	FW IN ANG
84-1239	F16C	457FS	TF AFRes
84-1240	F16C	457FS	TF AFRes
84-1241	F16C	366W	MO
84-1242	F16C	457FS	TF AFRes
84-1243	F16C	366W	MO
84-1244	F16C	163FS	FW IN ANG
84-1245	F16C	457FS	TF AFRes
84-1246	F16C	366W	MO
84-1247	F16C	112FS	OH OH ANG
84-1248	F16C	457FS	TF AFRes
84-1249	F16C		
84-1250	F16C	112FS	OH OH ANG
84-1251	F16C	366W	MO
84-1252	F16C	366W	MO
84-1253	F16C	457FS	TF AFRes
84-1254	F16C	366W	MO
84-1255	F16C	184FG	KS ANG
84-1256	F16C	366W	MO
84-1257	F16C		
84-1258	F16C	184FG	KS ANG
84-1259	F16C		
84-1260	F16C	184FG	KS ANG
84-1261	F16C	366W	MO
84-1262	F16C	112FS	OH OH ANG
84-1263	F16C	w.o.18.12.89	
84-1264	F16C	163FS	FW IN ANG
84-1265	F16C	184FG	KS ANG
84-1266	F16C	163FS	FW IN ANG
84-1267	F16C		
84-1268	F16C	457FS	TF AFRes
84-1269	F16C	85TES	OT
84-1270	F16C		
84-1271	F16C	184FG	KS ANG
84-1272	F16C	457FS	TF AFRes
84-1273	F16C	457FS	TF AFRes
84-1274	F16C	113FS	TH IN ANG
84-1275	F16C	184FG	KS ANG
84-1276	F16C	184FG	KS ANG
84-1277	F16C	366W	MO
84-1278	F16C	113FS	TH IN ANG
84-1279	F16C	184FG	KS ANG
84-1280	F16C	85TES	OT
84-1281	F16C	112FS	OH OH ANG
84-1282	F16C	113FS	TH IN ANG
84-1283	F16C	112FS	OH OH ANG
84-1284	F16C	184FG	KS ANG
84-1285	F16C	85TES	OT
84-1286	F16C	184FG	KS ANG
84-1287	F16C	184FG	KS ANG
84-1288	F16C	184FG	KS ANG
84-1289	F16C	w.o.12.10.88	
84-1290	F16C	184FG	KS ANG
84-1291	F16C	113FS	TH IN ANG
84-1292	F16C	184FG	KS ANG
84-1293	F16C	w.o.18.12.89	
84-1294	F16C	184FG	KS ANG
84-1295	F16C	184FG	KS ANG
84-1296	F16C	112FS	OH OH ANG
84-1297	F16C	112FS	OH OH ANG
84-1298	F16C	163FS	FW IN ANG
84-1299	F16C	184FG	KS ANG
84-1300	F16C	163FS	FW IN ANG
84-1301	F16C	457FS	TF AFRes
84-1302	F16C	184FG	KS ANG
84-1303	F16C	113FS	TH IN ANG
84-1304	F16C	112FS	OH OH ANG
84-1305	F16C	113FS	TH IN ANG
84-1306	F16C	184FG	KS ANG
84-1307	F16C	113FS	TH IN ANG
84-1308	F16C	112FS	OH OH ANG
84-1309	F16C	113FS	TH IN ANG
84-1310	F16C	163FS	FW IN ANG
84-1311	F16C	112FS	OH OH ANG
84-1312	F16C	184FG	KS ANG
84-1313	F16C	113FS	TH IN ANG
84-1314	F16C	58FW	LF
84-1315	F16C	163FS	FW IN ANG
84-1316	F16C	113FS	TH IN ANG
84-1317	F16C	113FS	TH IN ANG
84-1318	F16C	184FG	KS ANG
84-1319	F16D	457FS	TF AFRes
84-1320	F16D	457FS	TF AFRes
84-1321	F16D	w.o.7.8.90	
84-1322	F16D	366W	MO
84-1323	F16D	58FW	LF
84-1324	F16D	184FG	KS ANG
84-1325	F16D	184FG	KS ANG
84-1326	F16D	113FS	TH IN ANG
84-1327	F16D	184FG	KS ANG
84-1328	F16D	184FG	KS ANG
84-1329	F16D	58FW	LF
84-1330	F16D	58FW	LF
84-1331	F16D	163FS	FW IN ANG
84-1374	F16C	163FS	FW IN ANG
84-1375	F16C	184FG	KS ANG
84-1376	F16C	184FG	KS ANG
84-1377	F16C	163FS	FW IN ANG
84-1378	F16C	366W	MO
84-1379	F16C	w.o.16.2.91	
84-1380	F16C	184FG	KS ANG
84-1381	F16C	112FS	OHOH ANG
84-1382	F16C	dbf 27.6.90	
84-1383	F16C	184FG	KS ANG
84-1384	F16C	113FS	TH IN ANG
84-1385	F16C	184FG	KS ANG
84-1386	F16C	184FG	KS ANG
84-1387	F16C	184FG	KS ANG
84-1388	F16C	163FS	FW IN ANG
84-1389	F16C	w.o.31.3.88	
84-1390	F16C	w.o.27.2.91	
84-1391	F16C	184FG	KS ANG
84-1392	F16C	184FG	KS ANG
84-1393	F16C	113FS	TH IN ANG
84-1394	F16C	163FS	FW IN ANG
84-1395	F16C	w.o.29.6.88	
84-1396	F16D	184FG	KS ANG
84-1397	F16D	58FW	LF
85-1398	F16C	175FS	SD SD ANG
85-1399	F16C	w.o.17.9.87	
85-1400	F16C	86W	RS
85-1401	F16C	w.o.29.6.88	
85-1402	F16C	706FS	NOAFRes
85-1403	F16C	163FS	FW IN ANG
85-1404	F16C	184FG	KS ANG
85-1405	F16C	113FS	TH IN ANG
85-1406	F16C	113FS	TH IN ANG
85-1407	F16C	113FS	TH IN ANG
85-1408	F16C	706FS	NO AFRes
85-1409	F16C	113FS	TH IN ANG
85-1410	F16C	175FS	SD SD ANG
85-1411	F16C	113FS	TH IN ANG
85-1412	F16C	w.o.11.87	
85-1413	F16C	163FS	FW IN ANG
85-1414	F16C	w.o.1.9.88	
85-1415	F16C	113FS	TH IN ANG
85-1416	F16C	163FS	FW IN ANG
85-1417	F16C	163FS	FW IN ANG

Serial	Type	Unit	Code/Notes
85-1418	F16C	163FS	FWIN ANG
85-1419	F16C	112FS	OH OH ANG
85-1420	F16C	457FS	TF AFRes
85-1421	F16C	457FS	TF AFRes
85-1422	F16C	86W	RS
85-1423	F16C		
85-1424	F16C	w.o.23.6.87	
85-1425	F16C	58FW	LF
85-1426	F16C	86W	RS
85-1427	F16C	58FW	LF
85-1428	F16C	86W	RS
85-1429	F16C	58FW	LF
85-1430	F16C	58FW	LF
85-1431	F16C	58FW	LF
85-1432	F16C	174FS	IA ANG
85-1433	F16C	58FW	LF
85-1434	F16C	175FS	SD SD ANG
85-1435	F16C	58FW	LF
85-1436	F16C	162FS	OH OH ANG
85-1437	F16C	58FW	LF
85-1438	F16C	86W	RS
85-1439	F16C	58FW	LF
85-1440	F16C	175FS	SD SD ANG
85-1441	F16C	58FW	LF
85-1442	F16C	175FS	SD SD ANG
85-1443	F16C	58FW	LF
85-1444	F16C	706FS	NO AFRes
85-1445	F16C	58FW	LF
85-1446	F16C	175FS	SD SD ANG
85-1447	F16C	58FW	LF
85-1448	F16C	175FS	SD SD ANG
85-1449	F16C	706FS	NO AFRes
85-1450	F16C	86W	RS
85-1451	F16C	w.o.8.9.92	
85-1452	F16C	58FW	LF
85-1453	F16C	706FS	NO AFRes
85-1454	F16C	175FS	SD SD ANG
85-1455	F16C	86W	RS
85-1456	F16C		
85-1457	F16C	162FS	OH OH ANG
85-1458	F16C	162FS	OH OH ANG
85-1459	F16C	706FS	NO AFRes
85-1460	F16C	149FS	VA VA ANG
85-1461	F16C	86W	RS
85-1462	F16C	w.o.18.4.88	
85-1463	F16C	w.o.19.10.87	
85-1464	F16C	175FS	SD SD ANG
85-1465	F16C	86W	RS
85-1466	F16C	175FS	SD SD ANG
85-1467	F16C	706FS	NO AFRes
85-1468	F16C		
85-1469	F16C	175FS	SD SD ANG
85-1470	F16C	175FS	SD SD ANG
85-1471	F16C	86W	RS
85-1472	F16C	175FS	SD SD ANG
85-1473	F16C	175FS	SD SD ANG
85-1474	F16C	86W	RS
85-1475	F16C	706FS	NO AFRes
85-1476	F16C	86W	RS
85-1477	F16C	86W	RS
85-1478	F16C	175FS	SD SD ANG
85-1479	F16C	706FS	NO AFRes
85-1480	F16C	86W	RS
85-1481	F16C	86W	RS
85-1482	F16C	174FS	IA ANG
85-1483	F16C	174FS	IA ANG
85-1484	F16C	706FS	NO AFRes
85-1485	F16C	w.o.22.10.92	
85-1486	F16C	706FS	NO AFRes
85-1487	F16C	432FW	MJ
85-1488	F16C	432FW	MJ
85-1489	F16C	432FW	MJ
85-1490	F16C	432FW	MJ
85-1491	F16C	432FW	MJ
85-1492	F16C	432FW	MJ
85-1493	F16C	432FW	MJ
85-1494	F16C	432FW	MJ
85-1495	F16C	432FW	MJ
85-1496	F16C	w.o.22.1.92	
85-1497	F16C	432FW	MJ
85-1498	F16C	432FW	MJ
85-1499	F16C	432FW	MJ
85-1500	F16C	432FW	MJ
85-1501	F16C	432FW	MJ
85-1502	F16C	432FW	MJ
85-1503	F16C	432FW	MJ
85-1504	F16C	432FW	MJ
85-1505	F16C	432FW	MJ
85-1506	F16D	58FW	LF
85-1507	F16D	58FW	LF
85-1508	F16D	58FW	LF
85-1509	F16D		
85-1510	F16D	w.o.20.9.90	
85-1511	F16D	162FS	OH OH ANG
85-1512	F16D	58FW	LF
85-1513	F16D	706FS	NO AFRes
85-1514	F16D	58FW	LF
85-1515	F16D	58FW	LF
85-1516	F16D	58FW	LF
85-1517	F16D	w.o.27.8.87	
85-1544	F16C	432FW	MJ
85-1545	F16C	174FS	IA ANG
85-1546	F16C	162FS	OH OH ANG
85-1547	F16C	174FS	IA ANG
85-1548	F16C	174FS	IA ANG
85-1549	F16C	706FS	NO AFRes
85-1550	F16C	174FS	IA ANG
85-1551	F16C	174FS	IA ANG
85-1552	F16C	149FS	VA VA ANG
85-1553	F16C	706FS	NO AFRes
85-1554	F16C	174FS	IA ANG
85-1555	F16C	175FS	SD SD ANG
85-1556	F16C	174FS	IA ANG
85-1557	F16C	174FS	IA ANG
85-1558	F16C	706FS	NO AFRes
85-1559	F16C	175FS	SD SD ANG
85-1560	F16C	174FS	IA ANG
85-1561	F16C	174FS	IA ANG
85-1562	F16C	432FW	MJ
85-1563	F16C	174FS	IA ANG
85-1564	F16C	706FS	NO AFRes
85-1565	F16C	174FS	IA ANG
85-1566	F16C	174FS	IA ANG
85-1567	F16C	706FS	NO AFRes
85-1568	F16C	174FS	IA ANG
85-1569	F16C	174FS	IA ANG
85-1570	F16C	174FS	IA ANG
85-1571	F16D	706FS	NO AFRes
85-1572	F16D	149FS	VA VA ANG
85-1573	F16D	432FW	MJ
86-0039	F16D	85TES	OT
86-0040	F16D	302FS	LR AFRes
86-0041	F16D	ADS	Thunderbirds
86-0042	F16D	174FS	IA ANG
86-0043	F16D	175FS	SD SD ANG
86-0044	F16D	52FW	SP
86-0045	F16D	8FW	WP
86-0046	F16D	8FW	WP
86-0047	F16D	162FS	OH OH ANG
86-0048	F16D	Gen. Dyn.GD	'VISTA'
86-0049	F16D	86W	RS
86-0050	F16D	120FS	CO CO ANG
86-0051	F16D	176FS	WI WI ANG
86-0052	F16D		
86-0053	F16D		
86-0207	F16C	432FW	MJ
86-0208	F16C	432FW	MJ
86-0209	F16C	86W	RS
86-0210	F16C	302FS	LR AFRes
86-0211	F16C	302FS	LR AFRes
86-0212	F16C	302FS	LR AFRes
86-0213	F16C	w.o.20.2.88	
86-0214	F16C	302FS	LR AFRes
86-0215	F16C	302FS	LR AFRes
86-0216	F16C	149FS	VA VA ANG
86-0217	F16C	302FS	LR AFRes
86-0218	F16C	302FS	LR AFRes
86-0219	F16C	149FS	VA VA ANG
86-0220	F16C	57FW	WA
86-0221	F16C	432FW	MJ
86-0222	F16C	149FS	VA VA ANG
86-0223	F16C	149FS	VA VA ANG
86-0224	F16C	175FS	SD SD ANG
86-0225	F16C	149FS	VA VA ANG
86-0226	F16C	149FS	VA VA ANG
86-0227	F16C	149FS	VA VA ANG

Serial	Type	Unit	Code
86-0228	F16C	149FS	VA VA ANG
86-0229	F16C	149FS	VA VA ANG
86-0230	F16C	149FS	VA VA ANG
86-0231	F16C	149FS	VA VA ANG
86-0232	F16C	149FS	VA VA ANG
86-0233	F16C	432FW	MJ
86-0234	F16C	432FW	MJ
86-0235	F16C	432FW	MJ
86-0236	F16C	302FS	LR AFRes
86-0237	F16C	175FS	SD SD ANG
86-0238	F16C	302FS	LR AFRes
86-0239	F16C	302FS	LR AFRes
86-0240	F16C	302FS	LR AFRes
86-0241	F16C	302FS	LR AFRes
86-0242	F16C	149FS	VA VA ANG
86-0243	F16C	149FS	VA VA ANG
86-0244	F16C	149FS	VA VA ANG
86-0245	F16C	149FS	VA VA ANG
86-0246	F16C	149FS	VA VA ANG
86-0247	F16C		
86-0248	F16C	175FS	SD SD ANG
86-0249	F16C	149FS	VA VA ANG
86-0250	F16C	57FW	WA
86-0251	F16C	57FW	WA
86-0252	F16C	302FS	LR AFRes
86-0253	F16C	302FS	LR AFRes
86-0254	F16C	149FS	VA VA ANG
86-0255	F16C	86W	RS
86-0256	F16C	302FS	LR AFRes
86-0257	F16C	302FS	LR AFRes
86-0258	F16C	149FS	VA VA ANG
86-0259	F16C	149FS	VA VA ANG
86-0260	F16C	149FS	VA VA ANG
86-0261	F16C	149FS	VA VA ANG
86-0262	F16C	162FS	OH OH ANG
86-0263	F16C	120FS	CO CO ANG
86-0264	F16C	8FW	WP
86-0265	F16C	8FW	WP
86-0266	F16C	8FW	WP
86-0267	F16C	8FW	WP
86-0268	F16C	8FW	WP
86-0269	F16C	57FW	WA
86-0270	F16C	86W	RS
86-0271	F16C	57FW	WA
86-0272	F16C	57FW	WA
86-0273	F16C	302FS	LR AFRes
86-0274	F16C	w.o.14.2.89	
86-0275	F16C	8FW	WP
86-0276	F16C	w.o.13.2.89	
86-0277	F16C	8FW	WP
86-0278	F16C	8FW	WP
86-0279	F16C	302FS	LR AFRes
86-0280	F16C	302FS	LR AFRes
86-0281	F16C	57FW	WA
86-0282	F16C	8FW	WP
86-0283	F16C	302FS	LR AFRes
86-0284	F16C	8FW	WP
86-0285	F16C	302FS	LR AFRes
86-0286	F16C	8FW	WP
86-0287	F16C	162FS	OH OH ANG
86-0288	F16C	52FW	SP
86-0289	F16C	120FS	CO CO ANG
86-0290	F16C	8FW	WP
86-0291	F16C	302FS	LR AFRes
86-0292	F16C	302FS	LR AFRes
86-0293	F16C	8FW	WP
86-0294	F16C	8FW	WP
86-0295	F16C	8FW	WP
86-0296	F16C	302FS	R AFRes
86-0297	F16C	w.o.29.1.89	
86-0298	F16C	8FW	WP
86-0299	F16C	302FS	LR AFRes
86-0300	F16C	8FW	WP
86-0301	F16C	8FW	WP
86-0302	F16C	162FS	OHOH ANG
86-0303	F16C	86W	RS
86-0304	F16C	8FW	WP
86-0305	F16C	8FW	WP
86-0306	F16C	8FW	WP
86-0307	F16C	8FW	WP
86-0308	F16C	8FW	WP
86-0309	F16C	8FW	WP
86-0310	F16C	8FW	WP
86-0311	F16C	w.o.14.3.89	
86-0312	F16C	w.o.14.3.89	
86-0313	F16C	86W	RS
86-0314	F16C	8FW	WP
86-0315	F16C		
86-0316	F16C	w.o.5.12.88	
86-0317	F16C	8FW	WP
86-0318	F16C	8FW	WP
86-0319	F16C	8FW	WP
86-0320	F16C	8FW	WP
86-0321	F16C	8FW	WP
86-0322	F16C	8FW	WP
86-0323	F16C	8FW	WP
86-0324	F16C	120FS	CO CO ANG
86-0325	F16C	120FS	CO CO ANG
86-0326	F16C	52FW	SP
86-0327	F16C	52FW	SP
86-0328	F16C	120FS	CO CO ANG
86-0329	F16C	w.o.20.2.91	
86-0330	F16C	120FS	CO CO ANG
86-0331	F16C	8FW	WP
86-0332	F16C	8FW	WP
86-0333	F16C	8FW	WP
86-0334	F16C	8FW	WP
86-0335	F16C	8FW	WP
86-0336	F16C	8FW	WP
86-0337	F16C	8FW	WP
86-0338	F16C	120FS	CO CO ANG
86-0339	F16C	120FS	CO CO ANG
86-0340	F16C	120FS	CO CO ANG
86-0341	F16C	52FW	SP
86-0342	F16C	86W	RS
86-0343	F16C	w.o.11.8.93	
86-0344	F16C	w.o.18.10.88	
86-0345	F16C	120FS	CO CO ANG
86-0346	F16C	52FW	SP
86-0347	F16C	162FS	OHOH ANG
86-0348	F16C	52FW	SP
86-0349	F16C	52FW	SP
86-0350	F16C	52FW	SP
86-0351	F16C	8FW	WP
86-0352	F16C	8FW	WP
86-0353	F16C	8FW	WP
86-0354	F16C		
86-0355	F16C	432FW	MJ
86-0356	F16C		
86-0357	F16C	51FW	OS
86-0358	F16C	120FS	CO CO ANG
86-0359	F16C		
86-0360	F16C	120FS	CO CO ANG
86-0361	F16C	52FW	SP
86-0362	F16C	162FS	OH OH ANG
86-0363	F16C	52FW	SP
86-0364	F16C	52FW	SP
86-0365	F16C	52FW	SP
86-0366	F16C	52FW	SP
86-0367	F16C	120FS	CO CO ANG
86-0368	F16C	120FS	CO CO ANG
86-0369	F16C	52FW	SP
86-0370	F16C	120FS	CO CO ANG
86-0371	F16C	120FS	CO CO ANG
87-0217	F16C	162FS	OH OH ANG
87-0218	F16C	52FW	SP
87-0219	F16C	52FW	SP
87-0220	F16C	86W	RS
87-0221	F16C	86W	RS
87-0222	F16C	162FS	OH OH ANG
87-0223	F16C	52FW	SP
87-0224	F16C	w.o.21.1.91	
87-0225	F16C		
87-0226	F16C		
87-0227	F16C	86W	RS
87-0228	F16C	w.o.19.1.91	
87-0229	F16C	120FS	CO CO ANG
87-0230	F16C		
87-0231	F16C	120FS	CO CO ANG
87-0232	F16C	86W	RS
87-0233	F16C	52FW	SP
87-0234	F16C		
87-0235	F16C		
87-0236	F16C		
87-0237	F16C	120FS	CO CO ANG
87-0238	F16C		

87-0239	F16C		
87-0240	F16C		
87-0241	F16C	120FS	CO CO ANG
87-0242	F16C	86W	RS
87-0243	F16C	162FS	OH OH ANG
87-0244	F16C		
87-0245	F16C	162FS	OH OH ANG
87-0246	F16C	120FS	CO CO ANG
87-0247	F16C		
87-0248	F16C	86W	RS
87-0249	F16C	162FS	OH OH ANG
87-0250	F16C	86W	RS
87-0251	F16C	51FW	OS
87-0252	F16C		
87-0253	F16C		
87-0254	F16C	120FS	CO CO ANG
87-0255	F16C		
87-0256	F16C		
87-0257	F16C	w.o.19.1.91	
87-0258	F16C		
87-0259	F16C	86W	RS
87-0260	F16C	52FW	SP
87-0261	F16C		
87-0262	F16C		
87-0263	F16C		
87-0264	F16C		
87-0265	F16C		
87-0266	F16C		
87-0267	F16C	57FW	WA
87-0268	F16C	52FW	SP
87-0269	F16C	57FW	WA
87-0270	F16C	52FW	SP
87-0271	F16C	162FS	OH OH ANG
87-0272	F16C		
87-0273	F16C		
87-0274	F16C		
87-0275	F16C		
87-0276	F16C	86W	RS
87-0277	F16C		
87-0278	F16C	176FS	WI WI ANG
87-0279	F16C		
87-0280	F16C	176FS	WI WI ANG
87-0281	F16C	52FW	SP
87-0282	F16C	52FW	SP
87-0283	F16C	52FW	SP
87-0284	F16C	120FS	CO CO ANG
87-0285	F16C		
87-0286	F16C		
87-0287	F16C		
87-0288	F16C		
87-0289	F16C	176FS	WI WI ANG
87-0290	F16C		
87-0291	F16C	432FW	MJ
87-0292	F16C	432FW	MJ
87-0293	F16C	ALC	Ogden (repairs)
87-0294	F16C	51FW	OS
87-0295	GF16C	TTC	Lowry
87-0296	F16C	51FW	OS
87-0297	F16C	57FW	WA
87-0298	F16C		
87-0299	F16C	57FW	WA
87-0300	F16C		
87-0301	F16C	57FW	WA
87-0302	F16C	w.o.7.5.91	
87-0303	F16C	ADS	Thunderbirds
87-0304	F16C	51FW	OS
87-0305	F16C	57FW	WA
87-0306	F16C	51FW	OS
87-0307	F16C	57FW	WA
87-0308	F16C	432FW	MJ
87-0309	F16C	ADS	Thunderbirds
87-0310	F16C	51FW	OS
87-0311	F16C	302FS	LR AFRes
87-0312	F16C	432FW	MJ
87-0313	F16C	ADS	Thunderbirds
87-0314	F16C	432FW	MJ
87-0315	F16C	302FS	LR AFRes
87-0316	F16C	51FW	OS
87-0317	F16C	302FS	LR AFRes
87-0318	F16C	51FW	OS
87-0319	F16C	ADS	Thunderbirds
87-0320	F16C	51FW	OS
87-0321	F16C	57FW	WA
87-0322	F16C	432FW	MJ
87-0323	F16C	ADS	Thunderbirds
87-0324	F16C	432FW	MJ
87-0325	F16C	ADS	Thunderbirds
87-0326	F16C	51FW	OS
87-0327	F16C	ADS	Thunderbirds
87-0328	F16C	51FW	OS
87-0329	F16C	ADS	Thunderbirds
87-0330	F16C	432FW	MJ
87-0331	F16C	ADS	Thunderbirds
87-0332	F16C	51FW	OS
87-0333	F16C	302FS	LR AFRes
87-0334	F16C	432FW	MJ
87-0335	F16C		
87-0336	F16C	52FW	SP
87-0337	F16C	120FS	CO CO ANG
87-0338	F16C		
87-0339	F16C		
87-0340	F16C		
87-0341	F16C		
87-0342	F16C	52FW	SP
87-0343	F16C		
87-0344	F16C		
87-0345	F16C		
87-0346	F16C		
87-0347	F16C	86W	RS
87-0348	F16C		
87-0349	F16C		
87-0350	F16C	188FS	NMN MANG
87-0351	F16C	388FW	HL
87-0352	F16C	412TW	ED AFFTC
87-0353	F16C	46TW	ET AFMC\MSD
87-0354	F16C	388FW	HL
87-0355	F16C	388FW	HL
87-0356	F16C	58FW	LF
87-0357	F16C	388FW	HL
87-0358	F16C	58FW	LF
87-0359	F16C	388FW	HL
87-0360	F16C	58FW	LF
87-0361	F16C	363FW	SW
87-0362	F16C	363FW	SW
87-0363	F16D	w.o.15.3.89	
87-0364	F16D		
87-0365	F16D		
87-0366	F16D		
87-0367	F16D		
87-0368	F16D		
87-0369	F16D	w.o.15.10.89	
87-0370	F16D		
87-0371	F16D	138FS	NY NY ANG
87-0372	F16D	52FW	SP
87-0373	F16D		
87-0374	F16D	176FS	WI WI ANG
87-0375	F16D		
87-0376	F16D	176FS	WI WI ANG
87-0377	F16D		
87-0378	F16D		
87-0379	F16D		
87-0380	F16D	51FW	OS
87-0381	F16D	ADS	Thunderbirds
87-0382	F16D	138FS	NY NY ANG
87-0383	F16D	86W	RS
87-0384	F16D	138FS	NY NY ANG
87-0385	F16D	86W	RS
87-0386	F16D	138FS	NY NY ANG
87-0387	F16D		
87-0388	F16D		
87-0389	F16D	52FW	SP
87-0390	F16D		
87-0391	F16D	388FW	HL
87-0392	F16D	412TW	ED AFFTC
87-0393	F16D	388FW	HL
87-0394	F16D	58FW	LF
87-0395	F16D	58FW	LF
87-0396	F16D	58FW	LF
88-0150	F16D		
88-0151	F16D		
88-0152	F16D	86W	RS
88-0153	F16D	58FW	LF
88-0154	F16D	58FW	LF
88-0155	F16D	58FW	LF
88-0156	F16D	58FW	LF
88-0157	F16D	58FW	LF

88-0158	F16D	58FW	LF
88-0159	F16D	58FW	LF
88-0160	F16D	58FW	LF
88-0161	F16D	58FW	LF
88-0162	F16D	58FW	LF
88-0163	F16D	58FW	LF
88-0164	F16D	58FW	LF
88-0165	F16D	58FW	LF
88-0166	F16D	388FW	HL
88-0167	F16D	58FW	LF
88-0168	F16D	w.o.30.7.91	
88-0169	F16D	58FW	LF
88-0170	F16D	23W	FT
88-0171	F16D	23W	FT
88-0172	F16D	58FW	LF
88-0173	F16D	58FW	LF
88-0174	F16D	86W	RS
88-0175	F16D	58FW	LF
88-0397	F16C		
88-0398	F16C	52FW	SP
88-0399	F16C	52FW	SP
88-0400	F16C	52FW	SP
88-0401	F16C	120FS	CO CO ANG
88-0402	F16C	8FW	WP
88-0403	F16C	8FW	WP
88-0404	F16C	8FW	WP
88-0405	F16C	8FW	WP
88-0406	F16C	8FW	WP
88-0407	F16C	8FW	WP
88-0408	F16C	w.o.3.4.90	
88-0409	F16C		
88-0410	F16C	86W	RS
88-0411	F16C	176FS	WI WI ANG
88-0412	F16C	58FW	LF
88-0413	F16C	388FW	HL
88-0414	F16C	58FW	LF
88-0415	F16C	188FS	NM NM ANG
88-0416	F16C	388FW	HL
88-0417	F16C	363FW	SW
88-0418	F16C	388FW	HL
88-0419	F16C	388FW	HL
88-0420	F16C	57FW	WA
88-0421	F16C	388FW	HL
88-0422	F16C	388FW	HL
88-0423	F16C	57FW	WA
88-0424	F16C	388FW	HL
88-0425	F16C	388FW	HL
88-0426	F16C	388FW	HL
88-0427	F16C	58FW	LF
88-0428	F16C	388FW	HL
88-0429	F16C	388FW	HL
88-0430	F16C	388FW	HL
88-0431	F16C	388FW	HL
88-0432	F16C	388FW	HL
88-0433	F16C	388FW	HL
88-0434	F16C	58FW	LF
88-0435	F16C	188FS	NM NM ANG
88-0436	F16C	388FW	HL
88-0437	F16C	388FW	HL
88-0438	F16C	388FW	HL
88-0439	F16C	388FW	HL
88-0440	F16C	363FW	SW
88-0441	F16C	46TW	ET AFMC\MSD
88-0442	F16C	57FW	WA
88-0443	F16C	388FW	HL
88-0444	F16C	388FW	HL
88-0445	F16C	412TW	ED AFFTC
88-0446	F16C	388FW	HL
88-0447	F16C	388FW	HL
88-0448	F16C	58FW	LF
88-0449	F16C	388FW	HL
88-0450	F16C	388FW	HL
88-0451	F16C	58FW	LF
88-0452	F16C	388FW	HL
88-0453	F16C	388FW	HL
88-0454	F16C	388FW	HL
88-0455	F16C	58FW	LF
88-0456	F16C		
88-0457	F16C	188FS	NM NM ANG
88-0458	F16C	58FW	LF
88-0459	F16C	388FW	HL
88-0460	F16C	388FW	HL
88-0461	F16C	w.o.1.12.90	
88-0462	F16C	388FW	HL
88-0463	F16C	388FW	HL
88-0464	F16C	58FW	LF
88-0465	F16C	w.o.24.8.92	
88-0466	F16C	388FW	HL
88-0467	F16C	388FW	HL
88-0468	F16C	347FW	MY
88-0469	F16C	388FW	HL
88-0470	F16C	388FW	HL
88-0471	F16C	388FW	HL
88-0472	F16C	58FW	LF
88-0473	F16C	188FS	NM NM ANG
88-0474	F16C	388FW	HL
88-0475	F16C	58FW	LF
88-0476	F16C	188FS	NM NM ANG
88-0477	F16C	188FS	NM NM ANG
88-0478	F16C	58FW	LF
88-0479	F16C	188FS	NM NM ANG
88-0480	F16C	188FS	NM NM ANG
88-0481	F16C	363FW	SW
88-0482	F16C	188FS	NM NM ANG
88-0483	F16C		
88-0484	F16C	58FW	LF
88-0485	F16C	388FW	HL
88-0486	F16C	388FW	HL
88-0487	F16C	58FW	LF
88-0488	F16C	388FW	HL
88-0489	F16C	188FS	NMNM ANG
88-0490	F16C	58FW	LF
88-0491	F16C	388FW	HL
88-0492	F16C	188FS	NM NM ANG
88-0493	F16C	58FW	LF
88-0494	F16C	363FW	SW
88-0495	F16C	388FW	HL
88-0496	F16C	58FW	LF
88-0497	F16C	388FW	HL
88-0498	F16C	388FW	HL
88-0499	F16C	58FW	LF
88-0500	F16C	388FW	HL
88-0501	F16C	388FW	HL
88-0502	F16C	363FW	SW
88-0503	F16C	188FS	NM NM ANG
88-0504	F16C	188FS	NM NM ANG
88-0505	F16C	58FW	LF
88-0506	F16C	188FS	NMNM ANG
88-0507	F16C	388FW	HL
88-0508	F16C	363FW	SW
88-0509	F16C	388FW	HL
88-0510	F16C	388FW	HL
88-0511	F16C	58FW	LF
88-0512	F16C	388FW	HL
88-0513	F16C	188FS	NM NM ANG
88-0514	F16C	58FW	LF
88-0515	F16C	347FW	MY
88-0516	F16C	363FW	SW
88-0517	F16C	58FW	LF
88-0518	F16C	363FW	SW
88-0519	F16C	347FW	MY
88-0520	F16C	51FW	OS
88-0521	F16C	31FW	HS
88-0522	F16C	347FW	MY
88-0523	F16C	347FW	MY
88-0524	F16C	58FW	LF
88-0525	F16C		
88-0526	F16C	347FW	MY
88-0527	F16C	347FW	MY
88-0528	F16C	347FW	MY
88-0529	F16C	347FW	MY
88-0530	F16C	51FW	OS
88-0531	F16C	347FW	MY
88-0532	F16C	347FW	MY
88-0533	F16C	31FW	HS
88-0534	F16C	51FW	OS
88-0535	F16C	347FW	MY
88-0536	F16C	347FW	MY
88-0537	F16C	347FW	MY
88-0538	F16C	31FW	HS
88-0539	F16C	51FW	OS
88-0540	F16C	347FW	MY
88-0541	F16C	347FW	MY
88-0542	F16C	51FW	OS
88-0543	F16C	188FS	NM NM ANG
88-0544	F16C	23W	FT

88-0545	F16C	51FW	OS		89-2077	F16C	343W	AK
88-0546	F16C	347FW	MY		89-2078	F16C	343W	AK
88-0547	F16C	31FW	HS		89-2079	F16C	363FW	SW
88-0548	F16C	51FW	OS		89-2080	F16C	343W	AK
88-0549	F16C	347FW	MY		89-2081	F16C	347FW	MY
88-0550	F16C	347FW	MY		89-2082	F16C	363FW	SW
89-2000	F16C	188FS	NM NM ANG		89-2083	F16C	347FW	MY
89-2001	F16C	347FW	MY		89-2084	F16C	347FW	MY
89-2002	F16C				89-2085	F16C	363FW	SW
89-2003	F16C	347FW	MY		89-2086	F16C	347FW	MY
89-2004	F16C	363FW	SW		89-2087	F16C	347FW	MY
89-2005	F16C	347FW	MY		89-2088	F16C	363FW	SW
89-2006	F16C	347FW	MY		89-2089	F16C	363FW	SW
89-2007	F16C	363FW	SW		89-2090	F16C	347FW	MY
89-2008	F16C	347FW	MY		89-2091	F16C	363FW	SW
89-2009	F16C	347FW	MY		89-2092	F16C	31FW	HS
89-2010	F16C				89-2093	F16C	31FW	HS
89-2011	F16C	347FW	MY		89-2094	F16C	363FW	SW
89-2012	F16C				89-2095	F16C	31FW	HS
89-2013	F16C	347FW	MY		89-2096	F16C	31FW	HS
89-2014	F16C	347FW	MY		89-2097	F16C	63FS	LF
89-2015	F16C	188FS	NM NM ANG		89-2098	F16C	363FW	SW
89-2016	F16C	31FW	HS		89-2099	F16C	31FW	HS
89-2017	F16C				89-2100	F16C	57FW	WA
89-2018	F16C	347FW	MY		89-2101	F16C	31FW	HS
89-2019	F16C				89-2102	F16C	31FW	HS
89-2020	F16C	347FW	MY		89-2103	F16C	363FW	SW
89-2021	F16C	188FS	NM NM ANG		89-2104	F16C	31FW	HS
89-2022	F16C				89-2105	F16C	31FW	HS
89-2023	F16C	347FW	MY		89-2106	F16C	363FW	SW
89-2024	F16C	347FW	MY		89-2107	F16C	363FW	SW
89-2025	F16C	347FW	MY		89-2108	F16C	31FW	HS
89-2026	F16C	347FW	MY		89-2109	F16C	363FW	SW
89-2027	F16C				89-2110	F16C	w.o.24.4.92	
89-2028	F16C	347FW	MY		89-2111	F16C	31FW	HS
89-2029	F16C	31FW	HS		89-2112	F16C	363FW	SW
89-2030	F16C	347FW	MY		89-2113	F16C	343W	AK
89-2031	F16C				89-2114	F16C	363FW	SW
89-2032	F16C	86W	RS		89-2115	F16C	31FW	HS
89-2033	F16C	347FW	MY		89-2116	F16C	31FW	HS
89-2034	F16C				89-2117	F16C	363FW	SW
89-2035	F16C	31FW	HS		89-2118	F16C	188FS	NM NM ANG
89-2036	F16C	347FW	MY		89-2119	F16C	31FW	HS
89-2037	F16C	347FW	MY		89-2120	F16C	363FW	SW
89-2038	F16C	347FW	MY		89-2121	F16C	343W	AK
89-2039	F16C	347FW	MY		89-2122	F16C	363FW	SW
89-2040	F16C				89-2123	F16C	57FW	WA
89-2041	F16C	188FS	NM NM ANG		89-2124	F16C	188FS	NM NM ANG
89-2042	F16C	347FW	MY		89-2125	F16C	363FW	SW
89-2043	F16C	31FW	HS		89-2126	F16C	57FW	WA
89-2044	F16C	31FW	HS		89-2127	F16C	363FW	SW
89-2045	F16C	347FW	MY		89-2128	F16C	363FW	SW
89-2046	F16C	31FW	HS		89-2129	F16C	GD	Fort Worth
89-2047	F16C	343W	AK		89-2130	F16C	363FW	SW
89-2048	F16C	363FW	SW		89-2131	F16C	31FW	HS
89-2049	F16C	347FW	MY		89-2132	F16C	363FW	SW
89-2050	F16C	347FW	MY		89-2133	F16C	363FW	SW
89-2051	F16C				89-2134	F16C	363FW	SW
89-2052	F16C	347FW	MY		89-2135	F16C	363FW	SW
89-2053	F16C	363FW	SW		89-2136	F16C	363FW	SW
89-2054	F16C	23W	FT		89-2137	F16C	363FW	SW
89-2055	F16C	347FW	MY		89-2138	F16C	363FW	SW
89-2056	F16C	363FW	SW		89-2139	F16C	363FW	SW
89-2057	F16C	347FW	MY		89-2140	F16C	363FW	SW
89-2058	F16C	347FW	MY		89-2141	F16C	363FW	SW
89-2059	F16C	57FW	WA		89-2142	F16C	363FW	SW
89-2060	F16C	347FW	MY		89-2143	F16C	31FW	HS
89-2061	F16C	w.o.4.4.91			89-2144	F16C	31FW	HS
89-2062	F16C	347FW	MY		89-2145	F16C	363FW	SW
89-2063	F16C	347FW	MY		89-2146	F16C	31FW	HS
89-2064	F16C	347FW	MY		89-2147	F16C	31FW	HS
89-2065	F16C	347FW	MY		89-2148	F16C	363FW	SW
89-2066	F16C	343W	AK		89-2149	F16C	31FW	HS
89-2067	F16C	343W	AK		89-2150	F16C	31FW	HS
89-2068	F16C	347FW	MY		89-2151	F16C	363FW	SW
89-2069	F16C	343W	AK		89-2152	F16C	31FW	HS
89-2070	F16C	85TES	OT		89-2153	F16C	363FW	SW
89-2071	F16C	347FW	MY		89-2154	F16C	363FW	SW
89-2072	F16C	347FW	MY		89-2155	F16D	58FW	LF
89-2073	F16C	363FW	SW		89-2156	F16D		
89-2074	F16C	343W	AK		89-2157	F16D	58FW	LF
89-2075	F16C	347FW	MY		89-2158	F16D	58FW	LF
89-2076	F16C	363FW	SW		89-2159	F16D	58FW	LF

Serial	Type	Unit	Code
89-2160	F16D	58FW	LF
89-2161	F16D	58FW	LF
89-2162	F16D	58FW	LF
89-2163	F16D	58FW	LF
89-2164	F16D	58FW	LF
89-2165	F16D	363FW	SW
89-2166	F16D	31FW	HS
89-2167	F16D	58FW	LF
89-2168	F16D	363FW	SW
89-2169	F16D	188FS	NM NM ANG
89-2170	F16D	58FW	LF
89-2171	F16D	347FW	MY
89-2172	F16D	343W	AK
89-2173	F16D	188FS	NM NM ANG
89-2174	F16D	388FW	HL
89-2175	F16D	363FW	SW
89-2176	F16D	388FW	HL
89-2177	F16D	58FW	LF
89-2178	F16D	347FW	MY
89-2179	F16D	58FW	LF
90-0700	F16C	363FW	SW
90-0701	F16C	363FW	SW
90-0702	F16C	363FW	SW
90-0703	F16C	31FW	HS
90-0704	F16C	363FW	SW
90-0705	F16C	363FW	SW
90-0706	F16C	363FW	SW
90-0707	F16C	57FW	WA
90-0708	F16C	57FW	WA
90-0709	F16C	31FW	HS
90-0710	F16C	363FW	SW
90-0711	F16C	343W	AK
90-0712	F16C	63FS	LF
90-0713	F16C	57FW	WA
90-0714	F16C	343W	AK
90-0715	F16C	57FW	WA
90-0716	F16C	57FW	WA
90-0717	F16C	343W	AK
90-0718	F16C	343W	AK
90-0719	F16C	363FW	SW
90-0720	F16C	57FW	WA
90-0721	F16C	57FW	WA
90-0722	F16C	363FW	SW
90-0723	F16C	343W	AK
90-0724	F16C	363FW	SW
90-0725	F16C	31FW	HS
90-0726	F16C	85TES	OT
90-0727	F16C	57FW	WA
90-0728	F16C	57FW	WA
90-0729	F16C	57FW	WA
90-0730	F16C	363FW	SW
90-0731	F16C	363FW	SW
90-0732	F16C	343W	AK
90-0733	F16C	343W	AK
90-0734	F16C	343W	AK
90-0735	F16C	343W	AK
90-0736	F16C	343W	AK
90-0737	F16C	363FW	SW
90-0738	F16C	363FW	SW
90-0739	F16C	57FW	WA
90-0740	F16C	57FW	WA
90-0741	F16C	363FW	SW
90-0742	F16C	343W	AK
90-0743	F16C	343W	AK
90-0744	F16C	343W	AK
90-0745	F16C	343W	AK
90-0746	F16C	57FW	WA
90-0747	F16C	57FW	WA
90-0748	F16C	363FW	SW
90-0749	F16C	363FW	SW
90-0750	F16C	57FW	WA
90-0751	F16C	57FW	WA
90-0752	F16C	57FW	WA
90-0753	F16C	31FW	HS
90-0754	F16C	363FW	SW
90-0755	F16C	363FW	SW
90-0756	F16C	347FW	MY
90-0757	F16C	363FW	SW
90-0758	F16C	363FW	SW
90-0759	F16C	363FW	SW
90-0760	F16C	363FW	SW
90-0761	F16C	w.o.27.10.92	
90-0762	F16C	363FW	SW
90-0763	F16C	31FW	HS
90-0764	F16C	363FW	SW
90-0765	F16C	363FW	SW
90-0766	F16C	363FW	SW
90-0767	F16C	363FW	SW
90-0768	F16C	363FW	SW
90-0769	F16C	363FW	SW
90-0770	F16C	363FW	SW
90-0771	F16C		
90-0772	F16C		
90-0773	F16C	347FW	MY
90-0774	F16C		
90-0775	F16C		
90-0776	F16C		
90-0777	F16D	188FS	NM NM ANG
90-0778	F16D	363FW	SW
90-0779	F16D	347FW	MY
90-0780	F16D	343W	AK
90-0781	F16D	363FW	SW
90-0782	F16D	31FW	HS
90-0783	F16D	363FW	SW
90-0784	F16D		
90-0785	F16D	57FW	WA
90-0786	F16D	363FW	SW
90-0787	F16D	363FW	SW
90-0788	F16D	363FW	SW
90-0789	F16D	57FW	WA
90-0790	F16D	57FW	WA
90-0791	F16D	343W	AK
90-0792	F16D	363FW	SW
90-0793	F16D	363FW	SW
90-0794	F16D	31FW	HS
90-0795	F16D	347FW	MY
90-0796	F16D	363FW	SW
90-0797	F16D	347FW	MY
90-0798	F16D		
90-0799	F16D	31FW	HS
90-0800	F16D		
90-0801	F16C	388FW	HL
90-0802	F16C	388FW	HL
90-0803	F16C	388FW	HL
90-0804	F16C	388FW	HL
90-0805	F16C	388FW	HL
90-0806	F16C	388FW	HL
90-0807	F16C	388FW	HL
90-0808	F16C	388FW	HL
90-0809	F16C	57FW	WA
90-0810	F16C	388FW	HL
90-0811	F16C	388FW	HL
90-0812	F16C	388FW	HL
90-0813	F16C	52FW	SP
90-0814	F16C	388FW	HL
90-0815	F16C		
90-0816	F16C		
90-0817	F16C		
90-0818	F16C		
90-0819	F16C	388FW	HL
90-0820	F16C		
90-0821	F16C	388FW	HL
90-0822	F16C		
90-0823	F16C		
90-0824	F16C		
90-0825	F16C		
90-0826	F16C		
90-0827	F16C	52FW	SP
90-0828	F16C	52FW	SP
90-0829	F16C	52FW	SP
90-0830	F16C		
90-0831	F16C	52FW	SP
90-0832	F16C	w.o.24.5.93	
90-0833	F16C	52FW	SP
90-0834	F16D	388FW	HL
90-0835	F16D		
90-0836	F16D	46TW	ET AFMC\MSD
90-0837	F16D	388FW	HL
90-0838	F16D		
90-0839	F16D	57FW	WA
90-0840	F16D	388FW	HL
90-0841	F16D		
90-0842	F16D		
90-0843	F16D		
90-0844	F16D	85TES	OT
90-0845	F16D		

Serial	Type	Unit	
90-0846	F16D	52FW	SP
90-0847	F16D	52FW	SP
90-0848	F16D		
90-0849	F16D	52FW	SP
91-0336	F16C	52FW	SP
91-0337	F16C	52FW	SP
91-0338	F16C	52FW	SP
91-0339	F16C	52FW	SP
91-0340	F16C	52FW	SP
91-0341	F16C	52FW	SP
91-0342	F16C	52FW	SP
91-0343	F16C	52FW	SP
91-0344	F16C	52FW	SP
91-0345	F16C	52FW	SP
91-0346	F16C		
91-0347	F16C		
91-0348	F16C	52FW	SP
91-0349	F16C	52FW	SP
91-0350	F16C		
91-0351	F16C	52FW	SP
91-0352	F16C		
91-0353	F16C		
91-0354	F16C	52FW	SP
91-0355	F16C		
91-0356	F16C		
91-0357	F16C		
91-0358	F16C		
91-0359	F16C		
91-0360	F16C		
91-0361	F16C		
91-0362	F16C		
91-0363	F16C		
91-0364	F16C		
91-0365	F16C		
91-0366	F16C		
91-0367	F16C		
91-0368	F16C		
91-0369	F16C		
91-0370	F16C		
91-0371	F16C		
91-0372	F16C		
91-0373	F16C		
91-0374	F16C		
91-0375	F16C		
91-0376	F16C		
91-0377	F16C		
91-0378	F16C		
91-0379	F16C		
91-0380	F16C		
91-0381	F16C		
91-0382	F16C		
91-0383	F16C		
91-0384	F16C		
91-0385	F16C		
91-0386	F16C		
91-0387	F16C		
91-0388	F16C		
91-0389	F16C		
91-0390	F16C		
91-0391	F16C		
91-0392	F16C		
91-0393	F16C		
91-0394	F16C		
91-0395	F16C		
91-0396	F16C		
91-0397	F16C		
91-0398	F16C		
91-0399	F16C		
91-0400	F16C		
91-0401	F16C		
91-0402	F16C		
91-0403	F16C		
91-0404	F16C		
91-0405	F16C		
91-0406	F16C		
91-0407	F16C		
91-0408	F16C		
91-0409	F16C		
91-0410	F16C		
91-0411	F16C		
91-0412	F16C		
91-0413	F16C		
91-0414	F16C		
91-0415	F16C		
91-0416	F16C		
91-0417	F16C		
91-0418	F16C		
91-0419	F16C		
91-0420	F16C		
91-0421	F16C		
91-0422	F16C		
91-0423	F16C		
91-0462	F16D		
91-0463	F16D		
91-0464	F16D		
91-0465	F16D	52FW	SP
91-0466	F16D		
91-0467	F16D		
91-0468	F16D		
91-0469	F16D		
91-0470	F16D		
91-0471	F16D		
91-0472	F16D		
91-0473	F16D		
91-0474	F16D		
91-0475	F16D		
91-0476	F16D		
91-0477	F16D		
91-0478	F16D		
91-0479	F16D		
91-0480	F16D		
91-0481	F16D		
92-3880	F16C		
92-3881	F16C		
92-3882	F16C		
92-3883	F16C		
92-3884	F16C		
92-3885	F16C		
92-3886	F16C		
92-3887	F16C		
92-3888	F16C		
92-3889	F16C		
92-3890	F16C		
92-3891	F16C		
92-3892	F16C		
92-3893	F16C		
92-3894	F16C		
92-3895	F16C		
92-3896	F16C		
92-3897	F16C		
92-3898	F16C		
92-3899	F16C		
92-3900	F16C		
92-3901	F16C		
92-3902	F16C		
92-3903	F16C		
92-3904	F16C		
92-3905	F16C		
92-3906	F16C		
92-3907	F16C		
92-3908	F16C		
92-3909	F16C		
92-3910	F16C		
92-3911	F16C		
92-3912	F16C		
92-3913	F16C		
92-3914	F16C		
92-3915	F16C		
92-3916	F16C		
92-3917	F16C		
92-3918	F16C		
92-3919	F16C		
92-3920	F16C		
92-3921	F16C		
92-3922	F16C		
92-3923	F16C		
92-3924	F16C		
92-3925	F16D		
92-3926	F16D		
92-3927	F16D		
93-0531	F16C		
93-0532	F16C		
93-0533	F16C		
93-0534	F16C		
93-0535	F16C		
93-0536	F16C		

93-0537	F16C		93-0545	F16C
93-0538	F16C		93-0546	F16C
93-0539	F16C		93-0547	F16C
93-0540	F16C		93-0548	F16C
93-0541	F16C		93-0549	F16C
93-0542	F16C		93-0550	F16C
93-0543	F16C		93-0551	F16C
93-0544	F16C		93-0552	F16C
			93-0553	F16C
			93-0554	F16C

F17 COBRA

72-1569	YF17A	to USN as	201569	72-1570	YF17A preserved	NAS Pensacola

F20 TIGERSHARK

c\n   GG.1001, GI.1001 plus 2 unknown.

82-0062	F20A	AFFTC		82-0064	F20A
82-0063	F20A			82-0065	F20A

F22 (Lockheed-ATF)

c\n   3997, 3998

87-0700	YF22A to N22YF w.o.25.4.92		87-0701	YF22A to N22YX

F23 (MDD\Northrop - ATF)

c\n   1001, 1002.

87-0800	YF23A to N231YF		87-0801	YF23A to N232YF

F106 DELTA DART

Serial	Type	Notes	Code
56-0451	F106A	USAF Museum	
56-0452	F106A	SoC	
56-0453	QF106A	6585TS	AD111
56-0454	QF106A	6585TS	AD102
56-0455	F106A	w.o.19.5.67	
56-0456	F106A	w.o.14.6.65	
56-0457	F106A	AMARC	FN084
56-0458	F106A	AMARC	FN051
56-0459	F106A	preserved McChord AFB	
56-0460	F106A	preserved Minot AFB	
56-0461	F106A	AMARC	FN074
56-0462	F106A	w.o.6.6.75	
56-0463	F106A	AMARC	FN060
56-0464	F106A	w.o.4.8.74	
56-0465	QF106A	6585TS	
56-0466	F106A	AMARC	FN091
56-0467	F106A	w.o.14.8.61	
57-0229	F106A	w.o.8.4.59	
57-0230	F106A	preserved Charleston AFB	
57-0231	F106A	w.o.28.5.81	
57-0232	F106A	FSI	
57-0233	F106A	w.o.12.4.63	
57-0234	QF106A	475WEG	AD175
57-0235	F106A		
57-0236	F106A		
57-0237	F106A	w.o.8.5.74	
57-0238	F106A	w.o.6.12.65	
57-0239	F106A	SoC	
57-0240	F106A		
57-0241	F106A		
57-0242	F106A	w.o.27.10.58	
57-0243	QF106A	w.o.10.3.93	
57-0244	F106A	AMARC	FN015
57-0245	F106A	AMARC	FN052
57-0246	QF106A	475WEG	AD161
57-2453	F106A	AMARC	FN135
57-2454	F106A	w.o.29.7.64	
57-2455	F106A		
57-2456	QF106A	475WEG	AD189
57-2457	F106A	w.o.22.4.69	
57-2458	F106A	w.o.3.9.79	
57-2459	F106A	FSI	AD103
57-2460	F106A	w.o.9.6.77	
57-2461	QF106A	475WEG	AD190
57-2462	F106A	w.o.21.12.61	
57-2463	F106A		
57-2464	F106A	w.o.1.7.83	
57-2465	QF106A	6585TS	AD132
57-2466	F106A		
57-2467	F106A		
57-2468	F106A	w.o.20.7.64	
57-2469	F106A	w.o.7.12.67	
57-2470	QF106A	475WED	AD...
57-2471	F106A	w.o.4.5.65	
57-2472	F106A	w.o.19.6.69	
57-2473	F106A	w.o.6.6.83	
57-2474	F106A	w.o.9.6.61	
57-2475	QF106A	6585TS	AD203
57-2476	F106A	AMARC	FN148
57-2477	QF106A	475WEG	AD212
57-2478	QF106A	6585TS	AD195
57-2479	F106A	w.o.4.2.63	
57-2480	F106A		
57-2481	QF106A	475WEG	AD185
57-2482	QF106A	475WEG	AD191
57-2483	QF106A	w.o.18.12.92	
57-2484	F106A	w.o.19.9.61	
57-2485	QF106A	475WEG	AD192
57-2486	F106A	w.o.14.12.72	
57-2487	QF106A	475WEG	AD188
57-2488	F106A	w.o.10.3.61	
57-2489	F106A	w.o.19.3.63	
57-2490	QF106A	475WEG	AD163
57-2491	F106A	w.o.11.7.73	
57-2492	QF106A	6585TS	
57-2493	QF106A	6585TS	
57-2494	QF106A	6585TS	
57-2495	QF106A	475WEG	AD165
57-2496	QF106A	475WEG	AD186
57-2497	QF106A	475WEG	AD194
57-2498	F106A	w.o.5.5.64	
57-2499	QF106A	6585TS	AD202
57-2500	F106A	w.o.8.1.70	
57-2501	F106A		
57-2502	F106A	SoC	
57-2503	QF106A	475WEG	AD...
57-2504	QF106A	475WEG	AD205
57-2505	F106A		
57-2506	F106A		
57-2507	NF106B to NASA		N607NA
57-2508	F106B	AMARC	FN042
57-2509	F106B	AMARC	FN...
57-2510	F106B	AMARC	FN048
57-2511	F106B	w.o.5.1.59	
57-2512	F106B	AMARC	FN007
57-2513	F106B	475WEG	
57-2514	F106B		
57-2515	F106B		
57-2516	NF106B to NASA N816NA		
57-2517	F106B	MASDC	
57-2518	F106B		
57-2519	F106B	SoC	
57-2520	F106B	w.o.3.10.72	
57-2521	F106B	w.o.2.4.85	
57-2522	F106B	AMARC	FN147
57-2523	F106B	preserved Andrews AFB	
57-2524	F106B		
57-2525	F106B	w.o.1.11.60	
57-2526	F106B	w.o.16.10.68	
57-2527	F106B	w.o.9.10.73	
57-2528	F106B	w.o.16.10.72	
57-2529	F106B	SoC	
57-2530	F106B	AMARC	FN...
57-2531	F106B	w.o. .2.66	
57-2532	F106B	AMARC	FN040
57-2533	F106B	preserved Kelly AFB	
57-2534	F106B	w.o.27.1.64	
57-2535	QF106B	475WEG	AD109
57-2536	QF106B	475WEG	AD107
57-2537	F106B		
57-2538	F106B	w.o.19.10.72	
57-2539	F106B	AMARC	FN034
57-2540	F106B		
57-2541	F106B	AMARC	FN041
57-2542	F106B	w.o.27.8.64	
57-2543	F106B		
57-2544	F106B	w.o.29.3.71	
57-2545	F106B	475WEG	
57-2546	F106B	AMARC	FN021
57-2547	F106B	AMARC	FN097
58-0759	F106A	w.o.11.4.67	
58-0760	QF106A	w.o.2.3.93	
58-0761	F106A	w.o.9.1.63	
58-0762	F106A	w.o.9.1.63	
58-0763	F106A	SoC	
58-0764	QF106A	475WEG	AD179
58-0765	F106A	w.o.18.1.71	
58-0766	QF106A	475WEG	AD129
58-0767	QF106A	475WEG	AD...
58-0768	F106A	w.o.2.9.69	
58-0769	F106A	w.o.10.3.66	
58-0770	F106A	w.o.9.1.61	
58-0771	F106A	w.o.8.1.62	
58-0772	QF106A	6585TS	

Serial	Type	Status	Code
58-0773	F106A	AMARC	FN057
58-0774	QF106A	6585TS	
58-0775	QF106A	6585TS	
58-0776	F106A	w.o.2.9.80	
58-0777	F106A	w.o.18.7.79	
58-0778	F106A	w.o.10.7.81	
58-0779	QF106A	475WEG	AD141
58-0780	QF106A	w.o.9.3.93	
58-0781	F106A	w.o.5.4.77	
58-0782	QF106A	475WEG	AD147
58-0783	QF106A	475WEG	AD108
58-0784	F106A	MASDC	
58-0785	F106A	w.o.7.2.77	
58-0786	QF106A	6585TS	
58-0787	F106A	USAF Museum	
58-0788	QF106A	6585TS	
58-0789	F106A	w.o.17.3.66	
58-0790	F106A	AMARC	FN080
58-0791	QF106A	6585TS	
58-0792	QF106A	475WEG	AD153
58-0793	F106A		
58-0794	F106A	w.o.4.11.64	
58-0795	QF106A	475WEG	AD180
58-0796	F106A	w.o.23.8.61	
58-0797	F106A	FSI	AD104
58-0798	F106A	w.o.13.6.66	
58-0900	F106B	AMARC	FN056
58-0901	F106B		
58-0902	F106B	AMARC	FN055
58-0903	F106B	AMARC	FN046
58-0904	F106B	AMARC	FN044
59-0001	F106A	w.o.12.10.66	
59-0002	QF106A	475WEG	AD113
59-0003	F106A	AMARC	FN070
59-0004	F106A	w.o.24.6.80	
59-0005	F106A		
59-0006	QF106A	475WEG	AD110
59-0007	QF106A	475WEG	AD215
59-0008	QF106A	6585TS	
59-0009	F106A	w.o.13.2.78	
59-0010	F106A	AMARC	FN069
59-0011	QF106A	475WEG	AD159
59-0012	F106A	AMARC	FN077
59-0013	F106A	w.o.3.10.63	
59-0014	F106A	w.o.10.3.69	
59-0015	QF106A	475WEG	AD115
59-0016	F106A	AMARC	FN066
59-0017	F106A	w.o.19.12.63	
59-0018	F106A	w.o.16.9.74	
59-0019	F106A	w.o.2.12.76	
59-0020	QF106A	475WEG	AD172
59-0021	F106A	w.o.15.5.73	
59-0022	F106A	w.o.3.10.67	
59-0023	QF106A	475WEG	AD...
59-0024	QF106A	475WEG	AD182
59-0025	F106A	AMARC	FN072
59-0026	QF106A	475WEG	AD139
59-0027	QF106A	6585TS	AD183
59-0028	F106A	w.o.19.3.74	
59-0029	F106A	w.o.28.7.61	
59-0030	F106A	w.o.16.9.74	
59-0031	QF106A	475WEG	AD...
59-0032	F106A	AMARC	FN088
59-0033	F106A		
59-0034	QF106A	475WEG	AD223
59-0035	QF106A	w.o.1.9.92	
59-0036	F106A	w.o.9.6.70	
59-0037	F106A		
59-0038	QF106A	475WEG	AD171
59-0039	F106A	w.o.8.10.63	
59-0040	F106A		
59-0041	F106A	w.o.7.9.65	
59-0042	QF106A	w.o.15.9.92	
59-0043	QF106A	475WEG	AD227
59-0044	F106A		
59-0045	F106A	w.o.7.6.61	
59-0046	F106A		
59-0047	F106A		
59-0048	F106A		
59-0049	F106A		
59-0050	F106A	w.o.8.4.64	
59-0051	F106A	AMARC	FN098
59-0052	F106A	w.o.11.12.73	
59-0053	QF106A	475WEG	AD173
59-0054	QF106A	475WEG	AD201
59-0055	F106A	w.o.30.11.65	
59-0056	QF106A	475WEG	AD164
59-0057	QF106A	475WEG	AD176
59-0058	QF106A	475WEG	AD184
59-0059	QF106A	475WEG	AD105
59-0060	QF106A	6585TS	
59-0061	QF106A	475WEG	AD156
59-0062	QF106A	475WEG	AD168
59-0063	F106A	AMARC	FN081
59-0064	QF106A	475WEG	AD155
59-0065	F106A	AMARC	FN003
59-0066	QF106A	475WEG	AD187
59-0067	F106A	w.o.20.2.73	
59-0068	F106A	w.o.19.3.67	
59-0069	F106A	preserved Griffiss AFB	
59-0070	F106A	w.o.22.9.61	
59-0071	F106A	w.o.4.6.81	
59-0072	QF106A		
59-0073	F106A	w.o.19.11.60	
59-0074	QF106A	w.o.2.2.92	
59-0075	F106A	w.o.22.8.75	
59-0076	QF106A	475WEG	AD169
59-0077	QF106A	475WEG	AD140
59-0078	F106A	SoC	
59-0079	F106A	AMARC	FN013
59-0080	F106A	AMARC	FN027
59-0081	QF106A	475WEG	AD177
59-0082	QF106A	w.o.15.5.92	
59-0083	QF106A	475WEG	AD143
59-0084	F106A	w.o.25.4.73	
59-0085	QF106A	6585TS	
59-0086	F106A		
59-0087	F106A	w.o.11.10.61	
59-0088	F106A	w.o.5.10.72	
59-0089	F106A	w.o.26.11.72	
59-0090	F106A	AMARC	FN067
59-0091	QF106A	475WEG	AD181
59-0092	QF106A	w.o.22.9.92	
59-0093	F106A	AMARC	FN073
59-0094	F106A	AMARC	FN038
59-0095	F106A	AMARC	FN089
59-0096	QF106A	475WEG	AD160
59-0097	QF106A	475WEG	AD...
59-0098	F106A	w.o.19.11.62	
59-0099	QF106A	475WEG	AD137
59-0100	QF106A	475WEG	AD174
59-0101	F106A	w.o.18.7.79	
59-0102	F106A		
59-0103	F106A	w.o.27.10.80	
59-0104	QF106A	6585TS	
59-0105	F106A	AMARC	FN059
59-0106	F106A	AMARC	FN058
59-0107	F106A	w.o.2.1.63	
59-0108	F106A		
59-0109	F106A	AMARC	FN104
59-0110	QF106A	w.o.19.2.93	
59-0111	F106A	w.o.12.10.65	
59-0112	F106A	w.o.9.10.73	
59-0113	F106A	w.o.29.1.64	
59-0114	F106A	w.o.4.4.63	
59-0115	F106A		
59-0116	F106A	MASDC	
59-0117	F106A	w.o.3.9.63	
59-0118	F106A	w.o.11.6.68	
59-0119	QF106A	475WEG	AD106
59-0120	F106A	w.o.14.12.61	
59-0121	F106A	w.o.24.4.69	
59-0122	F106A	MASDC	
59-0123	F106A	BDRT	Langley AFB
59-0124	F106A	w.o.9.3.64	
59-0125	F106A	w.o.7.11.71	
59-0126	QF106A	475WEG	AD158
59-0127	QF106A	w.o.25.6.92	
59-0128	QF106A	6585TS	
59-0129	QF106A	475WEG	AD162
59-0130	QF106A	6585TS	
59-0131	F106A	w.o.18.10.71	
59-0132	QF106A	6585TS	
59-0133	F106A	AMARC	FN064
59-0134	GF106A	preserved Lowry AFB	
59-0135	QF106A	475WEG	AD170

59-0136	QF106A	475WEG	AD178
59-0137	F106A	AMARC	FN062
59-0138	QF106A	475WEG	AD157
59-0139	F106A	w.o.17.5.61	
59-0140	QF106A	475WEG	AD...
59-0141	QF106A	6585TS	
59-0142	F106A	w.o.27.8.63	
59-0143	F106A		
59-0144	F106A	w.o.29.5.78	
59-0145	F106A	preserved	Tyndall AFB
59-0146	F106A	preserved	Fresno, Ca
59-0147	F106A	AMARC	FN017
59-0148	F106A	w.o.22.4.69	
59-0149	F106B	Rockwell	Palmdale
59-0150	F106B	AMARC	FN030
59-0151	F106B		
59-0152	QF106B	475WEG	AD101
59-0153	F106B	AMARC	FN130
59-0154	F106B	w.o.18.12.62	
59-0155	F106B		
59-0156	F106B	w.o.1.3.64	
59-0157	F106B	w.o.22.1.77	
59-0158	F106B	AMARC	FN019
59-0159	QF106B	6585TS	
59-0160	F106B	w.o.12.6.74	
59-0161	F106B	AMARC	FN038
59-0162	F106B	w.o.9.2.71	
59-0163	F106B	w.o.30.9.71	
59-0164	F106B	AMARC	FN050
59-0165	F106B	dump	McClellan AFB

F111 AARDVARK

c\n			
	F111A\F111E:	1 to 253.	
	FB111A\F111G:	1 to 76.	
	F111D:	1 to 96.	
	F111F:	1 to 106.	

63-9766	F111A	preserved	Edwards AFB
63-9767	F111A		
63-9768	GF111A	TTC	Sheppard
63-9769	F111A	w.o.18.5.68	
63-9770	GF111A	preserved	McClellan AFB
63-9771	F111A	preserved	Cannon AFB
63-9772	GF111A	TTC	Sheppard
63-9773	F111A	preserved	Sheppard AFB
63-9774	F111A	w.o.19.1.67AFB	
63-9777	F111A	AMARC	FV008
63-9778	F111A	NASA	Dryden center
63-9779	F111A	preserved	Carswell AFB
63-9780	F111A	w.o.19.10.67	
63-9781	F111A	AMARC	FV012
63-9782	F111A	ADC	Rome
63-9783	YFB111A	AMARC	FV016
65-5701	F111A	w.o.2.1.68	
65-5702	F111A	AMARC	FV...
65-5703	F111A	w.o.	
65-5704	F111A	AMARC	FV006
65-5705	F111A	AMARC	FV005
65-5706	F111A	cannibalised for 80082	
65-5707	F111A	AMARC	FV009
65-5708	F111A	AMARC	FV007
65-5709	F111A	preserved	Kirtland AFB
65-5710	F111A	AMARC	FV028
66-0011	F111A	MASDC	
66-0012	GF111A	TTC	Lowry
66-0013	EF111A	85TES	OT
66-0014	EF111A	366W	MO
66-0015	EF111A	27FW	CC
66-0016	EF111A	366W	MO
66-0017	F111A	w.o.30.3.68	
66-0018	EF111A	ALC	SM Sacramento
66-0019	EF111A	27FW	CC
66-0020	EF111A	27FW	CC
66-0021	EF111A	27FW	CC
66-0022	F111A	w.o.28.3.68	
66-0023	EF111A	w.o.14.2.91	
66-0024	F111A	w.o.22.4.68	
66-0025	F111A	w.o.20.6.75	
66-0026	F111A	w.o.13.3.84	
66-0027	EF111A	366W	MO
66-0028	EF111A	366W	MO
66-0029	F111A	w.o.	
66-0030	EF111A	366W	MO
66-0031	EF111A	27FW	CC
66-0032	F111A	w.o.	
66-0033	EF111A	366W	MO
66-0034	F111A	w.o.6.6.75	
66-0035	EF111A	366W	MO
66-0036	EF111A	366W	MO
66-0037	EF111A	366W	MO
66-0038	EF111A		
66-0039	EF111A	366W	MO
66-0040	F111A	w.o.	
66-0041	EF111A	366W	MO
66-0042	F111A	w.o.	
66-0043	F111A	w.o.4.3.69	
66-0044	EF111A	27FW	CC
66-0045	F111A	w.o.13.5.82	
66-0046	EF111A	366W	MO
66-0047	EF111A	27FW	CC
66-0048	EF111A	366W	MO
66-0049	EF111A	366W	MO
66-0050	EF111A	366W	MO
66-0051	EF111A	366W	MO
66-0052	F111A	w.o.31.7.79	
66-0053	F111A	AMARC	FV037
66-0054	F111A	w.o.13.4.83	
66-0055	EF111A	366W	MO
66-0056	EF111A	w.o.2.4.92	
66-0057	EF111A	366W	MO
66-0058	F111A	w.o.7.10.75	
67-0032	EF111A	366W	MO
67-0033	EF111A	366W	MO
67-0034	EF111A	366W	MO
67-0035	EF111A	366W	MO
67-0036	F111A	w.o.	
67-0037	EF111A	366W	MO
67-0038	EF111A	366W	MO
67-0039	EF111A	27FW	CC
67-0040	F111A	w.o.	
67-0041	EF111A	366W	MO
67-0042	EF111A	366W	MO
67-0043	F111A	w.o.	
67-0044	EF111A	366W	MO
67-0045	F111A	AMARC	FV...
67-0046	GF111A	TTC	Sheppard
67-0047	GF111A	TTC	Sheppard
67-0048	EF111A	366W	MO
67-0049	F111A	w.o.	
67-0050	F111A	ALC	SM Sacramento
67-0051	GF111A	TTC	Sheppard
67-0052	EF111A	366W	MO
67-0053	F111A	AMARC	FV070
67-0054	F111A		
67-0055	F111A	w.o.	
67-0056	GF111A	TTC	Sheppard
67-0057	GF111A	TTC	Sheppard
67-0058	F111A		
67-0059	F111A	w.o.	

Serial	Type	Status	
67-0060	F111A	w.o.16.10.72	
67-0061	F111A	AMARC	FV062
67-0062	F111A	AMARC	FV033
67-0063	F111A	w.o.6.11.72	
67-0064	F111A		
67-0065	F111A	AMARC	FV067
67-0066	F111A	w.o.	
67-0067	F111A	USAF	Museum
67-0068	F111A	w.o.22.12.72	
67-0069	F111A		
67-0070	F111A		
67-0071	F111A	AMARC	FV026
67-0072	F111A	w.o.20.2.73	
67-0073	F111A	w.o.19.1.82	
67-0074	F111A	AMARC	FV068
67-0075	F111A		
67-0076	F111A	AMARC	FV077
67-0077	F111A	AMARC	FV...
67-0078	F111A	w.o.28.9.72	
67-0079	F111A	AMARC	FV029
67-0080	F111A	w.o.	
67-0081	F111A	AMARC	FV...
67-0082	F111A	w.o.18.6.72	
67-0083	F111A	w.o.30.11.77	
67-0084	F111A	AMARC	FV065
67-0085	F111A	AMARC	FV027
67-0086	F111A	AMARC	FV...
67-0087	F111A	AMARC	FV...
67-0088	F111A	AMARC	FV058
67-0089	F111A		
67-0090	F111A	AMARC	FV069
67-0091	F111A	AMARC	FV075
67-0092	F111A	w.o.20.11.72	
67-0093	F111A	w.o.9.11.82	
67-0094	F111A	AMARC	FV...
67-0095	F111A	AMARC	FV072
67-0096	F111A	AMARC	FV...
67-0097	F111A	w.o.26.3.80	
67-0098	F111A	w.o.8.10.82	
67-0099	F111A	w.o.18.12.72	
67-0100	F111A		
67-0101	F111A	AMARC	FV032
67-0102	F111A	w.o.12.1.88	
67-0103	F111A		
67-0104	F111A	AMARC	FV...
67-0105	F111A	w.o.	
67-0106	F111A	AMARC	FV076
67-0107	F111A	AMARC	FV066
67-0108	F111A	AMARC	FV073
67-0109	F111A	to Australia	A8-109
67-0110	F111A	AMARC	FV031
67-0111	F111A	w.o.16.6.73	
67-0112	F111A	to Australia	A8-112
67-0113	F111A	to Australia	A8-113
67-0114	F111A	to Australia	A8-114
67-0115	F111E	AMARC	FV166
67-0116	F111E	w.o.6.10.76	
67-0117	F111E	w.o.23.4.71	
67-0118	F111E	46TW	ET AFMC\MSD
67-0119	F111E	AMARC	FV...
67-0120	F111E	20FW	UH
67-0121	F111E	AMARC	FV...
67-0122	F111E	AMARC	FV...
67-0123	F111E	20FW	UH
67-0124	F111E		
67-0159	FB111A	ALC	Sacramento
67-0160	FB111A	cannibalised for 77194	
67-0161	FB111A	AMARC	BF003
67-0162	F111G	AMARC	FV168
67-0163	FB111A	AMARC	BF009
67-7192	FB111A	AMARC	BF010
67-7193	F111G	27FW	CC
67-7194	F111G	27FW	CC
67-7195	FB111A	AMARC	BF004
67-7196	F111G	AMARC	FV156
68-0001	F111E	w.o.5.2.90	
68-0002	F111E		
68-0003	F111E	w.o.19.12.79	
68-0004	F111E	20FW	UH
68-0005	F111E	20FW	UH
68-0006	F111E	20FW	UH
68-0007	F111E	20FW	UH
68-0008	F111E	w.o.16.5.73	
68-0009	F111E	AMARC	FV...
68-0010	F111E	AMARC	FV143
68-0011	F111E	preserved RAFLakenheath,UK	
68-0012	F111E	w.o.30.10.79	
68-0013	F111E		
68-0014	F111E	AMARC	FV...
68-0015	F111E		
68-0016	F111E		
68-0017	F111E		
68-0018	F111E	w.o.18.1.72	
68-0019	F111E	w.o.9.8.84	
68-0020	F111E	20FW	UH
68-0021	F111E	AMARC	FV169
68-0022	F111E	27FW	CC
68-0023	F111E	20FW	UH
68-0024	F111E	w.o.11.1.73	
68-0025	F111E	20FW	UH
68-0026	F111E	20FW	UH
68-0027	GF111E	TTC	Sheppard
68-0028	F111E	AMARC	FV...
68-0029	F111E	20FW	UH
68-0030	F111E	20FW	UH
68-0031	F111E	AMARC	FV172
68-0032	F111E	27FW	CC
68-0033	F111E	AMARC	FV...
68-0034	F111E	20FW	UH
68-0035	F111E	AMARC	FV...
68-0036	F111E		
68-0037	F111E	20FW	UH
68-0038	F111E	AMARC	FV170
68-0039	F111E	AMARC	FV...
68-0040	F111E	27FW	CC
68-0041	F111E		
68-0042	F111E	w.o.24.7.79	
68-0043	F111E	20FW	UH
68-0044	F111E	27FW	CC
68-0045	F111E	w.o.12.12.79	
68-0046	F111E	AMARC	FV171
68-0047	F111E	27FW	CC
68-0048	F111E	20FW	UH
68-0049	F111E	20FW	UH
68-0050	F111E	27FW	CC
68-0051	F111E	AMARC	FV173
68-0052	F111E	w.o.17.9.92	
68-0053	F111E	AMARC	FV...
68-0054	F111E		
68-0055	F111E	20FW	UH
68-0056	F111E	20FW	UH
68-0057	F111E	w.o.29.4.80	
68-0058	F111E	preserved Eglin AFB	
68-0059	F111E	20FW	UH
68-0060	F111E	w.o.5.11.75	
68-0061	F111E	20FW	UH
68-0062	F111E	AMARC	FV...
68-0063	F111E	27FW	CC
68-0064	F111E	20FW	UH
68-0065	F111E	20FW	UH
68-0066	F111E	w.o.20.7.90	
68-0067	F111E		
68-0068	F111E	27FW	CC
68-0069	F111E	20FW	UH
68-0070	F111E	w.o.31.10.77	
68-0071	F111E	27FW	CC
68-0072	F111E		
68-0073	F111E		
68-0074	F111E	27FW	CC
68-0075	F111E	27FW	CC
68-0076	F111E	27FW	CC
68-0077	F111E	27FW	CC
68-0078	F111E	ALC	SM Sacramento
68-0079	F111E	27FW	CC
68-0080	F111E	27FW	CC
68-0081	F111E	w.o.5.3.75	
68-0082	F111E		
68-0083	F111E	27FW	CC
68-0084	F111E		
68-0085	F111D	AMARC	FV040
68-0086	F111D		
68-0087	F111D	AMARC	FV039
68-0088	F111D		
68-0089	F111D	AMARC	FV038
68-0090	F111D	AMARC	FV149
68-0091	F111D		

Serial	Type		
68-0092	F111D		
68-0093	F111D	w.o.3.10.77	
68-0094	F111D		
68-0095	F111D		
68-0096	F111D	AMARC	FV128
68-0097	F111D		
68-0098	F111D	w.o.8.6.88	
68-0099	F111D		
68-0100	F111D	AMARC	FV126
68-0101	F111D	AMARC	FV043
68-0102	F111D		
68-0103	F111D		
68-0104	F111D	AMARC	FV154
68-0105	F111D	w.o.30.3.73	
68-0106	F111D		
68-0107	F111D	AMARC	FV147
68-0108	F111D	AMARC	FV125
68-0109	F111D	w.o.	
68-0110	F111D	w.o.27.1.82	
68-0111	F111D	AMARC	FV...
68-0112	F111D	AMARC	FV114
68-0113	F111D	w.o.3.12.73	
68-0114	F111D		
68-0115	F111D		
68-0116	F111D		
68-0117	F111D	AMARC	FV042
68-0118	F111D	57FW	WA
68-0119	F111D		
68-0120	F111D	AMARC	FV080
68-0121	F111D	AMARC	FV135
68-0122	F111D	AMARC	FV138
68-0123	F111D	AMARC	FV148
68-0124	F111D	AMARC	FV046
68-0125	F111D		
68-0126	F111D	AMARC	FV045
68-0127	F111D	AMARC	FV150
68-0128	F111D	AMARC	FV063
68-0129	F111D	AMARC	FV044
68-0130	F111D		
68-0131	F111D		
68-0132	F111D	w.o.17.3.88	
68-0133	F111D		
68-0134	F111D	AMARC	FV...
68-0135	F111D		
68-0136	F111D		
68-0137	F111D	AMARC	FV142
68-0138	F111D		
68-0139	F111D	w.o.14.7.80	
68-0140	F111D		
68-0141	F111D		
68-0142	F111D		
68-0143	F111D	AMARC	FV131
68-0144	F111D		
68-0145	F111D		
68-0146	F111D	w.o.2.9.77	
68-0147	F111D		
68-0148	F111D		
68-0149	F111D	AMARC	FV144
68-0150	F111D	AMARC	FV145
68-0151	F111D	AMARC	FV119
68-0152	F111D		
68-0153	F111D		
68-0154	F111D		
68-0155	F111D		
68-0156	F111D		
68-0157	F111D		
68-0158	F111D	w.o.30.3.73	
68-0159	F111D		
68-0160	F111D	w.o.15.9.82	
68-0161	F111D		
68-0162	FB111D	AMARC	FV074
68-0163	F111D		
68-0164	F111D	w.o.17.9.84	
68-0165	F111D	AMARC	FV146
68-0166	F111D		
68-0167	F111D	w.o.	
68-0168	F111D		
68-0169	F111D		
68-0170	F111D		
68-0171	F111D	AMARC	FV081
68-0172	F111D	AMARC	FV129
68-0173	F111D	w.o.	
68-0174	F111D	AMARC	FV...
68-0175	F111D	AMARC	FV155
68-0176	F111D	AMARC	FV141
68-0177	F111D	AMARC	FV134
68-0178	F111D		
68-0179	F111D	AMARC	FV132
68-0180	F111D		
68-0239	FB111A		
68-0240	F111G	AMARC	BF012
68-0241	F111G	AMARC	FV167
68-0242	FB111A	w.o.8.6.83	
68-0243	FB111A	w.o.2.2.89	
68-0244	F111G	AMARC	FV078
68-0245	FB111A		
68-0246	FB111A	AMARC	BF008
68-0247	FB111A	ALC	SM Sacramano
68-0248	FB111A	AMARC	BF...
68-0249	FB111A	AMARC	BF018
68-0250	FB111A	AMARC	BF013
68-0251	FB111A		
68-0252	F111G	AMARC	FV079
68-0253	FB111A	w.o.10.7.70	
68-0254	FB111A	AMARC	FV165
68-0255	F111G	to Australia A8-255	
68-0256	FB111A	AMARC	BF017
68-0257	F111G	AMARC	FV164
68-0258	FB111A	AMARC	BF014
68-0259	F111G	to Australia A8-259	
68-0260	F111G	AMARC	FV161
68-0261	FB111A	w.o.8.9.79	
68-0262	FB111A	AMARC	BF016
68-0263	FB111A	w.o.30.1.81	
68-0264	F111G	to Australia A8-264	
68-0265	F111G	to Australia A8-265	
68-0266	FB111A	w.o.	
68-0267	FB111A	preserved Offutt AFB	
68-0268	FB111A	w.o.6.10.80	
68-0269	FB111A	AMARC	BF015
68-0270	F111G	to Australia A8-270	
68-0271	F111G	to Australia A8-271	
68-0272	F111G	to Australia A8-272	
68-0273	F111G		
68-0274	F117G	to Australia A8-274	
68-0275	FB111A	preserved Kelly AFB	
68-0276	F111G		
68-0277	F111G	to Australia A8-277	
68-0278	F111G	to Australia A8-278	
68-0279	FB111A	w.o.	
68-0280	FB111A	w.o.	
68-0281	F111G	to Australia A8-281	
68-0282	F111G	to Australia A8-282	
68-0283	FB111A	w.o.8.1.72	
68-0284	FB111A	stored	Barksdale AFB
68-0285	FB111A	w.o.	
68-0286	FB111A		
68-0287	GFB111	ATTC	Lowry
68-0288	FB111A	AMARC	BF011
68-0289	F111G		
68-0290	FB111A	w.o.	
68-0291	F111G	to Australia A8-291	
68-0292	FB111A		
69-6503	F111G		
69-6504	F111G	AMARC	FV050
69-6505	FB111A	w.o.	
69-6506	F111G	to Australia A8-506	
69-6507	FB111A		
69-6508	F111G		
69-6509	F111G		
69-6510	F111G		
69-6511	FB111A	w.o.	
69-6512	F111G	to Australia A8-512	
69-6513	FB111A	AMARC	BF007
69-6514	F111G	to Australia A8-514	
70-2362	F111F	27FW	CC
70-2363	F111F		
70-2364	F111F	27FW	CC
70-2365	F111F	27FW	CC
70-2366	F111F	w.o.21.12.83	
70-2367	F111F	w.o.20.4.79	
70-2368	F111F	w.o.2.5.90	
70-2369	F111F	27FW	CC
70-2370	F111F		
70-2371	F111F	27FW	CC
70-2372	F111F	27FW	CC

Serial	Type	Unit	Code		Serial	Type	Unit	Code
70-2373	F111F	27FW	CC		71-0883	F111F		
70-2374	F111F				71-0884	F111F	27FW	CC
70-2375	F111F	w.o.28.7.87			71-0885	F111F	27FW	CC
70-2376	F111F	27FW	CC		71-0886	F111F		
70-2377	F111F	w.o.7.12.82			71-0887	F111F		
70-2378	F111F	27FW	CC		71-0888	F111F	27FW	CC
70-2379	F111F	27FW	CC		71-0889	F111F	27FW	CC
70-2380	F111F	w.o.15.12.77			71-0890	F111F	27FW	CC
70-2381	F111F	27FW	CC		71-0891	F111F		
70-2382	F111F	27FW	CC		71-0892	F111F	27FW	CC
70-2383	F111F	27FW	CC		71-0893	F111F		
70-2384	F111F				71-0894	F111F	27FW	CC
70-2385	F111F	27FW	CC		72-1441	F111F	w.o.4.2.81	
70-2386	F111F	27FW	CC		72-1442	F111F	27FW	CC
70-2387	F111F	27FW	CC		72-1443	F111F		
70-2388	F111F	w.o.			72-1444	F111F	46TW	ET AFMC\MSD
70-2389	F111F	w.o.15.4.86			72-1445	F111F		
70-2390	F111F				72-1446	F111F		
70-2391	F111F	27FW	CC		72-1447	F111F	w.o.23.6.82	
70-2392	F111F				72-1448	F111F	27FW	CC
70-2393	F111F	w.o.			72-1449	F111F		
70-2394	F111F				72-1450	F111F		
70-2395	F111F	w.o.			72-1451	F111F	27FW	CC
70-2396	F111F	27FW	CC		72-1452	F111F		
70-2397	F111F	w.o.5.4.89			73-0707	F111F		
70-2398	F111F	27FW	CC		73-0708	F111F		
70-2399	F111F				73-0709	F111F	w.o.21.4.77	
70-2400	F111F	27FW	CC		73-0710	F111F	27FW	CC
70-2401	F111F	27FW	CC		73-0711	F111F		
70-2402	F111F	27FW	CC		73-0712	F111F	27FW	CC
70-2403	F111F				73-0713	F111F		
70-2404	F111F	27FW	CC		73-0714	F111F	w.o.20.4.79	
70-2405	F111F				73-0715	F111F		
70-2406	F111F				73-0716	F111F	w.o.1.11.82	
70-2407	F111F	w.o.3.72			73-0717	F111F	w.o.29.3.78	
70-2408	F111F	27FW	CC		73-0718	F111F	w.o.4.10.77	
70-2409	F111F	27FW	CC		74-0177	F111F	27FW	CC
70-2410	F111F	w.o.10.73			74-0178	F111F		
70-2411	F111F				74-0179	F111F	w.o.16.9.82	
70-2412	F111F	27FW	CC		74-0180	F111F		
70-2413	F111F	27FW	CC		74-0181	F111F	27FW	CC
70-2414	F111F	27FW	CC		74-0182	F111F		
70-2415	F111F				74-0183	F111F	w.o.9.10.90	
70-2416	F111F				74-0184	F111F	27FW	CC
70-2417	F111F	27FW	CC		74-0185	F111F	27FW	CC
70-2418	F111F	w.o.23.2.87			74-0186	F111F	27FW	CC
70-2419	F111F	27FW	CC		74-0187	F111F	57FW	WA
					74-0188	F111F	w.o.26.4.83	

F117 'Stealth Fighter'

note: FY for 782\783\784\785\792 remain unconfirmed.

Serial	Type	Unit	Code		Serial	Type	Unit	Code
79-10780	YF117A	preserved	Nellis AFB		82-0801	F117A	49FW	HO
79-10781	YF117A	USAF Museum			82-0802	F117A	w.o.4.8.92	
79-10782	F117A				82-0803	F117A	49FW	HO
79-10783	F117A	46TW	ED AFFTC		82-0804	F117A	57FW	WA
79-10784	F117A	46TW	ED AFFTC		82-0805	F117A	49FW	HO
80-0785	F117A	w.o.20.4.82			82-0806	F117A	49FW	HO
80-0786	F117A	49FW	HO		83-0807	F117A	49FW	HO
80-0787	F117A	49FW	HO		83-0808	F117A	49FW	HO
80-0788	F117A	49FW	HO		84-0809	F117A	49FW	HO
80-0789	F117A	49FW	HO		84-0810	F117A	49FW	HO
80-0790	F117A	49FW	HO		84-0811	F117A	49FW	HO
80-0791	F117A	49FW	HO		84-0812	F117A	49FW	HO
80-0792	F117A	w.o.11.7.86			84-0824	F117A	49FW	HO
81-10793	F117A	49FW	HO		84-0825	F117A	49FW	HO
81-10794	F117A	49FW	HO		84-0826	F117A	49FW	HO
81-10795	F117A	49FW	HO		84-0827	F117A	49FW	HO
81-10796	F117A	49FW	HO		84-0828	F117A	49FW	HO
81-10797	F117A	49FW	HO		85-0813	F117A	49FW	HO
81-10798	F117A	49FW	HO		85-0814	F117A	49FW	HO
82-0799	F117A	49FW	HO		85-0815	F117A	w.o.14.10.87	
82-0800	F117A	49FW	HO		85-0816	F117A	49FW	HO

85-0817	F117A	49FW	HO	85-0836	F117A	49FW	HO
85-0818	F117A	49FW	HO	86-0821	F117A	49FW	HO
85-0819	F117A	49FW	HO	86-0822	F117A	49FW	HO
85-0820	F117A	49FW	HO	86-0823	F117A	49FW	HO
85-0829	F117A	49FW	HO	86-0837	F117A	49FW	HO
85-0830	F117A	49FW	HO	86-0838	F117A	49FW	HO
85-0831	F117A	46TW	ED AFFTC	86-0839	F117A	49FW	HO
85-0832	F117A	49FW	HO	86-0840	F117A	49FW	HO
85-0833	F117A	49FW	HO	88-0841	F117A	49FW	HO
85-0834	F117A	49FW	HO	88-0842	F117A	49FW	HO
85-0835	F117A	49FW	HO	88-0843	F117A	49FW	HO

GLIDERS

81-0886	TG7A	USAF Academy		86-0404	RG8A	to USCG 8102
81-0887	TG7A	USAF Academy		87-0761	TG7A	USAF Academy
81-0888	TG7A	USAF		87-0762	TG7A	USAF Academy
81-0889	TG7A	USAF		87-0763	TG7A	USAF Academy
81-0890	TG7A	USAF		87-0764	TG7A	USAF Academy
82-0039	TG7A	USAF		88-0268	TG4A	USAF
82-0040	TG7A	USAF Academy		89-0461	TG3A	USAF
82-0041	TG7A	USAF Academy		89-0462	TG3A	USAF
82-0042	TG7A	USAF		89-0463	TG3A	USAF
82-0043	TG7A	USAF Academy		89-0464	TG4A	USAF Academy
85-0047	RG8A	to USCG 8101		91-0515	TG3A	USAF Academy
85-0048	RG8A	to USCG w.o.				

note: The serial 91-0515 is compromised by a C26B Metro.

UH1 IROQUOIS

c\n	UH1F:	7001 to 7120.
	THIF:	7301 to 7326 (conversions back to UH1F standard).
	UH1N:	31001 to 31079.
	HH1H:	17101 to 17130.
	UH1E:	ex USN (unknown).

63-13141	UH1F	Pima Museum		64-15477	UH1F	preserved	Davis-Monthan AFB
63-13142	UH1F	MASDC		64-15478	UH1F	w.o.	
63-13143	UH1F	Pima Museum		64-15479	UH1F	to N91471	
63-13144	UH1F	to N45770		64-15480	UH1F	preserved	March AFB
63-13145	UH1F	MASDC		64-15481	UH1F	to N64709	
63-13146	UH1P	AMARC	HF032	64-15482	UH1F	BDRT	Hill AFB
63-13147	UH1F	MASDC		64-15483	UH1F	scrapyard	Desert Air Parts
63-13148	UH1F	MASDC		64-15484	UH1F	w.o.	
63-13149	UH1P			64-15485	UH1F	MASDC	
63-13150	UH1P	MASDC		64-15486	UH1F	MASDC	
63-13151	UH1F	w.o.		64-15487	UH1F	to N48158	
63-13152	UH1F	w.o.		64-15488	UH1F	to N45774	
63-13153	UH1F	scrapyard	Desert Air Parts	64-15489	UH1F	MASDC	
63-13154	UH1F	to N45773		64-15490	UH1F	scrapyard	Desert Air Parts
63-13155	UH1F	w.o.		64-15491	UH1F	w.o.	
63-13156	UH1P	MASDC		64-15492	UH1F	w.o.	
63-13157	UH1F	w.o.		64-15493	UH1P	preserved	Hurlburt Field
63-13158	UH1F	w.o.		64-15494	UH1F	MASDC	
63-13159	UH1F	w.o.		64-15495	UH1F	preserved	Kirtland AFB
63-13160	UH1P	MASDC		64-15496	UH1F	scrapyard	Desert Air Parts
63-13161	UH1P	MASDC		64-15497	UH1F	to N45776	
63-13162	UH1P	MASDC		64-15498	UH1F	scrapyard	Desert Air Parts
63-13163	UH1F			64-15499	UH1F	BDRT	Hill AFB
63-13164	UH1F	w.o.		64-15500	UH1F	MASDC	
63-13165	GUH1P			64-15501	UH1F	to N24684	
64-15476	UH1P	USAF	Museum	65-7911	UH1F	to N4040P	

Serial	Type	Disposition	Notes
65-7912	UH1F	to N17LA	
65-7913	UH1F	w.o.	
65-7914	UH1F	MASDC	
65-7915	UH1F	37RQS	det.5
65-7916	UH1F	MASDC	
65-7917	UH1F	to USFS	N480DF
65-7918	UH1F	AMARC	HF146
65-7919	UH1F	w.o.	
65-7920	UH1F	w.o.	
65-7921	UH1F	to USFS	N481DF
65-7922	GUH1F	TTC	Sheppard
65-7923	UH1F	w.o.	
65-7924	UH1F	w.o.	
65-7925	UH1F	AMARC	HF...
65-7926	UH1P	MASDC	
65-7927	UH1F	w.o.	
65-7928	UH1F		
65-7929	UH1P	to N.....	
65-7930	UH1F	w.o.	
65-7931	UH1F	w.o.	
65-7932	UH1F	w.o.	
65-7933	UH1F	to USFS	N482DF
65-7934	UH1F	w.o.	
65-7935	UH1F	w.o.	
65-7936	UH1F	to N981	
65-7937	UH1F	w.o.	
65-7938	UH1F	to USFS	N483DF
65-7939	UH1F	w.o.	
65-7940	UH1F	to USFS	N484DF
65-7941	UH1F	preserved	Whiteman AFB
65-7942	UH1F	w.o.	
65-7943	UH1F	w.o.	
65-7944	UH1F	w.o.	
65-7945	UH1F	w.o.	
65-7946	UH1F	preserved	Grand Forks AFB
65-7947	UH1F	to USFS	N485DF
65-7948	UH1F		
65-7949	UH1F	to USFS	N486DF
65-7950	UH1F	w.o.	
65-7951	UH1F	preserved	Ellsworth AFB
65-7952	UH1F	to USFS	N487DF
65-7953	UH1F	preserved	F.E. Warren AFB
65-7954	UH1F	to USFS	N488DF
65-7955	UH1F	to N23021	
65-7956	UH1F	preserved	Malstrom AFB
65-7957	UH1F	w.o.	
65-7958	UH1F		
65-7959	UH1F	preserved	Robins AFB
65-7960	UH1F	to N.....	
65-7961	UH1F	to N7403	
65-7962	UH1F	AMARC	HF174
65-7963	UH1F	to USFS	N489DF
65-7964	UH1F	w.o.19.4.71	
65-7965	UH1F	37RQS	det..10
66-1211	UH1P		
66-1212	UH1F	to USFS	N490DF
66-1213	UH1F	to USFS	N491DF
66-1214	UH1F	w.o.	
66-1215	UH1F	preserved Minot AFB	
66-1216	UH1F	to N90413	
66-1217	UH1F	w.o.	
66-1218	UH1F	w.o.	
66-1219	GUH1F	GIA	Kirtland AFB
66-1220	UH1P	AMARC	HF...
66-1221	UH1F		
66-1222	UH1F	to N18LA	
66-1223	UH1F	37RQS	det.10
66-1224	UH1F	AMARC	HF150
66-1225	TH1F	AMARC	HF154
66-1226	UH1F	w.o.	
66-1227	TH1F	AMARC	HF165
66-1228	UH1F	AMARC	HF151
66-1229	UH1F	AMARC	HF167
66-1230	TH1F		
66-1231	TH1F	AMARC	HF169
66-1232	TH1F	AMARC	HF172
66-1233	TH1F	AMARC	HF155
66-1234	TH1F	AMARC	HF...
66-1235	UH1F	AMARC	HF...
66-1236	TH1F	AMARC	HF...
66-1237	TH1F	AMARC	HF173
66-1238	UH1F	AMARC	HF...
66-1239	UH1F	to N38921	
66-1240	TH1F	AMARC	HF152
66-1241	TH1F	MASDC	
66-1242	UH1F	w.o.	
66-1243	UH1F	w.o.	
66-1244	UH1F	AMARC	HF...
66-1245	TH1F	MASDC	
66-1246	UH1F	AMARC	HF...
66-1247	UH1F	AMARC	HF153
66-1248	UH1F	to N67163	
66-1249	UH1F	AMARC	HF156
66-1250	UH1F	to N55781	
68-10772	GUH1N	TTC	Sheppard
68-10773	HH1N		
68-10774	HH1N		
68-10775	HH1N		
68-10776	HH1N	412TW	ED AFFTC
69-6600	HH1N		
69-6601	HH1N		
69-6602	HH1N		
69-6603	UH1N	542CTW	
69-6604	HH1N	89AW	
69-6605	HH1N	1FW	FF
69-6606	UH1N	86W	
69-6607	UH1N	86W	
69-6608	UH1N	86W	
69-6609	UH1N	86W	
69-6610	UH1N	542CTW	
69-6611	HH1N		
69-6612	UH1N	1FW	FF
69-6613	HH1N	57FW	WAdet.1
69-6614	HH1N	1FW	FF
69-6615	HH1N		
69-6616	HH1N	37RQS	det.24
69-6617	HH1N	46TW	ET AFMC\MSD
69-6618	HH1N	89AW	
69-6619	HH1N	86W	
69-6620	HH1N		
69-6621	HH1N	w.o.	
69-6622	HH1N		
69-6623	HH1N	76RQS	
69-6624	HH1N	w.o.	
69-6625	HH1N		
69-6626	HH1N	46TW	ET AFMC\MSD
69-6627	HH1N	76RQS	
69-6628	HH1N		
69-6629	HH1N		
69-6630	HH1N		
69-6631	HH1N	w.o.	
69-6632	HH1N	374AW	YJ
69-6633	HH1N	89AW	
69-6634	HH1N	412TW	ED AFFTC
69-6635	HH1N	412TW	ED AFFTC
69-6636	HH1N	89AW	
69-6637	VH1N	89AW	
69-6638	HH1N	76RQS	
69-6639	HH1N	374AW	YJ
69-6640	HH1N		
69-6641	HH1N	57FW WA	det.1
69-6642	HH1N	89AW	
69-6643	HH1N	w.o.	
69-6644	HH1N	w.o.	
69-6645	HH1N	374AW	YJ
69-6646	HH1N	374AW	YJ
69-6647	HH1N	ATC Survey School	
69-6648	HH1N		
69-6649	UH1N	542CTW	
69-6650	UH1N	542CTW	
69-6651	HH1N	w.o.	
69-6652	HH1N	76RQS	
69-6653	HH1N	412TW	ED AFFTC
69-6654	HH1N	89AW	
69-6655	VH1N	89AW	
69-6656	VH1N	89AW	
69-6657	VH1N	89AW	
69-6658	VH1N	89AW	
69-6659	HH1N		
69-6660	HH1N	76RQS	
69-6661	HH1N	89AW	
69-6662	HH1N	1FW	FF
69-6663	HH1N	89AW	
69-6664	HH1N		
69-6665	UH1N	542CTW	
69-6666	HH1N	76RQS	

69-6667	VH1N	89AW	
69-6668	VH1N	89AW	
69-6669	VH1N	89AW	
69-6670	HH1N		
69-7536	HH1N	89AW	
69-7537	HH1N	89AW	
69-7538	VH1N	89AW	
70-2457	HH1H	37RQS	det.9
70-2458	HH1H	37RQS	det.9
70-2459	HH1H		
70-2460	HH1H		
70-2461	HH1H	37RQS	det.3
70-2462	HH1H	37RQS	det.3
70-2463	HH1H	37RQS	det.3
70-2464	HH1H		
70-2465	HH1H	37RQS	det.3
70-2466	HH1H	37RQS	det.7
70-2467	HH1H	37RQS	det.2
70-2468	HH1H	412TW	ED AFFTC
70-2469	HH1H	37RQS	det.3
70-2470	HH1H	545TG	
70-2471	HH1H	37RQS	det.9
70-2472	HH1H		
70-2473	HH1H	37RQS	det.2
70-2474	HH1H	37RQS	det.2
70-2475	HH1H	37RQS	det.2
70-2476	HH1H	37RQS	det.7
70-2477	HH1H	37RQS	det.7
70-2478	HH1H		
70-2479	HH1H		
70-2480	HH1H		
70-2481	HH1H		
70-2482	HH1H		
70-2483	HH1H	37RQS	det.7
70-2484	HH1H	545TG	
70-2485	HH1H		
70-2486	HH1H	37RQS	det.7
84-0474	GUH1E	ex USN	

H3 SEA KING

62-12571	CH3B	to USN	
62-12572	CH3B	w.o.	
62-12573	CH3B	to N1049G	
62-12574	CH3B	to USN	
62-12575	CH3B	to USN	
62-12576	CH3B	to N1048Y	
62-12577	JCH3C	w.o.	
62-12578	CH3E	to USCG	2578
62-12579	CH3E		
62-12580	JCH3E	to USCG	
62-12581	JCH3E		
62-12582	CH3C	w.o.	
63-9676	CH3E	USAF	Museum
63-9677	GCH3C	TTC	Sheppard
63-9678	CH3C	w.o.	
63-9679	CH3E	to USCG	
63-9680	CH3C	w.o.	
63-9681	CH3C	w.o.	
63-9682	CH3C	w.o.	
63-9683	CH3E	stored	Perkasie, Pa
63-9684	CH3C	w.o.	
63-9685	CH3C	w.o.	
63-9686	CH3E	stored	Perkasie, Pa
63-9687	CH3C	AMARC	HH027
63-9688	CH3C	scrapped	
63-9689	CH3C	w.o.	
63-9690	JCH3E	USArmy	USAATC
63-9691	CH3E	to USCG	9691
64-14221	JCH3E	US Army	USAATC
64-14222	CH3C	w.o.	
64-14223	CH3E		
64-14224	CH3E	to USCG	
64-14225	CH3E		
64-14226	CH3E	301RQS	AFRes
64-14227	CH3C	w.o.	
64-14228	GCH3E	TTC	Sheppard
64-14229	CH3B	w.o.	
64-14230	JHH3E	US Army	USAATC
64-14231	CH3B	w.o.	
64-14232	HH3E		
64-14233	CH3B	w.o.	
64-14234	CH3E	to USCG	
64-14235	CH3E	to USCG	
64-14236	CH3B	w.o.	
64-14237	CH3E	w.o.	
65-5690	HH3E	preserved	McClellan AFB
65-5691	CH3B	w.o.	
65-5692	CH3E	preserved	Davis Monthan AFB
65-5693	CH3E	AMARC	HH022
65-5694	CH3B	w.o.	
65-5695	CH3C	to USCG	
65-5696	CH3E	AMARC	HH021
65-5697	CH3E	to USCG	
65-5698	JCH3E	US Army	USAATC
65-5699	JCH3E	US Army	USAATC
65-5700	CH3E	301RQS	AFRes
65-12777	JHH3E	US Army	USAATC
65-12778	CH3C	w.o.	
65-12779	CH3C	w.o.	
65-12780	HH3E		
65-12781	HH3E	w.o.1.9.88	
65-12782	HH3E		
65-12783	CH3E	301RQS	AFRes
65-12784	HH3E		
65-12785	HH3E	w.o.	
65-12786	CH3C	w.o.	
65-12787	JHH3E	US Army	USAATC
65-12788	CH3E	to USCG	
65-12789	CH3E	to USCG	
65-12790	CH3E		
65-12791	CH3E	to USCG	2791
65-12792	JCH3E	US Army	USAATC
65-12793	CH3E	to USCG	2793
65-12794	HH3E	w.o.	
65-12795	JCH3E	US Army	USAATC
65-12796	JCH3E	US Army	USAATC
65-12797	CH3E		
65-12798	CH3E		
65-12799	CH3E		
65-12800	JCH3E	US Army	USAATC
66-13284	HH3E		
66-13285	JHH3E	US Army	USAATC
66-13286	JHH3E	US Army	USAATC
66-13287	CH3E	w.o.	
66-13288	CH3E	w.o.	
66-13289	CH3E	w.o.	
66-13290	CH3E		
66-13291	HH3E		
66-13292	HH3E		
66-13293	CH3E	w.o.	
66-13294	CH3E	w.o.	
66-13295	CH3E	w.o.	
66-13296	JHH3E	US Army	USAATC
67-14702	CH3E	w.o.	
67-14703	CH3E		
67-14704	HH3E		
67-14705	HH3E	AMARC	HH...
67-14706	HH3E		
67-14707	CH3E		
67-14708	CH3E	w.o.	

Serial	Type	Unit/Notes	Base
67-14709	HH3E		
67-14710	CH3E	w.o.	
67-14711	HH3E		
67-14712	GHH3E	TTC	Sheppard
67-14713	HH3E		
67-14714	HH3E	542CTW	
67-14715	JHH3E	US Army	USAATC
67-14716	HH3E	w.o.18.12.79	
67-14717	HH3E		
67-14718	CH3E		
67-14719	HH3E		
67-14720	HH3E		
67-14721	CH3E	w.o.	
67-14722	HH3E		
67-14723	HH3E	542CTW	
67-14724	HH3E		
67-14725	HH3E	dump	Mather AFB
68-8282	HH3E	w.o.	
69-5798	HH3		
69-5799	JHH3E	US Army	USAATC
69-5800	HH3E	w.o.13.6.78	
69-5801	JHH3E	US Army	USAATC
69-5802	HH3E		
69-5803	HH3E		
69-5804	HH3E		
69-5805	JHH3E	US Army	USAATC
69-5806	HH3E	542CTW	
69-5807	HH3E	w.o.	
69-5808	HH3E		
69-5809	HH3E		
69-5810	HH3E		
69-5811	CH3E		
69-5812	JHH3E	US Army	USAATC

H53 'JOLLY GREEN GIANT'

Serial	Type	Unit/Notes	Base
63-13693	NCH53A	545TG	
63-13694	NCH53A	545TG	
66-14428	MH53J	542CTW	
66-14429	MH53J	542CTW	
66-14430	HH53B	w.o.18.7.80	
66-14431	MH53J	21SOS	
66-14432	MH53J	20SOS	
66-14433	MH53J	542CTW	
66-14434	HH53B	w.o.	
66-14435	HH53B	dump	Kirtland AFB
66-14468	TH53A	542CTW	
66-14469	TH53A		
66-14470	TH53A	542CTW	
66-14471	TH53A	542CTW	
66-14472	TH53A	542CTW	
66-14473	TH53A		
67-14993	HH53C	545TG	
67-14994	HH53C		
67-14995	MH53J	20SOS	
67-14996	HH53C	w.o.8.10.68	
68-8283	HH53C	w.o.	
68-8284	MH53J		
68-8285	HH53C	w.o.	
68-8286	MH53J	20SOS	
68-10354	HH53C	545TG	
68-10355	HH53C	w.o.15.1.85	
68-10356	HH53C		
68-10357	MH53J	21SOS	
68-10358	MH53J	20SOS	
68-10359	HH53C	w.o.	
68-10360	HH53C		
68-10361	HH53C	w.o.	
68-10362	HH53C	w.o.	
68-10363	MH53J	20SOS	
68-10364	MH53J	20SOS	
68-10365	HH53C	w.o.	
68-10366	HH53C	w.o.	
68-10367	MH53J	20SOS	
68-10368	HH53C	w.o.24.6.77	
68-10369	HH53C		
68-10922	CH53C	w.o.	
68-10923	MH53J	20SOS	
68-10924	CH53C		
68-10925	CH53C	w.o.14.5.75	
68-10926	CH53C	w.o.14.5.75	
68-10927	CH53C	w.o.17.3.76	
68-10928	CH53C		
68-10929	CH53C	w.o.	
68-10930	MH53J	20SOS	
68-10931	CH53C	w.o.	
68-10932	CH53C		
68-10933	CH53C	w.o.13.5.75	
69-5784	MH53J	21SOS	
69-5785	HH53C		
69-5786	HH53C	w.o.	
69-5787	HH53C	w.o.	
69-5788	HH53C	w.o.	
69-5789	MH53J	21SOS	
69-5790	MH53J	20SOS	
69-5791	MH53J	20SOS	
69-5792	HH53C	w.o.28.7.82	
69-5793	MH53J	542CTW	
69-5794	HH53C		
69-5795	MH53J	20SOS	
69-5796	MH53J	20SOS	
69-5797	MH53J	20SOS	
70-1625	CH53C		
70-1626	MH53J	21SOS	
70-1627	CH53C	w.o.14.5.75	
70-1628	CH53C	w.o.	
70-1629	MH53J	21SOS	
70-1630	MH53J	20SOS	
70-1631	MH53J	20SOS	
70-1632	CH53C	w.o.26.9.75	
73-1647	HH53H	w.o.	
73-1648	HH53C	w.o.25.9.87	
73-1649	CH53C	20SOS	
73-1650	CH53C	w.o.17.10.84	
73-1651	CH53C	20SOS	
73-1652	MH53J	542CTW	

H60 BLACKHAWK

81-23643	HH60G	542CTW		89-26208	HH60G		
81-23644	HH60G	55SOS		89-26209	HH60G		
81-23645	HH60G	55SOS		89-26210	HH60G		
81-23646	HH60G	542CTW		89-26211	HH60G	39RQS	
82-23671	HH60G	542CTW		89-26212	HH60G		
82-23680	HH60G	55SOS		90-26222	HH60G	304RQS	AFRes
82-23689	HH60G	55SOS		90-26223	HH60G	301RQS	AFRes
82-23708	HH60G	542CTW		90-26224	HH60G	71SOS	AFRes
82-23718	HH60G	55SOS		90-26225	HH60G	71SOS	AFRes
82-23728	HH60G	55SOS		90-26226	HH60G	71SOS	AFRes
87-26006	HH60G	55SOS		90-26227	HH60G	71SOS	AFRes
87-26007	HH60G	55SOS		90-26228	HH60G	71SOS	AFRes
87-26008	HH60G	55SOS		90-26229	HH60G	542CTW	
87-26009	HH60G	55SOS		90-26230	HH60G	301RQS	AFRes
87-26010	HH60G	55SOS		90-26231	HH60G	301RQS	AFRes
87-26011	HH60G	55SOS		90-26232	HH60G	301RQS	AFRes
87-26012	HH60G	55SOS		90-26233	HH60G	542CTW	
87-26013	HH60G	55SOS		90-26234	HH60G	41RQS	
87-26014	HH60G	55SOS		90-26235	HH60G		
87-26018	HH60G	542CTW		90-26236	HH60G	71SOS	AFRes
88-26105	HH60G	210RQS	AK ANG	90-26237	HH60G	41RQS	
88-26106	HH60G	210RQS	AK ANG	90-26238	HH60G	41RQS	
88-26107	HH60G	210RQS	AK ANG	90-26239	HH60G	542CTW	
88-26108	HH60G	102RQS	NY ANG	90-26309	HH60G	66RQS	
88-26109	HH60G	210RQS	AK ANG	90-26310	HH60G	66RQS	
88-26110	HH60G	w.o.31.10.91		90-26311	HH60G	66RQS	
88-26111	HH60G			90-26312	HH60G	66RQS	
88-26112	HH60G	102RQS	NY ANG	91-26352	HH60G		
88-26113	HH60G			91-26353	HH60G		
88-26114	HH60G			91-26354	HH60G		
88-26115	HH60G	129RQS	CA ANG	91-26355	HH60G	55SOS	
88-26116	HH60G	w.o.29.10.92		91-26356	HH60G	55SOS	
88-26117	HH60G			91-26357	HH60G	66RQS	
88-26118	HH60G			91-26358	HH60G	66RQS	
88-26119	HH60G			91-26359	HH60G	66RQS	
89-26120	HH60G			92-26401	HH60G		
89-26195	HH60G	304RQS	AFRes	92-26402	HH60G	66RQS	
89-26196	HH60G	304RQS	AFRes	92-26403	HH60G	66RQS	
89-26197	HH60G	304RQS	AFRes	92-26404	HH60G		
89-26198	HH60G			92-26405	HH60G		
89-26199	HH60G	304RQS	AFRes	92-26406	HH60G		
89-26200	HH60G	301RQS	AFRes	92-26407	HH60G		
89-26201	HH60G			92-26460	HH60G	71SOS	AFRes
89-26202	HH60G	301RQS	AFRes	92-26461	HH60G	71SOS	AFRes
89-26203	HH60G			92-26462	HH60G	55SOS	
89-26204	HH60G	56RQS		xx-26463	HH60G	304RQS	AFRes
89-26205	HH60G	56RQS		xx-26464	HH60G	304RQS	AFRes
89-26206	HH60G	56RQS		xx-26465	HH60G	41RQS	
89-26207	HH60G						

O2 SUPER SKYMASTER

c\n	O2A:	337M-0001to337M-0467.				
	O2B:	31 purchased from civil stocks.				
67-21295	O2A			67-21300	O2A	to N590D
67-21296	O2A			67-21301	O2A	w.o.20.1.69
67-21297	O2A	MASDC		67-21302	O2A	w.o.24.3.68
67-21298	O2A			67-21303	O2A	AMARC	HV178
67-21299	O2A	AMARC	HV150	67-21304	O2A	

Serial	Type	Disposition	Notes
67-21305	O2A	w.o.2.5.69	
67-21306	O2A	w.o.20.5.71	
67-21307	O2A	to N1102A	
67-21308	O2A	w.o.14.12.68	
67-21309	O2A	AMARC	HV156
67-21310	O2A	AMARC	HV116
67-21311	O2A	to N311J	
67-21312	O2A		
67-21313	O2A	AMARC	HV120
67-21314	O2A	w.o.14.9.67	
67-21315	O2A	w.o.1.9.68	
67-21316	O2A	w.o.2.6.69	
67-21317	O2A	w.o.6.8.67	
67-21318	O2A	to USN	
67-21319	O2A	w.o.1.6.69	
67-21320	O2A		
67-21321	O2A	AMARC	HV179
67-21322	O2A	w.o.6.8.67	
67-21323	O2A		
67-21324	O2A	AMARC	HV158
67-21325	O2A	w.o.6.6.69	
67-21326	O2A	preserved	Dyess AFB
67-21327	O2A	w.o.17.1.68	
67-21328	O2A	w.o.3.11.67	
67-21329	O2A	w.o.9.5.69	
67-21330	O2A	AMARC	HV143
67-21331	O2A	USAF	Museum
67-21332	O2A	w.o.2.8.69	
67-21333	O2A	w.o.12.5.68	
67-21334	O2A	to N256Z	
67-21335	O2A	w.o.24.5.68	
67-21336	O2A	w.o.12.5.68	
67-21337	O2A	w.o.25.12.67	
67-21338	O2A	w.o.20.3.68	
67-21339	O2A	w.o.1.2.68	
67-21340	O2A	preserved	Selfridge ANGB
67-21341	O2A	w.o.11.5.70	
67-21342	O2A	AMARC	HV159
67-21343	O2A	w.o.8.10.69	
67-21344	O2A		
67-21345	O2A	AMARC	HV216
67-21346	O2A	AMARC	HV127
67-21347	O2A	w.o.1.2.68	
67-21348	O2A	w.o.23.8.68	
67-21349	O2A	to USAICS	
67-21350	O2A	w.o.18.1.69	
67-21351	O2A	w.o.30.1.78	
67-21352	O2A	AMARC	HV126
67-21353	O2A	AMARC	HV138
67-21354	O2A	AMARC	HV111
67-21355	O2A	AMARC	HV164
67-21356	O2A	AMARC	HV133
67-21357	O2A	w.o.1.2.68	
67-21358	O2A	w.o.13.2.70	
67-21359	O2A	w.o.1.2.68	
67-21360	O2A	AMARC	HV115
67-21361	O2A	w.o.25.5.69	
67-21362	O2A	w.o.11.2.68	
67-21363	O2A	AMARC	HV224
67-21364	O2A		
67-21365	O2A	to USN	
67-21366	O2A	AMARC	HV176
67-21367	O2A	w.o.28.4.68	
67-21368	O2A	preserved	Hurlburt Field
67-21369	O2A	w.o.12.12.67	
67-21370	O2A	w.o.10.5.68	
67-21371	O2A	AMARC	HV177
67-21372	O2A	AMARC	HV225
67-21373	O2A	AMARC	HV119
67-21374	O2A	to N19GH	
67-21375	O2A	to N16GH	
67-21376	O2A		
67-21377	O2A	to N58BB	
67-21378	O2A	w.o.17.11.68	
67-21379	O2A		
67-21380	O2A	AMARC	HV...
67-21381	O2A	to Vietnam	
67-21382	O2A	w.o.21.5.68	
67-21383	O2A	AMARC	HV160
67-21384	O2A	w.o.17.7.68	
67-21385	GO2A	GIA	Shaw AFB
67-21386	O2A	w.o.6.3.69	
67-21387	O2A	AMARC	HV130
67-21388	O2A	w.o.26.4.71	
67-21389	O2A	w.o.20.5.71	
67-21390	O2A	w.o.25.12.67	
67-21391	O2A		
67-21392	O2A	AMARC	HV180
67-21393	O2A	AMARC	HV217
67-21394	O2A	w.o.23.10.68	
67-21395	GO2A	TTC	Sheppard
67-21396	O2A		
67-21397	O2A	AMARC	HV183
67-21398	O2A	AMARC	HV144
67-21399	O2A	w.o.9.5.68	
67-21400	O2A		
67-21401	O2A	w.o.11.2.68	
67-21402	O2A	to N202A	
67-21403	O2A	AMARC	HV112
67-21404	O2A	to USN	
67-21405	O2A	w.o.30.5.68	
67-21406	O2A	w.o.1.4.68	
67-21407	O2A	AMARC	HV128
67-21408	O2A	w.o.16.3.68	
67-21409	O2A	w.o.28.9.70	
67-21410	O2A	AMARC	HV161
67-21411	O2A	preserved	Chanute AFB
67-21412	O2A	w.o.28.2.69	
67-21413	O2A	preserved	Castle AFB
67-21414	O2A	to USAICS	
67-21415	O2A	w.o.12.6.68	
67-21416	O2A	AMARC	HV219
67-21417	O2A	AMARC	HV173
67-21418	O2A		
67-21419	O2A	w.o.26.11.68	
67-21420	O2A	w.o.21.4.68	
67-21421	O2A	AMARC	HV114
67-21422	O2A	preserved	Ellsworth AFB
67-21423	O2A		
67-21424	O2A	AMARC	HV154
67-21425	O2A	w.o.9.3.69	
67-21426	O2A	preserved	Volk Field ANGB
67-21427	O2A	to N427A	
67-21428	O2A	w.o.12.12.70	
67-21429	O2A	w.o.3.2.75	
67-21430	O2A		
67-21431	O2A	to N3371C	
67-21432	O2A	w.o.2.2.69	
67-21433	O2A	AMARC	HV131
67-21434	O2A		
67-21435	O2A	AMARC	HV118
67-21436	O2A	to N436Z	
67-21437	O2A	w.o.16.4.69	
67-21438	GO2A	TTC	Sheppard
67-21439	O2A	w.o.30.10.71	
67-21440	O2B	preserved	Barksdale AFB
67-21441	O2B	to N25896	
67-21442	O2B		
67-21443	O2B	MASDC	
67-21444	O2B	scrapped	10.77
67-21445	O2B	to N47784	
67-21446	O2B	MASDC	
67-21447	O2B	to N14536	
67-21448	O2B	dump	Fresno, Ca
67-21449	O2B	to N67136	
67-21450	O2B	to N87755	
67-21451	O2B		
67-21452	O2B		
67-21453	O2B	to N55591	
67-21454	O2B	AMARC	HV162
67-21455	O2B	w.o.22.6.69	
67-21456	O2B	stored	Fresno, Ca
67-21457	O2B	w.o.8.10.68	
67-21458	O2B		
67-21459	O2B	to N67153	
67-21460	O2B		
67-21461	O2B		
67-21462	O2B	stored	March AFB
67-21463	O2B	stored	Fresno, Ca
67-21464	O2B		
67-21465	O2B	preserved	March AFB
67-21466	O2B	MASDC	
67-21467	O2B	to N31416	
67-21468	O2B	stored	Fresno, Ca
67-21469	GO2B	TTC	Sheppard
67-21470	O2B	AMARC	HV152

68-6857	O2A		
68-6858	O2A		
68-6859	O2A	w.o.22.11.68	
68-6860	O2A	w.o.16.3.71	
68-6861	O2A	to USCS	
68-6862	O2A		
68-6863	O2A	w.o.10.1.70	
68-6864	O2A	preserved	Eglin AFB
68-6865	O2A	preserved	Kelly AFB
68-6866	O2A		
68-6867	O2A		
68-6868	O2A		
68-6869	O2A	w.o.19.6.69	
68-6870	O2A	w.o.19.10.68	
68-6871	O2A	preserved	Grissom AFB
68-6872	O2A	to N902A	
68-6873	O2A	AMARC	HV117
68-6874	O2A	AMARC	HV186
68-6875	O2A	AMARC	HV123
68-6876	O2A	AMARC	HV147
68-6877	O2A	AMARC	HV153
68-6878	O2A		
68-6879	O2A	w.o.26.6.68	
68-6880	O2A	AMARC	HV121
68-6881	O2A	AMARC	HV122
68-6882	O2A	to USCS	
68-6883	O2A	w.o.23.10.68	
68-6884	O2A	w.o.11.3.70	
68-6885	O2A	to N802A	
68-6886	O2A		
68-6887	O2A		
68-6888	O2A		
68-6889	O2A	AMARC	HV188
68-6890	O2A		
68-6891	O2A	w.o.30.7.68	
68-6892	O2A	AMARC	HV149
68-6893	O2A	AMARC	HV137
68-6894	O2A		
68-6895	O2A	AMARC	HV140
68-6896	O2A	AMARC	HV133
68-6897	O2A	AMARC	HV144
68-6898	O2A	GIA	Howard AB
68-6899	O2A	to N5169Y	
68-6900	O2A	AMARC	HV191
68-6901	O2A	GIA	Howard AB
68-6902	O2A	w.o.17.3.81	
68-6903	O2A	AMARC	HV189
68-10828	O2A		
68-10829	O2A	AMARC	HV276
68-10830	O2A	AMARC	HV195
68-10831	O2A	AMARC	HV155
68-10832	O2A	AMARC	HV163
68-10833	O2A	AMARC	HV124
68-10834	O2A		
68-10835	O2A		
68-10836	O2A		
68-10837	O2A		
68-10838	O2A	AMARC	HV...
68-10839	O2A	w.o.1.10.70	
68-10840	O2A	AMARC	HV207
68-10841	O2A	w.o.31.1.71	
68-10842	O2A	w.o.2.4.72	
68-10843	O2A	to Vietnam	
68-10844	O2A	to USFS	N468DF
68-10845	O2A	to Vietnam	
68-10846	O2A	to Vietnam	
68-10847	O2A	to Vietnam	
68-10848	O2A	preserved	Castle AFB
68-10849	O2A	AMARC	HV277
68-10850	O2A	to Vietnam	
68-10851	O2A	w.o.23.7.69	
68-10852	O2A	w.o.7.11.69	
68-10853	O2A	preserved	Hill AFB
68-10854	O2A	w.o.14.9.69	
68-10855	O2A	to Vietnam	
68-10856	O2A	to Vietnam	
68-10857	O2A	AMARC	HV211
68-10858	O2A		
68-10859	O2A	to Vietnam	
68-10860	O2A		
68-10861	O2A	to USFS	N465DF
68-10862	O2A		
68-10863	O2A	to Vietnam	
68-10864	O2A	w.o.18.1.70	
68-10865	O2A	AMARC	HV212
68-10866	O2A		
68-10867	O2A		
68-10868	O2A	AMARC	HV279
68-10869	O2A	AMARC	HV255
68-10870	O2A	AMARC	HV256
68-10871	O2A	w.o.	
68-10872	O2A	AMARC	HV248
68-10962	O2A	preserved	Shaw AFB
68-10963	O2A	stored	Wheeler AFB
68-10964	O2A	to N326MT	
68-10965	O2A	w.o.5.11.70	
68-10966	O2A	stored	Wheeler AFB
68-10967	O2A	AMARC	HV165
68-10968	O2A	to N325MT	
68-10969	O2A	w.o.7.10.70	
68-10970	O2A	to USFS	N471DF
68-10971	O2A	AMARC	HV187
68-10972	O2A	stored	Wheeler AFB
68-10973	O2A	to Vietnam	
68-10974	O2A		
68-10975	O2A	w.o.21.10.69	
68-10976	O2A	AMARC	HV145
68-10977	O2A	AMARC	HV171
68-10978	O2A	to USFS	N470DF
68-10979	O2A		
68-10980	O2A	w.o.17.8.72	
68-10981	O2A	w.o.5.6.72	
68-10982	O2A	w.o.26.6.72	
68-10983	O2A	AMARC	HV220
68-10984	O2A		
68-10985	O2A		
68-10986	O2A		
68-10987	O2A	w.o.12.2.69	
68-10988	O2A	to Vietnam	
68-10989	O2A	AMARC	HV181
68-10990	O2A		
68-10991	O2A	to Vietnam	
68-10992	O2A	w.o.5.9.70	
68-10993	O2A	to N48292	
68-10994	O2A	to Vietnam	
68-10995	O2A	w.o.6.10.69	
68-10996	O2A	to Vietnam	
68-10997	O2A		
68-10998	O2A		
68-10999	O2A	w.o.24.12.69	
68-11000	O2A	w.o.11.5.72	
68-11001	O2A	w.o.19.2.71	
68-11002	O2A	w.o.22.5.70	
68-11003	O2A	AMARC	HV230
68-11004	O2A	w.o.11.5.72	
68-11005	O2A	preserved	Eielson AFB
68-11006	O2A		
68-11007	O2A	to Vietnam	
68-11008	O2A	to Vietnam	
68-11009	O2A	to Vietnam	
68-11010	O2A	to Vietnam	
68-11011	O2A	to Vietnam	
68-11012	O2A		
68-11013	O2A	to N64675	
68-11014	O2A	to USFS	N467DF
68-11015	O2A	w.o.25.6.72	
68-11016	O2A		
68-11017	O2A	to USFS	N470DF
68-11018	O2A	to N48233	
68-11019	O2A	to Vietnam	
68-11020	O2A	to USFS	N473DF
68-11021	O2A	to Vietnam	
68-11022	O2A		
68-11023	O2A	to Vietnam	
68-11024	O2A	to Vietnam	
68-11025	O2A	to Vietnam	
68-11026	O2A	w.o.27.6.72	
68-11027	O2A		
68-11028	O2A	AMARC	HV184
68-11029	O2A	to N256Y	
68-11030	O2A	w.o.16.8.71	
68-11031	O2A	stored	Fresno, Ca
68-11032	O2A	to USAICS	(spares)
68-11033	O2A	w.o.29.5.70	
68-11034	O2A	to N48330	
68-11035	O2A	to Vietnam	

Serial	Type	Status	Notes
68-11036	O2A		
68-11037	O2A	to Vietnam	
68-11038	O2A	w.o.9.9.71	
68-11039	O2A	to Vietnam	
68-11040	O2A	w.o.1.7.72	
68-11041	O2A		
68-11042	O2A	to USFS	N474DF
68-11043	O2A	to Vietnam	
68-11044	O2A	to USFS	N466DF
68-11045	O2A	to USFS	N472DF
68-11046	O2A	AMARC	HV172
68-11047	O2A	to USFS	N469DF
68-11048	O2A		
68-11049	O2A	to USFS	N463DF
68-11050	O2A	AMARC	HV170
68-11051	O2A		
68-11052	O2A		
68-11053	O2A		
68-11054	O2A	to N45715	
68-11055	O2A	to N5539F	
68-11056	O2A	w.o.24.5.72	
68-11057	O2A	AMARC	HV166
68-11058	O2A	w.o.4.4.70	
68-11059	O2A		
68-11060	O2A	to Vietnam	
68-11061	O2A	stored	Fresno, Ca
68-11062	O2A	to Vietnam	
68-11063	O2A	to USFS	N464DF
68-11064	O2A	w.o.2.5.70	
68-11065	O2A	w.o.23.11.72	
68-11066	O2A	to Vietnam	
68-11067	O2A	to USFS	N475DF
68-11068	O2A	AMARC	HV268
68-11069	O2A	to Vietnam	
68-11070	O2A		
68-11122	O2A		
68-11123	O2A	AMARC	HV263
68-11124	O2A	to N57BB	
68-11125	O2A	AMARC	HV262
68-11126	O2A	AMARC	HV267
68-11127	O2A		
68-11128	O2A		
68-11129	O2A		
68-11130	O2A		
68-11131	O2A		
68-11132	O2A		
68-11133	O2A		
68-11134	O2A	stored	Wheeler AFB
68-11135	O2A	AMARC	HV265
68-11136	O2A		
68-11137	O2A		
68-11138	O2A		
68-11139	O2A		
68-11140	O2A		
68-11141	O2A		
68-11142	O2A	stored	Wheeler AFB
68-11143	O2A		
68-11144	O2A	preserved	Wheeler AFB
68-11145	O2A	stored	Wheeler AFB
68-11146	O2A	stored	Wheeler AFB
68-11147	O2A		
68-11148	O2A	to N59327	
68-11149	O2A	to N8481L	
68-11150	O2A	to N400TH	
68-11151	O2A	w.o.6.9.74	
68-11152	O2A	AMARC	HV168
68-11153	O2A	to N21GH	
68-11154	O2A		
68-11155	O2A	AMARC	HV196
68-11156	O2A	preserved	Hickam AFB
68-11157	O2A	AMARC	HV203
68-11158	O2A	AMARC	HV266
68-11159	O2A		
68-11160	O2A	preserved	Gtr.PeoriaAP, Il
68-11161	O2A		
68-11162	O2A		
68-11163	O2A	AMARC	HV269
68-11164	O2A	preserved	Lackland AFB
68-11165	O2A	to USFS	N454DF
68-11166	O2A	preserved	Hickam AFB
68-11167	O2A	AMARC	HV264
68-11168	O2A	AMARC	HV208
68-11169	O2A	AMARC	HV174
68-11170	O2A		
68-11171	O2A	AMARC	HV198
68-11172	O2A	AMARC	HV199
68-11173	O2A	AMARC	HV193
69-7601	O2A	AMARC	HV287
69-7602	O2A	preserved	Kelly AFB
69-7603	O2A	to USFS	N455DF
69-7604	O2A	AMARC	HV251
69-7605	O2A	AMARC	HV243
69-7606	O2A	AMARC	HV252
69-7607	O2A	AMARC	HV244
69-7608	O2A	AMARC	HV190
69-7609	O2A	AMARC	HV245
69-7610	O2A	w.o.26.1.78	
69-7611	O2A	AMARC	HV232
69-7612	O2A	AMARC	HV246
69-7613	O2A		
69-7614	O2A		
69-7615	O2A	AMARC	HV258
69-7616	O2A	w.o.2.11.82	
69-7617	O2A	to N502A	
69-7618	O2A	to USCS	
69-7619	O2A		
69-7620	O2A		
69-7621	O2A		
69-7622	O2A	to N7654A	
69-7623	O2A	AMARC	HV226
69-7624	O2A	AMARC	HV234
69-7625	O2A	AMARC	HV221
69-7626	O2A	AMARC	HV239
69-7627	O2A	AMARC	HV240
69-7628	O2A	AMARC	HV259
69-7629	O2A		
69-7630	O2A	preserved	Rick'backerANGB
69-7631	O2A	AMARC	HV175
69-7632	O2A	AMARC	HV223
69-7633	O2A		
69-7634	O2A		
69-7635	O2A	AMARC	HV...
69-7636	O2A	AMARC	HV257
69-7637	O2A	AMARC	HV182
69-7638	O2A	AMARC	HV286
69-7639	O2A	AMARC	HV272
69-7640	O2A	AMARC	HV254
69-7641	O2A	AMARC	HV270
69-7642	O2A	AMARC	HV278
69-7643	O2A	AMARC	HV235
69-7644	O2A	AMARC	HV229
69-7645	O2A	AMARC	HV275
69-7646	O2A	AMARC	HV271
69-7647	O2A		
69-7648	O2A		
69-7649	O2A	AMARC	HV236
69-7650	O2A	AMARC	HV241
69-7651	O2A	AMARC	HV260
69-7652	O2A		
69-7653	O2A	AMARC	HV261
69-7654	O2A	AMARC	HV...
69-7655	O2A	AMARC	HV237
69-7656	O2A	AMARC	HV110
69-7657	O2A	AMARC	HV285
69-7658	O2A		
69-7659	O2A	to N370SD	
69-7660	O2A		
69-7661	O2A	to USCS	
69-7662	O2A	AMARC	HV247
69-7663	O2A	AMARC	HV242
69-7664	O2A	AMARC	HV...
69-7665	O2A	AMARC	HV233
69-7666	O2A	AMARC	HV238
69-7667	O2A		
69-7668	O2A		
69-7669	O2A	to N51680	

SR71 'BLACKBIRD'

c\n A11A: all unknown.
     YF12A: all unknown.
     SR71A: 2001 to 2032.

60-6924	A11A	preserved	Palmdale, Ca
60-6925	A11A	stored	Palmdale, Ca
60-6926	A11A	w.o.24.5.63	
60-6927	A11A	stored	Palmdale, Ca
60-6928	A11A	w.o.1.5.67	
60-6929	A11A	w.o. . .65	
60-6930	A11A	stored	Palmdale, Ca
60-6931	A11A	preserved	Minneapolis AP, Mn
60-6932	A11A	w.o.5.6.63	
60-6933	A11A	preserved	San Diego, Ca
60-6934	YF12A	to SR71C	64-17981
60-6935	YF12A	USAF	Museum
60-6936	YF12A	w.o.24.6.71	
60-6937	YF12A	to SR71A	64-17951
60-6938	A11A	stored	Palmdale, Ca
60-6939	A11A	w.o.1965	
60-6940	A11A	preserved	Boeing Field, Wa
60-6941	A11A	w.o.30.7.66	
64-17950	SR71A	w.o.10.1.67	
64-17951	SR71A	stored	Palmdale, Ca
64-17952	SR71A	w.o.25.1.66	
64-17953	SR71A	w.o.18.12.69	
64-17954	SR71A	w.o.11.4.69	
64-17955	SR71A	Lockheed	
64-17956	SR71B		

64-17957	SR71B	w.o.11.1.68	
64-17958	SR71A		
64-17959	SR71A	preserved	Eglin AFB
64-17960	SR71A		
64-17961	SR71A		
64-17962	SR71A		
64-17963	SR71A		
64-17964	SR71A	preserved	Offutt AFB
64-17965	SR71A	w.o.25.10.67	
64-17966	SR71A	w.o.13.4.67	
64-17967	SR71A		
64-17968	SR71A		
64-17969	SR71A	w.o. . .70	
64-17970	SR71A	w.o.17.6.70	
64-17971	SR71A		
64-17972	SR71A	Lockheed	Palmdale
64-17973	SR71A	preserved	Palmdale, Ca
64-17974	SR71A		
64-17975	SR71A		
64-17976	SR71A	USAF	Museum
64-17977	SR71A	w.o.10.10.68	
64-17978	SR71A	w.o. .5.73	
64-17979	SR71A		
64-17980	SR71A		
64-17981	SR71C		

T1 JAYHAWK

c\n TT-5, TT-3, TT-7 to TT-9, TT-6, TT-4, TT-11, TT-10, TT-12 to TT-15, TT-2, TT-16, TT-18, TT-17, TT-1, TT-19 to TT-77.

89-0284	T1A	64FTW	
90-0400	T1A	64FTW	
90-0401	T1A	64FTW	
90-0402	T1A	64FTW	
90-0403	T1A	64FTW	
90-0404	T1A	64FTW	
90-0405	T1A	64FTW	
90-0406	T1A	64FTW	
90-0407	T1A	64FTW	
90-0408	T1A	64FTW	
90-0409	T1A	64FTW	
90-0410	T1A	64FTW	
90-0411	T1A	64FTW	
90-0412	T1A	412TW	ED AFFTC
90-0413	T1A	64FTW	
91-0075	T1A	64FTW	
91-0076	T1A	64FTW	
91-0077	T1A		
91-0078	T1A	64FTW	
91-0079	T1A	64FTW	
91-0080	T1A	64FTW	
91-0081	T1A	64FTW	
91-0082	T1A	64FTW	
91-0083	T1A		
91-0084	T1A		
91-0085	T1A		

91-0086	T1A	64FTW
91-0087	T1A	
91-0088	T1A	
91-0089	T1A	
91-0090	T1A	
91-0091	T1A	
91-0092	T1A	
91-0093	T1A	
91-0094	T1A	
91-0095	T1A	
91-0096	T1A	
91-0097	T1A	
91-0098	T1A	
91-0099	T1A	
91-0100	T1A	
91-0101	T1A	
91-0102	T1A	
92-0330	T1A	
92-0331	T1A	
92-0332	T1A	
92-0333	T1A	
92-0334	T1A	
92-0335	T1A	
92-0336	T1A	
92-0337	T1A	
92-0338	T1A	

92-0339	T1A
92-0340	T1A
92-0341	T1A
92-0342	T1A
92-0343	T1A
92-0344	T1A
92-0345	T1A
92-0346	T1A
92-0347	T1A
92-0348	T1A
92-0349	T1A
92-0350	T1A

92-0351	T1A
92-0352	T1A
92-0353	T1A
92-0354	T1A
92-0355	T1A
92-0356	T1A
92-0357	T1A
92-0358	T1A
92-0359	T1A
92-0360	T1A
92-0361	T1A
92-0362	T1A
92-0363	T1A

note: A further 133 aircraft of this type are on order.

T3 FIREFLY

113 Slingsby T67M-260 Firefly\T3A on order for USAF Academy

T37 TWEETY BIRD

c\n   40001 to 40004, 40006 to 40015, 40005, 40016 to 40051, 40053 to 40064, 40052, 40065 to 40585, 40602, 40587 to 40601, 40603 to 40609, 40611, 40613, 40615, 40617, 40619 to 40640, 40642, 40643, 40645, 40647, 40649, 40651, 40653, 40655 to 40658, 40660, 40661, 40663 to 40724, 40824 to 40843, 40848 to 40885, 40902 to 40905, 40920 to 40960, 40966, 40978 to 40990, 40992, 40993, 40995 to 41000, 41002, 41004 to 41010, 41012, 41013, 41015, 41017, 41019, 41021, 41023, 41024, 41026, 41027, 41029, 41031, 41033, 41035, 41037, 41039, 41041, 41044, 41046, 41047, 41050, 41052 to 41054, 41056, 41058 to 41060, 41062 to 41064, 41066, 41067, 41069, 41071 to 41073, 41075 to 41087, 41089, 41091, 41092, 41094 to 41096, 41098 to 41100, 41102, 41104, 41106, 41107, 41109, 41111, 41113 to 41173, 41175, 41181, 41186, 41189, 41192, 41197, 41198, 41200, 41202, 41204 41206 to 41211.

54-0716	XT37	SoC		56-3464	T37B	to A37A 67-14518	
54-0717	XT37	preserved	Wichita, Ks	56-3465	T37B	to A37A 67-14519	
54-0718	XT37	preserved	Lackland AFB	56-3466	T37B	to A37A 67-14520	
54-2729	T37B	USAF	Museum	56-3467	T37B	to A37A 67-14521	
54-2730	T37B	preserved	Battle Creek AP	56-3468	T37B	to A37A 67-14522	
54-2731	T37B	static test rig		56-3469	T37B	to A37A 67-14523	
54-2732	T37B	preserved	Clearwater IAP, Fl	56-3470	T37B	to A37A 67-14524	
54-2733	GT37B	TTC	Sheppard	56-3471	T37B	to A37A 67-14525	
54-2734	GT37B	TTC	Sheppard	56-3472	T37B	to A37A 67-14526	
54-2735	T37B	SoC		56-3473	T37B	to A37A 67-14527	
54-2736	GT37B	TTC	Sheppard	56-3474	T37B	to A37A 67-14528	
54-2737	T37B	preserved	Columbus AFB	56-3475	T37B	to A37A 67-14529	
54-2738	GT37B	TTC	Sheppard	56-3476	T37B	SoC	
54-2739	GT37B	TTC	Chanute	56-3477	T37B	to A37A 67-14530	
55-2972	T37B	SoC		56-3478	T37B	to A37A 67-14531	
55-4302	T37B	to Vietnam		56-3479	T37B	to A37A 67-14532	
55-4303	T37B	to A37A 67-14503		56-3480	GT37B	SoC	
55-4304	T37B	SoC		56-3481	T37B	to A37A 67-14533	
55-4305	T37B	to A37A 67-14504		56-3482	T37B	to A37A 67-14534	
55-4306	T37B	to A37A 67-14505		56-3483	T37B	to A37A 67-14535	
55-4307	T37B	SoC		56-3484	T37B	to A37A 67-14536	
55-4308	T37B	to A37A 67-14506		56-3485	T37B	SoC	
55-4309	T37B	to A37A 67-14507		56-3486	T37B	to A37A 67-14537	
55-4310	T37B	to A37A 67-14508		56-3487	T37B	to A37A 67-14538	
55-4311	T37B	to A37A 67-14509		56-3488	T37B	SoC	
55-4312	T37B	to A37A 67-14510		56-3489	T37B	to A37A 67-14539	
55-4313	T37B	to A37A 67-14511		56-3490	T37B	to A37A 67-14540	
55-4314	T37B	to A37A 67-14512		56-3491	T37B	to A37A 67-14541	
55-4315	T37B	to A37A 67-14513		56-3492	T37B	14FTW	CB
55-4316	T37B	to A37A 67-14514		56-3493	T37B	AMARC	TE011
55-4317	T37B	SoC		56-3494	T37B		
55-4318	T37B	to A37A 67-14515		56-3495	T37B	to Thailand	
55-4319	T37B	to A37A 67-14516		56-3496	T37B	to Vietnam	
55-4320	T37B	to A37A 67-14517		56-3497	T37B	64FTW	
55-4321	T37B	SoC		56-3498	T37B	preserved	Vance AFB

Serial	Type	Disposition	Code
56-3499	T37B	47FTW	
56-3500	T37B	SoC	
56-3501	T37B	SoC	
56-3502	T37B	SoC	
56-3503	T37B	SoC	
56-3504	T37B	to Peru	
56-3505	T37B	to Vietnam	
56-3506	T37B		
56-3507	T37B	SoC	
56-3508	T37B	64FTW	
56-3509	T37B	71FTW	VN
56-3510	T37B	AMARC	TE...
56-3511	T37B	71FTW	VN
56-3512	T37B	64FTW	
56-3513	T37B	64FTW	
56-3514	T37B	47FTW	
56-3515	T37B	to Peru	
56-3516	T37B	to Vietnam	
56-3517	T37B	AMARC	TE014
56-3518	T37B	to Thailand	
56-3519	T37B	71FTW	VN
56-3520	T37B	80FTW	
56-3521	T37B	64FTW	
56-3522	T37B	SoC	
56-3523	T37B	to Vietnam	
56-3524	T37B	14FTW	CB
56-3525	T37B	SoC	
56-3526	T37B	71FTW	VN
56-3527	T37B	SoC	
56-3528	T37B	SoC	
56-3529	T37B	64FTW	
56-3530	T37B	64FTW	
56-3531	T37B	14FTW	CB
56-3532	T37B	to Thailand	
56-3533	T37B	64FTW	
56-3534	T37B	SoC	
56-3535	T37B	preserved	Kelly AFB
56-3536	T37B	64FTW	
56-3537	T37B	AMARC	TE031
56-3538	T37B	to Thailand	
56-3539	T37B	71FTW	VN
56-3540	T37B	14TFW	
56-3541	T37B	71TFW	VN
56-3542	T37B	47TFW	
56-3543	T37B	to Vietnam	
56-3544	T37B	80FTW	
56-3545	T37B	AMARC	TE053
56-3546	T37B	64FTW	
56-3547	T37B	12FTW	
56-3548	T37B	14FTW	CB
56-3549	T37B	SoC	
56-3550	T37B		
56-3551	T37B	SoC	
56-3552	T37B	to Thailand	
56-3553	T37B	64FTW	
56-3554	T37B	12FTW	
56-3555	T37B	12FTW	
56-3556	T37B	to Vietnam	
56-3557	T37B	AMARC	TE...
56-3558	T37B	64FTW	
56-3559	T37B	SoC	
56-3560	T37B	SoC	
56-3561	T37B		
56-3562	T37B	14FTW	CB
56-3563	T37B	64FTW	
56-3564	T37B	47FTW	
56-3565	T37B	SoC	
56-3566	T37B	AMARC	TE...
56-3567	T37B	to Vietnam	
56-3568	T37B	47FTW	
56-3569	T37B	SoC	
56-3570	T37B	71FTW	VN
56-3571	T37B	to Vietnam	
56-3572	T37B	SoC	
56-3573	T37B	to Vietnam	
56-3574	T37B	SoC	
56-3575	T37B	SoC	
56-3576	T37B		
56-3577	T37B	47FTW	
56-3578	T37B	to Vietnam	
56-3579	T37B	AMARC	TE017
56-3580	T37B	to Peru	
56-3581	T37B	AMARC	TE005
56-3582	T37B	to Vietnam	
56-3583	T37B	AMARC	TE015
56-3584	T37B	47FTW	
56-3585	T37B	to Thailand	
56-3586	T37B	64FTW	
56-3587	T37B	AMARC	TE...
56-3588	T37B	SoC	
56-3589	T37B	64FTW	
56-3590	T37B	64FTW	
57-2230	T37B	to Pakistan	
57-2231	T37B	AMARC	TE058
57-2232	T37B	SoC	
57-2233	T37B	AMARC	TE...
57-2234	T37B	64FTW	
57-2235	T37B	47FTW	
57-2236	T37B	AMARC	TE013
57-2237	T37B	47FTW	
57-2238	T37B	2W	LA
57-2239	T37B	AMARC	TE018
57-2240	T37B	80FTW	
57-2241	T37B	47FTW	
57-2242	T37B	SoC	
57-2243	T37B	AMARC	TE...
57-2244	T37B	14FTW	CB
57-2245	T37B	SoC	
57-2246	T37B	71FTW	VN
57-2247	T37B	AMARC	TE019
57-2248	T37B	80FTW	
57-2249	T37B	71FTW	VN
57-2250	T37B	12FTW	
57-2251	T37B	AMARC	TE024
57-2252	T37B	w.o.	
57-2253	T37B	AMARC	TE024
57-2254	T37B	14FTW	CB
57-2255	T37B	SoC	
57-2256	T37B	64FTW	
57-2257	T37B	to Pakistan	
57-2258	T37B	12FTW	
57-2259	T37B	64FTW	
57-2260	T37B	to Pakistan	
57-2261	T37B	14FTW	CB
57-2262	T37B	71FTW	VN
57-2263	T37B	47FTW	
57-2264	T37B	AMARC	TE...
57-2265	T37B	80FTW	EN
57-2266	T37B	SoC	
57-2267	T37B	AMARC	TE009
57-2268	T37B	71FTW	VN
57-2269	T37B	AMARC	TE023
57-2270	T37B	14FTW	CB
57-2271	T37B	SoC	
57-2272	T37B	47FTW	
57-2273	T37B	AMARC	TE...
57-2274	T37B	to Thailand	
57-2275	T37B	to Vietnam	
57-2276	T37B	47FTW	
57-2277	T37B	SoC	
57-2278	T37B	AMARC	TE025
57-2279	T37B	71FTW	VN
57-2280	T37B	71FTW	VN
57-2281	T37B	AMARC	TE062
57-2282	T37B	SoC	
57-2283	T37B	64FTW	
57-2284	T37B	71FTW	VN
57-2285	T37B	47FTW	
57-2286	T37B	SoC	
57-2287	T37B	80FTW	
57-2288	T37B	14FTW	CB
57-2289	T37B	USAF	Museum
57-2290	T37B	SoC	
57-2291	T37B	AMARC	TE029
57-2292	T37B	AMARC	TE...
57-2293	T37B	SoC	
57-2294	T37B	w.o.	
57-2295	T37B	AMARC	TE016
57-2296	T37B	AMARC	TE021
57-2297	T37B	14FTW	CB
57-2298	T37B	w.o.14.8.83	
57-2299	T37B	71FTW	VN
57-2300	T37B	SoC	
57-2301	T37B	SoC	
57-2302	T37B	to Vietnam	
57-2303	T37B	to Vietnam	

Serial	Type	Disposition	Code
57-2304	T37B	64FTW	
57-2305	T37B	AMARC	TE..
57-2306	T37B	SoC	
57-2307	T37B	AMARC	TE...
57-2308	T37B	47FTW	
57-2309	T37B	AMARC	TE002
57-2310	T37B	64FTW	
57-2311	T37B	AMARC	TE...
57-2312	T37B	SoC	
57-2313	T37B	AMARC	TE026
57-2314	T37B	SoC	
57-2315	T37B	SoC	
57-2316	T37B	AMARC	TE007
57-2317	T37B	AMARC	TE028
57-2318	T37B	47FTW	
57-2319	T37B	47FTW	
57-2320	T37B	80FTW	
57-2321	T37B	SoC	
57-2322	T37B	80FTW	
57-2323	T37B	SoC	
57-2324	T37B	SoC	
57-2325	T37B	SoC	
57-2326	T37B	SoC	
57-2327	T37B	SoC	
57-2328	T37B	47FTW	
57-2329	T37B	14FTW	CB
57-2330	T37B	80FTW	
57-2331	T37B	AMARC	TE032
57-2332	T37B	71FTW	VN
57-2333	T37B	SoC	
57-2334	T37B	SoC	
57-2335	T37B		
57-2336	T37B	AMARC	TE061
57-2337	T37B	AMARC	TE...
57-2338	T37B	64FTW	
57-2339	T37B	14FTW	CB
57-2340	T37B	80FTW	
57-2341	T37B	AMARC	TE012
57-2342	T37B	to Cambodia	
57-2343	T37B	AMARC	TE022
57-2344	T37B	64FTW	
57-2345	T37B	AMARC	TE...
57-2346	T37B	71FTW	VN
57-2347	T37B	to Pakistan	
57-2348	T37B	14FTW	CB
57-2349	T37B	12FTW	
57-2350	T37B	AMARC	TE027
57-2351	T37B	to Vietnam	
57-2352	T37B		
58-1861	T37B	SoC	
58-1862	T37B	AMARC	TE...
58-1863	T37B	12FTW	
58-1864	T37B	64FTW	
58-1865	T37B	SoC	
58-1866	T37B	71FTW	VN
58-1867	T37B	to Vietnam	
58-1868	T37B	71FTW	VN
58-1869	T37B	71FTW	VN
58-1870	T37B	SoC	
58-1871	T37B	to Pakistan	
58-1872	T37B	AMARC	TE001
58-1873	T37B	14FTW	CB
58-1874	T37B	AMARC	TE...
58-1875	T37B	SoC	
58-1876	T37B	SoC	
58-1877	T37B	71FTW	VN
58-1878	T37B	80FTW	
58-1879	T37B	71FTW	VN
58-1880	T37B	to Vietnam	
58-1881	T37B	14FTW	CB
58-1882	T37B	12FTW	
58-1883	T37B		
58-1884	T37B	64FTW	
58-1885	T37B	to Chile	J391
58-1886	T37B	14FTW	CB
58-1887	T37B	AMARC	TE064
58-1888	T37B	14FTW	CB
58-1889	T37B	64FTW	
58-1890	T37B	64FTW	
58-1891	T37B	80FTW	
58-1892	T37B	to Chile	J371
58-1893	T37B	to Colombia	
58-1894	T37B	to Vietnam	
58-1895	T37B	71FTW	VN
58-1896	T37B	SoC	
58-1897	T37B	71FTW	VN
58-1898	T37B	80FTW	
58-1899	T37B	64FTW	
58-1900	T37B	SoC	
58-1901	T37B	416BW	GR
58-1902	T37B	64FTW	
58-1903	T37B	to Pakistan	
58-1904	T37B	SoC	
58-1905	T37B	71FTW	VN
58-1906	T37B	12FTW	
58-1907	T37B	SoC	
58-1908	T37B	SoC	
58-1909	T37B	SoC	
58-1910	T37B	64FTW	
58-1911	T37B	SoC	
58-1912	T37B	w.o.	
58-1913	T37B	SoC	
58-1914	T37B	14FTW	CB
58-1915	T37B	47FTW	
58-1916	T37B	SoC	
58-1917	T37B	w.o.26.11.85	
58-1918	T37B	to Pakistan	
58-1919	T37B	14FTW	CB
58-1920	T37B	71FTW	VN
58-1921	T37B		
58-1922	T37B	47FTW	
58-1923	T37B	71FTW	VN
58-1924	T37B	47FTW	
58-1925	T37B	47FTW	
58-1926	T37B	64FTW	
58-1927	T37B	to Pakistan	
58-1928	T37B		
58-1929	T37B	14FTW	CB
58-1930	T37B	14FTW	CB
58-1931	T37B	SoC	
58-1932	T37B	SoC	
58-1933	T37B	64FTW	
58-1934	T37B	to Chile	J390
58-1935	T37B	80FTW	
58-1936	T37B	47FTW	
58-1937	T37B	SoC	
58-1938	T37B	AMARC	TE...
58-1939	T37B	AMARC	TE...
58-1940	T37B	47FTW	
58-1941	T37B	71FTW	VN
58-1942	T37B	14FTW	CB
58-1943	T37B		
58-1944	T37B	w.o.	
58-1945	T37B	12FTW	
58-1946	T37B	SoC	
58-1947	T37B	64FTW	
58-1948	T37B	12FTW	
58-1949	T37B	47FTW	
58-1950	T37B	410BW	KI
58-1951	T37B	14FTW	CB
58-1952	T37B	47FTW	
58-1953	T37B	71FTW	VN
58-1954	T37B	14FTW	CB
58-1955	T37B	14FTW	CB
58-1956	T37B	64FTW	
58-1957	T37B	12FTW	
58-1958	T37B	12FTW	
58-1959	T37B	47FTW	
58-1960	T37B	12FTW	
58-1961	T37B	71FTW	VN
58-1962	T37B	AMARC	TE065
58-1963	T37B	80FTW	
58-1964	T37B	64FTW	
58-1965	T37B		
58-1966	T37B	to Chile	J380
58-1967	T37B	71FTW	VN
58-1968	T37B	64FTW	
58-1969	T37B	to Pakistan	
58-1970	T37B	SoC	
58-1971	T37B	14FTW	CB
58-1972	T37B	71FTW	VN
58-1973	T37B	80FTW	
58-1974	T37B	SoC	
58-1975	T37B		
58-1976	T37B	14FTW	CB
58-1977	T37B	AMARC	TE047

59-0241	T37B	SoC	
59-0242	T37B	80FTW	
59-0243	T37B	SoC	
59-0244	T37B	SoC	
59-0245	T37B	to Pakistan	
59-0246	T37B	80FTW	
59-0247	T37B	64FTW	
59-0248	T37B	to Turkey	90248
59-0249	T37B	64FTW	
59-0250	T37B	to Turkey	90250
59-0251	T37B	to Turkey	90251
59-0252	T37B	SoC	
59-0253	T37B	71FTW	VN
59-0254	T37B	71FTW	VN
59-0255	T37B	47FTW	
59-0256	T37B	to Colombia	
59-0257	T37B	64FTW	
59-0258	T37B	to Chile	J386
59-0259	T37B	to Chile	J381
59-0260	T37B	to Pakistan	
59-0261	T37B	416BW	GR
59-0262	T37B	to Turkey	90262
59-0263	T37B	14FTW	CB
59-0264	T37B	SoC	
59-0265	T37B	47FTW	
59-0266	T37B	SoC	
59-0267	T37B	71FTW	VN
59-0268	T37B	14FTW	CB
59-0269	T37B	71FTW	VN
59-0270	T37B	47FTW	
59-0271	T37B	SoC	
59-0272	T37B	12FTW	
59-0273	T37B	47FTW	
59-0274	T37B	14FTW	CB
59-0275	T37B	64FTW	
59-0276	T37B	64FTW	
59-0277	T37B	to Chile	J387
59-0278	T37B	to Pakistan	
59-0279	T37B	to Pakistan	
59-0280	T37B	80FTW	
59-0281	T37B	SoC	
59-0282	T37B	416BW	GR
59-0283	T37B	47FTW	
59-0284	T37B	to Pakistan	
59-0285	T37B	80FTW	
59-0286	T37B	80FTW	
59-0287	T37B	12FTW	
59-0288	T37B	to Vietnam	
59-0289	T37B	AMARC	TE006
59-0290	T37B	to Pakistan	
59-0291	T37B		
59-0292	T37B	80FTW	
59-0293	T37B	71FTW	VN
59-0294	T37B	71FTW	VN
59-0295	T37B	12FTW	
59-0296	T37B	64FTW	
59-0297	T37B	SoC	
59-0298	T37B	SoC	
59-0299	T37B	47FTW	
59-0300	T37B	to Chile	J394
59-0301	T37B	SoC	
59-0302	T37B	to Chile	J397
59-0303	T37B	12FTW	
59-0304	T37B	71FTW	VN
59-0305	T37B	to Turkey	90305
59-0306	T37B	to Turkey	90306
59-0307	T37B	SoC	
59-0308	T37B	64FTW	
59-0309	T37B	SoC	
59-0310	T37B	to Turkey	90310
59-0311	T37B	47FTW	
59-0312	T37B	to Turkey	90312
59-0313	T37B	to Vietnam	
59-0314	T37B	to Turkey	90314
59-0315	T37B	SoC	
59-0316	T37B		
59-0317	T37B	14FTW	CB
59-0318	T37B	to Turkey	90318
59-0319	T37B	FMS	
59-0320	T37B	64FTW	
59-0321	T37B	47FTW	
59-0322	T37B	SoC	
59-0323	T37B	to Turkey	90323
59-0324	T37B		
59-0325	T37B	71FTW	VN
59-0326	T37B		
59-0327	T37B	to Turkey	90327
59-0328	T37B	to Turkey	90328
59-0329	T37B	47FTW	
59-0330	T37B		
59-0331	T37B	to Turkey	90331
59-0332	T37B	14FTW	CB
59-0333	T37B	71FTW	VN
59-0334	T37B	64FTW	
59-0335	T37B	80FTW	
59-0336	T37B	SoC	
59-0337	T37B	47FTW	
59-0338	T37B	AMARC	TE008
59-0339	T37B		
59-0340	T37B	14FTW	CB
59-0341	T37B	64FTW	
59-0342	T37B	47FTW	
59-0343	T37B	14FTW	CB
59-0344	T37B	to Vietnam	
59-0345	T37B	71FTW	VN
59-0346	T37B	12FTW	
59-0347	T37B	71FTW	VN
59-0348	T37B	80FTW	
59-0349	T37B	w.o.20.2.80	
59-0350	T37B	AMARC	TE004
59-0351	T37B	47FTW	
59-0352	T37B	to Pakistan	
59-0353	T37B	SoC	
59-0354	T37B	64FTW	
59-0355	T37B	64FTW	
59-0356	T37B	71FTW	VN
59-0357	T37B	14FTW	CB
59-0358	T37B	14FTW	CB
59-0359	T37B	to Chile	J395
59-0360	T37B	71FTW	VN
59-0361	T37B	14FTW	CB
59-0362	T37B	SoC	
59-0363	T37B	71FTW	VN
59-0364	T37B	to Chile	J393
59-0365	T37B	71FTW	VN
59-0366	T37B	71FTW	VN
59-0367	T37B	SoC	
59-0368	T37B	to Pakistan	
59-0369	T37B	12FTW	
59-0370	T37B	71FTW	VN
59-0371	T37B	to Turkey	90371
59-0372	T37B	SoC	
59-0373	T37B	SoC	
59-0374	T37B	to Pakistan	
59-0375	T37B	47FTW	
59-0376	T37B	64FTW	
59-0377	T37B	71FTW	VN
59-0378	T37B	47FTW	
59-0379	T37B	14FTW	CB
59-0380	T37B	12FTW	
59-0381	T37B	SoC	
59-0382	T37B	12FTW	
59-0383	T37B	12FTW	
59-0384	T37B	to Turkey	90384
59-0385	T37B	to Turkey	90385
59-0386	T37B	71FTW	VN
59-0387	T37B	80FTW	
59-0388	T37B	to Turkey	90388
59-0389	T37B	to Turkey	90389
59-0390	T37B	AMARC	TE...
60-0071	T37B	71FTW	VN
60-0072	T37B	AMARC	TE030
60-0073	T37B	SoC	
60-0074	T37B	2W	LA
60-0075	T37B	71FTW	VN
60-0076	T37B		
60-0077	T37B	71FTW	VN
60-0078	T37B	14FTW	CB
60-0079	T37B	to Chile	J381(2)
60-0080	T37B	71FTW	VN
60-0081	T37B	47FTW	
60-0082	T37B	64FTW	
60-0083	T37B	64FTW	
60-0084	T37B	NASA	
60-0085	T37B	to Chile	J388
60-0086	T37B	80FTW	

Serial	Type	Unit	Code
60-0087	T37B	AMARC	TE...
60-0088	T37B	SoC	
60-0089	T37B	SoC	
60-0090	T37B	14FTW	CB
60-0091	T37B	71FTW	VN
60-0092	T37B	80FTW	
60-0093	T37B	to Pakistan	
60-0094	T37B	71FTW	VN
60-0095	T37B	64FTW	
60-0096	T37B	14FTW	CB
60-0097	T37B	64FTW	
60-0098	T37B	12FTW	
60-0099	T37B	SoC	
60-0100	T37B	71FTW	VN
60-0101	T37B	14FTW	CB
60-0102	T37B	to Chile	J392
60-0103	T37B	AMARC	TE037
60-0104	T37B	12FTW	
60-0105	T37B	64FTW	
60-0106	T37B	to Pakistan	
60-0107	T37B	47FTW	
60-0108	T37B	71FTW	VN
60-0109	T37B	to Pakistan	
60-0110	T37B	80FTW	
60-0111	T37B	71FTW	VN
60-0112	T37B	71FTW	VN
60-0113	T37B	71FTW	VN
60-0114	T37B	71FTW	VN
60-0115	T37B	to Pakistan	
60-0116	T37B	SoC	
60-0117	T37B	64FTW	
60-0118	T37B	47FTW	
60-0119	T37B	14FTW	CB
60-0120	T37B	SoC	
60-0121	T37B	71FTW	VN
60-0122	T37B	71FTW	VN
60-0123	T37B	80FTW	
60-0124	T37B	12FTW	
60-0125	T37B	71FTW	VN
60-0126	T37B	to Chile	J375(2)
60-0127	T37B		
60-0128	T37B	47FTW	
60-0129	T37B	12FTW	
60-0130	T37B	71FTW	VN
60-0131	T37B		
60-0132	T37B	to Chile	J389
60-0133	T37B	to Thailand	
60-0134	T37B	64FTW	
60-0135	T37B	to Thailand	
60-0136	T37B	AMARC	TE...
60-0137	T37B	to Thailand	
60-0138	T37B	to Thailand	
60-0139	T37B	to Thailand	
60-0140	T37B	to Thailand	
60-0141	T37B	71FTW	VN
60-0142	T37B	to Thailand	
60-0143	T37B	AMARC	TE063
60-0144	T37B	64FTW	
60-0145	T37B	to Cambodia	
60-0146	T37B	to Thailand	
60-0147	T37B	14FTW	CB
60-0148	T37B	to Cambodia	
60-0149	T37B	to Cambodia	
60-0150	T37B	to Cambodia	
60-0151	T37B	SoC	
60-0152	T37B	64FTW	
60-0153	T37B	47FTW	
60-0154	T37B	80FTW	
60-0155	T37B	to Chile	J370
60-0156	T37B	to Chile	J371
60-0157	T37B	to Chile	J372
60-0158	T37B	to Chile	J373
60-0159	T37B	to Cambodia	
60-0160	T37B	14FTW	CB
60-0161	T37B	14FTW	CB
60-0162	T37B	14FTW	CB
60-0163	T37B	47FTW	
60-0164	T37B	to Peru	
60-0165	T37B	to Peru	
60-0166	T37B	to Peru	
60-0167	T37B	to Peru	
60-0168	T37B	to Peru	
60-0169	T37B	to Pakistan	
60-0170	T37B	to Pakistan	
60-0171	T37B	to Pakistan	
60-0172	T37B	to Pakistan	
60-0173	T37B	to Pakistan	
60-0174	T37B	to Pakistan	
60-0175	T37B	71FTW	VN
60-0176	T37B	14FTW	CB
60-0177	T37B	71FTW	VN
60-0178	T37B	71FTW	VN
60-0179	T37B	71FTW	VN
60-0180	T37B	71FTW	VN
60-0181	T37B	80FTW	
60-0182	T37B	80FTW	EN
60-0183	T37B	71FTW	VN
60-0184	T37B	to Chile	J396
60-0185	T37B	14FTW	CB
60-0186	T37B	14FTW	CB
60-0187	T37B	64FTW	
60-0188	T37B	80FTW	
60-0189	T37B	71FTW	VN
60-0190	T37B	AMARC	TE...
60-0191	T37B	71FTW	VN
60-0192	T37B	SoC	
60-0193	T37B	71FTW	VN
60-0194	T37B	to Pakistan	
60-0195	T37B	71FTW	VN
60-0196	T37B	to Pakistan	
60-0197	T37B	71FTW	VN
60-0198	T37B	71FTW	VN
60-0199	T37B	71FTW	VN
60-0200	T37B	to Pakistan	
61-2494	T37B	71FTW	VN
61-2495	T37B	71FTW	VN
61-2496	T37B	80FTW	
61-2497	T37B	w.o.22.9.82	
61-2498	T37B	71FTW	VN
61-2499	T37B	71FTW	VN
61-2500	T37B	SoC	
61-2501	T37B	12FTW	
61-2502	T37B	71FTW	VN
61-2503	T37B	71FTW	VN
61-2504	T37B	71FTW	VN
61-2505	T37B	71FTW	VN
61-2506	T37B	w.o.28.6.88	
61-2507	T37B	71FTW	VN
61-2508	T37B	71FTW	VN
61-2915	T37B	71FTW	VN
61-2916	T37B	71FTW	VN
61-2917	T37B	71FTW	VN
61-2918	T37B	71FTW	VN
61-2919	T37B	71FTW	VN
62-5950	GYA37A	preserved	Sheppard AFB
62-5951	YA37A	USAF	Museum
62-5952	T37B	71FTW	VN
62-5953	T37B	14FTW	CB
62-5954	T37B	71FTW	VN
62-5955	T37B	14FTW	CB
62-5956	T37B	71FTW	VN
64-13409	T37B	47FTW	
64-13410	T37B	47FTW	
64-13411	T37B	47FTW	
64-13412	T37B	47FTW	
64-13413	T37B	71FTW	VN
64-13414	T37B	71FTW	VN
64-13415	T37B	64FTW	
64-13416	T37B	SoC	
64-13417	T37B		
64-13418	T37B	71FTW	VN
64-13419	T37B	71FTW	VN
64-13420	T37B	47FTW	
64-13421	T37B	47FTW	
64-13422	T37B	47FTW	
64-13423	T37B	71FTW	VN
64-13424	T37B	47FTW	
64-13425	T37B	12FTW	
64-13426	T37B	AMARC	TE066
64-13427	T37B	47FTW	
64-13428	T37B	47FTW	
64-13433	T37B	47FTW	
64-13434	T37B	47FTW	
64-13435	T37B	71FTW	VN
64-13436	T37B	416BW	GR
64-13437	T37B	SoC	

Serial	Type	Unit	Notes
64-13438	T37B	12FTW	
64-13439	T37B	71FTW	VN
64-13440	T37B	71FTW	VN
64-13441	T37B	14FTW	CB
64-13442	T37B	14FTW	CB
64-13443	T37B	14FTW	CB
64-13444	T37B	SoC	
64-13445	T37B	12FTW	
64-13446	T37B	80FTW	
64-13447	T37B	71FTW	VN
64-13448	T37B	80FTW	
64-13449	T37B	SoC	
64-13450	T37B	80FTW	
64-13451	T37B	80FTW	
64-13452	T37B	80FTW	EN
64-13453	T37B	80FTW	
64-13454	T37B	80FTW	
64-13455	T37B	80FTW	
64-13456	T37B	80FTW	
64-13457	T37B	80FTW	
64-13458	T37B	14FTW	CB
64-13459	T37B	80FTW	
64-13460	T37B	80FTW	EN
64-13461	T37B	80FTW	EN
64-13462	T37B	80FTW	
64-13463	T37B	80FTW	
64-13464	T37B	80FTW	
64-13465	T37B	80FTW	
64-13466	T37B	80FTW	
64-13467	T37B	80FTW	
64-13468	T37B	80FTW	
64-13469	T37B	80FTW	EN
64-13470	T37B	80FTW	
65-10823	T37B	SoC	
65-10824	T37B	80FTW	
65-10825	T37B	80FTW	
65-10826	T37B	80FTW	
66-7960	T37B	80FTW	
66-7961	T37B	80FTW	
66-7962	T37B	80FTW	
66-7963	T37B	80FTW	
66-7964	T37B	80FTW	
66-7965	T37B	80FTW	
66-7966	T37B	80FTW	
66-7967	T37B	80FTW	
66-7968	T37B	80FTW	
66-7969	T37B	80FTW	
66-7970	T37B	80FTW	
66-7971	T37B	80FTW	
66-7972	T37B	80FTW	
66-7973	T37B	SoC	
66-7974	T37B	80FTW	
66-7975	T37B	80FTW	
66-7976	T37B	80FTW	
66-7977	T37B	SoC	
66-7978	T37B	64FTW	
66-7979	T37B	SoC	
66-7980	T37B	47FTW	
66-7981	T37B	to Greece	67981
66-7982	T37B	12FTW	
66-7983	T37B	14FTW	CB
66-7984	T37B	to Greece	67984
66-7985	T37B	14FTW	CB
66-7986	T37B	80FTW	
66-7987	T37B	12FTW	
66-7988	T37B	12FTW	
66-7989	T37B	12FTW	
66-7990	T37B	14FTW	CB
66-7991	T37B	80FTW	
66-7992	T37B	14FTW	CB
66-7993	T37B	12FTW	
66-7994	T37B	12FTW	
66-7995	T37B	12FTW	
66-7996	T37B	47FTW	
66-7997	T37B	14FTW	CB
66-7998	T37B	12FTW	
66-7999	T37B	12FTW	
66-8000	T37B	12FTW	
66-8001	T37B	12FTW	
66-8002	T37B	80FTW	EN
66-8003	T37B	80FTW	
66-8004	T37B	80FTW	
66-8005	T37B	80FTW	
66-8006	T37B	80FTW	
67-14730	T37B	80FTW	
67-14731	T37B	to Greece	14731
67-14732	T37B	64FTW	
67-14733	T37B	SoC	
67-14734	T37B	64FTW	
67-14735	T37B	64FTW	
67-14736	T37B	12FTW	
67-14737	T37B	12FTW	
67-14738	T37B	to Greece	14738
67-14739	T37B	80FTW	
67-14740	T37B	71FTW	VN
67-14741	T37B	80FTW	
67-14742	T37B	to Greece	14742
67-14743	T37B	47FTW	
67-14744	T37B	64FTW	
67-14745	T37B	12FTW	
67-14746	T37B	47FTW	
67-14747	T37B	SoC	
67-14748	T37B	14FTW	CB
67-14749	T37B	71FTW	VN
67-14750	T37B	12FTW	
67-14751	T37B	to Greece	14751
67-14752	T37B	80FTW	
67-14753	T37B	47FTW	
67-14754	T37B	47FTW	
67-14755	T37B	12FTW	
67-14756	T37B		
67-14757	T37B	12FTW	
67-14758	T37B	to Greece	14758
67-14759	T37B	47FTW	
67-14760	T37B	12FTW	
67-14761	T37B	80FTW	
67-14762	T37B	47FTW	
67-14763	T37B	47FTW	
67-14764	T37B	71FTW	VN
67-14765	T37B	14FTW	CB
67-14766	T37B	64FTW	
67-14767	T37B	14FTW	CB
67-14768	T37B	to Greece	14768
67-22240	T37B	SoC	
67-22241	T37B	71FTW	VN
67-22242	T37B	12FTW	
67-22243	T37B	12FTW	
67-22244	T37B	71FTW	VN
67-22245	T37B	80FTW	EN
67-22246	T37B	64FTW	
67-22247	T37B	12FTW	
67-22248	T37B	12FTW	
67-22249	T37B	64FTW	
67-22250	T37B	12FTW	
67-22251	T37B	64FTW	
67-22252	T37B	12FTW	
67-22253	T37B	71FTW	VN
67-22254	T37B	12FTW	
67-22255	T37B	12FTW	
67-22256	T37B	12FTW	
67-22257	T37B	12FTW	
67-22258	T37B	71FTW	VN
67-22259	T37B		
67-22260	T37B	SoC	
67-22261	T37B	71FTW	VN
67-22262	T37B		
68-7981	T37B	14FTW	CB
68-7982	T37B	12FTW	
68-7983	T37B		
68-7984	T37B	71FTW	VN
68-7985	T37B	to Jordan	
68-7986	T37B		
68-7987	T37B	47FTW	
68-7988	T37B	64FTW	
68-7989	T37B	64FTW	
68-7990	T37B	71FTW	VN
68-7991	T37B	64FTW	
68-7992	T37B	71FTW	VN
68-7993	T37B	80FTW	
68-7994	T37B	64FTW	
68-7995	T37B	12FTW	
68-7996	T37B	64FTW	
68-7997	T37B	SoC	
68-7998	T37B	to Jordan	
68-7999	T37B		
68-8000	T37B	71FTW	VN

Serial	Type	Unit	Code		Serial	Type	Unit	Code
68-8001	T37B	71FTW	VN		68-8044	T37B	to Jordan	
68-8002	T37B	71FTW	VN		68-8045	T37B	47FTW	
68-8003	T37B	80FTW			68-8046	T37B	14FTW	CB
68-8004	T37B	14FTW	CB		68-8047	T37B	14FTW	CB
68-8005	T37B	64FTW			68-8048	T37B	to Jordan	
68-8006	T37B	47FTW			68-8049	T37B	71FTW	VN
68-8007	T37B	47FTW			68-8050	T37B	80FTW	
68-8008	T37B	64FTW			68-8051	T37B	to Jordan	
68-8009	T37B	64FTW			68-8052	T37B	to Jordan	
68-8010	T37B	SoC			68-8053	T37B	14FTW	CB
68-8011	T37B	71FTW	VN		68-8054	T37B	71FTW	VN
68-8012	T37B	47FTW			68-8055	T37B	71FTW	VN
68-8013	T37B	71FTW	VN		68-8056	T37B	14FTW	CB
68-8014	T37B	80FTW			68-8057	T37B	80FTW	
68-8015	T37B	47FTW			68-8058	T37B	14FTW	CB
68-8016	T37B	to Jordan			68-8059	T37B	80FTW	
68-8017	T37B	12FTW			68-8060	T37B	47FTW	
68-8018	T37B	to Jordan			68-8061	T37B	80FTW	
68-8019	T37B	80FTW			68-8062	T37B	12FTW	
68-8020	T37B	64FTW			68-8063	T37B	to Jordan	
68-8021	T37B	64FTW			68-8064	T37B	71FTW	VN
68-8022	T37B	47FTW			68-8065	T37B	80FTW	
68-8023	T37B	to Jordan			68-8066	T37B	to Jordan	
68-8024	T37B	12FTW			68-8067	T37B	14FTW	CB
68-8025	T37B	12FTW			68-8068	T37B	14FTW	CB
68-8026	T37B	47FTW			68-8069	T37B	14FTW	CB
68-8027	T37B	80FTW			68-8070	T37B	12FTW	
68-8028	T37B	14FTW	CB		68-8071	T37B	14FTW	CB
68-8029	T37B	14FTW	CB		68-8072	T37B	to Jordan	
68-8030	T37B				68-8073	T37B	80FTW	
68-8031	T37B	64FTW			68-8074	T37B	to Jordan	
68-8032	T37B				68-8075	T37B	14FTW	CB
68-8033	T37B	SoC			68-8076	T37B	to Jordan	
68-8034	T37B				68-8077	T37B	71FTW	VN
68-8035	T37B	to Jordan			68-8078	T37B	to Jordan	
68-8036	T37B	14FTW	CB		68-8079	T37B	to Jordan	
68-8037	T37B	SoC			68-8080	T37B	80FTW	
68-8038	T37B				68-8081	T37B	to Jordan	
68-8039	T37B	47FTW			68-8082	T37B	SoC	
68-8040	T37B	80FTW	EN		68-8083	T37B	12FTW	
68-8041	T37B	71FTW	VN		68-8084	T37B	14FTW	CB
68-8042	T37B	14FTW	CB					
68-8043	T37B	w.o.9.7.82						

T38 TALON

c\n   N.5001, N.5002, N.5100, N5103 to N.5904, N.5906 to N.5916, N.5918 to N.5929, N.5978 to N.5983, N.5990 to N.5999, T.6000 to T.6020, N.5905, N.5966 to N.5974, N.5917, N.5930 to N.5965, N.5975 to N.5977, N.5984 to N.5989, T.6027, T.6021 to T.6026, T.6028 to T.6279, T.6282, T.6283, T.6280, T.6281, T.6284 to T.6289.

Serial	Type	Status	Location		Serial	Type	Status	Location
58-1191	YT38A	SoC			60-0549	GT38A	TTC	Chanute
58-1192	YT38A	preserved	Ellsworth AFB		60-0550	AT38B	w.o.24.5.83	
58-1193	YT38A	static test airframe			60-0551	T38A	ALC	SM Sacramento
58-1194	T38A	to US Navy			60-0552	T38A	MASDC	
58-1195	T38A	to US Navy			60-0553	AT38B	49FW	HO
58-1196	T38A	preserved	Los Angeles		60-0554	T38A	12FTW	
58-1197	T38A	w.o.66			60-0555	GT38A	TTC	Chanute
59-1594	QT38A	to US Navy			60-0556	T38A	AMARC	TF022
59-1595	QT38A	to US Navy			60-0557	GT38A	TTC	Chanute
59-1596	QT38A	to US Navy			60-0558	GT38A	TTC	Chanute
59-1597	QT38A	to US Navy			60-0559	T38A	dump	Chino, Ca
59-1598	QT38A	to US Navy			60-0560	T38A	w.o.	
59-1599	T38A	w.o.66			60-0561	AT38B	49FW	HO
59-1600	QT38A	to US Navy			60-0562	T38A	AMARC	TF020
59-1601	T38A	preserved	Maxwell AFB		60-0563	T38A	w.o.62	
59-1602	T38A	Air Force Academy			60-0564	T38A	AMARC	TF023
59-1603	QT38A	to US Navy			60-0565	T38A	AMARC	TF015
59-1604	QT38A	to US Navy			60-0566	T38A	scrapped	
59-1605	T38A	preserved	Lackland AFB		60-0567	GT38A	TTC	Chanute
59-1606	T38A	w.o.72			60-0568	T38A	w.o.62	
60-0547	T38A	scrapped			60-0569	AT38B	w.o.22.6.88	
60-0548	T38A	AMARC	TF001		60-0570	T38A	preserved	Beale AFB

Serial	Type	Status	Notes		Serial	Type	Status	Notes
60-0571	T38A	AMARC	TF017		61-0861	T38A		
60-0572	T38A				61-0862	T38A	stored	Chino, Ca
60-0573	AT38B	49FW	HO		61-0863	T38B		
60-0574	T38A	MASDC			61-0864	AT38B	AMARC	TF...
60-0575	T38A	AMARC	TF018		61-0865	T38A	scrapped	
60-0576	AT38B	preserved	Holloman AFB		61-0866	AT38B	AMARC	TF...
60-0577	T38A	AMARC	TF014		61-0867	T38A	to Portugal	2610
60-0578	T38A	9RW	BB		61-0868	T38A	to Portugal	2603
60-0579	T38A	AMARC	TF026		61-0869	T38A	scrapped	
60-0580	GT38A	TTC	Chanute		61-0870	T38A		
60-0581	T38A	9RW	BB		61-0871	T38A	w.o.65	
60-0582	NT38A	to US Navy	'600582'		61-0872	T38A	to Portugal	2611
60-0583	GT38A	TTC	Sheppard		61-0873	T38A	w.o.63	
60-0584	T38A	w.o.66			61-0874	T38A	46TW	ET AFMC\MSD
60-0585	GT38A	GIA	Holloman AFB		61-0875	AT38B	AMARC	TF103
60-0586	GT38A	TTC	Chanute		61-0876	AT38B	AMARC	TF...
60-0587	T38A	w.o.62			61-0877	T38A	w.o.	
60-0588	GT38A	GIA	Holloman AFB		61-0878	AT38B	w.o.	
60-0589	AT38B	AMARC	TF...		61-0879	T38A	64FTW	
60-0590	GT38A	TTC	Sheppard		61-0880	AT38B	AMARC	TF062
60-0591	AT38B	AMARC	TF...		61-0881	T38A	w.o.71	
60-0592	GT38A	TTC	Sheppard		61-0882	T38A	64FTW	
60-0593	GT38A	TTC	Sheppard		61-0883	T38A	12FTW	
60-0594	AT38B	AMARC	TF...		61-0884	T38A	w.o.	
60-0595	AT38B	AMARC	TF080		61-0885	T38A	w.o.67	
60-0596	T38A	w.o.70			61-0886	AT38B	AMARC	TF...
61-0804	AT38B	AMARC	TF...		61-0887	T38A	to Taiwan	
61-0805	T38A	w.o.70			61-0888	GAT38B	TTC	Sheppard
61-0806	AT38B	AMARC	TF...		61-0889	T38A	to US Navy	
61-0807	AT38B	AMARC	TF071		61-0890	T38A	to Portugal	2604
61-0808	T38A	w.o.			61-0891	AT38B	AMARC	TF...
61-0809	AT38B	AMARC	TF...		61-0892	T38A	scrapped	
61-0810	T38A	preserved	Edwards AFB		61-0893	T38A	to Taiwan	
61-0811	T38A	w.o.			61-0894	T38A		
61-0812	AT38B	AMARC	TF...		61-0895	GT38A	TTC	Sheppard
61-0813	T38A	scrapped			61-0896	T38A		
61-0814	GT38A	TTC	Sheppard		61-0897	T38A	to Portugal	2605
61-0815	T38A	to Portugal	2607		61-0898	AT38B	TTC	Sheppard
61-0816	T38A	64FTW			61-0899	AT38B	AMARC	TF074
61-0817	GAT38B	TTC	Sheppard		61-0900	T38A	14FTW	CB
61-0818	AT38B	AMARC	TF...		61-0901	T38A	w.o.70	
61-0819	T38A	47FTW			61-0902	T38A	scrapped	
61-0820	AT38B	49FW	HO		61-0903	T38A	to Portugal	2612
61-0821	T38B	w.o.64			61-0904	AT38B	AMARC	TF...
61-0822	AT38B	w.o.			61-0905	T38A	to Taiwan	
61-0823	T38A	14FTW	CB		61-0906	T38A	w.o.68	
61-0824	GT38A	TTC	Sheppard		61-0907	AT38B	stored	El Paso, Tx
61-0825	T38A	412TW	ED AFFTC		61-0908	T38A	47FTW	
61-0826	T38A	80FTW			61-0909	T38A	14FTW	CB
61-0827	T38A	14FTW	CB		61-0910	T38A		
61-0828	GAT38B	TTC	Sheppard		61-0911	AT38B	AMAR	TF...
61-0829	T38A				61-0912	T38A	to NASA	N905NA
61-0830	T38A	scrapped			61-0913	T38A		
61-0831	AT38B	AMARC	TF...		61-0914	T38A	scrapped	
61-0832	T38A	scrapped			61-0915	T38A	to Portugal	2606
61-0833	T38A	w.o.73			61-0916	T38A	to Taiwan	
61-0834	T38A	14FTW	CB		61-0917	AT38B	AMARC	TF101
61-0835	AT38B	AMARC	TF...		61-0918	T38A		
61-0836	AT38B	AMARC	TF...		61-0919	T38A	scrapped	
61-0837	T38A	to Portugal	2608		61-0920	T38A	80FTW	
61-0838	T38A	preserved	Randolph AFB		61-0921	T38A	w.o.71	
61-0839	T38A	w.o.62			61-0922	T38A	w.o.63	
61-0840	T38A	to Portugal	2609		61-0923	GAT38B	TTC	Sheppard
61-0841	T38A	to Taiwan			61-0924	T38A	80FTW	
61-0842	AT38B	AMARC	TF...		61-0925	T38A	80FTW	
61-0843	T38A	to Portugal	2601		61-0926	GT38A	TTC	Sheppard
61-0844	T38A				61-0927	T38A	14FTW	CB
61-0845	AT38B	AMARC	TF105		61-0928	T38A	w.o.67	
61-0846	T38A	w.o.			61-0929	T38A	71FTW	VN
61-0847	AT38B	AMARC	TF083		61-0930	T38A	14FTW	CB
61-0848	AT38B	49FW	HO		61-0931	T38A	w.o.	
61-0849	T38A	14FTW	CB		61-0932	T38A	w.o.66	
61-0850	T38A	stored	Chino, Ca		61-0933	T38A		
61-0851	AT38B	49FW	HO		61-0934	T38A	w.o.66	
61-0852	AT38B	AMARC	TF...		61-0935	T38B	preserved	Chino, Ca
61-0853	T38A	to Portugal	2602		61-0936	T38A		
61-0854	T38A	Pima Museum			61-0937	T38A	scrapped	
61-0855	T38A	14FTW	CB		61-0938	AT38B	49FW	HO
61-0856	T38A	scrapped			61-0939	T38A	w.o.63	
61-0857	AT38B	AMARC	TF...		61-0940	AT38B	AMARC	TF060
61-0858	GT38A	preserved	Sheppard AFB		61-0941	GT38A	TTC	Sheppard
61-0859	AT38B				61-0942	T38A	14FTW	CB
61-0860	AT38B	AMARC	TF...		61-0943	T38A	w.o.64	

Serial	Type	Status	Notes
61-0944	T38A	to Taiwan	
61-0945	T38A	64FTW	
61-0946	T38A	w.o.	
61-0947	AT38B	AMARC	TF...
62-3609	T38A	w.o.65	
62-3610	T38A	12FTW	
62-3611	T38A	to Turkey	23611
62-3612	T38A	w.o.65	
62-3613	T38A	w.o.63	
62-3614	AT38B	AMARC	TF...
62-3615	T38A	80FTW	
62-3616	T38A		
62-3617	T38A	to Turkey	23617
62-3618	T38A	80FTW	
62-3619	T38A		
62-3620	T38A		
62-3621	T38A	to Turkey	23621
62-3622	T38A	w.o.71	
62-3623	T38A	71FTW	VN
62-3624	T38A	to Turkey	23624
62-3625	T38A	22ARW	
62-3626	T38A	64FTW	
62-3627	AT38B	AMARC	TF...
62-3628	T38A	64FTW	
62-3629	T38A	64FTW	
62-3630	T38A	64FTW	
62-3631	T38A	64FTW	
62-3632	AT38B	49FW	HO
62-3633	T38A	80FTW	
62-3634	T38A	64FTW	
62-3635	T38A	w.o.63	
62-3636	T38A	14FTW	CB
62-3637	T38A	64FTW	
62-3638	T38A	64FTW	
62-3639	T38A		
62-3640	T38A	64FTW	
62-3641	AT38B	AMARC	TF...
62-3642	T38A	w.o.	
62-3643	T38A	14FTW	CB
62-3644	T38A	64FTW	
62-3645	T38A	64FTW	
62-3646	T38A	22ARW	
62-3647	T38A	w.o.66	
62-3648	T38A	71FTW	VN
62-3649	T38A	to Turkey	23649
62-3650	T38A	64FTW	
62-3651	T38A	64FTW	
62-3652	T38A	64FTW	
62-3653	T38A	64FTW	
62-3654	T38A	64FTW	
62-3655	T38A	w.o.	
62-3656	T38A	64FTW	
62-3657	T38A	80FTW	
62-3658	T38A	w.o.66	
62-3659	T38A	80FTW	
62-3660	AT38B	AMARC	TF...
62-3661	T38A	64FTW	
62-3662	T38A	64FTW	
62-3663	T38A		
62-3664	T38A	w.o.66	
62-3665	T38A	64FTW	
62-3666	T38A	w.o.73	
62-3667	T38A		
62-3668	T38A	64FTW	
62-3669	T38A	64FTW	
62-3670	T38A	w.o.66	
62-3671	T38A	14FTW	CB
62-3672	T38A	14FTW	CB
62-3673	AT38B	49FW	HO
62-3674	T38A	64FTW	
62-3675	T38A	80FTW	
62-3676	T38A	w.o.65	
62-3677	T38A	80FTW	
62-3678	AT38B	AMARC	TF...
62-3679	T38A	64FTW	
62-3680	T38A	64FTW	
62-3681	T38A	64FTW	
62-3682	T38A	w.o.72	
62-3683	T38A	12FTW	
62-3684	T38A	w.o.64	
62-3685	T38A	64FTW	
62-3686	T38A	64FTW	
62-3687	T38A	w.o.64	
62-3688	T38A	to Turkey	23688
62-3689	T38A	64FTW	
62-3690	T38A	64FTW	
62-3691	T38A	64FTW	
62-3692	T38A	64FTW	
62-3693	T38A	71FTW	VN
62-3694	T38A	w.o.67	
62-3695	T38A	w.o.	
62-3696	T38A	to Turkey	23696
62-3697	T38A	64FTW	
62-3698	T38A		
62-3699	T38A	64FTW	
62-3700	T38A	w.o.72	
62-3701	T38A	64FTW	
62-3702	T38A	64FTW	
62-3703	AT38B	49FW	HO
62-3704	T38A	w.o.	
62-3705	T38A	64FTW	
62-3706	T38A	64FTW	
62-3707	T38A	w.o.65	
62-3708	T38A	to Turkey	23708
62-3709	T38A	71FTW	VN
62-3710	T38A	w.o.70	
62-3711	T38A	to Turkey	23711
62-3712	T38A	w.o.	
62-3713	T38A	71FTW	VN
62-3714	T38A	71FTW	VN
62-3715	AT38B	stored	Edwards AFB
62-3716	T38A	71FTW	VN
62-3717	T38A	71FTW	VN
62-3718	T38A	71FTW	VN
62-3719	T38A	71FTW	VN
62-3720	T38A	80FTW	
62-3721	T38A	71FTW	VN
62-3722	T38A	71FTW	VN
62-3723	T38A	71FTW	VN
62-3724	T38A	71FTW	VN
62-3725	T38A	71FTW	VN
62-3726	T38A	w.o.71	
62-3727	T38A	to Turkey	63727
62-3728	T38A	to Turkey	23728
62-3729	T38A	71FTW	VN
62-3730	T38A	71FTW	VN
62-3731	T38A	w.o.69	
62-3732	T38A	14FTW	CB
62-3733	T38A	71FTW	VN
62-3734	T38A	71FTW	VN
62-3735	T38A	w.o.67	
62-3736	T38A	71FTW	VN
62-3737	T38A	to Turkey	23737
62-3738	AT38B	AMARC	TF112
62-3739	T38A	71FTW	VN
62-3740	T38A	71FTW	VN
62-3741	T38A		
62-3742	T38A	71FTW	VN
62-3743	T38A	71FTW	VN
62-3744	T38A	71FTW	VN
62-3745	T38A	71FTW	VN
62-3746	AT38B	stored	Edwards AFB
62-3747	T38A	71FTW	VN
62-3748	T38A	71FTW	VN
62-3749	T38A	to Turkey	23749
62-3750	T38A	80FTW	
62-3751	T38A	80FTW	
62-3752	AT38B	AMARC	TF...
63-8111	T38A	71FTW	VN
63-8112	AT38B	49FW	HO
63-8113	T38A	71FTW	VN
63-8114	T38A	w.o.69	
63-8115	T38A	71FTW	VN
63-8116	T38A	71FTW	VN
63-8117	AT38B	to NASA	Langley
63-8118	T38A	71FTW	VN
63-8119	T38A	47FTW	
63-8120	T38A	47FTW	
63-8121	T38A	71FTW	VN
63-8122	T38A	w.o.67	
63-8123	T38A	47FTW	
63-8124	T38A	64FTW	
63-8125	T38A	TTC	Sheppard
63-8126	T38A	71FTW	VN
63-8127	T38A	71FTW	VN
63-8128	T38A	w.o.16.10.80	

Serial	Type	Unit	Code
63-8129	T38A	80FTW	EN
63-8130	T38A	w.o.72	
63-8131	T38A	71FTW	VN
63-8132	T38A	71FTW	VN
63-8133	T38A	71FTW	VN
63-8134	T38A	w.o.70	
63-8135	T38A	412TW	ED AFFTC
63-8136	T38A	w.o.	
63-8137	T38A	71FTW	VN
63-8138	T38A	71FTW	VN
63-8139	T38A	47FTW	
63-8140	T38A	w.o.70	
63-8141	T38A	47FTW	
63-8142	T38A	71FTW	VN
63-8143	T38A	71FTW	VN
63-8144	T38A	14FTW	CB
63-8145	T38A	to Turkey	38145
63-8146	T38A	71FTW	VN
63-8147	T38A	14FTW	CB
63-8148	T38A	80FTW	
63-8149	AT38B	AMARC	TF...
63-8150	T38A	47FTW	
63-8151	T38A	47FTW	
63-8152	T38A	71FTW	VN
63-8153	AT38B	w.o.	
63-8154	T38A	64FTW	
63-8155	T38A	71FTW	VN
63-8156	T38A	64FTW	
63-8157	T38A	71FTW	VN
63-8158	T38A	w.o.	
63-8159	T38A	to Turkey	38159
63-8160	T38A	w.o.	
63-8161	T38A	to Turkey	38161
63-8162	AT38B	49FW	HO
63-8163	T38A	80FTW	
63-8164	AT38B	AMARC	TF106
63-8165	T38A	w.o.65	
63-8166	AT38B	49FW	HO
63-8167	T38A	80FTW	
63-8168	T38A	14FTW	CB
63-8169	T38B	w.o.	
63-8170	T38A	w.o.71	
63-8171	T38A	w.o.	
63-8172	AT38B	AMARC	TF...
63-8173	T38A	to Turkey	38173
63-8174	T38A	14FTW	CB
63-8175	AT38B	49FW	HO
63-8176	T38A		
63-8177	T38A	14FTW	CB
63-8178	T38A	80FTW	
63-8179	T38A	64FTW	
63-8180	T38A	64FTW	
63-8181	T38A	to NASA	
63-8182	T38A	64FTW	
63-8183	T38A	to Turkey	38183
63-8184	T38A	14FTW	CB
63-8185	T38A	64FTW	
63-8186	T38A	w.o.67	
63-8187	AT38B	49FW	HO
63-8188	T38A	to NASA	
63-8189	T38A	64FTW	
63-8190	T38A	80FTW	
63-8191	T38A	to Turkey	38191
63-8192	T38A	64FTW	
63-8193	T38A	to NASA	N902NA
63-8194	T38A	w.o.	
63-8195	T38A	to Turkey	38195
63-8196	T38A	64FTW	
63-8197	T38A	64FTW	
63-8198	T38A	47FTW	
63-8199	T38A	w.o.65	
63-8200	T38A	to NASA	N903NA
63-8201	T38A	to Turkey	38201
63-8202	T38A	47FTW	
63-8203	T38A	to Turkey	38203
63-8204	T38A	to NASA	N904NA
63-8205	T38A	to Turkey	38205
63-8206	T38A	to Turkey	38206
63-8207	AT38B	49FW	HO
63-8208	T38A	to Turkey	38208
63-8209	T38A	47FTW	
63-8210	T38A	to Turkey	38210
63-8211	AT38B	AMARC	TF...
63-8212	T38A	w.o.71	
63-8213	T38A	w.o.66	
63-8214	AT38B	AMARC	TF...
63-8215	AT38B	AMARC	TF...
63-8216	T38A	to Turkey	38216
63-8217	T38A	w.o.23.2.77	
63-8218	T38A	80FTW	
63-8219	T38A		
63-8220	T38A	to Turkey	38220
63-8221	T38A		
63-8222	T38A	47FTW	
63-8223	T38A	w.o.70	
63-8224	T38A	64FTW	
63-8225	T38A	22ARW	
63-8226	T38A	64FTW	
63-8227	T38A		
63-8228	T38A		
63-8229	T38A	71FTW	VN
63-8230	T38A		
63-8231	T38A	to Turkey	38231
63-8232	T38A	14FTW	CB
63-8233	T38A	to Turkey	38233
63-8234	T38A		
63-8235	T38A	64FTW	
63-8236	T38A	47FTW	
63-8237	T38A		
63-8238	T38A	64FTW	
63-8239	T38A	47FTW	
63-8240	T38A	80FTW	
63-8241	T38A	47FTW	
63-8242	T38A	w.o.	
63-8243	T38A		
63-8244	T38A	w.o.65	
63-8245	T38A	w.o.70	
63-8246	T38A	47FTW	
63-8247	AT38B		
64-13166	T38A	w.o.	
64-13167	T38A	47FTW	
64-13168	T38A	12FTW	
64-13169	AT38B		
64-13170	T38A	w.o.72	
64-13171	T38A	47FTW	
64-13172	AT38B	49FW	HO
64-13173	T38A	47FTW	
64-13174	T38A	47FTW	
64-13175	T38A	64FTW	
64-13176	T38A	47FTW	
64-13177	T38A	w.o.65	
64-13178	T38A		
64-13179	T38A	96W	DY
64-13180	T38A		
64-13181	T38A	12FTW	
64-13182	T38A	80FTW	
64-13183	T38A	47FTW	
64-13184	T38A	w.o.70	
64-13185	T38A	47FTW	
64-13186	T38A	47FTW	
64-13187	T38A	12FTW	
64-13188	AT38B	AMARC	TF...
64-13189	T38A	80FTW	
64-13190	T38A	9RW	BB
64-13191	T38A	47FTW	
64-13192	T38A	w.o.	
64-13193	AT38B	49FW	HO
64-13194	T38A	12FTW	
64-13195	T38A	47FTW	
64-13196	T38A	47FTW	
64-13197	T38A	12FTW	
64-13198	T38A		
64-13199	T38A	71FTW	VN
64-13200	T38A		
64-13201	T38A	80FTW	
64-13202	T38A	14FTW	CB
64-13203	AT38B	49FW	HO
64-13204	T38A	80FTW	
64-13205	T38A	w.o.	
64-13206	T38A	12FTW	
64-13207	T38A		
64-13208	T38A	47FTW	
64-13209	T38A	12FTW	
64-13210	T38A	64FTW	
64-13211	AT38B	49FW	HO
64-13212	T38A	9RW	BB

Serial	Type	Unit	Code
64-13213	T38A	64FTW	
64-13214	T38A	47FTW	
64-13215	AT38B	AMARC	TF...
64-13216	T38A	47FTW	
64-13217	T38A	9RW	BB
64-13218	T38A	w.o.67	
64-13219	T38A	47FTW	
64-13220	T38A	47FTW	
64-13221	T38A	47FTW	
64-13222	T38A	47FTW	
64-13223	T38A		
64-13224	T38A	47FTW	
64-13225	T38A	71FTW	VN
64-13226	T38A	w.o.65	
64-13227	T38A	47FTW	
64-13228	T38A	47FTW	
64-13229	T38A	w.o.70	
64-13230	T38A	47FTW	
64-13231	T38A	64FTW	
64-13232	AT38B	49FW	HO
64-13233	T38A	80FTW	
64-13234	T38A	80FTW	
64-13235	T38A	71FTW	VN
64-13236	T38A	47FTW	
64-13237	T38A	47FTW	
64-13238	T38A	80FTW	
64-13239	T38A	47FTW	
64-13240	T38A	9RW	BB
64-13241	T38A	47FTW	
64-13242	T38A	47FTW	
64-13243	T38A	47FTW	
64-13244	T38A	47FTW	
64-13245	AT38B	AMARC	TF...
64-13246	T38A	12FTW	
64-13247	T38A	64FTW	
64-13248	T38A	w.o.68	
64-13249	T38A	47FTW	
64-13250	T38A	w.o.	
64-13251	T38A	47FTW	
64-13252	T38A	AMARC	TF...
64-13253	T38A		
64-13254	T38A	47FTW	
64-13255	T38A	47FTW	
64-13256	T38A	47FTW	
64-13257	T38A	47FTW	
64-13258	T38A	47FTW	
64-13259	T38A	47FTW	
64-13260	T38A	w.o.72	
64-13261	AT38B	49FW	HO
64-13262	T38A	64FTW	
64-13263	T38A	14FTW	CB
64-13264	AT38B	49FW	HO
64-13265	T38A	64FTW	
64-13266	T38A	80FTW	EN
64-13267	AT38B	stored	El Paso, Tx
64-13268	T38A	64FTW	
64-13269	AT38B	49FW	HO
64-13270	T38A	9RW	BB
64-13271	T38A	9RW	BB
64-13272	T38A	47FTW	
64-13273	T38A	47FTW	
64-13274	T38A	14FTW	CB
64-13275	T38A	71FTW	VN
64-13276	AT38B	AMARC	TF...
64-13277	T38A	80FTW	
64-13278	T38A	64FTW	
64-13279	T38A		
64-13280	AT38B	AMARC	TF...
64-13281	T38A	9RW	BB
64-13282	T38A	64FTW	
64-13283	T38A	12FTW	
64-13284	T38A	47FTW	
64-13285	T38A	64FTW	
64-13286	T38A	80FTW	
64-13287	T38A	64FTW	
64-13288	AT38B	49FW	HO
64-13289	T38A	22ARW	
64-13290	T38A	80FTW	
64-13291	T38A	w.o..66	
64-13292	AT38B	AMARC	TF...
64-13293	T38A	71FTW	VN
64-13294	T38A	SoC	
64-13295	T38A	80FTW	EN
64-13296	T38A		
64-13297	T38A	9RW	BB
64-13298	AT38B	49FW	HO
64-13299	T38A	71FTW	VN
64-13300	T38A	71FTW	VN
64-13301	T38A	9RW	BB
64-13302	T38A	9RW	BB
64-13303	T38A	w.o.70	
64-13304	T38A	9RW	BB
64-13305	T38A	47FTW	
65-10316	T38A	w.o.68	
65-10317	T38A		
65-10318	T38A	w.o.15.3.77	
65-10319	T38A	w.o.73	
65-10320	T38A		
65-10321	GAT38B	TTC	Sheppard
65-10322	T38A	80FTW	
65-10323	T38A	w.o.69	
65-10324	T38A		
65-10325	T38A	412TW	ED AFFTC
65-10326	T38A	to NASA	N906NA
65-10327	T38A	to USN\TPS	510327
65-10328	T38A	to NASA	N908NA
65-10329	T38A	to NASA	N511NA
65-10330	T38A	AMARC	TF052
65-10331	T38A	47FTW	
65-10332	T38A	80FTW	
65-10333	T38A	47FTW	
65-10334	T38A	w.o.66	
65-10335	T38A	71FTW	VN
65-10336	T38A	47FTW	
65-10337	AT38B	49FW	HO
65-10338	T38A	47FTW	
65-10339	T38A	47FTW	
65-10340	T38A	71FTW	VN
65-10341	AT38B		
65-10342	T38A	384BW	OZ
65-10343	T38A	12FTW	
65-10344	T38A	w.o.23.1.81	
65-10345	T38A	47FTW	
65-10346	AT38B		
65-10347	T38A		
65-10348	T38A	64FTW	
65-10349	T38A	71FTW	VN
65-10350	AT38B	49FW	HO
65-10351	T38A	to NASA	N909NA
65-10352	T38A	to NASA	N910NA
65-10353	T38A	to NASA	N821NA
65-10354	T38A	to NASA	N912NA
65-10355	T38A	to NASA	N913NA
65-10356	T38A	to NASA	N914NA
65-10357	T38A	to NASA	N915NA
65-10358	T38A	71FTW	VN
65-10359	T38A	80FTW	
65-10360	T38A	w.o.68	
65-10361	T38A	12FTW	
65-10362	T38A	to NASA	
65-10363	T38A	Lockheed	Palmdale
65-10364	T38A		
65-10365	T38A	AMARC	TF 053
65-10366	T38A	64FTW	
65-10367	AT38B	49FW	HO
65-10368	T38A	w.o.18.5.85	
65-10369	T38A	71FTW	VN
65-10370	AT38B	49FW	HO
65-10371	AT38B	49FW	HO
65-10372	T38A	14FTW	CB
65-10373	T38A	12FTW	
65-10374	T38A	14FTW	CB
65-10375	T38A	412TW	ED AFFTC
65-10376	T38A	12FTW	
65-10377	T38A	71FTW	VN
65-10378	T38A	47FTW	
65-10379	T38A	64FTW	
65-10380	T38A	64FTW	
65-10381	AT38B	49FW	HO
65-10382	AT38B	49FW	HO
65-10383	T38A	w.o.	
65-10384	T38A	47FTW	
65-10385	T38A	71FTW	VN
65-10386	T38A	14FTW	CB
65-10387	T38A		
65-10388	T38A	47FTW	

Serial	Type	Unit/Status	Notes
65-10389	T38A	71FTW	VN
65-10390	T38A	47FTW	
65-10391	T38A	w.o.	
65-10392	T38A	47FTW	
65-10393	T38A		
65-10394	T38A	64FTW	
65-10395	T38A	71FTW	VN
65-10396	T38A	71FTW	VN
65-10397	T38A	w.o.5.4.83	
65-10398	T38A	w.o.	
65-10399	AT38B	49FW	HO
65-10400	T38A	14FTW	CB
65-10401	T38A	47FTW	
65-10402	T38A	412TW	ED AFFTC
65-10403	AT38B	49FW	HO
65-10404	T38A	14FTW	CB
65-10405	T38A	w.o.	
65-10406	T38A	47FTW	
65-10407	T38A	64FTW	
65-10408	T38A	412TW	ED AFFTC
65-10409	T38A	71FTW	VN
65-10410	T38A	w.o.66	
65-10411	T38A	dump	Edwards AFB
65-10412	T38A		
65-10413	T38A	47FTW	
65-10414	T38A	64FTW	
65-10415	T38A	w.o.	
65-10416	T38A	64FTW	
65-10417	T38A	14FTW	CB
65-10418	T38A	12FTW	
65-10419	T38A	64FTW	
65-10420	T38A	47FTW	
65-10421	T38A	47FTW	
65-10422	T38A	12FTW	
65-10423	T38A	w.o.69	
65-10424	T38A	47FTW	
65-10425	AT38B	49FW	HO
65-10426	T38B	preserved	Vance AFB
65-10427	T38A	80FTW	
65-10428	T38A	12FTW	
65-10429	T38A	12FTW	
65-10430	T38A	71FTW	VN
65-10431	T38A	71FTW	VN
65-10432	AT38B	49FW	HO
65-10433	T38A	80FTW	
65-10434	T38A	71FTW	VN
65-10435	T38A	64FTW	
65-10436	T38A	w.o.73	
65-10437	AT38B	49FW	HO
65-10438	T38A		
65-10439	AT38B	AMARC	TF...
65-10440	T38A	12FTW	
65-10441	T38A	USAF	Museum
65-10442	T38A	64FTW	
65-10443	T38A	SoC	
65-10444	T38A	w.o.14.1.83	
65-10445	T38A	47FTW	
65-10446	T38A	w.o.68	
65-10447	T38A		
65-10448	T38A	dump	Randolph AFB
65-10449	T38A	64FTW	
65-10450	AT38B	stored	El Paso, Tx
65-10451	T38A	14FTW	CB
65-10452	AT38B	49FW	HO
65-10453	T38A	80FTW	
65-10454	T38A	12FTW	
65-10455	T38A	12FTW	
65-10456	AT38B	AMARC	TF...
65-10457	AT38B	49FW	HO
65-10458	T38A	12FTW	
65-10459	T38A	80FTW	
65-10460	T38A	12FTW	
65-10461	T38A	47FTW	
65-10462	T38A	w.o..70	
65-10463	T38A	12FTW	
65-10464	T38A	47FTW	
65-10465	T38A	w.o.67	
65-10466	AT38B	49FW	HO
65-10467	T38A	64FTW	
65-10468	T38A	64FTW	
65-10469	T38A	12FTW	
65-10470	T38A		
65-10471	T38A	64FTW	
65-10472	AT38B	49FW	HO
65-10473	T38A	71FTW	VN
65-10474	T38A		
65-10475	T38A	w.o..67	
66-4320	T38A	80FTW	
66-4321	T38A	80FTW	
66-4322	T38A	w.o.70	
66-4323	T38A	80FTW	
66-4324	T38A	80FTW	
66-4325	T38A	80FTW	
66-4326	T38A	47FTW	
66-4327	T38A	14FTW	CB
66-4328	T38A	12FTW	
66-4329	T38A	w.o.21.3.84	
66-4330	T38A	64FTW	
66-4331	T38A	w.o.21.3.84	
66-4332	T38A	64FTW	
66-4333	T38A	12FTW	
66-4334	T38A		
66-4335	T38A	71FTW	VN
66-4336	T38A	w.o.67	
66-4337	T38A	47FTW	
66-4338	T38A	64FTW	
66-4339	T38A	14FTW	CB
66-4340	T38A	80FTW	
66-4341	T38A	80FTW	
66-4342	T38A	80FTW	
66-4343	T38A	80FTW	
66-4344	T38A	80FTW	
66-4345	T38A	80FTW	
66-4346	T38A	80FTW	
66-4347	T38A	80FTW	
66-4348	T38A	80FTW	
66-4349	T38A	80FTW	
66-4350	T38A	80FTW	
66-4351	T38A	80FTW	
66-4352	T38A	80FTW	
66-4353	T38A	w.o.70	
66-4354	T38A	64FTW	
66-4355	T38A	80FTW	
66-4356	T38A	w.o.	
66-4357	T38A		
66-4358	T38A	14FTW	CB
66-4359	T38A	47FTW	
66-4360	T38A	preserved	Williams AFB
66-4361	T38A	80FTW	
66-4362	T38A	14FTW	CB
66-4363	T38A	80FTW	
66-4364	T38A	w.o.31.8.78	
66-4365	T38A	12FTW	
66-4366	T38A	80FTW	
66-4367	T38A	80FTW	
66-4368	T38A	80FTW	
66-4369	T38A	80FTW	
66-4370	T38A	71FTW	VN
66-4371	T38A	w.o.69	
66-4372	T38A	80FTW	
66-4373	T38A	w.o.	
66-4374	T38A	80FTW	EN
66-4375	T38A	80FTW	
66-4376	T38A	w.o.	
66-4377	T38A	w.o.68	
66-4378	T38A	14FTW	CB
66-4379	T38A		
66-4380	T38A	64FTW	
66-4381	T38A	14FTW	CB
66-4382	T38A	12FTW	
66-4383	T38A	12FTW	
66-4384	T38A	80FTW	
66-4385	T38A		
66-4386	T38A	80FTW	
66-4387	T38A	80FTW	
66-4388	T38A	12FTW	
66-4389	T38A	12FTW	
66-8349	T38A	w.o.67	
66-8350	T38A	80FTW	
66-8351	T38A	w.o.	
66-8352	T38A	80FTW	
66-8353	T38A	80FTW	
66-8354	T38A	to NASA	
66-8355	T38A	to NASA	N923NA
66-8356	T38A	14FTW	CB
66-8357	T38A	12FTW	

Serial	Type	Unit	Code
66-8358	T38A	80FTW	
66-8359	T38A	80FTW	
66-8360	T38A	80FTW	
66-8361	T38A	80FTW	
66-8362	T38A	80FTW	
66-8363	T38A	w.o.	
66-8364	T38A	80FTW	
66-8365	T38A	80FTW	
66-8366	T38A	80FTW	
66-8367	T38A	80FTW	
66-8368	T38A	80FTW	
66-8369	T38A	80FTW	
66-8370	T38A	w.o.67	
66-8371	T38A	w.o.68	
66-8372	T38A	80FTW	
66-8373	T38A	80FTW	
66-8374	T38A	80FTW	
66-8375	T38A	80FTW	
66-8376	T38A	w.o.68	
66-8377	T38A	80FTW	
66-8378	T38A	80FTW	
66-8379	T38A	80FTW	
66-8380	T38A	80FTW	
66-8381	T38A	to NASA	N901NA
66-8382	T38A	to NASA	N916NA
66-8383	T38A	to NASA	N917NA
66-8384	T38A	to NASA	N918NA
66-8385	T38A	to NASA	N919NA
66-8386	T38A	to NASA	N920NS
66-8387	T38A	to NASA	N921NS
66-8388	T38A	80FTW	
66-8389	T38A	80FTW	
66-8390	T38A	80FTW	
66-8391	T38A	80FTW	
66-8392	T38A	80FTW	
66-8393	T38A	80FTW	
66-8394	T38A	80FTW	EN
66-8395	T38A	80FTW	EN
66-8396	T38A	w.o.69	
66-8397	T38A	14FTW	CB
66-8398	T38A	12FTW	
66-8399	T38A	12FTW	
66-8400	T38A	64FTW	
66-8401	T38A	12FTW	
66-8402	T38A	384BW	OZ
66-8403	T38A	12FTW	
66-8404	T38A	384BW	OZ
67-14825	T38A	to NASA	N924NA
67-14826	T38A	12FTW	
67-14827	T38A	12FTW	
67-14828	T38A	55W	OF
67-14829	T38A	12FTW	
67-14830	T38A	w.o.	
67-14831	T38A	384BW	OZ
67-14832	T38A	71FTW	VN
67-14833	T38A	12FTW	
67-14834	T38A		
67-14835	T38A	w.o.	
67-14836	T38A		
67-14837	T38A	14FTW	CB
67-14838	T38A		
67-14839	T38A	w.o.69	
67-14840	T38A		
67-14841	T38A	80FTW	
67-14842	AT38B	49FW	HO
67-14843	T38A	w.o.	
67-14844	T38A	47FTW	
67-14845	T38A	22ARW	
67-14846	T38A	12FTW	
67-14847	T38A	w.o.	
67-14848	T38A		
67-14849	T38A	71FTW	VN
67-14850	T38A	64FTW	
67-14851	T38A	71FTW	VN
67-14852	T38A	47FTW	
67-14853	T38A	12FTW	
67-14854	T38A	14FTW	CB
67-14855	T38A	14FTW	CB
67-14856	T38A	412TW	ED AFFTC
67-14857	T38A		
67-14858	T38A	71FTW	VN
67-14859	T38A	14FTW	CB
67-14915	T38A	71FTW	VN
67-14916	T38A	47FTW	
67-14917	T38A	12FTW	
67-14918	T38A	w.o.	
67-14919	T38A		
67-14920	T38A	28BW	EL
67-14921	T38A	71FTW	VN
67-14922	T38A	71FTW	VN
67-14923	T38A	12FTW	
67-14924	T38A	47FTW	
67-14925	T38A	14FTW	CB
67-14926	T38A	47FTW	
67-14927	T38A		
67-14928	T38A		
67-14929	T38A	80FTW	
67-14930	T38A	14FTW	CB
67-14931	T38A		
67-14932	T38A	47FTW	
67-14933	T38A	14FTW	CB
67-14934	T38A	14FTW	CB
67-14935	T38A	12FTW	
67-14936	T38A	71FTW	VN
67-14937	T38A	14FTW	CB
67-14938	T38A		
67-14939	T38A	12FTW	
67-14940	T38A		
67-14941	T38A	71FTW	VN
67-14942	T38A	12FTW	
67-14943	T38A	412TW	ED AFFTC
67-14944	T38A	71FTW	VN
67-14945	T38A	71FTW	VN
67-14946	T38A	384BW	OZ
67-14947	T38A	12FTW	
67-14948	T38A	w.o.15.5.81	
67-14949	T38A	71FTW	VN
67-14950	T38A	12FTW	
67-14951	T38A	47FTW	
67-14952	T38A	96W	DY
67-14953	T38A	12FTW	
67-14954	T38A	w.o.6.82	
67-14955	T38A	47FTW	
67-14956	T38A	412TW	ED AFFTC
67-14957	T38A	14FTW	CB
67-14958	T38A	71FTW	VN
68-8095	T38A	w.o.20.6.93	
68-8096	T38A	71FTW	VN
68-8097	T38A	14FTW	CB
68-8098	T38A	14FTW	CB
68-8099	T38A	14FTW	CB
68-8100	T38A	w.o.25.7.77	
68-8101	T38A	64FTW	
68-8102	T38A	71FTW	VN
68-8103	T38A	w.o.	
68-8104	T38A	14FTW	CB
68-8105	T38A	14FTW	CB
68-8106	AT38B	49FW	HO
68-8107	T38A	12FTW	
68-8108	T38A	w.o.72	
68-8109	AT38B	49FW	HO
68-8110	T38A	71FTW	VN
68-8111	T38A	w.o.	
68-8112	T38A	12FTW	
68-8113	AT38B	stored	El Paso, Tx
68-8114	T38A	14FTW	CB
68-8115	T38A	w.o.70	
68-8116	AT38B		
68-8117	T38A	71FTW	VN
68-8118	T38A	47FTW	
68-8119	T38A	14FTW	CB
68-8120	T38A	14FTW	CB
68-8121	T38A	47FTW	
68-8122	T38A	14FTW	CB
68-8123	AT38B		
68-8124	T38A	12FTW	
68-8125	T38A	14FTW	CB
68-8126	T38A	w.o.72	
68-8127	T38A	14FTW	CB
68-8128	T38A		
68-8129	T38A	14FTW	CB
68-8130	T38A	71FTW	VN
68-8131	T38A	w.o.9.5.81	
68-8132	T38A	12FTW	
68-8133	AT38B	stored	El Paso, Tx
68-8134	T38A	12FTW	

68-8135	T38A	12FTW	
68-8136	T38A	12FTW	
68-8137	T38A		
68-8138	AT38B	49FW	HO
68-8139	T38A	12FTW	
68-8140	AT38B	AMARC	TF...
68-8141	T38A	12FTW	
68-8142	AT38B	49FW	HO
68-8143	T38A	12FTW	
68-8144	T38A	71FTW	VN
68-8145	T38A	14FTW	CB
68-8146	T38A	14FTW	CB
68-8147	T38A	47FTW	
68-8148	T38A	14FTW	CB
68-8149	T38A	w.o.	
68-8150	T38A	14FTW	CB
68-8151	T38A	14FTW	CB
68-8152	T38A		
68-8153	T38A	412TW	ED AFFTC
68-8154	T38A	412TW	ED AFFTC
68-8155	T38A	14FTW	CB
68-8156	T38A	w.o.18.1.82	
68-8157	T38A	14FTW	CB
68-8158	T38A	412TW	ED AFFTC
68-8159	T38A	71FTW	VN
68-8160	T38A	71FTW	VN
68-8161	T38A	80FTW	
68-8162	T38A	14FTW	CB
68-8163	T38A	14FTW	CB
68-8164	T38A	14FTW	CB
68-8165	T38A	14FTW	CB
68-8166	T38A	14FTW	CB
68-8167	T38A	14FTW	CB
68-8168	AT38B	49FW	HO
68-8169	T38A	14FTW	CB
68-8170	T38A	14FTW	CB
68-8171	T38A	71FTW	VN
68-8172	T38A	12FTW	
68-8173	T38A	71FTW	VN
68-8174	T38B	w.o.	
68-8175	T38A	w.o.18.1.82	
68-8176	T38A	w.o.18.1.82	
68-8177	T38A	12FTW	
68-8178	T38A	w.o.73	
68-8179	T38A	12FTW	
68-8180	T38A	14FTW	CB
68-8181	T38A	14FTW	CB
68-8182	T38A	w.o.8.9.81	
68-8183	T38B	dump	Norton AFB
68-8184	T38A	w.o.18.1.82	
68-8185	T38A	12FTW	
68-8186	T38A	12FTW	
68-8187	T38A	14FTW	CB
68-8188	T38A	14FTW	CB
68-8189	T38A	14FTW	CB
68-8190	T38A	12FTW	
68-8191	T38A	12FTW	
68-8192	T38A	47FTW	
68-8193	T38A	71FTW	VN
68-8194	T38A	to US Navy	Bu158197
68-8195	T38A	14FTW	CB
68-8196	T38A	71FTW	VN
68-8197	T38A	80FTW	
68-8198	T38A	14FTW	CB
68-8199	T38A	14FTW	CB
68-8200	T38A	14FTW	CB
68-8201	T38A	14FTW	CB
68-8202	T38A	14FTW	CB
68-8203	T38A	80FTW	
68-8204	T38A	14FTW	CB
68-8205	T38A	412TW	ED AFFTC
68-8206	T38A	71FTW	VN
68-8207	T38A	14FTW	CB
68-8208	T38A	14FTW	CB
68-8209	T38A	to US Navy	Bu158198
68-8210	T38A	71FTW	VN
68-8211	T38A	14FTW	CB
68-8212	T38A	to US Navy	Bu158199
68-8213	T38A	12FTW	
68-8214	T38A	to US Navy	Bu158200
68-8215	T38A	14FTW	CB
68-8216	T38A	to US Navy	Bu158201
68-8217	T38A		
69-7073	T38A	80FTW	
69-7074	T38A	14FTW	CB
69-7075	T38A	14FTW	CB
69-7076	T38A	14FTW	CB
69-7077	T38A	80FTW	
69-7078	T38A		
69-7079	T38A	71FTW	VN
69-7080	T38A	64FTW	
69-7081	T38A		
69-7082	T38A	to NASA	N955NA
69-7083	T38A	12FTW	
69-7084	T38A	to NASA	N956NA
69-7085	T38A	12FTW	
69-7086	T38A	to NASA	N957NA
69-7087	T38A	12FTW	
69-7088	T38A	to NASA	N958NA
70-1549	T38A	12FTW	
70-1550	T38A	to NASA	N959NA
70-1551	T38A	64FTW	
70-1552	T38A	to NASA	N960NA
70-1553	T38A	64FTW	
70-1554	T38A	12FTW	
70-1555	T38A	to NASA	N961NA
70-1556	T38A	to NASA	N962NA
70-1557	T38A	71FTW	VN
70-1558	T38A	412TW	ED AFFTC
70-1559	T38A	412TW	ED AFFTC
70-1560	T38A		
70-1561	T38A	47FTW	
70-1562	T38A	64FTW	
70-1563	T38A	14FTW	CB
70-1564	T38A	64FTW	
70-1565	T38A	71FTW	VN
70-1566	T38A	14FTW	CB
70-1567	T38A	64FTW	
70-1568	T38A	47FTW	
70-1569	T38A	14FTW	CB
70-1570	T38A	80FTW	
70-1571	T38A	80FTW	
70-1572	T38A	to NASA	N963NA
70-1573	T38A	71FTW	VN
70-1574	T38A	412TW	ED AFFTC
70-1575	T38A	412TW	ED AFFTC
70-1576	T38A	14FTW	CB
70-1577	T38A	64FTW	
70-1578	T38A		
70-1579	T38A	412TW	ED AFFTC
70-1580	T38A	80FTW	
70-1581	T38A	71FTW	VN
70-1582	T38A		
70-1583	T38A	14FTW	CB
70-1584	T38A	64FTW	
70-1585	T38A	47FTW	
70-1586	T38A	47FTW	
70-1587	T38A	71FTW	VN
70-1588	T38A	64FTW	
70-1589	T38A	71FTW	VN
70-1590	T38A	80FTW	
70-1591	T38A	SoC	
70-1949	T38A	12FTW	
70-1950	T38A	71FTW	VN
70-1951	T38A	12FTW	
70-1952	T38A	14FTW	CB
70-1953	T38A	47FTW	
70-1954	T38A	47FTW	
70-1955	T38A		
70-1956	T38A	w.o.73	

T39 SABRELINER

| c\n | T39A: | 265-1 to 265-88, 276-1 to 276-55. |
| | T39B: | 270-1 to 270-6. |

59-2868	CT39A	preserved	Kirtland AFB
59-2869	CT39A	AMARC	TG033
59-2870	NT39A	4950TW	AFMC\ASD
59-2871	T39A	w.o.13.11.69	
59-2872	CT39A	AMARC	TG015
59-2873	CT39B	4950TW	AFMC\ASD
59-2874	NT39B	4950TW	AFMC\ASD
60-3474	CT39B	4950TW	AFMC\ASD
60-3475	CT39B	4950TW	AFMC\ASD
60-3476	NT39B	4950TW	AFMC\ASD
60-3477	CT39B	4950TW	AFMC\ASD
60-3478	NT39A	12FTW	USAF\IFC
60-3479	CT39A	AMARC	TG082
60-3480	CT39A	scrapyard	Bobs
60-3481	CT39A	AMARC	TG085
60-3482	CT39A	AMARC	TG016
60-3483	CT39A	preserved	Travis AFB
60-3484	CT39A	AMARC	TG024
60-3485	CT39A	AMARC	TG003
60-3486	CT39A	AMARC	TG008
60-3487	CT39A	AMARC	TG021
60-3488	CT39A	NASA	
60-3489	CT39A	AMARC	TG058
60-3490	CT39A	AMARC	TG062
60-3491	CT39A	AMARC	TG009
60-3492	CT39A	AMARC	TG007
60-3493	CT39A	AMARC	TG057
60-3494	CT39A	AMARC	TG094
60-3495	CT39A	preserved	Scott AFB
60-3496	CT39A	AMARC	TG072
60-3497	CT39A	AMARC	TG066
60-3498	CT39A	stored	Chandler MAP, Az
60-3499	CT39A	AMARC	TG037
60-3500	CT39A	AMARC	TG030
60-3501	CT39A	AMARC	TG093
60-3502	CT39A	scrapyard	Bobs
60-3503	GCT39A	TTC	Chanute
60-3504	CT39A	GIA	Bi States Col. Il
60-3505	CT39A	preserved	Edwards AFB
60-3506	T39A	w.o.9.2.74	
60-3507	CT39A	GIA	West LA Col. Ca
60-3508	CT39A	AMARC	TG042
61-0634	CT39A	preserved	Dyess AFB
61-0635	CT39A	stored	Layfayette RAP, La
61-0636	CT39A	AMARC	TG089
61-0637	CT39A	AMARC	TG035
61-0638	CT39A	AMARC	TG096
61-0639	CT39A	GIA	Janesville, Wi
61-0640	T39A	w.o.16.4.70	
61-0641	CT39A	AMARC	TG036
61-0642	CT39A	AMARC	TG045
61-0643	CT39A	AMARC	TG022
61-0644	T39A	w.o.7.5.63	
61-0645	CT39A	scrapyard	Bobs
61-0646	T39A	w.o.14.5.75	
61-0647	CT39A	scrapyard	Bobs
61-0648	CT39A	AMARC	TG017
61-0649	T39A	stored	Portland, Or
61-0650	CT39A	AMARC	TG043
61-0651	CT39A	AMARC	TG040
61-0652	CT39A	AMARC	TG087
61-0653	CT39A	AMARC	TG071
61-0654	CT39A	GIA	Daytona Beach, Fl
61-0655	CT39A	scrapyard	Bobs
61-0656	CT39A	AMARC	TG010
61-0657	CT39A	AMARC	TG023
61-0658	CT39A	AMARC	TG034
61-0659	CT39A	to N6581K	

61-0660	CT39A	preserved	McClellan AFB
61-0661	T39A	w.o.9.2.69	
61-0662	CT39A	stored	St. Louis, Mo
61-0663	CT39A	AMARC	TG067
61-0664	T39A	AMARC	TG063
61-0665	CT39A	AMARC	TG028
61-0666	CT39A	AMARC	TG014
61-0667	CT39A	AMARC	TG088
61-0668	CT39A	AMARC	TG051
61-0669	CT39A	AMARC	TG075
61-0670	CT39A		
61-0671	CT39A	stored	Keesler AFB
61-0672	T39A	w.o.13.3.79	
61-0673	CT39A	scrapyard Bobs	
61-0674	CT39A	preserved	Norton AFB
61-0675	CT39A	preserved	Yokota AB, Japan
61-0676	CT39A	AMARC	TG049
61-0677	CT39A	to N9166Y	
61-0678	CT39A	AMARC	TG012
61-0679	CT39A	to N6581E	
61-0680	CT39A	AMARC	TG039
61-0681	CT39A	preserved	Willow Run AP, Mi
61-0682	CT39A	AMARC	TG032
61-0683	CT39A	AMARC	TG025
61-0684	CT39A	scrapyard	Bobs
61-0685	CT39A	preserved	USAAM Ft. Rucker
62-4448	T39A	w.o.29.1.64	
62-4449	CT39A	Pima Museum	
62-4450	CT39A	scrapyard	Bobs
62-4451	CT39A	AMARC	TG060
62-4452	T39A	preserved	Travis AFB
62-4453	CT39A	to N6552R	
62-4454	CT39A	AMARC	TG018
62-4455	CT39A	AMARC	TG065
62-4456	CT39A	scrapyard	Bobs
62-4457	CT39A	AMARC	TG002
62-4458	T39A	w.o.25.3.65	
62-4459	CT39A	AMARC	TG041
62-4460	T39A	w.o.28.2.70	
62-4461	CT39A	preserved	Warner-Robins AFB
62-4462	CT39A	AMARC	TG046
62-4463	CT39A		
62-4464	CT39A	AMARC	TG004
62-4465	CT39A	preserved	March AFB
62-4466	CT39A	GIA	Detroit, Mi
62-4467	CT39A	scrapyard	Bobs
62-4468	CT39A	AMARC	TG020
62-4469	CT39A	AMARC	TG064
62-4470	CT39A		
62-4471	CT39A	preserved	Ramstein AB, FRG
62-4472	CT39A	scrapyard	Bobs
62-4473	CT39A	AMARC	TG029
62-4474	CT39A	AMARC	TG027
62-4475	CT39A	AMARC	TG054
62-4476	NT39A	12FTW	USAF\IFC
62-4477	CT39A	AMARC	TG048
62-4478	T39A	USAF	Museum
62-4479	CT39A	to N988MT	
62-4480	CT39A	scrapyard	Bobs
62-4481	CT39A	to N33UT	
62-4482	CT39A	preserved	Lackland AFB
62-4483	CT39A	AMARC	TG055
62-4484	CT39A	preserved	Kadena AB, Japan
62-4485	CT39A	dump	Yokota AB, Japan
62-4486	CT39A	AMARC	TG050
62-4487	CT39A	preserved	Offutt AFB
62-4488	CT39A		
62-4489	CT39A	to N65618	

62-4490	CT39A	AMARC	TG079
62-4491	CT39A	AMARC	TG081
62-4492	CT39A		
62-4493	CT39A	AMARC	TG076
62-4494	CT39A	preserved	Chanute AFB
62-4495	CT39A	to	N6612S
62-4496	CT39A	w.o.20.4.85	

62-4497	CT39A	AMARC	TG053
62-4498	CT39A	preserved	Salt Lake City, Ut
62-4499	T39A	w.o.24.6.69	
62-4500	CT39A	AMARC	TG070
62-4501	CT39A	AMARC	TG073
62-4502	T39A	w.o.31.12.68	

T41 MESCALERO

65-5100	T41A		
65-5101	T41A		
65-5102	T41A	Randolph AFB	
65-5103	T41A	Beale AFB	
65-5104	T41A	Randolph AFB	
65-5105	T41A	OTS	Hondo simulator
65-5106	T41A	Randolph AFB simulator	
65-5107	T41A	Kirtland AFB	
65-5108	T41A		
65-5109	T41A	OTS	Hondo
65-5110	T41A		
65-5111	T41A		
65-5112	T41A		
65-5113	T41A		
65-5114	T41A	AFMTC	Lackland AFB
65-5115	T41A		
65-5116	T41A		
65-5117	T41A		
65-5118	T41A	Patrick AFB	
65-5119	T41A	Moody AFB	
65-5120	T41A	Davis Monthan AFB	
65-5121	T41A	Burlington, Vt	
65-5122	T41A		
65-5123	T41A		
65-5124	T41A	dump	Kirtland AFB
65-5125	T41A		
65-5126	T41A		
65-5127	T41A		
65-5128	T41A		
65-5129	T41A		
65-5130	T41A		
65-5131	T41A		
65-5132	T41A	to El Salvador 92	
65-5133	T41A		
65-5134	T41A	OTS	Hondo
65-5135	T41A		
65-5136	T41A		
65-5137	T41A	TTC	Sheppard
65-5138	T41A		
65-5139	T41A		
65-5140	T41A	Holloman AFB	
65-5141	T41A		
65-5142	T41A		
65-5143	T41A		
65-5144	T41A	Bergstrom AFB	
65-5145	T41A		
65-5146	T41A		
65-5147	T41A		
65-5148	T41A		
65-5149	T41A	Travis AFB	
65-5150	T41A	March AFB	
65-5151	T41A		
65-5152	T41A	Myrtle Beach AFB	
65-5153	T41A	Edwards AFB	
65-5154	T41A		
65-5155	T41A	GIA	Helena
65-5156	T41A	Grissom AFB	
65-5157	T41A		
65-5158	T41A		
65-5159	T41A		
65-5160	T41A		
65-5161	T41A	Charleston AFB	
65-5162	T41A	w.o.7.7.74	

65-5163	T41A		
65-5164	T41A		
65-5165	T41A	OTS	Hondo
65-5166	T41A		
65-5167	T41A	OTS	Hondo
65-5168	T41A	preserved	Vance AFB
65-5169	T41A		
65-5170	T41A		
65-5171	T41A		
65-5172	T41A		
65-5173	T41A		
65-5174	T41A		
65-5175	T41A		
65-5176	T41A		
65-5177	T41A		
65-5178	T41A		
65-5179	T41A		
65-5180	T41A	OTS	Hondo
65-5181	T41A	Eglin AFB	
65-5182	T41A	Peterson AFB	
65-5183	T41A		
65-5184	T41A		
65-5185	T41A		
65-5186	T41A	SoC	
65-5187	T41A	SoC	
65-5188	T41A	Myrtle Beach AFB	
65-5189	T41A	Kirtland AFB	
65-5190	T41A		
65-5191	T41A	OTS	Hondo
65-5192	T41A		
65-5193	T41A		
65-5194	T41A		
65-5195	T41A		
65-5196	T41A		
65-5197	T41A		
65-5198	T41A	OTS	Hondo
65-5199	T41A	OTS	Hondo
65-5200	T41A		
65-5201	T41A		
65-5202	T41A	OTS	Hondo
65-5203	T41A	OTS	Hondo
65-5204	T41A	Cannon AFB	
65-5205	T41A		
65-5206	T41A	w.o.10.73	
65-5207	T41A	OTS	Hondo
65-5208	T41A	OTS	Hondo
65-5209	T41A	Scott AFB	
65-5210	T41A		
65-5211	T41A		
65-5212	T41A	Charleston AFB	
65-5213	T41A		
65-5214	T41A		
65-5215	T41A	March AFB	
65-5216	T41A	OTS	Hondo
65-5217	T41A		
65-5218	T41A		
65-5219	T41A	Eglin AFB	
65-5220	T41A	Charleston AFB	
65-5221	T41A	Myrtle Beach AFB	
65-5222	T41A	Bergstrom AFB	
65-5223	T41A	Charleston AFB	
65-5224	T41A		
65-5225	T41A	Osan AB, S.Korea	

Serial	Type	Unit	Base
65-5226	T41A		Holloman AFB
65-5227	T41A	OTS	Hondo
65-5228	T41A	OTS	Hondo
65-5229	T41A	TTC	Lowry
65-5230	T41A		
65-5231	T41A		
65-5232	T41A		
65-5233	T41A		
65-5234	T41A		Myrtle Beach AFB
65-5235	T41A	OTS	Hondo
65-5236	T41A		to El Salvador 91
65-5237	T41A		
65-5238	T41A	OTS	Hondo
65-5239	T41A		
65-5240	T41A		
65-5241	T41A		Wright-Patterson AFB
65-5242	T41A		Peterson AFB
65-5243	T41A		Davis Monthan AFB
65-5244	T41A		Myrtle Beach AFB
65-5245	T41A		
65-5246	T41A		
65-5247	T41A		
65-5248	T41A		
65-5249	T41A		
65-5250	T41A	OTS	Hondo
65-5251	T41A	OTS	Hondo
65-5252	T41A	SoC	
65-5253	T41A	OTS	Hondo
65-5254	T41A		Peterson AFB
65-5255	T41A		Beale AFB
65-5256	T41A		
65-5257	T41A		
65-5258	T41A	OTS	Hondo
65-5259	T41A	SoC	
65-5260	T41A		
65-5261	T41A		Moody AFB
65-5262	T41A	SoC	
65-5263	T41A		
65-5264	T41A		
65-5265	T41A		
65-5266	T41A		
65-5267	T41A		Cannon AFB
65-5268	T41A		
65-5269	T41A		Edwards AFB
67-14959	T41A	OTS	Hondo
67-14960	T41A	OTS	Hondo
67-14961	T41A	OTS	Hondo
67-14962	T41A	OTS	Hondo
67-14963	T41A	w.o.	
67-14964	T41A	OTS	Hondo
67-14965	T41A		
67-14966	T41A	OTS	Hondo
67-14967	T41A		
67-14968	T41A		
67-14969	T41A	OTS	Hondo
67-14970	T41A	OTS	Hondo
67-14971	T41A		
67-14972	T41A	OTS	Hondo
67-14973	T41A	OTS	Hondo
67-14974	T41A	OTS	Hondo
67-14975	T41A	OTS	Hondo
67-14976	T41A	OTS	Hondo
67-14977	T41A	OTS	Hondo
67-14978	T41A	OTS	Hondo
67-14979	T41A	OTS	Hondo
67-14980	T41A	OTS	Hondo
67-14981	T41A	OTS	Hondo
67-14982	T41A	OTS	Hondo
67-14983	T41A	OTS	Hondo
67-14984	T41A	OTS	Hondo
67-14985	T41A	OTS	Hondo
67-14986	T41A	OTS	Hondo
67-14987	T41A	OTS	Hondo
67-14988	T41A	OTS	Hondo
67-14989	T41A	OTS	Hondo
67-14990	T41A	OTS	Hondo
67-14991	T41A		
67-14992	T41A	OTS	Hondo
67-15079	GT41B	IAAFA	Homestead AFB
67-15134	T41B		
67-15201	T41B		Kirtland AFB
67-15217	T41B		Edwards AFB
68-7866	T41C	557FTS	USAF Academy
68-7867	T41C	557FTS	USAF Academy
68-7868	T41C	557FTS	USAF Academy
68-7869	T41C	557FTS	USAF Academy
68-7870	T41C	557FTS	USAF Academy
68-7871	T41C	557FTS	USAF Academy
68-7872	T41C	557FTS	USAF Academy
68-7873	T41C	557FTS	USAF Academy
68-7874	T41C	557FTS	USAF Academy
68-7875	T41C	557FTS	USAF Academy
68-7876	T41C	557FTS	USAF Academy
68-7877	T41C		
68-7878	T41C	557FTS	USAF Academy
68-7879	T41C	557FTS	USAF Academy
68-7880	T41C	557FTS	USAF Academy
68-7881	T41C	557FTS	USAF Academy
68-7882	T41C	557FTS	USAF Academy
68-7883	T41C	557FTS	USAF Academy
68-7884	T41C	557FTS	USAF Academy
68-7885	T41C	557FTS	USAF Academy
68-7886	T41C	557FTS	USAF Academy
68-7887	T41C	557FTS	USAF Academy
68-7888	T41C	557FTS	USAF Academy
68-7889	T41C	557FTS	USAF Academy
68-7890	T41C	557FTS	USAF Academy
68-7891	T41C	557FTS	USAF Academy
68-7892	T41C		to El Salvador 94
68-7893	T41C	557FTS	USAF Academy
68-7894	T41C	557FTS	USAF Academy
68-7895	T41C	557FTS	USAF Academy
68-7896	T41C	557FTS	USAF Academy
68-7897	T41C	557FTS	USAF Academy
68-7898	T41C	557FTS	USAF Academy
68-7899	T41C	557FTS	USAF Academy
68-7900	T41C	557FTS	USAF Academy
68-7901	T41C	557FTS	USAF Academy
68-7902	T41C	557FTS	USAF Academy
68-7903	T41C	557FTS	USAF Academy
68-7904	T41C	557FTS	USAF Academy
68-7905	T41C	557FTS	USAF Academy
68-7906	T41C	557FTS	USAF Academy
68-7907	T41C	557FTS	USAF Academy
68-7908	T41C	557FTS	USAF Academy
68-7909	T41C	557FTS	USAF Academy
68-7910	T41C	557FTS	USAF Academy
69-7743	T41A		
69-7744	T41A		Scott AFB
69-7745	T41A		
69-7746	T41A		
69-7747	T41A		Barksdale AFB
69-7748	T41A		Tyndall AFB
69-7749	T41A		Tyndall AFB
69-7750	T41C		w.o.31.3.81
69-7751	T41C	557FTS	USAF Academy
69-7752	T41C	557FTS	USAF Academy
69-7753	T41C	557FTS	USAF Academy
69-7754	T41C	557FTS	USAF Academy
69-7755	T41C	557FTS	USAF Academy
69-7756	T41C	557FTS	USAF Academy

Note: These aircraft also carry civilian registrations: N5100F to N5269F, N4959R to N4992R, N7866N to N7910N and N7743L to N7756L respectively.

T43

c\n	20685 to 20703.		
71-1403	CT43A	12FTW	
71-1404	CT43A	12FTW	
71-1405	CT43A	12FTW	
71-1406	CT43A	12FTW	
72-0282	CT43A	to N5175U	
72-0283	CT43A	N5175U	
72-0284	CT43A	200ALS	HQ West-ANG
72-0285	CT43A	to N5176Y	
72-0286	CT43A	to N5177C	
72-0287	CT43A	200ALS	HQ West-ANG
72-0288	CT43A	200ALS	HQ West-ANG
73-1149	CT43A	86W	RS
73-1150	CT43A	12FTW	
73-1151	CT43A	12FTW	
73-1152	CT43A	12FTW	
73-1153	CT43A	12FTW	
73-1154	CT43A	200ALS	HQ West-ANG
73-1155	CT43A	12FTW	
73-1156	CT43A	12FTW	

T46

84-0492	YT46A	6512TS	ED AFFTC
84-0493	YT46A	AMARC	TM002
85-1596	T46A	AMARC	TM001

U2

c\n	'Article' 341 to 395 built. FY80 aircraft unknown.
Note:	341 (w.o.4.4.57) and 390 were never allotted US military serials. FY80 aircraft originally referred to as TR1A, TR1B and ER2.

56-6675	U2A	'55741'	
56-6676	U2A	w.o.27.10.62	
56-6677	U2A	w.o.	
56-6678	U2C	w.o.	
56-6679	U2B	w.o.	
56-6680	U2F	preserved	NASM
56-6681	U2C	preserved	Moffett Field
56-6682	U2C	preserved	Warner Robins AFB
56-6683	U2B	w.o.20.11.63	
56-6684	U2B	w.o.	
56-6685	U2B	w.o.	
56-6686	U2C	w.o.	
56-6687	U2F	w.o.	
56-6688	U2F	w.o.	
56-6689	U2C	w.o.8.10.66	
56-6690	U2C	w.o.26.9.64	
56-6691	U2C	w.o.10.1.65	
56-6692	U2CT	preserved	IWM Duxford, UK
56-6693	U2C	w.o.1.5.60	
56-6694	U2C	w.o.	
56-6695	U2C	w.o.	
56-6696	WU2A		
56-6697	U2C	w.o.	
56-6698	U2C	stored	Beale AFB
56-6699	U2C	w.o.57	
56-6700	U2C	w.o.29.5.75	
56-6701	U2C	preserved	Offutt AFB
56-6702	U2C	stored	Beale AFB
56-6703	U2C	w.o.	
56-6704	U2C	w.o.	
56-6705	U2C	w.o.	
56-6706	U2C	w.o.	
56-6707	U2C	preserved	Laughlin AFB
56-6708	U2C	w.o.	
56-6709	U2C	w.o.	
56-6710	U2C	w.o.	
56-6711	U2C	w.o.	
56-6712	U2C	w.o.	
56-6713	U2C	w.o.	
56-6714	U2C	preserved	Beale AFB
56-6715	WU2A	w.o.	
56-6716	U2C	preserved	Davis-Monthan AFB
56-6717	WU2A	w.o.	
56-6718	U2C	w.o.	
56-6719	U2C	w.o.	
56-6720	U2C	w.o.	
56-6721	U2D	preserved	March AFB
56-6722	U2C	USAF Museum	
56-6951	U2C	w.o.	
56-6952	U2C	w.o.	
56-6953	U2CT	preserved	Edwards AFB
56-6954	U2D	w.o.	
56-6955	U2D	w.o.14.8.64	
68-10329	U2R	9RW	

68-10330	U2R	w.o.7.12.77		80-1076	U2R	9RW	
68-10331	U2R	9RW		80-1077	U2R	9RW	
68-10332	U2R	w.o.15.1.92		80-1078	U2R		
68-10333	U2R			80-1079	U2R	95RS	
68-10334	U2R	w.o.8.75		80-1080	U2R	95RS	
68-10335	U2R			80-1081	U2R		
68-10336	U2R			80-1082	U2R	9RW	
68-10337	U2R	9RW		80-1083	U2R	9RW	
68-10338	U2R	9RW		80-1084	U2R	9RW	
68-10339	U2R	9RW		80-1085	U2R	9RW	
68-10340	U2R			80-1086	U2R		
80-1063	ER2	to NASA	N706NA	80-1087	U2R	9RW	
80-1064	U2RT	9RW		80-1088	U2R		
80-1065	U2RT	9RW		80-1089	U2R	9RW	
80-1066	U2R	9RW		80-1090	U2R	9RW	
80-1067	U2R			80-1091	U2RT	9RW	
80-1068	U2R	9RW		80-1092	U2R	9RW	
80-1069	U2R	to NASA	NASA708	80-1093	U2R	9RW	
80-1070	U2R	9RW		80-1094	U2R	9RW	
80-1071	U2R	9RW		80-1095	U2R		
80-1072	U2R	9RW		80-1096	U2R	9RW	
80-1073	U2R	9RW		80-1097	ER2	to NASA	N709NA
80-1074	U2R	9RW		80-1098	U2R	9RW	
80-1075	U2R			80-1099	U2R	9RW	

U3 BLUE CANOE

c\n   38001, 38005, 38025, 38047, 38059, 38084, 38132, 0007, 0010.

57-5846	U3A	preserved	Lowry AFB	58-2110	U3A	Buckley ANGB	
57-5850	U3A	Vandenberg AFB		58-2158	U3A	GIA	San Francisco IAP
57-5870	U3A	Randolph AFB		60-6052	U3B	Eglin AFB	
57-5892	U3A	Vandenberg AFB		60-6055	U3B	Sheppard AFB	
57-5904	U3A	Warner Robins AFB					

U6 BEAVER

c\n   563.

53-2781	U6A	412TW	ED AFFTC	

U26

82-0667	U26A to USCS

OV10 BRONCO

c\n   305-1 to 305-157, plus 2 USMC aircraft.

66-13552	OV10A		
66-13553	OV10A	w.o.17.9.81	
66-13554	OV10A		
66-13555	OV10A	to Philippine	
66-13556	OV10A		
66-13557	OV10A		
66-13558	OV10A		
66-13559	OV10A	to Philippine	
66-13560	OV10A		
66-13561	OV10A		
66-13562	OV10A		
67-14604	OV10A		
67-14605	OV10A	to Philippine	
67-14606	GOV10A	scrapped	
67-14607	OV10A	w.o.	
67-14608	OV10A		
67-14609	OV10A		
67-14610	OV10A	to Philippine	
67-14611	OV10A	to Philippine	
67-14612	OV10A	to N95LM	
67-14613	OV10A	to Philippine	
67-14614	OV10A		
67-14615	OV10A	to N93LM	
67-14616	OV10A	to N97LM	
67-14617	OV10A	w.o.8.70	
67-14618	OV10A		
67-14619	OV10A	w.o.	
67-14620	OV10A	w.o.24.2.72	
67-14621	OV10A		
67-14622	OV10A	w.o.19.5.72	
67-14623	OV10A	to USMC	
67-14624	OV10A		
67-14625	OV10A		
67-14626	OV10A	to USMC	
67-14627	OV10A	w.o.13.12.68	
67-14628	OV10A	w.o.5.12.69	
67-14629	OV10A	to Philippine	
67-14630	OV10A	to Philippine	
67-14631	OV10A	w.o.11.11.68	
67-14632	OV10A	w.o.23.11.72	
67-14633	OV10A	w.o.26.7.69	
67-14634	OV10A	w.o.5.7.71	
67-14635	OV10A	w.o.8.10.69	
67-14636	OV10A	preserved	Hickam AFB
67-14637	OV10A		
67-14638	OV10A	w.o.28.1.71	
67-14639	OV10A	to Philippine	
67-14640	OV10A	w.o.16.9.72	
67-14641	OV10A	to Philippine	
67-14642	OV10A	w.o.30.1.69	
67-14643	OV10A		
67-14644	OV10A	w.o.16.11.69	
67-14645	OV10A	w.o.21.4.71	
67-14646	OV10A	w.o.8.2.70	
67-14647	OV10A		
67-14648	OV10A	w.o.26.12.71	
67-14649	OV10A		
67-14650	OV10A		
67-14651	OV10A	w.o.	
67-14652	OV10A	to NASA	
67-14653	OV10A		
67-14654	OV10A		
67-14655	OV10A	w.o.82	
67-14656	OV10A	w.o.8.4.70	
67-14657	OV10A	w.o.26.12.79	
67-14658	OV10A	w.o.2.3.72	
67-14659	OV10A	w.o.7.4.73	
67-14660	OV10A	w.o.13.8.70	
67-14661	OV10A	w.o.28.12.70	
67-14662	OV10A	w.o.12.8.71	
67-14663	OV10A	w.o.30.4.82	
67-14664	OV10A	w.o.26.9.71	
67-14665	OV10A	w.o.6.6.69	
67-14666	OV10A	to Philippine	
67-14667	OV10A	w.o.24.12.71	
67-14668	OV10A	w.o.20.6.71	
67-14669	OV10A	to Philippine	
67-14670	OV10A		
67-14671	OV10A	w.o.	
67-14672	OV10A	w.o.24.7.70	
67-14673	OV10A	w.o.6.10.72	
67-14674	OV10A		
67-14675	OV10A		
67-14676	OV10A	w.o.4.11.69	
67-14677	OV10A		
67-14678	OV10A	w.o.25.5.72	
67-14679	OV10A		
67-14680	OV10A	w.o.27.1.70	
67-14681	OV10A		
67-14682	OV10A	w.o.15.3.70	
67-14683	OV10A	w.o.25.7.80	
67-14684	OV10A		
67-14685	OV10A	w.o.26.9.72	
67-14686	OV10A	w.o.6.5.72	
67-14687	OV10A	to NASA	N524NA
67-14688	OV10A	w.o.23.11.70	
67-14689	OV10A		
67-14690	OV10A		
67-14691	OV10A	w.o.16.4.69	
67-14692	OV10A	w.o.4.6.71	
67-14693	OV10A	w.o.24.3.71	
67-14694	OV10A		
67-14695	OV10A	w.o.83	
67-14696	OV10A	w.o.11.10.70	
67-14697	OV10A		
67-14698	OV10A		
67-14699	OV10A		
67-14700	OV10A	w.o.30.6.72	
67-14701	OV10A	w.o.8.9.82	
68-3784	OV10A	w.o.	
68-3785	OV10A		
68-3786	OV10A	w.o.19.10.69	
68-3787	OV10A	USAF Museum	
68-3788	OV10A	w.o.3.7.70	
68-3789	OV10A	w.o.3.4.72	
68-3790	OV10A		
68-3791	OV10A		
68-3792	OV10A	to Philippine	
68-3793	OV10A		
68-3794	OV10A	w.o.21.3.70	
68-3795	OV10A		
68-3796	GOV10A	IAAFA	
68-3797	OV10A		
68-3798	OV10A	w.o.30.8.70	
68-3799	OV10A	to NASA	(spares)
68-3800	OV10A	w.o.17.9.70	
68-3801	OV10A	to Philippine	
68-3802	OV10A	w.o.20.6.70	
68-3803	OV10A	w.o.8.6.72	
68-3804	OV10A	w.o.29.6.72	
68-3805	OV10A		
68-3806	OV10A	w.o.27.3.73	
68-3807	OV10A	w.o.30.6.70	
68-3808	OV10A		
68-3809	GOV10A	IAAFA	
68-3810	OV10A	w.o.17.9.82	
68-3811	OV10A	to N91LM	
68-3812	OV10A	w.o.31.12.69	

68-3813	OV10A	w.o.29.4.75
68-3814	OV10A	to N....
68-3815	OV10A	w.o.17.6.70
68-3816	OV10A	to N94LM
68-3817	OV10A	w.o.
68-3818	OV10A	w.o.15.6.78
68-3819	OV10A	w.o.21.5.77
68-3820	OV10A	w.o.7.4.72
68-3821	OV10A	w.o.18.12.70
68-3822	OV10A	
68-3823	OV10A	w.o.18.3.72
68-3824	OV10A	w.o.29.4.70
68-3825	OV10A	
68-3826	OV10A	
68-3827	OV10A	w.o.7.5.70
68-3828	OV10A	w.o.30.4.82
68-3829	OV10A	
68-3830	OV10A	to Philippine
68-3831	OV10A	w.o.3.2.78
87-0405	OV10A	from\to USMC
87-0406	OV10A	from\to USMC

UV18 TWIN OTTER

c\n   554, 555, 437.

77-0464	UV18B	USAF Academy\N70464		87-0802	UV18B
77-0465	UV18B	USAF Academy\N70465			

X29

82-0003	X29A	NASA	AFFTC	82-0049	X29A	NASA	AFFTC

BEECH 200 KING AIR

c\n   BL-54, BL-61, BL-62.

N654BA	B200C	Dept. of AF	N662BA	B200C	Dept. of AF
N661BA	B200C	Dept. of AF			

BEECH 1900

N20RA	1900	Dept. of AF

CASA 212C AVIOCAR

c\n   320, 328, 336, 348.

87-0158 C212   317AW det.6
87-0159 C212   317AW

90-0168 C212   317AW
90-0169 C212

CESSNA 150

84-0483 C150   USAF Academy

DASSAULT FALCON 20

c\n   61.

N20NY   F20C   AFMC

E9 'DASH 8'

c\n   037, 045

N801AP   E9A   475WEG

N802AP   E9A   475WEG

UNITED STATES ARMY

C6 KING AIR A90

c\n   LJ-153

66-15361 VC6A ex N901R

C11 GULFSTREAM II

89-0266   VC11A Corps of Eng. N51741

C12 KING AIR 200\1900

c\n   BB-3 to 5, BD-5, BC-1 to 33, BC-35 to 75, BP-1 to 6, BP-12 to 39, GR-14 to 19, BP-46 to 51, FE-1 to 9, BP-52
to 71, FE-10 to FE-26.

note: BP-1 to 6 and BP-12 to 23 were changed as follows:-
BP-1\GR-6,7,8,9,10,2\FC-3\GR-4,12,5,11,3,13\FC-1,2\BP-22\GR-1.

71-21058	C12L	Moffett	ex RU21D		76-22558	C12C	to USCS	N283B
71-21059	C12L	Moffett	ex RU21J		76-22559	C12C	Dharan	Saudi Arabia
71-21060	C12L	DAC\MDW	ex RU21J		76-22560	C12C	FORSCOM	
73-1209	C12C	STARC	SC NG		76-22561	C12C	Lawson AAF	
73-22250	C12C	TRADOC	Langley		76-22562	C12C	US Army	Japan
73-22251	C12C	to USCS	N7064B		76-22563	C12C	to USCS	N7068D
73-22252	C12C	Lawson AAF			76-22564	C12C	56AvCo	
73-22253	C12C				77-22931	C12C	LANDSOUTHEAST	
73-22254	C12C	207AvCo	HQ\USAREUR		77-22932	C12C	6AvDet	SETAF
73-22255	C12C	5-158AVN			77-22933	C12C		
73-22256	C12C	Ft. Bliss			77-22934	C12C	Ft. Lewis	
73-22257	C12C	US Army Signal Center			77-22935	C12C	AATC	Cairns AAF
73-22258	C12C	Ft. Lewis			77-22936	C12C	Ft. Rucker	
73-22259	C12C	78AVN			77-22937	C12C	Libby AAF	
73-22260	C12C	7ATC			77-22938	C12C	Ft. Riley	
73-22261	C12C	5-158AVN			77-22939	C12C	FORSCOM	
73-22262	C12C				77-22940	C12C	Fort Lewis	
73-22263	C12C	Libby AAF			77-22941	C12C	1-159AVN	
73-22264	C12C		IL NG		77-22942	C12C	CTF	Hamilton AAF
73-22265	C12C	to USCS	N7068B		77-22943	C12C	USCS	
73-22266	C12C	to USCS	N7074G		77-22944	C12C	56AvCo	
73-22267	C12C	TRADOC	Langley		77-22945	C12C	USSOCOM	
73-22268	C12C	DAC\MDW			77-22946	C12C	A\1-501AVN	
73-22269	C12C	TRADOC	Langley		77-22947	C12C	A\1-501AVN	
76-22545	JC12C	Bryant AAF			77-22948	C12C	US Army	Korea
76-22546	C12C	Pat Henry AAF			77-22949	C12C	Ft. Sill	
76-22547	C12C	US Army Korea			77-22950	C12C	207AvCo	HQ\USAREUR
76-22548	C12C	A\1-501AVN			78-23126	C12C	207AvCo	HQ\USAREUR
76-22549	C12C	1AvDet	HQ\USEUCOM		78-23127	C12C	207AvCo	HQ\USAREUR
76-22550	C12C	1AvDet	HQ\USEUCOM		78-23128	C12C	207AvCo	HQ\USAREUR
76-22551	C12C	HQ\REDCOM			78-23129	C12C	Davison AAF	
76-22552	C12C	Randolph AFB			78-23130	C12C	DAC\MDW	
76-22553	C12C	Buckley ANGB			78-23131	C12C	Washington	
76-22554	C12C	FORSCOM			78-23132	C12C	LANDSOUTHEAST	
76-22555	C12C	USSOCOM			78-23133	JC12C	AATC	Cairns AAF
76-22556	C12C	6AvDet	SETAF		78-23134	C12C	A\1-501AVN	
76-22557	C12C	207AvCo	HQ\USAREUR		78-23135	C12C	Wheeler AFB	

Serial	Type	Unit	Location
78-23136	C12C	NAS	Glenview
78-23137	C12C	FORSCOM	
78-23138	C12C	Miami	
78-23139	C12C		
78-23140	JC12D	AATC	Cairns AAF
78-23141	RC12D	1MIBtn	
78-23142	RC12D	1MIBtn	
78-23143	RC12D	1MIBtn	
78-23144	RC12D	2MIBtn	
78-23145	RC12D	1MIBtn	
80-23371	RC12D	2MIBtn	
80-23372	RC12G	1MIBtn	
80-23373	RC12D	2MIBtn	
80-23374	RC12D	2MIBtn	
80-23375	RC12D	1MIBtn	
80-23376	RC12D	224MIBtn	
80-23377	RC12D	1MIBtn	
80-23378	RC12D	AATC	Cairns AAF
80-23379	RC12G	1MIBtn	
80-23380	RC12G	1MIBtn	
81-23541	C12D	HHC\1-159AVN	
81-23542	RC12D	1MIBtn	
81-23543	C12D	Elmendorf AFB	
81-23544	C12D	Wheeler AFB	
81-23545	C12D	USSOCOM	
81-23546	C12D		
82-23780	C12D		KS NG
82-23781	C12D		
82-23782	C12D		MT NG
82-23783	C12D	STARC	AZ NG
82-23784	C12D	STAR	LA NG
82-23785	C12D	STARC	MO NG
83-24145	C12D		ND NG
83-24146	C12D		UT NG
83-24147	C12D	STARC	NM NG
83-24148	C12D		WA NG
83-24149	C12D	STARC	OK NG
83-24150	C12D		VA NG
83-24313	RC12H	B\3MIB	
83-24314	RC12H	B\3MIB	
83-24315	RC12H	B\3MIB	
83-24316	RC12H	B\3MIB	
83-24317	RC12H	B\3MIB	
83-24318	RC12H	B\3MIB	
84-24375	C12D	TRADOC	Langley
84-24376	C12D	A\1-501AVN	
84-24377	C12D		
84-24378	C12D		
84-24379	C12D	USSOCOM	
84-24380	C12D	207AvCo	HQ\USAREUR
85-0147	RC12K	1MIBtn	
85-0148	RC12K	1MIBtn	
85-0149	RC12K		
85-0150	RC12K	1MIBtn	
85-0151	RC12K	1MIBtn	
85-0152	RC12K	1MIBtn	
85-0153	RC12K	1MIBtn	
85-0154	RC12K	1MIBtn	
85-0155	RC12K	1MIBtn	
85-1261	C12F	w.o.12.11.92	
85-1262	C12F		
85-1263	C12F		
85-1264	C12F	Sinop AAF	
85-1265	C12F	Co.B\1-228AVN	
85-1266	C12F	USSOCOM	
85-1267	C12F	AATC	Cairns AAF
85-1268	C12F	Sinop AAF	
85-1269	C12F	Co.B\1-228AVN	
85-1270	C12F		
85-1271	C12F		DC NG
85-1272	C12F		
86-0084	C12F	STARC	VA NG
86-0085	C12F		
86-0086	C12F	STARC	CA NG
86-0087	C12F		
86-0088	C12F		FL NG
86-0089	C12F		
87-0160	C12F		
87-0161	C12F	STARC	AL NG
88-0325	RC12K		
88-0326	RC12K		
88-0327	RC12K		
89-0267	RC12K		
89-0268	RC12K		
89-0269	RC12K		
89-0270	RC12N		
89-0271	RC12N		
89-0272	RC12N		
89-0273	RC12N		
89-0274	RC12N		
89-0275	RC12N		
89-0276	RC12N		
91-0516	RC12N		
91-0517	RC12N		
91-0518	RC12N		
91-0519	RC12N		

C20 GULFSTREAM III\IV

c\n   456, 458, 497, 498, 1162.

Serial	Type	Location
85-0049	C20C	Andrews AFB
85-0050	C20C	Andrews AFB
87-0139	C20E	Andrews AFB
87-0140	C20E	Andrews AFB
91-0108	C20F	Andrews AFB

C21A LEARJET 35A

c\n   35A-280.

87-0026   C21A

C23 SHERPA

c\n   SH.3107, SH.3110, SH.3111, SH.3113, SH.3114, SH.3115, SH.3117, SH.3118, SH.2120, SH.3010, SH.3011, SH.3019, SH.3027, SH.3201 to SH.3216.

84-0461	C23A			88-1863	C23B		
84-0463	C23A			88-1864	C23B	AVCRAD	CA NG
84-0464	C23A	ERADCOM		88-1865	C23B	AVCRAD	MS NG
84-0466	C23A			88-1866	C23B	STARC	AL NG
84-0467	C23A			88-1867	C23B		OR NG
84-0468	C23A			88-1868	C23B	AVCRAD	MD NG
84-0470	C23A			88-1869	C23B	STARC	UT NG
84-0471	JC23A			88-1870	C23B		
84-0473	C23A			90-7011	C23B		
85-25342	C23C	stored	NAS Lakehurst	90-7012	C23B		
85-25343	C23C	stored	NAS Lakehurst	90-7013	C23B		
85-25344	C23C	stored	NAS Lakehurst	90-7014	C23B		
85-25345	C23C	stored	NAS Lakehurst	90-7015	C23B	AVCRAD	MS NG
88-1861	C23B			90-7016	C23B		
88-1862	C23B						

C26 METRO III

87-1000	C26A		87-1001	C26A

C31 FRIENDSHIP

c\n   10652, 10668.

85-1607	C31A	Golden Knights	85-1608	C31A	Golden Knights

AH1 HUEY COBRA

c\n   200002 to 21049, 21054 to 21123, 22001 to 22347.

66-15246	AH1G	preserved	Ft. Rucker	66-15253	AH1S	1-285AVN	AZ NG
66-15247	AH1G	SoC		66-15254	AH1S	w.o.7.2.91	
66-15248	AH1G	to NASA	N736NA	66-15255	AH1F		
66-15249	AH1F	308AHB		66-15256	AH1G	SoC	
66-15250	AH1F	2ACR		66-15257	AH1F	1-224AVN	MD NG
66-15251	AH1F	11ACR		66-15258	AH1G	SoC	
66-15252	AH1F	11ACR		66-15259	AH1F	1-224AVN	MD NG

Serial	Type	Unit/Status	Code	Location
66-15260	AH1G	SoC		
66-15261	AH1F	2ACR		
66-15262	AH1F			
66-15263	AH1F	1\17Cav		
66-15264	AH1F			
66-15265	AH1G	w.o.15.1.70		
66-15266	AH1F	2ACR		
66-15267	AH1G	SoC		
66-15268	AH1S	3ACR		
66-15269	AH1G	w.o.1.7.70		
66-15270	AH1S			
66-15271	AH1F	1-224AVN		MD NG
66-15272	AH1G	SoC		
66-15273	AH1F	SoC		
66-15274	AH1G	SoC		
66-15275	GAH1F	GIA		Fort Eustis
66-15276	AH1G	SoC		
66-15277	AH1F	1-224AVN		MD NG
66-15278	AH1S	3ACR		
66-15279	AH1S			
66-15280	AH1S	26AVN		
66-15281	AH1S	1-150AVN		NJ NG
66-15282	AH1S	163ACR		
66-15283	AH1F			
66-15284	AH1F	1-224AVN		MD NG
66-15285	AH1F			
66-15286	AH1F	2ACR		
66-15287	AH1G	SoC		
66-15288	TH1G	w.o.		
66-15289	AH1F			
66-15290	TH1G	2ACR		
66-15291	AH1S	24AVN		
66-15292	AH1F	4AvRgt		
66-15293	AH1F	10AHB		
66-15294	AH1G	SoC		
66-15295	AH1F	2ACR		
66-15296	AH1F	1-224AVN		MD NG
66-15297	AH1S			
66-15298	AH1S	193AVN		
66-15299	AH1F			
66-15300	AH1F	AATC	00J	Hanchey AHP
66-15301	AH1G	SoC		
66-15302	AH1S	47AVN		
66-15303	AH1S	47AVN		
66-15304	AH1G	SoC		
66-15305	AH1G	w.o.28.7.70		
66-15306	AH1F			
66-15307	AH1F			
66-15308	AH1G	SoC		
66-15309	AH1S	AATC	09P	Hanchey AHP
66-15310	AH1G	w.o.8.6.70		
66-15311	AH1G	SoC		
66-15312	AH1F	w.o.9.87		
66-15313	AH1F	AATC	13H	Hanchey AHP
66-15314	AH1G	SoC		
66-15315	AH1F	308AHB		
66-15316	AH1F	3AVN		
66-15317	AH1S	126ACS		
66-15318	AH1F	AATC	18L	Hanchey AHP
66-15319	AH1G	w.o.3.3.82		
66-15320	AH1G	SoC		
66-15321	AH1F	4AVN		
66-15322	AH1F	10AHB		
66-15323	AH1G	SoC		
66-15324	AH1F	10AHB		
66-15325	AH1F			
66-15326	AH1G	SoC		
66-15327	AH1F			
66-15328	AH1F	SoC		
66-15329	AH1G	SoC		
66-15330	AH1F	24AVN		
66-15331	AH1S	42AVN		
66-15332	AH1G	SoC		
66-15333	AH1G	SoC		
66-15334	AH1G	SoC		
66-15335	AH1F	11ACR		
66-15336	AH1G	SoC		
66-15337	AH1G	w.o.8.5.70		
66-15338	AH1F	1 Inf Div		
66-15339	JAH1F	w.o.16.5.88		
66-15340	AH1G	w.o.5.2.71		
66-15341	TAH1F	w.o.23.9.81		
66-15342	AH1G	SoC		
66-15343	AH1F			
66-15344	AH1G	w.o.8.7.75		
66-15345	AH1G	USN\TPS 53		
66-15346	AH1G	w.o.8.5.78		
66-15347	AH1G	SoC		
66-15348	AH1F	1\17Cav		
66-15349	AH1G	SoC		
66-15350	AH1F	21AvCo		
66-15351	AH1G	SoC		
66-15352	AH1F			
66-15353	AH1F			
66-15354	AH1F	AATC	54G	Hanchey AHP
66-15355	AH1G	SoC		
66-15356	AH1F	11ACR		
66-15357	AH1S			
67-15450	AH1F	2ACR		
67-15451	AH1G	SoC		
67-15452	AH1F	8AVN		
67-15453	AH1G	SoC		
67-15454	AH1G	SoC		
67-15455	AH1F	2ACR		
67-15456	AH1F	10AHB		
67-15457	AH1F	308AHB		
67-15458	AH1G	SoC		
67-15459	AH1F			
07-15460	AH1F	13AHB		
67-15461	AH1G	SoC		
67-15462	AH1F	1-285AVN		AZ NG
67-15463	AH1S	126ACS		
67-15464	AH1G	w.o.6.3.71		
67-15465	AH1F	AATC	65B	Hanchey AHP
67-15466	AH1G	w.o.21.7.70		
67-15467	AH1S			
67-15468	AH1G	SoC		
67-15469	AH1F			
67-15470	AH1F	11ACR		
67-15471	AH1S	1-285AVN		AZ NG
67-15472	AH1F	1 Inf Div		
67-15473	AH1F	11ACR		
67-15474	AH1F	9Cav		
67-15475	AH1F	11ACR		
67-15476	AH1F			
67-15477	AH1F	3AVN		
67-15478	AH1G	w.o.27.9.71		
67-15479	AH1F			
67-15480	AH1F	308AHB		
67-15481	AH1F			
67-15482	AH1S			
67-15483	AH1G	w.o.26.10.73		
67-15484	AH1G	SoC		
67-15485	AH1F			
67-15486	AH1S	124ACS		
67-15487	AH1F			
67-15488	AH1G	SoC		
67-15489	AH1F	4AVN		
67-15490	AH1F	308AHB		
67-15491	AH1F	4AVN		
67-15492	AH1F	477AVN		
67-15493	AH1G	SoC		
67-15494	AH1S	3ACR		
67-15495	AH1F	SoC		
67-15496	AH1F	1-285AVN		AZ NG
67-15497	AH1F	2ACR		
67-15498	AH1S	1-285AVN		AZ NG
67-15499	AH1S	1-285AVN		AZ NG
67-15500	AH1G	SoC		
67-15501	AH1G	w.o.28.8.70		
67-15502	AH1F			
67-15503	AH1S	3ACR		
67-15504	AH1F			
67-15505	AH1F	1-285AVN		AZ NG
67-15506	AH1F	10AHB		
67-15507	AH1F	24AVN		
67-15508	AH1F	10AHB		
67-15509	AH1S	1-150AVN		NJ NG
67-15510	AH1G	SoC		
67-15511	AH1G	SoC		
67-15512	AH1F			
67-15513	AH1F	1-285AVN		AZ NG
67-15514	AH1G	SoC		
67-15515	AH1G	w.o.14.2.71		
67-15516	AH1S	w.o.26.5.82		
67-15517	AH1G	w.o.		

Serial	Type	Unit/Status	Location
67-15518	AH1G	SoC	
67-15519	AH1G	w.o.1.8.70	
67-15520	AH1F	10AHB	
67-15521	AH1F	24AVN	
67-15522	AH1F	3AVN	
67-15523	AH1G	SoC	
67-15524	AH1F		
67-15525	AH1G	SoC	
67-15526	AH1G	SoC	
67-15527	AH1G	SoC	
67-15528	AH1F	3AVN	
67-15529	AH1G	SoC	
67-15530	AH1F	4AVN	
67-15531	AH1G	SoC	
67-15532	AH1S		
67-15533	AH1F		
67-15534	AH1G	w.o.8.10.70	
67-15535	AH1F	3AVN	
67-15536	AH1S		
67-15537	AH1F	AATC	37N Hanchey AHP
67-15538	AH1G	SoC	
67-15539	AH1G	SoC	
67-15540	AH1F	4AVN	
67-15541	AH1S	10Cav	
67-15542	AH1G	SoC	
67-15543	AH1S	193AVN	
67-15544	AH1S	spares	NAD Pensacola
67-15545	AH1G	SoC	
67-15546	AH1S	40AVN	
67-15547	AH1S		
67-15548	AH1F	4AVN	
67-15549	AH1G	w.o.19.5.80	
67-15550	AH1S	1-285AVN	AZ NG
67-15551	AH1F		
67-15552	AH1G	SoC	
67-15553	AH1S	spares	NAD Pensacola
67-15554	AH1G	w.o.2.5.70	
67-15555	AH1S	124ACS	
67-15556	AH1G	SoC	
67-15557	AH1G	SoC	
67-15558	AH1G	SoC	
67-15559	AH1G	SoC	
67-15560	AH1S	spares	NAD Pensacola
67-15561	AH1G	SoC	
67-15562	GAH1S	GIA	Ft. Eustis
67-15563	AH1S	7AVN	
67-15564	AH1S	7AVN	
67-15565	AH1F	3AVN	
67-15566	AH1G	SoC	
67-15567	AH 1F	AATC	67B Hanchey AHP
67-15568	AH1F	477AVN	
67-15569	AH1G	SoC	
67-15570	AH1F	AATC	70J Hanchey AHP
67-15571	AH1S	8AVN	
67-15572	AH1F	11ACR	
67-15573	AH1G	SoC	
67-15574	AH1G	w.o.14.4.76	
67-15575	AH1G	SoC	
67-15576	AH1G	w.o.18.1.70	
67-15577	AH1F	24AVN	
67-15578	AH1G	SoC	
67-15579	AH1G	SoC	
67-15580	AH1G	SoC	
67-15581	AH1G	SoC	
67-15582	AH1G	SoC	
67-15583	AH1G	SoC	
67-15584	AH1G	SoC	
67-15585	AH1G	SoC	
67-15586	AH1F	26AVN	
67-15587	AH1F	10AHB	
67-15588	AH1G	SoC	
67-15589	AH1F		
67-15590	AH1G	Soc	
67-15591	AH1G	w.o.8.9.70	
67-15592	AH1G	SoC	
67-15593	AH1F	11ACR	
67-15594	AH1G	SoC	
67-15595	AH1F	AATC	95C Hanchey AHP
67-15596	AH1G	SoC	
67-15597	AH1G	w.o.20.3.71	
67-15598	AH1G	SoC	
67-15599	AH1F	AATC	99E Hanchey AHP
67-15600	AH1F		
67-15601	AH1S	1-150AVN	NJ NG
67-15602	AH1G	SoC	
67-15603	AH1F	229AVN	
67-15604	AH1G	w.o.15.7.70	
67-15605	AH1G	SoC	
67-15606	AH1G	SoC	
67-15607	JAH1F		
67-15608	AH1F		
67-15609	AH1G	SoC	
67-15610	AH1F	2ACR	
67-15611	AH1S	SoC	
67-15612	AH1G	SoC	
67-15613	AH1F	SoC	
67-15614	AH1F	4AVN	
67-15615	AH1G	SoC	
67-15616	AH1F		
67-15617	AH1F	308AHB	
67-15618	AH1S	10Cav	
67-15619	AH1S		
67-15620	AH1G	w.o.4.5.70	
67-15621	AH1F	2ACR	
67-15622	AH1F		
67-15623	TH1G	w.o.	
67-15624	AH1G	10AHB	
67-15625	AH1G	SoC	
67-15626	AH1G	w.o.16.1.73	
67-15627	AH1F	AATC27A	Han'y AHP
67-15628	AH1G	SoC	
67-15629	AH1S		
67-15630	AH1G	w.o.10.9.78	
67-15631	AH1G	SoC	
67-15632	AH1G	SoC	
67-15633	AH1F	10AHB	
67-15634	AH1G	SoC	
67-15635	AH1G	SoC	
67-15636	AH1G	w.o.30.4.70	
67-15637	AH1G	SoC	
67-15638	AH1G	SoC	
67-15639	AH1F		
67-15640	AH1S		
67-15641	AH1S	spares	NAD Pensacola
67-15642	AH1F	8AVN	
67-15643	AH1F	2ACR	
67-15644	AH1G	w.o.11.7.71	
67-15645	AH1S	7AVN	
67-15646	AH1F	1-224AVN	MD NG
67-15647	AH1S	spares	NAD Pensacola
67-15648	AH1S		
67-15649	AH1F	AATC	49F Hanchey AHP
67-15650	AH1F	308AHB	
67-15651	AH1G	SoC	
67-15652	AH1F	10AHB	
67-15653	AH1F	4AVN	
67-15654	AH1S		
67-15655	AH1S	193AVN	
67-15656	AH1S	47AVN	
67-15657	AH1S		
67-15658	AH1F	w.o.17.1.86	
67-15659	AH1F	2ACR	
67-15660	AH1S	3AVN	
67-15661	AH1S	47AVN	
67-15662	AH1F	11ACR	
67-15663	AH1S	1-150AVN	NJ NG
67-15664	AH1F	3AVN	
67-15665	AH1F	4AVN	
67-15666	AH1F	11ACR	
67-15667	AH1G	SoC	
67-15668	AH1G	SoC	
67-15669	AH1G	w.o.27.3.74	
67-15670	AH1G	SoC	
67-15671	AH1G	w.o.11.3.70	
67-15672	AH1G	SoC	
67-15673	AH1G	SoC	
67-15674	AH1G	w.o.15.1.70	
67-15675	AH1F	11ACR	
67-15676	AH1G	w.o.17.3.71	
67-15677	AH1G		
67-15678	AH1S	SoC	
67-15679	AH1F	4AVN	
67-15680	AH1S	SoC	
67-15681	JAH1G	w.o.29.8.84	
67-15682	AH1F	8AVN	
67-15683	AH1F	2ACR	

Serial	Type	Unit		
67-15684	AH1F	AATC	84G	Hanchey AHP
67-15685	AH1S			
67-15686	AH1G	SoC		
67-15687	AH1S			
67-15688	AH1F	477AVN		
67-15689	AH1F	2ACR		
67-15690	AH1F			
67-15691	AH1G	w.o.3.6.75		
67-15692	AH1G	w.o.20.2.74		
67-15693	AH1S	1-150AVN	NJ NG	
67-15694	AH1G	SoC		
67-15695	AH1G	SoC		
67-15696	AH1G	SoC		
67-15697	FAH1S	w.o.9.4.86		
67-15698	AH1G	SoC		
67-15699	AH1G	SoC		
67-15700	AH1F			
67-15701	AH1F	2ACR		
67-15702	AH1S	90ARCOM		
67-15703	AH1F	163ACR		
67-15704	AH1G	SoC		
67-15705	AH1G	24AVN		
67-15706	AH1G	SoC		
67-15707	AH1G	SoC		
67-15708	AH1S	1-285AVN	AZ NG	
67-15709	AH1G	SoC		
67-15710	AH1F	1\17Cav		
67-15711	AH1G	Soc		
67-15712	AH1G	SoC		
67-15713	AH1G	SoC		
67-15714	AH1G	SoC		
67-15715	AH1F			
67-15716	AH1F	308AHB		
67-15717	AH1F	2ACR		
67-15718	AH1G	SoC		
67-15719	AH1G	w.o.25.8.70		
67-15720	AH1F	229AVN		
67-15721	AH1F	308AHB		
67-15722	AH1F			
67-15723	AH1G	SoC		
67-15724	AH1G	w.o.6.10.72		
67-15725	AH1G	SoC		
67-15726	AH1F	17Cav		
67-15727	AH1G	SoC		
67-15728	AH1G	SoC		
67-15729	AH1G	w.o.15.7.70		
67-15730	AH1G	SoC		
67-15731	AH1S			
67-15732	AH1S	1-150AVN	NJ NG	
67-15733	AH1G	SoC		
67-15734	AH1G	SoC		
67-15735	AH1G	SoC		
67-15736	AH1F	2AVN		
67-15737	AH1S	w.o.15.9.85		
67-15738	AH1S			
67-15739	AH1G	SoC		
67-15740	AH1S	3ACR		
67-15741	AH1F	308AHB		
67-15742	AH1S	SoC		
67-15743	AH1G	SoC		
67-15744	AH1S			
67-15745	AH1F	3ACR		
67-15746	AH1G	SoC		
67-15747	AH1G	SoC		
67-15748	AH1G	SoC		
67-15749	AH1S			
67-15750	AH1S	193AVN		
67-15751	AH1G	SoC		
67-15752	AH1G	w.o.22.10.71		
67-15753	AH1G	SoC		
67-15754	AH1S	477AVN		
67-15755	AH1G	SoC		
67-15756	AH1G	SoC		
67-15757	AH1F	SoC		
67-15758	AH1S	1Cav		
67-15759	AH1F	AATC	59D	Hanchey AHP
67-15760	AH1G	w.o.3.7.71		
67-15761	AH1S	1-150AVN	NJ NG	
67-15762	AH1F	308AHB		
67-15763	AH1G	SoC		
67-15764	AH1F	3AVN		
67-15765	AH1G	SoC		
67-15766	AH1F			
67-15767	AH1G	SoC		
67-15768	AH1F			
67-15769	AH1F	1\17Cav		
67-15770	AH1G	SoC		
67-15771	AH1F	308AHB		
67-15772	AH1F	8AVN		
67-15773	AH1G	w.o.1.11.79		
67-15774	AH1S	10Cav		
67-15775	AH1F	308AHB		
67-15776	AH1F	308AHB		
67-15777	AH1S	SoC		
67-15778	AH1G	SoC		
67-15779	AH1G	w.o.		
67-15780	AH1G	SoC		
67-15781	AH1F	229AVN		
67-15782	AH1S	3AVN		
67-15783	AH1G	w.o.3.3.70		
67-15784	AH1F	8AVN		
67-15785	AH1S	126ACS		
67-15786	AH1G	SoC		
67-15787	AH1S	10Cav		
67-15788	AH1G	SoC		
67-15789	AH1F	4\9Cav		
67-15790	AH1F	3\3AVN		
67-15791	AH1S	126ACS		
67-15792	AH1S	1-150AVN	NJ NG	
67-15793	AH1G	SoC		
67-15794	AH1S			
67-15795	AH1F	477AVN		
67-15796	AH1S	7AVN		
67-15797	AH1F	1-224AVN	MD NG	
67-15798	AH1G	SoC		
67-15799	AH1G	SoC		
67-15800	AH1S	10Cav		
67-15801	AH1G	w.o.3.9.72		
67-15802	AH1G	SoC		
67-15803	AH1F	17Cav		
67-15804	AH1F			
67-15805	AH1F	10AHB		
67-15806	AH1G	w.o.14.12.70		
67-15807	AH1G	SoC		
67-15808	AH1F			
67-15809	AH1F	1-285AVN	AZ NG	
67-15810	AH1F	24AVN		
67-15811	AH1F	1Cav		
67-15812	AH1G	SoC		
67-15813	AH1F			
67-15814	AH1G	SoC		
67-15815	AH1F	10AHB		
67-15816	AH1G	w.o.11.9.71		
67-15817	AH1S	124ACS		
67-15818	AH1G	SoC		
67-15819	AH1F			
67-15820	AH1G	SoC		
67-15821	AH1G	SoC		
67-15822	AH1F	2ACR		
67-15823	AH1F			
67-15824	AH1F	AATC	24E	Hanchey AHP
67-15825	AH1F			
67-15826	AH1F	42AVN		
67-15827	AH1S	w.o.		
67-15828	AH1S	42AVN		
67-15829	AH1F	10AHB		
67-15830	AH1S	SoC		
67-15831	AH1F			
67-15832	AH1G	SoC		
67-15833	AH1F	8AVN		
67-15834	AH1S			
67-15835	AH1G	SoC		
67-15836	AH1G	SoC		
67-15837	AH1S	SoC		
67-15838	AH1F	8AVN		
67-15839	AH1G	SoC		
67-15840	AH1G	SoC		
67-15841	AH1G	SoC		
67-15842	AH1F	SoC		
67-15843	AH1F	1-224AVN	MD NG	
67-15844	AH1F	1Cav		
67-15845	AH1G	SoC		
67-15846	AH1G	w.o.21.6.70		
67-15847	AH1G	w.o.		
67-15848	AH1G	SoC		
67-15849	AH1G	SoC		

Serial	Type	Unit/Notes	Location
67-15850	AH1G	SoC	
67-15851	AH1S	126ACS	
67-15852	AH1F	308AHB	
67-15853	AH1G	SoC	
67-15854	AH1S	8AVN	
67-15855	AH1G	w.o.27.9.93	
67-15856	AH1G	w.o.4.4.73	
67-15857	AH1S	spares	NAD Pensacola
67-15858	AH1G	SoC	
67-15859	AH1F		
67-15860	AH1F	4AVN	
67-15861	AH1G	SoC	
67-15862	AH1G	SoC	
67-15863	AH1F	4AVN	
67-15864	AH1G	SoC	
67-15865	AH1G	SoC	
67-15866	AH1S		
67-15867	AH1F		
67-15868	AH1G	SoC	
67-15869	AH1G	SoC	
68-15000	AH1F	26AVN	
68-15001	AH1F	1-224AVN	MD NG
68-15002	AH1G	SoC	
68-15003	AH1F		
68-15004	AH1G	SoC	
68-15005	AH1G	SoC	
68-15006	AH1G	SoC	
68-15007	AH1F	8AVN	
68-15008	AH1G	w.o.10.5.70	
68-15009	AH1G	SoC	
68-15010	AH1G	SoC	
68-15011	AH1S	1Cav	
68-15012	AH1S		
68-15013	AH1G	SoC	
68-15014	AH1S	SoC	
68-15015	AH1F	122AHB	
68-15016	AH1F	4AVN	
68-15017	AH1S	6Cav	
68-15018	AH1S	1AVN	
68-15019	AH1G	w.o.6.3.70	
68-15020	AH1G	w.o.9.3.70	
68-15021	AH1S	8AVN	
68-15022	AH1S	10Cav	
68-15023	AH1S		
68-15024	AH1G	SoC	
68-15025	AH1G	SoC	
68-15026	AH1G	SoC	
68-15027	AH1F	24AVN	
68-15028	AH1S	SoC	
68-15029	AH1G	SoC	
68-15030	AH1S		
68-15031	AH1F	1Cav	
68-15032	AH1G	SoC	
68-15033	AH1G	w.o.	
68-15034	AH1G	w.o.	
68-15035	AH1S	1-285AVN	AZ NG
68-15036	AH1F	308AHB	
68-15037	AH1G	to USMC	
68-15038	AH1F	10AHB	
68-15039	AH1G	to USMC	
68-15040	AH1S	SoC	
68-15041	AH1G	SoC	
68-15042	AH1F	1-224AVN	MD NG
68-15043	AH1G	w.o.6.5.70	
68-15044	AH1G	SoC	
68-15045	FAH1S	w.o.11.3.85	
68-15046	AH1F	10AHB	
68-15047	AH1G	SoC	
68-15048	AH1G	SoC	
68-15049	AH1S		
68-15050	AH1G	SoC	
68-15051	AH1S	spares	NAD Pensacola
68-15052	AH1S	1-150AVN	NJ NG
68-15053	AH1S		
68-15054	AH1S	10Cav	
68-15055	AH1G	SoC	
68-15056	AH1G	SoC	
68-15057	AH1F	10AHB	
68-15058	AH1S	1-150AVN	NJ NG
68-15059	AH1G	w.o.10.9.70	
68-15060	AH1G	SoC	
68-15061	AH1G	SoC	
68-15062	AH1G	SoC	
68-15063	AH1G	w.o.17.11.84	
68-15064	AH1S	1-285AVN	AZ NG
68-15065	AH1G	SoC	
68-15066	AH1S		
68-15067	AH1S	10Cav	
68-15068	AH1F		
68-15069	AH1F	10AHB	
68-15070	AH1G	SoC	
68-15071	AH1S	40AVN	
68-15072	AH1G	to USMC	
68-15073	AH1G	to USMC	
68-15074	AH1S	SoC	
68-15075	AH1G	SoC	
68-15076	AH1S	10Cav	
68-15077	AH1G	SoC	
68-15078	AH1G	w.o.15.6.70	
68-15079	AH1G	to USMC	
68-15080	AH1G	to USMC	
68-15081	AH1G	SoC	
68-15082	AH1G	SoC	
68-15083	AH1S	193AVN	
68-15084	AH1F	308AHB	
68-15085	AH1F	2ACR	
68-15086	AH1S	7AVN	
68-15087	AH1G	SoC	
68-15088	AH1S	40AVN	
68-15089	AH1S	224AVN	
68-15090	AH1S	1-285AVN	AZ NG
68-15091	AH1S	8AVN	
68-15092	AH1F		
68-15093	AH1F	8AVN	
68-15094	AH1S	193AVN	
68-15095	AH1S	3ACR	
68-15096	AH1S	1Cav	
68-15097	AH1F		
68-15098	AH1G	SoC	
68-15099	AH1G	SoC	
68-15100	AH1F		
68-15101	AH1F	101AbDiv	
68-15102	AH1F		
68-15103	AH1G	SoC	
68-15104	AH1F	4AVN	
68-15105	AH1F	2ACR	
68-15106	AH1F	3AVN	
68-15107	AH1G	SoC	
68-15108	AH1G	SoC	
68-15109	AH1G	SoC	
68-15110	AH1F	308AHB	
68-15111	AH1G	SoC	
68-15112	AH1F	308AHB	
68-15113	AH1F	308AHB	
68-15114	AH1G	SoC	
68-15115	AH1F	24AVN	
68-15116	AH1F	2ACR	
68-15117	AH1F		
68-15118	AH1F	2ACR	
68-15119	AH1G	SoC	
68-15120	AH1G	SoC	
68-15121	AH1G	w.o.2.5.70	
68-15122	AH1S		
68-15123	AH1S	47AVN	
68-15124	AH1G	SoC	
68-15125	AH1G	SoC	
68-15126	AH1S		
68-15127	AH1S	SoC	
68-15128	GAH1S	GIA	Ft. Eustis
68-15129	AH1S		
68-15130	AH1S	40AVN	
68-15131	AH1F	4AVN	
68-15132	AH1G	SoC	
68-15133	AH1F	4AVN	
68-15134	AH1F	10AHB	
68-15135	AH1G	w.o.7.8.73	
68-15136	AH1G	SoC	
68-15137	AH1G	SoC	
68-15138	AH1F		
68-15139	AH1S		
68-15140	AH1G	preserved	NAS New Orleans
68-15141	AH1G	SoC	
68-15142	AH1F	8AVN	
68-15143	AH1S		
68-15144	AH1S	spares	NAD Pensacola
68-15145	AH1G	SoC	

Serial	Type	Unit	Notes
68-15146	AH1S	40AVN	
68-15147	AH1F	155AvCo	
68-15148	AH1F	1Cav	
68-15149	AH1S	107ACR	OH NG
68-15150	AH1G	SoC	
68-15151	AH1S	47AVN	
68-15152	AH1F	10AHB	
68-15153	AH1S	3ACR	
68-15154	AH1G	SoC	
68-15155	AH1S	10Cav	
68-15156	AH1S	10AHB	
68-15157	AH1S	1-114AVN	AR NG
68-15158	AH1S	3AVN	
68-15159	AH1S	40AVN	
68-15160	AH1G		
68-15161	AH1G	w.o.10.10.70	
68-15162	AH1G	SoC	
68-15163	AH1G	SoC	
68-15164	AH1G	SoC	
68-15165	AH1G	SoC	
68-15166	AH1S	8AVN	
68-15167	AH1F	308AHB	
68-15168	AH1G	SoC	
68-15169	AH1S	47AVN	
68-15170	AH1G	to USMC	
68-15171	AH1G	SoC	
68-15172	AH1S	10Cav	
68-15173	AH1F	10AHB	
68-15174	AH1F	107ACR	OH NG
68-15175	AH1G	SoC	
68-15176	AH1G	SoC	
68-15177	AH1S		
68-15178	GAH1S	GIA	Ft. Eustis
68-15179	AH1S	224AVN	
68-15180	AH1F	308AHB	
68-15181	AH1S	spares	NAD Pensacola
68-15182	AH1G	SoC	
68-15183	AH1S	3ACR	
68-15184	AH1G	SoC	
68-15185	AH1G	w.o.3.7.71	
68-15186	AH1S		
68-15187	AH1G	SoC	
68-15188	AH1G	SoC	
68-15189	AH1G	SoC	
68-15190	AH1G	to USMC	
68-15191	JAH1G	w.o.1.11.75	
68-15192	GAH1S	GIA	Ft. Eustis
68-15193	AH1G	w.o.13.9.70	
68-15194	AH1G	to USMC	
68-15195	AH1S	40AVN	
68-15196	AH1S	42AVN	
68-15197	AH1S	6Cav	
68-15198	AH1G	to USMC	
68-15199	AH1G	SoC	
68-15200	AH1S	3ACR	
68-15201	AH1G	SoC	
68-15202	AH1S	8AVN	
68-15203	AH1S		
68-15204	AH1S	8AVN	
68-15205	AH1G	42AVN	
68-15206	AH1G	w.o.24.6.72	
68-15207	AH1G	SoC	
68-15208	AH1F	308AHB	
68-15209	GAH1S	GIA	Ft. Eustis
68-15210	AH1G	w.o.12.4.70	
68-15211	AH1G	w.o.14.10.70	
68-15212	AH1S	193AVN	
68-15213	AH1G	to USMC	
68-17020	AH1G	SoC	
68-17021	AH1G	w.o.11.2.70	
68-17022	AH1S	40AVN	
68-17023	AH1F	308AHB	
68-17024	AH1S	3AVN	
68-17025	AH1S	124ACS	
68-17026	AH1S	47AVN	
68-17027	AH1G	to USMC	
68-17028	GAH1F	GIA	Ft. Eustis
68-17029	AH1G	SoC	
68-17030	AH1G	w.o.9.4.70	
68-17031	AH1F	1-285AVN	AZ NG
68-17032	AH1G	SoC	
68-17033	AH1G	SoC	
68-17034	AH1S	1Cav	
68-17035	AH1G	SoC	
68-17036	AH1S	107ACR	OH NG
68-17037	AH1S	4-107ACR	OH NG
68-17038	AH1G	SoC	
68-17039	AH1G	w.o.27.1.70	
68-17040	AH1F	1Cav	
68-17041	AH1G	to USMC	
68-17042	AH1S	224AVN	
68-17043	AH1F		
68-17044	AH1S		
68-17045	AH1G	to USMC	
68-17046	AH1S	224AVN	
68-17047	AH1F	11ACR	
68-17048	AH1S	SoC	
68-17049	AH1G	8AVN	
68-17050	AH1G	SoC	
68-17051	AH1S	107ACR	OH NG
68-17052	AH1F		
68-17053	AH1G	SoC	
68-17054	AH1S	10Cav	
68-17055	AH1S		
68-17056	AH1S	1-285AVN	AZ NG
68-17057	AH1G	SoC	
68-17058	GAH1S	GIA	Ft. Eustis
68-17059	AH1G	SoC	
68-17060	AH1S		
68-17061	AH1S	42AVN	
68-17062	AH1F	308AHB	
68-17063	AH1F	308AHB	
68-17064	AH1G	w.o.	
68-17065	AH1F	w.o.13.11.91	
68-17066	AH1F	308AHB	
68-17067	AH1F	308AHB	
68-17068	AH1G	SoC	
68-17069	AH1S		
68-17070	AH1F	10AHB	
68-17071	AH1S	3AVN	
68-17072	AH1F	9Cav	
68-17073	AH1G	SoC	
68-17074	AH1F		
68-17075	AH1F		
68-17076	AH1F	2ACR	
68-17077	AH1G	SoC	
68-17078	AH1F	4AVN	
68-17079	AH1F	308AHB	
68-17080	AH1G	AATC	80A Hanchey AHP
68-17081	AH1G	SoC	
68-17082	AH1F	2ACR	
68-17083	AH1G	SoC	
68-17084	AH1F	3ACR	
68-17085	AH1F	308AHB	
68-17086	AH1G	SoC	
68-17087	AH1F	10AHB	
68-17088	AH1F	11ACR	
68-17089	AH1G	SoC	
68-17090	AH1G	SoC	
68-17091	AH1F	w.o.12.6.86	
68-17092	FAH1S	SoC	
68-17093	AH1S		
68-17094	AH1G	SoC	
68-17095	AH1F	4AVN	
68-17096	AH1F		
68-17097	AH1G	SoC	
68-17098	AH1S	42AVN	
68-17099	AH1S	47AVN	
68-17100	AH1F	2ACR	
68-17101	AH1F	2ACR	
68-17102	AH1G	SoC	
68-17103	AH1G	w.o.12.74	
68-17104	AH1F		
68-17105	AH1F	2ACR	
68-17106	AH1G	SoC	
68-17107	AH1G	SoC	
68-17108	AH1F	308AHB	
68-17109	AH1S	126ACS	
68-17110	AH1G	SoC	
68-17111	AH1F	4AVN	
68-17112	AH1F		
68-17113	AH1S	3ACR	
69-16410	AH1F		
69-16411	AH1F	3AVN	
69-16412	AH1S	1-285AVN	AZ NG
69-16413	AH1S	5InfDiv	

Serial	Type	Unit/Status		
69-16414	AH1G	w.o.8.6.71		
69-16415	AH1F			
69-16416	AH1F			
69-16417	AH1S	w.o.16.12.81		
69-16418	AH1G	SoC		
69-16419	AH1G	w.o.14.1.71		
69-16420	AH1G	w.o.9.2.75		
69-16421	AH1G	SoC		
69-16422	AH1F	10AHB		
69-16423	JAH1F	USAAEFA		
69-16424	AH1F	1Cav		
69-16425	AH1G	w.o.19.5.71		
69-16426	AH1F	2ACR		
69-16427	AH1F			
69-16428	AH1F	AATC	28B	Hanchey AHP
69-16429	AH1F	3AVN		
69-16430	AH1F	4\9Cav		
69-16431	AH1F	3AVN		
69-16432	AH1F	308AHB		
69-16433	AH1F	4AVN		
69-16434	GAH1F	GIA	Ft. Eustis	
69-16435	AH1F			
69-16436	AH1F	308AHB		
69-16437	AH1S	1-150AVN	NJ NG	
69-16438	AH1G	w.o.7.3.71		
69-16439	AH1F	3ACR		
69-16440	AH1F	11ACR		
69-16441	AH1F			
69-16442	AH1F			
69-16443	AH1S			
69-16444	AH1S			
69-16445	AH1F	8AVN		
69-16446	AH1F	1-285AVN	AZ NG	
69-16447	AH1F	SoC		
70-15936	AH1S			
70-15937	AH1S	8AVN		
70-15938	AH1S			
70-15939	AH1S	1Cav		
70-15940	AH1S	12Cav		
70-15941	AH1S	1-150AVN	NJ NG	
70-15942	AH1S	477AVN		
70-15943	AH1S	42AVN		
70-15944	AH1S	10Cav		
70-15945	AH1S	124ACS		
70-15946	AH1S	SoC		
70-15947	AH1G	spares	NAD Pensacola	
70-15948	AH1S	47AVN		
70-15949	AH1S			
70-15950	AH1F	SoC		
70-15951	AH1F	308AHB		
70-15952	AH1F	11ACR		
70-15953	AH1F	AATC	53F	Hanchey AHP
70-15954	AH1F	308AHB		
70-15955	AH1G	w.o.26.11.73		
70-15956	AH1S			
70-15957	AH1S	spares	NAD Pensacola	
70-15958	AH1F	11ACR		
70-15959	AH1F	8AVN		
70-15960	AH1G	w.o.12.3.76		
70-15961	AH1G	8AVN		
70-15962	AH1S	SoC		
70-15963	AH1F			
70-15964	AH1S			
70-15965	AH1S	spares	NAD Pensacola	
70-15966	AH1S	1-285AVN	AZ NG	
70-15967	AH1F	AATC	67D	Hanchey AHP
70-15968	AH1S	1-285AVN	AZ NG	
70-15969	AH1F			
70-15970	AH1F	w.o.14.8.80		
70-15971	AH1F	3AVN		
70-15972	AH1G	w.o.		
70-15973	AH1S			
70-15974	AH1S	8AVN		
70-15975	AH1S			
70-15976	AH1S			
70-15977	AH1F			
70-15978	AH1S	1-285AVN	AZ NG	
70-15979	AH1S			
70-15980	AH1F	229AVN		
70-15981	AH1S			
70-15982	AH1S	SoC		
70-15983	AH1S			
70-15984	AH1S			
70-15985	AH1S	1-285AVN	AZ NG	
70-15986	AH1F	AATC	86K	Hanchey AHP
70-15987	AH1S	163ACR		
70-15988	AH1S	8AVN		
70-15989	AH1F	229AVN		
70-15990	AH1F			
70-15991	AH1F	229AVN		
70-15992	AH1G	SoC		
70-15993	AH1F	1-285AVN	AZ NG	
70-15994	AH1F			
70-15995	AH1F	2ACR		
70-15996	AH1F	1-285AVN	AZ NG	
70-15997	AH1S	126ACS		
70-15998	AH1S	47AVN		
70-15999	AH1F			
70-16000	AH1S	47AVN		
70-16001	AH1F	AATC	01J	Hanchey AHP
70-16002	AH1G	w.o.6.10.80		
70-16003	AH1F	SoC		
70-16004	AH1S	1-285AVN	AZ NG	
70-16005	AH1S	4AVN		
70-16006	AH1S	1-285AVN	AZ NG	
70-16007	AH1S	SoC		
70-16008	AH1S	SoC		
70-16009	AH1S			
70-16010	AH1S	107ACR	OH NG	
70-16011	AH1F	24AVN		
70-16012	AH1F	3AVN		
70-16013	AH1S	3AVN		
70-16014	AH1G	SoC		
70-16015	AH1S	SoC		
70-16016	AH1F	3AVN		
70-16017	AH1S	10Cav		
70-16018	AH1S	8AVN		
70-16019	YAH1S	Textron Bell		
70-16020	AH1F			
70-16021	AH1S	126ACS		
70-16022	AH1S	47AVN		
70-16023	AH1S	SoC		
70-16024	AH1S	spares	NAD Pensacola	
70-16025	AH1S	SoC		
70-16026	AH1S			
70-16027	AH1G	SoC		
70-16028	AH1S			
70-16029	AH1S	10Cav		
70-16030	AH1G	SoC		
70-16031	AH1S	3AVN		
70-16032	AH1F	AATC	32E	Hanchey AHP
70-16033	AH1F	24AVN		
70-16034	AH1F	4AVN		
70-16035	AH1S	3AVN		
70-16036	AH1G	SoC		
70-16037	AH1G	w.o.6.8.74		
70-16038	AH1S			
70-16039	AH1S	8AVN		
70-16040	AH1S			
70-16041	AH1S	82AVN		
70-16042	AH1S	1-285AVN	AZ NG	
70-16043	AH1F	163ACR		
70-16044	AH1F	47AVN		
70-16045	AH1F	229AVN		
70-16046	AH1F	1-285AVN	AZ NG	
70-16047	AH1S	47AVN		
70-16048	AH1F	3AVN		
70-16049	AH1G	w.o.11.6.73		
70-16050	AH1S			
70-16051	AH1S	224AVN		
70-16052	AH1S	126ACS		
70-16053	AH1S	229AVN		
70-16054	AH1F	10AHB		
70-16055	AH1S	1-150AVN	NJ NG	
70-16056	AH1S			
70-16057	AH1S	1 Inf Div		
70-16058	AH1S	7 Inf Div		
70-16059	AH1S	spares	NAD Pensacola	
70-16060	AH1S			
70-16061	AH1S			
70-16062	AH1S			
70-16063	AH1S			
70-16064	AH1S	126ACS		
70-16065	AH1S	3AVN		
70-16066	AH1S			
70-16067	AH1S	10Cav		

Serial	Type	Unit	Notes
70-16068	AH1G	SoC	
70-16069	AH1S		
70-16070	AH1S	42AVN	
70-16071	AH1S	1 Inf Div	
70-16072	AH1S		
70-16073	AH1S		
70-16074	AH1S	1-285AVN	AZ NG
70-16075	AH1S	1-285AVN	AZ NG
70-16076	AH1S		
70-16077	AH1S	126ACS	
70-16078	AH1S	10Cav	
70-16079	AH1S	SoC	
70-16080	AH1S		
70-16081	AH1S	10Cav	
70-16082	AH1S		
70-16083	AH1S	10Cav	
70-16084	AH1S		CA NG
70-16085	AH1S	126ACS	
70-16086	AH1S	3AVN	
70-16087	AH1S	8AVN	
70-16088	AH1S	SoC	
70-16089	AH1S	1-285AVN	AZ NG
70-16090	AH1S	1-285AVN	AZ NG
70-16091	AH1F	10AHB	
70-16092	AH1S	1-285AVN	AZ NG
70-16093	AH1S		
70-16094	AH1S	SoC	
70-16095	AH1S	8AVN	
70-16096	AH1S		
70-16097	AH1S	8AVN	
70-16098	AH1S	193AVN	
70-16099	AH1S	3AVN	
70-16100	AH1S	997AvCo	
70-16101	AH1S	1-150AVN	NJ NG
70-16102	AH1S	SoC	
70-16103	AH1S		
70-16104	AH1S		
70-16105	AH1S	25AvCo	
71-20983	AH1F	1\17Cav	
71-20984	AH1F	3ACT	
71-20985	AH1F	AATC	85B Hanchey AHP
71-20986	AH1S	1-150AVN	NJ NG
71-20987	AH1S		
71-20988	AH1S	w.o.8.4.82	
71-20989	AH1S	SoC	
71-20990	AH1S		
71-20991	AH1S	1-285AVN	AZ NG
71-20992	AH1S		
71-20993	AH1S	SoC	
71-20994	AH1S	193AVN	
71-20995	AH1S	spares	NAD Pensacola
71-20996	AH1S	SoC	
71-20997	AH1S	7AVN	
71-20998	AH1F	11ACR	
71-20999	AH1S	4AVN	
71-21000	AH1S	8AVN	
71-21001	AH1G	SoC	
71-21002	AH1S		
71-21003	AH1F	2ACR	
71-21004	AH1S		
71-21005	AH1S		
71-21006	AH1S	SoC	
71-21007	AH1S	10Cav	
71-21008	AH1S	7 Inf Div	
71-21009	AH1S	8AVN	
71-21010	AH1S	3AVN	
71-21011	AH1S		
71-21012	AH1F		
71-21013	AH1G	w.o.10.3.77	
71-21014	AH1F	308AHB	
71-21015	AH1S		
71-21016	AH1S	26AVN	
71-21017	AH1S	1AVN	
71-21018	AH1S	193AVN	
71-21019	AH1S	193AVN	
71-21020	AH1S		
71-21021	AH1S	8AVN	
71-21022	AH1S	126ACS	
71-21023	AH1S	3AVN	
71-21024	AH1S	25AVN	
71-21025	AH1S	7 Inf Div	
71-21026	AH1S	1AVN	
71-21027	AH1S	3AVN	
71-21028	AH1F		
71-21029	AH1F	308AHB	
71-21030	AH1S	7AVN	
71-21031	AH1F	1-285AVN	AZ NG
71-21032	AH1G	w.o.28.3.72	
71-21033	AH1F	477AVN	
71-21034	AH1S		
71-21035	AH1F	2ACR	
71-21036	AH1G	SoC	
71-21037	AH1S	26AVN	
71-21038	AH1S	SoC	
71-21039	AH1S		
71-21040	AH1S		
71-21041	AH1S	3AVN	
71-21042	AH1S		
71-21043	AH1S	224AVN	
71-21044	AH1F		
71-21045	AH1S	SoC	
71-21046	AH1S	4 Inf Div	
71-21047	AH1S	SoC	
71-21048	AH1S	4 Inf Div	
71-21049	AH1G	SoC	
71-21050	AH1F	4 Inf Div	
71-21051	AH1F	25AvCo	
71-21052	AH1S	1AVN	
76-22567	AH1F	4 Inf Div	
76-22568	AH1P	1-167Cav	NE NG
76-22569	AH1P	4InfDiv	
76-22570	AH1P	AATC	70K Hanchey AHP
76-22571	AH1P	1-167Cav	NE NG
76-22572	AH1P	AATC	72C Hanchey AHP
76-22573	AH1P	USAAEFA	
76-22574	AH1P	SoC	
76-22575	AH1P	AATC	75H Hanchey AHP
76-22576	JPAH1S	w.o.15.2.85	
76-22577	AH1P	4 Inf Div	
76-22578	AH1P	229AVN	
76-22579	AH1P	229AVN	
76-22580	AH1S	w.o.20.11.81	
76-22581	AH1P		
76-22582	AH1P		
76-22583	AH1P		
76-22584	AH1P		
76-22585	AH1P		
76-22586	AH1P		
76-22587	AH1P		
76-22588	AH1P		
76-22589	AH1P		
76-22590	AH1P		
76-22591	AH1P	1 Inf Div	
76-22592	AH1P		
76-22593	AH1P	1Cav	
76-22594	AH1P	SoC	
76-22595	AH1P	9AVN	
76-22596	AH1P	SoC	
76-22597	AH1P	229AVN	
76-22598	AH1P		
76-22599	AH1P	AATC	99B Hanchey AHP
76-22600	JAH1F	229AVN	
76-22601	AH1P	AATC	01B Hanchey AHP
76-22602	AH1P	AATC	02F Hanchey AHP
76-22603	AH1P	4 Inf Div	
76-22604	AH1P	SoC	
76-22605	AH1P	AATC	05A Hanchey AHP
76-22606	AH1P	229AVN	
76-22607	AH1P	AATC	07G Hanchey AHP
76-22608	AH1P	229AVN	
76-22609	AH1P	229AVN	
76-22610	AH1P	1-167Cav	NE NG
76-22692	AH1P	AATC	92F Hanchey AHP
76-22693	AH1P	SoC	
76-22694	AH1P	AATC	94G Hanchey AHP
76-22695	AH1P	AATC	95A Hanchey AHP
76-22696	AH1P	1-167Cav	NE NG
76-22697	AH1P	w.o.	
76-22698	AH1P	SoC	
76-22699	AH1P	4 Inf Div	
76-22700	AH1P	4Cav	
76-22701	AH1P		
76-22702	AH1P		
76-22703	AH1P		
76-22704	AH1P		
76-22705	AH1P	SoC	

Serial	Type	Unit/Status	Code	Location
76-22706	AH1P	AATC	06L	Hanchey AHP
76-22707	AH1P	AATC	07.	Hanchey AHP
76-22708	AH1P	4Cav		
76-22709	AH1P	SoC		
76-22710	AH1P	SoC		
76-22711	AH1P	4Cav		
76-22712	AH1P	AATC	12B	Hanchey AHP
76-22713	AH1P			
77-22729	AH1P	w.o.		
77-22730	AH1P	SoC		
77-22731	AH1P			
77-22732	AH1P	4 Inf Div		
77-22733	AH1P	4 Inf Div		
77-22734	AH1P	229AVN		
77-22735	AH1P	4InfDiv		
77-22736	AH1S	w.o.1.8.78		
77-22737	AH1P	4 Inf Div		
77-22738	AH1P	229AVN		
77-22739	AH1P	AATC	39K	Hanchey AHP
77-22740	AH1P			
77-22741	AH1P	229AVN		
77-22742	AH1P	1-167Cav		NE NG
77-22743	AH1P	4 Inf Div		
77-22744	AH1P	4 Inf Div		
77-22745	AH1P	1-167Cav		NE NG
77-22746	AH1P	w.o.20.6.83		
77-22747	AH1P	4 Inf Div		
77-22748	AH1P	AATC	48.	Hanchey AHP
77-22749	AH1P	AATC	49.	Hanchey AHP
77-22750	AH1P	1-167Cav		NE NG
77-22751	AH1P	229AVN		
77-22752	AH1P	AATC	52.	Hanchey AHP
77-22753	AH1P	AATC	53B	Hanchey AHP
77-22754	AH1P	AATC	54.	Hanchey AHP
77-22755	AH1P	w.o.		
77-22756	AH1P	3ACT		
77-22757	AH1P	229AVN		
77-22758	AH1P	AATC	58.	Hanchey AHP
77-22759	AH1P	AATC	59F	Hanchey AHP
77-22760	AH1P	1-167Cav		NE NG
77-22761	AH1P	3ACT		
77-22762	AH1P	3ACT		
77-22763	AH1E	7Cav		
77-22764	AH1E	1 Inf Div		
77-22765	AH1E	derelict		Hunter AAF
77-22766	AH1E			
77-22767	AH1E	SoC		
77-22768	AH1E	to NASA	N730NA	
77-22769	AH1E			
77-22770	AH1E	4AHB		
77-22771	AH1E	SoC		
77-22772	AH1E			
77-22773	AH1E	1-135AVN		MO NG
77-22774	AH1E	1-135AVN		MO NG
77-22775	AH1E			
77-22776	AH1E	SoC		
77-22777	AH1E			
77-22778	AH1E	SoC		
77-22779	AH1E	7Cav		
77-22780	AH1E			
77-22781	AH1E			
77-22782	AH1E	derelict		Hunter AAF
77-22783	AH1E			
77-22784	AH1E	derelict		Hunter AAF
77-22785	AH1E	USAICS		
77-22786	AH1E	224AVN		
77-22787	AH1E	1-135AVN		MO NG
77-22788	AH1E	224AVN		
77-22789	AH1E	1-135AVN		MO NG
77-22790	AH1E	w.o.		
77-22791	AH1E			
77-22792	AH1E	derelict		Hunter AAF
77-22793	AH1E			
77-22794	AH1E	w.o.21.1.90		
77-22795	AH1E			
77-22796	AH1E	SoC		
77-22797	AH1E	224AVN		
77-22798	AH1E	derelict		Hunter AAF
77-22799	AH1E	derelict		Hunter AAF
77-22800	AH1E	1-135AVN		MO NG
77-22801	EAH1S	w.o.26.4.83		
77-22802	AH1E	w.o.30.7.82		
77-22803	AH1E	1-135AVN		MO NG
77-22804	AH1E	11ACR		
77-22805	AH1E			
77-22806	AH1E	derelict		Hunter AAF
77-22807	AH1E	derelict		Hunter AAF
77-22808	AH1E			
77-22809	AH1E	17Cav		
77-22810	AH1E	w.o.		
78-23043	AH1E	1-135AVN		MO NG
78-23044	AH1E	3ACT		
78-23045	AH1E	3ACT		
78-23046	AH1E	derelict		Hunter AAF
78-23047	AH1E	3ACT		
78-23048	AH1E	224AVN		
78-23049	AH1E	1-135AVN		MO NG
78-23050	AH1E	224AVN		
78-23051	AH1E	101AbDiv		
78-23052	AH1E	82AVN		
78-23053	AH1E	1-135AVN		MO NG
78-23054	AH1E	derelict		Hunter AAF
78-23055	AH1E	1-135AVN		MO NG
78-23056	AH1E	w.o.1.2.89		
78-23057	AH1E			
78-23058	AH1E			
78-23059	AH1E	82AVN		
78-23060	AH1E	w.o.4.10.88		
78-23061	AH1E			
78-23062	AH1E	1-135AVN		MO NG
78-23063	AH1E	82AVN		
78-23064	AH1E	17Cav		
78-23065	AH1E	1-135AVN		MO NG
78-23066	AH1E	17Cav		
78-23067	AH1E	1-135AVN		MO NG
78-23068	AH1E	derelict		Hunter AAF
78-23069	AH1E	1-135AVN		MO NG
78-23070	AH1E	derelict		Hunter AAF
78-23071	AH1E			
78-23072	AH1E	224AVN		
78-23073	AH1E	224AVN		
78-23074	AH1E			
78-23075	AH1E			
78-23076	AH1E	w.o.23.3.90		
78-23077	AH1E	4 Inf Div		
78-23078	AH1E	224AVN		
78-23079	AH1E	SoC		
78-23080	AH1E	224AVN		
78-23081	AH1E			
78-23082	AH1E	SoC		
78-23083	AH1E	224AVN		
78-23084	AH1E			
78-23085	AH1E			
78-23086	AH1E	4 Inf Div		
78-23087	AH1E	derelict		Hunter AAF
78-23088	AH1E	224AVN		
78-23089	AH1E	w.o.23.4.82		
78-23090	AH1E	1-135AVN		MO NG
78-23091	EAH1S	w.o.6.8.85		
78-23092	AH1E	224AVN		
78-23093	AH1F			
78-23094	AH1F	AATC	94E	Hanchey AHP
78-23095	AH1F	preserved		Ft. Rucker
78-23096	AH1F			
78-23097	AH1F	45AVN		
78-23098	AH1F	7 Inf Div		
78-23099	AH1F			
78-23100	AH1F			
78-23101	AH1F			
78-23102	AH1F			
78-23103	AH1F			
78-23104	AH1F			
78-23105	AH1F	9AVN		
78-23106	AH1F			
78-23107	AH1F			
78-23108	AH1F	4\9Cav		
78-23109	AH1F			
78-23110	AH1F			
78-23111	AH1F			
78-23112	AH1F			
78-23113	AH1F	9AVN		
78-23114	AH1F	9AVN		
78-23115	AH1F			WA NG
78-23116	AH1F			
78-23117	AH1F	24AVN		
78-23118	AH1F	3AVN		

78-23119	AH1F	24AVN	
78-23120	AH1F	4\9Cav	
78-23121	AH1F	1AHB	
78-23122	AH1F	9Cav	
78-23123	AH1F	24AVN	
78-23124	AH1F	9Cav	
78-23125	AH1S	w.o.27.10.80	
79-23187	AH1F	w.o.113.11.91	
79-23188	AH1F	w.o.6.8.81	
79-23189	AH1F	7Cav	
79-23190	AH1F	3AVN	
79-23191	AH1F	24AVN	
79-23192	AH1F	24AVN	
79-23193	FAH1S	w.o.4.8.83	
79-23194	AH1F	1\17Cav	
79-23195	AH1F		
79-23196	AH1F		
79-23197	AH1F	1\17Cav	
79-23198	AH1F	1 Inf Div	
79-23199	AH1F	3AVN	
79-23200	AH1F	3AVN	
79-23201	AH1F	4\9Cav	
79-23202	AH1F		
79-23203	AH1F		
79-23204	FAH1S	w.o.18.5.85	
79-23205	AH1F		
79-23206	AH1F		
79-23207	AH1F	101AbDiv	
79-23208	AH1F		
79-23209	AH1F		
79-23210	AH1F		
79-23211	AH1F	101AbDiv	
79-23212	AH1F		
79-23213	AH1F	9AVN	
79-23214	AH1F		
79-23215	AH1F	4\9Cav	
79-23216	AH1F	268AvDet	
79-23217	AH1F		
79-23218	AH1F		
79-23219	GAH1F		
79-23220	AH1F		
79-23221	AH1F	9AVN	
79-23222	AH1F	w.o.5.2.91	
79-23223	AH1F		
79-23224	AH1F		
79-23225	AH1F	SoC	
79-23226	AH1F	9AVN	
79-23227	AH1F		
79-23228	AH1F		
79-23229	AH1F		
79-23230	AH1F		
79-23231	AH1F		
79-23232	AH1F		
79-23233	AH1F		
79-23234	AH1F	1\17Cav	
79-23235	AH1F	3AVN	
79-23236	AH1F	9Cav	
79-23237	AH1F	w.o.8.5.88	
79-23238	AH1F		
79-23239	AH1F		
79-23240	GAH1F		
79-23241	AH1F	4\9Cav	
79-23242	AH1F		
79-23243	AH1F		
79-23244	GAH1F		
79-23245	AH1F		
79-23246	AH1F	4\9Cav	
79-23247	AH1F		
79-23248	AH1F		
79-23249	AH1F		
79-23250	AH1F		
79-23251	AH1F		
79-23252	AH1F	w.o.21.1.91	
80-23510	AH1F	163ACR	
80-23511	AH1F	163ACR	
80-23512	AH1F	2-135AVN	CO NG
80-23513	AH1F	163ACR	
80-23514	AH1F	2-135AVN	CO NG
80-23515	AH1F	163ACR	
80-23516	AH1F	2-135AVN	CO NG
80-23517	AH1F	116AHB	
80-23518	AH1F	116AHB	
80-23519	AH1F	116AHB	
80-23520	AH1F	116AHB	
80-23521	AH1F	8AVN	
81-23526	AH1F	11ACR	
81-23527	AH1F	11ACR	
81-23528	AH1F	8AVN	
81-23529	AH1F	4AVN	
81-23530	AH1F	4AVN	
81-23531	AH1F	8AVN	
81-23532	AH1F	w.o.11.5.88	
81-23533	AH1F	4AVN	
81-23534	AH1F	4AVN	
81-23535	AH1F	1 Inf Div	
81-23536	AH1F	8AVN	
81-23537	AH1F	4AVN	
81-23538	AH1F	4AVN	
81-23539	AH1F	3AVN	
81-23540	AH1F	3AVN	
82-24065	AH1F	2-135AVN	CO NG
82-24066	AH1F		
82-24067	AH1F		
82-24068	AH1F		WA NG
82-24069	AH1F	2-135AVN	CO NG
82-24070	AH1F		WA NG
82-24071	AH1F		
82-24072	AH1F		
82-24073	AH1F		
82-24074	AH1F	1-224AVN	MD NG
82-24075	AH1F	1-224AVN	MD NG
82-24076	AH1F	1-224AVN	MD NG
83-24189	AH1F		
83-24190	AH1F	2-135AVN	CO NG
83-24191	AH1F	2-135AVN	CO NG
83-24192	AH1F	w.o.18.5.86	
83-24193	AH1F	1-224AVN	MD NG
83-24194	AH1F	1-224AVN	MD NG
83-24195	AH1F	1-224AVN	MD NG
83-24196	AH1F	1-224AVN	MD NG
83-24197	AH1F	1-224AVN	MD NG
83-24198	AH1F	2-135AVN	CO NG
83-24199	AH1F	1-224AVN	MD NG

UH1 IROQUOIS

c\n

X\YH40:	1 to 9.
UH1A:	10 to 191.
UH1B:	192 to 383, 392 to 700, 708 to 713, 722 to 880, 888 to 1013, 1026 to 1045, 1048 to 1069,1072 to 1084, 1088, 1092 to 1224, 2048.
UH1C:	1225 to 1471, 1473 to 1973.
UH1D\H:	701 to 707, 4001 to 4508, 4513 to 4548, 4551 to 4807, 4814 to 4934, 4939 to 5153, 5161 to 5693, 9339 to 9342, 5694 to 6000, 8501 to 9338, 9343 to 10057, 10067 to 10080,

55-4459	XH40	preserved	Ft. Rucker
55-4460	XH40	SoC	
55-4461	XH40	SoC	
56-6723	YH40	preserved	Ft. Rucker
56-6724	YH40	SoC	
56-6725	YH40	SoC	
56-6726	YH40	SoC	
56-6727	YH40	SoC	
56-6728	YH40	SoC	
57-6095	UH1A	SoC	
57-6096	UH1A	SoC	
57-6097	UH1A	SoC	
57-6098	UH1A	SoC	
57-6099	UH1A	SoC	
57-6100	UH1A	SoC	
57-6101	UH1A	SoC	
57-6102	UH1A	SoC	
57-6103	UH1A	SoC	
58-2078	UH1A	SoC	
58-2079	UH1A	to N93066	
58-2080	UH1A	SoC	
58-2081	UH1A	SoC	
58-2082	UH1A	SoC	
58-2083	UH1A	SoC	
58-2084	UH1A	SoC	
58-2085	UH1A	SoC	
58-2086	UH1A	scrapped	
58-2087	UH1A	SoC	
58-2088	UH1A	SoC	
58-2089	UH1A	SoC	
58-2090	UH1A	SoC	
58-2091	UH1A	preserved	Ft. Campbell
58-2092	UH1A	SoC	
58-2093	UH1A	SoC	
58-3017	UH1A	SoC	
58-3018	UH1A	SoC	
58-3019	UH1A	SoC	
58-3020	UH1A	SoC	
58-3021	UH1A	SoC	
58-3022	UH1A	SoC	
58-3023	UH1A	SoC	
58-3024	UH1A	SoC	
58-3025	UH1A	SoC	
58-3026	UH1A	SoC	
58-3027	UH1A	SoC	
58-3028	UH1A	SoC	
58-3029	UH1A	SoC	
58-3030	UH1A	SoC	
58-3031	UH1A	SoC	
58-3032	UH1A	SoC	
58-3033	UH1A	SoC	
58-3034	UH1A	SoC	
58-3035	UH1A	ranges	Aberdeen, Md.
58-3036	UH1A	SoC	
58-3037	UH1A	SoC	
58-3038	UH1A	SoC	
58-3039	UH1A	to N707FW	
58-3040	UH1A	SoC	
58-3041	UH1A	SoC	
58-3042	UH1A	SoC	
58-3043	UH1A	SoC	
58-3044	UH1A	SoC	
58-3045	UH1A	w.o.24.4.70	
58-3046	UH1A	SoC	
58-3047	UH1A	SoC	
59-1607	UH1A	SoC	
59-1608	UH1A	SoC	
59-1609	UH1A	scrapped	
59-1610	UH1A	SoC	
59-1611	UH1A	ranges	Aberdeen, Md
59-1612	UH1A	SoC	
59-1613	UH1A	SoC	
59-1614	UH1A	SoC	
59-1615	UH1A	SoC	
59-1616	UH1A	SoC	
59-1617	UH1A	SoC	
59-1618	UH1A	SoC	
59-1619	UH1A	SoC	
59-1620	UH1A	ranges	Aberdeen, Md.
59-1621	UH1A	SoC	
59-1622	UH1A	SoC	
59-1623	UH1A	SoC	
59-1624	UH1A	SoC	
59-1625	UH1A	preserved	Ft. Hood
59-1626	UH1A	SoC	
59-1627	UH1A	SoC	
59-1628	UH1A	SoC	
59-1629	UH1A	SoC	
59-1630	UH1A	SoC	
59-1631	UH1A	SoC	
59-1632	UH1A	SoC	
59-1633	UH1A	SoC	
59-1634	UH1A	SoC	
59-1635	UH1A	SoC	
59-1636	UH1A	w.o.20.2.70	
59-1637	UH1A	SoC	
59-1638	UH1A	SoC	
59-1639	UH1A	SoC	
59-1640	UH1A	SoC	
59-1641	UH1A	SoC	
59-1642	UH1A	SoC	
59-1643	UH1A	ranges	Aberdeen, Md.
59-1644	UH1A	to N64286	
59-1645	UH1A	SoC	
59-1646	UH1A	SoC	
59-1647	UH1A	SoC	
59-1648	UH1A	SoC	
59-1649	UH1A	SoC	
59-1650	UH1A	SoC	
59-1651	UH1A	SoC	
59-1652	UH1A	SoC	
59-1653	UH1A	SoC	
59-1654	UH1A	SoC	
59-1655	UH1A	SoC	
59-1656	UH1A	ranges	Aberdeen, Md.
59-1657	UH1A	SoC	
59-1658	UH1A	SoC	
59-1659	UH1A	SoC	
59-1660	UH1A	w.o.24.8.70	
59-1661	UH1A	SoC	
59-1662	UH1A	SoC	
59-1663	UH1A	SoC	
59-1664	UH1A	SoC	
59-1665	UH1A	w.o.14.1.70	
59-1666	UH1A	SoC	
59-1667	UH1A	w.o.17.3.71	
59-1668	UH1A	w.o.25.3.70	
59-1669	UH1A	SoC	
59-1670	UH1A	SoC	
59-1671	UH1A	SoC	
59-1672	UH1A	SoC	
59-1673	UH1A	SoC	
59-1674	UH1A	SoC	
59-1675	UH1A	SoC	
59-1676	UH1A	SoC	
59-1677	UH1A	SoC	
59-1678	UH1A	SoC	
59-1679	UH1A	SoC	
59-1680	UH1A	SoC	
59-1681	UH1A	ranges	Aberdeen, Md.
59-1682	UH1A	SoC	
59-1683	UH1A	SoC	
59-1684	UH1A	SoC	
59-1685	UH1A	SoC	
59-1686	UH1A	preserved	Ft. Rucker, Al
59-1687	UH1A	SoC	
59-1688	UH1A	SoC	
59-1689	UH1A	SoC	
59-1690	UH1A	SoC	
59-1691	UH1A	SoC	
59-1692	UH1A	SoC	
59-1693	UH1A	SoC	
59-1694	UH1A	SoC	
59-1695	UH1A	preserved	Ft. Rucker, Al
59-1696	UH1A	w.o.25.11.70	
59-1697	UH1A	SoC	
59-1698	UH1A	SoC	
59-1699	UH1A	SoC	
59-1700	UH1A	SoC	
59-1701	UH1A	SoC	

59-1702	UH1A	SoC	
59-1703	UH1A	SoC	
59-1704	UH1A	SoC	
59-1705	UH1A	SoC	
59-1706	UH1A	ranges	Aberdeen, Md.
59-1707	UH1A	ranges	Aberdeen, Md.
59-1708	UH1A	SoC	
59-1709	UH1A	w.o.2.11.70	
59-1710	UH1A	SoC	
59-1711	UH1A	preserved	Ft.Bragg, NC
59-1712	UH1A	SoC	
59-1713	UH1A	SoC	
59-1714	UH1A	SoC	
59-1715	UH1A	SoC	
59-1716	UH1A	SoC	
60-3530	UH1A	SoC	
60-3531	UH1A	SoC	
60-3532	UH1A	SoC	
60-3533	UH1A	SoC	
60-3534	UH1A	SoC	
60-3535	UH1A	ranges	Aberdeen, Md.
60-3536	UH1A	SoC	
60-3537	UH1A	SoC	
60-3538	UH1A	SoC	
60-3539	UH1A	SoC	
60-3540	UH1A	SoC	
60-3541	UH1A	ranges	Aberdeen, Md.
60-3542	UH1A	SoC	
60-3543	UH1A	SoC	
60-3544	UH1A	SoC	
60-3545	UH1A	SoC	
60-3546	UH1B	SoC	
60-3547	UH1B	w.o.18.9.70	
60-3548	UH1B	SoC	
60-3549	UH1B	SoC	
60-3550	UH1B	to Singapore	
60-3551	UH1B	dumped	Tucson, Az
60-3552	UH1B	to N8043Z	
60-3553	UH1B	preserved	Ft. Rucker, Al
60-3554	UH1B	preserved	Ft. Knox, Ky
60-3555	UH1B	SoC	
60-3556	UH1B	to N3145F	
60-3557	UH1B	SoC	
60-3558	UH1B	preserved	Cuatro Vientos
60-3559	UH1B	to N37995	
60-3560	UH1B	SoC	
60-3561	UH1B	to N92826	
60-3562	UH1B	to N22949	
60-3563	UH1B	SoC	
60-3564	UH1B	SoC	
60-3565	UH1B	to Uruguay 060	
60-3566	UH1B	GIA	OtisANGB
60-3567	UH1B	to Panama 103	
60-3568	UH1B	to N9970F	
60-3569	UH1B	SoC	
60-3570	UH1B	to N3830F	
60-3571	UH1B	SoC	
60-3572	UH1B	SoC	
60-3573	UH1B	to N98163	
60-3574	UH1B	SoC	
60-3575	UH1B	to Uruguay	
60-3576	UH1B	SoC	
60-3577	UH1B	SoC	
60-3578	UH1B	SoC	
60-3579	UH1B	SoC	
60-3580	UH1B	to Norway 580	
60-3581	UH1B	SoC	
60-3582	UH1B	SoC	
60-3583	UH1B	to N59367	
60-3584	UH1B	to Norway 584	
60-3585	UH1B	to Norway 585	
60-3586	UH1B	to Norway 586	
60-3587	UH1B	to N1901R	
60-3588	UH1B	to N91348	
60-3589	UH1B	SoC	
60-3590	UH1B	w.o.5.12.70	
60-3591	UH1B	to Norway 591	
60-3592	UH1B	SoC	
60-3593	UH1B	SoC	
60-3594	UH1B	to N96142	
60-3595	UH1B	to N1902R	
60-3596	UH1B	to N1903R	
60-3597	UH1B	to Norway 597	
60-3598	UH1B	to N330WN	
60-3599	UH1B	SoC	
60-3600	UH1B	to N1904R	
60-3601	UH1B	preserved	Lackland AFB
60-3602	UH1B	SoC	
60-3603	UH1B	SoC	
60-3604	UH1B	to N5022Q	
60-3605	UH1B	to Singapore 258	
60-3606	UH1B	to N394HP	
60-3607	UH1B	to N3231F	
60-3608	UH1B	SoC	
60-3609	UH1B	SoC	
60-3610	UH1B	SoC	
60-3611	UH1B	SoC	
60-3612	UH1B	to N70512	
60-3613	UH1B	to Norway 613	
60-3614	UH1B	to N70264	
60-3615	UH1B	to N4237V	
60-3616	UH1B	SoC	
60-3617	UH1B	to N.....	
60-3618	UH1B	SoC	
60-3619	UH1B	to N2296J	
60-6028	UH1D	SoC	
60-6029	UH1D	SoC	
60-6030	YUH1D	preserved	Ft Rucker, Al.
60-6031	UH1D	preserved	Ft. Sam Houston
60-6032	UH1D	w.o.8.12.70	
60-6033	UH1D	SoC	
60-6034	UH1D	SoC	
61-0686	UH1B	preserved	Chanute AFB
61-0687	UH1B	to N57HP	
61-0688	UH1B	to Norway 688	
61-0689	UH1B	to Singapore 273	
61-0690	UH1B	to N2298E	
61-0691	UH1B	SoC	
61-0692	UH1B	SoC	
61-0693	UH1B	preserved	Ft. Hood, Tx
61-0694	UH1B	to N3361F	
61-0695	UH1B	to Singapore 269	
61-0696	UH1B	SoC	
61-0697	UH1B	SoC	
61-0698	UH1B	to N70544	
61-0699	UH1B	to Norway 699	
61-0700	UH1B	to Singapore 259	
61-0701	UH1B	to N46968	
61-0702	UH1B	to N404PD	
61-0703	UH1B	to Uruguay 061	
61-0704	UH1B	SoC	
61-0705	UH1B	to N80479	
61-0706	UH1B	SoC	
61-0707	UH1B	SoC	
61-0708	UH1B	SoC	
61-0709	UH1B	SoC	
61-0710	UH1B	SoC	
61-0711	UH1B	SoC	
61-0712	UH1B	to N46928	
61-0713	UH1B	to Uruguay 062	
61-0714	UH1B	SoC	
61-0715	UH1B	SoC	
61-0716	UH1B	SoC	
61-0717	UH1B	SoC	
61-0718	UH1B	to N88992	
61-0719	UH1B	SoC	
61-0720	UH1B	to N103MF	
61-0721	UH1B	SoC	
61-0722	UH1B	SoC	
61-0723	UH1B	to N9645A	
61-0724	UH1B	to N88987	
61-0725	UH1B	to Korea	
61-0726	UH1B	SoC	
61-0727	UH1B	SoC	
61-0728	UH1B	to Honduras	
61-0729	UH1B	SoC	
61-0730	UH1B	SoC	
61-0731	UH1B	SoC	
61-0732	UH1B	SoC	
61-0733	UH1B	to N9050Q	
61-0734	UH1B	SoC	
61-0735	UH1B	to Korea	
61-0736	UH1B	SoC	
61-0737	UH1B	SoC	
61-0738	UH1B	to N331WN	
61-0739	UH1B	to N96261	

61-0740	UH1B	to N88983		62-1891	UH1B	to Singapore	
61-0741	UH1B	to N91525		62-1892	UH1B	to Paraguay	
61-0742	UH1B	SoC		62-1893	UH1B	to N96017	
61-0743	UH1B	SoC		62-1894	UH1B	to N64CC	
61-0744	UH1B	to N98154		62-1895	UH1B	to Panama 116	
61-0745	UH1B	w.o.15.9.71		62-1896	UH1B	to Honduras	
61-0746	UH1B	SoC		62-1897	UH1B	SoC	
61-0747	UH1B	to N3032F		62-1898	UH1B	SoC	
61-0748	UH1B	to N31341		62-1899	UH1B	w.o.27.3.72	
61-0749	UH1B	SoC		62-1900	UH1B	to N9066D	
61-0750	UH1B	to N45731		62-1901	UH1B	SoC	
61-0751	UH1B	SoC		62-1902	UH1B	to N10SP	
61-0752	UH1B	SoC		62-1903	UH1B	SoC	
61-0753	UH1B	to N98102		62-1904	UH1B	SoC	
61-0754	UH1B	SoC		62-1905	UH1B	to N65SP	
61-0755	UH1B	SoC		62-1906	UH1B	to Panama 111	
61-0756	UH1B	to N96142		62-1907	UH1B	SoC	
61-0757	UH1B	SoC		62-1908	UH1B	to NASA	N732NA(1)
61-0758	UH1B	to N91307		62-1909	UH1B	SoC	
61-0759	UH1B	SoC		62-1910	UH1B	to N105MF	
61-0760	UH1B	SoC		62-1911	UH1B	SoC	
61-0761	UH1B	to N91348		62-1912	UH1B	to N70547	
61-0762	UH1B	to N98126		62-1913	UH1B	SoC	
61-0763	UH1B	to N842M		62-1914	UH1B	SoC	
61-0764	UH1B	SoC		62-1915	UH1B	to N31213	
61-0765	UH1B	to N4491D		62-1916	UH1B	SoC	
61-0766	UH1B	SoC		62-1917	UH1B	SoC	
61-0767	UH1B	to N87845		62-1918	UH1B	to Singapore	
61-0768	UH1B	w.o.3.6.70		62-1919	UH1B	SoC	
61-0769	UH1B	SoC		62-1920	UH1B	preserved	NAS Willow Grove
61-0770	UH1B	to N88979		62-1921	UH1B	dump	Tucson, Az
61-0771	UH1B	to N98049		62-1922	UH1B	to N17765	
61-0772	UH1B	SoC		62-1923	UH1B	to N22961	
61-0773	UH1B	SoC		62-1924	UH1B	to N6NJ	
61-0774	UH1B	to Singapore 268		62-1925	UH1B	to HP-999	
61-0775	UH1B	to Honduras		62-1926	UH1B	to N1564F	
61-0776	UH1B	to Singapore (GIA)		62-1927	UH1B	SoC	
61-0777	UH1B	SoC		62-1928	UH1B	to N59368	
61-0778	UH1B	to N81958		62-1929	UH1B	SoC	
61-0779	UH1B	SoC		62-1930	UH1B	SoC	
61-0780	UH1B	SoC		62-1931	UH1B	to N91284	
61-0781	UH1B	SoC		62-1932	UH1B	SoC	
61-0782	UH1B	SoC		62-1933	UH1B	to N59HP	
61-0783	UH1B	SoC		62-1934	UH1B	to N8NJ	
61-0784	UH1B	SoC		62-1935	UH1B	to Panama	
61-0785	UH1B	SoC		62-1936	UH1B	to N1564F	
61-0786	UH1B	to N9846F		62-1937	UH1B	to Norway 937	
61-0787	UH1B	to N9993Q		62-1938	UH1B	SoC	
61-0788	UH1B	preserved	Ft. Eustis	62-1939	UH1B	SoC	
61-0789	UH1B	SoC		62-1940	UH1B	to Colombia	
61-0790	UH1B	w.o.20.2.70		62-1941	UH1B	SoC	
61-0791	UH1B	to N68077		62-1942	UH1B	SoC	
61-0792	UH1B	SoC		62-1943	UH1B	SoC	
61-0793	UH1B	SoC		62-1944	UH1B	to N83985	
61-0794	UH1B	SoC		62-1945	UH1B	SoC	
61-0795	UH1B	SoC		62-1946	UH1B	SoC	
61-0796	UH1B	to Singapore 257		62-1947	UH1B	w.o.17.1.72	
61-0797	UH1B	to N96206		62-1948	UH1B	to N4028F	
61-0798	UH1B	to N2252A		62-1949	UH1B	to N8503K	
61-0799	UH1B	SoC		62-1950	UH1B	SoC	
61-0800	UH1B	SoC		62-1951	UH1B	SoC	
61-0801	UH1B	w.o.23.4.72		62-1952	UH1B	SoC	
61-0802	UH1B	SoC		62-1953	UH1B	to N50023	
61-0803	UH1B	SoC		62-1954	UH1B	to N910PD	
62-1872	UH1B	SoC		62-1955	UH1B	SoC	
62-1873	UH1B	SoC		62-1956	UH1B	to N841M	
62-1874	UH1B	SoC		62-1957	UH1B	to Norway 957	
62-1875	UH1B	to N85290		62-1958	UH1B	SoC	
62-1876	UH1B	SoC		62-1959	UH1B	SoC	
62-1877	UH1B	SoC		62-1960	UH1B	to N1565F	
62-1878	UH1B	SoC		62-1961	UH1B	SoC	
62-1879	UH1B	SoC		62-1962	UH1B	w.o.20.10.70	
62-1880	UH1B	SoC		62-1963	UH1B	SoC	
62-1881	UH1B	to N58HP		62-1964	UH1B	SoC	
62-1882	UH1B	SoC		62-1965	UH1B	SoC	
62-1883	UH1B	SoC		62-1966	UH1B	SoC	
62-1884	UH1B	preserved	Ft. Rucker, Al	62-1967	UH1B	SoC	
62-1885	UH1B	SoC		62-1968	UH1B	to N98F	
62-1886	UH1B	w.o.27.4.70		62-1969	UH1B	SoC	
62-1887	UH1B	to N91320		62-1970	UH1B	dump	Tucson, Az
62-1888	UH1B	to N5023U		62-1971	UH1B	SoC	
62-1889	UH1B	SoC		62-1972	UH1B	SoC	
62-1890	UH1B	to Panama 102		62-1973	UH1B	SoC	

Serial	Type	Status	Note
62-1974	UH1B	SoC	
62-1975	UH1B	SoC	
62-1976	UH1B	SoC	
62-1977	UH1B	to Korea	
62-1978	UH1B	SoC	
62-1979	UH1B	SoC	
62-1980	UH1B	to N63CD	
62-1981	UH1B	SoC	
62-1982	UH1B	SoC	
62-1983	UH1B	SoC	
62-1984	UH1B	to N332WN	
62-1985	UH1B	to Singapore 255	
62-1986	UH1B	preserved	Mobile, Al
62-1987	UH1B	SoC	
62-1988	UH1B	SoC	
62-1989	UH1B	SoC	
62-1990	UH1B	SoC	
62-1991	UH1B	to N2266T	
62-1992	UH1B	to Norway 992	
62-1993	UH1B	to Norway 993	
62-1994	UH1B	to Norway 994	
62-1995	UH1B	to Norway 995	
62-1996	UH1B	to Norway 996	
62-1997	UH1B	SoC	
62-1998	UH1B	SoC	
62-1999	UH1B	SoC	
62-2000	UH1B	to N4284N	
62-2001	UH1B	SoC	
62-2002	UH1B	SoC	
62-2003	UH1B	SoC	
62-2004	UH1B	SoC	
62-2005	UH1B	SoC	
62-2006	UH1B	SoC	
62-2007	UH1B	to Korea	
62-2008	UH1B	SoC	
62-2009	UH1B	SoC	
62-2010	UH1B	preserved	Ft. Campbell, Ky
62-2011	UH1B	SoC	
62-2012	UH1B	SoC	
62-2013	UH1B	SoC	
62-2014	UH1B	SoC	
62-2015	UH1B	SoC	
62-2016	UH1B	SoC	
62-2017	UH1B	SoC	
62-2018	UH1B	preserved	
62-2019	UH1B	SoC	
62-2020	UH1B	SoC	
62-2021	UH1B	SoC	
62-2022	UH1B	SoC	
62-2023	UH1B	w.o.15.2.73	
62-2024	UH1B	SoC	
62-2025	UH1B	to Norway 025	
62-2026	UH1B	to N328WN	
62-2027	UH1B	SoC	
62-2028	UH1B	SoC	
62-2029	UH1B	SoC	
62-2030	UH1B	SoC	
62-2031	UH1B	to N91350	
62-2032	UH1B	SoC	
62-2033	UH1B	SoC	
62-2034	UH1B	to N87729	
62-2035	UH1B	SoC	
62-2036	UH1B	SoC	
62-2037	UH1B	SoC	
62-2038	UH1B	to N106MF	
62-2039	UH1B	SoC	
62-2040	UH1B	SoC	
62-2041	UH1B	SoC	
62-2042	UH1B	SoC	
62-2043	UH1B	to N31379	
62-2044	UH1B	SoC	
62-2045	UH1B	to Taiwan	
62-2046	UH1B	SoC	
62-2047	UH1B	SoC	
62-2048	UH1B	to N3979C	
62-2049	UH1B	SoC	
62-2050	UH1B	SoC	
62-2051	UH1B	SoC	
62-2052	UH1B	SoC	
62-2053	UH1B	to N3879G	
62-2054	UH1B	to Honduras	
62-2055	UH1B	SoC	
62-2056	UH1B	SoC	
62-2057	UH1B	SoC	
62-2058	UH1B	SoC	
62-2059	UH1B	SoC	
62-2060	UH1B	SoC	
62-2061	UH1B	SoC	
62-2062	UH1B	w.o.27.7.70	
62-2063	UH1B	SoC	
62-2064	UH1B	to NASA	N416NA
62-2065	UH1B	SoC	
62-2066	UH1B	dump	Tucson, Az
62-2067	UH1B	SoC	
62-2068	UH1B	SoC	
62-2069	UH1B	SoC	
62-2070	UH1B	SoC	
62-2071	UH1B	SoC	
62-2072	UH1B	to N1385W	
62-2073	UH1B	w.o.22.10.70	
62-2074	UH1B	SoC	
62-2075	UH1B	SoC	
62-2076	UH1B	to Panama 112	
62-2077	UH1B	SoC	
62-2078	UH1B	to N50330	
62-2079	UH1B	SoC	
62-2080	UH1B	w.o.4.2.70	
62-2081	UH1B	SoC	
62-2082	UH1B	to N333WN	
62-2083	UH1B	SoC	
62-2084	UH1B	to N832M	
62-2085	UH1B	to N83980	
62-2086	UH1B	to Norway 086	
62-2087	UH1B	SoC	
62-2088	UH1B	to Singapore 272	
62-2089	UH1B	to N1525T	
62-2090	UH1B	to N80WF	
62-2091	UH1B	SoC	
62-2092	UH1B	SoC	
62-2093	UH1B	SoC	
62-2094	UH1B	to Singapore 270	
62-2095	UH1B	SoC	
62-2096	UH1B	to N87765	
62-2097	UH1B	to N334WN	
62-2098	UH1B	SoC	
62-2099	UH1B	preserved	Ft. Rucker, Al
62-2100	UH1B	SoC	
62-2101	UH1B	SoC	
62-2102	UH1B	SoC	
62-2103	UH1B	to Korea	
62-2104	UH1B	dump	Tucson, Az
62-2105	UH1B	to Uruguay 063	
62-2106	UH1H	AMARC	XA315
62-2107	UH1H	SoC	
62-2108	UH1H	SoC	
62-2109	UH1H		
62-2110	UH1D	SoC	
62-2111	UH1D	SoC	
62-2112	UH1H		
62-2113	GUH1H	GIA	USATS
62-4566	UH1B	to N61589	
62-4567	UH1B	SoC	
62-4568	UH1B	to N87948	
62-4569	UH1B	SoC	
62-4570	UH1B	SoC	
62-4571	UH1B	SoC	
62-4572	UH1B	w.o.28.4.71	
62-4573	UH1B	SoC	
62-4574	UH1B	SoC	
62-4575	UH1B	SoC	
62-4576	UH1B	to N61589	
62-4577	UH1B	SoC	
62-4578	UH1B	SoC	
62-4579	UH1B	SoC	
62-4580	UH1B	SoC	
62-4581	UH1B	SoC	
62-4582	UH1B	SoC	
62-4583	UH1B	to Singapore 262	
62-4584	UH1B	SoC	
62-4585	UH1B	SoC	
62-4586	UH1B	SoC	
62-4587	UH1B	SoC	
62-4588	UH1B	preserved	Oklahoma City
62-4589	UH1B	SoC	
62-4590	UH1B	to N70105	
62-4591	UH1B	to N335WN	

Serial	Type	Status/Disposition	Location
62-4592	UH1B	to N1386L	
62-4593	UH1B	SoC	
62-4594	UH1B	SoC	
62-4595	UH1B	to N3950Z	
62-4596	UH1B	SoC	
62-4597	UH1B	to Singapore 254	
62-4598	UH1B	to N64789	
62-4599	UH1B	SoC	
62-4600	UH1B	SoC	
62-4601	UH1B	to N99675	
62-4602	UH1B	to N84402	
62-4603	UH1B	to N911CD	
62-4604	UH1B	dump	Tucson, Az
62-4605	UH1B	dump	Tucson, Az
62-12351	UH1D	SoC	
62-12352	UH1D	SoC	
62-12353	UH1D	SoC	
62-12354	UH1D	SoC	
62-12355	UH1H		
62-12356	UH1D	SoC	
62-12357	UH1H	to N81523	
62-12358	UH1D	SoC	
62-12359	UH1H		
62-12360	UH1H		
62-12361	UH1H	1-185AVN	MS NG
62-12362	UH1H		
62-12363	UH1D	SoC	
62-12364	UH1D	SoC	
62-12365	UH1D	SoC	
62-12366	GUH1H	GIA	USATS
62-12367	UH1D	SoC	
62-12368	GUH1H		
62-12369	UH1H		
62-12370	UH1D	SoC	
62-12371	UH1D	SoC	
62-12372	UH1H	SoC	
62-12515	UH1B	SoC	
62-12516	UH1B	SoC	
62-12517	UH1B	SoC	
62-12518	UH1B	SoC	
62-12519	UH1B	SoC	
62-12520	UH1B	w.o.30.7.73	
62-12521	UH1B	SoC	
62-12522	UH1B	SoC	
62-12523	UH1B	SoC	
62-12524	UH1B	SoC	
62-12525	UH1B	to Singapore 271	
62-12526	UH1B	w.o.14.4.70	
62-12527	UH1B	SoC	
62-12528	UH1B	SoC	
62-12529	UH1B	to N9687R	
62-12530	UH1B	stored	Tucson IAP
62-12531	UH1B	SoC	
62-12532	UH1B	w.o.23.9.70	
62-12533	UH1B	to N394HP	
62-12534	UH1B	to N747FW	
62-12535	UH1B	SoC	
62-12536	UH1B	to N336WN	
62-12537	UH1B	to N.....	
62-12538	UH1B	SoC	
62-12539	UH1B	SoC	
62-12540	UH1B	SoC	
62-12541	UH1B	SoC	
62-12542	UH1B	to Korea	
62-12543	UH1B	to N96113	
62-12544	UH1B	SoC	
62-12545	UH1B	SoC	
62-12546	UH1B	SoC	
62-12547	UH1B	w.o.25.9.71	
62-12548	UH1B	SoC	
62-12549	UH1B	SoC	
62-12550	UH1B	preserved	Bradley ANGB
62-12551	UH1B	SoC	
62-12552	UH1B	SoC	
62-12553	UH1B	to Paraguay	
62-12554	UH1B	preserved	Andrews AFB
62-12555	UH1B	to N80477	
63-8500	UH1B	to N337WN	
63-8501	UH1B	to Norway 501	
63-8502	UH1B	preserved	Donnally Field
63-8503	UH1B	to N96119	
63-8504	UH1B	scrapyard	Dross Metals
63-8505	UH1B	preserved	Ft. Rucker
63-8506	UH1B	w.o.16.3.72	
63-8507	UH1B	SoC	
63-8508	UH1B	to N3080W	
63-8509	UH1B	to N2770N	
63-8510	UH1B	preserved	Ft. Sill, Ok
63-8511	UH1B	to N96250	
63-8512	UH1B	SoC	
63-8513	UH1B	to N99676	
63-8514	UH1B	to Singapore	
63-8515	UH1B	SoC	
63-8516	UH1B	SoC	
63-8517	UH1B	SoC	
63-8518	UH1B	to N2580V	
63-8519	UH1B	SoC	
63-8520	UH1B	dump	Tucson, Az
63-8521	UH1B	to N46942	
63-8522	UH1B	w.o.30.1.73	
63-8523	UH1B	to N3121A	
63-8524	UH1B	to Korea	
63-8525	UH1B	SoC	
63-8526	UH1B	to N104MF	
63-8527	UH1B	SoC	
63-8528	UH1B	to N844M	
63-8529	UH1B	to N2295B	
63-8530	UH1B	to Singapore	
63-8531	UH1B	to Singapore	
63-8532	UH1B	to Honduras	
63-8533	UH1B	to Singapore 266	
63-8534	UH1B	to Singapore 265	
63-8535	UH1B	to N91281	
63-8536	UH1B	SoC	
63-8537	UH1B	SoC	
63-8538	UH1B	SoC	
63-8539	UH1B	to Singapore	
63-8540	UH1B	to Korea	
63-8541	UH1B	to N567SJ	
63-8542	UH1B	SoC	
63-8543	UH1B	SoC	
63-8544	UH1B	to Singapore 256	
63-8545	UH1B	to N45267	
63-8546	UH1B	to Panama	
63-8547	UH1B	to N96036	
63-8548	UH1B	to N46969	
63-8549	UH1B	SoC	
63-8550	UH1B	SoC	
63-8551	UH1B	SoC	
63-8552	UH1B	SoC	
63-8553	UH1B	SoC	
63-8554	UH1B	to N39502	
63-8555	UH1B	SoC	
63-8556	UH1B	to N98024	
63-8557	UH1B	SoC	
63-8558	UH1B	SoC	
63-8559	UH1B	SoC	
63-8560	UH1B	to Korea	
63-8561	UH1B	dump	Tucson, Az
63-8562	UH1B	SoC	
63-8563	UH1B	GIA	Dannelly Field, Al
63-8564	UH1B	SoC	
63-8565	UH1B	SoC	
63-8566	UH1B	to N8142W	
63-8567	UH1B	SoC	
63-8568	UH1B	SoC	
63-8569	UH1B	SoC	
63-8570	UH1B	SoC	
63-8571	UH1B	SoC	
63-8572	UH1B	SoC	
63-8573	UH1B	SoC	
63-8574	UH1B	SoC	
63-8575	UH1B	SoC	
63-8576	UH1B	SoC	
63-8577	UH1B	SoC	
63-8578	UH1B	to N2142V	
63-8579	UH1B	SoC	
63-8580	UH1B	SoC	
63-8581	UH1B	to N.....	
63-8582	UH1B	to Korea	
63-8583	UH1B	SoC	
63-8584	UH1B	SoC	
63-8585	UH1B	SoC	
63-8586	UH1B	SoC	
63-8587	UH1B	SoC	
63-8588	UH1B	to Norway 588	

63-8589	UH1B	SoC	
63-8590	UH1B	SoC	
63-8591	UH1B	SoC	
63-8592	UH1B	SoC	
63-8593	UH1B	SoC	
63-8594	UH1B	SoC	
63-8595	UH1B	SoC	
63-8596	UH1B	SoC	
63-8597	UH1B	SoC	
63-8598	UH1B	SoC	
63-8599	UH1B	SoC	
63-8600	UH1B	SoC	
63-8601	UH1B	SoC	
63-8602	UH1B	SoC	
63-8603	UH1B	SoC	
63-8604	UH1B	SoC	
63-8605	UH1B	SoC	
63-8606	UH1B	to N15SP	
63-8607	UH1B	SoC	
63-8608	UH1B	SoC	
63-8609	UH1B	SoC	
63-8610	UH1B	to N5HF	
63-8611	UH1B	dump	Tucson, Az
63-8612	UH1B	w.o.12.2.70	
63-8613	UH1B	to Korea	
63-8614	UH1B	to Korea	
63-8615	UH1B	to N15SD	
63-8616	UH1B	SoC	
63-8617	UH1B	SoC	
63-8618	UH1B	SoC	
63-8619	UH1B	SoC	
63-8620	UH1B	SoC	
63-8621	UH1B	SoC	
63-8622	UH1B	to Panama 117	
63-8623	UH1B	SoC	
63-8624	UH1B	SoC	
63-8625	UH1B	SoC	
63-8626	UH1B	SoC	
63-8627	UH1B	SoC	
63-8628	UH1B	SoC	
63-8629	UH1B	SoC	
63-8630	UH1B	SoC	
63-8631	UH1B	SoC	
63-8632	UH1B	SoC	
63-8633	UH1B	SoC	
63-8634	UH1B	SoC	
63-8635	UH1B	SoC	
63-8636	UH1B	to N99478	
63-8637	UH1B	SoC	
63-8638	UH1B	SoC	
63-8639	UH1B	SoC	
63-8640	UH1B	SoC	
63-8641	UH1B	SoC	
63-8642	UH1B	SoC	
63-8643	UH1B	SoC	
63-8644	UH1B	SoC	
63-8645	UH1B	SoC	
63-8646	UH1B	to N99634	
63-8647	UH1B	SoC	
63-8648	UH1B	SoC	
63-8649	UH1B	SoC	
63-8650	UH1B	SoC	
63-8651	UH1B	SoC	
63-8652	UH1B	SoC	
63-8653	UH1B	SoC	
63-8654	UH1B	SoC	
63-8655	UH1B	SoC	
63-8656	UH1B	SoC	
63-8657	UH1B	SoC	
63-8658	UH1B	to N2581E	
63-8663	UH1B	SoC	
63-8664	UH1B	to N9434A	
63-8665	UH1B	SoC	
63-8666	UH1B	SoC	
63-8667	UH1B	SoC	
63-8668	UH1B	to N9378A	
63-8669	UH1B	SoC	
63-8670	UH1B	SoC	
63-8671	UH1B	SoC	
63-8672	UH1B	SoC	
63-8673	UH1B	to Korea	
63-8674	UH1B	to N9650N	
63-8675	UH1B	SoC	
63-8676	UH1B	to N843M	
63-8677	UH1B	SoC	
63-8678	UH1B	SoC	
63-8679	UH1B	to Singapore	
63-8680	UH1B	SoC	
63-8681	UH1B	SoC	
63-8682	UH1B	SoC	
63-8683	UH1B	SoC	
63-8684	NUH1M	NOAA	Miami IAP
63-8685	UH1B	w.o.19.2.70	
63-8686	UH1B	SoC	
63-8687	UH1B	SoC	
63-8688	UH1B	SoC	
63-8689	UH1B	to NASA	N415NA
63-8690	UH1B	SoC	
63-8691	UH1B	SoC	
63-8692	UH1B	to Singapore 264	
63-8693	UH1B	SoC	
63-8694	UH1B	to N847MC	
63-8695	UH1B	to N1363J	
63-8696	UH1B	SoC	
63-8697	UH1B	SoC	
63-8698	UH1B	to Korea	
63-8699	UH1B	to N39SD	
63-8700	UH1B	SoC	
63-8701	UH1B	SoC	
63-8702	UH1B	SoC	
63-8703	UH1B	to '0927'	
63-8704	UH1B	to N64771	
63-8705	UH1B	SoC	
63-8706	UH1B	SoC	
63-8707	UH1B	SoC	
63-8708	UH1B	to Singapore 267	
63-8709	UH1B	SoC	
63-8710	UH1B	SoC	
63-8711	UH1B	to Uruguay 064	
63-8712	UH1B	SoC	
63-8713	UH1B	SoC	
63-8714	UH1B	SoC	
63-8715	UH1B	SoC	
63-8716	UH1B	SoC	
63-8717	UH1B	SoC	
63-8718	UH1B	SoC	
63-8719	UH1B	SoC	
63-8720	UH1B	SoC	
63-8721	UH1B	to N64770	
63-8722	UH1B	SoC	
63-8723	UH1B	SoC	
63-8724	UH1B	SoC	
63-8725	UH1B	dump	Tucson, Az
63-8726	UH1B	SoC	
63-8727	UH1B	SoC	
63-8728	UH1B	to N88389	
63-8729	UH1B	SoC	
63-8730	UH1B	to Uruguay	
63-8731	UH1B	to N3295F	
63-8732	UH1B	w.o.16.8.70	
63-8733	UH1B	dump	Tucson, Az
63-8734	UH1B	to Korea	
63-8735	UH1B	to Uruguay 065	
63-8736	UH1B	SoC	
63-8737	UH1B	SoC	
63-8738	UH1B	SoC	
63-8739	GUH1H	GIA	USATS
63-8740	UH1H	to N8148C	
63-8741	UH1H		
63-8742	UH1H		
63-8743	UH1H	SoC	
63-8744	UH1D	w.o.	
63-8745	UH1D	w.o.	
63-8746	UH1D	31AVN	
63-8747	UH1D	w.o.	
63-8748	UH1H	26AVN	
63-8749	UH1D	to Israel	
63-8750	UH1D	w.o.	
63-8751	UH1D	w.o.	
63-8752	UH1D	SoC	
63-8753	UH1H	AMARC	XA317
63-8754	UH1D	w.o.	
63-8755	UH1D	w.o.	
63-8756	UH1H	455\457MedDet	MO NG
63-8757	UH1D	w.o.	
63-8758	UH1H	AATC	58K Lowe AHP

63-8759	UH1D	w.o.	
63-8760	UH1D	SoC	
63-8761	UH1D	w.o.	
63-8762	UH1H	w.o.20.3.76	
63-8763	UH1H	w.o.3.7.86	
63-8764	UH1D	w.o.17.6.71	
63-8765	UH1H		
63-8766	UH1H		
63-8767	UH1D	w.o.	
63-8768	UH1H		
63-8769	UH1D	SoC	
63-8770	UH1H	AATC	70A Lowe AHP
63-8771	UH1D	w.o.	
63-8772	UH1D	SoC	
63-8773	UH1D	SoC	
63-8774	UH1D	SoC	
63-8775	UH1H	to N81526	
63-8776	UH1D	w.o.	
63-8777	UH1D	w.o.	
63-8778	UH1H	AMARC	XA368
63-8779	UH1D	SoC	
63-8780	UH1D	SoC	
63-8781	UH1H		
63-8782	UH1H	AMARC	XA344
63-8783	UH1D	SoC	
63-8784	UH1H	AATC	84. Lowe AHP
63-8785	UH1H	to N8149P	
63-8786	UH1D	SoC	
63-8787	UH1D	SoC	
63-8788	UH1D	w.o.	
63-8789	UH1D	w.o.	
63-8790	UH1D	w.o.	
63-8791	UH1D	w.o.	
63-8792	UH1D	SoC	
63-8793	UH1D	SoC	
63-8794	UH1H		
63-8795	UH1D	w.o.	
63-8796	UH1D	SoC	
63-8797	UH1D	SoC	
63-8798	UH1D	w.o.	
63-8799	UH1D	w.o.	
63-8800	UH1D	w.o.	
63-8801	UH1H		
63-8802	UH1D	SoC	
63-8803	UH1H		
63-8804	UH1D	SoC	
63-8805	UH1H		
63-8806	UH1D	w.o.10.8.72	
63-8807	UH1H		
63-8808	UH1D	w.o.	
63-8809	UH1H		IL NG
63-8810	UH1H	SoC	
63-8811	UH1D	w.o.	
63-8812	UH1H	SoC	
63-8813	UH1D	w.o.	
63-8814	UH1D	w.o.	
63-8815	UH1H	w.o.19.7.83	
63-8816	UH1H	to N482DF	
63-8817	UH1H	SoC	
63-8818	UH1D	w.o.	
63-8819	UH1H	AMARC	XA343
63-8820	UH1D	SoC	
63-8821	UH1H	42AVN	
63-8822	UH1D	w.o.	
63-8823	UH1H		
63-8824	UH1D	w.o.13.5.75	
63-8825	UH1H		
63-8826	UH1H		
63-8827	UH1D	w.o.	
63-8828	UH1H	w.o.12.11.74	
63-8829	UH1H	AATC	29A Lowe AHP
63-8830	UH1D	w.o.	
63-8831	UH1H		
63-8832	UH1D	w.o.	
63-8833	UH1D	SoC	
63-8834	UH1D	w.o.	
63-8835	UH1D	w.o.	
63-8836	UH1H	AMARC	XA345
63-8837	UH1D	SoC	
63-8838	UH1H		
63-8839	UH1D	SoC	
63-8840	UH1D	SoC	
63-8841	UH1H	31AVN	
63-8842	UH1D	w.o.	
63-8843	UH1D	SoC	
63-8844	UH1D	w.o.	
63-8845	GUH1H	GIA	USATS
63-8846	UH1H	to N338WN	
63-8847	UH1H	198AvCo	
63-8848	UH1H	to N81785	
63-8849	UH1D	w.o.	
63-8850	UH1H	AMARC	XA369
63-8851	UH1H	w.o.	
63-8852	UH1D	SoC	
63-8853	UH1H	31AVN	
63-8854	UH1H	31AVN	
63-8855	UH1D	w.o.	
63-8856	UH1D	w.o.	
63-8857	UH1H	40AVN	
63-8858	UH1H	AATC	58L Lowe AHP
63-8859	UH1H		
63-12903	UH1B	SoC	
63-12904	UH1B	SoC	
63-12905	UH1B	SoC	
63-12906	UH1B	to N106MF	
63-12907	UH1B	SoC	
63-12907	UH1B	SoC	
63-12908	UH1B	SoC	
63-12909	UH1B	SoC	
63-12910	UH1B	to N106MF	
63-12911	UH1B	dump	Tucson, Az
63-12912	UH1B	SoC	
63-12913	UH1B	SoC	
63-12914	UH1B	SoC	
63-12915	UH1B	SoC	
63-12916	UH1B	to N5598G	
63-12917	UH1B	SoC	
63-12918	UH1B	SoC	
63-12919	UH1B	SoC	
63-12920	UH1B	SoC	
63-12921	UH1B	to Korea	
63-12922	UH1B	SoC	
63-12923	UH1B	SoC	
63-12924	UH1B	SoC	
63-12925	UH1B	SoC	
63-12926	UH1B	SoC	
63-12927	UH1B	SoC	
63-12928	UH1B	SoC	
63-12929	UH1B	SoC	
63-12930	UH1B	SoC	
63-12931	UH1B	SoC	
63-12932	UH1B	SoC	
63-12933	UH1B	SoC	
63-12934	UH1B	SoC	
63-12935	UH1B	SoC	
63-12936	UH1B	SoC	
63-12937	UH1B	SoC	
63-12938	UH1B	SoC	
63-12939	UH1B	SoC	
63-12940	UH1B	SoC	
63-12941	UH1B	SoC	
63-12942	UH1B	SoC	
63-12943	UH1B	SoC	
63-12944	UH1B	SoC	
63-12945	UH1B	SoC	
63-12946	UH1B	SoC	
63-12947	UH1B	SoC	
63-12948	UH1B	SoC	
63-12949	UH1B	w.o.1.6.70	
63-12950	UH1B	to N2956F	
63-12951	UH1B	to Singapore	
63-12952	UH1B	SoC	
63-12956	UH1D	w.o.	
63-12957	UH1D	w.o.	
63-12958	UH1D	w.o.	
63-12959	UH1H		
63-12960	GUH1H	GIA	USATS
63-12961	UH1D	SoC	
63-12962	UH1H	to N81500	
63-12963	UH1H	83ARCOM	
63-12964	UH1D	w.o.	
63-12965	UH1D	SoC	
63-12966	UH1D	w.o.	
63-12967	UH1D	w.o.	
63-12968	UH1D	w.o.	
63-12969	UH1D	w.o.	

Serial	Type	Note	Extra
63-12970	UH1H	26AVN	
63-12971	UH1D	w.o.	
63-12972	UH1D	SoC	
63-12973	UH1H		
63-12974	UH1H	AMARC	XA365
63-12975	UH1H	26AVN	
63-12976	UH1H	AMARC	XA366
63-12977	UH1D	w.o.	
63-12978	UH1D	w.o.	
63-12979	UH1H	to N339WN	
63-12980	UH1D	w.o.	
63-12981	UH1D	w.o.	
63-12982	UH1H	42AVN	
63-12983	UH1D	SoC	
63-12984	UH1D	w.o.	
63-12985	UH1D	SoC	
63-12986	UH1H	AMARC	XA340
63-12987	UH1D	w.o.	
63-12988	UH1D	w.o.	
63-12989	UH1H	198AvCo	
63-12990	UH1H	1-185AVN	MS NG
63-12991	GUH1H		
63-12992	UH1D	SoC	
63-12993	UH1H	AATC	93. Lowe AHP
63-12994	UH1D	w.o.	
63-12995	UH1D	w.o.	
63-12996	UH1H	w.o.29.8.00	
63-12997	UH1H	26AVN	
63-12998	UH1H	AMARC	XA319
63-12999	UH1D	w.o.	
63-13000	UH1D	w.o.	
63-13001	UH1D	SoC	
63-13002	UH1D	SoC	
64-13492	UH1H	150AVN	
64-13493	UH1H		
64-13494	UH1H	w.o.11.9.82	
64-13495	UH1H		
64-13496	UH1D	SoC	
64-13497	UH1H	126MedCo	
64-13498	UH1H		
64-13499	UH1D	w.o.	
64-13500	UH1D	w.o.	
64-13501	UH1D	w.o.	
64-13502	UH1H		
64-13503	UH1D	w.o.	
64-13504	UH1H	A\1-244AVN	LA NG
64-13505	UH1D	w.o.	
64-13506	GUH1H		
64-13507	UH1H	SoC	
64-13508	UH1D	w.o.	
64-13509	GUH1H		
64-13510	UH1H	w.o.22.3.75	
64-13511	UH1H	44AVN	
64-13512	UH1D	w.o.11.1.71	
64-13513	UH1H	44AVN	
64-13514	GUH1H	GIA	USATS
64-13515	UH1D	w.o.	
64-13516	UH1D	w.o.	
64-13517	UH1H		
64-13518	UH1D	w.o.	
64-13519	UH1D	w.o.	
64-13520	UH1D	SoC	
64-13521	UH1D	w.o.	
64-13522	UH1D	w.o.	
64-13523	UH1D	SoC	
64-13524	UH1D	SoC	
64-13525	UH1D	w.o.	
64-13526	UH1D	w.o.2.10.70	
64-13527	UH1D	w.o.	
64-13528	UH1H		
64-13529	UH1D	w.o.	
64-13530	UH1D	w.o.	
64-13531	UH1D	w.o.	
64-13532	UH1D	w.o.	
64-13533	UH1H	to N81522	
64-13534	UH1D	w.o.	
64-13535	UH1D	w.o.	
64-13536	UH1H		
64-13537	UH1D	w.o.	
64-13538	UH1H		
64-13539	UH1H	SoC	
64-13540	GUH1H	GIA	USATS
64-13541	UH1D	w.o.	
64-13542	UH1D	w.o.	
64-13543	UH1D	w.o	
64-13544	UH1H	166ACR	
64-13545	UH1D	w.o.	
64-13546	UH1H	w.o.25.1.72	
64-13547	UH1D	w.o.	
64-13548	UH1D	w.o.	
64-13549	UH1D	w.o.	
64-13550	UH1D	w.o.	
64-13551	UH1D	w.o.	
64-13552	UH1D	w.o.	
64-13553	UH1H		
64-13554	UH1D	w.o.	
64-13555	UH1D	w.o.	
64-13556	UH1D	preserved	
64-13557	UH1H	AATC	57F Lowe AHP
64-13558	UH1H	w.o.24.10.89	
64-13559	UH1D	w.o.	
64-13560	GUH1H	GIA	USATS
64-13561	UH1H		Kansas NG
64-13562	UH1H		
64-13563	UH1D	w.o.	
64-13564	UH1H	w.o.30.1.81	
64-13565	UH1H	70Div	
64-13566	UH1D	SoC	
64-13567	UH1D	w.o.	
64-13568	UH1D	w.o.	
64-13569	UH1H	891EngBtn	
64-13570	UH1D	SoC	
64-13571	UH1D	w.o.	
64-13572	UH1H	to N8152Q	
64-13573	UH1D	SoC	
64-13574	UH1H		
64-13575	UH1D	w.o.	
64-13576	UH1D	w.o.	
64-13577	UH1H	to N81569	
64-13578	UH1D	w.o.	
64-13579	UH1H	SoC	
64-13580	UH1H	SoC	
64-13581	UH1H	w.o.4.8.72	
64-13582	UH1D	w.o.	
64-13583	UH1D	w.o.	
64-13584	UH1H		
64-13585	UH1H	w.o.	
64-13586	UH1H		
64-13587	UH1D	SoC	
64-13588	UH1D	SoC	
64-13589	UH1D	SoC	
64-13590	UH1D	SoC	
64-13591	GUH1H		
64-13592	GUH1H		
64-13593	UH1H	SoC	
64-13594	UH1D	w.o.	
64-13595	UH1D	w.o.	
64-13596	UH1D	w.o.	
64-13597	UH1D	w.o.	
64-13598	UH1H		
64-13599	UH1D	w.o.	
64-13600	UH1D	w.o.	
64-13601	UH1H	163ACR	
64-13602	UH1D	w.o.	
64-13603	UH1H		NY NG
64-13604	UH1D	w.o.	
64-13605	UH1H	AATC	05F Lowe AHP
64-13606	UH1D	SoC	
64-13607	UH1D	w.o.	
64-13608	UH1D	w.o.	
64-13609	UH1H	to N8158Q	
64-13610	UH1D	w.o.	
64-13611	UH1H	163ACR	
64-13612	UH1D	w.o.	
64-13613	UH1D	SoC	
64-13614	GUH1H	GIA	USATS
64-13615	UH1D	w.o.	
64-13616	UH1D	w.o.	
64-13617	UH1H	AATC	17E Lowe AHP
64-13618	UH1D	SoC	
64-13619	UH1H		
64-13620	UH1D	w.o.	
64-13621	UH1H		
64-13622	UH1D	w.o.	
64-13623	UH1H	198AvCo	
64-13624	GUH1H	GIA	USATS

64-13625	UH1D	w.o.	
64-13626	UH1H	AMARC	XA326
64-13627	UH1D	w.o.	
64-13628	UH1H	to NASA	N734NA
64-13629	UH1H	AATC	29N Lowe AHP
64-13630	UH1D	SoC	
64-13631	UH1D	w.o.	
64-13632	UH1H		
64-13633	UH1D	w.o.	
64-13634	UH1D	SoC	
64-13635	UH1D	w.o.	
64-13636	UH1D	w.o.	
64-13637	UH1D	w.o.	
64-13638	UH1H		
64-13639	UH1H	42AVN	
64-13640	UH1D	w.o.	
64-13641	UH1D	w.o.	
64-13642	UH1D	w.o.	
64-13643	UH1D	SoC	
64-13644	GUH1H		
64-13645	UH1D	w.o.	
64-13646	UH1H	AMARC	XA371
64-13647	UH1D	SoC	
64-13648	UH1D	w.o.	
64-13649	UH1D	SoC	
64-13650	UH1H	26AVN	
64-13651	UH1D	w.o.	
64-13652	UH1H	2-135AVN	CO NG
64-13653	UH1D	w.o.	
64-13654	UH1H	AMARC	XA357
64-13655	UH1D	w.o.	
64-13656	GUH1H	GIA	USATS
64-13657	UH1D	w.o.	
64-13658	UH1D	w.o.	
64-13659	UH1D	SoC	
64-13660	UH1H	AATC	60M Lowe AHP
64-13661	UH1D	to Germany 7037	
64-13662	UH1D	to Germany 7038	
64-13663	UH1D	w.o.	
64-13664	UH1D	w.o.22.9.70	
64-13665	UH1D	SoC	
64-13666	UH1H		
64-13667	UH1H	to N7042N	
64-13668	UH1D	w.o.	
64-13669	UH1H		
64-13670	UH1H	461AVN	NM NG
64-13671	UH1D	w.o.	
64-13672	UH1H		
64-13673	UH1H		
64-13674	UH1H	2-135AVN	CO NG
64-13675	UH1H	AATC	75B Lowe AHP
64-13676	UH1D	w.o.	
64-13677	UH1D	w.o.	
64-13678	UH1H	44AVN	
64-13679	UH1H	AATC	79B Lowe AHP
64-13680	UH1D	w.o.	
64-13681	UH1D	w.o.	
64-13682	UH1H		
64-13683	UH1H	AATC	83F Lowe AHP
64-13684	UH1D	SoC	
64-13685	UH1H	to N394M	
64-13686	GUH1H		
64-13687	UH1D	w.o.	
64-13688	UH1H		
64-13689	UH1H	79ARCOM	
64-13690	UH1H		
64-13691	UH1H		
64-13692	UH1D	w.o.	
64-13693	UH1D	w.o.	
64-13694	UH1H	GIA	USATS
64-13695	UH1H	AATC	95G Lowe AHP
64-13696	UH1H	w.o.16.12.80	
64-13697	UH1H	to N4396H	
64-13698	UH1D	w.o.	
64-13699	UH1H	AATC	99D Lowe AHP
64-13700	UH1D	w.o.	
64-13701	UH1D		
64-13702	UH1D	w.o.	
64-13703	UH1D	w.o.	
64-13704	UH1D	w.o.	
64-13705	UH1D	w.o.	
64-13706	UH1D	w.o.	
64-13707	UH1D	SoC	
64-13708	UH1D	w.o.	
64-13709	UH1H	GIA	Lakefront AP, La
64-13710	UH1H	40AVN	
64-13711	UH1H		
64-13712	UH1D	w.o.	
64-13713	UH1H	SoC	
64-13714	UH1D	w.o.	
64-13715	UH1D	w.o.	
64-13716	UH1D	SoC	
64-13717	UH1D	w.o.	
64-13718	UH1H	w.o.4.90	
64-13719	UH1D	w.o.	
64-13720	UH1D	w.o.	
64-13721	UH1H	AATC	21H Lowe AHP
64-13722	UH1H	SoC	
64-13723	UH1H	to N8160V	
64-13724	UH1D	w.o.	
64-13725	UH1D	w.o.	
64-13726	UH1D	w.o.	
64-13727	UH1D	w.o.	
64-13728	UH1D	SoC	
64-13729	UH1H		
64-13730	UH1H		
64-13731	UH1H	31AVN	
64-13732	UH1H	42AVN	
64-13733	UH1D	w.o.	
64-13734	UH1D	w.o.	
64-13735	UH1D	w.o.	
64-13736	UH1H	81ARCOM	
64-13737	UH1D	w.o.	
64-13738	UH1D	w.o.	
64-13739	UH1H	62AvCo	
64-13740	UH1H	AMARC	XA318
64-13741	UH1D	w.o.	
64-13742	UH1D	w.o.	
64-13743	UH1D	SoC	
64-13744	UH1D	w.o.	
64-13745	UH1H	31AVN	
64-13746	UH1D	w.o.	
64-13747	UH1H		
64-13748	UH1D	w.o.	
64-13749	UH1D	w.o.	
64-13750	UH1D	SoC	
64-13751	UH1H		
64-13752	UH1H	SoC	
64-13753	UH1H	AATC	53H Lowe AHP
64-13754	UH1H		
64-13755	UH1D	w.o.	
64-13756	UH1D	w.o.	
64-13757	UH1D	w.o.	
64-13758	UH1D	SoC	
64-13759	UH1H		
64-13760	UH1H	SoC	
64-13761	UH1H	SoC	
64-13762	UH1D	w.o.	
64-13763	UH1D	w.o.	
64-13764	UH1D	w.o.	
64-13765	UH1H	w.o.26.8.81	
64-13766	UH1H	D\1-228AVN	
64-13767	UH1H		AL NG
64-13768	UH1H	AATC	68G Lowe AHP
64-13769	UH1D	SoC	
64-13770	UH1H	86ARCOM	
64-13771	UH1H	SoC	
64-13772	UH1D	w.o.	
64-13773	UH1H		
64-13774	UH1H	31AVN	
64-13775	UH1D	w.o.	
64-13776	UH1D	w.o.	
64-13777	UH1H	42AVN	
64-13778	UH1D	w.o.	
64-13779	UH1D	SoC	
64-13780	UH1D	w.o.	
64-13781	UH1D	w.o.	
64-13782	UH1D	SoC	
64-13783	UH1H	SoC	
64-13784	UH1H		
64-13785	UH1D	w.o.	
64-13786	UH1H	AATC	86. Lowe AHP
64-13787	UH1H	w.o.	
64-13788	UH1H	w.o.	
64-13789	UH1H	w.o.	
64-13790	UH1H	D\135AVN	KS NG

64-13791	UH1H	SoC	
64-13792	UH1D	w.o.	
64-13793	UH1D	w.o.	
64-13794	UH1H	SoC	
64-13795	UH1D	w.o.	
64-13796	UH1H	79ARCOM	
64-13797	UH1D	w.o.	
64-13798	UH1H	86ARCOM	
64-13799	UH1D	w.o.	
64-13800	UH1D	w.o.	
64-13801	UH1H		
64-13806	UH1D	to Germany 7038	
64-13807	UH1H	SoC	
64-13808	UH1H	26AVN	
64-13809	UH1D	w.o.	
64-13810	UH1H	to N945MF	
64-13811	UH1H		
64-13812	UH1D	w.o.	
64-13813	UH1H		
64-13814	UH1H	AMARC	XA325
64-13815	UH1D	w.o.	
64-13816	UH1H	AMARC	XA348
64-13817	UH1D	w.o.	
64-13818	UH1H	45AVN	
64-13819	UH1H	AATC	19N Lowe AHP
64-13820	UH1H		
64-13821	UH1D	w.o	
64-13822	UH1H		
64-13823	UH1H		
64-13824	UH1D	w.o.	
64-13825	UH1D	SoC	
64-13826	UH1D	w.o.	
64-13827	UH1H	SoC	
64-13828	UH1D	w.o.	
64-13829	UH1D	w.o.	
64-13830	UH1D	w.o.	
64-13831	UH1D	SoC	
64-13832	UH1D	w.o.	
64-13833	UH1D	w.o.	
64-13834	UH1D	w.o.	
64-13835	UH1D	w.o.	
64-13836	UH1D	w.o.	
64-13837	UH1D	w.o.	
64-13838	UH1D	SoC	
64-13839	UH1D	w.o.	
64-13840	UH1H	42AVN	
64-13841	UH1H		
64-13844	GUH1H	2AASF	
64-13845	UH1D	w.o.	
64-13846	UH1D	w.o.	
64-13847	UH1D	SoC	
64-13848	UH1D	w.o.	
64-13849	UH1H	AATC	49A Lowe AHP
64-13850	UH1D	w.o.	
64-13851	UH1H		
64-13852	UH1H	44AVN	
64-13853	UH1D	w.o.	
64-13854	UH1D	w.o.	
64-13855	UH1H		
64-13856	UH1H	42AVN	
64-13857	UH1D	w.o.	
64-13858	UH1H		
64-13859	UH1H		
64-13860	UH1D	SoC	
64-13861	UH1H	163ACR	
64-13862	UH1D	w.o.	
64-13863	UH1D	SoC	
64-13864	UH1H	31AVN	
64-13865	UH1D	SoC	
64-13866	UH1H		
64-13867	UH1D	SoC	
64-13868	UH1H		
64-13869	UH1H	AATC	69J Lowe AHP
64-13870	UH1D	w.o.	
64-13871	UH1D	w.o.	
64-13872	UH1H	26AVN	
64-13873	GUH1H	GIA	USATS
64-13874	UH1D	w.o.	
64-13875	UH1H	to N8152J	
64-13876	UH1H	AATC	76L Lowe AHP
64-13877	UH1H		
64-13878	UH1D	w.o.	
64-13879	UH1H	40AVN	

64-13880	UH1H		
64-13881	UH1D	to Germany 7040	
64-13882	UH1H	26AVN	
64-13883	UH1D	w.o.	
64-13884	UH1D	w.o.	
64-13885	UH1H	AATC	85A Lowe AHP
64-13886	UH1D	w.o.	
64-13887	UH1H		
64-13888	UH1H		
64-13889	UH1H	SoC	
64-13890	UH1H		
64-13891	UH1D	w.o.	
64-13892	UH1D		
64-13893	UH1D	w.o.	
64-13894	UH1D	w.o.	
64-13895	UH1H	AATC	95D Lowe AHP
64-13896	UH1H	AMARC	XA359
64-13897	GUH1H		
64-13898	UH1H	AATC	98D Lowe AHP
64-13899	UH1H	166ACR	
64-13900	UH1H	w.o.25.4.72	
64-13901	UH1H		
64-13902	UH1B	SoC	
64-13903	UH1B	SoC	
64-13904	UH1B	SoC	
64-13905	UH1B	SoC	
64-13906	UH1B	to N406PD	
64-13907	UH1B	to NASA	N424NA
64-13908	UH1B	SoC	
64-13909	UH1B	SoC	
64-13910	UH1B	SoC	
64-13911	UH1B	to Honduras	
64-13912	UH1B	SoC	
64-13913	UH1B	SoC	
64-13914	UH1B	to N87944	
64-13915	UH1B	SoC	
64-13916	UH1B	SoC	
64-13917	UH1B	SoC	
64-13918	UH1B	w.o.30.6.74	
64-13919	UH1B	SoC	
64-13920	UH1B	to Panama	
64-13921	UH1B	SoC	
64-13924	UH1B	SoC	
64-13925	UH1B	to N781CC	
64-13926	UH1B	to Korea	
64-13927	UH1B	SoC	
64-13928	UH1B	SoC	
64-13929	UH1B	SoC	
64-13930	UH1B	to N405PD	
64-13931	UH1B	SoC	
64-13932	UH1B	SoC	
64-13933	UH1B	dump	Tucson, Az
64-13934	UH1B	SoC	
64-13935	UH1B	SoC	
64-13936	UH1B	SoC	
64-13937	UH1B	SoC	
64-13938	UH1B	SoC	
64-13939	UH1B	to Korea	
64-13940	UH1B	to N91259	
64-13941	UH1B	SoC	
64-13942	UH1B	SoC	
64-13943	UH1B	to Honduras	
64-13944	UH1B	SoC	
64-13945	UH1B	to N5598E	
64-13948	UH1B	to N9044N	
64-13949	UH1B	to N99021	
64-13950	UH1B	SoC	
64-13951	UH1B	SoC	
64-13952	UH1B	SoC	
64-13953	UH1B	SoC	
64-13954	UH1B	preserved	Huntsville
64-13955	UH1B	SoC	
64-13956	UH1B	SoC	
64-13957	UH1B	to Singapore	
64-13958	UH1B	SoC	
64-13959	UH1B	SoC	
64-13960	GUH1B	GIA	IAAFA
64-13964	UH1B	SoC	
64-13968	UH1B	to N87966	
64-13969	UH1B	SoC	
64-13970	UH1B	to N87923	
64-13971	UH1B	SoC	
64-13972	UH1B	SoC	

64-13973	UH1B	SoC	
64-13974	UH1B	SoC	
64-13975	UH1B	SoC	
64-13976	UH1B	dump	Tucson, Az
64-13977	UH1B	SoC	
64-13978	UH1B	SoC	
64-13979	UH1B	SoC	
64-13980	UH1B	SoC	
64-13981	UH1B	SoC	
64-13982	UH1B	to Singapore	
64-13983	UH1B	SoC	
64-13984	UH1B	SoC	
64-13985	UH1B	to N22753	
64-13986	UH1B	SoC	
64-13987	UH1B	SoC	
64-13988	UH1B	SoC	
64-13989	UH1B	SoC	
64-13990	UH1B	SoC	
64-13991	UH1B	SoC	
64-13992	UH1B	SoC	
64-13993	UH1B	SoC	
64-13994	UH1B	SoC	
64-13995	UH1B	SoC	
64-13996	UH1B	SoC	
64-13997	UH1B	SoC	
64-13998	UH1B	SoC	
64-13999	UH1B	SoC	
64-14000	UH1B	to N65380	
64-14001	UH1B	SoC	
64-14002	UH1B	SoC	
64-14003	UH1B	to Korea	
64-14004	UH1B	w.o.9.1.70	
64-14005	UH1B	preserved	Ft. Knox, Ky
64-14006	UH1B	SoC	
64-14007	UH1B	derelict	Salina MAP, Ks
64-14008	UH1B	SoC	
64-14009	UH1B	to N88976	
64-14010	UH1B	to Korea	
64-14011	UH1B	SoC	
64-14012	UH1B	SoC	
64-14013	UH1B	SoC	
64-14014	UH1B	SoC	
64-14015	UH1B	SoC	
64-14016	UH1B	SoC	
64-14017	UH1B	SoC	
64-14018	UH1B	SoC	
64-14019	UH1B	SoC	
64-14020	UH1B	SoC	
64-14021	UH1B	SoC	
64-14022	UH1B	dump	Tucson, Az
64-14023	UH1B	to N266F	
64-14024	UH1B	SoC	
64-14025	UH1B	SoC	
64-14026	UH1B	SoC	
64-14027	UH1B	SoC	
64-14028	UH1B	SoC	
64-14029	UH1B	SoC	
64-14030	UH1B	SoC	
64-14031	UH1B	SoC	
64-14032	UH1B	SoC	
64-14033	UH1B	to N3231F	
64-14034	UH1B	SoC	
64-14035	UH1B	SoC	
64-14036	UH1B	SoC	
64-14037	UH1B	SoC	
64-14038	UH1B	to N87701	
64-14039	UH1B	SoC	
64-14040	UH1B	to Korea	
64-14041	UH1B	SoC	
64-14042	UH1B	SoC	
64-14043	UH1B	SoC	
64-14044	UH1B	SoC	
64-14045	UH1B	SoC	
64-14046	UH1B	to N92376	
64-14047	UH1B	SoC	
64-14048	UH1B	SoC	
64-14049	UH1B	mobile display unit	
64-14050	UH1B	to Panama 113	
64-14051	UH1B	SoC	
64-14052	UH1B	SoC	
64-14053	UH1B	SoC	
64-14054	UH1B	to Korea	
64-14055	UH1B	SoC	

64-14056	UH1B	SoC	
64-14057	UH1B	to N46884	
64-14058	UH1B		
64-14059	UH1B	SoC	
64-14060	UH1B	SoC	
64-14061	UH1B	to FMS	
64-14062	UH1B	to FMS	
64-14063	UH1B	SoC	
64-14064	UH1B	w.o.26.6.70	
64-14065	UH1B	SoC	
64-14066	UH1B	SoC	
64-14067	UH1B	SoC	
64-14068	UH1B	SoC	
64-14069	UH1B	SoC	
64-14070	UH1B	to N51929	
64-14071	UH1B	SoC	
64-14072	UH1B	SoC	
64-14073	UH1B	to Panama 114	
64-14074	UH1B	SoC	
64-14075	UH1B	dump	Tucson, Az
64-14076	UH1B	to N845M	
64-14077	UH1B	to Singapore 263	
64-14078	UH1B	SoC	
64-14079	UH1B	to Norway 079	
64-14080	UH1B	SoC	
64-14081	UH1B	to Korea	
64-14082	UH1B	to Norway 082	
64-14083	UH1B	SoC	
64-14084	UH1B	SoC	
64-14085	UH1B	SoC	
64-14086	UH1B	SoC	
64-14087	UH1B	to Korea	
64-14088	UH1B	SoC	
64-14089	UH1B	SoC	
64-14090	UH1B	to N90632	
64-14091	UH1B	SoC	
64-14092	UH1B	SoC	
64-14093	UH1B	SoC	
64-14094	UH1B	preserved	Quebec
64-14095	UH1B	SoC	
64-14096	UH1B	to Korea	
64-14097	UH1B	SoC	
64-14098	UH1B	SoC	
64-14099	UH1B	SoC	
64-14100	UH1B	to Panama 115	
64-14101	UH1M		
64-14102	UH1M		
64-14103	UH1C	w.o.	
64-14104	UH1C	SoC	
64-14105	UH1C	SoC	
64-14106	UH1C	SoC	
64-14107	QUH1M	Foster Aviation	
64-14108	UH1C	SoC	
64-14109	UH1C	w.o.	
64-14110	UH1M		
64-14111	UH1M		
64-14112	UH1C	w.o.	
64-14113	UH1C	w.o.	
64-14114	UH1C	SoC	
64-14115	UH1M		
64-14116	UH1M		
64-14117	UH1M		
64-14118	UH1M		
64-14119	QUH1M	Foster Aviation	
64-14120	UH1C	w.o.	
64-14121	UH1M		
64-14122	UH1C	SoC	
64-14123	UH1M		
64-14124	UH1C	SoC	
64-14125	UH1M		
64-14126	UH1C	w.o.	
64-14127	UH1M		
64-14128	UH1C	SoC	
64-14129	UH1M		
64-14130	UH1C	w.o.21.2.70	
64-14131	UH1M		
64-14132	UH1C	SoC	
64-14133	UH1C	SoC	
64-14134	UH1C	w.o.9.4.71	
64-14135	QUH1M		
64-14136	QUH1M		
64-14137	UH1M		
64-14138	UH1M	preserved	Salt Lake Cty

Serial	Type	Status	Location
64-14139	UH1M		
64-14140	QUH1M		
64-14141	UH1C	SoC	
64-14142	UH1M		
64-14143	UH1C	w.o.	
64-14144	UH1M	w.o.15.8.81	
64-14145	UH1M	to N57RF	
64-14146	UH1C	w.o.	
64-14147	UH1C	w.o.	
64-14148	UH1M		
64-14149	UH1C	w.o.	
64-14150	UH1M		
64-14151	UH1C	w.o.	
64-14152	UH1M		
64-14153	UH1C	w.o.	
64-14154	UH1M		
64-14155	UH1C	SoC	
64-14156	UH1M	preserved	Papago AAF
64-14157	UH1M		
64-14158	UH1C	w.o.	
64-14159	UH1C	w.o.	
64-14160	UH1C	w.o.	
64-14161	UH1C	w.o.	
64-14162	UH1C	w.o.	
64-14163	UH1C	w.o.	
64-14164	UH1C	w.o.	
64-14165	UH1C	w.o.	
64-14166	QUH1M		
64-14167	UH1C	w.o.4.2.71	
64-14168	UH1C	w.o.	
64-14169	UH1C	SoC	
64-14170	UH1C	w.o.	
64-14171	UH1C	w.o.	
64-14172	UH1C	w.o.	
64-14173	UH1C	w.o.	
64-14174	UH1C	w.o.	
64-14175	UH1M		
64-14176	UH1M		
64-14177	UH1C	w.o.	
64-14178	UH1C	w.o.	
64-14179	UH1M		
64-14180	UH1C	w.o.7.6.71	
64-14181	UH1C	SoC	
64-14182	UH1M		
64-14183	UH1C	w.o.	
64-14184	UH1C	w.o.	
64-14185	UH1M	pr. AVSCOM St. Louis	
64-14186	UH1C	w.o.	
64-14187	UH1C	w.o.	
64-14188	UH1C	w.o.	
64-14189	UH1C	w.o.	
64-14190	UH1M		
64-14191	UH1C	w.o.	
64-18261	NUH1B	SoC	
65-9416	UH1C	w.o.	
65-9417	UH1C	w.o.	
65-9418	QUH1M	Foster Aviation	
65-9419	UH1C	w.o.	
65-9420	QUH1M		
65-9421	UH1C	w.o.	
65-9422	UH1C	w.o.	
65-9423	JUH1M	SoC	
65-9424	UH1M	w.o.11.5.77	
65-9425	UH1C	w.o.	
65-9426	QUH1M	Foster Aviation	
65-9427	UH1C	w.o.	
65-9428	UH1C	w.o.	
65-9429	UH1M		
65-9430	UH1M		
65-9431	UH1C	w.o.	
65-9432	UH1C	w.o.	
65-9433	UH1C	w.o.7.4.70	
65-9434	UH1C	w.o.	
65-9435	UH1M		
65-9436	UH1M	w.o.8.6.84	
65-9437	UH1C	w.o.	
65-9438	UH1C	SoC	
65-9439	UH1C	w.o.13.4.70	
65-9440	QUH1M		
65-9441	UH1C	w.o.	
65-9442	UH1M		
65-9443	UH1C	w.o.	
65-9444	QUH1M		
65-9445	UH1M		
65-9446	UH1M	preserved	NASM
65-9447	UH1C	SoC	
65-9448	UH1C	w.o.	
65-9449	UH1C	w.o.	
65-9450	UH1C	SoC	
65-9451	UH1C	w.o.	
65-9452	UH1C	w.o.	
65-9453	UH1C	w.o.	
65-9454	UH1C	w.o.20.6.70	
65-9455	UH1M	w.o.	
65-9456	QUH1M		
65-9457	UH1C	SoC	
65-9458	UH1C	w.o.	
65-9459	UH1C	w.o.	
65-9460	UH1M		
65-9461	UH1C	SoC	
65-9462	UH1M		
65-9463	UH1M		
65-9464	UH1C	w.o.	
65-9465	UH1C	w.o.	
65-9466	UH1C	w.o.	
65-9467	UH1C	SoC	
65-9468	UH1C	w.o.	
65-9469	UH1C	w.o.	
65-9470	UH1M	GIA	Mercer Cty MAP
65-9471	UH1C	w.o.	
65-9472	UH1M		
65-9473	QUH1M		
65-9474	UH1C	w.o.	
65-9475	UH1M		
65-9476	QUH1M		
65-9477	UH1C	w.o.	
65-9478	UH1C	w.o.	
65-9479	UH1C	w.o.	
65-9480	UH1C	w.o.	
65-9481	UH1C	w.o.	
65-9482	UH1C	SoC	
65-9483	UH1M		
65-9484	UH1M	stored	Buckley ANGB
65-9485	UH1C	w.o.	
65-9486	UH1C	w.o.	
65-9487	UH1C	w.o.	
65-9488	QUH1M		
65-9489	UH1C	w.o.	
65-9490	UH1M		
65-9491	UH1C	SoC	
65-9492	UH1C	w.o.	
65-9493	UH1C	w.o.	
65-9494	UH1C	w.o.6.6.70	
65-9495	UH1M		
65-9496	UH1M		
65-9497	UH1M	preserved	Wright AAF
65-9498	UH1M		
65-9499	UH1M		
65-9500	UH1M		
65-9501	UH1M		
65-9502	UH1C	dump Miami, Fl	
65-9503	UH1C	w.o.23.2.71	
65-9504	UH1M	preserved	Avra Valley AP
65-9505	UH1C	w.o.	
65-9506	UH1C	SoC	
65-9507	UH1M		
65-9508	UH1C	w.o.	
65-9509	UH1C	w.o.	
65-9510	QUH1M		
65-9511	UH1C	w.o.	
65-9512	UH1C	w.o.30.3.70	
65-9513	UH1M		
65-9514	UH1C	w.o.	
65-9515	UH1C	w.o.	
65-9516	UH1M		
65-9517	UH1M		
65-9518	UH1C	w.o.	
65-9519	UH1M		
65-9520	UH1M		
65-9521	UH1M	to N59RF	
65-9522	UH1C	w.o.	
65-9523	UH1C	w.o.	
65-9524	UH1C	w.o.21.4.70	
65-9525	UH1M		
65-9526	UH1C	w.o.	
65-9527	UH1C	w.o.	

Serial	Type	Notes	Location
65-9528	UH1M	w.o.29.1.82	
65-9529	UH1C	w.o.	
65-9530	UH1C	w.o.	
65-9531	UH1M	w.o.20.5.71	
65-9532	UH1M	stored	Davison AAF
65-9533	QUH1M		
65-9534	UH1M		
65-9535	UH1C	SoC	
65-9536	UH1M		
65-9537	UH1C	SoC	
65-9538	UH1C	w.o.	
65-9539	UH1C	w.o.	
65-9540	UH1M	stored	Buckley ANGB
65-9541	QUH1M		
65-9542	UH1C	w.o.11.6.70	
65-9543	UH1C	w.o.	
65-9544	UH1C	w.o.	
65-9545	UH1C	w.o.	
65-9546	UH1C	w.o.	
65-9547	UH1M		
65-9548	UH1M		
65-9549	UH1C	SoC	
65-9550	UH1C	SoC	
65-9551	UH1C	w.o.	
65-9552	QUH1M		
65-9553	UH1C	w.o.	
65-9554	UH1C	w.o.	
65-9555	QUH1M		
65-9556	UH1M		
65-9557	UH1C	w.o.	
65-9558	UH1M		
65-9559	UH1M		
65-9560	UH1M	preserved	Quonset Pt. RI
65-9561	UH1C	w.o.	
65-9562	UH1C	w.o.	
65-9563	UH1C	w.o.11.7.70	
65-9564	UH1C	w.o.	
65-9565	UH1H		
65-9566	UH1D	SoC	
65-9567	UH1H		
65-9568	UH1D	SoC	
65-9569	UH1H		
65-9570	UH1D	to Philippine	
65-9571	GUH1H		
65-9572	UH1H		
65-9573	UH1D	SoC	
65-9574	UH1D	SoC	
65-9575	UH1D	w.o.	
65-9576	UH1H	455\457MedDet	MO NG
65-9577	UH1H	163ACR	
65-9578	UH1D	w.o.	
65-9579	GUH1D	GIA	IAAFA
65-9580	UH1H		
65-9581	UH1D	SoC	
65-9582	UH1D	SoC	
65-9583	UH1H	86ARCOM	
65-9584	UH1H	26AVN	
65-9585	UH1H	524ASA Co	
65-9586	UH1D	w.o.	
65-9587	UH1H	107AvCo	
65-9588	UH1D	w.o.	
65-9589	UH1H		
65-9590	UH1H	AATC	90. Lowe AHP
65-9591	UH1D	w.o.	
65-9592	UH1H	AMARC	XA370
65-9593	UH1H		
65-9594	UH1D	w.o.	
65-9595	UH1D	w.o.	
65-9596	UH1H	to N8151G	
65-9597	UH1D	w.o.	
65-9598	UH1H	AATC	98A Lowe AHP
65-9599	UH1D	SoC	
65-9600	UH1H		
65-9601	UH1H	w.o.19.2.75	
65-9602	UH1H	149AHB	
65-9603	UH1H	AMARC	XA346
65-9604	UH1H	198AvCo	
65-9605	UH1D	w.o.	
65-9606	UH1H	AATC	06A Lowe AHP
65-9607	UH1D	w.o.	
65-9608	UH1D	w.o.	
65-9609	UH1H	AATC	09N Lowe AHP
65-9610	UH1H	B\26AVN	
65-9611	UH1D	w.o.	
65-9612	UH1D	w.o.	
65-9613	UH1H	150AVN	
65-9614	GUH1H		
65-9615	UH1H	44AVN	
65-9616	UH1H	SoC	
65-9617	UH1H		
65-9618	UH1H	to N106SW	
65-9619	UH1H		
65-9620	UH1D	w.o.	
65-9621	UH1H		
65-9622	UH1H		
65-9623	UH1D	w.o.	
65-9624	UH1D	w.o.	
65-9625	UH1D	w.o.	
65-9626	UH1D	w.o.	
65-9627	UH1D	SoC	
65-9628	UH1H		
65-9629	UH1H	w.o.	
65-9630	UH1H	AMARC	XA332
65-9631	UH1D	w.o.	
65-9632	UH1H		
65-9633	UH1D	w.o.	
65-9634	UH1H		
65-9635	UH1D	w.o.	
65-9636	UH1H		
65-9637	UH1H	86ARCOM	
65-9638	UH1D	w.o.22.6.67	
65-9639	GUH1H		
65-9640	GUH1H	GIA	USATS
65-9641	UH1D	SoC	
65-9642	UH1H	SoC	
65-9643	UH1H	AATC	43F Lowe AHP
65-9644	UH1H	to N8157G	
65-9645	UH1D	w.o.	
65-9646	UH1H	6Cav	
65-9647	UH1D	w.o.	
65-9648	UH1H	AMARC	XA327
65-9649	UH1H		
65-9650	UH1D	SoC	
65-9651	UH1H		
65-9652	UH1H	AATC	52A Lowe AHP
65-9653	UH1H	SoC	
65-9654	UH1H	150AVN	DE NG
65-9655	UH1H	w.o.22.6.81	
65-9656	UH1D	w.o.	
65-9657	UH1H		
65-9658	UH1D	SoC	
65-9659	UH1D	w.o.	
65-9660	UH1H	w.o.15.4.77	
65-9661	UH1H		
65-9662	UH1H	to N395M	
65-9663	UH1H		
65-9664	UH1D	w.o.	
65-9665	UH1H	w.o.14.7.78	
65-9666	UH1H	278ACR	
65-9667	UH1H	31AVN	
65-9668	GUH1H		
65-9669	UH1H	w.o.21.7.72	
65-9670	UH1D	SoC	
65-9671	UH1H	SoC	
65-9672	UH1H	w.o.	
65-9673	UH1H	w.o.	
65-9674	UH1D	SoC	
65-9675	UH1D	w.o.	
65-9676	UH1D	SoC	
65-9677	UH1D	w.o.	
65-9678	UH1D	w.o.	
65-9679	UH1H	150AVN	
65-9680	UH1D	w.o.	
65-9681	UH1D	SoC	
65-9682	GUH1H		
65-9683	UH1D	w.o.	
65-9684	UH1D	SoC	
65-9685	GUH1H		
65-9686	UH1D	w.o.	
65-9687	UH1H	w.o.9.7.74	
65-9688	UH1H		
65-9689	UH1H	C\1-111AVN	FL NG
65-9690	UH1H	26AVN	
65-9691	UH1H		
65-9692	UH1H	69AvBde	
65-9693	UH1D	SoC	

158

Serial	Type	Notes	
65-9694	UH1H	SoC	
65-9695	UH1H	AATC	95. Lowe AHP
65-9696	UH1H	107AvCo	
65-9697	UH1H		
65-9698	UH1H	108AvCo	
65-9699	UH1H		
65-9700	UH1H	AATC	00B Lowe AHP
65-9701	UH1D	SoC	
65-9702	UH1H	26AVN	
65-9703	UH1H	SoC	
65-9704	UH1D	SoC	
65-9705	UH1H	to N122FC	
65-9706	UH1D	w.o.13.3.70	
65-9707	UH1D	w.o.	
65-9708	UH1H	374MedDet	
65-9709	UH1D	w.o.	
65-9710	UH1H		
65-9711	UH1D	w.o.	
65-9712	UH1H	w.o.3.5.86	
65-9713	UH1D	w.o.	
65-9714	UH1H	31AVN	
65-9715	UH1D	SoC	
65-9716	UH1D	w.o.	
65-9717	UH1D	w.o.	
65-9718	UH1H	AATC	18A Lowe AHP
65-9719	UH1H	AATC	19P Lowe AHP
65-9720	UH1D	w.o.	
65-9721	UH1D	w.o.	
65-9722	UH1H	AATC	22. Lowe AHP
65-9723	UH1H	C\1-111AVN	FL NG
65-9724	UH1H	w.o.5.5.80	
65-9725	UH1D	w.o.	
65-9726	UH1D	w.o.	
65-9727	UH1H		
65-9728	UH1D	w.o.	
65-9729	UH1D	w.o.	
65-9730	UH1D	w.o.	
65-9731	UH1H	to N8152G	
65-9732	UH1D	w.o.6.2.70	
65-9733	UH1D	w.o.	
65-9734	UH1H		
65-9735	GUH1H	GIA	USATS
65-9736	UH1H	AMARC	XA320
65-9737	UH1D	SoC	
65-9738	UH1D	w.o.	
65-9739	UH1H	26AVN	
65-9740	UH1D	SoC	
65-9741	UH1D	SoC	
65-9742	UH1H	149AHB	
65-9743	UH1H		
65-9744	UH1H	166ACR	
65-9745	UH1D	SoC	
65-9746	UH1D	SoC	
65-9747	UH1H	AATC	47G Lowe AHP
65-9748	UH1H	198AvCo	
65-9749	UH1H		
65-9750	UH1H	SoC	
65-9751	UH1H		
65-9752	UH1H		
65-9753	UH1H	AMARC	XA322
65-9754	UH1H	42AVN	
65-9755	UH1D	SoC	
65-9756	UH1H		
65-9757	UH1H		
65-9758	UH1H	1-131AVN	AL ANG
65-9759	GUH1H	GIA	USATS
65-9760	UH1D	SoC	
65-9761	UH1H		
65-9762	UH1H		
65-9763	UH1H		
65-9770	UH1H	AATC	70E Lowe AHP
65-9771	UH1H	92AvCo	
65-9772	UH1D	w.o.26.8.72	
65-9773	GUH1H		
65-9774	UH1H	163ACR	
65-9775	UH1H	AATC	75D Lowe AHP
65-9776	UH1H	247MedDet	
65-9777	UH1H		DC NG
65-9778	UH1D	w.o.	
65-9779	UH1H	812MedDet	
65-9780	UH1D	w.o.	
65-9781	UH1H	42AVN	
65-9782	UH1D	SoC	
65-9783	UH1D	w.o.	
65-9784	UH1H	AATC	84B Lowe AHP
65-9785	UH1H	B\26AVN	
65-9786	UH1H		
65-9787	UH1D	SoC	
65-9788	UH1H		
65-9789	UH1H	SoC	
65-9790	UH1H	AATC	90H Lowe AHP
65-9791	UH1H	to N943SM	
65-9792	UH1H	26AVN	
65-9793	UH1D	w.o.	
65-9794	UH1H	AATC	94. Lowe AHP
65-9795	UH1H	1\18ACS	
65-9796	UH1D	w.o.	
65-9797	UH1D	w.o.	
65-9798	UH1D	w.o.	
65-9799	UH1D	SoC	
65-9800	UH1D	w.o.	
65-9801	UH1H		
65-9802	UH1H	w.o.	
65-9803	UH1H	163ACR	
65-9804	UH1H		
65-9805	UH1H		
65-9806	UH1D	w.o.	
65-9807	UH1D	w.o.	
65-9808	UH1H	w.o.14.4.82	
65-9809	UH1D	w.o.	
65-9810	UH1D	w.o.	
65-9811	UH1D	w.o.	
65-9812	UH1D	w.o.	
65-9813	UH1D	w.o.	
65-9814	UH1H		
65-9815	UH1H	40AVN	
65-9816	UH1H	AMARC	XA352
65-9817	UH1D	w.o.	
65-9818	UH1H	w.o.25.11.74	
65-9819	UH1H	AATC	19F Lowe AHP
65-9820	UH1D	SoC	
65-9821	UH1D	w.o.	
65-9822	UH1H	w.o.17.12.76	
65-9823	UH1D	10SFG	
65-9824	UH1D	SoC	
65-9825	UH1H	AMARC	XA353
65-9826	GUH1H		
65-9827	UH1D	SoC	
65-9828	UH1D	w.o.	
65-9829	UH1D	SoC	
65-9830	UH1D	w.o.	
65-9831	UH1H	AATC	31J Lowe AHP
65-9832	UH1H	AMARC	XA314
65-9833	UH1H	7InfDiv	
65-9834	UH1H	AATC	34A Lowe AHP
65-9835	UH1H	AATC	35A Lowe AHP
65-9836	UH1H	SoC	
65-9837	UH1H	AATC	37F Lowe AHP
65-9838	UH1D	w.o.	
65-9839	UH1H	AATC	39B Lowe AHP
65-9840	UH1H	w.o.11.2.75	
65-9841	UH1D	w.o.	
65-9842	UH1D	w.o.	
65-9843	UH1D	w.o.	
65-9844	UH1H	AMARC	XA...
65-9845	UH1D	w.o.	
65-9846	UH1H		
65-9847	UH1H	120AvCo	
65-9848	UH1H		
65-9849	UH1D	w.o.	
65-9850	UH1D	w.o.	
65-9851	UH1D	w.o.	
65-9852	UH1H		
65-9853	UH1H	AMARC	XA...
65-9854	UH1H	w.o.	
65-9855	UH1D	w.o.20.1.71	
65-9856	UH1H		
65-9857	UH1H	AMARC	XA354
65-9858	UH1H	AATC	58E Lowe AHP
65-9859	UH1D	SoC	
65-9860	UH1D	w.o.	
65-9861	UH1D	w.o.	
65-9862	GUH1H	GIA	USATS
65-9863	UH1H	40AVN	
65-9864	UH1H		
65-9865	UH1H	B\26AVN	

Serial	Type	Notes	Extra
65-9866	UH1D	SoC	
65-9867	UH1D	w.o.	
65-9868	UH1D	w.o.	
65-9869	UH1D	SoC	
65-9870	UH1H	AATC	70G Lowe AHP
65-9871	UH1H	SoC	
65-9872	UH1H		
65-9873	UH1D	w.o.	
65-9874	UH1H	53InfDiv	
65-9875	UH1H	w.o.12.7.84	
65-9876	UH1D	w.o.	
65-9877	UH1H	9AVN	
65-9878	UH1H	AMARC	XA358
65-9879	UH1H	AATC	79ELowe AHP
65-9880	UH1D	w.o.	
65-9881	UH1H		
65-9882	UH1D	w.o.	
65-9883	UH1H	to Iran	
65-9884	UH1H	104ACS	
65-9885	UH1D	w.o.	
65-9886	UH1H	AATC	86H Lowe AHP
65-9887	UH1D	w.o.	
65-9888	UH1H		
65-9889	GUH1H	GIA	USATS
65-9890	UH1D	w.o.	
65-9895	UH1D	SoC	
65-9896	UH1D	SoC	
65-9897	UH1H	AATC	97M Lowe AHP
65-9898	UH1H		
65-9899	UH1H	A\3-158AVN	Reserve
65-9900	UH1D	SoC	
65-9901	UH1D	SoC	
65-9902	UH1H	Ft. Eustis	
65-9903	UH1D	w.o.	
65-9904	UH1D	w.o.	
65-9905	UH1D	w.o.	
65-9906	UH1H	42AVN	
65-9907	UH1D	w.o.	
65-9908	UH1H	B\26AVN	
65-9909	UH1H	SoC	
65-9910	UH1H	SoC	
65-9911	UH1H	SoC	
65-9912	UH1D	w.o.	
65-9913	UH1H	w.o.18.5.81	
65-9914	UH1D	w.o.	
65-9915	UH1H	2-135AVN	CO NG
65-9916	UH1D	w.o.	
65-9917	UH1D	w.o.	
65-9918	UH1D	w.o.	
65-9919	UH1D	w.o.	
65-9920	UH1H	SoC	
65-9921	UH1D	w.o.	
65-9922	UH1H	198AvCo	
65-9923	UH1D	w.o.	
65-9924	UH1D	w.o.	
65-9925	UH1D	w.o.12.3.70	
65-9926	UH1H		
65-9927	UH1D	w.o.	
65-9928	UH1D	w.o.	
65-9929	UH1H	42AVN	
65-9930	UH1D	w.o.	
65-9931	UH1H	AATC	31E Lowe AHP
65-9932	UH1D	w.o.	
65-9933	UH1D	w.o.	
65-9934	UH1H		
65-9935	UH1D	SoC	
65-9936	UH1D	w.o.	
65-9937	UH1H	A\1-244AVN	LA NG
65-9938	UH1D	w.o.	
65-9939	UH1H		
65-9940	UH1D	w.o.	
65-9941	UH1D	SoC	
65-9942	UH1D	w.o.	
65-9943	UH1H	31AVN	
65-9944	UH1H	AATC	44E Lowe AHP
65-9945	UH1H	SoC	
65-9946	UH1D	w.o.	
65-9947	UH1H	AATC	47J Lowe AHP
65-9948	UH1D	w.o.	
65-9949	UH1H	31AVN	
65-9950	UH1D	w.o.	
65-9951	UH1D	w.o.	
65-9952	UH1H	42AVN	
65-9953	UH1H	w.o.8.90	
65-9954	UH1D	w.o.	
65-9955	UH1D	w.o.	
65-9956	UH1H	SoC	
65-9957	UH1D	w.o.	
65-9958	UH1H	AATC	58E Lowe AHP
65-9959	UH1H	w.o.23.11.88	
65-9960	UH1H	SoC	
65-9961	UH1H	126MedDet	
65-9962	UH1H		
65-9963	UH1D	w.o.	
65-9964	UH1H	SoC	
65-9965	UH1H		
65-9966	UH1D	w.o.	
65-9967	UH1H	w.o.25.2.87	
65-9968	GUH1H	GIA	USATS
65-9969	UH1H	1-159AVN	
65-9970	UH1D	w.o.	
65-9971	UH1H	AATC	70G Lowe AHP
65-9972	UH1D	w.o.	
65-9973	UH1H		
65-9974	GUH1H	GIA	USATS
65-9975	UH1D	w.o.14.11.83	
65-9976	UH1D	w.o.	
65-9977	UH1H	AMARC	XA349
65-9978	UH1H	AATC	78E Lowe AHP
65-9979	UH1H	SoC	
65-9980	UH1H	SoC	
65-9981	UH1D	SoC	
65-9982	UH1D	w.o.	
65-9983	UH1D	w.o.	
65-9984	UH1H	to N398M	
65-9985	UH1D	w.o.	
65-9986	UH1H	1-131AVN	AL ANG
65-9987	UH1D	w.o.	
65-9988	UH1H		
65-9989	UH1D	w.o.	
65-9990	UH1D	w.o.	
65-9991	UH1D	w.o.	
65-9992	UH1H	to NASA	Langley
65-9993	UH1D	w.o.	
65-9994	UH1D	w.o.	
65-9995	UH1D	w.o.	
65-9996	UH1H	164AvCo	
65-9997	UH1D	w.o.	
65-9998	UH1H	NM NG	
65-9999	UH1H	40AVN	
65-10000	UH1H	B\26AVN	
65-10001	UH1D	w.o.	
65-10002	UH1D	w.o.	
65-10003	UH1D	w.o.	
65-10004	UH1D	w.o.	
65-10005	UH1H	AATC	05L Lowe AHP
65-10006	UH1H	AATC	06. Lowe AHP
65-10007	UH1H		
65-10008	UH1H		
65-10009	UH1H	A\3-158AVN	Reserve
65-10010	UH1H	D\135AVN	KS NG
65-10011	UH1H		
65-10012	UH1H	AMARC	XA321
65-10013	UH1H		
65-10014	GUH1H	GIA	USATS
65-10015	UH1H	SoC	
65-10016	UH1H		
65-10017	UH1H		
65-10018	UH1D	w.o.	
65-10019	UH1D	w.o.	
65-10020	UH1D	w.o.	
65-10021	UH1H	44AVN	
65-10022	UH1H	42AVN	
65-10023	UH1H	to N81518	
65-10024	UH1D	w.o.16.2.70	
65-10025	UH1H		
65-10026	UH1H	1-131AVN	AL ANG
65-10027	UH1H		
65-10028	UH1H	6AvCo	
65-10029	UH1H		
65-10030	UH1H	44AVN	
65-10031	UH1H	31AVN	
65-10032	UH1H	1-131AVN	AL ANG
65-10033	UH1D	w.o.8.6.70	
65-10034	UH1H		
65-10035	UH1H	164AvCo	

Serial	Type	Notes	
65-10036	UH1H		
65-10037	UH1D	w.o.	
65-10038	UH1D	w.o.	
65-10039	UH1D	w.o.	
65-10040	UH1D	w.o.	
65-10041	UH1D	to Australia A2-041	
65-10042	UH1H	w.o.19.1.79	
65-10043	UH1D	w.o.	
65-10044	UH1D	w.o.	
65-10045	UH1D	w.o.	
65-10046	UH1D	w.o.	
65-10047	UH1D	w.o.	
65-10048	UH1D	SoC	
65-10049	UH1H	w.o.28.1.74	
65-10050	UH1H		
65-10051	UH1D	w.o.	
65-10052	GUH1D	GIA	USATS
65-10053	UH1D	w.o.	
65-10054	UH1H		
65-10055	UH1H	w.o.11.12.78	
65-10056	UH1D	w.o.	
65-10057	UH1D	w.o.	
65-10058	UH1D	w.o.	
65-10059	UH1H	126MedCo	
65-10060	UH1H	104ACS	
65-10061	UH1H		
65-10062	UH1D	w.o.	
65-10063	UH1H	SoC	
65-10064	UH1D	w.o.7.11.71	
65-10065	UH1H	AATC	65A Lowe AHP
65-10066	UH1D	w.o.	
65-10067	UH1H	AATC	67. Lowe AHP
65-10068	UH1H	AATC	68B Lowe AHP
65-10069	UH1H		
65-10070	UH1D	SoC	
65-10071	UH1H		
65-10072	UH1H	SoC	
65-10073	GUH1H		
65-10074	GUH1H	GIA	USATS
65-10075	UH1H	Muir AAF	
65-10076	UH1D	w.o.	
65-10077	UH1H		
65-10078	UH1D	w.o.	
65-10079	UH1D	w.o.	
65-10080	UH1H		
65-10081	UH1H	to N340WN	
65-10082	UH1D	w.o.	
65-10083	UH1H	w.o.	
65-10084	UH1D	w.o.	
65-10085	UH1H	to NASA	N415NA (2)
65-10086	UH1H	w.o.21.1.81	
65-10087	UH1H	w.o.	
65-10088	UH1D	w.o.	
65-10089	UH1D	w.o.	
65-10090	GUH1H	GIA	USATS
65-10091	UH1H	AATC	91E Lowe AHP
65-10092	UH1D	SoC	
65-10093	UH1D	SoC	
65-10094	UH1H		
65-10095	UH1H		
65-10096	UH1H	63ARCOM	
65-10097	UH1D	w.o.8.1.70	
65-10098	UH1H		
65-10099	UH1H	w.o.26.2.77	
65-10100	UH1H	166ACR	
65-10101	UH1H	44AVN	
65-10102	UH1H	278ACR	
65-10103	UH1H	w.o.5.11.75	
65-10104	UH1H	to NASA	N420NA
65-10105	UH1H		
65-10106	UH1H	AMARC	XA350
65-10107	UH1D	w.o.	
65-10108	UH1H		
65-10109	UH1H	w.o.10.7.86	
65-10117	UH1H	26AVN	
65-10118	JUH1H		
65-10119	UH1D	w.o.	
65-10120	UH1D	SoC	
65-10121	UH1D	w.o.	
65-10122	UH1D	w.o.	
65-10123	UH1D	w.o.	
65-10124	UH1H	26AVN	
65-10125	UH1H	Ft. Eustis	
65-10126	UH1H	498MedCo	
65-10127	UH1D	SoC	
65-10128	UH1H	AATC	28. Lowe AHP
65-10129	UH1H	C\1-111AVN	FL ANG
65-10130	UH1H	to N341WN	
65-10131	UH1H	B\26AVN	
65-10132	GUH1H	GIA	USATS
65-10133	UH1H		
65-10134	UH1D	SoC	
65-10135	UH1H	w.o.9.90	
65-12738	UH1C	SoC	
65-12739	QUH1M		
65-12740	UH1M	preserved	Ft. Rucker, Al
65-12741	UH1M	preserved	Wright AAF, Ga
65-12742	UH1M		
65-12743	QUH1M		
65-12744	UH1M		
65-12773	GUH1H		
65-12774	UH1H	86ARCOM	
65-12775	UH1D	SoC	
65-12776	UH1H	45AVN	
65-12847	UH1D	SoC	
65-12848	UH1D	w.o.	
65-12849	UH1H	1 Cav	
65-12850	UH1D	w.o.	
65-12851	UH1H		
65-12852	UH1H	AATC	52. Lowe AHP
65-12857	UH1H	w.o.6.11.82	
65-12858	UH1H	SoC	
65-12859	UH1H	335AvCo	
65-12860	UH1H	62AvCo	
65-12861	UH1D	w.o.	
65-12862	UH1D	w.o.	
65-12863	UH1D	w.o.	
65-12864	UH1D	SoC	
65-12865	UH1D	w.o.	
65-12866	UH1H	w.o.1.8.80	
65-12867	UH1D	w.o.	
65-12868	UH1H	AATC	68K Lowe AHP
65-12869	UH1D	w.o.	
65-12870	UH1D	w.o.	
65-12871	UH1H	79AvCo	
65-12872	UH1H	w.o.9.90	
65-12873	UH1H		
65-12874	UH1H	79AvCo	
65-12875	UH1D	w.o.	
65-12876	UH1H	C\1-111AVN	FL NG
65-12877	UH1H	w.o.	
65-12878	UH1H		
65-12879	UH1H	w.o.25.3.85	
65 12880	UH1H	42AVN	
65-12881	UH1H	w.o.	
65-12882	UH1H	AATC	82A Lowe AHP
65-12883	UH1H	w.o.	
65-12884	UH1H	AATC	84E Lowe AHP
65-12885	UH1D	w.o.	
65-12886	UH1H	AMARC	XA361
65-12887	UH1D	SoC	
65-12888	UH1D	w.o.	
65-12889	UH1H	SoC	
65-12890	UH1D	w.o.	
65-12891	UH1D	w.o.	
65-12892	UH1D	w.o.	
65-12893	UH1D	w.o.	
65-12894	UH1D	w.o.	
65-12895	UH1H	42AVN	
66-0491	QUH1M	Foster Aviation	
66-0492	UH1M		
66-0493	UH1C	SoC	
66-0494	UH1M		
66-0495	UH1C	SoC	
66-0496	UH1C	w.o.	
66-0497	UH1M		
66-0498	UH1M		
66-0499	UH1M		
66-0500	UH1C	w.o.	
66-0501	UH1C	w.o.	
66-0502	UH1C	w.o.	
66-0503	UH1C	w.o.	
66-0504	UH1C	w.o.	
66-0505	UH1C	w.o.	
66-0506	QUH1M		
66-0507	UH1C	w.o.	

66-0508	UH1C	w.o.	
66-0509	UH1C	w.o.	
66-0510	UH1C	w.o.	
66-0511	UH1M		
66-0512	UH1M		
66-0513	UH1M		
66-0514	UH1C	w.o.	
66-0515	UH1C	w.o.	
66-0516	UH1C	w.o.	
66-0517	UH1C	w.o.	
66-0518	UH1C	w.o.	
66-0519	UH1C	to N204L	
66-0520	UH1M		
66-0521	UH1C	w.o.	
66-0522	UH1M		
66-0523	UH1C	w.o.	
66-0524	QUH1M	Foster Aviation	
66-0525	UH1C	w.o.	
66-0526	UH1C	w.o.	
66-0527	UH1C	w.o.	
66-0528	UH1M		
66-0529	UH1C	SoC	
66-0530	UH1C	w.o.	
66-0531	UH1C	SoC	
66-0532	UH1C	w.o.	
66-0533	UH1C	w.o.	
66-0534	UH1C	SoC	
66-0535	UH1M		
66-0536	UH1C	w.o.	
66-0537	UH1M		
66-0538	UH1C	w.o.	
66-0539	UH1M		
66-0540	QUH1M		
66-0541	UH1C	w.o.	
66-0542	UH1C	w.o.	
66-0543	UH1C	w.o.	
66-0544	QUH1M		
66-0545	UH1C	w.o.	
66-0546	UH1M	166ACR	
66-0547	UH1M		
66-0548	UH1C	SoC	
66-0549	UH1C	w.o.	
66-0550	UH1C	w.o.	
66-0551	UH1M		
66-0552	UH1C	w.o.	
66-0553	UH1C	w.o.	
66-0554	UH1C	w.o.	
66-0555	QUH1M	40AVN	
66-0556	UH1C	w.o.23.3.71	
66-0557	UH1M		
66-0558	QUH1M	107AvCo	
66-0559	UH1C	w.o.	
66-0560	UH1C	SoC	
66-0561	UH1C	w.o.	
66-0562	UH1C	w.o.	
66-0563	UH1M		
66-0564	UH1C	w.o.	
66-0565	UH1C	w.o.	
66-0566	UH1C	w.o.	
66-0567	UH1M	SoC	
66-0568	QUH1M		
66-0569	QUH1M		
66-0570	UH1C	w.o.	
66-0571	UH1C	to N204VC	
66-0572	UH1M	w.o.24.3.78	
66-0573	UH1C	w.o.	
66-0574	UH1C	w.o.	
66-0575	UH1C	w.o.	
66-0576	UH1M		
66-0577	UH1C	w.o.	
66-0578	UH1C	w.o.	
66-0579	UH1C	w.o.	
66-0580	UH1C	w.o.	
66-0581	UH1C	w.o.	
66-0582	UH1C	w.o.	
66-0583	UH1C	w.o.	
66-0584	QUH1M	Foster Aviation	
66-0585	UH1C	w.o.	
66-0586	UH1M		
66-0587	UH1C	w.o.	
66-0588	UH1M		
66-0589	UH1C	w.o.	
66-0590	UH1C	w.o.	
66-0591	UH1C	w.o.21.1.70	
66-0592	UH1C	w.o.	
66-0593	UH1C	w.o.	
66-0594	UH1C	w.o.	
66-0595	UH1M		
66-0596	UH1C	w.o.	
66-0597	UH1C	w.o.	
66-0598	UH1C	w.o.	
66-0599	UH1M		
66-0600	UH1C	w.o.	
66-0601	UH1C	w.o.	
66-0602	UH1M		
66-0603	UH1C	SoC	
66-0604	UH1C	w.o.	
66-0605	QUH1M		
66-0606	UH1C	w.o.	
66-0607	UH1M		
66-0608	UH1M		
66-0609	UH1M		
66-0610	UH1C	w.o.	
66-0611	UH1C	w.o.	
66-0612	UH1C	w.o.	
66-0613	UH1C	w.o.	
66-0614	UH1C	w.o.	
66-0615	UH1C	w.o.	
66-0616	UH1M		
66-0617	UH1C	w.o.	
66-0618	UH1M		
66-0619	UH1C	w.o.	
66-0620	UH1C	w.o.	
66-0621	UH1C	SoC	
66-0622	UH1C	w.o.	
66-0623	QUH1M		
66-0624	UH1C	w.o.	
66-0625	QUH1M		
66-0626	UH1C	w.o.2.10.70	
66-0627	QUH1M		
66-0628	UH1C	w.o.	
66-0629	QUH1M		
66-0630	UH1M		
66-0631	UH1C	w.o.	
66-0632	UH1M		
66-0633	UH1C	w.o.	
66-0634	UH1C	w.o.	
66-0635	UH1M		
66-0636	UH1M	w.o.7.10.71	
66-0637	UH1C	w.o.	
66-0638	UH1C	w.o.	
66-0639	UH1C	w.o.	
66-0640	UH1C	w.o.	
66-0641	UH1M	w.o.2.11.79	
66-0642	UH1C	w.o.	
66-0643	UH1M		
66-0644	UH1M		
66-0645	QUH1M		
66-0646	UH1C	w.o.	
66-0647	UH1C	w.o.24.1.70	
66-0648	UH1M	preserved	Hampto n, Va
66-0649	UH1M		
66-0650	UH1C	w.o.	
66-0651	UH1C	w.o.	
66-0652	UH1C	w.o.	
66-0653	UH1C	w.o.	
66-0654	UH1C	w.o.	
66-0655	UH1M		
66-0656	UH1C	w.o.5.5.70	
66-0657	UH1C	w.o.28.6.70	
66-0658	UH1C	SoC	
66-0659	UH1M		
66-0660	UH1C	w.o.7.3.70	
66-0661	UH1C	w.o.	
66-0662	UH1C	w.o.	
66-0663	UH1M		
66-0664	JUHIM	to NASA	N418NA
66-0665	UH1C	w.o.	
66-0666	UH1C	w.o.	
66-0667	UH1C	SoC	
66-0668	UH1C	w.o.	
66-0669	UH1M		
66-0670	UH1C	w.o.	
66-0671	UH1C	w.o.	
66-0672	QUH1M	Foster Aviation	
66-0673	UH1C	w.o.	

Serial	Type	Notes	
66-0674	UH1C	w.o.	
66-0675	UH1C	w.o.	
66-0676	UH1M		
66-0677	UH1C	w.o.	
66-0678	UH1C	to N58RF	
66-0679	QUH1M	51AvCo	
66-0680	UH1C	w.o.	
66-0681	UH1C	w.o.	
66-0682	UH1C	w.o.	
66-0683	UH1M		
66-0684	UH1M		
66-0685	UH1M		
66-0686	UH1C	w.o.	
66-0687	UH1M		
66-0688	UH1M		
66-0689	UH1M		
66-0690	UH1M		
66-0691	UH1C	w.o.	
66-0692	UH1C	w.o.	
66-0693	UH1C	w.o.	
66-0694	UH1C	w.o.	
66-0695	UH1C	w.o.	
66-0696	UH1C	w.o.9.5.71	
66-0697	UH1C	w.o.	
66-0698	UH1M		
66-0699	UH1C	w.o.	
66-0700	UH1C	w.o.	
66-0701	UH1C	w.o.	
66-0702	UH1C	w.o.	
66-0703	QUH1M		
66-0704	UH1C	w.o.	
66-0705	UH1C	w.o.	
66-0706	UH1C	w.o.	
66-0707	UH1C	w.o.	
66-0708	UH1M		
66-0709	UH1M		
66-0710	UH1C	w.o.	
66-0711	UH1C	w.o.	
66-0712	UH1C	SoC	
66-0713	UH1C	w.o.	
66-0714	UH1C	w.o.	
66-0715	UH1C	SoC	
66-0716	UH1C	w.o.	
66-0717	UH1M	dump	Burlington IAP, Vt
66-0718	UH1C	SoC	
66-0719	UH1M		
66-0720	UH1C	w.o.	
66-0721	UH1M		
66-0722	UH1M		
66-0723	UH1C	w.o.	
66-0724	UH1C	w.o.	
66-0725	UH1C	w.o.2.1.70	
66-0726	UH1M		
66-0727	UH1C	w.o.	
66-0728	UH1C	w.o.	
66-0729	UH1C	w.o.	
66-0730	UH1C	w.o.	
66-0731	UH1M		
66-0732	UH1C	w.o.	
66-0733	QUH1M		
66-0734	QUH1M		
66-0735	UH1C	w.o.	
66-0736	UH1C	w.o.	
66-0737	UH1C	w.o.	
66-0738	UH1C	w.o.	
66-0739	UH1C	w.o.	
66-0740	UH1C	SoC	
66-0741	UH1C	w.o.	
66-0742	UH1C	w.o.	
66-0743	UH1C	w.o.	
66-0744	UH1C	w.o.	
66-0745	UH1C	w.o.	
66-0746	UH1H		
66-0747	UH1H		
66-0748	UH1H		
66-0749	UH1H		
66-0750	UH1H	166ACR	
66-0751	UH1H	AATC	51. Lowe AHP
66-0752	UH1D	w.o.	
66-0753	UH1D	SoC	
66-0754	UH1D	w.o.	
66-0755	UH1D	SoC	
66-0756	UH1D	w.o.	
66-0757	UH1D	w.o.	
66-0758	UH1H		
66-0759	UH1H		
66-0760	UH1H	VA NG	
66-0761	UH1H		
66-0762	UH1H		
66-0763	UH1D	w.o.	
66-0764	UH1H	163ACR	
66-0765	GUH1H	GIA	USATS
66-0766	UH1D	w.o.	
66-0767	UH1H	63ARCOM	
66-0768	UH1H	44AVN	
66-0769	UH1H	D\1-245AVN	OK NG
66-0770	UH1H		
66-0771	UH1D	w.o.	
66-0772	UH1M		
66-0773	UH1D	w.o.	
66-0774	UH1D	SoC	
66-0775	UH1H		
66-0776	UH1H	w.o.29.3.82	
66-0777	UH1H	to Korea	
66-0778	UH1D	w.o.	
66-0779	UH1D	w.o.	
66-0780	UH1H	1133MedDet	
66-0781	UH1H	224AVN	
66-0782	UH1H		
66-0783	UH1H	w.o.21.1.81	
66-0784	UH1H	SoC	
66-0785	UH1H	1-185AVN	MS NG
66-0786	UH1H	47AVN	
66-0787	UH1D	w.o.7.4.83	
66-0788	UH1H	w.o.10.6.71	
66-0789	UH1H	86ARCOM	
66-0790	UH1D	w.o.	
66-0791	UH1H		
66-0792	UH1H		
66-0793	UH1D	w.o.	
66-0794	UH1H	w.o.	
66-0795	UH1H		
66-0796	UH1H		
66-0797	UH1D	w.o.	
66-0798	UH1H		
66-0799	UH1D	w.o.	
66-0800	UH1D	w.o.	
66-0801	UH1H	C\1-245AVN	OK NG
66-0802	UH1D	w.o.	
66-0803	UH1D	SoC	
66-0804	UH1D	w.o.	
66-0805	UH1H	w.o.9.9.90	
66-0806	UH1D	w.o.	
66-0807	UH1D	w.o.	
66-0808	UH1D	w.o.	
66-0809	UH1H		
66-0810	UH1H	AATC	10A Cairns AAF
66-0811	UH1D	w.o.	
66-0812	UH1H	107AvCo	
66-0813	UH1H	to Korea	
66-0814	UH1H	SoC	
66-0815	UH1H		
66-0816	UH1H		
66-0817	UH1D	SoC	
66-0818	UH1H	w.o.15.9.84	
66-0819	UH1D	w.o.	
66-0820	UH1D	w.o.	
66-0821	UH1H		
66-0822	UH1H		
66-0823	UH1D	w.o.	
66-0824	UH1H	w.o.6.12.81	
66-0825	UH1H	81ARCOM	
66-0826	UH1H	193AVN	
66-0827	UH1H	SoC	
66-0828	UH1H	AATC	28C Lowe AHP
66-0829	UH1D	w.o.	
66-0830	UH1D	w.o.	
66-0831	UH1M		
66-0832	UH1D	SoC	
66-0833	UH1D	SoC	
66-0834	UH1D	SoC	
66-0835	UH1H		
66-0836	UH1H		
66-0837	UH1H	AATC	37R Lowe AHP
66-0838	UH1D	SoC	
66-0839	UH1D	w.o.	

Serial	Type	Notes	
66-0840	UH1D	w.o.	
66-0841	UH1H		
66-0842	UH1H	SoC	
66-0843	UH1H		
66-0844	UH1H	w.o.13.6.90	
66-0845	UH1D	w.o.	
66-0846	UH1D	w.o.	
66-0847	UH1D	w.o.	
66-0848	UH1H	AMARC	XA356
66-0849	UH1H	40AVN	
66-0850	UH1H	AMARC	XA329
66-0851	UH1H	w.o.28.2.84	
66-0852	UH1H	SoC	
66-0853	UH1H	40AVN	
66-0854	UH1H	w.o.1.3.70	
66-0855	UH1H	w.o.5.89	
66-0856	UH1H	to Korea	
66-0857	UH1H		
66-0858	UH1H	1-185AVN	MS NG
66-0859	UH1H	1-185AVN	MS NG
66-0860	UH1H	SoC	
66-0861	UH1D	SoC	
66-0862	UH1H		
66-0863	UH1H	1-131AVN	AL ANG
66-0864	UH1D	w.o.	
66-0865	UH1H	SoC	
66-0866	UH1D	SoC	
66-0867	UH1D	SoC	
66-0868	UH1H		
66-0869	NUH1H		
66-0870	UH1D	w.o.	
66-0871	UH1H	461AVN	NM NG
66-0872	UH1H	C\1-245AVN	OK NG
66-0873	UH1H		
66-0874	UH1D	w.o.	
66-0875	UH1H	79ARCOM	
66-0876	UH1H		
66-0877	UH1H	1-185AVN	MS NG
66-0878	UH1H	to Korea	
66-0879	UH1D	w.o.	
66-0880	UH1D	w.o.	
66-0881	UH1D	w.o.	
66-0882	UH1H		
66-0883	UH1H	335AvCo	
66-0884	UH1H	SoC	
66-0885	UH1H		
66-0886	UH1H	to Korea	
66-0887	UH1D	w.o.	
66-0888	UH1H	AMARC	XA360
66-0889	UH1D	w.o.	
66-0890	UH1H	SoC	
66-0891	UH1H		
66-0892	UH1D	w.o.	
66-0893	UH1H	31AVN	
66-0894	JUH1H	ERADCOM	
66-0895	UH1D	SoC	
66-0896	UH1H	AATC	96B Lowe AHP
66-0897	UH1D	w.o.	
66-0898	UH1D	w.o.	
66-0899	UH1H	to Korea	
66-0900	UH1D	w.o.	
66-0901	UH1H	SoC	
66-0902	UH1M		
66-0903	UH1H	C\1-108AVN	AZ NG 32
66-0904	UH1H		
66-0905	UH1D	SoC	
66-0906	UH1H	40AVN	
66-0907	UH1D	SoC	
66-0908	UH1D	SoC	
66-0909	UH1D	w.o.	
66-0910	UH1D	w.o.	
66-0911	UH1H	193AVN	
66-0912	UH1H	B\26AVN	
66-0913	UH1H	193AVN	
66-0914	UH1H		
66-0915	UH1H	AATC	15D Lowe AHP
66-0916	UH1D	w.o.	
66-0917	UH1D	w.o.	
66-0918	UH1D	w.o.	
66-0919	UH1D	w.o.	
66-0920	GUH1H	GIA	USATS
66-0921	UH1D	w.o.	
66-0922	UH1D	w.o.22.7.73	
66-0923	UH1D	SoC	
66-0924	UH1D	SoC	
66-0925	UH1D	81ARCOM	
66-0926	UH1H	163ACR	
66-0927	UH1D	w.o.	
66-0928	UH1H	159AHB 01	
66-0929	UH1H	163ACR	
66-0930	UH1D		
66-0931	UH1D	w.o.	
66-0932	UH1H	G\1-149AVN	TX NG
66-0933	UH1D	SoC	
66-0934	UH1D	1Cav	
66-0935	UH1H		
66-0936	UH1D	w.o.	
66-0937	UH1D	w.o.	
66-0938	UH1H	2-135AVN	CO NG
66-0939	UH1H	323AVN	
66-0940	UH1D	w.o.	
66-0941	UH1H		
66-0942	UH1D	w.o.2.8.72	
66-0943	UH1D	w.o.	
66-0944	UH1H		
66-0945	UH1M		
66-0946	UH1D	w.o.	
66-0947	UH1D	SoC	
66-0948	UH1D	SoC	
66-0949	UH1H	1-159AVN	
66-0950	UH1D	w.o.	
66-0951	UH1D	w.o.	
66-0952	UH1D	SoC	
66-0953	UH1H		NY NG
66-0954	UH1H	SoC	
66-0955	UH1H	SoC	
66-0956	UH1H	SoC	
66-0957	UH1H	to Korea	
66-0958	UH1H	A\3-158AVN	Reserve
66-0959	UH1D	SoC	
66-0960	UH1H	B\26AVN	
66-0961	UH1H	w.o.16.12.84	
66-0962	UH1H	AATC	62D Lowe AHP
66-0963	UH1H	w.o.12.2.76	
66-0964	UH1H	w.o.3.12.73	
66-0965	UH1H		
66-0966	UH1H	79ARCOM	
66-0967	UH1H		
66-0968	UH1D	w.o.	
66-0969	UH1D	w.o.	
66-0970	UH1H		
66-0971	UH1H		
66-0972	UH1H	SoC	
66-0973	UH1H	SoC	
66-0974	UH1H	3ACR	
66-0975	UH1D	SoC	
66-0976	UH1D	SoC	
66-0977	UH1D	SoC	
66-0978	UH1H		
66-0979	UH1H	to Korea	
66-0980	UH1H		
66-0981	UH1H	SoC	
66-0982	UH1H		
66-0983	UH1H		
66-0984	UH1H	SoC	
66-0985	UH1D	SoC	
66-0986	UH1H	SoC	
66-0987	UH1H	SoC	
66-0988	UH1H	6ACR	
66-0989	UH1H		
66-0990	UH1D	SoC	
66-0991	UH1H	SoC	
66-0992	UH1H	44AVN	
66-0993	UH1D	SoC	
66-0994	UH1H	150AVN	
66-0995	UH1H	w.o.15.11.71	
66-0996	UH1H	SoC	
66-0997	UH1H	w.o.28.2.70	
66-0998	UH1H	SoC	
66-0999	UH1H		
66-1000	UH1H	B\1-132AVN	DC NG
66-1001	UH1H	AMARC	XA338
66-1002	UH1D	SoC	
66-1003	UH1H	to Vietnam	
66-1004	UH1D	SoC	
66-1005	UH1H	to Korea	

Serial	Type	Notes	
66-1006	UH1D	SoC	
66-1007	UH1H		
66-1008	UH1D	dump	Rosamund, Ca
66-1009	UH1D	Berlin Brigade	
66-1010	UH1H		
66-1011	UH1H		
66-1012	UH1D	GIA	USN
66-1013	UH1H	w.o.11.3.75	
66-1014	UH1D	SoC	
66-1015	UH1H	A\1-244AVN	LA NG
66-1016	UH1D	SoC	
66-1017	UH1D	SoC	
66-1018	UH1H	240EngGrp	
66-1019	UH1D	w.o.4.5.70	
66-1020	UH1D	SoC	
66-1021	UH1D	w.o.7.8.70	
66-1022	UH1D	SoC	
66-1023	UH1H		
66-1024	UH1H		
66-1025	UH1H	to Iran	
66-1026	UH1H	44AVN	
66-1027	UH1D	SoC	
66-1028	UH1D	SoC	
66-1029	UH1D	SoC	
66-1030	UH1H	to Iran	
66-1031	UH1D	SoC	
66-1032	UH1D	SoC	
66-1033	UH1H	2-135AVN	CO NG
66-1034	UH1D	SoC	
66-1035	UH1D	SoC	
66-1036	UH1H	to Iran	
66-1037	UH1D	SoC	
66-1038	UH1H	w.o.5.6.89	
66-1039	UH1H	47AVN	
66-1040	UH1H	1898AvCo	
66-1041	UH1H	SoC	
66-1042	UH1H	1-158AVN	
66-1043	UH1H	AMARC	XA324
66-1044	UH1H	150AVN	DE NG
66-1045	UH1H	92InfDiv	
66-1046	UH1M		
66-1047	UH1H	AATC	47K Lowe AHP
66-1048	UH1H	47AVN	
66-1049	UH1D	SoC	
66-1050	UH1H		
66-1051	UH1H	to Iran	
66-1052	UH1H	812MedDet	
66-1053	UH1D	SoC	
66-1054	UH1D	SoC	
66-1055	UH1D	SoC	
66-1056	UH1H	w.o.28.7.78	
66-1057	UH1H	AATC	57E Lowe AHP
66-1058	UH1H		
66-1059	UH1H	SoC	
66-1060	UH1D	SoC	
66-1061	UH1D	w.o.	
66-1062	UH1H	1Cav	
66-1063	UH1D	w.o.	
66-1064	UH1D	w.o.	
66-1065	UH1H	812MedDet	
66-1066	UH1D	w.o.	
66-1067	UH1D	w.o.	
66-1068	UH1H	A\1-244AVN	LA NG
66-1069	UH1D	w.o.	
66-1070	UH1H	to Vietnam	
66-1071	UH1H	198AvCo	
66-1072	UH1D	w.o.	
66-1073	UH1H	42AVN	
66-1074	UH1D	SoC	
66-1075	UH1H		
66-1076	UH1H	AATC	76. Cairns AAF
66-1077	UH1H	to Korea	
66-1078	UH1H	C\1-245AVN	OK NG
66-1079	UH1H	w.o.25.9.71	
66-1080	UH1H		
66-1081	UH1H		
66-1082	UH1H		
66-1083	UH1D	SoC	
66-1084	UH1H	w.o.16.6.81	
66-1085	UH1H	to Vietnam	
66-1086	UH1H	1255MedCo	NV NG
66-1087	UH1H	126MedCo	
66-1088	UH1H	to Korea	
66-1089	UH1D	w.o.	
66-1090	UH1D	w.o.	
66-1091	UH1H		
66-1092	UH1H	26AVN	
66-1093	UH1H	166ACR	
66-1094	UH1D	SoC	
66-1095	UH1H		
66-1096	UH1H		
66-1097	UH1H	1159MedCo	
66-1098	UH1D	w.o.	
66-1099	UH1D	w.o.	
66-1100	UH1D	SoC	
66-1101	UH1H	26AVN	
66-1102	UH1D	w.o.	
66-1103	UH1H	40AVN	
66-1104	UH1M		
66-1105	UH1M		
66-1106	UH1D	w.o.	
66-1107	UH1H		
66-1108	UH1H		
66-1109	UH1H		
66-1110	UH1H	352ASA Co	
66-1111	UH1H	w.o.6.8.79	
66-1112	UH1H	w.o.	
66-1113	UH1H	SoC	
66-1114	UH1H	w.o.	
66-1115	UH1D	w.o.	
66-1116	UH1H	AMARC	XA347
66-1117	UH1H		
66-1118	UH1D	w.o.	
66-1119	UH1D	w.o.	
66-1120	UH1H	AATC	20B Lowe AHP
66-1121	UH1H	C\1-111AVN	FL NG
66-1122	UH1D	w.o.	
66-1123	UH1H	SoC	
66-1124	UH1H	w.o.25.6.74	
66-1125	UH1H	26AVN	
66-1126	UH1H	26AVN	
66-1127	UH1H	4InfDiv	
66-1128	UH1D	w.o.	
66-1129	UH1H		
66-1130	UH1H		IL NG
66-1131	UH1H	w.o.4.9.70	
66-1132	UH1H	w.o.10.9.84	
66-1133	UH1H	w.o.	
66-1134	UH1H	to Iran	
66-1135	UH1H		
66-1136	UH1H	79ARCOM	
66-1137	UH1H		
66-1138	UH1D	w.o.	
66-1139	UH1H	AATC	39J Lowe AHP
66-1140	UH1H	45AVN	
66-1141	UH1H		
66-1142	UH1D	w.o.	
66-1143	UH1D	w.o.	
66-1144	UH1H	26AVN	
66-1145	UH1H	AATC	45. Lowe AHP
66-1146	UH1D	w.o.	
66-1147	UH1H	42AVN	
66-1148	UH1H		
66-1149	UH1D	w.o.	
66-1150	UH1D	w.o.	
66-1151	UH1H	6ACR	
66-1152	UH1D	w.o.	
66-1153	UH1D	w.o.	
66-1154	UH1H		
66-1155	UH1H	4ARC	
66-1156	UH1H	26AVN	
66-1157	UH1D	w.o.	
66-1158	UH1H	335AvCo	
66-1159	UH1D	w.o.	
66-1160	UH1H		
66-1161	UH1H		
66-1162	UH1H	164AvCo	
66-1163	UH1H	to Iran	
66-1164	JUH1H	White Sands	
66-1165	UH1H		
66-1166	UH1D	to Australia A2-166	
66-1167	UH1H	2-135AVN	CO NG
66-1168	UH1H	31AVN	
66-1169	UH1H	62AvCo	
66-1170	UH1H	AATC	70B Lowe AHP
66-1171	UH1H	1-149AVN	TX NG

66-1172	UH1D	w.o.
66-1173	UH1H	86ARCOM
66-1174	UH1H	
66-1175	UH1D	w.o.
66-1176	UH1D	w.o.
66-1177	UH1H	to Korea
66-1178	UH1D	SoC
66-1179	UH1D	SoC
66-1180	UH1H	
66-1181	UH1H	B\26AVN
66-1182	UH1H	163ACR
66-1183	UH1H	26AVN
66-1184	UH1H	26AVN
66-1185	UH1D	w.o.
66-1186	UH1D	w.o.
66-1187	UH1D	w.o.
66-1188	UH1H	w.o.21.8.70
66-1189	UH1D	w.o.
66-1190	UH1M	70AVN
66-1191	UH1D	w.o.
66-1192	UH1H	to N81568
66-1193	UH1H	
66-1194	UH1H	
66-1195	UH1D	w.o.
66-1196	UH1H	to Vietnam
66-1197	UH1H	
66-1198	UH1D	w.o.
66-1199	UH1D	w.o.
66-1200	UH1D	w.o.
66-1201	UH1D	w.o.
66-1202	UH1D	w.o.
66-1203	UH1D	w.o.
66-1204	UH1H	
66-1205	UH1H	w.o.15.7.75
66-1206	UH1D	w.o.
66-1207	UH1H	w.o.12.3.70
66-1208	UH1H	51AvCo
66-1209	UH1D	w.o.
66-1210	UH1D	SoC
66-8574	UH1H	w.o.
66-8575	UH1H	SoC
66-8576	UH1H	
66-8577	UH1H	SoC
66-15000	UH1M	
66-15001	UH1C	w.o.
66-15002	UH1C	w.o.
66-15003	UH1M	
66-15004	UH1C	w.o.
66-15005	UH1M	preserved Charleston, SC
66-15006	UH1C	w.o.
66-15007	UH1C	w.o.
66-15008	UH1M	92InfBde
66-15009	UH1C	w.o.
66-15010	UH1C	w.o.
66-15011	UH1C	SoC
66-15012	UH1C	w.o.
66-15013	UH1C	w.o.
66-15014	UH1C	w.o.
66-15015	UH1C	w.o.
66-15016	UH1M	107AvCo
66-15017	QUH1M	
66-15018	UH1C	w.o.
66-15019	UH1C	w.o.
66-15020	UH1C	w.o.
66-15021	UH1M	47AVN
66-15022	UH1C	w.o.
66-15023	QUH1M	
66-15024	UH1M	163ACR
66-15025	UH1C	w.o.
66-15026	UH1C	w.o.
66-15027	UH1C	w.o.26.2.71
66-15028	UH1M	
66-15029	UH1C	w.o.
66-15030	UH1C	w.o.
66-15031	UH1C	w.o.
66-15032	UH1C	w.o.
66-15033	UH1C	w.o.
66-15034	UH1M	w.o.29.9.83
66-15035	UH1C	w.o.
66-15036	UH1M	
66-15037	UH1C	184AHC
66-15038	UH1M	
66-15039	UH1C	w.o.
66-15040	UH1C	w.o.5.1.70
66-15041	UH1C	w.o.
66-15042	UH1C	w.o.
66-15043	UH1C	w.o.
66-15044	UH1C	w.o.
66-15045	UH1C	w.o.
66-15046	QUH1M	
66-15047	UH1M	116AHB
66-15048	UH1C	w.o.
66-15049	UH1C	w.o.28.5.70
66-15050	UH1M	
66-15051	UH1C	w.o.
66-15052	UH1C	w.o.
66-15053	UH1C	w.o.
66-15054	QUH1M	
66-15055	UH1C	w.o.
66-15056	QUH1M	
66-15057	UH1M	
66-15058	UH1M	26AVN
66-15059	UH1C	w.o.
66-15060	UH1C	SoC
66-15061	UH1C	w.o.
66-15062	UH1M	1-183AVN ID NG
66-15063	QUH1M	Foster Aviation
66-15064	UH1C	w.o.
66-15065	UH1C	w.o.17.1.70
66-15066	UH1C	SoC
66-15067	UH1C	w.o.
66-15068	UH1M	
66-15069	UH1M	
66-15070	UH1M	
66-15071	UH1M	
66-15072	UH1C	SoC
66-15073	UH1C	w.o.
66-15074	UH1C	w.o.
66-15075	UH1C	w.o.
66-15076	UH1M	preserved USS Intrepid
66-15077	UH1M	
66-15078	UH1M	
66-15079	UH1C	w.o.
66-15080	UH1M	116AHB
66-15081	UH1C	w.o.
66-15082	QUH1M	Foster Aviation
66-15083	UH1M	
66-15084	UH1M	
66-15085	UH1C	w.o.
66-15086	UH1C	w.o.
66-15087	UH1C	w.o.
66-15088	UH1C	w.o.
66-15089	UH1M	51AvCo
66-15090	UH1C	w.o.
66-15091	UH1C	SoC
66-15092	UH1C	w.o.
66-15093	UH1C	w.o.
66-15094	UH1C	w.o.
66-15095	UH1C	w.o.
66-15096	UH1C	w.o.
66-15097	UH1C	w.o.
66-15098	UH1M	40AVN
66-15099	QUH1M	
66-15100	UH1C	w.o.
66-15101	UH1M	
66-15102	UH1M	
66-15103	UH1C	w.o.
66-15104	UH1C	w.o.
66-15105	UH1C	w.o.
66-15106	UH1M	
66-15107	UH1M	278ACR
66-15108	UH1M	
66-15109	UH1C	w.o.
66-15110	UH1C	w.o.
66-15111	QUH1M	w.o.5.6.70
66-15112	UH1C	w.o.
66-15113	UH1C	w.o.
66-15114	UH1C	w.o.
66-15115	UH1C	w.o.1.6.70
66-15116	QUH1M	Foster Aviation
66-15117	UH1C	w.o.
66-15118	UH1M	
66-15119	UH1C	w.o.
66-15120	UH1C	w.o.
66-15121	UH1C	w.o.
66-15122	QUH1M	

66-15123	UH1C	w.o.	
66-15124	QUH1M	AMARC	XA308
66-15125	UH1C	w.o.	
66-15126	UH1C	w.o.	
66-15127	UH1M		
66-15128	UH1M	to N56RF	
66-15129	UH1C	w.o.	
66-15130	UH1C	w.o.	
66-15131	UH1M		
66-15132	UH1C	SoC	
66-15133	UH1C	w.o.	
66-15134	UH1M	w.o.17.1.84	
66-15135	UH1C	w.o.24.2.70	
66-15136	UH1C	w.o.	
66-15137	QUH1M		
66-15138	UH1M	1-183AVN	ID NG
66-15139	UH1C	w.o.	
66-15140	UH1C	w.o.	
66-15141	UH1C	w.o.	
66-15142	UH1C	SoC	
66-15143	UH1M	derelict	Rosamund, Ca
66-15144	UH1M	26AVN	
66-15145	QUH1M	163ACR	
66-15146	QUH1M		
66-15147	QUH1M		
66-15148	UH1C	w.o.	
66-15149	UH1C	w.o.	
66-15150	UH1M	116AHB	
66-15151	UH1C	w.o.11.1.70	
66-15152	UH1C	w.o.	
66-15153	UH1C	w.o.	
66-15154	UH1C	w.o.	
66-15155	UH1C	w.o.	
66-15156	UH1M	116AHB	
66-15157	UH1C	w.o.	
66-15158	UH1C	w.o.	
66-15159	UH1C	w.o.	
66-15160	UH1M		
66-15161	UH1C	w.o.	
66-15162	UH1C	w.o.	
66-15163	UH1C	SoC	
66-15164	UH1C	w.o.	
66-15165	UH1C	w.o.	
66-15166	UH1C	w.o.	
66-15167	UH1C	w.o.	
66-15168	UH1C	w.o.	
66-15169	UH1C	w.o.	
66-15170	UH1M		
66-15171	UH1C	w.o.	
66-15172	UH1C	w.o.	
66-15173	UH1C	w.o.	
66-15174	UH1M	40AVN	
66-15175	UH1C	w.o.	
66-15176	UH1M		
66-15177	UH1C	w.o.	
66-15178	UH1C	w.o.	
66-15179	UH1M	w.o.10.2.81	
66-15180	UH1C	w.o.18.2.71	
66-15181	QUH1M		
66-15182	UH1C	w.o.	
66-15183	UH1M		
66-15184	UH1C	w.o.	
66-15185	UH1M	396AvCo	
66-15186	UH1M		
66-15187	UH1M	396AvCo	
66-15188	UH1M	278ACR	
66-15189	UH1C	w.o.	
66-15190	QUH1M		TN NG
66-15191	UH1M	51AvCo	
66-15192	UH1C	SoC	
66-15193	UH1M		
66-15194	UH1M		
66-15195	UH1M		
66-15196	UH1M		
66-15197	UH1M	51AvCo	
66-15198	UH1C	w.o.	
66-15199	UH1C	w.o.	
66-15200	UH1M		
66-15201	UH1C	w.o.9.8.70	
66-15202	UH1C	w.o.	
66-15203	UH1C	w.o.	
66-15204	UH1C	w.o.	
66-15205	QUH1M		

66-15206	UH1C	w.o.	
66-15207	UH1C	w.o.	
66-15208	UH1C	w.o.	
66-15209	UH1C	w.o.	
66-15210	UH1C	w.o.	
66-15211	UH1M	116AHB	
66-15212	UH1M		
66-15213	UH1C	w.o.	
66-15214	UH1C	w.o.26.4.70	
66-15215	UH1C	w.o.	
66-15216	UH1M	116AHB	
66-15217	UH1C	SoC	
66-15218	QUH1M	AMARC	XA306
66-15219	UH1C	w.o.2.11.70	
66-15220	QUH1M		
66-15221	UH1M	116AHB	
66-15222	UH1M	1-183AVN	ID NG
66-15223	UH1C	w.o.	
66-15224	UH1C	w.o.	
66-15225	UH1M	116AHB	
66-15226	UH1C	w.o.	
66-15227	UH1C	w.o.	
66-15228	UH1C	w.o.	
66-15229	UH1M	1-183AVN	ID NG
66-15230	UH1M	w.o.2.4.72	
66-15231	QUH1M		
66-15232	UH1C	w.o.	
66-15233	UH1C	w.o.	
66-15234	UH1M	40AVN	
66-15235	UH1C	w.o.	
66-15236	UH1M	166ACR	
66-15237	UH1C	w.o.	
66-15238	UH1M		
66-15239	UH1C	w.o.	
66-15240	UH1C	w.o.	
66-15241	UH1C	w.o.	
66-15242	QUH1M	Foster Aviation	
66-15243	UH1M		
66-15244	UH1C	w.o.	
66-15245	UH1M	w.o.30.5.84	
66-16000	GUH1H	GIA	USATS
66-16001	UH1D	w.o.	
66-16002	UH1D	w.o.	
66-16003	UH1D	SoC	
66-16004	UH1H	25AvCo	
66-16005	UH1V	812Medco	det.1
66-16006	UH1H	A\3-158AVN	Reserve
66-16007	UH1H	126ACS	
66-16008	UH1D	w.o.	
66-16009	UH1H	w.o.20.5.70	
66-16010	UH1H	B\1-132AVN	DC NG
66-16011	UH1H	to Vietnam	
66-16012	UH1H	198AvCo	
66-16013	UH1H	112MedDet	
66-16014	UH1H	224AVN	
66-16015	UH1H	w.o.24.4.73	
66-16016	UH1H	26AVN	
66-16017	UH1H	to Iran	
66-16018	UH1D	SoC	
66-16019	UH1H	112MedDet	
66-16020	UH1H	to Vietnam	
66-16021	UH1H	w.o.9.10.71	
66-16022	UH1D	w.o.	
66-16023	UH1H	6ACR	
66-16024	UH1H		
66-16025	UH1D	w.o.	
66-16026	UH1D	w.o.	
66-16027	UH1H		
66-16028	UH1H	w.o.19.7.71	
66-16029	UH1D	w.o.	
66-16030	UH1H		
66-16031	UH1H	w.o.4.10.72	
66-16032	UH1D	SoC	
66-16033	UH1D	w.o.	
66-16034	UH1H		
66-16035	UH1H	1267MedCo	
66-16036	UH1H	w.o.10.11.71	
66-16037	UH1D	to El Salvador	
66-16038	UH1H	w.o.24.2.71	
66-16039	UH1D	SoC	
66-16040	UH1H		
66-16041	UH1H		
66-16042	UH1D	w.o.	

66-16043	UH1D	w.o.	
66-16044	UH1D	w.o.	
66-16045	GUH1H		
66-16046	UH1D	w.o.	
66-16047	UH1H	323AvCo	
66-16048	UH1H	88ARCOM	
66-16049	UH1H	42AVN	
66-16050	UH1D	w.o.13.1.71	
66-16051	UH1D	SoC	
66-16052	UH1H	D\1-245AVN	OK NG
66-16053	UH1D	SoC	
66-16054	UH1H	44AVN	
66-16055	UH1D	SoC	
66-16056	UH1H	31AVN	
66-16057	UH1H	107AvCo	
66-16058	UH1H	to Vietnam	
66-16059	UH1D	w.o.	
66-16060	UH1D	SoC	
66-16061	UH1D	SoC	
66-16062	UH1D	w.o.	
66-16063	UH1H	w.o.12.4.71	
66-16064	UH1D	w.o.	
66-16065	UH1H	to Iran	
66-16066	UH1H		
66-16067	UH1D	w.o.	
66-16068	UH1D	w.o.	
66-16069	UH1H	2\10Cav	
66-16070	UH1H		
66-16071	UH1H		
66-16072	UH1H	207AvCo	
66-16073	UH1D		
66-16074	UH1H		
66-16075	UH1H		
66-16076	UH1H	SoC	
66-16077	UH1D	w.o.	
66-16078	UH1H	to N8146H	
66-16079	UH1D	w.o.	
66-16080	UH1D	w.o.	
66-16081	UH1D	w.o.	
66-16082	UH1D	w.o.	
66-16083	UH1H	24Medco	
66-16084	UH1D	w.o.	
66-16085	UH1D	w.o.	
66-16086	UH1H	1-131AVN	AL NG
66-16087	UH1H	3ACT	
66-16088	UH1D	w.o.	
66-16089	UH1H	to Vietnam	
66-16090	UH1D	SoC	
66-16091	UH1D	w.o.	
66-16092	UH1H	w.o.14.3.79	
66-16093	UH1H		
66-16094	UH1H	5InfDiv	
66-16095	UH1D	w.o.	
66-16096	UH1D	w.o.	
66-16097	UH1D	w.o.	
66-16098	UH1D	w.o.	
66-16099	UH1D	w.o.	
66-16100	UH1D	w.o.	
66-16101	UH1H	150AVN	DE NG
66-16102	UH1D	w.o.	
66-16103	UH1H	56AvCo	
66-16104	UH1D	w.o.	
66-16105	UH1D	w.o.	
66-16106	UH1H	155ArmBde	
66-16107	UH1H		
66-16108	UH1H		
66-16109	UH1H	AATC	09. Lowe AHP
66-16110	UH1D	w.o.	
66-16111	UH1D	w.o.	
66-16112	UH1D	w.o.	
66-16113	UH1D	w.o.	
66-16114	UH1V	190AvCo	
66-16115	UH1H	w.o.14.11.71	
66-16116	UH1D	w.o.	
66-16117	UH1D	w.o.	
66-16118	UH1H		
66-16119	UH1H	461AVN	NM NG
66-16120	UH1D	w.o.	
66-16121	UH1D	w.o.	
66-16122	UH1H		
66-16123	UH1H		
66-16124	UH1D	SoC	
66-16125	UH1D	w.o.	
66-16126	UH1H	47AvBde	
66-16127	UH1H	498MedCo	
66-16128	UH1H	40AVN	
66-16129	UH1H		
66-16130	UH1D	SoC	
66-16131	UH1H	812MedDet	
66-16132	UH1H		
66-16133	UH1D	SoC	
66-16134	UH1H	w.o.	
66-16135	UH1H	D\1-245AVN	OK NG
66-16136	UH1D	w.o.	
66-16137	UH1H	7InfDiv	
66-16138	UH1H	SoC	
66-16139	UH1H	SoC	
66-16140	UH1D	w.o.	
66-16141	UH1H		
66-16142	UH1D	w.o.	
66-16143	UH1V	7SigBde	
66-16144	UH1D	SoC	
66-16145	UH1H	163ACR	
66-16146	UH1H		
66-16147	UH1D	w.o.	
66-16148	UH1H	D\135AVN	KS NG
66-16149	UH1H	40AVN	
66-16150	UH1H	150AVN	
66-16151	UH1D	w.o.	
66-16152	UH1D	w.o.	
66-16153	UH1D	w.o.	
66-16154	UH1D	SoC	
66-16155	UH1V	1159MedCo	
66-16156	UH1D	w.o.	
66-16157	UH1D	SoC	
66-16158	UH1D	SoC	
66-16159	UH1D	w.o.	
66-16160	UH1D	w.o.	
66-16161	UH1H	1-131AVN	AL NG
66-16162	UH1D	to N57RF	
66-16163	UH1H	w.o.20.9.77	
66-16164	UH1H	244AvCo	
66-16165	UH1D	SoC	
66-16166	UH1D	w.o.	
66-16167	UH1D	SoC	
66-16168	UH1H	to Vietnam	
66-16169	UH1D	w.o.	
66-16170	UH1D	w.o.	
66-16171	UH1H	C\1-245AVN	OK NG
66-16172	UH1D	w.o.	
66-16173	UH1H		
66-16174	UH1H	to Vietnam	
66-16175	UH1H	to Vietnam	
66-16176	UH1D	w.o.	
66-16177	UH1H	to Vietnam	
66-16178	UH1D	w.o.	
66-16179	UH1H		
66-16180	UH1H	to Vietnam	
66-16181	UH1D	w.o.	
66-16182	UH1D	w.o.24.3.70	
66-16183	UH1D	w.o.	
66-16184	UH1H	to Vietnam	
66-16185	UH1D	w.o.	
66-16186	UH1H	SoC	
66-16187	UH1H	1133MedDet	
66-16188	UH1D	w.o.	
66-16189	UH1D	w.o.	
66-16190	UH1D	w.o.	
66-16191	UH1D	SoC	
66-16192	UH1D	w.o.	
66-16193	UH1D	w.o.	
66-16194	UH1D	SoC	
66-16195	UH1D	w.o.	
66-16196	UH1D	SoC	
66-16197	UH1H		
66-16198	UH1H	222AVN	
66-16199	UH1D	w.o.	
66-16200	UH1D		
66-16201	UH1H	116AHB	
66-16202	UH1H		SC NG
66-16203	UH1H		
66-16204	UH1D	w.o.	
66-16205	UH1D	w.o.	
66-16206	UH1D	w.o.	
66-16207	UH1D	w.o.	
66-16208	UH1D	SoC	

66-16209	UH1H	A\1-207AVN AK NG	
66-16210	UH1H	92InfBde	
66-16211	UH1H	w.o.	
66-16212	UH1D	w.o.	
66-16213	UH1D	w.o.	
66-16214	UH1H		
66-16215	UH1H		
66-16216	UH1D	w.o.	
66-16217	UH1D	w.o.	
66-16218	UH1H		
66-16219	UH1D	w.o.	
66-16220	UH1H	498MedCo	
66-16221	UH1H	w.o.19.5.70	
66-16222	UH1H	SoC	
66-16223	UH1H		
66-16224	UH1D	SoC	
66-16225	UH1D	SoC	
66-16226	UH1H	to Iran	
66-16227	UH1H		
66-16228	UH1V	w.o.15.9.85	
66-16229	UH1D	SoC	
66-16230	UH1D	w.o.	
66-16231	UH1D	w.o.	
66-16232	UH1D	w.o.	
66-16233	UH1D	w.o.	
66-16234	UH1H	6Cav	
66-16235	UH1D	w.o.	
66-16236	UH1H		
66-16237	UH1D	w.o.	
66-16238	UH1H	278ACR	
66-16239	UH1V	1159MedCo	
66-16240	UH1H	SoC	
66-16241	UH1D	w.o.	
66-16242	UH1D	w.o.	
66-16243	UH1V	w.o.11.9.91	
66-16244	UH1D	w.o.	
66-16245	UH1D	w.o.	
66-16246	UH1D	w.o.	
66-16247	UH1D	w.o.	
66-16248	UH1D	w.o.	
66-16249	UH1D	w.o.	
66-16250	UH1D	w.o.	
66-16251	UH1D	w.o.	
66-16252	UH1D	SoC	
66-16253	UH1D	w.o.	
66-16254	UH1D	w.o.	
66-16255	UH1H		
66-16256	UH1D	SoC	
66-16257	UH1D	SoC	
66-16258	UH1H		
66-16259	UH1D	SoC	
66-16260	UH1D	w.o.	
66-16261	UH1H		
66-16262	UH1D	SoC	
66-16263	UH1D	SoC	
66-16264	UH1D	w.o.	
66-16265	UH1D	SoC	
66-16266	UH1H	w.o.16.1.72	
66-16267	UH1D	w.o.	
66-16268	UH1D	w.o.	
66-16269	UH1H	w.o.23.6.86	
66-16270	UH1D	SoC	
66-16271	UH1D	SoC	
66-16272	UH1V	w.o.13.10.81	
66-16273	UH1H	w.o.2.3.70	
66-16274	UH1H	w.o.18.1.71	
66-16275	UH1D	w.o.	
66-16276	UH1H	w.o.3.2.71	
66-16277	UH1H		
66-16278	UH1H		
66-16279	UH1D	w.o.	
66-16280	UH1H	to Vietnam	
66-16281	UH1H	6ACR	
66-16282	UH1D	w.o.	
66-16283	UH1D	w.o.29.7.70	
66-16284	UH1D	w.o.	
66-16285	UH1D	w.o.	
66-16286	UH1H	to Vietnam	
66-16287	UH1H		
66-16288	UH1H	w.o.12.2.70	
66-16289	UH1H	1Cav	
66-16290	UH1H	1-149AVN TX NG	
66-16291	UH1D	SoC	
66-16292	UH1D	w.o.	
66-16293	UH1H	w.o.23.1.76	
66-16294	UH1H	1267MedCo	
66-16295	UH1D	SoC	
66-16296	UH1V	w.o.7.5.81	
66-16297	UH1D	w.o.	
66-16298	UH1H	124ARCOM	
66-16299	UH1H	w.o.15.6.70	
66-16300	UH1H		
66-16301	UH1H		
66-16302	UH1D	w.o.	
66-16303	UH1D	w.o.	
66-16304	UH1H		
66-16305	UH1D	42AVN	
66-16306	UH1D	w.o.	
66-16307	UH1D	w.o.	
66-16308	UH1H	SoC	
66-16309	UH1H	to Vietnam	
66-16310	UH1D	w.o.	
66-16311	UH1D	w.o.	
66-16312	UH1H	1-158AVN	
66-16313	UH1H	SoC	
66-16314	UH1D	w.o.	
66-16315	UH1D	w.o.	
66-16316	UH1D	SoC	
66-16317	UH1H	to Vietnam	
66-16318	UH1H	SoC	
66-16319	UH1H		
66-16320	UH1D	w.o.	
66-16321	UH1H	1-183AVN ID NG	
66-16322	UH1D	SoC	
66-16323	UH1D	w.o.	
66-16324	UH1D	w.o.	
66-16325	UH1H		
66-16326	UH1H	w.o.	
66-16327	UH1H	w.o.17.6.85	
66-16328	UH1D	SoC	
66-16329	UH1D	w.o.	
66-16330	UH1D	w.o.	
66-16331	UH1D	w.o.	
66-16332	UH1D	w.o.	
66-16333	UH1D	SoC	
66-16334	UH1D	w.o.19.8.71	
66-16335	UH1H	63ARCOM	
66-16336	UH1H	w.o.	
66-16337	UH1D	w.o.	
66-16338	UH1D	w.o.6.6.70	
66-16339	UH1D	SoC	
66-16340	UH1H	to Vietnam	
66-16341	UH1H	AATC 41A Cairns AAF	
66-16342	UH1H	to Vietnam	
66-16343	UH1D	w.o.	
66-16344	UH1D	w.o.	
66-16345	UH1H	A\1-207AVN AK NG	
66-16346	UH1H	to Vietnam	
66-16347	UH1D	w.o.	
66-16348	UH1D	w.o.13.12.78	
66-16349	UH1H		
66-16350	UH1H	w.o.6.5.74	
66-16351	UH1H	to Iran	
66-16352	UH1H	to Vietnam	
66-16353	UH1D	w.o.	
66-16354	UH1H	AMARC XA328	
66-16355	UH1V	56AvCo	
66-16356	UH1D	w.o.	
66-16357	UH1D	w.o.	
66-16358	UH1H	to Vietnam	
66-16359	UH1H	to Vietnam	
66-16360	UH1D	w.o.	
66-16361	UH1D	w.o.	
66-16362	UH1D	w.o.	
66-16363	UH1D	w.o.13.4.70	
66-16364	UH1D	w.o.	
66-16365	UH1D	w.o.	
66-16366	UH1D	SoC	
66-16367	UH1D	w.o.	
66-16368	UH1D	w.o.	
66-16369	UH1D	SoC	
66-16370	UH1H	to Vietnam	
66-16371	UH1H	to Vietnam	
66-16372	UH1V	1159MedCo	
66-16373	UH1D	w.o.	
66-16374	UH1H	C\1-108AVN AZ NG 34	

Serial	Type	Notes		
66-16375	UH1D	w.o.		
66-16376	UH1D	SoC		
66-16377	UH1D	SoC		
66-16378	UH1D	w.o.		
66-16379	UH1H			
66-16380	UH1H	w.o.20.7.75		
66-16381	UH1H	to Vietnam		
66-16382	UH1D	w.o.		
66-16383	UH1D	w.o.		
66-16384	UH1H			
66-16385	UH1H	SoC		
66-16386	UH1D	w.o.		
66-16387	UH1H	1-131AVN	AL NG	
66-16388	UH1D	SoC		
66-16389	UH1D	SoC		
66-16390	UH1D	SoC		
66-16391	UH1D	w.o.		
66-16392	UH1D	w.o.		
66-16393	UH1D	w.o.		
66-16394	UH1H	w.o.7.3.71		
66-16395	UH1D	SoC		
66-16396	UH1D	w.o.		
66-16397	UH1V			
66-16398	UH1D	SoC		
66-16399	UH1H	w.o.		
66-16400	UH1D	SoC		
66-16401	UH1H	63ARCOM		
66-16402	UH1D	w.o.		
66-16403	UH1H	126MedDet		
66-16404	UH1D	SoC		
66-16405	UH1H	812MedDet		
66-16406	UH1H	to Vietnam		
66-16407	UH1H	to Korea		
66-16408	UH1H	w.o.13.12.70		
66-16409	UH1H	w.o.10.6.70		
66-16410	UH1D	w.o.		
66-16411	UH1H	w.o.3.12.76		
66-16412	UH1H	w.o.3.10.86		
66-16413	UH1H	w.o.13.2.70		
66-16414	UH1D	w.o.		
66-16415	UH1H			
66-16416	UH1H	8AVN		
66-16417	UH1D	SoC		
66-16418	UH1H	to Iran		
66-16419	UH1V	1159MedCo		
66-16420	UH1H	to Norway 420		
66-16421	UH1H	166ACR		
66-16422	UH1H	SoC		
66-16423	UH1H	w.o.23.7.72		
66-16424	UH1H	AATC	24ALowe AHP	
66-16425	UH1D	w.o.		
66-16426	UH1H			
66-16427	UH1H	to Korea		
66-16428	UH1H	to Vietnam		
66-16429	UH1H	79ARCOM		
66-16430	UH1H			
66-16431	UH1H	1-131AVN	AL NG	
66-16432	UH1H	to Korea		
66-16433	UH1D	w.o.		
66-16434	UH1D	w.o.		
66-16435	UH1H	w.o.14.1.70		
66-16436	UH1D	w.o.		
66-16437	UH1H			
66-16438	UH1H	to Vietnam		
66-16439	UH1H	2\10Cav		
66-16440	UH1D	SoC		
66-16441	UH1D	w.o.		
66-16442	UH1D	SoC		
66-16443	UH1H	to Vietnam		
66-16444	UH1H			
66-16445	UH1D	SoC		
66-16446	UH1H	126MedCo	CA NG	
66-16447	UH1H	w.o.		
66-16448	UH1D	w.o.		
66-16449	UH1H	to Korea		
66-16450	UH1H	to N81463		
66-16451	UH1H	w.o.		
66-16452	UH1H	to Vietnam		
66-16453	UH1D	SoC		
66-16454	UH1H			
66-16455	UH1D	w.o.		
66-16456	UH1H	812MedDet		
66-16457	UH1D	w.o.		
66-16458	UH1H			
66-16459	UH1D	w.o.		
66-16460	UH1H	62AvCo		
66-16461	UH1D	w.o.		
66-16462	UH1D	w.o.		
66-16463	UH1H	to Korea		
66-16464	UH1H	w.o.1.3.70		
66-16465	UH1H	w.o.12.4.70		
66-16466	UH1H			
66-16467	UH1H	to Thailand		
66-16468	UH1D	w.o.		
66-16469	UH1D	w.o.		
66-16470	UH1D	SoC		
66-16471	UH1H	C\1-108AVN	AZ NG	19
66-16472	UH1D	w.o.		
66-16473	UH1D	w.o.		
66-16474	UH1D	w.o.		
66-16475	UH1V	w.o.15.3.86		
66-16476	UH1D	SoC		
66-16477	UH1D	w.o.		
66-16478	UH1H	126MedCo	CA NG	
66-16479	UH1D	SoC		
66-16480	UH1D	w.o.		
66-16481	UH1D	SoC		
66-16482	UH1H	461AVN	NM NG	
66-16483	UH1H	13AHB		
66-16484	UH1V			
66-16485	UH1D	SoC		
66-16486	UH1D	SoC		
66-16487	UH1H			
66-16488	UH1D	SoC		
66-16489	UH1H			
66-16490	UH1D	w.o.		
66-16491	UH1H	to Thailand		
66-16492	UH1D	w.o.		
66-16493	UH1H			
66-16494	UH1H	w.o.		
66-16495	GUH1H	GIA	Sheppard TTC	
66-16496	UH1H	to Vietnam		
66-16497	UH1D	w.o.		
66-16498	UH1D	SoC		
66-16499	UH1D	w.o.		
66-16500	UH1D	w.o.		
66-16501	UH1D	207AvCo		
66-16502	UH1D	w.o.		
66-16503	UH1D	to Vietnam		
66-16504	UH1D	w.o.		
66-16505	UH1D	w.o.		
66-16506	UH1D	SoC		
66-16507	UH1D	w.o.		
66-16508	UH1H	455\457MedDet	MO NG	
66-16509	UH1D	w.o.		
66-16510	UH1D	SoC		
66-16511	UH1D	w.o.		
66-16512	UH1D	SoC		
66-16513	UH1V	63MedDet		
66-16514	UH1H	6ACR		
66-16515	UH1H	163ACR		
66-16516	UH1D	w.o.		
66-16517	UH1H	w.o.		
66-16518	UH1H	to Vietnam		
66-16519	UH1D	w.o.		
66-16520	UH1H			
66-16521	UH1H	SoC		
66-16522	UH1H			
66-16523	UH1H	w.o.16.6.70		
66-16524	UH1D	SoC		
66-16525	UH1D	SoC		
66-16526	UH1D	w.o.		
66-16527	UH1D	w.o.		
66-16528	UH1D	w.o.		
66-16529	UH1D	w.o.		
66-16530	UH1D	SoC		
66-16531	UH1D	w.o.		
66-16532	UH1D	w.o.		
66-16533	UH1H	w.o.		
66-16534	UH1H	to Vietnam		
66-16535	UH1D	SoC		
66-16536	UH1H	26AVN		
66-16537	UH1H	to N81499		
66-16538	UH1V	273MedDet		
66-16539	UH1D	w.o.		
66-16540	UH1H	to Vietnam		

66-16541	UH1H	to Vietnam	
66-16542	UH1D	w.o.	
66-16543	UH1H		
66-16544	UH1D	SoC	
66-16545	UH1H	w.o.1.12.70	
66-16546	UH1D	SoC	
66-16547	UH1D	w.o.	
66-16548	UH1D	w.o.	
66-16549	UH1H	to Vietnam	
66-16550	UH1D	w.o.	
66-16551	UH1D	w.o.	
66-16552	UH1D		
66-16553	UH1D	SoC	
66-16554	UH1V	283MedCo	
66-16555	UH1D	w.o.	
66-16556	UH1D	w.o.	
66-16557	UH1D	SoC	
66-16558	UH1H	to Vietnam	
66-16559	UH1H	A\1-244AVN	LA NG
66-16560	UH1H	B\1-132AVN	DC NG
66-16561	UH1H	44AVN	
66-16562	UH1H	to Vietnam	
66-16563	UH1H	w.o.19.6.72	
66-16564	UH1D	w.o.	
66-16565	UH1D	w.o.	
66-16566	UH1H		
66-16567	UH1D	SoC	
66-16568	UH1D	SoC	
66-16569	UH1H		
66-16570	UH1H	to Vietnam	
66-16571	UH1H	to Vietnam	
66-16572	UH1H		
66-16573	UH1D	w.o.	
66-16574	UH1H	to Vietnam	
66-16575	UH1H	SoC	
66-16576	UH1D	w.o.	
66-16577	UH1D	w.o.	
66-16578	UH1D	w.o.	
66-16579	UH1H		
66-16580	UH1D	w.o.	
66-16581	UH1H	to Vietnam	
66-16582	UH1D	w.o.	
66-16583	UH1D	w.o.	
66-16584	UH1H		
66-16585	UH1D	w.o.	
66-16586	UH1D	w.o.	
66-16587	UH1D	w.o.	
66-16588	UH1D	w.o.	
66-16589	UH1D	w.o.	
66-16590	UH1D	w.o.	
66-16591	UH1H		
66-16592	UH1D	w.o.	
66-16593	UH1D	w.o.	
66-16594	UH1H	455\457MedDet	MO NG
66-16595	UH1D	w.o.	
66-16596	UH1D	w.o.	
66-16597	UH1D	w.o.	
66-16598	UH1D	w.o.	
66-16599	UH1D	SoC	
66-16600	UH1H		
66-16601	UH1D	SoC	
66-16602	UH1D	w.o.	
66-16603	UH1D	w.o.	
66-16604	UH1H		
66-16605	UH1D	w.o.	
66-16606	UH1H		
66-16607	UH1H	to Iran	
66-16608	UH1H	to Vietnam	
66-16609	UH1H		
66-16610	UH1H	6ACR	
66-16611	UH1H	1-131AVN	AL NG
66-16612	UH1D	w.o.	
66-16613	UH1H	10AHB	
66-16614	UH1D	w.o.	
66-16615	UH1H	42AVN	
66-16616	UH1D	SoC	
66-16617	UH1H	w.o.20.10.70	
66-16618	UH1H	AMARC	XA355
66-16619	UH1H	to Iran	
66-16620	UH1H	to Philippine	
66-16621	UH1D	SoC	
66-16622	UH1D	SoC	
66-16623	UH1D	w.o.	
66-16624	UH1H	1-131AVN	AL NG
66-16625	UH1D	SoC	
66-16626	UH1D	SoC	
66-16627	UH1D	w.o.	
66-16628	UH1D	w.o.	
66-16629	UH1D	w.o.	
66-16630	UH1D	w.o.	
66-16631	UH1D	w.o.	
66-16632	UH1H	w.o.26.7.70	
66-16633	UH1H	126MedCo	CA NG
66-16634	UH1H	224AVN	
66-16635	UH1H	SoC	
66-16636	UH1H	to Vietnam	
66-16637	UH1H	w.o.	
66-16638	UH1H	1-101AVN	NY NG
66-16639	UH1H	C\1-108AVN	AZ NG 39
66-16640	UH1D	SoC	
66-16641	UH1D	1InfDiv	
66-16642	UH1H	SoC	
66-16643	UH1H	26AVN	
66-16644	UH1H	w.o.	
66-16645	UH1H	SoC	
66-16646	UH1H	w.o.	
66-16647	UH1D	w.o.	
66-16648	UH1D	w.o.	
66-16649	UH1H	A\1-244AVN	LA NG
66-16650	UH1D	w.o.	
66-16651	UH1H	to Vietnam	
66-16652	UH1H	A\3-158AVN	Reserve
66-16653	UH1H	1898AvCo	
66-16654	UH1H		
66-16655	UH1D	w.o.	
66-16656	UH1H		
66-16657	UH1H		
66-16658	UH1H	812MedDet	
66-16659	UH1H	323AvCo	
66-16660	UH1D	w.o.	
66-16661	UH1H		
66-16662	UH1H	w.o.26.8.76	
66-16663	UH1H		
66-16664	UH1H	w.o.10.2.70	
66-16665	UH1H		
66-16666	UH1D	w.o.	
66-16667	UH1D	SoC	
66-16668	UH1H	w.o.30.10.75	
66-16669	UH1H	to Vietnam	
66-16670	UH1H	112MedDet	
66-16671	UH1H	to Vietnam	
66-16672	UH1D	w.o.	
66-16673	UH1D	SoC	
66-16674	UH1H	D\135AVN	KS NG
66-16675	UH1D	w.o.	
66-16676	UH1D	w.o.	
66-16677	UH1H		
66-16678	UH1H	126MedCo	CA NG
66-16679	UH1D	w.o.	
66-16680	UH1H	1133MedDet	
66-16681	UH1D	w.o.	
66-16682	UH1H	26AVN	
66-16683	UH1H	47AVN	
66-16684	UH1D	SoC	
66-16685	UH1D	w.o.	
66-16686	UH1D	w.o.	
66-16687	UH1H	to Korea	
66-16688	UH1H	to Thailand	
66-16693	UH1D	w.o.	
66-16694	UH1D	w.o.	
66-16695	UH1D	w.o.	
66-16696	UH1H		
66-16697	UH1H	w.o.10.3.70	
66-16698	UH1H	to Vietnam	
66-16699	UH1D	w.o.	
66-16700	UH1H		
66-16701	UH1D	w.o.	
66-16702	UH1D	w.o.	
66-16703	UH1D	SoC	
66-16704	UH1H	SoC	
66-16705	UH1D	w.o.	
66-16706	UH1H	to Korea	
66-16707	UH1D	w.o.	
66-16708	UH1H	to Vietnam	
66-16709	UH1D	w.o.	
66-16710	UH1D	w.o.	

Serial	Type	Notes	
66-16711	UH1H	w.o.8.3.70	
66-16712	UH1H	to Vietnam	
66-16713	UH1H	w.o.8.12.76	
66-16714	UH1D	w.o.	
66-16715	UH1H	w.o.20.3.76	
66-16716	UH1D	w.o.	
66-16717	UH1D	SoC	
66-16718	UH1D	w.o.	
66-16719	UH1H	to Vietnam	
66-16720	UH1H	25AvCo	
66-16721	UH1D	SoC	
66-16722	UH1D	w.o.	
66-16723	UH1D	w.o.	
66-16724	UH1D	w.o.	
66-16725	UH1H	44AVN	
66-16726	UH1H	w.o.4.9.70	
66-16727	UH1H	to Vietnam	
66-16728	UH1D	w.o.	
66-16729	UH1H	A\3-158AVN	Reserve
66-16730	UH1H	to Vietnam	
66-16731	UH1D	SoC	
66-16732	UH1D	w.o.	
66-16733	UH1D	w.o.	
66-16734	UH1H		
66-16735	UH1D	w.o.	
66-16736	UH1D	w.o.	
66-16737	UH1H	w.o.16.2.70	
66-16738	UH1H	to Iran	
66-16739	UH1H	to Vietnam	
66-16740	UH1V		
66-16741	UH1H	to N56RF	
66-16742	UH1H	26AVN	
66-16743	UH1D	w.o.	
66-16744	UH1H	w.o.2.9.71	
66-16745	UH1H	to Vietnam	
66-16746	UH1H	to Vietnam	
66-16747	UH1H		
66-16748	UH1H	Berlin Brigade	
66-16749	UH1H	to Vietnam	
66-16750	UH1H	203AvCo	
66-16751	UH1H	to Vietnam	
66-16752	UH1D	w.o.	
66-16753	UH1H	to Vietnam	
66-16754	UH1D	SoC	
66-16755	UH1D	w.o.	
66-16756	UH1D	w.o.	
66-16757	UH1H		
66-16758	UH1H	SoC	
66-16759	UH1H	w.o.16.1.70	
66-16760	UH1D	w.o.	
66-16761	UH1H	1-131AVN	AL NG
66-16762	UH1H	to Vietnam	
66-16763	UH1D	w.o.	
66-16764	UH1H	323AVN	
66-16765	UH1H	to Vietnam	
66-16766	UH1D	w.o.	
66-16767	UH1D	w.o.	
66-16768	UH1H		
66-16769	NUH1H		
66-16770	UH1D	w.o.	
66-16771	UH1H	to Vietnam	
66-16772	UH1D	SoC	
66-16773	UH1D	w.o.	
66-16774	UH1D	w.o.	
66-16775	UH1D	w.o.	
66-16776	UH1D	w.o.	
66-16777	UH1D	w.o.	
66-16778	UH1D	w.o.	
66-16779	UH1H	126MedCo	CA NG
66-16780	UH1D	w.o.	
66-16781	UH1H	w.o.8.2.71	
66-16782	UH1H	w.o.22.4.78	
66-16783	UH1H	to Vietnam	
66-16784	UH1D	w.o.	
66-16785	UH1D	w.o.	
66-16786	UH1D	w.o.	
66-16787	UH1D	w.o.	
66-16788	UH1D	w.o.	
66-16789	UH1D	w.o.	
66-16790	UH1D	SoC	
66-16791	UH1H	to Vietnam	
66-16792	UH1H	to Vietnam	
66-16793	UH1D	w.o.	
66-16794	UH1H	2ACR	
66-16795	UH1H	1898AvCo	
66-16796	UH1H	to Vietnam	
66-16797	UH1H	1InfDiv	
66-16798	UH1H	184AHC	
66-16799	UH1D	w.o.	
66-16800	UH1D	w.o.	
66-16801	UH1H	2ACR	
66-16802	UH1D	w.o.	
66-16803	UH1D	w.o.	
66-16804	UH1D	w.o.	
66-16805	UH1D	w.o.	
66-16806	UH1D	w.o.	
66-16807	UH1H	166ACR	
66-16808	UH1H	B\26AVN	
66-16809	UH1H	to Vietnam	
66-16810	UH1H	C\1-108AVN	AZ NG 31
66-16811	UH1D	w.o.	
66-16812	UH1H	to Vietnam	
66-16813	UH1H	to Vietnam	
66-16814	UH1H	w.o.	
66-16815	UH1H		
66-16816	UH1H		
66-16817	UH1D	SoC	
66-16818	UH1H		
66-16819	UH1H		
66-16820	UH1D	w.o.	
66-16821	UH1D	SoC	
66-16822	UH1D	SoC	
66-16823	UH1H	D\1-245AVN	OK NG
66-16824	UH1D	w.o.	
66-16825	UH1D	SoC	
66-16826	UH1H	47AVN	
66-16827	UH1H	A\1-207AVN	AK NG
66-16828	UH1D	w.o.	
66-16829	UH1H	A\3-158AVN	Reserve
66-16830	UH1D	w.o.	
66-16831	UH1H	to Iran	
66-16832	UH1H	to Vietnam	
66-16833	UH1D	w.o.	
66-16834	UH1D	w.o.	
66-16835	UH1D	w.o.	
66-16836	UH1V	1159MedCo	
66-16837	UH1H	SoC	
66-16838	UH1H	to Vietnam	
66-16839	UH1H		
66-16840	UH1H	w.o.25.2.72	
66-16841	UH1H	to Vietnam	
66-16842	UH1H		
66-16843	UH1H	to Vietnam	
66-16844	UH1D	w.o.	
66-16845	UH1D	w.o.	
66-16846	UH1D	w.o.	
66-16847	UH1H	6ACR	
66-16848	UH1H	w.o.8.2.70	
66-16849	UH1H	w.o.	
66-16850	UH1H		
66-16851	UH1H		
66-16852	UH1H	to Vietnam	
66-16853	UH1D	w.o.	
66-16854	UH1H	136AvCo	
66-16855	UH1H	to Korea	
66-16856	UH1H	to Vietnam	
66-16857	UH1H	1898AvCo	
66-16858	UH1H	to Vietnam	
66-16859	UH1H		
66-16860	UH1D	w.o.	
66-16861	UH1D	w.o.	
66-16862	UH1H	to Vietnam	
66-16863	UH1H	to N8152K	
66-16864	UH1H	to Vietnam	
66-16865	UH1D	SoC	
66-16866	UH1H	198AvCo	
66-16867	UH1D	SoC	
66-16868	UH1V	273MedDet	
66-16869	UH1D	w.o.	
66-16870	UH1H		
66-16871	UH1D	w.o.	
66-16872	UH1D	w.o.	
66-16873	UH1H	1-131AVN	AL NG
66-16874	UH1D	w.o.	
66-16875	UH1H		
66-16876	UH1H	w.o.10.8.70	

Serial	Type	Notes	Extra
66-16877	UH1V	1InfDiv	
66-16878	UH1H	to Vietnam	
66-16879	UH1H	198AvCo	
66-16880	UH1D	w.o.	
66-16881	UH1H	to Vietnam	
66-16882	UH1H		
66-16883	UH1D	SoC	
66-16894	UH1V	1InfDiv	
66-16895	UH1D	w.o.20.5.70	
66-16896	UH1V	348MedDet	
66-16897	UH1H		SD NG
66-16898	UH1D	SoC	
66-16899	UH1H		
66-16900	UH1H	166ACR	
66-16901	UH1H	SoC	
66-16902	UH1H	107AvCo	
66-16903	UH1H		
66-16904	UH1H	w.o.12.1.85	
66-16905	UH1D	SoC	
66-16906	UH1H	150AVN	
66-16907	UH1H		
66-16908	UH1D	SoC	
66-16909	UH1H		
66-16910	UH1D	SoC	
66-16911	UH1H	to Iran	
66-16912	UH1H	SoC	
66-16916	UH1D	w.o.	
66-16917	UH1H	w.o.22.10.71	
66-16918	UH1D	w.o.	
66-16919	UH1H		
66-16920	UH1H		
66-16921	UH1D	w.o.	
66-16922	UH1D	w.o.	
66-16923	UH1H	278ACR	
66-16924	UH1D	w.o.	
66-16925	UH1H	to Vietnam	
66-16926	UH1D	w.o.	
66-16927	UH1D	w.o.	
66-16928	UH1D	SoC	
66-16929	UH1H	25AvCo	
66-16930	UH1D	w.o.	
66-16931	UH1H	w.o.13.5.84	
66-16932	UH1H	w.o.1.4.70	
66-16933	UH1D	SoC	
66-16934	UH1H	to Panama	
66-16935	UH1H	163ACR	
66-16936	UH1D	SoC	
66-16937	UH1H	to Panama	
66-16938	UH1H	2ACR	
66-16939	UH1H		
66-16940	UH1D	w.o.	
66-16941	UH1H	to Vietnam	
66-16942	UH1H	to N8149H	
66-16943	UH1D	w.o.	
66-16944	UH1V	1267MedCo	MO NG
66-16945	UH1D	SoC	
66-16946	UH1D	w.o.	
66-16947	UH1D	w.o.	
66-16948	UH1H		
66-16949	UH1H	1Cav	
66-16950	UH1D	w.o.	
66-16951	UH1H	to Vietnam	
66-16952	UH1D	w.o.	
66-16953	UH1H	to Vietnam	
66-16954	UH1H	to Vietnam	
66-16955	UH1H	to Vietnam	
66-16956	UH1H	to Vietnam	
66-16957	UH1H		
66-16958	UH1D	w.o.	
66-16959	UH1D	SoC	
66-16960	UH1H	349MedDet	
66-16961	UH1D	SoC	
66-16962	UH1D	w.o.	
66-16963	UH1H	Berlin Brigade	
66-16964	UH1H	40AVN	
66-16965	UH1H	126MedCo	CA NG
66-16966	UH1D	w.o.	
66-16967	UH1H	173AvCo	
66-16968	UH1H	AATC	68L Lowe AHP
66-16969	UH1H	40AVN	
66-16970	UH1H	25AVN	
66-16971	UH1H	D\135AVN	KS NG
66-16972	UH1V		SD NG
66-16973	UH1H	3AVN	
66-16974	UH1H	SoC	
66-16975	UH1H	1-149AVN	TX NG
66-16976	UH1H	to Vietnam	
66-16977	JUH1H	ERADCOM	
66-16978	UH1D	SoC	
66-16979	UH1H	w.o.	
66-16980	UH1H	to Korea	
66-16981	UH1D	w.o.	
66-16982	UH1H	w.o.1.4.71	
66-16983	UH1D	w.o.	
66-16984	UH1H	w.o.21.5.70	
66-16985	UH1H	w.o.11.6.70	
66-16986	UH1H	2ACR	
66-16987	UH1H	163ACR	
66-16988	UH1H	98DivTrng	
66-16989	UH1D	w.o.22.6.70	
66-16990	UH1H		
66-16991	UH1H	w.o.	
66-16992	UH1H	308AHB	
66-16993	UH1H		
66-16994	UH1D	SoC	
66-16995	UH1D	SoC	
66-16996	UH1D	SoC	
66-16997	UH1H	207AvCo	
00-10998	UH1H		
66-16999	UH1H		
66-17000	UH1H	to Thailand	
66-17001	UH1H		
66-17002	UH1H	SoC	
66-17003	UH1H	to Thailand	
66-17004	UH1H	126MedCo	CA NG
66-17005	UH1H		
66-17006	UH1H	SoC	
66-17007	UH1H	SoC	
66-17008	UH1H	w.o.	
66-17009	UH1H	w.o.	
66-17010	UH1H	w.o.	
66-17011	UH1H	w.o.22.5.70	
66-17012	UH1H	w.o.	
66-17013	UH1H		
66-17014	UH1H	w.o.	
66-17015	UH1H	w.o.	
66-17016	UH1H	w.o.	
66-17017	UH1H	w.o.	
66-17018	UH1H	w.o.	
66-17019	UH1H	31AVN	
66-17020	JUH1H	ERADCOM	
66-17021	UH1H		
66-17022	UH1H	SoC	
66-17023	UH1H	SoC	
66-17024	UH1H	507MedDet	
66-17025	UH1H	w.o.	
66-17026	UH1H	w.o.28.8.70	
66-17027	UH1H	w.o.	
66-17028	UH1H	w.o.	
66-17029	UH1H	w.o.	
66-17030	UH1V	4InfDiv	
66-17031	UH1H		
66-17032	UH1H	w.o.	
66-17033	UH1H		
66-17034	UH1H	SoC	
66-17035	UH1H	w.o.	
66-17036	UH1H	w.o.	
66-17037	UH1H		
66-17038	UH1H	SoC	
66-17039	UH1H	w.o.	
66-17040	UH1H	to Vietnam	
66-17041	UH1H	SoC	
66-17042	UH1H	w.o.	
66-17043	UH1H	w.o.	
66-17044	UH1H	120AvCo	
66-17045	UH1H	w.o.	
66-17046	UH1H	w.o.	
66-17047	UH1H	SoC	
66-17048	UH1H	42AVN	
66-17049	UH1H	w.o.	
66-17050	UH1H	SoC	
66-17051	UH1H	w.o.17.3.72	
66-17052	UH1H	62AvCo	
66-17053	UH1H	244AvCo	
66-17054	UH1H		MN NG
66-17055	UH1H	w.o.20.5.70	

Serial	Type	Unit	Note
66-17056	UH1H	155ArmBde	
66-17057	UH1H	w.o.	
66-17058	UH1H	to Vietnam	
66-17059	UH1H		
66-17060	UH1H	278ACR	
66-17061	UH1H	47AVN	
66-17062	UH1H	to Vietnam	
66-17063	UH1H	to Vietnam	
66-17064	UH1H	SoC	
66-17065	UH1H	to Vietnam	
66-17066	UH1H	308AHB	
66-17067	UH1H	w.o.	
66-17068	UH1H		
66-17069	UH1H	SoC	
66-17070	UH1H	207AvCo	
66-17071	UH1H		
66-17072	UH1H		
66-17073	UH1H	158AVN	
66-17074	UH1H	SoC	
66-17075	UH1H	w.o.	
66-17076	UH1H	812MedDet	
66-17077	UH1H	w.o.	
66-17078	UH1H	1150MedCo	
66-17079	UH1H	C\1-245AVN OK NG	
66-17080	JUH1H		
66-17081	UH1H		
66-17082	UH1H	w.o.	
66-17083	UH1H	w.o.	
66-17084	UH1H	A\1-131AVN AL NG	
66-17085	UH1H	SoC	
66-17086	UH1H	SoC	
66-17087	UH1H		
66-17088	UH1H	1898AvCo	
66-17089	UH1V	68MedDet	
66-17090	UH1H	1-149AVN TX NG	
66-17091	UH1H	C\1-108AVN AZ NG 23	
66-17092	UH1H	w.o.	
66-17093	UH1H	w.o.	
66-17094	UH1H	SoC	
66-17095	UH1H	to Vietnam	
66-17096	UH1H	to Vietnam	
66-17097	UH1H	to Vietnam	
66-17098	UH1H	w.o.	
66-17099	UH1H		
66-17100	UH1V		
66-17101	UH1H	79ARCOM	
66-17102	UH1H		
66-17103	UH1H	to Vietnam	
66-17104	UH1H	26AVN	
66-17105	UH1H		
66-17106	UH1H	1267MedCo MO NG	
66-17107	UH1H	to Vietnam	
66-17108	UH1H	w.o.	
66-17109	UH1V	to Vietnam	
66-17110	UH1H	w.o.20.1.70	
66-17111	UH1H	126MedCo CA NG	
66-17112	UH1H	w.o.	
66-17113	UH1H	1-131AVN AL NG	
66-17114	UH1H	120AvCo	
66-17115	UH1H		
66-17116	UH1H	to Vietnam	
66-17117	UH1H	w.o.	
66-17118	UH1H		
66-17119	UH1H	SoC	
66-17120	UH1H		
66-17121	UH1H	w.o.	
66-17122	UH1H	w.o.	
66-17123	UH1H	150AVN DE NG	
66-17124	UH1H	to Vietnam	
66-17125	UH1H	w.o.	
66-17126	UH1H	w.o.	
66-17127	UH1H	w.o.14.2.70	
66-17128	UH1H	w.o.	
66-17129	UH1H	w.o.	
66-17130	UH1H		
66-17131	UH1H	w.o.	
66-17132	UH1H	w.o.13.5.70	
66-17133	UH1H	SoC	
66-17134	UH1H	to Vietnam	
66-17135	UH1H	w.o.	
66-17136	UH1H	to Vietnam	
66-17137	UH1H	w.o.26.4.78	
66-17138	UH1H		
66-17139	UH1H		
66-17140	UH1H	w.o.	
66-17141	UH1H	w.o.8.1.72	
66-17142	UH1H	to Iran	
66-17143	UH1H	SoC	
66-17144	UH1H	1267MedCo MO NG	
67-17145	UH1H	USAAEFA	
67-17146	UH1H	321MedDet	
67-17147	UH1H	A\1-244AVN LA NG	
67-17148	UH1H	w.o.15.11.70	
67-17149	UH1H	w.o.	
67-17150	UH1H	IL NG	
67-17151	UH1H	to Vietnam	
67-17152	UH1H	w.o.	
67-17153	UH1H		
67-17154	UH1H	w.o.	
67-17155	UH1H		
67-17156	UH1H	w.o.	
67-17157	UH1H	w.o.	
67-17158	UH1H	w.o.	
67-17159	UH1H	to Vietnam	
67-17160	UH1H	1AVN	
67-17161	UH1H	to Vietnam	
67-17162	UH1H	6ACR	
67-17163	UH1H	to Vietnam	
67-17164	UH1H	w.o.	
67-17165	UH1H	w.o.10.3.70	
67-17166	UH1H	w.o.	
67-17167	UH1H	w.o.	
67-17168	UH1H	w.o.20.2.70	
67-17169	UH1H	SoC	
67-17170	UH1H	to Vietnam	
67-17171	UH1H	A\1-131AVN AL NG	
67-17172	UH1H	w.o.	
67-17173	UH1H	to Panama	
67-17174	UH1H		
67-17175	UH1H	w.o.17.2.70	
67-17176	UH1H	w.o.	
67-17177	UH1H	SoC	
67-17186	UH1H	to Vietnam	
67-17187	UH1H	349MedDet	
67-17188	UH1H	to Vietnam	
67-17189	UH1V	498MedCo	
67-17190	UH1H	to Vietnam	
67-17191	UH1H	w.o.	
67-17192	UH1H	to N8159C	
67-17193	UH1H	w.o.	
67-17194	UH1H	26AVN	
67-17195	UH1H	6AvCo	
67-17196	UH1H	w.o.	
67-17197	UH1H	150AVN DE NG	
67-17198	UH1H		
67-17199	UH1H	63ARCOM	
67-17200	UH1V	507MedCo	
67-17201	UH1H	2\10Cav	
67-17202	UH1V	283MedCo	
67-17203	UH1H	w.o.	
67-17204	UH1H		
67-17205	UH1H	w.o.	
67-17206	UH1H	w.o.	
67-17207	UH1H	w.o.	
67-17208	UH1H	SoC	
67-17209	UH1H	w.o.	
67-17210	UH1H	44AVN	
67-17211	UH1H		
67-17212	UH1H	1AVN	
67-17213	UH1H	SoC	
67-17214	UH1H	w.o.	
67-17215	UH1H	w.o.	
67-17216	UH1H	w.o.17.8.82	
67-17217	UH1H	10AHB	
67-17218	UH1H	to Vietnam	
67-17219	UH1H	149AHB	
67-17220	UH1H		
67-17221	UH1H	SoC	
67-17222	UH1H	to Vietnam	
67-17223	UH1H	Foster Aviation	
67-17224	UH1H		
67-17225	UH1H	to Vietnam	
67-17226	UH1H	w.o.	
67-17227	UH1H	w.o.	
67-17228	UH1H	w.o.	
67-17229	UH1H	to Vietnam	

Serial	Type	Notes	Location
67-17230	UH1H	w.o.	
67-17231	UH1H	w.o.	
67-17232	UH1H	w.o.	
67-17233	UH1H	to N57RF	
67-17234	UH1H		
67-17235	UH1H	w.o.24.7.70	
67-17236	UH1H	69AvBde	
67-17237	UH1H	w.o.	
67-17238	UH1H	82AbDiv	
67-17239	UH1H		
67-17240	UH1H		
67-17241	UH1H	dump	Ansbach AB, FRG
67-17242	UH1H	w.o.23.6.70	
67-17243	UH1H	SoC	
67-17244	UH1H	to Vietnam	
67-17245	UH1H	w.o.	
67-17246	UH1H	w.o.	
67-17247	UH1H	158AVN	
67-17248	UH1H	w.o.26.7.70	
67-17249	UH1H	w.o.	
67-17250	UH1H		
67-17251	UH1H		
67-17252	UH1H	w.o.	
67-17253	UH1H	w.o.	
67-17254	UH1H	w.o.	
67-17255	UH1H	w.o.	
67-17256	UH1H	A\1-131AVN	AL NG
67-17257	UH1H	w.o.	
67-17258	UH1H	1-158AVN	
67-17259	UH1H	w.o.3.6.70	
67-17260	UH1H	w.o.	
67-17261	UH1H	w.o.	
67-17262	UH1H	w.o.	
67-17263	UH1H	stored	Ramstein AB,FRG
67-17264	UH1H	to Vietnam	
67-17265	UH1H	w.o.	
67-17266	UH1H	173AvCo	
67-17267	UH1H	45AVN	
67-17268	UH1V	1267MedCo	MO NG
67-17269	UH1H	w.o.	
67-17270	UH1H		
67-17271	UH1H	to Thailand	
67-17272	UH1H	w.o.14.12.86	
67-17273	UH1H	w.o.	
67-17274	UH1H	Berlin Brigade	
67-17275	UH1H	w.o.4.1.71	
67-17276	UH1H	to Vietnam	
67-17277	UH1H	to Iran	
67-17278	UH1H	w.o.3.5.72	
67-17279	UH1H	1-159AVN	
67-17280	UH1H	w.o.	
67-17281	UH1H	AATC	81. Cairns AAF
67-17282	UH1H	145MedDet	
67-17283	UH1H	w.o.	
67-17284	UH1H		
67-17285	UH1H		
67-17286	UH1H	278ACR	
67-17287	UH1H	w.o.	
67-17288	UH1H		
67-17289	UH1H	26AVN	
67-17290	UH1H		
67-17291	UH1H	w.o.	
67-17292	UH1H		
67-17293	UH1H	to Vietnam	
67-17294	UH1H	SoC	
67-17295	UH1H	SoC	
67-17296	UH1H		
67-17297	UH1H		
67-17298	UH1H	SoC	
67-17299	UH1H		
67-17300	UH1H	SoC	
67-17301	UH1H	SoC	
67-17302	UH1H		
67-17303	UH1H	to Vietnam	
67-17304	UH1H	A\1-207AVN	AK NG
67-17305	UH1H	Berlin Brigade	
67-17306	UH1H	to Vietnam	
67-17313	UH1H	to Vietnam	
67-17314	UH1H	to Vietnam	
67-17315	UH1H	w.o.5.3.90	
67-17316	UH1H	w.o.	
67-17317	UH1H	to Vietnam	
67-17318	UH1H	w.o.	
67-17319	UH1H	w.o.	
67-17320	UH1H	to Vietnam	
67-17321	UH1H	3ACR	
67-17322	UH1H	to Iran	
67-17323	UH1H	to Chile	
67-17324	UH1H	to Argentina	
67-17325	UH1H	to Argentina H-14	
67-17326	UH1H	to Argentina	
67-17327	UH1H	w.o.	
67-17328	UH1H		
67-17329	UH1H	w.o.	
67-17330	UH1H	w.o.	
67-17331	UH1H	w.o.21.9.77	
67-17332	UH1H		
67-17333	UH1H	w.o.	
67-17334	UH1V	126MedCo	CA NG
67-17335	UH1H	w.o.	
67-17336	UH1H	166ACR	
67-17337	UH1V	283MedCo	
67-17338	UH1H		
67-17339	UH1H		
67-17340	UH1H	to Vietnam	
67-17341	UH1H	w.o.	
67-17342	UH1H	w.o.	
67-17343	UH1H	w.o.	
67-17344	UH1H	SoC	
67-17345	UH1H	w.o.	
67-17346	UH1H	to Vietnam	
67-17347	UH1H	w.o.29.5.78	
67-17348	UH1H	w.o.	
67-17349	UH1H	w.o.	
67-17350	UH1H		
67-17351	UH1H	w.o.17.10.70	
67-17352	UH1H	SoC	
67-17353	UH1H	w.o.	
67-17354	UH1H	w.o.	
67-17355	UH1H	207AvCo	
67-17356	UH1H	AATC	56A Cairns AAF
67-17357	UH1H	w.o.	
67-17358	UH1H	40AVN	
67-17359	UH1H	SoC	
67-17360	UH1H	to Vietnam	
67-17361	UH1H	w.o.	
67-17362	UH1H	w.o.20.5.70	
67-17363	UH1H	SoC	
67-17364	UH1H		
67-17365	UH1H	w.o.	
67-17366	UH1H	w.o.	
67-17367	UH1H	w.o.	
67-17368	UH1H		
67-17369	UH1H	163ACR	
67-17370	UH1H	w.o.	
67-17371	UH1H	w.o.27.9.71	
67-17372	UH1H		
67-17373	UH1H		
67-17374	UH1H		
67-17375	UH1H	w.o.	
67-17376	UH1H	w.o.15.6.70	
67-17377	UH1H	w.o.	
67-17378	UH1H	w.o.	
67-17379	UH1H	SoC	
67-17380	UH1H	w.o.	
67-17381	UH1H	8AVN	
67-17382	UH1H	SoC	
67-17383	UH1H	to Vietnam	
67-17384	UH1H	to Thailand	
67-17385	UH1H	to Vietnam	
67-17386	UH1H	SoC	
67-17387	UH1H		
67-17388	UH1H		
67-17389	UH1H	w.o.27.4.76	
67-17390	UH1H	to Thailand	
67-17391	UH1H	79ARCOM	
67-17392	UH1H	to Vietnam	
67-17393	UH1H	w.o.	
67-17394	UH1H	to Vietnam	
67-17395	UH1H	w.o.	
67-17396	UH1H	A\1-244AVN	LA NG
67-17397	UH1H	w.o.	
67-17398	UH1H		
67-17399	UH1H	w.o.	
67-17400	UH1H	w.o.	
67-17401	UH1H		

Serial	Type	Notes			Serial	Type	Notes	
67-17402	UH1H	w.o.			67-17485	UH1H	w.o.	
67-17403	UH1H	SoC			67-17486	UH1H	SoC	
67-17404	UH1H	SoC			67-17487	UH1H	247MedDet	
67-17405	UH1H	SoC			67-17488	UH1H	w.o.	
67-17406	UH1H	507MedCo			67-17489	UH1H	to Vietnam	
67-17407	UH1H	w.o.			67-17490	UH1H	w.o.	
67-17408	UH1H	w.o.5.8.75			67-17491	UH1H	w.o.	
67-17409	UH1H	w.o.			67-17492	UH1H	w.o.5.4.70	
67-17410	UH1H	w.o.			67-17493	UH1H		
67-17411	UH1H	C\1-245AVN	OK NG		67-17494	UH1H	w.o.	
67-17412	UH1H	w.o.			67-17495	UH1H	347MedDet	
67-17413	UH1H	11ACR			67-17496	UH1H	w.o.1.2.70	
67-17414	UH1V	427MedCo			67-17497	UH1H	w.o.	
67-17415	UH1H	w.o.			67-17498	UH1H	SoC	
67-17416	UH1H	1133MedDet			67-17499	UH1H	w.o.	
67-17417	UH1H	w.o.5.12.70			67-17500	UH1H		
67-17418	UH1H	to Vietnam			67-17501	UH1H		
67-17419	UH1H	w.o.			67-17502	UH1H	26AVN	
67-17420	UH1H	200AvCo			67-17503	UH1H		
67-17421	UH1H	w.o.12.12.84			67-17504	UH1H	w.o.	
67-17422	UH1H				67-17505	UH1H		
67-17423	UH1H	w.o.10.4.78			67-17506	UH1H	to Vietnam	
67-17424	UH1H	A\3-158AVN	Reserve		67-17507	UH1H	to Vietnam	
67-17425	UH1H	10AHB			67-17508	UH1H	to Vietnam	
67-17426	UH1H	26AVN			67-17509	UH1H	to Vietnam	
67-17427	UH1H	to Thailand			67-17510	UH1H		
67-17428	UH1H	SoC			67-17511	UH1H	D\135AVN	KS NG
67-17429	UH1H	w.o.			67-17512	UH1H	w.o.	
67-17430	UH1H	to Vietnam			67-17513	UH1H	to Iran	
67-17431	UH1H	173AvCo			67-17514	UH1H		
67-17432	UH1H	to Vietnam			67-17515	UH1H	w.o.	
67-17433	UH1H	to Vietnam			67-17516	UH1H	w.o.24.3.70	
67-17434	UH1H	w.o.			67-17517	UH1H	w.o.	
67-17435	UH1H	w.o.			67-17518	UH1H		
67-17436	UH1H	SoC			67-17519	UH1V		
67-17437	UH1H	to Iran			67-17520	UH1H	150AVN	DE NG
67-17438	UH1H	to Vietnam			67-17521	UH1H	to Vietnam	
67-17439	UH1H	w.o.4.4.70			67-17522	UH1H	w.o.	
67-17440	UH1H	w.o.			67-17523	UH1V	507MedCo	
67-17441	UH1H				67-17524	UH1H	w.o.13.2.72	
67-17442	UH1H	2-135AVN	CO NG		67-17525	UH1H	w.o.	
67-17443	UH1H	w.o.			67-17526	UH1V		
67-17444	UH1H	to Vietnam			67-17527	UH1H	163ACR	
67-17445	UH1H				67-17528	UH1H	w.o.	
67-17446	UH1H	w.o.			67-17529	UH1H	w.o.	
67-17447	UH1H	w.o.			67-17530	UH1H	w.o.	
67-17448	JUH1H	AMARC	XA...		67-17531	UH1H	w.o.	
67-17449	UH1H	w.o.			67-17532	UH1H	w.o.16.12.70	
67-17450	UH1H	w.o.			67-17533	UH1H	w.o.	
67-17451	UH1H	79ARCOM			67-17534	UH1V	131AVN	
67-17452	UH1H				67-17535	UH1V		
67-17453	UH1H				67-17536	UH1H	w.o.	
67-17454	UH1H	1267MedCo	MO NG		67-17537	UH1H	7ATC	
67-17455	UH1V	54MedDet			67-17538	UH1H	w.o.2.9.72	
67-17456	UH1H				67-17539	UH1H		
67-17457	UH1H	to Thailand			67-17540	UH1H		
67-17458	UH1H	w.o.10.9.70			67-17541	UH1H		
67-17459	UH1V	w.o.8.2.74			67-17542	UH1H	to Korea	
67-17460	UH1H	w.o.			67-17543	UH1H	173AvCo	
67-17461	UH1H				67-17544	UH1H	w.o.	
67-17462	UH1H	w.o.			67-17545	UH1H	340EngGrp	
67-17463	UH1H	w.o.11.2.70			67-17546	UH1H	4ARC	
67-17464	UH1H	SoC			67-17547	UH1V	427MedCo	
67-17465	UH1H	w.o.			67-17548	UH1H	to Vietnam	
67-17466	UH1H	w.o.			67-17549	UH1H	10Cav	
67-17467	UH1H				67-17550	UH1H	w.o.	
67-17468	UH1H	w.o.			67-17551	UH1H	w.o.22.7.70	
67-17469	UH1H	107AvCo			67-17552	UH1H	w.o.	
67-17470	UH1H	w.o.			67-17553	UH1H	w.o.23.3.70	
67-17471	UH1H	SoC			67-17554	UH1H	SoC	
67-17472	UH1H	26AVN			67-17555	UH1H	to Vietnam	
67-17473	UH1H	w.o.			67-17556	UH1H	w.o.	
67-17474	UH1H	SoC			67-17557	UH1H	to Vietnam	
67-17475	UH1H	SoC			67-17558	UH1H		
67-17476	UH1H	w.o.18.2.70			67-17559	UH1H	w.o.	
67-17477	UH1H	to Vietnam			67-17560	UH1H	to Vietnam	
67-17478	UH1H	to Vietnam			67-17561	UH1H	to Vietnam	
67-17479	UH1H	to Vietnam			67-17562	UH1H		
67-17480	UH1H	to Vietnam			67-17563	UH1H	to Vietnam	
67-17481	UH1H	to Vietnam			67-17564	UH1H	to Vietnam	
67-17482	UH1H	to Vietnam			67-17565	UH1H		
67-17483	UH1H	w.o.			67-17566	UH1H	w.o.	
67-17484	UH1H	SoC			67-17567	UH1H	to Peru	

Serial	Type	Notes	
67-17586	UH1H		
67-17587	UH1H	w.o.17.8.70	
67-17588	UH1H	w.o.	
67-17589	UH1H		
67-17590	UH1H		
67-17591	UH1H		
67-17592	UH1H	w.o.	
67-17593	UH1H	w.o.	
67-17594	UH1H	w.o.23.6.70	
67-17595	UH1H	w.o.	
67-17596	UH1H		
67-17597	UH1H	w.o.	
67-17598	UH1H	92InfBde	
67-17599	UH1V		
67-17600	UH1H	to Thailand	
67-17601	UH1H		
67-17602	UH1H	derelict	NAS Dallas
67-17603	UH1H	w.o.	
67-17604	UH1H	to Thailand	
67-17605	UH1H	w.o.	
67-17606	UH1H	C\1-108AVN	AZ NG 16
67-17607	UH1H	w.o.	
67-17608	UH1H	to Vietnam	
67-17609	UH1H	w.o.	
67-17610	UH1H	to Vietnam	
67-17611	UH1H	1\104ACS	
67-17612	UH1H	to Vietnam	
67-17613	UH1H	w.o.	
67-17614	UH1H	w.o.	
67-17615	UH1H	SoC	
67-17616	UH1H	w.o.	
67-17617	UH1H	to Iran	
67-17618	UH1H	w.o.	
67-17619	UH1H	w.o.4.8.71	
67-17620	UH1H	to Vietnam	
67-17621	UH1H	w.o.	
67-17622	UH1H	w.o.	
67-17623	UH1H	62AvCo	
67-17624	UH1H	166ACR	
67-17625	UH1H	w.o.12.4.70	
67-17626	UH1H	w.o.27.4.70	
67-17627	UH1H	to Vietnam	
67-17628	UH1H	w.o.10.4.71	
67-17629	UH1H	104ACS	
67-17630	UH1H	w.o.	
67-17631	UH1H	w.o.	
67-17632	UH1H	w.o.26.7.70	
67-17633	UH1H	120AvCo	
67-17634	UH1H	w.o.	
67-17635	UH1H	to Vietnam	
67-17636	UH1H	w.o.	
67-17637	UH1H	SoC	
67-17638	UH1H	w.o.	
67-17639	UH1H	w.o.	
67-17640	UH1H	w.o.	
67-17641	UH1H	w.o.	
67-17642	UH1H	w.o.	
67-17643	UH1H	w.o.	
67-17644	UH1H	w.o.	
67-17645	UH1V	273Medco	
67-17646	UH1H	w.o.	
67-17647	UH1H	w.o.	
67-17648	UH1H	w.o.	
67-17649	UH1H	w.o.	
67-17650	UH1H	24MedDet	
67-17651	UH1H	to Vietnam	
67-17652	UH1H	D\135AVN	KS NG
67-17653	UH1H		
67-17654	UH1H	w.o.	
67-17655	UH1H		
67-17656	UH1H	w.o.	
67-17657	UH1H	to Korea	
67-17658	UH1H	107AvCo	
67-17659	UH1V	455\457MedDet	MO NG
67-17660	UH1H	w.o.	
67-17661	UH1V	126MedCo	CA NG
67-17662	UH1H	to Thailand	
67-17663	UH1H	w.o.	
67-17664	UH1H	to Vietnam	
67-17665	UH1H	w.o.18.5.70	
67-17666	UH1H		
67-17667	UH1H	to Thailand	
67-17668	UH1H	88ARCOM	
67-17669	UH1H		
67-17670	UH1H	62AvCo	
67-17671	UH1H	w.o.	
67-17672	UH1H	to Korea	
67-17673	UH1H	w.o	
67-17674	UH1H	to Vietnam	
67-17675	UH1H	to Vietnam	
67-17676	UH1H	w.o.	
67-17677	UH1H	w.o.	
67-17678	GUH1H		
67-17679	UH1H		
67-17680	UH1H	w.o.	
67-17681	UH1H	D\1-245AVN	OK NG
67-17682	UH1H	w.o.	
67-17683	UH1H	w.o.11.6.70	
67-17684	UH1H		
67-17685	UH1H	w.o.	
67-17686	UH1H	140AVN	CA NG
67-17687	UH1H	to Vietnam	
67-17688	UH1H	to Vietnam	
67-17689	UH1H	to Vietnam	
67-17690	UH1H	308AHB	
67-17691	JUH1H	AMARC	XA...
67-17692	UH1H	w.o.	
67-17693	UH1H	w.o.	
67-17094	UH1H	w.o.	
67-17695	UH1H	308AHB	
67-17696	UH1H	w.o.	
67-17697	UH1H	w.o.	
67-17698	UH1H	SoC	
67-17699	UH1H	w.o.	
67-17700	UH1H	to Vietnam	
67-17701	UH1V		
67-17702	UH1H	w.o.	
67-17703	UH1H	SC NG	
67-17704	UH1H		
67-17705	UH1H	14AvCo	
67-17706	UH1H	w.o.17.4.70	
67-17707	UH1H	6ACR	
67-17708	UH1H	w.o.	
67-17709	UH1H	to Vietnam	
67-17710	UH1H	w.o.	
67-17711	UH1H	A\1-244AVN	LA NG
67-17712	UH1H	SoC	
67-17713	UH1H	w.o.	
67-17714	UH1H	to Vietnam	
67-17715	UH1H	w.o.	
67-17716	UH1H	to Vietnam	
67-17717	UH1H	w.o.29.7.74	
67-17718	UH1H	to Vietnam	
67-17719	UH1H	SoC	
67-17720	UH1H	w.o.	
67-17721	UH1H	w.o.	
67-17722	UH1H	w.o.	
67-17723	UH1H	to Vietnam	
67-17724	UH1H	SoC	
67-17725	UH1H		
67-17726	UH1H	w.o.14.2.70	
67-17727	UH1H	w.o.	
67-17728	UH1H	150AVN	
67-17729	UH1H	w.o.	
67-17730	UH1H	to Vietnam	
67-17731	UH1H	to Vietnam	
67-17732	UH1H	SoC	
67-17733	UH1H		
67-17734	UH1H	w.o.18.6.70	
67-17735	UH1H	11ACR	
67-17736	UH1H		
67-17737	UH1H	w.o.	
67-17738	UH1H		
67-17739	UH1H	w.o.	
67-17740	UH1H	w.o.	
67-17741	UH1H	to Vietnam	
67-17742	UH1H	to Vietnam	
67-17743	UH1H	163ACR	
67-17744	UH1H	w.o.16.7.70	
67-17745	UH1H	to Vietnam	
67-17746	UH1H	w.o.	
67-17747	UH1H	w.o.	
67-17748	UH1H	to Vietnam	
67-17749	UH1H	308AHB	
67-17750	UH1H	w.o.30.8.70	
67-17751	UH1H		

Serial	Type	Notes	
67-17752	UH1H	to Vietnam	
67-17753	UH1H	w.o.	
67-17754	UH1H	w.o.	
67-17755	JUH1H	USAICS	
67-17756	UH1H	3AVN	
67-17757	UH1H	A\3-158AVN Reserve	
67-17758	UH1H	w.o.	
67-17759	UH1H	to Vietnam	
67-17760	UH1H	w.o.	
67-17761	UH1H	to Vietnam	
67-17762	UH1H	193AVN	
67-17763	UH1H	SoC	
67-17764	UH1H	SoC	
67-17765	UH1H	w.o.26.10.70	
67-17766	UH1H		
67-17767	UH1H	w.o.14.5.70	
67-17768	UH1H	w.o.	
67-17769	UH1H	w.o.	
67-17770	UH1H		
67-17771	UH1H	to Vietnam	
67-17772	UH1H	1-158AVN	
67-17773	UH1H	to Vietnam	
67-17774	UH1H	w.o.	
67-17775	UH1H	SoC	
67-17776	UH1H	w.o.	
67-17777	UH1H	SoC	
67-17778	UH1H		
67-17779	UH1H		
67-17780	UH1H	56AvCo	
67-17781	UH1H	SoC	
67-17782	UH1H	w.o.	
67-17783	UH1V		
67-17784	UH1H	w.o.	
67-17785	UH1H	w.o.14.3.70	
67-17786	UH1H	w.o.	
67-17787	UH1H	to Vietnam	
67-17788	UH1H		
67-17789	UH1H	to Vietnam	
67-17790	UH1H	w.o.	
67-17791	UH1H	w.o.	
67-17792	UH1H		
67-17793	UH1H	to Vietnam	
67-17794	UH1H	to Vietnam	
67-17795	UH1H		
67-17796	UH1H	to Vietnam	
67-17797	UH1H	w.o.	
67-17798	UH1H		
67-17799	UH1H	w.o.	
67-17800	UH1H	w.o.	
67-17801	UH1H	w.o.	
67-17802	UH1V		
67-17803	UH1H	w.o.	
67-17804	UH1H	308AHB	
67-17805	UH1H		
67-17806	UH1H	to Vietnam	
67-17807	UH1H	w.o.	
67-17808	UH1H	244AvCo	
67-17809	UH1H	SoC	
67-17810	UH1V		
67-17811	UH1H	to Vietnam	
67-17812	UH1H	to Panama	
67-17813	UH1V	412MedDet	
67-17814	UH1H	SoC	
67-17815	UH1H	to Vietnam	
67-17816	UH1H	to Vietnam	
67-17817	UH1H	w.o.	
67-17818	UH1H	w.o.	
67-17819	UH1H	to Vietnam	
67-17820	UH1H	w.o.17.3.77	
67-17821	UH1H	w.o.	
67-17822	UH1H	to Thailand	
67-17823	UH1H	w.o.	
67-17824	UH1H	w.o.	
67-17825	UH1H	SoC	
67-17826	UH1H	to Vietnam	
67-17827	UH1H	w.o.	
67-17828	UH1H	SoC	
67-17829	UH1H	to Vietnam	
67-17830	UH1H	to Vietnam	
67-17831	UH1H	AATC	31. Cairns AAF
67-17832	UH1H	44AVN	
67-17833	UH1H	6ACR	
67-17834	UH1H	w.o.	
67-17835	UH1H	79ARCOM	
67-17836	UH1H	w.o.	
67-17837	UH1H	w.o.	
67-17838	UH1H	w.o.	
67-17839	UH1H	to Vietnam	
67-17840	UH1H	to Vietnam	
67-17841	UH1H	w.o.	
67-17842	UH1H	42AVN	
67-17843	UH1H	to Vietnam	
67-17844	UH1H	w.o.	
67-17845	UH1H	to Vietnam	
67-17846	UH1H	278ACR	
67-17847	UH1H	to Vietnam	
67-17848	UH1H		
67-17849	UH1H	B\26AVN	
67-17850	UH1H	to N8159Z	
67-17851	UH1H	to Vietnam	
67-17852	UH1H		
67-17853	UH1H	SoC	
67-17854	UH1H	11ACR	
67-17855	UH1H	SoC	
67-17856	UH1H	to Vietnam	
67-17857	UH1H	to Vietnam	
67-17858	UH1H	w.o.26.9.70	
67-17859	GUH1H	GIA	USATS
67-18564	UH1H	SoC	
67-18565	UH1H	w.o.	
67-18566	UH1H	to Vietnam	
67-18567	UH1H	to Vietnam	
67-18568	UH1H		
67-18569	UH1H	1159MedDet	
67-18570	UH1H	to Vietnam	
67-18571	UH1H	w.o.28.9.70	
67-18572	UH1H	w.o.	
67-18573	UH1H	w.o.	
67-18574	UH1H	w.o.	
67-18575	UH1H	w.o.	
67-18576	UH1H	w.o.	
67-18577	UH1H	24MedDet	
67-19484	UH1H	w.o.15.2.70	
67-19485	UH1H		
67-19486	UH1H	w.o.	
67-19487	UH1H	SoC	
67-19488	UH1H	w.o.	
67-19489	UH1H	w.o.	
67-19490	UH1H	to Vietnam	
67-19491	UH1H	308AHB	
67-19492	UH1V		
67-19493	UH1H	63ARCOM	
67-19494	UH1V	126MedCo	CA NG
67-19495	UH1H	SoC	
67-19496	UH1H	w.o.	
67-19497	UH1H	w.o.17.10.71	
67-19498	UH1H	w.o.	
67-19499	UH1H	USAFSA	
67-19500	UH1H	w.o.	
67-19501	UH1H	SoC	
67-19502	UH1H	120AvCo	
67-19503	UH1H	w.o.	
67-19504	UH1H	to Vietnam	
67-19511	UH1H	SoC	
67-19512	UH1H	w.o.	
67-19513	UH1H	w.o.	
67-19514	UH1H	w.o.	
67-19515	UH1H	w.o.	
67-19516	UH1H	w.o.23.2.71	
67-19517	UH1H	w.o.	
67-19518	UH1H	to Vietnam	
67-19519	UH1H		
67-19520	UH1H	w.o.	
67-19521	UH1V	126MedCo	CA NG
67-19522	UH1H		
67-19523	UH1H		
67-19524	UH1H	w.o.	
67-19525	UH1H	to Vietnam	
67-19526	UH1H	A\1-244AVN	LA NG
67-19527	UH1H	SoC	
67-19528	UH1H	to Vietnam	
67-19529	UH1H	w.o.	
67-19530	UH1H	w.o.	
67-19531	UH1H	w.o.	
67-19532	UH1H		
67-19533	UH1H	w.o.	

Serial	Type	Unit/Status	Notes		Serial	Type	Unit/Status	Notes
67-19534	UH1V	507MedCo			68-15301	UH1H	to Vietnam	
67-19535	UH1H	to Thailand			68-15302	UH1H	to Vietnam	
67-19536	UH1H	to Vietnam			68-15303	UH1H	w.o.3.3.70	
67-19537	UH1H	159AVN			68-15304	UH1H	w.o.	
68-15214	UH1V	w.o.17.5.88			68-15305	UH1H	to Vietnam	
68-15215	UH1H	w.o.			68-15306	UH1H	w.o.	
68-15216	UH1H	to Vietnam			68-15307	UH1H		
68-15217	UH1H	SoC			68-15308	UH1H	w.o.	
68-15218	UH1H	to Vietnam			68-15309	UH1H	w.o.1.3.70	
68-15219	UH1H	to Australia			68-15310	UH1H	w.o.	
68-15220	UH1H	w.o.3.1.70			68-15311	UH1H	w.o.	
68-15221	UH1H	w.o.16.5.70			68-15312	UH1H	w.o.	
68-15222	UH1H	C\1-108AVN	AZ NG		68-15313	UH1H		
68-15223	UH1H				68-15314	UH1H	to Vietnam	
68-15224	UH1H	SoC			68-15315	UH1H	to Vietnam	
68-15225	UH1H	w.o.			68-15316	UH1H	w.o.25.2.70	
68-15226	UH1H	51AvCo			68-15317	UH1H		
68-15227	UH1V	126MedCo	CA NG		68-15318	UH1H	to Vietnam	
68-15228	UH1H	w.o.13.4.75			68-15319	UH1H	308AHB	
68-15229	UH1H	w.o.			68-15320	UH1H	to Thailand	
68-15230	UH1H	to Thailand			68-15321	UH1H	to Vietnam	
68-15231	UH1H	to Greece			68-15322	UH1H	26AVN	
68-15232	UH1H	to Greece			68-15323	UH1H	to Vietnam	
68-15233	UH1H	to Greece			68-15324	UH1H	to Vietnam	
68-15234	UH1H	to Greece			68-15325	UH1H	w.o.	
68-15235	UH1H	to Greece			68-15326	UH1H	w.o.	
68-15236	UH1H	w.o.			68-15327	UH1H	1\104ACS	
68-15237	UH1H	w.o.			68-15328	UH1H	SoC	
68-15238	UH1H	SoC			68-15329	UH1H	to Vietnam	
68-15239	JUH1H	AAOD			68-15330	UH1H	SoC	
68-15240	UH1H				68-15331	UH1H	A\1-207AVN	AK NG
68-15241	UH1H	w.o.6.2.70			68-15332	UH1H	26AVN	
68-15242	UH1H	to Vietnam			68-15333	UH1H	w.o.18.10.70	
68-15243	UH1H	to Vietnam			68-15334	UH1H	92AvCo	
68-15244	UH1H	w.o.28.3.70			68-15337	UH1H		
68-15245	UH1H	to Vietnam			68-15338	UH1H	to Vietnam	
68-15246	UH1H	w.o.			68-15339	UH1H	w.o.	
68-15247	UH1H	to Vietnam			68-15340	UH1H	w.o.	
68-15248	JUH1H				68-15341	UH1H	w.o.	
68-15249	UH1H	w.o.2.10.70			68-15345	UH1H	w.o.	
68-15250	UH1H	w.o.			68-15346	UH1H	w.o.	
68-15251	UH1H	w.o.			68-15347	UH1H	w.o.21.8.73	
68-15252	UH1V	w.o.20.10.84			68-15348	UH1H	w.o.	
68-15253	UH1H				68-15349	UH1H	w.o.	
68-15254	UH1H	to N9424A			68-15350	UH1H	w.o.	
68-15255	UH1H	w.o.			68-15351	UH1H		
68-15256	UH1H	w.o.20.2.70			68-15352	UH1H	321MedCo	
68-15257	UH1H	to Vietnam			68-15353	UH1H	to Vietnam	
68-15258	UH1H	w.o.			68-15354	UH1V	507MedDet	
68-15259	UH1H	B\4-228AVN			68-15355	UH1H	to Vietnam	
68-15260	UH1H	w.o.			68-15356	UH1H	w.o.	
68-15261	UH1H	278ACR			68-15357	UH1H	to Vietnam	
68-15262	UH1H	w.o.			68-15358	UH1H	to Vietnam	
68-15263	UH1H	SoC			68-15359	UH1H	w.o.27.9.70	
68-15264	UH1H	w.o.			68-15360	UH1H	42AVN	
68-15270	UH1H	w.o.			68-15361	UH1H	441MedDet	
68-15271	UH1H	126MedCo	CA NG		68-15362	UH1H	42AVN	
68-15272	UH1H	to Vietnam			68-15363	UH1H	to Vietnam	
68-15273	UH1H	AATC	73N Lowe AHP		68-15364	UH1H	w.o.	
68-15274	UH1H	w.o.			68-15365	UH1H	w.o.2.2.78	
68-15275	UH1H				68-15366	UH1H	w.o.9.1.75	
68-15276	UH1H				68-15367	UH1H		
68-15277	UH1H	352ASA Co			68-15368	UH1H	to Vietnam	
68-15278	UH1H	w.o.10.10.85			68-15369	UH1H		
68-15279	UH1H	w.o.			68-15370	UH1H	w.o.	
68-15280	UH1H				68-15371	UH1H	w.o.	
68-15281	UH1H	SoC			68-15372	UH1H	7InfDiv	
68-15282	UH1H	w.o.			68-15373	UH1H	w.o.10.9.70	
68-15283	UH1H	150AVN	DE NG		68-15374	UH1H	1898AvCo	
68-15284	UH1H	stored	Ramstein AB,FRG		68-15375	UH1H		
68-15285	UH1H	63ARCOM			68-15376	UH1H	w.o.	
68-15286	UH1H	w.o.			68-15377	UH1H	w.o.23.4.71	
68-15287	GUH1H	GIA	USATS		68-15378	UH1H	SoC	
68-15288	UH1H	w.o.10.9.75			68-15379	UH1H	to Vietnam	
68-15289	UH1H	to Chile			68-15380	UH1H	SoC	
68-15290	UH1H	SoC			68-15381	UH1H	w.o.	
68-15294	UH1H	to Argentina			68-15382	UH1H		
68-15295	UH1H				68-15383	UH1H	w.o.25.2.91	
68-15296	UH1H	to Thailand			68-15384	UH1H	to Vietnam	
68-15297	UH1H				68-15385	UH1H	w.o.	
68-15298	UH1H	to Vietnam			68-15386	UH1H	w.o.	
68-15299	UH1H	w.o.21.7.70			68-15387	UH1H	w.o.9.6.88	
68-15300	UH1H	to Vietnam			68-15388	UH1H	to Vietnam	

Serial	Type	Notes	Location
68-15389	UH1H	to Vietnam	
68-15390	UH1H	AATC	90B Cairns AAF
68-15391	UH1H	w.o.	
68-15392	UH1H	w.o.	
68-15393	UH1H	w.o.26.3.70	
68-15394	UH1H	w.o.22.7.70	
68-15395	UH1H	SoC	
68-15396	UH1H	w.o.	
68-15397	UH1H	w.o.	
68-15398	UH1H	to Vietnam	
68-15399	UH1H	w.o.	
68-15400	UH1H	104ACS	
68-15401	UH1H	w.o.	
68-15402	UH1H		SD NG
68-15403	UH1H	w.o.	
68-15404	UH1H	w.o.1.4.70	
68-15405	UH1H		
68-15406	UH1H	w.o.	
68-15407	UH1H	to Vietnam	
68-15408	UH1H	w.o.	
68-15409	UH1H	A\1-207AVN	AK NG
68-15410	UH1H	w.o.9.8.70	
68-15411	UH1H	AATC	11A Lowe AHP
68-15412	UH1H	w.o.	
68-15413	UH1H	to Thailand	
68-15414	UH1H	SoC	
68-15419	UH1H		
68-15420	UH1H	SoC	
68-15421	UH1H		
68-15422	UH1H	25AvCo	
68-15423	UH1H	to Korea	
68-15424	UH1H	to Vietnam	
68-15425	UH1H		
68-15426	UH1H	w.o.27.4.72	
68-15427	UH1H	w.o.	
68-15428	UH1H	to Vietnam	
68-15429	UH1H	to Vietnam	
68-15430	UH1H	to Thailand	
68-15431	UH1V	126MedCo	CA NG
68-15432	UH1H		
68-15433	UH1H	126MedCo	CA NG
68-15434	UH1H	5-158AVN	
68-15435	UH1H	w.o.2.10.70	
68-15436	UH1H	to Vietnam	
68-15437	UH1H		
68-15438	UH1H		
68-15439	UH1H	SoC	
68-15440	UH1H	SoC	
68-15441	UH1H	w.o.15.3.71	
68-15442	UH1H	to Thailand	
68-15443	UH1H	w.o.	
68-15444	UH1V		
68-15445	UH1H	to Thailand	
68-15446	UH1H	193AVN	
68-15447	UH1H	w.o.	
68-15448	UH1H	w.o.30.6.74	
68-15449	UH1H	w.o.	
68-15450	UH1H	24MedDet	
68-15451	UH1H		SD NG
68-15452	UH1H	D\135AVN	KS NG
68-15453	UH1H	to Vietnam	
68-15454	UH1H	to Vietnam	
68-15455	UH1H	to Vietnam	
68-15456	UH1H	to Vietnam	
68-15457	UH1H	to Vietnam	
68-15458	UH1H		
68-15459	UH1H	w.o.	
68-15460	UH1H	SoC	
68-15461	UH1V		
68-15462	UH1H	w.o.	
68-15463	UH1H		
68-15464	UH1H	SoC	
68-15465	UH1H	to Vietnam	
68-15466	UH1H	w.o.	
68-15467	UH1H	to Vietnam	
68-15468	UH1H	to Vietnam	
68-15469	UH1H	SoC	
68-15470	UH1H	1-158AVN	
68-15471	UH1H	w.o.	
68-15472	UH1H	stored	Ramstein AB,FRG
68-15473	UH1H	to Vietnam	
68-15474	UH1H		
68-15475	UH1H	SoC	
68-15476	UH1H	D\135AVN	KS NG
68-15477	UH1H	41InfBde	
68-15478	UH1H	1AvBde	
68-15479	UH1H	to Vietnam	
68-15480	JUH1H		
68-15481	UH1H	1-183AVN	ID NG
68-15482	UH1H	1-111AVN	FL NG
68-15483	UH1H	w.o.	
68-15484	UH1H		
68-15485	UH1H	w.o.	
68-15486	UH1H	to Vietnam	
68-15487	UH1H	to Vietnam	
68-15488	UH1H	w.o.	
68-15489	UH1H		
68-15490	UH1H	SoC	
68-15491	UH1H	w.o.	
68-15492	UH1V		
68-15493	UH1H		
68-15494	UH1H	w.o.10.3.77	
68-15495	UH1H	SoC	
68-15496	UH1H	w.o.21.9.72	
68-15497	UH1H	dump Van Nuys, Ca	
68-15498	UH1H	w.o.13.1.70	
68-15503	UH1H	to Vietnam	
68-15504	UH1H	to Vietnam	
68-15505	UH1H	to Vietnam	
68-15506	UH1H	to Vietnam	
68-15507	UH1H	to Vietnam	
68-15508	UH1H	to Vietnam	
68-15509	UH1H	to Vietnam	
68-15510	UH1H	to Vietnam	
68-15511	UH1H	to Vietnam	
68-15512	UH1H	to Vietnam	
68-15513	UH1H	to Vietnam	
68-15514	UH1H	to Vietnam	
68-15515	UH1H	to Vietnam	
68-15516	UH1H	to Vietnam	
68-15517	UH1H	to Vietnam	
68-15518	UH1H	to Vietnam	
68-15519	UH1H	to Vietnam	
68-15520	UH1H	B\4-228AVN	
68-15521	UH1H		
68-15522	UH1H	to Vietnam	
68-15523	UH1H	A\1-131AVN	AL NG
68-15524	UH1H	w.o.	
68-15525	UH1H	w.o.	
68-15526	UH1V	62AvCo	
68-15527	UH1H	to Vietnam	
68-15528	UH1H	to Vietnam	
68-15529	UH1V	81ARCOM	
68-15530	UH1H		
68-15531	UH1H	w.o.	
68-15532	UH1V	26AVN	
68-15533	UH1H	158AVN	
68-15534	UH1H	w.o.	
68-15535	UH1H	to Vietnam	
68-15536	UH1H	w.o.	
68-15537	UH1H	SoC	
68-15538	UH1H	to Vietnam	
68-15539	UH1H	to Vietnam	
68-15540	UH1H	w.o.8.3.70	
68-15541	UH1H	w.o.	
68-15542	UH1H	to Vietnam	
68-15543	UH1V	5-158AVN	
68-15544	UH1H	to Vietnam	
68-15545	UH1H	to Vietnam	
68-15546	UH1H	to Vietnam	
68-15547	UH1H	to Thailand	
68-15548	UH1H	to Vietnam	
68-15549	UH1H	w.o.	
68-15550	UH1H		MO NG
68-15551	UH1H	w.o.25.10.71	
68-15552	UH1H		
68-15553	UH1H	to Vietnam	
68-15554	UH1H		
68-15555	UH1H	to Thailand	
68-15556	UH1H		
68-15557	UH1H	w.o.2.6.71	
68-15558	UH1H	w.o.	
68-15559	UH1H	w.o.	
68-15560	UH1H	to Vietnam	
68-15561	UH1H	79ARCOM	
68-15562	UH1H	stored	Ramstein AB,FRG

68-15563	UH1H	w.o.	
68-15564	UH1H	w.o.	
68-15565	UH1H	w.o.	
68-15566	UH1H		
68-15567	UH1H	w.o.	
68-15568	UH1H	w.o.4.4.71	
68-15569	UH1H	w.o.	
68-15570	UH1H	25AvCo	
68-15571	UH1H	w.o.3.3.70	
68-15572	UH1H	w.o.	
68-15573	UH1H	w.o.7.11.80	
68-15574	UH1H		
68-15575	UH1H	to Vietnam	
68-15576	UH1H	to Vietnam	
68-15577	UH1H	to Thailand	
68-15578	UH1H	1-158AVN	
68-15579	UH1H	to Thailand	
68-15580	UH1H	w.o.	
68-15581	UH1H	w.o.	
68-15582	UH1H	to Vietnam	
68-15583	UH1H	to Vietnam	
68-15584	UH1H		
68-15585	UH1H	w.o.	
68-15586	UH1H		
68-15587	UH1H	w.o.	
68-15588	UH1H		
68-15589	UH1H	SoC	
00-15590	UH1H	w.o.11.3.89	
68-15591	UH1H	to Vietnam	
68-15592	UH1V	112MedDet	
68-15593	UH1V	157MedDet	
68-15594	UH1V		
68-15595	UH1H	8AVN	
68-15601	UH1H	to Vietnam	
68-15602	UH1H	to Vietnam	
68-15603	UH1H	to Vietnam	
68-15604	UH1H	to Vietnam	
68-15605	UH1H	to Vietnam	
68-15606	UH1H	to Vietnam	
68-15607	UH1H	to Vietnam	
68-15608	UH1H	to Vietnam	
68-15609	UH1H	to Vietnam	
68-15610	UH1H	to Vietnam	
68-15611	UH1H	to Vietnam	
68-15612	UH1H	to Vietnam	
68-15613	UH1H	to Vietnam	
68-15614	UH1H	to Vietnam	
68-15615	UH1H	to Vietnam	
68-15616	UH1H	to Vietnam	
68-15617	UH1H	to Vietnam	
68-15618	UH1H	to Vietnam	
68-15619	UH1H	to Vietnam	
68-15620	UH1H	w.o.2.5.70	
68-15621	UH1H	to Vietnam	
68-15622	UH1V		
68-15623	UH1H	w.o.	
68-15624	UH1H	to Korea	
68-15625	UH1H	SoC	
68-15626	UH1H	150AVN	
68-15627	UH1H	w.o.20.9.70	
68-15628	UH1H	w.o.	
68-15629	UH1H	w.o.	
68-15630	UH1H		
68-15631	UH1H	w.o.	
68-15632	UH1H	C\1-111AVN FL NG	
68-15633	UH1H	w.o.	
68-15634	UH1H	to Vietnam	
68-15635	UH1H		
68-15636	UH1H	to Vietnam	
68-15637	UH1H	308AHB	
68-15638	UH1H	79ARCOM	
68-15639	UH1H	w.o.	
68-15640	UH1H	SoC	
68-15641	UH1H	to Cambodia	
68-15642	UH1H	to Vietnam	
68-15643	UH1H	to Vietnam	
68-15644	UH1H	w.o.28.9.73	
68-15645	UH1H	9AVN	
68-15646	UH1H	w.o.4.10.83	
68-15647	UH1H	to Vietnam	
68-15648	UH1H	w.o.26.9.70	
68-15649	UH1H	w.o.14.5.91	
68-15650	UH1H	25AvCo	
68-15651	UH1H	to Vietnam	
68-15652	UH1H		
68-15653	UH1H		
68-15654	UH1H	w.o.1.9.70	
68-15655	UH1H		
68-15656	UH1H		
68-15657	UH1H	Berlin Brigade	
68-15658	UH1H	SoC	
68-15659	UH1H	145MedDet	
68-15660	UH1H	to Vietnam	
68-15661	UH1H	w.o.	
68-15662	UH1H	to Vietnam	
68-15663	UH1H	w.o.	
68-15664	UH1H	w.o.3.3.71	
68-15665	UH1H	w.o.8.9.72	
68-15666	UH1H	w.o.	
68-15667	UH1H		
68-15668	UH1H	to Vietnam	
68-15669	UH1H	6ACR	
68-15670	UH1H	w.o.	
68-15671	UH1H	w.o.	
68-15672	UH1H	w.o.14.1.71	
68-15673	UH1H	to Iran	
68-15674	UH1H	to Vietnam	
68-15675	UH1H	to Vietnam	
68-15676	UH1H	to Thailand	
68-15677	UH1H	166ACR	
68-15678	UH1V		
68-15679	UH1H		
68-15680	UH1H	w.o.	
68-15681	UH1H	w.o.1.3.70	
68-15682	UH1H	to Vietnam	
68-15683	UH1H	131AVN	
68-15684	UH1H	w.o.	
68-15685	UH1H	w.o.1.1.70	
68-15686	UH1H	w.o.7.1.70	
68-15687	UH1H	SoC	
68-15688	UH1V		
68-15689	UH1H	to Vietnam	
68-15690	UH1H	w.o.	
68-15691	UH1H		IL NG
68-15692	UH1H	w.o.31.7.71	
68-15693	UH1H	1-158AVN	
68-15694	UH1H	to Thailand	
68-15701	UH1H	to Vietnam	
68-15702	UH1H	308AHB	
68-15703	UH1H	to Vietnam	
68-15704	UH1H	to Vietnam	
68-15705	UH1H	to Vietnam	
68-15706	UH1H	to Vietnam	
68-15707	UH1H	to Vietnam	
68-15708	UH1H	to Vietnam	
68-15709	UH1H	to Vietnam	
68-15710	UH1H	to Vietnam	
68-15711	UH1H	to Vietnam	
68-15712	UH1H	to Vietnam	
68-15713	UH1H	to Vietnam	
68-15714	UH1H	to Vietnam	
68-15715	UH1H	to Vietnam	
68-15716	UH1H	to Vietnam	
68-15717	UH1H	to Vietnam	
68-15718	UH1H	to Vietnam	
68-15719	UH1H	to Thailand	
68-15720	UH1H	to Iran	
68-15721	UH1H	w.o.9.8.70	
68-15722	UH1H		
68-15723	UH1H	w.o.	
68-15724	UH1H	120AvCo	
68-15725	UH1H		
68-15726	UH1H	to Vietnam	
68-15727	UH1H	6ACR	
68-15728	UH1H	w.o.	
68-15729	UH1H	to Thailand	
68-15730	UH1H	to Thailand	
68-15731	UH1H		
68-15732	UH1H	to Vietnam	
68-15733	UH1H		
68-15734	UH1H	308AHB	
68-15735	UH1H	w.o.	
68-15736	UH1H	w.o.3.5.72	
68-15737	UH1H	1-149AVN	TX NG
68-15738	UH1H	to Thailand	
68-15739	UH1H	w.o.15.3.75	

Serial	Type	Unit/Status	Notes
68-15740	UH1H	w.o.	
68-15741	UH1H	w.o.	
68-15742	UH1H	w.o.	
68-15743	UH1V	236MedDet	
68-15744	UH1H	w.o.28.8.80	
68-15745	UH1H		
68-15746	UH1H	to Iran	
68-15747	UH1H		
68-15748	UH1H	SoC	
68-15749	UH1H		
68-15750	UH1H	to Vietnam	
68-15751	UH1H	to Thailand	
68-15752	UH1H	to Vietnam	
68-15753	UH1H	w.o.	
68-15754	UH1H	1133MedDet	
68-15755	UH1H	56AvCo	
68-15756	UH1H	to Vietnam	
68-15757	UH1H	w.o.	
68-15758	UH1H	to Vietnam	
68-15759	UH1H	w.o.	
68-15760	UH1H	C\1-108AVN AZ NG 37	
68-15761	UH1H	w.o.	
68-15762	UH1V		
68-15763	UH1H	w.o.	
68-15764	UH1H	to Vietnam	
68-15765	UH1H	to Thailand	
68-15766	UH1H	w.o.	
68-15767	UH1H	w.o.	
68-15768	UH1H	207AvCo	
68-15769	UH1H		
68-15770	UH1H	to Vietnam	
68-15771	UH1H	w.o.	
68-15772	UH1H	w.o.15.8.70	
68-15773	UH1H	A\1-244AVN LA NG	
68-15774	UH1H	69ADA Group	
68-15775	UH1H		
68-15776	UH1H	w.o.	
68-15777	UH1H	to Thailand	
68-15778	UH1H	to Vietnam	
68-15784	UH1H		
68-15785	UH1H		
68-15794	UH1H	SoC	
68-16050	UH1H	to Vietnam	
68-16051	UH1H	SoC	
68-16052	UH1H	to Vietnam	
68-16053	UH1H	w.o.	
68-16054	UH1H		
68-16055	UH1H	4InfDiv	
68-16056	UH1H	A\1-131AVN AL NG	
68-16057	UH1H	w.o.	
68-16058	UH1H	w.o.	
68-16059	UH1H	SoC	
68-16060	UH1H	to Vietnam	
68-16061	UH1H		
68-16062	UH1H	SoC	
68-16063	UH1H	w.o.	
68-16064	UH1H	w.o.28.9.70	
68-16065	UH1H	w.o.25.1.71	
68-16071	UH1H	SoC	
68-16072	UH1H	56AvCo	
68-16073	UH1H		
68-16074	UH1H	to Vietnam	
68-16075	UH1H	w.o.3.6.70	
68-16076	UH1H	w.o.	
68-16077	UH1H		
68-16078	UH1H		
68-16079	UH1H	AATC	79. Cairns AAF
68-16080	UH1H	to Vietnam	
68-16081	UH1H	w.o.24.7.70	
68-16082	UH1H		
68-16083	UH1H	120AvCo	
68-16084	UH1H	1150MedCo	
68-16085	UH1H	to Vietnam	
68-16086	UH1H	w.o.	
68-16087	UH1H		
68-16088	UH1H		
68-16089	UH1H		
68-16090	UH1H	to Vietnam	
68-16091	UH1H	w.o.	
68-16092	UH1H	6ACR	
68-16093	UH1H	w.o.	
68-16094	UH1H	SoC	
68-16095	UH1H	SoC	
68-16096	UH1H		
68-16097	UH1H	AATC	97. Cairns AAF
68-16098	UH1H	to Vietnam	
68-16099	UH1H	w.o.31.7.70	
68-16100	UH1H	SoC	
68-16101	UH1H	244AvCo	
68-16102	UH1H	to Thailand	
68-16103	UH1H	to Vietnam	
68-16104	UH1H	1-1131AVN AL ANG	
68-16114	UH1H	62AvCo	
68-16115	UH1H	A\3-158AVN Reserve	
68-16116	UH1H		
68-16117	UH1H	to Vietnam	
68-16118	UH1H	498MedCo	
68-16119	UH1H	to Vietnam	
68-16120	UH1H	w.o.	
68-16121	UH1H	to Vietnam	
68-16122	UH1H	to Vietnam	
68-16123	UH1H	w.o.26.9.70	
68-16124	UH1H	56AvCo	
68-16125	UH1H	SoC	
68-16126	UH1H	to Vietnam	
68-16127	UH1H	to Vietnam	
68-16128	UH1H	A\1-207AVN AK NG	
68-16129	UH1H	to Vietnam	
68-16130	UH1H	w.o.	
68-16131	UH1V	53InfBde	
68-16132	UH1H		
68-16133	UH1H	to Vietnam	
68-16134	GUH1H	GIA	USATS
68-16135	UH1H	90ARCOM	
68-16136	UH1H	90ARCOM	
68-16137	UH1H	to Vietnam	
68-16138	UH1V	347MedDet	
68-16139	UH1H	to Thailand	
68-16140	UH1H		
68-16141	UH1H	w.o.3.6.70	
68-16142	UH1H	SoC	
68-16143	UH1H	to Vietnam	
68-16144	UH1H	to N16144	
68-16145	UH1H	A\3-158AVN Reserve	
68-16146	UH1H	to Vietnam	
68-16147	UH1H	to Thailand	
68-16148	UH1H	to Vietnam	
68-16149	UH1H	to Vietnam	
68-16150	UH1H	to Vietnam	
68-16151	UH1H	166ACR	
68-16152	UH1H		
68-16153	UH1H	w.o.	
68-16154	UH1H	40AVN	
68-16155	UH1H	to Vietnam	
68-16156	JUH1H	278ACR	
68-16157	UH1H	w.o.	
68-16158	UH1H	to Vietnam	
68-16159	UH1H	to Vietnam	
68-16160	UH1H	w.o.	
68-16161	UH1H	79ARCOM	
68-16162	UH1H	w.o.	
68-16163	UH1H	to Vietnam	
68-16164	UH1H	SoC	
68-16165	UH1V		
68-16170	UH1H	w.o.	
68-16171	UH1H		145MedDet
68-16172	UH1H	w.o.23.11.70	
68-16173	UH1H	to Vietnam	
68-16174	UH1V	498MedDet	
68-16175	UH1H	to Vietnam	
68-16176	UH1H	w.o.	
68-16177	UH1H	w.o.	
68-16178	UH1H		
68-16179	UH1H		
68-16180	UH1H		872MedDet
68-16186	UH1H	to Vietnam	
68-16187	UH1H	1-207AVN	
68-16188	UH1H	to Vietnam	
68-16189	UH1H	1-149AVN TX NG	
68-16190	UH1H	w.o.26.7.70	
68-16191	UH1H	10AHB	
68-16192	UH1H	to Vietnam	
68-16193	UH1H	62AvCo	
68-16194	UH1H	w.o.28.11.71	
68-16195	UH1H	166ACR	
68-16196	UH1H	w.o.	

Serial	Type	Notes	Location
68-16197	UH1H		
68-16198	UH1H	to Korea	
68-16199	UH1H	w.o.	
68-16200	UH1H	w.o.9.5.74	
68-16201	UH1H	SoC	
68-16202	UH1H	120AvCo	
68-16203	UH1H	w.o.	
68-16204	UH1H	to Iran	
68-16205	UH1H	w.o.	
68-16206	UH1H	w.o.	
68-16207	UH1H	to Vietnam	
68-16208	UH1H	to Cambodia	
68-16209	UH1H	w.o.	
68-16210	UH1H		
68-16211	UH1H	1-158AVN	
68-16212	UH1H	w.o.7.7.70	
68-16213	UH1H	to Thailand	
68-16214	UH1H	to Vietnam	
68-16215	UH1H	193AVN	
68-16216	UH1H	w.o.4.5.83	
68-16217	UH1H	to Iran	
68-16218	UH1H	to Vietnam	
68-16219	UH1H	to Vietnam	
68-16220	UH1H	w.o.	
68-16221	UH1H	w.o.4.8.78	
68-16222	UH1H	to Vietnam	
68-16223	UH1H	w.o.2.7.81	
68-16224	UI1H		
68-16225	UH1H	w.o.14.5.70	
68-16226	UH1H	SoC	
68-16227	UH1H	w.o.29.1.72	
68-16228	JUH1H		
68-16229	UH1V		
68-16230	UH1H	to Vietnam	
68-16231	UH1H	to Vietnam	
68-16232	UH1H	w.o.	
68-16233	UH1H	to Vietnam	
68-16234	UH1H	w.o.	
68-16235	UH1V		
68-16236	UH1H	w.o.24.4.70	
68-16237	UH1H	AATC	37M Cairns AAF
68-16238	UH1H	w.o.	
68-16239	UH1H	w.o.	
68-16240	UH1H	w.o.28.5.70	
68-16241	UH1H	w.o.	
68-16242	UH1H	to Vietnam	
68-16243	UH1H	w.o.14.10.70	
68-16244	UH1H	w.o.4.5.70	
68-16245	UH1H		
68-16246	UH1H	w.o.14.3.70	
68-16247	UH1H		
68-16248	UH1H	to Vietnam	
68-16249	UH1H	w.o.	
68-16250	UH1H		
68-16251	UH1H		
68-16252	UH1H	w.o.20.10.84	
68-16253	UH1H	to Thailand	
68-16254	UH1H	1-159avn	
68-16255	UH1H	to Vietnam	
68-16256	UH1H		
68-16257	UH1H	w.o.	
68-16258	UH1H	to Thailand	
68-16259	UH1H	SoC	
68-16260	UH1H	to Vietnam	
68-16261	UH1H		
68-16262	UH1H	w.o.17.3.70	
68-16263	UH1V	112MedDet	
68-16264	UH1H	308AHB	
68-16265	UH1H	86ARCOM	
68-16270	UH1H	to Vietnam	
68-16271	UH1H	26AVN	
68-16272	UH1H	A\1-131AVN	AL NG
68-16273	UH1H		
68-16274	UH1H	w.o.	
68-16275	UH1H		
68-16276	UH1H	w.o.	
68-16277	UH1H	to Vietnam	
68-16278	UH1H	to Vietnam	
68-16279	UH1H	SoC	
68-16280	UH1H		
68-16281	UH1H	321MedDet	
68-16282	UH1H	166ACR	
68-16283	UH1H		
68-16284	UH1H	w.o.	
68-16285	UH1H	w.o.	
68-16286	UH1H	w.o.	
68-16287	UH1H	to Vietnam	
68-16288	UH1H	w.o.	
68-16289	UH1H	120AVN	
68-16290	UH1H	to Vietnam	
68-16291	UH1H	to Vietnam	
68-16292	UH1H	SoC	
68-16293	UH1H	SoC	
68-16294	UH1H	w.o.	
68-16295	UH1H	123ARCOM	
68-16296	UH1H		
68-16297	UH1H	to Vietnam	
68-16298	UH1H	w.o.	
68-16299	UH1H	w.o.	
68-16300	UH1H	w.o.	
68-16301	UH1V	157MedDet	
68-16302	UH1H	w.o.4.8.70	
68-16303	UH1H	w.o.	
68-16304	UH1H		
68-16305	UH1H	104ACS	
68-16306	UH1H	to Vietnam	
68-16307	UH1H	w.o.	
68-16308	UH1H	to Thailand	
68-16309	UH1H	SoC	
68 16310	UH1H	w.o.26.8.70	
68-16311	UH1H	w.o.19.8.82	
68-16312	UH1H	to Vietnam	
68-16313	UH1H	w.o.	
68-16314	UH1H	to Vietnam	
68-16315	UH1H	SoC	
68-16316	UH1H	w.o.	
68-16317	UH1H	to Vietnam	
68-16318	UH1H		
68-16319	UH1H	w.o.13.4.70	
68-16320	UH1H	to Vietnam	
68-16321	UH1V	1159MedCo	
68-16322	UH1H	26AVN	
68-16323	UH1H	w.o.	
68-16324	UH1H	A\1-207AVN	AK NG
68-16325	UH1H	to Vietnam	
68-16326	UH1H	3ACR	
68-16327	UH1H	w.o.	
68-16328	UH1H	w.o.	
68-16329	UH1H		
68-16330	UH1H	w.o.3.4.72	
68-16331	UH1H	to Vietnam	
68-16332	UH1H		
68-16333	UH1H	1-158AVN	
68-16334	UH1H	to Vietnam	
68-16335	UH1H	to Vietnam	
68-16336	UH1H	to Vietnam	
68-16337	UH1H	to Vietnam	
68-16338	UH1H	178AvCo	
68-16339	UH1H	to N58RF	
68-16340	UH1H	AATC	40B Lowe AHP
68-16341	UH1H		
68-16342	UH1H	w.o.	
68-16343	UH1H	to Vietnam	
68-16344	UH1H	w.o.	
68-16345	UH1H	w.o.	
68-16346	UH1H	w.o.20.2.80	
68-16347	UH1H	w.o.	
68-16348	UH1H	stored	Ramstein AB,FRG
68-16349	UH1V	131AVN	
68-16350	UH1H	w.o.	
68-16351	UH1H	to Vietnam	
68-16352	UH1H	to Vietnam	
68-16353	UH1H	to Vietnam	
68-16354	UH1H	w.o.	
68-16355	UH1V	1159MedCo	
68-16356	UH1H	to Vietnam	
68-16357	UH1H	to Vietnam	
68-16358	UH1H	derelict	NAS Dallas
68-16359	UH1H	w.o.	
68-16360	UH1H	w.o.13.2.71	
68-16361	UH1H		
68-16362	UH1H	w.o.	
68-16363	UH1H	to Vietnam	
68-16364	UH1H	to Vietnam	
68-16365	UH1H	w.o.10.7.70	
68-16366	UH1H	to Vietnam	

68-16367	UH1H	w.o.	
68-16368	UH1H	w.o.	
68-16373	UH1H	to Philippine	
68-16374	UH1H	to Vietnam	
68-16375	UH1H	w.o.	
68-16376	UH1H	AATC	76C Cairns AAF
68-16377	UH1H	to Vietnam	
68-16378	UH1H	A\1-131AVN	AL NG
68-16379	UH1H	to Vietnam	
68-16380	UH1H	to Vietnam	
68-16381	UH1H	to Thailand	
68-16382	UH1H	w.o.	
68-16383	UH1H		
68-16384	UH1H		
68-16385	UH1H	w.o.	
68-16386	UH1H	w.o.	
68-16387	UH1H	w.o.	
68-16388	UH1H	to Vietnam	
68-16389	UH1H	w.o.	
68-16390	UH1H	to Vietnam	
68-16391	UH1H	to Vietnam	
68-16392	UH1H		
68-16393	UH1H	to Vietnam	
68-16394	UH1H	1-149AVN	TX NG
68-16395	UH1H	to Vietnam	
68-16396	UH1H	w.o.	
68-16397	UH1H	to Vietnam	
68-16398	UH1H	to Thailand	
68-16399	UH1H	w.o.26.7.70	
68-16400	UH1H	to Vietnam	
68-16401	UH1H	25AVN	
68-16402	GUH1H		
68-16403	UH1H	SoC	
68-16404	UH1H	SoC	
68-16405	UH1H		
68-16406	UH1H	26AVN	
68-16407	UH1H	to Vietnam	
68-16408	UH1H	to Vietnam	
68-16409	UH1H	w.o.	
68-16410	UH1H	w.o.	
68-16411	UH1H	193AVN	
68-16412	UH1H	to Vietnam	
68-16415	UH1V		
68-16416	UH1H	to Vietnam	
68-16417	UH1H		
68-16418	UH1H		
68-16419	UH1H	SoC	
68-16420	UH1H	w.o.11.11.70	
68-16421	UH1V	126MedCo	CA NG
68-16422	UH1H	to Vietnam	
68-16423	UH1H	to Vietnam	
68-16424	UH1H	to Vietnam	
68-16425	UH1H	A\1-131AVN	AL NG
68-16426	UH1H		
68-16427	UH1H	to Vietnam	
68-16428	UH1H	to Vietnam	
68-16429	UH1H	w.o.25.4.70	
68-16430	UH1H	to Vietnam	
68-16431	UH1H	to Vietnam	
68-16432	UH1H	to Vietnam	
68-16433	UH1H	w.o.	
68-16434	UH1H	to Vietnam	
68-16435	UH1H	to Vietnam	
68-16436	UH1H	B\1-132AVN	DC NG
68-16437	UH1H		
68-16438	UH1H	w.o.	
68-16439	UH1H	SoC	
68-16440	UH1H	to Vietnam	
68-16441	UH1H	SoC	
68-16442	UH1H	to Vietnam	
68-16443	UH1H	to Cambodia	
68-16444	UH1H	to Vietnam	
68-16445	UH1H	SoC	
68-16446	UH1V	145MedDet	
68-16447	UH1H	w.o.23.7.70	
68-16448	UH1H	to Vietnam	
68-16449	UH1H	to Cambodia	
68-16450	UH1H		
68-16451	UH1H	to Australia	
68-16452	UH1H	to Vietnam	
68-16453	UH1H	SoC	
68-16454	UH1H	w.o.	
68-16455	UH1H	to Vietnam	
68-16456	UH1H	SoC	
68-16459	UH1H	to Korea	
68-16460	UH1H	w.o.10.10.71	
68-16461	UH1H	to Vietnam	
68-16462	UH1H	to Iran	
68-16463	UH1H	to Vietnam	
68-16464	UH1H	to Vietnam	
68-16465	UH1H	w.o.	
68-16466	UH1H	1-158AVN	
68-16467	UH1H	w.o.25.7.70	
68-16468	UH1H	3ACR	
68-16469	UH1H		
68-16470	UH1H	w.o.	
68-16471	UH1H	SoC	
68-16472	UH1H	to Vietnam	
68-16473	UH1H		
68-16474	UH1H		
68-16475	UH1H	24MedCo	
68-16476	UH1H	1Cav	
68-16477	UH1H	to Thailand	
68-16478	UH1H	w.o.	
68-16479	UH1H	to Vietnam	
68-16480	UH1H	w.o.30.10.72	
68-16481	UH1H	to Vietnam	
68-16482	UH1H	to Vietnam	
68-16483	UH1H	w.o.	
68-16484	UH1H	to Vietnam	
68-16485	UH1H	w.o.	
68-16486	UH1H		
68-16487	UH1H	SoC	
68-16488	UH1H	w.o.26.2.70	
68-16489	UH1H	to Vietnam	
68-16490	UH1H	to Vietnam	
68-16491	UH1H		
68-16492	UH1H	w.o.	
68-16493	UH1H	to Vietnam	
68-16494	UH1H	w.o.	
68-16495	UH1H	24MedCo	
68-16496	UH1H	to Vietnam	
68-16497	UH1H	to Vietnam	
68-16498	UH1H	SoC	
68-16499	UH1H	to Vietnam	
68-16500	UH1H	C\1-108AVN	AZ NG 38
68-16501	UH1H	w.o.5.9.71	
68-16502	UH1H	364EngCo	
68-16503	UH1H		
68-16504	UH1H		
68-16505	UH1H		
68-16506	UH1H	to Vietnam	
68-16507	UH1V	112MedDet	
68-16508	UH1H	to Vietnam	
68-16509	UH1H	w.o.	
68-16510	UH1H	13AHB	
68-16511	UH1H	to Iran	
68-16512	UH1H	w.o.	
68-16513	UH1H	308AHB	
68-16514	UH1H	SoC	
68-16515	UH1H	SoC	
68-16516	UH1H	to Thailand	
68-16519	UH1V	126MedCo	CA NG
68-16520	UH1H	w.o.	
68-16521	UH1H		
68-16522	UH1H	to Guatemala	
68-16523	UH1H	4InfDiv	
68-16524	UH1H	to Guatemala	
68-16525	UH1H	w.o.6.2.70	
68-16526	UH1H	62AvCo	
68-16527	UH1H	SoC	
68-16528	UH1H	w.o.31.5.70	
68-16529	UH1H	w.o.	
68-16530	UH1H	w.o.19.4.70	
68-16531	UH1H	24MedCo	
68-16532	UH1H	to Vietnam	
68-16533	UH1H	to Vietnam	
68-16534	UH1H	w.o.	
68-16535	UH1H	to Vietnam	
68-16536	UH1H		
68-16537	UH1H	w.o.	
68-16538	UH1H	145MedDet	
68-16539	UH1H	to Vietnam	
68-16540	UH1H	to Vietnam	
68-16541	UH1H	to Vietnam	
68-16542	UH1H	w.o.13.4.71	

Serial	Type	Notes		Serial	Type	Notes
68-16543	UH1H	to Vietnam		69-15000	JUH1H	
68-16544	UH1H	to Vietnam		69-15001	UH1H	
68-16545	UH1H	to Vietnam		69-15002	UH1V	A\2-224AVN
68-16546	UH1H	to Vietnam		69-15003	UH1H	
68-16547	UH1H	to Vietnam		69-15004	UH1H	
68-16548	UH1H	w.o.		69-15005	UH1H	327AVN
68-16549	UH1H	w.o.		69-15006	UH1H	w.o.30.11.70
68-16550	UH1H	w.o.		69-15007	UH1H	6ACR
68-16551	UH1H	to Vietnam		69-15008	UH1H	323AVN
68-16552	UH1H	to Vietnam		69-15009	UH1H	NTC\TPS
68-16553	UH1V	237MedCo		69-15010	UH1H	
68-16554	UH1H	w.o.15.2.71		69-15011	UH1H	
68-16555	UH1H	to Vietnam		69-15012	UH1V	
68-16556	UH1H	to Vietnam		69-15013	UH1H	w.o.
68-16557	UH1H	w.o.		69-15014	UH1H	207AvCo
68-16558	UH1H	w.o.		69-15015	EH1X	to USFS N494DF
68-16559	UH1H	6ACR		69-15016	UH1V	w.o.7.2.91
68-16560	UH1H	to Vietnam		69-15017	UH1H	
68-16561	UH1H	to Vietnam		69-15018	UH1H	to Vietnam
68-16562	UH1H	to Vietnam		69-15019	UH1H	to Vietnam
68-16563	UH1H			69-15020	UH1H	to Vietnam
68-16564	UH1H	to Vietnam		69-15021	UH1H	w.o.27.9.70
68-16565	UH1H	224AVN		69-15022	UH1H	
68-16566	UH1H	to Vietnam		69-15023	UH1H	to Vietnam
68-16567	UH1H	to Vietnam		69-15024	UH1H	to Vietnam
68-16568	UH1H			69-15025	UI1H	w.o.
68 16569	UH1H			69-15026	UH1H	A\1-244AVN LA NG
68-16570	UH1H	to Vietnam		69-15027	UH1H	229AVN
68-16571	UH1H	to Vietnam		69-15028	UH1H	to Vietnam
68-16572	UH1H			69-15029	EH1X	SoC
68-16573	UH1H	to Vietnam		69-15030	UH1H	278ACR
68-16574	UH1H	6ACR		69-15031	UH1H	w.o.17.2.72
68-16575	UH1V	126MedCo CA NG		69-15032	UH1H	to Vietnam
68-16576	UH1H	126ACS		69-15033	UH1H	A\1-244AVN LA NG
68-16577	UH1H	to Vietnam		69-15034	UH1H	to Vietnam
68-16578	UH1H	w.o.		69-15035	UH1H	
68-16579	UH1H	stored Ramstein AB,FRG		69-15036	UH1H	to Vietnam
68-16580	UH1H	to Vietnam		69-15037	UH1H	to Vietnam
68-16581	UH1H	SoC		69-15038	UH1H	w.o.
68-16582	UH1H	w.o.7.5.71		69-15039	UH1H	to Vietnam
68-16583	UH1H	w.o.2.12.71		69-15040	UH1H	to Vietnam
68-16584	UH1H	to Vietnam		69-15041	UH1H	w.o.22.8.75
68-16585	UH1H	w.o.31.5.71		69-15042	UH1H	to Vietnam
68-16586	UH1H	SoC		69-15043	UH1H	to Vietnam
68-16587	UH1H			69-15044	UH1H	w.o.
68-16588	UH1H	to Vietnam		69-15045	UH1H	
68-16589	UH1H	to Vietnam		69-15046	UH1H	to Vietnam
68-16590	UH1H	to Vietnam		69-15047	UH1H	3ACT
68-16591	UH1H	278ACR		69-15048	UH1H	to Vietnam
68-16592	UH1H	w.o.1.7.70		69-15049	UH1H	to Vietnam
68-16593	UH1H	to Vietnam		69-15050	UH1H	104ACS
68-16594	UH1H	455\457MedDet MO NG		69-15051	UH1H	D\1-245AVN OK NG
68-16595	UH1H	SoC		69-15052	UH1H	to Vietnam
68-16596	UH1H	6ACR		69-15053	UH1H	
68-16597	UH1H	SoC		69-15054	UH1H	
68-16598	UH1H			69-15055	UH1H	to Vietnam
68-16599	UH1H	w.o.		69-15056	UH1H	to Vietnam
68-16600	UH1H	to Vietnam		69-15057	UH1H	to Vietnam
68-16601	UH1H	A\1-244AVN LA NG		69-15058	UH1H	w.o.
68-16602	UH1H	to Vietnam		69-15059	UH1H	to Vietnam
68-16603	UH1H	222AVN		69-15060	UH1H	to Vietnam
68-16604	UH1H	SoC		69-15061	UH1H	to Vietnam
68-16608	UH1H			69-15062	UH1H	455\457MedDet MO NG
68-16609	UH1H	1898AvCo		69-15063	UH1H	w.o.
68-16610	UH1H			69-15064	UH1H	w.o.
68-16611	UH1H			69-15065	UH1H	w.o.22.3.72
68-16612	UH1H	56AvCo		69-15066	UH1H	w.o.22.7.75
68-16613	UH1H	SoC		69-15067	UH1H	
68-16614	UH1V	w.o.		69-15068	UH1H	to Vietnam
68-16615	UH1H			69-15069	UH1H	w.o.28.2.79
68-16616	UH1H	26AVN		69-15070	UH1H	
68-16617	UH1H	w.o.12.11.73		69-15071	UH1V	
68-16618	UH1H	193AVN		69-15072	UH1H	w.o.
68-16619	UH1H	163ACR		69-15073	UH1H	to Vietnam
68-16620	UH1H	308AHB		69-15074	UH1H	w.o.
68-16621	UH1H			69-15075	UH1H	
68-16622	UH1H			69-15076	UH1H	to Vietnam
68-16623	UH1V			69-15077	UH1H	25InfDiv
68-16624	UH1H	SoC		69-15078	UH1H	w.o.
68-16625	UH1H			69-15079	UH1H	to Vietnam
68-16626	UH1H	321MedDet		69-15080	UH1H	w.o.22.10.71
68-16627	UH1V			69-15081	UH1V	w.o.19.5.87
68-16628	UH1H	w.o.10.7.85		69-15082	UH1H	w.o.15.8.70

69-15083	UH1H	w.o.	
69-15084	UH1H	to Vietnam	
69-15085	EH1H	to USFS	
69-15086	UH1H		
69-15087	UH1H	w.o.	
69-15088	UH1H	w.o.	
69-15089	UH1H		
69-15090	UH1H	to Vietnam	
69-15091	UH1H	166ACR	
69-15092	UH1H	to N800NT	
69-15093	UH1H	w.o.	
69-15094	UH1H	SoC	
69-15095	UH1H	w.o.	
69-15096	UH1H	to Vietnam	
69-15097	UH1H	w.o.	
69-15098	UH1H	to Argentina	
69-15099	UH1H	to Argentina	
69-15100	UH1H	to Vietnam	
69-15101	UH1H	to Vietnam	
69-15102	UH1H	w.o.25.10.91	
69-15103	EH1H	to N318LD	
69-15104	UH1H		
69-15105	UH1H	to Vietnam	
69-15106	UH1H	w.o.	
69-15107	UH1H	412MedDet	
69-15108	UH1H	w.o.	
69-15109	UH1H	to Vietnam	
69-15113	UH1H		
69-15114	UH1H	w.o.	
69-15115	UH1H	w.o.23.5.70	
69-15116	UH1H	to Vietnam	
69-15117	UH1H	to Vietnam	
69-15118	UH1H	374MedDet	
69-15119	UH1H	w.o.29.7.70	
69-15120	UH1H	to Vietnam	
69-15121	UH1H	w.o.	
69-15122	UH1H	SoC	
69-15123	UH1H	USAICS	
69-15124	UH1H	40AVN	
69-15125	UH1H	w.o.	
69-15126	UH1H	158AVN	
69-15127	UH1H		
69-15128	UH1H	to Vietnam	
69-15129	UH1H	w.o.	
69-15130	UH1H	to Vietnam	
69-15131	UH1H	w.o.	
69-15132	UH1H		
69-15133	UH1H	to Vietnam	
69-15134	UH1H	to Vietnam	
69-15135	UH1H	w.o.	
69-15136	UH1H		
69-15137	UH1H	w.o.	
69-15138	UH1H	w.o.7.7.70	
69-15139	UH1H	w.o.	
69-15140	UH1H	A\1-207AVN	AK NG
69-15141	UH1H	w.o.	
69-15142	UH1H	w.o.27.10.70	
69-15143	UH1H	to Vietnam	
69-15144	UH1H	321MedDet	
69-15145	EH1H	to USFS	N429DF
69-15146	UH1H	w.o.17.6.72	
69-15147	UH1H	w.o.	
69-15148	UH1H	w.o.29.3.70	
69-15149	UH1H	w.o.8.5.70	
69-15150	UH1H	374MedDet	
69-15151	UH1H	w.o.	
69-15152	UH1H		
69-15153	UH1H	MDDH	Mesa Field
69-15154	UH1H	40AVN	
69-15155	JUH1H	1898AvCo	
69-15156	UH1H	to Vietnam	
69-15157	UH1H	to Philippine	
69-15158	UH1H	1-158AVN	
69-15159	UH1H	SoC	
69-15160	UH1H	to Vietnam	
69-15161	UH1H	to Vietnam	
69-15162	UH1H	24MedCo	
69-15163	UH1H	to Vietnam	
69-15164	UH1H	w.o.	
69-15165	UH1H	w.o.	
69-15166	UH1H		
69-15167	UH1H	to Philippine	
69-15168	UH1H	to Vietnam	
69-15169	UH1H	to Vietnam	
69-15170	UH1H	to Vietnam	
69-15171	UH1V	69AvBde	
69-15172	UH1H	to Vietnam	
69-15173	UH1H	207AvCo	
69-15174	UH1H	40AVN	
69-15175	UH1H	131AVN	
69-15178	UH1H	w.o.	
69-15179	UH1H	to Vietnam	
69-15180	UH1V	347MedDet	
69-15181	UH1H	to Vietnam	
69-15182	UH1H	to Vietnam	
69-15183	UH1H	to Vietnam	
69-15184	UH1H	w.o.6.12.70	
69-15185	UH1H		
69-15186	UH1H	3AVN	
69-15187	UH1H	w.o.28.10.70	
69-15188	UH1H	1898AvCo	
69-15189	UH1H	w.o.	
69-15190	UH1H	166ACR	
69-15191	UH1H		
69-15192	UH1H	166ACR	
69-15193	UH1H	to Vietnam	
69-15194	UH1H	SoC	
69-15195	UH1H	to Vietnam	
69-15196	UH1H	to Vietnam	
69-15197	UH1H	w.o.	
69-15198	UH1H		
69-15199	UH1H		
69-15200	UH1H	w.o.	
69-15201	UH1H	to Vietnam	
69-15202	UH1H	to Thailand	
69-15203	UH1H	w.o.19.8.70	
69-15204	UH1H	w.o.27.9.70	
69-15205	UH1H	to Vietnam	
69-15206	UH1H	to Vietnam	
69-15207	UH1H	to Vietnam	
69-15208	UH1V	348MedDet	
69-15209	UH1H	to Vietnam	
69-15210	UH1H	w.o.24.10.70	
69-15211	UH1H	to Vietnam	
69-15212	UH1H	w.o.	
69-15213	EH1X	to N955MF	
69-15214	UH1H	to Vietnam	
69-15215	UH1H	to Korea	
69-15216	UH1H	1Cav	
69-15217	UH1H	308AHB	
69-15218	UH1H	C\1-245AVN	OK NG
69-15219	UH1V	63MedCo	
69-15220	UH1H	w.o.	
69-15221	UH1H	104ACS	
69-15222	UH1H	to Vietnam	
69-15223	UH1H	to Vietnam	
69-15224	UH1H	to Vietnam	
69-15225	UH1H	to Vietnam	
69-15226	UH1H	w.o.18.8.70	
69-15227	UH1H	1-149AVN	TX NG
69-15228	UH1V	1259MedCo	
69-15229	UH1V	441MedDet	
69-15230	UH1H		
69-15231	UH1H	to NASA	N733NA
69-15232	UH1H	396AvCo	
69-15233	UH1H		
69-15234	EH1X	SoC	
69-15235	UH1H	to Iran	
69-15236	UH1H	A\1-131AVN	AL NG
69-15237	UH1H		
69-15238	UH1H	227AVN	
69-15239	EH1H	to USFS	
69-15240	UH1H		
69-15241	UH1H		
69-15242	UH1H	to Iran	
69-15243	UH1H	26AVN	
69-15244	UH1H	w.o.9.5.74	
69-15245	UH1H		
69-15246	UH1H		
69-15247	UH1V		
69-15248	UH1H		
69-15249	UH1V	441MedDet	
69-15250	UH1H		
69-15251	UH1H	w.o.12.8.88	
69-15252	UH1H	56AvCo	
69-15253	UH1V		

Serial	Type	Unit/Notes	Extra
69-15254	EH1X	to USFS	
69-15255	UH1H	335AvCo	
69-15256	UH1H	1AVN	
69-15257	UH1H		
69-15258	UH1H	1AVN	
69-15259	UH1H	1AVN	
69-15260	UH1H	w.o.17.7.75	
69-15261	UH1H	25AvCo	
69-15262	UH1H	w.o.15.6.84	
69-15263	UH1H		
69-15264	UH1H	4InfDiv	
69-15265	EH1H	to USFS	
69-15266	UH1H		
69-15267	UH1V		
69-15268	UH1V	507MedCo	
69-15269	UH1H		
69-15270	UH1H	to Philippine	
69-15271	UH1V		
69-15272	UH1H	w.o.	
69-15273	UH1H	w.o.20.2.71	
69-15274	UH1H	to Vietnam	
69-15275	UH1H	w.o.	
69-15276	UH1H	1-183AVN	ID NG
69-15277	UH1H	w.o.	
69-15278	UH1H	to Vietnam	
69-15279	UH1H	to Vietnam	
69-15280	UH1H	to Vietnam	
69-15281	UH1H	to Vietnam	
69-15282	UH1H		
69-15283	UH1H	w.o.20.5.73	
69-15284	UH1H	w.o.	
69-15285	UH1H	to Vietnam	
69-15286	UH1H	w.o.6.9.82	
69-15287	UH1H	w.o.23.8.70	
69-15288	UH1H	to Vietnam	
69-15289	UH1H	to Vietnam	
69-15292	UH1H	w.o.7.3.72	
69-15293	UH1H	to Vietnam	
69-15294	UH1H		
69-15295	UH1H	to Vietnam	
69-15296	UH1H	to Vietnam	
69-15297	UH1H	w.o.27.9.70	
69-15298	UH1V	AATC	Cairns AAF
69-15299	UH1H	to Vietnam	
69-15300	UH1H		
69-15301	UH1H	to Vietnam	
69-15302	UH1H		
69-15303	UH1H	C\1-108AVN	AZ NG
69-15304	UH1H		
69-15305	UH1H	w.o.30.3.79	
69-15306	UH1H		
69-15307	UH1H	CCAD	
69-15308	UH1H	A\1-207AVN	AK NG
69-15309	UH1H		
69-15310	UH1H		
69-15311	UH1H	w.o.11.11.80	
69-15312	UH1H	3-159AVN 04	
69-15313	UH1H	159AHB	
69-15314	UH1H	SoC	
69-15315	UH1H		
69-15316	UH1H	w.o.4.10.70	
69-15317	UH1H	to Vietnam	
69-15318	UH1H		
69-15319	UH1H	w.o.15.5.73	
69-15320	UH1H	w.o.19.7.78	
69-15321	UH1H		
69-15322	UH1V	441MedDet	
69-15323	UH1V	w.o.6.4.82	
69-15324	UH1H	w.o.7.10.75	
69-15325	UH1V	101AbDiv	
69-15326	UH1V	441MedDet	
69-15327	UH1H	SoC	
69-15328	UH1H		
69-15329	UH1H	w.o.9.6.83	
69-15330	UH1H	stored	Ramstein AB,FRG
69-15331	UH1V		
69-15332	UH1H	26AVN	
69-15333	UH1H	w.o.12.2.81	
69-15334	UH1H	7InfDiv	
69-15335	EH1X	SoC	
69-15336	UH1H		
69-15337	UH1H		
69-15338	EH1H	to USFS	
69-15339	UH1H		
69-15340	JEH1H	to USFS	
69-15341	UH1V	SoC	
69-15342	UH1H	10SFG	
69-15343	UH1H		
69-15344	UH1H	to Vietnam	
69-15345	UH1H	B\4-228AVN	
69-15346	UH1H	to Vietnam	
69-15347	UH1H	B\4-228AVN	
69-15348	UH1H	to Vietnam	
69-15349	EH1H	to USFS	
69-15350	UH1H	159AHB	
69-15351	UH1H	to Vietnam	
69-15352	UH1H	to Korea	
69-15353	UH1H	to Vietnam	
69-15354	UH1H	7InfDiv	
69-15355	UH1H	to Philippine	
69-15356	UH1H		
69-15357	UH1H	w.o.	
69-15358	UH1H	w.o.	
69-15359	UH1H	to Vietnam	
69-15360	UH1H	to Vietnam	
69-15361	UH1H	to Vietnam	
69-15362	UH1H	166ACR	
69-15363	UH1H	159AHB	
69-15364	UH1H		
69-15365	UH1H	455\457MedDet	MO NG
69-15366	UH1H	to Vietnam	
69-15367	UH1H	w.o.	
69-15368	UH1H	w.o.18.4.72	
69-15369	UH1H		
69-15370	UH1V	240EngGrp	
69-15371	UH1H	w.o.	
69-15372	UH1H	63ARCOM	
69-15373	UH1H	to Vietnam	
69-15374	UH1H	w.o.	
69-15375	UH1H	w.o.	
69-15376	UH1H	to Vietnam	
69-15377	UH1H	to Vietnam	
69-15378	UH1H	w.o.15.11.71	
69-15379	UH1H	w.o.	
69-15380	UH1H	w.o.	
69-15381	UH1H	w.o.	
69-15382	UH1H	to Vietnam	
69-15383	UH1H	to Vietnam	
69-15384	UH1H	to Vietnam	
69-15385	UH1H	to Vietnam	
69-15386	UH1H	to Vietnam	
69-15387	UH1H	to Vietnam	
69-15388	UH1H	to Vietnam	
69-15389	UH1H	166ACR	
69-15390	UH1H	to Vietnam	
69-15391	UH1H	to Vietnam	
69-15392	UH1H	to Vietnam	
69-15393	UH1H	to Vietnam	
69-15394	UH1H	40 AVN	
69-15395	UH1H		
69-15396	UH1H	w.o.30.12.72	
69-15397	UH1H	w.o.24.11.70	
69-15398	UH1H	to Vietnam	
69-15399	UH1H	to Vietnam	
69-15400	UH1H	to Vietnam	
69-15401	UH1H	to Vietnam	
69-15402	UH1H	to Vietnam	
69-15403	UH1H	w.o.	
69-15404	UH1H	C\1-245AVN	OK NG
69-15405	UH1H		
69-15406	UH1H	w.o.	
69-15407	UH1H	to Vietnam	
69-15408	UH1V	498MedCo	
69-15409	UH1H	to Vietnam	
69-15410	UH1H	to Vietnam	
69-15411	UH1H	A\1-207AVN	AK NG
69-15412	UH1H	1-158AVN	
69-15413	UH1H	to Vietnam	
69-15414	UH1H	26AVN	
69-15415	UH1H	to Australia	
69-15416	UH1H	to Vietnam	
69-15421	UH1H	7SigBde	
69-15422	UH1H	to Vietnam	
69-15423	UH1H	to Vietnam	
69-15424	UH1H	to Vietnam	
69-15425	UH1H		

69-15426	UH1H	to Vietnam	
69-15427	UH1H		
69-15428	UH1H	w.o.	
69-15429	UH1H	w.o.	
69-15430	UH1H	to Vietnam	
69-15431	UH1H	to Vietnam	
69-15432	UH1H	to Vietnam	
69-15433	UH1H	to Vietnam	
69-15434	UH1H	to Vietnam	
69-15435	UH1H	to Australia	
69-15436	UH1H	w.o.9.9.89	
69-15437	UH1H	to Vietnam	
69-15438	UH1H	to Vietnam	
69-15439	UH1V	347MedCo	
69-15440	UH1H		
69-15441	UH1H	w.o.	
69-15442	UH1H	w.o.9.11.71	
69-15443	UH1H	to Vietnam	
69-15444	UH1H	to Vietnam	
69-15445	UH1H	to Vietnam	
69-15446	UH1H		
69-15447	UH1H	193AVN	
69-15448	UH1H	to Cambodia	
69-15449	UH1H	w.o.3.6.73	
69-15450	UH1H	25AvCo	
69-15451	UH1H	to Vietnam	
69-15452	UH1H	to Vietnam	
69-15453	UH1H	w.o.17.6.71	
69-15454	UH1H	w.o.	
69-15455	EH1H	to N16TV	
69-15456	UH1H	to Vietnam	
69-15462	UH1H	to Vietnam	
69-15463	UH1H	to Vietnam	
69-15464	UH1H	to Vietnam	
69-15465	UH1H	to Vietnam	
69-15466	UH1H		
69-15467	UH1H	to Vietnam	
69-15468	UH1H	to Vietnam	
69-15469	EH1H	to USFS	
69-15470	UH1H	to Vietnam	
69-15471	UH1H	JRTC	Ft. Rucker
69-15472	UH1H	to Vietnam	
69-15473	UH1H	to Vietnam	
69-15474	UH1H	193AvCo	
69-15475	UH1H	335AvCo	
69-15476	UH1H	to Vietnam	
69-15477	UH1H	to Vietnam	
69-15478	UH1H	to Vietnam	
69-15479	UH1H	to Vietnam	
69-15480	UH1H	SoC	
69-15481	UH1H		
69-15482	UH1H		
69-15499	EH1H	to USFS	
69-15500	UH1H	SigCenter	
69-15501	UH1H	to Vietnam	
69-15502	UH1H	SoC	
69-15503	UH1H		
69-15504	UH1H	to Vietnam	
69-15505	UH1H	w.o.	
69-15506	UH1H	to Vietnam	
69-15507	UH1H	to Vietnam	
69-15508	UH1H	to Vietnam	
69-15509	UH1V	145MedDet	
69-15510	UH1H		
69-15511	UH1H	to Thailand	
69-15512	UH1H	to Vietnam	
69-15513	UH1H	to Vietnam	
69-15514	UH1H	w.o.13.10.71	
69-15515	UH1H	335AvCo	
69-15516	UH1H	to Vietnam	
69-15517	UH1H		
69-15518	UH1V	507MedCo	
69-15519	UH1H	to Cambodia	
69-15520	UH1H		
69-15521	UH1H	w.o.22.9.70	
69-15522	UH1H	to Vietnam	
69-15523	UH1H	to Vietnam	
69-15524	UH1H	w.o.19.3.71	
69-15525	UH1H	to Vietnam	
69-15526	UH1H	to Vietnam	
69-15527	UH1H	to Vietnam	
69-15528	UH1H	w.o.30.8.88	
69-15529	UH1H	to Vietnam	

69-15530	UH1H	SoC	
69-15531	UH1V		
69-15532	JUH1H	USAAEFA	
69-15533	UH1H		
69-15534	UH1H	to Vietnam	
69-15535	UH1H	to Vietnam	
69-15536	UH1H	to Vietnam	
69-15537	UH1H		
69-15538	UH1H	to Thailand	
69-15539	UH1H	w.o.20.10.71	
69-15540	UH1H	to Vietnam	
69-15541	UH1H	to Vietnam	
69-15542	UH1H	to Thailand	
69-15543	UH1H	5-158AVN	
69-15544	UH1H	to Cambodia	
69-15545	UH1H	1-159AVN	
69-15546	UH1H	to Vietnam	
69-15547	UH1H	to Vietnam	
69-15548	UH1H	w.o.25.11.70	
69-15549	UH1H	to Vietnam	
69-15550	UH1H	173AvCo	
69-15551	UH1H	A\1-131AVN	AL NG
69-15552	UH1H	to Vietnam	
69-15553	UH1H	to Vietnam	
69-15554	UH1H	to Vietnam	
69-15555	UH1H	to Vietnam	
69-15556	UH1H	to Vietnam	
69-15557	UH1H	w.o.	
69-15558	UH1H	to Vietnam	
69-15559	UH1H	to Vietnam	
69-15560	UH1H	to Vietnam	
69-15561	UH1H	to Vietnam	
69-15562	UH1H	to Vietnam	
69-15563	UH1H	to Vietnam	
69-15564	UH1H	to Vietnam	
69-15565	UH1H	w.o.13.11.70	
69-15566	UH1H	to Vietnam	
69-15567	UH1H	to Vietnam	
69-15568	UH1H	to Vietnam	
69-15569	UH1H	to Vietnam	
69-15574	UH1H	to Vietnam	
69-15575	UH1H	w.o.	
69-15576	UH1H	to Vietnam	
69-15577	UH1H		
69-15578	EH1H	to USFS	
69-15579	UH1H	DAC\MDW	
69-15580	UH1H	DAC\MDW	
69-15581	UH1H	1AVN	
69-15582	UH1H	B\1-132AVN	DC NG
69-15583	UH1H		
69-15584	UH1H	B\1-132AVN	DC NG
69-15585	UH1H	B\1-132AVN	DC NG
69-15586	UH1H	B\1-132AVN	DC NG
69-15587	UH1H		
69-15588	UH1H		
69-15589	UH1H		
69-15590	UH1H	B\1-132AVN	DC NG
69-15591	UH1H		
69-15592	UH1H		
69-15593	UH1H	AATC	93. Cairns AAF
69-15594	UH1H		
69-15595	UH1H	to Argentina	
69-15596	UH1H	to Argentina	
69-15597	UH1H	to Argentina H-16	
69-15598	UH1H		
69-15599	UH1H	B\1-132AVN	DC NG
69-15600	UH1H		
69-15601	UH1H		
69-15602	UH1H	DAC\MDW	
69-15603	UH1H	ERADCOM	
69-15604	UH1H	B\1-132AVN	DC NG
69-15605	UH1H	USEUCOM	
69-15606	UH1H	USEUCOM	
69-15607	UH1H		
69-15608	UH1H	56AvCo	
69-15609	UH1H	SoC	
69-15610	UH1H	308AHB	
69-15611	UH1H	56AvCo	
69-15612	UH1H	to Vietnam	
69-15613	UH1H	to Vietnam	
69-15614	UH1H	to Vietnam	
69-15615	UH1H	to Vietnam	
69-15616	UH1H	244AvCo	

69-15617	UH1H	159AHB 03	
69-15618	UH1H	to Vietnam	
69-15619	UH1H	SoC	
69-15620	UH1H	to Vietnam	
69-15621	UH1H	to Vietnam	
69-15622	UH1H	to Vietnam	
69-15623	UH1H	to Vietnam	
69-15624	UH1H	SoC	
69-15625	UH1H		
69-15626	UH1H	to Vietnam	
69-15627	UH1H	SoC	
69-15628	UH1H	w.o.13.7.71	
69-15629	UH1H		
69-15630	UH1H		
69-15631	UH1H	to Vietnam	
69-15632	UH1H	w.o.	
69-15633	UH1H	to Vietnam	
69-15634	UH1H	to Vietnam	
69-15635	UH1H	to Vietnam	
69-15636	UH1V	1InfDiv	
69-15637	UH1H	to Vietnam	
69-15638	UH1H	to Vietnam	
69-15639	UH1H	to Vietnam	
69-15640	UH1H	to Vietnam	
69-15641	UH1V	347MedDet	
69-15642	UH1H	w.o.22.8.71	
69-15643	UH1H	to Vietnam	
69-15644	UH1H	to Vietnam	
69-15660	UH1H	104ACS	
69-15661	UH1H	to Vietnam	
69-15662	UH1H	w.o.	
69-15663	UH1H	to Vietnam	
69-15664	UH1H	w.o.	
69-15665	UH1H	w.o.	
69-15666	UH1H	SoC	
69-15667	UH1H	to Vietnam	
69-15668	UH1H	to Vietnam	
69-15669	UH1H	to Vietnam	
69-15670	UH1H	to Vietnam	
69-15671	UH1H		
69-15672	UH1H	to Vietnam	
69-15673	UH1H	to Vietnam	
69-15674	UH1H	to Vietnam	
69-15675	UH1H	w.o.24.10.88	
69-15676	EH1H	to USFS	N496DF
69-15677	UH1H		
69-15678	UH1H	to Vietnam	
69-15679	UH1H	w.o.7.11.90	
69-15680	UH1H		
69-15681	UH1H	w.o.13.7.76	
69-15682	UH1H	w.o.26.1.71	
69-15683	UH1V	427MedCo	
69-15684	UH1H	w.o.	
69-15685	UH1H	to Vietnam	
69-15686	UH1H	to Vietnam	
69-15687	UH1H		
69-15688	UH1H		
69-15689	UH1H	B\1-245AVN	OK NG
69-15690	UH1H	1085Medco	Det.1
69-15691	UH1H	SoC	
69-15692	UH1H	w.o.24.4.71	
69-15693	UH1H		
69-15694	UH1H	to Vietnam	
69-15695	UH1H	w.o.	
69-15696	UH1H	to Vietnam	
69-15697	UH1H	to Vietnam	
69-15698	UH1H		
69-15699	UH1H		
69-15700	UH1H	to Vietnam	
69-15701	UH1H	to Vietnam	
69-15702	UH1H	w.o.24.2.71	
69-15703	UH1H	198AvCo	
69-15704	UH1H	SoC	
69-15705	UH1H		
69-15706	EH1H	to USFS	
69-15707	UH1H	to Vietnam	
69-15708	UH1H		
69-15709	UH1H	to Vietnam	
69-15710	UH1H		
69-15711	UH1H	to Vietnam	
69-15712	UH1H	to Vietnam	
69-15713	EH1H	to USFS	N413DF
69-15714	UH1H	to Vietnam	

69-15715	UH1H	w.o.	
69-15716	UH1H		
69-15717	UH1H		
69-15718	UH1H	w.o.	
69-15719	UH1H		
69-15729	UH1H		
69-15730	UH1H	1-1131AVN	AL NG
69-15731	UH1H	to Vietnam	
69-15732	UH1H	to Cambodia	
69-15733	UH1H	10Cav	
69-15734	UH1H	to Vietnam	
69-15735	UH1H	to Vietnam	
69-15736	UH1H	to Vietnam	
69-15737	UH1H	to Vietnam	
69-15738	UH1H	stored	Ramstein AB,FRG
69-15739	UH1H	to Vietnam	
69-15740	UH1H	to Vietnam	
69-15741	UH1H	to Vietnam	
69-15742	UH1H	to Vietnam	
69-15743	UH1H	to Vietnam	
69-15744	UH1H	to Vietnam	
69-15745	UH1H	to Vietnam	
69-15746	UH1H	to Vietnam	
69-15747	UH1H	to Vietnam	
69-15748	UH1H	to Vietnam	
60 15749	UH1H	to Vietnam	
69-15750	UH1H	to Vietnam	
69-15751	UH1H	to Vietnam	
69-15752	EH1X	SoC	
69-15753	UH1H	to Vietnam	
69-15754	UH1H		
69-15755	UH1H	to Vietnam	
69-15756	UH1V		
69-15757	UH1H		
69-15758	UH1H		
69-15759	UH1H	to Vietnam	
69-15760	UH1H	to Vietnam	
69-15761	UH1H	to Vietnam	
69-15762	UH1H	to N46905	
69-15763	UH1H	w.o.	
69-15764	UH1H	to Vietnam	
69-15765	UH1H	to N8146M	
69-15766	UH1H	3AVN	
69-15767	UH1H	w.o.	
69-15768	UH1H	to Vietnam	
69-15769	UH1H	to Vietnam	
69-15770	UH1H	to Vietnam	
69-15771	UH1H		
69-15772	UH1H	to Vietnam	
69-15773	UH1H	to Vietnam	
69-15774	UH1H	to Vietnam	
69-15775	UH1H	to Vietnam	
69-15776	UH1H	to Vietnam	
69-15777	UH1H	w.o.	
69-15778	UH1H	to Vietnam	
69-15779	UH1H	to Vietnam	
69-15780	UH1H	to Vietnam	
69-15781	UH1H	to Vietnam	
69-15782	UH1H	to Vietnam	
69-15783	UH1H	w.o.	
69-15784	UH1H	SoC	
69-15785	UH1H	w.o.	
69-15786	UH1H	to Vietnam	
69-15787	UH1H	w.o.7.11.79	
69-15788	UH1H	to Vietnam	
69-15789	UH1H	to Vietnam	
69-15790	UH1H		
69-15791	UH1H	to Vietnam	
69-15792	UH1H	w.o.	
69-15793	UH1H	to Korea	
69-15803	UH1H	to Vietnam	
69-15804	UH1H	to Vietnam	
69-15805	UH1H	to Vietnam	
69-15806	UH1H	to Vietnam	
69-15807	UH1H	to Vietnam	
69-15808	UH1H	to Vietnam	
69-15809	UH1H	to Vietnam	
69-15810	UH1H	to Vietnam	
69-15811	UH1H	to Vietnam	
69-15812	UH1H	to Vietnam	
69-15813	UH1H	8-158AVN	
69-15814	UH1H	352ASA Co	
69-15815	UH1H	to Vietnam	

69-15816	UH1H	to Vietnam
69-15817	UH1H	to Vietnam
69-15818	UH1V	MEDDAC
69-15819	UH1H	to Vietnam
69-15820	UH1H	to Vietnam
69-15821	UH1H	to Vietnam
69-15822	UH1H	to Uruguay
69-15823	UH1H	to Uruguay
69-15824	UH1H	w.o.
69-15825	UH1H	to Vietnam
69-15826	UH1H	w.o.
69-15827	UH1H	to Uruguay
69-15828	UH1H	to Vietnam
69-15829	UH1H	to Vietnam
69-15830	UH1H	to Vietnam
69-15831	UH1H	w.o.
69-15832	UH1H	to Vietnam
69-15833	UH1H	to Vietnam
69-15834	UH1H	to Vietnam
69-15835	UH1H	to Vietnam
69-15836	JUH1H	
69-15837	UH1H	to Vietnam
69-15838	UH1H	45AVN
69-15839	UH1H	
69-15840	UH1H	101AbDiv
69-15841	UH1H	455\457MedDet MO NG
69-15842	UH1H	w.o.16.8.71
69-15843	UH1H	AATC 43NLowe AHP
69-15844	UH1V	
69-15845	UH1V	157MedDet
69-15846	UH1H	1AVN
69-15847	UH1H	
69-15848	UH1V	507MedDet
69-15849	UH1H	193AVN
69-15850	UH1H	w.o.4.6.83
69-15851	UH1H	w.o.21.2.90
69-15852	UH1H	
69-15853	UH1H	224AVN
69-15854	UH1V	
69-15855	UH1H	NTC
69-15856	UH1H	45AVN
69-15857	UH1H	45AVN
69-15858	EH1H	to USFS
69-15859	UH1H	68MedDet
69-15860	UH1H	A\3-158AVN Reserve
69-15861	UH1H	
69-15862	UH1H	1267MedCo MO NG
69-15863	UH1H	to Iran
69-15864	UH1H	82AVN
69-15865	EH1H	to USFS
69-15866	UH1H	
69-15867	UH1H	
69-15877	UH1H	to Thailand
69-15878	UH1H	to Thailand
69-15879	UH1H	to Vietnam
69-15880	UH1H	
69-15881	UH1H	to Vietnam
69-15882	UH1H	
69-15883	UH1H	to Vietnam
69-15884	UH1H	
69-15885	UH1H	to Thailand
69-15886	UH1H	to Thailand
69-15887	UH1H	to Thailand
69-15888	UH1H	to Thailand
69-15889	UH1H	w.o.
69-15890	UH1H	to Vietnam
69-15891	UH1H	w.o.27.8.82
69-15892	UH1H	to Vietnam
69-15893	UH1H	w.o.23.4.71
69-15894	UH1H	w.o.
69-15895	UH1H	to Vietnam
69-15896	EH1H	82AVN
69-15897	UH1H	to Vietnam
69-15898	UH1H	w.o.
69-15899	UH1H	to Vietnam
69-15900	UH1H	26AVN
69-15901	UH1H	to Vietnam
69-15902	UH1H	to Vietnam
69-15903	UH1H	
69-15904	UH1H	to Vietnam
69-15905	EH1X	
69-15906	UH1H	to Vietnam
69-15907	UH1H	to Vietnam
69-15908	UH1H	to Vietnam
69-15909	UH1H	to Vietnam
69-15910	UH1H	
69-15911	UH1H	5-158AVN
69-15912	UH1H	
69-15913	UH1H	w.o.14.12.90
69-15914	UH1H	
69-15915	UH1H	3ACR
69-15916	UH1H	
69-15917	UH1V	145MedDet
69-15918	JUH1H	3ACR
69-15919	UH1H	47AVN
69-15920	EH1X	
69-15921	UH1H	
69-15922	UH1H	
69-15923	UH1H	5AvDet
69-15924	UH1H	
69-15925	UH1H	w.o.16.4.84
69-15926	UH1H	813MedDet
69-15927	UH1V	321MedDet
69-15928	UH1V	NTC\TPS 01
69-15929	UH1V	
69-15930	EH1H	to USFS
69-15931	UH1V	321MedDet
69-15932	UH1H	
69-15933	UH1H	
69-15934	UH1H	w.o.
69-15935	UH1H	7InfDiv
69-15936	EH1X	to USFS N489DF
69-15937	UH1H	10Cav
69-15938	UH1H	308AHB
69-15939	UH1H	10Cav
69-15940	UH1H	
69-15941	UH1V	
69-15946	UH1V	104ACS
69-15947	UH1H	to Vietnam
69-15948	UH1H	to Guatemala
69-15949	UH1H	
69-15950	UH1H	to Vietnam
69-15951	UH1H	SoC
69-15952	UH1H	
69-15953	UH1H	1898AvCo
69-15954	UH1H	
69-15955	UH1H	to Vietnam
69-15956	UH1H	to Vietnam
69-15957	UH1H	131AVN
69-15958	UH1H	to Vietnam
69-15959	UH1H	to Vietnam
69-16650	UH1H	to Vietnam
69-16651	UH1H	to Vietnam
69-16652	UH1H	
69-16653	UH1H	
69-16654	UH1H	SoC
69-16655	UH1H	to Vietnam
69-16656	UH1H	SoC
69-16657	UH1H	
69-16658	UH1H	to Vietnam
69-16659	UH1H	to Vietnam
69-16660	UH1H	to Vietnam
69-16661	UH1H	to Vietnam
69-16662	UH1H	to Vietnam
69-16663	UH1H	w.o.30.11.88
69-16664	UH1H	to Vietnam
69-16665	UH1H	to Vietnam
69-16666	UH1H	to Vietnam
69-16667	UH1H	
69-16668	UH1H	to Vietnam
69-16669	UH1H	to Vietnam
69-16670	UH1H	to Vietnam
69-16692	UH1H	to Vietnam
69-16693	UH1H	to Vietnam
69-16694	UH1H	to Vietnam
69-16695	UH1H	w.o.
69-16696	UH1H	to Vietnam
69-16697	UH1H	to Vietnam
69-16698	UH1H	to Vietnam
69-16699	UH1H	w.o.
69-16700	UH1H	w.o.
69-16701	UH1H	w.o.
69-16702	UH1H	w.o.
69-16703	UH1H	w.o.23.8.71
69-16704	UH1H	
69-16705	UH1H	to Vietnam

Serial	Type	Notes	
69-16706	UH1H	to Vietnam	
69-16707	UH1H	to Vietnam	
69-16708	UH1H	to Vietnam	
69-16709	UH1H	to Vietnam	
69-16710	UH1H	to Vietnam	
69-16711	UH1V	69AvBde	
69-16712	UH1H	3ACT	
69-16713	EH1H	to USFS	
69-16714	UH1H	to Vietnam	
69-16715	UH1H	to Philippine	
69-16716	UH1H	to Cambodia	
69-16717	UH1H	w.o.	
69-16718	UH1H	to Vietnam	
69-16719	UH1V	126MedCo	CA NG
69-16720	UH1H	to Vietnam	
69-16721	UH1H	to Vietnam	
69-16722	UH1H	to Colombia	
69-16723	UH1H		
69-16724	UH1H	to Vietnam	
69-16725	UH1H	to Vietnam	
69-16726	UH1H	to Vietnam	
69-16727	UH1V	283MedDet	
69-16728	UH1V	283MedDet	
69-16729	UH1H		
69-16730	UH1H		
69-16731	UH1H	w.o.13.9.71	
69-16732	UH1H	to Vietnam	
70-15700	UH1H	w.o.22.7.75	
70-15701	UH1H	SoC	
70-15702	UH1H	SoC	
70-15703	UH1H	SoC	
70-15704	UH1H	SoC	
70-15705	UH1H	to Vietnam	
70-15706	UH1H	6ACR	
70-15707	UH1H	163ACR	
70-15708	UH1H	1267MedCo	
70-15709	UH1H	w.o.19.7.76	
70-15710	UH1H	SoC	
70-15711	UH1H	5-158AVN	
70-15712	UH1H	w.o.16.9.71	
70-15713	UH1H		
70-15714	UH1H	to Vietnam	
70-15715	UH1H	to Vietnam	
70-15716	UH1H	to Vietnam	
70-15717	UH1H	to Vietnam	
70-15718	UH1H	40AVN	
70-15719	UH1H	SoC	
70-15720	UH1H		
70-15721	UH1H	to Vietnam	
70-15722	UH1H	to Vietnam	
70-15723	UH1H		
70-15724	UH1H	SoC	
70-15725	UH1H	to Vietnam	
70-15726	UH1H	SoC	
70-15727	UH1H	SoC	
70-15728	UH1H	SoC	
70-15729	UH1H		
70-15730	UH1H	122AHB	
70-15731	UH1H	SoC	
70-15732	UH1H	SoC	
70-15733	UH1H	SoC	
70-15734	UH1H	SoC	
70-15735	UH1H	SoC	
70-15736	UH1H		
70-15737	UH1H	to Vietnam	
70-15738	UH1H	3ACR	
70-15739	UH1H		
70-15740	UH1H	to Vietnam	
70-15741	UH1H	172InfBde	
70-15742	UH1H	SoC	
70-15743	UH1H	25AvCo	
70-15744	UH1H	SoC	
70-15745	UH1H	25AvCo	
70-15746	UH1H		
70-15747	UH1H	SoC	
70-15748	UH1H	to Vietnam	
70-15749	UH1H	SoC	
70-15750	UH1H	25AvCo	
70-15751	UH1H	to Vietnam	
70-15752	UH1H	SoC	
70-15753	UH1H	w.o.1.12.71	
70-15754	UH1H		
70-15755	UH1H	SoC	
70-15756	UH1H	SoC	
70-15757	UH1H		
70-15758	UH1H	SoC	
70-15759	UH1H	SoC	
70-15760	UH1H	SoC	
70-15761	UH1H	SoC	
70-15762	UH1H	SoC	
70-15763	UH1H	SoC	
70-15764	UH1H	USAFDA	
70-15765	UH1H		
70-15766	UH1H	1-158AVN	
70-15767	UH1H	w.o.28.11.85	
70-15768	UH1H	SoC	
70-15769	UH1H	SoC	
70-15770	UH1H	56AvCo	
70-15771	UH1H	25AvCo	
70-15772	UH1H	3ACR	
70-15773	UH1H	w.o.23.6.71	
70-15774	UH1H	SoC	
70-15775	UH1H	SoC	
70-15776	UH1H	SoC	
70-15777	UH1H	SoC	
70-15778	UH1H	SoC	
70-15779	UH1H	SoC	
70-15780	UH1H	SoC	
70-15781	UH1H	SoC	
70-15782	UH1H	SoC	
70-15783	UH1H	SoC	
70-15784	UH1H	184AHC	
70-15785	UH1H		
70-15786	UH1H	to Vietnam	
70-15787	UH1H	7InfDiv	
70-15788	UH1H	JRTC	Ft. Rucker
70-15789	UH1H	53InfBde	
70-15790	UH1H	SoC	
70-15791	UH1H	SoC	
70-15792	UH1H	SoC	
70-15793	UH1H	to Vietnam	
70-15794	UH1H	SoC	
70-15795	UH1H	SoC	
70-15796	UH1H	SoC	
70-15797	UH1H		
70-15798	UH1H		
70-15799	UH1H	3ACT	
70-15800	UH1H	to Vietnam	
70-15801	UH1H	SoC	
70-15802	UH1H	SoC	
70-15803	UH1H	to Vietnam	
70-15804	UH1H	to Vietnam	
70-15805	UH1H	SoC	
70-15806	UH1H	1AVN	
70-15807	UH1H	SoC	
70-15808	UH1H	to Vietnam	
70-15809	UH1H		
70-15810	UH1H	SoC	
70-15811	UH1H	SoC	
70-15812	UH1H	SoC	
70-15813	UH1H	to Vietnam	
70-15814	UH1H	to Vietnam	
70-15815	UH1H	1085Medco	Det.1
70-15816	UH1H		
70-15829	UH1H	82AVN	
70-15830	UH1H		
70-15831	UH1H	to Vietnam	
70-15832	UH1H		
70-15833	UH1H	SoC	
70-15834	UH1H	SoC	
70-15835	UH1H	SoC	
70-15836	UH1H	1InfDiv	
70-15837	UH1H	SoC	
70-15838	UH1H	SoC	
70-15839	UH1H		
70-15840	UH1H	to Vietnam	
70-15841	UH1H	SoC	
70-15842	UH1H	to Vietnam	
70-15843	UH1H	SoC	
70-15844	UH1H	4InfDiv	
70-15845	UH1H	SoC	
70-15846	UH1H	SoC	
70-15847	UH1H	SoC	
70-15848	UH1H	SoC	
70-15849	UH1H	SoC	
70-15850	UH1H	SoC	

Serial	Type	Notes	
70-15851	UH1H		
70-15852	UH1H	SoC	
70-15853	UH1H		
70-15854	UH1H	SoC	
70-15855	UH1H	NTC\TPS	04
70-15856	UH1H	SoC	
70-15857	UH1H	SoC	
70-15858	UH1H	SoC	
70-15859	UH1H	SoC	
70-15860	UH1H	349MedDet	
70-15861	UH1H	SoC	
70-15862	UH1H	SoC	
70-15863	UH1H	w.o.	
70-15864	UH1H	SoC	
70-15865	UH1H	SoC	
70-15866	UH1H		
70-15867	UH1H	SoC	
70-15868	UH1H	to Vietnam	
70-15869	UH1H	w.o.	
70-15870	UH1H	w.o.8.5.72	
70-15871	UH1H	56AvCo	
70-15872	UH1H	B\4-228AVN	
70-15873	UH1H	24MedCo	
70-15874	UH1H	to Vietnam	
70-15913	UH1H	6ACR	
70-15914	UH1H	SoC	
70-15915	UH1H	SoC	
70-15916	UH1H	SoC	
70-15917	UH1H	SoC	
70-15918	UH1H	SoC	
70-15919	UH1H	SoC	
70-15920	UH1H	SoC	
70-15921	UH1H	to Vietnam	
70-15922	UH1H	to Vietnam	
70-15923	UH1H	SoC	
70-15924	UH1H	to Vietnam	
70-15925	UH1H	SoC	
70-15926	UH1H	w.o.18.1.82	
70-15927	UH1H	6ACR	
70-15928	UH1H	159AHB	
70-15929	UH1H	24MedCo	
70-15930	UH1H	308AHB	
70-15931	UH1H	SoC	
70-15932	UH1H		
70-16200	UH1H	150AVN	
70-16201	UH1V	341MedDet	
70-16202	UH1H		
70-16203	UH1H	1-149AVN	TX NG
70-16204	UH1H		
70-16205	UH1H	166ACR	
70-16206	UH1H	SoC	
70-16207	UH1H		
70-16208	UH1H	SoC	
70-16209	UH1V	273MedCo	
70-16210	UH1H	51AvCo	
70-16211	UH1H	USAICS	
70-16212	UH1H		
70-16213	UH1H	6ACR	
70-16214	UH1H	10Cav	
70-16215	UH1H	SoC	
70-16216	UH1H	6ACR	
70-16217	UH1H	461AVN	NM NG
70-16218	UH1H	AATC	18J Cairns AAF
70-16219	UH1H	6ACR	
70-16220	UH1H		
70-16221	UH1H	A\1-244AVN	LA NG
70-16222	UH1H		
70-16223	JUH1H		
70-16224	UH1V	237MedCo	
70-16225	UH1H		
70-16226	UH1H	498MedDet	
70-16227	UH1H	1-183AVN	ID NG
70-16228	UH1H		
70-16229	UH1H	1-207AVN	
70-16230	UH1H	w.o.23.6.82	
70-16231	UH1H	40AVN	
70-16232	UH1H		
70-16233	UH1H	w.o.2.11.71	
70-16234	UH1H		
70-16235	JUH1H		
70-16236	UH1H		
70-16237	UH1H		SD NG
70-16238	UH1H	to N8147Q	
70-16239	UH1H		
70-16240	UH1V		
70-16241	UH1H		
70-16242	UH1H	w.o.18.9.80	
70-16243	UH1H		
70-16244	UH1H	A\1-207AVN	AK NG
70-16245	UH1H	SoC	
70-16246	UH1H	SoC	
70-16247	UH1H		
70-16248	UH1V	131AVN	
70-16249	UH1H		
70-16250	UH1H		
70-16251	UH1H	441MedDet	
70-16252	UH1H		
70-16253	UH1H		
70-16254	UH1H		
70-16255	UH1H		
70-16256	UH1H	w.o.17.7.78	
70-16257	UH1H	6ACR	
70-16258	UH1H	w.o.14.11.84	
70-16259	UH1V		
70-16260	UH1H	A\1-131AVN	AL NG
70-16261	UH1H		
70-16262	UH1V		
70-16263	UH1H	w.o.2.5.911	
70-16264	UH1H		
70-16265	UH1H		
70-16266	UH1V	273MedDet	
70-16267	UH1H	USAICS	
70-16268	UH1H		
70-16269	UH1H		
70-16270	UH1H		
70-16271	UH1H	79ARCOM	
70-16272	UH1H		
70-16273	UH1H	USAICS	
70-16274	UH1H	USAICS	
70-16275	UH1H		
70-16276	UH1H		
70-16277	UH1H		
70-16278	UH1H	USAICS	
70-16279	UH1H		SC NG
70-16280	UH1V	507MedCo	
70-16281	UH1H	158AVN	
70-16282	UH1H	158AVN	
70-16283	UH1H		
70-16284	UH1H	AATC	84. Lowe AHP
70-16285	UH1H	1085Medco	Det.1
70-16286	UH1H	B\1-132AVN	DC NG
70-16287	UH1H	308AHB	
70-16288	UH1H		
70-16289	UH1H		
70-16290	UH1H	6ACR	
70-16291	UH1H		
70-16292	UH1H		
70-16293	UH1H	SoC	
70-16294	UH1H		
70-16295	UH1H	USAAEFA	
70-16296	UH1H	USAAEFA	
70-16297	UH1H	w.o.12.10.85	
70-16298	UH1H		
70-16299	UH1H		
70-16300	UH1H		
70-16301	UH1H	82AbDiv	
70-16302	JUH1H		
70-16303	UH1H	SoC	
70-16304	UH1H	SoC	
70-16305	UH1H	SoC	
70-16306	UH1H	SoC	
70-16307	UH1H	SoC	
70-16308	UH1H		
70-16309	UH1H	C\1-108AVN	AZ NG 18
70-16310	UH1H	164AvCo	
70-16311	UH1H	SoC	
70-16312	UH1H	3AVN	
70-16313	UH1H	247MedDet	
70-16314	UH1H		
70-16315	UH1H		
70-16316	UH1H		
70-16317	UH1H	TECOM	
70-16318	UH1H	USAAEFA	
70-16319	UH1H	SoC	
70-16320	UH1H	SoC	
70-16321	UH1H	w.o.5.3.74	

Serial	Type	Unit	Notes
70-16322	UH1H	to Pakistan	
70-16323	UH1H	SoC	
70-16324	UH1H	SoC	
70-16325	UH1H	79ARCOM	
70-16326	UH1H		
70-16327	UH1H	SoC	
70-16328	UH1V		
70-16329	UH1H		
70-16330	JUH1H		
70-16331	UH1H	USAAEFA	
70-16332	UH1H	TECOM	
70-16333	JUH1H	w.o.18.1.88	
70-16334	UH1H	SoC	
70-16335	UH1H	SoC	
70-16336	UH1H		
70-16337	UH1H	w.o.5.3.75	
70-16338	UH1H		
70-16339	UH1H		
70-16340	UH1H	w.o.25.3.76	
70-16341	UH1H	w.o.22.6.87	
70-16342	UH1H		
70-16343	UH1H		
70-16344	UH1H	to N37	
70-16345	UH1H		
70-16346	UH1H	107AvCo	
70-16347	UH1H	107AvCo	
70-16348	UH1H	107AvCo	
70-16349	UH1H	107AvCo	
70 10350	UH1H		
70-16351	UH1H	107AvCo	
70-16352	UH1H		
70-16353	UH1H	107AvCo	
70-16354	UH1H		
70-16355	UH1H	150AVN	
70-16356	UH1H	150AVN	
70-16357	UH1H	193AVN	
70-16358	UH1H	193AVN	
70-16359	UH1H	159AHB	
70-16360	UH1V		
70-16361	UH1V		
70-16362	UH1H		
70-16363	UH1H	1267MedCo	
70-16364	UH1V	w.o.16.5.83	
70-16365	UH1H	w.o.30.9.77	
70-16366	UH1H	40AVN	
70-16367	UH1H		
70-16368	UH1V	812Medco	det.1
70-16369	UH1H	44AVN	
70-16370	UH1V	126MedCo	CA NG
70-16371	UH1V	126MedCo	CA NG
70-16372	UH1V	273MedDet	
70-16373	UH1V	498MedCo	
70-16374	UH1H	SoC	
70-16375	UH1H	SoC	
70-16376	UH1H	SoC	
70-16377	UH1H	SoC	
70-16378	UH1V	498MedCo	
70-16379	UH1V	498MedCo	
70-16380	UH1V	54MedDet	
70-16381	UH1V	54MedDet	
70-16382	UH1V		
70-16383	UH1V		
70-16384	UH1H		
70-16385	UH1V		
70-16386	UH1H	498MedCo	
70-16387	UH1V	498MedCo	
70-16388	UH1H	B\4-228AVN	
70-16389	UH1V		
70-16390	UH1V	498MedCo	
70-16391	UH1V		
70-16392	UH1V	498MedCo	
70-16393	UH1V	498MedCo	
70-16394	UH1V	498MedCo	
70-16395	UH1H	22AvDet	
70-16396	UH1V	1InfDiv	
70-16397	UH1H		
70-16398	UH1H	w.o.10.9.73	
70-16399	UH1H		
70-16400	UH1H	92InfBde	
70-16401	UH1H	107AvCo	
70-16402	UH1H		
70-16403	UH1H	44AVN	
70-16404	UH1H	107AvCo	
70-16405	UH1H		
70-16406	UH1H	278ACR	
70-16407	UH1H	81ARCOM	
70-16408	UH1H		
70-16409	UH1H		
70-16410	UH1H	w.o.20.8.75	
70-16411	UH1H		
70-16412	UH1H	SoC	
70-16413	UH1H		
70-16414	UH1H	6ACR	
70-16415	UH1H	104ACS	
70-16416	UH1H	A\1-244AVN	LA NG
70-16417	UH1H	HQ-Army West	
70-16418	UH1H	B\1-132AVN	DC NG
70-16419	UH1H	B\1-132AVN	DC NG
70-16420	UH1H		
70-16421	UH1H		
70-16422	UH1V	126MedCo	CA NG
70-16423	UH1H	150AVN	DE NG
70-16424	UH1H	183AVN	
70-16425	UH1H	DAC\MDW	
70-16426	UH1H		
70-16427	UH1H	104AvCo	
70-16428	UH1H	25AvCo	
70-16429	UH1H	w.o.29.11.85	
70-16430	UH1H	SoC	
/0-16431	UH1H		
70-16432	UH1H	107AvCo	
70-16433	UH1H		
70-16434	UH1H	A\1-131AVN	AL NG
70-16435	UH1H		
70-16436	UH1H		
70-16437	UH1V		
70-16438	UH1H		
70-16439	UH1H	1255MedCo	NV NG
70-16440	UH1H		
70-16441	UH1H	SoC	
70-16442	JUH1H	TECOM	
70-16443	UH1H	104ACS	
70-16444	UH1H		
70-16445	UH1H		
70-16446	UH1H	26AVN	
70-16447	UH1H		
70-16448	UH1H	207AvCo	
70-16449	UH1H	1-158AVN	
70-16450	UH1H		
70-16451	UH1H		
70-16452	UH1H	25AvCo	
70-16453	UH1H	SoC	
70-16454	UH1H	C\1-108AVN	AZ NG 30
70-16455	UH1H	150AVN	DE NG
70-16456	UH1H		
70-16457	UH1H	w.o.1.4.82	
70-16458	UH1H	SoC	
70-16459	UH1H	1InfDiv	
70-16460	UH1H		
70-16461	UH1H	1267MedCo	
70-16462	UH1H	1InfDiv	
70-16463	UH1H		
70-16464	UH1V		
70-16465	UH1V		
70-16466	UH1H		
70-16467	UH1H		
70-16468	UH1V	812Medco	
70-16469	UH1H	104ACS	
70-16470	UH1H	A\1-207AVN	AK NG
70-16471	UH1H	w.o.12.7.76	
70-16472	UH1H	SoC	
70-16473	UH1H		
70-16474	UH1H		
70-16475	UH1H	104ACS	
70-16476	UH1H	25AvCo	
70-16477	UH1H	B\1-132AVN	DC NG
70-16478	UH1H	126MedCo	CA NG
70-16479	UH1H	6ACR	
70-16480	UH1H	w.o.28.1.80	
70-16481	UH1H	116AHB	
70-16482	UH1H		
70-16483	UH1H	180AvCo	
70-16484	UH1H	SoC	
70-16485	UH1H	SnC	
70-16486	UH1H		
70-16487	UH1H		

Serial	Type	Notes	
70-16488	UH1H		
70-16489	UH1H	SoC	
70-16490	UH1H		
70-16491	JUH1H		
70-16492	UH1H	6ACR	
70-16493	UH1H		
70-16494	UH1H		
70-16495	UH1H	A\1-131AVN	AL NG
70-16496	UH1H	C\1-108AVN	AZ NG 35
70-16509	UH1H	SoC	
70-16510	UH1H	SoC	
70-16511	UH1H	SoC	
70-16512	UH1H	SoC	
70-16513	UH1H	SoC	
70-16514	UH1H	SoC	
70-16515	UH1H		
70-16516	UH1V	1259MedCo	
70-16517	UH1H	198AvCo	
70-16518	UH1H	198AvCo	
71-20000	UH1H	92InfBde	
71-20001	UH1V	5-158AVN	
71-20002	UH1V	51AvCo	
71-20003	UH1H	SoC	
71-20004	UH1H		
71-20005	UH1H	Berlin Brigade	
71-20006	UH1H	131AVN	
71-20007	UH1H	w.o.17.9.78	
71-20008	UH1H	w.o.15.7.75	
71-20009	UH1V	126MedCo	CA NG
71-20010	UH1H	40AVN	
71-20011	UH1H	40AVN	
71-20012	UH1H	150AVN	DE NG
71-20013	UH1H	150AVN	DE NG
71-20014	UH1H	44AVN	
71-20015	UH1H	44AVN	
71-20016	UH1V		
71-20017	UH1V	24MedCo	
71-20018	UH1V	421MedDet	
71-20019	UH1V	25AVN	
71-20020	UH1H		
71-20021	UH1H	DAC\MDW	
71-20022	UH1H	31ADB	
71-20023	UH1V	68MedDet	
71-20024	UH1V	68MedDet	
71-20025	UH1H	Berlin Brigade	
71-20026	UH1V	68MedDet	
71-20027	UH1H	6AvCo	
71-20028	UH1H	w.o.	
71-20029	UH1H	42AVN	
71-20030	UH1H	A\1-207AVN	AK NG
71-20031	UH1H		
71-20032	UH1H	104ACS	
71-20033	UH1H		
71-20034	UH1H		
71-20035	UH1H		
71-20036	UH1V		
71-20037	UH1H		
71-20038	UH1H	10	
71-20039	UH1H	25AVN	
71-20040	UH1H	1-159AVN	
71-20041	UH1H	25AVN	
71-20042	UH1H	US Army	Japan
71-20043	UH1H		
71-20044	UH1H		
71-20045	UH1H	scrapped	
71-20046	UH1H		
71-20047	UH1H	6ACR	
71-20048	UH1H	SoC	
71-20049	UH1H	w.o.24.2.91	
71-20050	UH1V		
71-20051	UH1H	w.o.4.10.84	
71-20052	UH1H	w.o.16.7.77	
71-20053	UH1H		
71-20054	UH1H	w.o.	
71-20055	UH1H	w.o.31.5.78	
71-20056	UH1H	1AVN	
71-20057	UH1H	4InfDiv	
71-20058	UH1H	2-158AVN Reserve	
71-20059	UH1H	163ACR	
71-20060	UH1H	163ACR	
71-20061	UH1H	40AVN	
71-20062	UH1H	9AVN	
71-20063	UH1V	63MedCo	
71-20064	UH1H	1InfDiv	
71-20065	UH1H		
71-20066	UH1H	47AVN	
71-20067	UH1H	SoC	
71-20068	UH1H	USMA	
71-20069	UH1H		
71-20070	UH1H	56AvCo	
71-20071	UH1H	SoC	
71-20072	UH1H	92InfBde	
71-20073	UH1H	B\1-132AVN	DC NG
71-20074	UH1H	B\1-132AVN	DC NG
71-20075	UH1H	SoC	
71-20076	UH1H	SoC	
71-20077	UH1H	SoC	
71-20078	UH1H		
71-20079	UH1H	w.o.	
71-20080	UH1H	w.o.	
71-20081	UH1H	198AvCo	
71-20082	UH1H	SoC	
71-20083	UH1H	SoC	
71-20084	UH1H	SoC	
71-20085	UH1H	SoC	
71-20086	UH1H	SoC	
71-20087	UH1H	SoC	
71-20088	UH1V	1259MedCo	
71-20089	UH1H	SoC	
71-20090	UH1H		
71-20091	UH1H	w.o.	
71-20092	UH1H	SoC	
71-20093	UH1H	B\26AVN	
71-20094	UH1H	193AVN	
71-20095	UH1H	SoC	
71-20096	UH1H	SoC	
71-20097	UH1H	SoC	
71-20098	UH1H	SoC	
71-20099	UH1H	SoC	
71-20100	UH1H	SoC	
71-20101	UH1H		
71-20102	UH1H	82AVN	
71-20103	UH1H	to N81477	
71-20104	UH1H		
71-20105	UH1V		
71-20106	UH1H	SoC	
71-20107	UH1H	SoC	
71-20108	UH1H	SoC	
71-20109	UH1H		
71-20110	UH1H	116AHB	
71-20111	UH1H	SoC	
71-20112	UH1V	131AVN	
71-20113	UH1H	SoC	
71-20114	UH1H	SoC	
71-20115	UH1H		
71-20116	UH1H	SoC	
71-20117	UH1H	207AvCo	
71-20118	UH1H	B\4-228AVN	
71-20119	UH1H	150AVN	DE NG
71-20120	UH1H		
71-20121	UH1H	193AVN	
71-20122	UH1H		
71-20123	UH1H	56AvCo	
71-20124	UH1H		
71-20127	UH1H	9AVN	
71-20128	UH1H	w.o.21.9.80	
71-20129	UH1H	26AVN	
71-20130	UH1H	12CAG	
71-20131	UH1H	SoC	
71-20132	UH1H	25AvCo	
71-20133	UH1H	w.o.28.5.75	
71-20134	UH1H		
71-20135	UH1H	SoC	
71-20138	UH1H	SoC	
71-20139	UH1H	207AvCo	
71-20140	UH1H	SoC	
71-20141	UH1H	SoC	
71-20142	UH1H	SoC	
71-20143	UH1H	SoC	
71-20144	UH1H	SoC	
71-20145	UH1H		
71-20146	UH1H	229AVN	
71-20147	UH1H	SoC	
71-20148	UH1H	SoC	
71-20149	UH1H	SoC	
71-20150	UH1H	SoC	

Serial	Type	Unit/Notes	
71-20151	UH1H	SoC	
71-20152	UH1H	SoC	
71-20153	UH1H	SoC	
71-20154	UH1H	SoC	
71-20155	UH1H	SoC	
71-20156	UH1V		SD NG
71-20157	UH1H	SoC	
71-20158	UH1H		
71-20159	UH1V	273MedDet	
71-20160	UH1H	207AvCo	
71-20161	UH1H	w.o.2.3.91	
71-20162	UH1H	308AHB	
71-20163	UH1H		
71-20164	UH1H	377TSARCOM	
71-20165	UH1H		
71-20166	UH1V	1159MedCo	
71-20167	UH1H	9AVN	
71-20168	UH1H		
71-20169	UH1H	9AVN	
71-20170	UH1H		
71-20171	JUH1H	TECOM	
71-20172	UH1H	166ACR	
71-20173	UH1H		
71-20174	UH1H		
71-20175	UH1H	to N8147G	
71-20176	UH1H	92InfBde	
71-20177	UH1V		
71-20178	UH1H	5-158AVN	
71-20179	UH1H	w.o.23.7.75	
71-20180	UH1H		
71-20181	UH1H	49AvCo	
71-20182	UH1H	w.o.9.6.78	
71-20183	UH1H		
71-20184	UH1H		
71-20185	UH1H	SoC	
71-20186	UH1H	SoC	
71-20187	UH1H	SoC	
71-20188	UH1H	SoC	
71-20189	UH1H	SoC	
71-20190	UH1H	184AHC	
71-20191	UH1H	92InfBde	
71-20192	UH1H		
71-20193	UH1H	SoC	
71-20194	UH1H	SoC	
71-20195	UH1H	SoC	
71-20196	UH1H	SoC	
71-20197	UH1H	SoC	
71-20198	UH1H	SoC	
71-20199	UH1H	SoC	
71-20200	UH1H	SoC	
71-20201	UH1H		
71-20202	UH1H		
71-20203	UH1H	B\4-228AVN	
71-20204	UH1H	SoC	
71-20205	UH1H	SoC	
71-20206	UH1V	w.o.14.6.84	
71-20207	UH1H	3ACR	
71-20208	UH1H		
71-20209	UH1H	w.o.16.8.76	
71-20210	UH1H	9AVN	
71-20211	UH1H		
71-20212	UH1H	131AVN	
71-20213	UH1V		
71-20214	UH1V	348MedDet	
71-20215	UH1H	6AvCo	
71-20216	UH1H	6AvCo	
71-20217	UH1H		
71-20218	UH1H	w.o.4.1.78	
71-20219	UH1H	A\1-207AVN	AK NG
71-20220	UH1V	812Medco	det.1
71-20221	UH1H		
71-20222	UH1H		
71-20223	UH1H	w.o.15.9.84	
71-20224	UH1H		
71-20225	UH1H	82AVN	
71-20226	UH1H	C\1-108AVN	AZ NG 17
71-20227	UH1H		
71-20228	UH1V		
71-20229	UH1H		
71-20230	UH1H	TF Viper	
71-20231	UH1H	w.o.	
71-20232	UH1H		
71-20233	UH1H		
71-20234	UH1H	23TF	
71-20235	UH1H		
71-20236	UH1H		
71-20237	UH1V	507MedCo	
71-20238	UH1H	SoC	
71-20239	UH1H	SoC	
71-20240	UH1H		
71-20241	UH1H		
71-20242	UH1H	9InfDiv	
71-20243	UH1H	to Thailand	
71-20244	UH1H		
71-20245	UH1H	6ACR	
71-20246	UH1H	A\1-207AVN	AK NG
71-20247	UH1H	498MedCo	
71-20248	UH1H		
71-20249	UH1H	SoC	
71-20250	UH1H	SoC	
71-20251	UH1H		
71-20252	UH1H		
71-20253	UH1H	6ACR	
71-20254	UH1H	3ACR	
71-20255	UH1H	7ATC	
71-20256	UH1H		
71-20257	UH1H		
71-20258	UH1H	1-183AVN	ID NG
71-20259	UH1V		
71-20260	UH1H	w.o.4.3.80	
71-20261	UH1H	SoC	
71-20262	UH1H	4-107ACR	OH NG
71-20263	UH1H	120AvCo	
71-20264	UH1H	SoC	
71-20265	UH1H	SoC	
71-20266	UH1H	SoC	
71-20267	UH1H	SoC	
71-20268	UH1H	SoC	
71-20269	UH1H	SoC	
71-20270	UH1H	225AvCo	
71-20271	UH1H	B\1-132AVN	DC NG
71-20272	UH1H	1-158AVN	
71-20273	UH1H	w.o.	
71-20274	UH1H	SoC	
71-20275	UH1H	SoC	
71-20276	UH1H	C\1-108AVN	AZ NG 20
71-20277	UH1H	SoC	
71-20278	UH1H	SoC	
71-20279	UH1H	SoC	
71-20280	UH1H	62AvCo	
71-20281	UH1H	62AvCo	
71-20282	UH1H	SoC	
71-20283	UH1H	w.o.	
71-20284	UH1V	w.o.4.9.82	
71-20285	UH1H	w.o.28.10.83	
71-20286	UH1V		
71-20287	UH1H		
71-20288	UH1H	26AVN	
71-20289	UH1H	SoC	
71-20290	UH1H	SoC	
71-20291	UH1H	SoC	
71-20292	UH1H	126MedCo	CA NG
71-20293	UH1H	SoC	
71-20294	UH1V	112MedDet	
71-20295	UH1H	126ACS	
71-20296	UH1H	w.o.27.6.78	
71-20297	UH1H	4ARC	
71-20298	UH1H		
71-20299	UH1H	SoC	
71-20300	UH1V	347MedDet	
71-20301	UH1V		
71-20302	UH1V	273MedDet	
71-20303	UH1V	6USACTF	
71-20304	UH1V	to N81464	
71-20305	UH1V	159MedDet	
71-20306	UH1V	10AHB	
71-20307	UH1V		
71-20308	UH1H	164AvCo	
71-20309	UH1H	1Cav	
71-20310	UH1H	to Bolivia	
71-20311	UH1V	145MedDet	
71-20312	UH1H		
71-20313	UH1H		
71-20314	UH1H		
71-20315	UH1V	w.o.	
71-20316	UH1H		

71-20317	UH1H	B\1-132AVN	DC NG
71-20318	UH1H	B\1-245AVN	OK NG
71-20319	UH1H	A\1-207AVN	AK NG
71-20320	UH1H		
71-20321	UH1H		
71-20322	UH1H	150AVN	DE NG
71-20323	UH1H	B\1-132AVN	DC NG
71-20324	UH1H	HQ\FORSCOM	
71-20325	UH1H		
71-20326	UH1H		
71-20327	UH1H	SoC	
71-20328	UH1H		
71-20329	UH1V		
71-20330	UH1H		
71-20331	UH1H	NTC\TPS	
71-20332	UH1V	347MedDet	
71-20333	UH1H	AATC	33BCairns AAF
72-21465	UH1H	w.o.21.3.88	
72-21467	UH1H	w.o.	
72-21469	UH1H	SoC	
72-21470	UH1H	SoC	
72-21473	UH1H	7InfDiv	
72-21474	UH1H	1-159AVN	
72-21475	UH1H		
72-21476	UH1H		
72-21477	UH1H	w.o.26.8.78	
72-21478	UH1H	to N81473	
72-21479	UH1H		
72-21480	UH1H	1AVN	
72-21481	UH1V	AATC	Cairns AAF
72-21482	UH1H		
72-21483	UH1H	92InfBde	
72-21484	UH1H	10Cav	
72-21485	UH1H	D\1-228AVN	
72-21486	UH1H		
72-21487	UH1H	26AVN	
72-21488	UH1V		
72-21489	UH1H	SoC	
72-21490	UH1H	SoC	
72-21491	UH1H	SoC	
72-21492	UH1H	47AVN	
72-21501	UH1H	SoC	
72-21502	UH1H	26AVN	
72-21503	UH1H		
72-21504	UH1H	SoC	
72-21505	UH1H	SoC	
72-21506	UH1H	SoC	
72-21507	UH1H		
72-21508	UH1H	82AVN	
72-21509	UH1H	86ARCOM	
72-21510	UH1H	HQ\FORSCOM	
72-21511	UH1H	SoC	
72-21512	UH1H	USAFDA	
72-21513	UH1H	323AVN	
72-21514	UH1H	USAFDA	
72-21515	UH1H	B\26AVN	
72-21516	UH1V	237MedCo	
72-21517	UH1H	A\1-131AVN	AL NG
72-21518	UH1H	1-158AVN	
72-21521	UH1H	SoC	
72-21522	UH1H	SoC	
72-21523	UH1H	42AVN	
72-21524	UH1H	1085Medco	Det.1
72-21533	UH1H	1-159AVN	
72-21534	UH1H	w.o.12.11.79	
72-21535	UH1H	12Cav	
72-21538	UH1H	SoC	
72-21539	UH1H	SoC	
72-21540	UH1H	SoC	
72-21541	UH1H	SoC	
72-21542	UH1H	92InfBde	
72-21547	UH1V	347MedDet	
72-21548	UH1H		
72-21549	UH1V	131AVN	
72-21550	UH1H		
72-21551	UH1H		
72-21552	UH1H	8AVN	
72-21553	UH1H		
72-21554	UH1H	102ARCOM	
72-21555	UH1H	207AvCo	
72-21556	UH1H		
72-21557	UH1H	SoC	
72-21558	UH1H	82AVN	

72-21559	UH1V	w.o.	
72-21560	UH1H		
72-21561	UH1H		
72-21562	UH1H	SoC	
72-21563	UH1H	SoC	
72-21564	UH1H		
72-21565	UH1V	w.o.31.8.90	
72-21566	UH1H	w.o.4.8.76	
72-21567	UH1H	56AvCo	
72-21568	UH1H	62AvCo	
72-21569	UH1H	56AvCo	
72-21570	UH1H		
72-21571	UH1H		
72-21572	UH1H		
72-21573	UH1H	SoC	
72-21574	UH1H		
72-21578	UH1H	SoC	
72-21579	UH1H	w.o.	
72-21580	UH1H	SoC	
72-21581	UH1H	56AvCo	
72-21582	UH1H	207AvCo	
72-21583	UH1H	25AvCo	
72-21584	UH1H	11ACR	
72-21585	UH1H	10AHB	
72-21586	UH1H		
72-21587	UH1H	323AvCo	
72-21588	UH1H	5-158AVN	
72-21589	UH1V	3ACT	
72-21590	UH1H		
72-21591	UH1H	SoC	
72-21592	UH1H	1-159AVN	
72-21596	UH1H	207AvCo	
72-21597	UH1H	SoC	
72-21598	UH1H	to Korea	
72-21599	UH1H	w.o.13.5.81	
72-21600	UH1H	198AvCo	
72-21601	UH1H	SoC	
72-21602	UH1H	SoC	
72-21603	UH1H	SoC	
72-21604	UH1H	SoC	
72-21605	UH1H	23TF	
72-21606	UH1H		
72-21607	UH1H	SoC	
72-21608	UH1H	SoC	
72-21609	UH1H	SoC	
72-21610	UH1H	SoC	
72-21611	UH1H	SoC	
72-21612	UH1H	SoC	
72-21613	UH1H	SoC	
72-21614	UH1H	SoC	
72-21615	UH1H	44AVN	
72-21616	UH1H	C\1-108AVN	AZ NG
72-21617	UH1H	SoC	
72-21618	UH1H	stored	Ramstein AB,FRG
72-21619	UH1H		
72-21620	UH1H	5-158AVN	
72-21621	UH1H		
72-21622	UH1H	308AHB	
72-21623	UH1H	SoC	
72-21624	UH1H	26AvBat	
72-21625	UH1H	3ACR	
72-21626	UH1H	82AVN	
72-21627	UH1H	1AVN	
72-21628	UH1H		VA NG
72-21629	UH1V	79ARCOM	
72-21630	UH1H	5-158AVN	
72-21631	UH1H	172InfBde	
72-21632	UH1H	56AvCo	
72-21633	UH1H	63ARCOM	
72-21634	UH1H	8AVN	
72-21635	UH1H	5-158AVN	
72-21636	UH1H	56AvCo	
72-21637	UH1H	6AvCo	
72-21638	UH1V		
72-21639	UH1H	92InfBde	
72-21640	UH1H	SoC	
72-21641	UH1H	HQ\REDCOM	
72-21642	UH1H	6ACR	
72-21643	UH1H	82AVN	
72-21644	UH1H	SoC	
72-21645	UH1H	120AvCo	
72-21646	UH1H	25AvCo	
72-21647	UH1H	HQ\REDCOM	

72-21648	UH1H	193AVN	
72-21649	UH1H	SoC	
73-21661	UH1H	8-158AVN	
73-21662	UH1H		
73-21663	UH1H	USAFDA	
73-21664	UH1V	131AVN	
73-21665	UH1H	813MedDet	
73-21666	UH1V	812MedCo	
73-21667	UH1H	Berlin Brigade	
73-21668	UH1H	62AvCo	
73-21669	UH1H		
73-21670	UH1H	6ACR	
73-21671	UH1H	6ACR	
73-21672	UH1H	w.o.28.10.85	
73-21673	UH1H	1AVN	
73-21674	UH1V		
73-21675	UH1H		
73-21676	UH1H		
73-21677	UH1H	SoC	
73-21678	UH1H	B\1-132AVN	DC NG
73-21679	UH1H	5-158AVN	
73-21680	UH1H	158AVN	
73-21681	UH1H	SoC	
73-21682	UH1H	5-158AVN	
73-21683	UH1H	158AVN	
73-21684	JUH1H	63MedDet	
73-21685	UH1H	ERADCOM	
73-21686	UH1H		
73-21687	UH1H	B\1-132AVN	DC NG
73-21688	UH1H	131AVN	
73-21689	UH1V	68MedDet	
73-21690	UH1H	USAICS	
73-21691	UH1V		
73-21692	UH1H		
73-21693	UH1H		
73-21694	UH1H	w.o.29.5.82	
73-21695	UH1H		
73-21696	UH1H	308AHB	
73-21697	UH1H	6AvCo	
73-21698	UH1H	5-158AVN	
73-21699	UH1H	SoC	
73-21700	UH1H	11ACR	
73-21701	UH1H	w.o.9.1.75	
73-21702	UH1H		
73-21703	UH1H	DAC\MDW	
73-21704	UH1H	SoC	
73-21705	UH1H	SoC	
73-21706	UH1H	2\10Cav	
73-21707	UH1H		
73-21708	UH1H	HQ-Army West	
73-21709	UH1H	25AVN	
73-21710	UH1H		
73-21711	UH1H	w.o.17.2.81	
73-21712	UH1H		
73-21713	UH1H	B\1-245AVN	OK NG
73-21714	UH1H	56AvCo	
73-21715	UH1H	207AvCo	
73-21716	UH1H		
73-21717	UH1V		
73-21718	UH1H	352ASA Co	
73-21719	UH1V	352ASA Co	
73-21720	UH1H	SoC	
73-21721	UH1H	6ACR	
73-21722	UH1H		
73-21723	UH1H	166ACR	
73-21724	JUH1H	82AVN	
73-21725	UH1H	5-158AVN	
73-21726	UH1H	5-158AVN	
73-21727	UH1V	63MedDet	
73-21728	UH1H	62AvCo	
73-21729	UH1H		
73-21730	UH1H	SoC	
73-21731	UH1H	SoC	
73-21732	UH1H		
73-21733	UH1H	SoC	
73-21734	UH1H		
73-21735	UH1H	347MedDet	
73-21736	UH1H		
73-21737	UH1H	3ACR	
73-21738	UH1H		
73-21739	UH1H	349MedDet	
73-21740	UH1H	82AVN	
73-21741	UH1H		

73-21742	UH1H	4InfDiv	
73-21743	UH1H		
73-21744	UH1H	1898AvCo	
73-21745	UH1V	812MedCo	
73-21746	UH1H	44AVN	
73-21747	UH1V		SD NG
73-21748	UH1V		
73-21749	UH1H	CCAD	
73-21750	UH1H		
73-21751	UH1H	205AvIm	
73-21752	UH1H	308AHB	
73-21753	UH1H		
73-21754	UH1H	308AHB	
73-21755	UH1H	SoC	
73-21756	UH1H	SoC	
73-21757	UH1H		
73-21758	UH1V		
73-21759	UH1H	SoC	
73-21760	UH1H		
73-21761	UH1V		
73-21762	UH1H		
73-21763	UH1V	352ASA Co	
73-21764	UH1H		
73-21765	UH1V	352ASA Co	
73-21766	UH1H	200AvCo	
73-21767	UH1V	236MedDet	
73-21768	UH1H	w.o.26.6.76	
73-21769	UH1H	8AVN	
73-21770	UH1H	25AVN	
73-21771	UH1H	1898AvCo	
73-21772	UH1H		
73-21773	UH1H		
73-21774	UH1H		
73-21775	UH1H	26AVN	
73-21776	UH1H	3ACT	
73-21777	UH1H	82AVN	
73-21778	UH1V	112MedDet	
73-21779	UH1H	A\1-207AVN	AK NG
73-21780	UH1H	92InfBde	
73-21781	UH1H	56AvCo	
73-21782	UH1H		
73-21783	UH1H	374MedDet	
73-21784	UH1H	56AvCo	
73-21785	UH1H	9AVN	
73-21786	UH1H		
73-21787	UH1H	193AVN	
73-21788	UH1H	3AVN	
73-21789	UH1H		
73-21790	UH1H	92InfDiv	
73-21791	UH1H		
73-21792	UH1H		
73-21793	UH1H		
73-21794	UH1H	SoC	
73-21795	UH1H	SoC	
73-21796	UH1H	SoC	
73-21797	UH1H	SoC	
73-21798	UH1H	SoC	
73-21799	UH1H	SoC	
73-21800	UH1H	SoC	
73-21801	UH1H		
73-21802	JUH1H	5AVN	
73-21803	UH1H	164AvCo	
73-21804	UH1H	131AVN	
73-21805	UH1H		
73-21806	UH1H	5-158AVN	
73-21807	UH1H	AATC	07E Lowe AHP
73-21808	UH1H		
73-21809	UH1H	150AVN	DE NG
73-21810	UH1H	ERADCOM	
73-21811	UH1H	SoC	
73-21812	UH1H	SoC	
73-21813	UH1H	SoC	
73-21814	UH1H	SoC	
73-21815	UH1H	SoC	
73-21816	UH1H	SoC	
73-21817	UH1H	SoC	
73-21818	UH1H	308AHB	
73-21819	UH1H	5-158AVN	
73-21820	UH1V	812MedCo	
73-21821	UH1H		
73-21822	UH1H	25AvCo	
73-21823	UH1H	308AHB	
73-21824	UH1H	26AVN	

Serial	Type	Notes	
73-21825	UH1H	w.o.8.7.82	
73-21826	UH1H	56AvCo	
73-21827	UH1H	SoC	
73-21828	UH1H	1898AvCo	
73-21829	UH1H	1AvBde	
73-21830	UH1H		
73-21831	UH1H	8-158AVN	
73-21832	UH1V	498MedCo	
73-21833	UH1H		GA NG
73-21834	UH1H	w.o.11.3.89	
73-21835	UH1H	222AVN	
73-21836	UH1H	w.o.19.11.75	
73-21837	UH1H	120AvCo	
73-21838	UH1H		
73-21839	UH1H		
73-21840	UH1H		
73-21841	UH1H	SoC	
73-21850	UH1H		
73-21851	UH1H	1ACR	
73-21852	UH1H	1Cav	
73-21853	UH1H	45AVN	
73-21854	UH1H	2-135AVN	CO NG
73-21855	UH1H	42AVN	
73-21856	UH1H	B\26AVN	
73-21857	UH1H		
73-21858	UH1H		
73-21859	UH1H		
73-21860	UH1H		
73-22066	UH1V		
73-22067	UH1V		
73-22072	UH1H		
73-22073	UH1H		
73-22074	UH1H	374MedDet	
73-22075	UH1H	25AvCo	
73-22076	UH1H	to Argentina	
73-22077	UH1H	SoC	
73-22078	UH1H	63MedDet	
73-22079	UH1H		
73-22080	UH1V	6USACTF	
73-22081	UH1H	w.o.12.2.76	
73-22082	UH1H	321MedDet	
73-22083	UH1H	SoC	
73-22084	UH1H	to Argentina	
73-22085	UH1H	SoC	
73-22086	UH1H	SoC	
73-22087	UH1H	SoC	
73-22088	UH1H	SoC	
73-22089	UH1H	SoC	
73-22090	JUH1H	ERADCOM	
73-22091	UH1V	131AVN	
73-22092	UH1H	56AvCo	
73-22093	UH1H		
73-22094	UH1H		
73-22095	UH1H	SoC	
73-22096	UH1H	SoC	
73-22097	UH1V	348MedDet	
73-22098	UH1H		
73-22099	UH1H	278ACR	
73-22100	UH1V		
73-22101	UH1H	B\1-132AVN	DC NG
73-22102	UH1V		
73-22122	UH1H	1Cav	
73-22123	UH1H	56AvCo	
73-22124	UH1H	56AvCo	
73-22125	UH1H	56AvCo	
73-22126	UH1H	421MedCo	
73-22127	UH1H	308AHB	
73-22128	UH1H	10AHB	
73-22129	UH1H	10AHB	
73-22130	UH1H	1-159AVN	
73-22131	UH1H	10AHB	
73-22132	UH1H	323AvCo	
73-22133	UH1H		
73-22134	UH1H	150ACR	
73-22135	UH1H		
74-22295	UH1H	412MedDet	
74-22296	UH1H	SoC	
74-22297	UH1H	120AvCo	
74-22298	UH1V	812MedCo	
74-22299	UH1V	63MedDet	
74-22300	UH1H	w.o.5.8.86	
74-22301	UH1H	6AvDet	
74-22302	UH1H		
74-22303	UH1H	207AvCo	
74-22304	UH1H		
74-22305	UH1H		
74-22306	UH1H		
74-22307	UH1H	B\1-245AVN	OK NG
74-22308	UH1H		
74-22309	UH1H	w.o.22.11.90	
74-22310	UH1H	7InfDiv	
74-22311	UH1V	412MedDet	
74-22312	UH1V	348MedDet	
74-22313	UH1H	25AVN	
74-22314	UH1H		
74-22315	UH1H	237MedDet	
74-22316	UH1H		
74-22317	UH1H		
74-22318	UH1H	56AvCo	
74-22324	UH1H		
74-22325	UH1H		
74-22326	UH1H	56AvCo	
74-22327	UH1H	11ACR	
74-22328	UH1H	308AHB	
74-22329	UH1H	205AvIm	
74-22330	UH1H	TF Viper	
74-22331	UH1H	321MedDet	
74-22332	UH1H	150AVN	
74-22333	UH1V	347MedDet	
74-22334	UH1H		
74-22335	UH1H	stored	Ramstein AB,FRG
74-22336	UH1H	B\26AVN	
74-22337	UH1H	w.o.2.12.77	
74-22338	UH1H	w.o.11.6.79	
74-22339	UH1H	SoC	
74-22340	UH1H		
74-22341	UH1V		
74-22342	UH1H		
74-22343	UH1H	B\1-245AVN	OK NG
74-22344	UH1H	1-159AVN	
74-22345	UH1H		
74-22346	UH1H	SoC	
74-22347	UH1H	5-158AVN	
74-22348	UH1H	w.o.12.7.77	
74-22349	UH1H	5-158AVN	
74-22350	UH1H	A\1-131AVN	AL NG
74-22351	UH1H	62AvCo	
74-22352	UH1H		
74-22353	UH1H		
74-22354	UH1V	w.o.24.6.90	
74-22355	UH1H	8AVN	
74-22356	UH1H	5-158AVN	
74-22357	UH1H		
74-22358	UH1H		GA NG
74-22359	UH1H		
74-22360	UH1H	1-158AVN	
74-22361	UH1H		
74-22362	UH1H	B\1-245AVN	OK NG
74-22363	UH1H		
74-22364	UH1H		
74-22365	UH1H		
74-22366	UH1V	68MedDet	
74-22367	JUH1H		
74-22368	UH1H		
74-22369	UH1H	1Cav	
74-22370	UH1H	10AHB	
74-22371	UH1H		
74-22372	UH1V	321MedDet	
74-22373	UH1H	to Philippine	
74-22374	UH1V	w.o.25.7.85	
74-22375	UH1H	116AHB	
74-22376	UH1H	124ARCOM	
74-22377	UH1H	124ARCOM	
74-22378	UH1H	1267MedCo	MO NG
74-22379	UH1H	AATC	79G Lowe AHP
74-22380	UH1V		
74-22381	UH1H	DAC\MDW	
74-22382	UH1H	DAC\MDW	
74-22383	UH1H		
74-22384	UH1H	w.o.5.11.77	
74-22385	UH1H	7InfDiv	
74-22386	UH1H	B\26AVN	
74-22387	UH1H		
74-22388	UH1H	26AVN	
74-22389	UH1H		
74-22390	UH1H		

Serial	Type	Unit	Note
74-22391	UH1H		
74-22392	UH1H		
74-22393	UH1H	SoC	
74-22394	UH1H	193AVN	
74-22395	UH1H	SoC	
74-22396	UH1H		
74-22397	UH1H	1-159AVN	
74-22398	UH1H	1267MedCo	MO NG
74-22399	UH1H		
74-22400	UH1H		
74-22401	UH1H		
74-22402	UH1H	SoC	
74-22403	UH1H	SoC	
74-22404	UH1H	SoC	
74-22405	UH1H	SoC	
74-22406	UH1H	SoC	
74-22407	UH1V		
74-22408	UH1H	3AVN	
74-22409	UH1H	6ACR	
74-22410	UH1H	11ACR	
74-22411	UH1H	25AvCo	
74-22412	UH1H	4ARC	
74-22413	UH1H	13AHB	
74-22414	UH1H	w.o.6.5.88	
74-22415	UH1V	SoC	
74-22416	UH1H		
74-22417	UH1H		
74-22418	UH1H		
74-22419	UH1H	1-183AVN	ID NG
74-22420	UH1H	B\1-245AVN	OK NG
74-22421	UH1V	812MedCo	
74-22422	UH1H	C\1-108AVN	AZ NG 29
74-22423	UH1H	278ACR	
74-22424	UH1H	278ACR	
74-22425	UH1H	B\1-132AVN	DC NG
74-22426	UH1H		
74-22427	UH1H	26AVN	
74-22428	UH1V	1255MedCo	NV NG
74-22429	UH1H	HQ\REDCOM	
74-22430	UH1V	1255MedCo	NV NG
74-22431	UH1V		NM NG
74-22432	UH1H	w.o.22.7.83	
74-22433	UH1V	1133MedDet	
74-22434	UH1H	107AvCo	
74-22435	UH1H	10SFG	
74-22436	JUH1H	10SFG	
74-22437	UH1H	123ARCOM	
74-22438	UH1H	123ARCOM	
74-22439	UH1H	123ARCOM	
74-22440	UH1H	98DivTrng	
74-22441	UH1H	98DivTrng	
74-22442	UH1V	63MedDet	
74-22443	UH1H		
74-22444	UH1H	2ACR	
74-22445	UH1H	1-159AVN	
74-22446	UH1H	B\4-228AVN	
74-22447	UH1V		
74-22448	UH1H	25AvCo	
74-22449	UH1H	SoC	
74-22450	UH1H	86ARCOM	
74-22451	UH1H	10Cav	
74-22452	UH1H	10Cav	
74-22453	UH1H		
74-22454	UH1H		
74-22455	UH1H		
74-22456	UH1H	158AVN	
74-22457	UH1H		
74-22458	UH1H		
74-22459	UH1H		
74-22460	UH1H	SoC	
74-22461	UH1H		
74-22462	UH1H	56AvCo	
74-22463	UH1V	131AVN	
74-22464	UH1H		
74-22465	UH1H	62AvCo	
74-22466	UH1H	126MedCo	CA NG
74-22467	UH1H		
74-22468	UH1V	126MedCo	CA NG
74-22469	UH1H		
74-22470	UH1H	7InfDiv	
74-22471	UH1H	24AVN	
74-22472	UH1V		
74-22473	UH1V	1150MedCo	
74-22474	UH1H	7\6Cav	
74-22475	UH1V	347MedDet	
74-22476	UH1H	26AVN	
74-22477	UH1H	7InfDiv	
74-22478	JUH1H		
74-22479	UH1H		
74-22480	UH1V		
74-22481	UH1H	w.o.15.9.82	
74-22482	UH1V		
74-22483	UH1H	56AvCo	
74-22484	UH1H	SoC	
74-22485	UH1H	SoC	
74-22486	UH1H	SoC	
74-22487	UH1H	SoC	
74-22488	UH1H	SoC	
74-22489	UH1H		
74-22490	UH1V	131AVN	
74-22491	UH1H	10AHB	
74-22492	UH1H	SoC	
74-22493	UH1H	SoC	
74-22494	UH1H	SoC	
74-22495	UH1H	HQ\REDCOM	
74-22496	UH1H		
74-22497	UH1H	2ACR	
74-22498	UH1V		
74-22499	UH1H	104ACS	
74-22500	UH1H	HQ\REDCOM	
74-22501	UH1H	158AVN	
74-22502	UH1H	207AvCo	
74-22503	UH1H		
74-22504	UH1H	207AvCo	
74-22505	UH1H	166ACR	
74-22506	UH1H	AFTC	
74-22507	UH1H	6ACR	
74-22508	UH1H	7InfDiv	
74-22509	UH1H	7InfDiv	
74-22510	UH1H	7InfDiv	
74-22511	UH1H		
74-22512	UH1H		
74-22513	UH1H	HQ\USEUCOM	
74-22514	UH1H	HQ\USEUCOM	
74-22515	UH1H		
74-22516	UH1H	7InfDiv	
74-22517	UH1H	SoC	
74-22518	UH1H	SoC	
74-22519	UH1H	SoC	
74-22520	UH1H	to Argentina	
74-22521	UH1H	159AHB	
74-22522	UH1H	7InfDiv	
74-22523	UH1H	SoC	
74-22524	UH1H	163ACR	
74-22526	UH1H		
74-22527	UH1H		
74-22528	UH1H		
74-22529	UH1H		
74-22530	UH1H	w.o.	
74-22531	UH1H	SoC	
74-22532	UH1H	SoC	
74-22533	UH1H		
74-22534	UH1H		
74-22535	UH1H	USAICS	
74-22536	UH1H		
74-22537	UH1H	2\10Cav	
74-22538	UH1H	2\10Cav	
74-22539	UH1H	166ACR	
74-22540	UH1H	7\6Cav	
74-22541	UH1V		
74-22542	UH1V		
74-22543	UH1H	1InfDiv	
74-22544	UH1V		
76-22651	UH1H	SoC	
76-22652	UH1H	SoC	
76-22653	UH1H	SoC	
76-22654	UH1H	SoC	
76-22655	UH1H	SoC	
76-22656	UH1H	SoC	
76-22659	UH1H	SoC	
76-22660	UH1H	SoC	
76-22663	UH1H	SoC	
76-22664	UH1H	SoC	
76-22665	UH1H	SoC	
76-22666	UH1H	SoC	
76-22667	UH1H	SoC	

76-22668	UH1H	to Philippine	
76-22669	UH1H	SoC	
76-22670	UH1H	8AVN	

76-22671	UH1H	w.o.12.3.84	
76-22672	UH1H	8AVN	

H6 CAYUSE

62-4212	YOH6A	ex N9696F	
62-4213	YOH6A	preserved	Ft. Rucker
62-4214	YOH6A		
62-4215	YOH6A		
62-4216	YOH6A		
62-12624	YOH6A	preserved	Gulfport RAP, Ms
65-12916	OH6A	SoC	
65-12917	OH6A	NOTAR demonstrator	
65-12918	OH6A	163ACR	
65-12919	OH6A	SoC	
65-12920	OH6A	SoC	
65-12921	OH6A	278ACR	
65-12922	OH6A		
65-12923	OH6A	SoC	
65-12924	OH6A	SoC	
65-12925	OH6A		VT NG
65-12926	OH6A	26AVN	
65-12927	OH6A	w.o.10.12.77	
65-12928	OH6A	SoC	
65-12929	OH6A	42AVN	
65-12930	OH6A	SoC	
65-12931	OH6A	104ACS	
65-12932	OH6A	SoC	
65-12933	OH6A	SoC	
65-12934	OH6A	SoC	
65-12935	OH6A	SoC	
65-12936	OH6A	w.o.9.4.70	
65-12937	OH6A	stored	Gulfport RAP, Ms
65-12938	OH6A	w.o.31.3.70	
65-12939	OH6A	SoC	
65-12940	OH6A	w.o.31.12.70	
65-12941	OH6A	SoC	
65-12942	OH6A	SoC	
65-12943	OH6A	SoC	
65-12944	OH6A	w.o.23.10.72	
65-12945	OH6A	SoC	
65-12946	OH6A	w.o.9.5.70	
65-12947	OH6A	SoC	
65-12948	OH6A	126ACS	
65-12949	OH6A	26AVN	
65-12950	OH6A	stored	Gulfport RAP, Ms
65-12951	OH6A	SoC	
65-12952	OH6A	SoC	
65-12953	OH6A	SoC	
65-12954	OH6A	SoC	
65-12955	OH6A	SoC	
65-12956	OH6A	SoC	
65-12957	OH6A	163ACR	
65-12958	OH6A	SoC	
65-12959	OH6A	SoC	
65-12960	OH6A	w.o.4.3.71	
65-12961	OH6A	SoC	
65-12962	OH6A	SoC	
65-12963	OH6A	SoC	
65-12964	OH6A		
65-12965	OH6A		
65-12966	OH6A	278ACR	
65-12967	OH6A	SoC	
65-12968	OH6A	NOTAR development	
65-12969	OH6A	SoC	
65-12970	OH6A	stored	Gulfport RAP, Ms
65-12971	OH6A	SoC	
65-12972	OH6A	76AvCo	
65-12973	OH6A	SoC	
65-12974	OH6A	42AVN	
65-12975	OH6A	SoC	
65-12976	OH6A	stored	Gulfport RAP, Ms
65-12977	OH6A	SoC	
65-12978	OH6A	323AVN	
65-12979	OH6A	SoC	
65-12980	OH6A	stored	Gulfport RAP, Ms
65-12981	OH6A	SoC	
65-12982	OH6A	stored	Gulfport RAP, Ms
65-12983	OH6A	SoC	
65-12984	OH6A	w.o.2.12.70	
65-12985	OH6A	SoC	
65-12986	OH6A	to N67BP	
65-12987	OH6A	SoC	
65-12988	OH6A	SoC	
65-12989	OH6A	SoC	
65-12990	OH6A	SoC	
65-12991	OH6A	SoC	
65-12992	OH6A	SoC	
65-12993	OH6A	SoC	
65-12994	OH6A	2-135AVN	CO NG
65-12995	OH6A	SoC	
65-12996	OH6A	SoC	
65-12997	OH6A	328AvCo	
65-12998	OH6A	SoC	
65-12999	OH6A	SoC	
65-13000	OH6A	SoC	
65-13001	OH6A	SoC	
65-13002	OH6A	SoC	
65-13003	OH6A	SoC	
66-7775	OH6A	104ACS	
66-7776	OH6A	SoC	
66-7777	OH6A	SoC	
66-7778	OH6A	SoC	
66-7779	OH6A	B\26AVN	
66-7780	OH6A	SoC	
66-7781	OH6A	SoC	
66-7782	OH6A	SoC	
66-7783	OH6A	SoC	
66-7784	OH6A	SoC	
66-7785	OH6A	SoC	
66-7786	OH6A	SoC	
66-7787	OH6A	SoC	
66-7788	OH6A	SoC	
66-7789	OH6A	42AVN	
66-7790	OH6A	SoC	
66-7791	OH6A	42AVN	
66-7792	OH6A	SoC	
66-7793	OH6A	SoC	
66-7794	OH6A	SoC	
66-7795	OH6A	SoC	
66-7796	OH6A	SoC	
66-7797	OH6A	SoC	
66-7798	OH6A	SoC	
66-7799	OH6A	SoC	
66-7800	OH6A	SoC	
66-7801	OH6A	SoC	
66-7802	OH6A	SoC	
66-7803	OH6A	SoC	
66-7804	OH6A	SoC	
66-7805	OH6A	SoC	
66-7806	OH6A	SoC	
66-7807	OH6A	w.o.18.12.70	
66-7808	OH6A	SoC	
66-7809	OH6A	SoC	
66-7810	OH6A	SoC	
66-7811	OH6A	SoC	
66-7812	OH6A	SoC	
66-7813	OH6A	w.o.6.10.70	
66-7814	OH6A	SoC	

66-7815	OH6A	SoC	
66-7816	OH6A	SoC	
66-7817	OH6A	42AVN	
66-7818	OH6A	SoC	
66-7819	OH6A	SoC	
66-7820	OH6A	to N6602N	
66-7821	OH6A	SoC	
66-7822	OH6A	SoC	
66-7823	OH6A	SoC	
66-7824	OH6A	SoC	
66-7825	OH6A	SoC	
66-7826	OH6A	SoC	
66-7827	OH6A	SoC	
66-7828	OH6A	SoC	
66-7829	OH6A	w.o.4.1.70	
66-7830	OH6A	SoC	
66-7831	OH6A	SoC	
66-7832	OH6A	SoC	
66-7833	OH6A	SoC	
66-7834	OH6A	SoC	
66-7835	OH6A	w.o.11.6.71	
66-7836	OH6A	SoC	
66-7837	OH6A	SoC	
66-7838	OH6A	104ACS	
66-7839	OH6A	SoC	
66-7840	OH6A	SoC	
66-7841	OH6A	SoC	
66-7842	OH6A	SoC	
66-7843	OH6A	SoC	
66-7844	OH6A	SoC	
66-7845	OH6A	SoC	
66-7846	OH6A	SoC	
66-7847	OH6A	SoC	
66-7848	OH6A	SoC	
66-7849	OH6A	SoC	
66-7850	OH6A	SoC	
66-7851	OH6A	w.o.4.9.91	
66-7852	OH6A	SoC	
66-7853	OH6A	SoC	
66-7854	OH6A	SoC	
66-7855	OH6A	SoC	
66-7856	OH6A	SoC	
66-7857	OH6A	SoC	
66-7858	OH6A	SoC	
66-7859	OH6A	SoC	
66-7860	OH6A	SoC	
66-7861	OH6A	SoC	
66-7862	OH6A	SoC	
66-7863	OH6A	SoC	
66-7864	OH6A	328AvCo	
66-7865	OH6A	SoC	
66-7866	OH6A	SoC	
66-7867	OH6A	SoC	
66-7868	OH6A	SoC	
66-7869	OH6A	SoC	
66-7870	OH6A	SoC	
66-7871	OH6A	SoC	
66-7872	OH6A	SoC	
66-7873	OH6A	SoC	
66-7874	OH6A	SoC	
66-7875	OH6A	SoC	
66-7876	OH6A	SoC	
66-7877	OH6A	SoC	
66-7878	OH6A	SoC	
66-7879	OH6A	SoC	
66-7880	OH6A	SoC	
66-7881	OH6A	SoC	
66-7882	OH6A	SoC	
66-7883	OH6A	SoC	
66-7884	OH6A	SoC	
66-7885	OH6A	SoC	
66-7886	OH6A	SoC	
66-7887	OH6A	w.o.16.10.72	
66-7888	OH6A	SoC	
66-7889	OH6A	w.o.23.5.70	
66-7890	OH6A	SoC	
66-7891	OH6A	SoC	
66-7892	OH6A	SoC	
66-7893	OH6A	SoC	
66-7894	OH6A	SoC	
66-7895	OH6A	SoC	
66-7896	OH6A	SoC	
66-7897	OH6A	SoC	
66-7898	OH6A	SoC	
66-7899	OH6A	SoC	
66-7900	OH6A	SoC	
66-7901	OH6A	SoC	
66-7902	OH6A	SoC	
66-7903	OH6A	SoC	
66-7904	OH6A	SoC	
66-7905	OH6A	SoC	
66-7906	OH6A	SoC	
66-7907	OH6A	SoC	
66-7908	OH6A	SoC	
66-7909	OH6A	SoC	
66-7910	OH6A	SoC	
66-7911	OH6A	SoC	
66-7912	OH6A	SoC	
66-7913	OH6A	SoC	
66-7914	OH6A	SoC	
66-7915	OH6A	163ACR	
66-7916	OH6A	SoC	
66-7917	OH6A	SoC	
66-7918	OH6A	SoC	
66-7919	OH6A	42AVN	
66-7920	OH6A	SoC	
66-7921	OH6A	SoC	
66-7922	OH6A	SoC	
66-7923	OH0A	SoC	
66-7924	OH6A	SoC	
66-7925	OH6A	SoC	
66-7926	OH6A	SoC	
66-7927	OH6A	SoC	
66-7928	OH6A	SoC	
66-7929	OH6A	SoC	
66-7930	OH6A	SoC	
66-7931	OH6A	w.o.6.10.70	
66-7932	OH6A	SoC	
66-7933	OH6A	SoC	
66-7934	OH6A	SoC	
66-7935	OH6A	SoC	
66-7936	OH6A	stored	Gulfport RAP, Ms
66-7937	OH6A	SoC	
66-7938	OH6A	SoC	
66-7939	OH6A	to Nicaragua 512	
66-7940	OH6A	SoC	
66-7941	OH6A	SoC	
66-7942	OH6A	SoC	
66-14376	OH6A	SoC	
66-14377	OH6A	SoC	
66-14378	OH6A	SoC	
66-14379	OH6A	SoC	
66-14380	OH6A	SoC	
66-14381	OH6A	SoC	
66-14382	OH6A	SoC	
66-14383	OH6A	SoC	
66-14384	OH6A	SoC	
66-14385	OH6A	SoC	
66-14386	OH6A	SoC	
66-14387	OH6A	SoC	
66-14388	OH6A	stored	Gulfport RAP, Ms
66-14389	OH6A	328AvCo	
66-14390	OH6A	SoC	
66-14391	OH6A	328AvCo	
66-14392	OH6A	SoC	
66-14393	OH6A	SoC	
66-14394	OH6A	SoC	
66-14395	OH6A	SoC	
66-14396	OH6A	SoC	
66-14397	OH6A	SoC	
66-14398	OH6A	SoC	
66-14399	OH6A	SoC	
66-14400	OH6A	SoC	
66-14401	OH6A	stored	Gulfport RAP, Ms
66-14402	OH6A	SoC	
66-14403	OH6A	w.o.1.3.70	
66-14404	OH6A	42AVN	
66-14405	OH6A	SoC	
66-14406	OH6A	SoC	
66-14407	OH6A	SoC	
66-14408	OH6A	SoC	
66-14409	OH6A	SoC	
66-14410	OH6A	SoC	
66-14411	OH6A		
66-14412	OH6A	SoC	
66-14413	OH6A	SoC	

66-14414	OH6A	SoC	
66-14415	OH6A	SoC	
66-14416	OH6A	SoC	
66-14417	OH6A	SoC	
66-14418	OH6A	SoC	
66-14419	OH6A	SoC	
66-17750	OH6A	to Nicaragua 511	
66-17751	OH6A	SoC	
66-17752	OH6A	SoC	
66-17753	OH6A	SoC	
66-17754	OH6A	278ACR	
66-17755	OH6A	SoC	
66-17756	OH6A	SoC	
66-17757	OH6A	SoC	
66-17758	OH6A	SoC	
66-17759	OH6A	w.o.18.12.70	
66-17760	OH6A	SoC	
66-17761	OH6A	SoC	
66-17762	OH6A	SoC	
66-17763	OH6A	1-131AVN AL NG	
66-17764	OH6A	SoC	
66-17765	OH6A	SoC	
66-17766	OH6A	SoC	
66-17767	OH6A	SoC	
66-17768	OH6A	SoC	
66-17769	OH6A	164AvCo	
66-17770	OH6A	SoC	
66-17771	OH6A	SoC	
66-17772	OH6A	SoC	
66-17773	OH6A	SoC	
66-17774	OH6A	stored	Gulfport RAP, Ms
66-17775	OH6A	SoC	
66-17776	OH6A	SoC	
66-17777	OH6A	SoC	
66-17778	OH6A	SoC	
66-17779	OH6A	SoC	
66-17780	OH6A	w.o.7.3.70	
66-17781	OH6A	1-150AVN	NJ NG
66-17782	OH6A	SoC	
66-17783	OH6A	SoC	
66-17784	OH6A	SoC	
66-17785	OH6A	SoC	
66-17786	OH6A	323AVN	
66-17787	OH6A	to N97BP	
66-17788	OH6A	SoC	
66-17789	OH6A	SoC	
66-17790	OH6A	SoC	
66-17791	OH6A	SoC	
66-17792	OH6A	126ACS	
66-17793	OH6A	SoC	
66-17794	OH6A	SoC	
66-17795	OH6A	stored	Gulfport RAP, Ms
66-17796	OH6A	SoC	
66-17797	OH6A	SoC	
66-17798	OH6A	SoC	
66-17799	OH6A	26AVN	
66-17800	OH6A	SoC	
66-17801	OH6A		
66-17802	OH6A	SoC	
66-17803	OH6A	SoC	
66-17804	OH6A	SoC	
66-17805	OH6A	SoC	
66-17806	OH6A	SoC	
66-17807	OH6A	SoC	
66-17808	OH6A		
66-17809	OH6A	SoC	
66-17810	OH6A	SoC	
66-17811	OH6A	SoC	
66-17812	OH6A	SoC	
66-17813	OH6A	SoC	
66-17814	OH6A	SoC	
66-17815	OH6A	SoC	
66-17816	OH6A	SoC	
66-17817	OH6A	SoC	
66-17818	OH6A	SoC	
66-17819	OH6A	SoC	
66-17820	OH6A	SoC	
66-17821	OH6A	SoC	
66-17822	OH6A	1-131AVN	AL NG
66-17823	OH6A	SoC	
66-17824	OH6A	stored	Gulfport RAP, Ms
66-17825	OH6A	SoC	
66-17826	OH6A	w.o.6.5.70	
66-17827	OH6A	SoC	
66-17828	OH6A	SoC	
66-17829	OH6A	104ACS	
66-17830	OH6A	SoC	
66-17831	OH6A	SoC	
66-17832	OH6A		
66-17833	OH6A	SoC	
66-17905	OH6A	w.o.2.6.71	
66-17918	OH6A	w.o.12.1.70	
67-16000	OH6A		
67-16001	OH6A	42AVN	
67-16002	OH6A	126ACS	
67-16003	OH6A	SoC	
67-16004	OH6A	SoC	
67-16005	OH6A	SoC	
67-16006	OH6A	w.o.16.2.70	
67-16007	OH6A	SoC	
67-16008	OH6A	SoC	
67-16009	OH6A	stored	Gulfport RAP, Ms
67-16010	OH6A	SoC	
67-16011	OH6A	SoC	
67-16012	OH6A	SoC	
67-16013	OH6A	SoC	
67-16014	OH6A	SoC	
67-16015	OH6A	SoC	
67-16016	OH6A	SoC	
67-16017	OH6A	42AVN	
67-16018	OH6A	SoC	
67-16019	OH6A		
67-16020	OH6A	stored	Gulfport RAP, Ms
67-16021	OH6A	GIA	Ft. Eustis
67-16022	OH6A	SoC	
67-16023	OH6A	SoC	
67-16024	OH6A	SoC	
67-16025	OH6A	SoC	
67-16026	OH6A		
67-16027	OH6A	SoC	
67-16028	OH6A	SoC	
67-16029	OH6A	SoC	
67-16030	OH6A	SoC	
67-16031	OH6A	SoC	
67-16032	OH6A	SoC	
67-16033	OH6A	SoC	
67-16034	OH6A	SoC	
67-16035	OH6A	stored	Gulfport RAP, Ms
67-16036	OH6A	126ACS	
67-16037	OH6A	SoC	
67-16038	OH6A	SoC	
67-16039	OH6A	SoC	
67-16040	OH6A	SoC	
67-16041	OH6B	42AVN	
67-16042	OH6A	SoC	
67-16043	OH6A	w.o.26.6.70	
67-16044	OH6A	SoC	
67-16045	OH6A	SoC	
67-16046	OH6A	SoC	
67-16047	OH6A	SoC	
67-16048	OH6A		
67-16049	OH6A	SoC	
67-16050	OH6A	26AVN	
67-16051	OH6A	SoC	
67-16052	OH6A	SoC	
67-16053	OH6A	SoC	
67-16054	OH6A	SoC	
67-16055	OH6A	SoC	
67-16056	OH6A	SoC	
67-16057	OH6A	SoC	
67-16058	OH6A	SoC	
67-16059	OH6A	SoC	
67-16060	OH6A	SoC	
67-16061	OH6A	150AVN	
67-16062	OH6A	SoC	
67-16063	OH6A	SoC	
67-16064	OH6A	104ACS	
67-16065	OH6A	SoC	
67-16066	OH6A	SoC	
67-16067	OH6A	SoC	
67-16068	OH6A	SoC	
67-16069	OH6A	SoC	
67-16070	OH6A	SoC	
67-16071	OH6A	26AVN	
67-16072	OH6A	SoC	
67-16073	OH6A	SoC	

Serial	Type	Notes	Operator
67-16074	OH6A	SoC	
67-16075	OH6A	SoC	
67-16076	OH6A	SoC	
67-16077	OH6A	SoC	
67-16078	OH6A	SoC	
67-16079	OH6A	SoC	
67-16080	OH6A	SoC	
67-16081	OH6A	SoC	
67-16082	OH6A	42AVN	
67-16083	OH6A	SoC	
67-16084	OH6A	SoC	
67-16085	OH6A	SoC	
67-16086	OH6A	SoC	
67-16087	OH6A	1-131AVN	AL NG
67-16088	OH6A	SoC	
67-16089	OH6A	SoC	
67-16090	OH6A	SoC	
67-16091	OH6A	SoC	
67-16092	OH6A		
67-16093	OH6A	SoC	
67-16094	OH6A	SoC	
67-16095	OH6A	SoC	
67-16096	OH6A	w.o.17.5.70	
67-16097	OH6A	SoC	
67-16098	OH6A	SoC	
67-16099	OH6A	SoC	
67-16100	OH6A	SoC	
67-16101	OH6A	SoC	
67-16102	OH6A	SoC	
67-16103	OH6A	w.o.13.12.70	
67-16104	OH6A	SoC	
67-16105	OH6A	SoC	
67-16106	OH6A	SoC	
67-16107	OH6A	SoC	
67-16108	OH6A	SoC	
67-16109	OH6A	SoC	
67-16110	OH6A	SoC	
67-16111	OH6A	SoC	
67-16112	OH6A		
67-16113	OH6A	stored	Gulfport RAP, Ms
67-16114	OH6A	w.o.7.3.70	
67-16115	OH6A	SoC	
67-16116	OH6A	w.o.28.8.70	
67-16117	OH6A		
67-16118	OH6A	SoC	
67-16119	OH6A	42AVN	
67-16120	OH6A	w.o.18.12.70	
67-16121	OH6A	2-135AVN	CO NG
67-16122	OH6A	SoC	
67-16123	OH6A	SoC	
67-16124	OH6A	w.o.1.7.70	
67-16125	OH6A	SoC	
67-16126	OH6A	SoC	
67-16127	OH6A	26AVN	
67-16128	OH6A	SoC	
67-16129	OH6A	SoC	
67-16130	OH6A	SoC	
67-16131	OH6A	SoC	
67-16132	OH6A	SoC	
67-16133	OH6A	GIA Ft. Eustis	
67-16134	OH6A	SoC	
67-16135	OH6A	SoC	
67-16136	OH6A	SoC	
67-16137	OH6A	SoC	
67-16138	OH6A	SoC	
67-16139	OH6A	SoC	
67-16140	OH6A	SoC	
67-16141	OH6A	SoC	
67-16142	OH6A	SoC	
67-16143	OH6A	SoC	
67-16144	OH6A	w.o.12.2.70	
67-16145	OH6A	SoC	
67-16146	OH6A	SoC	
67-16147	OH6A	SoC	
67-16148	OH6A	SoC	
67-16149	OH6A	SoC	
67-16150	OH6A	SoC	
67-16151	OH6A	SoC	
67-16152	OH6A	w.o.16.8.70	
67-16153	OH6A	SoC	
67-16154	OH6A	SoC	
67-16155	OH6A	26AVN	
67-16156	OH6A	SoC	
67-16157	OH6A	104ACS	
67-16158	OH6A	SoC	
67-16159	OH6A	SoC	
67-16160	OH6A	SoC	
67-16161	OH6A	SoC	
67-16162	OH6A	SoC	
67-16163	OH6A	SoC	
67-16164	OH6A		
67-16165	OH6A	SoC	
67-16166	OH6A	SoC	
67-16167	OH6A	SoC	
67-16168	OH6A	SoC	
67-16169	OH6A	SoC	
67-16170	OH6A	SoC	
67-16171	OH6A	SoC	
67-16172	OH6A	328AvCo	
67-16173	OH6A	SoC	
67-16174	OH6A	SoC	
67-16175	OH6A	SoC	
67-16176	OH6A	SoC	
67-16177	OH6A	w.o.22.3.71	
67-16178	OH6A	SoC	
67-16179	OH6A	w.o.26.7.70	
67-16180	OH6A	SoC	
67-16181	OH6A	126ACS	
67-16182	OH6A	SoC	
67-16183	OH6A	SoC	
67-16184	OH6A	SoC	
67-16185	OH6A	SoC	
67-16186	OH6A	SoC	
67-16187	OH6A	SoC	
67-16188	OH6A	1-150AVN	NJ NG
67-16189	OH6A	SoC	
67-16190	OH6A	SoC	
67-16191	OH6A	SoC	
67-16192	OH6A	SoC	
67-16193	OH6A	SoC	
67-16194	OH6A	SoC	
67-16195	OH6A	SoC	
67-16196	OH6A	SoC	
67-16197	OH6A	SoC	
67-16198	OH6A	SoC	
67-16199	OH6A	SoC	
67-16200	OH6A	SoC	
67-16201	OH6A	SoC	
67-16202	OH6A	SoC	
67-16203	OH6A	SoC	
67-16204	OH6A	SoC	
67-16205	OH6A	SoC	
67-16206	OH6A	SoC	
67-16207	OH6A	SoC	
67-16208	OH6A		SD NG
67-16209	OH6A	SoC	
67-16210	OH6A	SoC	
67-16211	OH6A	SoC	
67-16212	OH6A	SoC	
67-16213	OH6A	SoC	
67-16214	OH6A	1-150AVN	NJ NG
67-16215	OH6A	SoC	
67-16216	OH6A	SoC	
67-16217	OH6A	SoC	
67-16218	OH6A	SoC	
67-16219	OH6A	to NASA	N731NA
67-16220	OH6A	SoC	
67-16221	OH6A	SoC	
67-16222	OH6A	SoC	
67-16223	OH6A	SoC	
67-16224	OH6A	SoC	
67-16225	OH6A	w.o.25.2.70	
67-16226	OH6A	SoC	
67-16227	OH6A	SoC	
67-16228	OH6A	SoC	
67-16229	OH6A	SoC	
67-16230	OH6A	SoC	
67-16231	OH6A	SoC	
67-16232	OH6A	SoC	
67-16233	OH6A	SoC	
67-16234	OH6A	SoC	
67-16235	OH6A	126ACS	
67-16236	OH6A	w.o.18.5.70	
67-16237	OH6A		
67-16238	OH6A	SoC	
67-16239	OH6A	SoC	

Serial	Type	Unit/Status	Notes
67-16240	OH6A	w.o.28.8.71	
67-16241	OH6A	SoC	
67-16242	OH6A	1-131AVN	AL NG
67-16243	OH6A	1-131AVN	AL NG
67-16244	OH6A	SoC	
67-16245	OH6A	SoC	
67-16246	OH6A	SoC	
67-16247	OH6A	1-150AVN	NJ NG
67-16248	OH6A	42AVN	
67-16249	OH6A	SoC	
67-16250	OH6A	SoC	
67-16251	OH6A	SoC	
67-16252	OH6A	SoC	
67-16253	OH6A	w.o.24.5.71	
67-16254	OH6A	SoC	
67-16255	OH6A	SoC	
67-16256	OH6A	SoC	
67-16257	OH6A	SoC	
67-16258	OH6A	SoC	
67-16259	OH6A	SoC	
67-16260	OH6A	B\26AVN	
67-16261	OH6A	SoC	
67-16262	OH6A	SoC	
67-16263	OH6A	SoC	
67-16264	OH6A	SoC	
67-16265	OH6A	328AvCo	
67-16266	OH6A	SoC	
67-16267	OH6A	SoC	
67-16268	OH6A	SoC	
67-16269	OH6A	SoC	
67-16270	OH6A	SoC	
67-16271	OH6A	SoC	
67-16272	OH6A	SoC	
67-16273	OH6A	328AvCo	
67-16274	OH6A	w.o.14.8.72	
67-16275	OH6A	SoC	
67-16276	OH6A	SoC	
67-16277	OH6A	SoC	
67-16278	OH6A	SoC	
67-16279	OH6A	42AVN	
67-16280	OH6A	stored	Gulfport RAP, Ms
67-16281	OH6A	SoC	
67-16282	OH6A	SoC	
67-16283	OH6A		
67-16284	OH6A	SoC	
67-16285	OH6A	SoC	
67-16286	OH6A	1-150AVN	NJ NG
67-16287	OH6A	SoC	
67-16288	OH6A	SoC	
67-16289	OH6A	SoC	
67-16290	OH6A	w.o.13.7.70	
67-16291	OH6A	SoC	
67-16292	OH6A	SoC	
67-16293	OH6A	SoC	
67-16294	OH6A	SoC	
67-16295	OH6A	SoC	
67-16296	OH6A	SoC	
67-16297	OH6A	SoC	
67-16298	OH6A	SoC	
67-16299	OH6A	SoC	
67-16300	OH6A	SoC	
67-16301	OH6A	323AVN	
67-16302	OH6A	SoC	
67-16303	OH6A	SoC	
67-16304	OH6A	126ACS	
67-16305	OH6A	SoC	
67-16306	OH6A	SoC	
67-16307	OH6A	SoC	
67-16308	OH6A	SoC	
67-16309	OH6A	SoC	
67-16310	OH6A	SoC	
67-16311	OH6A	SoC	
67-16312	OH6A	SoC	
67-16313	OH6A	SoC	
67-16314	OH6A	SoC	
67-16315	OH6A	SoC	
67-16316	OH6A	SoC	
67-16317	OH6A	w.o.5.11.77	
67-16318	OH6A	SoC	
67-16319	OH6A	SoC	
67-16320	OH6A	SoC	
67-16321	OH6A	SoC	
67-16322	OH6A	SoC	
67-16323	OH6A	SoC	
67-16324	OH6A	SoC	
67-16325	OH6A	SoC	
67-16326	OH6A	104ACS	
67-16327	OH6A	42AVN	
67-16328	OH6A	SoC	
67-16329	OH6A	SoC	
67-16330	OH6A	SoC	
67-16331	OH6A	SoC	
67-16332	OH6A	to N8525G	
67-16333	OH6A	SoC	
67-16334	OH6A	126ACS	
67-16335	OH6A	SoC	
67-16336	OH6A	SoC	
67-16337	OH6A	SoC	
67-16338	OH6A		
67-16339	OH6A	w.o.10.5.71	
67-16340	OH6A	SoC	
67-16341	OH6A	SoC	
67-16342	OH6A	SoC	
67-16343	OH6A	SoC	
67-16344	OH6A	SoC	
67-16345	OH6A	SoC	
67-16346	OH6A	stored	Gulfport RAP, Ms
67-16347	OH6A	SoC	
67-16348	OH6A	SoC	
67-16349	OH6A	42AVN	
67-16350	OH6A	42AVN	
67-16351	OH6A	SoC	
67-16352	OH6A	SoC	
67-16353	OH6A	SoC	
67-16354	OH6A	SoC	
67-16355	OH6A	SoC	
67-16356	OH6A	w.o.17.3.71	
67-16357	OH6A	SoC	
67-16358	OH6A	40AVN	
67-16359	OH6A	SoC	
67-16360	OH6A	SoC	
67-16361	OH6A	1-131AVN	AL NG
67-16362	OH6A	SoC	
67-16363	OH6A	SoC	
67-16364	OH6A	SoC	
67-16365	OH6A	SoC	
67-16366	OH6A	SoC	
67-16367	OH6A	SoC	
67-16368	OH6A	SoC	
67-16369	OH6A	SoC	
67-16370	OH6A	SoC	
67-16371	OH6A	SoC	
67-16372	OH6A	42AVN	
67-16373	OH6A	SoC	
67-16374	OH6A	SoC	
67-16375	OH6A	SoC	
67-16376	OH6A		
67-16377	OH6A	SoC	
67-16378	OH6A	SoC	
67-16379	OH6A	SoC	
67-16380	OH6A	104ACS	
67-16381	OH6A	SoC	
67-16382	OH6A	SoC	
67-16383	OH6A	SoC	
67-16384	OH6A	SoC	
67-16385	OH6A	SoC	
67-16386	OH6A	SoC	
67-16387	OH6A	SoC	
67-16388	OH6A	SoC	
67-16389	OH6A	SoC	
67-16390	OH6A	SoC	
67-16391	OH6A	SoC	
67-16392	OH6A	w.o.5.9.70	
67-16393	OH6A	SoC	
67-16394	OH6A	SoC	
67-16395	OH6A	SoC	
67-16396	OH6A	SoC	
67-16397	OH6A	SoC	
67-16398	OH6A	stored	Gulfport RAP, Ms
67-16399	OH6A	SoC	
67-16400	OH6A	SoC	
67-16401	OH6A	SoC	
67-16402	OH6A	SoC	
67-16403	OH6A	w.o.8.4.70	
67-16404	OH6A	w.o.2.7.84	
67-16405	OH6A	SoC	

67-16406	OH6A	SoC	
67-16407	OH6A	SoC	
67-16408	OH6A	SoC	
67-16409	OH6A	SoC	
67-16410	OH6A	SoC	
67-16411	OH6A	SoC	
67-16412	OH6A	SoC	
67-16413	OH6A	SoC	
67-16414	OH6A	SoC	
67-16415	OH6A	SoC	
67-16416	OH6A	2-135AVN	CO NG
67-16417	OH6A	SoC	
67-16418	OH6A	SoC	
67-16419	OH6A	SoC	
67-16420	OH6A	SoC	
67-16421	OH6A	76AvCo	
67-16422	OH6A	SoC	
67-16423	OH6A	SoC	
67-16424	OH6A	SoC	
67-16425	OH6A	SoC	
67-16426	OH6A	150AVN	
67-16427	OH6A	SoC	
67-16428	OH6A	SoC	
67-16429	OH6A	SoC	
67-16430	OH6A	278ACR	
67-16431	OH6A	SoC	
67-16432	OH6A	SoC	
67-16433	OH6A	w.o.30.3.71	
67-16434	OH6A	SoC	
67-16435	OH6A		
67-16436	OH6A	SoC	
67-16437	OH6A	SoC	
67-16438	OH6A	SoC	
67-16439	OH6A	SoC	
67-16440	OH6A	42AVN	
67-16441	OH6A	SoC	
67-16442	OH6A	SoC	
67-16443	OH6A	42AVN	
67-16444	OH6A	SoC	
67-16445	OH6A	SoC	
67-16446	OH6A	SoC	
67-16447	OH6A	SoC	
67-16448	OH6A	SoC	
67-16449	OH6A		
67-16450	OH6A	SoC	
67-16451	OH6A	SoC	
67-16452	OH6A	SoC	
67-16453	OH6A	SoC	
67-16454	OH6A	SoC	
67-16455	OH6A	SoC	
67-16456	OH6A	SoC	
67-16457	OH6A	SoC	
67-16458	OH6A	SoC	
67-16459	OH6A	w.o.10.5.70	
67-16460	OH6A	SoC	
67-16461	OH6A	SoC	
67-16462	OH6A	104ACS	
67-16463	OH6A	SoC	
67-16464	OH6A	126ACS	
67-16465	OH6A	SoC	
67-16466	OH6A	w.o.22.3.71	
67-16467	OH6A	SoC	
67-16468	OH6A	SoC	
67-16469	OH6A	SoC	
67-16470	OH6A	1-131AVN	AL NG
67-16471	OH6A	SoC	
67-16472	OH6A	SoC	
67-16473	OH6A	SoC	
67-16474	OH6A	SoC	
67-16475	OH6A	1-131AVN	AL NG
67-16476	OH6A	SoC	
67-16477	OH6A		
67-16478	OH6A	SoC	
67-16479	OH6A	SoC	
67-16480	OH6A	SoC	
67-16481	OH6A	SoC	
67-16482	OH6A	SoC	
67-16483	OH6A	SoC	
67-16484	OH6A	SoC	
67-16485	OH6A	323AVN	
67-16486	OH6A	SoC	
67-16487	OH6A	SoC	
67-16488	OH6A	SoC	
67-16489	OH6A	SoC	
67-16490	OH6A	SoC	
67-16491	OH6A	SoC	
67-16492	OH6A	SoC	
67-16493	OH6A	w.o.12.12.70	
67-16494	OH6A	SoC	
67-16495	OH6A	42AVN	
67-16496	OH6A	1-131AVN	AL NG
67-16497	OH6A	w.o.9.7.70	
67-16498	OH6A	SoC	
67-16499	OH6A	SoC	
67-16500	OH6A	SoC	
67-16501	OH6A	SoC	
67-16502	OH6A	SoC	
67-16503	OH6A	SoC	
67-16504	OH6A	SoC	
67-16505	OH6A	SoC	
67-16506	OH6A	SoC	
67-16507	OH6A	SoC	
67-16508	OH6A	SoC	
67-16509	OH6A	SoC	
67-16510	OH6A	SoC	
67-16511	OH6A	SoC	
67-16512	OH6A	SoC	
67-16513	OH6A	SoC	
67-16514	OH6A	SoC	
67-16515	OH6A		
67-16516	OH6A	w.o.7.9.71	
67-16517	OH6A	SoC	
67-16518	OH6A	SoC	
67-16519	OH6A	SoC	
67-16520	OH6A	SoC	
67-16521	OH6A	SoC	
67-16522	OH6A	w.o.3.7.71	
67-16523	OH6A	SoC	
67-16524	OH6A	SoC	
67-16525	OH6A	1-131AVN	AL NG
67-16526	OH6A	SoC	
67-16527	OH6A	SoC	
67-16528	OH6A	SoC	
67-16529	OH6A	SoC	
67-16530	OH6A	SoC	
67-16531	OH6A	SoC	
67-16532	OH6A	w.o.1.6.70	
67-16533	OH6A	SoC	
67-16534	OH6A	SoC	
67-16535	OH6A	SoC	
67-16536	OH6A	SoC	
67-16537	OH6A	SoC	
67-16538	OH6A	1-131AVN	AL NG
67-16539	OH6A	SoC	
67-16540	OH6A	104ACS	
67-16541	OH6A	SoC	
67-16542	OH6A	SoC	
67-16543	OH6A	SoC	
67-16544	OH6A	to N6602M	
67-16545	OH6A	104ACS	
67-16546	OH6A	SoC	
67-16547	OH6A	SoC	
67-16548	OH6A	SoC	
67-16549	OH6A	stored	Gulfport RAP, Ms
67-16550	OH6A		
67-16551	OH6A	SoC	
67-16552	OH6A	SoC	
67-16553	OH6A	SoC	
67-16554	OH6A	SoC	
67-16555	OH6A	SoC	
67-16556	OH6A	SoC	
67-16557	OH6A	SoC	
67-16558	OH6A	SoC	
67-16559	OH6A	SoC	
67-16560	OH6A	SoC	
67-16561	OH6A	SoC	
67-16562	OH6A	SoC	
67-16563	OH6A	SoC	
67-16564	OH6A	SoC	
67-16565	OH6A		
67-16566	OH6A	SoC	
67-16567	OH6A	SoC	
67-16568	OH6A	SoC	
67-16569	OH6A	163ACR	
67-16570	OH6A	stored	Gulfport RAP, Ms
67-16571	OH6A	SoC	

67-16572	OH6A	SoC	
67-16573	OH6A	SoC	
67-16574	OH6A	SoC	
67-16575	OH6A	SoC	
67-16576	OH6A	SoC	
67-16577	OH6A	SoC	
67-16578	OH6A	SoC	
67-16579	OH6A		
67-16580	OH6A	SoC	
67-16581	OH6A	SoC	
67-16582	OH6A	SoC	
67-16583	OH6A	SoC	
67-16584	OH6A	126ACS	
67-16585	OH6A	SoC	
67-16586	OH6A	SoC	
67-16587	OH6A	stored	Gulfport RAP, Ms
67-16588	OH6A	SoC	
67-16589	OH6A	SoC	
67-16590	OH6A	SoC	
67-16591	OH6A	SoC	
67-16592	OH6A	SoC	
67-16593	OH6A	26AVN	
67-16594	OH6A	SoC	
67-16595	OH6A	SoC	
67-16596	OH6A	w.o.20.4.70	
67-16597	OH6A	SoC	
67-16598	OH6A	SoC	
67-16599	OH6A	SoC	
67-16600	OH6A	SoC	
67-16601	OH6A	SoC	
67-16602	OH6A	SoC	
67-16603	OH6A	SoC	
67-16604	OH6A	SoC	
67-16605	OH6A	SoC	
67-16606	OH6A	SoC	
67-16607	OH6A	SoC	
67-16608	OH6A	42AVN	
67-16609	OH6A	SoC	
67-16610	OH6A	SoC	
67-16611	OH6A	w.o.21.10.70	
67-16612	OH6A	SoC	
67-16613	OH6A	SoC	
67-16614	OH6A	SoC	
67-16615	OH6A	1-131AVN	AL NG
67-16616	OH6A	40AVN	
67-16617	OH6A	SoC	
67-16618	OH6A	SoC	
67-16619	OH6A	SoC	
67-16620	OH6A	SoC	
67-16621	OH6A	SoC	
67-16622	OH6A	SoC	
67-16623	OH6A	stored	Gulfport RAP, Ms
67-16624	OH6A	SoC	
67-16625	OH6A	SoC	
67-16626	OH6A	SoC	
67-16627	OH6A	126ACS	
67-16628	OH6A	SoC	
67-16629	OH6A	SoC	
67-16630	OH6A	SoC	
67-16631	OH6A	42AVN	
67-16632	OH6A	SoC	
67-16633	OH6A	26AVN	
67-16634	OH6A	SoC	
67-16635	OH6A	w.o.3.8.75	
67-16636	OH6A	1-150AVN	NJ NG
67-16637	OH6A	SoC	
67-16638	OH6A	B\26AVN	
67-16639	OH6A	SoC	
67-16640	OH6A	SoC	
67-16641	OH6A	SoC	
67-16642	OH6A	SoC	
67-16643	OH6A	SoC	
67-16644	OH6A	SoC	
67-16645	OH6A	SoC	
67-16646	OH6A	SoC	
67-16647	OH6A	42AVN	
67-16648	OH6A	SoC	
67-16649	OH6A	SoC	
67-16650	OH6A	w.o.22.10.70	
67-16651	OH6A	SoC	
67-16652	OH6A	SoC	
67-16653	OH6A	SoC	
67-16654	OH6A	SoC	
67-16655	OH6A	SoC	
67-16656	OH6A	w.o.2.11.71	
67-16657	OH6A	SoC	
67-16658	OH6A	SoC	
67-16659	OH6A	SoC	
67-16660	OH6A	stored	Gulfport RAP, Ms
67-16661	OH6A	328AvCo	
67-16662	OH6A	SoC	
67-16663	OH6A	150AVN	
67-16664	OH6A	SoC	
67-16665	OH6A	SoC	
67-16666	OH6A	SoC	
67-16667	OH6A	SoC	
67-16668	OH6A	SoC	
67-16669	OH6A	SoC	
67-16670	OH6A	SoC	
67-16671	OH6A	SoC	
67-16672	OH6A	SoC	
67-16673	OH6A	SoC	
67-16674	OH6A	SoC	
67-16675	OH6A	126ACS	
67-16676	OH6A	SoC	
67-16677	OH6A	w.o.27.2.70	
67-16678	OH6A	SoC	
67-16679	OH6A	SoC	
67-16680	OH6A	SoC	
67-16681	OH6A	SoC	
67-16682	OH6A	SoC	
67-16683	OH6A	SoC	
67-16684	OH6A	SoC	
67-16685	OH6A	SoC	
67-16686	OH6A	SoC	
68-17140	MH6B	SoC	
68-17141	OH6A	SoC	
68-17142	OH6A	w.o.17.2.70	
68-17143	OH6A	SoC	
68-17144	OH6A		
68-17145	OH6A		
68-17146	OH6A	1-150AVN	NJ NG
68-17147	OH6A	323AVN	
68-17148	OH6A	SoC	
68-17149	OH6A	SoC	
68-17150	OH6A		
68-17151	OH6A	SoC	
68-17152	OH6A	SoC	
68-17153	OH6A	SoC	
68-17154	OH6A	stored	Gulfport RAP, Ms
68-17155	MH6C		
68-17156	OH6A	SoC	
68-17157	OH6A	SoC	
68-17158	OH6A	SoC	
68-17159	OH6A	169AvRgt	
68-17160	OH6A	w.o.5.4.70	
68-17161	OH6A	SoC	
68-17162	OH6A	155ArmBde	
68-17163	OH6A	SoC	
68-17164	OH6A	104ACS	
68-17165	OH6A	SoC	
68-17166	OH6A	SoC	
68-17167	MH6B	to USBP	N51844
68-17168	AH6C		
68-17169	OH6A	SoC	
68-17170	OH6A	SoC	
68-17171	OH6A	SoC	
68-17172	OH6A	104ACS	
68-17173	OH6A	SoC	
68-17174	OH6A	SoC	
68-17175	MH6B	to USBP	N5185B
68-17176	OH6A	SoC	
68-17177	OH6A	w.o.24.11.70	
68-17178	OH6A	SoC	
68-17179	OH6A	SoC	
68-17180	OH6A	SoC	
68-17181	OH6A	SoC	
68-17182	OH6A	SoC	
68-17183	OH6A	SoC	
68-17184	OH6A	1-101AVN	NY NG
68-17185	OH6A	SoC	
68-17186	OH6A	SoC	
68-17187	OH6A	SoC	
68-17188	OH6A	1-131AVN	AL NG
68-17189	OH6A	SoC	
68-17190	OH6A	SoC	

Serial	Type	Notes	Extra
68-17191	AH6C		
68-17192	OH6A	SoC	
68-17193	MH6B	to USBP	N5185H
68-17194	OH6A	SoC	
68-17195	OH6A	SoC	
68-17196	OH6A	w.o.13.8.70	
68-17197	OH6A	1-131AVN	AL NG
68-17198	OH6A	SoC	
68-17199	OH6A	SoC	
68-17200	OH6A	SoC	
68-17201	OH6A	w.o.25.5.70	
68-17202	OH6A	1-131AVN	AL NG
68-17203	OH6A	SoC	
68-17204	OH6A	104ACS	
68-17205	OH6A	SoC	
68-17206	OH6A	150AVN	
68-17207	OH6A	to Taiwan 905	
68-17208	OH6A	w.o.12.2.74	
68-17209	OH6A	SoC	
68-17210	OH6A	w.o.31.1.70	
68-17211	OH6A	SoC	
68-17212	OH6A	to Nicaragua 515	
68-17213	OH6A	to Nicaragua 516	
68-17214	OH6A	to Nicaragua 517	
68-17215	OH6A	SoC	
68-17216	OH6A	B\26AVN	
68-17217	OH6A	1-131AVN	AL NG
68-17218	OH6A		
68-17219	OH6A	SoC	
68-17220	OH6A	w.o.1.1.70	
68-17221	OH6A	SoC	
68-17222	OH6A	w.o.12.9.72	
68-17223	OH6A	SoC	
68-17224	OH6A	SoC	
68-17225	MH6B	SoC	
68-17226	OH6A	w.o.8.6.74	
68-17227	OH6A	SoC	
68-17228	AH6C	w.o.15.9.88	
68-17229	OH6A	SoC	
68-17230	OH6A	NASA	
68-17231	OH6A	SoC	
68-17232	OH6A	SoC	
68-17233	OH6A	SoC	
68-17234	OH6A	1-131AVN	AL NG
68-17235	OH6A	SoC	
68-17236	OH6A	SoC	
68-17237	OH6A	SoC	
68-17238	OH6A	SoC	
68-17239	OH6A	w.o.29.5.70	
68-17240	OH6A	SoC	
68-17241	OH6A		
68-17242	AH6C	w.o.13.5.83	
68-17243	OH6A	1-131AVN	AL NG
68-17244	OH6A	stored	Gulfport RAP, Ms
68-17245	OH6A	w.o.24.6.70	
68-17246	OH6A	w.o.26.1.70	
68-17247	OH6A	SoC	
68-17248	OH6A	SoC	
68-17249	AH6C		OK NG
68-17250	OH6A	SoC	
68-17251	OH6A	1-150AVN	NJ NG
68-17252	OH6A	42AVN	
68-17253	OH6A	SoC	
68-17254	OH6A	SoC	
68-17255	OH6A	328AvCo	
68-17256	MH6B	SoC	
68-17257	OH6A	SoC	
68-17258	AH6C	328AvCo	
68-17259	OH6A	SoC	
68-17260	OH6A	SoC	
68-17261	OH6A	1-131AVN	AL NG
68-17262	OH6A	SoC	
68-17263	OH6A		
68-17264	OH6A	SoC	
68-17265	OH6A	to N.....	
68-17266	OH6A	SoC	
68-17267	OH6A	1-150AVN	NJ NG
68-17268	OH6A	SoC	
68-17269	OH6A	SoC	
68-17270	OH6A	w.o.14.11.70	
68-17271	OH6A		
68-17272	OH6A	stored	Gulfport RAP, Ms
68-17273	OH6A	SoC	
68-17274	OH6A	w.o.13.3.70	
68-17275	OH6A	SoC	
68-17276	AH6C	w.o.20.5.88	
68-17277	OH6A	w.o.12.6.72	
68-17278	OH6A	SoC	
68-17279	OH6A	164AvCo	
68-17280	OH6A	stored	Gulfport RAP, Ms
68-17281	OH6A	SoC	
68-17282	OH6A	42AVN	
68-17283	OH6A		
68-17284	OH6A	SoC	
68-17285	OH6A	SoC	
68-17286	OH6A		
68-17287	OH6A	w.o.19.5.72	
68-17288	OH6A	SoC	
68-17289	OH6A	SoC	
68-17290	MH6C	NLC	
68-17291	OH6A	SoC	
68-17292	OH6A	SoC	
68-17293	OH6A	SoC	
68-17294	OH6A	SoC	
68-17295	OH6A	SoC	
68-17296	OH6A	SoC	
68-17297	OH6A	1-131AVN	AL NG
68-17298	AH6C	SoC	
68-17299	OH6A	SoC	
68-17300	OH6A	SoC	
68-17301	AH6C	SoC	
68-17302	OH6A	w.o.10.7.70	
68-17303	OH6A	w.o.26.2.70	
68-17304	OH6A	SoC	
68-17305	OH6A	SoC	
68-17306	OH6A	SoC	
68-17307	AH6C		
68-17308	OH6A	104ACS	
68-17309	OH6A	SoC	
68-17310	OH6A	1-131AVN	AL NG
68-17311	OH6A		SD NG
68-17312	OH6A	SoC	
68-17313	OH6A	SoC	
68-17314	OH6A	SoC	
68-17315	OH6A	w.o.24.1.70	
68-17316	OH6A	SoC	
68-17317	OH6A	SoC	
68-17318	OH6A	1-131AVN	AL NG
68-17319	OH6A	SoC	
68-17320	MH6B	to USBP	N5185K
68-17321	OH6A	SoC	
68-17322	OH6A	SoC	
68-17323	OH6A	SoC	
68-17324	OH6A		
68-17325	OH6A	SoC	
68-17326	OH6A	SoC	
68-17327	OH6A	w.o.	
68-17328	OH6A	SoC	
68-17329	OH6A	SoC	
68-17330	OH6A	SoC	
68-17331	OH6A	42AVN	
68-17332	MH6C		
68-17333	OH6A	SoC	
68-17334	MH6B	to USBP	N5185L
68-17335	OH6A	SoC	
68-17336	OH6A	w.o.13.11.77	
68-17337	OH6A	SoC	
68-17338	OH6A	SoC	
68-17339	OH6A	42AVN	
68-17340	OH6A	preserved	Ft. Rucker
68-17341	MH6B	to USBP	N5185N
68-17342	OH6A	w.o.14.9.70	
68-17343	OH6A	SoC	
68-17344	OH6A	SoC	
68-17345	OH6A	150AVN	
68-17346	MH6B	to USBP	N5185S
68-17347	OH6A	150AVN	
68-17348	MH6B	to USBP	N5186J
68-17349	OH6A	SoC	
68-17350	OH6A	SoC	
68-17351	OH6A	SoC	
68-17352	OH6A	w.o.11.9.70	
68-17353	OH6A	SoC	
68-17354	OH6A	SoC	
68-17355	OH6A	SoC	
68-17356	OH6A	SoC	

Serial	Type	Notes	
68-17357	OH6A	SoC	
68-17358	MH6B	to USBP	N5187Y
68-17359	OH6A	SoC	
68-17360	OH6A	SoC	
68-17361	OH6A	SoC	
68-17362	OH6A	SoC	
68-17363	OH6A	SoC	
68-17364	OH6A	SoC	
68-17365	OH6A	328AvCo	
68-17366	OH6A	SoC	
68-17367	OH6A	w.o.21.8.70	
68-17368	OH6A	SoC	
68-17369	OH6A	B\26AVN	
69-15960	OH6A	w.o.24.10.70	
69-15961	OH6A	SoC	
69-15962	OH6A	w.o.14.5.70	
69-15963	OH6A	SoC	
69-15964	OH6A	SoC	
69-15965	OH6A	SoC	
69-15966	OH6A	to N.....	
69-15967	OH6A	SoC	
69-15968	OH6A	SoC	
69-15969	OH6A	SoC	
69-15970	OH6A	SoC	
69-15971	OH6A		
69-15972	OH6A	SoC	
69-15973	AH6C	76AvCo	
69-15974	OH6A	SoC	
69-15975	OH6A	SoC	
69-15976	OH6A	SoC	
69-15977	MH6B	to USBP	N5184C
69-15978	OH6A	w.o.17.4.70	
69-15979	OH6A	26AVN	
69-15980	OH6A	126ACS	
69-15981	OH6A	SoC	
69-15982	OH6A	w.o.27.4.74	
69-15983	OH6A	w.o.19.9.71	
69-15984	OH6A	SoC	
69-15985	OH6A	278ACR	
69-15986	OH6A		
69-15987	OH6A		
69-15988	OH6A		
69-15989	OH6A	SoC	
69-15990	OH6A	42AVN	
69-15991	OH6A	SoC	
69-15992	OH6A	SoC	
69-15993	OH6A	126ACS	
69-15994	OH6A	SoC	
69-15995	OH6A	328AvCo	
69-15996	OH6A	SoC	
69-15997	OH6A	1-131AVN	AL NG
69-15998	OH6A	SoC	
69-15999	OH6A	USN\TPS	
69-16000	OH6A	w.o.18.11.71	
69-16001	OH6A		
69-16002	OH6A	323AVN	
69-16003	OH6A	SoC	
69-16004	OH6A		
69-16005	OH6A	SoC	
69-16006	OH6A		
69-16007	OH6A	SoC	
69-16008	OH6A	42AVN	
69-16009	OH6A	SoC	
69-16010	OH6A	SoC	
69-16011	OH6A		
69-16012	OH6A	SoC	
69-16013	OH6A	SoC	
69-16014	OH6A	126ACS	
69-16015	MH6B	to USBP	N51841
69-16016	OH6A	SoC	
69-16017	OH6A	SoC	
69-16018	AH6C		
69-16019	OH6A		
69-16020	OH6A	SoC	
69-16021	OH6A	SoC	
69-16022	OH6A	1-150AVN	NJ NG
69-16023	OH6A	w.o.20.10.70	
69-16024	OH6A	w.o.24.12.70	
69-16025	OH6A	SoC	
69-16026	OH6A	42AVN	
69-16027	OH6A	w.o.17.10.70	
69-16028	OH6A	164AvCo	
69-16029	OH6A		
69-16030	OH6A	SoC	
69-16031	AH6C	SoC	
69-16032	OH6A	SoC	
69-16033	OH6A	SoC	
69-16034	OH6A	SoC	
69-16035	OH6A		
69-16036	OH6A	SoC	
69-16037	OH6A	SoC	
69-16038	OH6A		
69-16039	OH6A	w.o.7.10.81	
69-16040	OH6A	SoC	
69-16041	OH6A	SOC	
69-16042	OH6A	1-131AVN	AL NG
69-16043	OH6A	SoC	
69-16044	OH6A	SoC	
69-16045	OH6A	1-131AVN	AL NG
69-16046	OH6A		
69-16047	OH6A	SoC	
69-16048	OH6A	76AvCo	
69-16049	OH6A	SoC	
69-16050	OH6A	1-150AVN	NJ NG
69-16051	OH6A	SoC	
69-16052	AH6C		
69-16053	AH6C		
69-16054	AH6C	1-131AVN	AL NG
69-16055	OH6A	w.o.24.2.71	
69-16056	OH6A		
69-16057	MH6B	to USBP	N51843
69-16058	AH6C		
69-16059	OH6A		
69-16060	OH6A	1-131AVN	AL NG
69-16061	OH6A	SoC	
69-16062	OH6A		
69-16063	OH6A	SoC	
69-16064	OH6A	w.o.29.6.72	
69-16065	OH6A	stored	Gulfport RAP, Ms
69-16066	OH6A	164AvCo	
69-16067	OH6A	SoC	
69-16068	OH6A	1-131AVN	AL NG
69-16069	OH6A	SoC	
69-16070	OH6A	SoC	
69-16071	OH6A	w.o.15.5.72	
69-16072	AH6C		
69-16073	OH6A	328AvCo	
69-16074	OH6A	SoC	
69-16075	OH6A	SoC	
81-23629	MH6H		
81-23630	MH6H		
81-23631	MH6J		
81-23632	MH6H		
81-23633	MH6H	w.o.11.2.86	
81-23634	MH6H		
81-23635	MH6J		
81-23636	MH6J		
81-23637	MH6H	w.o.30.12.89	
81-23648	MH6H		
81-23649	MH6H	SoC	
81-23650	MH6J		
81-23651	MH6H		
81-23652	MH6H		
81-23653	MH6H		
81-23654	MH6H		
81-23655	MH6H	SoC	
81-23656	MH6H	w.o.27.4.87	
84-24319	AH6G		
84-24677	AH6G	SoC	
84-24678	AH6G	SoC	
84-24679	AH6G		
84-24680	AH6G		
84-24681	AH6G	w.o.23.6.88	
84-24682	AH6G	w.o.18.10.88	
84-24683	AH6G		
84-24684	AH6G		
85-25346	AH6G		
85-25347	AH6G	160SOAR	
85-25348	AH6G		
86-0041	AH6F	SoC	
86-0382	AH6F	SoC	
86-0383	AH6F	SoC	
86-0384	AH6F	SoC	
86-0385	AH6F	SoC	
86-0386	AH6F	SoC	
86-0387	AH6F	SoC	

88-25349	MH6H		89-25354	AH6J	
88-25350	MH6H		89-25355	AH6J	
88-25351	AH6J		89-25356	AH6J	
88-25352	AH6J		90-25357	AH6J	
88-25353	AH6J		90-25358	AH6J	
88-25354	AH6J		90-25359	AH6J	
88-25355	AH6J		90-25360	AH6J	
88-25356	AH6J		90-25361	MH6J	
88-25357	AH6J		90-25362	AH6J	
89-25351	AH6J		90-25363	AH6J	
89-25352	AH6J		91-25364	AH6J	
89-25353	AH6J		91-25366	MH6J	

H47 CHINOOK

c\n CH47A\B\C: B1 to B5, B7 to B744. (last 11 built by Meridionali)

note: All extant CH47A\B\C airframes have been involved in a conversion and rebuilding programme to CH47D or MH47E standard. A new FY serial and construction number has been issued (see below).

c\n YCH47D\CH47D: M3001 to M3464.
YMH47E\MH47E: M3258(2), M3701 to M3711.

note: serial 82-23780 is compromised on C12D Huron c\n BP-28.

59-4982	YCH47A		62-2134	CH47A	
59-4983	YCH47A		62-2135	CH47A	to CH47D 84-24182
59-4984	YCH47B		62-2136	CH47A	to CH47D 87-0088
59-4985	YCH47A		62-2137	CH47A	to CH47D 84-24184
59-4986	YCH47A	to CH47D 92-0291	63-7900	CH47A	to CH47D 92-0294
60-3448	GCH47A		63-7901	CH47A	
60-3449	CH47A	to CH47D 92-0292	63-7902	CH47A	to CH47D 87-0090
60-3450	JCH47A	preserved	63-7903	CH47A	to CH47D 87-0092
60-3451	CH47A	preserved Ft. Rucker	63-7904	CH47A	to CH47D 87-0094
60-3452	GCH47A	preserved	63-7905	CH47A	to CH47D 92-0285
61-2408	CH47A	preserved	63-7906	CH47A	to CH47D 87-0096
61-2409	CH47A	to CH47D 91-0264	63-7907	CH47A	to CH47D 87-0098
61-2410	CH47A	to CH47D 91-0265	63-7908	CH47A	to CH47D 92-0286
61-2411	CH47A		63-7909	CH47A	to CH47D 87-0100
61-2412	CH47A	to CH47D 92-0290	63-7910	CH47A	
61-2413	CH47A	to CH47D 92-0297	63-7911	CH47A	to CH47D 87-0102
61-2414	CH47A	GIA RAF Odiham	63-7912	CH47A	to CH47D 92-0296
61-2415	CH47A	to CH47D 91-0266	63-7913	CH47A	
61-2416	CH47A	to CH47D 91-0260	63-7914	CH47A	to CH47D 87-0104
61-2417	CH47A	to CH47D 91-0267	63-7915	CH47A	to CH47D 92-0295
61-2418	CH47A		63-7916	CH47A	to CH47D 92-0287
61-2419	CH47A	to CH47D 91-0268	63-7917	CH47A	to CH47D 87-0106
61-2420	CH47A	to CH47D 91-0261	63-7918	CH47A	to CH47D 84-24186
61-2421	CH47A	to CH47D 91-0269	63-7919	CH47A	to CH47D 87-0082
61-2422	CH47A	to CH47D 87-0074	63-7920	CH47A	to CH47D 87-0112
61-2423	CH47A	to CH47D 91-0270	63-7921	CH47A	to CH47D 87-0116
61-2424	CH47A	to CH47D 87-0070	63-7922	CH47A	to CH47D 83-24107
61-2425	CH47A		63-7923	CH47A	to CH47D 83-24111
62-2114	CH47A	to CH47D 91-0263	64-13106	CH47A	
62-2115	CH47A	to CH47D 91-0271	64-13107	CH47A	
62-2116	CH47A	to CH47D 92-0280	64-13108	CH47A	to CH47D 83-24145
62-2117	CH47A	to CH47D 87-0076	64-13109	CH47A	
62-2118	CH47A	to CH47D 87-0080	64-13110	CH47A	
62-2119	GCH47A		64-13111	CH47A	to CH47D 83-24110
62-2120	CH47A	w.o.29.7.70	64-13112	CH47A	to CH47D 83-24114
62-2121	CH47A		64-13113	CH47A	to CH47D 84-24156
62-2122	CH47A		64-13114	CH47A	
62-2123	CH47A	to CH47D 92-0301	64-13115	CH47A	to CH47D 83-24104
62-2124	CH47A	to CH47D 92-0302	64-13116	CH47A	w.o.13.5.71
62-2125	CH47A		64-13117	CH47A	to CH47D 84-24171
62-2126	CH47A		64-13118	CH47A	to CH47D 84-24117
62-2127	CH47A	to CH47D 92-0288	64-13119	CH47A	to CH47D 84-24178
62-2128	CH47A	to CH47D 92-0293	64-13120	CH47A	to CH47D 84-24153
62-2129	CH47A	to CH47D 87-0108	64-13121	CH47A	to CH47D 84-24174
62-2130	CH47A	to CH47D 92-0283	64-13122	CH47A	to CH47D 83-24119
62-2131	CH47A	to CH47D 92-0284	64-13123	CH47A	to CH47D 84-24168
62-2132	CH47A	to CH47D 87-0084	64-13124	CH47A	
62-2133	CH47A	to CH47D 87-0086	64-13125	CH47A	to CH47D 84-24159

64-13126	CH47A	to CH47D 84-24165		65-8008	CH47A	to YCH47D '76-8008'
64-13127	CH47A	to CH47D 83-24123		65-8009	CH47A	to CH47D 86-1657
64-13128	CH47A	w.o.		65-8010	CH47A	to CH47D 85-24367
64-13129	CH47A	to CH47D 84-24162		65-8011	CH47A	to CH47D 82-23764
64-13130	CH47A	to CH47D 84-24180		65-8012	CH47A	to CH47D 86-1681
64-13131	CH47A			65-8013	CH47A	to CH47D 86-1648
64-13132	CH47A	to CH47D 81-23388		65-8014	CH47A	to CH47D 85-24335
64-13133	CH47A	to CH47D 81-23389		65-8015	CH47A	to CH47D 86-1651
64-13134	CH47A	to CH47D 81-23387		65-8016	CH47A	to Vietnam
64-13135	CH47A	to CH47D 86-1676		65-8017	CH47A	to CH47D 85-24333
64-13136	CH47A	to Thailand		65-8018	CH47A	to CH47D 85-24353
64-13137	CH47A	to CH47D 86-1635		65-8019	CH47A	to CH47D 85-24339
64-13138	CH47A			65-8020	CH47A	to CH47D 86-1645
64-13139	CH47A			65-8021	CH47A	to Vietnam
64-13140	CH47A	to CH47D 84-24154		65-8022	CH47A	
64-13141	CH47A	to Vietnam		65-8023	CH47A	to CH47D 86-1636
64-13142	CH47A	to CH47D 86-1662		65-8024	CH47A	
64-13143	CH47A	to Vietnam		65-8025	CH47A	to Vietnam
64-13144	CH47A	to CH47D 82-23744		66-0066	CH47A	to Vietnam
64-13145	CH47A			66-0067	CH47A	
64-13146	CH47A			66-0068	CH47A	to Vietnam
64-13147	CH47A	to Vietnam		66-0069	CH47A	to Vietnam
64-13148	CH47A	to Thailand		66-0070	CH47A	to Vietnam
64-13149	CH47A			66-0071	CH47A	
64-13150	CH47A	to CH47D 84-24163		66-0072	CH47A	
64-13151	CH47A			66-0073	CH47A	to Vietnam
64-13152	CH47A			66-0074	CH47A	to CH47D 86-1647
64-13153	CH47A			66-0075	CH47A	to CH47D 86-1668
64-13154	CH47A			66-0076	CH47A	
64-13155	CH47A	to CH47D 86-1663		66-0077	CH47A	to CH47D 85-24341
64-13156	CH47A			66-0078	CH47A	to Vietnam
64-13157	CH47A	w.o.10.5.72		66-0079	CH47A	to Vietnam
64-13158	CH47A			66-0080	CH47A	to Vietnam
64-13159	CH47A	to CH47D 86-1638		66-0081	CH47A	to Vietnam
64-13160	CH47A	to CH47D 84-24166		66-0082	CH47A	to Vietnam
64-13161	CH47A			66-0083	CH47A	
64-13162	CH47A			66-0084	CH47A	to Vietnam
64-13163	CH47A	to Vietnam		66-0085	CH47A	to Vietnam
64-13164	CH47A	to CH47D 86-1641		66-0086	CH47A	to Vietnam
64-13165	CH47A	to CH47D 87-0072		66-0087	CH47A	to Vietnam
65-7966	CH47A	w.o.23.6.70		66-0088	CH47A	
65-7967	CH47A	to CH47D 86-1667		66-0089	CH47A	to CH47D 85-24347
65-7968	CH47A	to Vietnam		66-0090	CH47A	to CH47D 85-24345
65-7969	CH47A			66-0091	CH47A	to Vietnam
65-7970	CH47A	to Vietnam		66-0092	CH47A	to Vietnam
65-7971	CH47A	to CH47D 85-24351		66-0093	CH47A	to CH47D 85-24357
65-7972	CH47A	to Vietnam		66-0094	CH47A	to Vietnam
65-7973	CH47A	w.o.22.3.72		66-0095	CH47A	to Vietnam
65-7974	CH47A	to Vietnam		66-0096	CH47A	to Vietnam
65-7975	CH47A	to Vietnam		66-0097	CH47A	to CH47D 87-0069
65-7976	CH47A			66-0098	CH47A	to Vietnam
65-7977	CH47A	to CH47D 85-24366		66-0099	CH47A	to Vietnam
65-7978	CH47A	to CH47D 84-24157		66-0100	CH47A	to Vietnam
65-7979	CH47A	to CH47D 85-24338		66-0101	CH47A	to CH47D 85-24348
65-7980	CH47A	to CH47D 86-1656		66-0102	CH47A	to CH47D 86-1659
65-7981	CH47A	to CH47D 85-24323		66-0103	CH47A	to CH47D 86-1650
65-7982	CH47A	to CH47D 85-24326		66-0104	CH47A	to CH47D 86-1674
65-7983	CH47A	to CH47D 85-24322		66-0105	CH47A	to Vietnam
65-7984	CH47A	to CH47D 86-1639		66-0106	CH47A	to CH47D 82-23768
65-7985	CH47A	to Vietnam		66-0107	CH47A	to CH47D 85-24344
65 7986	CH47A	to Vietnam		66-0108	CH47A	to CH47D 82-23770
65-7987	CH47A	to Vietnam		66-0109	CH47A	w.o.6.74
65-7988	CH47A	w.o.18.1.70		66-0110	CH47A	to Vietnam
65-7989	CH47A			66-0111	CH47A	w.o.25.5.71
65-7990	CH47A	to CH47D 84-24169		66-0112	CH47A	to Vietnam
65-7991	CH47A	to CH47D 86-1679		66-0113	CH47A	to Vietnam
65-7992	CH47A	preserved Ft. Rucker		66-0114	CH47A	to CH47D 86-1665
65-7993	CH47A	to CH47D 86-1653		66-0115	CH47A	to CH47D 82-23776
65-7994	CH47A	to Vietnam		66-0116	CH47A	to CH47D 82-23772
65-7995	CH47A	to CH47D 86-1660		66-0117	CH47A	to Vietnam
65-7996	CH47A	to Vietnam		66-0118	CH47A	w.o.22.11.74
65-7997	CH47A	to Vietnam		66-0119	CH47A	to CH47D 82-23765
65-7998	CH47A	to Vietnam		66-0120	CH47A	
65-7999	CH47A			66-0121	CH47A	w.o.19.2.71
65-8000	CH47A	to Vietnam		66-0122	CH47A	to CH47D 82-23762
65-8001	CH47A	w.o.20.8.79		66-0123	CH47A	to CH47D 85-24332
65-8002	CH47A	to CH47D 85-24365		66-0124	CH47A	to CH47D 85-24336
65-8003	CH47A	to CH47D 86-1680		66-0125	CH47A	to Vietnam
65-8004	CH47A	to CH47D 86-1644		66-19000	CH47A	to CH47D 85-24356
65-8005	CH47A	to CH47D 82-23769		66-19001	CH47A	
65-8006	CH47A	to Vietnam		66-19002	CH47A	to Vietnam
65-8007	CH47A			66-19003	CH47A	to Vietnam

Serial	Type	Notes		Serial	Type	Notes
66-19004	CH47A	to Vietnam		66-19087	CH47A	to CH47D 86-1670
66-19005	CH47A			66-19088	CH47A	to CH47D 81-23386
66-19006	CH47A			66-19089	CH47A	to Vietnam
66-19007	CH47A	to Vietnam		66-19090	CH47A	
66-19008	CH47A	to CH47D 84-24176		66-19091	CH47A	
66-19009	CH47A	to CH47D 85-24354		66-19092	CH47A	to Vietnam
66-19010	CH47A	to Vietnam		66-19093	CH47A	to Vietnam
66-19011	CH47A			66-19094	CH47A	w.o.10.4.72
66-19012	CH47A			66-19095	CH47A	w.o.11.2.70
66-19013	CH47A			66-19096	CH47A	to CH47D 86-1671
66-19014	CH47A			66-19097	CH47A	to CH47D 86-1673
66-19015	CH47A			66-19098	CH47B	to CH47D 87-0109
66-19016	CH47A			66-19099	CH47B	to CH47D 88-0081
66-19017	CH47A	to CH47D 81-23383		66-19100	CH47B	to CH47D 88-0089
66-19018	CH47A	to CH47D 85-24360		66-19101	CH47B	
66-19019	CH47A			66-19102	CH47B	to CH47D 87-0075
66-19020	CH47A	to CH47D 84-24160		66-19103	CH47B	GIA RAF
66-19021	CH47A	to CH47D 85-24324		66-19104	CH47B	to CH47D 88-0092
66-19022	CH47A			66-19105	CH47B	to CH47D 88-0066
66-19023	CH47A	w.o.18.8.71		66-19106	CH47B	w.o.9.3.70
66-19024	CH47A	to CH47D 85-24362		66-19107	CH47B	to CH47D 86-1675
66-19025	CH47A	to CH47D 81-23381		66-19108	CH47B	to CH47D 88-0080
66-19026	CH47A	to CH47D 85-24327		66-19109	CH47B	to CH47D 88-0083
66-19027	CH47A			66-19110	CH47B	w.o.19.8.80
66-19028	CH47A	to CH47D 86-1654		66-19111	CH47B	to CH47D 87-0087
66-19029	CH47A			66-19112	CH47B	to CH47D 87-0103
66-19030	CH47A	to CH47D 86-1677		66-19113	CH47B	
66-19031	CH47A	to CH47D 82-23775		66-19114	CH47B	to CH47D 86-1655
66-19032	CH47A			66-19115	CH47B	to CH47D 86-1646
66-19033	CH47A	to Vietnam		66-19116	CH47B	to CH47D 88-0085
66-19034	CH47A	to Vietnam		66-19117	CH47B	w.o.4.10.84
66-19035	CH47A	to Vietnam		66-19118	CH47B	
66-19036	CH47A	to CH47D 85-24329		66-19119	CH47B	to CH47D 87-0101
66-19037	CH47A	to Vietnam		66-19120	CH47B	to CH47D 87-0079
66-19038	CH47A	to Vietnam		66-19121	CH47B	to CH47D 86-1643
66-19039	CH47A	w.o.		66-19122	CH47B	to CH47D 88-0073
66-19040	CH47A	to Vietnam		66-19123	CH47B	to CH47D 88-0088
66-19041	CH47A	w.o.		66-19124	CH47B	to CH47D 87-0111
66-19042	CH47A	to Vietnam		66-19125	CH47B	to CH47D 86-1658
66-19043	CH47A	to CH47D 82-23767		66-19126	CH47B	
66-19044	CH47A	to CH47D 85-24368		66-19127	CH47B	to CH47D 87-0105
66-19045	CH47A			66-19128	CH47B	
66-19046	CH47A	w.o.		66-19129	CH47B	to CH47D 88-0070
66-19047	CH47A			66-19130	CH47B	to CH47D 88-0075
66-19048	CH47A	to CH47D 86-1642		66-19131	CH47B	to CH47D 88-0072
66-19049	CH47A	to CH47D 82-23763		66-19132	CH47B	to CH47D 87-0077
66-19050	CH47A			66-19133	CH47B	to CH47D 86-1664
66-19051	CH47A	to CH47D 85-24330		66-19134	CH47B	to CH47D 87-0091
66-19052	CH47A	to CH47D 81-23382		66-19135	CH47B	to CH47D 87-0093
66-19053	CH4/A			66-19136	CH47B	to CH47D 88-0062
66-19054	CH47A	to CH47D 85-24369		66-19137	CH47B	to CH47D 83-24102
66-19055	CH47A	to CH47D 84-24172		66-19138	CH47B	to CH47D 89-0176
66-19056	CH47A	to Vietnam		66-19139	CH47B	
66-19057	CH47A	to CH47D 81-23384		66-19140	CH47B	to CH47D 86-1661
66-19058	CH47A	to CH47D 82-23766		66-19141	CH47B	to CH47D 88-0063
66-19059	CH47A			66-19142	CH47B	
66-19060	CH47A	to CH47D 85-24350		66-19143	CH47B	w.o.26.10.71
66-19061	CH47A			67-18432	CH47B	to CH47D 87-0107
66-19062	CH47A			67-18433	CH47B	w.o.21.12.70
66-19063	CH47A			67-18434	CH47B	to CH47D 87-0095
66-19064	CH47A	w.o.12.9.70		67-18435	CH47B	
66-19065	CH47A	to Vietnam		67-18436	CH47B	to CH47D 87-0078
66-19066	CH47A	to CH47D 85-24359		67-18437	CH47B	to CH47D 87-0097
66-19067	CH47A			67-18438	CH47B	to CH47D 86-1637
66-19068	CH47A	to Vietnam		67-18439	CH47B	to CH47D 86-1640
66-19069	CH47A			67-18440	CH47B	to CH47D 88-0079
66-19070	CH47A	w.o.24.10.70		67-18441	CH47B	to CH47D 88-0071
66-19071	CH47A	to CH47D 85-24342		67-18442	CH47B	w.o.17.3.70
66-19072	CH47A	to CH47D 82-23777		67-18443	CH47B	to CH47D 87-0113
66-19073	CH47A	to CH47D 81-23385		67-18444	CH47B	to CH47D 86-1669
66-19074	CH47A	to CH47D 82-23771		67-18445	CH47B	
66-19075	CH47A			67-18446	CH47B	to CH47D 87-0073
66-19076	CH47A			67-18447	CH47B	to CH47D 88-0064
66-19077	CH47A	to CH47D 85-24363		67-18448	CH47B	to CH47D 86-1672
66-19078	CH47A			67-18449	CH47B	
66-19079	CH47A			67-18450	CH47B	to CH47D 88-0067
66-19080	CH47A			67-18451	CH47B	to CH47D 86-1652
66-19081	CH47A			67-18452	CH47B	to CH47D 87-0114
66-19082	CH47A			67-18453	CH47B	to CH47D 87-0085
66-19083	CH47A			67-18454	CH47B	to CH47D 88-0082
66-19084	CH47A			67-18455	CH47B	
66-19085	CH47A	w.o.10.4.71		67-18456	CH47B	to CH47D 83-24115
66-19086	CH47A	w.o.4.73		67-18457	CH47B	

67-18458	CH47B	
67-18459	CH47B	to CH47D 87-0115
67-18460	CH47B	to CH47D 83-24112
67-18461	CH47B	
67-18462	CH47B	w.o.9.6.71
67-18463	CH47B	to CH47D 87-0110
67-18464	CH47B	to CH47D 83-24108
67-18465	CH47B	to CH47D 88-0090
67-18466	CH47B	to CH47D 87-0083
67-18467	CH47B	to CH47D 88-0091
67-18468	CH47B	w.o.15.2.71
67-18469	CH47B	
67-18470	CH47B	
67-18471	CH47B	to CH47D 87-0071
67-18472	CH47B	to CH47D 86-1649
67-18473	CH47B	to CH47D 88-0087
67-18474	CH47B	to CH47D 87-0099
67-18475	CH47B	
67-18476	CH47B	to CH47D 88-0069
67-18477	CH47B	to CH47D 87-0081
67-18478	CH47B	to CH47D 88-0078
67-18479	CH47B	to YCH47D '76-18479'
67-18480	CH47B	
67-18481	CH47B	
67-18482	CH47B	to CH47D 88-0077
67-18483	CH47B	to CH47D 83-24105
67-18484	CH47B	to CH47D 87-0089
67-18485	CH47B	w.o.31.5.70
67-18486	CH47B	to CH47D 88-0068
67-18487	CH47B	
67-18488	CH47B	to CH47D 88-0086
67-18489	CH47B	to CH47D 88-0074
67-18490	CH47B	
67-18491	CH47B	to CH47D 88-0065
67-18492	CH47B	to CH47D 88-0076
67-18493	CH47B	w.o.6.2.71
67-18494	CH47C	to CH47D 91-0240
67-18495	CH47C	to CH47D 91-0247
67-18496	CH47C	
67-18497	CH47C	
67-18498	CH47C	
67-18499	CH47C	
67-18500	CH47C	to CH47D 89-0160
67-18501	CH47C	
67-18502	CH47C	
67-18503	CH47C	to CH47D 89-0170
67-18504	CH47C	to CH47D 85-24346
67-18505	CH47C	to CH47D 89-0134
67-18506	CH47C	
67-18507	CH47C	
67-18508	CH47C	
67-18509	CH47C	to CH47D 85-24337
67-18510	CH47C	to CH47D 90-0189
67-18511	CH47C	w.o.2.3.77
67-18512	CH47C	w.o.13.6.70
67-18513	CH47C	
67-18514	CH47C	
67-18515	CH47C	to CH47D 90-0186
67-18516	CH47C	to CH47D 89-0142
67-18517	CH47C	to CH47D 89-0137
67-18518	CH47C	w.o.5.3.71
67-18519	CH47C	
67-18520	CH47C	to CH47D 89-0167
67-18521	CH47C	to CH47D 90-0222
67-18522	CH47C	to CH47D 86-1666
67-18523	CH47C	
67-18524	CH47C	w.o.10.1.70
67-18525	CH47C	to CH47D 91-0235
67-18526	CH47C	to CH47D 89-0159
67-18527	CH47C	w.o.20.3.83
67-18528	CH47C	to CH47D 89-0173
67-18529	CH47C	
67-18530	CH47C	to CH47D 90-0201
67-18531	CH47C	to CH47D 82-23780
67-18532	CH47C	to CH47D 89-0161
67-18533	CH47C	to CH47D 89-0138
67-18534	CH47C	
67-18535	CH47C	w.o.20.3.70
67-18536	CH47C	
67-18537	CH47C	to CH47D 84-24167
67-18538	CH47C	to YCH47D '76-18538'
67-18539	CH47C	w.o.
67-18540	CH47C	to CH47D 92-0281
68-15541	CH47C	to CH47D 83-24121
67-18542	CH47C	w.o.8.2.71
67-18543	CH47C	
67-18544	CH47C	
67-18545	CH47C	w.o.2.4.71
67-18546	CH47C	to CH47D 89-0166
67-18547	CH47C	to CH47D 91-0248
67-18548	CH47C	to CH47D 92-0282
67-18549	CH47C	to CH47D 89-0164
67-18550	CH47C	to CH47D 84-24181
67-18551	CH47C	to CH47D 89-0174
68-15810	CH47C	
68-15811	CH47C	to CH47D 91-0250
68-15812	CH47C	to CH47D 90-0190
68-15813	CH47C	to CH47D 83-24113
68-15814	CH47C	to CH47D 91-0234
68-15815	CH47C	to CH47D 89-0162
68-15816	CH47C	to CH47D 90-0183
68-15817	CH47C	to CH47D 90-0224
68-15818	CH47C	to CH47D 90-0209
68-15819	CH47C	to CH47D 89-0177
68-15820	CH47C	to CH47D 89-0135
68-15821	CH47C	to CH47D 89-0171
68-15822	CH47C	to CH47D 89-0154
68-15823	CH47C	
68-15824	CH47C	
68-15825	CH47C	to CH47D 89-0172
68-15826	CH47C	
68-15827	CH47C	to CH47D 88-0104
68-15828	CH47C	to CH47D 90-0184
68-15829	CH47C	to CH47D 89-0143
68-15830	CH47C	to CH47D 84-24164
68-15831	CH47C	to CH47D 88-0106
68-15832	CH47C	
68-15833	CH47C	to CH47D 91-0236
68-15834	CH47C	to CH47D 89-0141
68-15835	CH47C	w.o.15.2.71
68-15836	CH47C	to CH47D 89-0147
68-15837	CH47C	w.o.13.5.70
68-15838	CH47C	to CH47D 88-0084
68-15839	CH47C	to CH47D 85-24349
68-15840	CH47C	to CH47D 84-24173
68-15841	CH47C	
68-15842	CH47C	to CH47D 90-0215
68-15843	CH47C	to CH47D 85-24352
68-15844	CH47C	to CH47D 91-0251
68-15845	CH47C	w.o.10.7.83
68-15846	CH47C	to CH47D 88-0101
68-15847	CH47C	to CH47D 88-0095
68-15848	CH47C	to CH47D 89-0175
68-15849	CH47C	to CH47D 89-0139
68-15850	CH47C	to CH47D 85-24343
68-15851	CH47C	to CH47D 88-0094
68-15852	CH47C	to CH47D 89-0157
68-15853	CH47C	to CH47D 90-0210
68-15854	CH47C	
68-15855	CH47C	to CH47D 86-1678
68-15856	CH47C	to CH47D 90-0212
68-15857	CH47C	to CH47D 90-0219
68-15858	CH47C	to CH47D 90-0220
68-15859	CH47C	to CH47D 90-0193
68-15860	CH47C	to CH47D 89-0156
68-15861	CH47C	to CH47D 85-24355
68-15862	CH47C	to CH47D 89-0153
68-15863	CH47C	to CH47D 89-0148
68-15864	CH47C	to CH47D 90-0195
68-15865	CH47C	to CH47D 88-0099
68-15866	CH47C	w.o.28.11.71
68-15867	CH47C	to CH47D 88-0100
68-15868	CH47C	to CH47D 88-0093
68-15869	CH47C	w.o.30.1.70
68-15990	CH47C	to CH47D 89-0140
68-15991	CH47C	to CH47D 84-24152
68-15992	CH47C	to CH47D 90-0214
68-15993	CH47C	w.o.25.2.80
68-15994	CH47C	w.o.31.8.70
68-15995	CH47C	to CH47D 88-0105
68-15996	CH47C	to CH47D 83-24116
68-15997	CH47C	to CH47D 88-0102
68-15998	CH47C	to CH47D 90-0217
68-15999	CH47C	
68-16000	CH47C	
68-16001	CH47C	to CH47D 83-24122

Serial	Model	Disposition	Unit	Location
68-16002	CH47C	to CH47D 82-23773		
68-16003	CH47C	to MH47E 90-0414		
68-16004	CH47C	to CH47D 90-0203		
68-16005	CH47C	to MH47E 91-0499		
68-16006	CH47C	to CH47D 88-0103		
68-16007	CH47C	to CH47D 90-0221		
68-16008	CH47C	to CH47D 89-0144		
68-16009	CH47C	to CH47D 82-23779		
68-16010	CH47C	to CH47D 85-24331		
68-16011	CH47C	to CH47D 89-0155		
68-16012	CH47C	to CH47D 89-0169		
68-16013	CH47C	to CH47D 85-24364		
68-16014	CH47C	to CH47D 82-24119		
68-16015	CH47C	to CH47D 90-0216		
68-16016	CH47C	to CH47D 88-0096		
68-16017	CH47C	to CH47D 89-0130		
68-16018	CH47C	to CH47D 91-0249		
68-16019	CH47C	to CH47D 91-0233		
68-16020	CH47C	to CH47D 84-24187		
68-16021	CH47C	to CH47D 85-24361		
68-16022	CH47C			
69-17100	CH47C	w.o.16.9.70		
69-17101	CH47C	to CH47D 91-0256		
69-17102	CH47C	to CH47D 83-24106		
69-17103	CH47C	to CH47D 91-0259		
69-17104	CH47C	to CH47D 82-23778		
69-17105	CH47C	w.o.12.7.79		
69-17106	CH47C	to CH47D 89-0131		
69-17107	CH47C	to CH47D 90-0196		
69-17108	CH47C	w.o.5.8.76		
69-17109	CH47C	to CH47D 83-24124		
69-17110	CH47C	to CH47D 91-0246		
69-17111	CH47C	to CH47D 85-24328		
69-17112	CH47C	to CH47D 85-24334		
69-17113	CH47C	to CH47D 84-24175		
69-17114	CH47C	to CH47D 88-0108		
69-17115	CH47C	to CH47D 90-0223		
69-17116	CH47C	to CH47D 88-0098		
69-17117	CH47C	to CH47D 90-0197		
69-17118	CH47C	to CH47D 92-0311		
69-17119	CH47C			
69-17120	CH47C	w.o.21.7.71		
69-17121	CH47C	to CH47D 85-24340		
69-17122	CH47C	to CH47D 90-0208		
69-17123	CH47C	to CH47D 88-0109		
69-17124	CH47C	w.o.15.7.77		
69-17125	CH47C	to CH47D 90-0205		
69-17126	CH47C	to CH47D 83-24104		
70-15000	CH47C	to CH47D 90-0207		
70-15001	CH47C	to CH47D 84-24170		
70-15002	CH47C	to CH47D 89-0149		
70-15003	CH47C	to CH47D 91-0262		
70-15004	CH47C	to CH47D 90-0218		
70-15005	CH47C	to CH47D 89-0150		
70-15006	CH47C	to CH47D 84-24179		
70-15007	CH47C	to MH47E 92-0400		
70-15008	CH47C	to CH47D 89-0132		
70-15009	CH47C	to CH47D 90-0185		
70-15010	CH47C	to CH47D 83-24118		
70-15011	CH47C	to CH47D 85-24358		
70-15012	CH47C	to CH47D 89-0145		
70-15013	CH47C	to CH47D 84-24185		
70-15014	CH47C	to CH47D 84-24158		
70-15015	CH47C	to CH47D 92-0299		
70-15016	CH47C	to CH47D 84-24161		
70-15017	CH47C	to CH47D 84-24155		
70-15018	CH47C	to CH47D 90-0187		
70-15019	CH47C	to CH47D 84-24183		
70-15020	CH47C	to CH47D 91-0258		
70-15021	CH47C	to CH47D 90-0211		
70-15022	CH47C	to CH47D 86-1682		
70-15023	CH47C	to CH47D 89-0163		
70-15024	CH47C	to CH47D 91-0253		
70-15025	CH47C	to CH47D 85-24325		
70-15026	CH47C	to CH47D 91-0252		
70-15027	CH47C	to CH47D 89-0168		
70-15028	CH47C	to CH47D 90-0200		
70-15029	CH47C	to MH47E 92-0401		
70-15030	CH47C	to MH47E 92-0403		
70-15031	CH47C	to CH47D 89-0146		
70-15032	CH47C	to CH47D 91-0254		
70-15033	CH47C	to CH47D 91-0232		
70-15034	CH47C	to CH47D 83-24109		
70-15035	CH47C	w.o.13.6.75		
71-20944	CH47C	w.o.12.1.72		
71-20945	CH47C	to CH47D 89-0133		
71-20946	CH47C	to CH47D 92-0298		
71-20947	CH47C	to CH47D 84-24177		
71-20948	CH47C	to CH47D 91-0237		
71-20949	CH47C	to CH47D 90-0225		
71-20950	CH47C	to CH47D 92-0300		
71-20951	CH47C	to CH47D 92-0305		
71-20952	CH47C	to CH47D 88-0097		
71-20953	CH47C	to CH47D 91-0230		
71-20954	CH47C	to CH47D 92-0312		
71-20955	CH47C	to CH47D 90-0188		
74-22271	CH47C	to CH47D 90-0181		
74-22272	CH47C	to CH47D 90-0182		
74-22273	CH47C	to CH47D 90-0191		
74-22274	CH47C	to CH47D 90-0192		
74-22275	CH47C	w.o.16.2.84		
74-22276	CH47C	to MH47E 92-0402		
74-22277	CH47C	to MH47E 91-0500		
74-22278	CH47C	to CH47D 89-0165		
74-22279	CH47C	to CH47D 89-0158		
74-22280	CH47C	to CH47D 91-0231		
74-22281	CH47C	to MH47E 91-0501		
74-22282	CH47C	to CH47D 91-0241		
74-22283	CH47C	to CH47D 92-0289		
74-22284	CH47C	to CH47D 91-0255		
74-22285	CH47C	to CH47D 92-0310		
74-22286	CH47C	to CH47D 90-0180		
74-22287	CH47C	to CH47D 90-0204		
74-22288	CH47C	to MH47E 91-0497		
74-22289	CH47C	to MH47E 91-0498		
74-22290	CH47C	w.o.17.7.80		
74-22291	CH47C	to CH47D 91-0257		
74-22292	CH47C	w.o.11.9.82		
74-22293	CH47C	to CH47D 90-0202		
74-22294	CH47C	to CH47D 90-0194		
76-8008	YCH47D	to CH47D 92-0303		
76-18479	YCH47D			
76-18538	YCH47D	to CH47D 92-0304		
76-22673	CH47C	to CH47D 91-0238		
76-22674	CH47C	to CH47D 91-0239		
76-22675	CH47C	to CH47D 89-0136		
76-22676	CH47C	to CH47D 90-0199		
76-22677	CH47C	to CH47D 92-0308		
76-22678	CH47C	to CH47D 89-0151		
76-22679	CH47C	to CH47D 92-0309		
76-22680	CH47C	to CH47D 91-0245		
76-22681	CH47C	to MH47E 91-0496		
76-22682	CH47C	to CH47D 90-0198		
76-22683	CH47C	to CH47D 90-0206		
76-22684	CH47C	to CH47D 89-0152		
79-23394	CH47C	to CH47D 92-0306		
79-23395	CH47C	to CH47D 90-0213		
79-23396	CH47C	to CH47D 92-0307		
79-23397	CH47C	to CH47D 90-0226		
79-23398	CH47C	to CH47D 88-0107		
79-23399	CH47C	to CH47D 90-0242		
79-23400	CH47C	to CH47D 91-0244		
79-23401	CH47C	to CH47D 91-0243		
81-23381	GCH47D			
81-23382	GCH47D			
81-23383	JCH47D	USAADTA		
81-23384	GCH47D	AFTC		
81-23385	CH47D	355AvCo		
81-23386	CH47D			
81-23387	CH47D			
81-23388	GCH47D			
81-23389	CH47D			
82-23762	CH47D	AATC	62G Hanchey AHP	
82-23763	CH47D			
82-23764	CH47D	w.o.24.7.90		
82-23765	CH47D			
82-23766	CH47D			
82-23767	CH47D			
82-23768	CH47D	159AVN		
82-23769	CH47D			
82-23770	CH47D	w.o.21.4.85		
82-23771	CH47D	AATC	71H Hanchey AHP	
82-23772	CH47D			
82-23773	CH47D			
82-23774	CH47D			
82-23775	CH47D			

Serial	Type	Unit/Notes	Code	Location
82-23776	CH47D			
82-23777	CH47D			
82-23778	CH47D			
82-23779	CH47D			
82-23780	CH47D			
83-24102	CH47D			
83-24103	CH47D			
83-24104	CH47D			
83-24105	GCH47D			
83-24106	GCH47D			
83-24107	CH47D			
83-24108	GCH47D			
83-24109	CH47D			
83-24110	MH47D	w.o.12.6.90		
83-24111	CH47D			
83-24112	GCH47D			
83-24113	CH47D			
83-24114	CH47D			
83-24115	CH47D			
83-24116	CH47D	AATC	16G	Hanchey AHP
83-24117	GCH47D			
83-24118	MH47D	160AVN		
83-24119	CH47D	AATC	19J	Hanchey AHP
83-24120	CH47D	AATC	20F	Hanchey AHP
83-24121	CH47D			
83-24122	CH47D			
83-24123	CH47D			
83-24124	CH47D			
83-24125	CH47D			
84-24152	CH47D			
84-24153	CH47D			
84-24154	CH47D			
84-24155	CH47D			
84-24156	CH47D			
84-24157	CH47D			
84-24158	CH47D			
84-24159	JCH47D			
84-24160	CH47D	AATC	60E	Hanchey AHP
84-24161	CH47D	AATC	61B	Hanchey AHP
84-24162	CH47D			
84-24163	CH47D	AATC	63F	Hanchey AHP
84-24164	CH47D	AATC	64D	Hanchey AHP
84-24165	CH47D			
84-24166	CH47D	w.o.12.7.85		
84-24167	CH47D			
84-24168	CH47D			
84-24169	CH47D			
84-24170	CH47D			
84-24171	CH47D			
84-24172	CH47D			
84-24173	CH47D			
84-24174	CH47D			
84-24175	CH47D			
84-24176	CH47D	132AvCo		
84-24177	CH47D	w.o.1.3.91		
84-24178	CH47D			
84-24179	CH47D			
84-24180	CH47D			
84-24181	CH47D			
84-24182	CH47D			
84-24183	CH47D			
84-24184	CH47D	159AVN		
84-24185	CH47D			
84-24186	CH47D			
84-24187	CH47D			
85-24322	CH47D	2-15AVN		
85-24323	CH47D	2-15AVN		
85-24324	CH47D	132AvCo		
85-24325	CH47D	w.o.9.4.86		
85-24326	CH47D			
85-24327	CH47D			
85-24328	CH47D			
85-24329	CH47D			
85-24330	CH47D			
85-24331	CH47D	2-15AVN		
85-24332	CH47D	w.o.8.12.88		
85-24333	CH47D	2-15AVN		
85-24334	CH47D	2-15AVN		
85-24335	CH47D			
85-24336	CH47D			
85-24337	CH47D	2-15AVN		
85-24338	CH47D			
85-24339	CH47D	2-15AVN		
85-24340	CH47D	2-15AVN		
85-24341	CH47D			
85-24342	CH47D			
85-24343	CH47D	2-15AVN		
85-24344	CH47D	AATC	44F	Hanchey AHP
85-24345	CH47D			
85-24346	CH47D			
85-24347	CH47D			
85-24348	CH47D			
85-24349	CH47D	132AvCo		
85-24350	CH47D			
85-24351	CH47D			
85-24352	CH47D			
85-24353	CH47D			
85-24354	CH47D			
85-24355	CH47D			
85-24356	CH47D			
85-24357	CH47D			
85-24358	CH47D			
85-24359	CH47D	132AvCo		
85-24360	CH47D			
85-24361	CH47D			
85-24362	CH47D			
85-24363	CH47D			
85-24364	CH47D			
85-24365	CH47D			
85-24366	CH47D			
85-24367	CH47D	160AVN		
85-24368	CH47D	AATC	68F	Hanchey AHP
85-24369	CH47D			
85-24734	CH47C	to CH47D 92-0313		
85-24735	CH47C	to CH47D 92-0314		
85-24736	CH47C	to CH47D 92-0315		
85-24737	CH47C	to CH47D 92-0316		
85-24738	CH47C	to CH47D 92-0317		
85-24739	CH47C	to CH47D 92-0318		
85-24740	CH47C	to CH47D 92-0319		
85-24741	CH47C	to CH47D 92-0320		
85-24742	CH47C	to CH47D 92-0321		
85-24743	CH47C	to CH47D 92-0322		
85-24744	CH47C	to CH47D 92-0323		
86-1635	CH47D			
86-1636	CH47D			
86-1637	CH47D			
86-1638	CH47D			
86-1639	CH47D			
86-1640	CH47D			
86-1641	CH47D			
86-1642	CH47D	2-158AVN		
86-1643	CH47D	w.o.25.2.88		
86-1644	CH47D	AATC	44C	Hanchey AHP
86-1645	CH47D	C\4-228AVN		
86-1646	CH47D			
86-1647	CH47D	2-158AVN		
86-1648	CH47D	2-158AVN		
86-1649	CH47D			
86-1650	CH47D			
86-1651	CH47D			
86-1652	CH47D			
86-1653	CH47D			
86-1654	CH47D			
86-1655	CH47D			
86-1656	CH47D	C\4-228AVN		
86-1657	CH47D			
86-1658	CH47D			
86-1659	CH47D			
86-1660	CH47D	6Cav		
86-1661	CH47D	C\4-228AVN		
86-1662	CH47D			
86-1663	CH47D			
86-1664	CH47D			
86-1665	CH47D			
86-1666	CH47D	6Cav		
86-1667	CH47D			
86-1668	CH47D			
86-1669	CH47D			
86-1670	CH47D			
86-1671	CH47D			
86-1672	CH47D			
86-1673	CH47D			
86-1674	CH47D	stored		Coleman Barracks
86-1675	CH47D			
86-1676	CH47D	205AvCo		

Serial	Type	Unit	Notes
86-1677	CH47D		
86-1678	CH47D		
86-1679	CH47D	158AVN	
86-1680	CH47D	158AVN	
86-1681	CH47D	2-158AVN	
86-1682	CH47D	158AVN	
87-0069	CH47D	AATC	69K Hanchey AHP
87-0070	CH47D	AATC	70C Hanchey AHP
87-0071	CH47D		
87-0072	CH47D		
87-0073	CH47D	205AvCo	
87-0074	CH47D		
87-0075	CH47D	stored	Coleman Barracks
87-0076	CH47D	stored	Coleman Barracks
87-0077	CH47D	7-158AVN	
87-0078	CH47D	stored	Coleman Barracks
87-0079	CH47D	5-159AVN	
87-0080	CH47D	7-158AVN	
87-0081	CH47D	A\5-159AVN	
87-0082	CH47D	7-158AVN	
87-0083	CH47D	7-158AVN	
87-0084	CH47D	5-159AVN	
87-0085	CH47D	5-159AVN	
87-0086	CH47D	B\6-158AVN	
87-0087	CH47D	5-159AVN	
87 0088	CH47D	5 159AVN	
87-0089	CH47D	7-158AVN	
87-0090	CH47D	5-159AVN	
87-0091	CH47D	5-159AVN	
87-0092	CH47D	5-159AVN	
87-0093	CH47D	7-158AVN	
87-0094	CH47D	7-158AVN	
87-0095	CH47D	7-158AVN	
87-0096	CH47D	B\6-158AVN	
87-0097	CH47D	B\6-158AVN	
87-0098	CH47D	B\6-158AVN	
87-0099	CH47D	B\6-158AVN	
87-0100	CH47D	B\6-158AVN	
87-0101	CH47D	B\6-158AVN	
87-0102	CH47D		
87-0103	CH47D	B\6-158AVN	
87-0104	CH47D	B\6-158AVN	
87-0105	CH47D	B\6-158AVN	
87-0106	CH47D	B\6-158AVN	
87-0107	CH47D	B\6-158AVN	
87-0108	CH47D	G\1-149AVN	TX NG
87-0109	CH47D	B\6-158AVN	
87-0110	CH47D	B\6-158AVN	
87-0111	CH47D	B\6-158AVN	
87-0112	CH47D	B\6-158AVN	
87-0113	CH47D	G\1-149AVN	TX NG
87-0114	CH47D	AATC	14H Hanchey AHP
87-0115	CH47D	G\1-149AVN	TX NG
87-0116	CH47D	G\1-149AVN	TX NG
88-0062	CH47D		
88-0063	CH47D	G\1-149AVN	TX NG
88-0064	CH47D	G\1-149AVN	TX NG
88-0065	CH47D	G\1-149AVN	TX NG
88-0066	CH47D	G\1-149AVN	TX NG
88-0067	CH47D		
88-0068	CH47D		
88-0069	CH47D		
88-0070	CH47D		
88-0071	CH47D		
88-0072	CH47D		
88-0073	CH47D		
88-0074	CH47D		
88-0075	CH47D		
88-0076	CH47D		
88-0077	CH47D		
88-0078	CH47D		
88-0079	CH47D		
88-0080	CH47D		
88-0081	CH47D		
88-0082	CH47D		
88-0083	CH47D		
88-0084	CH47D	to YMH47E 88-0267	
88-0085	CH47D		
88-0086	CH47D		
88-0087	CH47D		
88-0088	CH47D		
88-0089	CH47D		
88-0090	CH47D		
88-0091	CH47D		
88-0092	CH47D	w.o.4.12.89	
88-0093	CH47D		
88-0094	CH47D		
88-0095	CH47D		
88-0096	CH47D		
88-0097	CH47D		
88-0098	CH47D	E\502AVN	
88-0099	CH47D	E\502AVN	
88-0100	CH47D	E\502AVN	
88-0101	CH47D	E\502AVN	
88-0102	CH47D	E\502AVN	
88-0103	CH47D	E\502AVN	
88-0104	CH47D	E\502AVN	
88-0105	CH47D		
88-0106	CH47D	E\502AVN	
88-0107	CH47D	AATC	07B Hanchey AHP
88-0108	CH47D		
88-0109	CH47D	AATC	09N Hanchey AHP
88-0267	MH47E		
89-0130	CH47D		
89-0131	CH47D		
89-0132	CH47D		
89-0133	CH47D		
89-0134	CH47D		
80 0135	CH47D		
89-0136	CH47D		
89-0137	CH47D	AATC	37. Hanchey AHP
89-0138	CH47D	502AVN	
89-0139	CH47D	502AVN	
89-0140	CH47D	502AVN	
89-0141	CH47D	502AVN	
89-0142	CH47D	502AVN	
89-0143	CH47D	502AVN	
89-0144	CH47D	502AVN	
89-0145	CH47D	502AVN	
89-0146	CH47D		
89-0147	CH47D		
89-0148	CH47D		
89-0149	CH47D		
89-0150	CH47D		
89-0151	CH47D		
89-0152	CH47D		
89-0153	CH47D		
89-0154	CH47D		
89-0155	CH47D		
89-0156	CH47D		
89-0157	CH47D		
89-0158	CH47D		
89-0159	CH47D		
89-0160	CH47D		
89-0161	MH47D	A-2\160SOAR	
89-0162	CH47D		
89-0163	CH47D		
89-0164	CH47D		
89-0165	CH47D	w.o.11.1.91	
89-0166	CH47D		
89-0167	CH47D	C\228AVN	
89-0168	CH47D	C\228AVN	
89-0169	CH47D	C\228AVN	
89-0170	CH47D	C\228AVN	
89-0171	CH47D	C\228AVN	
89-0172	CH47D	C\228AVN	
89-0173	CH47D	C\228AVN	
89-0174	CH47D		
89-0175	CH47D		
89-0176	CH47D	C\228AVN	
89-0177	CH47D	C\228AVN	
90-0180	JCH47D		
90-0181	CH47D	C\228AVN	
90-0182	CH47D		
90-0183	CH47D		
90-0184	CH47D	2-158AVN	Reserve
90-0185	CH47D	2-158AVN	Reserve
90-0186	CH47D	2-158AVN	Reserve
90-0187	CH47D	2-158AVN	Reserve
90-0188	CH47D	AATC	88. Hanchey AHP
90-0189	CH47D		
90-0190	CH47D	2-158AVN	Reserve
90-0191	CH47D	AATC	91. Hanchey AHP
90-0192	CH47D	C\228AVN	
90-0193	CH47D	2-158AVN	Reserve
90-0194	CH47D	2-158AVN	Reserve

90-0195	CH47D		
90-0196	CH47D		
90-0197	CH47D	3-140AVN	
90-0198	CH47D	6-158AVN	Reserve
90-0199	CH47D		
90-0200	CH47D		
90-0201	CH47D	2-158AVN	Reserve
90-0202	CH47D		
90-0203	CH47D	2-158AVN	Reserve
90-0204	CH47D	6-158AVN	Reserve
90-0205	CH47D	2-158AVN	Reserve
90-0206	CH47D		
90-0207	CH47D	6-158AVN	Reserve
90-0208	CH47D		
90-0209	CH47D	2-158AVN	Reserve
90-0210	CH47D	G\1-149AVN	TX NG
90-0211	CH47D	6-158AVN	Reserve
90-0212	CH47D	6-158AVN	Reserve
90-0213	CH47D	6-158AVN	Reserve
90-0214	CH47D	6-158AVN	Reserve
90-0215	CH47D		
90-0216	CH47D		
90-0217	CH47D		
90-0218	CH47D	2-158AVN	Reserve
90-0219	CH47D		
90-0220	CH47D	6-158AVN	Reserve
90-0221	CH47D	6-158AVN	Reserve
90-0222	CH47D	6-158AVN	Reserve
90-0223	CH47D	6-158AVN	Reserve
90-0224	CH47D	6-158AVN	Reserve
90-0225	CH47D	6-158AVN	Reserve
90-0226	CH47D	G\1-149AVN	TX NG
90-0414	MH47E		
91-0230	CH47D		
91-0231	CH47D		
91-0232	CH47D		
91-0233	CH47D		
91-0234	CH47D	3-140AVN	
91-0235	CH47D	G\1-149AVN	TX NG
91-0236	CH47D		
91-0237	CH47D		
91-0238	CH47D		
91-0239	CH47D		
91-0240	CH47D		
91-0241	CH47D		
91-0242	CH47D		
91-0243	CH47D		
91-0244	CH47D		
91-0245	CH47D	1-106AVN	IL NG
91-0246	CH47D	1-106AVN	IL NG
91-0247	CH47D	1-106AVN	IL NG
91-0248	CH47D	1-106AVN	IL NG
91-0249	CH47D		
91-0250	CH47D		
91-0251	CH47D		
91-0252	CH47D		
91-0253	CH47D		
91-0254	CH47D		
91-0255	CH47D		
91-0256	CH47D		
91-0257	CH47D		
91-0258	CH47D		
91-0259	CH47D		
91-0260	CH47D		
91-0261	CH47D		

91-0262	CH47D		
91-0263	CH47D		
91-0264	CH47D		
91-0265	CH47D	E\185AVN	MS NG
91-0266	CH47D	E\185AVN	MS NG
91-0267	CH47D		
91-0268	CH47D		
91-0269	CH47D		
91-0270	CH47D		
91-0271	CH47D		
91-0496	MH47E		
91-0497	MH47E		
91-0498	MH47E		
91-0499	MH47E		
91-0500	MH47E		
91-0501	MH47E		
92-0280	CH47D		
92-0281	CH47D		
92-0282	CH47D		
92-0283	CH47D		
92-0284	CH47D		
92-0285	CH47D		
92-0286	CH47D		
92-0287	CH47D		
92-0288	CH47D		
92-0289	CH47D		
92-0290	CH47D		
92-0291	CH47D		
92-0292	CH47D		
92-0293	CH47D		
92-0294	CH47D		
92-0295	CH47D		
92-0296	CH47D		
92-0297	CH47D		
92-0298	CH47D		
92-0299	CH47D		
92-0300	CH47D		
92-0301	CH47D		
92-0302	CH47D		
92-0303	CH47D		
92-0304	CH47D		
92-0305	CH47D		
92-0306	CH47D		
92-0307	CH47D		
92-0308	CH47D		
92-0309	CH47D		
92-0310	CH47D		
92-0311	CH47D		
92-0312	CH47D		
92-0313	CH47D		
92-0314	CH47D		
92-0315	CH47D		
92-0316	CH47D		
92-0317	CH47D		
92-0318	CH47D		
92-0319	CH47D		
92-0320	CH47D		
92-0321	CH47D		
92-0322	CH47D		
92-0323	CH47D		
92-0400	MH47E		
92-0401	MH47E		
92-0402	MH47E		
92-0403	MH47E		

H54 TARHE

64-14202	YCH54A	w.o.9.8.66		64-14207	YCH54A	w.o.23.6.67	
64-14203	YCH54A	preserved	Ft. Eustis	66-18408	CH54A	stored	Hunter AAF
64-14204	YCH54A	w.o.5.1.66		66-18409	CH54A	preserved	Meridian AP, Ms
64-14205	YCH54A	w.o.19.4.68		66-18410	CH54A	stored	Hunter AAF
64-14206	YCH54A	w.o.6.7.67		66-18411	CH54A	stored	Meridian AP, Ms

66-18412	CH54A	wfu	
66-18413	CH54A	wfu	
67-18414	CH54A	wfu	
67-18415	CH54A	stored	Reno-Stead AP
67-18416	CH54A	stored	Reno-Stead AP
67-18417	CH54A	stored	Reno-Stead AP
67-18418	CH54A	D\113AVN	NV NG
67-18419	CH54A	w.o.13.8.81	
67-18420	CH54A	stored	Reno-Stead AP
67-18421	CH54A	wfu	
67-18422	CH54A	w.o.1.9.89	
67-18423	GCH54A	GIA	
67-18424	CH54A	wfu	
67-18425	CH54A	stored	Reno-Stead AP
67-18426	CH54A	stored	Reno-Stead AP
67-18427	CH54A	stored	Hunter AAF
67-18428	CH54A	stored	Hunter AAF
67-18429	CH54A	stored	Meridian AP, Ms
67-18430	CH54A	stored	Meridian AP, Ms
67-18431	CH54A	stored	Meridian AP, Ms
68-18432	CH54A	w.o.9.1.70	
68-18433	CH54A	stored	Meridian AP, Ms
68-18434	CH54A	w.o.1981	
68-18435	CH54A	wfu	
68-18436	CH54A	stored	Reno-Stead AP
68-18437	CH54A	stored	Reno-Stead AP
68-18438	CH54A	wfu	
68-18439	CH54A	D\113AVN	NV NG
68-18440	CH54A	stored	Hunter AAF
68-18441	CH54A	wfu	
68-18442	CH54A	w.o.24.4.72	
68-18443	CH54A	w.o.17.3.69	
68-18444	CH54A	w.o.1.2.71	
68-18445	CH54A	w.o.24.7.70	
68-18446	CH54A	D\113AVN	NV NG
68-18447	CH54A	stored	Meridian AP, Ms
68-18448	CH54A	stored	Reno-Stead AP
68-18449	CH54A	stored	Meridian AP, Ms
68-18450	CH54A	stored	Reno-Stead AP
68-18451	CH54A	stored	Hunter AAF
68-18452	CH54A	w.o.28.7.70	
68-18453	CH54A	stored	Hunter AAF
68-18454	CH54A	stored	Meridian AP, Ms
68-18455	CH54A	stored	Meridian AP, Ms
68-18456	CH54A	stored	Meridian AP, Ms
68-18457	CH54A	w.o.29.9.79	
68-18458	CH54A	wfu	
68-18459	CH54A	stored	Reno-Stead AP
68-18460	CH54A	wfu	
68-18461	CH54A	stored	Hunter AAF
69-18462	CH54B	wfu	
69-18463	CH54B	stored	Birmingham IAP
69-18464	CH54B	stored	Birmingham IAP
69-18465	CH54B	wfu	
69-18466	CH54B	wfu	
69-18467	CH54B	wfu	
69-18468	CH54B	wfu	
69-18469	CH54B	stored	Birmingham IAP
69-18470	CH54B	wfu	
69-18471	CH54B	stored	Birmingham IAP
69-18472	CH54B	stored	Fort Richardson
69-18473	CH54B	wfu	
69-18474	CH54B	w.o.28.7.72	
69-18475	CH54B	w.o.6.4.76	
69-18476	CH54B	wfu	
69-18477	CH54B	stored	Birmingham IAP
69-18478	CH54B	stored	Birmingham IAP
69-18479	CH54B	stored	Birmingham IAP
69-18480	CH54B	stored	Birmingham IAP
69-18481	CH54B	stored	Birmingham IAP
69-18482	CH54B	stored	Birmingham IAP
69-18483	CH54B	stored	Fort Richardson
69-18484	CH54B	wfu	
69-18485	CH54B	wfu	
69-18486	CH54B	stored	Birmingham IAP
69-18487	CH54B	stored	Birmingham IAP
69-18488	CH54B	stored	Fort Richardson
69-18489	CH54B	stored	Fort Richardson
69-18490	CH54B	wfu	

H58 KIOWA

c\n	40001 to 42200
note:	Some OH58A airframes have been involved in a conversion and rebuilding programme to OH58D AHIP and Warrior standard. A new FY serial and construction number has been issued (see below).
c\n	OH58D: 43001 to 43005 (68-16748\54\69-16139\322\285), 43006 to 43320.

68-16687	OH58A	preserved	Ft. Rucker
68-16688	OH58CR	197AvSect	
68-16689	OH58CR	13AHB	
68-16690	OH581		AL NG
68-16691	OH58A	to OH58D 87-0726	
68-16692	OH58CF		
68-16693	OH58A		
68-16694	OH58A	79ARCOM	
68-16695	OH58A	to OH58D 88-0298	
68-16696	OH58CR	USN\NATC	
68-16697	OH58CF		
68-16698	OH58A	to OH58D 86-8902	
68-16699	OH58A	derelict Salina MAP, Ks	
68-16700	OH58A		
68-16701	OH58A	to OH58D 85-24709	
68-16702	OH58CR	224AVN	
68-16703	GOH58A		
68-16704	OH58A	to OH58D 85-24706	
68-16705	OH58CF	stored Ramstein AB, FRG	
68-16706	OH58A	to OH58D 88-0290	
68-16707	OH58A	12Cav	
68-16708	OH58CR		
68-16709	OH58A		
68-16710	OH58A	to OH58D 85-24713	
68-16711	OH58A	E\1-192AVN	AL NG
68-16712	OH58A		
68-16713	OH58A		
68-16714	GOH58A		
68-16715	OH58A	to OH58D 85-24710	
68-16716	OH58CF		
68-16717	OH58A	to OH58D 83-24137	
68-16718	OH58CF	CCAD	
68-16719	OH58A	to OH58D 91-0541	
68-16720	OH58A	81ARCOM	
68-16721	GOH58A		
68-16722	OH58A		
68-16723	OH58A	to OH58D 87-0727	
68-16724	OH58CF	USAAEFA	
68-16725	OH58A	to OH58D 91-0571	
68-16726	OH58CR	22AVN	
68-16727	OH58CR	3AVN	
68-16728	OH58CF	AATC 28A Shell AHP	
68-16729	OH58A	57MedDet	
68-16730	OH58A	13AHB	
68-16731	OH58A	to OH58D 88-0315	
68-16732	OH58CR	2ACR	

Serial	Type	Notes
68-16733	OH58A	
68-16734	OH58CR	24AVN
68-16735	OH58CR	
68-16736	OH58A	197AvSect
68-16737	OH58CF	
68-16738	OH58CR	62AvCo
68-16739	OH58CR	68-16740 OH58A
68-16741	OH58CF	17Cav
68-16742	OH58CF	24AVN
68-16743	OH58CR	8AVN
68-16744	GOH58A	
68-16745	OH58A	11ACR
68-16746	OH58CR	
68-16747	OH58CF	
68-16748	OH58D	
68-16749	OH58A	w.o.1.12.70
68-16750	OH58CF	
68-16751	OH58CF	
68-16752	OH58A	244AvCo
68-16753	OH58CR	w.o.4.4.91
68-16754	OH58D	
68-16755	OH58CR	4AVN
68-16756	OH58A	to OH58D 89-0082
68-16757	OH58CR	9AVN
68-16758	OH58CR	
68-16759	OH58A	
68-16760	OH58A	w.o.27.3.71
68-16761	OH58CR	224AVN
68-16762	OH58CF	
68-16763	OH58A	8AVN
68-16764	GOH58C	
68-16765	OH58A	10AHB
68-16766	OH58A	107AvCo
68-16767	OH58A	A\132AVN OK NG
68-16768	OH58A	2-135AVN CO NG
68-16769	OH58CR	3AVN
68-16770	OH58A	107AvCo
68-16771	OH58A	to OH58D 85-24708
68-16772	GOH58A	
68-16773	OH58A	82AVN
68-16774	OH58A	
68-16775	OH58CR	1InfDiv
68-16776	OH58A	HI NG
68-16777	OH58A	
68-16778	OH58A	w.o.16.1.84
68-16779	OH58A	107AvCo
68-16780	OH58A	w.o.12.9.70
68-16781	OH58A	
68-16782	OH58A	
68-16783	OH58CF	11ACR
68-16784	OH58A	to OH58D 92-0584
68-16785	OH58A	
68-16786	OH58A	to OH58D 87-0757
68-16787	OH58CF	8AVN
68-16788	GOH58A	
68-16789	OH58A	107AvCo
68-16790	OH58A	1-183AVN ID NG
68-16791	OH58A	to OH58D 90-0369
68-16792	OH58A	to OH58D 89-0117
68-16793	OH58CF	9Cav
68-16794	OH58CF	stored Ramstein AB
68-16795	OH58A	
68-16796	OH58A	AATC 96D Shell AHP
68-16797	OH58A	to OH58D 88-0289
68-16798	OH58CF	6Cav
68-16799	OH58A	to OH58D 92-0571
68-16800	OH58CF	
68-16801	GOH58A	
68-16802	OH58A	to OH58D 90-0367
68-16803	OH58A	w.o.4.12.70
68-16804	OH58A	AATC 04K Shell AHP
68-16805	OH58A	to OH58D 88-0318
68-16806	OH58A	to OH58D 88-0302
68-16807	OH58A	E\1-192AVN AL NG
68-16808	OH58CR	
68-16809	GOH58A	
68-16810	OH58CF	
68-16811	OH58CF	17Cav
68-16812	OH58CF	
68-16813	OH58A	
68-16814	OH58A	
68-16815	OH58CR	4AVN
68-16816	OH58A	
68-16817	OH58CR	
68-16818	OH58A	
68-16819	OH58CF	
68-16820	OH58A	to OH58D 85-24695
68-16821	OH58A	308AHB
68-16822	OH58A	
68-16823	OH58A	w.o.21.1.70
68-16824	OH58A	
68-16825	OH58CF	11ACR
68-16826	OH58CF	
68-16827	OH58CR	
68-16828	OH58A	
68-16829	OH58A	1-185AVN MS NG
68-16830	OH58CF	
68-16831	OH58A	w.o.26.3.85
68-16832	OH58A	to OH58D 86-8939
68-16833	OH58A	
68-16834	OH58A	A\132AVN OK NG
68-16835	OH58A	2ACR
68-16836	OH58A	w.o.31.7.82
68-16837	OH58A	to OH58D 90-0346
68-16838	OH58A	
68-16839	OH58CR	224AVN
68-16840	OH58CR	224AVN
68-16841	GOH58A	
68-16842	OH58A	HI NG
68-16843	OH58A	w.o.24.1.85
68-16844	OH58A	w.o.29.4.70
68-16845	OH58CF	
68-16846	OH58CR	278ACR
68-16847	OH58CR	4AVN
68-16848	OH58A	17Cav
68-16849	OH58CR	AATC 49K Shell AHP
68-16850	JOH58C	
68-16851	OH58CR	2ACR
68-16852	OH58A	w.o.23.2.72
68-16853	OH58CF	stored Ramstein AB
68-16854	OH58A	to OH58D 91-0565
68-16855	OH58CF	6Cav
68-16856	OH58CR	
68-16857	OH58CR	
68-16858	OH58A	D\1-228AVN
68-16859	OH58CF	
68-16860	OH58A	
68-16861	OH58A	to OH58D 83-24133
68-16862	OH58A	
68-16863	OH58A	to OH58D 86-8912
68-16864	OH58A	
68-16865	OH58A	
68-16866	OH58A	
68-16867	OH58CR	1InfDiv
68-16868	OH58A	to OH58D 83-24132
68-16869	OH58A	to OH58D 88-0293
68-16870	OH58CF	
68-16871	OH58A	to OH58D 89-0105
68-16872	OH58CR	w.o.
68-16873	OH58A	
68-16874	OH58A	
68-16875	OH58A	4InfDiv
68-16876	OH58A	w.o.13.4.76
68-16877	OH58CF	stored Ramstein AB
68-16878	OH58CR	
68-16879	OH58A	
68-16880	OH58A	to OH58D 90-0347
68-16881	OH58CF	82AVN
68-16882	OH58CF	11ACR
68-16883	OH58A	
68-16884	OH58A	w.o.17.3.71
68-16885	OH58CR	3AVN
68-16886	OH58CF	7InfDiv
68-16887	OH58A	
68-16888	OH58A	
68-16889	OH58CR	10AHB
68-16890	OH58CR	
68-16891	OH58A	
68-16892	OH58A	7InfDiv
68-16893	OH58A	7\6Cav
68-16894	OH58A	
68-16895	OH58CR	10AHB
68-16896	OH58CF	stored Ramstein AB
68-16897	OH58A	
68-16898	OH58CF	
68-16899	OH58CF	3ACT

Serial	Type	Notes	
68-16900	OH58CR		
68-16901	OH58CR	26AVN	
68-16902	OH58CF		
68-16903	GOH58C		
68-16904	OH58A	40AVN	
68-16905	OH58A	56AvCo	
68-16906	OH58A	7InfDiv	
68-16907	OH58CF	4InfDiv	
68-16908	OH58A	to OH58D 92-0573	
68-16909	OH58CF	101AbDiv	
68-16910	OH58A	1InfDiv	
68-16911	OH58CF		
68-16912	OH58A		
68-16913	OH58A		
68-16914	OH58A	A\132AVN	OK NG
68-16915	OH58CR	163ACR	
68-16916	GOH58A		
68-16917	OH58CF	2ACR	
68-16918	OH58A	A\132AVN	OK NG
68-16919	OH58CR	3AVN	
68-16920	OH58CR		
68-16921	OH58CR		
68-16922	OH58A	to OH58D 83-24134	
68-16923	OH58A	to OH58D 85-24705	
C0-16924	OH58A	w.o.11.2.74	
68-16925	OH58A	to OH58D 86-8927	
68-16926	OH58A		
68-16927	OH58A		
68-16928	OH58CF	3Cav	
68-16929	OH58A		
68-16930	OH58A	to OH58D 88-0319	
68-16931	OH58CF	11ACR	
68-16932	OH58A	w.o.18.3.71	
68-16933	OH58A	to OH58D 85-24711	
68-16934	OH58A		
68-16935	OH58CF	17Cav	
68-16936	OH58A	25AvCo	
68-16937	OH58A	w.o.4.11.71	
68-16938	OH58A	w.o.9.6.72	
68-16939	OH58CF	w.o.6.11.91	
68-16940	GOH58C		
68-16941	OH58A		
68-16942	OH58CR		
68-16943	OH58A		
68-16944	OH58C	w.o.25.9.85	
68-16945	OH58A		
68-16946	OH58A		
68-16947	OH58CF		
68-16948	OH58CF	1InfDiv	
68-16949	OH58CR	1AVN	
68-16950	OH58A		
68-16951	OH58CF	AATC 51K Shell AHP	
68-16952	OH58A	w.o.2.5.78	
68-16953	OH58CR	4AVN	
68-16954	OH58CR	224AVN	
68-16955	OH58CF		
68-16956	OH58A	3ACT	
68-16957	OH58CR	3-227AVN	
68-16958	OH58CR	10AHB	
68-16959	OH58CF		
68-16960	OH58A	AATC 60D Shell AHP	
68-16961	OH58CF	stored Ramstein AB	
68-16962	OH58CF		
68-16963	OH58CR	10AHB	
68-16964	OH58A	w.o.30.12.70	
68-16965	OH58CF	11ACR	
68-16966	OH58A		
68-16967	OH58A		
68-16968	OH58A	to OH58D 88-0291	
68-16969	OH58CR	224AVN	
68-16970	OH58CR	1AVN	
68-16971	OH58A	2ACR	
68-16972	OH58A	w.o.21.1.80	
68-16973	OH58A		
68-16974	OH58A	to OH58D 85-24724	
68-16975	OH58A	w.o.18.6.71	
68-16976	OH58A	to OH58D 89-0108	
68-16977	OH58A		
68-16978	OH58CR	10AHB	
68-16979	OH58CR	5InfDiv	
68-16980	OH58A	to OH58D 83-24142	
68-16981	GOH58A		
68-16982	OH58CR	224AVN	
68-16983	OH58A		
68-16984	OH58CF		
68-16985	OH58CF	w.o.11.11.85	
68-16986	OH58A		
69-16080	OH58A	40AVN	
69-16081	OH58CR		
69-16082	OH58CR	10AHB	
69-16083	OH58A	40AVN	
69-16084	OH58CF	56AvCo	
69-16085	OH58A	25AVN	
69-16086	OH58A	126MedCo	
69-16087	OH58A		
69-16088	OH58A		
69-16089	OH58A		
69-16090	OH58A	w.o.15.12.70	
69-16091	OH58A		
69-16092	OH58A		
69-16093	OH58A		
69-16094	OH58CF		
69-16095	OH58A	w.o.29.5.71	
69-16096	OH58CR	224AVN	
69-16097	OH58CF	9Cav	
69-16098	OH58A	w.o.4.7.70	
69-16099	OH58A		
69-16100	OH58A	A\132AVN	OK NG
69-16101	OH58CR		
69-16102	OH58A	w.o.28.10.72	
69-16103	OH58A	to OH58D 83-24131	
69-16104	OH58A	to OH58D 88-0294	
69-16105	OH58A	62AvCo	
69-16106	OH58CF		
69-16107	OH58A		
69-16108	OH58CF	17Cav	
69-16109	OH58A		
69-16110	OH58CF		
69-16111	OH58CR	224AVN	
69-16112	OH58A		
69-16113	OH58CR	w.o.18.2.91	
69-16114	OH58A	to OH58D 88-0287	
69-16115	OH58CR	4AVN	
69-16116	OH58A	to OH58D 89-0112	
69-16117	JOH58A		DC NG
69-16118	OH58CF	7InfDiv	
69-16119	OH58CR	10AHB	
69-16120	OH58A	7\6Cav	
69-16121	OH58CR		
69-16122	OH58CR		
69-16123	OH58A		
69-16124	OH58A	w.o.8.10.70	
69-16125	OH58A	AATC 25L Shell AHP	
69-16126	OH58CR		
69-16127	OH58CF		
69-16128	OH58A	AATC 28E Shell AHP	
69-16129	OH58A		
69-16130	OH58A		
69-16131	OH58A		
69-16132	OH58CR		
69-16133	OH58CF		
69-16134	GOH58A		
69-16135	OH58CF		VA NG
69-16136	OH58A		
69-16137	OH58A	2ACR	
69-16138	OH58A	to OH58D 86-8920	
69-16139	OH58D		
69-16140	OH58A	1AVN	
69-16141	OH58C	w.o.27.2.91	
69-16142	OH58CR	5\6Cav	
69-16143	OH58CR	6ACR	
69-16144	OH58A	stored Ramstein AB, FRG	
69-16145	OH58CR	4Bde	
69-16146	OH58A	1-285AVN	AZ NG
69-16147	OH58CR	13AHB	
69-16148	OH58CR	3-227AVN	
69-16149	OH58A	w.o.26.5.75	
69-16150	OH58A	to OH58D 89-0083	
69-16151	OH58A		
69-16152	OH58A	to OH58D 89-0084	
69-16153	OH58A	3AVN	
69-16154	OH58A	2-135AVN	CO NG
69-16155	OH58CF	w.o.29.1.91	
69-16156	OH58A	7SigBde	
69-16157	OH58A	1-158AVN	
69-16158	OH58A	7InfDiv	

69-16159	OH58A		
69-16160	OH58CR		
69-16161	OH58A	w.o.17.2.82	
69-16162	OH58CR	6ACR	
69-16163	OH58CR	to OH58D 85-24712	
69-16164	OH58CR	4AVN	
69-16165	OH58CR		
69-16166	OH58CF	3ACT	
69-16167	OH58A	to OH58D 91-0546	
69-16168	OH58A	1-158AVN	
69-16169	OH58A	to OH58D 83-24129	
69-16170	OH58A	w.o.15.12.81	
69-16171	OH58CF	7InfDiv	
69-16172	OH58A	w.o.	
69-16173	OH58CR	224AVN	
69-16174	OH58A	w.o.22.2.73	
69-16175	OH58A	10AHB	
69-16176	OH58CR		
69-16177	OH58A	to OH58D 86-8932	
69-16178	OH58CF	stored Ramstein AB	
69-16179	OH58A		
69-16180	OH58A		
69-16181	OH58CF	82AVN	
69-16182	OH58A	to OH58D 85-24731	
69-16183	OH58A	to OH58D 85-24699	
69-16184	OH58A	1-185AVN	MS NG
69-16185	OH58A	28AVN	
69-16186	OH58CR	4-11ACR	
69-16187	OH58A		
69-16188	OH58A	w.o.10.9.76	
69-16189	OH58CR	3AVN	
69-16190	OH58A	56AvCo	
69-16191	OH58A	w.o.6.10.73	
69-16192	OH58CR		
69-16193	OH58A	to OH58D 89-0085	
69-16194	OH58CR	10AHB	
69-16195	OH58A	w.o.27.9.71	
69-16196	OH58A		
69-16197	OH58A	w.o.28.4.80	
69-16198	OH58A	40AVN	
69-16199	OH58A	1-285AVN	AZ NG
69-16200	OH58CR	4AVN	
69-16201	OH58CR	62AvCo	
69-16202	OH58CR	10AHB	
69-16203	OH58A	to OH58D 88-0308	
69-16204	OH58CR	4AVN	
69-16205	OH58A		
69-16206	OH58CR	4-11ACR	
69-16207	OH58A	308AHB	
69-16208	OH58CF	stored Ramstein AB	
69-16209	OH58CR	4AVN	
69-16210	OH58A	308AHB	
69-16211	OH58A	to OH58D 86-8931	
69-16212	OH58A	w.o.5.4.83	
69-16213	OH58CR	10AHB	
69-16214	OH58CF	w.o.	
69-16215	OH58CR	56AvCo	
69-16216	OH58A	197AvSect	
69-16217	OH58A	62AvCo	
69-16218	OH58A	to OH58D 85-24691	
69-16219	OH58A	2-135AVN	CO NG
69-16220	OH58CR		
69-16221	OH58A		
69-16222	OH58A	to OH58D 89-0088	
69-16223	OH58CR	10AHB	
69-16224	OH58A	10AHB	
69-16225	OH58CR	2ACR	
69-16226	OH58CR	2ACR	
69-16227	OH58A	w.o.19.5.81	
69-16228	OH58A	to OH58D 88-0314	
69-16229	OH58A		
69-16230	OH58A	2-135AVN	CO NG
69-16231	OH58A	56AvCo	
69-16232	OH58CR	10AHB	
69-16233	OH58A	1-285AVN	AZ NG
69-16234	OH58A	1-285AVN	AZ NG
69-16235	OH58A		
69-16236	OH58CF	AATC 36. Shell AHP	
69-16237	OH58CF		
69-16238	OH58CR	10AHB	
69-16239	OH58CF	307AVN	
69-16240	OH58A	w.o.10.9.88	
69-16241	OH58CR	3-227AVN	
69-16242	OH58CR	227AVN	
69-16243	OH58A	to OH58D 88-0292	
69-16244	OH58CR	4AVN	
69-16245	OH58CR	3AVN	
69-16246	OH58CR	4AVN	
69-16247	OH58A	w.o.7.12.72	
69-16248	OH58CR	stored Ramstein AB	
69-16249	OH58CR	11ACR	
69-16250	OH58A	w.o.11.8.73	
69-16251	OH58A	w.o.28.1.70	
69-16252	OH58A	AATC 52B Shell AHP	
69-16253	OH58A	10Cav	
69-16254	OH58A	308AHB	
69-16255	OH58A	to OH58D 86-8909	
69-16256	OH58CF	25AVN	
69-16257	OH58A	w.o.3.6.72	
69-16258	OH58CF	4InfDiv	
69-16259	OH58A	w.o.21.8.71	
69-16260	OH58CR	3AVN	
69-16261	OH58A		
69-16262	OH58A	44AVN	
69-16263	OH58A	to OH58D 86-8903	
69-16264	OH58CF	229AVN	
69-16265	OH58A	w.o.13.9.77	
69-16266	OH58A	w.o.13.10.74	
69-16267	OH58A	to OH58D 86-8933	
69-16268	OH58CR	4-11ACR	
69-16269	OH58A	to OH58D 86-8921	
69-16270	OH58A	229AVN	
69-16271	OH58A	2-135AVN	CO NG
69-16272	OH58A	w.o.28.11.72	
69-16273	OH58A	1-167Cav	NE NG
69-16274	OH58CR	4-11ACR	
69-16275	OH58CR	11ACR	
69-16276	OH58A	w.o.21.10.74	
69-16277	OH58CR	11ACR	
69-16278	OH58A	w.o.21.4.71	
69-16279	OH58CR	11AVN	
69-16280	OH58A	11CAG\3ACR	
69-16281	OH58A		
69-16282	OH58A	w.o.7.9.83	
69-16283	OH58A	w.o.27.10.76	
69-16284	OH58A	11CAG\3ACR	
69-16285	OH58D		
69-16286	OH58A		
69-16287	OH58A	1-285AVN	AZ NG
69-16288	OH58CR	308AHB	
69-16289	OH58A	1-285AVN	AZ NG
69-16290	OH58A	w.o.19.1.77	
69-16291	OH58CR	4-11ACR	
69-16292	OH58A	1AVN	
69-16293	OH58A	44AVN	
69-16294	OH58A	w.o.15.9.80	
69-16295	OH58A	w.o.25.8.78	
69-16296	OH58A	w.o.5.12.75	
69-16297	OH58CF	1AVN	
69-16298	OH58CR		
69-16299	OH58A	3AVN	
69-16300	OH58A	w.o.31.1.79	
69-16301	OH58A	1InfDiv	
69-16302	OH58A	1-158AVN	
69-16303	OH58A		
69-16304	OH58A		
69-16305	OH58CF		
69-16306	OH58A	86ARCOM	
69-16307	OH58A	3ACR	
69-16308	OH58A		
69-16309	OH58A	to OH58D 89-0087	
69-16310	OH58A	w.o.3.3.78	
69-16311	OH58A		
69-16312	OH58CR	9AVN	
69-16313	OH58CR		
69-16314	OH58CR		
69-16315	OH58CR	4-11ACR	
69-16316	OH58A	to OH58D 87-0730	
69-16317	OH58A	to OH58D 83-24130	
69-16318	OH58A	to OH58D 83-24144	
69-16319	OH58A	229AVN	
69-16320	OH58D		
69-16321	OH58A	to OH58D 91-0552	
69-16322	OH58D		
69-16323	OH58CR	3AVN	
69-16324	OH58A	1-158AVN	

69-16325	OH58A	w.o.27.2.79
69-16326	OH58CR	stored Ramstein AB
69-16327	OH58CR	3AVN
69-16328	OH58CR	10Cav
69-16329	OH58A	to OH58D 85-24729
69-16330	OH58CR	7InfDiv
69-16331	OH58A	
69-16332	OH58A	to OH58D 85-24701
69-16333	OH58A	285AvCo
69-16334	OH58CR	285AvCo
69-16335	OH58A	w.o.1.4.91
69-16336	OH58A	4AVN
69-16337	OH58A	1-285AVN AZ NG
69-16338	OH58CR	308AHB
69-16339	OH58A	3AVN
69-16340	OH58A	to OH58D 83-24140
69-16341	OH58A	8AVN
69-16342	OH58A	w.o.10.4.76
69-16343	OH58A	8AVN
69-16344	OH58A	3AVN
69-16345	OH58A	w.o.17.9.75
69-16346	OH58CR	3AVN
69-16347	OH58CR	4-11ACR
69-16348	OH58A	to OH58D 83-24136
69-16349	OH58A	stored Ramstein AB, FRG
69-16350	OH58A	to OH58D 88-0310
69-16351	OH58A	to OH58D 85-24702
69-16352	OH58CF	w.o.2.5.89
69-16353	OH58A	163ACR
69-16354	OH58CF	3ACR
69-16355	OH58A	7InfDiv
69-16356	OH58CR	163ACR
69-16357	OH58CR	
69-16358	OH58A	w.o.25.10.83
69-16359	OH58CR	10AHB
69-16360	OH58CF	
69-16361	OH58A	to OH58D 85-24723
69-16362	OH58CR	9AVN
69-16363	OH58CF	
69-16364	OH58A	
69-16365	OH58CF	w.o.18.11.83
69-16366	OH58A	w.o.23.8.80
69-16367	OH58CR	2ACR
69-16368	OH58A	to OH58D 92-0574
69-16369	OH58CR	
69-16370	OH58A	28AVN
69-16371	OH58A	
69-16372	OH58A	229AVN
69-16373	OH58A	AATC 73D Shell AHP
69-16374	OH50A	
69-16375	OH58A	
69-16376	OH58CF	172InfBde
69-16377	OH58A	8AVN
69-16378	OH58A	to OH58D 89-0089
69-16379	OH58A	stored Ramstein AB, FRG
70-15050	OH58A	
70-15051	OH58A	to OH58D 83-24135
70-15052	OH58A	
70-15053	OH58CR	229AVN
70-15054	OH58A	to OH58D 86-8911
70-15055	OH58CR	
70-15056	OH58A	8AVN
70-15057	OH58A	308AHB
70-15058	OH58A	278ACR
70-15059	OH58A	
70-15060	OH58A	
70-15061	OH58CR	
70-15062	OH58A	w.o.5.11.80
70-15063	OH58A	3ACR
70-15064	OH58A	w.o.24.4.73
70-15065	OH58A	to OH58D 86-8924
70-15066	OH58CR	w.o.2.4.89
70-15067	OH58A	w.o.10.8.71
70-15068	OH58A	to OH58D 91-0566
70-15069	OH58A	w.o.23.4.73
70-15070	OH58CR	6ACR
70-15071	OH58A	to OH58D 92-0585
70-15072	OH58CR	10AHB
70-15073	OH58A	
70-15074	OH58CF	3ACR
70-15075	OH58A	44AVN
70-15076	OH58A	
70-15077	OH58A	w.o.26.10.79
70-15078	OH58A	w.o.21.7.83
70-15079	OH58CF	
70-15080	OH58A	w.o.15.11.85
70-15081	OH58A	
70-15082	OH58A	40AVN
70-15083	OH58CR	
70-15084	OH58A	to OH58D 90-0381
70-15085	OH58A	11ACR
70-15086	OH58CR	w.o.
70-15087	OH58CR	
70-15088	OH58A	
70-15089	OH58A	to OH58D 86-8934
70-15090	OH58CF	6ACR
70-15091	OH58CF	
70-15092	OH58CF	82AVN
70-15093	OH58CR	
70-15094	OH58A	
70-15095	OH58A	244AvCo
70-15096	OH58CR	
70-15097	OH58CF	
70-15098	OH58CR	4-11ACR
70-15099	OH58A	9AVN
70-15100	OH58A	9AVN
70-15101	JOH58A	
70-15102	OH58A	to OH58D 83-24138
70-15103	OH58A	to OH58D 83-24139
70-15104	OH58A	w.o.27.6.91
70-15105	OH58A	2ACR
70-15106	OH58A	8AVN
70-15107	OH58A	
70-15108	OH58A	w.o.14.6.88
70-15109	OH58A	2-135AVN CO NG
70-15110	OH58A	13AHB
70-15111	OH58A	to OH58D 83-24143
70-15112	OH58A	8AVN
70-15113	OH58A	13AHB
70-15114	OH58A	stored Ramstein AB
70-15115	OH58A	8AVN
70-15116	OH58A	244AvCo
70-15117	OH58A	2ACR
70-15118	OH58A	7\6Cav
70-15119	OH58A	13AHB
70-15120	OH58A	25AvCo
70-15121	OH58A	56AvCo
70-15122	OH58A	56AvCo
70-15123	OH58A	1-167Cav NE NG
70-15124	OH58A	w.o.2.4.82
70-15125	OH58A	25AvCo
70-15126	OH58CR	3AVN
70-15127	OH58CR	308AHB
70-15128	OH58CF	25AvCo
70-15129	OH58A	17Cav
70-15130	OH58A	82AVN
70-15131	OH58CR	53InfBde
70-15132	OH58A	2ACR
70-15133	OH58CF	82AVN
70-15134	OH58A	7InfDiv
70-15135	OH58CF	
70-15136	OH58CF	w.o.
70-15137	OH58CF	
70-15138	OH58CF	
70-15139	GOH58A	1AVN
70-15140	OH58CF	AATC 40A Shell AHP
70-15141	OH58A	to OH58D 85-24715
70-15142	OH58A	10AHB
70-15143	OH58A	
70-15144	OH58A	to OH58D 85-24720
70-15145	OH58CR	
70-15146	OH58A	w.o.28.9.71
70-15147	OH58A	
70-15148	OH58A	82AVN
70-15149	OH58A	82AVN
70-15150	OH58CR	
70-15151	OH58A	3AVN
70-15152	OH58A	7\6Cav
70-15153	OH58CR	
70-15154	OH58CR	224AVN
70-15155	OH58A	w.o.29.3.83
70-15156	OH58A	to OH58D 90-0349
70-15157	OH58A	
70-15158	OH58A	w.o.25.3.71
70-15159	OH58CF	
70-15160	OH58A	to OH58D 90-0363

70-15161	OH58CF	101AbDiv		
70-15162	OH58CF	2ACR		
70-15163	OH58A	w.o.28.5.911		
70-15164	OH58A			
70-15165	OH58A	3AVN		
70-15166	OH58CF	11ACR		
70-15167	OH58CR	4-11ACR		
70-15168	OH58A	w.o.21.5.71		
70-15169	OH58A	3AVN		
70-15170	OH58CF	17Cav		
70-15171	OH58A	10AHB		
70-15172	OH58CR	308AHB		
70-15173	OH58A	AATC	73S Shell AHP	
70-15174	OH58A	1InfDiv		
70-15175	OH58CR	62AvCo		
70-15176	OH58CR	62AvCo		
70-15177	OH58CR	2ACR		
70-15178	OH58CR	62AvCo		
70-15179	OH58A	203AvCo		
70-15180	OH58A	158AVN		
70-15181	OH58A	w.o.24.8.71		
70-15182	OH58A	8AVN		
70-15183	OH58CR			
70-15184	OH58A	to OH58D 89-0090		
70-15185	OH58A	to OH58D 89-0091		
70-15186	OH58A	56AvCo		
70-15187	OH58A			
70-15188	OH58CR	8AVN		
70-15189	OH58A	w.o.25.4.74		
70-15190	OH58A	10AHB		
70-15191	OH58A	56AvCo		
70-15192	OH58CR	25AvCo		
70-15193	OH58CF			
70-15194	OH58A	AATC	94F Shell AHP	
70-15195	OH58CF	AATC	95H Shell AHP	
70-15196	OH58A	1-158AVN		
70-15197	OH58A	to OH58D 87-0758		
70-15198	OH58A			
70-15199	OH58CR	2ACR		
70-15200	OH58CR	2ACR		
70-15201	OH58A	7\6Cav		
70-15202	OH58A	stored Ramstein AB, FRG		
70-15203	OH58A	56AvCo		
70-15204	OH58CF	AATC	04M Shell AHP	
70-15205	OH58A	3AVN		
70-15206	OH58A	to OH58D 89-0092		
70-15207	OH58A	308AHB		
70-15208	OH58CF	17Cav		
70-15209	OH58A	w.o.2.2.78		
70-15210	OH58A	82AVN		
70-15211	OH58A	25AvCo		
70-15212	OH58CR	25AvCo		
70-15213	OH58A	AATC	13. Shell AHP	
70-15214	OH58A	to OH58D 90-0359		
70-15215	OH58CF	101AbDiv		
70-15216	OH58A	AATC	16. Shell AHP	
70-15217	OH58A	w.o.12.10.76		
70-15218	OH58A	to OH58D 89-0093		
70-15219	OH58A	2ACR		
70-15220	OH58A	25AvCo		
70-15221	OH58A			
70-15222	OH58A	to OH58D 87-0731		
70-15223	OH58A	62AvCo		
70-15224	OH58A	w.o.1972		
70-15225	OH58CR	4AVN		
70-15226	OH58CR	4AVN		
70-15227	OH58A	25AvCo		
70-15228	OH58A	158AVN		
70-15229	OH58A	to OH58D 89-0094		
70-15230	GOH58A			
70-15231	OH58A	w.o.8.3.79		
70-15232	OH58A	to OH58D 83-24141		
70-15233	OH58A	w.o.4.10.72		
70-15234	OH58A	11ACR		
70-15235	OH58A			
70-15236	OH58A	w.o.		
70-15237	OH58A			
70-15238	OH58A	w.o.29.3.88		
70-15239	OH58A	w.o.15.6.84		
70-15240	OH58A	to OH58D 89-0109		
70-15241	OH58CF			
70-15242	OH58A		FL NG	
70-15243	OH58A	116AHB		
70-15244	GOH58A	1AVN		
70-15245	OH58CR	8AVN		
70-15246	OH58A	3AVN		
70-15247	OH58A	AATC	47E Shell AHP	
70-15248	OH58A	25AvCo		
70-15249	OH58CF			
70-15250	OH58A	158AVN		
70-15251	OH58CF	82AVN		
70-15252	OH58A			
70-15253	OH58A	2-135AVN	CO NG	
70-15254	OH58CF	11ACR		
70-15255	OH58A			
70-15256	OH58A			
70-15257	OH58A	w.o.10.1.80		
70-15258	OH58A	25AVN		
70-15259	OH58A	25AvCo		
70-15260	OH58A			
70-15261	OH58A	69ADA Grp		
70-15262	OH58A	to OH58D 86-8936		
70-15263	OH58A	1Cav		
70-15264	OH58A	w.o.20.7.72		
70-15265	OH58A	4\9Cav		
70-15266	OH58A	to OH58D 90-0371		
70-15267	OH58A	1Cav		
70-15268	OH58A	1Cav		
70-15269	OH58A	4\9Cav		
70-15270	OH58A	to OH58D 87-0733		
70-15271	OH58A	4\9Cav		
70-15272	OH58A	172InfBde		
70-15273	OH58A	1Cav		
70-15274	OH58A	4\9Cav		
70-15275	OH58A	4\9Cav		
70-15276	OH58A	1-185AVN	MS NG	
70-15277	OH58A	222AVN		
70-15278	GOH58A	222AVN		
70-15279	OH58CR			
70-15280	OH58CR	3ACT		
70-15281	OH58A	w.o.18.8.71		
70-15282	OH58A	to OH58D 85-24714		
70-15283	OH58A	AATC	83D Shell AHP	
70-15284	OH58A	to OH58D 87-0745		
70-15285	OH58CR			
70-15286	OH58A	AATC	86E Shell AHP	
70-15287	OH58CF			
70-15288	OH58A	E\1-192AVN AL NG		
70-15289	OH58A	w.o.12.4.83		
70-15290	OH58CF	w.o.7.11.84		
70-15291	OH58A	w.o.4.12.72		
70-15292	OH58A	A\132AVN OK NG		
70-15293	OH58A	51AvCo		
70-15294	OH58A	to OH58D 89-0099		
70-15295	OH58A	49AvCo		
70-15296	OH58A	to OH58D 91-0569		
70-15297	JOH58A			
70-15298	OH58A			
70-15299	OH58A			
70-15300	OH58CR	82AVN		
70-15301	OH58CR	224AVN		
70-15302	OH58CF			
70-15303	OH58A			
70-15304	OH58A	158AVN		
70-15305	OH58A	1-167Cav NE NG		
70-15306	OH58A			
70-15307	OH58CF	w.o.11.11.85		
70-15308	OH58A			
70-15309	OH58A			
70-15310	GOH58A	10Cav		
70-15311	OH58CF	w.o.2.10.90		
70-15312	OH58A	82AVN		
70-15313	OH58A	to OH58D 90-0372		
70-15314	OH58A	to OH58D 90-0357		
70-15315	OH58A	to OH58D 85-24719		
70-15316	OH58CF			
70-15317	OH58A	to OH58D 86-8922		
70-15318	OH58CF			
70-15319	OH58A			
70-15320	OH58A	to OH58D 90-0353		
70-15321	OH58A	to OH58D 89-0095		
70-15322	OH58CR	w.o.29.10.73		
70-15323	OH58A	to OH58D 90-0364		
70-15324	OH58A			
70-15325	OH58CF	24AVN		
70-15326	OH58A	8AVN		

Serial	Type	Details	
70-15327	OH58A		
70-15328	OH58A	to OH58D 89-0096	
70-15329	OH58A	AATC 29C Shell AHP	
70-15330	OH58A	to OH58D 85-24697	
70-15331	OH58A	10AHB	
70-15332	OH58CR	2ACR	
70-15333	OH58A	to OH58D 89-0097	
70-15334	OH58A		
70-15335	OH58A	w.o.7.6.72	
70-15336	OH58A	3ACR	
70-15337	OH58A	1-285AVN	AZ NG
70-15338	OH58A	AATC 38G Shell AHP	
70-15339	OH58CF	3-227AVN	
70-15340	OH58A	to OH58D 87-0744	
70-15341	OH58A	to OH58D 89-0110	
70-15342	OH58A		
70-15343	OH58A	to OH58D 90-0365	
70-15344	OH58A	to OH58D 88-0285	
70-15345	OH58CF	4InfDiv	
70-15346	OH58A		
70-15347	OH58CR	stored Avra Valley AP, Az	
70-15348	OH58A	to OH58D 88-0309	
70-15349	OH58CF	9AVN	
70-15350	OH58A		
70-15351	OH58A		
70-15352	OH58A	to OH58D 86-8904	
70-15353	OH58CF	172InfBde	
70-15354	OH58A		
70-15355	OH58A	1-158AVN	
70-15356	OH58CR	9AVN	
70-15357	OH58A		
70-15358	OH58A		
70-15359	OH58A	to OH58D 86-8901	
70-15360	OH58A	to OH58D 87-0738	
70-15361	OH58A		
70-15362	OH58A		
70-15363	OH58CF		
70-15364	OH58CF	11ACR	
70-15365	OH58CR	9AVN	
70-15366	OH58A	4InfDiv	
70-15367	OH58A		
70-15368	OH58A	to OH58D 85-24694	
70-15369	OH58CR	1AVN	
70-15370	OH58A	AATC 70H Shell AHP	
70-15371	OH58A	to OH58D 86-8913	
70-15372	OH58A		
70-15373	OH58A	to OH58D 88-0304	
70-15374	OH58CF		
70-15375	OH58A	w.o.22.6.83	
70-15376	OH58CR		
70-15377	OH58CF	w.o.23.7.84	
70-15378	OH58A	to OH58D 90-0374	
70-15379	OH58A	w.o.8.6.75	
70-15380	OH58A	to OH58D 91-0567	
70-15381	OH58CF	24AVN	
70-15382	OH58A	w.o.5.5.75	
70-15383	OH58A	101AbDiv	
70-15384	OH58A	57MedDet	
70-15385	OH58A	to OH58D 85-24730	
70-15386	OH58A	1-158AVN	
70-15387	OH58CF		
70-15388	OH58CF	11ACR	
70-15389	OH58A	to OH58D 86-8906	
70-15390	OH58CF	9CavRgt	
70-15391	OH58A		
70-15392	OH58A	28AVN	
70-15393	OH58A	to OH58D 85-24727	
70-15394	OH58A	7InfDiv	
70-15395	OH58A	3ACR	
70-15396	OH58A	to OH58D 89-0104	
70-15397	OH58CR	w.o.30.11.89	
70-15398	OH58CF	2ACR	
70-15399	OH58A		
70-15400	OH58A	to OH58D 89-0111	
70-15401	OH58A	to OH58D 89-0115	
70-15402	OH58A		
70-15403	OH58CR	9AVN	
70-15404	OH58A		
70-15405	OH58A	40AVN	
70-15406	OH58A	40AVN	
70-15407	OH58A	to OH58D 89-0107	
70-15408	OH58CR	3-227AVN	
70-15409	OH58CR	163ACR	
70-15410	OH58A	1-167Cav	NE NG
70-15411	GOH58A		
70-15412	OH58CR	6ACR	
70-15413	OH58CR		
70-15414	OH58CR	3AVN	
70-15415	OH58A	USAICS	
70-15416	OH58CF	3ACT	
70-15417	OH58CF	2ACR	
70-15418	OH58A	w.o.12.8.71	
70-15419	OH58A		
70-15420	OH58A	w.o.1.2.89	
70-15421	OH58A	1-167Cav	NE NG
70-15422	OH58A	1-167Cav	NE NG
70-15423	OH58A	1-167Cav	NE NG
70-15424	OH58A	1-185AVN	MS NG
70-15425	OH58A	1-185AVN	MS NG
70-15426	OH58A	1-185AVN	MS NG
70-15427	OH58A	1-185AVN	MS NG
70-15428	OH58A	to OH58D 90-0370	
70-15429	OH58A	to OH58D 87-0751	
70-15430	OH58A	to OH58D 86-8914	
70-15431	OH58A	to OH58D 87-0749	
70-15432	OH58CF	229AVN	
70-15433	OH58A		
70-15434	OH58CR		
70-15435	OH58A	to OH58D 88-0307	
70-15436	OH58CF		
70-15437	OH58A	1-158AVN	
70-15438	OH58A		
70-15439	OH58A		
70-15440	OH58A	A\132AVN	OK NG
70-15441	OH58A		HI NG
70-15442	OH58A		
70-15443	OH58CR	9AVN	
70-15444	OH58A	40AVN	
70-15445	OH58A	w.o.15.6.76	
70-15446	OH58A		
70-15447	OH58CR	6ACR	
70-15448	OH58A		
70-15449	OH58CR		
70 15450	OH58A	1-158AVN	
70-15451	OH58A	1-158AVN	
70-15452	OH58CR	w.o.	
70-15453	OH58A	40AVN	
70-15454	OH58A	88ARCOM	
70-15455	OH58A	w.o.13.6.79	
70-15456	OH58A		
70-15457	OH58A		
70-15458	OH58A	AATC 58J Shell AHP	
70-15459	OH58CF	163ACR	
70-15460	OH58CR		
70-15461	OH58A	to OH58D 91-0570	
70-15462	OH58CF	2ACR	
70-15463	OH58A		
70-15464	OH58A		HI NG
70-15465	OH58A	w.o.15.7.76	
70-15466	OH58A	1-183AVN	ID NG
70-15467	OH58A	to OH58D 90-0350	
70-15468	OH58A	to OH58D 89-0114	
70-15469	OH58A	1-158AVN	
70-15470	OH58A		
70-15471	OH58A	24AVN	
70-15472	OH58A	1-158AVN	
70-15473	OH58CR	6ACR	
70-15474	OH58CR	5\6Cav	
70-15475	OH58A		
70-15476	OH58A	to OH58D 85-24725	
70-15477	OH58CR	6ACR	
70-15478	OH58A	2ACR	
70-15479	OH58A	to OH58D 90-0360	
70-15480	OH58CR	3-227AVN	
70-15481	OH58CF	82AVN	
70-15482	OH58CR	308AHB	
70-15483	OH58A	to OH58D 87-0734	
70-15484	OH58A		
70-15485	OH58CF	24AVN	
70-15486	OH58A	to OH58D 89-0106	
70-15487	OH58CF	9Cav	
70-15488	OH58CF	1InfDiv	
70-15489	OH58A		
70-15490	OH58A	to OH58D 88-0286	
70-15491	OH58A	w.o.11.4.78	
70-15492	OH58CR	w.o.2.11.83	

70-15493	OH58CR			
70-15494	OH58CF	11ACR		
70-15495	OH58CR			
70-15496	OH58A	3ACR		
70-15497	OH58CF			
70-15498	OH58A	3ACR		
70-15499	OH58A	w.o.11.7.74		
70-15500	OH58CR			
70-15501	OH58A			
70-15502	OH58A	w.o.9.12.86		
70-15503	OH58A	USAFDA		
70-15504	OH58A	to OH58D 90-0373		
70-15505	OH58A	w.o.5.9.72		
70-15506	OH58A			
70-15507	OH58A			
70-15508	OH58CR	7InfDiv		
70-15509	OH58A	w.o.20.5.83		
70-15510	OH58CF	82AVN		
70-15511	OH58CR			
70-15512	OH58CR	9AVN		
70-15513	OH58CR			
70-15514	OH58A			
70-15515	OH58CF	2ACR		
70-15516	OH58A	E\1-192AVN AL NG		
70-15517	OH58A	1-285AVN AZ NG		
70-15518	OH58A			
70-15519	OH58A			
70-15520	OH58CF	2ACR		
70-15521	OH58A	31AVN		
70-15522	OH58A			
70-15523	OH58A	308AHB		
70-15524	OH58A	w.o.17.10.72		
70-15525	OH58A			
70-15526	OH58A			
70-15527	OH58A			
70-15528	OH58A	w.o.21.2.90		
70-15529	OH58A	88ARCOM		
70-15530	OH58A			
70-15531	OH58A	w.o.21.1.84		
70-15532	OH58CR			
70-15533	OH58A			
70-15534	OH58A			
70-15535	OH58A			
70-15536	OH58A		HI NG	
70-15537	OH58A		HI NG	
70-15538	OH58A	25AVN		
70-15539	OH58A	A\132AVN OK NG		
70-15540	OH58A			
70-15541	OH58A			
70-15542	OH58A	to OH58D 90-0366		
70-15543	OH58CF	4InfDiv		
70-15544	OH58A	w.o.15.8.80		
70-15545	OH58A	1-185AVN MS NG		
70-15546	OH58A	to OH58D 88-0306		
70-15547	OH58A	w.o.17.11.81		
70-15548	OH58A	to OH58D 86-8935		
70-15549	OH58A			
70-15550	OH58CR			
70-15551	OH58CF	10AHB		
70-15552	OH58A	to OH58D 90-0354		
70-15553	GOH58A			
70-15554	OH58A	88ARCOM		
70-15555	OH58A	88ARCOM		
70-15556	OH58A	88ARCOM		
70-15557	OH58A	to OH58D 87-0742		
70-15558	OH58A	TECOM		
70-15559	OH58A	TECOM		
70-15560	OH58A	w.o.5.4.88		
70-15561	OH58CR	10AHB		
70-15562	OH58A	w.o.14.3.81		
70-15563	OH58CR	9AVN		
70-15564	OH58A	107AvCo		
70-15565	OH58A	107AvCo		
70-15566	OH58A	w.o.16.7.72		
70-15567	OH58A	4-107ACR	OH NG	
70-15568	OH58A	CCAD		
70-15569	OH58A	AATC	69G Shell AHP	
70-15570	OH58A	TECOM		
70-15571	OH58A	w.o.		
70-15572	OH58A	3AVN		
70-15573	OH58A			
70-15574	OH58A	w.o.20.9.78		
70-15575	OH58A	to OH58D 90-0361		
70-15576	OH58A	to OH58D 90-0380		
70-15577	OH58A	to OH58D 90-0368		
70-15578	OH58A	40AVN		
70-15579	OH58CF	CCAD		
70-15580	OH58A	to OH58D 87-0755		
70-15581	OH58A	1-158AVN		
70-15582	OH58A			
70-15583	OH58A	w.o.21.11.74		
70-15584	OH58A	1-167Cav	NE NG	
70-15585	OH58CF	11ACR		
70-15586	OH58CR	10AHB		
70-15587	OH58A			
70-15588	GOH58C			
70-15589	OH58A	1-130AVN	NC NG	
70-15590	OH58A		CA NG	
70-15591	OH58A	to OH58D 87-0741		
70-15592	OH58CF	4InfDiv		
70-15593	OH58A	40AVN		
70-15594	OH58A	w.o.23.2.73		
70-15595	OH58A	10Cav		
70-15596	OH58A	to OH58D 90-0375		
70-15597	OH58A	to OH58D 88-0301		
70-15598	OH58CR			
70-15599	OH58A			
70-15600	OH58A	w.o.19.12.91		
70-15601	OH58A			
70-15602	OH58A			
70-15603	OH58CF	w.o.21.3.85		
70-15604	OH58A			
70-15605	OH58A			
70-15606	OH58A	107AvCo		
70-15607	OH58A			
70-15608	OH58CR	1InfDiv		
70-15609	OH58CF	AATC	09. Shell AHP	
70-15610	OH58A			
70-15611	OH58A			
70-15612	JOH58A			
70-15613	OH58A			
70-15614	OH58A	to OH58D 89-0113		
70-15615	OH58A			
70-15616	OH58A	1-158AVN		
70-15617	OH58A	to OH58D 86-8907		
70-15618	OH58A			
70-15619	OH58CF			
70-15620	OH58A			
70-15621	OH58CF			
70-15622	OH58CR			
70-15623	OH58A	to OH58D 87-0725		
70-15624	OH58A	w.o.15.3.90		
70-15625	OH58CR	4AVN		
70-15626	OH58A			
70-15627	OH58A	w.o.6.5.72		
70-15628	OH58A	to OH58D 88-0295		
70-15629	OH58A	AATC	29D Shell AHP	
70-15630	OH58A	AATC	30B Shell AHP	
70-15631	OH58A	1-158AVN		
70-15632	OH58A	w.o.2.6.72		
70-15633	OH58CF			
70-15634	OH58CF	1InfDiv		
70-15635	OH58A	w.o.18.2.72		
70-15636	OH58A			
70-15637	OH58A	AATC	37E Shell AHP	
70-15638	OH58A			
70-15639	OH58CF			
70-15640	OH58A	to OH58D 91-0545		
70-15641	OH58A	to OH58D 86-8918		
70-15642	OH58A	1-183AVN	ID NG	
70-15643	OH58A	E\1-192AVN	AL NG	
70-15644	OH58A			
70-15645	OH58A	A\132AVN	OK NG	
70-15646	OH58A			
70-15647	GOH58C			
70-15648	OH58A	to OH58D 85-24721		
70-15649	OH58A			
71-20340	OH58A			
71-20341	OH58A	1-158AVN		
71-20342	OH58A	to OH58D 86-8923		
71-20343	OH58CR	224AVN		
71-20344	OH58A	1-158AVN		
71-20345	OH58CF	2ACR		
71-20346	OH58A	244AvCo		
71-20347	OH58A	to OH58D 91-0536		
71-20348	OH58A	to OH58D 87-0728		

Serial	Type	Unit	Notes
71-20349	OH58CR	1InfDiv	
71-20350	OH58A	3ACR	
71-20351	OH58A		
71-20352	OH58CF	3ACR	
71-20353	OH58CF		
71-20354	OH58CF		
71-20355	OH58A	w.o.28.3.90	
71-20356	OH58CF		
71-20357	OH58A	FL NG	
71-20358	OH58A	E\1-192AVN	AL NG
71-20359	OH58A	w.o.6.4.83	
71-20360	OH58A	229AVN	
71-20361	OH58A	1-158AVN	
71-20362	OH58A	57MedDet	
71-20363	OH58A	40AVN	
71-20364	OH58CF	AATC 64E Shell AHP	
71-20365	OH58CF	4InfDiv	
71-20366	OH58A	8AVN	
71-20367	OH58A	8AVN	
71-20368	OH58A	to OH58D 88-0311	
71-20369	OH58CF		
71-20370	OH58A	stored Ramstein AB, FRG	
71-20371	OH58A	to OH58D 86-8938	
71-20372	OH58A	w.o.10.10.76	
71-20373	OH58A	to OH58D 85-24704	
71-20374	OH58A	200AvCo	
71-20375	OH58A		
71-20376	OH58CR	1InfDiv	
71-20377	OH58A	AATC 77C Shell AHP	
71-20378	GOH58A	1AVN	
71-20379	OH58A	to OH58D 86-8925	
71-20380	OH58A	to OH58D 87-0760	
71-20381	OH58CF		
71-20382	OH58A	to OH58D 89-0101	
71-20383	OH58A		
71-20384	OH58CF	AATC 84D Shell AHP	
71-20385	OH58A	to OH58D 87-0732	
71-20386	OH58CF		
71-20387	OH58A		
71-20388	JOH58A	17Cav	
71-20389	OH58A		
71-20390	OH58A	116AHB	
71-20391	OH58A	A\132AVN	OK NG
71-20392	OH58A		
71-20393	OH58A	1-167Cav	NE NG
71-20394	OH58A	1-167Cav	NE NG
71-20395	OH58A		
71-20396	OH58CF	17Cav	
71-20397	OH58A		
71-20398	OH58A	1-185AVN	MS NG
71-20399	OH58A	1-185AVN	MS NG
71-20400	OH58A	to OH58D 91-0554	
71-20401	OH58A	150AVN	DE NG
71-20402	OH58A		
71-20403	OH58CF	3ACT	
71-20404	OH58A	116AHB	
71-20405	OH58A	116AHB	
71-20406	OH58A	to OH58D 91-0539	
71-20407	OH58A	to OH58D 91-0547	
71-20408	OH58A	to OH58D 90-0362	
71-20409	OH58A	to OH58D 89-0086	
71-20410	OH58A	44AVN	
71-20411	OH58CR		
71-20412	OH58A		
71-20413	OH58A	to OH58D 88-0296	
71-20414	OH58CR	AATC 14E Shell AHP	
71-20415	OH58CR	3-227AVN	
71-20416	OH58A		
71-20417	OH58A	107AvCo	
71-20418	OH58A	w.o.20.6.73	
71-20419	OH58A	1-151AVN	SC NG
71-20420	OH58A	1-185AVN	MS NG
71-20421	OH58CF		
71-20422	OH58A	w.o.20.6.79	
71-20423	OH58CR	8AVN	
71-20424	OH58A	13AHB	
71-20425	OH58A	to OH58D 85-24718	
71-20426	OH58CR	6ACR	
71-20427	OH58CF		
71-20428	OH58A	82AVN	
71-20429	OH58A		
71-20430	OH58A	7InfDiv	
71-20431	OH58A	to OH58D 88-0320	
71-20432	OH58A	w.o.12.9.82	
71-20433	OH58CR	11ADA Grp	
71-20434	OH58A	57MedDet	
71-20435	OH58CF		
71-20436	OH58A	to OH58D 86-8916	
71-20437	OH58A	to OH58D 86-8917	
71-20438	OH58A	w.o.28.6.76	
71-20439	OH58A	57MedDet	
71-20440	OH58A		
71-20441	OH58A	4InfDiv	
71-20442	OH58A	w.o.23.6.76	
71-20443	OH58CR	1InfDiv	
71-20444	OH58CF	2ACR	
71-20445	OH58CR	5\6Cav	
71-20446	OH58CR	62AvCo	
71-20447	OH58A	w.o.8.12.81	
71-20448	OH58CR	3-227AVN	
71-20449	OH58A	158AVN	
71-20450	OH58A	10Cav	
71-20451	OH58A	278ACR	
71-20452	OH58A	47AVN	
71-20453	OH58A	to OH58D 89-0100	
71-20454	OH58A	to OH58D 88-0316	
71-20455	OH58A		
71-20456	OH58A		
71-20457	OH58A	to OH58D 88-0299	
71-20458	OH58A	107AvCo	
71-20459	OH58A		
71-20460	OH58A	1-185AVN	MS NG
71-20461	OH58A		
71-20462	OH58A	1-185AVN	MS NG
71-20463	OH58A	w.o.18.5.73	
71-20464	OH58A	AATC 64C Shell AHP	
71-20465	OH58A	1-285AVN	AZ NG
71-20466	OH58A	116AHB	
71-20467	OH58A	116AHB	
71-20468	OH58A		
71-20469	OH58A	E\1-192AVN	AL NG
71-20470	OH58A	150AVN	DE NG
71-20471	OH58A	2ACR	
71-20472	OH58A	1-158AVN	
71-20473	OH58A		
71-20474	OH58CR		
71-20475	OH58CR	9AVN	
71-20476	OH58CF		
71-20477	OH58A	7\6Cav	
71-20478	OH58A	7InfDiv	
71-20479	OH58A	7\6Cav	
71-20480	OH58A	to OH58D 88-0300	
71-20481	OH58CF	1AHB	
71-20482	OH58A		
71-20483	OH58A	224AVN	
71-20484	OH58A	w.o.6.4.72	
71-20485	OH58A	w.o.6.4.72	
71-20486	OH58A	1-149AVN	TX NG
71-20487	OH58A	AATC 87K Shell AHP	
71-20488	OH58A	AATC 88D Shell AHP	
71-20489	OH58A	AATC 89G Shell AHP	
71-20490	OH58A	to OH58D 87-0740	
71-20491	OH58A	278ACR	
71-20492	OH58CF		
71-20493	OH58A	1-151AVN	SC NG
71-20494	OH58A	to OH58D 91-0563	
71-20495	OH58CF		
71-20496	OH58A		
71-20497	OH58A	w.o.16.9.88	
71-20498	OH58A	E\1-192AVN	AL NG
71-20499	OH58A	150AVN	DE NG
71-20500	OH58A	51AvCo	
71-20501	OH58A	D\1-137AVN	OH NG
71-20502	OH58A	E\1-192AVN	AL NG
71-20503	OH58A	E\1-192AVN	AL NG
71-20504	OH58A		
71-20505	OH58A	w.o.17.6.81	
71-20506	OH58A		
71-20507	OH58A	83ARCOM	
71-20508	OH58C	w.o.31.8.81	
71-20509	OH58A	w.o.11.12.77	
71-20510	OH58A	49AvCo	
71-20511	OH58A	49AvCo	
71-20512	OH58A	49AvCo	
71-20513	OH58CR		
71-20514	OH58A	107AvCo	

71-20515	OH58A	107AvCo	
71-20516	OH58A	51AvCo	
71-20517	OH58CF	9Cav	
71-20518	OH58A	28AVN	
71-20519	OH58A		
71-20520	OH58A	1-285AVN	AZ NG
71-20521	OH58A		FL NG
71-20522	OH58A	47AVN	
71-20523	OH58A	1-158AVN	
71-20524	OH58A	to OH58D 92-0576	
71-20525	OH58A	to OH58D 92-0578	
71-20526	OH58A	1-158AVN	
71-20527	OH58A	86ARCOM	
71-20528	OH58A	47AVN	
71-20529	OH58A	to OH58D 90-0355	
71-20530	OH58A	40AVN	
71-20531	OH58A	40AVN	
71-20532	OH58A	53InfBde	
71-20533	OH58A	53InfBde	
71-20534	OH58A	81ARCOM	
71-20535	OH58A	1-151AVN	SC NG
71-20536	OH58A	1-183AVN	ID NG
71-20537	OH58A	1-183AVN	ID NG
71-20538	OH58A	308AHB	
71-20539	OH58A	to OH58D 90-0348	
71-20540	OH58A		
71-20541	OH58A	28AVN	
71-20542	OH58A		
71-20543	OH58A	to OH58D 87-0747	
71-20544	OH58CF		
71-20545	OH58A	47AVN	
71-20546	OH58CR		
71-20547	OH58A	1-285AVN	AZ NG
71-20548	OH58A	1-285AVN	AZ NG
71-20549	OH58CR	9AVN	
71-20550	OH58CF	82AVN	
71-20551	OH58A	to OH58D 86-8926	
71-20552	OH58A	toOH58D87-0759	
71-20553	OH58A	to OH58D 86-8915	
71-20554	OH58A	USN\TPS	
71-20555	OH58A		
71-20556	OH58CF		
71-20557	OH58A	to OH58D 91-0537	
71-20558	OH58A		
71-20559	OH58A		
71-20560	OH58A		
71-20561	OH58CR	2ACR	
71-20562	OH58A	to OH58D 90-0351	
71-20563	OH58CF	9Cav	
71-20564	OH58CR	2ACR	
71-20565	OH58A		
71-20566	OH58CR		
71-20567	OH58A	10AHB	
71-20568	OH58A		
71-20569	OH58A		
71-20570	OH58A	4Cav	
71-20571	OH58A		
71-20572	OH58CR	229AVN	
71-20573	OH58CF		
71-20574	OH58CF		
71-20575	OH58A	225AvCo	
71-20576	OH58CR	4AVN	
71-20577	OH58CF	229AVN	
71-20578	OH58A	1InfDiv	
71-20579	OH58A	to OH58D 90-0356	
71-20580	OH58A	AATC 80J Shell AHP	
71-20581	OH58A	to OH58D 92-0580	
71-20582	OH58A	3ACT	
71-20583	OH58A	9AVN	
71-20584	OH58CR		
71-20585	OH58A		
71-20586	OH58A		
71-20587	OH58CF		
71-20588	OH58A	w.o.6.12.72	
71-20589	OH58A		
71-20590	OH58CF	w.o.19.10.72	
71-20591	OH58A		
71-20592	OH58CF		
71-20593	OH58CR		
71-20594	OH58CF		
71-20595	OH58A		
71-20596	OH58CF		
71-20597	OH58CF		

71-20598	OH58A	40AVN	
71-20599	OH58A		
71-20600	OH58A	126MedCo	
71-20601	OH58A	to OH58D 92-0572	
71-20602	OH58A	1-183AVN	ID NG
71-20603	OH58A	1-183AVN	ID NG
71-20604	OH58C	w.o.29.4.83	
71-20605	OH58A	150AVN	DE NG
71-20606	OH58A	1-151AVN	SC NG
71-20607	OH58CR	163ACR	
71-20608	OH58A	w.o.16.7.73	
71-20609	OH58CR	3-227AVN	
71-20610	OH58A	to OH58D 91-0557	
71-20611	OH58A	E\1-192AVN	AL NG
71-20612	OH58A		
71-20613	OH58A	3ACR	
71-20614	OH58A	1-285AVN	AZ NG
71-20615	OH58A	47AVN	
71-20616	OH58A	49AvCo	
71-20617	OH58A	40AVN	
71-20618	OH58A	44AVN	
71-20619	OH58A	1-183AVN	ID NG
71-20620	OH58A	1-183AVN	ID NG
71-20621	OH58A	1-135AVN	MO NG
71-20622	OH58CR	1AVN	
71-20623	OH58A	1InfDiv	
71-20624	OH58CF	28AVN	
71-20625	OH58CF	w.o.20.3.90	
71-20626	OH58CF		
71-20627	OH58CR	6ACR	
71-20628	OH58A	E\1-192AVN	AL NG
71-20629	OH58A	44AVN	
71-20630	OH58A	47AVN	
71-20631	OH58A	1-285AVN	AZ NG
71-20632	OH58A		
71-20633	OH58A	to OH58D 88-0305	
71-20634	OH58A	to OH58D 85-24703	
71-20635	OH58A	w.o.6.5.77	
71-20636	OH58CF	1Cav	
71-20637	OH58A	to OH58D 90-0376	
71-20638	OH58A	to OH58D 91-0542	
71-20639	OH58A	to OH58D 87-0735	
71-20640	OH58A	w.o.25.10.79	
71-20641	OH58A	w.o.14.4.82	
71-20642	OH58A	82AVN	
71-20643	OH58A	44AVN	
71-20644	OH58A	w.o.23.8.73	
71-20645	OH58A		
71-20646	OH58CF		
71-20647	OH58A	w.o.15.5.73	
71-20648	OH58CF		
71-20649	OH58A	to OH58D 85-24722	
71-20650	OH58A	7InfDiv	
71-20651	OH58A	7InfDiv	
71-20652	OH58A	w.o.18.10.72	
71-20653	OH58A	4InfDiv	
71-20654	OH58A	28AVN	
71-20655	OH58A	1-149AVN	TX NG
71-20656	OH58A		
71-20657	OH58A		
71-20658	OH58A		
71-20659	OH58A	w.o.15.3.80	
71-20660	OH58A	47AVN	
71-20661	OH58A	1-158AVN	
71-20662	OH58A	244AvCo	
71-20663	OH58CF	stored Ramstein AB, FRG	
71-20664	OH58A		
71-20665	OH58A	w.o.9.3.85	
71-20666	OH58CF		
71-20667	OH58CR	3ACT	
71-20668	OH58A	to OH58D 85-24707	
71-20669	OH58A	to OH58D 85-24700	
71-20670	OH58CF	4InfDiv	
71-20671	OH58CF	11ACR	
71-20672	OH58A		
71-20673	OH58CF		
71-20674	OH58CR	56AvCo	
71-20675	OH58CR	1Cav	
71-20676	OH58A	82AVN	
71-20677	OH58A	229AVN	
71-20678	OH58A		
71-20679	OH58CF		
71-20680	OH58CR	9AVN	

Serial	Type	Notes	
71-20681	OH58A	to OH58D 85-24698	
71-20682	OH58A	1-158AVN	
71-20683	OH58CR	AATC	83E Shell AHP
71-20684	OH58A		
71-20685	OH58CR	308AHB	
71-20686	OH58A	to OH58D 89-0098	
71-20687	OH58A	166ACR	
71-20688	OH58A	47AVN	
71-20689	OH58A		
71-20690	OH58A		
71-20691	OH58A		
71-20692	OH58A	47AVN	
71-20693	OH58A		
71-20694	OH58A	to OH58D 91-0559	
71-20695	OH58A	to OH58D 85-24693	
71-20696	OH58A	to OH58D 88-0288	
71-20697	OH58A	1-158AVN	
71-20698	OH58A		
71-20699	OH58A		WI NG
71-20700	OH58A		WI NG
71-20701	OH58A		
71-20702	OH58A	w.o.26.12.72	
71-20703	OH58A	to OH58D 92-0581	
71-20704	OH58CF	stored Ramstein AB	
71-20705	OH58A	86ARCOM	
71-20706	OH58A	278ACR	
71-20707	OH58A	278ACR	
71-20708	OH58A	40AVN	
71-20709	OH58A	40AVN	
71-20710	OH58A		
71-20711	OH58A	to OH58D 91-0564	
71-20712	OH58A	166ACR	
71-20713	OH58CF		
71-20714	OH58A	11CAG	
71-20715	OH58A	47AVN	
71-20716	OH58A	1-135AVN	MO NG
71-20717	OH58A	1-285AVN	AZ NG
71-20718	OH58A	28AVN	
71-20719	OH58A	28AVN	
71-20720	OH58A	28AVN	
71-20721	OH58A	47AVN	
71-20722	OH58A	1-285AVN	AZ NG
71-20723	OH58A	w.o.14.9.73	
71-20724	JOH58C	9AVN	
71-20725	OH58A	w.o.18.5.88	
71-20726	OH58CR		
71-20727	OH58A	4AVN	
71-20728	OH58CR		
71-20729	OH58A		
71-20730	OH58A	3ACR	
71-20731	OH58A	to OH58D 88-0303	
71-20732	OH58A	224AVN	
71-20733	OH58A	w.o.14.12.78	
71-20734	OH58A		
71-20735	OH58A	224AVN	
71-20736	GOH58A	45AVN	
71-20737	OH58A	25AvCo	
71-20738	OH58A	to OH58D 92-0575	
71-20739	OH58CF	9Cav	
71-20740	OH58A	82AVN	
71-20741	OH58A		
71-20742	OH58A	to OH58D 91-0538	
71-20743	OH58A	to OH58D 86-8908	
71-20744	OH58A	to OH58D 86-8919	
71-20745	OH58A	to OH58D 87-0729	
71-20746	OH58CF		
71-20747	OH58A	to OH58D 85-24692	
71-20748	OH58A	1-158AVN	
71-20749	OH58CF	stored Ramstein AB	
71-20750	OH58CR		
71-20751	OH58CF		
71-20752	OH58A	w.o.14.1.81	
71-20753	OH58CF	8AVN	
71-20754	OH58A		
71-20755	OH58A	11ADA Grp	
71-20756	OH58CR	4AVN	
71-20757	OH58A	26AVN	
71-20758	OH58A	to OH58D 87-0736	
71-20759	JOH58C	w.o.20.3.85	
71-20760	OH58CR	AATC	60B Shell AHP
71-20761	OH58CF	11ACR	
71-20762	OH58CR		
71-20763	OH58A	9AVN	
71-20764	OH58A	9AVN	
71-20765	OH58A	9AVN	
71-20766	OH58CF		
71-20767	OH58A	AATC	67C Shell AHP
71-20768	OH58A	9AVN	
71-20769	OH58CR	1AVN	
71-20770	OH58A	w.o.19.5.76	
71-20771	OH58CF	stored Ramstein AB	
71-20772	OH58A	w.o.7.3.85	
71-20773	OH58CF	11ACR	
71-20774	OH58A		
71-20775	OH58CF	AATC	75J Shell AHP
71-20776	OH58A	7\6Cav	
71-20777	OH58A		
71-20778	JOH58A	1AvBde	
71-20779	OH58CF		
71-20780	OH58A	w.o.16.10.90	
71-20781	OH58A	to OH58D 92-0577	
71-20782	OH58A		
71-20783	OH58CR		
71-20784	OH58A		
71-20785	OH58A	45AVN	
71-20786	OH58A		
71-20787	OH58A	40AVN	
71-20788	OH58A		
71-20789	OH58A	w.o.15.8.89	
71-20790	OH58A		
71-20791	OH58A	126MedCo	
71-20792	OH58A		FL NG
71-20793	OH58A	24AVN	
71-20794	OH58A		
71-20795	OH58A	to OH58D 85-24726	
71-20796	OH58A	229AVN	
71-20797	OH58A	229AVN	
71-20798	OH58CR	9AVN	
71-20799	OH58A	USN\TPS	
71-20800	OH58CF	3ACT	
71-20801	OH58A		
71-20802	OH58CR		
71-20803	OH58A	40AVN	
71-20804	OH58A	AATC	04L Shell AHP
71-20805	OH58A	to OH58D 89-0116	
71-20806	OH58CR	1InfDiv	
71-20807	OH58CR	6ACR	
71-20808	OH58A	w.o.6.5.83	
71-20809	OH58A	7InfDiv	
71-20810	OH58A	9AVN	
71-20811	OH58A	3ACR	
71-20812	OH58A	10Cav	
71-20813	OH58CF	9Cav	
71-20814	OH58A	USN\TPS 53	
71-20815	OH58A		
71-20816	OH58A		
71-20817	OH58CF	9Cav	
71-20818	OH58A	193AVN	
71-20819	OH58A	47AVN	
71-20820	OH58A	25AvCo	
71-20821	OH58A	116AHB	
71-20822	OH58A		
71-20823	OH58CR	163ACR	
71-20824	OH58A	1-285AVN	AZ NG
71-20825	OH58A	1-151AVN	SC NG
71-20826	OH58A	1-151AVN	SC NG
71-20827	OH58A	to OH58D 87-0746	
71-20828	OH58A		
71-20829	GOH58A		
71-20830	OH58A	1-149AVN	TX NG
71-20831	OH58A		
71-20832	OH58A	to OH58D 86-8930	
71-20833	OH58CR	3-227AVN	
71-20834	OH58A	to OH58D 85-24733	
71-20835	OH58CF	w.o.7.9.90	
71-20836	OH58A	51AvCo	
71-20837	OH58CF	w.o.20.3.89	
71-20838	OH58CF	17Cav	
71-20839	OH58CR	163ACR	
71-20840	OH58A	116AHB	
71-20841	OH58A		
71-20842	OH58A		
71-20843	OH58A	AATC	43G Shell AHP
71-20844	OH58A	57MedDet	
71-20845	GOH58A		
71-20846	OH58CF	1InfDiv	

Serial	Type	Unit / Notes	
71-20847	OH58A		HI NG
71-20848	OH58A	to OH58D 85-24696	
71-20849	OH58A	1-158AVN	
71-20850	OH58A	E\1-192AVN	AL NG
71-20851	OH58A	3ACT	
71-20852	OH58A		
71-20853	OH58A	57MedDet	
71-20854	OH58A	7\6Cav	
71-20855	OH58A	278ACR	
71-20856	OH58A		DC NG
71-20857	OH58A		FL NG
71-20858	OH58A		
71-20859	OH58A		
71-20860	OH58A		
71-20861	OH58A	116AHB	
71-20862	OH58A	116AHB	
71-20863	OH58A	116AHB	
71-20864	OH58A	1-185AVN	MS NG
71-20865	OH58A	1-285AVN	AZ NG
72-21061	OH58A	to OH58D 92-0582	
72-21062	OH58A		
72-21063	OH58A		
72-21064	OH58A		
72-21065	OH58A	285AvCo	
72-21066	OH58CF	6ACR	
72-21067	OH58CF	163ACR	
72-21068	OH58A	to OH58D 90-0352	
72-21069	OH58A	82AVN	
72-21070	OH58A		
72-21071	OH58A		
72-21072	OH58CF	11ADA Grp	
72-21073	OH58CF		
72-21074	OH58A	to OH58D 91-0556	
72-21075	OH58A	w.o.10.12.84	
72-21076	OH58A		
72-21077	OH58CR		
72-21078	OH58A	101AbDiv	
72-21079	OH58A		
72-21080	OH58CF	2ACR	
72-21081	OH58A		HI NG
72-21082	OH58A	D\1-228AVN	
72-21083	OH58A	308AHB	
72-21084	OH58CR	1AVN	
72-21085	OH58CR	6ACR	
72-21086	OH58A	9AVN	
72-21087	OH58A	w.o.31.3.80	
72-21088	OH58CR	9AVN	
72-21089	OH58CF	9Cav	
72-21090	OH58CF		
72-21091	OH58CF		
72-21092	OH58CR	1InfDiv	
72-21093	OH58CF		
72-21094	OH58A		
72-21095	OH58A	to OH58D 90-0379	
72-21096	OH58A		
72-21097	OH58A	AATC 97N Shell AHP	
72-21098	OH58A	w.o.15.12.81	
72-21099	OH58A	w.o.4.4.77	
72-21100	OH58A	3ACR	
72-21101	OH58A	to OH58D 90-0378	
72-21102	OH58CR		
72-21103	OH58CR		
72-21104	OH58CR	w.o.	
72-21105	OH58CR	9AVN	
72-21106	OH58CF	11ACR	
72-21107	OH58A	to OH58D 85-24717	
72-21108	OH58A	9AVN	
72-21109	OH58A	3ACR	
72-21110	OH58CR	1InfDiv	
72-21111	OH58A	to OH58D 87-0737	
72-21112	OH58A	82AVN	
72-21113	OH58A		
72-21114	OH58A	3ACT	
72-21115	OH58CR	163ACR	
72-21116	OH58CR		
72-21117	OH58A	to OH58D 90-0358	
72-21118	OH58A	197AvSect	
72-21119	OH58A	AATC 19C Shell AHP	
72-21120	OH58A		
72-21121	OH58A	to OH58D 91-0562	
72-21122	OH58A	1-183AVN	ID NG
72-21123	OH58CF		
72-21124	OH58A	47AVN	
72-21125	OH58A		
72-21126	OH58A		
72-21127	OH58A	w.o.25.2.86	
72-21128	OH58A	D\135AVN	KS NG
72-21129	OH58A		HI NG
72-21130	OH58A		NC NG
72-21131	OH58A	47AVN	
72-21132	OH58A	1-285AVN	AZ NG
72-21133	OH58A	to OH58D 91-0543	
72-21134	OH58A	1-135AVN	MO NG
72-21135	OH58A		
72-21136	OH58A	28AVN	
72-21137	OH58CF		
72-21138	OH58A		
72-21139	OH58A	to OH58D 87-0753	
72-21140	OH58A	7InfDiv	
72-21141	OH58A	56AvCo	
72-21142	OH58A	7\6Cav	
72-21143	OH58A	w.o.23.8.85	
72-21144	OH58A	1-151AVN	SC NG
72-21145	OH58A	49AvCo	
72-21146	OH58A	53InfBde	
72-21147	OH58A	53InfBde	
72-21148	OH58A		
72-21149	OH58A		
72-21150	OH58CF		
72-21151	OH58A		
72-21152	OH58A	w.o.6.3.79	
72-21153	OH58A	1-285AVN	AZ NG
72-21154	OH58A		
72-21155	OH58CR	163ACR	
72-21156	OH58A	107AvCo	
72-21157	OH58A	107AvCo	
72-21158	OH58A	47AVN	
72-21159	OH58A	49AvCo	
72-21160	OH58A		
72-21161	OH58A	82AVN	
72-21162	OH58A	224AVN	
72-21163	OH58A		
72-21164	OH58A	1AVN	
72-21165	OH58CF	24AVN	
72-21166	OH58A	w.o.6.11.75	
72-21167	OH58A	1-151AVN	SC NG
72-21168	OH58A	10Cav	
72-21169	OH58CF	40AVN	
72-21170	OH58A		
72-21171	OH58A	to OH58D 89-0102	
72-21172	OH58CF		
72-21173	OH58A		
72-21174	OH58CF	7InfDiv	
72-21175	OH58A		
72-21176	OH58CR		
72-21177	OH58A	w.o.10.10.74	
72-21178	OH58A	4InfDiv	
72-21179	OH58CF	9Cav	
72-21180	OH58CF		
72-21181	OH58CF	49AVN	
72-21182	OH58A	1-151AVN	SC NG
72-21183	OH58A	to OH58D 85-24728	
72-21184	OH58A	1-167Cav	NE NG
72-21185	OH58A	240EngGrp	
72-21186	OH58A	USAICS	
72-21187	OH58A	A\132AVN	OK NG
72-21188	OH58A	USAICS	
72-21189	OH58A		
72-21190	OH58A	to OH58D 87-0748	
72-21191	OH58A	to OH58D 87-0752	
72-21192	OH58A	to OH58D 87-0750	
72-21193	OH58A	USN\TPS	
72-21194	OH58A	82AVN	
72-21195	OH58A		
72-21196	OH58A	to OH58D 85-24690	
72-21197	OH58A		
72-21198	OH58A	w.o.22.10.81	
72-21199	OH58CF	2ACR	
72-21200	OH58A	79ARCOM	
72-21201	OH58A		
72-21202	OH58A	w.o.11.1.84	
72-21203	OH58A	w.o.2.4.89	
72-21204	OH58CF		
72-21205	OH58A	to OH58D 91-0544	
72-21206	OH58A	AATC 06A Shell AHP	
72-21207	OH58A	24EngGrp	

Serial	Type	Notes	Code
72-21208	OH58A		
72-21209	OH58A	7SigBde	
72-21210	OH58A		HI NG
72-21211	OH58A	3ACR	
72-21212	OH58A	308AHB	
72-21213	OH58A	A\132AVN	OK NG
72-21214	OH58A	308AHB	
72-21215	OH58CR	3-227AVN	
72-21216	OH58A	193AVN	
72-21217	OH58A	AATC 17C Shell AHP	
72-21218	OH58A	w.o.14.5.85	
72-21219	OH58A	193AVN	
72-21220	GOH58A		
72-21221	OH58A	7\6Cav	
72-21222	OH58A		
72-21223	OH58A		HI NG
72-21224	OH58A	1-285AVN	AZ NG
72-21225	OH58A	to OH58D 88-0312	
72-21226	OH58A	278ACR	
72-21227	OH58A	81ARCOM	
72-21228	OH58A	44AVN	
72-21229	OH58A	229AVN	
72-21230	OH58A		HI NG
72-21231	OH58CR		
72-21232	OH58CF	w.o.29.5.80	
72-21233	OH58A	AATC 33. Shell AHP	
72-21234	OH58A	to OH58D 87-0756	
72-21235	OH58CF	101AbDiv	
72-21236	OH58CR		
72-21237	OH58A	AATC 37H Shell AHP	
72-21238	OH58A	to OH58D 90-0377	
72-21239	OH58CR	1Cav	
72-21240	OH58A	to OH58D 86-8905	
72-21241	OH58A		
72-21242	OH58A	w.o.14.5.83	
72-21243	OH58A	w.o.26.10.81	
72-21244	OH58A	to OH58D 87-0739	
72-21245	OH58A		
72-21246	OH58A	to OH58D 86-8910	
72-21247	OH58CF		
72-21248	OH58A		
72-21249	OH58A		
72-21250	OH58A		
72-21251	OH58A	w.o.29.6.80	
72-21252	OH58A	USN\NATC	
72-21253	OH58A	USN\TPS 52	
72-21254	OH58A	97ARCOM	
72-21255	OH58A	26AVN	
72-21256	OH58A	44AVN	
72-21257	OH58A		
72-21258	OH58A	116AHB	
72-21259	OH58A	116AHB	
72-21260	OH58A	D\135AVN	KS NG
72-21261	OH58A	1-135AVN	MO NG
72-21262	OH58A	w.o.14.7.73	
72-21263	OH58A		
72-21264	OH58A	AATC 64J Shell AHP	
72-21265	OH58A	107AvCo	
72-21266	OH58A		
72-21267	OH58A		
72-21268	OH58A	1-285AVN	AZ NG
72-21269	GOH58A		AZ NG
72-21270	OH58A	1-151AVN	SC NG
72-21271	OH58A	51AvCo	
72-21272	OH58A	to OH58D 91-0540	
72-21273	OH58A		
72-21274	OH58A	9AVN	
72-21275	OH58CF	stored Ramstein AB	
72-21276	OH58A	1-285AVN	AZ NG
72-21277	OH58A	1-167Cav	NE NG
72-21278	OH58A	81ARCOM	
72-21279	OH58A		
72-21280	OH58A		
72-21281	OH58A		
72-21282	OH58CF	AATC 82E Shell AHP	
72-21283	OH58A	A\132AVN	OK NG
72-21284	OH58CR	4AVN	
72-21285	OH58A	D\135AVN	KS NG
72-21286	OH58A	D\135AVN	KS NG
72-21287	OH58A		
72-21288	OH58A	preserved Shell AHP	
72-21289	OH58A		
72-21290	OH58A		
72-21291	OH58A	1-135AVN	MO NG
72-21292	OH58A	1-285AVN	AZ NG
72-21293	OH58A	to OH58D 91-0550	
72-21294	OH58A		
72-21295	OH58A		
72-21296	OH58CR	10AHB	
72-21297	GOH58A	(w.o.18.3.85)	
72-21298	OH58A	26AVN	
72-21299	OH58A		
72-21300	OH58A	18AvCo	
72-21301	OH58CF	11ACR	
72-21302	OH58CF		
72-21303	OH58CF	229AVN	
72-21304	OH58CF	11ACR	
72-21305	OH58A	to OH58D 86-8928	
72-21306	OH58A	AATC 06M Shell AHP	
72-21307	OH58A	to OH58D 85-24716	
72-21308	OH58CF	229AVN	
72-21309	OH58A	47AVN	
72-21310	OH58A	47AVN	
72-21311	OH58CR	w.o.15.6.89	
72-21312	OH58CR	6ACR	
72-21313	OH58A	to OH58D 88-0297	
72-21314	OH58A		
72-21315	OH58A		
72-21316	OH58A	to OH58D 91-0558	
72-21317	OH58A	to OH58D 91-0561	
72-21318	OH58A	51AvCo	
72-21319	OH58A	to OH58D 91-0548	
72-21320	OH58A	47AVN	
72-21321	OH58A		
72-21322	OH58A	to OH58D 86-8929	
72-21323	OH58A	to OH58D 85-24732	
72-21324	OH58CF	9Cav	
72-21325	OH58A	166ACR	
72-21326	OH58A		
72-21327	OH58CF		
72-21328	OH58A		
72-21329	OH58A		
72-21330	OH58A	1-151AVN	SC NG
72-21331	OH58A		
72-21332	OH58A		
72-21333	OH58A		
72-21334	OH58A	w.o.17.2.90	
72-21335	OH58A		
72-21336	OH58A		
72-21337	OH58A	7\6Cav	
72-21338	OH58A		
72-21339	OH58A	AATC 39E Shell AHP	
72-21340	OH58A		
72-21341	OH58A		
72-21342	OH58A		
72-21343	OH58A	1-285AVN	AZ NG
72-21344	OH58A	w.o.8.9.79	
72-21345	OH58A		
72-21346	OH58A		
72-21347	OH58A		
72-21348	OH58A	1-285AVN	AZ NG
72-21349	OH58A	to OH58D 91-0551	
72-21350	OH58A	to OH58D 91-0560	
72-21351	OH58A		
72-21352	OH58A	to OH58D 92-0579	
72-21353	OH58A	108DivTrng	
72-21354	OH58A	1-158AVN	
72-21355	OH58A	193AvCo	
72-21356	OH58A	193AvCo	
72-21357	OH58A	193AvCo	
72-21358	OH58A	193AvCo	
72-21359	OH58A		
72-21360	OH58A	w.o.10.11.74	
72-21361	OH58A	40AVN	
72-21362	OH58A	40AVN	
72-21363	OH58A	40AVN	
72-21364	OH58A	40AVN	
72-21365	OH58A	53InfBde	
72-21366	OH58A	1-285AVN	AZ NG
72-21367	OH58A	53InfBde	
72-21368	OH58A	53InfBde	
72-21369	OH58A		
72-21370	OH58A		
72-21371	OH58A	1-285AVN	AZ NG
72-21372	OH58A	1-135AVN	MO NG
72-21373	OH58A	w.o.19.7.76	

72-21374	OH58A		
72-21375	OH58A	D\135AVN KS NG	
72-21376	OH58CR		
72-21377	OH58A		
72-21378	OH58A	86ARCOM	
72-21379	OH58A	derelict Rosamund, Ca	
72-21380	OH58CF		
72-21381	OH58A	w.o.30.7.79	
72-21382	OH58A		
72-21383	OH58A	13AHB	
72-21384	OH58A	w.o.17.3.81	
72-21385	OH58A	25AVN	
72-21386	OH58A	97ARCOM	
72-21387	OH58A	346MedDet	
72-21388	OH58A	2ACR	
72-21389	OH58A	1-224AVN	MD NG
72-21390	OH58A	1-183AVN	ID NG
72-21391	OH58A	AATC 91D Shell AHP	
72-21392	OH58A	AATC 92. Shell AHP	
72-21393	OH58A	w.o.10.11.74	
72-21394	OH58A	w.o.14.10.74	
72-21395	OH58CR	2ACR	
72-21396	OH58CR	3-227AVN	
72-21397	OH58CR	2ACR	
72-21398	OH58CR	2ACR	
72-21399	OH58CR	308AHB	
72-21400	OH58A	to OH58D 86-8937	
72-21401	OH58A	2ACR	
72-21402	OH58CR	2ACR	
72-21403	OH58A	308AHB	
72-21404	OH58CR	2ACR	
72-21405	OH58CR	3-227AVN	
72-21406	OH58A	2ACR	
72-21407	OH58A	8AVN	
72-21408	OH58A	25AvCo	
72-21409	OH58CR		
72-21410	OH58A	to OH58D 89-0103	
72-21411	OH58A		
72-21412	OH58A	62AvCo	
72-21413	OH58A	56AvCo	
72-21414	OH58A	3ACR	
72-21415	OH58A	AATC 15A Shell AHP	
72-21416	OH58A	1-158AVN	
72-21417	OH58A		
72-21418	OH58A		
72-21419	OH58CF		
72-21420	OH58CF	11ACR	
72-21421	OH58A	158AVN	
72-21422	OH58A		OH NG
72-21423	OH58A		
72-21424	OH58A		
72-21425	OH58A		
72-21426	OH58CR	AATC 26F Shell AHP	
72-21427	OH58A	to OH58D 87-0754	
72-21428	OH58A		
72-21429	OH58A		
72-21430	OH58A	7\6Cav	
72-21431	OH58A		
72-21432	OH58A	83ARCOM	
72-21433	OH58A	40AVN	
72-21434	OH58A	to OH58D 87-0743	
72-21435	OH58A		
72-21436	OH58A		
72-21437	OH58A		
72-21438	OH58A		
72-21439	OH58A		
72-21440	OH58A		
72-21441	OH58A	1\17Cav	
72-21442	OH58A		OH NG
72-21443	OH58A	1-149AVN	TX NG
72-21444	OH58CF	17Cav	
72-21445	OH58A		NM NG
72-21446	OH58A	193AvCo	
72-21447	OH58A	193AvCo	
72-21448	OH58CR	10AHB	
72-21449	OH58A	25AvCo	
72-21450	OH58A	1-285AVN	AZ NG
72-21451	OH58A	278ACR	
72-21452	OH58A	1-114AVN	AR NG
72-21453	OH58A	40AVN	
72-21454	OH58A	44AVN	
72-21455	OH58A	26AVN	
72-21456	OH58A	26AVN	
72-21457	OH58A	to OH58D 91-0568	
72-21458	OH58A	to OH58D 92-0583	
72-21459	OH58A	47AVN	
72-21460	OH58CF	1AHB	
73-21861	OH58A	1-183AVN	ID NG
73-21862	OH58A	1-183AVN	ID NG
73-21863	OH58A	D\135AVN	KS NG
73-21864	OH58A		
73-21865	OH58A	1-185AVN	MS NG
73-21866	OH58A	A\132AVN	OK NG
73-21867	OH58A		
73-21868	OH58A		
73-21869	OH58A	47AVN	
73-21870	OH58CR	6Cav	
73-21871	OH58CR	47AVN	
73-21872	OH58A		
73-21873	OH58A		NH NG
73-21874	OH58A	to OH58D 88-0313	
73-21875	OH58A	1-185AVN	MS NG
73-21876	OH58A	to OH58D 91-0553	
73-21877	OH58A	1-151AVN	SC NG
73-21878	OH58CF		
73-21879	OH58A	116AHB	
73-21880	OH58A	3ACT	
73-21881	OH58A	to OH58D 91-0549	
73-21882	OH58A		
73-21883	OH58A	w.o.1.9.77	
73-21884	OH58A	1-285AVN	AZ NG
73-21885	OH58A	997AvCo 64	
73-21886	OH58A	49AvCo	
73-21887	OH58A	49AvCo	
73-21888	OH58A		OH NG
73-21889	OH58A	1-151AVN	SC NG
73-21890	GOH58A		
73-21891	OH58A	1-149AVN	TX NG
73-21892	OH58A	1-149AVN	TX NG
73-21893	OH58A	1-149AVN	TX NG
73-21894	OH58A	1-149AVN	TX NG
73-21895	OH58A	1-149AVN	TX NG
73-21896	OH58A	1-149AVN	TX NG
73-21897	OH58A	1-149AVN	TX NG
73-21898	OH58A	1-149AVN	TX NG
73-21899	OH58A		
73-21900	OH58A		
73-21901	OH58A		
73-21902	OH58A	1-149AVN	TX NG
73-21903	OH58A	1-149AVN	TX NG
73-21904	OH58A	1-149AVN	TX NG
73-21905	OH58A	1-149AVN	TX NG
73-21906	OH58A	1-149AVN	TX NG
73-21907	OH58A	1-149AVN	TX NG
73-21908	OH58A	1-149AVN	TX NG
73-21909	OH58A	53InfDiv	
73-21910	OH58A		
73-21911	OH58A		
73-21912	OH58A	112MedDet	
73-21913	OH58A	to OH58D 88-0317	
73-21914	OH58A	1-135AVN	MO NG
73-21915	OH58A	1-135AVN	MO NG
73-21916	OH58A	1-135AVN	MO NG
73-21917	OH58A	1-135AVN	MO NG
73-21918	OH58A		
73-21919	OH58A		
73-21920	OH58A		
73-21921	OH58A		
73-21922	OH58A	51AvCo	
73-21923	OH58A	4\9Cav	
73-21924	OH58A	w.o.14.9.83	
73-21925	OH58A	w.o.19.3.91	
73-21926	OH58A	4\9Cav	
73-21927	OH58A	222AVN	
73-21928	OH58A		
73-21929	OH58A	4\9Cav	
73-21930	OH58A	4\9Cav	
73-21931	OH58A	222AVN	
73-21932	OH58CF		
73-21933	OH58A	to OH58D 91-0555	
73-21934	OH58A	4\9Cav	
83-24129	OH58D	w.o.21.8.90	
83-24130	OH58D	4-159AVN	
83-24131	OH58D		
83-24132	OH58D		
83-24133	OH58D	AATC 33G Hanchey AHP	

Serial	Type	Unit	Code	Location
83-24134	OH58D			
83-24135	OH58D	AATC	35E	Hanchey AHP
83-24136	OH58D	AATC	36C	Hanchey AHP
83-24137	OH58D			
83-24138	OH58D	AATC	38J	Hanchey AHP
83-24139	OH58D	AATC	39N	Hanchey AHP
83-24140	OH58D	AATC	40M	Hanchey AHP
83-24141	OH58D	AATC	41G	Hanchey AHP
83-24142	OH58D			
83-24143	OH58D			
83-24144	OH58D			
85-24690	OH58D			
85-24691	OH58D			
85-24692	OH58D			
85-24693	OH58D			
85-24694	OH58D			
85-24695	OH58D			
85-24696	OH58D			
85-24697	OH58D			
85-24698	OH58D			
85-24699	OH58D			
85-24700	OH58D			
85-24701	OH58D	AATC	01.	Hanchey AHP
85-24702	OH58D			
85-24703	OH58D	w.o.19.10.88		
85 24704	OH58D	2AD		
85-24705	OH58D	2AD		
85-24706	OH58D	2AD		
85-24707	OH58D	2AD		
85-24708	OH58D	2AD		
85-24709	OH58D	2AD		
85-24710	OH58D			
85-24711	OH58D			
85-24712	OH58D	AATC	12A	Hanchey AHP
85-24713	OH58D	AATC	13D	Hanchey AHP
85-24714	OH58D			
85-24715	OH58D	AATC	15M	Hanchey AHP
85-24716	OH58D			
85-24717	OH58D			
85-24718	OH58D			
85-24719	OH58D			
85-24720	OH58D			
85-24721	OH58D			
85-24722	OH58D			
85-24723	OH58D	23TF		
85-24724	OH58D			
85-24725	OH58D	23TF		
85-24726	OH58D			
85-24727	OH58D	23TF		
85-24728	OH58D			
85-24729	OH58D	TF Viper		
85-24730	OH58D			
85-24731	OH58D	TF Viper		
85-24732	OH58D			
85-24733	OH58D	TF Viper		
86-8901	OH58D	TF Viper		
86-8902	OH58D	4-11ACR		
86-8903	OH58D	240AvCo		
86-8904	OH58D	4-11ACR		
86-8905	OH58D	240AvCo		
86-8906	OH58D	240AvCo		
86-8907	OH58D	240AvCo		
86-8908	OH58D	240AvCo		
86-8909	OH58D			
86-8910	OH58D	w.o.20.9.91		
86-8911	OH58D			
86-8912	OH58D			
86-8913	OH58D			
86-8914	OH58D			
86-8915	OH58D			
86-8916	OH58D			
86-8917	OH58D			
86-8918	OH58D			
86-8919	OH58D			
86-8920	OH58D			
86-8921	OH58D			
86-8922	OH58D	4-159AVN		
86-8923	OH58D			
86-8924	OH58D			
86-8925	OH58D			
86-8926	OH58D			
86-8927	OH58D			
86-8928	OH58D			
86-8929	OH58D	w.o.24.2.91		
86-8930	OH58D			
86-8931	OH58D			
86-8932	OH58D			
86-8933	OH58D			
86-8934	OH58D			
86-8935	OH58D			
86-8936	OH58D			
86-8937	OH58D			
86-8938	OH58D			
86-8939	OH58D			
87-0725	OH58D			
87-0726	OH58D			
87-0727	OH58D			
87-0728	OH58D	w.o.1.4.91		
87-0729	OH58D			
87-0730	OH58D			
87-0731	OH58D			
87-0732	OH58D			
87-0733	OH58D			
87-0734	OH58D			
87-0735	OH58D			
87-0736	OH58D			
87-0737	OH58D			
87-0738	OH58D			
87-0739	OH50D	w.o.23.11.89		
87-0740	OH58D			
87-0741	OH58D			
87-0742	OH58D			
87-0743	OH58D			
87-0744	OH58D	4-159AVN		
87-0745	OH58D	4-159AVN		
87-0746	OH58D	4-159AVN		
87-0747	OH58D	4-159AVN		
87-0748	OH58D	4-159AVN		
87-0749	OH58D	5-158AVN		
87-0750	OH58D	5-158AVN		
87-0751	OH58D	5-158AVN		
87-0752	OH58D	5-158AVN		
87-0753	OH58D	5-158AVN		
87-0754	OH58D	5-158AVN		
87-0755	OH58D	5-158AVN		
87-0756	OH58D	5-158AVN		
87-0757	OH58D	5-158AVN		
87-0758	OH58D	5-158AVN		
87-0759	OH58D	5-158AVN		
87-0760	OH58D	5-158AVN		
88-0285	OH58D	5-158AVN		
88-0286	OH58D	5-158AVN		
88-0287	OH58D	5-158AVN		
88-0288	OH58D	TF Viper		
88-0289	OH58D	TF Viper		
88-0290	OH58D	TF Viper		
88-0291	OH58D	TF Viper		
88-0292	OH58D	4-159AVN		
88-0293	OH58D	w.o.16.8.91		
88-0294	OH58D	4-11ACR		
88-0295	OH58D			
88-0296	OH58D	4-11ACR		
88-0297	OH58D			
88-0298	OH58D	4-11ACR		
88-0299	OH58D	4-11ACR		
88-0300	OH58D	AATC	00C	Hanchey AHP
88-0301	OH58D	4-11ACR		
88-0302	OH58D	w.o.13.9.90		
88-0303	OH58D			
88-0304	OH58D			
88-0305	OH58D	w.o.20.8.90		
88-0306	OH58D			
88-0307	OH58D			
88-0308	OH58D			
88-0309	OH58D			
88-0310	OH58D			
88-0311	OH58D	5-158AVN		
88-0312	OH58D	5-158AVN		
88-0313	OH58D			
88-0314	OH58D	4-11ACR		
88-0315	OH58D	4-11ACR		
88-0316	OH58D	5-158AVN		
88-0317	OH58D			
88-0318	OH58D			
88-0319	OH58D	4-11ACR		
88-0320	OH58D	4-11ACR		

Serial	Type	Unit				Serial	Type
89-0082	OH58D	4-11ACR				91-0536	OH58D
89-0083	OH58D					91-0537	OH58D
89-0084	OH58D					91-0538	OH58D
89-0085	OH58D					91-0539	OH58D
89-0086	OH58D					91-0540	OH58D
89-0087	OH58D	5-158AVN				91-0541	OH58D
89-0088	OH58D	4-11ACR				91-0542	OH58D
89-0089	OH58D					91-0543	OH58D
89-0090	OH58D					91-0544	OH58D
89-0091	OH58D					91-0545	OH58D
89-0092	OH58D					91-0546	OH58D
89-0093	OH58D					91-0547	OH58D
89-0094	OH58D					91-0548	OH58D
89-0095	OH58D					91-0549	OH58D
89-0096	OH58D	1\17Cav				91-0550	OH58D
89-0097	OH58D					91-0551	OH58D
89-0098	OH58D					91-0552	OH58D
89-0099	OH58D					91-0553	OH58D
89-0100	OH58D					91-0554	OH58D
89-0101	OH58D					91-0555	OH58D
89-0102	OH58D					91-0556	OH58D
89-0103	OH58D					91-0557	OH58D
89-0104	OH58D					91-0558	OH58D
89-0105	OH58D					91-0559	OH58D
89-0106	OH58D					91-0560	OH58D
89-0107	OH58D					91-0561	OH58D
89-0108	OH58D					91-0562	OH58D
89-0109	OH58D					91-0563	OH58D
89-0110	OH58D					91-0564	OH58D
89-0111	OH58D					91-0565	OH58D
89-0112	OH58D					91-0566	OH58D
89-0113	OH58D					91-0567	OH58D
89-0114	OH58D					91-0568	OH58D
89-0115	OH58D					91-0569	OH58D
89-0116	OH58D	AATC	16A	Hanchey AHP		91-0570	OH58D
89-0117	OH58D	AATC	17D	Hanchey AHP		91-0571	OH58D
90-0346	OH58D	AATC	46A	Hanchey AHP		92-0571	OH58D
90-0347	OH58D					92-0572	OH58D
90-0348	OH58D	AATC	48C	Hanchey AHP		92-0573	OH58D
90-0349	OH58D					92-0574	OH58D
90-0350	OH58D					92-0575	OH58D
90-0351	OH58D					92-0576	OH58D
90-0352	OH58D					92-0577	OH58D
90-0353	OH58D					92-0578	OH58D
90-0354	OH58D					92-0579	OH58D
90-0355	OH58D					92-0580	OH58D
90-0356	OH58D	AATC	56C	Hanchey AHP		92-0581	OH58D
90-0357	OH58D	AATC	57A	Hanchey AHP		92-0582	OH58D
90-0358	OH58D	AATC	58C	Hanchey AHP		92-0583	OH58D
90-0359	OH58D	AATC	59A	Hanchey AHP		92-0584	OH58D
90-0360	OH58D					92-0585	OH58D
90-0361	OH58D	AATC	61A	Hanchey AHP		92-0586	OH58D
90-0362	OH58D					92-0587	OH58D
90-0363	OH58D	57MedDet				92-0588	OH58D
90-0364	OH58D					92-0589	OH58D
90-0365	OH58D	57MedDet				92-0590	OH58D
90-0366	OH58D	57MedDet				92-0591	OH58D
90-0367	OH58D					92-0592	OH58D
90-0368	OH58D	AATC	68C	Hanchey AHP		92-0593	OH58D
90-0369	OH58D					92-0594	OH58D
90-0370	OH58D	AATC	70D	Hanchey AHP		92-0595	OH58D
90-0371	OH58D	AATC	71A	Hanchey AHP		92-0596	OH58D
90-0372	OH58D					92-0597	OH58D
90-0373	OH58D					92-0598	OH58D
90-0374	OH58D					92-0599	OH58D
90-0375	OH58D					92-0600	OH58D
90-0376	OH58D					92-0601	OH58D
90-0377	OH58D					92-0602	OH58D
90-0378	OH58D					92-0603	OH58D
90-0379	OH58D					92-0604	OH58D
90-0380	OH58D					92-0605	OH58D
90-0381	OH58D					92-0606	OH58D

H60 BLACK HAWK

c\n   70-001 to 70-004, 70-006 to 70-311, 70-313 to 70-347, 70-312, 70-352 to 70-363, one unknown, 70-495
      to 70-584, 70-662 to 70-757, 70-760 to 70-855, plus deliveries from FY85 onwards.

73-21650	YUH60A	w.o.19.5.78		78-23008	UH60A	101AbDiv
73-21651	GYUH60A			78-23009	UH60A	w.o.
73-21652	YUH60A	not completed		78-23010	UH60A	to USCS
73-21653	YUH60A	not completed		78-23011	UH60A	101AbDiv
73-21654	UH60A			78-23012	JUH60A	ERADCOM
73-21655	UH60A			78-23013	YEH60B	101AbDiv
77-22714	UH60A	w.o.3.10.86		78-23014	UH60A	
77-22715	GUH60A	GIA		78-23015	UH60A	101AbDiv
77-22716	UH60A	USN\TPS 60		78-23016	UH60A	
77-22717	NUH60A	ERADCOM		79-23263	UH60A	101AbDiv
77-22718	UH60A	to USCS		79-23264	UH60A	101AbDiv
77-22719	GUH60A			79-23265	GUH60A	101AbDiv
77-22720	UH60A			79-23266	UH60A	101AbDiv
77-22721	GUH60A			79-23267	UH60A	w.o.5.8.82
77-22722	UH60A	MASDC		79-23268	UH60A	w.o.2.2.84
77-22723	UH60A	159MedCo		79-23269	UH60A	101AbDiv
77-22724	UH60A			79-23270	UH60A	101AbDiv
77-22725	UH60A	USN\TPS 62		79-23271	UH60A	236MedCo
77-22726	GUH60A			79-23272	UH60A	101AbDiv
77-22727	UH60A	63MedDet		79-23273	UH60A	101AbDiv
77-22728	UH60A			79-23274	UH60A	40AVN
78-22960	UH60A			79-23275	UH60A	101AbDiv
78-22961	UH60A	9AVN		79-23276	UH60A	101AbDiv
78-22962	UH60A	9AVN		79-23277	UH60A	101AbDiv
78-22963	GUH60A			79-23278	UH60A	101AbDiv
78-22964	GUH60A			79-23279	UH60A	101AbDiv
78-22965	GUH60A			79-23280	UH60A	101AbDiv
78-22966	UH60A	A\4-228AVN		79-23281	UH60A	101AbDiv
78-22967	UH60A			79-23282	UH60A	
78-22968	UH60A			79-23283	UH60A	101AbDiv
78-22969	UH60A	7MEDCOM		79-23284	UH60A	101AbDiv
78-22970	UH60A	w.o.		79-23285	UH60A	101AbDiv
78-22971	UH60A			79-23286	UH60A	101AbDiv
78-22972	GUH60A			79-23287	UH60A	101AbDiv
78-22973	UH60A			79-23288	UH60A	101AbDiv
78-22974	UH60A			79-23289	UH60A	101AbDiv
78-22975	UH60A	w.o.		79-23290	UH60A	
78-22976	UH60A			79-23291	UH60A	2-82AVN
78-22977	UH60A	63MedDet		79-23292	UH60A	2-82AVN
78-22978	UH60A	421MedCo		79-23293	UH60A	
78-22979	UH60A	9AVN		79-23294	UH60A	
78-22980	UH60A			79-23295	UH60A	2-82AVN
78-22981	GUH60A			79-23296	UH60A	2-82AVN
78-22982	UH60A	to USCS		79-23297	UH60A	to USCS
78-22983	UH60A			79-23298	UH60A	
78-22984	UH60A	w.o.29.3.82		79-23299	UH60A	to USCS
78-22985	UH60A			79-23300	UH60A	2-82AVN
78-22986	UH60A	236MedDet		79-23301	UH60A	USAICS
78-22987	UH60A			79-23302	UH60A	2-82AVN
78-22988	GUH60A	101AbDiv		79-23303	UH60A	2-82AVN
78-22989	UH60A	101AbDiv		79-23304	UH60A	2-82AVN
78-22990	UH60A	7-1AVN		79-23305	UH60A	2-82AVN
78-22991	UH60A	421MedCo		79-23306	UH60A	2-82AVN
78-22992	GUH60A	101AbDiv		79-23307	UH60A	6ACR
78-22993	UH60A	w.o.26.8.83		79-23308	UH60A	2-82AVN
78-22994	UH60A	w.o.25.2.91		79-23309	UH60A	2-82AVN
78-22995	UH60A	159MedCo		79-23310	UH60A	2-82AVN
78-22996	UH60A	236MedCo		79-23311	UH60A	2-82AVN
78-22997	UH60A	159MedCo		79-23312	UH60A	2-82AVN
78-22998	UH60A	101AbDiv		79-23313	UH60A	2-82AVN
78-22999	UH60A	w.o.4.10.83		79-23314	UH60A	w.o.13.3.85
78-23000	UH60A	159MedCo		79-23315	UH60A	2-82AVN
78-23001	UH60A	236MedCo		79-23316	UH60A	2-82AVN
78-23002	UH60A	101AbDiv		79-23317	UH60A	2-82AVN
78-23003	UH60A	159MedCo		79-23318	UH60A	2-82AVN
78-23004	UH60A	441MedDet		79-23319	UH60A	2-82AVN
78-23005	UH60A	101AbDiv		79-23320	UH60A	to USCS
78-23006	UH60A	101AbDiv		79-23321	UH60A	to USCS
78-23007	UH60A	101AbDiv		79-23322	UH60A	w.o.25.8.81

Serial	Type	Unit		Serial	Type	Unit	
79-23323	UH60A	B\2-224AVN	VA NG	80-23465	UH60A	to USCS	
79-23324	UH60A	w.o.4.3.82		80-23466	GUH60A	2-82AVN	
79-23325	UH60A	2-82AVN		80-23467	UH60A		
79-23326	UH60A	2-82AVN		80-23468	UH60A	2-82AVN	
79-23327	UH60A	2-82AVN		80-23469	UH60A	24InfDiv	
79-23328	UH60A			80-23470	UH60A	24InfDiv	
79-23329	UH60A	9AVN		80-23471	UH60A	w.o.21.3.84	
79-23330	UH60A	2-82AVN		80-23472	UH60A		
79-23331	UH60A	AATC	31. Cairns AAF	80-23473	UH60A		
79-23332	UH60A	2-82AVN		80-23474	UH60A		
79-23333	UH60A	2-82AVN		80-23475	UH60A		
79-23334	UH60A	2-82AVN		80-23476	UH60A		
79-23335	UH60A			80-23477	UH60A		
79-23336	UH60A	w.o.2.3.84		80-23478	UH60A	A\4-228AVN	
79-23337	UH60A	A\4-228AVN		80-23479	UH60A		
79-23338	UH60A	247MedDet		80-23480	UH60A	w.o.26.2.91	
79-23339	UH60A	A\4-228AVN		80-23481	UH60A		
79-23340	UH60A	2-82AVN		80-23482	UH60A		
79-23341	UH60A	A\4-228AVN		80-23483	UH60A	24InfDiv	
79-23342	UH60A	A\4-228AVN		80-23484	UH60A		
79-23343	UH60A	A\4-228AVN		80-23485	UH60A	AATC	85G Cairns AAF
79-23344	UH60A	to USCS		80-23486	UH60A		
79-23345	GUH60A	24InfDiv		80-23487	UH60A		
79-23346	UH60A	2-82AVN		80-23488	UH60A		
79-23347	UH60A	247MedDet		80-23489	UH60A	203AvCo	
79-23348	UH60A	AFTC		80-23490	UH60A	21AvCo	
79-23349	UH60A	247MedDet		80-23491	UH60A		
79-23350	UH60A	to USCS		80-23492	UH60A		
79-23351	UH60A	247MedDet		80-23493	UH60A	21AvCo	
79-23352	GUH60A	9AVN		80-23494	UH60A	21AvCo	
79-23353	UH60A	24InfDiv		80-23495	UH60A	21AvCo	
79-23354	UH60A	9AVN		80-23496	UH60A	TF Phoenix	
79-23369	UH60A	24InfDiv		80-23497	UH60A		
79-23370	UH60A	24InfDiv		80-23498	UH60A	w.o.26.2.85	
80-23416	UH60A	24InfDiv		80-23499	UH60A		
80-23417	UH60A	24InfDiv		80-23500	UH60A	B\2-224AVN	VA NG
80-23418	GUH60A	24InfDiv		80-23501	UH60A		
80-23419	UH60A	24InfDiv		80-23502	UH60A		
80-23420	UH60A	24InfDiv		80-23503	UH60A		
80-23421	UH60A	24InfDiv		80-23504	UH60A		
80-23422	UH60A	w.o.16.2.91		80-23505	UH60A		
80-23423	UH60A	to USCS		80-23506	UH60A	9AVN	
80-23424	UH60A	24InfDiv		80-23507	UH60A	9AVN	
80-23425	UH60A	21AvCo		80-23508	UH60A	17Cav	
80-23426	UH60A	w.o.2.7.83		80-23509	UH60A	9AVN	
80-23427	UH60A	21AvCo		81-23547	UH60A	A\4-228AVN	
80-23428	UH60A			81-23548	UH60A	2ACR	
80-23429	UH60A	17Cav		81-23549	JUH60A		
80-23430	UH60A	2-82AVN		81-23550	UH60A		
80-23431	UH60A	23TF		81-23551	UH60A	236MedCo	
80-23432	UH60A	21AvCo		81-23552	UH60A	2-82AVN	
80-23433	UH60A	21AvCo		81-23553	UH60A	A\4-228AVN	
80-23434	UH60A	236MedCo		81-23554	UH60A	2-82AVN	
80-23435	UH60A	21AvCo		81-23555	UH60A	A\4-228AVN	
80-23436	UH60A	21AvCo		81-23556	UH60A	2-82AVN	
80-23437	UH60A	21AvCo		81-23557	UH60A	w.o.18.10.83	
80-23438	UH60A			81-23558	UH60A	9AVN	
80-23439	UH60A	21AvCo		81-23559	UH60A	A\4-228AVN	
80-23440	UH60A	21AvCo		81-23560	UH60A	2-82AVN	
80-23441	UH60A	w.o.28.1.87		81-23561	UH60A	A\4-228AVN	
80-23442	UH60A	21AvCo		81-23562	UH60A	9AVN	
80-23443	UH60A			81-23563	UH60A	A\4-228AVN	
80-23444	UH60A	4-11ACR		81-23564	UH60A	9AVN	
80-23445	UH60A	2-82AVN		81-23565	GUH60A		
80-23446	UH60A	B\2-224AVN	VA NG	81-23566	UH60A	2-82AVN	
80-23447	UH60A	17Cav		81-23567	UH60A	57MedDet	
80-23448	UH60A	17Cav		81-23568	UH60A	158AVN	
80-23449	UH60A	17Cav		81-23569	UH60A	2-82AVN	
80-23450	UH60A	9AVN		81-23570	UH60A	57MedDet	
80-23451	UH60A	w.o.17.1.90		81-23571	UH60A	21AvCo	
80-23452	UH60A	17Cav		81-23572	UH60A	5-6Cav	
80-23453	UH60A	17Cav		81-23573	UH60A	4-11ACR	
80-23454	UH60A	2-82AVN		81-23574	UH60A	101AbDiv	
80-23455	UH60A	2-82AVN		81-23575	UH60A	stored Ramstein AB, FRG	
80-23456	UH60A	2-82AVN		81-23576	UH60A	to USCS	
80-23457	UH60A	2-82AVN		81-23577	UH60A	to USCS	
80-23458	UH60A	57MedDet		81-23578	UH60A	48AvCo	
80-23459	UH60A	2-82AVN		81-23579	UH60A	w.o.19.1.91	
80-23460	UH60A	9AVN		81-23580	UH60A		
80-23461	UH60A	24InfDiv		81-23581	UH60A	57AvCo	
80-23462	UH60A	2-82AVN		81-23582	UH60A	H\3AVN	TF Warrior
80-23463	UH60A	17Cav		81-23583	UH60A	57AvCo	
80-23464	UH60A	2-82AVN		81-23584	UH60A	308AHB	

81-23585	UH60A	375AvDet	
81-23586	UH60A	308AHB	
81-23587	UH60A	H\3AVN	TF Warrior
81-23588	UH60A	48AvCo	
81-23589	UH60A	48AvCo	
81-23590	UH60A	TF Viper	
81-23591	UH60A		
81-23592	UH60A	236MedCo	
81-23593	UH60A		
81-23594	UH60A	TF Phoenix	
81-23595	UH60A	2ACR	
81-23596	UH60A	236MedCo	
81-23597	UH60A	159MedCo	
81-23598	UH60A	4-11ACR	
81-23599	UH60A	48AvCo	
81-23600	UH60A	48AvCo	
81-23601	UH60A	15MedDet	
81-23602	UH60A	4-11ACR	
81-23603	UH60A	308AHB	
81-23604	UH60A	4-11ACR	
81-23605	UH60A	57AvCo	
81-23606	UH60A	4-11ACR	
81-23607	UH60A	4-11ACR	
81-23608	UH60A	2ACR	
81-23609	UH60A	57AvCo	
81-23610	UH60A	308AHB	
81-23611	JUH60A	ERADCOM	
81-23612	JUH60A	USAICS	
81-23613	UH60A	7-1AVN	
81-23614	UH60A	308AHB	
81-23615	UH60A	57AvCo	
81-23616	UH60A	5-6Cav	
81-23617	UH60A	308AHB	
81-23618	UH60A	stored Ramstein AB, FRG	
81-23619	UH60A	227AVN	
81-23620	UH60A	227AVN	
81-23621	UH60A	48AvCo	
81-23622	UH60A	21AvCo	
81-23623	UH60A	w.o.3.10.86	
81-23624	UH60A	4AvCo	
81-23625	UH60A	4-11ACR	
81-23626	UH60A	5-6Cav	
81-23647	UH60A	w.o.14.8.87	
82-23660	UH60A		
82-23661	UH60A	TF Phoenix	
82-23662	UH60A	10AHB	
82-23663	UH60A	7-1AVN	
82-23664	UH60A	10AHB	
82-23665	UH60A	4-11ACR	
82-23666	UH60A	10AHB	
82-23667	UH60A	4-11ACR	
82-23668	UH60A	10AHB	
82-23669	UH60A	10AHB	
82-23670	UH60A	to USCS Homstead AFB	
82-23671	UH60A	to USAF	
82-23672	UH60A	203AvCo	
82-23673	UH60A		
82-23674	UH60A	E\140AVN	CA NG
82-23675	UH60A	421MedCo	
82-23676	UH60A	236MedCo	
82-23677	UH60A	107ACR	
82-23678	UH60A	9AVN	
82-23679	UH60A	9AVN	
82-23680	UH60A	to USAF	
82-23681	UH60A	8AVN	
82-23682	UH60A	H\3AVN TF Warrior	
82-23683	UH60A	stored Ramstein AB, FRG	
82-23684	UH60A	4AvCo	
82-23685	UH60A	4AvCo	
82-23686	UH60A	TF Viper	
82-23687	UH60A	4AvCo	
82-23688	UH60A	203AvCo	
82-23689	UH60A	to USAF	
82-23690	UH60A	2ACR	
82-23691	UH60A	2ACR	
82-23692	UH60A	4-11ACR	
82-23693	UH60A	8-158AVN	
82-23694	UH60A	2ACR	
82-23695	UH60A	203AvCo	
82-23696	UH60A	3ACR	
82-23697	UH60A	203AvCo	
82-23698	UH60A	203AvCo	
82-23699	UH60A		

82-23700	UH60A	236MedCo	
82-23701	UH60A	203AvCo	
82-23702	UH60A	7-1AVN	
82-23703	UH60A	203AvCo	
82-23704	UH60A	2ACR	
82-23705	UH60A	40AVN	
82-23706	UH60A	203AvCo	
82-23707	UH60A	w.o.25.1.85	
82-23708	UH60A	to USAF	
82-23709	UH60A	AATC	09. Cairns AAF
82-23710	UH60A		
82-23711	UH60A		
82-23712	UH60A		
82-23713	UH60A		
82-23714	UH60A	w.o.18.4.85	
82-23715	UH60A	AATC	15. Cairns AAF
82-23716	UH60A		
82-23717	UH60A		
82-23718	HH60A	412TW	ED AFFTC
82-23719	UH60A	w.o.11.3.86	
82-23720	UH60A		
82-23721	UH60A	203AvCo	
82-23722	UH60A	8-158AVN	
82-23723	UH60A	159MedCo	
82-23724	UH60A	421MedCo	
82-23725	UI100A	421MedCo	
82-23726	UH60A	63MedDet	
82-23727	UH60A	236MedCo	
82-23728	UH60A	to USAF	
82-23729	UH60A	7-1AVN	
82-23730	UH60A	236MedCo	
82-23731	UH60A	421MedCo	
82-23732	UH60A	B\1-132AVN	DC NG
82-23733	UH60A	7-1AVN	
82-23734	UH60A	TF Viper	
82-23735	UH60A	236MedCo	
82-23736	UH60A	159MedCo	
82-23737	UH60A	159MedCo	
82-23738	UH60A	159MedCo	
82-23739	UH60A	stored Ramstein AB, FRG	
82-23740	UH60A	63MedDet	
82-23741	UH60A	63MedDet	
82-23742	UH60A	B\1-132AVN DC NG	
82-23743	UH60A	8-158AVN	
82-23744	UH60A	TF Phoenix	
82-23745	UH60A	7-1AVN	
82-23746	UH60A	159MedCo	
82-23747	UH60A	to USCS	
82-23748	UH60A	7-1AVN	
82-23749	UH60A	159MedCo	
82-23750	UH60A	159MedCo	
82-23751	UH60A	159MedCo	
82-23752	UH60A	w.o.	
82-23753	UH60A	159MedCo	
82-23754	UH60A	7-1AVN	
82-23755	UH60A	7-1AVN	
82-23756	UH60A	421MedCo	
82-23757	UH60A	201AvCo	
82-23758	UH60A	w.o.9.12.85	
82-23759	UH60A		
82-23760	UH60A	to N3124B\G-RRTM	
82-23761	UH60A	TF Viper	
83-23837	UH60A		
83-23838	UH60A	w.o.22.12.86	
83-23839	UH60A		
83-23840	UH60A		
83-23841	UH60A		
83-23842	UH60A	to Philippine	
83-23843	UH60A		
83-23844	UH60A	USAK	
83-23845	UH60A		
83-23846	UH60A		
83-23847	UH60A		
83-23848	UH60A	1-285AVN	AZ NG
83-23849	UH60A	w.o.4.8.91	
83-23850	UH60A		
83-23851	UH60A	A\1-245AVN	OK NG
83-23852	UH60A		
83-23853	UH60A		
83-23854	UH60A	H\3AVN TF Warrior	
83-23855	UH60A	4-11ACR	
83-23856	UH60A	201AvCo	
83-23857	UH60A	w.o.22.12.86	

Serial	Type	Unit	Notes
83-23858	UH60A		
83-23859	UH60A		
83-23860	UH60A	USAK	
83-23861	UH60A		
83-23862	UH60A		
83-23863	UH60A		
83-23864	UH60A		
83-23865	UH60A	w.o.6.7.89	
83-23866	UH60A		
83-23867	UH60A	201AvCo	
83-23868	UH60A		
83-23869	UH60A		
83-23870	UH60A	USAK	
83-23871	UH60A		
83-23872	UH60A		
83-23873	UH60A		
83-23874	UH60A		
83-23875	UH60A	B\1-132AVN	DC NG
83-23876	UH60A		
83-23877	UH60A	1InfDiv	
83-23878	UH60A	USAK	
83-23879	UH60A	USAK	
83-23880	UH60A	6ACR	
83-23881	UH60A	USAK	
83-23882	UH60A	57AvCo	
83-23883	UH60A		
83-23884	UH60A	USAK	
83-23885	UH60A		
83-23886	UH60A	USAK	
83-23887	UH60A		
83-23888	UH60A	1InfDiv	
83-23889	UH60A	w.o.8.1.85	
83-23890	UH60A		
83-23891	UH60A		
83-23892	UH60A	3ACR	
83-23893	UH60A	USAK	
83-23894	UH60A		
83-23895	UH60A		
83-23896	UH60A		
83-23897	UH60A		
83-23898	UH60A		
83-23899	UH60A		
83-23900	UH60A		
83-23901	UH60A		
83-23902	UH60A		
83-23903	UH60A		
83-23904	UH60A	USAK	
83-23905	UH60A		
83-23906	UH60A		
83-23907	UH60A		
83-23908	MH60L	160SOAR	
83-23909	UH60A		
83-23910	UH60A		
83-23911	UH60A		
83-23912	UH60A		
83-23913	UH60A		
83-23914	UH60A		
83-23915	UH60A		
83-23916	UH60A		
83-23917	UH60A	101AbDiv	
83-23918	UH60A	USAK	
83-23919	UH60A		
83-23920	UH60A	101AbDiv	
83-23921	UH60A	101AbDiv	
83-23922	UH60A		
83-23923	UH60A		
83-23924	UH60A		
83-23925	UH60A		
83-23926	UH60A	78AVN	
83-23927	UH60A	USCS	
83-23928	UH60A		
83-23929	UH60A	377MedDet	
83-23930	UH60A		
83-23931	UH60A	377MedDet	
83-23932	UH60A		
83-23933	UH60A		
84-23934	UH60A	AATC	34. Cairns AAF
84-23935	UH60A		
84-23936	UH60A		
84-23937	UH60A		
84-23938	UH60A		
84-23939	UH60A	377MedDet	
84-23940	UH60A		
84-23941	UH60A		
84-23942	UH60A		
84-23943	UH60A	201AvCo	
84-23944	UH60A		
84-23945	UH60A	377MedDet	
84-23946	UH60A		
84-23947	UH60A		
84-23948	UH60A		
84-23949	UH60A		
84-23950	UH60A	AATC	50M Cairns AAF
84-23951	UH60A	AATC	51. Cairns AAF
84-23952	UH60A		
84-23953	UH60A		
84-23954	UH60A		
84-23955	UH60A		
84-23956	UH60A		
84-23957	UH60A		
84-23958	UH60A		
84-23959	UH60A	377MedDet	
84-23960	UH60A		
84-23961	UH60A		
84-23962	UH60A		
84-23963	UH60A		
84-23964	UH60A		
84-23965	UH60A	USAK	
84-23966	UH60A		
84-23967	UH60A		
84-23968	UH60A		
84-23969	UH60A		
84-23970	UH60A		
84-23971	UH60A	1-285AVN	AZ NG
84-23972	UH60A		
84-23973	UH60A		
84-23974	UH60A		
84-23975	UH60A		
84-23976	UH60A		
84-23977	UH60A		
84-23978	UH60A		
84-23979	UH60A		
84-23980	UH60A		
84-23981	UH60A		
84-23982	UH60A		
84-23983	UH60A		
84-23984	UH60A		
84-23985	UH60A		
84-23986	UH60A		
84-23987	UH60A	w.o.30.9.89	
84-23988	UH60A		
84-23989	UH60A		
84-23990	UH60A		
84-23991	UH60A		
84-23992	UH60A	17AHC	
84-23993	UH60A	17AHC	
84-23994	UH60A	17AHC	
84-23995	UH60A	17AHC	
84-23996	UH60A		
84-23997	MH60A	w.o.13.5.90	
84-23998	UH60A	184AvCo	
84-23999	UH60A		
84-24000	UH60A		
84-24001	UH60A	17AHC	
84-24002	UH60A	17AHC	
84-24003	UH60A	17AHC	
84-24004	UH60A	17AHC	
84-24005	UH60A	17AHC	
84-24006	UH60A	17AHC	
84-24007	UH60A	17AHC	
84-24008	UH60A	17AHC	
84-24009	UH60A	17AHC	
84-24010	UH60A	17AHC	
84-24011	UH60A	17AHC	
84-24012	UH60A	17AHC	
84-24013	UH60A	17AHC	
84-24014	UH60A		
84-24015	UH60A	A\1-245AVN	OK NG
84-24016	UH60A		
84-24017	YEH60A		
84-24018	EH60A		
84-24019	EH60A	23TF	
84-24020	EH60A		
84-24021	EH60A		
84-24022	EH60A		
84-24023	EH60C	TF Viper	

Serial	Type	Unit		Location
84-24024	EH60A			
84-24025	EH60A			
84-24026	EH60A			
84-24027	EH60A			
84-24028	EH60A			
85-24387	UH60A			
85-24388	UH60A			
85-24389	UH60A	A\1-245AVN		OK NG
85-24390	UH60A			
85-24391	UH60A			
85-24392	UH60A			
85-24393	UH60A			
85-24394	UH60A			
85-24395	UH60A			
85-24396	UH60A	AATC	96A	Cairns AAF
85-24397	UH60A			
85-24398	UH60A			
85-24399	UH60A			
85-24400	UH60A	2-82AVN		
85-24401	UH60A			
85-24402	UH60A	7InfDiv		
85-24403	UH60A	A\1-245AVN		OK NG
85-24404	UH60A	AATC	04F	Cairns AAF
85-24405	UH60A	7InfDiv		
85-24406	UH60A			
85-24407	UH60A			
85-24408	UH60A			
85-24409	UH60A	1AHB		
85-24410	UH60A			
85-24411	UH60A	1-285AVN		AZ NG
85-24412	UH60A	57MedDet		
85-24413	UH60A	A\1-245AVN		OK NG
85-24414	UH60A	A\1-245AVN		OK NG
85-24415	UH60A			
85-24416	UH60A	A\1-245AVN		OK NG
85-24417	UH60A	A\1-245AVN		OK NG
85-24418	UH60A	A\1-245AVN		OK NG
85-24419	UH60A	A\1-245AVN		OK NG
85-24420	UH60A			
85-24421	UH60A	A\4-228AVN		
85-24422	UH60A			
85-24423	UH60A			
85-24424	UH60A	7InfDiv		
85-24425	UH60A			
85-24426	UH60A	7InfDiv		
85-24427	UH60A			
85-24428	UH60A			
85-24429	UH60A	7InfDiv		
85-24430	UH60A	7InfDiv		
85-24431	UH60A	7InfDiv		
85-24432	UH60A	7InfDiv		
85-24433	UH60A			
85-24434	UH60A	7InfDiv		
85-24435	UH60A			
85-24436	UH60A	7InfDiv		
85-24437	UH60A	7InfDiv		
85-24438	UH60A	441MedDet		
85-24439	UH60A	7InfDiv		
85-24440	UH60A	AATC	40N	Cairns AAF
85-24441	UH60A	AATC	41F	Cairns AAF
85-24442	UH60A	7InfDiv		
85-24443	UH60A			
85-24444	UH60A			
85-24445	UH60A			
85-24446	UH60A	7InfDiv		
85-24447	UH60A			
85-24448	UH60A	78AVN		
85-24449	UH60A	78AVN		
85-24450	UH60A	78AVN		
85-24451	UH60A	78AVN		
85-24452	UH60A			
85-24453	UH60A			
85-24454	UH60A			
85-24455	UH60A			
85-24456	UH60A			
85-24457	UH60A			
85-24458	UH60A			
85-24459	UH60A			
85-24460	UH60A			
85-24461	UH60A			
85-24462	UH60A	w.o.8.3.88		
85-24463	UH60A			
85-24464	UH60A			

Serial	Type	Unit		Location
85-24465	EH60A			
85-24466	EH60A			
85-24467	EH60C	TF Phoenix		
85-24468	EH60C	TF Viper		
85-24469	EH60A	7InfDiv		
85-24470	EH60A			
85-24471	EH60A			
85-24472	EH60A			
85-24473	EH60A			
85-24474	EH60C	4-11ACR		
85-24475	EH60C	4-11ACR		
85-24476	EH60A			
85-24477	EH60A			
85-24478	EH60C	TF Phoenix		
85-24479	EH60A			
85-24480	EH60A			
85-24481	EH60A			
85-24482	EH60A			
86-24483	UH60A			
86-24484	UH60A			
86-24485	UH60A			
86-24486	UH60A			
86-24487	UH60A	7InfDiv		
86-24488	UH60A	7InfDiv		
86-24489	UH60A	7InfDiv		
86-24490	UH60A	7InfDiv		
86-24491	UH60A	6ACR		
86-24492	UH60A			
86-24493	UH60A	6ACR		
86-24494	UH60A			
86-24495	UH60A			
86-24496	UH60A			
86-24497	UH60A			
86-24498	UH60A	TF Viper		
86-24499	UH60A			
86-24500	UH60A	w.o.27.7.90		
86-24501	UH60A	A\1-245AVN		OK NG
86-24502	UH60A	7InfDiv		
86-24503	UH60A	w.o.15.9.87		
86-24504	UH60A	6ACR		
86-24505	UH60A	7InfDiv		
86-24506	UH60A	6ACR		
86-24507	JUH60A	USAATDA		
86-24508	UH60A			
86-24509	UH60A	A\1-245AVN		OK NG
86-24510	UH60A			
86-24511	UH60A			
86-24512	UH60A			
86-24513	UH60A	2\158AVN		
86-24514	UH60A			
86-24515	UH60A			
86-24516	UH60A	AATC	16F	Cairns AAF
86-24517	UH60A	AATC	17.	Cairns AAF
86-24518	UH60A			
86-24519	UH60A			
86-24520	UH60A			
86-24521	UH60A	28AVN		
86-24522	UH60A	B\2-224AVN		VA NG
86-24523	UH60A			
86-24524	UH60A	A\1-245AVN		OK NG
86-24525	UH60A	2-82AVN		
86-24526	UH60A			
86-24527	UH60A	A\1-245AVN		OK NG
86-24528	UH60A	A\1-245AVN		OK NG
86-24529	UH60A			
86-24530	UH60A	2-6Cav		
86-24531	UH60A	stored	Ramstein AB, FRG	
86-24532	UH60A	2-6Cav		
86-24533	UH60A			
86-24534	UH60A			
86-24535	UH60A			
86-24536	UH60A			
86-24537	UH60A	207AvCo		
86-24538	UH60A	207AvCo		
86-24539	UH60A			
86-24540	UH60A			
86-24541	UH60A			
86-24542	UH60A	w.o.29.9.90		
86-24543	JUH60A			
86-24544	UH60A	207AvCo		
86-24545	UH60A			
86-24546	UH60A	w.o.12.3.91		
86-24547	UH60A			

Serial	Type	Unit/Notes		
86-24548	UH60A	to USCS		
86-24549	UH60A	w.o.22.7.89		
86-24550	UH60A	236MedCo		
86-24551	UH60A	236MedCo		
86-24552	UH60A	4-11ACR		
86-24553	UH60A	48AvCo		
86-24554	UH60A	TF Viper		
86-24555	UH60A			
86-24556	UH60A			
86-24557	UH60A	to USCS		
86-24558	UH60A	to USCS		
86-24559	UH60A			
86-24560	EH60C	3ACR		
86-24561	EH60A			
86-24562	EH60A			
86-24563	EH60A			
86-24564	EH60A	1InfDiv		
86-24565	EH60C	23TF		
86-24566	EH60C	TF Viper		
86-24567	EH60A	USAK		
86-24568	EH60A	USAICS		
86-24569	EH60A			
86-24570	EH60C	23TF		
86-24571	EH60A			
86-24572	EH60A			
86-24573	EH60A	7InfDiv		
86-24574	EH60A	USAK		
86-24575	EH60A			
86-24576	EH60A			
86-24577	EH60C	4-2ACR		
86-24578	EH60A			
87-24579	UH60A			
87-24580	UH60A	3ACR		
87-24581	UH60A	236MedCo		
87-24582	UH60A	2-82AVN		
87-24583	UH60A	357AvDet		
87-24584	UH60A	357AvDet		
87-24585	UH60A	2-82AVN		
87-24586	UH60A			
87-24587	UH60A	2-82AVN		
87-24588	UH60A			
87-24589	UH60A			
87-24590	UH60A			
87-24591	UH60A	USAAEFA		
87-24592	UH60A			
87-24593	UH60A			
87-24594	UH60A	28AVN		
87-24595	UH60A			
87-24596	UH60A			
87-24597	UH60A			
87-24598	UH60A			
87-24599	UH60A	2-82AVN		
87-24600	UH60A	7InfDiv		
87-24601	UH60A			
87-24602	UH60A	7InfDiv		
87-24603	UH60A	USAK		
87-24604	UH60A			
87-24605	UH60A	w.o.8.3.88		
87-24606	UH60A			
87-24607	UH60A			
87-24608	UH60A			
87-24609	UH60A	AATC	09.	Cairns AAF
87-24610	UH60A	AATC	10H	Cairns AAF
87-24611	UH60A	AATC	11.	Cairns AAF
87-24612	UH60A			
87-24613	UH60A			
87-24614	UH60A			
87-24615	UH60A			
87-24616	UH60A			
87-24617	UH60A			
87-24618	UH60A			
87-24619	UH60A			
87-24620	UH60A			
87-24621	UH60A	3-1AVN		
87-24622	UH60A			
87-24623	UH60A	2-82AVN		
87-24624	UH60A	2-82AVN		
87-24625	UH60A	2-82AVN		
87-24626	UH60A	2-82AVN		
87-24627	UH60A	2-82AVN		
87-24628	UH60A	2-82AVN		
87-24629	UH60A	2-82AVN		
87-24630	UH60A	2-82AVN		
87-24631	UH60A	2-82AVN		
87-24632	UH60A			
87-24633	UH60A	2-82AVN		
87-24634	UH60A			
87-24635	UH60A	4-228AVN Medevac Det.		
87-24636	UH60A	4-228AVN Medevac Det.		
87-24637	UH60A			
87-24638	UH60A	4-228AVN Medevac Det.		
87-24639	UH60A	1-285AVN	AZ NG	
87-24640	UH60A	1-285AVN	AZ NG	
87-24641	UH60A	to USCS		
87-24642	UH60A	2ACR		
87-24643	UH60A	2ACR		
87-24644	UH60A	2ACR		
87-24645	UH60A	2ACR		
87-24646	UH60A	7-1AVN		
87-24647	UH60A	2ACR		
87-24648	UH60A			
87-24649	UH60A			
87-24650	UH60A	AATC	50.	Cairns AAF
87-24651	UH60A			
87-24652	UH60A	AATC	52F	Cairns AAF
87-24653	UH60A	AATC	53K	Cairns AAF
87-24654	UH60A			
87-24655	UH60A	4\6Cav		
87-24656	UH60A	2ACR		
87-24657	EH60C	4-2ACR		
87-24658	EH60A			
87-24659	EH60A	3-24AVN		
87-24660	EH60C	TF Viper		
87-24661	EH60A			
87-24662	EH60A			
87-24663	EH60C	TF Viper		
87-24664	EH60C	TF		
87-24665	EH60C	TF Viper		
87-24666	EH60C	4-11ACR		
87-24667	EH60A	4-11ACR		
87-24668	EH60A	7InfDiv		
87-24669	EH60A	7InfDiv		
87-24670	EH60A	1InfDiv		
87-24671	EH60A	2-82AVN		
87-24672	EH60A			
87-24673	EH60A			
87-24674	EH60A			
85-24745	UH60A	441MedCo		
85-24746	UH60A			
85-24747	UH60A			
85-24748	UH60A	7InfDiv		
85-24749	UH60A			
85-24750	UH60A			
85-25511	UH60A			
85-25512	UH60A			
87-26000	UH60A			
87-26001	UH60A	C\6-159AVN		
87-26002	UH60A	159MedCo		
87-26003	UH60A			
87-26004	UH60A			
87-26005	JUH60A			
88-26015	UH60A			
88-26016	UH60A	AATC	16G	Cairns AAF
88-26017	UH60A			
88-26018	UH60A			
88-26019	UH60A	TF Viper		
88-26020	UH60A	4-229AVN		
88-26021	UH60A			
88-26022	UH60A			
88-26023	UH60A	158AVN		
88-26024	UH60A	H\3AVN	TF Warrior	
88-26025	UH60A	H\3AVN	TF Warrior	
88-26026	UH60A	H\3AVN	TF Warrior	
88-26027	UH60A	H\3AVN	TF Warrior	
88-26028	UH60A	H\3AVN	TF Warrior	
88-26029	UH60A	w.o.12.5.88		
88-26030	UH60A	ALS		
88-26031	UH60A	4-11ACR		
88-26032	UH60A	ATS		
88-26033	UH60A	AATC	33N	Cairns AAF
88-26034	UH60A	4-229AVN		
88-26035	UH60A			
88-26036	UH60A			
88-26037	UH60A	stored Ramstein AB, FRG		
88-26038	UH60A	4-11ACR		
88-26039	UH60A	4-11ACR		

Serial	Type	Unit	
88-26040	UH60A	4-11ACR	
88-26041	UH60A	4-11ACR	
88-26042	UH60A	stored Ramstein AB, FRG	
88-26043	UH60A	USAK	
88-26044	UH60A	USAK	
88-26045	UH60A	4AVN	
88-26046	UH60A	B\2-224AVN	VA NG
88-26047	UH60A		SC NG
88-26048	UH60A		
88-26049	UH60A		SC NG
88-26050	UH60A	stored Ramstein AB, FRG	
88-26051	UH60A	H\3AVN TF Warrior	
88-26052	UH60A	TF Viper	
88-26053	UH60A	stored Ramstein AB, FRG	
88-26054	UH60A	236MedCo	
88-26055	UH60A	stored Ramstein AB, FRG	
88-26056	UH60A		
88-26057	UH60A	6ACR	
88-26058	UH60A	4AVN	
88-26059	UH60A	H\3AVN TF Warrior	
88-26060	UH60A		
88-26061	UH60A	7InfDiv	
88-26062	UH60A	7InfDiv	
88-26063	UH60A	TF Viper	
88-26064	UH60A	A\4-228AVN	
88-26065	UH60A	USAK	
88-26066	UH60A		
88-26067	UH60A		
88-26068	UH60A	TF Viper	
88-26069	JUH60A		
88-26070	UH60A		
88-26071	UH60A	TF Viper	
88-26072	UH60A	TF Viper	
88-26073	UH60A		
88-26074	UH60A	STARC	DC NG
88-26075	UH60A	3-227AVN	
88-26076	UH60A		
88-26077	UH60A	stored Ramstein AB, FRG	
88-26078	UH60A		
88-26079	UH60A	111AVN	
88-26080	UH60A	1AVN	
88-26081	UH60A		
88-26082	UH60A		
88-26083	UH60A		
88-26084	UH60A		
88-26085	UH60A		
88-26086	UH60A	3-227AVN	
89-26123	UH60A		
89-26124	UH60A		
89-26125	UH60A		VA NG
89-26126	UH60A		
89-26127	UH60A		
89-26128	UH60A	24AVN	
89-26129	UH60A		
89-26130	UH60A		
89-26131	UH60A		
89-26132	UH60A		
89-26133	UH60A		
89-26134	UH60A		
89-26135	UH60A	5\229AVN	
89-26136	UH60A		
89-26137	UH60A		
89-26138	UH60A	5\229AVN	
89-26139	UH60A	5\229AVN	
89-26140	UH60A		
89-26141	UH60A		
89-26142	UH60A	stored Ramstein AB, FRG	
89-26143	UH60A		
89-26144	UH60A	w.o.9.1.90	
89-26145	UH60A	3-227AVN	
89-26146	UH60A	stored Ramstein AB, FRG	
89-26147	UH60A		
89-26148	UH60A		
89-26149	JUH60L		
89-26150	UH60A	3ACR	
89-26151	UH60A	1AVN	
89-26152	UH60A		
89-26153	UH60A	1AVN	
89-26154	UH60L		
89-26155	UH60A		
89-26156	UH60A	3ACR	
89-26157	UH60A	3ACR	
89-26158	UH60A		
89-26159	UH60A		
89-26160	UH60A		
89-26161	UH60A		
89-26162	UH60A	5\229AVN	
89-26163	UH60A	1InfDiv	
89-26164	UH60A	1AVN	
89-26165	UH60A	1AVN	
89-26166	UH60A		
89-26167	UH60A		
89-26168	UH60A	1InfDiv	
89-26169	UH60A		
89-26170	UH60A		
89-26171	UH60A		
89-26172	UH60A		
89-26173	UH60A		
89-26179	UH60L		
89-26180	UH60L		
89-26181	UH60L		
89-26182	UH60L		
89-26183	UH60L	E\1-149AVN	TX NG
89-26184	UH60L		
89-26185	UH60L		
89-26186	UH60L		
89-26187	UH60L		
89-26188	UH60L		
89-26189	UH60L		
89-26190	UH60L	E\1-149AVN	TX NG
89-26191	UH60L	E\1-149AVN	TX NG
89-26192	UH60L		
89-26193	UH60L		
89-26194	MH60K		
89-26212	UH60L		
89-26213	UH60L		
89-26214	UH60L		
89-26215	UH60L		
89-26216	UH60L	E\1-149AVN	TX NG
89-26217	UH60L		
90-26218	UH60L		
90-26219	UH60L		
90-26220	UH60L		
90-26221	UH60L		
89-26242	UH60L		
90-26243	UH60L	3ACR	
90-26244	UH60L	3ACR	
90-26245	UH60L	3ACR	
90-26246	UH60L	3ACR	
90-26247	UH60L	3ACR	
90-26248	UH60L		
90-26249	UH60L		
90-26250	UH60L		
90-26251	MH60L	w.o.21.2.91	
90-26252	UH60L		
90-26253	UH60L		
90-26254	UH60L		
90-26255	UH60L	B\1-132AVN	DC NG
90-26256	UH60L	3ACR	
90-26257	UH60L		
90-26258	UH60L		
90-26259	UH60L		
90-26260	UH60L		
90-26261	UH60L	3ACR	
90-26262	UH60L		
90-26263	UH60L	3ACR	
90-26264	UH60L	3ACR	
90-26265	UH60L	3ACR	
90-26266	UH60L		
90-26267	UH60L		
90-26268	UH60L		
90-26269	UH60L		
90-26270	UH60L		
90-26271	UH60L	3ACR	
90-26272	UH60L		
90-26273	UH60L		
90-26274	UH60L		
90-26275	UH60L		
90-26276	UH60L		
90-26277	UH60L	1-130AVN	NC NG
90-26278	UH60L	28AVN	
90-26279	UH60L	28AVN	
90-26280	UH60L	3ACR	
90-26281	UH60L	3ACR	
90-26282	UH60L	3ACR	
90-26283	UH60L		
90-26284	UH60L		

90-26285	UH60L		
90-26286	UH60L		
90-26287	UH60L		
90-26288	UH60L		
90-26289	UH60L		
90-26290	UH60L	2-82AVN	
90-26291	UH60L		
90-26292	UH60L	2-82AVN	
90-26293	UH60L		
90-26294	UH60L	2-82AVN	
90-26295	UH60L	2-82AVN	
90-26296	UH60L	2-82AVN	
90-26297	UH60L	2-82AVN	
90-26298	UH60L		
90-26299	UH60L		
90-26300	UH60L	2-82AVN	
90-26301	UH60L	2-82AVN	
90-26302	UH60L		
90-26303	UH60L		
90-26304	UH60L	2-82AVN	
90-26305	UH60L		
90-26306	UH60L		
90-26307	UH60L		
90-26308	UH60L		
90-26313	UH60L		
90-26314	UH60L		
90-26315	UH60L		
90-26316	UH60L		
90-26317	UH60L		
90-26318	UH60L		
91-26319	UH60L	1-149AVN	TX NG
91-26320	UH60L		
91-26321	UH60L		
91-26322	UH60L		
91-26323	UH60L		
91-26324	UH60L		
91-26325	UH60L		
91-26326	UH60L	1-151AVN	SC NG
91-26327	UH60L		
91-26328	UH60L	1-151AVN	SC NG
91-26329	UH60L		
91-26330	UH60L	1-151AVN	SC NG
91-26331	UH60L		
91-26332	UH60L		
91-26333	UH60L		
91-26334	UH60L		
91-26335	UH60L		
91-26336	UH60L	E\1-149AVN	TX NG
91-26337	UH60L		
91-26338	UH60L	B\1-207AVN	AK NG
91-26339	UH60L		
91-26340	UH60L	B\1-207AVN	AK NG
91-26341	UH60L		
91-26342	UH60L		
91-26343	UH60L		
91-26344	UH60L	B\1-207AVN	AK NG
91-26345	UH60L	B\1-207AVN	AK NG
91-26346	UH60L	B\1-207AVN	AK NG
91-26347	UH60L	B\1-207AVN	AK NG
91-26348	UH60L	B\1-207AVN	AK NG
91-26349	UH60L	B\1-207AVN	AK NG
91-26350	UH60L		
91-26351	UH60L	B\1-132AVN	DC NG
91-26360	UH60L		
91-26361	UH60L		
91-26362	UH60L		
91-26363	UH60L		
91-26364	UH60L		
91-26365	UH60L		
91-26366	UH60L		
91-26367	UH60L		
91-26368	MH60K		
91-26369	MH60K		
91-26370	MH60K		
91-26371	MH60K		
91-26372	MH60K		
91-26373	MH60K		
91-26374	MH60K		
91-26375	MH60K		
91-26376	MH60K		
91-26377	MH60K		
91-26378	MH60K		
91-26379	MH60K		
91-26380	MH60K		
91-26381	MH60K		
91-26382	MH60K		
91-26383	MH60K		
91-26384	MH60K		
91-26385	MH60K		
91-26386	MH60K		
91-26387	MH60K		
91-26388	MH60K		
91-26389	MH60K		
91-26390	UH60L	1-149AVN	TX NG
91-26391	UH60L	1-149AVN	TX NG
91-26392	UH60L	1-149AVN	TX NG
92-26408	UH60L		
92-26409	UH60L		
92-26410	UH60L		
92-26411	UH60L		
92-26412	UH60L		
92-26413	UH60L		
92-26414	UH60L		
92-26415	UH60L		
92-26416	UH60L		
92-26417	UH60L		
92-26418	UH60L		
92-26419	UH60L		
92-26420	UH60L		
92-26421	UH60L		
92-26422	UH60L		
92-26423	UH60L		
92-26424	UH60L		
92-26425	UH60L		
92-26426	UH60L		
92-26427	UH60L		
92-26428	UH60L		
92-26429	UH60L		
92-26430	UH60L		
92-26431	UH60L		
92-26432	UH60L		
92-26433	UH60L		
92-26434	UH60L		
92-26435	UH60L		
92-26436	UH60L		
92-26437	UH60L		
92-26438	UH60L		
92-26439	UH60L		
92-26440	UH60L		
92-26441	UH60L		
92-26442	UH60L		
92-26443	UH60L		
92-26444	UH60L		
92-26445	UH60L		
92-26446	UH60L		
92-26447	UH60L		
92-26448	UH60L		
92-26449	UH60L		
92-26450	UH60L		
92-26451	UH60L		
92-26452	UH60L		
92-26453	UH60L		
92-26454	UH60L		
92-26455	UH60L		
92-26456	UH60L		
92-26457	UH60L		
92-26458	UH60L		
92-26459	UH60L		

H64 APACHE

c\n   AV.01 to AV.06, PV.01 to PV.807.

73-22247	YAH64	preserved	Ft. Rucker	84-24204	AH64	3-227AVN	
73-22248	YAH64	preserved	Ft. Rucker	84-24205	AH64	6Cav	
73-22249	YAH64	preserved		84-24206	AH64	6Cav	
79-23257	YAH64	w.o.28.11.80		84-24207	AH64	6Cav	
79-23258	YAH64	MDH		84-24208	AH64	6Cav	
79-23259	YAH64A			84-24209	AH64	w.o.19.7.91	
82-23355	AH64	MDH		84-24210	AH64	1-149AVN	TX NG
82-23356	AH64	MDH		84-24211	AH64	AATC	11. Hanchey AHP
82-23357	GAH64	MDH		84-24212	AH64	6Cav	
82-23358	GAH64A			84-24213	AH64	6Cav	
82-23359	AH64	AATC	59E Hanchey AHP	84-24214	AH64	w.o.5.11.88	
82-23300	AGH64A			84-24215	AH64	1-149AVN	TX NG
82-23361	GAH64			84-24216	JAH64	USADTA	
82-23362	AH64	AATC	62B Hanchey AHP	84-24217	AH64	5-229AVN	
82-23363	GAH64A			84-24218	AH64	2-6Cav	
82-23364	AH64	AATC	64B Hanchey AHP	84-24219	AH64	5-229AVN	
82-23365	AH64A			84-24220	AH64	6Cav	
83-23787	AH64	AATC	87H Hanchey AHP	84-24221	AH64	AATC	21K Hanchey AHP
83-23788	AH64A			84-24222	GAH64	6Cav	
83-23789	AH64A			84-24223	GAH64	6Cav	
83-23790	AH64	AATC	90. Hanchey AHP	84-24224	AH64	6Cav	
83-23791	GAH64			84-24225	AH64	AATC	25J Hanchey AHP
83-23792	GAH64A			84-24226	AH64	6Cav	
83-23793	AH64A			84-24227	GAH64	6Cav	
83-23794	AH64	AATC	94. Hanchey AHP	84-24228	AH64	6Cav	
83-23795	AH64	AATC	95D Hanchey AHP	84-24229	AH64	AATC	29M Hanchey AHP
83-23796	AH64	AATC	96F Hanchey AHP	84-24230	AH64	AATC	30H Hanchey AHP
83-23797	AH64	AATC	97A Hanchey AHP	84-24231	AH64	6Cav	
83-23798	GAH64A			84-24232	AH64	6Cav	
83-23799	GAH64A			84-24233	AH64	1-149AVN	TX NG
83-23800	AH64	AATC	00. Hanchey AHP	84-24234	AH64	6Cav	
83-23801	AH64	AATC	01F Hanchey AHP	84-24235	AH64	6Cav	
83-23802	AH64	AATC	02H Hanchey AHP	84-24236	AH64	1\6Cav	
83-23803	AH64	AATC	03C Hanchey AHP	84-24237	AH64	6Cav	
83-23804	AH64	AATC	04D Hanchey AHP	84-24238	JAH64	6Cav	
83-23805	AH64	AATC	05. Hanchey AHP	84-24239	AH64	6Cav	
83-23806	AH64A			84-24240	AH64	6Cav	
83-23807	GAH64			84-24241	AH64	6Cav	
83-23808	AH64A			84-24242	AH64	6Cav	
83-23809	AH64	AATC	09M Hanchey AHP	84-24243	AH64	6Cav	
83-23810	AH64	AATC	10. Hanchey AHP	84-24244	AH64	6Cav	
83-23811	AH64A			84-24245	AH64	6Cav	
83-23812	GAH64A			84-24246	AH64	6Cav	
83-23813	GAH64A			84-24247	AH64	6Cav	
83-23814	AH64	AATC	14F Hanchey AHP	84-24248	AH64	6Cav	
83-23815	AH64	AATC	15. Hanchey AHP	84-24249	AH64	6Cav	
83-23816	AH64	AATC	16. Hanchey AHP	84-24250	AH64	6Cav	
83-23817	AH64A			84-24251	AH64	6Cav	
83-23818	AH64	AATC	18D Hanchey AHP	84-24252	AH64	3-227AVN	
83-23819	AH64A			84-24253	AH64	6Cav	
83-23820	AH64	AATC	20. Hanchey AHP	84-24254	AH64	5-229AVN	
83-23821	AH64	AATC	21. Hanchey AHP	84-24255	AH64	6Cav	
83-23822	AH64	AATC	22E Hanchey AHP	84-24256	JAH64	6Cav	
83-23823	AH64	w.o.20.2.86		84-24257	AH64	4-229AVN	
83-23824	AH64A			84-24258	AH64	2-6Cav	
83-23825	AH64	w.o.14.7.86		84-24259	AH64	w.o.19.10.89	
83-23826	AH64	AATC	26G Hanchey AHP	84-24260	AH64	2-6Cav	
83-23827	AH64	MDH		84-24261	AH64	151AVN	
83-23828	AH64A			84-24262	AH64	2-6Cav	
83-23829	AH64	1-149AVN	TX NG	84-24263	AH64	2-6Cav	
83-23830	AH64	1-149AVN	TX NG	84-24264	AH64	AATC	64. Hanchey AHP
83-23831	AH64A			84-24265	AH64	5-229AVN	
83-23832	AH64A			84-24266	AH64	6Cav	
83-23833	AH64A			84-24267	AH64	6Cav	
83-23834	AH64A			84-24268	AH64	AATC	68. Hanchey AHP
84-24200	AH64	6Cav		84-24269	AH64	6Cav	
84-24201	AH64	4\6Cav		84-24270	AH64	6Cav	
84-24202	AH64	6Cav		84-24271	AH64	1-149AVN	TX NG
84-24203	AH64	6Cav		84-24272	AH64	1-149AVN	TX NG

Serial	Type	Unit		Note
84-24273	AH64	AATC	73T	Hanchey AHP
84-24274	AH64	6Cav		
84-24275	GAH64	6Cav		
84-24276	AH64	AATC	76H	Hanchey AHP
84-24277	AH64	6Cav		
84-24278	AH64	1-149AVN		TX NG
84-24279	AH64	AATC	79.	Hanchey AHP
84-24280	AH64	6Cav		
84-24281	AH64	6Cav		
84-24282	AH64	1-149AVN		TX NG
84-24283	AH64	6Cav		
84-24284	AH64	6Cav		
84-24285	AH64	6Cav		
84-24286	AH64	6Cav		
84-24287	AH64	6Cav		
84-24288	AH64	6Cav		
84-24289	AH64	6Cav		
84-24290	AH64	2-6Cav		
84-24291	AH64	2-6Cav		
84-24292	AH64	2-6Cav		
84-24293	AH64	2-6Cav		
84-24294	AH64	5-6Cav		
84-24295	AH64	2-6Cav		
84-24296	AH64	2-6Cav		
84-24297	AH64	2-6Cav		
84-24298	AH64	2-6Cav		
84-24299	AH64	2-6Cav		
84-24300	AH64	6Cav		
84-24301	AH64	6Cav		
84-24302	AH64	2-6Cav		
84-24303	AH64	2-6Cav		
84-24304	AH64	2-6Cav		
84-24305	AH64	6Cav		
84-24306	AH64	6Cav		
84-24307	AH64	6Cav		
84-24308	AH64	6Cav		
84-24309	AH64	6Cav		
84-24310	AH64	1-149AVN		TX NG
84-24311	AH64	6Cav		
85-25351	AH64	5-229AVN		
85-25352	AH64			
85-25353	AH64A			
85-25354	AH64A			
85-25355	AH64A			
85-25356	AH64A			
85-25357	AH64	6Cav		
85-25358	AH64	1-149AVN		TX NG
85-25359	AH64A			
85-25360	AH64A			
85-25361	AH64A			
85-25362	AH64	1Cav		
85-25363	AH64A			
85-25364	AH64A			
85-25365	AH64A			
85-25366	AH64	1Cav		
85-25367	AH64A			
85-25368	AH64A			
85-25369	AH64A			
85-25370	AH64	1-149AVN		TX NG
85-25371	AH64A			
85-25372	AH64A			
85-25373	AH64A			
85-25374	AH64	ATS		
85-25375	AH64A			
85-25376	AH64A			
85-25377	AH64			
85-25378	AH64	1Cav		
85-25379	AH64A			
85-25380	AH64A			
85-25381	AH64	28AVN		
85-25382	AH64	w.o.18.11.91		
85-25383	AH64A			
85-25384	AH64	1-82AVN		
85-25385	AH64	28AVN		
85-25386	AH64	28AVN		
85-25387	AH64A			
85-25388	AH64	1-149AVN		TX NG
85-25389	AH64	1-82AVN		
85-25390	AH64A			
85-25391	AH64A			
85-25392	AH64A			
85-25393	AH64A			
85-25394	AH64	1-82AVN		
85-25395	AH64A			
85-25396	AH64	1-82AVN		
85-25397	AH64	5-6Cav		
85-25398	AH64	2-6Cav		
85-25399	AH64	1-82AVN		
85-25400	AH64	28AVN		
85-25401	AH64	w.o.17.7.91		
85-25402	AH64	28AVN		
85-25403	AH64	28AVN		
85-25404	AH64	28AVN		
85-25405	AH64	1-149AVN		TX NG
85-25406	AH64	28AVN		
85-25407	AH64	1-149AVN		TX NG
85-25408	AH64C	28AVN		
85-25409	AH64	28AVN		
85-25410	AH64A			
85-25411	AH64	28AVN		
85-25412	AH64	28AVN		
85-25413	AH64	28AVN		
85-25414	AH64	28AVN		
85-25415	AH64	28AVN		
85-25416	AH64	28AVN		
85-25417	AH64	1-130AVN		NC NG
85-25418	AH64A			
85-25419	AH64	28AVN		
85-25420	AH64	28AVN		
85-25421	AH64	28AVN		
85-25422	AH64	I\3AVN		
85-25423	AH64	1-149AVN		TX NG
85-25424	AH64	1-82AVN		
85-25425	AH64	1-82AVN		
85-25426	AH64	1-82AVN		
85-25427	AH64	1-82AVN		
85-25428	AH64	1-82AVN		
85-25429	AH64A			
85-25430	AH64	I\3AVN		
85-25431	AH64A			
85-25432	AH64A			
85-25433	AH64			
85-25434	AH64	1-82AVN		
85-25435	AH64A			
85-25436	AH64	1-82AVN		
85-25437	AH64	1-82AVN		
85-25438	AH64A			
85-25439	AH64	1-82AVN		
85-25440	AH64	1-82AVN		
85-25441	AH64	1-82AVN		
85-25442	AH64	AATC	42.	Hanchey AHP
85-25443	AH64	AATC	43H	Hanchey AHP
85-25444	AH64	I\3AVN		
85-25445	AH64	AATC	45C	Hanchey AHP
85-25446	AH64A			
85-25447	AH64A			
85-25448	AH64A			
85-25449	AH64A			
85-25450	AH64A			
85-25451	AH64A			
85-25452	AH64	5-229AVN		
85-25453	AH64A			
85-25454	AH64A			
85-25455	AH64A			
85-25456	AH64A			
85-25457	AH64A			
85-25458	AH64A			
85-25459	AH64A			
85-25460	AH64	1AVN		
85-25461	AH64	1-149AVN		TX NG
85-25462	AH64	I\3AVN		
85-25463	AH64A			
85-25464	AH64	I\3AVN		
85-25465	AH64A			
85-25466	AH64	1-82AVN		
85-25467	AH64A			
85-25468	AH64A			
85-25469	AH64	5-6Cav		
85-25470	AH64	5-6Cav		
85-25471	AH64	5-6Cav		
85-25472	AH64	5-6Cav		
85-25473	AH64			
85-25474	AH64	5-6Cav		
85-25475	AH64	5-6Cav		
85-25476	AH64	5-6Cav		
85-25477	AH64C	227AVN		

85-25478	AH64	1AVN	
85-25479	AH64	1AVN	
85-25480	AH64	5-6Cav	
85-25481	AH64	w.o.26.9.89	
85-25482	AH64	5-6Cav	
85-25483	AH64	1-149AVN	TX NG
85-25484	AH64	5-6Cav	
85-25485	AH64	5-6Cav	
85-25486	AH64	5-6Cav	
85-25487	AH64	5-6Cav	
85-25488	AH64	5-6Cav	
86-8940	AH64	4-229AVN	
86-8941	AH64	4-229AVN	
86-8942	AH64	4-229AVN	
86-8943	AH64	4-229AVN	
86-8944	AH64	4-229AVN	
86-8945	JAH64	4-229AVN	
86-8946	AH64	4-229AVN	
86-8947	AH64	B\4-229AVN	
86-8948	AH64	4-229AVN	
86-8949	AH64	4-229AVN	
86-8950	AH64	4-229AVN	
86-8951	AH64	4-229AVN	
86-8952	AH64	4-229AVN	
86-8953	AH64	w.o.7.1.91	
86-8954	AH64	w.o.29.10.90	
86-8955	AH64	4-229AVN	
86-8956	AH64	3-1AVN	
86-8957	AH64	4-229AVN	
86-8958	AH64	1-149AVN	TX NG
86-8959	AH64	4-229AVN	
86-8960	AH64	4-229AVN	
86-8961	AH64	4-229AVN	
86-8962	AH64	1-151AVN	SC NG
86-8963	AH64	1-151AVN	SC NG
86-8964	AH64	1-151AVN	SC NG
86-8965	AH64A		
86-8966	AH64	151AVN	
86-8967	AH64A		
86-8968	AH64A		
86-8969	AH64A		
86-8970	AH64	3-227AVN	
86-8971	AH64A		
86-8972	AH64A		
86-8973	AH64A		
86-8974	AH64A		
86-8975	AH64A		
86-8976	AH64A		
86-8977	AH64A		
86-8978	AH64	3ACR	
86-8979	AH64	101AbDiv	
86-8980	AH64A		
86-8981	AH64	5-6Cav	
86-8982	AH64	w.o.26.4.90	
86-8983	AH64	5-6Cav	
86-8984	AH64	w.o.26.4.90	
86-8985	AH64A		
86-8986	AH64	1-151AVN	SC NG
86-8987	AH64A		
86-8988	AH64		
86-8989	JAH64A		
86-8990	AH64		
86-8991	AH64		
86-8992	AH64	1-149AVN	TX NG
86-8993	AH64	1-151AVN	SC NG
86-8994	AH64A		
86-8995	AH64		
86-8996	AH64	1-151AVN	SC NG
86-8997	AH64	1-151AVN	SC NG
86-8998	AH64		
86-8999	AH64A		
86-9000	AH64	1-151AVN	SC NG
86-9001	AH64		
86-9002	AH64	1-151AVN	SC NG
86-9003	AH64A		
86-9004	AH64		
86-9005	AH64A		
86-9006	AH64	1-151AVN	SC NG
86-9007	AH64	1-151AVN	SC NG
86-9008	AH64	1-151AVN	SC NG
86-9009	AH64A		
86-9010	AH64	1AVN	
86-9011	AH64	1AVN	
86-9012	AH64	w.o.10.7.89	
86-9013	AH64	1-151AVN	SC NG
86-9014	AH64A		
86-9015	AH64A		
86-9016	AH64		
86-9017	AH64		
86-9018	AH64A		
86-9019	AH64	1AVN	
86-9020	AH64A		
86-9021	AH64	1-111AVN	FL NG
86-9022	AH64	1-111AVN	FL NG
86-9023	AH64	1-111AVN	FL NG
86-9024	AH64	224AVN	
86-9025	AH64	5-229AVN	
86-9026	AH64	6Cav	
86-9027	AH64	3ACR	
86-9028	AH64A		
86-9029	AH64		
86-9030	AH64	3-1AVN	
86-9031	AH64	1-151AVN	SC NG
86-9032	AH64	1AVN	
86-9033	AH64A		
86-9034	AH64	3ACR	
86-9035	AH64	1-151AVN	SC NG
86-9036	AH64	1-151AVN	SC NG
86-9037	AH64	6Cav	
86-9038	AH64A		
86-9039	AH64	1AVN	
86-9040	AH64	3ACR	
86-9041	AH64	227AVN	
86-9042	AH64A		
86-9043	AH64	5-229AVN	
86-9044	AH64	1AVN	
86-9045	AH64A		
86-9046	AH64A		
86-9047	AH64	5-229AVN	
86-9048	AH64	1AVN	
86-9049	AH64	1-111AVN	FL NG
86-9050	AH64	1-111AVN	FL NG
86-9051	AH64	224AVN	
86-9052	AH64	w.o.29.6.89	
86-9053	AH64A		
86-9054	AH64	3ACR	
86-9055	AH64A		
87-0407	AH64	3ACR	
87-0408	AH64	1AVN	
87-0409	AH64	1AVN	
87-0410	AH64A		
87-0411	AH64	4-229AVN	
87-0412	AH64	1AVN	
87-0413	AH64	1AVN	
87-0414	AH64	3ACR	
87-0415	AH64A		
87-0416	AH64	1-111AVN	FL NG
87-0417	AH64	1AVN	
87-0418	AH64	6Cav	
87-0419	AH64	1AVN	
87-0420	AH64	1AVN	
87-0421	AH64	3ACR	
87-0422	AH64	5-229AVN	
87-0423	AH64	1AVN	
87-0424	AH64		
87-0425	AH64	1-111AVN	FL NG
87-0426	AH64	1-111AVN	FL NG
87-0427	AH64	5-229AVN	
87-0428	AH64	1AVN	
87-0429	AH64	1-111AVN	FL NG
87-0430	AH64	1-111AVN	FL NG
87-0431	AH64A		
87-0432	AH64	1AVN	
87-0433	AH64	3-227AVN	
87-0434	AH64		
87-0435	AH64A		
87-0436	AH64	1AVN	
87-0437	AH64	1AVN	
87-0438	AH64	3-227AVN	
87-0439	AH64	1AVN	
87-0440	AH64	3-227AVN	
87-0441	AH64	3-227AVN	
87-0442	AH64	1AVN	
87-0443	AH64	3-227AVN	
87-0444	AH64	3-227AVN	
87-0445	AH64	3-227AVN	

Serial	Type	Unit	Notes		Serial	Type	Unit	Notes
87-0446	AH64	3-227AVN			88-0218	AH64	5-229AVN	
87-0447	AH64	3-227AVN			88-0219	AH64	5-229AVN	
87-0448	AH64A				88-0220	AH64	5-229AVN	
87-0449	AH64	3-227AVN			88-0221	AH64A		
87-0450	AH64	w.o.20.1.91			88-0222	AH64	5-229AVN	
87-0451	AH64	3-227AVN			88-0223	AH64	5-229AVN	
87-0452	AH64	w.o.24.2.91			88-0224	AH64	5-229AVN	
87-0453	AH64A				88-0225	AH64	5-229AVN	
87-0454	AH64	3-227AVN			88-0226	AH64	5-229AVN	
87-0455	AH64	3-227AVN			88-0227	AH64A		
87-0456	AH64	1-151AVN	SC NG		88-0228	AH64	5-229AVN	
87-0457	AH64	3-227AVN			88-0229	AH64	5-229AVN	
87-0458	AH64	1AVN			88-0230	AH64	5-229AVN	
87-0459	AH64	1AVN			88-0231	AH64	5-229AVN	
87-0460	AH64	1-151AVN	SC NG		88-0232	AH64	5-229AVN	
87-0461	AH64A				88-0233	AH64	5-229AVN	
87-0462	AH64A				88-0234	AH64	6-6Cav	
87-0463	AH64A				88-0235	AH64	5-229AVN	
87-0464	AH64	1AVN			88-0236	AH64A		
87-0465	AH64A				88-0237	AH64	5-229AVN	
87-0466	AH64	1-111AVN	FL NG		88-0238	AH64	6-6Cav	
87-0467	AH64A				88-0239	AH64	5-229AVN	
87-0468	AH64	1-111AVN	FL NG		88-0240	AH64	A-1\6Cav	
87-0469	AH64	1-111AVN	FL NG		88-0241	AH64	5-229AVN	
87-0470	AH64	1AVN			88-0242	AH64	5-229AVN	
87-0471	AH64	1AVN			88-0243	AH64	6-6Cav	
87-0472	AH64	1AVN			88-0244	AH64	5-229AVN	
87-0473	AH64	1AVN			88-0245	AH64	5-229AVN	
87-0474	AH64	1AVN			88-0246	AH64	5-229AVN	
87-0475	AH64	6Cav			88-0247	AH64	5-229AVN	
87-0476	AH64	1AVN			88-0248	AH64	5-229AVN	
87-0477	AH64	1AVN			88-0249	AH64	5-229AVN	
87-0478	AH64	1AVN			88-0250	AH64A		
87-0479	AH64	1AVN			88-0251	AH64A		
87-0480	AH64	3ACR			88-0252	AH64A		
87-0481	AH64	1AVN			88-0253	AH64A		
87-0482	AH64	1AVN			88-0254	AH64A		
87-0483	AH64A				88-0255	AH64A		
87-0484	AH64A				88-0256	AH64A		
87-0485	AH64A				88-0257	AH64A		
87-0486	AH64A				88-0258	AH64A		
87-0487	AH64	227AVN			88-0259	AH64A		
87-0488	AH64A				88-0260	AH64A		
87-0489	AH64A				88-0261	AH64A		
87-0490	AH64A				88-0262	AH64A		
87-0491	AH64A				88-0263	AH64A		
87-0492	AH64	3ACR			88-0275	AH64A		
87-0493	AH64	3ACR			88-0276	AH64A		
87-0494	AH64	3ACR			88-0277	AH64	w.o.15.11.90	
87-0495	AH64	3ACR			88-0278	AH64A		
87-0496	AH64	227AVN			88-0279	AH64A		
87-0497	AH64	3ACR			88-0280	AH64A		
87-0498	AH64	6Cav			88-0281	AH64A		
87-0499	AH64	3ACR			88-0282	AH64A		
87-0500	AH64	3ACR			88-0283	AH64A		
87-0501	AH64	3ACR			88-0284	AH64A		
87-0502	AH64	3ACR			89-0192	AH64D		
87-0503	AH64	227AVN			89-0193	AH64A		
87-0504	AH64	227AVN			89-0194	AH64	to Israel	
87-0505	AH64	6Cav			89-0195	AH64	to Israel	
87-0506	AH64	227AVN			89-0196	AH64	to Israel	
87-0507	AH64	227AVN			89-0197	AH64	to Israel	
88-0197	AH64	6Cav			89-0198	AH64	to Israel	
88-0198	AH64	227AVN			89-0199	AH64A		
88-0199	AH64	227AVN			89-0200	AH64	to Israel	
88-0200	AH64A				89-0201	AH64A		
88-0201	AH64	5-229AVN			89-0202	AH64A		
88-0202	AH64	5-229AVN			89-0203	AH64A		
88-0203	AH64	5-229AVN			89-0204	AH64A		
88-0204	AH64	4AVN			89-0205	AH64A		
88-0205	AH64	5-229AVN			89-0206	AH64	to Israel	
88-0206	AH64	5-229AVN			89-0207	AH64	to Israel	
88-0207	AH64	4AVN			89-0208	AH64A		
88-0208	AH64	5-229AVN			89-0209	AH64A		
88-0209	AH64	4AVN			89-0210	AH64A		
88-0210	AH64	4AVN			89-0211	AH64	229AHR	
88-0211	AH64	4AVN			89-0212	AH64	to Israel	
88-0212	AH64	5-229AVN			89-0213	AH64	to Israel	
88-0213	AH64	5-229AVN			89-0214	AH64A		
88-0214	AH64	5-229AVN			89-0215	AH64A		
88-0215	AH64	5-229AVN			89-0216	AH64A		
88-0216	AH64	5-229AVN			89-0217	AH64A		
88-0217	AH64	6-6Cav			89-0218	AH64	to Israel	

Serial	Type	Notes
89-0219	AH64	to Israel
89-0220	AH64A	
89-0221	AH64A	
89-0222	AH64A	
89-0223	AH64A	
89-0224	AH64	to Israel
89-0225	AH64	to Israel
89-0226	AH64A	
89-0227	AH64A	
89-0228	AH64D	
89-0229	AH64A	
89-0230	AH64	to Israel
89-0231	AH64	to Israel
89-0232	AH64A	
89-0233	AH64A	
89-0234	AH64A	
89-0235	AH64A	
89-0236	AH64	to Israel
89-0237	AH64	to Israel
89-0238	AH64A	
89-0239	AH64A	
89-0240	AH64A	
89-0241	AH64A	
89-0242	AH64A	
89-0243	AH64A	
89-0244	AH64A	
89-0245	AH64A	
89-0246	AH64A	
89-0247	AH64A	
89-0248	AH64A	
89-0249	AH64A	
89-0250	AH64A	
89-0251	AH64A	
89-0252	AH64A	
89-0253	AH64A	
89-0254	AH64A	
89-0255	AH64A	
89-0256	AH64A	
89-0257	AH64A	
89-0258	AH64A	
89-0259	AH64A	
89-0260	AH64A	
89-0261	AH64A	
89-0262	AH64A	
89-0263	AH64A	
90-0280	AH64A	
90-0281	AH64A	
90-0282	AH64A	
90-0283	AH64A	
90-0284	AH64A	
90-0285	AH64	7\6Cav
90-0286	AH64A	
90-0287	AH64A	
90-0288	AH64A	
90-0289	AH64A	
90-0290	AH64A	
90-0291	AH64	to S. Arabia
90-0292	AH64A	
90-0293	AH64A	
90-0294	AH64A	
90-0295	AH64A	
90-0296	AH64A	
90-0297	AH64A	
90-0298	AH64A	
90-0299	AH64A	
90-0300	AH64A	
90-0301	AH64A	
90-0302	AH64A	
90-0303	AH64A	
90-0304	AH64A	
90-0305	AH64A	
90-0306	AH64A	
90-0307	AH64A	
90-0308	AH64A	
90-0309	AH64A	
90-0310	AH64A	
90-0311	AH64	1-285AVN AZ NG
90-0312	AH64A	
90-0313	AH64	1-285AVN AZ NG
90-0314	AH64	1-285AVN AZ NG
90-0315	AH64A	
90-0316	AH64A	
90-0317	AH64A	
90-0318	AH64A	
90-0319	AH64A	
90-0320	AH64A	
90-0321	AH64	7\6Cav
90-0322	AH64	1-285AVN AZ NG
90-0323	AH64	1-285AVN AZ NG
90-0324	AH64D	
90-0325	AH64	1-285AVN AZ NG
90-0326	AH64	1-285AVN AZ NG
90-0327	AH64A	
90-0328	AH64A	
90-0329	AH64A	
90-0330	AH64A	
90-0331	AH64	1-285AVN AZ NG
90-0332	AH64A	
90-0333	AH64A	
90-0334	AH64	1-285AVN AZ NG
90-0335	AH64	1-285AVN AZ NG
90-0336	AH64	1-285AVN AZ NG
90-0337	AH64	1-285AVN AZ NG
90-0338	AH64A	
90-0339	AH64A	
90-0340	AH64A	
90-0341	AH64	to S. Arabia
90-0342	AH64	to S. Arabia
90-0343	AH64	to S. Arabia
90-0344	AH64	to S. Arabia
90-0345	AH64	to S. Arabia
90-0415	AH64A	
90-0416	AH64A	
90-0417	AH64A	
90-0418	AH64A	
90-0419	AH64A	
90-0420	AH64A	
90-0421	AH64A	
90-0422	AH64A	
90-0423	AH64D	
90-0424	AH64A	
90-0425	AH64A	
90-0426	AH64A	
90-0427	AH64A	
90-0428	AH64A	
90-0429	AH64	to S. Arabia
90-0430	AH64	to S. Arabia
90-0431	AH64	to S. Arabia
90-0432	AH64	to S. Arabia
90-0433	AH64	to S. Arabia
90-0434	AH64	to S. Arabia
90-0435	AH64A	
90-0436	AH64A	
90-0437	AH64A	
90-0438	AH64A	
90-0439	AH64A	
90-0440	AH64A	
90-0441	AH64A	
90-0442	AH64A	
90-0443	AH64A	
90-0444	AH64A	
90-0445	AH64A	
90-0446	AH64A	
90-0447	AH64A	
90-0448	AH64A	
90-0449	AH64A	
90-0450	AH64A	
90-0451	AH64A	
90-0452	AH64A	
90-0453	AH64A	
90-0454	AH64A	
90-0455	AH64A	
90-0456	AH64A	
90-0457	AH64A	
90-0458	AH64A	
90-0459	AH64A	
90-0460	AH64A	
90-0461	AH64A	
90-0462	AH64A	
90-0463	AH64A	
90-0464	AH64A	
90-0465	AH64A	
90-0466	AH64A	
90-0467	AH64A	
90-0468	AH64A	
90-0469	AH64A	

90-0470	AH64A		
90-0471	AH64A		
90-0472	AH64A		
90-0473	AH64A		
90-0474	AH64A		
90-0475	AH64A		

90-0476	AH64A		
90-0477	AH64A		
90-0478	AH64A		
90-0479	AH64A		
90-0480	AH64A		

T41 MESCALERO

c\n   0147, 0162, 228, 1 unknown.

67-15146	T41B	2AvnDet	
67-15161	T41B	preserved	Ft. Rucker

67-15227	T41B	w.o.24.9.86
85-24372	T41B	2AvnDet

T42 COCHISE

c\n   TF-1 to TF-65.

65-12679	T42A	SoC	
65-12680	T42A	w.o.25.1.83	
65-12681	T42A	to N56AF	
65-12682	T42A	to N9768B	
65-12683	T42A	to N55NE	
65-12684	T42A		CA NG
65-12685	T42A	SoC	
65-12686	T42A	to N5148Z	
65-12687	T42A	SoC	
65-12688	T42A	to N65KT	
65-12689	T42A		
65-12690	T42A	SoC	
65-12691	T42A	SoC	
65-12692	T42A		MS NG
65-12693	T42A	to N95RG	
65-12694	T42A	to N9775Q	
65-12695	T42A	SoC	
65-12696	T42A	192AVN	AL NG
65-12697	T42A		
65-12698	T42A	to N97RG	
65-12699	T42A		
65-12700	T42A	to N9177Y	
65-12701	T42A	SoC	
65-12702	T42A	to N5085Q	
65-12703	T42A	w.o.23.12.87	
65-12704	T42A	to N7042H	
65-12705	T42A		
65-12706	T42A	w.o.26.5.72	
65-12707	JT42A	to N7042R	
65-12708	T42A	to N7041M	
65-12709	T42A	w.o.4.3.70	
65-12710	T42A	to N43657	
65-12711	T42A	w.o.	

65-12712	T42A	192AVN	AL NG
65-12713	T42A	SoC	
65-12714	T42A	to N5069J	
65-12715	T42A	w.o.20.7.77	
65-12716	T42A	SoC	
65-12717	T42A	to N19SJ	
65-12718	T42A	STARC	DE NG
65-12719	T42A	SoC	
65-12720	T42A	to N42WR	
65-12721	T42A		
65-12722	T42A	SoC	
65-12723	T42A		
65-12724	T42A	to N9114S	
65-12725	T42A	w.o.9.10.70	
65-12726	T42A	SoC	
65-12727	T42A	to N727AF	
65-12728	T42A	to N7041L	
65-12729	T42A	w.o.	
65-12730	T42A	STARC	CA NG
65-12731	T42A	SoC	
65-12732	T42A		NY NG
65-12733	T42A	STARC	CO NG
66-4300	T42A	to N820M	
66-4301	T42A	to N5150M	
66-4302	T42A	192AVN	AL NG
66-4303	T42A		
66-4304	T42A	w.o.25.5.77	
66-4305	T42A	to N9110Y	
66-4306	T42A	w.o.2.7.86	
66-4307	T42A	to N54339	
66-4308	T42A	to N.....	
66-4309	T42A	to N522SP	

U3 BLUE CANOE

c\n	U3A:	38001 to 38160
	U3B:	0002 to 0036

c\n	type		
57-5846	U3A	to N4303	
57-5847	U3A	79ARCOM	
57-5848	U3A		
57-5849	U3A	preserved	Castle AFB
57-5850	U3A	to N5795X	
57-5851	U3A		
57-5852	U3A	to N133Z	
57-5853	U3A		
57-5854	U3A		
57-5855	U3A		
57-5856	U3A		
57-5857	U3A	to N134Z	
57-5858	U3A		
57-5859	U3A		
57-5860	U3A	Propeller Co.	Santee, Ca
57-5861	U3A		
57-5862	U3A		
57-5863	U3A	preserved	Ft. Rucker
57-5864	U3A		
57-5865	U3A		
57-5866	U3A		
57-5867	U3A	to N145Z	
57-5868	U3A		
57-5869	U3A	preserved	Ft. Eustis
57-5870	U3A		
57-5871	U3A		MA NG
57-5872	U3A	AFRES	
57-5873	U3A		
57-5874	U3A	to N65386	
57-5875	U3A		
57-5876	U3A	to N5275G	
57-5877	U3A	to N115ZL	
57-5878	U3A		CT NG
57-5879	U3A	NAS	Glenview
57-5880	U3A		IA NG
57-5881	U3A	w.o.3.10.79	
57-5882	U3A		
57-5883	U3A	Propeller Co.	Santee, Ca
57-5884	U3A		
57-5885	U3A		
57-5886	U3A		
57-5887	U3A		
57-5888	U3A	MASDC	
57-5889	U3A		
57-5890	U3A	to N12NP	
57-5891	U3A	NWC	China Lake
57-5892	U3A	to N3979A	
57-5893	U3A	to N154Z	
57-5894	U3A	to N4406	
57-5895	U3A		
57-5896	U3A	NAS	Glenview
57-5897	U3A		
57-5898	U3A	to N155F	
57-5899	U3A		
57-5900	U3A		
57-5901	U3A		
57-5902	U3A		
57-5903	U3A		
57-5904	U3A	NAS	Glenview
57-5905	U3A	to N5272G	
57-5906	U3A		
57-5907	U3A		PA NG
57-5908	U3A		
57-5909	U3A	Randolph AFB	
57-5910	U3A		
57-5911	U3A		
57-5912	U3A		
57-5913	U3A		
57-5914	U3A	MASDC	
57-5915	U3A		
57-5916	U3A	to N99639	
57-5917	U3A		FL NG
57-5918	U3A		
57-5919	U3A	to N7410N	
57-5920	U3A		
57-5921	U3A	to NASA	N505NA
57-5922	U3A	MASDC	
57-5923	U3A	Selfridge ANGB	
57-5924	U3A		
57-5925	U3A		IL NG
58-2107	U3A	Pima Museum	
58-2108	U3A	to N1363P	
58-2109	U3A		PR NG
58-2110	U3A		
58-2111	U3A	to N9062Y	
58-2112	U3A		
58-2113	U3A	to N2273A	
58-2114	U3A	to N4266	
58-2115	U3A		
58-2116	U3A		
58-2117	U3A	to N505Z	
58-2118	U3A	to N65387	
58-2119	U3A	79ARCOM	
58-2120	U3A	to N107Z	
58-2121	U3A		
58-2122	U3A	to N6468	
58-2123	U3A	NWC	China Lake
58-2124	U3A	USAF	Museum
58-2125	U3A	to N5273G	
58-2126	U3A		
58-2127	U3A		
58-2128	U3A		
58-2129	U3A		
58-2130	U3A		SC NG
58-2131	U3A		
58-2132	U3A	to N6747	
58-2133	U3A		
58-2134	U3A		
58-2135	U3A		
58-2136	U3A		
58-2137	U3A		
58-2138	U3A		
58-2139	U3A		
58-2140	U3A		
58-2141	U3A	to N83637	
58-2142	U3A		GA NG
58-2143	U3A		
58-2144	U3A	79ARCOM	
58-2145	U3A		
58-2146	U3A		KS NG
58-2147	U3A	to N487LA	
58-2148	U3A		
58-2149	U3A		
58-2150	U3A	to N308S	
58-2151	U3A		KS NG
58-2152	U3A		
58-2153	U3A	to N5975	
58-2154	U3A		
58-2155	U3A		
58-2156	U3A	to N7349	
58-2157	U3A		TX NG
58-2158	U3A	San Francisco IAP	
58-2159	U3A		
58-2160	U3A	Selfridge ANGB	
58-2161	U3A		
58-2162	U3A	to N75CA	

58-2163	U3A	to N5000W	
58-2164	U3A	to N4568	
58-2165	U3A		
58-2166	U3A	to N885M	
58-2167	U3A		
58-2168	U3A		
58-2169	U3A	to N5070F	
58-2170	U3A	to N110Z	
58-2171	U3A	to N7274N	
58-2172	U3A	to N79MU	
58-2173	U3A		
58-2174	U3A	to N20982	
58-2175	U3A	to N1032	
58-2176	U3A	Lakehurst Flying Club	
58-2177	U3A		
58-2178	U3A		
58-2179	U3A	to N3161Z	
58-2180	U3A		
58-2181	U3A		
58-2182	U3A		
58-2183	U3A		
58-2184	U3A		
58-2185	U3A		
58-2186	U3A		
60-6047	U3B	US Navy	Andrews AFB
60-6048	U3B		SD NG
60-6049	U3B	to N3AF	
60-6050	U3B	to N5331G	
60-6051	U3B	Ft. Bliss	
60-6052	U3B	Elmendorf AFB	
60-6053	U3B		
60-6054	U3B	to N605AF	
60-6055	U3B	Sheppard AFB	
60-6056	U3B		WA NG
60-6057	U3B		NV NG
60-6058	U3B		
60-6059	U3B		UT NG
60-6060	U3B		OH NG
60-6061	U3B		NV NG
60-6062	U3B		PA NG
60-6063	U3B		
60-6064	U3B		ND NG
60-6065	U3B		
60-6066	U3B		
60-6067	U3B		
60-6068	U3B	to N5215G	
60-6069	U3B		
60-6070	U3B		
60-6071	U3B		
60-6072	U3B		NY NG
60-6073	U3B		KS NG
60-6074	U3B		
60-6075	U3B		
60-6076	U3B		
60-6077	U3B		IL NG
60-6078	U3B		
60-6079	U3B		
60-6080	U3B		
60-6081	U3B		

U8 SEMINOLE

not listed

serial | U8A: | 52-1800 to 52-1803, 52-6162 to 52-6216.
| U8B: | 53-6153 TO 53-6192.

c\n | U8A: | 4 unknown, LH-1 to LH-55 (most rebuilt as U8D with RLH- c\n)
| U8B: | LH-56 to LH-95 (most rebuilt as U8D with RLH- c\n).

listed

serial | all: | as shown below

c\n | U8G: | 1 unknown.
| U8D: | LH-96 to LH-137, RLH-1 to RLH-48, LH-138 to LH-180, 8 unknown, RLH-49 to RLH-93, 9 unknown, LH-196 to LH-198.
| U8E: | 6 unknown, JH-163.
| U8F: | L-3 to L-5, 68 unknown (believed in range LF-7, LF-9 to LF-76, not necessarily in order), LF-8.
| NU8F: | LG-1.

55-3465	U8G	MASDC		
56-3695	U8D	NAS	Glenview	
56-3696	U8D	MASDC		
56-3697	U8D			
56-3698	U8D	MASDC		
56-3699	U8D			
56-3700	U8D	to N2874A		
56-3701	U8D	Pima	Museum	
56-3702	U8D			
56-3703	U8D			
56-3704	U8D	MASDC		
56-3705	U8D	MASDC		
56-3706	U8D	MASDC		
56-3707	U8D	MASDC		
56-3708	U8D	MASDC		
56-3709	U8D			
56-3710	U8D	Felker AAF		
56-3711	U8D			
56-3712	U8D	preserved	Ft. Rucker	
56-3713	U8D			
56-3714	U8D	AMARC	UB068	
56-3715	U8D			
56-3716	U8D	MASDC		
56-3717	U8D			
56-3718	U8D	MASDC		
56-4039	U8E	MASDC		
56-4040	U8E			
56-4041	U8G			
56-4042	U8E			
56-4043	U8E	Felker AAF		
56-4044	U8G	MASDC		
57-3084	U8D	102ARCOM		
57-3085	U8D	MASDC		
57-3086	U8D			
57-3087	U8D			
57-3088	U8D	MASDC		
57-3089	U8D			
57-3090	U8D			
57-3091	U8D	Felker AAF		
57-3092	U8D	stored	Salina MAP, Ks	

Serial	Type	Status	Location
57-3093	U8D	MASDC	
57-3094	U8D		
57-3095	U8D	MASDC	
57-3096	U8D		
57-3097	U8D	MASDC	
57-3098	U8D	MASDC	
57-3099	U8D		
57-3100	U8D		
57-3101	U8D	MASDC	
57-6029	U8D		
57-6030	U8D	MASDC	
57-6031	U8D		
57-6032	U8D	MASDC	
57-6033	U8D		
57-6034	U8D	to N925TB	
57-6035	U8D		
57-6036	U8D		
57-6037	U8D	MASDC	
57-6038	U8D	MASDC	
57-6039	U8D		
57-6040	U8D		
57-6041	U8D	MASDC	
57-6042	U8D	MASDC	
57-6043	U8D	MASDC	
57-6044	U8D	MASDC	
57-6045	U8D	MASDC	
57-6046	RU8D		
57-6047	U8D	MASDC	
57-6048	U8D	MASDC	
57-6049	U8D	MASDC	
57-6050	U8D	MASDC	
57-6051	U8D	MASDC	
57-6052	U8D		IN NG
57-6053	U8D		NJ NG
57-6054	U8D	89ARCOM	
57-6055	U8D		
57-6056	U8D		
57-6057	U8D		
57-6058	U8D		
57-6059	U8D	MASDC	
57-6060	U8D		VT NG
57-6061	U8D		
57-6062	U8D		
57-6063	U8D	MASDC	
57-6064	U8D		
57-6065	U8D	MASDC	
57-6066	U8D		
57-6067	U8D		
57-6068	U8D	MASDC	
57-6069	U8D		
57-6070	U8D	MASDC	
57-6071	U8D		
57-6072	U8D		
57-6073	U8D		SC NG
57-6074	U8D	MASDC	
57-6075	U8D	MASDC	
57-6076	U8D	MASDC	
57-6077	U8D	MASDC	
57-6078	U8D		
57-6079	U8D		
57-6080	U8D	MASDC	
57-6081	U8D	MASDC	
57-6082	U8D	MASDC	
57-6083	U8D	MASDC	
57-6084	U8D	MASDC	
57-6085	U8D	stored	NAEC
57-6086	U8D	to N99128	
57-6087	U8D		CA NG
57-6088	U8D		
57-6089	U8D	MASDC	
57-6090	U8D		
57-6091	U8D		
57-6092	U8D	MASDC	
57-6093	U8D	MASDC	
57-6094	U8D	MASDC	
58-1329	U8D		
58-1330	U8D		
58-1331	U8D	MASDC	
58-1332	U8D		
58-1333	U8D	MASDC	
58-1334	U8D	MASDC	
58-1335	U8D		
58-1336	U8D	to N.....	
58-1337	U8D		
58-1338	U8D	MASDC	
58-1339	U8D	stored	NAEC
58-1340	U8D	MASDC	
58-1341	U8D		OH NG
58-1342	U8D	MASDC	
58-1343	U8D		
58-1344	U8D	to N1373K	
58-1345	U8D	MASDC	
58-1346	U8D	to N1373W	
58-1347	U8D	MASDC	
58-1348	U8D	MASDC	
58-1349	U8D		
58-1350	U8D	MASDC	
58-1351	U8D		
58-1352	U8D		
58-1353	U8D		
58-1354	U8F	SoC	
58-1355	U8F		
58-1356	U8F	SoC	
58-1357	RU8D	MASDC	
58-1358	RU8D	MASDC	
58-1359	RU8D	preserved	Ft. Rucker
58-1360	RU8D	to N50KT	
58-1361	RU8D	MASDC	
50-1362	RU8D	MASDC	
58-1363	U8G	MASDC	
58-1364	RU8D		
58-3048	U8D	MASDC	
58-3049	U8D		
58-3050	U8D	MASDC	
58-3051	U8D	preserved	Ft. Eustis
58-3052	U8D	to N28173	
58-3053	U8D		
58-3054	U8D	MASDC	
58-3055	U8D	stored	NAEC
58-3056	U8D		
58-3057	U8D	MASDC	
58-3058	U8D	MASDC	
58-3059	U8D	MASDC	
58-3060	RU8D	MASDC	
58-3061	U8D	MASDC	
58-3062	U8D	MASDC	
58-3063	U8D	MASDC	
58-3064	U8D	MASDC	
58-3065	U8D	MASDC	
58-3066	U8D	MASDC	
58-3067	U8D		
58-3068	U8D	MASDC	
58-3069	U8D		
58-3070	U8D	MASDC	
58-3071	U8D	Felker AAF	
58-3072	U8D	MASDC	
58-3073	U8D		NY NG
58-3074	U8D	MASDC	
58-3075	U8D	MASDC	
58-3076	U8D		
58-3077	U8D	MASDC	
58-3078	U8D	MASDC	
58-3079	U8D		
58-3080	U8D	MASDC	
58-3081	U8D		
58-3082	U8D		
58-3083	U8D	MASDC	
58-3084	U8D	MASDC	
58-3085	U8D		
58-3086	U8D	stored	Mesa Field, Az
58-3087	U8D	MASDC	
58-3088	U8D		
58-3089	U8D	MASDC	
58-3090	U8D		
58-3091	U8D	MASDC	
58-3092	U8D	MASDC	
59-2535	RU8D		
59-2536	RU8D	to N5071N	
59-2537	RU8D	to N50736	
59-2538	RU8D	MASDC	
59-2539	RU8D		
59-2540	RU8D		
59-2541	RU8D		
59-2542	RU8D		
59-2543	RU8D		
59-4990	U8D	MASDC	

Serial	Type	Note	Location
59-4991	U8D	preserved	Ft. Eustis
59-4992	U8D		
60-3453	U8F		
60-3454	U8F	SoC	
60-3455	U8F	SoC	
60-3456	U8F		
60-3457	U8F	SoC	
60-3458	U8F	SoC	
60-3459	U8F		
60-3460	U8F		MT NG
60-3461	U8F		CA NG
60-3462	U8F	stored	Lakefront AP, La
60-3463	U8F		
60-5386	U8F		
60-5387	U8F	STARC	TX NG
60-5388	U8F		AZ NG
60-5389	U8F	SoC	
60-5390	U8F		
61-2426	U8F	SoC	
61-2427	U8F		
61-2428	U8F		
61-2429	U8F	SoC	
61-2430	U8F	SoC	
62-3832	U8F	SoC	
62-3833	U8F		GA NG
62-3834	U8F		
62-3835	U8F		
62-3836	U8F		DE NG
62-3837	U8F	AATC	Cairns AAF
62-3838	U8F		
62-3839	U8F	SoC	
62-3840	U8F		OK NG
62-3841	U8F		
62-3842	U8F	SoC	
62-3843	U8F		
62-3844	U8F		
62-3845	U8F	SoC	
62-3846	U8F	SoC	
62-3847	U8F	Felker AAF	
62-3848	U8F	SoC	
62-3849	U8F	SoC	
62-3850	U8F		CA NG
62-3851	U8F		
62-3852	U8F		SC NG
62-3853	U8F		
62-3854	U8F		FL NG
62-3855	U8F		
62-3856	U8F		
62-3857	U8F		AR NG
62-3858	U8F		UT NG
62-3859	U8F	SoC	
62-3860	U8F	SoC	
62-3861	U8F		
62-3862	U8F	Ft. Bliss	
62-3863	U8F	Felker AAF	
62-3864	U8F	NAS	Glenview
62-3865	U8F	SoC	
62-3866	U8F		
62-3867	U8F	SoC	
62-3868	U8F		
62-3869	U8F		
62-3870	U8F	HHC\1-132AVN OK NG	
62-3871	U8F		
62-3872	U8F		
62-3873	U8F	HHC\1-132AVN OK NG	
62-3874	U8F	SoC	
62-3875	U8F	STARC	KS NG
63-7975	U8F	Macdill AFB	
63-12902	NU8F	preserved	Ft. Rucker
63-13636	U8F	SoC	
63-13637	U8F	stored	Lakefront AP, La
66-15360	NU8E		
66-15365	U8F		

U21 UTE

c\n	U21A\F:	LM-1 to LM-146.
	U21B:	LS-1 to LS-3.
	U21C:	LT-1 to LT-2.
	U21E\H:	LU-1 to LU-16.

Serial	Type	Note	Location
66-18000	U21A	56AvCo	
66-18001	U21A	TRADOC	Langley
66-18002	U21A	w.o.14.4.77	
66-18003	U21A	SoC	
66-18004	U21A	USN\TPS	
66-18005	U21A	Ft. Ord	
66-18006	GU21A	6AvDet	
66-18007	U21A	SoC	
66-18008	JU21A	USAICS	
66-18009	JU21A	Ft. Rucker	
66-18010	U21A	56AvCo	
66-18011	U21A	w.o.20.1.78	
66-18012	GU21A	5-158AVN	V Corps
66-18013	EU21A	Berlin Brigade	
66-18014	U21A	56AvCo	
66-18015	U21A	Golden Knights	
66-18016	U21A	DAC\MDW	
66-18017	U21A	HQ\FORSCOM	
66-18018	U21A	Ft. Meade	
66-18019	U21A	STARC	DE NG
66-18020	U21A	TRADOC	Langley
66-18021	U21A	56AvCo	
66-18022	U21A	SoC	
66-18023	U21A	HQ\FORSCOM	
66-18024	U21A	56AvCo	
66-18025	U21A	224MIBtn	
66-18026	U21A	w.o.25.5.70	
66-18027	U21A	56AvCo	
66-18028	U21A	SoC	
66-18029	U21A	136AvCo	
66-18030	U21A	56AvCo	
66-18031	U21A	HHC\1-159AVN	
66-18032	U21A		
66-18033	U21A	136AvCo	
66-18034	U21A	224MIBtn	
66-18035	U21A	SoC	
66-18036	JU21A	3SUPCOM	
66-18037	U21A	V Corps	
66-18038	U21A	HHC\1-159AVN	
66-18039	U21A	w.o.26.4.75	
66-18040	U21A		FL NG
66-18041	U21A	w.o.14.12.71	
66-18042	U21A	Ft. Meade	
66-18043	U21A	AATC	Cairns AAF
66-18044	U21A	Ft. Sill	
66-18045	U21A	SoC	
66-18046	U21A		MD NG
66-18047	U21A	136AvCo	
67-18048	U21A	Ft. Indian Gap	
67-18049	U21A		
67-18050	U21A	207AvCo	
67-18051	U21A	Presidio of San Francisco	
67-18052	U21A	NTC	
67-18053	U21A	Ft. Ord	
67-18054	U21A	AATC	Cairns AAF
67-18055	U21A		

Serial	Type	Unit/Notes	
67-18056	U21A		
67-18057	U21A	Anniston Army Depot	
67-18058	U21A	5-158AVN	V Corps
67-18059	RU21A	200AvCo	
67-18060	U21A	Ft. Polk	
67-18061	U21A	w.o.20.1.87	
67-18062	U21A	w.o.12.12.84	
67-18063	RU21A	HQ\FORSCOM	
67-18064	U21A	AATC	Cairns AAF
67-18065	JU21A	SoC	
67-18066	U21A	Ft. Sill	
67-18067	U21A	US Army Signal Center	
67-18068	U21A	TRADOC	Langley
67-18069	U21A		NV NG
67-18070	U21A	Ft. Bliss	
67-18071	U21A	56AvCo	
67-18072	U21A		TX NG
67-18073	EU21A	TECOM	
67-18074	EU21A	101AbDiv	
67-18075	U21A	TRADOC	Langley
67-18076	U21A	Golden Knights	
67-18077	RU21B	138AvCo	
67-18078	U21A	5-158AVN	V Corps
67-18079	U21A	AATC	Cairns AAF
67-18080	U21A	56AvCo	
67-18081	U21A	w.o.10,2,77	
67-18082	U21A	USAICS	
67-18083	U21A	HQ\FORSCOM	
67-18084	U21A	63ARCOM	
67-18085	RU21C	USAICS	
67-18086	U21A	to Korea	
67-18087	RU21B	138AvCo	
67-18088	U21A		
67-18089	RU21C	USAICS	
67-18090	U21A	Ft. Sam Houston	
67-18091	U21A	DAC\MDW	
67-18092	U21A		VA NG
67-18093	RU21B	138AvCo	
67-18094	U21A	Ft. Knox	
67-18095	U21A		
67-18096	U21A	USN\TPS 41	
67-18097	U21A		NM NG
67-18098	U21A		IA NG
67-18099	U21A		MN NG
67-18100	U21A		NH NG
67-18101	U21A	136AvCo	
67-18102	U21A	NTC	
67-18103	U21A	AATC	Cairns AAF
67-18104	U21D	Hunter AAF	
67-18105	RU21H	224MIBtn	
67-18106	U21D	w.o.30.3.81	
67-18107	RU21D	Selfridge ANGB	
67-18108	RU21D	63ARCOM	
67-18109	RU21D	96ARCOM	
67-18110	U21D	3-58AVN	ATC
67-18111	RU21H	224MIBtn	
67-18112	RU21A	w.o.24.7.85	
67-18113	RU21A	AATC	Cairns AAF
67-18114	RU21A	138AvCo	
67-18115	RU21A	138AvCo	
67-18116	U21A	56AvCo	
67-18117	U21A	HHC\1-159AVN	
67-18118	U21A	Simmons AAF	
67-18119	RU21H	Libby AAF	
67-18120	U21A	HHC\1-159AVN	
67-18121	U21A	w.o.17.1.91	
67-18122	U21D	TRADOC	Langley
67-18123	U21A	AATC	Cairns AAF
67-18124	U21A	124ARCOM	
67-18125	U21A	w.o.30.3.78	
67-18126	U21A	STARC	MD NG
67-18127	U21A	HQ\Army West Command	
67-18128	U21D	HQ\DESCOM	
70-15875	U21H	ERADCOM	
70-15876	RU21H	224MIBtn	
70-15877	U21H	Holloman AFB	
70-15878	U21H	Nellis AFB	
70-15879	RU21H	2MIBtn	
70-15880	RU21H	224MIBtn	
70-15881	U21H	to Korea	
70-15882	U21H		NJ NG
70-15883	U21H	224MIBtn	
70-15884	RU21H	Hunter AAF	
70-15885	RU21H	3MIBtn	
70-15886	RU21H	2MIBtn	
70-15887	RU21H	2MIBtn	
70-15888	U21H	US Army Missile Command	
70-15889	RU21H	2MIBtn	
70-15890	RU21E	w.o.19.10.75	
70-15891	RU21H	2MIBtn	
70-15892	U21G	USMA	West Point
70-15893	RU21H	15MIBtn	
70-15894	RU21H		
70-15895	RU21H		
70-15896	U21G	REDCOM	
70-15897	U21G	CCAD	
70-15898	RU21H		
70-15899	RU21H	2MIBtn	
70-15900	U21G	REDCOM	
70-15901	JU21G	AATC	Cairns AAF
70-15902	RU21H		
70-15903	RU21H	2MIBtn	
70-15904	RU21H		
70-15905	U21G	Ft. Jackson	
70-15906	JU21G	ERADCOM	
70-15907	U21G	HQ\USAREUR	
70-15908	U21F	DAC\MDW	
70-15909	U21F	DAC\MDW	
70-15910	U21F	DAC\MDW	
70-15911	U21F	AATC	Cairns AAF
70-15912	U21F	DAC\MDW	

OV1 MOHAWK

Serial	Type	Notes	
57-6463	YOV1A		
57-6464	YOV1A		
57-6465	YOV1A		
57-6466	YOV1A		
57-6467	YOV1A		
57-6538	YOV1A		
57-6539	YOV1A	preserved	Ft. Rucker
57-6540	YOV1A		
57-6541	YOV1A		
59-2603	OV1A		
59-2604	OV1A	GIA	Felker AAF
59-2605	OV1A		
59-2606	OV1A	MASDC	
59-2607	OV1A		
59-2608	OV1A	MASDC	
59-2609	OV1A		
59-2610	OV1A	MASDC	
59-2611	OV1A		
59-2612	OV1A		
59-2613	OV1A		
59-2614	OV1A		
59-2615	OV1A		
59-2616	OV1A		
59-2617	OV1A		
59-2618	OV1A	Fort Huachuca	
59-2619	OV1A	derelict	Felker AAF

59-2620	OV1A	Fort Huachuca	
59-2621	OV1B	AMARC	YA040
59-2622	OV1B	AMARC	YA104
59-2623	OV1B	AMARC	YA054
59-2624	OV1B	SoC	
59-2625	OV1B	w.o.28.4.83	
59-2626	OV1B	MASDC	
59-2627	OV1B	AMARC	YA087
59-2628	OV1B	AMARC	YA027
59-2629	OV1B	AMARC	YA...
59-2630	OV1B	AMARC	YA078
59-2631	OV1B	preserved	Ft. Rucker, Al
59-2632	OV1B	SoC	
59-2633	OV1B	AMARC	YA...
59-2634	OV1B	w.o.26.11.71	
59-2635	OV1B	w.o.5.3.70	
59-2636	OV1B	AMARC	YA055
59-2637	OV1B	USN\TPS	35
60-3720	OV1A		
60-3721	OV1A	Felker AAF	
60-3722	OV1A	scrapyard	Bobs Airpark
60-3723	OV1A		
60-3724	OV1A	MASDC	
60-3725	OV1A		
60-3726	OV1A	scrapyard	Allied
60-3727	OV1A	scrapyard	Allied
60-3728	OV1A		
60-3729	OV1A	scrapyard	Allied
60-3730	OV1A		
60-3731	OV1A		
60-3732	OV1A	GIA	Felker AAF
60-3733	OV1A		
60-3734	OV1A	stored	Felker AAF
60-3735	OV1A		
60-3736	OV1A	scrapyard	Bobs Airpark
60-3737	OV1A	scrapyard	Allied
60-3738	OV1A		
60-3739	OV1A	w.o.24.2.70	
60-3740	OV1A	stored	Felker AAF
60-3741	OV1A	scrapyard	Allied
60-3742	OV1A		
60-3743	GOV1A	GIA	Felker AAF
60-3744	OV1A		
60-3745	OV1C	w.o.	
60-3746	OV1C		
60-3747	OV1C	GIA	Felker AAF
60-3748	OV1C	Edwards AFB	
60-3749	OV1C	SoC	
60-3750	GOV1C	GIA	Libby AAF
60-3751	OV1C	SoC	
60-3752	OV1C	SoC	
60-3753	OV1C	SoC	
60-3754	OV1C	SoC	
60-3755	OV1C	SoC	
60-3756	OV1C	SoC	
60-3757	OV1C	AMARC	YA069
60-3758	OV1C	SoC	
60-3759	OV1C	AMARC	YA079
60-3760	OV1C	AMARC	YA089
60-3761	OV1C	w.o.17.3.75	
61-2675	OV1C	SoC	
61-2676	OV1C	SoC	
61-2677	OV1C	AMARC	YA012
61-2678	OV1C	SoC	
61-2679	OV1C	SoC	
61-2680	OV1C	SoC	
61-2681	OV1C	SoC	
61-2682	OV1C	SoC	
61-2683	OV1C	SoC	
61-2684	OV1C	SoC	
61-2685	OV1C	AMARC	YA016
61-2686	OV1C	w.o.13.7.85	
61-2687	OV1C	SoC	
61-2688	OV1C	SoC	
61-2689	OV1C	AMARC	YA098
61-2690	OV1C	SoC	
61-2691	OV1C	SoC	
61-2692	OV1C	stored	Cairns AAF
61-2693	OV1C	w.o.24.5.72	
61-2694	OV1C	SoC	
61-2695	OV1C	AMARC	YA101
61-2696	OV1C	AMARC	YA...
61-2697	OV1C	AMARC	YA090

61-2698	OV1C	w.o.12.2.78	
61-2699	OV1C	SoC	
61-2700	OV1C	SoC	
61-2701	OV1C	SoC	
61-2702	OV1C	SoC	
61-2703	OV1C	SoC	
61-2704	OV1C	SoC	
61-2705	OV1C	SoC	
61-2706	OV1C	SoC	
61-2707	OV1C	SoC	
61-2708	OV1C	SoC	
61-2709	OV1C	SoC	
61-2710	OV1C	AMARC	YA099
61-2711	OV1C	SoC	
61-2712	OV1C	SoC	
61-2713	OV1C	SoC	
61-2714	OV1C	AMARC	YA081
61-2715	OV1C	SoC	
61-2716	OV1C	SoC	
61-2717	OV1C	AMARC	YA021
61-2718	OV1C	AMARC	YA059
61-2719	OV1C	w.o.26.1.79	
61-2720	OV1C	w.o.9.11.79	
61-2721	OV1C	AMARC	YA102
61-2722	OV1C	SoC	
61-2723	OV1C	SoC	
61-2724	OV1C	Pima	Museum
61-2725	OV1C	SoC	
61-2726	OV1C	Edwards AFB	
61-2727	OV1C	SoC	
61-2728	OV1C	AMARC	YA100
62-5849	OV1C	w.o.	
62-5850	OV1C	SoC	
62-5851	OV1C	SoC	
62-5852	OV1C	SoC	
62-5853	OV1C	SoC	
62-5854	OV1C	SoC	
62-5855	OV1C	SoC	
62-5856	OV1C	SoC	
62-5857	JOV1C	AMARC	YA077
62-5858	OV1C	SoC	
62-5859	OV1B	Seattle Community College	
62-5860	OV1B	preserved	Ft. Rucker
62-5861	OV1B	AMARC	YA022
62-5862	OV1B	AMARC	YA109
62-5863	OV1B	SoC	
62-5864	OV1B	Seattle Community College	
62-5865	OV1D	1MIBtn	
62-5866	OV1B	USN\TPS	34
62-5867	OV1D	SoC	
62-5868	OV1B	w.o.8.6.71	
62-5869	OV1B	AMARC	YA106
62-5870	OV1B	AMARC	YA014
62-5871	OV1B	AMARC	YA063
62-5872	OV1D	AMARC	YA...
62-5873	OV1D		GA NG
62-5874	OV1D		
62-5875	OV1D		
62-5876	OV1D	stored	Gulfport RAP, Ms
62-5877	OV1B	SoC	
62-5878	OV1D	A\151AEB	GA NG
62-5879	OV1B	SoC	
62-5880	OV1B	to NASA	N512NA
62-5881	OV1B	SoC	
62-5882	OV1B	AMARC	YA110
62-5883	OV1B	AMARC	YA023
62-5884	OV1B	AMARC	YA105
62-5885	OV1D	w.o.16.12.88	
62-5886	OV1D		
62-5887	OV1D		
62-5888	OV1D		
62-5889	OV1D		
62-5890	OV1D	224MIBtn	
62-5891	RV1D	stored	Gulfport RAP, Ms
62-5892	OV1B	AMARC	YA031
62-5893	OV1B	AMARC	YA107
62-5894	OV1B	SoC	
62-5895	OV1B	AMARC	YA108
62-5896	OV1B	w.o.15.7.86	
62-5897	RV1D	SoC	
62-5898	OV1D		
62-5899	OV1B	w.o.10.5.85	
62-5900	OV1B	derelict	Winder

Serial	Type	Status	Note
62-5901	OV1B	AMARC	YA053
62-5902	OV1D	stored	Gulfport RAP, Ms
62-5903	JOV1B	Pima	Museum
62-5904	OV1B	SoC	
62-5905	OV1B	AMARC	YA017
62-5906	OV1B	MASDC	
63-13114	OV1A		
63-13115	OV1A		
63-13116	OV1A		
63-13117	OV1A		
63-13118	OV1A	MASDC	
63-13119	OV1A	to USN	
63-13120	OV1A		
63-13121	OV1A		
63-13122	OV1A		
63-13123	OV1A		
63-13124	OV1A		
63-13125	OV1A		
63-13126	OV1A	scrapyard	Allied
63-13127	OV1A		
63-13128	OV1A	to 87864	
63-13129	OV1A	scrapyard	Allied
63-13130	OV1A		
63-13131	OV1A		
63-13132	OV1A	scrapyard	Allied
63 13133	OV1A		
63-13134	OV1A		
64-14238	RV1D	w.o.8.9.84	
64-14239	RV1D	1MIBtn	
64-14240	OV1B	w.o.4.3.71	
64-14241	OV1B	SoC	
64-14242	OV1B	SoC	
64-14243	RV1D	SoC	
64-14244	RV1D	to NASA	N637NA
64-14245	RV1D	SoC	
64-14246	RV1D	stored	Gulfport RAP, Ms
64-14247	RV1D	SoC	
64-14248	RV1D	SoC	
64-14249	OV1B	AMARC	YA009
64-14250	RV1D	SoC	
64-14251	OV1B	AMARC	YA066
64-14252	RV1D	224MIBtn	
64-14253	RV1D	w.o.10.12.84	
64-14254	RV1D	w.o.7.3.79	
64-14255	RV1D	to Korea	
64-14256	RV1D	2MIBtn	
64-14257	OV1B	SoC	
64-14258	RV1D	3MIBtn	
64-14259	RV1D	224MIBtn	
64-14260	RV1D	Hunter AAF	
64-14261	RV1D	224MIBtn	
64-14262	RV1D	w.o.29.10.79	
64-14263	RV1D	3MIBtn	
64-14264	OV1B	SoC	
64-14265	RV1D	SoC	
64-14266	OV1B	SoC	
64-14267	RV1D	1MIBtn	
64-14268	RV1D	SoC	
64-14269	RV1D	1MIBtn	
64-14270	RV1D	2MIBtn	
64-14271	RV1D	2MIBtn	
64-14272	RV1D	SoC	
64-14273	RV1D	w.o.8.2.91	
66-18881	OV1C	SoC	
66-18882	OV1C	w.o.13.11.73	
66-18883	OV1C	AMARC	YA064
66-18884	OV1C	AMARC	YA103
66-18885	OV1C	SoC	
66-18886	OV1C	SoC	
66-18887	OV1C	AMARC	YA035
66-18888	OV1C	SoC	
66-18889	OV1C	AMARC	YA091
66-18890	OV1C	AMARC	YA097
66-18891	OV1C	AMARC	YA010
66-18892	OV1C	AMARC	YA085
66-18893	OV1C	AMARC	YA044
66-18894	OV1C	AMARC	YA014
66-18895	OV1C	SoC	
66-18896	OV1C	SoC	
67-18897	OV1C	w.o.2.11.70	
67-18898	OV1D	E\304MIBtn	
67-18899	OV1D	A\151AEB	GA NG
67-18900	OV1D	stored	Gulfport RAP, Ms
67-18901	OV1C	w.o.	
67-18902	OV1D	SoC	
67-18903	OV1D	to Grumman	
67-18904	OV1D	SoC	
67-18905	OV1D	1MIBtn	
67-18906	OV1D	SoC	
67-18907	OV1D	15MIBtn	
67-18908	OV1D	1MIBtn	
67-18909	OV1D		
67-18910	OV1D	1MIBtn	
67-18911	OV1D	224MIBtn	
67-18912	OV1D	Hunter AAF	
67-18913	OV1C	Edwards AFB	
67-18914	OV1C	SoC	
67-18915	OV1C	SoC	
67-18916	OV1D	w.o.26.1.79	
67-18917	OV1D	w.o.29.7.78	
67-18918	OV1D	rework	Grumman
67-18919	OV1D	1MIBtn	
67-18920	OV1D	w.o.26.1.82	
67-18921	OV1D	USAICS	
67-18922	OV1D	SoC	
67-18923	OV1D	rework	Grumman
67-18924	OV1D	OR NG	
67-18925	OV1D	1MIBtn	
67 18026	OV1D	OR NG	
67-18927	OV1D	1MIBtn	
67-18928	OV1C	w.o.3.9.70	
67-18929	OV1D	3MIBtn	
67-18930	RV1D	to Korea	
67-18931	OV1D	224MIBtn	
67-18932	OV1D	224MIBtn	
68-15930	OV1D	2MIBtn	
68-15931	OV1D	A\151AEB	GA NG
68-15932	OV1D	to Argentina	
68-15933	OV1D		OR NG
68-15934	OV1D	SoC	
68-15935	OV1D	w.o.14.1.81	
68-15936	OV1C	GIA	Felker AAF
68-15937	OV1D	w.o.	
68-15938	OV1D		
68-15939	GOV1D		
68-15940	OV1D	AMARC	YA...
68-15941	OV1D	224MIBtn	
68-15942	OV1D		
68-15943	OV1D	1MIBtn	
68-15944	OV1C	w.o.19.8.71	
68-15945	OV1D	SoC	
68-15946	OV1D	A\151AEB	GA NG
68-15947	OV1D	rework	Grumman
68-15948	OV1D	w.o.15.11.82	
68-15949	OV1C	w.o.7.7.70	
68-15950	OV1D	1MIBtn	
68-15951	OV1D	USAICS	
68-15952	OV1D	AATC	Cairns AAF
68-15953	OV1D	1MIBtn	
68-15954	OV1D		GA NG
68-15955	OV1D	224MIBtn	
68-15956	OV1D	stored	Gulfport RAP, Ms
68-15957	1V1D	1MIBtn	
68-15958	GOV1D		
68-15959	OV1D	rework	Grumman
68-15960	OV1D	AMARC	YA...
68-15961	OV1D	A\151AEB	GA NG
68-15962	OV1D	SoC	
68-15963	OV1D	rework	Grumman
68-15964	OV1D	2MIBtn	
68-15965	OV1D	USAICS	
68-16990	OV1D		
68-16991	OV1D	USAICS	
68-16992	JOV1D	stored	Gulfport RAP, Ms
68-16993	OV1D		
68-16994	OV1D	224MIBtn	
68-16995	OV1D	USAICS	
68-16996	OV1D	1MIBtn	
69-16997	OV1D	AATC	Cairns AAF
69-16998	OV1D	A\151AEB	GA NG
69-16999	OV1D	w.o.27.2.80	
69-17000	JOV1D	w.o.5.4.79	
69-17001	OV1D	w.o.	
69-17002	OV1D	AATC	Cairns AAF
69-17003	OV1D	w.o.28.10.81	
69-17004	OV1D		

69-17005	OV1D	15MIBtn		69-17017	OV1D		
69-17006	OV1D	A\151AEB	GA NG	69-17018	OV1D	w.o.22.2.74	
69-17007	OV1D	SoC		69-17019	JOV1D	stored	Gulfport RAP, Ms
69-17008	OV1D	224MIBtn		69-17020	OV1D	w.o.10.10.83	
69-17009	OV1D	Libby AAF		69-17021	OV1D	SoC	
69-17010	JOV1D	SoC		69-17022	OV1D	to Korea	
69-17011	OV1D	15MIBtn		69-17023	OV1D	15MIBtn	
69-17012	OV1D	USAICS		69-17024	OV1D	SoC	
69-17013	OV1D	w.o.6.12.73		69-17025	OV1D	w.o.12.8.72	
69-17014	OV1D	SoC		69-17026	OV1D	USAICS	
69-17015	OV1D	w.o.22.1.74					
69-17016	OV1D	AATC	Cairns AAF				

UV18 TWIN OTTER

c\n   495, 496, 680, 681, 800, 801.

76-22565	UV18A	B\1-207AVN det.1	AK NG	79-23256	UV18A		AK NG
76-22566	UV18A		AK NG	82-23835	UV18A		AK NG
79-23255	UV18A	B\1-207AVN	AK NG	82-23836	UV18A	B\1-207AVN	AK NG

V20 TURBO-PORTER

c\n   802, 803.

79-23253	UV20A	Golden Knights	79-23254	UV20A	Golden Knights

ANTONOV AN-2 PTCHELKA

91-0074	An-2	USAASOTB

BEECH D18S

89-0126	D18S

BEECH 65 QUEENAIR

c\n   LC-40, 2 unknown, LC-62, LC-70, 2 unknown, LC-165, 3 unknown.

81-23658	B65			85-23525	B65	SoC
82-23659	B65			85-24370	B65	
82-24029	B65	SoC		85-24373	B65	
82-24054	B65	D\1-228AVN		85-24374	B65-80	SoC
83-24188	B65	D\1-228AVN		85-25349	B65	
84-24320	B65-80	SOC				

BEECH 90 KINGAIR

c\n   1 unknown, LJ-129, 1 unknown.

86-0092	A90	
86-1683	A90	
87-0142	A90	SoC

BEECH A100 KINGAIR

90-0060 A100

CASA 212 AVIOCAR

c\n   296.

88-0321 C212   w.o.1.12.89

CESSNA 182 SKYLANE II

89-0264   C182   89-0265   C182

CESSNA 185

86-0142   C185

CESSNA 310

82-23786 C310   84-24312 C310
83-24151 C310   STARC   OK NG 85-25350 C310

CESSNA 402B

82-24101 C402B

FAIRCHILD SA227AT MERLIN IV

c\n   AT-004

N55CE   227AT   Corps of Eng.

GRUMMAN GULFSTREAM I

c\n   2

86-0402   G159

HP\CENTURY JETSTREAM

c\n   224, 218, 240.

85-24687 HP137   85-24689 HP137
85-24688 HP137

MIL Mi-17

5733   Mi17   Ft.Bliss   ex Iraq AF

MIL Mi-24

4493   Mi24   Ft. Bliss   ex Iraq AF

PBN BN2B-21 ISLANDER

88-0196   BN2B   USANVL   Davison AAF

PILATUS PC-9

c\n   180 to 182.

91-0071	PC9	rtd. to Manufacturer	91-0073	PC9	rtd. to Manufacturer
91-0072	PC9	rtd. to Manufacturer			

PIPER PA31T CHEYENNE II

c\n   31T-7720051

85-1609   PA31T

TURBO COMMANDER 680W

c\n   1772-10.

83-24126   R680W to NASA

SHORT SC7 SKYVAN

90-0042   SC7   SoC

UNITED STATES NAVY

A3 SKYWARRIOR

c\n   7588, 7589, 12 unknown, 10300 to 10337, 10763 to 10837, 78 unknown, 12071 to 12113, 35 unknown.

c/n	Type	Status	Location
125412	XA3D-1	scrapped	
125413	XA3D-1	preserved	USS Intrepid
130352	NA3A	w.o.6.65	
130353	A3A	w.o.10.64	
130354	A3A	scrapped	Kolar
130355	A3A	scrapped	
130356	YEA3A	w.o.16.10.58	
130357	A3A	scrapped	
130358	NRA3A	scrapped	
130359	A3A	w.o.29.10.55	
130360	EA3A	w.o.	
130361	YEA3A	Pima Museum	
130362	EA3A	w.o.28.5.59	
130363	EA3A	scrapped	
135407	A3A	scrapped	Kolar
135408	A3A	scrapped	
135409	NA3A	MASDC	
135410	A3A	w.o.1.59	
135411	NRA3A	MASDC	
135412	A3A	scrapped	
135413	A3A	w.o.9.63	
135414	A3A	w.o.11.65	
135415	A3A	scrapped	
135416	A3A	scrapped	
135417	A3A	w.o.9.57	
135418	A3A	preserved	NAS Pensacola
135419	A3A	scrapped	
135420	A3A	scrapped	
135421	A3A	scrapped	
135422	A3A	w.o.5.60	
135423	A3A	scrapped	
135424	A3A	scrapped	
135425	A3A	scrapped	
135426	A3A	w.o.11.63	
135427	A3A	scrapped	
135428	A3A	scrapped	
135429	A3A	w.o.6.59	
135430	A3A	scrapped	
135431	A3A	scrapped	
135432	A3A	scrapped	
135433	A3A	scrapped	
135434	A3A	dump	Edwards AFB
135435	A3A	scrapped	
135436	A3A	scrapped	
135437	A3A	scrapped	
135438	A3A	scrapped	
135439	A3A	scrapped	
135440	A3A	w.o.8.59	
135441	A3A	w.o.8.57	
135442	A3A	w.o.3.63	
135443	A3A	scrapped	
135444	A3A	scrapyard	Consolidated
138902	A3A	w.o.8.60	
138903	A3A	w.o.1.58	
138904	KA3B	MASDC	
138905	KA3B	MASDC	
138906	KA3B	MASDC	
138907	A3B	w.o.8.63	
138908	A3B	w.o.1.58	
138909	A3B	w.o.2.12.68	
138910	A3B	w.o.8.57	
138911	KA3B	MASDC	
138912	A3B	w.o.8.60	
138913	A3B	w.o.5.59	
138914	KA3B	MASDC	
138915	EKA3B	scrapped	
138916	A3B	scrapped	
138917	A3B	scrapped	
138918	EKA3B	dest. 14.1.69	
138919	A3B	w.o.5.57	
138920	A3B	scrapped	
138921	KA3B	w.o.4.68	
138922	A3B	scrapped	
138923	A3B	scrapped	
138924	A3B	w.o.7.61	
138925	KA3B	w.o.16.11.87	
138926	A3B	scrapped	
138927	A3B	w.o.2.63	
138928	A3B	scrapped	
138929	EKA3B	w.o.26.1.80	
138930	A3B	w.o.12.2.65	
138931	KA3B	MASDC	
138932	KA3B	AMARC	2A109
138933	KA3B	MASDC	
138934	A3B	w.o.3.63	
138935	A3B	scrapped	
138936	A3B	w.o.10.57	
138937	KA3B	SoC	
138938	NA3A	AMARC	2A...
138939	KA3B	MASDC	
138940	KA3B	scrapped	
138941	KA3B	scrapped	
138942	A3B	MASDC	
138943	KA3B	w.o.17.2.69	
138944	NA3A		
138945	A3B	scrapped	Consolidated
138946	A3B	scrapped	Consolidated
138947	EKA3B	w.o.25.5.65	
138948	A3B	w.o.8.60	
138949	A3B	w.o.8.58	
138950	A3B	w.o.3.59	
138951	KA3B	scrapped	
138952	A3B	scrapped	
138953	KA3B	dump	NAS Alameda
138954	A3B	w.o.2.58	
138955	KA3B	scrapped	
138956	A3B	scrapped	
138957	KA3B	scrapped	
138958	A3B	w.o.4.60	
138959	A3B	scrapped	Consolidated
138960	A3B	w.o.1.61	
138961	KA3B	scrapped	Consolidated
138962	A3B	w.o.6.62	
138963	KA3B	scrapped	
138964	KA3B	dump	NAS Alameda
138965	KA3B	AMARC	2A110
138966	KA3B	scrapped	
138967	KA3B	MASDC	
138968	A3B	w.o.10.10.72	
138969	KA3B	AMARC	2A068
138970	A3B	scrapped	
138971	A3B	scrapped	Consolidated
138972	KA3B	w.o.10.68	
138973	KA3B	AMARC	2A052
138974	KA3B	scrapped	
138975	A3B	scrapped	
138976	A3B	w.o.3.61	
142236	A3B	w.o.11.62	
142237	EKA3B	AMARC	2A071
142238	KA3B	MASDC	
142239	EKA3B	scrapped	
142240	A3B	preserved	USS Yorktown
142241	KA3B	scrapped	
142242	A3B	scrapped	
142243	A3B	w.o.1.62	
142244	KA3B	dump	NAS Alameda

142245	A3B	w.o.5.61	
142246	A3B	preserved	Bradley ANGB
142247	KA3B	AMARC	2A050
142248	KA3B	dump	NAS Alameda
142249	KA3B	AMARC	2A072
142250	A3B	w.o.27.12.64	
142251	EKA3B	AMARC	2A073
142252	KA3B	w.o.8.70	
142253	KA3B	AMARC	2A067
142254	A3B	w.o.3.59	
142255	EKA3B	AMARC	2A081
142256	YRA3B	Westinghouse Corp.	
142257	KA3B	w.o.8.3.74	
142400	EKA3B	w.o.4.7.70	
142401	KA3B	AMARC	2A051
142402	EKA3B	w.o.24.9.62	
142403	EKA3B	MASDC	
142404	EKA3B	AMARC	2A087
142405	A3B	scrapped	
142406	KA3B	scrapped	Kolar
142407	A3B	scrapped	
142630	NA3B	AMARC	2A132
142631	A3B	w.o.7.64	
142632	EKA3B	AMARC	2A082
142633	A3B	w.o.2.10.60	
142634	EKA3B	w.o.21.1.73	
142635	KA3B	scrapped	Kolar
142636	A3B	w.o.12.58	
142637	A3B	w.o.10.61	
142638	EKA3B	AMARC	2A088
142639	A3B	w.o.7.59	
142640	A3B	w.o.18.9.59	
142641	A3B	w.o.7.59	
142642	EKA3B	w.o.3.60	
142643	A3B	w.o.4.59	
142644	KA3B	AMARC	2A058
142645	A3B	w.o.10.58	
142646	EKA3B	AMARC	2A083
142647	EKA3B	AMARC	2A086
142648	A3B	w.o.10.61	
142649	KA3B	w.o.10.72	
142650	EKA3B	AMARC	2A124
142651	EKA3B	AMARC	2A076
142652	EKA3B	AMARC	2A077
142653	KA3B	w.o.12.4.66	
142654	EKA3B	preserved	D-M AFB
142655	KA3B	w.o.21.10.67	
142656	EKA3B	AMARC	2A092
142657	EKA3B	w.o.16.5.70	
142658	KA3B	w.o.28.7.67	
142659	EKA3B	w.o.10.7.77	
142660	EKA3B	MASDC	
142661	EKA3B	MASDC	
142662	EKA3B		
142663	A3B	w.o.10.61	
142664	EKA3B	VAK-208	AF-520
142665	A3B	w.o.1.4.66	
142666	RA3B	AMARC	2A040
142667	NRA3B	NAWC-WD	71
142668	RA3B		
142669	RA3B	AMARC	2A060
142670	EA3B	w.o.4.6.68	
142671	EA3B	AMARC	2A117
142672	VA3B	w.o.23.1.85	
142673	EA3B	dump	NAS Alameda
144626	A3B	w.o.11.62	
144627	A3B	w.o.8.3.67	
144628	EKA3B	AMARC	2A100
144629	A3B	w.o.10.60	
144825	NRA3B	NAWC-WD	75
144826	RA3B	w.o.8.8.69	
144827	ERA3B	w.o.16.11.87	
144828	RA3B	w.o.16.6.67	
144829	RA3B	w.o.6.64	
144830	RA3B	w.o.16.5.63	
144831	RA3B	scrapped	
144832	ERA3B	to N162TB	
144833	RA3B	NAWC-WD	73
144834	URA3B	AMARC	2A125
144835	RA3B	w.o.25.8.67	
144836	RA3B	w.o.3.60	
144837	RA3B	w.o.4.69	
144838	ERA3B		
144839	RA3B	scrapped	
144840	NRA3B	NAWC-WD	78
144841	ERA3B		
144842	RA3B	w.o.13.6.66	
144843	RA3B	Raytheon	
144844	RA3B	w.o.14.10.67	
144845	RA3B	w.o.9.60	
144846	ERA3B		
144847	RA3B	w.o.1.1.68	
144848	EA3B	scrapped	
144849	EA3B	AMARC	2A123
144850	EA3B	w.o.25.1.87	
144851	EA3B	w.o.26.2.70	
144852	EA3B		
144853	EA3B	w.o.11.59	
144854	EA3B	w.o.26.6.87	
144855	EA3B	w.o.19.9.73	
144856	TA3B	to N160TB	
144857	TA3B	AMARC	2A121
144858	TA3B		
144859	TA3B	AMARC	2A122
144860	TA3B	AMARC	2A119
144861	TA3B	w.o.27.8.73	
144862	TA3B		
144863	TA3B	w.o.9.7.74	
144864	TA3B	dump	NAS Alameda
144865	TA3B	NARF	Alameda
144866	TA3B	AMARC	2A128
144867	TA3B	to N14867	
146446	ERA3B	to N161TB	
146447	ERA3B	AMARC	2A120
146448	EA3B		
146449	EA3B	AMARC	2A131
146450	EA3B	w.o.4.8.82	
146451	EA3B		
146452	EA3B	AMARC	2A130
146453	EA3B	AMARC	2A126
146454	EA3B		
146455	EA3B	AMARC	2A129
146456	EA3B	w.o.1.61	
146457	EA3B		
146458	EA3B	w.o.3.11.66	
146459	EA3B		
147648	EKA3B		
147649	EKA3B	w.o.18.6.71	
147650	A3B	w.o.6.10.66	
147651	A3B	scrapped	
147652	EKA3B	MASDC	
147653	KA3B	w.o.3.11.67	
147654	A3B	w.o.12.63	
147655	KA3B		
147656	EKA3B		
147657	KA3B	AMARC	2A118
147658	EKA3B	AMARC	2A090
147659	EKA3B	AMARC	2A079
147660	EKA3B	dump	NAS Alameda
147661	A3B	w.o.9.61	
147662	A3B	w.o.11.62	
147663	EKA3B		
147664	EKA3B	w.o.24.2.65	
147665	EKA3B	w.o.13.1.88	
147666	KA3B	preserved	Oakland, Ca
147667	EKA3B		

A4 SKYHAWK

c\n   10709 to 10728, 11284 to 11489, 11601 to 11608, 11736 to 12015, 12114 to 12392, 12433 to 12807,
          12812 to 12991, 13012 to 13191, 12992 to 13011, 13192 to 13524 plus later deliveries.

137812	YA4A				139965	A4A		
137813	A4A	preserved	NAS Pensacola		139966	A4A		
137814	A4A	preserved	NAWC China Lake		139967	A4A		
137815	A4A				139968	A4A	preserved	Annapolis, Md
137816	A4A				139969	A4A		
137817	A4A				139970	A4A		
137818	A4A	stored	NAWC China Lake		142082	A4B		
137819	A4A				142083	A4B		
137820	A4A				142084	A4B		
137821	A4A				142085	A4B		
137822	A4A				142086	A4B		
137823	A4A				142087	A4B		
137824	A4A				142088	A4B		
137825	A4A				142089	NA4B	scrapyard	Consolidated
137826	A4A				142090	A4B		
137827	A4A				142091	A4B		
137828	A4A				142092	A4B		
137829	A4A				142093	A4B		
137830	A4A	preserved	NAS Pensacola		142094	A4B	preserved	NAS Lemoore
137831	A4A				142095	A4B		
139919	A4A				142096	A4B		
139920	A4A				142097	A4B		
139921	A4A				142098	A4B		
139922	A4A	dump	NAS Memphis		142099	A4B		
139923	A4A				142100	A4B		
139924	A4A				142101	A4S	to Singapore	
139925	A4A				142102	A4B		
139926	A4A				142103	A4B		
139927	A4A				142104	A4B		
139928	A4A				142105	A4B		
139929	A4A				142106	A4B	preserved	NAS Lakehurst
139930	A4A				142107	A4B		
139931	A4A				142108	A4B		
139932	A4A				142109	A4B		
139933	A4A				142110	A4B		
139934	A4A				142111	A4B		
139935	A4A				142112	A4B		
139936	A4A				142113	A4B		
139937	A4A				142114	A4B	w.o.3.10.67	
139938	A4A	dump	NAS Memphis		142115	A4B		
139939	A4A	NATTC			142116	A4B	to N116MD	
139940	A4A				142117	A4B		
139941	A4A				142118	A4B		
139942	A4A				142119	A4B		
139943	TA4B				142120	TA4B	MASDC	
139944	A4A				142121	A4B		
139945	A4A				142122	A4B		
139946	A4A				142123	A4B		
139947	A4A	NATTC			142124	A4B		
139948	A4A				142125	A4S	to Singapore	619
139949	A4A				142126	A4B		
139950	A4A				142127	A4B		
139951	A4A				142128	A4B		
139952	A4A				142129	A4B		
139953	A4A				142130	A4B		
139954	A4A				142131	A4S	to Singapore	604
139955	A4A				142132	A4B		
139956	A4A				142133	A4B		
139957	A4A				142134	A4B		
139958	A4A				142135	A4B		
139959	A4A				142136	A4B		
139960	A4A				142137	A4B		
139961	A4A				142138	A4B		
139962	A4A				142139	A4B		
139963	A4A				142140	TA4S	to Singapore	
139964	A4A				142141	A4B		

Serial	Type	Status	Location
142142	A4A		
142143	A4A		
142144	A4A		
142145	A4A		
142146	A4A		
142147	A4A		
142148	A4A		
142149	A4A		
142150	A4A	preserved	NAS Mayport
142151	A4A		
142152	A4A		
142153	A4A		
142154	A4A		
142155	A4A		
142156	A4A		
142157	A4A		
142158	A4A		
142159	A4A		
142160	A4A		
142161	A4A		
142162	A4A		
142163	A4A		
142164	A4A		
142165	A4A	dump	NAS Memphis
142166	A4A	to N5548	
142167	A4A		
142168	A4A		
142169	A4A		
142170	A4A		
142171	A4A		
142172	A4A		
142173	A4A		
142174	A4A		
142175	A4A		
142176	A4A		
142177	A4A	stored	NAWC China Lake
142178	A4A		
142179	A4A		
142180	A4A		
142181	A4A		
142182	A4A		
142183	A4A		
142184	A4A		
142185	A4A		
142186	A4A		
142187	A4A		
142188	A4A		
142189	A4A		
142190	A4A		
142191	A4A		
142192	A4A		
142193	A4A		
142194	A4A		
142195	A4A		
142196	A4A		
142197	A4A		
142198	A4A		
142199	A4A		
142200	A4A		
142201	A4A		
142202	A4A		
142203	A4A		
142204	A4A		
142205	A4A		
142206	A4A		
142207	A4A		
142208	A4A		
142209	A4A		
142210	A4A		
142211	A4A		
142212	A4A		
142213	A4A		
142214	A4A		
142215	A4A		
142216	A4A		
142217	A4A		
142218	A4A		
142219	A4A	preserved	Bradley ANGB
142220	A4A		
142221	A4A		
142222	A4A		
142223	A4A		
142224	A4A		
142225	A4A		
142226	A4A	preserved	MCAS Quantico
142227	A4A	stored	NAWC China Lake
142228	A4A		
142229	A4A		
142230	A4A		
142231	A4A		
142232	A4A		
142233	A4A		
142234	A4A		
142235	A4A	stored	NAWC China Lake
142416	A4B		
142417	A4B		
142418	A4B		
142419	A4B		
142420	A4B		
142421	A4B		
142422	A4B		
142423	A4B		
142674	A4B		
142675	A4B		
142676	TA4A	AMARC	3A113
142677	A4B		
142678	A4B		
142679	A4B	ranges	Eglin, Fl
142680	A4B		
142681	A4B		
142682	A4B		
142683	A4B		
142684	A4B		
142685	A4B		
142686	A4B		
142687	A4B		
142688	A4B		
142689	A4B	scrapyard	AirMet
142690	TA4A	AMARC	3A136
142691	A4B		
142692	A4B		
142693	A4B	to NASA	Ames
142694	A4B		
142695	A4B		
142696	A4B		
142697	A4B		
142698	A4B		
142799	A4B		
142700	A4B		
142701	A4B		
142702	A4B		
142703	A4B	ranges	Eglin, Fl
142704	TA4B	AMARC	3A104
142705	A4B		
142706	A4B		
142707	A4B		
142708	TA4B	AMARC	3A103
142709	A4B		
142710	A4B		
142711	A4S	to Singapore	621
142712	A4B		
142713	TA4S	to Singapore	
142714	A4B		
142715	A4B		
142716	A4B		
142717	A4B		
142718	A4B	stored	NAD Pensacola
142719	A4B		
142720	A4B		
142721	A4B		
142722	A4B	dump	NAS Memphis
142723	A4B		
142724	A4B		
142725	A4B		
142726	A4B	scrapyard	Consolidated
142727	A4B		
142728	A4B		
142729	A4B		
142730	A4B		
142731	A4B		
142732	A4B		
142733	A4B		
142734	A4B		
142735	A4B		
142736	A4B		
142737	A4B		

Serial	Type	Status	Location
142738	A4B		
142739	A4B		
142740	A4B		
142741	A4B		
142742	A4B	w.o.26.11.67	
142743	A4B		
142744	A4B		
142745	TA4B	AMARC	3A195
142746	A4S	to Singapore	
142747	A4B		
142748	A4B		
142749	A4S		
142750	A4B		
142751	A4S	to Singapore	
142752	A4B		
142753	A4B		
142754	TA4B	AMARC	3A269
142755	A4B		
142756	A4B		
142757	A4B		
142758	A4B		
142759	A4B		
142760	A4B		
142761	A4B	preserved	Selfridge ANGB
142762	A4B		
142763	A4B		
142764	A4B	stored	Ontario, Ca
142765	A4B		
142766	A4B		
142767	A4B		
142768	TA4S	to Singapore	
142769	A4B		
142770	A4S	to Singapore	633
142771	A4S	to Singapore	602
142772	TA4B	AMARC	3A240
142773	A4B		
142774	A4B		
142775	TA4B	AMARC	3A106
142776	A4B		
142777	A4B		
142778	TA4B	to Singapore	
142779	A4B		
142780	A4B		
142781	TA4B	AMARC	3A158
142782	A4B		
142783	A4B		
142784	A4B		
142785	A4B		
142786	A4B		
142787	A4B		
142788	A4B		
142789	A4B		
142790	A4B		
142791	A4B		
142792	A4B		
142793	A4B		
142794	A4B		
142795	A4B		
142796	A4B		
142797	A4B		
142798	A4B	stored	Ontario, Ca
142799	A4S		
142800	A4S	to Singapore	
142801	A4B		
142802	A4B		
142803	A4B		
142804	A4B		
142805	A4B		
142806	A4B		
142807	A4B		
142808	A4B		
142809	A4B		
142810	A4B		
142811	A4B		
142812	A4B		
142813	A4B		
142814	TA4B	AMARC	3A116
142815	A4B		
142816	A4B		
142817	A4B		
142818	A4B		
142819	A4S	to Singapore	636
142820	A4B		
142821	A4B		
142822	A4B		
142823	A4B		
142824	A4B	ranges	Eglin, Fl
142825	A4B	AMARC	3A197
142826	A4B		
142827	A4B		
142828	A4B		
142829	A4B		
142830	A4B		
142831	A4B		
142832	A4S	to Singapore	601
142833	TA4B	preserved	USS Intrepid
142834	A4B		
142835	A4B	stored	Ontario, Ca
142836	A4B		
142837	A4B		
142838	A4B		
142839	A4B		
142840	A4S	to Singapore	
142841	A4B		
142842	A4B	MASDC	
142843	TA4B	AMARC	3A131
142844	TA4B	AMARC	3A159
142845	A4B		
142846	A4B	stored	Ontario, Ca
142847	TA4B	AMARC	3A165
142848	A4B	GIA	NAWC Lakehurst
142849	TA4B	AMARC	3A236
142850	A4S	to Singapore	600
142851	A4B		
142852	A4B		
142853	A4B		
142854	TA4S	to Singapore	
142855	A4B		
142856	A4B		
142857	A4B		
142858	A4B		
142859	A4B		
142860	TA4S	to Singapore	
142861	A4B		
142862	A4B		
142863	A4B	ranges	Eglin, Fl
142864	A4B		
142865	A4B		
142866	A4B		
142867	A4B		
142868	A4B		
142869	A4B		
142870	A4S	to Singapore	645
142871	TA4B	MASDC	
142872	A4B		
142873	A4B		
142874	A4B	preserved	NAS Chase Field
142875	TA4B	AMARC	3A128
142876	A4S	to Singapore	646
142877	A4B		
142878	A4B		
142879	A4B	dump	MCAS El Toro
142880	A4B		
142881	TA4S	to Singapore	
142882	A4S	to Singapore	647
142883	A4B		
142884	TA4B		
142885	A4B		
142886	A4B		
142887	A4B		
142888	A4B		
142889	A4B		
142890	A4B		
142891	TA4B	to Singapore	
142892	A4B		
142893	A4B		
142894	A4B		
142895	A4B		
142896	A4B		
142897	A4B		
142898	A4B		
142899	TA4B	AMARC	3A112
142900	A4B		
142901	A4B		
142902	A4B		
142903	A4B		

142904	TA4B	MASDC	
142905	A4B	to N905MD	
142906	A4B		
142907	A4B		
142908	A4S	to Singapore	603
142909	TA4B	AMARC	3A282
142910	A4B		
142911	A4B		
142912	A4B		
142913	A4B		
142914	A4B		
142915	A4B		
142916	A4B		
142917	TA4B	AMARC	3A107
142918	A4B		
142919	A4B		
142920	A4B		
142921	TA4B	AMARC	3A133
142922	A4B		
142923	A4B		
142924	A4B		
142925	A4B		
142926	A4B		
142927	A4B		
142928	TA4B	Pima Museum	
142929	A4B		
142930	A4B		
142931	A4B		
142932	A4B		
142933	A4B		
142934	A4B		
142935	A4B		
142936	TA4B	AMARC	3A163
142937	A4B		
142938	A4B	MASDC	
142939	A4B		
142940	A4B	preserved	NAS S.Weymouth
142941	A4B		
142942	A4S	to Singapore	650
142943	A4B		
142944	A4B	dump	NAS S.Weymouth
142945	A4B		
142946	A4B		
142947	A4B		
142948	TA4B	Pima Museum	
142949	A4B		
142950	A4B		
142951	A4B		
142952	A4B		
142953	A4B		
144868	A4B		
144869	A4B		
144870	A4B		
144871	A4B		
144872	A4B		
144873	A4B		
144874	A4S	to Singapore	605
144875	TA4B	AMARC	3A259
144876	A4B		
144877	A4B		
144878	A4B		
144879	A4B		
144880	A4B		
144881	A4B		
144882	A4B		
144883	A4B		
144884	TA4B	AMARC	3A117
144885	TA4B	to Singapore	
144886	A4B		
144887	A4B		
144888	A4B	ranges	Eglin, Fl
144889	A4B		
144890	A4B		
144891	A4B	ranges	Eglin, Fl
144892	A4B		
144893	A4B		
144894	A4B		
144895	A4B		
144896	TA4S	to Singapore	
144897	A4B		
144898	A4B		
144899	A4B		
144900	A4B		

144901	A4B		
144902	A4B		
144903	TA4B	AMARC	3A160
144904	TA4B	to Singapore	
144905	A4B		
144906	A4B	to NASA	Ames
144907	A4B		
144908	A4B		
144909	A4B	stored	Ontario, Ca
144910	A4B		
144911	A4B		
144912	A4B		
144913	A4B		
144914	A4B	ranges	Eglin, Fl
144915	A4B		
144916	TA4S	to Singapore	
144917	TA4B	AMARC	3A188
144918	A4B		
144919	TA4B	AMARC	3A233
144920	A4B		
144921	A4B		
144922	TA4S	to Singapore	
144923	TA4B	scrapyard	Allied
144924	A4B		
144925	TA4B	dump	NAS Memphis
144926	A4S	to Singapore	656
144927	A4B		
144928	A4B		
144929	A4B	preserved	NAS Pensacola
144930	A4B		
144931	A4B		
144932	A4B		
144933	A4B		
144934	A4B		
144935	A4B		
144936	A4B		
144937	A4B	stored	Ontario, Ca
144938	A4B	w.o.21.5.65	
144939	A4B		
144940	A4B		
144941	A4B		
144942	A4B		
144943	A4B		
144944	TA4S	to Singapore	
144945	A4B		
144946	A4B		
144947	A4B	stored	Ontario, Ca
144948	A4B		
144949	TA4B	scrapyard	Consolidated
144950	A4B		
144951	A4B		
144952	A4B		
144953	A4B		
144954	A4B		
144955	A4B		
144956	A4B		
144957	A4B		
144958	A4B		
144959	A4B		
144960	TA4B	AMARC	3A205
144961	TA4S	to Singapore	
144962	TA4B		
144963	A4B		
144964	A4B	dump	NAS Memphis
144965	TA4B	AMARC	3A227
144966	A4B		
144967	TA4B	scrapyard	Consolidated
144968	A4B		
144969	A4B		
144970	A4B		
144971	A4B		
144972	A4B	stored	Ontario, Ca
144973	A4B		
144974	A4S	to Singapore	660
144975	A4B		
144976	A4B		
144977	TA4B	AMARC	3A123
144978	A4B		
144979	A4B		
144980	A4S	to Singapore	606
144981	A4B		
144982	A4B		
144983	A4B		

Serial	Type	Fate	Location
144984	A4B		
144985	A4B		
144986	A4B		
144987	A4B		
144988	A4B		
144989	A4B		
144990	A4B	ranges	Eglin, Fl
144991	A4B		
144992	A4B		
144993	A4B		
144994	A4B		
144995	A4B		
144996	A4B		
144997	A4B		
144998	A4B		
144999	A4B		
145000	A4B	NATTC	
145001	A4B		
145002	A4B	w.o.23.7.67	
145003	A4B	ranges	Eglin, Fl
145004	A4B		
145005	A4B		
145006	A4B		
145007	A4B		
145008	A4B		
145009	A4B		
145010	A4B		
145011	A4B		
145012	A4B		
145013	A4S	to Singapore	607
145014	A4B		
145015	A4B		
145016	A4B		
145017	A4B		
145018	A4B		
145019	A4B		
145020	A4B		
145021	TA4S	to Singapore	
145022	A4B		
145023	A4B		
145024	A4B		
145025	A4B		
145026	A4B		
145027	A4B		
145028	A4B		
145029	A4B		
145030	TA4S	to Singapore	679
145031	A4B		
145032	A4B		
145033	TA4B	AMARC	3A115
145034	A4B	dump	Memphis NAS
145035	A4B		
145036	A4B	scrapyard	AirMet
145037	TA4S	to Singapore	651
145038	TA4S	to Singapore	680
145039	A4B		
145040	A4B	w.o.7.8.66	
145041	TA4S	to Singapore	
145042	A4B		
145043	TA4S	to Singapore	653
145044	A4B		
145045	A4B		
145046	A4S	to Singapore	681
145047	A4S	to Singapore	
145048	A4B		
145049	A4B		
145050	A4B		
145051	A4B		
145052	A4B		
145053	A4B		
145054	A4B		
145055	A4B		
145056	A4S	to Singapore	682
145057	A4B		
145058	A4B		
145059	A4S	to Singapore	683
145060	A4B		
145061	A4B		
145062	NA4C		
145063	NA4C	to Singapore	
145064	A4C	scrapyard	AirMet
145065	A4C	to Malaysia	
145066	A4C		
145067	YA4S	to Singapore	
145068	A4C	to Singapore	
145069	A4B	dump	NAS Lakehurst
145070	A4B		
145071	A4S	to Singapore	
145072	A4B		
145073	A4S	to Singapore	
145074	A4B		
145075	A4B		
145076	A4L	scrapyard	AirMet
145077	A4L	NATTC	010
145078	A4L	to Malaysia	
145079	A4C		
145080	A4C		
145081	A4C	w.o.20.3.66	
145082	A4C	NATTC	Memphis
145083	A4C	MASDC	
145084	A4C		
145085	A4C		
145086	A4C		
145087	A4C	w.o.14.1.67	
145088	A4C		
145089	A4C		
145090	A4C		
145091	A4C		
145092	A4L	to Malaysia	
145093	A4C		
145094	A4C		
145095	A4C		
145096	A4C	w.o.6.5.68	
145097	A4C		
145098	A4C		
145099	A4C	MASDC	
145100	A4C		
145101	A4L	to Malaysia	
145102	A4C		
145103	A4L	to Malaysia	
145104	A4C		
145105	A4C		
145106	A4S	to Singapore	
145107	A4C		
145108	A4S	to Singapore	
145109	A4C	w.o.28.3.66	
145110	A4S	to Singapore	
145111	A4C		
145112	A4C	decoy	Gila Bend, Az
145113	A4C	MASDC	
145114	A4L	to Malaysia	
145115	A4C		
145116	A4C	w.o.2.12.66	
145117	A4C		
145118	A4S	to Singapore	
145119	A4L	to Malaysia	
145120	A4C		
145121	A4L	to Malaysia	
145122	A4L	NATTC	
145123	A4C		
145124	A4C		
145125	A4C		
145126	A4C		
145127	A4C	MASDC	
145128	A4L	scrapyard	AirMet
145129	A4C	MASDC	
145130	A4C		
145131	A4C	to Malaysia	
145132	A4S	to Singapore	
145133	A4C		
145134	A4C		
145135	A4C		
145136	A4C		
145137	A4C		
145138	A4C		
145139	A4C	MASDC	
145140	A4C		
145141	A4L	to Malaysia	
145142	A4C		
145143	A4C	w.o.2.12.66	
145144	A4C		
145145	A4C	w.o.20.1.67	
145146	A4C		
147669	A4L	scrapyard	AirMet
147670	A4C	w.o.1.8.67	
147671	A4L	scrapyard	AirMet

Serial	Type	Status	Location
147672	A4C	scrapyard	AirMet
147673	A4C	to Malaysia	
147674	A4C		
147675	A4C	MASDC	
147676	A4L		
147677	A4C	w.o.31.7.66	
147678	A4L	derelict	NAS Memphis
147679	A4L		
147680	A4L	w.o.23.7.66	
147681	A4C	MASDC	
147682	A4C	w.o.14.9.65	
147683	A4C		
147684	A4C		
147685	A4C	scrapyard	AirMet
147686	A4C		
147687	A4C	MASDC	
147688	A4C		
147689	A4C		
147690	A4L	scrapyard	AirMet
147691	A4C		
147692	A4C		
147693	A4C		
147694	A4C		
147695	A4C		
147696	A4C	scrapyard	AirMet
147697	A4C		
147698	A4C	MASDC	
147699	A4C		
147700	A4C		
147701	A4C	dump	NAS Lakehurst
147702	A4C	preserved	Pueblo, Co
147703	A4L	to Malaysia	
147704	A4C	w.o.2.1.66	
147705	A4C		
147706	A4L	MASDC	
147707	A4C		
147708	A4C	NATTC	400
147709	A4C	w.o.14.7.67	
147710	A4C	dump	NAS Lakehurst
147711	A4C		
147712	A4C	w.o.30.6.67	
147713	A4C		
147714	A4C	stored	Ontario, Ca
147715	A4L	NATTC	002
147716	A4C		
147717	A4C		
147718	A4C	w.o.11.11.66	
147719	A4C	w.o.8.8.67	
147720	A4C		
147721	A4C	MASDC	
147722	A4C	to Malaysia	
147723	A4C		
147724	A4C	w.o.14.1.67	
147725	A4C		
147726	A4C	to Malaysia	
147727	A4C		
147728	A4C	NPRO	Long Beach, Ca
147729	A4C		
147730	A4C		
147731	A4S	to Singapore	
147732	A4C	w.o.10.7.66	
147733	A4C		
147734	A4C	MASDC	
147735	A4C		
147736	A4L	to Malaysia	
147737	A4C	w.o.4.10.66	
147738	A4C	w.o.23.3.66	
147739	A4C		
147740	A4C	w.o.17.3.66	
147741	A4C	stored	Ontario, Ca
147742	A4C	to Singapore	
147743	A4C	to Singapore	
147744	A4C	MASDC	
147745	A4C	to Singapore	
147746	A4C		
147747	A4C	to Argentina	
147748	A4C	MASDC	
147749	A4C	MASDC	
147750	A4L	NATTC	7
147751	A4C		
147752	A4S	to Singapore	
147753	A4C	w.o.14.1.66	
147754	A4L	scrapyard	AirMet
147755	A4C		
147756	A4C		
147757	A4C		
147758	A4C		
147759	A4C	w.o.14.7.67	
147760	A4C		
147761	A4C	scrapyard	AirMet
147762	A4C	w.o.23.5.66	
147763	A4C		
147764	A4C	w.o.3.1.69	
147765	A4C	to Argentina	
147766	A4C		
147767	A4C	w.o.1.10.68	
147768	A4L	scrapyard	AirMet
147769	A4C		
147770	A4C		
147771	A4C		
147772	A4C	dump	MCAS Beaufort
147773	A4C	NATTC	Memphis
147774	A4C		
147775	A4C	w.o.20.10.66	
147776	A4C	w.o.13.12.66	
147777	A4C		
147778	A4C		
147779	A4S	to Singapore	927
147780	A4L	to Malaysia	
147781	A4C	MASDC	
147782	A4C	to Malaysia	
147783	A4S	to Singapore	
147784	A4C		
147785	A4S	to Singapore	
147786	A4S	to Singapore	
147787	A4C	stored	NAD Pensacola
147788	A4C	preserved	NAS Pensacola
147789	A4C		
147790	A4C		
147791	A4C		
147792	A4S	to Singapore	
147793	A4L	scrapyard	AirMet
147794	A4C		
147795	A4C	scrapyard	AirMet
147796	A4L	scrapyard	AirMet
147797	A4S	scrapyard	AirMet
147798	A4L	to Malaysia	
147799	A4C	w.o.25.4.67	
147800	A4C		
147801	A4C		
147802	A4L	to Malaysia	
147803	A4C	w.o.28.4.70	
147804	A4C	w.o.25.6.68	
147805	A4C		
147806	A4C	stored	Ontario, Ca
147807	A4L	to Malaysia	
147808	A4C	w.o.5.5.66	
147809	A4S	to Singapore	
147810	A4C		
147811	A4C		
147812	A4C		
147813	A4C		
147814	A4C	MASDC	
147815	A4L	scrapyard	AirMet
147816	A4C	w.o.18.5.67	
147817	A4C		
147818	A4C		
147819	A4C	w.o.13.12.66	
147820	A4C		
147821	A4S	to Singapore	
147822	A4C		
147823	A4C	to Singapore	
147824	A4C	w.o.14.7.72	
147825	A4C	NATTC	012
147826	A4C	MASDC	
147827	A4L	to Malaysia	
147828	A4C		
147829	A4C		
147830	A4C	stored	Ontario, Ca
147831	A4C		
147832	A4C		
147833	A4C	w.o.26.7.69	
147834	A4C		
147835	A4S	to Singapore	
147836	A4L	scrapyard	AirMet
147837	A4C		

Serial	Type	Status	Location/Code
147838	A4S	to Singapore	
147839	A4C		
147840	A4C		
147841	A4S	to Singapore	
147842	A4C	w.o.18.5.67	
147843	A4C		
147844	A4C	w.o.30.3.67	
147845	A4S	to Singapore	
147846	A4C		
147847	A4C	MASDC	
147848	A4C		
147849	A4C		
148304	A4S	to Singapore	
148305	A4C	w.o.22.12.65	
148306	A4L	to Malaysia	
148307	A4L	scrapyard	AirMet
148308	A4C		
148309	A4C		
148310	A4C	w.o.22.7.69	
148311	A4S	to Singapore	
148312	A4C		
148313	A4C	w.o.20.3.66	
148314	A4C	preserved NASM	
148315	A4C	scrapyard	AirMet
148316	A4L	scrapyard	AirMet
148317	A4C	w.o.7.4.65	
148435	A4C	to Argentina	
148436	A4L	to Malaysia	
148437	A4L		
148438	A4L	stored	Ontario, Ca
148439	A4S	to Singapore	
148440	A4L	w.o.25.8.67	
148441	A4L		
148442	A4L		
148443	A4L		
148444	A4L	w.o.25.3.66	
148445	A4C		
148446	A4L	scrapyard	AirMet
148447	A4C		
148448	A4C		
148449	A4S	to Singapore	
148450	A4C	to Argentina	
148451	A4C		
148452	A4C	scrapyard	Ontario, Ca
148453	A4C		
148454	A4C		
148455	A4C		
148456	A4C	w.o.7.7.66	
148457	A4C		
148458	A4S	to Singapore	
148459	A4C		
148460	A4C		
148461	A4C		
148462	A4S	to Singapore	
148463	A4S	to Singapore	
148464	A4C		
148465	A4C	MASDC	
148466	A4C	w.o.30.6.67	
148467	A4C	scrapyard	Ontario, Ca
148468	A4C	MASDC	
148469	A4S	to Singapore	
148470	A4C	w.o.20.8.68	
148471	A4C		
148472	A4C		
148473	A4C	w.o.21.5.66	
148474	A4C		
148475	A4C	w.o.13.8.65	
148476	A4C		
148477	A4C		
148478	A4C		
148479	A4L	to Malaysia	
148480	A4C		
148481	A4C	w.o.9.4.65	
148482	A4S	to Singapore	
148483	A4S	to Singapore	
148484	A4C	w.o.21.4.70	
148485	A4C	NATTC	007
148486	A4C	w.o.3.1.68	
148487	A4C		
148488	A4C	w.o.17.9.66	
148489	A4C	w.o.11.5.65	
148490	A4L	NATTC	403
148491	A4C	NATTC	006
148492	A4F	preserved	MCAS El Toro
148493	A4S	to Singapore	
148494	A4C	MASDC	
148495	A4C	w.o.22.6.70	
148496	A4C	w.o.17.11.66	
148497	A4S	to Singapore	
148498	A4C	scrapyard	AirMet
148499	A4C	w.o.21.3.66	
148500	A4C	to Malaysia	
148501	A4C		
148502	A4C	scrapyard	AirMet
148503	A4C		
148504	A4S	to Singapore	
148505	A4C	NATTC	011
148506	A4C	MASDC	
148507	A4C	w.o.21.12.66	
148508	A4C		
148509	A4C	to Malaysia	
148510	A4C	w.o.17.12.65	
148511	A4C		
148512	A4C	w.o.20.4.66	
148513	A4S	to Singapore	
148514	A4C	w.o.4.5.67	
148515	A4C		
148516	A4C		
148517	A4C	scrapyard	AirMet
148518	A4C	w.o.9.3.66	
148519	A4C	w.o.27.3.67	
148520	A4C		
148521	A4S	to Singapore	
148522	A4C		
148523	A4C		
148524	A4C		
148525	A4S	to Singapore	
148526	A4S	to Singapore	
148527	A4C		
148528	A4S	to Singapore	
148529	A4S	to Singapore	
148530	A4C		
148531	A4C	scrapyard	AirMet
148532	A4C		
148533	A4C	scrapyard	AirMet
148534	A4S	to Singapore	
148535	A4C	w.o.12.5.66	
148536	A4C	MASDC	
148537	A4C		
148538	A4C	NATTC	001
148539	A4C		
148540	A4C		
148541	A4C	to Singapore	
148542	A4C		
148543	A4C		
148544	A4C	w.o.4.7.67	
148545	A4C	MASDC	
148546	A4C		
148547	A4C	w.o.14.2.69	
148548	A4S	to Singapore	
148549	A4C	w.o.24.5.68	
148550	A4C	MASDC	
148551	A4C		
148552	A4S	to Singapore	
148553	A4C		
148554	A4C		
148555	A4L	to Malaysia	
148556	A4C	stored	Ontario, Ca
148557	A4C	w.o.3.4.65	
148558	A4S	to Singapore	
148559	A4C	stored	Ontario, Ca
148560	A4C		
148561	A4C		
148562	A4C	to Argentina	
148563	A4C		
148564	A4C	w.o.13.8.65	
148565	A4C		
148566	A4C	w.o.7.11.67	
148567	A4C		
148568	A4C		
148569	A4C	NATTC	009
148570	A4C		
148571	A4C	Pima Museum	
148572	A4C	NATTC	08
148573	A4C	scrapyard	AirMet
148574	A4C		

Serial	Type	Status	Notes
148575	A4C	to Malaysia	
148576	A4C	to Malaysia	
148577	A4C	w.o.3.6.65	
148578	A4C		
148579	A4C		
148580	A4C		
148581	A4L	scrapyard	AirMet
148582	A4C		
148583	A4C	w.o.17.4.66	
148584	A4C	w.o.19.10.65	
148585	A4C	w.o.17.3.67	
148586	A4C	dump	NATTC
148587	A4C	scrapyard	Consolidated
148588	A4L	to Malaysia	
148589	A4C	MASDC	
148590	A4C	MASDC	
148591	A4S	to Singapore	
148592	A4C	preserved	St. Petersburg
148593	A4C	to Malaysia	
148594	A4C		
148595	A4C	w.o.28.10.65	
148596	A4C	w.o.28.10.65	
148597	A4C	scrapyard	AirMet
148598	A4S	to Singapore	
148599	A4C	w.o.1.8.68	
148600	A4C		
148601	A4C		
148602	A4C	scrapyard	AirMet
148603	A4S	to Singapore	
148604	A4C	scrapyard	AirMet
148605	A4S	to Singapore	
148606	A4C	dump	NAS N. Island
148607	A4C	w.o.27.2.67	
148608	A4C	w.o.21.11.68	
148609	A4C		
148610	A4C	preserved	Alameda, Ca
148611	A4L	to Malaysia	
148612	A4C	stored	Ontario, Ca
148613	NA4E	preserved	Lowry AFB
148614	NA4E		
149487	A4C		
149488	A4C		
149489	A4C	NPRO	Long Beach, Ca
149490	A4C	w.o.24.8.65	
149491	A4C		
149492	A4S	to Singapore	
149493	A4S	to Singapore	902
149494	A4C	w.o.8.7.66	
149495	A4C	w.o.20.4.66	
149496	A4C		
149497	A4L	to Malaysia	
149498	A4S	to Singapore	
149499	A4C		
149500	A4L	scrapyard	AirMet
149501	A4C		
149502	A4L	scrapyard	AirMet
149503	A4C	Boeing Co. Seattle, Wa	
149504	A4C		
149505	A4S	NATTC	004
149506	A4L	to Malaysia	
149507	A4C	w.o.20.4.65	
149508	A4L	NATTC	22
149509	A4C	w.o.10.5.67	
149510	A4C		
149511	A4C		
149512	A4C		
149513	A4C		
149514	A4C	stored	Ontario, Ca
149515	A4C	w.o.21.3.66	
149516	A4L	FSI	N402FS
149517	A4C		
149518	A4L	to Malaysia	
149519	A4C		
149520	A4S	to Singapore	
149521	A4C	w.o.22.12.65	
149522	A4S	to Singapore	904
149523	A4C		
149524	A4C		
149525	A4C	w.o.22.9.70	
149526	A4C	stored	Ontario, Ca
149527	A4C	w.o.1.2.66	
149528	A4C	w.o.17.6.66	
149529	A4C	w.o.14.2.69	
149530	A4S	to Singapore	
149531	A4L	to Malaysia	
149532	A4C	NATTC	008
149533	A4E		
149534	A4C		
149535	A4C		
149536	A4L	to Malaysia	
149537	A4S	to Singapore	
149538	A4C		
149539	A4C		
149540	A4L	scrapyard	AirMet
149541	A4C	MASDC	
149542	A4C	w.o.9.7.67	
149543	A4C		
149544	A4S	to Singapore	
149545	A4C		
149546	A4C	w.o.17.11.67	
149547	A4C	to Malaysia	
149548	A4C		
149549	A4C		
149550	A4C	scrapyard	AirMet
149551	A4L	to Malaysia	
149552	A4C	w.o.14.2.66	
149553	A4C	w.o.6.8.70	
149554	A4C		
149555	A4C	scrapyard	AirMet
149556	A4L	NATTC	
149557	A4C	w.o.9.2.66	
149558	A4C		
149559	A4C	MASDC	
149560	A4C	w.o.1.12.65	
149561	A4C		
149562	A4C		
149563	A4C	NATTC	003
149564	A4C	to Argentina	
149565	A4C		
149566	A4C		
149567	A4C	w.o.25.6.66	
149568	A4C		
149569	A4C		
149570	A4C	w.o.13.11.64	
149571	A4C	w.o.5.5.66	
149572	A4C	w.o.11.2.65	
149573	A4L	to Malaysia	
149574	A4C	w.o.25.6.65	
149575	A4C	scrapyard	AirMet
149576	A4C	w.o.15.7.65	
149577	A4C	dump	NAS Fallon
149578	A4C	w.o.5.8.64	
149579	A4C		
149580	A4C	to Malaysia	
149581	A4C	scrapyard	AirMet
149582	A4C		
149583	A4L	to Malaysia	
149584	A4C		
149585	A4C	to Argentina	
149586	A4C		
149587	A4S	to Singapore	
149588	A4S	to Singapore	
149589	A4C		
149590	A4C	w.o.18.9.67	
149591	A4L	scrapyard	AirMet
149592	A4C		
149593	A4C	NATTC	415
149594	A4L	to Malaysia	
149595	A4C	scrapyard	AirMet
149596	A4C		
149597	A4C		
149598	A4C		
149599	A4C	scrapyard	Allied
149600	A4C	MASDC	
149601	A4C		
149602	A4C		
149603	A4C	w.o.9.7.67	
149604	A4L	to Malaysia	
149605	A4C		
149606	A4C	to Malaysia	
149607	A4C	to Malaysia	
149608	A4C	to Malaysia	
149609	A4C		
149610	A4C		
149611	A4C		
149612	A4C	w.o.3.3.65	

149613	NA4E		
149614	A4S	to Singapore	
149615	A4C		
149616	A4C	w.o.4.7.66	
149617	A4S	to Singapore	
149618	A4C	stored	Ontario, Ca
149619	A4C	w.o.4.10.67	
149620	A4L	scrapyard	AirMet
149621	A4C	scrapyard	Allied
149622	A4C		
149623	A4L	NATTC	
149624	A4C		
149625	A4C	w.o.30.12.64	
149626	A4C	to Malaysia	
149627	A4C		
149628	A4S	to Singapore	
149629	A4S	to Singapore	
149630	A4L	to Malaysia	
149631	A4C		
149632	A4C	w.o.2.8.67	
149633	A4L	to Malaysia	
149634	A4C		
149635	A4L	NATTC	09
149636	A4C	scrapyard	AirMet
149637	A4C		
149638	A4C	MASDC	
149639	A4C	w.o.7.4.67	
149640	A4C		
149641	A4C	w.o.27.12.69	
149642	A4C	stored	Ontario, Ca
149643	A4C		
149644	A4C		
149645	A4C	NPRO	Long Beach, Ca
149646	A4C		
149647	A4E		
149648	A4E		
149649	A4E		
149650	A4E		
149651	A4E	VF-45	AD-617
149652	A4E	w.o.20.5.67	
149653	A4E	VF-45	AD-612
149654	A4E	w.o.5.5.84	
149655	A4E	w.o.22.5.85	
149656	A4E	NFWS	554
149657	A4E		
149658	A4E		
149659	A4E		
149660	A4E	w.o.11.11.68	
149661	A4E		
149662	A4E		
149663	A4E		
149664	A4E		
149665	A4E	w.o.15.6.68	
149666	A4E		
149959	A4E	w.o.26.10.67	
149960	A4E		
149961	A4E	w.o.25.7.67	
149962	A4E	w.o.28.7.65	
149963	A4E	w.o.24.10.67	
149964	A4E		
149965	A4E		
149966	A4E		
149967	A4E		
149968	A4E		
149969	A4E	w.o.14.1.83	
149970	A4E	w.o.17.5.68	
149971	A4E	to Israel	
149972	A4E		
149973	A4E		
149974	A4E		
149975	A4E	w.o.31.8.67	
149976	A4E	w.o.29.1.68	
149977	A4E	VF-43	AD-615
149978	A4E		
149979	A4E		
149980	A4E		
149981	A4E	MASDC	
149982	A4E		
149983	A4E		
149984	A4E		
149985	A4E	VF-43	AD-329
149986	A4E		
149987	A4E		
149988	A4E		
149989	A4E	w.o.26.8.79	
149990	A4E	w.o.26.8.79	
149991	A4E	w.o.10.9.65	
149992	A4E	w.o.22.8.66	
149993	A4E	w.o.26.5.70	
149994	A4E		
149995	A4E		
149996	A4E	w.o.29.7.67	
149997	A4E		
149998	A4E		
149999	A4E	w.o.13.9.65	
150000	A4E	w.o.27.6.66	
150001	A4E	VF-43	AD-328
150002	A4E		
150003	A4E		
150004	A4E		
150005	A4E		
150006	A4E		
150007	A4E	w.o.25.8.87	
150008	A4E		
150009	A4E		
150010	A4E		
150011	A4E	w.o.11.5.67	
150012	A4E		
150013	A4E		
150014	A4E		
150015	A4E		
150016	A4E		
150017	A4E	w.o.1.7.66	
150018	A4E	w.o.13.8.64	
150019	A4E	w.o.29.12.65	
150020	A4E	w.o.8.9.66	
150021	A4E		
150022	A4E		
150023	A4E		
150024	A4E	w.o.6.9.64	
150025	A4E		
150026	A4E	to Israel	
150027	A4E		
150028	A4E	w.o.1.9.70	
150029	A4E		
150030	A4E		
150031	A4E		
150032	A4E	w.o.30.5.67	
150033	A4E	w.o.19.8.64	
150034	A4E	w.o.27.1.66	
150035	A4E		
150036	A4E	w.o.6.1.67	
150037	A4E	w.o.25.11.67	
150038	A4E	w.o.28.8.67	
150039	A4E		
150040	A4E	w.o.24.7.66	
150041	A4E	w.o.21.3.68	
150042	A4E		
150043	A4E	NFWS	556
150044	A4E	NFWS	555
150045	A4E		
150046	A4E		
150047	A4E	w.o.10.9.67	
150048	A4E	w.o.12.11.66	
150049	A4E		
150050	YA4F		
150051	A4E	w.o.12.11.66	
150052	A4E	w.o.4.8.67	
150053	A4E	w.o.23.1.68	
150054	A4E	w.o.20.9.66	
150055	A4E		
150056	A4E	w.o.8.6.84	
150057	A4E	w.o.25.1.68	
150058	A4E	MASDC	
150059	A4E	w.o.26.10.67	
150060	A4E		
150061	A4E		
150062	A4E	w.o.7.6.70	
150063	A4E	w.o.19.4.68	
150064	A4E	w.o.29.7.67	
150065	A4E		
150066	A4E		
150067	A4E	w.o.12.8.65	
150068	A4E	w.o.29.7.67	
150069	A4E		
150070	A4E		

Serial	Type	Notes	Code
150071	A4E	w.o.7.11.65	
150072	A4E	w.o.23.10.66	
150073	A4E	VF-126	NJ-631
150074	A4E		
150075	A4E	w.o.7.2.65	
150076	A4E	VFA-127	NJ-705
150077	A4E		
150078	A4E	w.o.29.3.65	
150079	A4E	w.o.27.8.66	
150080	A4E	w.o.6.12.69	
150081	A4E		
150082	A4E	to Israel	
150083	A4E		
150084	A4E	w.o.29.7.67	
150085	A4E		
150086	A4E		
150087	A4E		
150088	A4E	w.o.12.11.66	
150089	A4E		
150090	A4E		
150091	A4E		
150092	A4E		
150093	A4E		
150094	A4E	w.o.26.9.68	
150095	A4E		
150096	A4E	w.o.20.7.67	
150097	A4E	w.o.20.7.67	
150098	A4E		
150099	A4E		
150100	A4E	w.o.2.9.68	
150101	A4E	w.o.21.7.66	
150102	A4E	w.o.12.7.67	
150103	A4E		
150104	A4E	w.o.25.2.68	
150105	A4E		
150106	A4E	w.o.14.1.67	
150107	A4E	w.o.17.10.69	
150108	A4E		
150109	A4E		
150110	A4E	VF-126	NJ-626
150111	A4E	w.o.27.11.68	
150112	A4E		
150113	A4E		
150114	A4E		
150115	A4E	w.o.29.7.67	
150116	A4E	w.o.22.10.67	
150117	A4E		
150118	A4E	w.o.29.7.67	
150119	A4E	w.o.23.5.69	
150120	A4E		
150121	A4E		
150122	A4E	w.o.14.8.67	
150123	A4E		
150124	A4E		
150125	A4E		
150126	A4E		
150127	A4E		
150128	A4E	w.o.27.3.66	
150129	A4E	w.o.29.7.67	
150130	A4E	w.o.26.3.65	
150131	A4E	w.o.5.1.68	
150132	A4E	VF-43	AD-327
150133	A4E		
150134	A4E		
150135	A4E		
150136	A4E		
150137	A4E		
150138	A4E		
150581	A4C	scrapyard	AirMet
150582	A4C		
150583	A4C		
150584	A4C	w.o.4.1.67	
150585	A4C	MASDC	
150586	A4E	preserved	MCAS Yuma
150587	A4C		
150588	A4C	MASDC	
150589	A4C		
150590	A4C		
150591	A4C	scrapyard	Consolidated
150592	A4C	to Malaysia	
150593	A4L	to Malaysia	
150594	A4C		
150595	A4C	stored	Ontario, Ca
150596	A4C		
150597	A4C		
150598	A4C		
150599	A4C		
150600	A4C		
151022	A4E		
151023	A4E		
151024	A4E	w.o.15.7.66	
151025	A4E	w.o.29.8.67	
151026	A4E	w.o.4.7.66	
151027	A4E		
151028	A4E		
151029	A4E	w.o.26.12.67	
151030	A4E	VMA-131	QG-17
151031	A4E		
151032	A4E	w.o.6.7.67	
151033	A4E	NFWS	552
151034	A4E	w.o.2.5.66	
151035	A4E		
151036	A4E	VMA-131	QG-20
151037	A4E	w.o.23.11.64	
151038	A4E	VF-45	AD-616
151039	A4E	w.o.12.4.67	
151040	A4E		
151041	A4E		
151042	A4E	w.o.21.2.67	
151043	A4E	w.o.21.3.68	
151044	A4E	VF-126	NJ-633
151045	A4E	w.o.20.3.69	
151046	A4E	MASDC	
151047	A4E	w.o.29.4.66	
151048	A4E		
151049	A4E	w.o.30.5.67	
151050	A4E	VF-45	AD-614
151051	A4E		
151052	A4E		
151053	A4E		
151054	A4E	w.o.24.7.67	
151055	A4E	w.o.29.1.68	
151056	A4E		
151057	A4E	w.o.1.6.66	
151058	A4E	w.o.17.4.66	
151059	A4E	VF-126	NJ-632
151060	A4E	w.o.21.3.68	
151061	A4E		
151062	A4E		
151063	A4E		
151064	A4E	VF-45	AD-610
151065	A4E	w.o.10.8.66	
151066	A4E		
151067	A4E	w.o.13.11.65	
151068	A4E	w.o.14.12.66	
151069	A4E		
151070	A4E	w.o.27.4.68	
151071	A4E		
151072	A4E		
151073	A4E	w.o.26.4.67	
151074	A4E		
151075	A4E	w.o.26.10.66	
151076	A4E	w.o.24.5.67	
151077	A4E	w.o.21.3.68	
151078	A4E	VMA-131	QG-
151079	A4E		
151080	A4E	w.o.9.6.68	
151081	A4E		
151082	A4E	w.o.6.5.67	
151083	A4E	w.o.17.11.65	
151084	A4E		
151085	A4E		
151086	A4E		
151087	A4E		
151088	A4E	w.o.11.8.67	
151089	A4E	w.o.18.7.65	
151090	A4E		
151091	A4E	w.o.21.6.67	
151092	A4E		
151093	A4E	w.o.15.1.67	
151094	A4E		
151095	A4E	NFWS	55
151096	A4E		
151097	A4E	w.o.24.1.69	
151098	A4E		
151099	A4E	w.o.26.9.72	

151100	A4E		
151101	A4E		
151102	A4E	w.o.25.4.67	
151103	A4E	w.o.8.7.69	
151104	A4E		
151105	A4E	w.o.5.5.68	
151106	A4E	w.o.22.6.67	
151107	A4E		
151108	A4E	w.o.11.3.67	
151109	A4E	w.o.21.8.66	
151110	A4E		
151111	A4E		
151112	A4E		
151113	A4E	w.o.31.5.67	
151114	A4E		
151115	A4E	w.o.20.9.65	
151116	A4E	w.o.25.4.67	
151117	A4E		
151118	A4E	w.o.20.6.88	
151119	A4E	w.o.20.7.67	
151120	A4E		
151121	A4E		
151122	A4E	w.o.21.10.68	
151123	A4E	w.o.24.11.66	
151124	A4E	w.o.21.3.68	
151125	A4E		
151126	A4E	w.o.2.10.68	
151127	A4E	w.o.21.3.68	
151128	A4E		
151129	A4E		
151130	A4E		
151131	A4E	w.o.10.8.69	
151132	A4E		
151133	A4E	w.o.1.1.68	
151134	A4E	w.o.9.9.65	
151135	A4E	w.o.26.2.70	
151136	A4E	w.o.5.1.67	
151137	A4E		
151138	A4E	w.o.1.11.66	
151139	A4E		
151140	A4E	w.o.21.1.68	
151141	A4E		
151142	A4E	w.o.1.11.65	
151143	A4E	w.o.27.7.68	
151144	A4E	w.o.2.6.65	
151145	A4E	w.o.30.4.66	
151146	A4E		
151147	A4E		
151148	A4E		
151149	A4E	w.o.21.3.68	
151150	A4E	w.o.10.10.66	
151151	A4E		
151152	A4E	w.o.11.1.68	
151153	A4E		
151154	A4E		
151155	A4E		
151156	A4E		
151157	A4E		
151158	A4E	w.o.13.1.67	
151159	A4E	w.o.1.10.68	
151160	A4E	w.o.21.10.68	
151161	A4E	w.o.2.6.65	
151162	A4E	w.o.18.7.89	
151163	A4E		
151164	A4E		
151165	A4E	w.o.10.9.70	
151166	A4E		
151167	A4E	VF-43	AD-328
151168	A4E	w.o.15.1.67	
151169	A4E		
151170	A4E		
151171	A4E		
151172	A4E	w.o.23.11.66	
151173	A4E	w.o.31.10.65	
151174	A4E		
151175	A4E	w.o.18.7.67	
151176	A4E		
151177	A4E		
151178	A4E	w.o.15.11.87	
151179	A4E	w.o.1.5.66	
151180	A4E		
151181	A4E	w.o.12.7.67	
151182	A4E		
151183	A4E	w.o.31.5.67	
151184	A4E		
151185	A4E	w.o.11.8.65	
151186	A4E	VF-43	AD-325
151187	A4E		
151188	A4E	w.o.15.3.67	
151189	A4E		
151190	A4E		
151191	A4E	w.o.9.11.72	
151192	A4E	w.o.7.10.65	
151193	A4E		
151194	A4E	AMARC	3A538
151195	A4E		
151196	A4E	w.o.6.5.68	
151197	A4E		
151198	A4E	w.o.6.5.68	
151199	A4E		
151200	A4E		
151201	A4E	w.o.20.6.66	
151984	A4E		
151985	A4E	w.o.2.11.67	
151986	A4E	w.o.18.7.67	
151987	A4E		
151988	A4E	VF-126	NJ-630
151989	A4E		
151990	A4E	w.o.21.10.81	
151991	A4E	w.o.31.8.67	
151992	A4E	w.o.9.12.65	
151993	A4E	w.o.4.4.69	
151994	A4E		
151995	A4E	w.o.11.5.66	
151996	A4E		
151997	A4E	w.o.10.5.67	
151998	A4E	w.o.27.1.66	
151999	A4E		
152000	A4E	w.o.2.9.72	
152001	A4E		
152002	A4E		
152003	A4E	w.o.13.8.69	
152004	A4E	w.o.11.2.91	
152005	A4E	w.o.8.5.68	
152006	A4E		
152007	A4E		
152008	A4E		
152009	A4E		
152010	A4E		
152011	A4E		
152012	A4E	VF-43	AD-323
152013	A4E		
152014	A4E		
152015	A4E		
152016	A4E	w.o.11.10.65	
152017	A4E		
152018	A4E	w.o.29.7.67	
152019	A4E		
152020	A4E	w.o.16.9.66	
152021	A4E	w.o.25.1.66	
152022	A4E	w.o.26.5.67	
152023	A4E		
152024	A4E	w.o.29.7.67	
152025	A4E	w.o.23.4.66	
152026	A4E		
152027	A4E	w.o.7.2.66	
152028	A4E	w.o.31.8.67	
152029	A4E	w.o.22.6.69	
152030	A4E		
152031	A4E		
152032	A4E		
152033	A4E		
152034	A4E	w.o.18.7.67	
152035	A4E		
152036	A4E	w.o.29.7.67	
152037	A4E		
152038	A4E	w.o.17.10.67	
152039	A4E		
152040	A4E	w.o.29.7.67	
152041	A4E		
152042	A4E	w.o.6.9.65	
152043	A4E	w.o.6.5.69	
152044	A4E		
152045	A4E	w.o.29.7.66	
152046	A4E	w.o.5.11.68	
152047	A4E	w.o.3.6.66	

Serial	Type	Unit	Code
152048	A4E	w.o.18.10.67	
152049	A4E	w.o.14.7.67	
152050	A4E		
152051	A4E	w.o.27.2.67	
152052	A4E	w.o.6.4.66	
152053	A4E		
152054	A4E	w.o.13.8.67	
152055	A4E	w.o.3.5.67	
152056	A4E	w.o.27.9.67	
152057	A4E	w.o.1.3.66	
152058	A4E		
152059	A4E		
152060	A4E	w.o.13.5.67	
152061	A4E	NFWS	57
152062	A4E	w.o.7.6.70	
152063	A4E	w.o.15.6.66	
152064	A4E		
152065	A4E		
152066	A4E	w.o.31.1.66	
152067	A4E	w.o.28.3.87	
152068	A4E	w.o.24.3.68	
152069	A4E		
152070	A4E	VF-45	AD-613
152071	A4E	w.o.22.12.67	
152072	A4C	VMA-131	QG-15
152073	A4E	w.o.27.6.66	
152074	A4E	w.o.5.1.68	
152075	A4E	w.o.12.10.66	
152076	A4E	w.o.26.4.67	
152077	A4E	w.o.28.7.66	
152078	A4E		
152079	A4E		
152080	A4E	VMA-131	QG-16
152081	A4E	VMA-131	QG-
152082	A4E	w.o.9.1.70	
152083	A4E		
152084	A4E	w.o.25.8.66	
152085	A4E	w.o.9.10.67	
152086	A4E	w.o.7.10.67	
152087	A4E	w.o.7.3.67	
152088	A4E	w.o.13.3.68	
152089	A4E		
152090	A4E		
152091	A4E	w.o.23.9.68	
152092	A4E	w.o.9.5.68	
152093	A4E	w.o.26.8.66	
152094	A4E		
152095	A4E	w.o.29.7.66	
152096	A4E		
152097	A4E		
152098	A4E	w.o.15.9.83	
152099	A4E	w.o.11.4.70	
152100	A4E	w.o.23.7.66	
152101	NA4F	w.o.30.11.89	
152102	NTA4F	NAWC-WD	042
152103	TA4E		
152846	TA4F		
152847	TA4J	w.o.17.11.89	
152848	TA4J	NAWC-WD	042
152849	TA4F		
152850	TA4J	VT-7	A-701
152851	TA4F		
152852	EA4F	VAQ-33	GD-105
152853	TA4J	VX-5	XE-14
152854	TA4F		
152855	TA4J	VT-22	B-145
152856	OA4M	MALS-12	WA-00
152857	TA4F		
152858	TA4J	VT-22	B-155
152859	TA4F		
152860	TA4F	MALS-31	EX-
152861	TA4J	VT-7	A-706
152862	TA4J	VX-5	XE-12
152863	TA4J		
152864	TA4J	w.o.2.6.77	
152865	TA4F		
152866	TA4F		
152867	TA4F	VC-8	GF-4
152868	TA4J	NAWC\TPS	06
152869	EA4F	w.o.14.4.80	
152870	TA4F		
152871	TA4F	VC-8	GF-3
152872	TA4J	VT-7	A-746
152873	TA4F		
152874	OA4M	MALS-12	WA-01
152875	TA4F		
152876	TA4F		
152877	TA4J	NFWS	57
152878	TA4J	NAWC\TPS	07
153459	TA4J		
153460	TA4F	w.o.	
153461	TA4F	VF-45	AD-600
153462	TA4J	VT-86	F-237
153463	TA4J	VT-7	A-732
153464	TA4F		
153465	TA4F		
153466	TA4F		
153467	TA4J		
153468	TA4F		
153469	TA4J	VT-7	A-756
153470	TA4F		
153471	TA4J	VT-7	A-739
153472	TA4F		
153473	TA4F		
153474	TA4J	AMARC	3A623
153475	TA4F		
153476	TA4F		
153477	TA4J	w.o.24.5.84	
153478	TA4J	AMARC	3A649
153479	TA4J		
153480	TA4F		
153481	EA4F	VC-10	JH-00
153482	TA4J	VT-86	F-236
153483	TA4J	MALS-49	QZ-21
153484	TA4J	VMA-124	QP-21
153485	TA4F		
153486	TA4F		
153487	TA4F		
153488	TA4J	AMARC	3A...
153489	TA4J	VT-7	A-766
153490	TA4F		
153491	TA4F	MALS-24	EW-04
153492	TA4J		
153493	TA4F		
153494	TA4F		
153495	TA4J		
153496	TA4F	VF-45	AD-606
153497	TA4J	w.o.22.10.84	
153498	TA4J	VT-22	B-109
153499	TA4F	w.o.29.9.67	
153500	TA4J	AMARC	3A614
153501	TA4F	w.o.13.2.68	
153502	TA4J	VC-8	GF-2
153503	TA4F		
153504	TA4J	w.o.7.8.87	
153505	TA4J	VT-7	A-753
153506	TA4J	MALS-31	EX-03
153507	OA4M	MALS-32	DA-00
153508	TA4F	w.o.26.5.72	
153509	TA4J		
153510	OA4M	MALS-32	DA-02
153511	TA4F	w.o.11.4.68	
153512	TA4F	w.o.13.12.68	
153513	TA4J	VF-126	NJ-612
153514	TA4F		
153515	TA4J	VT-7	A-762
153516	TA4J	VF-43	AD-340
153517	TA4J	VT-7	A-712
153518	TA4J	dump	NAS Meridian
153519	TA4F		
153520	TA4F		
153521	TA4J	w.o.10.7.89	
153522	TA4J	VT-7	A-752
153523	TA4F	w.o.1.10.68	
153524	TA4J	NAWC\TPS	08
153525	TA4F	preserved	NAS Pensacola
153526	TA4J	VC-10	JH-5
153527	OA4M	NAD	Cherry Point
153528	TA4J	w.o.15.2.84	
153529	TA4F	AMARC	3A566
153530	TA4F	w.o.25.5.88	
153531	OA4M	AMARC	3A567
153660	TA4F		
153661	TA4J	w.o.11.4.83	
153662	TA4F		
153663	TA4J	AMARC	3A...

Serial	Type	Unit/Notes	Code
153664	TA4F		
153665	TA4F		
153666	TA4F		
153667	TA4J	VT-22	B-135
153668	TA4F		
153669	TA4J	AMARC	3A632
153670	TA4J		
153671	TA4J	VT-7	A-716
153672	TA4J	AMARC	3A610
153673	TA4M		
153674	TA4J	VT-22	B-166
153675	TA4J	AMARC	3A607
153676	TA4J	VT-22	B-148
153677	TA4F	VT-22	B-163
153678	TA4J	VT-7	A-725
153679	TA4J	VF-126	NJ-613
153680	TA4J	VT-22	B-119
153681	TA4J	w.o.7.9.87	
153682	TA4F		
153683	TA4J		
153684	TA4F		
153685	TA4J	NFWS	51
153686	TA4F		
153687	TA4F		
153688	TA4F		
153689	TA4J	VT-7	A-747
153690	TA4J	VF-45	AD-604
154172	A4F	NFWS	54
154173	A4F	NFWS	
154174	A4F	w.o.30.5.68	
154175	A4F		
154176	A4F		
154177	A4F	dump	NAS Pensacola
154178	A4F		
154179	A4F	AMARC	3A532
154180	A4F	MASDC	
154181	A4F	VF-126	NJ-620
154182	A4F	w.o.26.7.68	
154183	A4F	VF-126	NJ-600
154184	A4F	w.o.29.7.69	
154185	A4F	w.o.15.9.75	
154186	A4F		
154187	A4F	w.o.6.9.68	
154188	A4F		
154189	A4F	w.o.23.7.68	
154190	A4F		
154191	A4F		
154192	A4F		
154193	A4F	VF-43	AD-321
154194	A4F	w.o.22.6.83	
154195	A4F		
154196	A4F		
154197	A4F	w.o.27.5.72	
154198	A4F	w.o.14.5.68	
154199	A4F	w.o.21.7.69	
154200	A4F		
154201	A4F		
154202	A4F		
154203	A4F		
154204	A4F		
154205	A4F	w.o.8.3.72	
154206	A4F	w.o.15.2.69	
154207	A4F		
154208	A4F	dump	NAS Pensacola
154209	A4F	NFWS	56
154210	A4F		
154211	A4F	VF-43	AD-
154212	A4F		
154213	A4F		
154214	A4F	w.o.7.5.68	
154215	A4F	w.o.7.6.70	
154216	A4F	w.o.23.6.68	
154217	A4F	preserved	NAS Pensacola
154287	TA4J	VT-22	B-103
154288	TA4F		
154289	TA4J	VF-126	NJ-606
154290	TA4J	VF-126	NJ-607
154291	TA4J	VT-22	B-156
154292	TA4J	VT-22	B-161
154293	TA4J		
154294	OA4M	AMARC	3A598
154295	TA4F	MALS-32	DA-000
154296	TA4J	VF-43	AD-342
154297	TA4J	AMARC	3A650
154298	TA4J	VT-7	A-754
154299	TA4F	w.o.9.4.69	
154300	TA4J	VT-22	B-102
154301	TA4F		
154302	TA4F	w.o.8.9.70	
154303	TA4J		
154304	TA4J	w.o.28.10.68	
154305	TA4J	AMARC	3A...
154306	OA4M	AMARC	3A600
154307	OA4M	AMARC	3A615
154308	TA4F		
154309	TA4F		
154310	TA4J	AMARC	3A...
154311	TA4J	MALS-31	EX-04
154312	TA4J	VF-45	AD-
154313	TA4J	VT-22	B-152
154314	TA4F		
154315	TA4J	VC-10	JH-
154316	TA4F		
154317	TA4F		
154318	TA4J	VF-45	AD-600
154319	TA4J	AMARC	3A609
154320	TA4F		
154321	TA4F		
154322	TA4J	w.o.29.9.86	
154323	TA4J		
154324	TA4F		
154325	TA4J	VF-126	NJ-617
154326	TA4F		
154327	TA4J	VF-45	AD-603
154328	OA4M	AMARC	3A568
154329	TA4F		
154330	TA4J	VF-126	NJ-610
154331	TA4J	VMA-124	QP-22
154332	TA4J	NAWC-WD	040
154333	OA4M	MALS-12	WA-04
154334	TA4J	VFA-126	NJ-615
154335	OA4M	AMARC	3A565
154336	TA4F	dump	NAS Pensacola
154337	TA4J	MALS-42	MW-18
154338	TA4J	VT-22	B-129
154339	TA4F	w.o.5.10.89	
154340	OA4M	MALS-12	WA-03
154341	TA4F		
154342	TA4J		
154343	TA4J	VF-43	AD-343
154614	TA4J	VT-22	B-171
154615	TA4F		
154616	TA4F	VC-8	GF-00
154617	TA4F		
154618	TA4J	w.o.26.4.85	
154619	TA4J	VT-7	A-750
154620	TA4F		
154621	TA4F	w.o.27.12.69	
154622	TA4F	w.o.	
154623	TA4F	AMARC	3A569
154624	OA4M	w.o.15.3.86	
154625	TA4F		
154626	TA4J		
154627	TA4F		
154628	OA4M	MALS-32	DA-04
154629	TA4F		
154630	OA4M	AMARC	3A562
154631	TA4J	AMARC	3A612
154632	TA4J	VF-126	NJ-614
154633	OA4M	MALS-12	WA-
154634	TA4J	VT-22	B-115
154635	TA4F		
154636	TA4F		
154637	TA4F		
154638	OA4M	MALS-12	WA-
154639	TA4J	MALS-49	QZ-20
154640	TA4F	w.o.17.3.87	
154641	TA4F	MALS-42	MW-17
154642	TA4F		
154643	TA4F	MALS-24	EW-
154644	TA4F		
154645	TA4F	AMARC	3A599
154646	TA4F	w.o.11.7.70	
154647	TA4G	to Australia	N13-154647
154648	TA4G	to Australia	N13-154648
154649	TA4F	VC-8	GF-1

Serial	Type	Unit/Status	Code
154650	TA4J	VT-22	B-122
154651	OA4M	MALS-12	WA-07
154652	TA4F	w.o.18.1.84	
154653	TA4F	w.o.3.8.84	
154654	TA4F		
154655	EA4F	w.o.25.5.85	
154656	TA4J	AMARC	3A613
154657	TA4J	VF-126	NJ-613
154970	A4F		
154971	A4F	w.o.27.6.91	
154972	A4F	w.o.9.7.72	
154973	A4F	VF-43	AD-322
154974	A4F	w.o.22.5.68	
154975	A4F		
154976	A4F		
154977	A4F		
154978	A4F		
154979	A4F	VMA-142	MB-09
154980	A4F	w.o.26.1.71	
154981	A4F	w.o.30.8.68	
154982	A4F	w.o.28.5.68	
154983	A4F	preserved	NAS Pensacola
154984	A4F	dump	NAS Pensacola
154985	A4F	w.o.24.1.71	
154986	A4F		
154987	A4F		
154988	A4F	w.o.21.5.68	
154989	A4F	NARF	Pensacola
154990	A4F		
154991	A4F	VF-43	AD-323
154992	A4F	w.o.13.7.85	
154993	A4F	w.o.20.7.69	
154994	A4F	w.o.8.3.70	
154995	A4F	w.o.11.4.68	
154996	A4F		
154997	A4F		
154998	A4F		
154999	A4F		
155000	A4F	preserved	NAS ensacola
155001	A4F		
155002	A4F	w.o.4.7.68	
155003	A4F	w.o.21.7.69	
155004	A4F		
155005	A4F	w.o.9.2.70	
155006	A4F		
155007	A4F		
155008	A4F		
155009	A4F		
155010	A4F	w.o.14.3.70	
155011	A4F	w.o.28.9.68	
155012	A4F		
155013	A4F	VMA-142	MB-
155014	A4F	dump	NAS Pensacola
155015	A4F	w.o.23.9.68	
155016	A4F		
155017	A4F	VF-126	NJ-621
155018	A4F	VF-126	NJ-627
155019	A4F		
155020	A4F		
155021	A4F	w.o.6.9.72	
155022	A4F		
155023	A4F	w.o.10.2.70	
155024	A4F		
155025	A4F	VF-126	NJ-626
155026	A4F	VF-43	AD-324
155027	A4F	NFWS	
155028	A4F	VMA-142	MB-
155029	A4F	w.o.13.7.85	
155030	A4F	VMA-142	MB-04
155031	A4F	VF-126	NJ-622
155032	A4F		
155033	A4F	NARF	Pensacola
155034	A4F	VFC-12	UX-21
155035	A4F		
155036	A4F	VF-126	NJ-623
155037	A4F		
155038	A4F		
155039	A4F		
155040	A4F	w.o.15.3.70	
155041	A4F		
155042	A4F	dump	NAS Pensacola
155043	A4F		
155044	A4F		
155045	A4F	w.o.25.5.72	
155046	A4F	w.o.11.7.72	
155047	A4F	VMA-142	MB-
155048	A4F	w.o.27.5.72	
155049	NA4M	NAWC-AD	SATD SD-300
155050	A4F		
155051	A4G	to Australia	N13-155051
155052	A4G	to Australia	N13-155052
155053	A4F		
155054	A4F		
155055	A4G	to Australia	N13-155055
155056	A4F		
155057	A4F	w.o.24.6.85	
155058	A4F		
155059	A4F	w.o.16.1.68	
155060	A4G	to Australia	N13-155060
155061	A4G	to Australia	N13-155061
155062	A4G	to Australia	N13-155062
155063	A4G	to Australia	N13-155063
155064	A4F	VF-126	NJ-625
155065	A4F		
155066	A4F		
155067	A4F		
155068	A4F		
155069	A4G	to Australia	N13-155069
155070	TA4J	AMARC	3A608
155071	TA4F		
155072	TA4J	VF-43	AD-341
155073	TA4J		
155074	TA4J	AMARC	3A683
155075	TA4J		
155076	TA4J	VT-22	B-106
155077	TA4J	w.o.25.8.86	
155078	TA4J	BT-22	B-000
155079	TA4J	w.o.25.11.86	
155080	TA4J		
155081	TA4J	VT-22	B-125
155082	TA4J	VT-22	B-159
155083	TA4J	VT-22	B-151
155084	TA4J		
155085	TA4J		
155086	TA4J		
155087	TA4J	VT-7	A-744
155088	TA4J	VT-22	B-101
155089	TA4J	AMARC	3A634
155090	TA4J		
155091	TA4J		
155092	TA4J		
155093	TA4J		
155094	TA4J		
155095	TA4J		
155096	TA4J	AMARC	3A611
155097	TA4J		
155098	TA4J		
155099	TA4J	AMARC	3A...
155100	TA4J	VT-7	A-741
155101	TA4J		
155102	TA4J	VT-22	B-124
155103	TA4J		
155104	TA4J		
155105	TA4J	VT-7	A-738
155106	TA4J	AMARC	3A...
155107	TA4J	VT-22	B-131
155108	TA4J		
155109	TA4J	w.o.26.4.85	
155110	TA4J	VF-45	AD-602
155111	TA4J	VT-22	B-153
155112	TA4J	VT-22	B-111
155113	TA4J	VT-22	B-143
155114	TA4J		
155115	TA4J		
155116	TA4J		
155117	TA4J		
155118	TA4J	VT-22	B-147
155119	TA4J	NAWC\TPS	04
156891	TA4J	VT-7	A-748
156892	TA4J	w.o.16.12.83	
156893	TA4J		
156894	TA4J	VT-22	B-117
156895	TA4J		
156896	TA4J	NAWC-AD	SATD SD-323
156897	TA4J	VT-86	F-234
156898	TA4J		

156899	TA4J	VT-22	B-112		158104	TA4J	VT-7	A-711
156900	TA4J	VT-22	B-113		158105	TA4J	VT-7	A-700
156901	TA4J	VT-22	B-114		158106	TA4J	VT-22	B-137
156902	TA4J	w.o.13.6.84			158107	TA4J	VT-22	B-110
156903	TA4J	VT-22	B-127		158108	TA4J	AMARC	3A...
156904	TA4J	VT-22	B-162		158109	TA4J	VT-7	A-733
156905	TA4J	VT-22	B-157		158110	TA4J	VT-7	A-717
156906	TA4J				158111	TA4J	w.o.17.12.75	
156907	TA4J				158112	TA4J		
156908	TA4J	VT-22	B-116		158113	TA4J	VF-43	AD-310
156909	TA4J	AMARC	3A...		158114	TA4J		
156910	TA4J				158115	TA4J		
156911	TA4J	AMARC	3A606		158116	TA4J	w.o.10.2.84	
156912	TA4J	VT-22	B-136		158117	TA4J	VT-22	B-164
156913	TA4J				158118	TA4J	VT-7	A-718
156914	TA4J	VT-7	A-715		158119	TA4J	VT-22	B-121
156915	TA4J				158120	TA4J	dump	MCAS Cherry Pt.
156916	TA4J				158121	TA4J	w.o.6.8.86	
156917	TA4J	VT-22	B-118		158122	TA4J	AMARC	3A605
156918	TA4J				158123	TA4J		
156919	TA4J				158124	TA4J		
156920	TA4J	AMARC	3A...		158125	TA4J	VT-7	A-725
156921	TA4J	AMARC	3A...		158126	TA4J	NAWC\TPS	05
156922	TA4J	AMARC	3A687		158127	TA4J	w.o.6.9.78	
156923	TA4J	VT-22	B-125		158128	TA4J	w.o.6.9.78	
156924	TA4J	AMARC	3A...		158129	TA4J	VT-7	A-729
156925	TA4J	VT-7	A-708		158130	TA4J		
156926	TA4J	VT-22	B-120		158131	TA4J	VT-7	A-721
156927	TA4J	VT-22	B-138		158132	TA4J		
156928	TA4J	AMARC	3A...		158133	TA4J	NARF	Pensacola
156929	TA4J	VT-7	A-720		158134	TA4J	VT-7	A-719
156930	TA4J				158135	TA4J	dump	NAS Pensacola
156931	TA4J				158136	TA4J	VT-22	B-146
156932	TA4J	VT-7	A-727		158137	TA4J	VT-7	A-760
156933	TA4J	VT-22	B-123		158138	TA4J	VT-7	A-759
156934	TA4J	AMARC	3A...		158139	TA4J		
156935	TA4J	VT-7	A-757		158140	TA4J		
156936	TA4J	AMARC	3A...		158141	TA4J	VT-7	A-714
156937	TA4J	AMARC	3A666		158142	TA4J	VT-7	A-755
156938	TA4J				158143	TA4J	VT-7	A-746
156939	TA4J	VT-7	A-764		158144	TA4J	VT-7	A-724
156940	TA4J	AMARC	3A604		158145	TA4J	VT-7	A-
156941	TA4J	VT-22	B-158		158146	TA4J	VT-7	A-722
156942	TA4J	AMARC	3A622		158147	TA4J	AMARC	3A624
156943	TA4J	VT-22	B-126		158148	A4M	NAWC-AD	SATD SD-304
156944	TA4J	VT-7	A-705		158149	A4M	VMA-131	QG-06
156945	TA4J	VT-22	B-128		158150	A4M		
156946	TA4J	w.o.16.9.84			158151	A4M	w.o.9.9.81	
156947	TA4J	AMARC	3A...		158152	A4M		
156948	TA4J				158153	A4M		
156949	TA4J				158154	A4M		
156950	TA4J	w.o.11.4.91			158155	A4M		
158073	TA4J	VT-22	B-130		158156	A4M	AMARC	3A542
158074	TA4J	w.o.			158157	A4M		
158075	TA4J	VT-22	B-132		158158	A4M	w.o.30.5.86	
158076	TA4J				158159	A4M	NAWC-WD	
158077	TA4J				158160	A4M		
158078	TA4J	VT-7	A-740		158161	A4M	VMA-131	QG-04
158079	TA4J	AMARC	3A690		158162	A4M		
158080	TA4J				158163	A4M		
158081	TA4J	AMARC	3A630		158164	A4M	VMA-131	QG-02
158082	TA4J	VT-22	B-167		158165	A4M	VMA-131	QG-02
158083	TA4J	VF-45	AD-601		158166	A4M	dump	NAS Memphis
158084	TA4J				158167	A4M	AMARC	3A547
158085	TA4J	VT-22	B-133		158168	A4M	AMARC	3A546
158086	TA4J				158169	A4M	NFWS	
158087	TA4J	VT-22	B-134		158170	A4M	VMA-131	QG-03
158088	TA4J				158171	A4M	NFWS	51
158089	TA4J				158172	A4M		
158090	TA4J	VT-7	A-740		158173	A4M		
158091	TA4J	w.o.21.9.86			158174	A4M		
158092	TA4J	VF-43	AD-311		158175	A4M		
158093	TA4J	VT-22	B-131		158176	A4M	VMA-142	MB-03
158094	TA4J	VT-7	A-745		158177	A4M	AMARC	3A576
158095	TA4J	AMARC	3A...		158178	A4M		
158096	TA4J				158179	A4M	derelict	NAS Willow Grove
158097	TA4J				158180	A4M	VX-5	XE-15
158098	TA4J	VT-7	A-739		158181	A4M	AMARC	3A563
158099	TA4J	AMARC	3A...		158182	A4M		
158100	TA4J	AMARC	3A...		158183	A4M	NARF	Pensacola
158101	TA4J	AMARC	3A...		158184	A4M	AMARC	3A539
158102	TA4J				158185	A4M	AMARC	3A543
158103	TA4J				158186	A4M	VMA-142	MB-07

Bu.No.	Type	Unit	Code
158187	A4M	w.o.12.6.85	
158188	A4M	w.o.30.11.74	
158189	A4M		
158190	A4M		
158191	A4M	NAD	Cherry Point
158192	A4M	VMA-142	MB-08
158193	A4M	AMARC	3A558
158194	A4M	AMARC	3A560
158195	A4M	NAWC-AD	SATD SD-301
158196	A4M	AMARC	3A561
158412	A4M		
158413	A4M	w.o.28.1.84	
158414	A4M	VMA-131	QG-10
158415	A4M		
158416	A4M	AMARC	3A545
158417	A4M	VMA-131	QG-11
158418	A4M	w.o.7.7.80	
158419	A4M	VMA-131	QG-09
158420	A4M		
158421	A4M	w.o.4.11.81	
158422	A4M	VMA-131	QG-01
158423	A4M		
158424	A4M	AMARC	3A575
158425	A4M	NAWC-WD	043
158426	A4M	VMA-131	QG-12
158427	A4M	AMARC	3A557
158428	A4M		
158429	A4M	AMARC	3A544
158430	A4M	dumped	NAS Memphis
158431	A4M		
158432	A4M		
158433	A4M		
158434	A4M		
158435	A4M	AMARC	3A573
158453	TA4J	VT-22	B-105
158454	TA4J		
158455	TA4J	AMARC	3A...
158456	TA4J	VT-22	B-150
158457	TA4J	w.o.26.5.88	
158458	TA4J	VT-7	A-710
158459	TA4J	VT-22	B-140
158460	TA4J	w.o.17.11.72	
158461	TA4J	w.o.27.12.89	
158462	TA4J	VT-7	A-712
158463	TA4J	NAD	Cherry Point
158464	TA4J	VT-22	B-104
158465	TA4J	VT-7	A-765
158466	TA4J	VT-7	A-707
158467	TA4J	VT-7	A-734
158468	TA4J	VT-7	A-726
158469	TA4J	AMARC	3A658
158470	TA4J		
158471	TA4J		
158472	TA4J		
158473	TA4J	VT-7	A-735
158474	TA4J	VT-7	A-702
158475	TA4J	VT-7	A-736
158476	TA4J	w.o.24.5.84	
158477	TA4J	VX-5	XE-15
158478	TA4J	VT-22	B-149
158479	TA4J	VT-7	A-730
158480	TA4J		
158481	TA4J	VT-7	A-761
158482	TA4J	VT-7	A-709
158483	TA4J	w.o.13.2.85	
158484	TA4J	dump	NAS Pensacola
158485	TA4J		
158486	TA4J		
158487	TA4J		
158488	TA4J		
158489	TA4J	VT-86	F-230
158490	TA4J	VT-22	B-141
158491	TA4J	VT-22	B-142
158492	TA4J		
158493	TA4J		
158494	TA4J	VT-22	B-143
158495	TA4J	VT-22	B-144
158496	TA4J	AMARC	3A...
158497	TA4J		
158498	TA4J	VAQ-33	GD-108
158499	TA4J		
158500	TA4J	VT-7	A-723
158501	TA4J	VT-7	A-744
158502	TA4J		
158503	TA4J	to Israel	
158504	TA4J	VT-22	B-
158505	TA4J	AMARC	3A...
158506	TA4J	AMARC	3A...
158507	TA4J	VT-22	B-149
158508	TA4J		
158509	TA4J	AMARC	3A684
158510	TA4J		
158511	TA4J		
158512	TA4J	AMARC	3A616
158513	TA4J	VT-7	A-737
158514	TA4J	VT-7	A-757
158515	TA4J		
158516	TA4J	AMARC	3A638
158517	TA4J		
158518	TA4J		
158519	TA4J	AMARC	3A...
158520	TA4J	VT-7	A-736
158521	TA4J		
158522	TA4J		
158523	TA4J	VT-7	A-731
158524	TA4J	AMARC	3A641
158525	TA4J	AMARC	3A686
158526	TA4J	VT 7	A-704
158527	TA4J	VT-7	A-763
158712	TA4J	VT-22	B-142
158713	TA4J	AMARC	3A688
158714	TA4J	VT-7	A-742
158715	TA4J	AMARC	3A...
158716	TA4J	VT-22	B-139
158717	TA4J		
158718	TA4J	VT-22	B-154
158719	TA4J		
158720	TA4J		
158721	TA4J	AMARC	3A665
158722	TA4J	VT-22	B-160
158723	TA4J		
159099	TA4J		
159100	TA4J	w.o.14.9.89	
159101	TA4J		
159102	TA4J		
159103	TA4J		
159104	TA4J		
159470	A4M		
159471	A4M	AMARC	3A578
159472	A4M	VMA-131	QG-03
159473	A4M	VMA-131	QG-07
159474	A4M	w.o.29.1.85	
159475	A4M	NFWS	
159476	A4M		
159477	A4M	OMD	7D-04
159478	A4M	VMA-131	QG-08
159479	A4M		
159480	A4M	w.o.29.8.80	
159481	A4M	w.o.9.6.81	
159482	A4M	VMA-142	MB-12
159483	A4M		
159484	A4M	OMD	7D-02
159485	A4M	AMARC	3A559
159486	A4M		
159487	A4M	VMA-131	QG-09
159488	A4M	w.o.11.6.84	
159489	A4M		
159490	A4M	OMD	7D-03
159491	A4M		
159492	A4M	AMARC	3A552
159493	A4M	AMARC	3A553
159778	A4M	AMARC	3A556
159779	A4M		
159780	A4M	VMA-131	QG-12
159781	A4M		
159782	A4M	w.o.13.6.84	
159783	A4M	AMARC	3A541
159784	A4M	w.o.3.11.89	
159785	A4M		
159786	A4M	AMARC	3A540
159787	A4M		
159788	A4M	OMD	7D-01
159789	A4M		
159790	A4M	AMARC	3A564
159791	A4M		
159792	A4M		

159793	A4M		
159794	A4M		
159795	TA4J	VT-7	A-743
159796	TA4J	VT-22	B-100
159797	TA4J	VT-22	B-156
159798	TA4J	VT-7	A-728
160022	A4M	AMARC	3A554
160023	A4M	w.o.9.7.85	
160024	A4M	VMA-131	QG-06
160025	A4M	AMARC	3A548
160026	A4M	AMARC	3A551
160027	A4M	NARU	7D-03
160028	A4M		
160029	A4M	AMARC	3A549
160030	A4M		
160031	A4M	VMA-142	MB-15
160032	A4M		
160033	A4M	w.o.30.5.86	
160034	A4M		
160035	A4M	AMARC	3A570
160036	A4M	VMA-124	QP-17
160037	A4M		
160038	A4M		
160039	A4M		
160040	A4M	AMARC	3A550
160041	A4M		
160042	A4M	AMARC	3A572
160043	A4M	VMA-131	QG-05
160044	A4M		
160045	A4M	AMARC	3A555
160241	A4M		
160242	A4M		
160243	A4M	VMA-124	QP-09
160244	A4M	VMA-124	QP-11
160245	A4M	NAWC-WD	048
160246	A4M		
160247	A4M	dump	NAS Memphis
160248	A4M		
160249	A4M	dump	NAS Memphis
160250	A4M	w.o.	
160251	A4M	VMA-124	QP-
160252	A4M	VMA-124	QP-
160253	A4M	VMA-124	QP-05
160254	A4M	VMA-124	QP-02
160255	A4M	VMA-124	QP-16
160256	A4M	VMA-124	QP-00
160257	A4M	w.o.14.4.85	
160258	A4M	AMARC	3A571
160259	A4M	VMA-124	QP-07
160260	A4M	VMA-124	QP-08
160261	A4M	AMARC	3A574
160262	A4M	VMA-124	QP-06
160263	A4M	VMA-124	QP-03
160264	A4M	VMA-124	QP-04

A6 INTRUDER\PROWLER

147864	YA6A		
147865	EA6A		
147866	NA6A	dump	NATTC
147867	NA6A	preserved	USS Intrepid
148615	EA6B	GIA	
148616	EA6A	AMARC	5A004
148617	NA6A	to Grumman	
148618	EA6A	AMARC	5A028
149475	EA6A		
149476	A6A		
149477	EA6A		
149478	EA6A	AMARC	5A006
149479	NEA6B		
149480	A6A	w.o.	
149481	NEA6B		
149482	KA6D	VA-52	NL-523
149483	A6A		
149484	KA6D	VA-35	AA-524
149485	KA6D	VA-115	NF-514
149486	KA6D	VA-304	ND-420
149935	NEA6A	dump	NAS Lakehurst
149936	KA6D	AMARC	5A053
149937	KA6D	VA-304	ND-421
149938	A6A		
149939	KA6D		
149940	KA6D	VAK-208	AF-523
149941	A6E	w.o.30.9.83	
149942	KA6D	VA-196	NK-514
149943	A6E	VA-145	NE-514
149944	A6E	VA-145	NE-505
149945	KA6D	w.o.22.2.78	
149946	A6E	preserved	NAS Jacksonville
149947	A6A		
149948	A6E		
149949	A6E	VA-34	AG-504
149950	A6E		
149951	KA6D	VA-52	NL-516
149952	KA6D	VA-304	ND-422
149953	A6E		
149954	KA6D	w.o.1.8.85	
149955	A6E	VA-85	AB-506
149956	A6E	VA-304	ND-405
149957	A6E	VA-65	AJ-504
149958	A6A		
151558	A6B	MASDC	
151559	A6B		
151560	A6B	w.o.20.8.68	
151561	A6B	w.o.28.8.68	
151562	A6E	VA-205	AF-506
151563	A6B		
151564	A6E	MASDC	
151565	A6E	VA-304	ND-401
151566	KA6D		
151567	KA6D	w.o.28.1.66	
151568	KA6D	AMARC	5A054
151569	A6A		
151570	KA6D		
151571	A6A		
151572	A6A	VA-35	AA-523
151573	A6E		
151574	A6A		
151575	KA6D		
151576	KA6D	VA-165	NG-514
151577	A6A	w.o.18.7.65	
151578	A6A	w.o.1.5.68	
151579	A6A	preserved NAS Oceana	
151580	KA6D		
151581	KA6D	VA-304	ND-423
151582	KA6D	w.o.22.2.79	
151583	KA6D	VA-165	NG-
151584	A6A	w.o.14.7.65	
151585	A6A	w.o.24.7.65	
151586	A6A		
151587	A6E	w.o.	
151588	A6A	w.o.17.9.65	
151589	KA6D	VA-34	AG-522
151590	A6A	w.o.19.1.67	
151591	A6E	VA-52	NL-502
151592	KA6D	VA-65	AJ-536
151593	A6E	VA-75	AC-504
151594	A6A		
151595	EA6A		
151596	EA6A	VAQ-209	AF-604
151597	EA6A		

Serial	Type	Unit	Code
151598	EA6A		
151599	EA6A		
151600	EA6A		
151780	A6A		
151781	A6A	w.o.21.12.65	
151782	A6E	VA-34	AG-506
151783	KA6D	w.o.18.2.86	
151784	A6E	VA-196	NK-512
151785	A6A	w.o.22.4.66	
151786	A6A		
151787	KA6D	VA-196	NK-
151788	A6A	w.o.27.4.66	
151789	KA6D	VA-95	NH-515
151790	A6E	VA-128	NJ-819
151791	KA6D	stored	NAD Alameda
151792	KA6D	VA-34	AG-521
151793	KA6D		
151794	KA6D	w.o.17.4.66	
151795	KA6D	w.o.7.12.77	
151796	KA6D	VA-196	NK-521
151797	A6A	w.o.18.2.66	
151798	A6A	w.o.21.4.66	
151799	A6A		
151800	A6A	w.o.15.5.66	
151801	KA6D	AMARC	5A056
151802	A6E	MASDC	
151803	A6A		
151804	A6E	w.o.27.7.87	
151805	A6A		
151806	KA6D		
151807	A6E	VA-304	ND-400
151808	KA6D	VAK-208	AF-520
151809	KA6D	VA-35	AA-515
151810	KA6D	w.o.	
151811	A6E	MASDC	
151812	A6E	VX-5	XE-17
151813	KA6D	VA-95	NH-514
151814	KA6D	VA-85	AB-522
151815	A6A		
151816	A6A		
151817	A6A		
151818	KA6D	VAK-208	AF-
151819	KA6D	VA-165	NG-517
151820	KA6D	VAK-208	AF-524
151821	KA6D	w.o.5.11.80	
151822	A6A	w.o.27.8.66	
151823	KA6D		
151824	KA6D		
151825	A6A		
151826	KA6D	VA-52	NL-514
151827	KA6D	VA-34	AG-521
152583	A6E	MASDC	
152584	A6E	VA-145	NE-500
152585	A6E	w.o.13.7.84	
152586	A6A	w.o.17.1.69	
152587	A6E	VA-176	AE-518
152588	A6A	w.o.30.1.68	
152589	A6A	w.o.24.4.67	
152590	KA6D		
152591	A6E	VA-128	NJ-829
152592	KA6D	AMARC	5A...
152593	A6E		
152594	A6A	w.o.19.5.67	
152595	A6A	w.o.27.7.68	
152596	A6E		
152597	KA6D	w.o.2.5.72	
152598	KA6D	w.o.12.8.71	
152599	A6E		
152600	A6E		
152601	A6A	w.o.30.10.67	
152602	A6A		
152603	A6E	MASDC	
152604	A6A		
152605	A6A		
152606	KA6D	VA-35	AA-522
152607	A6E	NSWC	03
152608	A6A	w.o.23.3.67	
152609	A6A	w.o.17.4.67	
152610	A6E	VA-128	NJ-825
152611	KA6D	VA-85	AA-521
152612	A6A	w.o.25.11.67	
152613	A6A		
152614	A6E	VA-196	NK-503
152615	A6A		
152616	A6B		
152617	A6E	NARF	Norfolk
152618	KA6D		
152619	KA6D	VA-196	NK-516
152620	A6E	VA-85	AB-503
152621	A6E	VA-42	AD-556
152622	A6A		
152623	A6E	VA-128	NJ-837
152624	KA6D	VA-205	AF-521
152625	A6A	w.o.21.8.67	
152626	KA6D	VA-165	NG-515
152627	A6A	w.o.21.8.67	
152628	KA6D		
152629	A6A	w.o.2.11.67	
152630	A6E		
152631	A6A	MASDC	
152632	KA6D		
152633	A6A		
152634	A6E	VA-128	NJ-807
152635	A6E		
152636	A6A	w.o.18.1.68	
152637	KA6D		
152638	A6A	w.o.21.8.67	
152639	A6A	w.o.26.8.67	
152640	A6E		
152641	A6E		
152642	A6E	VA-65	AJ-531
152643	A6A		
152644	A6A		
152645	A6E		
152646	A6A		
152891	A6A	w.o.26.12.69	
152892	KA6D		
152893	KA6D	VA-196	NK-521
152894	KA6D	w.o.17.7.84	
152895	A6E	VA-128	NJ-836
152896	KA6D	VA-35	AA-521
152897	A6A		
152898	A6A		
152899	A6A		
152900	A6A		
152901	A6A	w.o.21.1.68	
152902	A6E	VA-42	AD-503
152903	A6A		
152904	A6E	NATC	
152905	A6E	VA-145	NE-514
152906	KA6D	VA-52	NL-514
152907	A6E	VA-128	NJ-840
152908	A6E	w.o.12.2.82	
152909	A6A		
152910	KA6D	VA-165	NG-516
152911	KA6D	VA-95	NH-513
152912	A6E	VA-205	AF-512
152913	KA6D	VA-165	NG-516
152914	KA6D		
152915	A6E	w.o.4.12.83	
152916	A6E		
152917	A6A	w.o.31.12.67	
152918	A6E		
152919	KA6D	VA-205	AF-522
152920	KA6D		
152921	KA6D	VA-304	ND-422
152922	A6A	w.o.6.3.68	
152923	A6E	VA-34	AG-500
152924	A6E	VA-205	AF-502
152925	A6E	VA-145	NE-511
152926	A6A		
152927	KA6D	VA-34	AG-523
152928	A6E	w.o.18.1.91	
152929	A6E	w.o.14.1.80	
152930	A6E	VA-145	NE-512
152931	A6E	VA-34	AG-510
152932	A6A	w.o.23.1.68	
152933	A6E		
152934	KA6D	w.o.12.7.84	
152935	A6E	MASDC	
152936	A6E	VA-75	AC-505
152937	A6A	w.o.2.1.70	
152938	A6A	w.o.28.2.68	
152939	KA6D	VA-34	AG-515
152940	A6A	w.o.16.3.68	
152941	A6E	MASDC	

Serial	Type	Unit	Code
152942	A6E	NARF	NAS Norfolk
152943	A6A	w.o.12.3.68	
152944	A6A	w.o.1.3.68	
152945	A6E	VA-85	AB-510
152946	A6A	w.o.21.12.72	
152947	A6E	w.o.16.11.78	
152948	A6E	VA-52	NL-500
152949	A6A	w.o.24.6.68	
152950	A6E		
152951	A6A	w.o.13.5.68	
152952	KA6D	AMARC	5A055
152953	A6E	VA-128	NJ-804
152954	A6E	VA-145	NE-504
154124	A6E	VX-5	XE-24
154125	A6A		
154126	A6E	VA-52	NL-503
154127	A6A	w.o.6.9.68	
154128	A6E	VA-128	NJ-808
154129	A6E	VA-95	NH-505
154130	A6A		
154131	A6E	VA-145	NE-506
154132	A6E	w.o.11.1.84	
154133	KA6D		
154134	A6E	VA-52	NL-505
154135	A6E		
154136	A6E		
154137	A6E	VA-128	NJ-815
154138	A6A		
154139	A6A		
154140	A6E	w.o.10.5.85	
154141	A6A	w.o.13.10.68	
154142	A6E	VA-42	AD-550
154143	A6A		
154144	A6E	w.o.22.5.83	
154145	A6A	w.o.11.6.72	
154146	A6E	VA-145	NE-505
154147	KA6D	VA-34	AG-520
154148	A6E		
154149	A6A	w.o.30.9.68	
154150	A6A		
154151	A6E	w.o.1.2.78	
154152	A6A	w.o.19.12.68	
154153	A6E		
154154	KA6D		
154155	A6A		
154156	A6E		
154157	A6A		
154158	A6E		
154159	A6E	VA-65	AJ-503
154160	A6A	w.o.17.3.69	
154161	A6E	VA-165	NG-502
154162	A6E	VA-145	NE-504
154163	A6E	VA-304	ND-402
154164	A6A	w.o.2.5.68	
154165	A6A		
154166	A6A	w.o.25.7.68	
154167	A6E		
154168	A6A	w.o.27.6.83	
154169	A6E		
154170	A6E	NAWC	
154171	A6E		
155581	A6E	VA-42	AD-521
155582	KA6D	VA-75	AC-515
155583	KA6D	VA-65	AG-521
155584	KA6D	AMARC	5A...
155585	A6E	VA-304	ND-404
155586	A6E	VA-95	NH-500
155587	A6A	w.o.30.4.69	
155588	A6E		
155589	A6A	VA-65	AJ-510
155590	A6E	VA-42	AD-546
155591	A6E		
155592	A6E	VA-34	AG-505
155593	A6A		
155594	A6A	w.o.20.12.72	
155595	A6E		
155596	A6E	NAWC-WD	57
155597	KA6D	AMARC	5A059
155598	A6E	VA-85	AB-523
155599	A6E	VA-34	AG-504
155600	A6E	VA-42	AD-602
155601	A6A		
155602	A6E	w.o.15.2.91	
155603	A6E		
155604	A6E	VA-85	AB-520
155605	A6A	w.o.28.2.70	
155606	A6E		
155607	A6E	w.o.22.11.69	
155608	A6E	w.o.2.10.83	
155609	A6E	w.o.	
155610	A6E	VA-128	NJ-803
155611	A6A	w.o.21.9.69	
155612	A6E		
155613	A6A	w.o.22.11.69	
155614	A6E		
155615	A6E		
155616	A6E	VA-52	NL-507
155617	A6A		
155618	A6E	w.o.6.2.70	
155619	KA6D	VA-35	AA-520
155620	A6E	VA-42	AD-505
155621	A6E		
155622	A6A	w.o.28.11.72	
155623	A6E	VA-128	NJ-
155624	A6E	w.o.23.7.80	
155625	A6E	w.o.4.8.87	
155626	A6A	w.o.6.9.72	
155627	A6E		
155628	A6E	Grumman	
155629	A6E	VA-34	AG-507
155630	A6E	VA-128	NJ-824
155631	A6A		
155632	A6E	w.o.2.2.91	
155633	A6E	w.o.2.1.81	
155634	A6A		
155635	A6E		
155636	A6E		
155637	A6E	w.o.5.5.88	
155638	A6E	VA-95	NH-516
155639	A6A		
155640	A6E		
155641	A6A		
155642	A6E		
155643	A6E	VA-128	NJ-535
155644	A6E		
155645	A6A	w.o.19.9.78	
155646	A6E	VA-42	AD-522
155647	A6C	w.o.8.1.71	
155648	A6E	VA-52	NL-506
155649	A6E	w.o.23.10.87	
155650	A6A	w.o.29.5.72	
155651	A6E	VA-128	NJ-805
155652	A6A	w.o.9.4.72	
155653	A6E	VA-35	AA-501
155654	A6E	VA-95	NH-501
155655	A6E	MASDC	
155656	A6E	VA-42	AD-500
155657	A6E	w.o.12.5.87	
155658	A6E	w.o.	
155659	A6E	MASDC	
155660	A6E	w.o.18.10.83	
155661	A6E	VA-304	ND-406
155662	A6E		
155663	A6A		
155664	A6E	VA-115	NF-512
155665	A6E	VA-95	NH-511
155666	A6A	w.o.27.12.72	
155667	A6E	w.o.21.11.84	
155668	A6E	w.o.11.12.85	
155669	A6E	VA-42	AD-551
155670	A6E	VA-304	ND-411
155671	A6A	w.o.7.4.77	
155672	A6E		
155673	A6E	VA-196	NK-510
155674	A6E	w.o.23.9.87	
155675	A6A	w.o.8.2.80	
155676	A6E		
155677	A6A	w.o.30.12.71	
155678	A6E	VA-85	AB-500
155679	A6E	VA-42	AD-512
155680	A6E	w.o.26.8.83	
155681	A6E	VA-34	AG-500
155682	A6E	NAWC-WD	602
155683	A6A	VA-35	AA-511
155684	A6E		
155685	A6E	VA-185	NF-400

Serial	Type	Unit	Code
155686	A6E		
155687	A6E	VA-35	AA-510
155688	A6E	w.o.14.1.87	
155689	A6E		
155690	A6A	w.o.7.7.72	
155691	KA6D	AMARC	5A057
155692	A6A	w.o.20.7.78	
155693	A6A	w.o.9.1.73	
155694	A6E	VA-304	ND-407
155695	A6E	VA-34	AG-502
155696	A6A	w.o.29.9.69	
155697	A6E	VA-128	NJ-531
155698	A6E	NAWC-WD	608
155699	A6E		
155700	A6A	w.o.11.10.72	
155701	A6A	w.o.15.11.69	
155702	A6E		
155703	A6E	VA-42	AD-501
155704	A6E		
155705	A6A	w.o.29.10.72	
155706	A6E		
155707	A6E	NAWC-WD	053
155708	A6E		
155709	A6A	w.o.3.5.72	
155710	A6E	VA-85	AB-514
155711	A6E	VA-145	NE-506
155712	A6E	VA-34	AG-501
155713	A6E	VA-128	NJ-553
155714	A6E	VX-5	XE-22
155715	A6E	VA-52	NL-511
155716	A6E	VA-36	AJ-530
155717	A6E	VA-128	NJ-814
155718	A6E	VA-42	AD-506
155719	A6E		
155720	A6A		
155721	A6E		
156478	EA6B	AMARC	5A025
156479	EA6B	AMARC	5A...
156480	EA6B	w.o.30.3.71	
156481	EA6B	VAQ-137	AB-625
156482	EA6B		
156979	EA6A	w.o.12.4.72	
156980	EA6A	w.o.23.10.72	
156981	EA6A	VAQ-309	ND-610
156982	EA6A	w.o.29.12.72	
156983	EA6A		
156984	EA6A		
156985	EA6A	VAQ-209	AF-...
156986	EA6A	VAQ-309	ND-613
156987	EA6A		
156988	EA6A	w.o.	
156989	EA6A	VAQ-309	ND-605
156990	EA6A		
156991	EA6A	VAQ-309	ND-607
156992	EA6A		
156993	EA6A		
156994	A6A	w.o.24.2.71	
156995	A6E		
156996	A6E	VA-128	NJ-551
156997	A6E	NAWC\SATD	SD-501
156998	EA6A		
156999	A6A		
157000	A6E		
157001	A6E	VA-36	AJ-502
157002	A6E	VA-34	AG-503
157003	A6E	VA-42	AD-500
157004	A6E	VA-95	NH-504
157005	A6E		
157006	A6E	VA-35	AA-504
157007	A6A	w.o.24.1.73	
157008	A6A		
157009	A6A		
157010	A6E	VA-34	AG-512
157011	A6E	w.o.6.12.79	
157012	A6A	w.o.16.3.79	
157013	A6A	VA-128	NJ-556
157014	A6E	NAWC	
157015	A6A		
157016	A6E	VA-304	ND-410
157017	A6E	VA-205	AF-510
157018	A6A	w.o.19.8.72	
157019	A6E	VA-65	AJ-503
157020	A6A		
157021	A6E	VA-42	AD-552
157022	A6A		
157023	A6E		
157024	A6E	VA-115	NF-515
157025	A6E		
157026	A6E	VA-75	AC-520
157027	A6E	VA-75	AC-521
157028	A6A		
157029	A6E	VA-35	AA-502
158029	EA6B	VAQ-209	AF-620
158030	EA6B	VMAQ-3	MD-01
158031	EA6B		
158032	EA6B	VMAQ-3	MD-02
158033	EA6B		
158034	EA6B	VAQ-129	NJ-908
158035	EA6B	VMAQ-2	CY-002
158036	EA6B	VMAQ-2	CY-003
158037	EA6B	w.o.16.12.79	
158038	EA6B	w.o.16.11.71	
158039	EA6B	VAQ-129	NJ-930
158040	EA6B	VMAQ-2	CY-004
158041	A6E		
158042	A6E	VA-165	NG-507
158043	A6E	VA-42	AD-570
158044	A6E		
158045	A6E	VA-196	NK-507
158046	A6E	VA-196	NK-517
158047	A6E	w.o.29.4.78	
158048	A6E	w.o.20.4.75	
158049	A6E	w.o.15.8.77	
158050	A6E		
158051	A6E	VA-205	AF-501
158052	A6E	VA-196	NK-522
158528	A6E		
158529	A6E	VA-205	AF-503
158530	A6E	w.o.6.7.81	
158531	A6E	VA-304	ND-403
158532	A6E		
158533	A6E	VA-196	NK-511
158534	A6E	stored	NAD Alameda
158535	A6E	w.o.18.12.84	
158536	A6E	VAQ-129	NJ-901
158537	A6E	NAWC-WD	608
158538	A6E	VA-205	AF-507
158539	A6E	w.o.17.1.91	
158540	EA6B	VMAQ-2	CY-001
158541	EA6B	w.o.28.11.79	
158542	EA6B	VAQ-129	NJ-920
158543	EA6B	w.o.27.6.86	
158544	EA6B	VMAQ-4	RM-00
158545	EA6B	w.o.11.8.76	
158546	EA6B	w.o.13.4.82	
158547	EA6B	VAQ-132	AE-605
158649	EA6B	VAQ-209	AF-604
158650	EA6B	VAQ-129	NJ-915
158651	EA6B	w.o.10.7.84	
158787	A6E	VA-34	AG-511
158788	A6E	VA-128	NJ-802
158789	A6E	w.o.4.12.89	
158790	A6E	w.o.27.12.84	
158791	A6E	VA-115	NF-503
158792	A6E	VA-196	NK-505
158793	A6E		
158794	A6E	VA-52	NL-503
158795	A6E		
158796	A6E	w.o.14.4.88	
158797	A6E	VA-205	AF-505
158798	A6E		
158799	EA6B	w.o.20.5.80	
158800	EA6B	w.o.24.10.85	
158801	EA6B	VAQ-134	AA-620
158802	EA6B		
158803	EA6B	w.o.30.7.78	
158804	EA6B		
158805	EA6B		
158806	EA6B	w.o.19.87.81	
158807	EA6B	VAQ-129	NJ-900
158808	EA6B	w.o.28.12.82	
158809	EA6B	w.o.11.11.77	
158810	EA6B		
158811	EA6B	VAQ-129	NJ-652
158812	EA6B	w.o.13.1.75	
158813	EA6B	w.o.23.7.81	

158814	EA6B	w.o.25.6.75		160705	EA6B	w.o.28.4.83	
158815	EA6B	VAQ-129	NJ-905	160706	EA6B	VAQ-134	NL-622
158816	EA6B			160707	EA6B	VMAQ-2	CY-14
158817	EA6B	w.o.28.9.78		160708	EA6B	w.o.22.1.82	
159174	A6E			160709	EA6B	VAQ-134	NL-624
159175	A6E	VA-65	AJ-512	160786	EA6B	VMAQ-3	MD-05
159176	A6E	VA-205	AF-500	160787	EA6B		
159177	A6E	VA-145	NE-512	160788	EA6B	VMAQ-2	CY-05
159178	A6E	VA-42	AD-540	160789	EA6B	w.o.7.4.86	
159179	A6E	VA-35	AA-507	160790	EA6B	w.o.24.7.89	
159180	A6E	VA-145	NE-515	160791	EA6B	VAQ-129	NJ-936
159181	A6E	NSWC	04	160993	A6E		
159182	A6E			160994	A6E	w.o.15.12.86	
159183	A6E	VA-42	AD-501	160995	A6E	VA-42	AD-530
159184	A6E	VA-128	NJ-533	160996	A6E	VA-65	AJ-502
159185	A6E	w.o.2.10.80		160997	A6E	VA-128	NJ-821
159309	A6E	w.o.3.3.77		160998	A6E		
159310	A6E	VA-145	NE-506	161082	A6E	VA-128	NJ-532
159311	A6E	VA-35	AA-503	161083	A6E	VA-42	AD-534
159312	A6E			161084	A6E	VA-42	AD-535
159313	A6E	w.o.10.10.78		161085	A6E	w.o.3.2.86	
159314	A6E	VA-35	AA-541	161086	A6E	VA-36	AJ-533
159315	A6E			161087	A6E	VA-205	AF-510
159316	A6E	VA-42	AD-542	161088	A6E	VA-128	NJ-555
159317	A6E			161089	A6E	VA-52	NL-511
159567	A6E			161090	A6E		
159568	A6E	VX-5	XE-	161091	A6E	VA-128	NJ-810
159569	A6E	VA-42	AD-537	161092	A6E	VA-65	AJ-506
159570	A6E	VA-128	NJ-833	161093	A6E	VA-36	AJ-534
159571	A6E	VA-35	AA-504	161100	A6E	VA-115	NF-506
159572	A6E	w.o.6.11.89		161101	A6E	NSWC	05
159573	A6E	w.o.		161102	A6E	VA-128	NJ-807
159574	A6E	VA-36	AJ-542	161103	A6E	VA-205	AF-505
159575	A6E	VA-65	AJ-500	161104	A6E	VA-128	NJ-823
159576	A6E			161105	A6E	w.o.18.9.87	
159577	A6E			161106	A6E	VA-85	AB-501
159578	A6E	VA-128	NJ-534	161107	A6E	VA-165	NG-502
159579	A6E	VA-145	NE-514	161108	A6E	VA-128	NJ-830
159580	A6E			161109	A6E	w.o.8.8.89	
159581	A6E			161110	A6E		
159582	EA6B	w.o.29.10.81		161111	A6E		
159583	EA6B	VAQ-129	NJ-600	161115	EA6E	VAQ-139	NK-620
159584	EA6B	VAQ-129	NJ-910	161116	EA6E	VAQ-129	NJ-932
159585	EA6B	VAQ-129	NJ-603	161117	EA6E	w.o.25.7.81	
159586	EA6B	NAWC\SATD	SD-536	161118	EA6E	VAQ-209	AF-605
159587	EA6B	VAQ-131	NE-605	161119	EA6E	VAQ-139	NK-621
159895	A6E	VA-165	NG-507	161120	EA6E	VAQ-134	AA-604
159896	A6E	VA-85	AA-510	161230	A6E		
159897	A6E	w.o.24.10.86		161231	A6E	VA-65	AJ-511
159898	A6E	VA-52	NL-504	161232	A6E	VA-85	AB-504
159899	A6E	VA-52	NL-510	161233	A6E		
159900	A6E			161234	A6E		
159901	A6E	VA-36	AJ-540	161235	A6E	VA-95	NH-502
159902	A6E	w.o.		161242	EA6B		
159903	A6E	VA-128	NJ-824	161243	EA6B	VAQ-139	NK-622
159904	A6E			161244	EA6B	VAQ-129	NJ-928
159905	A6E	w.o.17.1.87		161245	EA6B	VMAQ-4	RM-03
159906	A6E	VA-165	NG-504	161246	EA6B	w.o.30.7.82	
159907	EA6B	VMAQ-2	CY-02	161247	EA6B	VMAQ-4	RM-04
159908	EA6B	VMAQ-4	RM-01	161347	EA6B		
159909	EA6B			161348	EA6B	VAQ-137	AB-622
159910	EA6B	w.o.26.5.81		161349	EA6B	VMAQ-4	RM-05
159911	EA6B	VAQ-129	NJ-934	161350	EA6B	VAQ-309	ND-607
159912	EA6B	VAQ-141	AJ-621	161351	EA6B	w.o.28.10.84	
160421	A6E	w.o.17.5.79		161352	EA6B	VAQ-129	NJ-935
160422	A6E	VA-165	NG-506	161659	A6E	VA-85	AB-511
160423	A6E	VA-196	NK-500	161660	A6E	VA-36	AJ-540
160424	A6E	VA-205	AF-500	161661	A6E	VA-115	NF-501
160425	A6E			161662	A6E	VA-36	AJ-532
160426	A6E			161663	A6E		
160427	A6E	VA-196	NK-	161664	A6E	VA-128	NJ-810
160428	A6E	w.o.29.8.79		161665	A6E		
160429	A6E	VA-128	NJ-803	161666	A6E		
160430	A6E	VA-145	NE-510	161667	A6E	VA-42	AD-511
160431	A6E			161668	A6E	w.o.17.1.91	
160432	EA6B	VMAQ-2	CY-07	161669	A6E	VA-128	NJ-813
160433	EA6B	VAQ-309	ND-604	161670	A6E		
160434	EA6B	VAQ-129	NJ-627	161671	A6E		
160435	EA6B	VMAQ-2	CY-03	161672	A6E	VA-196	NK-504
160436	EA6B	VMAQ-2	CY-04	161673	A6E	w.o.1.4.85	
160437	EA6B	VAQ-134	NL-623	161674	A6E	VA-196	NK-512
160609	EA6B	VAQ-129	NJ-612	161675	A6E	VA-42	AD-507
160704	EA6B	w.o.25.8.83		161676	A6E	VA-42	AD-501

Serial	Type	Unit	Code
161677	A6E	VA-85	AB-512
161678	A6E		
161679	A6E	VA-42	AD-504
161680	A6E		
161681	A6E		
161682	A6E	VA-42	AD-531
161683	A6E	VA-196	NK-506
161684	A6E	w.o.22.4.85	
161685	A6E	w.o.20.5.87	
161686	A6E		
161687	A6E		
161688	A6E		
161689	A6E		
161774	EA6B	VAQ-129	NJ-929
161775	EA6B	VAQ-137	AB-620
161776	EA6B	VAQ-129	NJ-910
161777	EA6B	w.o.13.12.84	
161778	EA6B	w.o.13.7.86	
161779	EA6B	VAQ-137	AB-
161880	EA6B	VAQ-137	AB-621
161881	EA6B	VAQ-137	AB-606
161882	EA6B	VAQ-137	AB-623
161883	EA6B		
161884	EA6B	VAQ-129	NJ-901
161885	EA6B	VAQ-129	NJ-908
162179	A6E	VA-75	AC-510
162180	A6E	VA-128	NJ-801
162181	A6E	w.o.22.5.86	
162182	A6E	VA-145	NE-500
162183	A6F		
162184	A6F		
162185	A6F		
162186	A6F		
162187	A6F		
162188	A6E	VA-128	NJ-805
162189	A6E	VA-128	NJ-554
162190	A6E	VA-75	AC-500
162191	A6E	VA-75	AC-501
162192	A6E	VA-75	AC-502
162193	A6E	VA-75	AC-503
162194	A6E	VA-75	AC-504
162195	A6E	VA-75	AC-505
162196	A6E	VA-75	AC-506
162197	A6E	VA-145	NE-501
162198	A6E	VA-75	AC-523
162199	A6E	VA-75	AC-524
162200	A6E	VA-145	NE-502
162201	A6E	VA-145	NE-503
162202	A6E	VA-34	AG-507
162203	A6E		
162204	A6E	VA-165	NG-501
162205	A6E	VA-165	NG-502
162206	A6E	NSWC	06
162207	A6E	VA-165	NG-503
162208	A6E	VA-95	NH-503
162209	A6E	VA-128	NJ-530
162210	A6E	VA-95	NH-507
162211	A6E	VA-42	AD-510
162212	A6E	VA-115	NF-516
162213	A6E		
162214	A6E		
162215	A6E		
162216	A6E		
162217	A6E		
162218	A6E		
162219	A6E		
162220	A6E		
162221	A6E		
162222	A6E		
162223	EA6B	VAQ-131	NE-607
162224	EA6B	VAQ-132	AA-622
162225	EA6B	VAQ-129	NJ-907
162226	EA6B	w.o.19.11.87	
162227	EA6B	VAQ-129	NJ-909
162228	EA6B	VAQ-134	NL-620
162229	EA6B	VAQ-136	NF-604
162230	EA6B	VAQ-137	AB-606
162231	EA6B		
162232	EA6B		
162233	EA6B	w.o.15.4.87	
162234	EA6B		
162235	EA6B		
162236	EA6B		
162237	EA6B		
162238	EA6B		
162239	EA6B		
162240	EA6B		
162241	EA6B		
162242	EA6B		
162243	EA6B		
162244	EA6B		
162245	EA6B		
162246	EA6B		
162934	EA6B	VAQ-132	AA-623
162935	EA6B	VMAQ-2	CY-...
162936	EA6B	VAQ-134	NL-621
162937	EA6B	VMAQ-2	CY-03
162938	EA6B	VAQ-137	AB-623
162939	EA6B	VMAQ-2	CY-16
162940	EA6B		
162941	EA6B		
163030	EA6B	VAQ-129	NJ-905
163031	EA6B	VMAQ-2	CY-17
163032	EA6B	VAQ-138	NG-607
163033	EA6B	VMAQ-2	CY-20
163034	EA6B	VAQ-141	AJ-623
163035	EA6B	VMAQ-1	CB-06
163044	EA6B	w.o.5.12.88	
163045	EA6B	VAQ-136	NF-621
163046	EA6B	VAQ-136	NF-622
163047	EA6B	VAQ-136	NF-623
163048	EA6B	VAQ-136	NF-624
163049	EA6B	VAQ-137	AB-620
163395	EA6B	VAQ-140	AG-620
163396	EA6B	VAQ-138	NG-620
163397	EA6B	VAQ-209	AF-607
163398	EA6B	VAQ-130	AC-620
163399	EA6B	VAQ-130	AC-621
163400	EA6B	VAQ-129	NJ-931
163401	EA6B	VX-5	XE-60
163402	EA6B	VAQ-130	AC-622
163403	EA6B	VAQ-130	AC-623
163404	EA6B	VAQ-130	AC-624
163405	EA6B	VAQ-141	AJ-620
163406	EA6B		
163520	EA6B	VAQ-138	NG-622
163521	EA6B	VAQ-138	NG-623
163522	EA6B	VAQ-138	NG-624
163523	EA6B	VAQ-139	NK-624
163524	EA6B	VAQ-131	NE-605
163525	EA6B	VAQ-131	NE-606
163526	EA6B	VAQ-131	NE-607
163527	EA6B	VAQ-141	AJ-621
163528	EA6B	VAQ-141	AJ-622
163529	EA6B	VAQ-141	AJ-623
163530	EA6B	VAQ-141	AJ-624
163531	EA6B	VAQ-135	NH-620
163884	EA6B	VAQ-140	AG-621
163885	EA6B	VAQ-135	NH-621
163886	EA6B	VAQ-135	NH-622
162887	EA6B	VAQ-135	NH-623
163888	EA6B	VAQ-140	AG-622
163889	EA6B	VAQ-140	AG-623
163890	EA6B	VAQ-135	NH-624
163891	EA6B	VAQ-129	NJ-933
163892	EA6B		
163955	A6F		
163956	A6F		
163957	A6F		
163958	A6F		
163959	A6F		
163960	A6F		
163961	A6F		
163962	A6F		
163963	A6F		
163964	A6F		
163965	A6F		
163966	A6F		
163967	A6F		
163968	A6F		
163969	A6F		
163970	A6F		
163971	A6F		
163972	A6F		
163973	A6F		
163974	A6F		

163975	A6F			164190	EA6B		
163976	A6F			164191	EA6B		
163977	A6F			164192	EA6B		
163978	A6F			164193	EA6B		
163979	A6F			164376	A6E	VA-35	AA-500
163980	A6F			164377	A6E	VA-52	NL-501
163981	A6F			164378	A6E	VA-35	AA-501
163982	A6F			164379	A6E	VA-52	NL-506
163983	A6F			164380	A6E		
163984	A6F			164381	A6E	VA-65	AJ-505
164182	EA6B			164382	A6E	VA-36	AJ-530
164183	EA6B			164383	A6E		
164184	EA6B			164384	A6E	VA-42	AD-505
164185	EA6B			164385	A6E		
164186	EA6B			164401	EA6B	VAQ-140	AG-624
164187	EA6B			164402	EA6B	VAQ-131	NE-604
164188	EA6B			164403	EA6B		
164189	EA6B						

A7 CORSAIR II

152580	YA7A			153143	A7A	w.o.28.2.70	
152581	YA7A			153144	A7P	LTV	
152582	YA7A			153145	A7A	MASDC	
152647	A7A	NARTC	Orlando, Fl	153146	A7A	w.o.29.5.70	
152648	A7A	stored	NAD N. Island	153147	A7A	w.o.6.8.72	
152649	A7A	stored	NAD N. Island	153148	A7A		
152650	A7A			153149	A7A		
152651	A7A	LTV		153150	A7A	NATTC	
152652	A7A	MASDC		153151	A7A	LTV	
152653	A7A			153152	A7P	to Portugal	5510
152654	A7A			153153	A7A	w.o.8.4.70	
152655	A7A	LTV		153154	A7A		
152656	A7A	LTV		153155	A7P	to Portugal	5522
152657	A7A	LTV		153156	A7A	w.o.15.11.69	
152658	A7A	preserved	NAS Patuxent Riv.	153157	A7P	to Portugal	5511
152659	A7P	LTV		153158	A7A	w.o.13.11.76	
152660	A7A	MASDC		153159	A7P	to Portugal	5523
152661	A7A			153160	A7A		
152662	A7A			153161	A7A	MASDC	
152663	A7A			153162	A7P	to Portugal	5524
152664	A7A	w.o.1.8.67		153163	A7A	NATTC	Memphis
152665	A7A			153164	A7A	w.o.28.4.69	
152666	A7A			153165	A7A		
152667	A7A	LTV		153166	A7A		
152668	A7A	NATTC		153167	A7A		
152669	A7A	LTV		153168	A7A		
152670	A7A	preserved	NAS Cecil Field	153169	A7A		
152671	A7A			153170	A7P	to Portugal	5512
152672	A7A			153171	A7P	to Portugal	5525
152673	A7A	LTV		153172	A7A		
152674	A7A	w.o.1.8.67		153173	A7P	to Portugal	
152675	A7A	scrapyard	Consolidated	153174	A7A		
152676	A7A	LTV		153175	A7A	w.o.31.10.68	
152677	A7A	AMARC	6A122	153176	A7A	w.o.28.6.70	
152678	A7A	LTV		153177	A7P	to Portugal	5526
152679	A7A	w.o.15.11.69		153178	A7A		
152680	A7A	w.o.1.5.70		153179	A7P	to Portugal	5527
152681	A7A			153180	A7A	w.o.2.5.69	
152682	A7P	LTV		153181	A7A	w.o.14.2.69	
152683	A7A			153182	A7A		
152684	A7A			153183	A7A		
152685	A7A			153184	A7P	to Portugal	5504
153134	A7P	to Portugal	5521	153185	A7A	w.o.1.8.69	
153135	A7A	NARTC	Orlando, Fl	153186	A7A		
153136	A7A	w.o.3.3.70		153187	A7P	to Portugal	5528
153137	A7A			153188	A7P	to Portugal	5513
153138	A7A	LTV		153189	A7A	w.o.1.11.71	
153139	A7A			153190	A7P	to Portugal	5505
153140	A7P	LTV		153191	A7A	w.o.30.3.77	
153141	A7A			153192	A7A		
153142	A7A	NATTC		153193	A7A	w.o.31.7.72	

153194	A7P	to Portugal	5507
153195	A7P	to Portugal	5529
153196	A7P	to Portugal	5530
153197	A7A	w.o.15.6.72	
153198	A7A		
153199	A7P	LTV	
153200	A7P	to Portugal	5502
153201	A7A	MASDC	
153202	A7A		
153203	A7A		
153204	A7A		
153205	A7A		
153206	A7A	w.o.13.6.72	
153207	A7A	w.o.17.8.72	
153208	A7P	to Portugal	5534
153209	A7A		
153210	A7A		
153211	A7A		
153212	TA7P	to Portugal	
153213	A7A	w.o.12.9.72	
153214	A7A		
153215	A7P	to Portugal	5514
153216	TA7P	to Portugal	
153217	TA7P	to Portugal	
153218	A7A		
153219	A7P	to Portugal	5508
153220	A7A	NARC	Great Lakes II
153221	A7P	to Portugal	5515
153222	A7A	w.o.12.4.69	
153223	A7A	w.o.8.9.71	
153224	TA7P	to Portugal	
153225	A7A	w.o.2.9.68	
153226	A7A	MASDC	
153227	A7P	to Portugal	5516
153228	A7P	to Portugal	5517
153229	A7P	to Portugal	5531
153230	A7A	w.o.17.6.72	
153231	A7A	w.o.7.1.70	
153232	A7A		
153233	A7A	w.o.7.4.70	
153234	A7A	MASDC	
153235	A7A	dump	NAS Alameda
153236	A7A	NARTC	San Diego, Ca
153237	A7P	to Portugal	5518
153238	A7A	NARTC	San Diego, Ca
153239	A7A	w.o.22.12.67	
153240	A7P	to Portugal	5532
153241	A7A	LTV	
153242	A7A	NATTC	Memphis
153243	A7A		
153244	A7P	to Portugal	5509
153245	A7A	LTV	
153246	A7A		
153247	A7A		
153248	A7P	to Portugal	5519
153249	A7P	to Portugal	5533
153250	A7P	to Portugal	5506
153251	A7A		
153252	A7A	w.o.1.10.69	
153253	A7A	w.o.24.7.68	
153254	A7P	to Portugal	5535
153255	A7A	w.o.31.5.68	
153256	A7A	w.o.12.11.73	
153257	A7A	w.o.22.6.68	
153258	A7A	w.o.31.5.68	
153259	A7A		
153260	A7P	to Portugal	5536
153261	A7P	to Portugal	5520
153262	A7A		
153263	A7A		
153264	A7A		
153265	A7A	w.o.10.6.68	
153266	A7P	to Portugal	5538
153267	A7A		
153268	A7P	to Portugal	5537
153269	A7A	w.o.21.6.68	
153270	A7A		
153271	A7A	w.o.25.6.68	
153272	A7P	to Portugal	5503
153273	A7A	w.o.6.10.68	
153344	A7A	w.o.14.9.68	
153345	A7A	NATTC	
154346	A7P	to Portugal	5539

154347	A7A		
154348	A7A	MASDC	
154349	A7P	to Portugal	5540
154350	A7A		
154351	A7P	to Portugal	5541
154352	A7P	to Portugal	5501
154353	A7A		
154354	A7P	to Portugal	5542
154355	A7P	to Portugal	5543
154356	A7P	to Portugal	5544
154357	A7A		
154358	A7A	w.o.3.4.70	
154359	A7A	w.o.24.8.68	
154360	A7B	LTV	
154361	TA7C	AMARC	6A368
154362	A7B		
154363	A7B	w.o.20.9.72	
154364	A7B		
154365	A7B		
154366	A7B	AMARC	6A130
154367	A7B		
154368	A7B	AMARC	6A129
154369	A7B		
154370	A7B		
154371	A7B	AMARC	6A131
154372	A7B	AMARC	6A151
154373	A7B	AMARC	6A206
154374	A7B		
154375	A7B	AMARC	6A153
154376	A7B		
154377	TA7C		
154378	A7B		
154379	TA7C	AMARC	6A...
154380	A7B		
154381	A7B	AMARC	6A147
154382	A7B		
154383	A7B	w.o.8.6.69	
154384	A7B		
154385	A7B		
154386	A7B		
154387	A7B		
154388	A7B	AMARC	6A157
154389	A7B		
154390	A7B	AMARC	6A169
154391	A7B	w.o.5.2.70	
154392	A7B		
154393	A7B	w.o.7.2.72	
154394	A7B		
154395	A7B		
154396	A7B	AMARC	6A125
154397	A7B	AMARC	6A138
154398	A7B		
154399	A7B	w.o.10.11.72	
154400	A7B		
154401	A7B		
154402	TA7C	AMARC	6A370
154403	A7B		
154404	TA7C	AMARC	6A369
154405	A7B	w.o.23.5.72	
154406	A7B	AMARC	6A181
154407	TA7C	NAWC-AD	SATD SD-422
154408	A7B		
154409	A7B	AMARC	6A148
154410	TA7C	w.o.17.1.84	
154411	A7B		
154412	TA7C		
154413	A7B	AMARC	6A149
154414	A7B		
154415	A7B	AMARC	6A155
154416	A7B	AMARC	6A186
154417	A7B		
154418	A7B		
154419	A7B		
154420	A7B		
154421	A7B		
154422	A7B		
154423	A7B	w.o.19.7.69	
154424	TA7C		
154425	TA7C	AMARC	6A320
154426	A7B		
154427	A7B		
154428	A7B		
154429	A7B		

154430	A7B	w.o.22.9.71		154461	A7B		
154431	A7B			154462	A7B	AMARC	6A170
154432	A7B			154463	A7B	AMARC	6A156
154433	A7B	AMARC	6A200	154464	TA7C	NAWC-WD	81
154434	A7B			154465	A7B	w.o.18.8.69	
154435	A7B			154466	A7B	AMARC	6A145
154436	A7B	w.o.24.9.72		154467	TA7C	w.o.2.6.86	
154437	TA7C	stored	NAS Cecil Field	154468	A7B	AMARC	6A134
154438	A7B	AMARC	6A174	154469	A7B	AMARC	6A144
154439	A7B	AMARC	6A165	154470	A7B		
154440	A7B	AMARC	6A155	154471	TA7C		
154441	A7B			154472	A7B	AMARC	6A142
154442	A7B	w.o.18.8.69		154473	A7B	w.o.9.3.69	
154443	A7B	GIA	NAWC Lakehurst	154474	A7B		
154444	A7B			154475	A7B	AMARC	6A152
154445	A7B	w.o.27.3.86		154476	A7B		
154446	A7B			154477	TA7C	NAWC-AD	SATD SD-420
154447	A7B			154478	A7B	AMARC	6A139
154448	A7B	AMARC	6A143	154479	A7B	AMARC	6A182
154449	A7B			154480	A7B		
154450	TA7C	w.o.29.11.84		154481	A7B	AMARC	6A150
154451	A7B	AMARC	6A127	154482	A7B	dump	NAS Jacksonville
154452	A7B	AMARC	6A136	154483	A7B		
154453	A7B	AMARC	6A163	154484	A7B	AMARC	6A140
154454	A7B	AMARC	6A203	154485	A7B		
154455	TA7C	AMARC	6A321	154486	A7B		
154456	A7B			154487	A7B		
154457	A7B			154488	A7B	AMARC	6A175
154458	TA7C			154489	TA7C	AMARC	6A392
154459	A7B	w.o.12.1.84		154490	AI		
154460	A7B	AMARC	6A161				

AV8 HARRIER

158384	AV8A	w.o.5.9.80		158957	AV8C		
158385	AV8A	to NASA	N716NA	158958	AV8A		
158386	AV8A	w.o.18.6.71		158959	AV8A	AMARC	7A002
158387	AV8C			158960	AV8A		
158388	AV8A	w.o.27.3.73		158961	AV8A		
158389	AV8A			158962	AV8C	w.o.24.9.82	
158390	AV8C			158963	AV8A	NARF	Cherry Point
158391	AV8C	AMARC	7A034	158964	AV8A	AMARC	7A005
158392	AV8C	AMARC	7A035	158965	AV8A	NARF	Cherry Point
158393	AV8C	AMARC	7A026	158966	AV8A	AMARC	7A001
158394	YAV8B	to NASA	N704NA	158967	AV8A	w.o.11.2.77	
158395	AV8A	w.o.15.11.79		158968	AV8A	w.o.26.1.82	
158694	AV8A			158969	AV8C	MASDC	
158695	AV8A	preserved	MCAS Quantico	158970	AV8A	w.o.6.9.77	
158696	AV8A			158971	AV8A	w.o.28.7.74	
158697	AV8C	AMARC	7A025	158972	AV8C	AMARC	7A011
158698	AV8C	preserved	NAF El Centro	158973	AV8C	AMARC	7A012
158699	AV8C	w.o.22.9.83		158974	AV8C	w.o.30.8.76	
158700	AV8C	AMARC	7A027	158975	AV8C	Pima Museum	
158701	AV8C	AMARC	7A010	158976	AV8A	preserved	Havelock, NC
158702	AV8A	AMARC	7A014	158977	AV8C	AMARC	7A013
158703	AV8A	w.o.26.6.81		159230	AV8A	w.o.6.12.76	
158704	AV8A			159231	AV8A	w.o.25.4.85	
158705	AV8C	AMARC	7A028	159232	AV8C	preserved	USS Intrepid
158706	AV8C	NARF	Cherry Point	159233	AV8A	preserved	RNAS Yeovilton
158707	AV8A			159234	AV8A		
158708	AV8A	w.o.29.11.77		159235	AV8A	MASDC	
158709	AV8A	w.o.10.1.76		159236	AV8A	w.o.3.7.75	
158710	AV8A	NARF	Cherry Point	159237	AV8A	w.o.16.6.76	
158711	AV8A	NARF	Cherry Point	159238	AV8C	AMARC	7A024
158948	AV8A	w.o.5.6.74		159239	AV8A	AMARC	7A016
158949	AV8C	AMARC	7A003	159240	AV8C	AMARC	7A033
158950	AV8A	w.o.28.6.84		159241	AV8C	Pima Museum	
158951	AV8C	AMARC	7A029	159242	AV8A		
158952	AV8A	w.o.3.2.78		159243	AV8C	AMARC	7A031
158953	AV8A	w.o.27.7.77		159244	AV8A	w.o.3.7.75	
158954	AV8C	w.o.3.4.85		159245	AV8A	w.o.9.10.74	
158955	AV8A	w.o.1.12.82		159246	AV8A		
158956	AV8C	w.o.4.3.82		159247	AV8C	AMARC	7A007

159248	AV8A		
159249	AV8A		
159250	AV8A	w.o.12.7.77	
159251	AV8A	w.o.13.8.80	
159252	AV8A	AMARC	7A022
159253	AV8A	w.o.2.2.83	
159254	AV8C		
159255	AV8A	AMARC	7A023
159256	AV8A		
159257	AV8C	AMARC	7A032
159258	AV8C	MASDC	
159259	AV8A	w.o.27.11.77	
159366	AV8A	AMARC	7A018
159367	AV8A	AMARC	7A019
159368	AV8A	w.o.19.1.81	
159369	AV8A		
159370	AV8C	AMARC	7A037
159371	AV8A	AMARC	7A008
159372	AV8A	w.o.6.4.77	
159373	AV8A	AMARC	7A015
159374	AV8A	AMARC	7A020
159375	AV8A	AMARC	7A017
159376	AV8A	AMARC	7A021
159377	AV8A	w.o.19.3.77	
159378	TAV8A	AMARC	7A043
159379	TAV8A	AMARC	7A042
159380	TAV8A	w.o.12.8.87	
159381	TAV8A	w.o.27.4.83	
159382	TAV8A	AMARC	7A040
159383	TAV8A	AMARC	7A039
159384	TAV8A	w.o.1.8.80	
159385	TAV8A	AMARC	7A041
161396	AV8B	NAWC-AD	7T-
161397	AV8B	MDD	
161398	AV8B	NAWC-AD	7T-
161399	AV8B	NAWC-AD	7T-624
161573	AV8B	w.o.23.2.91	
161574	AV8B	VMAT-203	KD-22
161575	AV8B	VMA-542	WH-13
161576	AV8B	NAWC-WD	008
161577	AV8B	VMAT-203	KD-23
161578	AV8B	w.o.31.3.85	
161579	AV8B	VMAT-203	KD-31
161580	AV8B	VMAT-203	KD-25
161581	AV8B	VMAT-203	KD-26
161582	AV8B	w.o.13.7.88	
161583	AV8B	VMAT-203	KD-29
161584	AV8B	VMAT-203	KD-30
162068	AV8B	VMAT-203	KD-39
162069	AV8B	VMA-542	WH-12
162070	AV8B		
162071	AV8B	w.o.11.2.88	
162072	AV8B		
162073	AV8B	w.o.5.6.87	
162074	AV8B		
162075	AV8B	w.o.24.7.87	
162076	AV8B		
162077	AV8B	VMAT-203	KD-21
162078	AV8B	w.o.5.90	
162079	AV8B	w.o.27.2.86	
162080	AV8B		
162081	AV8B	w.o.9.2.91	
162082	AV8B	VMA-231	CG-
162083	AV8B		
162084	AV8B	VMA-223	WP-31
162085	AV8B		
162086	AV8B	VMA-231	CG-03
162087	AV8B	VMAT-203	KD-24
162088	AV8B	VMA-223	WP-02
162721	AV8B	NAWC-WD	88
162722	AV8B		
162723	AV8B	VMAT-203	KD-46
162724	AV8B	w.o.17.1.86	
162725	AV8B	VMAT-203	KD-47
162726	AV8B		
162727	AV8B	VMA-223	WP-11
162728	AV8B		
162729	AV8B	VMA-231	CG-13
162730	AV8B		
162731	AV8B		
162732	AV8B	VMA-542	WH-14
162733	AV8B	VMA-223	WP-07
162734	AV8B	VMA-231	CG-16

162735	AV8B	w.o.	
162736	AV8B		
162737	AV8B		
162738	AV8B	VMA-231	CG-08
162739	AV8B	VMA-542	WH-05
162740	AV8B	w.o.27.2.91	
162741	AV8B	VMA-542	WH-01
162742	AV8B	VMAT-203	KD-33
162743	AV8B	VMA-542	WH-
162744	AV8B	VMA-231	CG-14
162745	AV8B	w.o.4.11.86	
162746	AV8B	w.o.12.1.87	
162747	TAV8B	damaged	NAD Cherry Point
162942	AV8B	VMA-542	WH-04
162943	AV8B	VMA-231	CG-05
162944	AV8B	VMA-231	CG-06
162945	AV8B	VMA-542	WH-24
162946	AV8B	VMA-542	WH-20
162947	AV8B	VMA-513	WF-00
162948	AV8B	VMAT-203	KD-30
162949	AV8B	VMAT-203	KD-34
162950	AV8B	VMAT-203	KD-35
162951	AV8B	VMAT-203	KD-36
162952	AV8B	w.o.8.10.88	
162953	AV8B	VMA-231	CG-19
162954	AV8B	w.o.22.1.91	
162955	AV8B	VMA-231	CG-17
162956	AV8B	VMA-513	WF-21
162957	AV8B	VMA-214	WE-15
162958	AV8B	VMA-513	WF-23
162959	AV8B	VMA-513	WF-24
162960	AV8B	VMA-513	WF-25
162961	AV8B	w.o.10.9.87	
162962	AV8B	VMA-231	CG-07
162963	TAV8B	VMAT-203	KD-01
162964	AV8B	VMA-231	CG-08
162965	AV8B	VMA-542	WH-09
162966	AV8B	NAWC-WD	
162967	AV8B	VMA-231	CG-10
162968	AV8B	VMA-513	WF-27
162969	AV8B		
162970	AV8B	VMA-513	WF-08
162971	TAV8B	VMAT-203	KD-02
162972	AV8B	VMA-231	CG-11
162973	AV8B	VMA-231	CG-12
163176	AV8B	VMA-311	WL-24
163177	AV8B		
163178	AV8B	VMA-513	WF-46
163179	AV8B	VMA-542	WH-08
163180	TAV8B	VMAT-203	KD-03
163181	AV8B		
163182	AV8B	w.o.1.3.88	
163183	AV8B	VMA-231	CG-15
163184	AV8B	w.o.5.11.88	
163185	AV8B	w.o.	
163186	TAV8B	VMAT-203	KD-04
163187	AV8B	VMA-223	WP-09
163188	AV8B	VMA-542	WH-10
163189	AV8B	VMA-223	WP-14
163190	AV8B	w.o.25.2.91	
163191	TAV8B	VMAT-203	KD-05
163192	AV8B	VMA-214	WE-18
163193	AV8B	VMA-214	WE-19
163194	AV8B	VMA-214	WE-20
163195	AV8B	VMA-223	WP-12
163196	TAV8B	VMAT-203	KD-06
163197	AV8B	VMA-542	WH-21
163198	AV8B	VMA-542	WH-04
163199	AV8B	VMA-542	WH-15
163200	AV8B	VMA-223	WP-01
163201	AV8B	VMA-542	WH-02
163202	TAV8B	w.o.28.8.89	
163203	AV8B	VMA-513	WF-14
163204	AV8B		
163205	AV8B	VMA-542	WH-01
163206	AV8B	VMA-542	WH-16
163207	TAV8B	damaged	NAD Cherry Point
163419	AV8B		
163420	AV8B	w.o.15.5.91	
163421	AV8B		
163422	AV8B	VMAT-203	KD-38
163423	AV8B	VMA-513	WF-16
163424	AV8B	VMA-542	WH-17

163425	AV8B	VMA-542	WH-05		163668	AV8B	VMA-211	CF-24
163426	AV8B	VMA-231	CG-13		163669	AV8B	VMA-311	WL-09
163514	AV8B	VMA-231	CG-20		163670	AV8B	VMA-231	CG-04
153515	AV8B				163671	AV8B	VMA-311	WL-10
163516	AV8B	VMA-542	WH-06		163672	AV8B	VMA-211	CF-22
163517	AV8B				163673	AV8B	VMA-231	CG-03
163518	AV8B	w.o.28.1.91			163674	AV8B	VMA-211	CF-23
163519	AV8B	VMA-311	WL-20		163675	AV8B	VMA-311	WL-13
163659	AV8B	VMA-223	WP-18		163676	AV8B	VMA-542	WH-03
163660	AV8B				163677	AV8B	VMA-311	WL-14
163661	AV8B	VMA-311	WL-05		163678	AV8B	VMA-211	CF-26
163662	AV8B	VMA-231	CG-01		163679	AV8B	VMAT-203	KD-32
163663	AV8B	VMA-542	WH-11		163680	AV8B	VMA-311	WL-16
163664	AV8B	VMA-211	CF-25		163681	AV8B	VMA-311	WL-17
163665	AV8B	VMA-231	CG-02		163682	AV8B	VMA-231	CG-17
163666	AV8B	w.o.31.10.89			163683	AV8B	VMA-311	WL-IO
163667	AV8B	w.o.16.12.89						

A12 AVENGER II

164519	A12A		164529	A12A	
164520	A12A		164530	A12A	
164521	A12A		164531	A12A	
164522	A12A		164532	A12A	
164523	A12A		164533	A12A	
164526	A12A		164534	A12A	
164527	A12A		164535	A12A	
164528	A12A				

C2 GREYHOUND

148147	YC2A				162150	C2A	VRC-50	RG-424
148148	YC2A	AMARC	1C005		162151	C2A	VAW-120	AD-022
152786	C2A	AMARC	1C002		162152	C2A		
152787	C2A				162153	C2A		
152788	C2A				162154	C2A	VRC-50	RG-425
152789	C2A				162155	C2A		
152790	C2A	AMARC	1C001		162156	C2A		
152791	C2A	AMARC	1C004		162157	C2A		
152792	C2A				162158	C2A		
152793	C2A	w.o.12.12.71			162159	C2A		
152794	C2A	AMARC	1C003		162160	C2A		
152795	C2A				162161	C2A	VRC-30	RW-32
152796	C2A	w.o.2.10.69			162162	C2A	VAW-110	NJ-312
152797	C2A				162163	C2A	VRC-30	RW-31
155120	C2A	w.o.15.12.70			162164	C2A		
155121	C2A				162165	C2A	VRC-30	RW-34
155122	C2A	w.o.29.1.72			162166	C2A	VAW-120	AD-023
155123	C2A	AMARC	1C006		162167	C2A	VRC-30	RW-33
155124	C2A				162168	C2A		
162140	C2A	VAW-120	AD-025		162169	C2A		
162141	C2A	VRC-30	RW-30		162170	C2A	VRC-30	RW-35
162142	C2A	VAW-110	NJ-311		162171	C2A		
162143	C2A				162172	C2A		
162144	C2A				162173	C2A		
162145	C2A				162174	C2A		
162146	C2A	VAW-110	NJ-310		162175	C2A		
162147	C2A	VRC-50	RG-421		162176	C2A		
162148	C2A	VRC-50	RG-422		162177	C2A	VRC-30	RW-36
162149	C2A	VRC-50	RG-423		162178	C2A		

C4 ACADEME

c\n   176, 178, 180, 182 to 187.

155722	TC4C	VMAT-202	KC-722		155727	TC4C	VA-42	AD-575
155723	TC4C	w.o.15.10.75			155728	TC4C	VA-42	AD-574
155724	TC4C	VA-42	AD-576		155729	TC4C	VA-42	AD-577
155725	TC4C	VA-128	NJ-852		155730	TC4C	VA-128	NJ-853
155726	TC4C	VA-128	NJ-851					

C8 BUFFALO

161546	UC8A	NAWC-WD

C9 SKYTRAIN II

c\n   47577, 47584, 47587, 47580, 47581, 47585, 47578, 47586, 47684, 47687, 47681, 47699, 47698, 47770, 48137, 48165, 48166, 47003, 47004, 47065, 47325, 47410, 47476, 47041, 47221, 47431, 47474, 47477, 47639, plus 4 other ex civil DC9-32.

159113	C9B	VR-55	RU-113	162390	C9B	SoC	
159114	C9B	VR-57	RX-114	162391	C9B	SoC	
159115	C9B	VR-57	RX-115	162392	C9B	SoC	
159116	C9B	VR-57	RX-116	162393	C9B	SoC	
159117	C9B	VR-57	RX-117	162753	C9B	VR-51	RV-753
159118	C9B	VR-56	JU-118	162754	C9B	VR-52	RV-754
159119	C9B	VR-56	JU-119	163036	C9B	VR-52	JT-036
159120	C9B	VR-55	RU-120	163037	C9B	VR-52	JT-037
160046	C9B	SOES	Cherry Point	163208	C9B	VR-46	JS-208
160047	C9B	SOES	Cherry Point	163511	C9B	VR-62	JW-511
160048	C9B	VR-58	JV-048	163512	C9B	VR-46	JS-512
160049	C9B	VR-58	JV-049	163513	C9B	VR-62	JW-513
160050	C9B	VR-55	RU-050	164605	C9B	VR-60	RT-605
160051	C9B	VR-55	RU-051	164606	C9B	VR-60	RT-606
161266	C9B	VR-59	RY-266	164607	C9B	VR-61	RS-607
161529	C9B	VR-59	RY-529	164608	C9B	VR-61	RS-608
161530	C9B	VR-59	RY-530				

C12 SUPER KING AIR 200

c\n   BJ-1 to BJ-66, BU-1 to BU-12, BV-1 to BV-12.

161185	UC12B	NATC			161187	UC12B	MCAS	5B Beaufort
161186	UC12B	NAS		7X New Orleans	161188	UC12B	NAS	7W Willow Grove

161189	UC12B	w.o.2.1.82			161500	UC12B	NAS	7E Jacksonville
161190	UC12B				161501	UC12B	NAS	8A Atsugi
161191	UC12B	NAS	7E Jacksonville		161502	UC12B	NAS	8E R'velt Roads
161192	UC12B	MCAS	5Y Yuma		161503	UC12B	NAF	8N El Centro
161193	UC12B	NAS	7H Fallon		161504	UC12B	NAS	7S Lemoore
161194	UC12B	MWHS-4	EZ New Orleans		161505	UC12B	NAS	F Pensacola
161195	UC12B	NAS	8F Guantanamo		161506	UC12B	NAS	A Meridian
161196	UC12B	MCAS	5Y Yuma		161507	UC12B	MCAS	5D New River
161197	UC12B	MCAS	5T El Toro		161508	UC12B	NAS	7V Glenview
161198	UC12B	NAS	7X New Orleans		161509	UC12B	NAS	7C Norfolk
161199	UC12B	MCAS	5A Andrews		161510	UC12B	NAS	7B Atlanta
161200	UC12B	MCAS	5T El Toro		161511	UC12B	NAS	7V Glenview
161201	UC12B	NAS	7M N. Island		161512	UC12B	NAS	G C. Christi
161202	UC12B	NAS	7L Pt. Mugu		161513	UC12B	NAS	7R Oceana
161203	UC12B	NAS	7C Norfolk		161514	UC12B	MWHS-4	EZ New Orleans
161204	UC12B	NAF	8N El Centro		161515	UC12B	NAS	5B Beaufort
161205	UC12B	NAS	8F Guantanamo		161516	UC12B	NAS	7W Willow Grove
161206	UC12B	NAS	7G Whidbey Is.		161517	UC12B	NAS	7U Cecil Field
161306	UC12B				161518	UC12B	NAF	5G Iwakuni
161307	UC12B	NAS	7F Brunswick		163553	UC12F	NAS	8M Misawa
161308	UC12B	NAS	7J Alameda		163554	UC12F	NAS	8H Kadena
161309	UC12B	NAS	8J Agana		163555	UC12F	VRC-30	RW N. Island
161310	UC12B	NAS	8E R'velt Roads		163556	UC12F	NAF	8A Atsugi
161311	UC12B	NAS	7M N. Island		163557	UC12F	NAS	8M Misawa
161312	UC12B	NAS	7J Alameda		163558	UC12F	NAF	5G Iwakuni
161313	UC12B	NAS	7G Whidbey Is.		163559	UC12F	NAF	5G Iwakuni
161314	UC12B	MCAS	5A Andrews		163560	UC12F	NAF	8A Atsugi
161315	UC12B	NAS	F Pensacola		163561	UC12F	NAF	5G Iwakuni
161316	UC12B	NAS	7Y Detroit		163562	UC12F	NAF	8A Atsugi
161317	UC12B	NAS	F Pensacola		163563	UC12F	NAWC-WD	RANSAC
161318	UC12B	NAS	7Q Key West		163564	UC12F	NAWC-WD	RANSAC
161319	UC12B	NAF	8M Misawa		163836	UC12M	NAS	7C Norfolk
161320	UC12B	NAF	8H Kadena		163837	UC12M	NAF	8G Mildenhall
161321	UC12B	NAS	7C Norfolk		163838	UC12M	NAF	8C Sigonella
161322	UC12B	NAS	7X New Orleans		163839	UC12M	NS	8D Rota
161323	UC12B				163840	UC12M	NAF	8G Mildenhall
161324	UC12B	NAS	7S Lemoore		163841	UC12M	NAF	8C Sigonella
161325	UC12B	MCAS	5D New River		163842	UC12M	NS	8D Rota
161326	UC12B	NAS	7G Whidbey Is.		163843	UC12M	NAF	8G Mildenhall
161327	UC12B	VRC-30	RW N. Island		163844	UC12M	NAF	8C Sigonella
161497	UC12B	NAS	7K Memphis		163845	UC12M	NAS	7C Norfolk
161498	UC12B	NAF	7N Washington		163846	UC12M		
161499	UC12B	NAS	7Z S. Weymouth		163847	UC12M	NAS	7L Point Mugu

C20 GULFSTREAM III

c\n   480, 481 plus later deliveries

163691	C20D	CFLSW		165151	C20G
163692	C20D	CFLSW		165152	C20G
165093	C20G	VR-48		165153	C20G
165094	C20G	VR-48			

C24 DOUGLAS DC8-54F

c\n   45881\276.

163050	EC24A	FTRG

C28 (Cessna 404)

163917	C28A	NAWC-WD		164761	C28A	

C130 HERCULES

c\n 3554, 3555, 3566, 3573, 3574, 3577, 3562, 3564, 3565, 3567, 3592, 3605 to 3608, 3619, 3623, 3627, 3631, 3632, 3636, 3640, 3644, 3645, 3657, 3658, 3660, 3661, 3664 to 3666, 3680, 3684 to 3686, 3693 to 3696, 3703 to 3705, 3709 to 3711, 3718, 3719, 3723, 3725 to 3728, 3733, 3734, 3740 to 3742, 3849, 3858, 3871, 3878, 4305, 4239, 4249, 4269, 4277 to 4281, 3048, 3099, 4508, 4516, 4522, 4601, 4595, 4615, 4626, 4629, 4635, 4677, 4683, 4689, 4696, 4702, 4712, 4781, 4768, 4770, 4773, 4776, 4725, 4731, 4867, 4896, 4901, 4904, 4932, 4972, 4974, 4978, 4981, 4984, 4988, 5009, 5011, 5040, 5045, 5085, 5087, 5143, 5145, 5147, 5149, 5174, 5176, 5219, 5222, 5260, 5263, 5302, 5303, 5255, 5258, 5298 to 5301, 5304, 5305, 5339 to 5344 plus 2, and ex USAF aircraft (3122, 3203\4).

147572	KC130F	VMGR-352	QB-572
147573	KC130F	VMGR-352	QB-573
148246	KC130F	VMGR-234	QH-246
148247	KC130F	VMGR-152	QD-247
148248	KC130F	VMGR-252	BH-248
148249	KC130F	VMGR-253	GR-249
148318	LC130F	w.o.15.2.71	
148319	LC130F	VXE-6	XD-07
148320	LC130F	VXE-6	XD-06
148321	LC130F		
148890	KC130F		
148891	KC130F	VMGR-252	BH-891
148892	KC130F	VMGR-234	QH-892
148893	KC130F	VMGR-234	QH-893
148894	KC130F	VMGR-252	BH-894
148895	KC130F	VMGR-234	QH-895
148896	KC130F	VMGR-252	BH-896
148897	KC130F	VMGR-252	BH-897
148898	KC130F	VMGR-252	BH-898
148899	KC130F	VMGR-252	BH-899
149787	C130F		
149788	KC130F	VMGR-234	QH-788
149789	KC130F	VMGR-252	BH-789
149790	C130F	AMARC	
149791	KC130F	Blue Angels	
149792	KC130F	VMGR-252	QH-792
149793	C130F		
149794	C130F		
149795	KC130F	VMGR-252	BH-795
149796	KC130F	VMGR-234	QH-796
149797	C130F	AMARC	
149798	KC130F	VMGR-252	BH-798
149799	KC130F	VMGR-252	BH-799
149800	KC130F	VMGR-234	QH-800
149801	C130F		
149802	KC130F	w.o.24.8.65	
149803	KC130F	VMGR-234	BH-803
149804	KC130F	VMGR-352	QB-804
149805	C130F		
149806	KC130F	NAS	Adaksupport
149807	KC130F	VMGR-152	QD-807
149808	KC130F	VMGR-252	BH-808
149809	KC130F	w.o.1.2.66	
149810	KC130F	w.o.2.72	
149811	KC130F	VMGRT-253	GR-811
149812	KC130F	VMGR-152	QD-812
149813	KC130F	w.o.10.2.68	
149814	KC130F	w.o.18.5.69	
149815	KC130F	VMGR-252	BH-815
149816	KC130F	VMGR-352	QB-816
150684	KC130F	VMGR-252	BH-684
150685	KC130F	w.o.7.70	
150686	KC130F	VMGR-234	QH-686
150687	KC130F	VMGR-252	BH-687
150688	KC130F	VMGRT-253	GR-688
150689	KC130F	VMGR-352	QB-689
150690	KC130F	VMGR-352	QB-690
151888	TC130G	AMARC	2G011
151889	EC130G	AMARC	2G...
151890	EC130G	w.o.1.72	
151891	TC130G	NAWC-AD	ASW
155917	LC130R	w.o.28.1.73	
156170	TC130Q	AMARC	2G014
156171	EC130Q	AMARC	2G012
156172	EC130Q	AMARC	2G010
156173	EC130Q	AMARC	2G009
156174	TC130Q	AMARC	2G...
156175	EC130Q	AMARC	2G007
156176	EC130Q	w.o.21.6.77	
156177	EC130Q	AMARC	2G008
158228	DC130A	AMARC	2G003
158229	DC130A	AMARC	to N9724V
159129	LC130R	VXE-6	XD-05
159130	LC130R	VXE-6	XD-04
159131	LC130R	w.o.9.12.87	
159348	TC130Q	VR-24	JM-07
159469	EC130Q	AMARC	2G...
160013	KC130R	VMGR-352	QB-013
160014	KC130R	VMGR-352	QB-014
160015	KC130R	VMGR-352	QB-015
160016	KC130R	VMGR-352	QB-016
160017	KC130R	VMGR-352	QB-017
160018	KC130R	VMGR-352	QB-018
160019	KC130R	VMGR-352	QB-019
160020	KC130R	VMGR-352	QB-020
160021	KC130R	VMGR-352	QB-021
160240	KC130R	VMGR-352	QB-240
160608	EC130Q	AMARC	2G013
160625	KC130R	VMGR-252	BH-625
160626	KC130R	VMGR-252	BH-626
160627	KC130R	VMGR-252	BH-627
160628	KC130R	VMGR-252	BH-628
160740	LC130R	VXE-6	XD-02
160741	LC130R	VXE-6	XD-01
161223	EC130Q	w.o.	
161494	EC130Q	to NASA	

161495	EC130Q	to NASA	N427NA		164597	KC130T	3VMGR-452	NY-597
161496	EC130Q	F.S.I	Mojave		164598	KC130T	3VMGR-452	NY-598
161531	EC130Q	to NOAA	N41RF		164759	KC130T	VMGR-452	NY-759
162308	KC130T	VMGR-234	QH-308		164760	KC130T	VMGR-234	QH-760
162309	KC130T	VMGR-234	QH-309		164762	C130T	VR-54	CW-762
162310	KC130T	VMGR-234	QH-310		164763	C130T	VR-54	CW-763
162311	KC130T	VMGR-234	QH-311		164993	C130T	VR-54	CW-993
162312	EC130Q	AMARC	2G...		164994	C130T	VR-48	WV-994
162313	EC130Q	AMARC	2G...		164995	C130T	VR-54	CW-995
162785	KC130T	VMGR-234	QH-785		164996	C130T	VR-48	WV-996
162786	KC130T	VMGR-234	QH-786		164997	C130T	VR-48	WV-997
163022	KC130T	VMGR-234	QH-022		164998	C130T	VR-48	WV-998
163023	KC130T	VMGR-234	QH-023		164999	KC130T		
163310	KC130T	VMGR-234	QH-310		165000	KC130T		
163311	KC130T	VMGR-452	NY-311		165158	C130T		
163591	KC130T	VMGR-452	NY-591		165159	C130T		
163592	KC130T	VMGR-452	NY-592		165160	C130T		
164105	KC130T	VMGR-452	NY-105		165161	C130T		
164106	KC130T	VMGR-452	NY-106		165162	KC130T		
164180	KC130T	VMGR-452	NY-180		165163	KC130T		
164181	KC130T	VMGR-234	QH-181		560514	DC130A	F.S.I.	Mojave
164441	KC130T	VMGR-452	NY-441		570496	DC130A	F.S.I.	Mojave
164442	KC130T	VMGR-452	NY-442		570497	DC130A	F.S.I.	Mojave

C135 STRATOTANKER

c\n   17250, 17345.

553134	NKC135A FTRG		ex USAF		563596	NKC135A FTRG	ex USAF

E2 HAWKEYE

note: Between 1987 and 1992 the USCG operated up to nine E2C as 3501 to 3509. The corresponding USN serials were 160698\159497\159112\ 160011\159502\161342\158641\160415\160419. The last two originally operated with the USCS using their USN serials.

147263	E2A				151707	E2B		
147264	E2A				151708	E2B		
147265	E2A				151709	E2B	AMARC	2E014
148149	E2A	preserved	NAS Miramar		151710	E2B	AMARC	2E015
148711	E2A				151711	E2B	w.o.8.4.70	
148712	TE2C	w.o.			151712	E2B		
148713	TE2C	dump	NAS N. Island		151713	E2B		
149817	TE2A	dump	NAS N. Island		151714	E2B		
149818	TE2A	stored	NAS N. Island		151715	E2B	AMARC	2E009
149819	E2A	AMARC	2E006		151716	E2B	AMARC	2E003
150530	E2B				151717	E2B	AMARC	2E013
150531	E2B				151718	E2B	stored	NAS N. Island
150532	E2B	AMARC	2E001		151719	E2B		
150533	TE2A	dump	NAS N. Island		151720	E2B	AMARC	2E017
150534	E2B				151721	E2B		
150535	E2B	AMARC	2E002		151722	E2B	AMARC	2E007
150536	E2B	gate	NAS Miramar		151723	E2B		
150537	E2B	w.o.13.11.79			151724	E2B	AMARC	2E020
150538	E2B	AMARC	2E019		151725	E2B	AMARC	2E004
150539	E2B				152476	E2B	MASDC	
150540	E2B	MASDC			152477	E2B	AMARC	2E005
150541	E2B	preserved	NAS Norfolk		152478	E2B	AMARC	2E018
151702	E2B	AMARC	2E012		152479	E2B	AMARC	2E016
151703	E2B				152480	E2B		
151704	E2B				152481	E2B		
151705	E2B	w.o.8.8.72			152482	E2B	AMARC	2E011
151706	E2B	AMARC	2E010		152483	E2B	dump	NAS N. Island

Serial	Type	Unit	Code
152484	E2B	GIA	NAWC Lakehurst
152485	TE2A	NARF	NAS N. Island
152486	E2B		
152487	E2B		
152488	E2B	dump	NAS N. Island
152489	E2B	w.o.25.5.83	
158638	E2C	VAW-125	AA-602
158639	E2C	rework	Grumman
158640	E2C	VAW-121	AG-600
158641	E2C	to N15941	
158642	E2C	w.o.	
158643	E2C	VAW-88	ND-
158644	E2C	VAW-110	NJ-604
158645	E2C	VAW-124	AJ-600
158646	E2C	VAW-112	NG-601
158647	E2C	VAW-120	AD-017
158648	E2C	VAW-120	AD-020
159105	E2C	VAW-110	NJ-331
159106	E2C	VAW-120	AD-010
159107	E2C	VAW-125	AA-600
159108	E2C	w.o.14.1.78	
159109	E2C	VAW-117	NH-600
159110	E2C	VAW-115	NF-604
159111	E2C	VAW-110	NJ-
159112	E2C		
159494	E2C	VAW-120	AD-014
159495	E2C	VAW-110	NJ-340
159496	E2C	VAW-110	NJ-321
159497	E2C		
159498	E2C	VAW-121	AG-601
159499	E2C	VAW-110	NJ-606
159500	E2C	VAW-88	ND-600
159501	E2C	VAW-113	NK-605
159502	E2C	VAW-110	NJ-606
160007	E2C	VAW-121	AG-
160008	E2C		
160009	E2C	VAW-112	NG-
160010	E2C	VAW-117	NH-602
160011	E2C		
160012	E2C	NAWC-AD	ASATD
160415	E2C		
160416	E2C		
160417	E2C		
160418	E2C		
160419	E2C		
160420	E2C	VAW-115	NF-601
160697	E2C	VAW-123	AB-602
160698	E2C	w.o.14.8.90	(as 3501)
160699	E2C	VAW-116	NG-603
160700	E2C	VAW-120	AD-017
160701	E2C	VAW-110	NJ-347
160702	E2C		
160703	E2C	VAW-121	AG-600
160987	E2C	VAW-121	AG-603
160988	E2C	VAW-117	NH-603
160989	E2C	VAW-78	AF-013
160990	E2C	VAW-110	NJ-
160991	E2C	VAW-115	NF-603
160992	E2C	VAW-115	NF-602
161094	E2C	w.o.17.8.85	
161095	E2C	VAW-127	AK-602
161096	E2C	VAW-121	AG-602
161097	E2C	VAW-123	AB-600
161098	E2C	VAW-123	AB-603
161099	E2C		
161224	E2C		
161225	E2C	VAW-120	AD-011
161226	E2C	VAW-110	NJ-341
161227	E2C	VAW-116	NE-
161228	E2C	VAW-110	NJ-325
161229	E2C	VAW-116	NE-
161341	E2C	VAW-116	NE-601
161342	E2C		
161343	E2C	VAW-116	NE-603
161344	E2C	rework	Grumman
161345	E2C	NAWC-AD	SATD
161346	E2C	VAW-120	AD-
161547	E2C	VAW-125	AA-603
161548	E2C	VAW-125	AA-602
161549	E2C	VAW-123	AB-601
161550	E2C	VAW-125	AA-600
161551	E2C	VAW-120	AD-012
161552	E2C	VAW-126	AJ-601
161780	E2C	VAW-126	AJ-600
161781	E2C	VAW-126	AJ-602
161782	E2C	VAW-126	AJ-603
161783	E2C	VAW-120	AD-016
161784	E2C		
161785	E2C	VAW-120	AD-015
162614	E2C	VAW-120	AD-010
162615	E2C	VAW-120	AD-011
162616	E2C	VAW-126	AC-601
162617	E2C	VAW-126	AC-602
162618	E2C	VAW-126	AC-603
162619	E2C	VAW-113	NK-600
162795	E2C		
162796	E2C		
162797	E2C	VAW-113	NK-601
162798	E2C	VAW-112	NG-602
162799	E2C		
162800	E2C		
162801	E2C		
162802	E2C		
163024	E2C		
163025	E2C		
163026	E2C		
163027	E2C		
163028	E2C		
163029	E2C		
163535	E2C		
163536	E2C		
163537	E2C		
163538	E2C		
163539	E2C		
163540	E2C	VAW-113	NK-...
163565	E2C		
163693	E2C		
163694	E2C		
163695	E2C		
163696	E2C		
163697	E2C		
163698	E2C		
163848	E2C		
163849	E2C		
163850	E2C		
163851	E2C		
164107	E2C		
164108	E2C		
164109	E2C		
164110	E2C		
164111	E2C		
164112	E2C		
164352	E2C		
164353	E2C		
164354	E2C		
164355	E2C		
164356	E2C		
164357	E2C		
164483	E2C		
164484	E2C		
164485	E2C		
164486	E2C		
164487	E2C		
164488	E2C		
164489	E2C		
164490	E2C		
164491	E2C		
164492	E2C		
164493	E2C		
164494	E2C		
164495	E2C		
164496	E2C		
164497	E2C		
164498	E2C		
164499	E2C		
164500	E2C		
164501	E2C		
164502	E2C		
164503	E2C		
164504	E2C		
164505	E2C		
164506	E2C		
164507	E2C		
164508	E2C		
164509	E2C		
164510	E2C		

164511	E2C	
164512	E2C	
164513	E2C	
164514	E2C	
164515	E2C	
164516	E2C	
164517	E2C	
164518	E2C	

E6 MERCURY

c\n   23430, 23889 to 23894, 24500 to 24509.

162782	E6A	VQ-4
162783	E6A	NAWC
162784	E6A	VQ-4
163532	E6A	VQ-3
163533	E6A	VQ-3
163534	E6A	VQ-3
163918	E6A	VQ-3
163919	E6A	VQ-3
163920	E6A	VQ-3
164386	E6A	VQ-3
164387	E6A	VQ-3
164388	E6A	VQ-3
164404	E6A	VQ-4
164405	E6A	VQ-4
164406	E6A	VQ-4
164407	E6A	VQ-4
164408	E6A	VQ-4
164409	E6A	VQ-4
164410	E6A	VQ-4

F4 PHANTOM II

142259	F4A		
142260	F4A		
143388	F4A	preserved	MCAS Quantico
143389	F4A		
143391	F4A		
143392	F4A		
145307	F4A	preserved	Silver Hill, Md
145308	F4A	preserved	St. Petersburg
145309	F4A		
145310	F4A	preserved	Bradley ANGB
145311	F4A		
145312	F4A		
145313	F4A		
145314	F4A		
145315	F4A	dump	NAS Memphis
145316	F4A		
145317	F4A		
146817	F4A		
146818	F4A		
146819	F4A		
146820	F4A	stored	Kirtland AFB
146821	F4A	w.o.16.2.63	
148252	F4A		
148253	F4A		
148254	F4A		
148255	F4A		
148256	F4A		
148257	F4A		
148258	F4A		
148259	F4A	NATTC	
148260	TF4A	dump	NAS Memphis
148261	F4A	preserved	NAS Oceana
148262	F4A		
148263	F4A		
148264	F4A	w.o.65	
148265	F4A		
148266	F4A		
148267	F4A	dump	NAS Lakehurst
148268	F4A		
148269	F4A	w.o.65	
148270	F4A	w.o.64	
148271	F4A		
148272	F4A		
148273	F4A	preserved	NAS Lakehurst
148274	F4A		
148275	F4A	preserved	Annapolis, Md
148363	F4A	NMIMT	Socorro
148364	F4A		
148365	QF4B		
148366	F4B		
148367	F4B	preserved	MCAS Yuma
148368	F4B		
148369	F4B	dump	NAS Memphis
148370	F4B	w.o.63	
148371	F4B		
148372	F4B	dump	St. Louis, Mi
148373	F4B	preserved	MCAS El Toro
148374	F4B		
148375	F4B	w.o.65	
148376	F4B		
148377	F4B	w.o.73	
148378	QF4B		
148379	F4B	w.o.66	
148380	F4B		
148381	F4B		
148382	F4B		
148383	F4B	NATTC	NAS Memphis
148384	F4B		
148385	F4B		
148386	QF4B		
148387	F4B	w.o.16.7.67	
148388	F4B	w.o.22.12.67	
148389	F4B		
148390	F4B		
148391	F4B		
148392	F4B	NATTC	NAS Memphis
148393	F4B		
148394	F4B	w.o.63	
148395	F4B	w.o.65	
148396	F4B	w.o.64	
148397	F4B		
148398	F4B		

Serial	Type	Status	Location
148399	F4B	w.o.65	
148400	F4B		
148401	F4B		
148402	F4B		
148403	F4B		
148404	F4B		
148405	F4B		
148406	F4B		
148407	F4B	NATTC	NAS Memphis
148408	F4B		
148409	QF4B	w.o.10.9.68	
148410	F4B	w.o.65	
148411	F4B	w.o.64	
148412	F4B	to USAF	
148413	F4B		
148414	F4B	w.o.5.70	
148415	QF4B	stored	China Lake
148416	F4B		
148417	F4B	w.o.64	
148418	F4B	NATTC	NAS Memphis
148419	F4B	w.o.63	
148420	F4B	w.o.8.8.68	
148421	F4B	GIA	Lackland AFB
148422	F4B	w.o.26.9.67	
148423	QF4B	dump	NAS Memphis
148424	QF4B	stored	China Lake
148425	F4B		
148426	F4B		
148427	F4B	w.o.28.4.69	
148428	QF4B	BDRT	Hill AFB
148429	F4B		
148430	F4B	BDRT	Hill AFB
148431	F4B		
148432	F4B	w.o.8.7.68	
148433	F4B	w.o.3.11.66	
148434	F4B		
149403	F4B	w.o.27.5.70	
149404	F4B	w.o.66	
149405	F4B	to USAF	62-12168
149406	F4B	to USAF	62-12169
149407	F4B	NMIMT	Socorro, NM
149408	F4B		
149409	QF4B		
149410	F4B	w.o.26.10.65	
149411	F4B	w.o.68	
149412	F4B		
149413	F4B		
149414	QF4B	stored	China Lake
149415	F4B	MASDC	
149416	F4B	w.o.19.8.69	
149417	F4B		
149418	F4B	MASDC	
149419	F4B		
149420	F4B		
149421	F4B	preserved	Lackland AFB
149422	F4B	w.o.71	
149423	F4B	NATTC	
149424	F4B	w.o.1.5.68	
149425	F4B	w.o.62	
149426	F4B	MASDC	
149427	F4B	BDRT	W-P AFB
149428	F4B	scrapped	
149429	F4B	w.o.28.12.69	
149430	F4B		
149431	QF4B	NAWC-WD	410
149432	QF4B	stored	China Lake
149433	QF4B	stored	China Lake
149434	QF4B	NAWC-WD	41
149435	F4B	NARF	North Island
149436	F4B	w.o.63	
149437	F4B		
149438	F4B	w.o.66	
149439	F4B	w.o.64	
149440	F4B	w.o.3.12.67	
149441	QF4B	stored	China Lake
149442	F4B		
149443	F4B	w.o.16.9.68	
149444	F4B	w.o.65	
149445	F4B	scrapyard	Allied
149446	QF4B	stored	China Lake
149447	F4B	NAWC-AD	SATD
149448	F4B	w.o.67	
149449	F4B	w.o.2.8.68	
149450	F4B	dump	NAS N. Island
149451	QF4B	stored	China Lake
149452	QF4B	stored	China Lake
149453	F4B	w.o.69	
149454	F4B	w.o.64	
149455	F4B	dump	NAS Lakehurst
149456	F4B	w.o.17.11.68	
149457	F4B	preserved	NAS Pensacola
149458	F4B	derelict	NAS Norfolk
149459	F4B		
149460	F4B	MASDC	
149461	F4B	NAWC-WD	43
149462	F4B		
149463	F4B	NMIMT	Socorro, NM
149464	F4B		
149465	F4B	stored	NAS N. Island
149466	QF4B		
149467	F4B	w.o.8.2.70	
149468	F4B	w.o.2.12.65	
149469	F4B		
149470	F4B	derelict	NAS Oceana
149471	QF4B	NAWC-WD	42
149472	F4B	NARF	North Island
149473	F4B	MASDC	
149474	F4B		
150406	F4B	w.o.73	
150407	F4N	MASDC	
150408	F4B		
150409	F4B		
150410	F4B		
150411	F4N		
150412	QF4N	stored	China Lake
150413	F4B	w.o.20.2.67	
150414	F4B	w.o.73	
150415	QF4N	stored	China Lake
150416	F4B	w.o.4.4.70	
150417	F4B	w.o.3.3.72	
150418	F4B	w.o.30.12.71	
150419	GF4N	stored	China Lake
150420	F4B		
150421	F4B	w.o.24.10.67	
150422	F4N	MASDC	
150423	F4N	stored	China Lake
150424	F4B	w.o.15.5.70	
150425	F4N	scrapyard	Bobs
150426	F4N	scrapyard	Bobs
150427	F4B	w.o.19.2.70	
150428	F4B		
150429	F4B	scrapyard	Bobs
150430	F4N	MASDC	
150431	F4B	w.o.68	
150432	QF4N	stored	China Lake
150433	F4B	w.o.28.1.64	
150434	F4B	w.o.24.8.68	
150435	F4N		
150436	F4N	scrapyard	Bobs
150437	F4B		
150438	F4N	AMARC	8F112
150439	F4B	w.o.19.6.67	
150440	F4N	AMARC	8F090
150441	F4N	scrapyard	Bobs
150442	F4N		
150443	F4B	w.o.1.3.66	
150444	F4N	NATTC	
150445	F4N	preserved	Hill AFB
150446	F4B	w.o.70	
150447	F4B	w.o.25.3.69	
150448	F4N	stored	Hill AFB
150449	F4B	w.o.29.12.67	
150450	F4N	w.o.73	
150451	F4N	w.o.65	
150452	F4N	scrapyard	Bobs
150453	F4B	w.o.2.6.68	
150454	F4B		
150455	F4B		
150456	QF4N	NAWC-WD	411
150457	F4B	w.o.64	
150458	F4B	w.o.64	
150459	F4B	w.o.21.9.68	
150460	F4N	scrapyard	Bobs
150461	F4B	w.o.63	
150462	F4B		
150463	F4B	w.o.5.4.68	

Serial	Type	Status	Location
150464	QF4N	stored	China Lake
150465	F4N	NAD	Cherry Point
150466	F4N	w.o.2.9.70	
150467	F4B		
150468	F4N	NAD	Cherry Point
150469	F4B	w.o.23.10.67	
150470	F4B	w.o.15.7.66	
150471	F4B	w.o.63	
150472	F4N		
150473	F4B	w.o.69	
150474	F4B	w.o.8.10.67	
150475	QF4N	stored	China Lake
150476	F4N		
150477	F4B	w.o.13.10.67	
150478	F4N	MASDC	
150479	F4N	scrapyard	Bobs
150480	F4N		
150481	F4N	scrapyard	Bobs
150482	F4N	MASDC	
150483	F4B	w.o.66	
150484	F4N	AMARC	8F091
150485	F4N	dump	NAS Lakehurst
150486	F4N	w.o.73	
150487	F4B	w.o.72	
150488	F4N	MASDC	
150489	QF4N	stored	China Lake
150490	F4N	AMARC	8F100
150491	F4N	MASDC	
150492	F4N	AMARC	8F154
150493	F4B		
150624	F4B		
150625	F4N	w.o.30.4.76	
150626	F4B	w.o.28.10.65	
150627	F4N	w.o.73	
150628	F4N	preserved	USS Intrepid
150629	F4B	w.o.30.10.67	
150630	QF4N	NAWC-WD	42
150631	F4B	w.o.17.10.65	
150632	F4N		
150633	F4B	w.o.66	
150634	F4N	MASDC	
150635	F4N	MASDC	
150636	F4B	w.o.71	
150637	F4B		
150638	F4N	scrapyard	Bobs
150639	F4N	preserved	D-M AFB
150640	F4N		
150641	F4B		
150642	F4N		
150643	F4N	AMARC	8F121
150644	F4B	w.o.14.4.68	
150645	F4B	w.o.28.4.66	
150646	F4B	w.o.28.7.65	
150647	F4B	w.o.64	
150648	F4N	scrapyard	Bobs
150649	F4N	w.o.17.12.70	
150650	F4B	w.o.13.7.68	
150651	F4N	AMARC	8F089
150652	F4N	MASDC	
150653	F4B	w.o.19.6.70	
150993	QF4N	NAWC-WD	407
150994	F4B	w.o.7.6.68	
150995	F4B	w.o.24.10.67	
150996	F4N	MASDC	
150997	F4B	w.o.19.11.67	
150998	F4B	w.o.26.10.65	
150999	F4B	w.o.66	
151000	F4N		
151001	F4B	w.o.18.5.69	
151002	QF4N	stored	China Lake
151003	F4N	AMARC	8F086
151004	QF4N	stored	China Lake
151005	F4B	w.o.63	
151006	F4N	scrapyard	Bobs
151007	F4N	stored	China Lake
151008	F4N	w.o.20.11.78	
151009	F4B	w.o.22.10.66	
151010	F4B	w.o.21.4.66	
151011	F4N	GIA	NAS Alameda
151012	F4B	w.o.19.8.69	
151013	F4B	w.o.27.8.72	
151014	F4B	w.o.2.12.66	
151015	F4N	MASDC	
151016	F4N	scrapyard	Bobs
151017	F4B	w.o.64	
151018	F4B	w.o.3.11.66	
151019	F4B		
151020	F4B	w.o.30.9.67	
151021	F4B		
151397	F4N	w.o.71	
151398	F4N	MASDC	
151399	F4B		
151400	F4N	MASDC	
151401	F4N	AMARC	8F055
151402	F4B	w.o.13.11.64	
151403	F4B	w.o.9.4.65	
151404	F4B	w.o.17.8.68	
151405	F4B	w.o.19.9.69	
151406	F4N		
151407	F4B	w.o.30.1.68	
151408	F4B	w.o.67	
151409	F4B	w.o.2.12.65	
151410	F4B	w.o.20.3.66	
151411	F4B	w.o.24.2.66	
151412	F4B	w.o.28.12.65	
151413	F4N	w.o.27.7.79	
151414	F4B	w.o.22.1.68	
151415	F4N	MASDC	
151416	F4B	w.o.25.6.69	
151417	F4N		
151418	F4B	w.o.24.5.67	
151419	F4B	w.o.69	
151420	F4B		
151421	F4B	w.o.2.7.67	
151422	F4N		
151423	F4N	w.o.29.10.67	
151424	F4N	scrapyard	Bobs
151425	F4B	w.o.65	
151426	F4B		
151427	F4B	w.o.73	
151428	F4B		
151429	F4B	w.o.29.9.67	
151430	F4N	stored	China Lake
151431	F4N	AMARC	8F141
151432	F4N	w.o.65	
151433	F4N	MASDC	
151434	F4N	MASDC	
151435	QF4N	stored	China Lake
151436	F4N		
151437	F4B		
151438	F4B	w.o.28.12.65	
151439	F4N	MASDC	
151440	QF4N	stored	China Lake
151441	F4B	w.o.12.1.69	
151442	F4N		
151443	F4B		
151444	F4N	AMARC	8F111
151445	F4B	dump	NAS N. Island
151446	F4N	AMARC	8F143
151447	F4B	w.o.3.1.68	
151448	F4N	w.o.77	
151449	QF4N	stored	China Lake
151450	F4B	w.o.18.5.69	
151451	F4N	w.o.9.81	
151452	F4N	MASDC	
151453	F4B	w.o.3.3.66	
151454	F4B	w.o.12.2.70	
151455	QF4N	stored	China Lake
151456	F4N	stored	NAS Norfolk
151457	F4B	w.o.19.10.67	
151458	F4B	w.o.6.6.71	
151459	F4B	w.o.3.11.78	
151460	F4B	w.o.64	
151461	QF4N	stored	China Lake
151462	F4B	w.o.73	
151463	F4N	NAD	Cherry Point
151464	F4N		
151465	QF4N	stored	China Lake
151466	F4B		
151467	F4B	w.o.4.7.68	
151468	F4N	w.o.75	
151469	F4N	MASDC	
151470	F4B		
151471	QF4N	stored	China Lake
151472	F4B	w.o.27.4.72	
151473	YF4J	NAWC-WD	473

151474	F4B				152235	QF4N	stored	China Lake
151475	QF4N	stored	China Lake		152236	F4N		
151476	F4N	MASDC			152237	F4N	scrapyard	Bobs
151477	F4N				152238	F4B	w.o.8.9.67	
151478	F4B	w.o.7.6.70			152239	F4B	w.o.17.5.70	
151479	F4B				152240	F4B	w.o.73	
151480	F4N				152241	F4N		
151481	F4B				152242	F4B	w.o.28.6.67	
151482	F4N	AMARC	8F120		152243	QF4N	stored	China Lake
151483	F4B				152244	F4N	w.o.9.7.85	
151484	F4N	MASDC			152245	F4B	w.o.72	
151485	F4B				152246	F4N	NAD	Cherry Point
151486	F4B	w.o.18.4.67			152247	F4B	w.o.21.8.67	
151487	F4N	scrapyard	Bobs		152248	F4B	w.o.65	
151488	F4B	w.o.17.11.67			152249	F4B	dump	NAS Norfolk
151489	F4N				152250	F4N	NAD	Cherry Point
151490	F4B				152251	F4B	w.o.15.6.66	
151491	F4N	scrapyard	Bobs		152252	F4N	stored	NAS N. Island
151492	F4B	w.o.16.12.67			152253	QF4N	stored	China Lake
151493	F4B	w.o.4.4.67			152254	F4N	dump	Ft. Chaffee
151494	F4B	w.o.17.10.65			152255	F4B	w.o.26.4.66	
151495	F4B	scrapped			152256	F4B	NATTC	
151496	YF4J	stored	China Lake		152257	F4B	w.o.18.5.66	
151497	YF4J	Pima Museum			152258	QF4N	stored	China Lake
151498	F4N				152259	F4N		
151499	F4B	w.o.10.1.68			152260	F4B	w.o.22.1.69	
151500	F4B	w.o.23.5.68			152261	F4B	w.o.7.12.65	
151501	F4B				152262	F4B	w.o.11.7.66	
151502	F4N	AMARC	8F085		152263	F4N		
151503	F4N	NAD	Cherry Point		152264	F4N	w.o.19.5.67	
151504	F4N	NAD	Cherry Point		152265	F4B	w.o.24.1.65	
151505	F4B	w.o.25.10.65			152266	F4B	w.o.16.5.67	
151506	F4B	w.o.10.1.68			152267	F4N	preserved	NAS Dallas
151507	F4B	w.o.23.5.68			152268	F4B	w.o.66	
151508	F4B	w.o.31.1.68			152269	F4N	NAD	Cherry Point
151509	F4B	w.o.66			152270	F4N	preserved	Beaufort MCAS
151510	F4N	preserved	Luke AFB		152271	F4B	w.o.67	
151511	F4N				152272	F4N	NAWC-WD	409
151512	F4B	w.o.67			152273	F4B	w.o.68	
151513	F4N				152274	F4B	w.o.66	
151514	F4N	AMARC	8F136		152275	F4N	AMARC	8F094
151515	F4B	w.o.17.10.65			152276	F4B	w.o.65	
151516	F4B				152277	F4N	MASDC	
151517	F4B	w.o.65			152278	F4N	AMARC	8F144
151518	F4B				152279	QF4N	stored	China Lake
151519	F4N	w.o.78			152280	F4N		
151975	RF4B	NAWC-AD	SATD		152281	QF4N	stored	China Lake
151976	RF4B	w.o.18.10.68			152282	F4N		
151977	RF4B	AMARC	8F325		152283	F4B	w.o.69	
151978	RF4B	AMARC	8F328		152284	F4N	w.o.27.7.79	
151979	RF4B				152285	F4B	w.o.66	
151980	RF4B	AMARC	8F333		152286	F4B	w.o.70	
151981	RF4B				152287	F4B	w.o.68	
151982	RF4B	w.o.70			152288	F4N	AMARC	8F107
151983	RF4B	AMARC	8F330		152289	F4B	w.o.68	
152207	F4B				152290	F4N	MASDC	
152208	F4N	AMARC	8F101		152291	F4N	preserved	NAS Barbers Pt.
152209	F4B	w.o.9.11.68			152292	F4B	dump	NAS N. Island
152210	F4N	scrapyard	Bobs		152293	F4N	AMARC	8F138
152211	F4B	w.o.73			152294	F4N	dump	NAS Norfolk
152212	F4N				152295	F4N	AMARC	FP482
152213	F4B	w.o.18.9.69			152296	F4B		
152214	F4N				152297	F4B	w.o.66	
152215	F4B	w.o.24.8.65			152298	F4N		
152216	F4B	w.o.65			152299	F4B	w.o.69	
152217	QF4N	stored	China Lake		152300	F4N	AMARC	8F171
152218	F4B	w.o.10.1.66			152301	F4B	w.o.67	
152219	F4B	w.o.12.2.67			152302	F4N	MASDC	
152220	F4B	w.o.2.12.65			152303	QF4N	NAWC-WD	408
152221	F4N	stored	China Lake		152304	F4B	w.o.19.11.67	
152222	QF4N	stored	China Lake		152305	F4B		
152223	QF4N	stored	China Lake		152306	F4N	AMARC	8F170
152224	F4B	w.o.5.3.66			152307	F4N	AMARC	8F148
152225	F4N				152308	F4B		
152226	QF4N	stored	China Lake		152309	F4B	w.o.66	
152227	F4N	w.o.75			152310	F4N	NAD	Cherry Point
152228	F4N	MASDC			152311	F4B	w.o.65	
152229	QF4N	stored	China Lake		152312	F4B	MASDC	
152230	QF4N	stored	China Lake		152313	F4N		
152231	F4B	w.o.69			152314	F4B		
152232	F4B	w.o.19.9.68			152315	F4B	w.o.66	
152233	F4B	w.o.31.1.66			152316	F4B	w.o.74	
152234	F4B	w.o.23.6.66			152317	F4N	AMARC	8F116

Serial	Type	Status	Notes
152318	F4N	AMARC	8F132
152319	F4B	w.o.70	
152320	F4B	w.o.67	
152321	QF4N	stored	China Lake
152322	F4B	w.o.70	
152323	QF4N	stored	China Lake
152324	F4B	w.o.66	
152325	F4B	w.o.70	
152326	QF4N	stored	China Lake
152327	F4N		
152328	F4B	w.o.66	
152329	F4B	w.o.68	
152330	F4B	w.o.66	
152331	F4B		
152965	F4N	AMARC	8F114
152966	F4B		
152967	F4N		
152968	F4N	NAWC-WD	412
152969	F4N	MASDC	
152970	F4N	AMARC	8F125
152971	F4N	NAD	Cherry Point
152972	F4B		
152973	F4B	w.o.20.9.66	
152974	F4B	w.o.66	
152975	F4N	AMARC	8F129
152976	F4B	w.o.66	
152977	F4N		
152978	F4B	w.o.67	
152979	F4B	w.o.67	
152980	F4B		
152981	F4N	AMARC	8F131
152982	F4N	MASDC	
152983	F4N	MASDC	
152984	F4B	w.o.67	
152985	F4B	w.o.66	
152986	F4N		
152987	F4B	w.o.67	
152988	F4B	w.o.69	
152989	F4B	w.o.67	
152990	F4N	MASDC	
152991	F4N	AMARC	8F087
152992	F4N	preserved	NAS Alameda
152993	F4B	w.o.66	
152994	F4B	w.o.68	
152995	F4B	w.o.66	
152996	F4N	preserved	Birmingham, Al
152997	F4B	w.o.67	
152998	F4B	w.o.74	
152999	F4B	w.o.67	
153000	F4B	w.o.67	
153001	F4B	w.o.67	
153002	F4B	w.o.68	
153003	F4B	w.o.68	
153004	F4B	w.o.67	
153005	F4B	w.o.67	
153006	F4N	dump	Hill AFB
153007	F4B	AMARC	8F175
153008	F4N	to USAF	84-0494
153009	F4B	w.o.72	
153010	F4N		
153011	QF4N	stored	China Lake
153012	F4N	AMARC	8F150
153013	F4B	w.o.68	
153014	F4B	w.o.68	
153015	F4B	w.o.69	
153016	F4N		
153017	F4N	stored	NAS N. Island
153018	F4B	w.o.69	
153019	F4N	preserved	NAS Key West
153020	F4B	w.o.72	
153021	F4B	w.o.70	
153022	F4B	w.o.70	
153023	F4N		
153024	F4N	MASDC	
153025	F4N	w.o.72	
153026	F4B	w.o.15.4.81	
153027	F4N	AMARC	8F142
153028	F4B		
153029	F4B	w.o.1.67	
153030	QF4N	stored	China Lake
153031	F4B	w.o.72	
153032	F4B	w.o.72	
153033	F4B	w.o.67	
153034	QF4N	NAWC-WD	
153035	F4B	w.o.66	
153036	F4N	dump	MCAS Yuma
153037	F4B	w.o.69	
153038	F4B	w.o.66	
153039	F4N	AMARC	8F145
153040	F4B	w.o.67	
153041	F4B	w.o.69	
153042	F4B		
153043	F4B	w.o.68	
153044	F4B	w.o.67	
153045	F4N	w.o.29.7.67	
153046	F4B	MASDC	
153047	F4N	MASDC	
153048	F4B		
153049	F4B		
153050	F4N	AMARC	8F127
153051	F4B	w.o.68	
153052	F4B		
153053	QF4N	stored	China Lake
153054	F4B	w.o.29.7.67	
153055	F4B	w.o.67	
153056	F4N	NAD	Cherry Point
153057	F4N	dump	NAS N. Island
153058	F4N	MASDC	
153059	F4N	NAD	Cherry Point
153060	F4N	w.o.29.7.67	
153061	F4B	w.o.29.7.67	
153062	F4N	dump	Hill AFB
153063	F4B	w.o.1967	
153064	QF4N	stored	China Lake
153065	QF4N	stored	China Lake
153066	F4B	w.o.29.7.67	
153067	F4N		
153068	F4B	w.o.72	
153069	F4B	w.o.29.7.67	
153070	EF4B	MASDC	
153071	F4J	preserved	Patuxent River
153072	F4J	w.o.73	
153073	F4J	NAWC-WD 93	
153074	F4J	NAWC-WD 90	
153075	F4J	w.o.69	
153076	EF4J	stored	China Lake
153077	F4J	NAWC-AD	SATD 7T-101
153078	F4J	w.o.69	
153079	F4J	w.o.73	
153080	F4J	w.o.73	
153081	F4J	w.o.73	
153082	F4J		
153083	F4J	w.o.73	
153084	DF4J	AMARC	8F199
153085	F4J		
153086	F4J	w.o.72	
153087	F4J	w.o.70	
153088	F4J	NAWC-AD	SATD
153089	RF4B	w.o.68	
153090	RF4B	w.o.66	
153091	RF4B	AMARC	8F337
153092	RF4B	AMARC	8F339
153093	RF4B	AMARC	8F338
153094	RF4B	AMARC	8F335
153095	RF4B	AMARC	8F327
153096	RF4B	AMARC	8F329
153097	RF4B		
153098	RF4B	w.o.22.10.75	
153099	RF4B	MASDC	
153100	RF4B		
153101	RF4B		
153102	RF4B	AMARC	8F336
153103	RF4B	AMARC	8F332
153104	RF4B		
153105	RF4B	AMARC	8F334
153106	RF4B	AMARC	8F326
153107	RF4B	AMARC	8F331
153108	RF4B	AMARC	8F324
153109	RF4B	w.o.27.10.87	
153110	RF4B		
153111	RF4B	w.o.3.11.74	
153112	RF4B	w.o.72	
153113	RF4B		
153114	RF4B	w.o.67	
153115	RF4B	w.o.68	
153768	F4J	to RAF	ZE350

Serial	Type	Status	Location/Code
153769	F4J		
153770	F4J		
153771	F4S		
153772	F4J		
153773	F4J	to RAF	ZE351
153774	F4J		
153775	F4J	w.o.7.11.77	
153776	F4J	AMARC	8F128
153777	F4J	AMARC	8F164
153778	F4J	w.o.7.11.77	
153779	F4S	AMARC	8F256
153780	F4S	AMARC	8F300
153781	F4J	w.o.5.2.78	
153782	F4J	w.o.69	
153783	F4J	to RAF	ZE352
153784	F4S	AMARC	8F284
153785	F4J	to RAF	ZE353
153786	F4J	w.o.67	
153787	F4S		
153788	F4J	w.o.68	
153789	F4J	w.o.72	
153790	F4J		
153791	F4S		
153792	F4S		
153793	F4J	w.o.74	
153794	F4J	w.o.69	
153795	F4J	to RAF	ZE354
153796	F4J	AMARC	8F184
153797	F4J	dump	MCAS Yuma
153798	F4S	w.o.8.10.82	
153799	F4J	w.o.72	
153800	F4S	AMARC	8F259
153801	F4J	w.o.71	
153802	F4J	w.o.67	
153803	F4J	to RAF	ZE355
153804	F4J		
153805	F4S	AMARC	8F270
153806	F4J	w.o.69	
153807	F4S	AMARC	8F188
153808	F4S	dump	NAS N. Island
153809	F4S		
153810	F4S	dump	MCAS Cherry Pt.
153811	F4J		
153812	F4J	AMARC	8F104
153813	F4J		
153814	F4S		
153815	F4J	w.o.68	
153816	F4J	w.o.73	
153817	F4J		
153818	F4S	w.o.2.1.80	
153819	F4S	w.o.13.7.82	
153820	F4S		
153821	F4S		
153822	F4S	w.o.72	
153823	F4S		
153824	F4S		
153825	F4S	AMARC	8F195
153826	F4S	AMARC	8F249
153827	F4S	AMARC	8F214
153828	F4S	AMARC	8F244
153829	F4J		
153830	F4J	dump	NAS Barbers Pt.
153831	F4J		
153832	F4S		
153833	F4S		
153834	F4J	w.o.73	
153835	F4S		
153836	F4J	w.o.74	
153837	F4J		
153838	F4J		
153839	F4J	NAWC-AD	SATD SD-104
153840	F4S	w.o.20.11.83	
153841	F4J	AMARC	8F182
153842	F4S	AMARC	8F212
153843	F4S		
153844	F4J		
153845	F4S	AMARC	8F314
153846	F4J	w.o.73	
153847	F4S	AMARC	8F228
153848	F4S	AMARC	8F190
153849	F4J	w.o.20.11.72	
153850	F4J	to RAF	ZE356
153851	F4S		
153852	F4J	AMARC	8F106
153853	F4S	w.o.2.1.80	
153854	F4J	w.o.13.9.72	
153855	F4S	AMARC	8F169
153856	F4S		
153857	F4S	AMARC	8F298
153858	F4S		
153859	F4S	AMARC	8F243
153860	F4S		
153861	F4S	AMARC	8F165
153862	F4S	AMARC	8F268
153863	F4J	w.o.69	
153864	F4S	AMARC	8F230
153865	F4J	w.o.70	
153866	F4S	w.o.16.12.80	
153867	F4J		
153868	F4S	stored	NAS N. Island
153869	F4S	w.o.4.6.83	
153870	F4J	w.o.71	
153871	F4J		
153872	F4S	AMARC	8F172
153873	F4S	AMARC	8F174
153874	F4S	w.o.19.6.86	
153875	F4J		
153876	F4J		
153877	F4S		
153878	F4S	w.o.19.6.86	
153879	F4S	stored	NAS N. Island
153880	F4S		
153881	F4S	AMARC	8F218
153882	F4S		
153883	F4J		
153884	F4S		
153885	F4J	w.o.23.12.72	
153886	F4J		
153887	F4S		
153888	F4J		
153889	F4S		
153890	F4S	AMARC	8F239
153891	F4S	AMARC	8F303
153892	F4J	to RAF	ZE357
153893	F4S	AMARC	8F294
153894	F4J		
153895	F4J		
153896	F4S	AMARC	8F191
153897	F4J		
153898	F4S	AMARC	8F231
153899	F4S	NAWC-WD	40
153900	F4S	AMARC	8F192
153901	F4J		
153902	F4S		
153903	F4S	AMARC	8F263
153904	F4S		
153905	F4J		
153906	F4J	w.o.68	
153907	F4S	AMARC	8F175
153908	F4S	AMARC	8F166
153909	F4S		
153910	F4S	AMARC	8F299
153911	F4S		
153912	F4B	w.o.29.7.67	
153913	F4B	w.o.8.70	
153914	F4N	NAD	Cherry Point
153915	F4N	preserved	NAS Pensacola
154781	F4S	AMARC	8F307
154782	F4S	AMARC	8F236
154783	F4J		
154784	F4J	w.o.11.9.72	
154785	F4J	AMARC	8F180
154786	F4S		
154787	F4J		
154788	F4S		
155504	F4S	AMARC	8F219
155505	F4J	AMARC	8F183
155506	F4J		
155507	F4J		
155508	F4J	w.o.4.11.83	
155509	F4J	stored	NAS N. Island
155510	F4J	to RAF	ZE358
155511	F4S	dump	NAS N. Island
155512	F4J		
155513	F4J		
155514	F4J		

155515	F4S	AMARC	8F265	155748	F4J	AMARC	8F163
155516	F4J	scrapped		155749	F4S		
155517	F4S	w.o.14.1.87		155750	F4J	w.o.69	
155518	F4S			155751	F4J	w.o.68	
155519	F4S	AMARC	8F209	155752	F4J		
155520	F4J			155753	F4S	AMARC	8F167
155521	F4S			155754	F4S	AMARC	8F283
155522	F4S	AMARC	8F255	155755	F4J	to RAF	ZE362
155523	F4J			155756	F4J	w.o.68	
155524	F4S			155757	F4S	w.o.2.4.84	
155525	F4S	AMARC	8F226	155758	F4J	w.o.68	
155526	F4J	w.o.11.9.72		155759	F4S		
155527	F4S			155760	F4J	w.o.17.2.69	
155528	F4S	AMARC	8F267	155761	F4S	AMARC	8F261
155529	F4J	to RAF	ZE359	155762	F4J	w.o.7.2.69	
155530	F4S	AMARC	8F203	155763	F4J	w.o.20.2.69	
155531	F4S	w.o.19.6.86		155764	F4S		
155532	F4S	AMARC	8F193	155765	F4S	AMARC	8F272
155533	F4J			155766	F4S	AMARC	8F202
155534	F4J			155767	F4S	AMARC	8F229
155535	F4J			155768	F4J	w.o.73	
155536	F4J			155769	F4S	w.o.8.4.83	
155537	F4J			155770	F4J	w.o.69	
155538	F4J			155771	F4J	w.o.69	
155539	F4S	AMARC	8F222	155772	F4S		
155540	F4J	w.o.25.7.68		155773	F4S	AMARC	8F168
155541	F4S	AMARC	8F194	155774	F4J	w.o.73	
155542	F4S	AMARC	8F297	155775	F4J	w.o.9.3.70	
155543	F4S			155776	F4J	w.o.68	
155544	F4S			155777	F4J	w.o.5.1.71	
155545	F4S	AMARC	8F241	155778	F4J	w.o.70	
155546	F4J	w.o.68		155779	F4S	AMARC	8F210
155547	F4J			155780	F4J	w.o.72	
155548	F4J	w.o.68		155781	F4S	AMARC	8F252
155549	F4S	stored	MCAS Cherry Pt.	155782	F4J	w.o.68	
155550	F4S	AMARC	8F286	155783	F4S	AMARC	8F248
155551	F4J	w.o.24.7.68		155784	F4S	AMARC	8F260
155552	F4S			155785	F4J		
155553	F4J			155786	F4S	stored	MCAS Cherry Pt.
155554	F4J	w.o.68		155787	F4S	AMARC	8F285
155555	F4S	AMARC	8F291	155788	F4J	w.o.69	
155556	F4J	w.o.20.11.79		155789	F4J	w.o.70	
155557	F4J			155790	F4J	w.o.70	
155558	F4S	AMARC	8F233	155791	F4J	w.o.68	
155559	F4S	AMARC	8F277	155792	F4S	AMARC	8F208
155560	F4S	AMARC	8F273	155793	F4J	w.o.68	
155561	F4S			155794	F4S	AMARC	8F305
155562	F4S	w.o.2.9.79		155795	F4J	w.o.70	
155563	F4J	NAWC-WD	92	155796	F4J	w.o.68	
155564	F4J	w.o.68		155797	F4J	w.o.72	
155565	F4S	VX-4	XF-4	155798	F4J		
155566	F4S	AMARC	8F311	155799	F4S	AMARC	8F186
155567	F4S	AMARC	8F247	155800	F4J	w.o.10.5.72	
155568	F4S			155801	F4S	AMARC	8F213
155569	F4S			155802	F4J	w.o.69	
155570	F4S	AMARC	8F223	155803	F4J	w.o.10.7.72	
155571	F4J			155804	F4J	w.o.68	
155572	F4S			155805	F4S	AMARC	8F302
155573	F4S			155806	F4S		
155574	F4S	to RAF	ZE360	155807	F4S		
155575	F4S	AMARC	8F309	155808	F4S	w.o.4.11.83	
155576	F4J	w.o.7.5.72		155809	F4J	w.o.25.7.69	
155577	F4J	w.o.5.1.71		155810	F4S		
155578	F4J			155811	F4J	w.o.72	
155579	F4S	w.o.1.8.84		155812	F4S	AMARC	8F204
155580	F4S			155813	F4S	w.o.11.12.84	
155731	F4S	AMARC	8F196	155814	F4J	w.o.70	
155732	F4S	AMARC	8F257	155815	F4J	w.o.74	
155733	F4S	AMARC	8F198	155816	F4J		
155734	F4J	to RAF	ZE361	155817	F4J	w.o.2.8.72	
155735	F4S	AMARC	8F173	155818	F4S	AMARC	8F240
155736	F4S	AMARC	8F312	155819	F4J	w.o.4.6.72	
155737	F4J	w.o.72		155820	F4S	AMARC	8F197
155738	F4J	w.o.20.5.70		155821	F4S	AMARC	8F271
155739	F4S	AMARC	8F323	155822	F4S	AMARC	8F306
155740	F4J			155823	F4J	AMARC	8F279
155741	F4J			155824	F4J		
155742	F4J	w.o.68		155825	F4S	AMARC	8F278
155743	F4S	AMARC	8F200	155826	F4J		
155744	F4J	w.o.68		155827	F4S	AMARC	8F237
155745	F4S			155828	F4S	w.o.1.6.79	
155746	F4S	AMARC	8F220	155829	F4S		
155747	F4S	AMARC	8F177	155830	F4J	AMARC	8F262

Serial	Type	Disposition	Notes
155831	F4J		
155832	F4S		
155833	F4S	w.o.23.9.86	
155834	F4S		
155835	F4J	AMARC	8F189
155836	F4S	w.o.30.4.85	
155837	F4J		
155838	F4S	AMARC	8F266
155839	F4S	AMARC	8F250
155840	F4S		
155841	F4S		
155842	F4J		
155843	F4J		
155844	F4S	AMARC	8F317
155845	F4S		
155846	F4J	w.o.73	
155847	F4S		
155848	F4S	preserved	RNAS Yeovilton
155849	F4S	AMARC	8F290
155850	F4J		
155851	F4S	AMARC	8F296
155852	F4S	AMARC	8F215
155853	F4J		
155854	F4J	AMARC	8F206
155855	F4S		
155856	F4J		
155857	F4J	w.o.72	
155858	F4S	AMARC	8F253
155859	F4S	AMARC	8F287
155860	F4J		
155861	F4S		
155862	F4S	AMARC	8F235
155863	F4S	AMARC	8F292
155864	F4S		
155865	F4S	w.o.20.1.81	
155866	F4S		
155867	F4J		
155868	F4J	to RAF	ZE363
155869	F4S	NARF	Cherry Point
155870	F4J		
155871	F4S	AMARC	8F254
155872	F4S		
155873	F4J		
155874	F4S	AMARC	8F281
155875	F4J	w.o.73	
155876	F4S	AMARC	8F280
155877	F4J	w.o.69	
155878	F4S		
155879	F4S	AMARC	8F288
155880	F4S	AMARC	8F224
155881	F4S	AMARC	8F319
155882	F4S		
155883	F4S	AMARC	8F293
155884	F4J	w.o.27.2.71	
155885	F4J		
155886	F4J	w.o.73	
155887	F4S	stored	MCAS Cherry Pt.
155888	F4S	AMARC	8F301
155889	F4J	w.o.69	
155890	F4S		
155891	F4S	AMARC	8F289
155892	F4S		
155893	F4S	AMARC	8F246
155894	F4J	to RAF	ZE364
155895	F4J	w.o.74	
155896	F4S	AMARC	8F304
155897	F4S	AMARC	8F221
155898	F4J		
155899	F4J		
155900	F4J		
155901	F4S	w.o.23.11.79	
155902	F4J		
155903	F4J		
157242	F4S		
157243	F4S		
157244	F4J	w.o.70	
157245	F4S	AMARC	8F205
157246	F4S		
157247	F4J		
157248	F4S	AMARC	8F201
157249	F4S	AMARC	8F242
157250	F4S	w.o.8.8.84	
157251	F4S		
157252	F4J	w.o.14.4.72	
157253	F4J	w.o.72	
157254	F4S	AMARC	8F282
157255	F4S	w.o.11.8.86	
157256	F4J	w.o.73	
157257	F4S	AMARC	8F178
157258	F4J	AMARC	8F295
157259	F4S	PMTC	94
157260	F4S	AMARC	8F258
157261	F4S		
157262	F4J	w.o.72	
157263	F4J	w.o.70	
157264	F4J	AMARC	8F216
157265	F4J		
157266	F4J	w.o.72	
157267	F4S	preserved	San Diego
157268	F4S	AMARC	8F264
157269	F4S	AMARC	8F211
157270	F4J		
157271	F4J	w.o.18.12.79	
157272	F4S	AMARC	8F313
157273	F4J	w.o.72	
157274	F4S		
157275	F4J	w.o.13.4.80	
157276	F4S	AMARC	8F275
157277	F4J		
157278	F4S	AMARC	8F318
157279	F4S	w.o.6.7.82	
157280	F4J		
157281	F4J	AMARC	8F316
157282	F4J	NARF	North Island
157283	F4S	AMARC	8F274
157284	F4J		
157285	F4J		
157286	F4J	AMARC	8F308
157287	F4S	AMARC	8F276
157288	F4J	w.o.20.11.72	
157289	F4J		
157290	F4S		
157291	F4S	AMARC	8F225
157292	F4J	AMARC	8F245
157293	F4S		
157294	F4J		
157295	F4J		
157296	F4S	AMARC	8F251
157297	F4S		
157298	F4S		
157299	F4J		
157300	F4J	AMARC	8F217
157301	F4J	AMARC	8F234
157302	F4J	w.o.8.9.72	
157303	F4J		
157304	F4J		
157305	F4S		
157306	F4S		
157307	F4S		
157308	F4S	AMARC	8F232
157309	F4S	AMARC	8F238
157342	RF4B		
157343	RF4B		
157344	RF4B		
157345	RF4B	AMARC	8F340
157346	RF4B		
157347	RF4B		
157348	RF4B		
157349	RF4B		
157350	RF4B		
157351	RF4B		
158346	F4S	w.o.14.9.82	
158347	F4J	dump	NAS N. Island
158348	F4S	AMARC	8F207
158349	F4J	w.o.5.8.80	
158350	F4S	AMARC	8F227
158351	F4S		
158352	F4S	AMARC	8F320
158353	F4S		
158354	F4S	AMARC	8F269
158355	F4J	preserved	USS Intrepid
158356	F4J	AMARC	8F181
158357	F4J	w.o.74	
158358	F4J	VX-4	XF-1
158359	F4J		
158360	F4S	VX-4	XF-7

158361	F4J	w.o.29.1.73			158371	F4J	w.o.7.1.80	
158362	F4S				158372	F4J	AMARC	8F315
158363	F4J	AMARC	8F187		158373	F4S	AMARC	8F310
158364	F4J	w.o.72			158374	F4S	w.o.23.9.86	
158365	F4J	AMARC	8F185		158375	F4J	w.o.73	
158366	F4J	w.o.73			158376	F4S	AMARC	8F322
158367	F4J	preserved	MCAS Yuma		158377	F4S	w.o.13.7.82	
158368	F4J				158378	F4S	AMARC	8F179
158369	F4J	AMARC	8F321		158379	F4J	w.o.72	
158370	F4S							

F5 TIGER II

159878	F5E				741528	F5E	VMFT-401	05
159879	F5E	w.o.13.5.81			741529	F5E	VMFT-401	06
159880	F5E	VF-43	14		741530	F5E	VFA-127	NJ-720
159881	F5E	VFA-127	NJ-727		741536	F5E	VFA-127	NJ-745
159882	F5E	w.o.21.9.84			741537	F5E	VMFT-401	08
160792	F5E	VFA-127	NJ-723		741539	F5E	VFA-127	NJ-724
160793	F5E	VFA-127	NJ-726		741540	F5E	VMFT-401	09
160794	F5E	VF-43	12		741541	F5E	VMFT-401	10
160795	F5E	VFA-127	NJ-721		741544	F5E	VFA-127	NJ-740
160796	F5E	VF-43	11		741545	F5E	VFA-127	NJ-741
160964	F5F	VMFT-401	00		741547	F5E	VFA-127	NJ-742
160965	F5F	VF-43	15		741554	F5E		
160966	F5F	FWS	550		741556	F5E		
162307	F5F	VF-43	10		741558	F5E	VFA-127	NJ-725
721387	F5E				741563	F5E		
730855	F5E	VMFT-401	01		741564	F5E	VFA-127	NJ-744
730865	F5E	VFA-127	NJ-743		741568	F5E	VFA-127	NJ-746
730879	F5E	VFA-127	NJ-722		741570	F5E	VMFT-401	11
730885	F5E	VMFT-401	02		741572	F5E		
731635	F5E	VMFT-401	03		840456	F5F	VFA-127	NJ-730
741519	F5E	VMFT-401	04					

F14 TOMCAT

157980	F14A	w.o.30.12.70			158625	F14A	NAWC-WD	226
157981	F14A	w.o.13.5.74			158626	F14A	VF-202	AF-213
157982	F14A	Grumman			158627	F14A	VF-201	AF-102
157983	YF14A	stored	NAS Memphis		158628	F14A	VF-202	AF-203
157984	F14A	w.o.			158629	F14A	VF-202	AF-204
157985	F14A	w.o.20.6.73			158630	F14A	VF-201	AF-104
157986	F14B				158631	F14A	VF-201	AF-113
157987	F14A	w.o.19.9.74			158632	F14A	VF-201	AF-105
157988	F14A	gate	NAS Oceana		158633	F14A	VF-201	AF-111
157989	F14A	w.o.30.6.72			158634	F14A	VF-202	AF-210
157990	F14A	w.o.			158635	F14A	VF-202	AF-205
157991	F14A	NASA	Dryden		158636	F14A	VF-202	AF-202
158612	F14A	VF-202	AF-206		158637	F14A	VF-202	AF-216
158613	F14A	stored	NAS Norfolk		158978	F14A	VX-4	XF-33
158614	F14A	VF-202	AF-217		158979	F14A	AMARC	1K...
158615	F14A	VF-202	AF-207		158980	F14A	AMARC	1K...
158616	F14A	VF-201	AF-107		158981	F14A	VF-302	ND-201
158617	F14A	VF-201	AF-114		158982	F14A	w.o.2.1.75	
158618	F14A	VF-201	AF-112		158983	F14A	w.o.20.6.86	
158619	F14A	w.o.22.2.77			158984	F14A	VF-302	ND-203
158620	F14A	VF-202	AF-215		158985	F14A		
158621	F14A	NAWC-WD	222		158986	F14A	AMARC	1K...
158622	F14A	AMARK	1K...		158987	F14A	AMARC	1K...
158623	F14A	NAWC-WD	224		158988	F14A	AMARC	1K...
158624	F14A	VF-201	AF-106		158989	F14A	AMARC	1K...

Serial	Type	Unit/Status	Code
158990	F14A	VF-301	ND-101
158991	F14A	AMARC	1K...
158992	F14A	VF-301	ND-106
158993	F14A	VF-302	ND-205
158994	F14A	AMARC	1K...
158995	F14A	w.o.27.3.78	
158996	F14A	w.o.26.6.77	
158997	F14A	VF-301	ND-110
158998	F14A	NADC	
158999	F14A	VF-202	AF-212
159000	F14A	AMARC	1K...
159001	F14A	w.o.13.1.75	
159002	F14A	AMARC	1K...
159003	F14A	AMARC	1K...
159004	F14A	AMARC	1K...
159005	F14A	VF-301	ND-112
159006	F14A	AMARC	1K...
159007	F14A	w.o.5.8.75	
159008	F14A	w.o.31.10.77	
159009	F14A	AMARC	1K...
159010	F14A	AMARC	1K...
159011	F14A	w.o.6.2.82	
159012	F14A	w.o.20.3.78	
159013	F14A	VF-33	AB-217
159014	F14A	AMARC	1K...
159015	F14A	AMARC	1K...
159016	F14A		
159017	F14A	AMARC	1K...
159018	F14A		
159019	F14A		
159020	F14A	AMARC	1K...
159021	F14A	AMARC	1K...
159022	F14A	w.o.5.12.79	
159023	F14A	VF-101	AD-134
159024	F14A	w.o.8.11.83	
159025	F14A	VF-302	ND-207
159421	F14A	w.o.18.6.87	
159422	F14A	AMARC	1K...
159423	F14A	VF-302	ND-202
159424	F14A	VX-4	XF-34
159425	F14A	AMARC	1K...
159426	F14A	NAWC-AD	
159427	F14A	AMARC	1K...
159428	F14A	VF-202	AF-
159429	F14A	AMARC	1K...
159430	F14A	w.o.5.10.78	
159431	F14A	w.o.3.1.87	
159432	F14A	w.o.24.6.75	
159433	F14A		
159434	F14A	AMARC	1K...
159435	F14A	AMARC	1K...
159436	F14A	AMARC	1K...
159437	F14A	VF-101	AD-130
159438	F14A	VF-101	AD-132
159439	F14A	w.o.15.7.84	
159440	F14A	VF-33	AB-
159441	F14A	w.o.25.3.88	
159442	F14A	AMARC	1K...
159443	F14A	w.o.28.3.77	
159444	F14A	VF-32	AC-203
159445	F14A	VF-33	AB-211
159446	F14A	stored	NAS Norfolk
159447	F14A	AMARC	1K...
159448	F14A	VF-33	AB-204
159449	F14A	VF-14	AC-112
159450	F14A	VF-33	AB-205
159451	F14A	w.o.10.11.77	
159452	F14A	VF-32	AC-214
159453	F14A	VF-14	AC-107
159454	F14A	VF-201	AF-
159455	F14A	VF-143	AG-
159456	F14A	w.o.21.5.79	
159457	F14A	VF-32	AC-215
159458	F14A	VF-102	AB-104
159459	F14A	VF-101	AD-111
159460	F14A	NAWC-WD	
159461	F14A	w.o.23.3.76	
159462	F14A	VF-143	AG-103
159463	F14A	VF-14	AC-115
159464	F14A	w.o.19.12.76	
159465	F14A	VF-302	ND-204
159466	F14A	VF-102	AB-106
159467	F14A	VF-124	NJ-407
159468	F14A	VF-101	AD-122
159588	F14A	w.o.14.9.76	
159589	F14A	AMARC	1K...
159590	F14A	w.o.29.10.75	
159591	F14A	VF-302	ND-211
159592	F14A		
159593	F14A	stored	NAS Norfolk
159594	F14A	w.o.21.6.77	
159595	F14A	VF-143	AG-
159596	F14A	VF-101	AD-136
159597	F14A	VF-101	AD-127
159598	F14A		
159599	F14A	w.o.6.3.82	
159600	F14A	VF-142	AG-206
159601	F14A	w.o.6.3.80	
159602	F14A	stored	NAS Norfolk
159603	F14A	VF-32	AC-206
159604	F14A	VF-201	AF-103
159605	F14A	w.o.13.9.80	
159606	F14A	VF-302	ND-212
159607	F14A	FWS	
159608	F14A	VF-211	NG-110
159609	F14A	stored	NAS Norfolk
159610	F14A	VF-21	NF-206
159611	F14A	VF-124	NJ-413
159612	F14A	stored	NAS Norfolk
159613	F14A	VF-11	NK-106
159614	F14A	AMARC	1K...
159615	F14A	NFWS	33
159616	F14A	VF-101	AD-134
159617	F14A	w.o.14.6.82	
159618	F14A	VF-14	AC-101
159619	F14D(R)	VX-4	XF-52
159620	F14A	VF-124	NJ-102
159621	F14A	VF-302	ND-200
159622	F14A	w.o.16.7.78	
159623	F14A	w.o.19.12.81	
159624	F14A	VF-211	NG-
159625	F14A	VF-24	NG-206
159626	F14A	VF-211	NG-106
159627	F14A	VF-124	NJ-116
159628	F14A	VF-301	ND-100
159629	F14A	VF-24	NG-207
159630	F14A	VF-301	ND-
159631	F14A	VF-124	NJ-103
159632	F14A	w.o.25.8.78	
159633	F14A	VF-211	NG-
159634	F14A	VF-11	NK-113
159635	F14A	VF-24	NG-
159636	F14A	w.o.25.11.78	
159637	F14A	VF-211	NG-107
159825	F14A	NFWS	30
159826	F14A	w.o.5.3.76	
159827	F14A	VF-213	NH-201
159828	F14A	VF-301	ND-100
159829	F14A	NAWC-AD	SATD SD-205
159830	F14A	VF-124	NJ-112
159831	F14A	VF-201	AF-110
159832	F14A		
159833	F14A	VF-301	ND-104
159834	F14A	VF-24	NG-224
159835	F14A		
159836	F14A	VF-124	NJ-124
159837	F14A	VF-1	NE-106
159838	F14A	w.o.25.3.78	
159839	F14A	w.o.28.6.77	
159840	F14A	w.o.18.1.84	
159841	F14A	VF-1	NE-105
159842	F14A	w.o.19.4.77	
159843	F14A	VF-124	NJ-
159844	F14A	VF-51	NL-102
159845	F14A	VF-124	NJ-
159846	F14A	w.o.14.2.86	
159847	F14A		
159848	F14A	VF-302	ND-210
159849	F14A	VF-302	ND-203
159850	F14A	VF-1	NE-110
159851	F14A	w.o.25.3.78	
159852	F14A	VF-124	NJ-412
159853	F14A	stored	NAS Norfolk
159854	F14A	w.o.28.6.77	
159855	F14A	VF-2	NE-202
159856	F14A	VF-301	ND-113

Serial	Type	Unit	Code
159857	F14A	VF-301	ND-
159858	F14A	VF-124	NJ-104
159859	F14A	w.o.8.8.84	
159860	F14A	VF-302	ND-202
159861	F14A	w.o.3.9.86	
159862	F14A	VF-51	NL-103
159863	F14A	VF-124	NJ-
159864	F14A	VF-51	NL-104
159865	F14A	w.o.16.12.88	
159866	F14A	VF-301	ND-105
159867	F14A	VF-2	NE-206
159868	F14A	VF-201	AF-104
159869	F14A	VF-124	NJ-414
159870	F14A	VF-124	NJ-
159871	F14A	VF-201	AF-101
159872	F14A	w.o.27.1.89	
159873	F14A	VF-2	NE-203
159874	F14A	VF-124	NJ-105
160379	F14A	VF-101	AD-123
160380	F14A	w.o.3.5.80	
160381	F14A	VF-33	AB-200
160382	F14A	VF-84	AJ-203
160383	F14A	w.o.3.11.79	
160384	F14A	VF-32	AC-202
160385	F14A	w.o.26.5.81	
160386	F14A	VF-142	AG-211
160387	F14A	VF-41	AJ-104
160388	F14A	w.o.1.4.80	
160389	F14A	VF-211	NG-112
160390	F14A	VF-103	AA-204
160391	F14A	VF-84	AJ-202
160392	F14A	w.o.3.10.77	
160393	F14A	VF-14	AC-110
160394	F14A	VF-41	AJ-106
160395	F14A	VF-33	AB-204
160396	F14A	VF-14	AC-104
160397	F14A	VF-102	AB-104
160398	F14A	VF-101	AD-
160399	F14A	VF-33	AB-203
160400	F14A	w.o.30.8.83	
160401	F14A	VF-102	AB-105
160402	F14A	VF-14	AC-105
160403	F14A	VF-101	AD-143
160404	F14A	stored	NAS Norfolk
160405	F14A	VF-84	AJ-210
160406	F14A	VF-33	AB-207
160407	F14A	VF-41	AJ-103
160408	F14A	VF-101	AD-143
160409	F14A	w.o.12.9.88	
160410	F14A	VF-211	NG-104
160411	F14A	VF-1	NE-101
160412	F14A	w.o.21.3.78	
160413	F14A	VF-14	AC-102
160414	F14A	VF-101	AD-146
160652	F14A	VF-124	NJ-
160653	F14A	w.o.15.6.78	
160654	F14A	VF-2	NE-207
160655	F14A		
160656	F14A	w.o.31.3.85	
160657	F14A	VF-24	NG-204
160658	F14A	NNAWC-AD	SATD SD-207
160659	F14A	w.o.13.9.78	
160660	F14A	w.o.16.7.91	
160661	F14A	VF-124	NJ-111
160662	F14A	w.o.9.4.83	
160663	F14A	w.o.2.9.86	
160664	F14A	VF-21	NF-203
160665	F14A	VF-154	NF-112
160666	F14A	VF-111	NL-201
160667	F14A	VF-154	NF-111
160668	F14A	VF-111	NL-213
160669	F14A	VF-111	NL-210
160670	F14A	w.o.12.80	
160671	F14A	VF-51-	NL-
160672	F14A	w.o.8.9.79	
160673	F14A	VF-1	NE-112
160674	F14A	w.o.27.6.81	
160675	F14A	w.o.	
160676	F14A	VF-111	NL-202
160677	F14A	w.o.7.9.81	
160678	F14A	VF-11	NK-111
160679	F14A	VF-51	NL-105
160680	F14A	VF-21	NF-210
160681	F14A	VF-301	ND-107
160682	F14A	VF-51	NL-107
160683	F14A	VF-51	NL-111
160684	F14A	VF-124	NJ-110
160685	F14A	w.o.26.3.86	
160686	F14A	VF-51	NL-106
160687	F14A	VF-51	NL-115
160688	F14A	VF-154	NF-107
160689	F14A	VF-111	NL-201
160690	F14A	VF-111	NL-206
160691	F14A	VF-124	NJ-120
160692	F14A	VF-21	NF-206
160693	F14A	VF-302	ND-206
160694	F14A	VF-1	NE-107
160695	F14A	VF-2	NE-205
160696	F14A	VF-124	NJ-
160887	F14A	w.o.22.1.92	
160888	F14A	w.o.	
160889	F14A	VF-124	NJ-417
160890	F14A	w.o.12.10.79	
160891	F14A	VF-14	AC-105
160892	F14A	VF-51	NL-110
160893	F14A	VF-101	AD-142
160894	F14A	VF-154	NF-106
160895	F14A	w.o.29.9.81	
160896	F14A	VF-32	AC-205
160897	F14A	VF-14	AC-106
160898	F14A	VF-41	AJ-111
160899	F14A	VF-101	AD-140
160900	F14A	VF-32	AC-206
160901	F14A	VF-32	AC-207
160902	F14A	VF-84	AJ-207
160903	F14A	VF-41	AJ-107
160904	F14A	VF-101	AD-147
160905	F14A	VF-74	AA-105
160906	F14A	VF-14	AC-107
160907	F14A	w.o.20.9.82	
160908	F14A	VF-84	AJ-204
160909	F14A	stored	NAS Norfolk
160910	F14A	VF-111	NL-204
160911	F14A	VF-302	ND-215
160912	F14A	stored	NAS Norfolk
160913	F14A	VF-103	AA-203
160914	F14A	VF-213	NH-113
160915	F14A	VF-124	NJ-126
160916	F14A	w.o.3.3.80	
160917	F14A	VF-14	AC-111
160918	F14A	VF-102	AB-102
160919	F14A	VF-124	NJ-106
160920	F14A	VF-213	NH-215
160921	F14A	w.o.17.1.85	
160922	F14A	VF-124	NJ-127
160923	F14A	stored	NAS Norfolk
160924	F14A	VF-102	AB-102
160925	F14A	VF-124	NJ-121
160926	F14A	VF-84	AJ-212
160927	F14A	VF-101	AD-145
160928	F14A	VF-1	NE-113
160929	F14A	VF-41	AJ-114
160930	F14A	VF-2	NE-213
161133	F14A	VF-101	AD-111
161134	F14A	VF-101	AD-
161135	F14A	VF-101	AD-152
161136	F14A	VF-124	NJ-122
161137	F14A	VF-84	AJ-214
161138	F14A	w.o.26.5.81	
161139	F14A	VF-101	AD-115
161140	F14A	VF-32	AC-210
161141	F14A	VF-84	AJ-
161142	F14A	VF-33	AB-210
161143	F14A	w.o.29.7.82	
161144	F14A	w.o.26.9.88	
161145	F14A	stored	NAS Norfolk
161146	F14A	VF-211	NG-106
161147	F14A	VF-102	AB-111
161148	F14A	w.o.23.8.86	
161149	F14A	w.o.11.11.83	
161150	F14A	VF-14	AC-212
161151	F14A	VF-33	AB-206
161152	F14A	VF-111	NL-212
161153	F14A	w.o.20.9.87	
161154	F14A	VF-101	AD-112
161155	F14A	VF-102	AB-113

161156	F14A	VF-32	AC-211		161607	F14A	VF-21	NF-202
161157	F14A	w.o.24.4.81			161608	F14B	VF-103	AA-213
161158	F14A(R)	VF-11	NK-111		161609	F14A	VF-21	NF-204
161159	F14A	VF-21	NF-211		161610	F14B	VF-211	NG-
161160	F14A	VF-101	AD-116		161611	F14A	VF-111	NL-207
161161	F14A	stored	NAS Norfolk		161612	F14A	VF-1	NE-103
161162	F14A	VF-32	AC-212		161613	F14A	w.o.12.3.86	
161163	F14A	NNAWC-AD	SATD SD-204		161614	F14A	w.o.20.3.87	
161164	F14A	VF-84	AJ-211		161615	F14A	VF-111	NL-212
161165	F14A	w.o.4.9.84			161616	F14A	VF-154	NF-103
161166	F14A	VF-143	AG-110		161617	F14A	VF-111	NL-211
161167	F14A	w.o.13.8.86			161618	F14A	VF-11	NK-100
161168	F14A	VF-2	NE-210		161619	F14A	VF-24	NG-205
161270	F14A	VF-154	NF-104		161620	F14A	VF-124	NJ-477
161271	F14A	VF-2	NE-211		161621	F14A	VF-154	NF-
161272	F14A	VF-154	NF-101		161622	F14A	VF-154	NF-106
161273	F14A	VF-2	NE-		161623	F14A	NAWC-AD	SATD SD-210
161274	F14A	VF-11	NK-112		161624	F14A	VF-124	NJ-472
161275	F14A	VF-154	NF-102		161625	F14A	w.o.2.8.89	
161276	F14A	VF-2	NE-213		161626	F14A	VF-154	NF-105
161277	F14A	VF-101	AD-157		161850	F14A	VF-211	NG-113
161278	F14A	VF-21	NF-205		161851	F14B	VF-101	AD-110
161279	F14A	VF-1	NE-		161852	F14A	VF-41	AJ-105
161280	F14A	VF-102	AB-112		161853	F14A	VF-24	NG-203
161281	F14A	VF-101	AD-155		161854	F14A	w.o.18.6.86	
161282	F14A	VF-211	NG-103		161855	F14A	VF-14	AC-100
161283	F14A	w.o.20.6.84			161856	F14A		
161284	F14A	VF-51	NL-112		161857	F14A	VX-4	XF-35
161285	F14A	VF-101	AD-154		161858	F14A	VF-14	AC-103
161286	F14A	w.o.14.11.89			161859	F14A	VF-14	AC-101
161287	F14B	VX-4	XF-46		161860	F14A	VF-41	AJ-110
161288	F14A	VF-1	NE-114		161861	F14A	w.o.15.12.92	
161289	F14A	w.o.28.2.83			161862	F14A		
161290	F14A	w.o.11.2.86			161863	F14A	VF-11	NK-
161291	F14A	VF-124	NJ-		161864	F14A		
161292	F14A	VF-21	NF-204		161865	F14A	NAWC	203
161293	F14A				161866	F14A	VF-102	AB-110
161294	F14A	VF-1	NE-111		161867	NF14D	NAWC	202
161295	F14A	VF-111	NL-206		161868	F14A		
161296	F14A	VF-2	NE-201		161869	F14A	VF-124	NJ-157
161297	F14A				161870	F14B	VF-74	AA-110
161298	F14A	w.o.			161871	F14B	VF-101	AD-107
161299	F14A	VF-2	NE-204		161872	F14A	w.o.12.8.87	
161416	F14B	VF-143	AG-100		161873	F14B	VF-103	AA-202
161417	F14B	VF-143	AG-101		162588	F14A	VF-24	NG-200
161418	F14B	VF-142	AG-211		162589	F14A	VF-154	NF-100
161419	F14B	VF-101	AD-104		162590	F14A	w.o.25.1.93	
161420	F14A	w.o.30.8.83			162591	F14A	VF-51	NL-100
161421	F14B	VF-142	AG-207		162592	F14A	VF-1	NE-113
161422	F14B	VF-143	AG-103		162593	F14A	w.o.17.8.87	
161423	F14A	w.o.			162594	F14A	VF-111	NL-200
161424	F14B	VF-103	AA-212		162595	F14A	NAWC-	SATD SD-221
161425	F14B	VF-101	AD-000		162596	F14A	w.o.22.9.88	
161426	F14B	VF-143	AG-104		162597	F14A	VF-1	NE-
161427	F14B	VF-142	AG-206		162598	F14A	VF-2	NE-202
161428	F14B	VF-103	AA-211		162599	F14A	VF-1	NE-101
161429	F14B	VF-143	AG-100		162600	F14A	VF-211	NG-106
161430		w.o.21.1.91			162601	F14A	VF-1	NE-100
161431	F14A	w.o.17.3.83			162602	F14A	VF-2	NE-204
161432	F14B	VF-74	AA-107		162603	F14A	VF-1	NE-
161433	F14B	w.o.13.1.92			162604	F14A	VF-51	NL-101
161434	F14B	VF-74	AA-102		162605	F14A	w.o.15.7.87	
161435	F14B	VF-74	AA-106		162606	F14A	VF-2	NE-206
161436	F14A	w.o.17.3.83			162607	F14A	VF-124	NJ-100
161437	F14B	stored	NAS Norfolk		162608	F14A	VF-2	NE-200
161438	F14B	stored	NAS Norfolk		162609	F14A	w.o.8.9.88	
161439	F14A	w.o.17.3.83			162610	F14A	stored	NAS Norfolk
161440	F14B	VF-124	NJ-425		162611	F14A	VF-1	NE-100
161441	F14B	VF-143	AG-102		162688	F14A	VF-84	AJ-200
161442	F14B	stored	NAS Norfolk		162689	F14A	VF-41	AJ-101
161443	F14A	VF-111	NL-213		162690	F14A	w.o.6.10.89	
161444	F14B	VX-4	XF-47		162691	F14A	VF-14	AC-101
161445	F14A	VF-21	NF-203		162692	F14A	VF-84	AJ-201
161597	F14A	w.o.29.6.91			162693	F14A	VF-101	AD-164
161598	F14A	VF-154	NF-101		162694	F14A	VF-32	AC-201
161599	F14B	VF-213	NH-101		162695	F14A	VF-102	AB-102
161600	F14A	VF-21	NF-		162696	F14A	VF-102	AB-103
161601	F14B	VF-24	NG-201		162697	F14A	VF-101	AD-162
161602	F14A	w.o.24.7.89			162698	F14A	VF-33	AB-202
161603	F14A	VF-21	NF-200		162699	F14A	VF-101	AD-163
161604	F14A	VF-84	AJ-215		162700	F14A	VF-102	AB-
161605	F14A	VF-101	AD-157		162701	F14A	VF-32	AC-200
161606	F14A	VF-21	NF-201		162702	F14A	w.o.	

162703	F14A	VF-41	AJ-100		163407	F14B	VF-143	AG-107
162704	F14A	VF-102	AB-102		163408	F14B	VF-211	NG-110
162705	F14A	VF-101	AD-160		163409	F14B	VF-143	AG-112
162706	F14A	w.o.18.4.89			163410	F14B	VF-101	
162707	F14A	w.o.18.1.93			163411	F14B	w.o.15.3.93	
162708	F14A	w.o.23.12.92			163412	F14D		
162709	F14A	VF-201	AF-100		163413	F14D	VX-4	XF-50
162710	F14A	VF-202	AF-200		163414	F14D	VX-4	XF-51
162711	F14A	VF-202	AF-201		163415	F14D	NAWC	201
162910	F14B	NAWC			163416	F14D		
162911	F14B	VX-4	XF-51		163417	F14D	NAWC-AD	SATD SD-231
162912	F14B	VF-101	AD-105		163418	F14D	VF-124	NJ-130
162913	F14B	VF-101	AD-114		163893	F14D	VF-124	NJ-131
162914	F14B	VF-142	AG-202		163894	F14D	VF-124	NJ-132
162915	F14B	VF-142	AG-204		163895	F14D	VF-124	NJ-133
162916	F14B	VF-143	AG-105		163896	F14D	VF-124	NJ-134
162917	F14B	VF-143	AG-106		163897	F14D	VF-124	NJ-135
162918	F14B	VF-101	AD-102		163898	F14D	VF-124	NJ-136
162919	F14B	VF-74	AA-101		163899	F14D	VF-124	NJ-137
162920	F14B	VF-103	AA-203		163900	F14D	VF-124	NJ-140
162921	F14B	VF-103	AA-200		163901	F14D	VF-124	NJ-141
162922	F14B	VF-142	AG-203		163902	F14D	VF-124	NJ-142
162923	F14B	VF-74	AA-103		163903	F14D	VF-124	NJ-143
162924	F14B	VF-74	AA-104		163904	F14D		
162925	F14B	VF-74	AA-100		164340	F14D		
162926	F14B	VF-142	AG-200		164341	F14D	VF-11	NK-101
162927	F14B	VF-101	AD-106		164342	F14D		
163215	F14B	VF-103	AA-204		164343	F14D	VF-11	NK-102
163216	F14B	VF-142	AG-210		164344	F14D		
163217	F14B	VF-142	AG-201		164345	F14D	VF-11	NK-103
163218	F14B	VF-143	AG-110		164346	F14D		
163219	F14B	VF-103	AA-201		164347	F14D	VF-11	NK-104
163220	F14B	VF-143	AG-111		164348	F14D		
163221	F14B	VF-143	AA-105		164349	F14D	VF-11	NK-105
163222	F14B	VF-101	AD-111		164350	F14D		
163223	F14B	VX-4	XF-45		164351	F14D	500	
163224	F14B	VF-142	AG-212		164599	F14D	600	
163225	F14B	VF-101	AD-112		164600	F14D	501	
163226	F14B	VF-101	AD-115		164601	F14D	601	
163227	F14B	VF-101	AD-103		164602	F14D	VF-124	NJ-162
163228	F14B	VF-101	AD-110		164603	F14D	VF-124	NJ-167
163229	F14B	VF-101	AD-100		164604	F14D		

F16 FIGHTING FALCON

ex   85-1369 to 1382 (14), 86-1684 to 1695 (12).

c\n   3M-1 to 3M-10, 3N-1 to 3N-4, 3M-11 to 3M-22.

163268	F16N	NFWS	41		163281	TF16N	VF-45	AD-625
163269	F16N	NFWS	42		163566	F16N	VF-126	NJ-604
163270	F16N	NFWS	43		163567	F16N	VF-126	NJ-605
163271	F16N	NFWS	44		163568	F16N	w.o.17.12.92	
163272	F16N	NFWS	45		163569	F16N	VF-45	AD-620
163273	F16N	NFWS	40		163570	F16N	VF-45	AD-621
163274	F16N	VF-43	04		163571	F16N	VF-45	AD-622
163275	F16N	VF-126	NJ-601		163572	F16N	VF-45	AD-623
163276	F16N	VF-126	NJ-602		163573	F16N	VF-45	AD-624
163277	F16N	VF-126	NJ-603		163574	F16N	VF-43	00
163278	TF16N	VF-43	05		163575	F16N	VF-43	01
163279	TF16N	NFWS	46		163576	F16N	VF-43	02
163280	TF16N	NFWS	47		163577	F16N	VF-43	03

FA18 HORNET

Serial	Type	Unit/Status	Code
160775	YFA18A	NASA	Dryden\spares
160776	YFA18A	derelict	NAS Memphis
160777	YFA18A	w.o.16.3.81	
160778	YFA18A	SoC 1992	
160779	YFA18A	SoC 1986	
160780	YFA18A	NASA	Dryden\N840NA
160781	FA18B	NASA	Dryden\N845NA
160782	FA18A	SoC 1987	
160783	FA18A	SoC 1984	
160784	FA18B	w.o.8.9.80	
160785	FA18A	SoC 1986	
161213	FA18A	w.o.7.10.88	
161214	FA18A	NASA	Dryden\N843NA
161215	FA18A	w.o.14.11.80	
161216	FA18A	NASA	Dryden\N841NA
161217	FA18B	NAWC\TPS	
161248	FA18A	w.o.2.5.83	
161249	FA18B	NAWC\TPS	01
161250	FA18A	NASA	Dryden\N842NA
161251	FA18A	NASA	Dryden
161353	FA18A	NAS	Pensacola
161354	FA18B	NAWC-WD	62
161355	FA18B	NASA	Dryden\N846NA
161356	FA18B	NAWC\TPS	03
161357	FA18B	NAWC-WD	
161358	FA18A	NAWC-AD	
161359	FA18A	NAWC-AD	
161360	FA18B	NAWC\TPS	04
161361	FA18A	NAWC-WD	65
161362	FA18A	NAWC-WD	
161363	FA18A	w.o.16.3.85	
161364	FA18A	w.o.11.2.91	
161365	FA18A	NAWC-WD	
161366	FA18A	stored	NAD N. Island
161367	FA18A	NAWC-AD	
161519	FA18A	NASA	
161520	FA18A	NASA	Dryden\N847NA
161521	FA18A	stored	NAS Moffett Field
161522	FA18A	w.o.12.2.87	
161523	FA18A		
161524	FA18A	w.o.23.1.90	
161525	FA18A	stored	NAD Jacksonville
161526	FA18A	NAWC	
161527	FA18A	NWEF	
161528	FA18A	SoC 1993	
161702	FA18A	NNAWC-AD	SATD SD-154
161703	FA18A	NAWC-WD	60
161704	FA18B	VFA-204	AF-415
161705	FA18A	NAWC-WD	
161706	FA18A	VMFA-134	MF-03
161707	FA18B	NSWC	17
161708	FA18A	VFA-303	ND-304
161709	FA18A	NAWC-AD	
161710	FA18A	VFA-303	ND-311
161711	FA18B	VFA-305	ND-414
161712	FA18A	VFA-305	ND-501
161713	FA18A	VFA-305	ND-410
161714	FA18B	NSWC	16
161715	FA18A	VFA-305	ND-404
161716	FA18A	VFA-303	ND-310
161717	FA18A	VFA-305	ND-403
161718	FA18A	VFA-305	ND-411
161719	FA18B	w.o.1.10.92	
161720	FA18A	VFA-305	ND-412
161721	FA18A	SoC 1993	
161722	FA18A	VFA-305	ND-400
161723	FA18B	VMFA-112	MA-00
161724	FA18A	VMFA-134	MF-07
161725	FA18A	VFA-204	AF-411
161726	FA18A	VFA-305	ND-402
161727	FA18B	w.o.20.5.87	
161728	FA18A	VAQ-34	04
161729	FA18A	SoC 1990	
161730	FA18A	VFA-303	ND-312
161731	FA18A	VFA-303	ND-313
161732	FA18A	VFA-303	ND-300
161733	FA18B	VAQ-34	00
161734	FA18A	VMFA-134	MF-00
161735	FA18A	VFA-303	ND-301
161736	FA18A	VAQ-34	05
161737	FA18A	VFA-303	ND-302
161738	FA18A	VFA-303	ND-306
161739	FA18A	VFA-305	ND-413
161740	FA18B	VAQ-34	03
161741	FA18A	w.o.17.11.83	
161742	FA18A	VMFA-134	MF-15
161743	FA18A	VFA-303	ND-303
161744	FA18A	NAWC-AD	
161745	FA18A	VMFA-134	MF-06
161746	FA18B	VFA-106	AD-360
161747	FA18A	VMFA-134	MF-12
161748	FA18A	VMFA-134	MF-10
161749	FA18A	VMFA-134	MF-05
161750	FA18A	VMFA-134	MF-02
161751	FA18A	SoC 1990	
161752	FA18A	VFA-303	ND-305
161753	FA18A	VMFA-134	MF-04
161754	FA18A	w.o.10.7.87	
161755	FA18A	VMFA-134	MF-11
161756	FA18A	VFA-305	ND-405
161757	FA18A	VMFA-134	MF-14
161758	FA18A	VMFA-134	MF-01
161759	FA18A	VFA-305	ND-406
161760	FA18A	VFA-303	ND-307
161761	FA18A	VFA-305	ND-407
161924	FA18B	VFA-106	AD-362
161925	FA18A	NAWC-AD	SATDSD-156
161926	FA18A	VFA-203	AF-300
161927	FA18A	VFA-203	AF-306
161928	FA18A		
161929	FA18A	VFA-203	AF-310
161930	FA18A	VFA-203	AF-313
161931	FA18A		
161932	FA18B	Blue Angels 7	
161933	FA18A	w.o.22.10.86	
161934	FA18A	VMFA-251	DW-
161935	FA18A		
161936	FA18A	VX-5	XE-31
161937	FA18A	NAWC-WD	104
161938	FA18B	VFA-125	NJ-542
161939	FA18A	VFA-203	AF-312
161940	FA18A	VFA-203	AF-314
161941	FA18A	NSWC	13
161942	FA18A	VFA-204	AF-401
161943	FA18B		
161944	FA18A	VFA-203	AF-311
161945	FA18A	Blue Angels 6	
161946	FA18A	VFA-204	AF-400
161947	FA18B	NAWC-AD	SATDSD-161
161948	FA18A	VFA-203	AF-302
161949	FA18A	VFA-203	AF-304
161950	FA18A	VFA-204	AF-407
161951	FA18A	VFA-204	AF-401
161952	FA18A	Blue Angels	4
161953	FA18A	VFA-203	AF-301
161954	FA18A	VFA-204	AF-410
161955	FA18A	Blue Angels	4
161956	FA18A	VFA-203	AF-305
161957	FA18A	Blue Angels	5
161958	FA18A		
161959	FA18A	stored	NAD N. Island
161960	FA18A	VMFA-321	MG-00

Serial	Type	Unit	Code
161961	FA18A	stored	NAS Pensacola
161962	FA18A	stored	NAD Jacksonville
161963	FA18A	VAQ-34	06
161964	FA18A	VFA-204	AF-403
161965	FA18A	VMFA-142	MB-03
161966	FA18A	w.o.19.6.84	
161967	FA18A	VFA-204	AF-406
161968	FA18A	VFA-204	AF-404
161969	FA18A		
161970	FA18A	VMFA-321	MG-01
161971	FA18A	w.o.7.2.87	
161972	FA18A	VMFA-142	MB-06
161973	FA18A	Blue Angels	1
161974	FA18A	VMFA-321	MG-02
161975	FA18A	NSWC	10
161976	FA18A		
161977	FA18A	VMFA-321	MG-03
161978	FA18A	Blue Angels	2
161979	FA18A	VMFA-312	DR-08
161980	FA18A	w.o.12.6.89	
161981	FA18A	VMFA-321	MG-05
161982	FA18A	VMFA-321	MG-06
161983	FA18A	VMFA-312	DR-10
161984	FA18A	Blue Angels	3
161985	FA18A	VMFA-142	MB-05
161986	FA18A		
161987	FA18A	w.o.4.6.87	
162394	FA18A	VMFA-323	WS-08
162395	FA18A	SoC 1992	
162396	FA18A	NAWC-WD	105
162397	FA18A	SoC 1990	
162398	FA18A	VMFA-323	WS-14
162399	FA18A	w.o.8.5.89	
162400	FA18A	VX-5	XE-33
162401	FA18A	VMFA-323	WS-00
162402	FA18B	VFA-125	NJ-543
162403	FA18A	VMFA-323	WS-04
162404	FA18A	w.o.17.7.85	
162405	FA18A	w.o.24.4.88	
162406	FA18A	VMFA-323	WS-
162407	FA18A	VMFA-314	VW-12
162408	FA18B	VMFAT-101	SH-106
162409	FA18A	VMFA-323	WS-03
162410	FA18A	stored	NAD N. Island
162411	FA18A	VFA-27	NL-404
162412	FA18A	VMFA-122	DC-01
162413	FA18B	w.o.8.9.87	
162414	FA18A		
162415	FA18A	VFA-127	NJ-01
162416	FA18A	VFA-127	NJ-02
162417	FA18A	VFA-127	NJ-03
162418	FA18A	NSWC	13
162419	FA18B	VMFAT-101	SH-111
162420	FA18A	NSWC	10
162421	FA18A	VFA-131	AG-401
162422	FA18A	damaged	NAS Lemoore
162423	FA18A	VMFA-323	WS-02
162424	FA18A	VMFA-115	VE-
162425	FA18A	NSWC	12
162426	FA18A	VFA-27	NL-
162427	FA18B	VX-5	XE-04
162428	FA18A	VMFA-142	MB-10
162429	FA18A	VMFA-323	WS-04
162430	FA18A	VMFA-142	MB-07
162431	FA18A	VX-4	XF-20
162432	FA18A	VFA-131	AG-403
162433	FA18A	VMFA-323	WS-05
162434	FA18A	VMFA-323	WS-11
162435	FA18A	VMFA-314	VW-07
162436	FA18A	VMFA-323	WS-09
162437	FA18A	stored	NAD Jacksonville
162438	FA18A	VFA-27	NL-
162439	FA18A	VMFA-323	WS-10
162440	FA18A	VFA-97	NL-303
162441	FA18A		
162442	FA18A	VFA-27	NL-402
162443	FA18A	VMFA-321	MG-09
162444	FA18A	NSWC	14
162445	FA18A		
162446	FA18A	VFA-127	NJ-04
162447	FA18A	w.o.8.1.86	
162448	FA18A	VFA-127	NJ-05
162449	FA18A		
162450	FA18A	w.o.8.12.87	
162451	FA18A	VFA-27	NL-403
162452	FA18A		
162453	FA18A	VFA-106	AD-333
162454	FA18A	VMFA-251	DW-04
162455	FA18A	VMFA-122	DC-05
162456	FA18A	VMFA-251	DW-06
162457	FA18A	VMFA-251	DW-07
162458	FA18A	VMFA-122	DC-08
162459	FA18A	VMFA-115	VE-
162460	FA18A	VMFA-122	DC-10
162461	FA18A	VMFA-115	VE-
162462	FA18A	stored	NAD Jacksonville
162463	FA18A	VMFA-142	MB-14
162464	FA18A	VMFA-321	MG-11
162465	FA18A	VMFA-251	DW-16
162466	FA18A	VMFA-321	MG-10
162467	FA18A	VMFA-323	WS-05
162468	FA18A	VMFA-321	MG-07
162469	FA18A	VMFA-314	VW-09
162470	FA18A	VMFA-314	VW-04
162471	FA18A	VMFA-314	VW-05
162472	FA18A	VMFA-323	WS-11
162473	FA18A	stored	NAD N. Island
162474	FA18A	NSWC	11
162475	FA18A	VMFA-314	VW-11
162476	FA18A	w.o.16.11.87	
162477	FA18A	w.o.9.10.87	
162826	FA18A	VX-5	XE-05
162827	FA18A	VFA-27	NL-
162828	FA18A	VMFAT-101	SH-220
162829	FA18A	VFA-97	NL-
162830	FA18A	VFA-113	NK-305
162831	FA18A	VFA-113	NK-
162832	FA18A	VFA-113	NK-
162833	FA18A	w.o.22.11.92	
162834	FA18A	VX-4	XF-22
162835	FA18A	VFA-125	NJ-515
162836	FA18B	VMFAT-101	SH-216
162837	FA18A	VFA-113	NK-
162838	FA18A	VFA-97	NL-300
162839	FA18A	VFA-151	NE-212
162840	FA18A	VFA-125	NJ-514
162841	FA18A	VFA-137	NE-403
162842	FA18B	VMFAT-101	SH-103
162843	FA18A	VFA-137	NE-402
162844	FA18A	NSWC	15
162845	FA18A	w.o.15.3.92	
162846	FA18A	VMFA-112	MA-05
162847	FA18A	w.o.14.4.87	
162848	FA18A	VFA-137	NE-400
162849	FA18A		
162850	FA18B	VFA-125	NJ-325
162851	FA18A	NAWC-WD	103
162852	FA18A	w.o.6.10.91	
162853	FA18A		
162854	FA18A	VFA-137	NE-404
162855	FA18A	w.o.14.4.87	
162856	FA18A	VFA-137	NE-405
162857	FA18B	VMFAT-101	SH-
162858	FA18A	w.o.31.3.90	
162859	FA18A	VFA-137	NE-406
162860	FA18A	VFA-136	AG-307
162861	FA18A	VFA-137	NE-407
162862	FA18A	VFA-106	AD-
162863	FA18A	VFA-137	NE-401
162864	FA18B	VFA-125	NJ-526
162865	FA18A	VFA-106	AD-330
162866	FA18A	VFA-137	NE-410
162867	FA18A		
162868	FA18A	VFA-106	AD-327
162869	FA18A	stored	NAD Jacksonville
162870	FA18B	VMFAT-101	SH-210
162871	FA18A	VFA-27	NL-411
162872	FA18A	VFA-106	AD-
162873	FA18A	VFA-27	NL-407
162874	FA18A		
162875	FA18A	VFA-137	NE-404
162876	FA18B	VFA-125	NJ-327
162877	FA18A	VFA-192	NF-
162878	FA18A	VFA-125	NJ-304
162879	FA18A	VFA-27	NL-403
162880	FA18A	VFA-27	NL-406

Serial	Type	Unit	Code
162881	FA18A	VFA-113	NK-
162882	FA18A	VMFA-112	MA-08
162883	FA18A	VFA-27	NL-405
162884	FA18A	VMFA-112	MA-09
162885	FA18B	VMFAT-101	SH-
162886	FA18A	VFA-125	NJ-301
162887	FA18A	VFA-113	NK-
162888	FA18A	VFA-151	NE-315
162889	FA18A	VFA-151	NE-
162890	FA18A	VFA-151	NE-312
162891	FA18A	VFA-151	NE-
162892	FA18A	stored	NAD N. Island
162893	FA18A	VFA-27	NL-
162894	FA18A	VFA-27	NL-405
162895	FA18A	SoC 1990	
162896	FA18A	VFA-151	NE-
162897	FA18A	VFA-195	NF-412
162898	FA18A	VX-5	XE-07
162899	FA18A	VFA-127	NJ-06
162900	FA18A	VFA-151	NE-
162901	FA18A	VMFA-323	WS-06
162902	FA18A	VFA-97	
162903	FA18A	VMFA-112	MA-12
162904	FA18A	VFA-151	NE-
162905	FA18A	VFA-97	
162906	FA18A	VFA-27	NL-400
162907	FA18A	stored	NAD N. Island
162908	FA18A	w.o.22.6.89	
162909	FA18A	VFA-113	NK-305
163092	FA18A	VMFAT-101	SH-22
163093	FA18A	NAWC-AD	
163094	FA18A	VMFA-314	NH-200
163095	FA18A	VFA-87	AJ-402
163096	FA18A	w.o.5.2.91	
163097	FA18A	VMFA-122	DC-01
163098	FA18A	stored	N. Island
163099	FA18A	VFA-106	AD-332
163100	FA18A	VFA-87	AJ-404
163101	FA18A	VFA-106	AD-326
163102	FA18A	VMFA-314	NH-202
163103	FA18A	VMFA-314	NH-212
163104	FA18B	VMFAT-101	SH-212
163105	FA18A	VFA-106	AD-300
163106	FA18A	VMFA-314	NH-204
163107	FA18A	VMFA-314	NH-205
163108	FA18A	VMFA-122	DC-02
163109	FA18A	w.o.28.1.88	
163110	FA18B	VMFAT-101	SH-215
163111	FA18A	VFA-106	AD-334
163112	FA18A	w.o.10.5.88	
163113	FA18A	VFA-15	AJ-300
163114	FA18A	VMFA-451	VM-
163115	FA18B	VFA-106	AD-364
163116	FA18A	stored	NAD Jacksonville
163117	FA18A	VFA-106	AD-301
163118	FA18A	VMFA-451	VM-00
163119	FA18A	VFA-106	AD-300
163120	FA18A	stored	NAD N. Island
163121	FA18A	w.o.24.1.91	
163122	FA18A	VMFA-314	NH-206
163123	FA18B	VFA-106	AD-370
163124	FA18A	VMFA-323	WS-16
163125	FA18A	w.o.28.5.92	
163126	FA18A	VMFAT-101	SH-224
163127	FA18A	VFA-106	AD-336
163128	FA18A	VMFA-314	NH-211
163129	FA18A	VFA-106	AD-
163130	FA18A	VMFAT-101	SH-226
163131	FA18A	VMFA-451	VM-14
163132	FA18A	VMFA-451	VM-
163133	FA18A	VMFA-451	VM-07
163134	FA18A	VMFA-451	VM-03
163135	FA18A	VFA-106	AD-304
163136	FA18A	w.o.28.1.88	
163137	FA18A	VMFA-451	VM-16
163138	FA18A	VMFA-314	NH-207
163139	FA18A	VFA-106	AD-335
163140	FA18A	VMFA-451	VM-01
163141	FA18A	VMFA-451	VM-06
163142	FA18A	VMFA-115	VE-
163143	FA18A	VMFA-323	WS-15
163144	FA18A	VMFA-314	NH-210
163145	FA18A	VMFA-451	VM-10
163146	FA18A	VMFA-451	VM-11
163147	FA18A	VMFA-115	VE-03
163148	FA18A	stored	NAD N. Island
163149	FA18A	VMFA-451	VM-00
163150	FA18A	VMFA-314	VW-08
163151	FA18A	VMFA-122	DC-04
163152	FA18A	VMFA-314	NH-201
163153	FA18A	VMFA-314	VW-
163154	FA18A	VMFA-115	VE-
163155	FA18A	VMFA-115	VE-05
163156	FA18A	VMFA-451	VM-12
163157	FA18A	VMFA-451	VM-05
163158	FA18A	VMFA-451	VM-13
163159	FA18A	VMFA-122	DC-06
163160	FA18A	VMFA-451	VM-02
163161	FA18A		
163162	FA18A	VMFA-115	VE-09
163163	FA18A	VMFA-122	DC-00
163164	FA18A	VMFA-122	DC-07
163165	FA18A	VMFA-115	VE-
163166	FA18A	VMFA-122	DC-
163167	FA18A	VMFA-312	DR-
163168	FA18A	VMFA-451	VM-13
163169	FA18A	VMFA-115	VE-
163170	FA18A	w.o.10.11.88	
163171	FA18A	VMFA-115	VE-
163172	FA18A	stored	NAD N. Island
163173	FA18A	VMFA-122	DC-10
163174	FA18A	VMFA-122	DC-11
163175	FA18A	VMFA-122	DC-
163427	FA18C	w.o.2.10.89	
163428	FA18C	w.o.15.2.90	
163429	FA18C	NAWC-WD	108
163430	FA18C	VFA-81	AA-411
163431	FA18C	VFA-86	AB-406
163432	FA18C	VFA-106	AD-311
163433	FA18C	VFA-106	AD-302
163434	FA18D	MDD	
163435	FA18C	stored	NAD N. Island
163436	FA18D	stored	NAD N. Island
163437	FA18C	VFA-86	AB-401
163438	FA18C	VFA-106	AD-
163439	FA18C	VFA-86	AB-402
163440	FA18C	VFA-82	AB-301
163441	FA18D	stored	NAD N. Island
163442	FA18C	VFA-82	AB-303
163443	FA18C	VFA-86	AB-400
163444	FA18C	VFA-83	AA-300
163445	FA18D	VFA-106	AD-352
163446	FA18C	VFA-106	AD-312
163447	FA18D	VMFAT-101	SH-241
163448	FA18C	VFA-82	AB-305
163449	FA18C	VFA-82	AB-306
163450	FA18C	VFA-86	AB-403
163451	FA18C	VFA-86	AB-404
163452	FA18D	VFA-106	AD-340
163453	FA18C	w.o.12.1.91	
163454	FA18D	w.o.4.3.92	
163455	FA18C	VFA-83	AA-304
163456	FA18C	VFA-82	AB-310
163457	FA18D	VFA-106	AD-347
163458	FA18C	VFA-86	AB-405
163459	FA18C	VFA-82	AB-311
163460	FA18D	VFA-106	AD-355
163461	FA18C	VFA-86	AB-410
163462	FA18C	VFA-106	AD-338
163463	FA18C	VFA-106	AD-
163464	FA18D	VFA-106	AD-000
163465	FA18C	VFA-106	AD-314
163466	FA18C	VFA-86	AB-411
163467	FA18C	VFA-82	AB-300
163468	FA18D	VFA-106	AD-346
163469	FA18C	VFA-106	AD-315
163470	FA18C	VFA-81	AA-400
163471	FA18C	VFA-82	AB-301
163472	FA18D	VFA-106	AD-343
163473	FA18C	VFA-86	AB-302
163474	FA18D	VFA-106	AD-344
163475	FA18C	w.o.6.10.90	
163476	FA18C	NAWC-AD	
163477	FA18C	VFA-81	AA-403
163478	FA18C	VFA-86	AB-412
163479	FA18D	VFA-106	AD-342

Serial	Type	Unit	Code
163480	FA18C	VFA-81	AA-402
163481	FA18C	VFA-83	AA-303
163482	FA18D	VFA-106	AD-341
163483	FA18C	NAWC-WD	
163484	FA18C	w.o.17.1.91	
163485	FA18C	VFA-83	AA-304
163486	FA18D	VFA-125	NJ-520
163487	FA18C	VFA-81	AA-404
163488	FA18D	VFA-125	NJ-522
163489	FA18C	VFA-83	AA-
163490	FA18C	NAWC-WD	66
163491	FA18C	VFA-81	AA-405
163492	FA18D	VFA-125	NJ-323
163493	FA18C	VFA-83	AA-306
163494	FA18C	VFA-82	AB-302
163495	FA18C	VFA-83	AA-307
163496	FA18C	VFA-125	NJ-
163497	FA18D	VFA-125	NJ-524
163498	FA18C	stored	NAD N. Island
163499	FA18C	VFA-83	AA-310
163500	FA18D	VFA-125	MJ-
163501	FA18D	VFA-125	NJ-552
163502	FA18C	VFA-81	AA-410
163503	FA18C	VFA-83	AA-302
163504	FA18C	VFA-125	NJ-306
163505	FA18C	VFA-81	AA-406
163506	FA18C	VFA-83	AA-301
163507	FA18D	VFA-125	NJ-
163508	FA18C	VFA-86	AB-405
163509	FA18C	VFA-106	AD-
163510	FA18D	VFA-125	NJ-334
163699	FA18C	VMFA-235	DB-00
163700	FA18D	VFA-125	NJ-535
163701	FA18C	VFA-125	NJ-505
163702	FA18C	VMFA-235	DB-01
163703	FA18C	VFA-195	NF-
163704	FA18C	w.o.8.5.89	
163705	FA18C	VFA-192	NF-300
163706	FA18C	NAWC	109
163707	FA18D	w.o.14.8.90	
163708	FA18C	VFA-195	NF-410
163709	FA18C	VMFA-235	DB-02
163710	FA18C	w.o.20.8.91	
163711	FA18C	SoC 1992	
163712	FA18C	w.o.13.5.92	
163713	FA18C	w.o.16.10.90	
163714	FA18C	VX-4	XF-11
163715	FA18C	VMFA-235	DB-03
163716	FA18C	VMFA-235	DB-05
163717	FA18C	stored	NAD N. Island
163718	FA18C	VFA-125	NJ-354
163719	FA18C	VFA-125	NJ-302
163720	FA18D	VFA-125	NJ-330
163721	FA18C	SoC 1992	
163722	FA18C	VMFA-235	DB-06
163723	FA18C	VMFA-235	DB-07
163724	FA18C	stored	NAD N. Island
163725	FA18C	VMFA-235	DB-10
163726	FA18C	VMFA-235	DB-11
163727	FA18C	VMFA-235	DB-12
163728	FA18C	w.o.8.3.91	
163729	FA18C	w.o.8.3.91	
163730	FA18C	VMFA-235	DB-14
163731	FA18C	VMFA-235	DB-15
163732	FA18C	w.o.23.2.92	
163733	FA18C	VMFA-232	WT-
163734	FA18D	VMFAT-101	SH-240
163735	FA18C	VMFA-212	WD-01
163736	FA18C	VMFA-212	WD-02
163737	FA18C	VMFA-212	WD-03
163738	FA18C	VMFA-212	WD-04
163739	FA18C	w.o.9.5.89	
163740	FA18C	VFA-192	NF-
163741	FA18C	VFA-192	NF-312
163742	FA18C	VMFA-212	WD-05
163743	FA18C	VMFA-212	WD-06
163744	FA18C	VFA-195	NF-
163745	FA18C	VMFA-212	WD-07
163746	FA18C	VFA-195	NF-404
163747	FA18C	VMFA-212	WD-10
163748	FA18C	VFA-192	NF-305
163749	FA18D	VMFAT-101	SH-243
163750	FA18C	VMFA-212	WD-11
163751	FA18C	VMFA-212	WD-12
163752	FA18C	VMFA-212	WD-13
163753	FA18C	VMFA-212	WD-14
163754	FA18C	VFA-192	NF-301
163755	FA18C	VFA-192	NF-302
163756	FA18C	damaged	NAS Atsugi
163757	FA18C	VFA-125	NJ-
163758	FA18C	VFA-195	NF-400
163759	FA18C	VFA-192	NF-
163760	FA18C	VFA-195	NF-
163761	FA18C	VFA-195	NF-403
163762	FA18C	VFA-195	NF-
163763	FA18D	VFA-125	NJ-340
163764	FA18C	VFA-192	NF-307
163765	FA18C	VFA-195	NF-405
163766	FA18C	VFA-192	NF-306
163767	FA18C	VFA-195	NF-406
163768	FA18C	VFA-192	NF-310
163769	FA18C	VMFA-235	DB-00
163770	FA18C	VMFA-235	DB-01
163771	FA18D	VMFAT-101	SH-244
163772	FA18C	VMFA-235	DB-02
163773	FA18C	VMFA-235	DB-03
163774	FA18C	w.o.9.10.91	
163775	FA18C	VMFA-235	DB-05
163776	FA18C	VMFA-235	DB-06
163777	FA18C	VMFA-235	DB-07
163778	FA18D	VMFAT-101	SH-245
163779	FA18C	VMFA-235	DB-14
163780	FA18C	VMFA-235	DB-10
163781	FA18C	VMFA-235	DB-11
163782	FA18C	VMFA-235	DB-12
163985	FA18C	NAWC-AD	
163986	FA18D	NAWC-AD	
163987	FA18C	NAWC-WD	107
163988	FA18C	VX-5	XE-01
163989	FA18D	NAWC-WD	112
163990	FA18C	VX-5	XE-02
163991	FA18D	VX-4	XF-14
163992	FA18C	VFA-146	NG-301
163993	FA18C	VFA-125	NJ-512
163994	FA18D	VX-5	XE-03
163995	FA18C	VFA-22	NH-304
163996	FA18C	VFA-22	NH-307
163997	FA18D	VMFAT-101	SH-150
163998	FA18C	VFA-147	NG-401
163999	FA18C	VFA-146	NG-302
164000	FA18C	w.o.12.7.91	
164001	FA18D	VMFAT-101	SH-257
164002	FA18C	VFA-146	NG-300
164003	FA18C	VFA-147	NG-404
164004	FA18C	w.o.24.8.91	
164005	FA18D	VFA-125	NJ-341
164006	FA18C	VFA-146	NG-306
164007	FA18C	VFA-147	NG-410
164008	FA18C	VFA-146	NG-303
164009	FA18D	VMFAT-101	SH-152
164010	FA18C	VFA-147	NG-402
164011	FA18D	VFA-125	NJ-321
164012	FA18C	VFA-146	NG-304
164013	FA18C	VFA-146	NG-405
164014	FA18D	VMFAT-101	SH-254
164015	FA18C	w.o.	
164016	FA18C	VFA-147	NG-406
164017	FA18D	VMFAT-101	SH-253
164018	FA18C	VFA-146	NG-305
164019	FA18D	VMFAT-101	SH-251
164020	FA18C	VFA-147	NG-407
164021	FA18C	VFA-146	NG-307
164022	FA18D	VMFA-242	DT-00
164023	FA18C	VFA-146	NG-
164024	FA18D	VMFA-242	DT-01
164025	FA18C	VFA-146	NG-311
164026	FA18D	VMFA-242	DT-02
164027	FA18C	VFA-125	NJ-311
164028	FA18D	VMFAT-101	SH-260
164029	FA18C	VFA-147	NG-411
164030	FA18C	VFA-147	NG-400
164031	FA18C	w.o.7.5.92	
164032	FA18D	VMFA-242	DT-04
164033	FA18C	VFA-147	NG-403
164034	FA18C	VFA-125	NJ-510
164035	FA18D	w.o.29.5.92	

Serial	Type	Unit	Code
164036	FA18C	VFA-125	NJ-351
164037	FA18C	VFA-125	NJ-
164038	FA18D	VFA-106	AD-351
164039	FA18C	VFA-22	NH-301
164040	FA18D	SoC 1992	
164041	FA18C	VFA-22	NH-302
164042	FA18C	VFA-94	NH-400
164043	FA18D	VFA-106	AD-355
164044	FA18C	VFA-22	NH-303
164045	FA18C	VFA-94	NH-401
164046	FA18D	VMFA-242	DT-05
164047	FA18C	VFA-94	NH-402
164048	FA18C	VFA-94	NH-403
164049	FA18D	VMFA-242	DT-06
164050	FA18C	VFA-94	NH-407
164051	FA18D	VMFA-242	DT-07
164052	FA18C	VFA-94	NH-404
164053	FA18D	VMFAT-101	SH-264
164054	FA18C	VFA-22	NH-305
164055	FA18C	VFA-94	NH-405
164056	FA18D	VMFAT-101	SH-263
164057	FA18C	VFA-22	NH-306
164058	FA18D	VFA-125	NJ-544
164059	FA18C	VFA-94	NH-406
164060	FA18C	VFA-22	NH-300
164061	FA18D	VMFA-242	DT-12
164062	FA18C	VFA-22	NH-310
164063	FA18C	VFA-22	NH-311
164064	FA18D	VMFAT-101	SH-271
164065	FA18C	w.o.4.12.91	
164066	FA18C	VFA-94	NH-410
164067	FA18C	VFA-94	NH-411
164068	FA18D	VFA-125	NJ-
164196	FA18D	w.o.15.5.92	
164197	FA18C	VFA-105	AC-406
164198	FA18D	VFA-106	AD-354
164199	FA18C	VFA-37	AC-311
164200	FA18C	VFA-105	AC-400
164201	FA18C	VFA-131	AG-401
164202	FA18C	VFA-136	AG-303
164203	FA18D	VMFA-242	DT-03
164204	FA18C	VFA-131	AG-404
164205	FA18C	VFA-131	AG-405
164206	FA18C	VFA-136	AG-307
164207	FA18D	VMFAT-101	SH-261
164208	FA18C	VFA-136	AG-304
164209	FA18C	VFA-136	AG-300
164210	FA18C	VFA-131	AG-410
164211	FA18D	VMFAT-101	SH-262
164212	FA18C	VFA-131	AG-400
164213	FA18C	VFA-136	AG-310
164214	FA18C	VFA-136	AG 311
164215	FA18C	VFA-37	AC-302
164216	FA18D	VMFAT-101	SH-263
164217	FA18C	VFA-136	AG-302
164218	FA18C	VFA-105	AC-407
164219	FA18D	VMFA-242	DT-10
164220	FA18C	VMFAT-101	SH-204
164221	FA18C	VX-4	XF-16
164222	FA18C	VFA-131	AG-403
164223	FA18C	VFA-136	AG-306
164224	FA18D	VMFA-121	VK-00
164225	FA18C	VFA-131	AG-406
164226	FA18C	VFA-131	AG-402
164227	FA18C	VFA-136	AG-
164228	FA18D	VMFA-121	VK-01
164229	FA18C	VFA-131	AG-407
164230	FA18C	VFA-136	AG-301
164231	FA18C	VFA-131	AG-411
164232	FA18C	VFA-37	AC-303
164233	FA18D	VMFA-121	VK-02
164234	FA18C	VFA-105	AC-412
164235	FA18C	VFA-37	AC-306
164236	FA18C	VFA-105	AC-411
164237	FA18D	VMFA-121	VK-03
164238	FA18C	VFA-37	AC-305
164239	FA18C	VFA-105	AC-403
164240	FA18C	VFA-37	AC-300
164241	FA18D	VMFA-121	VK-04
164242	FA18C	VFA-105	AC-402
164243	FA18C	VFA-37	AC-312
164244	FA18C	VFA-105	AC-404
164245	FA18D	VMFA-121	VK-05
164246	FA18C	VFA-37	AC-310
164247	FA18C	VFA-105	AC-405
164248	FA18C	w.o.12.8.91	
164249	FA18D	VMFA-121	VK-06
164250	FA18C	VFA-87	AJ-412
164251	FA18C	VFA-37	AC-301
164252	FA18C	VFA-37	AC-307
164253	FA18C	VFA-37	AC-304
164254	FA18D	VMFA-121	VK-07
164255	FA18C	VFA-105	AC-410
164256	FA18C	VMFAT-101	SH-100
164257	FA18C	VMFAT-101	SH-101
164258	FA18C	VMFAT-101	SH-136
164259	FA18D	VMFA-121	VK-10
164260	FA18C	w.o.24.7.92	
164261	FA18C	VFA-105	AC-401
164262	FA18C	VMFAT-101	SH-137
164263	FA18D	VMFA-121	VK-11
164264	FA18C	VMFA-312	DR-01
164265	FA18C	VMFA-312	DR-02
164266	FA18C	VMFAT-101	SH-202
164267	FA18D	VMFA-121	VK-12
164268	FA18C	VMFA-312	DR-03
164269	FA18C	VMFA-312	DR-04
164270	FA18C	VMFA-312	DR-05
164271	FA18C	VMFA-312	DR-06
164272	FA18C	VMFA-121	VK-13
164273	FA18C	VMFA-312	DR-07
164274	FA18C	VMFA-312	DR-08
164275	FA18C	VMFA-312	DR-09
164276	FA18C	VMFA-312	DR-10
164277	FA18C	VMFA-312	DR-11
164278	FA18C	VMFA-312	DR-12
164279	FA18D	NAWC-WD	113
164627	FA18C	NAWC-AD	
164628	FA18C	NAWC-WD	111
164629	FA18C	VX-4	XF-10
164630	FA18C	VX-5	XE-00
164631	FA18C	VFA-15	AJ-301
164632	FA18C	VFA-87	AJ-400
164633	FA18C	VFA-25	NK-400
164634	FA18C	VFA-113	NK-306
164635	FA18C	VFA-25	NK-401
164636	FA18C	VFA-113	NK-301
164637	FA18C	VFA-25	NK-402
164638	FA18C	VFA-113	NK-304
164639	FA18C	VFA-25	NK-403
164640	FA18C	VFA-113	NK-300
164641	FA18C	VFA-113	NK-305
164642	FA18C	VFA-25	NK-404
164643	FA18C	VFA-15	AJ-302
164644	FA18C	VFA-87	AJ-406
164645	FA18C	VFA-25	NK-405
164646	FA18C	VFA-15	AJ-303
164647	FA18C	VFA-87	AJ-402
164648	FA18C	VFA-113	NK-302
164649	FA18D	damaged	NAD N. Island
164650	FA18D	VMFA-225	CE-01
164651	FA18D	VMFA-225	CE-02
164652	FA18D	VMFA-225	CE-03
164653	FA18D	VMFA-225	CE-04
164654	FA18C	VFA-25	NK-406
164655	FA18C	VFA-15	AJ-304
164656	FA18D	VMFA-225	CE-05
164657	FA18C	VFA-87	AJ-405
164658	FA18C	VFA-113	NK-
164659	FA18D	VMFA-225	CE-06
164660	FA18C	VFA-25	NK-407
164661	FA18C	VFA-15	AJ-
164662	FA18D	VMFA-225	CE-07
164663	FA18C	VFA-87	AJ-403
164664	FA18C	VFA-113	NK-307
164665	FA18D	VMFA-225	CE-
164666	FA18C	VFA-15	AJ-
164667	FA18D	VMFA-225	CE-
164668	FA18C	VFA-87	AJ-407
164669	FA18C	VFA-15	AJ-
164670	FA18D		
164671	FA18C		
164672	FA18D	VMFA-225	CE-11
164673	FA18C	VFA-15	AJ-310
164674	FA18D	VMFA-242	DT-14
164675	FA18C		

164676	FA18C	VFA-25	NK-410
164677	FA18D	VMFAT-101	SH-272
164678	FA18C	VFA-106	AD-307
164679	FA18D	VMFAT-101	SH-270
164680	FA18C	VFA-113	NK-310
164681	FA18C	VFA-25	NK-410
164682	FA18C		
164683	FA18D	VMFAT-101	SH-273
164684	FA18C	w.o.9.9.92	
164685	FA18D	VMFAT-101	SH-274
164686	FA18C		
164687	FA18C		
164688	FA18D	VMFAT-101	SH-275
164689	FA18C	VFA-15	AJ-300
164690	FA18D		
164691	FA18C		
164692	FA18D	VMFAT-101	SH-276
164693	FA18C	VFA-137	NE-405
164694	FA18D	VMFA-533	ED-400
164695	FA18C	VFA-151	NE-301
164696	FA18C	NAWC-AD	104
164697	FA18C	VFA-151	NE-302
164698	FA18C	VFA-137	NE-401
164699	FA18D	VMFA-533	ED-401
164700	FA18C	VFA-151	NE-303
164701	FA18C	VFA-137	NE-402
164702	FA18D		
164703	FA18C	VFA-151	NE-300
164704	FA18C	VFA-137	NE-403
164705	FA18D		
164706	FA18C	VFA-151	NE-301
164707	FA18C		
164708	FA18C	VFA-151	NE-304
164709	FA18C	VFA-137	NE-404
164710	FA18C	VFA-151	NE-305
164711	FA18D		
164712	FA18C	VFA-137	NE-400
164713	FA18C	VFA-151	NE-306
164714	FA18D		
164715	FA18C	VFA-137	NE-406
164716	FA18C		
164717	FA18D		
164718	FA18C	VFA-151	NE-307
164719	FA18C		
164720	FA18C		
164721	FA18C		
164722	FA18C	VMFA-323	WS-01
164723	FA18D		
164724	FA18C		
164725	FA18C		
164726	FA18D		
164727	FA18C		
164728	FA18C		
164729	FA18D		
164730	FA18C		
164731	FA18C		
164732	FA18C		
164733	FA18C		
164734	FA18C		
164735	FA18D		
164736	FA18C		
164737	FA18C		
164738	FA18D		
164739	FA18C		
164740	FA18C		
164865	FA18C		
164866	FA18D		
164867	FA18C		
164868	FA18C		
164869	FA18D		
164870	FA18C		
164871	FA18C		
164872	FA18C		
164873	FA18D		
164874	FA18C		
164875	FA18C		
164876	FA18D		
164877	FA18C		
164878	FA18C		
164879	FA18C		
164880	FA18D		
164881	FA18C		
164882	FA18C		
164883	FA18D		
164884	FA18C		
164885	FA18C		
164886	FA18D		
164887	FA18C		
164888	FA18C		
164889	FA18D		
164890	FA18C		
164891	FA18C		
164892	FA18C		
164893	FA18D		
164894	FA18C		
164895	FA18C		
164896	FA18C		
164897	FA18D		
164898	FA18C		
164899	FA18C		
164900	FA18C		
164901	FA18D		
164902	FA18C		
164903	FA18C		
164904	FA18D		
164905	FA18C		
164906	FA18C		
164907	FA18D		
164908	FA18C		
164909	FA18C		
164910	FA18D		
164911	FA18C		
164912	FA18C		
164945	FA18C\D		
164946	FA18C\D		
164947	FA18C\D		
164948	FA18C\D		
164949	FA18C\D		
164950	FA18C\D		
164951	FA18C\D		
164952	FA18C\D		
164953	FA18C\D		
164954	FA18C\D		
164955	FA18C\D		
164956	FA18C\D		
164957	FA18C\D		
164958	FA18C\D		
164959	FA18C\D		
164960	FA18C\D		
164961	FA18C\D		
164962	FA18C\D		
164963	FA18C\D		
164964	FA18C\D		
164965	FA18C\D		
164966	FA18C\D		
164967	FA18C\D		
164968	FA18C\D		
164969	FA18C\D		
164970	FA18C\D		
164971	FA18C\D		
164972	FA18C\D		
164973	FA18C\D		
164974	FA18C\D		
164975	FA18C\D		
164976	FA18C\D		
164977	FA18C\D		
164978	FA18C\D		
164979	FA18C\D		
164980	FA18C\D		
164981	FA18C\D		
164982	FA18C\D		
164983	FA18C\D		
164984	FA18C\D		
164985	FA18C\D		
164986	FA18C\D		
164987	FA18C\D		
164988	FA18C\D		
164989	FA18C\D		
164990	FA18C\D		
164991	FA18C\D		
164992	FA18C\D		

F21 KFIR

999703	F21A	returned to IDF\AF
999708	F21A	returned to IDF\AF
999709	F21A	returned to IDF\AF
999710	F21A	returned to IDF\AF
999716	F21A	returned to IDF\AF
999724	F21A	returned to IDF\AF
999725	F21A	returned to IDF\AF
999726	F21A	returned to IDF\AF
999727	F21A	returned to IDF\AF
999728	F21A	returned to IDF\AF
999731	F21A	returned to IDF\AF
999732	F21A	returned to IDF\AF
999734	F21A	returned to IDF\AF
999735	F21A	returned to IDF\AF
999739	F21A	returned to IDF\AF
999747	F21A	returned to IDF\AF
999749	F21A	returned to IDF\AF
999750	F21A	returned to IDF\AF
999764	F21A	returned to IDF\AF
999785	F21A	returned to IDF\AF
999786	F21A	returned to IDF\AF
999787	F21A	returned to IDF\AF
999791	F21A	returned to IDF\AF
999794	F21A	returned to IDF\AF

F86 SABRE

513278	QF86F	stored	China Lake
524450	QF86F	NAWC-WD	832
531403	QF86F	NAWC-WD	
531409	QF86F	NAWC-WD	409
553432	QF86F	NAWC-WD	817
553822	QF86F	stored	China Lake
553823	QF86F	NAWC-WD	812
553824	QF86F		
553829	QF86F	stored	China Lake
553838	QF86F	NAWC-WD	818
553846	QF86F	NAWC-WD	824
553852	QF86F	NAWC-WD	820
553863	QF86F		
553864	QF86F	stored	China Lake
553865	QF86F	NAWC-WD	
553868	QF86F	stored	China Lake
553869	QF86F		
553872	QF86F	NAWC-WD	
553874	QF86F	stored	China Lake
553875	QF86F	NAWC-WD	31
553878	QF86F	NAWC-WD	885
553881	QF86F		
553882	QF86F		
553883	QF86F	NAWC-WD	806
553890	QF86F		
553891	F86F	stored	China Lake
553892	QF86F	NAWC-WD	
553895	QF86F	stored	China Lake
553898	QF86F	NAWC-WD	824
553900	QF86F	NAWC-WD	801
553903	QF86F	NAWC-WD	
553905	QF86F	NAWC-WD	829\stored
553906	QF86F	stored	China Lake
553908	QF86F		
553912	QF86F	NAWC-WD	
553913	QF86F	NAWC-WD	826\stored
553915	QF86F	stored	China Lake
553916	QF86F		
553919	QF86F	NAWC-WD	
553926	QF86F	NAWC-WD	830
553932	QF86F	NAWC-WD	880
553935	QF86F	NAWC-WD	802
553936	QF86F	NAWC-WD	863
553937	QF86F		
553939	QF86F	NAWC-WD	
553942	QF86F	stored	China Lake
553945	QF86F	NAWC-WD	834\stored
553948	F86F	stored	China Lake
553973	F86F	stored	China Lake
553998	QF86F	stored	China Lake
555011	QF86F	NAWC-WD	811
555014	F86F	stored	China Lake
555017	QF86F	NAWC-WD	808
555019	F86F	stored	China Lake
555021	QF86F		
555026	F86F	stored	China Lake
555032	F86F	stored	China Lake
555035	F86F	stored	China Lake
555047	QF86F	NAWC-WD	840
555048	QF86F	NAWC-WD	818
555052	QF86F	NAWC-WD	820
555053	QF86F	stored	China Lake
555056	QF86F	NAWC-WD	844
555057	QF86F		
555058	QF86F		
555069	QF86F		
555072	QF86F	NAWC-WD	898
555078	QF86F	NAWC-WD	882
555082	QF86F	NAWC-WD	
555084	F86F	stored	China Lake
555087	QF86F	NAWC-WD	35
555091	QF86F	NAWC-WD	899\stored
555095	QF86F		
555097	QF86F	NAWC-WD	816
555098	QF86F	NAWC-WD	867
555099	QF86F	NAWC-WD	871
555101	QF86F		
555102	QF86F	NAWC-WD	33
555103	QF86F		
555110	QF86F	NAWC-WD	849
555111	QF86F	NAWC-WD	821
555112	QF86F	NAWC-WD	813
555114	QF86F	NAWC-WD	
556412	QF86F	NAWC-WD	817
562781	QF86F		
562782	QF86F	NAWC-WD	833\stored

562783	QF86F	NAWC-WD			562858	QF86F	NAWC-WD	814
562784	QF86F	NAWC-WD	32		562865	QF86F	NAWC-WD	
562786	QF86F	NAWC-WD			562873	QF86F		
562787	QF86F	NAWC-WD			562874	QF86F	stored	China Lake
562792	QF86F	NAWC-WD	893		562875	F86F	stored	China Lake
562795	QF86F				562879	QF86F	NAWC-WD	877
562797	QF86F	NAWC-WD			562884	QF86F	NAWC-WD	828\stored
562811	QF86F	NAWC-WD			562896	QF86F	NAWC-WD	842
562813	QF86F	NAWC-WD	879		576346	QF86F	NAWC-WD	840
562814	QF86F	NAWC-WD			576352	QF86F	NAWC-WD	
562815	QF86F				576363	QF86F	NAWC-WD	835
562818	QF86F	NAWC-WD	836		576388	QF86F	NAWC-WD	839
562819	QF86F				576414	QF86F	NAWC-WD	30
562823	QF86F				576420	QF86F		
562825	QF86F	NAWC-WD	875		576422	QF86F	NAWC-WD	832
562826	QF86F	NAWC-WD	39		576424	QF86F	NAWC-WD	
562827	QF86F	NAWC-WD	881		576425	QF86F	NAWC-WD	825
562829	QF86F				576436	QF86F		
562830	QF86F	NAWC-WD	855		576440	QF86F	NAWC-WD	
562831	QF86F	NAWC-WD	828		576442	QF86F		
562832	QF86F	NAWC-WD			576444	QF86F		
562836	QF86F	NAWC-WD	861		576445	QF86F	stored	China Lake
562837	QF86F	NAWC-WD	878		576447	QF86F	NAWC-WD	837\stored
562838	QF86F	stored	China Lake		576449	QF86F	NAWC-WD	894
562839	QF86F	NAWC-WD	34		576450	QF86F	NAWC-WD	890
562840	QF86F	stored	China Lake		576457	QF86F		
562842	QF86F	NAWC-WD	853		576459	QF86F		
562845	QF86F				626416	RF86F	NAWC-WD	
562846	QF86F	stored	China Lake		627479	QF86F	NAWC-WD	830
562848	QF86F				727709	QF86F	NAWC-WD	844
562849	QF86F	NAWC-WD	865		727711	QF86F	NAWC-WD	891
562852	QF86F				827806	QF86F	NAWC-WD	869 (56-2806)
562854	QF86F	NAWC-WD	892		158436	F86H	NAWC-WD	01 ex USAF
562855	QF86F	NAWC-WD	838		158437	F86H	NAWC-WD	02 ex USAF

F111

151970	F111B				152714	F111B	preserved	McClellan AFB
151971	F111B				152715	F111B	stored	China Lake
151972	F111B				152716	F111B	
151973	F111B				152717	F111B	
151974	F111B						

AH1 SEA-COBRA

c\n   USN procurement 26001 to 26049

note: serials Bu157204 to 157241 were allotted to 38 ex US Army AH1G but were not used. Five AH1S were transferred also, but no BuA serals were allotted.

66-15345	AH1S	NAWC\TPS	53		68-15112	AH1G	MASDC	
67-15850	AH1G				68-15113	AH1G	MASDC	
68-15037	AH1G				68-15134	AH1G	MASDC	
68-15038	AH1G	MASDC			68-15140	AH1G	preserved	NAS Atlanta
68-15039	AH1G				68-15165	AH1G		
68-15045	AH1G	NAWC\TPS			68-15170	AH1G		
68-15046	AH1G	MASDC-			68-15190	AH1G		
68-15072	AH1G				68-15194	AH1G		
68-15073	AH1G				68-15198	AH1G		
68-15079	AH1G				68-15213	AH1G		
68-15080	AH1G				68-17023	AH1G	MASDC	
68-15085	AH1G	NAWC\TPS			68-17027	AH1G		
68-15104	AH1G				68-17041	AH1G		
68-15105	AH1G	MASDC			68-17045	AH1G		

Serial	Type	Unit	Code
68-17049	AH1G	NAWC\TPS	49
68-17062	AH1G		
68-17066	AH1G	MASDC	
68-17070	AH1G	MASDC	
68-17082	AH1G		
68-17086	AH1G		
68-17090	AH1G		
68-17101	AH1G		
68-17105	AH1G	MASDC	
68-17108	AH1G		
68-17113	AH1G		
69-16430	AH1S	NAWC\TPS	54
71-20983	AH1S	NAWC\TPS	55
71-22791	AH1S	NAWC\TPS	51
71-22792	AH1S	NAWC\TPS	52
157757	AH1J		
157758	AH1J	HMT-303	QT-420
157759	AH1J	AMARC	7H222
157760	AH1J	AMARC	7H216
157761	AH1J	AMARC	7H214
157762	AH1J	HMT-303	QT-
157763	AH1J		
157764	AH1J		
157765	AH1J	NARF	Pensacola
157766	AH1J		
157767	AH1J		
157768	AH1J	AMARC	7H219
157769	AH1J	HMT-303	QT-434
157770	AH1J	AMARC	7H213
157771	AH1J	NATTC	
157772	AH1J		
157773	AH1J		
157774	AH1J		
157775	AH1J	AMARC	7H217
157776	AH1J		
157777	AH1J	MASDC	
157778	AH1J	AMARC	7H221
157779	AH1J		
157780	AH1J		
157781	AH1J	HMLA-367	VT-215
157782	AH1J		
157783	AH1J	HMLA-369	SM-305
157784	AH1J	HMT-303	QT-404
157785	AH1J	AMARC	7H212
157786	AH1J	w.o.11.3.84	
157787	AH1J	AMARC	7H220
157788	AH1J	HMT-303	QT-
157789	AH1J		
157790	AH1J	AMARC	7H218
157791	AH1J		
157792	AH1J	w.o.29.4./6	
157793	AH1J	w.o.2.2.91	
157794	AH1J		
157795	AH1J		
157796	AH1J	HMT-303	QT-426
157797	AH1J	HMT-303	QT-432
157798	AH1J	AMARC	7H215
157799	AH1J	VX-5	XE-40
157800	AH1J	HMLA-269	HF-800
157801	AH1J	AMARC	7H225
157802	AH1J		
157803	AH1J	HMLA-269	HF-803
157804	AH1J	AMARC	7H223
157805	AH1J	AMARC	7H224
159210	AH1J	HMA-773	MP-700
159211	AH1J	HMA-773	MP-701
159212	AH1J		
159213	AH1J		
159214	AH1T	AMARC	7H231
159215	AH1J	HMA-773	MP-703
159216	AH1J	stored	MCAS Cherry Pt.
159217	AH1J		
159218	AH1J	HMA-773	MP-704
159219	AH1J	HMA-775	WR-705
159220	AH1J	HMA-773	MP-706
159221	AH1J	AMARC	7H233
159222	AH1J	HMA-773	MP-710
159223	AH1J	HMA-773	MP-711
159224	AH1T	HMA-773	MP-713
159225	AH1J	AMARC	7H232
159226	AH1J	HMLA-267	UV-306
159227	AH1J		
159228	YAH1W	VX-5	XE-43
159229	AH1J		
160105	AH1T		
160106	AH1T		
160107	AH1W	HMLA-269	HF-00
160108	AH1W	HMM-164	YT-43
160109	AH1T	w.o.16.9.85	
160110	AH1T		
160111	AH1W	HMLA-167	TV-22
160112	AH1W	HMLA-269	HF-112
160113	AH1W	HMLA-367	VT-47
160114	AH1T	w.o.22.5.81	
160742	AH1W	HMLA-167	TV-34
160743	AH1W	MEU	41
160744	AH1W	HMM-261	EM-32
160745	AH1W	HMLA-167	TV-26
160746	AH1T		
160747	AH1W	HMLA-269	HF-
160748	AH1T		
160797	AH1T	w.o.14.11.86	
160798	AH1T		
160799	AH1W	HMM-164	YT-464
160800	AH1W	HMLA-369	-
160801	AH1W	HMLA-169	SN-316
160802	AH1T		
160803	AH1W	HMM-161	YR-120
160804	AH1W	HMM-261	EM-33
160805	AH1W	HMM-265	EP-31
160806	AH1W	HMM-261	EM-34
160807	AH1T	w.o.22.11.89	
160808	AH1T	w.o.	
160809	AH1W	HMLA-369	SM-134
160810	AH1W	HMLA-269	HF-14
160811	AH1W	HMLA-167	TV-36
160812	AH1W	HMLA-269	HF-812
160813	AH1W	MEU	42
160814	AH1W	HMLA-269	HF-05
160815	AH1W	HMLA-269	HF-06
160816	AH1W	HMLA-167	TV-34
160817	AH1W	HMLA-269	HF-07
160818	AH1W	HMM-262	ET-33
160819	AH1W	HMLA-267	UV-
160820	AH1W	HMM-162	YS-23
160821	AH1W	HMLA-269	HF-821
160822	AH1W	HMLA-367	VT-122
160823	AH1T		
160824	AH1T		
160825	AH1W	MEU	43
160826	AH1W	HMLA-269	HF-11
161015	AH1W	HMLA-169	SN-123
161016	AH1W	MEU	44
161017	AH1W	HMLA-269	HF-
161018	AH1T	w.o.18.4.88	
161019	AH1W	HMLA-167	TV-35
161020	AH1W	HMLA-169	SN-306
161021	AH1W	HMM-261	EM-31
161022	AH1W	HMM-261	EM-32
162532	AH1W	NAWC-AD	RWATD
162533	AH1W	HMLA-369	SM-
162534	AH1W	HMLA-369	SM-
162535	AH1W	HMLA-169	SN-120
162536	AH1W	HMT-303	QT-
162537	AH1W	HMT-303	QT-435
162538	AH1W	HMLA-369	SM-
162539	AH1W	HMT-303	QT-434
162540	AH1W	w.o.20.6.89	
162541	AH1W		
162542	AH1W	HMLA-169	SN-303
162543	AH1W	HMT-303	QT-427
162544	AH1W	HMT-303	QT-440
162545	AH1W		
162546	AH1W	HMLA-367	VT-54
162547	AH1W		
162548	AH1W	HMT-303	QT-433
162549	AH1W	HMT-303	QT-445
162550	AH1W	HMLA-169	SN-310
162551	AH1W	HMLA-369	SM-311
162552	AH1W	HMLA-367	VT-55
162553	AH1W		
162554	AH1W	HMLA-267	UV-40
162555	AH1W	HMLA-169	SN-311
162556	AH1W	HMM-268	YQ-32
162557	AH1W	HMT-303	QT-444
162558	AH1W	HMT-303	QT-441

162559	AH1W				164586	AH1W	HMA-775	WR-722
162560	AH1W	HMLA-369	-		164587	AH1W	HMA-775	WR-723
162561	AH1W	HMLA-169	SN-305		164588	AH1W	HMA-775	WR-724
162562	AH1W	HMLA-267	UV-		164589	AH1W	HMA-775	WR-725
162563	AH1W	HMLA-169	SN-317		164590	AH1W	HMA-775	WR-726
162564	AH1W	HMLA-367	VT-42		164591	AH1W	HMA-775	WR-727
162565	AH1W	HMLA-367	VT-50		164592	AH1W		
162566	AH1W	HMLA-367	VT-51		164593	AH1W		
162567	AH1W	HMLA-367	VT-52		164594	AH1W		
162568	AH1W	HMLA-367	VT-53		164595	AH1W		
162569	AH1W	HMLA-367	VT-43		164596	AH1W		
162570	AH1W	HMT-303	QT-446		164913	AH1W		
162571	AH1W	HMLA-369	-		164914	AH1W		
162572	AH1W				164915	AH1W		
162573	AH1W	HMT-303	QT-437		164916	AH1W		
162574	AH1W				164917	AH1W		
162575	AH1W	HMT-303	QT-447		164918	AH1W		
163921	AH1W				164919	AH1W		
163922	AH1W				164920	AH1W		
163923	AH1W				164921	AH1W		
163924	AH1W	HMLA-167	TV-24		164922	AH1W		
163925	AH1W				164923	AH1W		
163926	AH1W				164924	AH1W		
163927	AH1W				164925	AH1W		
163928	AH1W				164926	AH1W		
163929	AH1W	HMLA-167	TV-22		164927	AH1W		
163930	AH1W	HMLA-167	TV-23		164928	AH1W		
163931	AH1W	HMLA-167	TV-25		164929	AH1W		
163932	AH1W	HMLA-167	TV-36		164930	AH1W		
163933	AH1W	NAWC-WD	001		164931	AH1W		
163934	AH1W				164932	AH1W		
163935	AH1W				164933	AH1W		
163936	AH1W				164934	AH1W		
163937	AH1W	VX-5	XE-42		164935	AH1W		
163938	AH1W	HMT-303	QT-442		164936	AH1W		
163939	AH1W	HMT-303	QT-443		164937	AH1W		
163940	AH1W	HMLA-169	SN-301		164938	AH1W		
163941	AH1W	HMT-303	QT-447		165037	AH1W		
163942	AH1W	HMLA-367	VT-131		165038	AH1W		
163943	AH1W	HMLA-369	SM-123		165039	AH1W		
163944	AH1W	HMLA-367	VT-44		165040	AH1W		
163945	AH1W	HMLA-367	VT-41		165041	AH1W		
163946	AH1W	HMLA-169	SN-304		165042	AH1W		
163947	AH1W	HMT-303	QT-436		165043	AH1W		
163948	AH1W	HMLA-169	SN-314		165044	AH1W		
163949	AH1W	HMLA-169	SN-315		165045	AH1W		
163950	AH1W	HMLA-367	VT-		165046	AH1W		
163951	AH1W				165047	AH1W		
163952	AH1W	HMLA-267	UV-44		165048	AH1W		
163953	AH1W	HMLA-267	UV-45		165049	AH1W		
163954	AH1W	HMLA-367	VT-46		165050	AH1W		
164572	AH1W	HMT-303	QT-432		165051	AH1W		
164573	AH1W	HMT-303	QT-431		165052	AH1W		
164574	AH1W	HMLA-169	SN-307		165053	AH1W		
164575	AH1W	HMA-775	WR-720		165054	AH1W		
164576	AH1W	HMA-576	WR-721		165055	AH1W		
164577	AH1W				165056	AH1W		
164578	AH1W	HMA-773	MP-08		165097	AH1W		

UH1 IROQUOIS

c\n	UH1E:	6001 to 6217.
	TH1E:	unknown (20).
	HHIK:	6301 to 6327.
	TH1L:	6401 to 6445.
	UH1N:	31401 to 31430, 31601 to 31647 plus later deliveries

151266	UH1E	to N540GH			151272	UH1E
151267	UH1E	MASDC			151273	UH1E
151268	UH1E	stored	NAS C. Christi		151274	UH1E
151269	UH1E				151275	UH1E
151270	UH1E				151276	UH1E
151271	UH1E	MASDC			151277	UH1E

Serial	Type	Status	Notes
151278	UH1E		
151279	UH1E		
151280	UH1E		
151281	UH1E		
151282	UH1E		
151283	UH1E	MASDC	
151284	UH1E		
151285	UH1E		
151286	UH1E		
151287	UH1E		
151288	UH1N	MCAS	El Toro
151289	UH1E		
151290	UH1E		
151291	UH1E		
151292	UH1E	MASDC	
151293	UH1E		
151294	UH1E		
151295	UH1E		
151296	UH1E		
151297	UH1E		
151298	UH1E		
151299	UH1E		
151840	UH1E		
151841	UH1E		
151842	UH1E		
151843	UH1E		
151844	UH1E	HML-771	QK-415
151845	UH1E		
151846	UH1E	MASDC	
151847	UH1E		
151848	UH1E	AMARC	7H172
151849	UH1E	w.o.23.10.68	
151850	UH1E		
151851	UH1E	to N9770N	
151852	UH1E		
151853	UH1E	AMARC	7H168
151854	UH1E		
151855	UH1E	MASDC	
151856	UH1E		
151857	UH1E	dump	Tallahasee AP
151858	UH1E	preserved	Tallahasee AP
151859	UH1E	MASDC	
151860	UH1E	MASDC	
151861	UH1E		
151862	UH1E		
151863	UH1E		
151864	UH1E		
151865	UH1E		
151866	UH1E	to N5363G	
151867	UH1E	dump	Tallahasee Apt, Fl
151868	UH1E		
151869	UH1E		
151870	UH1E		
151871	UH1E		
151872	UH1E		
151873	UH1E		
151874	UH1E	to N39AM	
151875	UH1E	to N151LC	
151876	UH1E	dump	NATTC
151877	UH1E		
151878	UH1E		
151879	UH1E	ranges	Aberdeen, Md
151880	UH1E		
151881	UH1E		
151882	UH1E		
151883	UH1E		
151884	UH1E		
151885	UH1E		
151886	UH1E		
151887	UH1E	MASDC	
152416	UH1E		
152417	UH1E		
152418	UH1E		
152419	UH1E		
152420	UH1E		
152421	UH1E		
152422	UH1E		
152423	UH1E		
152424	UH1E		
152425	UH1E		
152426	UH1E	AMARC	7H174
152427	UH1E		
152428	UH1E	to N120FC	
152429	UH1E		
152430	UH1E	MASDC	
152431	UH1E	AMARC	7H170
152432	UH1E		
152433	UH1E	MASDC	
152434	UH1E		
152435	UH1E		
152436	UH1E		
152437	UH1E		
152438	UH1E		
152439	UH1E		
153740	UH1E		
153741	UH1E		
153742	UH1E		
153743	UH1E		
153744	UH1E		
153745	UH1E	to N160RR	
153746	UH1E		
153747	UH1E		
153748	UH1E		
153749	UH1E		
153750	UH1E		
153751	UH1E		
153752	UH1E		
153753	UH1E		
153754	UH1E		
153755	UH1E		
153756	UH1E		
153757	UH1E		
153758	UH1E		
153759	UH1E		
153760	UH1E	MASDC	
153761	UH1E		
153762	UH1E	to N139US	
153763	UH1E		
153764	UH1E		
153765	UH1E	to N67RF	
153766	UH1E		
153767	UH1E		
154730	TH1E		
154731	TH1E		
154732	TH1E		
154733	TH1E		
154734	TH1E		
154735	TH1E		
154736	TH1E		
154737	TH1E		
154738	TH1E		
154739	TH1E		
154740	TH1E		
154741	TH1E		
154742	TH1E		
154743	TH1E		
154744	TH1E		
154745	TH1E		
154746	TH1E		
154747	TH1E		
154748	TH1E		
154749	TH1E		
154750	UH1E		
154751	UH1E		
154752	UH1E		
154753	UH1E		
154754	UH1E	MASDC	
154755	UH1E		
154756	UH1E		
154757	UH1E	AMARC	7H163
154758	UH1E		
154759	UH1E	preserved	NAS Whiting F'd
154760	UH1E	preserved	MCAS Quantico
154761	UH1E		
154762	UH1E		
154763	UH1E		
154764	UH1E	AMARC	7H169
154765	UH1E	to N98F	
154766	UH1E		
154767	UH1E		
154768	UH1E		
154769	UH1E	to N9678Z	
154770	UH1E	to N821DH	
154771	UH1E		
154772	UH1E		
154773	UH1E		

Serial	Type	Status/Unit	Code/Location
154774	UH1E	AMARC	7H173
154775	UH1E	to N38AM	
154776	UH1E		
154777	UH1E	MASDC	
154778	UH1E	preserved	Chanute AFB
154779	UH1E	MASDC	
154780	UH1E		
154943	UH1E	to N909KK	
154944	UH1E		
154945	UH1E	MASDC	
154946	UH1E		
154947	UH1E		
154948	UH1E	MASDC	
154949	UH1E	to N156RR	
154950	UH1E		
154951	UH1E	to N4692Z	
154952	UH1E		
154953	UH1E	MASDC	
154954	UH1E		
154955	UH1E		
154956	UH1E	to N5088J	
154957	UH1E	to N156RR	
154958	UH1E	GIA	NAS Norfolk
154959	UH1E	MASDC	
154960	UH1E	to N498RR	
154961	UH1E		
154962	UH1E		
154963	UH1E		
154964	UH1E		
154965	UH1E		
154966	UH1E		
154967	UH1E		
154968	UH1E		
154969	UH1E		
155337	UH1E		
155338	UH1E		
155339	UH1E		
155340	UH1E	MASDC	
155341	UH1E	to N118HS	
155342	UH1E	to N9173N	
155343	UH1E		
155344	UH1E	MASDC	
155345	UH1E	derelict	NAS Memphis
155346	UH1E	to N7160J	
155347	UH1E		
155348	UH1E	MASDC	
155349	UH1E	MASDC	
155350	UH1E	MASDC	
155351	UH1E	to N5089Q	
155352	UH1E	MASDC	
155353	UH1E	MASDC	
155354	UH1E		
157177	HH1K	HAL-5	NW-303
157178	HH1K	HAL-5	NW-305
157179	HH1K	AMARC	7H185
157180	HH1K	AMARC	7H186
157181	HH1K	AMARC	7H187
157182	HH1K	AMARC	7H191
157183	HH1K	AMARC	7H193
157184	HH1K	AMARC	7H194
157185	HH1K	MASDC	
157186	HH1K		
157187	HH1K		
157188	HH1K	HAL-5	NW-304
157189	HH1K	MASDC	
157190	HH1K	AMARC	7H196
157191	HH1K	HAL-5	NW-306
157192	HH1K		
157193	HH1K		
157194	HH1K	AMARC	7H192
157195	HH1K		
157196	HH1K		
157197	HH1K	preserved	Selfridge ANGB
157198	HH1K	AMARC	7H195
157199	HH1K	AMARC	7H197
157200	HH1K		
157201	HH1K		
157202	HH1K		
157203	HH1K		
157806	TH1L	to N55201	
157807	TH1L	preserved	NAS Whiting Field
157808	TH1L		
157809	TH1L	to N4963F	
157810	TH1L		
157811	TH1L	ranges	Aberdeen, Md
157812	TH1L		
157813	TH1L		
157814	TH1L		
157815	TH1L	to N154RR	
157816	TH1L	to N4962E	
157817	TH1L	NAWC	Lakehurst
157818	TH1L		
157819	TH1L		
157820	TH1L		
157821	TH1L		
157822	TH1L	to N155RR	
157823	TH1L		
157824	TH1L		
157825	TH1L	to N483RR	
157826	TH1L	to N153RR	
157827	TH1L	NAWC\RWATD	
157828	TH1L	to N465RR	
157829	TH1L	NAWC-WD	
157830	TH1L		
157831	TH1L	to N474RR	
157832	TH1L	to N5820X	
157833	TH1L	NWC	
157834	TH1L	stored	China Lake
157835	TH1L	stored	China Lake
157836	TH1L	stored	China Lake
157837	TH1L		
157838	TH1L	MASDC	
157839	TH1L	MASDC	
157840	TH1L	to N151RR	
157841	TH1L	AMARC	7H159
157842	TH1L		
157843	TH1L		
157844	TH1L	scrapyard	Bobs Air Park
157845	TH1L	MASDC	
157846	TH1L		
157847	TH1L		
157848	TH1L	to N5530U	
157849	TH1L	to N58205	
157850	TH1L		
157851	UH1E		
157852	UH1L		
157853	UH1E		
157854	UH1L	to N4964N	
157855	UH1L		
157856	UH1L	to N157LC	
157857	UH1E		
157858	UH1L	AMARC	7H158
158230	UH1N	USS Guadal Canal	
158231	UH1N	HC-16	BF-106
158232	UH1N	HML-767	MM-415
158233	UH1N		
158234	UH1N	VXE-6	XD-11
158235	UH1N	VXE-6	XD-12
158236	UH1N		
158237	UH1N	w.o.17.1.86	
158238	UH1N	VXE-6	XD-15
158239	UH1N		
158240	UH1N	HC-16	BF-100
158241	UH1N		
158242	UH1N	NAS	7H Fallon
158243	UH1N	NAS	7H Fallon
158244	UH1N	NAS	7H Fallon
158245	UH1N	HC-16	BF-104
158246	UH1N	NAS	7S-1 Lemoore
158247	UH1N		
158248	UH1N	NAS	G-201 C. Christi
158249	UH1N	NAS	7S Lemoore
158250	UH1N	HC-16	BF-102
158251	UH1N	MCAS	Yuma
158252	UH1N	NAS	A-02 Meridian
158253	UH1N	NAS	A-03 Meridian
158254	UH1N	NAS	G-204 C. Christi
158255	UH1N	w.o.	
158256	UH1N	HC-16	BF-108
158257	UH1N		
158258	UH1N	NAS	G-258 C. Christi
158259	UH1N	MCAS	El Toro SAR
158260	UH1N		
158261	UH1N	NAWC	
158262	UH1N	HML-767	MM-401
158263	UH1N	HML-776	QL-400

Serial	Type	Unit	Code
158264	UH1N	USS Iwo Jima	
158265	UH1N	HMT-303	QT-404
158266	UH1N	HML-767	MM-403
158267	UH1N	HML-767	MM-404
158268	UH1N	HMT-303	QT-405
158269	UH1N	HML-767	MM-416
158270	UH1N	HMT-303	QT-402
158271	UH1N	HMT-303	QT-406
158272	UH1N	VXE-6	XD-16
158273	UH1N		
158274	UH1N	HML-767	MM-406
158275	UH1N	HC-16	BF-105
158276	UH1N	MCAS	El Toro SAR
158277	UH1N	HML-767	MM-402
158278	UH1N	HC-16	BF-103
158279	UH1N		
158280	UH1N	HMLA-167	TV-
158281	UH1N	HMT-303	QT-407
158282	UH1N	HML-767	MM-409
158283	UH1N	VXE-6	XD-10
158284	UH1N	HNT-303	QT-414
158285	UH1N		
158286	UH1N	MCAS	El Toro SAR
158287	UH1N	HML-767	MM-411
158288	UH1N	MCAS	El Toro SAR
158289	UH1N	HML-767	MM-417
158290	UH1N	MCAS	El Toro SAR
158291	UH1N		
158548	UH1N		
158549	UH1N	HML-767	MM-414
158550	UH1N		
158551	VH1N	AMARC	7H198
158552	UH1N	AMARC	7H199
158553	VH1N	HC-16	BF-101
158554	VH1N	MASDC	
158555	UH1N		
158556	VH1N	AMARC	7H200
158557	UH1N	AMARC	7H201
158558	UH1N	HML-767	MM-400
158559	UH1N	NAS	7H Fallon
158560	UH1N	HML-771	QK-450
158561	UH1N	HML-771	QK-403
158562	UH1N	HML-771	QK-410
158762	UH1N	NAS	7S-3 Leeemore
158763	UH1N	USS Nassau	
158764	UH1N	NAS	Brunswick
158765	UH1N		
158766	UH1N	NAS	Brunswick
158767	UH1N	MASDC	
158768	UH1N	MCAS	Yuma
158769	UH1N	AMARC	7H055
158770	UH1N	NAWC-WD	018
158771	UH1N	NAWC-WD	017
158772	UH1N	HML-776	QL-403
158773	UH1N	HML-771	QK-453
158774	UH1N	HML-771	QK-454
158775	UH1N	HML-771	QK-455
158776	UH1N	HMM-164	YT-460
158777	UH1N	HML-771	QK-457
158778	UH1N	HML-771	QK-456
158779	UH1N		
158780	UH1N	w.o.19.7.81	
158781	UH1N	HML-771	QK-411
158782	UH1N	HML-771	QK-412
158783	UH1N		
158784	UH1N		
158785	UH1N		
159186	UH1N	HMLA-167	TV-15
159187	UH1N	HMLA-269	HF-30
159188	UH1N	HMM-261	EM-43
159189	UH1N	HMM-266	ES-14
159190	UH1N	HMM-264	EH-30
159191	UH1N	HMLA-269	HF-23
159192	UH1N	HMLA-269	HF-33
159193	UH1N	HMLA-167	TV-13
159194	UH1N	HMLA-269	HF-30
159195	UH1N	HMM-261	EM-42
159196	UH1N	HMM-261	EM-41
159197	UH1N	HMLA-169	SN-332
159198	UH1N	HMLA-169	SN-335
159199	UH1N	w.o.	
159200	UH1N	HMLA-169	SN-173
159201	UH1N		
159202	UH1N	HMLA-367	VT-
159203	UH1N		
159204	UH1N	HMLA-169	SN-334
159205	UH1N	HMLA-169	SN-325
159206	UH1N	HC-16	BF-99
159207	UH1N		
159208	UH1N	HMT-303	QT-400
159209	UH1N	HMLA-169	SN-340
159565	UH1N		
159680	UH1N	HMLA-367	VT-102
159681	UH1N	HMLA-169	SN-333
159682	UH1N		
159683	UH1N	HMLA-167	TV-5
159684	UH1N	HMLA-367	VT-104
159685	UH1N	HMT-303	QT-411
159686	UH1N		
159687	UH1N	w.o.25.10.88	
159688	UH1N	HMLA-367	VT-103
159689	UH1N	HMLA-367	VT-106
159690	UH1N	HMLA-367	VT-04
159691	UH1N		
159692	UH1N	HMLA-169	SN-320
159693	UH1N	HMT-303	QT-413
159694	UH1N		
159695	UH1N	HMLA-367	VT-101
159696	UH1N	HMLA-367	VT-06
159697	UH1N	HMT-303	QT-401
159698	UH1N	HMLA-367	VT-07
159699	UH1N	damaged	NAS C. Christi
159700	UH1N	HMLA-367	VT-510
159701	UH1N	HMLA-367	VT-11
159702	UH1N	HMLA-367	VT-
159703	UH1N	HMM-165	YW-
159774	UH1N		
159775	UH1N	HMLA-169	SN-336
159776	UH1N	HMLA-367	VT-
159777	UH1N	HMT-303	QT-403
160165	UH1N	HMLA-167	UV-16
160166	UH1N	HMT-303	QT-
160167	UH1N	HMLA-167	UV-10
160168	UH1N	HMM-161	YR-30
160169	UH1N	HMLA-169	SN-321
160170	UH1N	HMLA-367	VT-14
160171	UH1N	HMLA-369	SM-214
160172	UH1N	HMT-303	QT-412
160173	UH1N	HMLA-367	VT-01
160174	UH1N	HMT-303	QT-416
160175	UH1N	HMLA-169	SN-322
160176	UH1N	HMM-164	YT-461
160177	UH1N	HMLA-169	SN-331
160178	UH1N	w.o.	
160179	UH1N	HMLA-369	SM-
160438	UH1N	HMLA-369	SM-213
160439	UH1N	HMLA-269	HF-34
160440	UH1N	MEU	YS-35
160441	UH1N	HMLA-169	SN-323
160442	UH1N	HMLA-167	TV-36
160443	UH1N	HMM-164	YT-32
160444	UH1N	HMLA-269	HF-46
160445	UH1N	MEU	YS-37
160446	UH1N	HMLA-169	SN-324
160447	UH1N	w.o.4.10.87	
160448	UH1N	HMLA-269	HF-40
160449	UH1N	HMLA-167	TV-00
160450	UH1N	HMM-266	ES-03
160451	UH1N		
160452	UH1N	HMM-162	YS-32
160453	UH1N	HMH-262	ET-41
160454	UH1N	HMLA-269	HF-41
160455	UH1N	HMM-266	ES-06
160456	UH1N	HMLA-167	TV-01
160457	UH1N	HMLA-167	TV-04
160458	UH1N	w.o.3.5.89	
160459	UH1N	MEU	YS-36
160460	UH1N		
160461	UH1N	HMLA-367	VT-111
160619	UH1N	HMLA-169	SN-112
160620	UH1N	HMLA-267	UV-330
160621	UH1N	HMLA-367	VT-
160622	UH1N	w.o.	
160623	UH1N	HMLA-267	UV-337
160624	UH1N	w.o.3.2.89	
160827	UH1N	USS Nassau	

160828	UH1N	w.o.	
160829	UH1N	HC-16	BF-107
160830	UH1N	USS Tarawa	
160831	UH1N	USS Iwo Jima	
160832	UH1N	USS Tripoli	
160833	UH1N	USS Guam	
160834	UH1N	USS Bellau Wood	
160835	UH1N	w.o.12.8.85	
160836	UH1N	USS New Orleans	
160837	UH1N	USS Okinawa	
160838	UH1N	USS Inchcon	

H2 SEASPRITE

c\n   1 to 190 plus later deliveries.

147202	YUH2A		
147203	YUH2A		
147204	YUH2A		
147205	YUH2A		
147972	UH2A		
147973	UH2A		
147974	UH2A		
147975	UH2A		
147976	UH2A		
147977	UH2A		
147978	UH2A	to US Army	
147979	UH2A	w.o.8.4.68	
147980	SH2F		
147981	SH2F		
147982	UH2A		
147983	UH2A		
149013	SH2F	HSL-94	NW-27
149014	SH2F		
149015	SH2F		
149016	SH2F		
149017	SH2F	HSL-84	NW-03
149018	UH2A	w.o.1.7.64	
149019	UH2A		
149020	UH2A		
149021	SH2F	NAWC-AD	RWATD
149022	SH2F	HSL-35	TG-30
149023	SH2F		
149024	SH2F	HSL-94	NW-22
149025	UH2A		
149026	SH2F	preserved	NAS Norfolk
149027	UH2A		
149028	UH2A		
149029	UH2A		
149030	SH2F	HSL-32	HV-136
149031	SH2F	HSL-31	TD-14
149032	UH2A		
149033	SH2F	HSL-36	HY-
149034	UH2A		
149035	UH2A		
149036	SH2F	HSL-84	NW-04
149739	UH2A		
149740	UH2A		
149741	UH2A	w.o.4.12.67	
149742	UH2A		
149743	UH2A		
149744	SH2F	w.o.17.8.84	
149745	UH2A		
149746	UH2A		
149747	SH2F	HSL-94	NW-26
149748	SH2F		
149749	UH2A		
149750	SH2F	VX-1	JA-31
149751	UH2A	w.o.10.1.66	
149752	UH2A		
149753	SH2F	AMARC	8H023
149754	UH2A		
149755	SH2F		
149756	UH2A		
149757	UH2A	w.o.19.6.66	
149758	SH2F	HSL-84	NW-00
149759	UH2A		
149760	UH2A		
149761	SH2F	HSL-33	TF-15
149762	UH2A	w.o.18.3.67	
149763	UH2A		
149764	UH2A	w.o.7.1.69	
149765	UH2A	w.o.24.6.77	
149766	SH2F	HSL-37	TH-56
149767	UH2A	w.o.10.8.69	
149768	SH2F	HSL-34	HX-231
149769	SH2F	w.o.10.8.84	
149770	SH2F		
149771	SH2F		
149772	SH2F		
149773	SH2F	HSL-34	HX-237
149774	UH2A	w.o.26.10.66	
149775	UH2A		
149776	UH2A		
149777	UH2A		
149778	UH2A		
149779	SH2F	HSL-37	TH-51
149780	SH2F	HSL-94	NW-25
149781	UH2A		
149782	UH2A	w.o.17.9.64	
149783	UH2A	w.o.14.6.67	
149784	UH2A		
149785	UH2A	to US Army	
149786	UH2A	w.o.9.8.64	
150139	SH2F	stored	NAS Pensacola
150140	SH2F		
150141	SH2F		
150142	SH2F		
150143	SH2F	AMARC	8H019
150144	UH2B	w.o.21.2.69	
150145	UH2B		
150146	SH2F	HSL-30	HT-032
150147	UH2B		
150148	SH2F	damaged	NAS Pensacola
150149	SH2F	HSL-30	HT-
150150	SH2F	HSL-37	TH-54
150151	SH2F		
150152	SH2F	HSL-84	NW-05
150153	UH2B	w.o.4.10.67	
150154	SH2F	HSL-36	HY-337
150155	SH2F	HSL-31	TD-08
150156	SH2F	HSL-94	NW-24
150157	SH2F	HSL-37	TH-57
150158	SH2F	HSL-94	NW-24
150159	SH2F	Kaman	
150160	SH2F	HSL-84	NW-01
150161	SH2F		
150162	UH2B	w.o.15.4.66	
150163	SH2F	HSL-94	NW-20
150164	SH2F	w.o.19.11.85	
150165	SH2F	HSL-30	HT-033
150166	SH2F	w.o.11.4.84	
150167	SH2F	w.o.10.3.86	
150168	UH2B		
150169	SH2F	HSL-31	TD-04
150170	UH2B		
150171	SH2F	HSL-84	NW-06
150172	UH2B		
150173	SH2F	AMARC	8H024
150174	SH2F	HSL-31	TD-09

150175	SH2F	HSL-84	NW-04		161644	SH2F	HSL-34	HX-232
150176	UH2B				161645	SH2F	HSL-33	TF-22
150177	UH2B				161646	SH2F	w.o.30.3.88	
150178	SH2F	HSL-94	NW-22		161647	SH2F		
150179	SH2F	AMARC	8H022		161648	SH2F	w.o.3.3.89	
150180	UH2B				161649	SH2F	w.o.21.1.85	
150181	SH2F	HSL-31	TD-00		161650	SH2F	HSL-32	HV-140
150182	UH2B				161651	SH2F	w.o.11.11.88	
150183	UH2B	w.o.26.10.66			161652	SH2F	HSL-32	HV-133
150184	UH2B				161653	SH2F	Kaman	
150185	SH2F	HSL-31	TD-02		161654	SH2F	w.o.11.1.87	
150186	UH2B				161655	SH2F	w.o.	
151300	SH2F	w.o.30.7.84			161656	SH2F	HSL-34	HX-230
151301	SH2F				161657	SH2F	HSL-37	TH-59
151302	UH2B				161658	SH2F		
151303	SH2F	HSL-34	HX-235		161898	SH2F	w.o.29.11.89	
151304	SH2F	HSL-31	TD-03		161899	SH2F		
151305	UH2B				161900	SH2F	HSL-33	TF-19
151306	SH2F	HSL-33	TF-20		161901	SH2F	HSL-32	HV-141
151307	UH2B				161902	SH2F	w.o.	
151308	SH2F	HSL-33	TF-11		161903	SH2F	w.o.8.5.88	
151309	SH2F	HSL-31	TD-01		161904	SH2F	HSL-33	TF-16
151310	SH2F	HSL-30	HT-035		161905	SH2F	HSL-32	HV-
151311	SH2F	AMARC	8H020		161906	SH2F	HSL-33	TF-10
151312	SH2F	HSL-33	TF-14		161907	SH2F	HSL-34	HX-241
151313	SH2F	damaged			161908	SH2F	HSL-37	TH-61
151314	SH2F	HSL-33	TF-12		161909	SH2F		
151315	UH2B	w.o.23.1.68			161910	SH2F	w.o.5.3.88	
151316	SH2F	HSL-32	HV-132		161911	SH2F	HSL-32	HV-130
151317	UH2B	w.o.27.2.67			161912	SH2F	HSL-33	TF-14
151318	UH2B				161913	SH2F	HSL-34	HX-232
151319	SH2F	w.o.17.9.83			161914	SH2F	HSL-37	TH-54
151320	UH2B				161915	SH2F	w.o.2.10.86	
151321	SH2F	HSL-32	HV-135		162576	SH2F		
151322	SH2F				162577	SH2F	HSL-32	HV-132
151323	SH2F				162578	SH2F	HSL-33	TF-13
151324	SH2F				162579	SH2F	w.o.16.9.87	
151325	SH2F	w.o.25.11.85			162580	SH2F		
151326	SH2F				162581	SH2F	HSL-32	HV-141
151327	SH2F	w.o.18.8.84			162582	SH2F	HSL-33	TF-
151328	SH2F				162583	SH2F	HSL-36	HY-342
151329	SH2F	HSL-30	HT-037		162584	SH2F		
151330	SH2F	stored	NAS Pensacola		162585	SH2F	HSL-33	TF-21
151331	SH2F	HSL-32	HV-140		162586	SH2F	HSL-36	HY-341
151332	SH2F	HSL-94	NW-23		162587	SH2F		
151333	SH2F	HSL-36	HY-		162650	SH2F		
151334	SH2F	HSL-30	HT-041		162651	SH2F		
151335	SH2F				162652	SH2F		
152189	SH2G	HSL-31	TD-06		162653	SH2F		
152190	SH2F	HSL-94	NW-21		162654	SH2F		
152191	SH2F				162655	SH2F		
152192	SH2F	HSL-37	TH-60		163209	SH2G	HSL-36	HY-344
152193	UH2B				163210	SH2G	HSL-32	HV-142
152194	UH2B				163211	SH2G	HSL-37	TH-52
152195	UH2B	w.o.30.4.68			163212	SH2G		
152196	UH2B	w.o.16.9.66			163213	SH2G	HSL-33	TF-12
152197	UH2B	w.o.15.2.68			163214	SH2G		
152198	SH2F				163541	SH2F	NAWC-AD	RWATD
152199	SH2F				163542	SH2F		
152200	SH2F	HSL-30	HT-043		163543	SH2F		
152201	SH2F	HSL-30	HT-042		163544	SH2F		
152202	UH2B	w.o.17.8.68			163545	SH2F		
152203	SH2F	HSL-94	NW-21		163546	SH2F		
152204	SH2F	HSL-35	TG-41		163547	SH2F		
152205	SH2F				163548	SH2F		
152206	SH2F	HSL-30	HT-040		163549	SH2F		
161641	JSH2F	NAWC\RWATD			163550	SH2F		
161642	SH2F	HSL-32	HV-133		163551	SH2F		
161643	SH2F	HSL-37	TH-55		163552	SH2F		

H3 SEA KING

c\n   61-001 to 61-030 and later deliveries.

Serial	Type	Unit	Code
147137	SH3A		
147138	SH3A		
147139	SH3A		
147140	UH3A	NAWC-WD	5
147141	UH3A	NAWC-WD	7
147142	UH3A	NAWC-WD	4
147143	SH3A		
147144	SH3A		
147145	SH3A		
147146	SH3A	NAWC-WD	
148033	SH3A		
148034	SH3G	w.o.22.7.74	
148035	SH3H	NSWC	
148036	SH3H	NAD	Pensacola
148037	SH3G	w.o.30.7.87	
148038	UH3A	NAWC-WD	8
148039	SH3H	HS-75	NW-613
148040	UH3A	NAWC-WD 9	
148041	SH3A		
148042	SH3H	HS-1	AR-442
148043	SH3H	HS-5	AG-614
148044	SH3G		
148045	SH3H	HS-1	AR-450
148046	SH3G	w.o.22.1.85	
148047	SH3G	HC-2	HU-744
148048	SH3G	HC-2	HU-
148049	SH3H	HC-16	BF-
148050	SH3G	derelict	NAD Pensacola
148051	SH3G	AMARC	
148052	SH3H		
148964	SH3H	HS-1	AR-438
148965	SH3H	HS-1	AR-402
148966	SH3H	HS-14	NE-613
148967	SH3H		
148968	SH3A		
148969	SH3H	HC-1	UP-764
148970	SH3G	HC-2	HU-740
148971	SH3H	NSWC	
148972	SH3H		
148973	SH3H	HC-1	UP-01
148974	SH3H		
148975	SH3A		
148976	SH3H	w.o.19.10.85	
148977	SH3H	HS-1	AR-445
148978	SH3A		
148979	SH3G	HC-1	UP-
148980	SH3H	HS-1	AR-449
148981	SH3H		
148982	SH3A	w.o.26.8.67	
148983	SH3H		
148984	SH3A		
148985	SH3A	w.o.23.5.67	
148986	SH3H	HC-16	BF-
148987	SH3G	w.o.27.4.87	
148988	SH3H	stored	NAD Pensacola
148989	SH3G	derelict	NAD Pensacola
148990	SH3H	HS-7	AC-614
148991	SH3A		
148992	SH3A	w.o.23.5.67	
148993	SH3A	w.o.2.11.65	
148994	SH3A		
148995	SH3H	w.o.12.1.87	
148996	SH3G	HC-1	UP-732
148997	SH3H	stored	NAS North Island
148998	SH3H	HS-1	AR-400
148999	SH3H	HS-75	NW-611
149000	SH3G	NAD	Pensacola
149001	SH3A		
149002	SH3A		
149003	SH3G	AMARC	9H039
149004	SH3H	HS-8	NK-613
149005	SH3H	HS-75	NW-610
149006	SH3H	HS-4	NL-612
149007	SH3A		
149008	SH3A		
149009	SH3A	to USAF	62-12571
149010	SH3H	HS-1	AR-405
149011	SH3A	to USAF	62-12572
149012	SH3A	to USAF	62-12573
149679	SH3G	HC-1	UP-723
149680	SH3A		
149681	SH3A		
149682	HH3A	AMARC	
149683	SH3G	HC-1	UP-724
149684	SH3H		
149685	SH3A	w.o.20.6.67	
149686	SH3A		
149687	SH3H	HS-11	
149688	SH3G	w.o.18.1.86	
149689	SH3A		
149690	SH3H	dump	NAS Pensacola
149691	SH3A		
149692	SH3A		
149693	SH3H		
149694	SH3G	NAS	7R Oceana 694
149695	SH3G	NAS	Jacksonville
149696	SH3A		
149697	SH3G	w.o.	
149698	SH3H	stored	NAS North Island
149699	SH3G	w.o.8.5.72	
149700	SH3G	w.o.15.6.80	
149701	SH3H	NAF	Sigonella
149702	SH3H	HC-2	HU-724
149703	SH3A		
149704	SH3A	w.o.26.11.73	
149705	SH3H	HS-3	AA-612
149706	SH3H	VC-8	210
149707	SH3A	w.o.8.2.70	
149708	SH3H	HS-3	AA-611
149709	SH3A		
149710	SH3G	w.o.10.4.81	
149711	SH3H	HS-5	AG-615
149712	SH3H	stored	NAS Pensacola
149713	SH3H	HS-14	NE-614
149714	SH3A		
149715	SH3A		
149716	SH3A		
149717	SH3H	HS-1	AR-454
149718	SH3H	HC-16	BF-401
149719	SH3H	HC-1	UP-763
149720	SH3G		
149721	SH3A		
149722	SH3H	HC-16	BF-405
149723	SH3G	to NASA	N735NA
149724	SH3H	HC-1	UP-763
149725	SH3H	HS-5	AG-615
149726	SH3H	w.o.18.11.85	
149727	SH3H	HS-7	AC-611
149728	SH3G	NAD	Pensacola
149729	SH3G	NAWC\TPS	22
149730	SH3H	HS-11	AB-
149731	SH3G	HC-2	HU-723
149732	SH3A		
149733	SH3G	HC-2	HU-746
149734	SH3A		
149735	SH3H		
149736	SH3G	w.o.19.9.81	
149737	SH3G	HC-2	HU-721

Serial	Type	Unit	Code
149738	SH3H	HS-5	AG-611
149893	SH3G	w.o.23.10.73	
149894	SH3H	HS-8	NK-612
149895	SH3A	w.o.23.3.67	
149896	SH3A		
149897	SH3H	HS-14	NE-617
149898	SH3H	HS-85	NW-614
149899	SH3H	HS-4	NL-610
149900	SH3H	w.o.21.5.86	
149901	SH3A	w.o.23.6.68	
149902	SH3H	HS-75	NW-615
149903	HH3A		
149904	SH3H	HS-5	AG-616
149905	SH3H	HS-8	NK-611
149906	SH3H	NAD	Pensacola
149907	SH3A		
149908	SH3H	w.o.20.2.70	
149909	SH3A	w.o.12.1.67	
149910	SH3H	w.o.23.2.89	
149911	SH3A		
149912	SH3A	w.o.14.10.76	
149913	SH3H	HS-4	NL-613
149914	SH3G	HS-6	NH-613
149915	SH3G	stored	NAS North Island
149916	SH3H	HC-2	HU-721
149917	SH3H	HS-7	AC-610
149918	SH3H	HS-6	NH-614
149919	SH3G	HC-1	UP-734
149920	SH3A		
149921	SH3H	HS-5	AG-
149922	HH3A		
149923	SH3H	HS-12	NF-614
149924	SH3A	w.o.6.7.67	
149925	SH3G	stored	NAS North Island
149926	SH3A	w.o.5.2.66	
149927	SH3H	HS-1	AR-
149928	SH3A		
149929	SH3H	HS-2	NG-615
149930	SH3G	HC-1	UP-730
149931	SH3H	w.o.9.11.87	
149932	SH3G	HS-1	AR-420
149933	HH3A		
149934	SH3H	HS-11	AB-612
150610	VH3A	HC-6	HW-30
150611	VH3A	HC-6	HW-31
150612	VH3A	w.o.26.5.73	
150613	VH3A	HC-6	HW-32
150614	VH3A	MASDC	
150615	VH3A		
150616	VH3A	HC-2	HU-733
150617	VH3A	MASDC	
150618	SH3A	w.o.16.10.66	
150619	SH3A	w.o.5.2.66	
150620	SH3G	stored	NAS North Island
151522	SH3A	w.o.23.10.67	
151523	SH3H	VC-8	GF-22
151524	SH3H	w.o.26.3.82	
151525	SH3H	HS-7	AC-616
151526	SH3A		
151527	SH3G	HC-2	HU-741
151528	SH3H	HC-2	HU-725
151529	SH3G	w.o.	
151530	SH3A	w.o.21.5.67	
151531	HH3A	HS-10	RA-00
151532	SH3G	VC-8	GF-25
151533	SH3G	HC-1	UP-725
151534	SH3A	w.o.8.2.69	
151535	SH3H	HS-2	NG-611
151536	SH3G	HC-1	UP-722
151537	SH3A		
151538	SH3A	w.o.19.7.67	
151539	SH3G	HC-2	HU-723
151540	SH3A		
151541	SH3H	HS-12	NF-616
151542	SH3A		
151543	SH3H	w.o.30.5.86	
151544	SH3H		
151545	SH3G	HC-1	UP-726
151546	SH3H	NAD	Pensacola
151547	SH3G	NAS	7R Oceana 547
151548	SH3A		
151549	SH3H	stored	NAS North Island
151550	SH3H	HC-1	UP-761
151551	SH3H	HC-1	UP-760
151552	HH3A	w.o.23.12.72	
151553	HH3A	AMARC	
151554	SH3G	HC-1	UP-722
151555	SH3H	VC-8	GF-24
151556	HH3A		
151557	SH3A		
152104	SH3H	HC-16	BF-400
152105	SH3A		
152106	SH3D		
152107	SH3H	HS-7	AC-616
152108	SH3H	HS-75	NW-613
152109	SH3H	HS-7	AC-615
152110	SH3H	stored	NAS North Island
152111	SH3A		
152112	SH3H	HS-7	AC-615
152113	SH3H	HS-1	AR-404
152114	SH3A		
152115	SH3H	NARF	Pensacola
152116	SH3A	VX-1	JA-24
152117	SH3A		
152118	SH3A		
152119	SH3H	NSWC	72
152120	SH3A		
152121	SH3H	HS-1	AR-403
152122	SH3H	HS-7	AC-612
152123	SH3H	NAD	Pensacola
152124	SH3H	w.o.22.8.85	
152125	SH3H	w.o.	
152126	SH3A		
152127	SH3A		
152128	SH3H	HS-11	AB-610
152129	SH3H	HS-1	AR-401
152130	SH3H	NSWC	70
152131	SH3G	HC-2	HU-724
152132	SH3H	w.o.12.12.84	
152133	SH3H		
152134	SH3H		
152135	SH3H	HS-5	AG-612
152136	SH3H	w.o.12.2.86	
152137	SH3H	HS-1	AR-446
152138	SH3H	HS-1	AR-406
152139	SH3H	NAS	Oceana SAR
152690	SH3D		
152691	SH3D		
152692	SH3D		
152693	SH3D	NAWC\TPS	73
152694	SH3H	HS-1	AR-409
152695	SH3D		
152696	SH3G	NAS	7Q Key West
152697	SH3D	NAS	7G Whidbey I. 3
152698	SH3D		
156299	SH3H	HS-1	AR-401
152700	SH3H	VX-1	JA-23
152701	SH3H	HS-12	NF-611
152702	SH3H	HS-14	NE-615
152703	SH3H	NSWC	71
152704	SH3H	HS-12	NF-612
152705	SH3G	HS-75	NW-610
152706	SH3D		
152707	SH3H	HS-14	NE-613
152708	SH3H	HS-1	AR-402
152709	SH3H	HS-5	AG-610
152710	SH3H	HC-16	BF-404
152711	SH3D		
152712	SH3D	HS-1	AR-408
152713	SH3D		
154100	SH3D	NAS	7G Whidbey I. 2
154101	SH3H		
154102	SH3H	w.o.9.10.91	
154103	SH3D	NAS	7G Whidbey I. 3
154104	SH3D		
154105	SH3D	NAWC\TPS	72
154106	SH3D	HS-12	NH-613
154107	SH3D	AMARC	
154108	SH3D	NAS	7G Whidbey I. 4
154109	SH3D		
154110	SH3D		
154111	SH3D		
154112	SH3H	HS-75	NW-555
154113	SH3H	HS-75	NW-554
154114	SH3H	HC-16	BF-403
154115	SH3D		

154116	SH3D	NAWC\TPS	26
154117	SH3D		
154118	SH3D		
154119	SH3D	NAWC\TPS	71
154120	SH3D		
154121	SH3H	HS-14	NE-616
154122	SH3D	NAWC\TPS	
154123	SH3D		
156483	SH3H	NSWC	72
156484	SH3H	HC-16	BF-402
156485	SH3D	HS-75	NW-506
156486	SH3H	HC-16	BF-404
156487	SH3D	HC-16	BF-401
156488	SH3H		
156489	SH3D	NAWC\TPS	75
156490	SH3D	NAS	7G Whidbey I. 2
156491	SH3D	w.o.4.10.88	
156492	SH3D	NAS	7Q Key West
156493	SH3D	stored	NAS North Island
156494	SH3D	w.o.31.12.72	
156495	SH3H	HS-14	NE-617
156496	SH3D	NAS	7G Whidbey I. 4
156497	SH3D		
156498	SH3H	HS-12	NF-615
156499	SH3H	HC-16	BF-402
156500	SH3D	NAS	7G Whidbey I. 5
156501	SH3H	HS-1	AR-453
156502	SH3D		
156503	SH3D	NAS	7Q Key West
156504	SH3D		
156505	SH3H	HS-14	NE-610
156506	SH3H	HS-5	AG-613
159350	VH3D	HMX-1	
159351	VH3D	HMX-1	
159352	VH3D	HMX-1	
159353	VH3D	HMX-1	
159354	VH3D	HMX-1	
159355	VH3D	HMX-1	
159356	VH3D	HMX-1	
159357	VH3D	HMX-1	
159358	VH3D	HMX-1	
159359	VH3D	HMX-1	
159360	VH3D	HMX-1	
212574	SH3H	w.o.3.8.84	
212575	SH3H		

H46 SEA KNIGHT

c\n 2001 to 2003, 2005 to 2025, 2028 to 2037, 2039 to 2050, 2026, 2027, 2038, 2051, 2052, 2057, 2062, 2067, 2053 to 2056, 2058 to 2061, 2063 to 2066, 2068 to 2112, 2118, 2124, 2130, 2137, 2144, 2113 to 2117, 2119 to 2123, 2125 to 2129, 2131 to 2136, 2138 to 2143, 2145 to 2222, 2225 to 2234, 2237 to 2246, 2249 to 2258, 2261 to 2269, 2272 to 2301, 2223, 2224, 2235, 2236, 2247, 2248, 2259, 2260, 2270, 2271, 2302 to 2626.

150265	CH46A		
150266	CH46A	stored	Memphis
150267	CH46A	stored	Memphis
150268	CH46A		
150269	CH46A		
150270	CH46A		
150271	HH46A		
150272	CH46A		
150273	CH46A		
150274	CH46A		
150275	CH46A		
150276	CH46D	HC-6	HW-01
150277	CH46A		
150278	CH46A		
150933	CH46A		
150934	CH46A		
150935	CH46A		
150936	CH46A		
150937	CH46A		
150938	HH46A		
150939	CH46A		
150940	CH46A		
150941	HH46A	MCAS	Beaufort
150942	CH46D	HC-3	SA-20
150943	CH46A		
150944	CH46A		
150945	CH46A		
150946	CH46A	MASDC	
150947	HH46A		
150948	CH46A		
150949	HH46A		
150950	CH46A	w.o.1.3.78	
150951	HH46A	MCAS	Iwakuni
150952	HH46A	NAWC\TPS	63
150953	CH46A		
150954	HH46A	HC-11	VR-53
150955	CH46A		
150956	CH46A		
150957	CH46A	NAWC\TPS	41
150958	HH46A		
150959	CH46A		
150960	CH46A		
150961	CH46A		
150962	HH46A	HC-3	SA-
150963	HH46D	MCAS	Cherry Point
150964	HH46D	MCAS	Cherry Point
150965	UH46A		
150966	UH46A	HC-8	BR-40
150967	UH46A		
150968	UH46D	HC-11	VR-63
151902	UH46D	HC-11	VR-65
151903	UH46D	HC-11	VR-51
151904	UH46D	HC-8	BR-55
151905	UH46D	HC-11	VR-
151906	CH46A		
151907	CH46A		
151908	HH46A	HC-3	SA-02
151909	CH46A		
151910	HH46A	NAWC-AD	SATD
151911	HH46A		
151912	HH46A	MCAS	Beaufort
151913	HH46A	w.o.23.3.87	
151914	HH46A	NAD	Cherry Point
151915	UH46D	HC-11	VR-60
151916	CH46A		
151917	CH46A		
151918	HH46A	w.o.1.9.85	
151919	CH46A		
151920	HH46A	HC-3	SA-
151921	HH46A	MCAS	KB Kaneohe Bay
151922	CH46A		
151923	CH46A		
151924	HH46A	MCAS	KB Kaneohe Bay
151925	CH46A		
151926	HH46A		
151927	CH46A	HC-11	VR-54
151928	CH46A		
151929	CH46A		
151930	CH46A		
151931	CH46A		
151932	CH46A	HC-11	VR-75
151933	HH46A	HC-11	VR-64

151934	HH46A	HC-11	VR-73
151935	CH46A		
151936	HH46A	NAS	Oceana
151937	HH46D	HC-3	SA-01
151938	CH46A		
151939	HH46A	NAWC-WD	14
151940	CH46A		
151941	CH46A	HC-8	BR-54
151942	CH46D	HC-11	VR-71
151943	CH46A		
151944	HH46A	w.o.12.6.79	
151945	CH46A		
151946	CH46A		
151947	CH46D	HC-3	SA-04
151948	HH46D	MCAS	Cherry Point
151949	HH46D	HC-6	HW-05
151950	HH46A	MCAS	KB Kaneohe Bay
151951	HH46A	HC-3	SA-06
151952	UH46D	HC-6	HW-16
151953	HH46D	HC-6	HW-06
151954	CH46A		
151955	HH46A	HC-5	RB-06
151956	HH46A	HC-3	SA-15
151957	CH46A	HC-8	BR-41
151958	CH46A		
151959	HH46A	dump	Sanford, Fl
151960	CH46A		
151961	CH46A		
152490	UH46A		
152491	CH46A	NATTC	
152492	UH46A	MASDC	
152493	UH46A	HC-6	HW-21
152494	UH46A		
152495	CH46A	HC-11	VR-62
152496	CH46A	MCAS	KB Kaneohe Bay
152497	CH46A		
152498	HH46D	NAD	Cherry Point
152499	CH46E	w.o.1.11.84	
152500	CH46A		
152501	HH46A	NAWC-WD	13
152502	CH46A		
152503	HH46A	Cubi Point	NAS
152504	CH46A		
152505	CH46A		
152506	CH46A		
152507	CH46A		
152508	CH46D	w.o.28.2.83	
152509	CH46A		
152510	HH46A	dump	Norfolk NAS
152511	HH46A	w.o.13.5.82	
152512	HH46D	MCAS	Beaufort
152513	HH46A		
152514	CH46A		
152515	CH46A		
152516	CH46A		
152517	CH46A		
152518	HH46A	w.o.22.10.87	
152519	CH46A		
152520	HH46A	HC-8	BR-40
152521	CH46A		
152522	HH46A	NAWC-WD	15
152523	CH46A		
152524	CH46A		
152525	CH46A		
152526	CH46A		
152527	CH46A		
152528	CH46A	Iwakuni	MCAS
152529	CH46A		
152530	CH46A	Iwakuni	MCAS
152531	CH46A		
152532	CH46A		
152533	CH46A		
152534	CH46A		
152535	HH46D	HC-11	VR-67
152536	CH46A		
152537	CH46A		
152538	CH46E	HC-3	SA-03
152539	HH46A	HC-11	VR-70
152540	HH46A	w.o.27.8.87	
152541	CH46A		
152542	HH46A	w.o.12.7.84	
152543	HH46A	NARF	Alameda
152544	CH46A		
152545	CH46A		
152546	CH46A		
152547	CH46A		
152548	CH46A		
152549	CH46A		
152550	CH46A		
152551	CH46A		
152552	CH46A		
152553	HH46A	HC-11	VR-52
152554	CH46D	w.o.17.9.75	
152555	CH46D	HC-6	HW-03
152556	CH46D		
152557	CH46D		
152558	CH46D		
152559	CH46D		
152560	CH46D		
152561	CH46D		
152562	CH46E	HMM-164	YT-02
152563	CH46D	HC-8	BR-45
152564	CH46D		
152565	CH46D	w.o.17.11.85	
152566	CH46D		
152567	CH46D	HC-6	HW-04
152568	CH46D		
152569	CH46D		
152570	CH46D		
152571	CH46D		
152572	CH46D		
152573	CH46D	w.o.10.2.77	
152574	CH46E	HMM-166	YX-602
152575	CH46D	w.o.1.4.86	
152576	CH46D		
152577	CH46D		
152578	CH46E	HMM-774	MQ-402
152579	CH46D	HMM-774	MQ-
153314	CH46D	w.o.8.11.92	
153315	CH46D		
153316	CH46E	HMM-166	YX-603
153317	CH46D		
153318	CH46E	HMM-266	ES-53
153319	CH46D	HC-6	HW-03
153320	CH46D		
153321	CH46E	HMM-764	ML-425
153322	CH46E	HMM-268	YQ-04
153323	CH46D	w.o.21.6.82	
153324	CH46E	HMM-261	EM-
153325	CH46D	HC-3	SA-10
153326	CH46D	HC-3	SA-11
153327	CH46D		
153328	CH46E	HMM-161	YR-07
153329	CH46E	HC-11	VR-77
153330	CH46D	HMM-263	EG-08
153331	CH46E	w.o.2.3.88	
153332	CH46D		
153333	CH46E	HMM-365	YM-12
153334	CH46D		
153335	CH46D	HC-6	HW-18
153336	CH46D		
153337	CH46D	w.o.17.12.76	
153338	CH46D	HMM-164	YT-452
153339	CH46D	HC-8	BR-56
153340	CH46D		
153341	CH46D	HC-6	HW-11
153342	CH46D		
153343	CH46D		
153344	CH46D	NAWC\RWATD	
153345	CH46E	HC-3	SA-13
153346	CH46E	HMM-166	YX-605
153347	CH46E	HMT-204	GX-02
153348	CH46D		
153349	CH46D		
153350	CH46E	HMT-301	SU-06
153351	CH46D		
153352	CH46D	HC-8	BR-54
153353	CH46E	HMT-301	SU-04
153354	CH46D	w.o.20.8.73	
153355	CH46D	NAWC-AD	RWATD
153356	CH46D	HMM-364	PF-
153357	CH46E	HMM-162	YS-12
153358	CH46D	HC-3	SA-
153359	CH46D	HMM-364	PF-
153360	CH46D		
153361	CH46D		

Serial	Type	Unit	Code
153362	CH46E		
153363	CH46E	w.o.11.5.88	
153364	CH46D		
153365	CH46E	HMH-262	ET-
153366	CH46E	HMM-364	PF-
153367	CH46D		
153368	CH46D	HMM-774	MQ-
153369	CH46E	HMM-266	ES-01
153370	CH46D	w.o.11.6.81	
153371	CH46D	HC-6	HW-
153372	CH46E	HMM-263	EG-02
153373	CH46D	HMM-364	PF-
153374	CH46D	MASDC	
153375	CH46D		
153376	CH46D		
153377	CH46E	HMM-163	YP-07
153378	CH46D	w.o.18.1.78	
153379	CH46E	HMM-161J	YR-
153380	CH46E	HMM-265	EP-04
153381	CH46D		
153382	CH46E	HMH-264	EH-03
153383	CH46E	w.o.15.2.86	
153384	CH46D		
153385	CH46D		
153386	CH46D		
153387	CH46D	w.o.16.7.80	
153388	CH46D	HMT-301	SU-
153389	CH46E	HMM-268	YQ-11
153390	CH46D		
153391	CH46E	HMM-165	YW-
153392	CH46E	NAWC\RWATD	
153393	CH46E	HMM-265	EP-03
153394	CH46D		
153395	CH46E	HMM-365	YM-17
153396	CH46D		
153397	CH46D	w.o.7.7.69	
153398	CH46E	w.o.22.11.87	
153399	CH46E	HMM-162	YS-
153400	CH46E	HMM-165	YW-
153401	CH46D		
153402	CH46E	NAD	Cherry Point
153403	CH46D		
153404	UH46D	NAWC-WD	17
153405	UH46D	HC-8	BR-44
153406	UH46D	HC-6	HW-02
153407	UH46D		
153408	HH46D	NAD	Cherry Point
153409	UH46D	w.o.21.2.91	
153410	UH46D		
153411	UH46D	HC-6	HW-13
153412	UH46D	HC-8	BR-46
153413	UH46D	HC-8	BR-42
153951	CH46D		
153952	CH46E	HMM-165	YW-00
153953	CH46E	HMM-263	EG-03
153954	CH46E	NAD	Cherry Point
153955	CH46D		
153956	CH46E	HMM-164	YT-03
153957	CH46E	HMM-365	YM-15
153958	CH46E	HMM-162	YS-09
153959	CH46E	HMM-266	ES-12
153960	CH46E	HMM-365	YM-00
153961	CH46D	w.o.3.3.81	
153962	CH46E	HMM-774	MQ-409
153963	CH46D		
153964	CH46D		
153965	CH46E	HMM-266	ES-02
153966	CH46D		
153967	CH46D		
153968	CH46E	w.o.15.10.85	
153969	CH46E	HMM-163	YP-12
153970	CH46E	HMM-266	ES-24
153971	CH46D		
153972	HH46D	HC-8	BR-53
153973	CH46E	HMM-266	ES-23
153974	CH46D	HMT-204	GX-12
153975	CH46E	HMM-263	EG-11
153976	CH46D		
153977	CH46D	HMM-165	YW-
153978	CH46D	w.o.8.10.85	
153979	CH46E	HMM-163	YP-02
153980	CH46E	HMM-764	ML-433
153981	CH46E	HMM-263	EG-04
153982	CH46D		
153983	CH46E	HMM-365	YM-04
153984	CH46D		
153985	CH46E	w.o.9.10.86	
153986	CH46D		
153987	CH46D		
153988	CH46D		
153989	CH46D		
153990	CH46E	HMM-263	EG-07
153991	CH46D		
153992	CH46E	HMM-263	EG-05
153993	CH46E	HMM-163	YP-01
153994	CH46E	HMM-162	YS-05
153995	CH46D		
153996	CH46D		
153997	CH46D		
153998	CH46E	HMM-264	EH-04
153999	CH46E	HMM-365	YM-02
154000	CH46E	HMM-774	MQ-407
154001	CH46E	HMM-164	YT-
154002	CH46D	HC-6	HW-15
154003	CH46E	HMT-204	GX-23
154004	CH46E	HMM-774	MQ-401
154005	CH46E	HMM-774	MQ-
154006	CH46D		
154007	CH46D		
154008	CH46D	w.o.	
154009	CH46E	HMM-261	EM-03
154010	CH46E	NAD	Cherry Point
154011	CH46E	HMM-165	YW-01
154012	CH46E	HMM-364	PF-05
154013	CH46E		
154014	CH46E	HMM-774	MQ-405
154015	CH46E	HMM-265	EP-12
154016	CH46E	HMM-164	YT-13
154017	CH46D	HMM-165	YW-
154018	CH46D		
154019	CH46D		
154020	CH46E	HMT-304	GX-10
154021	CH46E	HMM-261	EM-05
154022	CH46D		
154023	CH46E	HMM-265	EP-02
154024	CH46D		
154025	CH46D		
154026	CH46D	w.o.5.12.73	
154027	CH46E	HMM-165	YW-02
154028	CH46D	w.o.17.11.82	
154029	CH46D		
154030	CH46D	HMM-164	YT-
154031	CH46E	HMM-261	EM-01
154032	CH46D	HC-6	HW-11
154033	CH46E	HMM-164	YT-00
154034	CH46E	HMM-162	YS-01
154035	CH46D	w.o.21.11.81	
154036	CH46E	HMM-161	YR-15
154037	CH46E	HMM-261	EM-02
154038	CH46E	HMM-266	ES-04
154039	CH46E	HMM-165	YW-04
154040	CH46E	HMM-364	PF-02
154041	CH46D		
154042	CH46D	w.o.	
154043	CH46D		
154044	CH46D		
154789	CH46E	HMT-301	SU-
154790	CH46E	HMT-301	SU-01
154791	CH46D		
154792	CH46E	HMM-165	YW-06
154793	CH46D		
154794	CH46D		
154795	CH46E	HMT-301	SU-
154796	CH46D		
154797	CH46D		
154798	CH46E	HMM-161	YR-01
154799	CH46E	HMM-166	YX-611
154800	CH46D		
154801	CH46E	HMM-162	YS-04
154802	CH46D		
154803	CH46E	HMM-764	ML-430
154804	CH46D		
154805	CH46E	HMM-268	YQ-
154806	CH46D	w.o.3.3.78	
154807	CH46E	HMM-165	YW-07
154808	CH46E	HMM-774	MQ-404

154809	CH46D		
154810	CH46E	HMM-774	MQ-403
154811	CH46D		
154812	CH46E	HMM-166	YX-613
154813	CH46E	HMM-164	YT-11
154814	CH46D		
154815	CH46E	HMM-764	ML-426
154816	CH46E	HMM-268	YQ-13
154817	CH46E	HMM-265	EP-14
154818	CH46D		
154819	CH46E	HMM-261	EM-13
154820	CH46D		
154821	CH46E	HMM-261	EM-
154822	CH46E	HMM-161	YR-03
154823	CH46D	HC-3	SA-12
154824	CH46D	HC-8	BR-43
154825	CH46D	HMM-268	YQ-05
154826	CH46D	HC-11	VR-72
154827	CH46E	HMM-163	YP-03
154828	CH46E	HMM-265	EP-05
154829	CH46E	HMM-764	ML-432
154830	CH46E		
154831	CH46E	NAD	Cherry Point
154832	CH46E	HMM-161	YR-10
154833	CH46E	HMM-163	YP-
154834	CH46E	HMM-163	YP-04
154835	CH46D		
154836	CH46D	HC-11	VR-67
154837	CH46D		
154838	CH46D		
154839	CH46D		
154840	CH46D		
154841	CH46D		
154842	CH46D		
154843	CH46D		
154844	CH46D	HMM-165	YW-16
154845	CH46F	w.o.24.2.91	
154846	CH46E	HMM-161	YR-14
154847	CH46E	HMM-774	MQ-405
154848	CH46F	HMM-164	YT-46
154849	CH46E	HMM-365	YM-06
154850	CH46E	HMM-265	EP-12
154851	CH46E	HMM-261	EM-11
154852	CH46F		
154853	CH46E	HMM-163	YP-00
154854	CH46E	HMM-365	YM-05
154855	CH46F	HMM-265	EP-4
154856	CH46E	HMM-162	YS-07
154857	CH46E	HMM-162	YS-11
154858	CH46F	w.o.4.2.77	
154859	CH46F		
154860	CH46F	HMT-301	SU-12
154861	CH46F	w.o.3.10.79	
154862	CH46F	El Centro	
155301	CH46E	HMT-204	GX-
155302	CH46E	HMM-774	MQ-400
155303	CH46E	HMM-161J	YR-103
155304	CH46E	HMT-204	GX-08
155305	CH46E	HMM-165	YW-
155306	CH46E	HMM-365	YM-07
155307	CH46E	HMM-264	EH-07
155308	CH46E	HMT-204	GX-04
155309	CH46E	HMM-164	YT-04
155310	CH46E	HMM-774	MQ-412
155311	CH46E	HMM-163	YP-05
155312	CH46E	HMM-774	MQ-406
155313	CH46E	HMM-265	EP-06
155314	CH46F		
155315	CH46E	HMM-161	YR-02
155316	CH46E	HMM-164	YT-05
155317	CH46F	w.o.14.2.86	
155318	CH46F	HMM-262	ET-5
156418	CH46E	HMM-266	ES-61
156419	CH46E	HMM-764	ML-429
156420	CH46E	HMM-166	YX-615
156421	CH46E	HMM-261	EM-04
156422	CH46E	HMM-163	YP-06
156423	CH46F	w.o.20.12.83	
156424	CH46E	HMM-161	YR-04
156425	CH46E		
156426	CH46E	HMM-266	ES-10
156427	CH46E	HMM-165	YW-10
156428	CH46F	w.o.13.5.76	
156429	CH46E	HMM-262	ET-
156430	CH46E	HMM-262	ET-1
156431	CH46E	HMM-774	MQ-411
156432	CH46F		
156433	CH46E	HMM-764	ML-434
156434	CH46E	HMM-165	YW-11
156435	CH46E	HMM-164	YT-07
156436	CH46E	HMM-264	EH-00
156437	CH46E	HMM-364	PF-12
156438	CH46E	HMM-166	YX-604
156439	CH46E	HMM-265	EP-01
156440	CH46E	HMM-265	EP-06
156441	CH46E	HMM-163	YP-11
156442	CH46E	HMM-165	YW-06
156443	CH46E	HMM-764	ML-423
156444	CH46E	HMM-764	ML-424
156445	CH46E	HMM-265	EP-07
156446	CH46E	HMM-268	YQ-
156447	CH46E	HMT-301	SU-05
156448	CH46F	w.o.7.11.83	
156449	CH46F	HMM-165	YW-12
156450	CH46E	HMM-162	YS-11
156451	CH46E	HMM-166	YX-606
156452	CH46E	HMT-301	SU-11
156453	CH46E	HMM-774	MQ-
156454	CH46E	HMM-264	EH-
156455	CH46E	HMM-161	YR-110
156456	CH46E	HMM-162	YS-3
156457	CH46E	HMM-161	YR-11
156458	CH46F	HMM-165	YW-
156459	CH46E	HMM-161	YR-06
156460	CH46E	HMM-364	PF-04
156461	CH46F	HMM-764	ML-417
156462	CH46E	HMM-774	MQ-408
156463	CH46E	w.o.20.2.91	
156464	CH46E	NARF	Cherry Point
156465	CH46E	HMM-364	PF-12
156466	CH46E	HMM-364	PF-
156467	CH46E	HMM-162	YS-03
156468	CH46F		
156469	CH46E		
156470	CH46E	HMT-301	SU-13
156471	CH46E	HMT-301	SU-02
156472	CH46E	HMT-204	GX-12
156473	CH46F	HMM-266	ES-63
156474	CH46F	HMM-262	ET-
156475	CH46F		
156476	CH46E	HMM-162	YS-08
156477	CH46E	HMM-266	ES-
157649	CH46E	HMM-163	YP-13
157650	CH46E	HMM-261	EM-10
157651	CH46E	HMM-365	YM-10
157652	CH46E	HMM-261	EM-07
157653	CH46E	HMM-264	EH-13
157654	CH46F	HMH-363	YZ-50
157655	CH46E	HMM-165	YW-13
157656	CH46E	HMM-264	EH-05
157657	CH46F	HMM-263	EG-
157658	CH46F		
157659	CH46E	w.o.13.6.86	
157660	CH46E	HMM-266	ES-07
157661	CH46E	HMM-265	EP-10
157662	CH46E	HMM-268	YQ-03
157663	CH46E	HMM-261	EM-14
157664	CH46E	HMT-301	SU-14
157665	CH46E	HMM-261	EM-12
157666	CH46F	w.o.10.2.81	
157667	CH46E	HMM-266	ES-08
157668	CH46E	HMM-263	EG-00
157669	CH46E	HMM-364	PF-13
157670	CH46E	HMM-164	YT-10
157671	CH46F	w.o.16.11.73	
157672	CH46E	HMM-266	ES-09
157673	CH46E	HMM-365	YM-01
157674	CH46F	HMM-161	YR-047
157675	CH46F	HMM-165	YW-14
157676	CH46F	w.o.3.4.74	
157677	CH46F	w.o.16.5.76	
157678	CH46E	HMM-266	ES-05
157679	CH46F	HMX-1	MX-
157680	CH46F	HMX-1	MX-18
157681	CH46F	HMX-1	MX-19
157682	CH46F	HMX-1	MX-20

157683	VH46F	HMX-1	MX-21
157684	CH46F	HMX-1	MX-14
157685	CH46E	HMT-301	SU-03
157686	CH46E	w.o.3.9.87	
157687	CH46E	HMM-264	EH-12
157688	CH46E	HMM-263	EG-
157689	CH46E	w.o.29.8.86	
157690	CH46E	HMM-261	EM-06
157691	CH46E	HMM-265	EP-13
157692	CH46E	HMM-161	YR-106
157693	CH46E	HMM-164	YT-12
157694	CH46E	HMT-301	SU-07
157695	CH46F		
157696	CH46E	HMM-166	YX-601
157697	CH46E	HMM-162	YS-06
157698	CH46E	HMM-266	ES-
157699	CH46E	HMM-262	ET-14
157700	CH46F		
157701	CH46E	HMM-162	YS-1
157702	CH46E	HMM-166	YX-607
157703	CH46E	HMM-166	YX-612
157704	CH46E	HMM-164	YT-01
157705	CH46E	damaged	MCAS Cherry Pt.
157706	CH46E	HMM-264	EH-
157707	CH46E	HMM-263	EG-
157708	CH46F	HMM-265	EP-14
157709	CH46F		
157710	CH46E	HMM-264	EH-06
157711	CH46F	HMM-261	EM-
157712	CH46F	HMM-364	PF-
157713	CH46E	HMM-365	YM-11
157714	CH46E	HMM-265	EP-03
157715	CH46E	HMM-161	YR-12
157716	CH46E	HMM-164	YT-443
157717	CH46F		
157718	CH46F	w.o.20.3.82	
157719	CH46F		
157720	CH46E	HMT-774	MQ-410
157721	CH46F	HMT-204	GX-11
157722	CH46E	HMM-165	YW-
157723	CH46E	HMM-268	YQ-37
157724	CH46E	HMM-161	YR-00
157725	CH46E	w.o.13.2.87	
157726	CH46E	HMM-263	EG-12

H53 SEA STALLION

151613	YCH53A		
151614	YCH53A		
151686	CH53A		
151687	CH53A	NARF	Pensacola
151688	CH53A	HMH-772	QM-401
151689	CH53A	NARF	Pensacola
151690	CH53A	VC-1	UA-14
151691	CH53A		
151692	CH53A	NARF	Pensacola
151693	CH53A		
151694	CH53A	HMH-772	MT-402
151695	CH53A	NATTC	
151696	CH53A	HMH-772	QM-402
151697	CH53A		
151698	CH53A	NARF	Pensacola
151699	CH53A	HMH-772	QM-404
151700	CH53A	NARF	North Island
151701	CH53A	HMT-204	GX-41
152392	CH53A	HMH-772	MT-405
152393	CH53A	w.o.	
152394	CH53A		
152395	CH53A	HMT-301	SU-21
152396	CH53A	HMH-772	QM-405
152397	CH53A		
152398	CH53A	NARF	Pensacola
152399	CH53A	to FAA	N39
152400	CH53A		
152401	CH53A	HMH-772	MS-486
152402	CH53A	HMT-302	UT-02
152403	CH53A	HMH-361	YN-68
152404	CH53A	MASDC	
152405	CH53A	HMH-361	YN-72
152406	CH53A	HMT-301	SU-26
152407	CH53A		
152408	CH53A	AMARC	2J031
152409	CH53A	HMH-363	YZ-
152410	CH53A		
152411	CH53A	AMARC	2J040
152412	CH53A	HMT-301	SU-27
152413	CH53A		
152414	CH53A	HMT-301	SU-
152415	CH53A	HMH-772	QM-
153274	CH53A	HMH-363	
153275	CH53A	HMT-301	SU-
153276	CH53A		
153277	CH53A	NARF	Pensacola
153278	CH53A		
153279	CH53A	AMARC	2J036
153280	CH53A		
153281	CH53A		
153282	CH53A		
153283	CH53A		
153284	CH53A		
153285	CH53A	HMH-772	MS-485
153286	CH53A	HMH-361	YN-67
153287	CH53A	AMARC	2J030
153288	CH53A		
153289	CH53A	HMH-772	QM-403
153290	CH53A	HMH-361	YN-03
153291	CH53A	to USAF	
153292	CH53A	to USAF	
153293	CH53A	HMH-361	YN-60
153294	CH53A	HMH-361	YN-1
153295	CH53A	w.o.29.8.86	
153296	CH53A		
153297	CH53A	w.o.2.8.78	
153298	CH53A	HMT-301	SU-25
153299	CH53A	HMH-772	MT-401
153300	CH53A	HMH-772	MT-404
153301	CH53A	HMH-772	QM-406
153302	CH53A	HMH-772	MS-484
153303	CH53A	HMH-772	MS-482
153304	CH53A		
153305	CH53A		
153306	CH53A		
153307	CH53A	HMH-361	YN-
153308	CH53A	HMT-301	SU-
153309	CH53A	HMH-363	YZ-76
153310	CH53A	HMH-772	MT-407
153311	CH53A	HMH-361	YN-20
153312	CH53A	w.o.14.9.82	
153313	CH53A		
153705	CH53A	HMH-361	YN-
153706	CH53A	HMT-301	SU-
153707	CH53A	VC-1	UA-15
153708	CH53A	AMARC	2J043
153709	CH53A		
153710	CH53A		
153711	CH53A	AMARC	2J037
153712	CH53A	VC-1	UA-16
153713	CH53A		
153714	CH53A		
153715	CH53A	NATTC	Memphis
153716	CH53A		
153717	CH53A	HMH-363	YZ-
153718	CH53A	GIA	NAS Norfolk

Serial	Type	Unit/Notes	Code
153719	CH53A	w.o.19.7.81	
153720	CH53A	AMARC	2J032
153721	CH53A	dump	Pensacola NAS
153722	CH53A		
153723	CH53A	w.o.25.2.73	
153724	CH53A		
153725	CH53A		
153726	CH53A	AMARC	2J033
153727	CH53A		
153728	CH53A	HMH-462	YF-
153729	CH53A	NARF	Pensacola
153730	CH53A		
153731	CH53A	NARF	Pensacola
153732	CH53A	AMARC	2J029
153733	CH53A		
153734	CH53A		
153735	CH53A		
153736	CH53A		
153737	CH53A		
153738	CH53A		
153739	CH53A	HMH-363	YZ-
154863	CH53A	AMARC	2J034
154864	CH53A		
154865	CH53A	NARF	Pensacola
154866	CH53A		
154867	CH53A	HMH-772	QM-400
154868	CH53A	HMH-363	YZ-00
154869	CH53A	HMH-361	YN-54
154870	CH53A	HMH-363	YZ-15
154871	CH53E	HMH-361	YN-
154872	CH53A		
154873	CH53A		
154874	CH53A		
154875	CH53A		
154876	CH53A	HMT-204	GX-30
154877	CH53A	AMARC	2J039
154878	CH53A	AMARC	2J035
154879	CH53A	HMT-301	SU-20
154880	CH53A		
154881	CH53A	NARF	Pensacola
154882	CH53A	HMT-301	SU-17
154883	CH53A	HMH-772	MT-406
154884	CH53A	HMH-772	MT-403
154887	CH53A	HMT-301	SU-
154888	CH53A		
156654	CH53D	HMH-772	MT-
156655	CH53D	HMH-463	YH-14
156656	CH53D	HMH-363	YZ-17
156657	CH53D		
156658	CH53D		
156659	CH53D	HMM-265	EP-22
156660	CH53D		
156661	CH53D	HMM-165	YW-21
156662	CH53D		
156663	CH53D	HMH-772	QM-400
156664	CH53D		
156665	CH53D		
156666	CH53D	HMH-463	YH-15
156667	CH53D		
156668	CH53D	HMH-772	MT-402
156669	CH53D	HMH-463	YH-17
156670	CH53D		
156671	CH53D	HMH-772	MT-71
156672	CH53D	HMH-772	MT-72
156673	CH53D	HMH-463	YH-22
156674	CH53D	HMH-772	QM-401
156675	CH53D		
156676	CH53D	HMH-772	QM-402
156677	CH53D	HMT-302	UT-
156951	CH53A		
156952	CH53A	w.o.18.5.80	
156953	CH53D		
156954	CH53D	HMT-302	UT-01
156955	CH53D	HMH-461	CJ-16
156956	CH53D		
156957	CH53D		
156958	CH53D		
156959	CH53D	HMH-362	YL-01
156960	CH53D	HMT-302	UT-05
156961	CH53D	HMH-362	YL-
156962	CH53D		
156963	CH53D	HMT-204	GX-5
156964	CH53D	HMH-362	YL-03
156965	CH53D	w.o.13.2.93	
156966	CH53D	w.o.15.12.79	
156967	CH53D	HMH-362	YL-15
156968	CH53D	w.o.8.1.75	
156969	CH53D	HMT-204	GX-3
156970	CH53D	w.o.12.7.85	
157127	CH53D		
157128	CH53D	HMH-463	YH-OO
157129	CH53D	HMH-361	YN-655
157130	CH53D	HMH-772	QM-403
157131	CH53D	HMH-463	YH-01
157132	CH53D	w.o.24.3.84	
157133	CH53D	NAS	Barbers Point
157134	CH53D	HMH-463	YH-03
157135	CH53D	HMH-463	YH-04
157136	CH53D	HMH-363	YZ-57
157137	CH53D	HMH-463	YH-05
157138	CH53D	w.o.6.5.85	
157139	CH53D		
157140	CH53D	w.o.20.3.89	
157141	CH53D	HMH-362	YL-
157142	CH53D	HMH-362	YL-14
157143	CH53D	NAD	Patuxent River
157144	CH53D	HMH-463	YH-06
157145	CH53D	HMH-361	YN-654
157146	CH53D	HMH-463	YH-07
157147	CH53D		
157148	CH53D		
157149	CH53D		
157150	CH53D	HMH-362	YL-00
157151	CH53D	HMM-162	YS-24
157152	CH53D	w.o.27.3.83	
157153	CH53D	w.o.14.3.74	
157154	CH53D	w.o.22.1.87	
157155	CH53D	HMH-362	YL-08
157156	CH53D	HMH-461	CJ-17
157157	CH53D	HMH-363	YZ-
157158	CH53D	HMH-463	YH-08
157159	CH53D	HMH-463	YH-09
157160	CH53D		
157161	CH53D		
157162	CH53D		
157163	CH53D	w.o.25.6.88	
157164	CH53D	HMH-463	YH-05
157165	CH53D	HMH-362	YL-
157166	CH53D	HMT-204	GX-00
157167	CH53D	HMT-302	UT-02
157168	CH53D	HMT-302	UT-
157169	CH53D	HMH-363	YZ-66
157170	CH53D	HMH-362	YL-05
157171	CH53D		
157172	CH53D	HMM-264	EH-
157173	CH53D	HMT-302	UT-06
157174	CH53D	HMH-462	YF-53
157175	CH53D	HMH-363	YZ-70
157176	CH53D	HMH-363	YZ-71
157727	CH53D	HMH-363	YZ-
157728	CH53D	HMH-362	YL-10
157729	CH53D	HMH-363	YZ-73
157730	CH53D		
157731	CH53D	w.o.17.9.81	
157732	CH53D	HMT-302	UT-03
157733	CH53D	NAS	Barbers Point
157734	CH53D	HMH-363	YZ-74
157735	CH53D		
157736	CH53D	HMT-302	UT-04
157737	CH53D	w.o.7.2.75	
157738	CH53D		
157739	CH53D	HMM-264	EH-26
157740	CH53D	HMH-363	YZ-75
157741	CH53D	HMH-463	YH-12
157742	CH53D	HMH-362	YL-19
157743	CH53D		
157744	CH53D		
157745	CH53D	HMH-461	CJ-
157746	CH53D	HMT-302	UT-07
157747	CH53D	HMT-302	UT-10
157748	CH53D	HMH-363	YZ-26
157749	CH53D	HMH-362	YL-
157750	CH53D		
157751	CH53D	HMH-463	YH-13
157752	CH53D		
157753	CH53D	HMH-463	YH-14

157754	VH53D	HMX-1			161541	CH53E	HC-2	HU-712
157755	VH53D	HMX-1			161542	CH53E	HC-1	UP-742
157756	VH53D	HMX-1			161543	CH53E	HC-2	HU-710
157930	VH53D	HMX-1	26		161544	CH53E		
157931	CH53D	HMH-363	YZ-57		161545	CH53E		
158682	RH53D	HM-12	DH-431		161988	CH53E		
158683	RH53D	HMH-772	MS-481		161989	CH53E		
158684	RH53D	AMARC	2J060		161990	CH53E	HMT-301	SU-
158685	RH53D	NCSC			161991	CH53E	HMH-465	YJ-55
158686	RH53D	w.o.24.4.80			161992	CH53E	HMH-466	YK-01
158687	RH53D	w.o.19.6.92			161993	CH53E	HMH-466	YK-04
158688	RH53D				161994	CH53E	HMH-464	EN-15
158689	RH53D	HMH-772	MS-482		161995	CH53E	HMH-464	EN-16
158690	RH53D	HM-18	NW-600		161996	CH53E		
158691	RH53D				161997	CH53E	HMT-302	UT-20
158692	RH53D	w.o.24.4.80			161998	CH53E	HMH-464	EN-53
158693	RH53D	HM-18	NW-602		161999	CH53E	w.o.7.2.85	
158744	RH53D	w.o.24.4.80			162000	CH53E	w.o.1.6.84	
158745	RH53D	HMH-772	MS-483		162001	CH53E	HMH-464	EN-17
158746	RH53D	HM-14	BJ-534		162002	CH53E	NARF	Pensacola
158747	RH53D	HM-12	DH-		162003	CH53E	HMH-465	YJ-56
158748	RH53D	HM-19	NW-624		162004	CH53E		
158749	RH53D	HMH-772	MS-484		162005	CH53E	HMH-466	YK-17
158750	RH53D	w.o.24.4.80			162006	CH53E	HMH-466	YK-06
158751	RH53D	HM-12	DH-430		162007	CH53E	HMM-268	YQ-20
158752	RH53D	HMH-772	MS-486		162008	CH53E	w.o.8.1.87	
158753	RH53D	w.o.24.4.80			162009	CH53E	HMT-301	SU-42
158754	RH53D	HM-19	NW-625		162010	CH53E		
158755	RH53D	HM-18	NW-602		162011	CH53E		
158756	RH53D	HM-19	NW-622		162012	CH53E		
158757	RH53D	HM-19	NW-627		162478	CH53E		
158758	RH53D	w.o.24.4.80			162479	CH53E	HMT-302	UT-24
158759	RH53D	HM-19	NW-623		162480	CH53E	HMH-465	YJ-62
158760	RH53D				162481	CH53E	HMH-464	EN-47
158761	RH53D	w.o.24.4.80			162482	CH53E	HMH-465	YJ-63
159121	YCH53E				162483	CH53E	HMH-466	YK-
159122	YCH53E				162484	CH53E	HMH-465	YJ-
159876	CH53E				162485	CH53E		
159877	CH53E				162486	CH53E	HMH-466	YK-07
161179	CH53E	HMH-466	YK-14		162487	CH53E		
161180	CH53E	MEU	YS-		162488	CH53E		
161181	CH53E	HMH-464	EN-20		162489	CH53E	HMH-466	YK-13
161182	CH53E	HMH-461	CJ-15		162490	CH53E	HMM-162	YS-44
161183	CH53E	HMH-461	CJ-01		162491	CH53E		
161184	CH53E	HMH-464	EN-05		162492	CH53E	HMH-461	CJ-23
161252	CH53E	HMH-464	EN-06		162493	CH53E	HMH-465	YJ-
161253	CH53E	w.o.18.11.84			162494	CH53E	HMM-466	YK-12
161254	CH53E	HMH-464	EN-04		162495	CH53E	HMH-464	EN-57
161255	CH53E	HMM-162	YS-21		162496	CH53E	HMH-462	YH-53
161256	CH53E	HMH-464	EN-08		162497	MH53E		
161257	CH53E	HMM-162	YS-41		162498	MH53E	HM-12	DH-444
161258	CH53E	NAD	Patuxent River		162499	MH53E		
161259	CH53E	HMH-464	EN-10		162500	MH53E		
161260	CH53E	HMH-464	EN-11		162501	MH53E	HMT-302	UT-27
161261	CH53E	MEU	YS-		162502	MH53E		
161262	CH53E	HMH-461	CJ-17		162503	MH53E	HM-12	DH-442
161263	CH53E	NAD	Patuxent River		162504	MH53E	HM-12	DH-440
161264	CH53E	w.o.9.5.86			162505	MH53E	HM-12	DH-441
161265	CH53E	HMH-462	YF-52		162506	MH53E	HM-12	DH-443
161381	CH53E	HMT-301	SU-		162507	MH53E	HM-15	TB-01
161382	CH53E	HMH-465	YJ-51		162508	MH53E	HM-15	TB-02
161383	CH53E	HMH-465	YJ-05		162509	MH53E	HM-15	TB-03
161384	CH53E	HMH-465	YJ-06		162510	MH53E	HM-15	TB-04
161385	CH53E	HMH-465	YJ-52		162511	MH53E	HM-15	TB-05
161386	CH53E	HMH-465	YJ-53		162512	MH53E	w.o.18.7.88	
161387	CH53E	HMH-465	YJ-11		162513	MH53E	NAWC-AD	RWATD
161388	CH53E	HMH-465	YJ-54		162514	MH53E	HM-15	TB-10
161389	CH53E	HMM-361	YN-11		162515	MH53E	HM-15	TB-11
161390	CH53E	HMM-268	YQ-22		162516	MH53E		
161391	CH53E	HMT-301	SU-		162517	MH53E	HMH-464	EN-50
161392	CH53E	HMT-301	SU-37		162518	MH53E	MEU	YS-
161393	CH53E	HMH-465	YJ-20		162519	MH53E	HMM-261	EM-22
161394	CH53E	MEU	YS-23		162520	MH53E	HMM-261	EM-23
161395	MH53E	HMH-461	CJ-22		162521	MH53E	HMM-261	EM-21
161532	CH53E	HC-4	HC-532		162522	MH53E		
161533	CH53E	HM-12	DH-435		162523	MH53E		
161534	CH53E	HM-12	DH-436		162524	MH53E		
161535	CH53E	HC-2	HU-711		162525	MH53E		
161536	CH53E	HC-4	HC-536		162526	MH53E		
161537	CH53E	HC-4	HC-537		162527	MH53E		
161538	CH53E	HC-4	HC-538		162528	MH53E		
161539	CH53E	HC-4	HC-539		162529	MH53E		
161540	CH53E	HC-4	HC-540		162530	MH53E		

162531	MH53E		
162687	MH53E		
162718	MH53E		
162719	MH53E		
162720	MH53E	HM-15	TB-14
163051	MH53E		
163052	MH53E	HM-15	TB-15
163053	MH53E	HM-14	BJ-540
163054	MH53E		
163055	MH53E	HM-12	DH-446
163056	MH53E	HM-14	BJ-543
163057	MH53E	HM-14	BJ-544
163058	MH53E		
163059	MH53E	HMM-261	EM-20
163060	MH53E	HMH-464	EN-10
163061	MH53E		
163062	MH53E	HMH-461	CJ-05
163063	MH53E	HMH-461	CJ-06
163064	MH53E	w.o.4.11.92	
163065	MH53E	HM-14	BJ-550
163066	MH53E	HM-14	BJ-551
163067	MH53E	HM-12	DH-445
163068	MH53E	HM-14	BJ-553
163069	MH53E	HM-14	BJ-554
163070	MH53E	HM-14	BJ-555
163071	MH53E		
163072	MH53E	HMT-302	UT-30
163073	MH53E	HMT-302	UT-31
163074	MH53E	HMH-465	YJ-60
163075	MH53E	HMT-302	UT-33
163076	MH53E	HMT-302	UT-34
163077	MH53E	HMT-302	UT-35
163078	MH53E	HMT-302	UT-36
163079	MH53E	HMH-466	YK-15
163080	MH53E		
163081	MH53E	HMH-466	YK-16
163082	MH53E	HMH-466	YK-17
163083	MH53E	HMH-466	YK-20
163084	MH53E	HMH-466	YK-10
163085	MH53E	HMT-302	UT-23
163086	MH53E	NAWC-AD	RWATD
163087	MH53E	HMH-466	YK-21
163088	MH53E		
163089	MH53E		
164358	CH53E	HMH-462	YF-58
164359	CH53E	HMH-462	YF-59
164360	CH53E	HMH-465	YJ-58
164361	CH53E	HMH-462	YF-61
164362	CH53E	HMH-462	YF-62
164363	CH53E	HMH-462	YF-63
164364	CH53E	HMH-462	YF-64
164365	CH53E		
164366	CH53E		
164367	CH53E		
164368	MH53E		
164369	MH53E		
164370	MH53E		
164371	MH53E		
164470	CH53E		
164471	CH53E		
164472	CH53E		
164473	CH53E		
164474	CH53E		
164475	CH53E		
164476	CH53E		
164477	CH53E		
164478	CH53E		
164479	CH53E		
164480	CH53E		
164481	CH53E		
164482	CH53E		
164536	CH53E		
164537	CH53E		
164538	CH53E		
164539	CH53E		
164859	CH53E		
164860	CH53E		
164861	MH53E		
164862	MH53E		
164863	MH53E		
164864	MH53E		

H57 SEA RANGER

c\n	TH57A:	5001 to 5040.	
157355	TH57A	MASDC	
157356	TH57A		
157357	TH57A		
157358	TH57A	MASDC	
157359	TH57A	MASDC	
157360	TH57A	MASDC	
157361	TH57A	AMARC	4H013
157362	TH57A	preserved	NAS Whiting Fd.
157363	TH57A	preserved	NAS Pensacola
157364	TH57A	AMARC	4H019
157365	TH57A	AMARC	4H003
157366	TH57A		
157367	TH57A	AMARC	4H004
157368	TH57A	MASDC	
157369	TH57A	MASDC	
157370	TH57A	AMARC	4H027
157371	TH57A	MASDC	
157372	TH57A		
157373	TH57A	AMARC	4H011
157374	TH57A	MASDC	
157375	TH57A	AMARC	4H005
157376	TH57A	AMARC	4H006
157377	TH57A	MASDC	
157378	TH57A	AMARC	4H029
157379	TH57A	AMARC	4H007
157380	TH57A	MASDC	
157381	TH57A	AMARC	4H015
157382	TH57A	AMARC	4H023
157383	TH57A	MASDC	
157384	TH57A		
157385	TH57A	AMARC	4H016
157386	TH57A		
157387	TH57A	AMARC	4H024
157388	TH57A	AMARC	4H017
157389	TH57A		
157390	TH57A		
157391	TH57A		
157392	TH57A	AMARC	4H018
157393	TH57A	MASDC	
157394	TH57A	AMARC	4H030
161695	TH57B	TW-5	E-146
161696	TH57B	TW-5	E-140
161697	TH57B	TW-5	E-141
161698	TH57B	TW-5	E-142
161699	TH57B	TW-5	E-143
161700	TH57B	TW-5	E-144
161701	TH57B	TW-5	E-145
162013	TH57C	TW-5	E-49
162014	TH57C	TW-5	E-50
162015	TH57C	TW-5	E-51
162016	TH57C	TW-5	E-52
162017	TH57C	TW-5	E-53
162018	TH57C	TW-5	E-54
162019	TH57C	TW-5	E-55

162020	TH57C	TW-5	E-56	162681	TH57C	TW-5	E-119
162021	TH57C	TW-5	E-57	162682	TH57C	TW-5	E-120
162022	TH57C	TW-5	E-58	162683	TH57C	TW-5	E-121
162023	TH57C	TW-5	E-59	162684	TH57C	TW-5	E-122
162024	TH57C	TW-5	E-60	162685	TH57C	TW-5	E-123
162025	TH57C	TW-5	E-61	162686	TH57C	TW-5	E-124
162026	TH57C	TW-5	E-62	162803	TH57B	TW-5	E-147
162027	TH57C	TW-5	E-63	162804	TH57B	TW-5	E-148
162028	TH57C	TW-5	E-64	162805	TH57B	TW-5	E-149
162029	TH57C	TW-5	E-65	162806	TH57B	TW-5	E-150
162030	TH57C	TW-5	E-66	162807	TH57B	TW-5	E-151
162031	TH57C	TW-5	E-67	162808	TH57B	TW-5	E-152
162032	TH57C	TW-5	E-68	162809	TH57B	TW-5	E-153
162033	TH57C	TW-5	E-69	162810	TH57B	w.o.6.88	
162034	TH57C	TW-5	E-70	162811	TH57C	TW-5	E-125
162035	TH57C	TW-5	E-71	162812	TH57C	TW-5	E-126
162036	TH57C	TW-5	E-72	162813	TH57C	TW-5	E-127
162037	TH57C	TW-5	E-73	162814	TH57C	TW-5	E-128
162038	TH57C	TW-5	E-74	162815	TH57C	TW-5	E-129
162039	TH57C	TW-5	E-75	162816	TH57C	TW-5	E-130
162040	TH57C	TW-5	E-76	162817	TH57C	TW-5	E-131
162041	TH57C	TW-5	E-77	162818	TH57C	TW-5	E-132
162042	TH57C	TW-5	E-78	162819	TH57C	TW-5	E-133
162043	TH57C	TW-5	E-79	162820	TH57C	TW-5	E-134
162044	TH57C	TW-5	E-80	162821	TH57C	TW-5	E-135
162045	TH57C	TW-5	E-81	162822	TH57C	TW-5	E-136
162046	TH57C	TW-5	E-82	162823	TH57C	TW-5	E-137
162047	TH57C	TW-5	E-83	163312	TH57B	TW-5	E-155
162048	TH57C	TW-5	E-84	163313	TH57B	TW-5	E-156
162049	TH57C	TW-5	E-85	163314	TH57B	TW-5	E-157
162050	TH57C	TW-5	E-86	163315	TH57B	TW-5	E-158
162051	TH57C	TW-5	E-87	163316	TH57B	TW-5	E-159
162052	TH57C	TW-5	E-88	163317	TH57B	TW-5	E-160
162053	TH57C	TW-5	E-89	163318	TH57B	TW-5	E-161
162054	TH57C	TW-5	E-90	163319	TH57B	TW-5	E-162
162055	TH57C	TW-5	E-91	163320	TH57B	TW-5	E-163
162056	TH57C	TW-5	E-92	163321	TH57B	TW-5	E-164
162057	TH57C	TW-5	E-93	163322	TH57B	TW-5	E-165
162058	TH57C	TW-5	E-94	163323	TH57B	TW-5	E-166
162059	TH57C	TW-5	E-95	163324	TH57B	TW-5	E-167
162060	TH57C	TW-5	E-96	163325	TH57B	TW-5	E-168
162061	TH57C	TW-5	E-97	163326	TH57B	TW-5	E-169
162062	TH57C	TW-5	E-98	163327	TH57B	TW-5	E-170
162063	TH57C	TW-5	E-99	163328	TH57B	TW-5	E-171
162064	TH57C	TW-5	E-100	163329	TH57B	TW-5	E-172
162065	TH57C	TW-5	E-101	163330	TH57B	w.o.6.88	
162066	TH57C	TW-5	E-102	163331	TH57B	TW-5	E-174
162067	TH57C	TW-5	E-103	163332	TH57B	TW-5	E-175
162666	TH57C	TW-5	E-104	163333	TH57B	TW-5	E-176
162667	TH57C	TW-5	E-105	163334	TH57B	TW-5	E-177
162668	TH57C	TW-5	E-106	163335	TH57B	TW-5	E-178
162669	TH57C	TW-5	E-107	163336	TH57B	TW-5	E-179
162670	TH57C	TW-5	E-108	163337	TH57B	TW-5	E-180
162671	TH57C	TW-5	E-109	163338	TH57B	TW-5	E-181
162672	TH57C	TW-5	E-110	163339	TH57B	TW-5	E-182
162673	TH57C	TW-5	E-111	163340	TH57B	TW-5	E-183
162674	TH57C	TW-5	E-112	163341	TH57B	TW-5	E-184
162675	TH57C	TW-5	E-113	163342	TH57B	TW-5	E-185
162676	TH57C	TW-5	E-114	163343	TH57B	TW-5	E-186
162677	TH57C	TW-5	E-115	163344	TH57B	TW-5	E-187
162678	TH57C	TW-5	E-116	163345	TH57B	TW-5	E-188
162679	TH57C	TW-5	E-117	163346	TH57B	TW-5	E-189
162680	TH57C	TW-5	E-118	163347	TH57B	TW-5	E-190

H58 KIOWA

71-20554	OH58A	NAWC\TPS	51	72-21193	OH58A	NAWC\TPS	53
71-20799	OH58A	NAWC\TPS	50	72-21253	OH58A	NAWC\TPS	52

H60 SEAHAWK

Serial	Type	Unit	Code
161169	SH60B	NAWC-AD	RWATD
161170	JSH60B	NAWC-AD	RWATD
161171	SH60B	NAWC-AD	RWATD
161172	SH60B	AMARC	1M001
161173	SH60B	preserved	Mayport
161553	SH60B	HSL-41	TS-00
161554	SH60B	HSL-41	TS-01
161555	SH60B	HSL-41	TS-02
161556	SH60B	HSL-41	TS-03
161557	SH60B	HSL-37	TH-
161558	SH60B	HSL-47	TY-66
161559	SH60B	HSL-41	TS-06
161560	SH60B	HSL-41	TS-07
161561	SH60B	VX-1	JA-41
161562	SH60B	HSL-49	TX-109
161563	SH60B	HSL-41	TS-10
161564	SH60B	HSL-41	TS-11
161565	SH60B	HSL-41	TS-12
161566	SH60B	HSL-41	TS-13
161567	SH60B	HSL-41	TS-14
161568	SH60B	HSL-41	TS-15
161569	SH60B	HSL-37	TH-51
161570	SH60B	HSL-43	TT-26
162092	SH60B	w.o.9.6.86	
162093	SH60B	HSL-49	TX-110
162094	SH60B	HSL-49	TX-108
162095	SH60B	HSL-42	HN-421
162096	SH60B	HSL-42	HN-422
162097	SH60B	w.o.23.2.85	
162098	SH60B		
162099	SH60B		
162100	SH60B	HSL-42	HN-424
162101	SH60B	HSL-42	HN-
162102	SH60B	HSL-37	TH-54
162103	SH60B	HSL-43	TT-23
162104	SH60B	HSL-42	HN-425
162105	SH60B	HSL-47	TY-64
162106	SH60B	HSL-42	HN-426
162107	SH60B	NAWC-AD	RWATD
162108	SH60B	HSL-43	TT-27
162109	SH60B		
162110	SH60B	w.o.26.6.86	
162111	SH60B	HSL-48	HR-
162112	SH60B	HSL-43	TT-28
162113	SH60B	w.o.2.8.89	
162114	SH60B	HSL-42	HN-
162115	SH60B	HSL-40	HK-402
162116	SH60B	HSL-40	HK-400
162117	SH60B	HSL-49	TX-104
162118	SH60B	HSL-43	TT-29
162119	SH60B	HSL-43	TT-30
162120	SH60B	HSL-43	TT-32
162121	SH60B	HSL-40	HK-401
162122	SH60B	HSL-42	HN-
162123	SH60B	HSL-42	HN-433
162124	SH60B	HSL-40	HK-404
162125	SH60B	HSL-40	HK-406
162126	SH60B	HSL-40	HK-405
162127	SH60B	HSL-40	HK-403
162128	SH60B	HSL-40	HK-410
162129	SH60B	HSL-40	HK-407
162130	SH60B	HSL-48	HR-504
162131	SH60B	HSL-40	HK-410
162132	SH60B	HSL-44	HP-441
162133	SH60B	HSL-40	HK-411
162134	SH60B	HSL-45	TZ-40
162135	SH60B	HSL-40	HK-
162136	SH60B	HSL-44	HP-451
162137	SH60B	HSL-44	HP-457
162138	SH60B	w.o.21.2.91	
162139	SH60B	HSL-40	HK-413
162326	SH60B	NAWC-AD	RWATD
162327	SH60B	HSL-45	TZ-44
162328	SH60B	HSL-44	HP-442
162329	SH60B	HSL-37	TH-62
162330	SH60B	HSL-49	TX-102
162331	SH60B	HSL-44	HP-455
162332	SH60B	HSL-47	TY-67
162333	SH60B	HSL-48	HR-
162334	SH60B	HSL-44	HP-445
162335	SH60B	HSL-44	HP-446
162336	SH60B	w.o.30.6.89	
162337	SH60B	NAWC-AD	RWATD
162338	SH60B	HSL-45	TZ-43
162339	SH60B	HSL-47	TY-73
162340	SH60B		
162341	SH60B	HSL-47	TY-61
162342	SH60B		
162343	SH60B	HSL-43	TT-26
162344	SH60B	HSL-46	HQ-460
162345	SH60B	HSL-49	TX-107
162346	SH60B	HSL-45	TZ-50
162347	SH60B	HSL-44	HP-453
162348	SH60B	HSL-47	TY-60
162349	SH60B	NAWC-AD	RWATD
162350	SH60B		
162351	SH60B		
162352	SH60B	HSL-45	TZ-51
162353	SH60B		
162354	SH60B		
162355	SH60B		
162356	SH60B		
162357	SH60B		
162358	SH60B		
162359	SH60B		
162360	SH60B		
162361	SH60B		
162362	SH60B		
162363	SH60B		
162364	SH60B		
162365	SH60B		
162366	SH60B	HSL-47	TY-65
162367	SH60B		
162368	SH60B		
162369	SH60B		
162370	SH60B		
162371	SH60B		
162372	SH60B		
162373	SH60B		
162374	SH60B		
162375	SH60B		
162376	SH60B		
162377	SH60B		
162378	SH60B		
162379	SH60B		
162380	SH60B		
162381	SH60B		
162382	SH60B	HSL-44	HP-
162383	SH60B		
162384	SH60B		
162385	SH60B		
162386	SH60B		
162387	SH60B		
162388	SH60B		
162389	SH60B	HSL-49	TX-103
162974	SH60B	NAWC\TPS	
162975	SH60B	HSL-51	TA-01
162976	SH60B	HSL-43	TT-24
162977	SH60B	HSL-46	HQ-460
162978	SH60B	w.o.19.1.88	
162979	SH60B	HSL-47	TY-62
162980	SH60B		

162981	SH60B	HSL-44	HP-455
162982	SH60B	HSL-46	HQ-461
162983	SH60B		
162984	SH60B	HSL-42	HN-435
162985	SH60B	HSL-47	TY-63
162986	SH60B	HSL-48	HR-511
162987	SH60B	HSL-37	TH-64
162988	SH60B	HSL-37	TH-50
162989	SH60B	HSL-46	HQ-464
162990	SH60B		
162991	SH60B	HSL-43	TT-22
162992	SH60B		
162993	SH60B		
162994	SH60B		
162995	SH60B	HSL-49	TX-101
162996	SH60B		
162997	SH60B		
163038	SH60B		
163039	SH60B		
163040	SH60B		
163041	SH60B		
163042	SH60B		
163043	SH60B		
163233	SH60B	HSL-46	HQ-460
163234	SH60B		
163235	SH60B		
163236	SH60B		
163237	SH60B	HSL-46	HQ-471
163238	SH60B	VX-1	JA-38
163239	SH60B	HSL-46	HQ-470
163240	SH60B	NAWC	
163241	SH60B	HSL-43	TT-32
163242	SH60B	VX-1	JA-42
163243	SH60B	VX-1	JA-43
163244	SH60B	HSL-46	HQ-472
163245	SH60B	HSL-48	HR-501
163246	SH60B	HSL-51	TA-111
163247	SH60B	HSL-47	TY-74
163248	SH60B	HSL-48	HR-502
163249	SH60B		
163250	SH60B		
163251	SH60B		
163252	SH60B		
163253	SH60B		
163254	SH60B		
163255	SH60B		
163256	SH60B		
163257	SH60F		
163258	SH60B		
163259	VH60N	HMX-1	
163260	VH60N	HMX-1	
163261	VH60N	HMX-1	
163262	VH60N	HMX-1	
163263	VH60N	HMX-1	
163264	VH60N	HMX-1	
163265	VH60N	HMX-1	
163266	VH60N	HMX-1	
163267	VH60N	HMX-1	
163282	SH60F	HS-10	RA-00
163283	SH60F		
163284	SH60F	HS-10	RA-10
163285	SH60F	HS-10	RA-11
163286	SH60F	HS-10	RA-12
163287	SH60F	HS-10	RA-14
163288	SH60F		
163593	SH60B	HSL-46	HQ-473
163594	SH60B	HSL-49	TX-101
163595	SH60B	HSL-48	HR-503
163596	SH60B	HSL-45	TZ-47
163597	SH60B		
163598	SH60B	HSL-47	TY-70
163783	HH60H	NAWC	
163784	HH60H	HCS-5	NW-302
163785	HH60H	HCS-5	NW-300
163786	HH60H	HCS-4	NW-202
163787	HH60H	HCS-5	NW-301
163788	HH60H	HCS-5	NW-303
163789	HH60H	HCS-5	NW-304
163790	HH60H	HCS-4	NW-200
163791	HH60H	HCS-4	NW-201
163792	HH60H	HS-4	NL-616
163793	HH60H		
163794	HH60H	HS-3	AJ-616
163795	HH60H		
163796	HH60H		
163797	HH60H	HCS-4	NW-204
163798	HH60H	HCS-5	NW-305
163799	HH60H	HCS-4	NW-205
163800	HH60H	HCS-5	NW-306
163905	SH60B	HSL-48	HR-510
163906	SH60B	HSL-51	TA-00
163907	SH60B	HSL-45	TZ-52
163908	SH60B	HSL-42	HN-436
163909	SH60B	HSL-43	TT-
163910	SH60B		
164069	SH60F	NAWC-AD	RWATD
164070	SH60F	VX-1	JA-44
164071	SH60F	HS-10	RA-16
164072	SH60F	HS-10	RA-20
164073	SH60F	HS-10	RA-17
164074	SH60F	HS-10	RA-18
164075	SH60F	HS-2	NG-611
164076	SH60F	HS-6	NH-614
164077	SH60F	HS-2	NG-613
164078	SH60F	HS-2	NG-614
164079	SH60F	HS-2	NG-615
164080	SH60F	HS-10	RA-19
164081	SH60F		
164082	SH60F	HS-2	NG-610
164083	SH60F	HS-6	NH-611
164084	SH60F	HS-6	NH-612
164085	SH60F	HS-6	NH-616
164086	SH60F	HS-2	NG-612
164087	SH60F	HS-6	NH-617
164088	SH60F		
164089	SH60F		
164090	SH60F	w.o.6.8.91	
164091	SH60F	HS-10	RA-17
164092	SH60F	HS-4	NL-614
164093	SH60F	HS-4	NL-610
164094	SH60F		
164095	SH60F	HS-10	RA-
164096	SH60F	HS-10	RA-21
164097	SH60F		
164098	SH60F	HS-10	RA-
164099	SH60F	HS-1	AR-105
164100	SH60F		
164101	SH60F	HS-1	AR-101
164102	SH60F	HS-1	AR-102
164103	SH60F	HS-15	AE-613
164104	SH60F		
164174	SH60B		
164175	SH60B	HSL-40	HK-411
164176	SH60B		
164177	SH60B	USN\IBM	
164178	SH60B		
164179	SH60B	HSL-40	HK-414
164443	SH60F	VX-1	JA-46
164444	SH60F	HS-8	NK-610
164445	SH60F	HS-8	NK-611
164446	SH60F		
164447	SH60F	HS-8	NK-612
164448	SH60F		
164449	SH60F	HS-8	NK-614
164450	SH60F		
164451	SH60F	HS-3	AJ-611
164452	SH60F	HS-3	AJ-612
164453	SH60F	HS-3	AJ-613
164454	SH60F	HS-3	AJ-614
164455	SH60F	HS-3	AJ-615
164456	SH60F	HS-2	NG-616
164457	SH60F	HS-6	NH-617
164458	SH60F	HS-2	NG-617
164459	SH60F		
164460	SH60F	HS-6	NH-618
164461	SH60B		
164462	SH60B		
164463	SH60B		
164464	SH60B		
164465	SH60B		
164466	SH60B		
164609	SH60F	HS-15	AE-616
164610	SH60F		
164611	SH60F		
164612	SH60F	VX-1	JA-47
164613	SH60F		

Serial	Type			
164614	SH60F			
164615	SH60F			
164616	SH60F			
164617	SH60F			
164618	SH60F			
164619	SH60F			
164620	SH60F			
164796	SH60F			
164797	SH60F			
164798	SH60F			
164799	SH60F			
164800	SH60F			
164801	SH60F			
164802	SH60F			
164803	SH60F			
164804	SH60F			
164805	SH60F			
164806	SH60F			
164807	SH60F			
164808	SH60B			
164809	SH60B			
164810	SH60B			
164811	SH60B			
164812	SH60B			
164813	SH60B			
164814	SH60B			
164815	SH60B			
164816	SH60B			
164817	SH60B			
164818	SH60B			
164819	SH60B			
164831	HH60H			
164832	HH60H			
164833	HH60H			
164834	HH60H			
164835	HH60H			
164836	HH60H			
164837	HH60H			
164838	HH60H			
164839	HH60H			
164840	HH60H			
164841	HH60H			
164842	HH60H			
164843	HH60H			
164844	HH60H			
164845	HH60H			
164846	HH60H			
164847	SH60B			
164848	SH60B			
164849	SH60B			
164850	SH60B			
164851	SH60B			
164852	SH60B			
164853	SH60B			
164854	SH60B			
164855	SH60B			
164856	SH60B			
164857	SH60B			
164858	SH60B			
165095	SH60B			
165106	SH60B			
165107	SH60B			
165108	SH60B			
165109	SH60B			
165110	SH60B			
165111	SH60B			
165112	SH60B			
165113	SH60F			
165114	SH60F			
165115	SH60F			
165116	SH60F			
165117	SH60F			
165118	SH60F			
165119	SH60F			
165120	HH60H			
165121	HH60H			
165122	HH60H			
165123	HH60H			
166128	SH60B			
165129	SH60B			
165130	SH60B			
165131	SH60B			
165132	SH60B			
165133	SH60B			
165134	SH60B			
165135	SH60F			
165136	SH60F			
165137	SH60F			
165138	SH60F			
165139	SH60F			
165140	SH60F			
165141	SH60F			
165142	HH60H			
165143	HH60H			
165144	HH60H			
165145	HH60H			
165154	SH60F			
165155	SH60F			
165156	SH60F			
165157	SH60F			
77-22716	UH60A	NAWC\TPS	61	
77-22725	UH60A	NAWC\TPS	62	
80-23507	UH60A	NAWC\TPS	60	

P3 ORION

c\n		
	P3A:	1003, 5001 to 5157.
	P3B:	5158 to 5189, 5191, 5193 to 5199, 5201 to 5207, 5209 to 5286, 5409 (ex RAAF).
	P3C:	5501 to 5550, 5552 to 5621, 5623 to 5632, 5634 to 5656, 5659, 5661, 5663, 5665, 5667, 5669, 5671, 5673, 5675, 5670, 5677 to 5681, 5683 to 5688, 5690 to 5692, 5694 to 5696, 5698 to 5703, 5705, 5707, 5710, 5713, 5716, 5718, 5721, 5724, 5726 to 5732, 5734 to 5736, 5738 to 5740, 5742 to 5744, 5746 to 5749, 5751 to 5753, 5755 to 5757, 5759 to 5761, 5763, 5764, 5766 to 5768, 5770 to 5772, 5775, 5777, 5779, 5781, 5783, 5786, 5788, 5790, 5792, 5794, 5796 to 5816, 5821 to 5824.
	P3D:	5551, 5622, 5633.

Serial	Type	Operator	Code			Serial	Type	Operator	Code
148276	NP3A	NASA	N428NA			149667	RP3A	scrapped	
148883	UP3A	NAWC-AD				149668	EP3E	VQ-2	JQ-21
148884	P3A	scrapped				149669	EP3B	scrapped	
148885	UP3A	scrapped				149670	RP3A	AMARC	2P...
148886	P3A	scrapyard	Kotz Metals			149671	EP3A	stored	China Lake
148887	EP3E	VQ-1	PR-33			149672	P3A	w.o.30.1.63	
148888	EP3E	VQ-2	JQ-23			149673	UP3A	AMARC	2P078
148889	UP3A	NAWC-AD				149674	RP3A	NRL	

Serial	Type	Status/Unit	Code/Location
149675	VP3A	ETD	CINCPAC
149676	VP3A	VP-30	CNO
149677	UP3A	to Chile	403
149678	EP3B	scrapped	
150494	EP3E	scrapped	
150495	UP3A	NAS	Keflavik
150496	VP3A	VP-30	CNO
150497	EP3E	stored	NAS Alameda
150498	EP3E	stored	NAS Alameda
150499	RP3A	NAWC-WD	37
150500	RP3A	derelict	NAS Jacksonville
150501	EP3E	VQ-1	PR-36
150502	EP3E	scrapped	
150503	EP3E	stored	China Lake
150504	UP3A	VQ-1	PR-00
150505	EP3E	VQ-2	JQ-24
150506	P3A	scrapyard	Kotz Metals
150507	P3A	to Chile	402
150508	P3A	w.o.4.12.64	
150509	P3A	preserved	NAS Moffett Field
150510	P3A	to Spain	P.3-5
150511	VP3A	VP-30	CINCLANT
150512	RP-3A	stored	NAD Alameda
150513	P3A	to Spain	P.3-6
150514	P3A	USCS	N18314
150515	VP3A	CINCAFSE	NAS Sigonella
150516	P3A	to Spain	P.3-7
150517	P3A	GIA	NAS Jacksonville
150518	UP3A	to Chile	401
150519	UP3A	AMARC	2P...
150520	RP3A	stored	NAD Alameda
150521	RP3A	NAWC-WD	41
150522	RP3A	NAWC-WD	40
150523	P3A	GIA	Waco, Tx
150524	RP3A	NAWC-WD	35
150525	RP3A	NAWC-WD	36
150526	UP3A	VRC-30	COMNAVAIRPAC
150527	UP3A	AMARC	2P041
150528	UP3A	AMARC	2P043
150529	EP3A	Hawkins & Powers	
150604	P3A	scrapped	
150605	UP3A	ETD	CINCPAC
150606	P3A	scrapyard	Kotz Metals
150607	UP3A	to Chile	406
150608	UP3A	scrapped	
150609	P3A	scrapyard	Kotz Metals
151349	P3A	scrapyard	Kotz Metaals
151350	P3A	w.o.5.4.68	
151351	P3A	scrapyard	Kotz Metal
151352	TP3A	VP-30	LL-
151353	UP3A	AMARC	2P060
151354	UP3A	to Chile	405
151355	P3A	Aero Union	
151356	UP3A	AMARC	2P025
151357	TP3A	VP-30	LL-26
151358	P3A	scrapped	
151359	P3A	w.o.17.10.91\N924AU	
151360	UP3A	AMARC	2P...
151361	P3A	USFS	N925AU
151362	P3A	w.o.17.11.64	
151363	P3A	w.o.2.6.69	
151364	TP3A	VP-30	LL-
151365	P3A	w.o.28.4.67	
151366	UP3A	AMARC	2P015
151367	UP3A	NAS	Bermuda
151368	UP3A	AMARC	2P073
151369	P3A	USFS	N927AU
151370	TP3A	VP-30	LL-28
151371	TP3A	VP-30	LL-
151372	UP3A	USFS	N923AU
151373	P3A	scrapped	
151374	P3A	preserved	NAS Jacksonville
151375	TP3A	VP-30	LL-
151376	TP3A	ETD	CINCPAC
151377	P3A	Aero Union	
151378	P3A	AMARC	2P028
151379	TP3A	VP-30	LL-
151380	P3A	w.o.27.7.65	
151381	P3A	w.o.23.3.78	
151382	TP3A	VP-30	LL-29
151383	UP3A	AMARC	2P023
151384	UP3A	to Chile	407
151385	P3A	USFS	N921AU
151386	UP3A	scrapped	
151387	UP3A	USFS	N181AU
151388	UP3A	scrapped	
151389	UP3A	AMARC	2P013
151390	P3A	USCS	N15390
151391	P3A	USFS	N900AU
151392	TP3A	VP-30	LL-30
151393	P3A	scrapped	
151394	TP3A	VP-30	LL-27
151395	P3A	USCS	N16295
151396	UP3A	scrapped	
152140	P3A	AMARC	2P051
152141	UP3A	to Chile	408
152142	P3A	AMARC	2P040
152143	P3A	NAD	Jacksonville
152144	P3A	w.o.16.1.68	
152145	P3A	to Spain	P.3-3
152146	P3A	AMARC	2P046
152147	P3A	AMARC	2P...
152148	P3A	AMARC	2P029
152149	P3A	to Spain	P.3-2
152150	UP3A	NAWC-AD	
152151	P3A	w.o.5.12.71	
152152	P3A	preserved	NAS Pensacola
152153	P3A	to Spain	P.3-1
152154	P3A	AMARC	2P031
152155	P3A	w.o.26.5.72	
152156	P3A	preserved	NAS Brunswick
152157	P3A	AMARC	2P052
152158	P3A	GIA	NAWC-23
152159	P3A	w.o.3.8.70	
152160	P3A	preserved	NAS Bermuda
152161	P3A	w.o.18.1.81	
152162	P3A	AMARC	2P053
152163	P3A	AMARC	2P055
152164	P3A	scrapped	
152165	P3A	to Chile	404
152166	P3A	derelict	Whidbey Is.
152167	P3A	AMARC	2P039
152168	P3A	AMARC	2P034
152169	UP3A	VPU-2	SP-
152170	P3A	USCS	N16370
152171	P3A	w.o.9.4.66	
152172	P3A	w.o.4.7.66	
152173	P3A	AMARC	2P038
152174	P3A	AMARC	2P048
152175	P3A	AMARC	2P033
152176	P3A	AMARC	2P044
152177	P3A	AMARC	2P049
152178	P3A	scrapyard	Kotz Metals
152179	UP3A	AMARC	2P009
152180	P3A	AMARC	2P037
152181	P3A	AMARC	2P047
152182	P3A	w.o.3.6.72	
152183	P3A	AMARC	2P030
152184	P3A	AMARC	2P050
152185	P3A	AMARC	2P057
152186	P3A	AMARC	2P032
152187	P3A	AMARC	2P035
152718	P3B	AMARC	2P095
152719	EP3J	VP-66	
152720	P3B	w.o.16.6.83	
152721	P3B	AMARC	2P069
152722	P3AEW	USCS	N147CS
152723	P3B	AMARC	2P063
152724	P3B	w.o.26.4.78	
152725	P3B	AMARC	2P...
152726	P3B	AMARC	2P091
152727	UP3B	VQ-1	PR-43
152728	P3B	VPU-1	
152729	P3B	AMARC	2P...
152730	P3B	AMARC	2P083
152731	P3B	VP-90	LX-7
152732	P3B	AMARC	2P...
152733	P3B	dump	Barbers Pt.
152734	P3B	AMARC	2P068
152735	P3B	NASA	N426NA
152736	P3B	AMARC	2P066
152737	P3B	AMARC	2P080
152738	RP3D	AMARC	2P...
152739	NP3B	GIA	NAWC-23
152740	UP3B	VQ-2	JQ-
152741	P3B	VP-94	LZ-7

152742	P3B	VP-60	LS-4	154588	P3B	VP-66	LV-66
152743	P3B	AMARC	2P079	154589	RP3D	NAD	Jacksonville
152744	P3B	AMARC	2P...	154590	P3B	VP-64	LU-00
152745	EP3J	VP-66	LV-	154591	P3B	w.o.5.9.80	
152746	P3B	AMARC	2P093	154592	P3B	VP-66	LV-00
152747	P3B	AMARC	2P...	154593	P3B	VP-64	LU-03
152748	P3B	VP-93	LH-	154594	P3B	VP-66	LV-7
152749	P3B	w.o.15.3.73		154595	P3B	VP-64	LU-05
152750	P3B	AMARC	2P086	154596	P3B	w.o.27.6.79	
152751	P3B	VP-60	LS-10	154597	P3B	VP-60	LS-1
152752	P3B	NAWC-AD	ASATD	154598	P3B	VP-64	LU-01
152753	P3B	AMARC	2P084	154599	P3B	VP-66	LV-8
152754	P3B	VP-66	LV-2	154600	RP3D	AMARC	2P...
152755	P3B	AMARC	2P...	154601	P3B	VP-94	LZ-3
152756	P3B	AMARC	2P064	154602	P3B	VP-93	LH-
152757	P3B	w.o.22.9.78		154603	P3B	VP-67	PL-00
152758	P3B	AMARC	2P094	154604	P3B	VP-67	PL-1
152759	P3B	AMARC	2P...	154605	P3AEW	USCS	N146CS
152760	P3B	AMARC	2P081	155299	P3AEW	USCS	N145CS
152761	P3B	AMARC	2P...	156507	EP3E	VQ-1	PR-31
152762	P3B	AMARC	2P071	156508	P3C	VP-65	PG-01
152763	P3B	VP-67	PL-6	156509	P3C	VP-65	PG-07
152764	P3B	VP-94	LZ-4	156510	P3C	VP-30	LL-44
152765	P3B	w.o.6.3.69		156511	EP3E	VQ-1	PR-32
153414	P3B	GIA	NAWC-23	156512	P3C	VP-65	PG-02
153415	P3B	VP-64	LU-	156513	P3C	VP-65	PG-03
153416	P3B	VP-94	LZ-1	156514	EP3E	stored	NAD Alameda
153417	P3B	AMARC	2P067	156515	P3C	VP-62	LT-
153418	P3B	AMARC	2P036	156516	P3C	VP-22	QA-
153419	P3B	VP-93	LH-1	156517	EP3E	VQ-1	PR-
153420	P3B	VP-90	LX-6	156518	P3C	NAWC-AD	
153421	P3B	VP-67	PL-4	156519	EP3E	stored	NAD Alameda
153422	P3B	VP-93	LH-	156520	P3C	VP-65	PG-00
153423	P3B	AMARC	2P065	156521	P3C	VP-91	PM-
153424	P3B	AMARC	2P...	156522	P3C	VP-31	RP-
153425	UP3B	scrapped		156523	P3C	VP-22	QA-
153426	P3B	AMARC	2P...	156524	P3C	VP-65	PG-04
153427	P3B	VP-66	LV-1	156525	P3C	VP-65	PG-05
153428	P3B	w.o.11.12.77		156526	P3C	VP-65	PG-06
153429	P3B	NASA		156527	P3C	VP-62	LT-6
153430	P3B	AMARC	2P087	156528	EP3E	stored	NAD Alameda
153431	P3B	VP-93	LH-3	156529	EP3E	stored	NAD Alameda
153432	P3B	AMARC	2P062	156530	P3C	VP-30	LL-45
153433	UP3B	VQ-1	PR-44	157310	P3C	VP-49	LP-
153434	P3B	AMARC	2P082	157311	P3C	VP-24	LR-3
153435	P3B	VP-66	LV-9	157312	P3C	VP-16	LF-7
153436	P3B	VP-93	LH-	157313	P3C	VP-49	LP-3
153437	P3B	AMARC	2P...	157314	P3C	VP-49	LP-4
153438	P3B	VP-90	LX-10	157315	P3C	VP-49	LP-5
153439	P3B	AMARC	2P090	157316	EP3E	stored	NAD Alameda
153440	P3B	w.o.6.2.68		157317	P3C	VP-22	QA-
153441	P3B	VP-60	LS-5	157318	EP3E	stored	NAD Alameda
153442	EP3B	NRL		157319	P3C	VP-45	LN-7
153443	RP3D	VXN-8	JB-01	157320	EP3E	VQ-2	JQ-26
153444	P3B	VP-60	LS-	157321	P3C	VP-16	LF-3
153445	P3B	w.o.1.4.68		157322	P3C	VP-46	RC-
153446	P3B	VP-93	LH-	157323	P3C	VP-91	PM-
153447	P3B	VP-90	LX-	157324	P3C	VP-4	YD-
153448	P3B	VP-67	PL-67	157325	EP3E	VQ-2	JQ-
153449	P3B	VP-90	LX-	157326	EP3E	stored	NAD Alameda
153450	P3B	VPU-1		157327	P3C	VP-22	QA-
153451	P3B	VP-94	LZ-2	157328	P3C	VP-30	LL-34
153452	P3B	VP-93	LH-6	157329	P3C	VP-1	YB-8
153453	P3B	VP-67	PL-3	157330	P3C	VP-31	RP-
153454	P3B	scrapped		157331	P3C	VP-30	LL-
153455	P3B	VP-90	LX-	157332	P3C	w.o.12.4.73	
153456	P3B	VP-94	LZ-6	158204	P3C	NAWC-AD	
153457	P3B	VP-60	LS-	158205	P3C	VP-46	RC-
153458	P3B	VP-90	LX-	158206	P3C	VX-1	JA-03
154574	P3B	stored	NAS Willow Grove	158207	P3C	VP-31	RP-
154575	P3AEW	USCS	N148CS	158208	P3C	VP-9	PD-
154576	P3B	to Norway	576	158209	P3C	VP-1	YB-1
154577	P3B	VPU-1		158210	P3C	VP-30	LL-43
154578	P3B	VP-60	LS-1	158211	P3C	VP-22	QA-
154579	P3B	VP-94	LZ-	158212	P3C	VP-22	QA-
154580	P3B	VP-67	PL-2	158213	P3C	w.o.17.4.80	
154581	P3B	VP-94	LZ-8	158214	P3C	VP-30	LL-
154582	P3B	VP-64	LU-02	158215	P3C	VP-46	RC-
154583	P3B	to Norway	583	158216	P3C	VP-1	YB-2
154584	P3B	AMARC	2P...	158217	P3C	VP-4	YD-
154585	P3B	VPU-2	SP-	158218	P3C	VP-4	YD-45
154586	P3B	VP-64	LU-06	158219	P3C	VP-45	LN-
154587	RP3D	VXN-8	JB-03	158220	P3C	VP-31	RP-11

158221	P3C	VP-46	RC-	160285	P3C	VP-17	ZE-
158222	P3C	VP-17	ZE-	160286	P3C	VP-46	RC-
158223	P3C	VP-1	YB-3	160287	P3C	VP-24	LR-
158224	P3C	VP-16	LF-	160288	P3C	VP-17	ZE-
158225	P3C	VP-9	PD-9	160289	P3C	VP-17	ZE-
158226	P3C	VP-1	YB-6	160290	P3C	NAWC-AD	ASATD
158227	RP3D	VXN-8	JB-02	160291	P3C	NAWC-AD	
158563	P3C	VP-1	YB-7	160292	P3C	NAD	Jacksonville
158564	P3C	VP-24	LR-4	160293	P3C	VX-1	JA-02
158565	P3C	VP-5	LA-1	160610	P3C	VP-23	LJ-00
158566	P3C	VP-5	LA-4	160611	P3C	VP-92	LY-1
158567	P3C	VP-5	LA-6	160612	P3C	VP-92	LY-2
158568	P3C	VP-49	LP-1	160761	P3C	VP-30	LL-50
158569	P3C	VP-45	LN-5	160762	P3C	VP-92	LY-3
158570	P3C	VP-45	LN-2	160763	P3C	VP-23	LJ-6
158571	P3C	VP-5	LA-8	160764	P3C	VP-11	LE-
158572	P3C	VP-45	LN-3	160765	P3C	VP-10	LD-
158573	P3C	VP-24	LR-2	160766	P3C	VP-92	LY-6
158574	P3C	VP-30	LL-40	160767	P3C	VP-92	LY-7
158912	P3C	NAWC-AD	ASATD	160768	P3C	VP-30	LL-56
158913	P3C	VP-46	RC-	160769	P3C	VP-92	LY-4
158914	P3C	VP-1	YB-4	160770	P3C	VP-30	LL-52
158915	P3C	VP-1	YB-	160999	P3C	VP-92	LY-5
158916	P3C	VP-30	LL-32	161000	P3C	VP-30	LL-
158917	P3C	VP-4	YD-	161001	P3C	VP-92	LY-8
158918	P3C	VP-4	YD-46	161002	P3C	VP-23	LJ-2
158919	P3C	VP-45	LN-4	161003	P3C	VP-11	LE-
158920	P3C	VP-16	LF-8	161004	P3C	VP-11	LE-
158921	P3C	VP-22	QA-	161005	P3C	VP-69	PJ-7
158922	P3C	VP-5	LA-7	161006	P3C	VP-30	LL-51
158923	P3C	VP-5	LA-2	161007	P3C	VP-26	LK-7
158924	P3C	VP-5	LA-5	161008	P3C	VP-26	LK-8
158925	P3C	VP-46	RC-	161009	P3C	VP-30	LL-
158926	P3C	VP-45	LN-1	161010	P3C	VP-10	LD-
158927	P3C	VP-49	LP-8	161011	P3C	VP-11	LE-
158928	P3C	VP-68	LW-	161012	P3C	VP-26	LK-2
158929	P3C	VP-49	LP-7	161013	P3C	VP-69	PJ-1
158930	P3C	w.o.21.3.91		161014	P3C	VX-1	JA-05
158931	P3C	VP-5	LA-3	161121	P3C	VP-30	LL-54
158932	P3C	VP-16	LF-4	161122	P3C	VP-69	PJ-3
158933	P3C	VP-45	LN-6	161123	P3C	VP-30	LL-
158934	P3C	VP-24	LR-5	161124	P3C	VP-11	LE-
158935	P3C	VP-24	LR-6	161125	P3C	VP-30	LL-53
159318	P3C	VP-16	LF-1	161126	P3C	VP-10	LD-6
159319	P3C	VP-45	LN-8	161127	P3C	VP-26	LK-4
159320	P3C	VP-24	LR-7	161128	P3C	VP-69	PJ-4
159321	P3C	VP-4	YD-43	161129	P3C	VP-23	LJ-
159322	P3C	VP-16	LF-5	161130	P3C	VP-69	PJ-5
159323	P3C	VP-9	PD-3	161131	P3C	VP-8	LC-
159324	P3C	VP-9	PD-4	161132	P3C	VP-8	LC-
159325	P3C	w.o.21.3.91		161329	P3C	VP-40	QE-
159326	P3C	VP-9	PD-5	161330	P3C	VP-11	LE-8
159327	P3C	VP-9	PD-2	161331	P3C	VP-26	LK-
159328	P3C	VP-4	YD-42	161332	P3C	VP-23	LJ-
159329	P3C	VP-9	PD-	161333	P3C	VP-11	LE-3
159503	P3C	VP-17	ZE-	161334	P3C	VP-23	LJ-
159504	P3C	VPU-2	SP-	161335	P3C	VP-26	LK-
159505	P3C	VP-69	PJ-69	161336	P3C	VP-23	LJ-
159506	P3C	VP-68	LW-01	161337	P3C	VP-10	LD-
159507	P3C	VP-4	YD-40	161338	P3C	VP-8	LC-11
159508	P3C	VP-17	ZE-	161339	P3C	VP-8	LC-2
159509	P3C	VP-69	PJ-2	161340	P3C	VP-8	LC-3
159510	P3C	VP-92	LY-	161404	P3C	VP-8	LC-4
159511	P3C	VP-68	LW-02	161405	P3C	VP-11	LE-9
159512	P3C	VP-68	LW-03	161406	P3C	VP-23	LJ-
159513	P3C	VP-68	LW-04	161407	P3C	VP-8	LC-7
159514	P3C	VP-68	LW-05	161408	P3C	VP-10	LD-
159773	WP3D	NOAA	N42RF	161409	P3C	VP-8	LC-9
159875	WP3D	NOAA	N43RF	161410	P3C	GIA	NAWC-23
159883	P3C	VP-69	PJ-	161411	P3C	VP-10	LD-
159884	P3C	VP-68	LW-06	161412	P3C	VP-40	QE-
159885	P3C	VP-17	ZE-	161413	P3C	VP-26	LK-
159886	P3C	VP-17	ZE-	161414	P3C	VP-26	LK-
159887	P3C	VX-1	JA-04	161415	P3C	VP-10	LD-
159888	P3C	VP-17	ZE-	161585	P3C	VP-26	LK-
159889	P3C	VP-30	LL-41	161586	P3C	VP-8	LC-
159890	P3C	VP-69	PJ-6	161587	P3C	VP-23	LJ-
159891	P3C	VP-4	YD-41	161588	P3C	VP-23	LJ-
159892	P3C	w.o.26.10.78		161589	P3C	VP-23	LJ-
159893	P3C	VP-47	RD-	161590	P3C	VP-40	QE-
159894	P3C	VP-17	ZE-	161591	P3C	VP-40	QE-
160283	P3C	VP-47	RD-	161592	P3C	VP-10	LD-
160284	P3C	VP-68	LW-07	161593	P3C	VP-10	LD-7

161594	P3C	VP-8	LC-	162776	P3C	VP-24	LR-8
161595	P3C	VP-8	LC-	162777	P3C	VP-47	RD-7
161596	P3C	VP-26	LK-	162778	P3C	VP-47	RD-
161762	P3C	w.o.25.9.90		162998	P3C	VP-47	RD-
161763	P3C	VP-40	QE-3	162999	P3C	VP-47	RD-9
161764	P3C	VP-40	QE-4	163000	P3C	VP-22	QA-
161765	P3C	VP-9	PD-	163001	P3C	VP-62	LT-1
161766	P3C	VP-91	PM-	163002	P3C	VP-62	LT-2
161767	P3C	VP-40	QE-2	163003	P3C	VP-62	LT-3
162314	P3C	VP-40	QE-8	163004	P3C	VP-62	LT-4
162315	P3C	VP-40	QE-9	163005	P3C	VP-62	LT-5
162316	P3C	VP-40	QE-8	163006	P3C	VX-1	JA-06
162317	P3C	VP-40	QE-7	163289	P3C	VP-62	LT-7
162318	P3C	VP-40	QE-	163290	P3C	VP-91	PM-2
162770	P3C	NAWC-AD		163291	P3C	VP-91	PM-3
162771	P3C	VP-22	QA-	163292	P3C	VP-16	LF-16
162772	P3C	VP-46	RC-	163293	P3C	VP-49	LP-2
162773	P3C	VP-47	RD-	163294	P3C	VP-91	PM-4
162774	P3C	VP-47	RD-4	163295	P3C	VP-91	PM-5
162775	P3C	VP-47	RD-5				

S3 VIKING

c\n   3001 to 3187.

157992	YS3A	stored	NAS Lakehurst	159413	S3B	VS-21	NF-704
157993	NS3A	gate	NAS Cecil Field	159414	S3A	VS-33	NG-
157994	US3A			159415	ES3A	Lockheed	NAS Cecil Field
157995	US3A	stored	NAD Alameda	159416	S3B	VS-27	AD-734
157996	US3A	w.o.20.1.89		159417	S3A	AMARC	2S...
157997	US3A			159418	S3B	w.o. 12.12.91	
157998	US3A			159419	S3A	VQ-6	ET-760
157999	YS3A	w.o.3.8.73		159420	ES3A	Lockheed	NAS Cecil Field
158861	S3B	VS-27	AD-712	159728	S3A	scrapped	
158862	ES3A	Lockheed	NAS Cecil Field	159729	S3B	VS-37	NL-
158863	S3B	VS-41	NJ-746	159730	S3A	w.o.1.11.83	
158864	S3B	stored	NAD Alameda	159731	S3B	VS-37	NL-
158865	S3B	VS-31	AG-706	159732	S3B	VS-31	AG-703
158866	S3B	VS-27	AD-726	159733	S3B	VS-30	AA-711
158867	S3A	VS-41	NJ-726	159734	S3B	VS-24	AJ-705
158868	US3A			159735	S3A	w.o.9.1.85	
158869	S3A	AMARC	2S...	159736	S3B	w.o.29.4.92	
158870	S3A	VS-29	NH-702	159737	S3B	VS-41	NJ-745
158871	S3A	VS-41	NJ-	159738	ES3A	VQ-6	ET-710
158872	S3A	stored	NAD Alameda	159739	S3A	VQ-5	SS-
158873	S3B	stored	NAD Alameda	159740	S3B	VS-33	NG-702
159386	S3A	w.o.19.2.92		159741	S3B	VS-24	AJ-704
159387	S3B	VS-33	NG-703	159742	S3B	NAWC-AD	
159388	S3A	VS-41	NJ-720	159743	S3B	VX-1	JA-
159389	S3B	VS-31	AG-701	159744	S3B	VS-27	AD-722
159390	S3B	VS-30	AA-710	159745	S3A	VS-35	NK-
159391	ES3A	NAWC-AD		159746	S3A	VS-41	NJ-725
159392	S3B	VS-33	NG-704	159747	S3B	VS-22	AC-705
159393	S3A	VQ-6	ET-767	159748	S3A	stored	NAD Alameda
159394	S3A	VQ-6	ET-	159749	S3A	AMARC	2S...
159395	S3A	AMARC	2S...	159750	S3A	VS-35	NK-
159396	S3A	AMARC	2S030	159751	S3B	VS-27	AD-715
159397	S3A	stored	NAD Alameda	159752	ES3A	Lockheed	NAS Cecil Field
159398	S3A	w.o.1.11.74		159753	S3B	VS-24	AJ-702
159399	S3A	VS-41	NJ-727	159754	S3A	w.o.9.9.86	
159400	ES3A	Lockheed	Cecil Field	159755	S3B	VS-24	AJ-703
159401	ES3A	NMAWC-AD	777	159756	S3B	VS-27	AD-720
159402	S3B	VS-31	AG-	159757	S3A	w.o.6.7.83	
159403	ES3A	VQ-5	SS-720	159758	S3B	VS-27	AD-
159404	ES3A	NAWC-AD	771	159759	S3B	w.o.7.10.89	
159405	S3A	VS-29	NH-711	159760	S3B	VS-22	AC-701
159406	S3A	VQ-5	SS-	159761	S3B	VS-24	AJ-707
159407	S3A	VS-29	NH-706	159762	S3B	VS-32	AB-702
159408	S3A	w.o.27.3.78		159763	S3B	VS-41	NJ-
159409	S3A	VS-38	NE-701	159764	S3B	VS-30	AA-
159410	S3A	AMARC	2S...	159765	S3B	VS-32	AB-705
159411	S3A	w.o.25.1.83		159766	S3B	VS-37	NL-
159412	S3A	scrapped		159767	S3B	VS-30	AA-

159768	S3B	VS-32	AB-704		160161	S3A	VS-35	NK-
159769	S3B	VS-31	AG-704		160162	S3B	VS-21	NF-702
159770	S3B	NAWC-AD			160163	S3B	VS-35	NK-701
159771	S3B	VS-30	AA-		160164	S3A	w.o.5.6.88	
159772	S3A	w.o.18.10.78			160567	S3A	VS-38	NE-704
160120	S3A	w.o.10.3.80			160568	S3A	AMARC	2S...
160121	S3B	stored	NAD Alameda		160569	S3A	VS-41	NJ-741
160122	S3B	VS-27	AD-732		160570	S3A	AMARC	2S...
160123	S3A	VS-29	NH-703		160571	S3A	VS-35	NK-703
160124	S3B	VS-41	NJ-724		160572	S3A	stored	NAD Alameda
160125	S3B	VS-24	AJ-		160573	S3A	VS-38	NE-700
160126	S3A	VS-38	NE-703		160574	S3A	AMARC	2S...
160127	S3B	VS-35	NK-700		160575	S3A	VS-41	NJ-736
160128	S3B	VS-41	NJ-722		160576	S3A	VS-38	NE-705
160129	S3B	VS-33	NG-		160577	S3B	VS-33	NG-
160130	S3B	VS-21	NF-701		160578	S3A	VS-33	NG-700
160131	S3B	VS-21	NF-		160579	S3A	w.o.21.3.87	
160132	S3A	VS-29	NH-705		160580	S3A	stored	NAD Alameda
160133	S3B	VS-21	NF-706		160581	S3B	VS-24	AJ-
160134	S3A	VS-41	NJ-		160582	S3A	VS-29	NH-712
160135	S3B	VS-33	NG-711		160583	S3A	VS-38	NE-702
160136	S3B	VS-29	NH-701		160584	S3A	VS-41	NJ-
160137	S3A	w.o.21.10.86			160585	S3A	AMARC	2S...
160138	S3B	stored	NAD Alameda		160586	S3A	AMARC	2S...
160139	S3B	VS-37	NL-		160587	S3A	stored	NAD Alameda
160140	S3B	VS-22	AC-703		160588	S3B	VS-27	AD-724
160141	S3B	VS-30	AA-704		160589	S3A	stored	NAD Alameda
160142	S3B	VS-31	AG-705		160590	S3A	w.o.2.11.78	
160143	S3B	VS-31	AG-700		160591	S3B	AMARC	2S...
160144	S3B	stored	NAD Alameda		160592	S3B	VX-1	JA-717
160145	S3B	VS-32	AB-703		160593	S3A	AMARC	2S...
160146	S3A	w.o.8.11.83			160594	S3A	AMARC	2S...
160147	S3B	stored	NAD Alameda		160595	S3A	AMARC	2S...
160148	S3B	VX-1	JA-715		160596	S3A	stored	NAD Alameda
160149	S3B	VS-22	AC-700		160597	S3A	AMARC	2S...
160150	S3A	w.o.17.11.81			160598	S3A	AMARC	2S...
160151	S3B	VS-22	AC-702		160599	S3A	VS-41	NJ-733
160152	S3B	VS-32	AB-		160600	S3B	VS-22	AC-704
160153	S3B	VS-27	AD-723		160601	S3B	VS-37	NL-
160154	S3A	w.o.8.12.79			160602	S3B	VQ-6	ET-
160155	S3B	VX-1	JA-		160603	S3B	VS-27	AD-716
160156	S3B	VS-32	AB-700		160604	S3B	VS-31	AG-702
160157	S3A	VS-33	NG-705		160605	S3B	VS-37	NL-
160158	S3A	VS-35	NK-702		160606	S3B	VS-32	AB-703
160159	S3B	VS-21	NF-705		160607	S3B	stored	NAD Alameda
160160	S3B	VS-21	NF-703					

T2 BUCKEYE

144217	T2A				146014	T2A	
144218	T2A				146015	T2A	
144219	T2A				147430	T2A	
144220	T2A				147431	T2A	
144221	T2A	scrapyard	Consilidated		147432	T2A	scrapyard	Kolar
144222	T2A				147433	T2A	
145996	T2A				147434	T2A	
145997	YT2B				147435	T2A	scrapyard	Kolar
145998	T2A				147436	T2A	preserved	Chino,Ca
145999	T2A	scrapyard	Consilidated		147437	T2A	scrapyard	Kolar
146000	T2A				147438	T2A	
146001	T2A				147439	T2A	
146002	T2A				147440	T2A	
146003	T2A				147441	T2A	
146004	T2A				147442	T2A	
146005	T2A				147443	T2A	NMIMT	Sorocco,NM
146006	T2A				147444	T2A	scrapyard	Consilidated
146007	T2A				147445	T2A	
146008	T2A				147446	T2A	
146009	T2A				147447	T2A	
146010	T2A				147448	T2A	scrapyard	Consilidated
146011	T2A				147449	T2A	
146012	T2A				147450	T2A	scrapyard	Consilidated
146013	T2A				147451	T2A	

147452	T2A	scrapyard	Consilidated
147453	T2A		
147454	T2A		
147455	T2A	scrapyard	Consilidated
147456	T2A		
147457	T2A		
147458	T2A		
147459	T2A		
147460	T2A		
147461	T2A		
147462	T2A		
147463	T2A		
147464	T2A		
147465	T2A	scrapyard	Consilidated
147466	T2A		
147467	T2A	scrapyard	Consilidated
147468	T2A		
147469	T2A	scrapyard	Consilidated
147470	T2A		
147471	T2A		
147472	T2A	scrapyard	Consilidated
147473	T2A	preserved	NASKingsville
147474	T2A	stored	Chino,Ca
147475	T2A		
147476	T2A		
147477	T2A		
147478	T2A		
147479	T2A		
147480	T2A	scrapyard	Consilidated
147481	T2A		
147482	T2A	scrapyard	Consilidated
147483	T2A		
147484	T2A		
147485	T2A		
147486	T2A	scrapyard	Consilidated
147487	T2A		
147488	T2A		
147489	T2A	scrapyard	Consilidated
147490	T2A	scrapyard	Consilidated
147491	T2A		
147492	T2A		
147493	T2A		
147494	T2A		
147495	T2A		
147496	T2A		
147497	T2A		
147498	T2A	scrapyard	Consilidated
147499	T2A	scrapyard	Consilidated
147500	T2A		
147501	T2A		
147502	T2A		
147503	T2A		
147504	T2A		
147505	T2A		
147506	T2A		
147507	T2A		
147508	T2A		
147509	T2A		
147510	T2A		
147511	T2A		
147512	T2A		
147513	T2A		
147514	T2A		
147515	T2A		
147516	T2A	preserved	Meridian,Ms
147517	T2A		
147518	T2A	scrapyard	Kolar
147519	T2A		
147520	T2A		
147521	T2A		
147522	T2A	preserved	NASMeridian
147523	T2A		
147524	T2A		
147525	T2A		
147526	T2A		
147527	T2A		
147528	T2A		
147529	T2A		
147530	T2A		
148150	T2A		
148151	T2A		
148152	T2A		
148153	T2A		
148154	T2A		
148155	T2A		
148156	T2A		
148157	T2A		
148158	T2A		
148159	T2A		
148160	T2A		
148161	T2A		
148162	T2A		
148163	T2A	scrapyard	Consilidated
148164	T2A		
148165	T2A		
148166	T2A		
148167	T2A		
148168	T2A	scrapyard	Consilidated
148169	T2A		
148170	T2A		
148171	T2A		
148172	T2A		
148173	T2A		
148174	T2A		
148175	T2A		
148176	T2A		
148177	T2A		
148178	T2A		
148179	T2A	scrapyard	Kolar
148180	T2A		
148181	T2A	NMIMT	Sorocco,NM
148182	T2A		
148183	T2A		
148184	T2A	scrapyard	Kolar
148185	T2A		
148186	T2A		
148187	T2A		
148188	T2A		
148189	T2A		
148190	T2A		
148191	T2A		
148192	T2A		
148193	T2A		
148194	T2A		
148195	T2A		
148196	T2A		
148197	T2A		
148198	T2A		
148199	T2A		
148200	T2A		
148201	T2A		
148202	T2A		
148203	T2A		
148204	T2A		
148205	T2A		
148206	T2A		
148207	T2A		
148208	T2A		
148209	T2A		
148210	T2A		
148211	T2A		
148212	T2A		
148213	T2A		
148214	T2A		
148215	T2A		
148216	T2A	scrapyard	Kolar
148217	T2A		
148218	T2A		
148219	T2A		
148220	T2A		
148221	T2A		
148222	T2A		
148223	T2A		
148224	T2A		
148225	T2A	scrapyard	Consilidated
148226	T2A		
148227	T2A		
148228	T2A		
148229	T2A		
148230	T2A	scrapyard	Kolar
148231	T2A		
148232	T2A	dump	NASPensacola
148233	T2A		
148234	T2A		
148235	T2A		
148236	T2A		

Serial	Type	Status	Location/Code
148237	T2A		
148238	T2A	stored MCAS	Cherry Point
148239	T2A		
152382	YT2C		
152383	T2B	NMIMT	Sorocco,NM
152384	T2B		
152385	T2B	derelict	NADJacksonville
152386	T2B	ranges	NellisAFB
152387	T2B	stored	Harrisburg,Pa
152388	T2B		
152389	T2B		
152390	T2B		
152391	T2B	to	N7139K
152440	T2B		
152441	T2B	ranges	NellisAFB
152442	T2B	ranges	NellisAFB
152443	T2B	NMIMT	Sorocco,NM
152444	T2B	NMIMT	Sorocco,NM
152445	T2B		
152446	T2B	NMIMT	Sorocco,NM
152447	T2B		
152448	T2B		
152449	T2B	stored	Harrisburg,Pa
152450	T2B	stored	Harrisburg,Pa
152451	T2B	NMIMT	Sorocco,NM
152452	T2B	ranges	NellisAFB
152453	T2B	stored	Harrisburg,Pa
152454	T2B	NMIMT	Sorocco,NM
152455	T2B	stored	Harrisburg,Pa
152456	T2B		
152457	T2B		
152458	T2B	stored	Harrisburg,Pa
152459	T2B	dump	PensacolaNAS
152460	T2B		
152461	T2B		
152462	T2B	ranges	IndianSprings
152463	T2B		
152464	T2B	NARF	Pensacola
152465	T2B		
152466	T2B	NMIMT	Sorocco,NM
152467	T2B		
152468	T2B	stored	Harrisburg,Pa
152469	T2B	ranges	NellisAFB
152470	T2B	ranges	NellisAFB
152471	T2B	stored	Harrisburg,Pa
152472	T2B	VT-10	F-862
152473	T2B	preserved	Seattle,Wa
152474	T2B	ranges	NellisAFB
152475	T2B	AMARC	2T107
153538	T2B	VT-10	F-863
153539	T2B	ranges	NellisAFB
153540	T2B	ranges	NellisAFB
153541	T2B	ranges	NellisAFB
153542	T2B	preserved	NASPensacola
153543	T2B	NMIMT	Sorocco,NM
153544	T2B	ranges	IndianSprings
153545	T2B	ranges	IndianSprings
153546	T2B	ranges	IndianSprings
153547	T2B		
153548	T2B	NMIMT	Sorocco,NM
153549	T2B	dump	PensacolaNAS
153550	T2B	NMIMT	Sorocco,NM
153551	DT2B	AMARC	2T166
153552	T2B	NMIMT	Sorocco,NM
153553	T2B	VT-10	F-861
153554	T2B	NMIMT	Sorocco,NM
153555	T2B	VT-10	F-864
155206	T2B		
155207	T2B	NMIMT	Sorocco,NM
155208	T2C	NARF	Pensacola
155209	T2B	NMIMT	Sorocco,NM
155210	T2B	NMIMT	Sorocco,NM
155211	T2B		
155212	T2B	NMIMT	Sorocco,NM
155213	T2C	VT-10	F-850
155214	T2B	NMIMT	Sorocco,NM
155215	T2B	scrapyard	SWAlloys
155216	T2B	NMIMT	Sorocco,NM
155217	T2B		
155218	T2B	NMIMT	Sorocco,NM
155219	T2B		
155220	T2B	derelict	NADJacksonville
155221	T2C	VT-10	F-851
155222	T2B	NMIMT	Sorocco,NM
155223	T2C	w.o.	22.10.86
155224	T2C	VT-10	F-853
155225	T2B		
155226	T2B	preserved	WillowRunAP
155227	T2C	VT-10	F-854
155228	T2B	scrapyard	SWAlloys
155229	T2B	VT-10	F-855
155230	T2B	VT-10	F-856
155231	T2B	VT-10	F-857
155232	T2B	VT-10	F-858
155233	T2B	VT-10	F-859
155234	T2B	VT-10	F-860
155235	T2B	to	N7139H
155236	T2B	stored	Harrisburg,Pa
155237	T2B		
155238	DT2B	AMARC	2T165
155239	T2C		
155240	T2C	AMARC	2T201
155241	T2C	dump	NASPensacola
156686	T2C	VT-23	B-325
156687	T2C	w.o.12.1.87	
156688	T2C	VT-23	B-
156689	T2C	VT-4	F-830
156690	T2C	w.o.11.1.85	
156691	T2C	VT-23	B-333
156692	T2C	dump	NASPensacola
156693	T2C	VT-23	B-337
156694	T2C	w.o.13.4.89	
156695	T2C	w.o.19.5.89	
156696	T2C	w.o.14.11.84	
156697	T2C	VT-19	A-974
156698	T2C	w.o.8.8.89	
156699	T2C	w.o.19.2.78	
156700	T2C	VT-19	A-956
156701	T2C		
156702	T2C	VT-4	F-810
156703	T2C	VT-19	A-959
156704	T2C	VT-19	A-993
156705	T2C	VT-19	A-992
156706	T2C		
156707	T2C	VT-23	B-365
156708	T2C		
156709	T2C	derelict	NAD Jacksonville
156710	T2C	VT-19	A-997
156711	T2C	VT-4	F-821
156712	T2C	VT-4	F-803
156713	T2C	VT-4	F-826
156714	T2C	AMARC	2T205
156715	T2C	VT-23	B-363
156716	T2C	VT-19	A-991
156717	T2C	AMARC	2T200
156718	T2C		
156719	T2C	VT-23	B-367
156720	T2C		
156721	T2C	AMARC	2T199
156722	T2C	AMARC	2T206
156723	T2C	VT-19	A-990
156724	T2C		
156725	T2C	VT-23	B-348
156726	T2C		
156727	T2C	VT-19	A-952
156728	T2C	VT-23	B-343
156729	T2C	VT-19	A-995
156730	T2C	VT-4	F-802
156731	T2C		
156732	T2C	VT-23	B-357
156733	T2C	VT-23	B-330
157030	T2C	VT-19	A-950
157031	T2C	VT-4	F-813
157032	T2C	NAWC\TPS	20
157033	T2C	VT-19	A-968
157034	T2C	VT-23	B-334
157035	T2C	AMARC	2T189
157036	T2C	VT-23	B-359
157037	T2C	AMARC	2T207
157038	T2C	VT-4	F-807
157039	T2C		
157040	T2C	w.o.13.7.89	
157041	T2C	VF-43	33
157042	T2C	AMARC	2T194
157043	T2C	VF-126	NJ-640
157044	T2C	VT-4	F-812

157045	T2C		
157046	T2C	VT-4	F-808
157047	T2C	AMARC	2T198
157048	T2C	w.o.30.8.84	
157049	T2C	VT-4	F-806
157050	T2C	VT-4	F-823
157051	T2C	VT-19	A-961
157052	T2C	VT-4	F-803
157053	T2C	VF-126	NJ-641
157054	T2C	VT-19	A-989
157055	T2C	VT-23	B-356
157056	T2C	VF-43	34
157057	T2C	NARF	Pensacola
157058	T2C	VT-19	A-967
157059	T2C	VT-23	B-360
157060	T2C	VT-19	A-970
157061	T2C	VT-23	B-301
157062	T2C	VT-19	A-962
157063	T2C	AMARC	2T195
157064	T2C	VT-19	A-988
157065	T2C	VT-23	B-302
158310	T2C	VT-19	A-999
158311	T2C	VT-23	B-340
158312	T2C	VT-4	F-822
158313	T2C	dump	NASMeridian
158314	T2C	VT-4	A-814
158315	T2C	VT-4	F-
158316	T2C	VT-19	A-986
158317	T2C	AMARC	2T183
158318	T2C	VT-4	F-814
158319	T2C	VT-23	B-328
158320	T2C	VT-19	A-969
158321	T2C	VT-19	A-985
158322	T2C		
158323	T2C	VT-23	B-366
158324	T2C	VT-19	A-966
158325	T2C	VT-4	F-804
158326	T2C	NAWC\TPS	22
158327	T2C	VT-4	F-827
158328	T2C	NAWC\TPS	23
158329	T2C		
158330	T2C		
158331	T2C	AMARC	2T203
158332	T2C	VT-23	B-311
158333	T2C	VT-23	B-349
158575	T2C	VT-19	A-971
158576	T2C	AMARC	2T190
158577	T2C	VT-23	B-335
158578	T2C	NAWC\TPS	25
158579	T2C	NAWC\TPS	26
158580	T2C	VT-19	A-951
158581	T2C	VT-23	B-320
158582	T2C		
158583	T2C	VT-4	F-829
158584	T2C	VT-4	F-815
158585	T2C	VT-19	A-984
158586	T2C	VT-4	F-805
158587	T2C	VT-23	B-368
158588	T2C	VT-23	B-362
158589	T2C		
158590	T2C	VT-4	F-819
158591	T2C	VT-19	A-948
158592	T2C		
158593	T2C		
158594	T2C	VT-4	F-800
158595	T2C	VF-126	NJ-643
158596	T2C	VT-23	B-344
158597	T2C	VT-23	B-345
158598	T2C	VT-4	F-801
158599	T2C	VT-23	B-353
158600	T2C		
158601	T2C	VT-23	B-355
158602	T2C	AMARC	2T202
158603	T2C	VT-23	B-318
158604	T2C		
158605	T2C	NAWC\TPS	27
158606	T2C	VT-23	B-319
158607	T2C	VT-23	B-300
158608	T2C	AMARC	2T184
158609	T2C	w.o.21.2.80	
158610	T2C	VT-23	B-327
158876	T2C	w.o.29.10.89	
158877	T2C	VT-4	F-807
158878	T2C	VT-19	A-960
158879	T2C	VT-19	A-983
158880	T2C	VT-23	B-338
158881	T2C	VT-19	A-994
158882	T2C	VT-23	B-
158883	T2C	VT-23	B-358
158884	T2C	VT-19	A-978
158885	T2C	AMARC	2T209
158886	T2C	VT-23	B-332
158887	T2C	stored	SilverHill,Md
158888	T2C	VT-19	A-958
158889	T2C	VF-126	NJ-642
158890	T2C	VT-19	A-955
158891	T2C	w.o.20.6.85	
158892	T2C	VT-23	B-323
158893	T2C	VT-4	F-818
158894	T2C	VT-23	B-346
158895	T2C	VT-19	A-983
158896	T2C	VT-19	A-981
158897	T2C	AMARC	2T204
158898	T2C	VT-19	A-987
158899	T2C	VT-4	F-809
158900	T2C	VT-23	B-309
158901	T2C	VT-19	A-949
158902	T2C		
158903	T2C	VT-23	B-307
158904	T2C	VF-43	32
158905	T2C	VT-19	A-996
158906	T2C	VT-23	B-336
158907	T2C	VT-23	B-310
158908	T2C	VT-4	F-828
158909	T2C	VT-19	A-982
158910	T2C	VF-43	31
158911	T2C	VT-19	A-957
159150	T2C	VT-23	B-321
159151	T2C	VT-19	A-963
159152	T2C	AMARC	2T181
159153	T2C	VT-4	F-816
159154	T2C	AMARC	2T191
159155	T2C		
159156	T2C	w.o.20.11.88	
159157	T2C	VT-19	A-980
159158	T2C	AMARC	2T182
159159	T2C	VT-19	A-979
159160	T2C		
159161	T2C	AMARC	2T193
159162	T2C		
159163	T2C	VT-19	A-953
159164	T2C	AMARC	2T192
159165	T2C	VT-4	F-820
159166	T2C		
159167	T2C		
159168	T2C	VT-23	B-315
159169	T2C	VT-23	B-303
159170	T2C	VT-23	B-301
159171	T2C	VT-19	A-975
159172	T2C	AMARC	2T208
159173	T2C	VT-19	A-973
159704	T2C	AMARC	2T188
159705	T2C	VT-19	A-977
159706	T2C	VT-19	A-965
159707	T2C	VT-23	B-322
159708	T2C	AMARC	2T197
159709	T2C	VT-23	B-317
159710	T2C	VT-23	B-304
159711	T2C		
159712	T2C	VT-19	A-998
159713	T2C	VT-19	A-964
159714	T2C	VT-23	B-324
159715	T2C	VT-23	B-369
159716	T2C	VT-19	A-972
159717	T2C	w.o.25.4.86	
159718	T2C	VT-19	A-976
159719	T2C	VT-23	B-331
159720	T2C	VT-19	A-954
159721	T2C	VT-23	B-352
159722	T2C	VT-23	B-329
159723	T2C	AMARC	2T186
159724	T2C	w.o.13.4.89	
159725	T2C		
159726	T2C		
159727	T2C	VT-4	F-825

T34 TURBO-MENTOR

c\n   BG-118, BG-195, GL-1 to GL-353.

140784	YT34C	preserved	NAS Pensacola		160511	T34C	TW-5	E-511
140861	YT34C	preserved	Whiting F'ld		160512	T34C	w.o.	
160265	T34C				160513	T34C	TW-5	E-513
160266	NT34C	NAWC-AD	ASATD		160514	T34C	VF-124	NJ-451
160267	T34C	w.o.15.7.77			160515	T34C	TW-5	E-515
160268	T34C	TW-5	E-268		160516	T34C	TW-5	E-516
160269	T34C	TW-5	E-269		160517	T34C	TW-5	E-517
160270	T34C	TW-5	E-270		160518	T34C	w.o.	
160271	T34C	VF-124	NJ-450		160519	T34C	TW-5	E-519
160272	T34C	TW-4	G-740		160520	T34C	TW-5	E-520
160273	T34C	VA-42	AD-560		160521	T34C	TW-4	G-767
160274	T34C	TW-5	E-274		160522	T34C	TW-5	E-522
160275	T34C	w.o.1.3.82			160523	T34C	TW-5	E-523
160276	T34C	TW-5	E-276		160524	T34C	TW-5	E-524
160277	T34C	TW-5	E-277		160525	T34C	w.o.	
160278	T34C	SFWSA	902		160526	T34C	TW-5	E-526
160279	T34C	TW-5	E-279		160527	T34C	TW-5	E-527
160280	T34C	TW-5	E-280		160528	T34C	TW-5	E-528
160281	T34C	TW-5	E-281		160529	T34C	TW-5	E-529
160282	T34C	TW-5	E-282		160530	T34C	VMFAT-101	SH-530
160462	T34C	TW-5	E-462		160531	T34C	TW-5	E-531
160463	T34C	TW-5	E-463		160532	T34C	VMFAT-101	SH-532
160464	T34C	TW-5	E-464		160533	T34C	TW-5	E-533
160465	T34C	TW-5	E-465		160534	T34C	VFA-125	NJ-534
160466	T34C	TW-5	E-466		160535	T34C	TW-4	G-771
160467	T34C	TW-5	E-467		160536	T34C	w.o.23.9.83	
160468	T34C	TW-5	E-468		160629	T34C	TW-5	E-629
160469	T34C	TW-5	E-469		160630	T34C	TW-5	E-630
160470	T34C	w.o.			160631	T34C	TW-5	E-631
160471	T34C	TW-5	E-471		160632	T34C	TW-5	E-632
160472	T34C	TW-5	E-472		160633	T34C	TW-5	E-633
160473	T34C	TW-5	E-473		160634	T34C	TW-5	E-634
160474	T34C	w.o.			160635	T34C	TW-5	E-635
160475	T34C	TW-5	E-475		160636	T34C	TW-4	G-756
160476	T34C	TW-5	E-476		160637	T34C	TW-5	E-637
160477	T34C	TW-5	E-477		160638	T34C	TW-5	E-638
160478	T34C	TW-5	E-478		160639	T34C	TW-5	E-639
160479	T34C	TW-5	E-479		160640	T34C	TW-5	E-640
160480	T34C	w.o.3.10.88			160641	T34C	TW-5	E-641
160481	T34C	TW-5	E-481		160642	T34C	TW-5	E-642
160482	T34C	TW-5	E-482		160643	T34C	TW-5	E-643
160483	T34C	TW-5	E-483		160644	T34C	TW-5	E-644
160484	T34C	TW-5	E-484		160645	T34C	TW-5	E-645
160485	T34C	TW-4	G-787		160646	T34C	TW-5	E-646
160486	T34C	VT-4	F-59		160647	T34C	TW-4	G-730
160487	T34C	TW-4	G-732		160648	T34C	VT-4	F-97
160488	T34C	TW-4	G-727		160649	T34C	TW-5	E-649
160489	T34C	TW-4	G-725		160650	T34C	TW-5	E-650
160490	T34C	TW-5	E-490		160651	T34C	TW-5	E-651
160491	T34C	VFA-125	NJ-491		160931	T34C	TW-5	E-931
160492	T34C	VFA-125	NJ-492		160932	T34C	TW-5	E-932
160493	T34C	TW-5	E-493		160933	T34C	TW-5	E-933
160494	T34C	SFWSA	901		160934	T34C	TW-5	E-934
160495	T34C	TW-5	E-495		160935	T34C	TW-5	E-935
160496	T34C	w.o.16.3.86			160936	T34C	TW-5	E-936
160497	T34C	TW-4	G-761		160937	T34C	TW-5	E-937
160498	T34C	w.o.			160938	T34C	TW-5	E-938
160499	T34C	NAWC-AD	ASATD		160939	T34C	TW-5	E-939
160500	T34C	TW-5 ·	E-500		160940	T34C	TW-5	E-940
160501	T34C	SFWSA			160941	T34C	TW-5	E-941
160502	T34C	TW-5	E-502		160942	T34C	TW-5	E-942
160503	T34C	w.o.9.1.84			160943	T34C	TW-5	E-943
160504	T34C	VMFAT-101	SH-504		160944	T34C	VT-4	F-96
160505	T34C	TW-4	G-766		160945	T34C	TW-5	E-945
160506	T34C	TW-5	E-506		160946	T34C	TW-5	E-946
160507	T34C	w.o.11.10.91			160947	T34C	TW-5	E-947
160508	T34C	TW-5	E-508		160948	T34C	TW-5	E-948
160509	T34C	TW-5	E-509		160949	T34C	VT-4	F-98
160510	T34C	TW-4	G-786		160950	T34C	TW-5	E-950

160951	T34C	TW-5	E-951	161826	T34C	TW-4	G-726
160952	T34C	TW-5	E-952	161827	T34C	TW-5	E-827
160953	T34C	TW-5	E-953	161828	T34C	TW-5	E-828
160954	T34C	TW-5	E-954	161829	T34C	TW-5	E-829
160955	T34C	w.o.13.7.82		161830	T34C	VT-4	F-60
160956	T34C	VFA-125	NJ-956	161831	T34C	TW-5	E-831
160957	T34C	TW-5	E-957	161832	T34C	TW-5	E-832
160958	T34C	TW-5	E-958	161833	T34C	TW-4	G-733
160959	T34C	TW-5	E-959	161834	T34C	TW-4	G-734
160960	T34C	TW-4	G-763	161835	T34C	TW-5	E-835
160961	T34C	TW-5	E-961	161836	T34C	TW-5	E-836
160962	T34C	TW-5	E-962	161837	T34C	TW-4	G-737
160963	T34C	TW-5	E-963	161838	T34C	TW-4	G-762
161023	T34C	TW-5	E-023	161839	T34C	TW-5	E-839
161024	T34C	TW-5	E-024	161840	T34C	TW-5	E-840
161025	T34C	TW-5	E-025	161841	T34C	TW-4	G-741
161026	T34C	TW-5	E-026	161842	T34C	TW-5	E-842
161027	T34C	TW-5	E-027	161843	T34C	TW-4	G-743
161028	T34C	TW-5	E-028	161844	T34C	US Army	
161029	T34C	TW-5	E-029	161845	T34C	TW-4	G-745
161030	T34C	TW-5	E-030	161846	T34C	TW-5	E-846
161031	T34C	TW-5	E-031	161847	T34C	TW-5	E-847
161032	T34C	TW-5	E-032	161848	T34C	TW-4	G-764
161033	T34C	TW-5	E-033	161849	T34C	TW-5	E-849
161034	T34C	TW-5	E-034	162247	T34C	w.o.5.3.87	
161035	T34C	TW-4	G-774	162248	T34C	VT-4	F-63
161036	T34C	TW-5	E-036	162249	T34C	TW-4	G-
161037	T34C	TW-5	E-037	162250	T34C	TW-5	E-250
161038	T34C	TW-5	E-038	162251	T34C	TW-5	E-251
161039	T34C	TW-5	E-039	162252	T34C	TW-5	E-252
161040	T34C	TW-4	G-781	162253	T34C	VT-4	F-62
161041	T34C	TW-5	E-041	162254	T34C	VT-4	F-99
161042	T34C	TW-5	E-042	162255	T34C	TW-5	E-255
161043	T34C	TW-4	G-731	162256	T34C	TW-5	E-256
161044	T34C	TW-5	E-044	162257	T34C	TW-5	E-257
161045	T34C	TW-5	E-045	162258	T34C	TW-5	E-258
161046	T34C	TW-5	E-046	162259	T34C	TW-5	E-259
161047	T34C	TW-5	E-047	162260	T34C	TW-4	G-780
161048	T34C	TW-5	E-048	162261	T34C	TW-5	E-261
161049	T34C	w.o.18.4.86		162262	T34C	VT-4	F-62
161050	T34C	TW-5	E-050	162263	T34C	TW-5	E-263
161051	T34C	TW-5	E-051	162264	T34C	USAAEFA	
161052	T34C	TW-5	E-052	162265	T34C	TW-5	E-265
161053	T34C	TW-5	E-053	162266	T34C	TW-5	E-266
161054	T34C	TW-5	E-054	162267	T34C	VA-42	AD-561
161055	T34C	TW-5	E-055	162268	T34C	VT-4	F-70
161056	T34C	w.o.1.3.82		162269	T34C	VT-4	F-71
161790	T34C	TW-5	E-790	162270	T34C	VT-4	F-72
161791	T34C	TW-5	E-791	162271	T34C	VT-4	F-73
161792	T34C	TW-5	E-792	162272	T34C	VT-4	F-74
161793	T34C	TW-5	E-793	162273	T34C	VT-4	F-75
161794	T34C	TW-4	G-704	162274	T34C	VT-4	F-76
161795	T34C	TW-4	G-705	162275	T34C	VT-4	F-77
161796	T34C	TW-4	G-706	162276	T34C	VT-4	F-78
161797	T34C	TW-4	G-707	162277	T34C	VT-4	F-79
161798	T34C	TW-4	G-754	162278	T34C	VT-4	F-80
161799	T34C	TW-4	G-799	162279	T34C	VT-4	F-81
161800	T34C	TW-5	E-800	162280	T34C	VT-4	F-82
161801	T34C	TW-5	E-801	162281	T34C	VT-4	F-83
161802	T34C	TW-5	E-802	162282	T34C	VT-4	F-84
161803	T34C	TW-5	E-803	162283	T34C	VT-4	F-85
161804	T34C	TW-5	E-804	162284	T34C	VT-4	F-86
161805	T34C	TW-4	G-752	162285	T34C	VT-4	F-87
161806	T34C	TW-4	G-703	162286	T34C	VT-4	F-88
161807	T34C	TW-5	E-807	162287	T34C	VT-4	F-89
161808	T34C	TW-4	G-750	162288	T34C	VT-4	F-61
161809	T34C	VFA-125	NJ-809	162289	T34C	TW-5	E-289
161810	T34C	TW-4	G-710	162290	T34C	TW-5	E-290
161811	T34C	TW-4	G-711	162291	T34C	TW-5	E-291
161812	T34C	w.o.14.1.92		162292	T34C	TW-5	E-292
161813	T34C	TW-4	G-713	162293	T34C	w.o.13.5.92	
161814	T34C	TW-4	G-714	162294	T34C	VT-4	F-64
161815	T34C	TW-4	G-715	162295	T34C	TW-5	E-295
161816	T34C	TW-4	G-716	162296	T34C	TW-5	E-296
161817	T34C	TW-4	G-717	162297	T34C	TW-5	E-297
161818	T34C	TW-5	E-818	162298	T34C	TW-5	E-298
161819	T34C	TW-4	G-757	162299	T34C	VA-42	AD-562
161820	T34C	TW-4	G-720	162300	T34C	TW-5	E-300
161821	T34C	TW-4	G-721	162301	T34C	TW-5	E-301
161822	T34C	TW-4	G-722	162302	T34C	TW-5	E-303
161823	T34C	TW-4	G-723	162303	T34C	TW-5	E-303
161824	T34C	TW-4	G-724	162304	T34C	TW-5	E-304
161825	T34C	TW-4	G-	162305	T34C	TW-5	E-305

162306	T34C	TW-5	E-306	162644	T34C	TW-5	E-285
162620	T34C	TW-5	E-620	162645	T34C	VT-4	F-94
162621	T34C	TW-5	E-621	162646	T34C	TW-5	E-287
162622	T34C	TW-5	E-622	162647	T34C	TW-5	E-537
162623	T34C	TW-5	E-623	162648	T34C	VT-4	F-92
162624	T34C	TW-5	E-624	162649	T34C	TW-5	E-539
162625	T34C	TW-5	E-625	164155	T34C	TW-5	E-155
162626	T34C	VT-4	F-66	164156	T34C	TW-5	E-156
162627	T34C	TW-5	E-627	164157	T34C	TW-5	E-157
162628	T34C	VT-4	F-65	164158	T34C	TW-5	E-158
162629	T34C	VT-4	F-91	164159	T34C		
162630	T34C	TW-5	E-468	164160	T34C	w.o.13.5.92	
162631	T34C	TW-5	E-470	164161	T34C	VT-4	F-95
162632	T34C	TW-5	E-474	164162	T34C	TW-5	E-162
162633	T34C	TW-5	E-498	164163	T34C	TW-5	E-163
162634	T34C	TW-5	E-503	164164	T34C	TW-5	E-164
162635	T34C	VT-4	F-90	164165	T34C	TW-5	E-165
162636	T34C	TW-5	E-518	164166	T34C	TW-5	E-166
162637	T34C	TW-5	E-525	164167	T34C	TW-5	E-167
162638	T34C	TW-5	E-	164168	T34C	TW-5	E-168
162639	T34C	VT-4	F-69	164169	T34C	TW-5	E-169
162640	T34C	TW-5	E-955	164170	T34C	TW-5	E-170
162641	T34C	TW-5	E-056	164171	T34C	TW-5	E-171
162642	T34C	VT-4	F-67	164172	T34C	TW-5	E-172
162643	T34C	TW-5	E-284	164173	T34C		

T39 SABRELINER

c\n	T39D:	277-1 to 277-10, 285-1 to 285-32. CT39E: 282-46, 282-84, 282-85, 282-95, 282-93, 282-92, 282-96, CT39G: 306-52, 306-55, 306-65 to 306-67, 306-69, 306-70, 306-104 to 306-108.
	T39N:	282-9, 282-81, 282-29, 282-2, 282-66, 282-30, 282-72, 282-90, 282-61, 282-77, 282-19, 282-94, 282-28, 282-20, 282-60, 282-100.

150542	T39D	stored	NAS China Lake	151341	T39D	AMARC	7T019
150543	T39D	NAS	F Pensacola	151342	T39D	AMARC	7T020
150544	T39D	AMARC	7T006	151343	T39D	dump	NAS Pensacola
150545	T39D	w.o.1.4.77		157352	CT39E	w.o.21.12.75	
150546	T39D	AMARC	7T014	157353	CT39E		
150547	T39D	AMARC	7T021	157354	CT39E		
150548	T39D	AMARC	7T008	158380	CT39E	to N425NA	NASA
150549	T39D	AMARC	7T011	158381	CT39E	w.o.12.7.88	
150550	T39D			158382	CT39E	SOES	
150551	T39D	dump	NAS Pensacola	158383	CT39E	w.o.	
150969	T39D	NAS	F Pensacola	158843	CT39G	CFSLW	
150970	T39D	to NASA	N431NA	158844	CT39G		
150971	T39D			159361	CT39G	VR-24	30
150972	T39D			159362	CT39G	VR-24	31
150973	T39D	AMARC	7T013	159363	CT39G	VR-24	32
150974	T39D	AMARC	7T015	159364	CT39G	HQ\Flt. Sec. USMC Iwakuni	
150975	T39D	AMARC	7T007	159365	CT39G	HQ\Flt. Sec. USMC El Toro	
150976	T39D	AMARC	7T016	160053	CT39G	CNR	USN
150977	T39D			160054	CT39G	SOES	MCAS Cherry Pt.
150978	T39D	AMARC	7T009	160055	CT39G	SOES	MCAS Cherry Pt.
150979	T39D	AMARC	7T017	160056	CT39G	SOES	MCAS Cherry Pt.
150980	T39D	AMARC	7T004	160057	CT39G	w.o.3.3.91	
150981	T39D	AMARC	7T012	N301NT	T39N	VT-86	F-01
150982	T39D	AMARC	7T022	N302NT	T39N	VT-86	F-02
150983	T39D	AMARC	7T023	N303NT	T39N	VT-86	F-03
150984	T39D	AMARC	7T010	N304NT	T39N	VT-86	F-04
150985	T39D	NAS	F Pensacola	N305NT	T39N	VT-86	F-05
150986	T39D	preserved	Robins AFB	N306NT	T39N	VT-86	F-06
150987	T39D	NAWC\TPS 39		N307NT	T39N	VT-86	F-07
150988	T39D	AMARC	7T005	N308NT	T39N	VT-86	F-08
150989	T39D	stored	NAS China Lake	N309NT	T39N	VT-86	F-09
150990	T39D	AMARC	7T024	N310NT	T39N	VT-86	F-10
150991	T39D	AMARC	7T003	N311NT	T39N	VT-86	F-11
150992	T39D	NAWC-WD		N313NT	T39N	VT-86	F-12
151336	T39D	AMARC	7T025	N314NT	T39N	VT-86	F-13
151337	T39D			N315NT	T39N	VT-86	F-14
151338	T39D			N316NT	T39N	VT-86	F-15
151339	T39D	preserved	NAS Pensacola	N317NT	T39N	VT-86	F-16
151340	T39D	AMARC	7T018				

T44 KING AIR

c\n   T44A\H90: LL-1 to LL-61.
      T44B   :   not known

160839	T44A	VT-31	G-839
160840	T44A	VT-31	G-462
160841	T44A	VT-31	G-471
160842	T44A	VT-31	G-842
160843	T44A	VT-31	G-843
160844	T44A	VT-31	G-844
160845	T44A	VT-31	G-945
160846	T44A	VT-31	G-846
160847	T44A	VT-31	G-967
160848	T44A	VT-31	G-448
160849	T44A	VT-31	G-849
160850	T44A	VT-31	G-800
160851	T44A	VT-31	G-801
160852	T44A	VT-31	G-032
160853	T44A	VT-31	G-553
160854	T44A	VT-31	G-854
160855	T44A	VT-31	G-055
160856	T44A	VT-31	G-956
160967	T44A	w.o.8.7.82	
160968	T44A	VT-31	G-827
160969	T44A	VT-31	G-459
160970	T44A	VT-31	G-110
160971	T44A	VT-31	G-111
160972	T44A	VT-31	G-522
160973	T44A	VT-31	G-113
160974	T44A	VT-31	G-834
160975	T44A	VT-31	G-425
160976	T44A	VT-31	G-437
160977	T44A	VT-31	G-577
160978	T44A	VT-31	G-118
160979	T44A	VT-31	G-519
160980	T44A	w.o.8.3.84	
160981	T44A	VT-31	G-020
160982	T44A	VT-31	G-122
160983	T44A	VT-31	G-403
160984	T44A	VT-31	G-484
160985	T44A	VT-31	G-595
160986	T44A	VT-31	G-414
161057	T44A	VT-31	G-857
161058	T44A	VT-31	G-858
161059	T44A	w.o.21.2.80	
161060	T44A	VT-31	G-900
161061	T44A	VT-31	G-901
161062	T44A	VT-31	G-862
161063	T44A	VT-31	G-026
161064	T44A	VT-31	G-064
161065	T44A	VT-31	G-865
161066	T44A	VT-31	G-566
161067	T44A	w.o.	
161068	T44A	VT-31	G-868
161069	T44A	VT-31	G-888
161070	T44A	VT-31	G-400
161071	T44A	VT-31	G-402
161072	T44A	VT-31	G-102
161073	T44A	VT-31	G-103
161074	T44A	VT-31	G-804
161075	T44A	VT-31	G-075
161076	T44A	VT-31	G-106
161077	T44A	VT-31	G-087
161078	T44A	VT-31	G-108
161079	T44A	VT-31	G-401
164579	T44B		
164580	T44B		
164581	T44B		
164582	T44B		
164583	T44B		

T45 GOSHAWK

162612	YT45A		
162613	YT45A		
162787	T45A	w.o.4.6.92	
162788	T45A	MDD	
162789	T45A		
162790	T45A		
163599	T45A		
163600	T45A	VT-21	B-200
163601	T45A	VT-21	B-201
163602	T45A	VT-21	B-202
163603	T45A	VT-21	B-203
163604	T45A	VT-21	B-204
163605	T45A	VT-21	B-205
163606	T45A	VT-21	B-206
163607	T45A	VT-21	B-207
163608	T45A	VT-21	B-208
163609	T45A	VT-21	B-209
163610	T45A		
163611	T45A		
163612	T45A		
163613	T45A		
163614	T45A		
163615	T45A		
163616	T45A		
163617	T45A		
163618	T45A		
163619	T45A		
163620	T45A		
163621	T45A		
163622	T45A		
163623	T45A		
163624	T45A		
163625	T45A		
163626	T45A		
163627	T45A		
163628	T45A		
163629	T45A		
163630	T45A		
163631	T45A		
163632	T45A		
163633	T45A		
163634	T45A		

163635	T45A		165063	T45A	
163636	T45A		165064	T45A	
163637	T45A		165065	T45A	
163638	T45A		165066	T45A	
163639	T45A		165067	T45A	
163640	T45A		165068	T45A	
163641	T45A		165069	T45A	
163642	T45A		165070	T45A	
163643	T45A		165071	T45A	
163644	T45A		165072	T45A	
163645	T45A		165073	T45A	
163646	T45A		165074	T45A	
163647	T45A		165075	T45A	
163648	T45A		165076	T45A	
163649	T45A		165077	T45A	
163650	T45A		165078	T45A	
163651	T45A		165079	T45A	
163652	T45A		165080	T45A	
163653	T45A		165081	T45A	
163654	T45A		165082	T45A	
163655	T45A		165083	T45A	
163656	T45A		165084	T45A	
163657	T45A		165085	T45A	
163658	T45A		165086	T45A	
165057	T45A		165087	T45A	
165058	T45A		165088	T45A	
165059	T45A		165089	T45A	
165060	T45A		165090	T45A	
165061	T45A		165091	T45A	
165062	T45A		165092	T45A	

T47 CITATION II

c\n   552-0001 to 0015.

N12855	T47A	w.o.22.7.93	N12763	T47A	w.o.22.7.93
N12756	T47A	w.o.22.7.93	N12564	T47A	w.o.22.7.93
N12557	T47A	w.o.22.7.93	N12065	T47A	w.o.22.7.93
N12058	T47A	w.o.22.7.93	N12566	T47A	returned to lessor
N12859	T47A	returned to lessor	N12967	T47A	w.o.22.7.93
N12660	T47A	w.o.22.7.93	N12568	T47A	returned to lessor
N12761	T47A	w.o.22.7.93	N12269	T47A	w.o.22.7.93
N12762	T47A	w.o.22.7.93			

note: US Navy Bureau numbers 162755 to 162769 were allocated but not carried by these leased aircraft. Most were destroyed in a hangar fire whilst stored after withdawal from use by the USN. They have now been replaced by reworked civil Sabreliner aircraft designated T-39N and bearing US civil markings.

U1 OTTER

c\n   76 to 79, 3 unknown, 148, 151, 158, 160, 163, 166, 126.

142424	U1B	w.o.22.12.55	144669	U1B	to OO-HAD	
142425	NU1B	to N1037G	144670	NU1B	NAWC\TPS	30
142426	U1B	w.o.30.8.57	144671	U1B	preserved	Calgary, Canada
142427	U1B	w.o.22.10.58	144672	NU1B	preserved	NAS Pensacola
144259	U1B	w.o.3.2.56	144673	U1B	w.o.3.7.59	
144260	U1B	w.o.3.2.56	144674	U1B	to N5348G	
144261	U1B	w.o.2.7.56	147574	U1B	to New Zealand	NZ6081

U3 BLUE CANOE

| c\n | U3A: | unknown |
| | U3B: | 0002 |

159073	U3A		606047	U3B	NavElexSysCom
159074	U3A				

U6 BEAVER

150191	U6A	NAWC\TPS 31	164524	U6A	
150192	U6A	to N99830	164525	U6A	NAWC\TPS
151348	U6A				

U9

c\n   466-136, 4771-141.

576183	RU9D	NAWC-WD	576184	RU9D	stored	NAS Pensacola

U11 AZTEC

149050	U11A			149060	U11A	
149051	U11A			149061	U11A	NAS	Mayport
149052	U11A	MASDC		149062	U11A	USMC
149053	U11A	NAS	Barbers Point	149063	U11A	
149054	U11A			149064	U11A	
149055	U11A	NAS	Barbers Point	149065	U11A	MASDC
149056	U11A			149066	U11A	
149057	U11A			149067	U11A	Pima Museum
149058	U11A	USMC		149068	U11A	MASDC
149059	U11A	to N8063		149069	U11A	NAS	North Island

OV1 MOHAWK

592625	OV1B	NAWC\TPS 33			625866	OV1B	NAWC\TPS 34
592637	OV1B	NAWC\TPS 35					

OV10 BRONCO

c\n   300-1 to 300-7, 305-1 to 305-114, 305A-31, 305A-34, 305A-60, 321-122, 321-125, 321-135. 152879 YOV10A derelict Rosamund, Ca

Serial	Type	Status	Code
152880	YOV10A	preserved	MCAS Quantico
152881	YOV10A	to NASA	N718NA
152882	YOV10A		
152883	YOV10A		
152884	YOV10A		
152885	YOV10A		
155390	OV10A	to NASA	N636NA
155391	OV10A		
155392	OV10A	w.o.6.8.81	
155393	OV10A	w.o.30.3.70	
155394	OV10A	w.o.29.10.71	
155395	OV10D	VMO-2	UU-06
155396	YOV10D	to NASA	N627NA
155397	OV10A	to Morocco	155397
155398	OV10A		
155399	OV10A		
155400	OV10A	VMO-4	MU-506
155401	OV10A	VMO-4	MU-507
155402	OV10A	VMO-4	MU-514
155403	OV10A	w.o.29.5.69	
155404	OV10A	to Morocco	155404
155405	OV10A	AMARC	1V033
155406	OV10A		
155407	OV10A	w.o.16.11.84	
155408	OV10A		
155409	OV10A	VMO-2	UU-01
155410	OV10A	VMO-4	MU-516
155411	OV10A	w.o.3.12.68	
155412	OV10A	w.o.25.7.68	
155413	OV10A		
155414	OV10A		
155415	OV10D	VMO-2	UU-21
155416	OV10A		
155417	OV10A	VMO-4	MU-515
155418	OV10D	VMO-2	UU-05
155419	OV10A	w.o.16.5.84	
155420	OV10A	w.o.8.4.69	
155421	OV10A	w.o.22.7.69	
155422	OV10A	w.o.22.10.68	
155423	OV10A	w.o.21.4.70	
155424	OV10A	w.o.25.2.91	
155425	OV10A	to Morocco	155425
155426	OV10A	VMO-2	UU-16
155427	OV10A	VMO-4	MU-505
155428	OV10A	AMARC	1V034
155429	OV10D	VMO-2	UU-25
155430	OV10A		
155431	OV10A	w.o.22.7.69	
155432	OV10A	w.o.	
155433	OV10A	to Morocco	155433
155434	OV10A	VMO-1	ER-10
155435	OV10A	w.o.18.1.91	
155436	OV10A	VMO-1	ER-14
155437	OV10A	w.o.31.7.85	
155438	OV10A	VMO-1	ER-11
155439	OV10A		
155440	OV10D	w.o.23.10.84	
155441	OV10A		
155442	OV10A	w.o.23.10.84	
155443	OV10A	VMO-1	ER-04
155444	OV10A	w.o.23.10.83	
155445	OV10A	VMO-4	MU-513
155446	OV10A	VMO-1	ER-08
155447	OV10D	VMO-1	ER-09
155448	OV10A	VMO-1	ER-07
155449	OV10A		
155450	OV10A	w.o.28.4.71	
155451	OV10D	VMO-2	UU-04
155452	OV10A		
155453	OV10A		
155454	OV10A	AMARC	1V026
155455	OV10A	w.o.9.8.69	
155456	OV10A	w.o.29.8.69	
155457	OV10A	AMARC	1V037
155458	OV10A		
155459	OV10A	AMARC	1V024
155460	OV10A	w.o.	
155461	OV10A		
155462	OV10A	to Morocco	155462
155463	OV10A	w.o.9.7.74	
155464	OV10A	VMO-1	ER-
155465	OV10D	VMO-1	ER-00
155466	OV10A	VMO-2	UU-22
155467	OV10A	VMO-4	MU-507
155468	OV10D	VMO-1	ER-06
155469	OV10D	VMO-2	UU-25
155470	OV10D	VMO-1	ER-03
155471	OV10A	AMARC	1V...
155472	OV10D	VMO-1	ER-02
155473	OV10D	VMO-2	UU-15
155474	OV10D	VMO-2	UU-12
155475	OV10A	AMARC	1V027
155476	OV10D	w.o.15.5.84	
155477	OV10D	VMO-2	UU-
155478	OV10A	VMO-1	ER-
155479	OV10A	VMO-2	UU-23
155480	OV10A	AMARC	1V025
155481	OV10D	VMO-1	ER-01
155482	OV10D	VMO-2	UU-01
155483	OV10D	VMO-2	UU-20
155484	OV10D	w.o.30.12.87	
155485	OV10A	w.o.24.8.89	
155486	OV10A	VMO-4	MU-511
155487	OV10D	VMO-1	ER-
155488	OV10A	VMO-4	MU-504
155489	OV10A	VMO-2	UU-26

155490	OV10A	w.o.19.7.69			155500	OV10A	VMO-	
155491	OV10A	to Morocco	155491		155501	OV10D	VMO-1	ER-02
155492	OV10A	VMO-2	UU-03		155502	OV10D		
155493	OV10D	VMO-1	ER-04		155503	OV10A	w.o.20.12.69	
155494	OV10D	VMO-2	UU-02		714623	OV10D		
155495	OV10A	w.o.7.6.70			714626	OV10D		
155496	OV10A	AMARC	1V023		714652	OV10D		
155497	OV10A				683796	OV10D		
155498	OV10A	VMO-4	MU-511		683799	OV10D		
155499	OV10D	VMO-1	ER-07		683809	OV10D		

V22 OSPREY

163911	XV22A			164395	V22A
163912	XV22A			164396	V22A
163913	XV22A			164397	V22A
163914	XV22A	w.o.21.7.92		164398	V22A
163915	XV22A	w.o.		164399	V22A
163916	XV22A			164400	V22A
164389	V22A			164939	V22A
164390	V22A			164940	V22A
164391	V22A			164941	V22A
164392	V22A			164942	V22A
164393	V22A			164943	V22A
164394	V22A			164944	V22A

X25 B8M ULTRALIGHT

68-10770 X25A   68-10771 X25B

X26 SCHWEIZER/LOCKHEED

c\n   67, 68, 74, last 3 unknown.

157932	X26A	w.o.9.80		159260	X26A	NAWC\TPS 32
157933	X26A			161571	X26A	
158818	X26A			715345	X26B	NAWC\TPS

X28 PEREIRA OSPREY

158786   X28A

X31

164584   X31A   NASA   Dryden   164585   X31A

FAIRCHILD F27

c\n   33
161628   F27   NAWC-AD

LEARJET 25B

c\n   25-082
N700FC   LJ25B   USN\NELO OPS

UNITED STATES COAST GUARD

C4 GULFSTREAM I

c\n 91

| 02 | VC4A | Elizabeth City | ex USCG 1380 |

C11 GULFSTREAM II

c\n 23

| 01 | VC11A | Washington, DC | ex USCG 1451 |

C130 HERCULES

ex USAF 58-5396, 58-5397, 58-6973, 58-6974, 60-0311, 60-0312, 61-2081 to 61-2083, 62-3753 to 62-3755, 66-4299, 67-7183 to 67-7185, 72-1300 to 72-1302, 73-0844, 73-0845, 77-0317 to 77-0320, 82-0081 to 0085, 83-0007, 83-0505 to 83-0508, 84-0479 to 84-0482, 85-0051, 85-0052, 85-1360, 86-0420 to 86-0422, 87-0156, 87-0157, 81-0999.

c\n 3529, 3533, 3542, 3548, 3594, 3595, 3638, 3641, 3650, 3745, 3763, 3773, 4158, 4255, 4260, 4265, 4501, 4507, 4513, 4528, 4529, 4757, 4760, 4762, 4764, 4947, 4958, 4966, 4967, 4969, 4993, 4996, 4999, 5002, 5005, 5028, 5031, 5033 to 5035, 5037, 5023, 5104, 5106, 5107, 5120, 5121, 4931.

1339	HC130B	AMARC	45009\CF037	1603	HC130H	Barbers Point
1340	HC130B	AMARC	45003\CF031	1700	HC130H	Sacramento
1341	HC130B	scrapped		1701	HC130H	Sacramento
1342	HC130B	AMARC	45002\CF032	1702	HC130H	Sacramento
1344	HC130B	AMARC	45001\CF033	1703	HC130H	Sacramento
1345	HC130B	AMARC	45007\CF038	1704	HC130H	Sacramento
1346	HC130B	GIA	Cherry Point	1705	HC130H	Sacramento
1347	HC130B	AMARC	45006\CF039	1706	HC130H	Kodiak
1348	HC130B	GIA	Davis Monthan	1707	HC130H	Barbers Point
1349	HC130B	AMARC	45004\CF034	1708	HC130H	Barbers Point
1350	HC130B	derelict	Westover AFB	1709	HC130H	Sacramento
1351	HC130B	scrapped		1710	HC130H	Borinquen
1414	EC130E	scrapyard	Bobs\CF049	1711	HC130H	Barbers Point
1452	HC130H	to USAF	67-7183	1712	HC130H	Elizabeth City
1453	HC130H	to USAF	67-7184	1713	HC130H	Clearwater
1454	HC130H	AMARC	67-7185\CF044	1714	HC130H	Borinquen
1500	HC130H	Elizabeth City		1715	HC130H	
1501	HC130H	Elizabeth City		1716	HC130H	Clearwater
1502	HC130H	Elizabeth City		1717	HC130H	Clearwater
1503	HC130H	Elizabeth City		1718	HC130H	Clearwater
1504	HC130H	Elizabeth City		1719	HC130H	Clearwater
1600	HC130H	w.o.30.7.82		1720	HC130H	Clearwater
1601	HC130H	Barbers Point		1721	EC130V	Clearwater
1602	HC130H	Barbers Point		1790	HC130H	Kodiak

E2 HAWKEYE

3501	E2C	w.o.14.8.90		3506	E2C	to USN	Bu161342
3502	E2C	to USN	Bu159497	3507	E2C	to USN	Bu158641
3503	E2C	to USN	Bu159112	3508	E2C	to USN	Bu160415
3504	E2C	to USN	Bu160011	3509	E2C	to USN	Bu160419
3505	E2C	to USN	Bu159502				

RG8 CONDOR

8101	RG8A	Miami	ex 85-0047	8103	RG8A	Miami	ex 86-0404
8102	RG8A	Miami	ex 85-0048 w.o.				

H3 SEA KING

1430	HH3F			1485	HH3F	Clearwater	
1431	HH3F	Clearwater		1486	HH3F	Clearwater	
1432	HH3F	New Orleans		1487	HH3F	Clearwater	
1433	HH3F	GIA	ElizabethCity	1488	HH3F		
1434	HH3F	GIA	ElizabethCity	1489	HH3F	Elizabeth City	
1435	HH3F			1490	HH3F	Mobile	
1436	HH3F	San Diego		1491	HH3F	Mobile	
1437	HH3F			1492	HH3F	Clearwater	
1438	HH3F	Clearwater		1493	HH3F	Clearwater	
1467	HH3F	Clearwater		1494	HH3F		
1468	HH3F	Traverse City		1495	HH3F	Cape Cod	
1469	HH3F	Elizabeth City		1496	HH3F		
1470	HH3F	Mobile		1497	HH3F	Clearwater	
1471	HH3F	San Francisco		2578	CH3E	ARSC	Elizabeth City
1472	HH3F	Cape Cod		2580	JCH3E		
1473	HH3F	w.o.2.11.86		2788	CH3E	ARSC	Elizabeth City
1474	HH3F			12789	CH3E	spares	ex USAF
1475	HH3F	Clearwater		2791	CH3E	Traverse City	
1476	HH3F			2793	CH3E	Traverse City	
1477	HH3F			4224	CH3E		
1478	HH3F			4234	CH3E		
1479	HH3F	Elizabeth City		4235	CH3E		
1480	HH3F	San Francisco		5695	CH3C		
1481	HH3F	Elizabeth City		5697	CH3E		
1482	HH3F	Clearwater		9679	CH3E		
1483	HH3F	Mobile		9691	CH3E	Traverse City	
1484	HH3F	Clearwater					

H52 SEAGUARD

1352	HH52A			1403	HH52A	Kodiak	
1353	HH52A			1404	HH52A	w.o.9.1.71	
1354	HH52A			1405	HH52A	Mobile	12
1355	HH52A	preserved	Pensacola	1406	HH52A	Mobile	9
1356	HH52A	rework	AFSC	1407	HH52A	Los Angeles	
1357	HH52A	Barbers Point		1408	HH52A	Mobile	
1358	HH52A			1409	HH52A	Mobile	10
1359	HH52A	MASDC		1410	HH52A		
1360	HH52A	Brooklyn		1411	HH52A	to TF-GNA	
1361	HH52A			1412	HH52A	w.o.	
1362	HH52A	scrapped		1413	HH52A	AMARC	
1363	HH52A			1415	HH52A	Port Angeles	
1364	HH52A	MASDC		1416	HH52A	Houston	
1365	HH52A			1417	HH52A	scrapped	
1366	HH52A	MASDC		1418	HH52A	Port Angeles	
1367	HH52A	Mobile	11	1419	HH52A	AMARC	46003
1368	HH52A	MASDC		1420	HH52A	Barbers Point	
1369	HH52A	GIA	ElizabethCity	1421	HH52A	AMARC	46009
1370	HH52A	GIA	ElizabethCity	1422	HH52A	MASDC	
1371	HH52A	MASDC		1423	HH52A	GIA	Mobile
1372	HH52A	Brooklyn		1424	HH52A	AMARC	46016
1373	HH52A	AMARC	46006	1425	HH52A	Houston	
1374	HH52A			1426	HH52A	Houston	
1375	HH52A	MASDC		1427	HH52A	w.o.22.10.81	
1376	HH52A	w.o.17.1.79		1428	HH52A	Detroit	
1377	HH52A	Los Angeles		1429	HH52A	preserved	USSIntrepid
1378	HH52A	Detroit		1439	HH52A	Houston	
1379	HH52A	AMARC	46002	1440	HH52A	MASDC	
1382	HH52A	Kodiak		1441	HH52A	MASDC	
1383	HH52A	Mobile	1	1442	HH52A	Port Angeles	
1384	HH52A	GIA	ElizabethCity	1443	HH52A	Houston	
1385	HH52A	Northwind	CG Cutter	1444	HH52A	North Bend	CG cutter
1386	HH52A	w.o.27.10.84		1445	HH52A	Savannah	
1387	HH52A			1446	HH52A	MASDC	
1388	HH52A	Los Angeles		1447	HH52A		
1389	HH52A	MASDC		1448	HH52A		
1390	HH52A	Mobile	6	1449	HH52A		
1391	HH52A	scrapped		1450	HH52A	Mobile	7
1392	HH52A	Mobile	2	1455	HH52A	Detroit	
1393	HH52A	Mobile	13	1456	HH52A	w.o.2.6.83	
1394	HH52A	to N1394		1457	HH52A		
1395	HH52A	MASDC		1458	HH52A		
1396	HH52A	GIA	ElizabethCity	1459	HH52A	Chicago	
1397	HH52A	Savannah		1460	HH52A	MASDC	
1398	HH52A	MASDC		1461	HH52A		
1399	HH52A	scrapped		1462	HH52A	Kodiak	
1400	HH52A	Houston		1463	HH52A	MASDC	
1401	HH52A	GIA	ElizabethCity	1464	HH52A	Barbers Point	
1402	HH52A	Los Angeles		1465	HH52A	AMARC	46015
				1466	HH52A	Detroit	

H60 JAYHAWK

ex   USN Bu163801 to 1673835, 164820 to 164830, 165096,165124 to 165127, 165146 to 165150.

6001	HH60J	Elizabeth City		6009	HH60J	Elizabeth City
6002	HH60J	Mobile		6010	HH60J	Traverse City
6003	HH60J	NATTC		6011	HH60J	Traverse City
6004	HH60J	Mobile		6012	HH60J	Traverse City
6005	HH60J	Mobile		6013	HH60J	San Francisco
6006	HH60J	Mobile		6014	HH60J	San Francisco
6007	HH60J	Mobile		6015	HH60J	San Francisco
6008	HH60J	Elizabeth City		6016	HH60J	Cape Cod

Reg	Type	Location		Reg	Type	Location
6017	HH60J	Cape Cod		6037	HH60J	
6018	HH60J	Cape Cod		6038	HH60J	
6019	HH60J	Cape Cod		6039	HH60J	
6020	HH60J			6040	HH60J	
6021	HH60J			6041	HH60J	
6022	HH60J			6042	HH60J	
6023	HH60J	Mobile		6043	HH60J	
6024	HH60J			6044	HH60J	
6025	HH60J			6045	HH60J	
6026	HH60J			6046	HH60J	
6027	HH60J			6047	HH60J	
6028	HH60J	Mobile		6048	HH60J	
6029	HH60J			6049	HH60J	
6030	HH60J	ARSC		6050	HH60J	
6031	HH60J	Elizabeth City		6051	HH60J	
6032	HH60J			6052	HH60J	
6033	HH60J			6053	HH60J	
6034	HH60J			6054	HH60J	
6035	HH60J			6055	HH60J	
6036	HH60J			6056	HH60J	

H65 DOLPHIN

Reg	Type	Location		Reg	Type	Location	
4101	HH65A	to N60035 thenIsrael		6547	HH65A	Miami	
4102	HH65A	GIA		6548	HH65A	Sacramento	
4103	HH65A	ret.to manufacturer		6549	HH65A	New Orleans	
4104	HH65A	to Israel		6550	HH65A	San Diego	
6501	HH65A	Houston		6551	HH65A	Astoria	1
6502	HH65A	San Diego		6552	HH65A	Brooklyn	
6503	HH65A	Savannah		6553	HH65A	Houston	
6504	HH65A	Corpus Christi		6554	HH65A	Port Angeles	
6505	HH65A	San Diego		6555	HH65A	Mobile	
6506	HH65A	Los Angeles		6556	HH65A	US Army	Phoenix
6507	HH65A	Barbers Point		6557	HH65A	Mobile	
6508	HH65A	Port Angeles		6558	HH65A	New Orleans	
6509	HH65A	Miami		6559	HH65A	Mobile	
6510	HH65A	Mobile		6560	HH65A		
6511	HH65A			6561	HH65A	San Diego	
6512	HH65A	San Diego		6562	HH65A	Mobile	
6513	HH65A			6563	HH65A	New Orleans	
6514	HH65A	North Bend		6564	HH65A	Mobile	
6515	HH65A			6565	HH65A	New Orleans	
6516	HH65A	Miami		6566	HH65A	New Orleans	
6517	HH65A	Los Angeles		6567	HH65A	Houston	
6518	HH65A	Elizabeth City ARSC		6568	HH65A		
6519	HH65A	New Orleans		6569	HH65A	Elizabeth City	
6520	HH65A	Miami		6570	HH65A	Mobile	
6521	HH65A	Barbers Point		6571	HH65A	Barbers Point	
6522	HH65A	North Bend		6572	HH65A	New Orleans	
6523	HH65A	Port Angeles		6573	HH65A	Houston	
6524	HH65A	Humboldt		6574	HH65A	Savannah	
6525	HH65A	North Bend		6575	HH65A	Savannah	
6526	HH65A	Brooklyn		6576	HH65A	San Diego	
6527	HH65A	Brooklyn		6577	HH65A	Miami	
6528	HH65A	Savannah		6578	HH65A	Barbers Point	
6529	HH65A	Port Angeles		6579	HH65A	New Orleans	
6530	HH65A			6580	HH65A	Detroit	
6531	HH65A	New Orleans		6581	HH65A	Chicago	
6532	HH65A	Miami		6582	HH65A	Miami	
6533	HH65A	Mobile		6583	HH65A	Savannah	
6534	HH65A	North Bend		6584	HH65A	Detroit	
6535	HH65A	Brooklyn		6585	HH65A	Houston	
6536	HH65A	w.o.1.4.92		6586	HH65A	Brooklyn	
6537	HH65A	Kodiak		6587	HH65A	Barbers Point	
6538	HH65A			6588	HH65A	Cape May	
6539	HH65A	North Bend		6589	HH65A	Los Angeles	
6540	HH65A	Miami		6590	HH65A	Detroit	
6541	HH65A	Mobile		6591	HH65A	Barbers Point	
6542	HH65A	Los Angeles		6592	HH65A	Mobile	
6543	HH65A	Los Angeles		6593	HH65A	Port Angeles	
6544	HH65A	Savannah		6594	HH65A	Detroit	
6545	HH65A	NAWC\TPS 67		6595	HH65A	Chicago	
6546	HH65A	w.o.		6596	HH65A		

note: 4105-4117 were reserialled as 6543\16\53\05\38\26\39\03\01\06\02\07\04

U25 GUARDIAN

c\n   374, 386, 394, 390, 398, 402, 409, 405, 407, 411, 413, 415, 417 to 421, 423 to 425, 431, 433, 435,
     437, 439, 441, 443, 445, 447, 450, 452, 454, 456, 458 to 460, 462, 464, 466, 467, 371.

2101	HU25A	Mobile			2122	HU25A	Miami	
2102	HU25A	Mobile			2123	HU25A	Mobile	
2103	HU25A				2124	HU25A	Mobile	
2104	HU25C	Miami			2125	HU25B	Miami	
2105	HU25A	Astoria	5		2126	HU25A	Miami	
2106	HU25A	Mobile			2127	HU25A	AMARC	41001
2107	HU25A	Miami			2128	HU25A	San Diego	
2108	HU25A	San Diego			2129	HU25A	Mobile	
2109	HU25A	Mobile			2130	HU25A	Cape Cod	
2110	HU25A	Elizabeth City ARSC			2131	HU25A	Miami	
2111	HU25A				2132	HU25A	Miami	
2112	HU25A	Miami			2133	HU25A	Mobile	
2113	HU25A	Cape Cod			2134	HU25B	Mobile	
2114	HU25A	Mobile			2135	HU25A	Miami	
2115	HU25A	San Diego			2136	HU25B	Miami	
2116	HU25A	Cape Cod			2137	HU25A	Astoria	
2117	HU25A	Astoria 7			2138	HU25A	Corpus Christi	
2118	HU25B	Cape Cod			2139	HU25C	Mobile	
2119	HU25A	Mobile			2140	HU25C	Miami	
2120	HU25A	Mobile			2141	HU25C	Miami	
2121	HU25A							

CASA 212-300 AVIOCAR

0393	C212	Miami

STATES OF THE USA

State and Territory abbreviations

Ak	Alaska	Ky	Kentucky	Oh	Ohio
Al	Alabama	La	Louisiana	Ok	Oklahoma
Ar	Arkansas	Ma	Massachusetts	Or	Oregon
Az	Arizona	Md	Maryland	Pa	Pennsylvania
Ca	California	Me	Maine	PR	Puerto Rico
Co	Colorado	Mi	Michigan	RI	Rhode Island
Ct	Connecticut	Mn	Minnesota	SC	South Carolina
CZ	Canal Zone	Mo	Missouri	SD	South Dakota
DC	District of Columbia	Ms	Mississippi	Tn	Tennessee
De	Delaware	Mt	Montana	Tx	Texas
Fl	Florida	Ne	Nebraska	Ut	Utah
Ga	Georgia	NC	North Carolina	Va	Virginia
Hi	Hawaii	ND	North Dakota	VI	Virgin Islands
Ia	Iowa	NH	New Hampshire	Vt	Vermont
Id	Idaho	NJ	New Jersey	Wa	Washington
Il	Illinois	NM	New Mexico	Wi	Wisconsin
In	Indiana	Nv	Nevada	WV	West Virginia
Ks	Kansas	NY	New York	Wy	Wyoming

AMARC - current residents

DOUGLAS A3 SKYWARRIOR (44)

138932	KA3B	2A109	142654	EKA3B	2A075
138938	NA3B	2A...	142656	EKA3B	2A092
138965	KA3B	2A110	142661	EKA3B	2A094
138969	KA3B	2A068	142666	RA3B	2A040
138973	KA3B	2A052	142669	RA3B	2A060
142237	EKA3B	2A071	142671	EA3B	2A117
142247	KA3B	2A050	144628	EKA3B	2A100
142249	KA3B	2A072	144834	UA3B	2A125
142251	EKA3B	2A073	144849	EA3B	2A123
142253	KA3B	2A067	144852	EA3B	2A127
142255	EKA3B	2A081	144857	TA3B	2A121
142401	KA3B	2A051	144859	TA3B	2A122
142404	EKA3B	2A087	144860	TA3B	2A119
142630	NA3B	2A132	144866	TA3B	2A128
142632	EKA3B	2A082	146447	ERA3B	2A120
142638	EKA3B	2A088	146448	EA3B	2A131
142644	KA3B	2A058	146452	EA3B	2A130
142646	EKA3B	2A083	146453	EA3B	2A126
142647	EKA3B	2A086	146455	EA3B	2A129
142650	KA3B	2A124	147657	KA3B	2A118
142651	EKA3B	2A076	147658	EKA3B	2A090
142652	EKA3B	2A077	147659	EKA3B	2A079

DOUGLAS A4 SKYHAWK (138)

142676	TA4B	3A113	154297	TA4J	3A650
142690	TA4A	3A136	154305	TA4J	3A...
142704	TA4B	3A104	154306	OA4M	3A600
142708	TA4B	3A103	154307	OA4M	3A615
142745	TA4B	3A195	154310	TA4J	3A...
142754	TA4B	3A269	154319	TA4J	3A609
142772	TA4B	3A240	154328	OA4M	3A568
142775	TA4B	3A106	154335	OA4M	3A565
142781	TA4B	3A158	154623	OA4M	3A569
142814	TA4B	3A116	154628	OA4M	3A...
142825	A4B	3A197	154630	OA4M	3A562
142843	TA4B	3A131	154631	TA4J	3A612
142844	TA4B	3A159	154633	OA4M	3A580
142847	TA4B	3A165	154645	OA4M	3A599
142849	TA4B	3A236	154656	TA4J	3A613
142875	TA4B	3A128	155074	TA4J	3A608
142899	TA4B	3A112	155074	TA4J	3A683
142909	TA4B	3A282	155089	TA4J	3A634
142917	TA4B	3A107	155097	TA4J	3A611
142921	TA4B	3A133	155099	TA4J	3A...
142936	TA4B	3A163	155106	TA4J	3A619
144875	TA4B	3A259	156897	TA4J	3A629
144884	TA4B	3A117	156909	TA4J	3A618
144903	TA4B	3A160	156911	TA4J	3A606
144917	TA4B	3A188	156920	TA4J	3A...
144919	TA4B	3A233	156921	TA4J	3A...
144960	TA4B	3A205	156922	TA4J	3A687
144965	TA4B	3A227	156924	TA4J	3A...
144977	TA4B	3A123	156928	TA4J	3A...
145033	TA4B	3A115	156934	TA4J	3A617
151194	A4E	3A538	156936	TA4J	3A...
152874	OA4M	3A...	156937	TA4J	3A666
153474	TA4J	3A623	156940	TA4J	3A604
153478	TA4J	3A649	156942	TA4J	3A622
143488	TA4F	3A...	156947	TA4J	3A621
153500	TA4J	3A614	158079	TA4J	3A690
153529	OA4M	3A566	158081	TA4J	3A630
153531	OA4M	3A567	158095	TA4J	3A...
153663	TA4J	3A...	158099	TA4J	3A...
153669	TA4J	3A632	158100	TA4J	3A...
153672	TA4J	3A610	158101	TA4J	3A...
153675	TA4J	3A607	158108	TA4J	3A...
154179	A4F	3A532	158122	TA4J	3A605
154294	OA4M	3A598	158143	TA4J	3A...

158147	TA4J	3A624	158519	TA4J	3A...
158156	A4M	3A542	158524	TA4J	3A641
158167	A4M	3A547	158525	TA4J	3A686
158168	A4M	3A546	158713	TA4J	3A688
158177	A4M	3A576	158715	TA4J	3A...
158181	A4M	3A563	158721	TA4J	3A665
158184	A4M	3A539	159471	A4M	3A578
158185	A4M	3A543	159482	A4M	3A590
158186	A4M	3A589	159485	A4M	3A559
158193	A4M	3A558	159492	A4M	3A552
158194	A4M	3A560	159493	A4M	3A553
158196	A4M	3A561	159778	A4M	3A556
158416	A4M	3A545	159783	A4M	3A541
158424	A4M	3A575	159786	A4M	3A540
158427	A4M	3A557	159790	A4M	3A564
158429	A4M	3A544	160022	A4M	3A554
158435	A4M	3A573	160025	A4M	3A548
158455	TA4J	3A...	160026	A4M	3A551
158469	TA4J	3A658	160029	A4M	3A549
158496	TA4J	3A...	160035	A4M	3A570
158505	TA4J	3A...	160040	A4M	3A550
158506	TA4J	3A...	160042	A4M	3A572
158509	TA4J	3A684	160045	A4M	3A555
158512	TA4J	3A616	160258	A4M	3A571
158516	TA4J	3A638	160261	A4M	3A574

GRUMMAN A6 INTRUDER\PROWLER (14)

148616	EA6A	5A004	152592	KA6D	5A055
148618	EA6B	5A028	152619	KA6D	5A...
149478	EA6A	5A006	155584	KA6D	5A...
149936	KA6D	5A053	155597	KA6D	5A059
151568	KA6D	5A054	155691	KA6D	5A057
151801	KA6D	5A056	156478	EA6B	5A025
151827	KA6D	5A...	156479	EA6B	5A...

LTV A7 CORSAIR II (497)

67-14582	YA7D	AE231	70-0943	A7D	AE167
67-14584	YA7D	AE230	70-0944	A7D	AE069
69-6194	A7D	AE018	70-0951	A7D	AE092
69-6195	A7D	AE008	70-0952	A7D	AE117
69-6199	A7D	AE125	70-0955	A7D	AE126
69-6203	A7D	AE103	70-0957	A7D	AE146
69-6209	A7D	AE009	70-0958	A7D	AE135
69-6210	A7D	AE127	70-0959	A7D	AE238
69-6212	A7D	AE162	70-0960	A7D	AE081
69-6213	A7D	AE039	70-0961	A7D	AE013
69-6214	A7D	AE169	70-0962	A7D	AE075
69-6215	A7D	AE122	70-0964	A7D	AE086
69-6217	A7D	AE227	70-0971	A7D	AE067
69-6218	A7D	AE112	70-0972	A7D	AE077
69-6219	A7D	AE116	70-0974	A7D	AE113
69-6223	A7D	AE049	70-0975	A7D	AE027
69-6225	A7D	AE089	70-0976	A7D	AE064
69-6226	A7D	AE026	70-0978	A7D	AE182
69-6227	A7D	AE110	70-0979	A7D	AE163
69-6229	A7D	AE087	70-0980	A7D	AE016
69-6232	A7D	AE140	70-0981	A7D	AE203
69-6233	A7D	AE049	70-0983	A7D	AE242
69-6236	A7D	AE124	70-0984	A7D	AE150
69-6243	A7D	AE050	70-0985	A7D	AE137
69-6244	A7D	AE171	70-0986	A7D	AE105
70-0929	A7D	AE076	70-0987	A7D	AE065
70-0932	A7D	AE178	70-0988	A7D	AE202
70-0933	A7D	AE188	70-0989	A7D	AE218
70-0934	A7D	AE082	70-0992	A7D	AE151
70-0935	A7D	AE174	70-0993	A7D	AE068
70-0936	A7D	AE226	70-0999	A7D	AE134
70-0939	A7D	AE118	70-1003	A7D	AE155
70-0940	A7D	AE164	70-1004	A7D	AE031
70-0941	A7D	AE145	70-1005	A7D	AE030
70-0942	A7D	AE107	70-1006	A7D	AE023

70-1007	A7D	AE017	72-0188	A7D	AE079
70-1009	A7D	AE111	72-0190	A7D	AE129
70-1010	A7D	AE025	72-0191	A7D	AE115
70-1011	A7D	AE106	72-0193	A7D	AE123
70-1013	A7D	AE011	72-0194	A7D	AE040
70-1016	A7D	AE204	72-0195	A7D	AE091
70-1017	A7D	AE015	72-0199	A7D	AE109
70-1021	A7D	AE217	72-0200	A7D	AE176
70-1022	A7D	AE108	72-0201	A7D	AE159
70-1023	A7D	AE090	72-0202	A7D	AE175
70-1025	A7D	AE201	72-0205	A7D	AE173
70-1026	A7D	AE080	72-0206	A7D	AE161
70-1027	A7D	AE100	72-0208	A7D	AE028
70-1029	A7D	AE084	72-0209	A7D	AE046
70-1030	A7D	AE193	72-0214	A7D	AE212
70-1031	A7D	AE190	72-0215	A7D	AE179
70-1033	A7D	AE045	72-0216	A7D	AE215
70-1034	A7D	AE154	72-0217	A7D	AE147
70-1036	A7D	AE136	72-0218	A7D	AE199
70-1038	A7D	AE232	72-0220	A7D	AE114
70-1041	A7D	AE029	72-0223	A7D	AE061
70-1042	A7D	AE138	72-0224	A7D	AE074
70-1044	A7D	AE208	72-0225	A7D	AE197
70-1047	A7D	AE172	72-0226	A7D	AE043
70-1048	A7D	AE060	72-0227	A7D	AE186
70-1049	A7D	AE035	72-0229	A7D	AE184
70-1056	A7D	AE133	72-0232	A7D	AE014
71-0292	A7D	AE012	72-0235	A7D	AE166
71-0293	A7D	AE185	72-0236	A7D	AE104
71-0294	A7D	AE194	72-0237	A7D	AE229
71-0295	A7D	AE207	72-0239	A7D	AE022
71-0298	A7D	AE048	72-0241	A7D	AE224
71-0301	A7D	AE210	72-0242	A7D	AE152
71-0303	A7D	AE038	72-0243	A7D	AE099
71-0304	A7D	AE041	72-0248	A7D	AE177
71-0307	A7D	AE098	72-0250	A7D	AE093
71-0308	A7D	AE191	72-0251	A7D	AE071
71-0309	A7D	AE206	72-0252	A7D	AE096
71-0311	A7D	AE094	72-0253	A7D	AE180
71-0314	A7D	AE141	72-0255	A7D	AE148
71-0317	A7D	AE214	72-0258	A7D	AE101
71-0318	A7D	AE225	72-0260	A7D	AE072
71-0321	A7D	AE187	73-0997	A7D	AE200
71-0325	A7D	AE078	73-1000	A7D	AE144
71-0327	A7D	AE097	73-1004	A7D	AE181
71-0330	A7D	AE006	73-1007	A7D	AE033
71-0332	A7D	AE211	73-1008	A7D	AE189
71-0334	A7D	AE183	73-1011	A7D	AE222
71-0335	A7D	AE034	73-1014	A7D	AE239
71-0338	A7D	AE102	74-1740	A7D	AE153
71-0341	A7D	AE085	74-1741	A7D	AE024
71-0343	A7D	AE143	74-1742	A7D	AE057
71-0349	A7D	AE235	74-1743	A7D	AE066
71-0350	A7D	AE006	74-1744	A7D	AE157
71-0352	A7D	AE168	74-1745	A7D	AE240
71-0353	A7D	AE223	74-1747	A7D	AE149
71-0354	A7D	AE160	74-1749	A7D	AE083
71-0358	A7D	AE120	74-1750	A7D	AE130
71-0359	A7D	AE059	74-1759	A7D	AE241
71-0362	A7D	AE142	75-0386	A7D	AE205
71-0365	A7D	AE221	75-0387	A7D	AE198
71-0369	A7D	AE073	75-0390	A7D	AE007
71-0370	A7D	AE234	75-0392	A7D	AE020
71-0371	A7D	AE044	75-0397	A7D	AE209
71-0374	A7D	AE042	75-0399	A7D	AE228
71-0376	A7D	AE063	75-0401	A7D	AE139
71-0377	A7D	AE170	75-0402	A7D	AE058
71-0379	A7D	AE158	75-0405	A7D	AE131
72-0169	A7D	AE047	75-0409	A7D	AE019
72-0170	A7D	AE088	79-0460	A7K	AE036
72-0171	A7D	AE192	79-0461	A7K	AE196
72-0176	A7D	AE219	79-0462	A7K	AE195
72-0180	A7D	AE132	79-0463	A7K	AE165
72-0182	A7D	AE128	79-0464	A7K	AE053
72-0186	A7D	AE037	79-0465	A7K	AE054

79-0466	A7K	AE021	154538	A7B	6A159
79-0467	A7K	AE055	154545	A7B	6A204
79-0468	A7K	AE070	154547	A7B	6A135
79-0469	A7K	AE119	154551	A7B	6A128
79-0470	A7K	AE121	154553	A7B	6A141
80-0284	A7K	AE052	154556	A7B	6A126
80-0285	A7K	AE051	156741	EA7L	6A404
80-0286	A7K	AE236	156745	TA7C	6A405
80-0287	A7K	AE056	156746	TA7C	6A357
80-0288	A7K	AE062	156757	EA7L	6A399
80-0290	A7K	AE213	156767	TA7C	6A323
80-0295	A7K	AE243	156773	TA7C	6A331
81-0072	A7K	AE156	156776	A7C	6A162
81-0074	A7K	AE220	156779	TA7C	6A327
81-0076	A7K	AE233	156784	TA7C	6A376
81-0077	A7K	AE237	156788	TA7C	6A361
152677	A7A	6A122	156805	A7E	6A264
154361	TA7C	6A368	156807	A7E	6A218
154366	A7B	6A130	156808	A7E	6A279
154368	A7B	6A129	156810	A7E	6A245
154371	A7B	6A131	156811	A7E	6A223
154372	A7B	6A151	156812	A7E	6A213
154373	A7B	6A206	156813	A7E	6A178
154375	A7B	6A153	156814	A7E	6A168
154377	TA7C	6A330	156815	A7E	6A179
154381	A7B	6A147	156817	A7E	6A187
154388	A7B	6A157	156818	A7E	6A164
154390	A7B	6A169	156819	A7E	6A221
154396	A7B	6A125	156821	A7E	6A209
154397	A7B	6A138	156822	A7E	6A324
154402	TA7C	6A370	156823	A7E	6A208
154404	TA7C	6A369	156827	A7E	6A281
154406	A7B	6A181	156828	A7E	6A180
154409	A7B	6A148	156829	A7E	6A280
154413	A7B	6A149	156830	A7E	6A237
154415	A7B	6A155	156831	A7E	6A192
154416	A7B	6A186	156832	A7E	6A220
154425	TA7C	6A320	156833	A7E	6A259
154433	A7B	6A200	156834	A7E	6A263
154438	A7B	6A174	156835	A7E	6A270
154439	A7B	6A165	156836	A7E	6A337
154440	A7B	6A155	156840	A7E	6A276
154448	A7B	6A143	156845	A7E	6A278
154451	A7B	6A127	156847	A7E	6A214
154452	A7B	6A136	156851	A7E	6A282
154453	A7B	6A163	156852	A7E	6A346
154454	A7B	6A203	156856	A7E	6A248
154455	TA7C	6A321	156857	A7E	6A366
154458	TA7C	6A322	156858	A7E	6A257
154460	A7B	6A161	156861	A7E	6A266
154462	A7B	6A170	156862	A7E	6A196
154463	A7B	6A156	156863	A7E	6A199
154466	A7B	6A145	156869	A7E	6A166
154468	A7B	6A134	156872	A7E	6A277
154469	A7B	6A144	156874	A7E	6A...
154472	A7B	6A142	156876	A7E	6A273
154475	A7B	6A152	156882	A7E	6A246
154478	A7B	6A139	156884	A7E	6A284
154479	A7B	6A182	156885	A7E	6A207
154481	A7B	6A150	156886	A7E	6A215
154484	A7B	6A140	156889	A7E	6A210
154488	A7B	6A175	156890	A7E	6A170
154489	TA7C	6A392	157435	A7E	6A222
154490	A7B	6A124	157440	A7E	6A268
154491	A7B	6A158	157441	A7E	6A304
154493	A7B	6A133	157444	A7E	6A286
154494	A7B	6A146	157445	A7E	6A239
154498	A7B	6A137	157450	A7E	6A190
154505	A7B	6A201	157453	A7E	6A243
154509	A7B	6A132	157460	A7E	6A212
154512	A7B	6A123	157463	A7E	6A256
154516	A7B	6A160	157466	A7E	6A194
154520	A7B	6A173	157471	A7E	6A258
154529	A7B	6A167	157472	A7E	6A247
154535	A7B	6A184		A7E	6A255

157475	A7E	6A185	158669	A7E	6A351
157476	A7E	6A364	158672	A7E	6A335
157477	A7E	6A296	158674	A7E	6A295
157480	A7E	6A191	158675	A7E	6A288
157481	A7E	6A269	158821	A7E	6A293
157485	A7E	6A233	158824	A7E	6A229
157486	A7E	6A328	158829	A7E	6A177
157489	A7E	6A316	158831	A7E	6A310
157490	A7E	6A224	158833	A7E	6A340
157491	A7E	6A244	158835	A7E	6A197
157492	A7E	6A367	158836	A7E	6A228
157493	A7E	6A365	158837	A7E	6A362
157494	A7E	6A312	158838	A7E	6A297
157496	A7E	6A217	159264	A74	6A235
157501	A7E	6A267	159270	A7E	6A...
157502	A7E	6A345	159272	A7E	6A300
157504	A7E	6A298	159278	A7E	6A...
157508	A7E	6A242	159280	A7E	6A313
157512	A7E	6A338	159283	A7E	6A271
157514	A7E	6A250	159284	A7E	6A314
157516	A7E	6A283	159286	A7E	6A363
157517	A7E	6A334	159290	A7E	6A308
157518	A7E	6A183	159293	A7E	6A358
157519	A7E	6A249	159294	A7E	6A...
157522	A7E	6A253	159296	A7E	6A...
157523	A7E	6A202	159300	A7E	6A290
157525	A7E	6A301	159302	A7E	6A311
157528	A7E	6A240	159304	A7E	6A289
157536	A7E	6A189	159306	A7E	6A374
157537	A7E	6A260	159308	A7E	6A341
157538	A7E	6A205	159638	A7E	6A231
157541	A7E	6A299	159651	A7E	6A359
157546	A7E	6A262	159653	A7E	6A230
157549	A7E	6A193	159660	A7E	6A292
157551	A7E	6A303	159661	A7E	6A226
157552	A7E	6A227	159969	A7E	6A326
157553	A7E	6A265	159970	A7E	6A319
157556	A7E	6A238	159976	A7E	6A305
157559	A7E	6A309	159981	A7E	6A325
157566	A7E	6A236	159985	A7E	6A406
157570	A7E	6A188	159986	A7E	6A241
157571	A7E	6A342	159988	A7E	6A232
157573	A7E	6A211	159989	A7E	6A317
157577	A7E	6A261	159992	A7E	6A355
157580	A7E	6A216	159997	A7E	6A302
157581	A7E	6A254	160001	A7E	6A329
157585	A7E	6A176	160538	A7E	6A356
157587	A7E	6A195	160539	A7E	6A349
157593	A7E	6A225	160544	A7E	6A347
158002	A7E	6A287	160545	A7E	6A348
158004	A7E	6A350	160547	A7E	6A272
158005	A7E	6A275	160549	A7E	6A343
158008	A7E	6A198	160555	A7E	6A353
158009	A7E	6A234	160556	A7E	6A388
158013	A7E	6A306	160558	A7E	6A352
158014	A7E	6A219	160565	A7E	6A...
158017	A7E	6A333	160712	A7E	6A332
158019	A7E	6A318	160729	A7E	6A315
158021	A7E	6A307	160732	A7E	6A339
158025	A7E	6A274	160736	A7E	6A387
158652	A7E	6A360	160859	A7E	6A251
158663	A7E	6A294	160863	A7E	6A372
158665	A7E	6A285	160874	A7E	6A344
158666	A7E	6A291			

McDONNELL DOUGLAS AV8 HARRIER (37)

158391	AV8C	7A034	158705	AV8C	7A028
158392	AV8C	7A035	158949	AV8C	7A003
158393	AV8C	7A026	158951	AV8C	7A029
158697	AV8C	7A025	158959	AV8A	7A002
158700	AV8C	7A027	158964	AV8C	7A005
158701	AV8C	7A010	158966	AV8A	7A001
158702	AV8A	7A014	158972	AV8C	7A011

158973	AV8C	7A012	159370	AV8C	7A037
158977	AV8C	7A013	159371	AV8A	7A008
159238	AV8C	7A024	159373	AV8A	7A015
159239	AV8A	7A016	159374	AV8A	7A020
159240	AV8C	7A033	159375	AV8A	7A017
159243	AV8C	7A031	159376	AV8A	7A021
159247	AV8C	7A007	159378	TAV8A	7A043
159252	AV8A	7A022	159379	TAV8A	7A042
159255	AV8A	7A023	159382	TAV8A	7A040
159257	AV8C	7A032	159383	TAV8A	7A039
159366	AV8A	7A018	159385	TAV8A	7A041
159367	AV8A	7A019			

FAIRCHILD A10 THUNDERBOLT II (169)

75-0258	A10A	AC020	77-0201	A10A	AC059
75-0261	A10A	AC001	77-0202	A10A	AC138
75-0264	A10A	AC064	77-0204	OA10A	AC072
75-0265	A10A	AC163	77-0207	OA10A	AC100
75-0267	A10A	AC153	77-0208	A10A	AC081
75-0269	A10A	AC071	77-0210	OA10A	AC074
75-0272	A10A	AC073	77-0211	A10A	AC147
75-0275	A10A	AC165	77-0212	OA10A	AC083
75-0276	A10A	AC133	77-0213	OA10A	AC087
75-0277	A10A	AC015	77-0214	A10A	AC095
75-0278	A10A	AC062	77-0217	A10A	AC121
75-0279	A10A	AC127	77-0219	OA10A	AC164
75-0280	A10A	AC042	77-0220	A10A	AC086
75-0281	A10A	AC146	77-0222	OA10A	AC101
75-0284	A10A	AC005	77-0224	OA10A	AC056
75-0285	A10A	AC149	77-0225	A10A	AC002
75-0287	A10A	AC150	77-0226	A10A	AC003
75-0288	A10A	AC135	77-0227	A10A	AC131
75-0290	A10A	AC129	77-0229	A10A	AC116
75-0291	A10A	AC036	77-0230	A10A	AC115
75-0292	A10A	AC161	77-0231	A10A	AC109
75-0293	A10A	AC077	77-0232	A10A	AC105
75-0296	A10A	AC155	77-0233	A10A	AC051
75-0297	A10A	AC063	77-0234	A10A	AC103
75-0298	A10A	AC151	77-0235	A10A	AC010
75-0299	A10A	AC162	77-0236	A10A	AC009
75-0300	A10A	AC037	77-0237	A10A	AC047
75-0303	A10A	AC139	77-0238	A10A	AC028
75-0304	A10A	AC125	77-0239	A10A	AC099
75-0306	A10A	AC092	77-0240	A10A	AC093
75-0307	A10A	AC006	77-0241	A10A	AC049
75-0309	OA10A	AC075	77-0242	A10A	AC048
76-0514	A10A	AC157	77-0245	A10A	AC096
76-0522	A10A	AC159	77-0246	A10A	AC034
76-0524	A10A	AC091	77-0247	A10A	AC031
76-0527	A10A	AC027	77-0249	A10A	AC097
76-0529	OA10A	AC166	77-0251	A10A	AC012
76-0531	A10A	AC084	77-0254	A10A	AC170
76-0532	OA10A	AC076	77-0256	A10A	AC029
76-0533	OA10A	AC104	77-0257	A10A	AC085
76-0534	OA10A	AC134	77-0260	A10A	AC152
76-0536	A10A	AC141	77-0262	A10A	AC132
76-0544	A10A	AC143	77-0263	A10A	AC030
76-0548	A10A	AC050	77-0266	A10A	AC046
76-0549	OA10A	AC044	77-0267	A10A	AC137
76-0554	A10A	AC089	77-0268	A10A	AC144
77-0177	OA10A	AC118	77-0269	A10A	AC148
77-0181	A10A	AC102	77-0271	A10A	AC088
77-0184	A10A	AC098	77-0272	A10A	AC094
77-0186	OA10A	AC128	77-0273	A10A	AC082
77-0187	OA10A	AC140	77-0274	A10A	AC145
77-0188	A10A	AC154	77-0275	A10A	AC130
77-0191	OA10A	AC065	77-0276	A10A	AC142
77-0192	A10A	AC117	78-0587	A10A	AC011
77-0193	A10A	AC114	78-0589	A10A	AC111
77-0194	A10A	AC169	78-0592	A10A	AC018
77-0195	A10A	AC004	78-0594	A10A	AC017
77-0198	OA10A	AC110	78-0594	A10A	AC025

78-0595	A10A	AC057	79-0098	A10A	AC035
78-0603	A10A	AC167	79-0099	A10A	AC022
78-0606	A10A	AC112	79-0101	A10A	AC053
78-0622	A10A	AC024	79-0102	A10A	AC106
78-0654	A10A	AC045	79-0112	A10A	AC014
78-0656	A10A	AC124	79-0115	A10A	AC019
78-0660	A10A	AC060	79-0124	A10A	AC058
78-0662	OA10A	AC090	79-0126	A10A	AC069
78-0664	A10A	AC052	79-0127	A10A	AC026
78-0665	A10A	AC016	79-0128	A10A	AC061
78-0666	A10A	AC120	79-0131	A10A	AC070
78-0667	A10A	AC119	79-0132	A10A	AC021
78-0668	A10A	AC023	79-0133	A10A	AC067
78-0672	A10A	AC079	79-0137	A10A	AC066
78-0675	A10A	AC108	79-0140	A10A	AC041
78-0677	A10A	AC055	79-0158	A10A	AC126
78-0678	A10A	AC168	79-0160	OA10A	AC136
78-0680	A10A	AC040	79-0163	A10A	AC122
78-0686	OA10A	AC080	79-0166	A10A	AC038
78-0698	A10A	AC054	79-0176	A10A	AC039
78-0699	A10A	AC078	79-0182	A10A	AC043
78-0710	A10A	AC013	79-0217	A10A	AC160
78-0713	A10A	AC107	79-0218	A10A	AC156
78-0714	A10A	AC068	79-0220	A10A	AC007
78-0715	A10A	AC123	79-0221	A10A	AC158
78-0724	A10A	AC033	79-0224	A10A	AC008
79-0096	A10A	AC032			

CESSNA A37 DRAGONFLY (15)

69-6397	OA37B	AB023	73-1066	OA37B	AB025
69-6438	OA37B	AB024	73-1073	OA37B	AB014
69-6443	OA37B	AB010	73-1077	OA37B	AB012
71-0859	OA37B	AB009	73-1098	OA37B	AB026
71-0860	OA37B	AB016	73-1101	OA37B	AB034
71-0863	OA37B	AB013	73-1102	OA37B	AB033
71-0873	OA37B	AB032	73-1111	OA37B	AB011
73-1061	OA37B	AB008			

BOEING B52 STRATOFORTRESS (307)

53-0400	B52C	BC205	55-0059	B52D	BC278
53-0401	B52C	BC156	55-0064	B52D	BC232
53-0402	B52C	BC206	55-0066	B52D	BC300
53-0403	B52C	BC109	55-0069	B52D	BC291
53-0404	B52C	BC170	55-0070	B52D	BC312
53-0405	B52C	BC180	55-0072	B52D	BC220
53-0407	B52C	BC176	55-0073	B52D	BC298
53-0408	B52C	BC178	55-0074	B52D	BC277
54-2665	B52C	BC159	55-0075	B52D	BC296
54-2668	B52C	BC191	55-0076	B52D	BC214
54-2669	B52C	BC158	55-0077	B52D	BC317
54-2671	B52C	BC194	55-0079	B52D	BC294
54-2672	B52C	BC185	55-0080	B52D	BC303
54-2673	B52C	BC204	55-0081	B52D	BC216
54-2674	B52C	BC160	55-0082	B52D	BC...
54-2675	B52C	BC168	55-0084	B52D	BC327
54-2677	B52C	BC174	55-0086	B52D	BC288
54-2678	B52C	BC184	55-0087	B52D	BC305
54-2679	B52C	BC163	55-0088	B52D	BC297
54-2681	B52C	BC200	55-0090	B52D	BC301
54-2683	B52C	BC196	55-0091	B52D	BC292
54-2684	B52C	BC198	55-0092	B52D	BC316
54-2685	B52C	BC172	55-0096	B52D	BC259
54-2686	B52C	BC192	55-0101	B52D	BC290
54-2687	B52C	BC166	55-0104	B52D	BC284
54-2688	B52C	BC182	55-0105	B52D	BC329
55-0049	B52D	BC245	55-0106	B52D	BC269
55-0051	B52D	BC263	55-0107	B52D	BC295
55-0053	B52D	BC219	55-0109	B52D	BC251
55-0054	B52D	BC268	55-0111	B52D	BC289

55-0113	B52D	BC314	57-0027	B52E	BC132
55-0117	B52D	BC244	57-0031	B52F	BC162
55-0673	B52D	BC308	57-0032	B52F	BC264
55-0674	B52D	BC330	57-0033	B52F	BC239
55-0675	B52D	BC302	57-0034	B52F	BC227
55-0680	B52D	BC234	57-0035	B52F	BC271
56-0580	B52D	BC299	57-0037	B52F	BC102
56-0581	B52D	BC218	57-0039	B52F	BC165
56-0582	B52D	BC235	57-0045	B52F	BC246
56-0583	B52D	BC223	57-0046	B52F	BC105
56-0587	B52D	BC307	57-0049	B52F	BC059
56-0588	B52D	BC280	57-0051	B52F	BC252
56-0590	B52D	BC261	57-0052	B52F	BC229
56-0592	B52D	BC273	57-0053	B52F	BC095
56-0596	B52D	BC285	57-0054	B52F	BC100
56-0598	B52D	BC213	57-0055	B52F	BC195
56-0600	B52D	BC310	57-0056	B52F	BC157
56-0602	B52D	BC325	57-0057	B52F	BC186
56-0604	B52D	BC240	57-0058	B52F	BC255
56-0606	B52D	BC281	57-0059	B52F	BC179
56-0609	B52D	BC267	57-0060	B52F	BC169
56-0611	B52D	BC221	57-0061	B52F	BC167
56-0613	B52D	BC228	57-0062	B52F	BC175
56-0614	B52D	BC324	57-0063	B52F	BC242
56-0615	B52D	BC246	57-0064	B52F	BC173
56-0617	B52D	BC326	57-0065	B52F	BC177
56-0619	B52D	BC222	57-0066	B52F	BC083
56-0620	B52D	BC207	57-0067	B52F	BC099
56-0621	B52D	BC286	57-0069	B52F	BC243
56-0623	B52D	BC260	57-0072	B52F	BC265
56-0624	B52D	BC226	57-0095	B52F	BC110
56-0626	B52D	BC253	57-0096	B52F	BC135
56-0631	B52E	BC085	57-0097	B52F	BC133
56-0632	B52E	BC211	57-0098	B52F	BC130
56-0634	B52E	BC116	57-0099	B52F	BC136
56-0636	B52E	BC275	57-0100	B52F	BC121
56-0639	B52E	BC149	57-0103	B52F	BC119
56-0640	B52E	BC140	57-0108	B52F	BC143
56-0644	B52E	BC131	57-0112	B52F	BC142
56-0645	B52E	BC115	57-0115	B52F	BC125
56-0646	B52E	BC113	57-0118	B52F	BC122
56-0650	B52E	BC129	57-0120	B52F	BC107
56-0651	B52E	BC127	57-0121	B52F	BC120
56-0653	B52E	BC118	57-0123	B52F	BC146
56-0656	B52E	BC144	57-0126	B52F	BC141
56-0658	B52D	BC276	57-0128	B52F	BC124
56-0660	B52D	BC320	57-0129	B52F	BC147
56-0663	B52D	BC282	57-0131	B52F	BC150
56-0666	B52D	BC321	57-0132	B52F	BC117
56-0667	B52D	BC323	57-0136	B52F	BC128
56-0668	B52D	BC306	57-0142	B52F	BC248
56-0670	B52D	BC279	57-0143	B52F	BC164
56-0671	B52D	BC287	57-0145	B52F	BC250
56-0672	B52D	BC313	57-0147	B52F	BC224
56-0673	B52D	BC233	57-0150	B52F	BC217
56-0675	B52D	BC247	57-0151	B52F	BC193
56-0678	B52D	BC252	57-0152	B52F	BC183
56-0679	B52D	BC293	57-0154	B52F	BC258
56-0682	B52D	BC256	57-0155	B52F	BC103
56-0684	B52D	BC315	57-0160	B52F	BC097
56-0686	B52D	BC318	57-0161	B52F	BC203
56-0690	B52D	BC304	57-0162	B52F	BC197
56-0691	B52D	BC249	57-0163	B52F	BC098
56-0693	B52D	BC230	57-0168	B52F	BC209
56-0694	B52D	BC319	57-0170	B52F	BC262
56-0697	B52D	BC328	57-0174	B52F	BC158
56-0698	B52D	BC311	57-0175	B52F	BC107
56-0704	B52E	BC145	57-0176	B52F	BC101
56-0705	B52E	BC114	57-0178	B52F	BC171
56-0706	B52E	BC123	57-0180	B52F	BC096
56-0707	B52E	BC152	57-0183	B52F	BC208
57-0015	B52E	BC109	57-6470	B52G	BC...
57-0017	B52E	BC148	57-6474	B52G	BC389
57-0020	B52E	BC134	57-6471	B52G	BC...
57-0024	B52E	BC112	57-6472	B52G	BC...

57-6473	B52G	BC439		58-0182	B52G	BC...
57-6474	B52G	BC...		58-0183	B52G	BC373
57-6475	B52G	BC378		58-0184	B52G	BC385
57-6477	B52G	BC...		58-0186	B52G	BC...
57-6478	B52G	BC334		58-0189	B52G	BC337
57-6480	B52G	BC...		58-0193	B52G	BC435
57-6483	B52G	BC...		58-0194	B52G	BC...
57-6484	B52G	BC333		58-0199	B52G	BC...
57-6485	B52G	BC...		58-0204	B52G	BC384
57-6486	B52G	BC377		58-0205	B52G	BC...
57-6487	B52G	BC...		58-0207	B52G	BC...
57-6489	B52G	BC343		58-0211	B52G	BC...
57-6490	B52G	BC433		58-0214	B52G	BC...
57-6491	B52G	BC...		58-0217	B52G	BC...
57-6492	B52G	BC437		58-0220	B52G	BC346
57-6495	B52G	BC...		58-0222	B52G	BC...
57-6499	B52G	BC...		58-0223	B52G	BC...
57-6500	B52G	BC331		58-0224	B52G	BC339
57-6501	B52G	BC...		58-0227	B52G	BC...
57-6502	B52G	BC340		58-0229	B52G	BC...
57-6504	B52G	BC...		58-0232	B52G	BC344
57-6505	B52G	BC...		58-0234	B52G	BC...
57-6506	B52G	BC342		58-0236	B52G	BC...
57-6508	B52G	BC404		58-0237	B52G	BC393
57-6510	B52G	BC...		58-0238	B52G	BC379
57-6511	B52G	BC...		58-0241	B52G	BC...
57-6512	B52G	BC...		58-0243	B52G	BC371
57-6513	B52G	BC335		58-0244	B52G	BC...
57-6514	B52G	BC...		58-0245	B52G	BC...
57-6516	B52G	BC387		58-0247	B52G	BC...
57-6517	B52G	BC347		58-0249	B52G	BC382
57-6518	B52G	BC...		58-0251	B52G	BC341
57-6519	B52G	BC...		58-0252	B52G	BC...
58-0158	B52G	BC...		58-0254	B52G	BC...
58-0159	B52G	BC...		59-2564	B52G	BC376
58-0160	B52G	BC...		59-2571	B52G	BC...
58-0162	B52G	BC...		59-2575	B52G	BC375
58-0166	B52G	BC438		59-2579	B52G	BC395
58-0167	B52G	BC...		59-2580	B52G	BC...
58-0168	B52G	BC...		59-2581	B52G	BC...
58-0171	B52G	BC...		59-2582	B52G	BC382
58-0172	B52G	BC332		59-2584	B52G	BC...
58-0173	B52G	BC...		59-2587	B52G	BC336
58-0175	B52G	BC...		59-2589	B52G	BC...
58-0176	B52G	BC...		59-2590	B52G	BC...
58-0177	B52G	BC381		59-2591	B52G	BC...
58-0178	B52G	BC...		59-2592	B52G	BC338
58-0179	B52G	BC436		59-2594	B52G	BC...
58-0181	B52G	BC...				

MARTIN B57 CANBERRA (35)

52-1506	EB57B	BM138		55-4293	EB57E	BM121
52-1511	EB57B	BM134		55-4295	EB57E	BM125
52-1521	EB57B	BM137		55-4298	EB57E	BM129
52-1545	EB57B	BM135		55-4300	EB57E	BM130
52-1564	EB57B	BM133		63-13286	WB57E	BM066
53-3831	B57C	BM139		63-13288	WB57F	BM105
53-3840	B57C	BM136		63-13289	WB57F	BM103
53-3856	B57C	BM140		63-13290	WB57F	BM078
55-4238	B57E	BM132		63-13291	WB57F	BM106
55-4239	B57E	BM101		63-13293	WB57F	BM065
55-4240	EB57E	BM128		63-13294	WB57F	BM104
55-4241	EB57E	BM131		63-13295	WB57F	BM074
55-4242	EB57E	BM126		63-13296	WB57F	BM107
55-4258	B57E	BM052		63-13299	WB57F	BM067
55-4265	B57E	BM118		63-13300	WB57F	BM083
55-4266	B57E	BM123		63-13301	WB57F	BM084
55-4278	EB57E	BM120		63-13302	WB57F	BM109
55-4285	B57E	BM113				

GRUMMAN C1 TRADER (24)

136748	C1A	7C015	146027	C1A	7C019
136756	C1A	7C031	146028	C1A	7C033
136763	C1A	7C004	146030	C1A	7C021
136766	C1A	7C022	146031	C1A	7C032
136773	C1A	7C008	146036	C1A	7C017
136778	C1A	7C034	146038	C1A	7C023
136780	C1A	7C002	146041	C1A	7C026
136783	C1A	7C006	146043	C1A	7C043
136786	C1A	7C024	146050	C1A	7C035
146024	C1A	7C025	146051	C1A	7C028
146025	C1A	7C029	146055	C1A	7C027
146026	C1A	7C003	146057	C1A	7C020

GRUMMAN C2 GREYHOUND (6)

148148	C2A	1C005	152791	C2A	1C004
152786	C2A	1C002	152794	C2A	1C003
152790	C2A	1C001	155123	C2A	1C006

BOEING C14 (1)

72-1874	YC14A	CW001

McDONNELL DOUGLAS C15 (1)

72-1876	YC15A	CX001

BOEING C22 (2)

N7004U	727-22	CU001	84-0193	C22A	CU002

DOUGLAS C47 SKYTRAIN (3)

42-100662	C47D	CB226	44-76642	C47B	CB334
43-48859	C47B	CB269			

BOEING 377 (C97) STRATOCRUISER (1)

N940NA	B377	CH626

DOUGLAS C118 LIFTMASTER (7)

53-3239	C118A	CG075	53-3300	C118A	CG083
53-3275	C118A	CG026	131568	C118B	8C030
53-3284	C118A	CG025	131591	C118B	8C028
53-3298	C118A	CG029			

LOCKHEED C121 CONSTELLATION (1)

54-0157	EC121S	CK201

FAIRCHILD C123 PROVIDER (29)

54-0565	C123K	CP084	54-0615	C123K	CP045
54-0583	C123K	CP086	54-0618	UC123K	CP071
54-0585	UC123K	CP091	54-0628	UC123K	CP076
54-0586	UC123K	CP088	54-0635	UC123K	CP087
54-0596	C123K	CP070	54-0658	UC123K	CP079
54-0605	UC123K	CP090	54-0659	C123K	CP060

54-0661	C123K	CP048		55-4532	UC123K	CP047
54-0674	C123K	CP054		55-4535	C123K	CP068
54-0693	C123K	CP081		55-4544	C123K	CP056
54-0695	C123K	CP069		55-4547	C123K	CP093
54-0701	UC123K	CP073		55-4571	UC123K	CP092
54-0706	C123K	CP072		55-4577	C123K	CP049
54-0709	C123K	CP064		56-4361	C123K	CP080
54-0711	C123K	CP046		56-4371	UC123K	CP082
55-4517	C123K	CP052				

LOCKHEED C130 HERCULES (77)

53-3133	NC130A	CF021		58-0725	C130B	CF095
53-3134	C130A	CF010		58-0744	C130B	CF...
54-1624	C130A	CF009		58-0746	C130B	CF...
54-1627	AC130A	CF015		58-0750	C130B	CF...
54-1634	C130A	CF...		58-0755	C130B	CF092
54-1640	C130A	CF088		58-0758	C130B	CF113
55-0004	C130A	CF085		59-5957	C130B	CF132
55-0018	C130A	CF081		60-0299	C130B	CF122
55-0024	C130A	CF062		61-0952	C130B	CF120
55-0026	C130A	CF086		61-0969	C130B	CF...
55-0033	C130A	CF...		61-2643	C130B	CF...
55-0036	C130A	CF073		67-7183	HC130H	CF119
55-0040	AC130A	CF018		67-7185	HC130H	CF044
55-0041	C130A	CF068		84-0454	HC130B	CF...
56-0470	C130A	CF071		149790	C130F	2G...
56-0471	C130A	CF087		149797	C130F	2G...
56-0481	C130A	CF069		151888	TC130G	2G011
56-0485	C130A	CF090		151889	EC130G	2G...
56-0486	C130A	CF075		156170	TC130Q	2G014
56-0495	C130A	CF097		156171	EC130Q	2G012
56-0503	C130A	CF...		156172	EC130Q	2G010
56-0513	C130A	CF066		156173	EC130Q	2G009
56-0523	C130A	CF082		156174	TC130Q	2G...
56-0527	DC130A	CF030		156175	EC130Q	2G007
56-0529	C130A	CF091		156177	EC130Q	2G008
56-0543	C130A	CF080		158228	DC130A	2G003
56-0544	C130A	CF084		158229	DC130A	2G004
57-0453	C130A	CF102		159469	EC130Q	2G015
57-0456	C130A	CF076		160608	EC130Q	2G013
57-0457	C130A	CF083		162312	EC130Q	2G...
57-0458	C130A	CF074		162313	EC130Q	2G016
57-0488	C130D	CF016		1339	HC130B 45009\CF037	
57-0492	C130D	CF046		1340	HC130B 45003\CF031	
57-0494	C130D	CF040		1342	HC130B 45002\CF032	
57-0529	C130B	CF115		1344	HC130B 45001\CF033	
58-0713	C130B	CF096		1345	HC130B 45007\CF038	
58-0715	C130B	CF...		1347	HC130B 45006\CF039	
58-0716	C130B	CF...		1349	HC130B 45004\CF034	
58-0720	C130B	CF...				

CONVAIR C131 SAMARITAN (46)

52-5781	HC131A	CS...		54-2823	C131D	CS083
52-5783	HC131A	CS068		54-2825	C131D	CS080
52-5791	HC131A	CS070		55-0296	C131D	CS001
52-5806	HC131A	CS...		55-0298	C131D	CS077
53-7788	C131B	CS056		57-2552	C131E	CS084
53-7792	C131B	CS081		140994	C131F	1G029
53-7795	C131B	CS048		140996	C131F	1G024
53-7796	C131B	CS019		140998	C131F	1G011
53-7801	C131B	CS054		140999	C131F	1G019
53-7802	C131B	CS046		141001	C131F	1G025
53-7803	C131B	CS036		141003	C131F	1G003
53-7819	C131B	CS050		141004	C131F	1G013
54-2805	C131D	CS014		141005	C131F	1G007
54-2809	C131D	CS078		141006	C131F	1G014
54-2813	C131D	CS086		141007	C131F	1G018
54-2818	C131D	CS079		141008	C131F	1G028
54-2819	C131D	CS085		141009	C131F	1G023

141011	C131F	1G015	141022	C131F	1G009
141013	C131F	1G010	141023	C131F	1G026
141016	C131F	1G022	141025	C131F	1G020
141017	C131F	1G030	141028	C131F	1G016
141018	C131F	1G031	145962	C131G	1G027
141020	C131F	1G017	145963	C131G	1G021

BOEING C135 STRATOTANKER (61)

55-3119	NKC135A	CA...	57-2592	KC135A	CA066
55-3127	NKC135A	CA040	58-0019	EC135P	CA009
55-3129	EC135P	CA008	58-0022	EC135P	CA011
55-3131	NKC135A	CA053	58-0029	KC135A	CA056
55-3137	KC135A	CA070	58-0033	KC135A	CA026
56-3594	KC135A	CA047	58-0081	KC135A	CA052
56-3603	KC135A	CA029	58-0097	KC135A	CA043
56-3608	KC135A	CA030	59-1454	KC135A	CA...
56-3601	KC135A	CA072	61-0261	EC135L	CA020
56-3610	KC135A	CA055	61-0263	EC135L	CA015
56-3615	KC135A	CA032	61-0274	EC135P	CA010
56-3619	KC135A	CA051	61-0278	EC135A	CA048
56-3627	KC135A	CA054	61-0279	EC135L	CA018
56-3633	KC135A	CA037	61-0283	EC135L	CA016
56-3634	KC135A	CA049	61-0285	EC135H	CA012
56-3635	KC135A	CA038	61-0289	EC135A	CA022
56-3636	KC135A	CA031	61-0291	EC135H	CA007
56-3637	KC135A	CA033	61-0297	EC135A	CA021
56-3644	KC135A	CA028	62-3497	KC135A	CA...
56-3646	KC135A	CA034	62-3501	KC135A	CA068
56-3647	KC135A	CA044	62-3532	KC135A	CA076
56-3649	KC135A	CA065	62-3570	EC135G	CA024
56-3651	KC135A	CA041	62-3574	KC135A	CA069
56 3653	KC135A	CA025	62-3579	EC135G	CA023
57-1420	KC135A	CA046	62-3583	EC135C	CA019
57-1467	KC135A	CA039	63-7994	EC135G	CA045
57-1476	KC135A	CA035	63-8001	EC135G	CA017
57-1477	KC135A	CA036	63-8051	EC135C	CA027
57-1490	KC135A	CA050	63-8056	EC135J	CA013
57-2590	KC135A	CA042	63-8057	EC135J	CA014
57-2591	KC135A	CA...			

LOCKHEED C140 JETSTAR (5)

61-2490	C140B	CL004	62-4199	C140B	CL002
61-2493	C140B	CL003	62-4200	C140B	CL005
62-4197	C140B	CL007			

LOCKHEED C141 STARLIFTER (9)

64-0636	C141B	CR004	65-9410	C141B	CR003
64-0648	C141B	CR005	66-0143	C141B	CR001
65-0233	C141B	CR009	66-0170	C141B	CR006
65-0262	C141B	CR007	66-0188	C141B	CR008
65-9398	C141B	CR002			

GRUMMAN E2 HAWKEYE (20)

149819	E2B	2E006	151717	E2B	2E013
150532	E2B	2E001	151720	E2B	2E017
150535	E2B	2E002	151722	E2B	2E007
150538	E2B	2E019	151724	E2B	2E020
151702	E2B	2E012	151725	E2B	2E004
151706	E2B	2E010	152477	E2B	2E005
151709	E2B	2E014	152478	E2B	2E018
151710	E2B	2E015	152479	E2B	2E016
151715	E2B	2E009	152482	E2B	2E011
151716	E2B	2E003	160417	E2C	2E...

BOEING E8\J-STARS (1)

88-0322   YE8B   Flight Line

McDONNELL F4 PHANTOM II (1059)

63-7409	NF4C	FP646		64-0726	F4C	FP039
63-7418	F4C	FP075		64-0727	NF4C	FP496
63-7420	F4C	FP295		64-0758	F4C	FP024
63-7428	F4C	FP320		64-0761	F4C	FP138
63-7436	F4C	FP335		64-0765	F4C	FP137
63-7452	F4C	FP089		64-0775	F4C	FP068
63-7454	F4C	FP079		64-0777	F4C	FP363
63-7460	F4C	FP179		64-0780	F4C	FP133
63-7468	F4C	FP321		64-0784	F4C	FP104
63-7475	F4C	FP365		64-0789	F4C	FP062
63-7484	F4C	FP081		64-0794	F4C	FP069
63-7490	F4C	FP174		64-0796	F4C	FP064
63-7516	F4C	FP028		64-0802	F4C	FP037
63-7517	F4C	FP334		64-0804	F4C	FP051
63-7529	F4C	FP180		64-0811	F4C	FP063
63-7530	F4C	FP294		64-0812	F4C	FP118
63-7542	F4C	FP136		64-0822	F4C	FP025
63-7543	F4C	FP140		64-0828	F4C	FP152
63-7545	F4C	FP785\ex N421FS		64-0831	F4C	FP181
63-7557	F4C	FP204		64-0836	F4C	FP026
63-7562	F4C	FP048		64-0840	F4C	FP088
63-7566	F4C	FP043		64-0852	F4C	FP208
63-7575	F4C	FP322		64-0860	F4C	FP029
63-7581	F4C	FP305		64-0865	F4C	FP070
63-7582	F4C	FP134		64-0869	NF4C	FP495
63-7585	F4C	FP182		64-0888	F4C	FP023
63-7591	F4C	FP027		64-0893	F4C	FP176
63-7595	F4C	FP247		64-0899	F4C	FP177
63-7598	F4C	FP034		64-0904	F4C	FP040
63-7605	F4C	FP102		64-0908	F4C	FP076
63-7607	F4C	FP733\ex N423FS		64-0918	F4C	FP119
63-7617	F4C	FP065		64-0926	F4C	FP351
63-7626	F4C	FP087		64-0928	F4C	FP077
63-7629	F4C	FP139		64-0930	F4D	FP530
63-7631	F4C	FP035		64-0937	F4D	FP485
63-7633	F4C	FP183		64-0938	F4D	FP486
63-7637	F4C	FP132		64-0939	F4D	FP130
63-7646	F4C	FP141		64-0942	F4D	FP484
63-7650	F4C	FP066		64-0945	F4D	FP292
63-7654	NF4C	FP853		64-0949	F4D	FP488
63-7655	F4C	FP073		64-0953	F4D	FP489
63-7662	F4C	FP293		64-0956	F4D	FP491
63-7683	F4C	FP357		64-0959	F4D	FP407
63-7689	F4C	FP842\N420FS		64-0963	F4D	FP444
63-7711	F4C	FP061		64-0968	F4D	FP490
63-7741	RF4C	FP338		64-0970	F4D	FP131
63-7744	NRF4C	FP871		64-0973	F4D	FP125
63-7747	RF4C	FP384		64-0975	F4D	FP217
63-7750	RF4C	FP673		64-0976	F4D	FP458
63-7752	RF4C	FP379		64-0977	F4D	FP487
63-7753	RF4C	FP261		64-0979	F4D	FP442
63-7754	RF4C	FP783		64-0980	F4D	FP544
63-7757	RF4C	FP739		64-0999	RF4C	FP812
63-7758	RF4C	FP297		64-1016	RF4C	FP850
63-7759	RF4C	FP362		64-1019	RF4C	FP810
63-7762	RF4C	FP609		64-1023	RF4C	FP818
64-0655	F4C	FP044		64-1024	RF4C	FP436
64-0661	F4C	FP030		64-1028	RF4C	FP623
64-0672	F4C	FP135		64-1029	RF4C	FP907
64-0677	F4C	FP060		64-1033	RF4C	FP905
64-0694	F4C	FP080		64-1035	RF4C	FP906
64-0711	F4C	FP358		64-1038	RF4C	FP814
64-0713	F4C	FP230		64-1046	RF4C	FP796
64-0725	F4C	FP052		64-1053	RF4C	FP854

64-1062	RF4C	FP821		65-0737	F4D	FP410
64-1066	RF4C	FP820		65-0738	F4D	FP515
64-1067	RF4C	FP729		65-0739	F4D	FP542
64-1071	RF4C	FP772		65-0740	F4D	FP231
64-1074	RF4C	FP773		65-0741	F4D	FP094
64-1080	RF4C	FP904		65-0742	F4D	FP508
64-1082	RF4C	FP732		65-0743	F4D	FP117
64-0581	F4D	FP382		65-0744	F4D	FP175
65-0583	F4D	FP239		65-0746	F4D	FP106
65-0584	F4D	FP251		65-0747	F4D	FP191
65-0586	F4D	FP541		65-0748	F4D	FP518
65-0590	F4D	FP492		65-0749	F4D	FP426
65-0595	F4D	FP126		65-0752	F4D	FP101
65-0596	F4D	FP090		65-0753	F4D	FP574
65-0597	F4D	FP548		65-0754	F4D	FP071
65-0598	F4D	FP521		65-0756	F4D	FP313
65-0609	F4D	FP467		65-0758	F4D	FP425
65-0611	F4D	FP536		65-0759	F4D	FP419
65-0613	F4D	FP123		65-0760	F4D	FP151
65-0614	F4D	FP083		65-0765	F4D	FP460
65-0615	F4D	FP084		65-0767	F4D	FP543
65-0617	F4D	FP525		65-0768	F4D	FP113
65-0629	F4D	FP529		65-0769	F4D	FP260
65-0631	F4D	FP504		65-0772	F4D	FP078
65-0635	F4D	FP301		65-0773	F4D	FP109
65-0638	F4D	FP252		65-0774	F4D	FP285
65-0639	F4D	FP409		65-0775	F4D	FP522
65-0641	F4D	FP396		65-0779	F4D	FP531
65-0643	F4D	FP085		65-0781	F4D	FP568
65-0644	F4D	FP427		65-0782	F4D	FP471
65-0646	F4D	FP528		65-0783	F4D	FP243
65-0647	F4D	FP551		65-0788	F4D	FP258
65-0648	F4D	FP178		65-0789	F4D	FP538
65-0652	F4D	FP509		65-0790	F4D	FP099
65-0654	F4D	FP550		65-0792	F4D	FP570
65-0655	F4D	FP234		65-0793	F4D	FP468
65-0658	F4D	FP120		65-0794	F4D	FP533
65-0659	F4D	FP577		65-0798	F4D	FP129
65-0661	F4D	FP114		65-0826	RF4C	FP730
65-0665	F4D	FP307		65-0830	RF4C	FP829
65-0666	F4D	FP414		65-0831	RF4C	FP771
65-0667	F4D	FP523		65-0837	RF4C	FP790
65-0670	NF4D	FP452		65-0838	RF4C	FP940
65-0671	F4D	FP107		65-0840	RF4C	FP819
65-0672	F4D	FP308		65-0843	RF4C	FP902
65-0676	F4D	FP235		65-0845	RF4C	FP700
65-0677	F4D	FP563		65-0849	RF4C	FP770
65-0680	F4D	FP464		65-0850	NRF4C	FP872
65-0682	F4D	FP244		65-0854	RF4C	FP9..
65-0683	F4D	FP122		65-0866	RF4C	FP795
65-0684	F4D	FP547		65-0868	RF4C	FP831
65-0687	F4D	FP545		65-0878	RF4C	FP804
65-0688	F4D	FP524		65-0881	RF4C	FP863
65-0692	F4D	FP253		65-0893	RF4C	FP815
65-0698	F4D	FP116		65-0896	RF4C	FP625
65-0699	F4D	FP250		65-0912	RF4C	FP778
65-0700	F4D	FP306		65-0920	RF4C	FP803
65-0701	F4D	FP086		65-0924	RF4C	FP689
65-0702	F4D	FP454		65-0925	RF4C	FP794
65-0703	F4D	FP093		65-0927	RF4C	FP637
65-0705	F4D	FP461		65-0935	RF4C	FP941
65-0707	F4D	FP423		65-0939	RF4C	FP9..
65-0710	F4D	FP463		65-0941	NRF4C	FP746
65-0711	F4D	FP219		66-0226	F4D	FP526
65-0714	F4D	FP433		66-0228	F4D	FP111
65-0716	F4D	FP112		66-0229	F4D	FP092
65-0718	F4D	FP300		66-0234	F4D	FP314
65-0719	F4D	FP115		66-0242	F4D	FP513
65-0720	F4D	FP110		66-0243	F4D	FP286
65-0721	F4D	FP415		66-0244	F4D	FP236
65-0729	F4D	FP108		66-0254	F4D	FP527
65-0730	F4D	FP095		66-0261	F4D	FP411
65-0731	F4D	FP124		66-0266	F4D	FP278
65-0734	F4D	FP438		66-0270	F4D	FP494
65-0736	F4D	FP443		66-0272	F4D	FP212

66-0276	F4D	FP210		66-7511	F4D	FP187
66-0277	F4D	FP417		66-7512	F4D	FP394
66-0279	F4D	FP532		66-7515	F4D	FP211
66-0283	F4D	FP480		66-7519	F4D	FP366
66-0284	NF4E	FP825		66-7520	F4D	FP148
66-0289	NF4E	FP627		66-7524	F4D	FP416
66-0291	NF4E	FP626		66-7525	F4D	FP097
66-0294	NF4E	FP747		66-7527	F4D	FP245
66-0306	F4E	FP390		66-7529	F4D	FP259
66-0315	NF4E	FP744		66-7531	F4D	FP312
66-0319	NF4E	FP748		66-7536	F4D	FP256
66-0329	NF4E	FP826		66-7537	F4D	FP198
66-0330	F4E	FP387		66-7538	F4D	FP207
66-0338	F4E	FP385		66-7539	F4D	FP344
66-0342	F4E	FP359		66-7544	F4D	FP317
66-0350	F4E	FP386		66-7545	F4D	FP348
66-0357	F4E	FP392		66-7547	F4D	FP404
66-0372	F4E	FP346		66-7548	F4D	FP163
66-0376	F4E	FP315		66-7549	F4D	FP373
66-0377	NF4E	FP658		66-7551	F4D	FP287
66-0378	F4E	FP309		66-7552	F4D	FP249
66-0384	RF4C	FP824		66-7553	F4D	FP470
66-0389	RF4C	FP736		66-7556	F4D	FP311
66-0393	RF4C	FP780		66-7558	F4D	FP273
66-0395	RF4C	FP502		66-7560	F4D	FP155
66-0410	RF4C	FP368		66-7561	F4D	FP224
66-0419	RF4C	FP340		66-7563	F4D	FP274
66-0421	RF4C	FP767		66-7566	F4D	FP408
66-0422	RF4C	FP903		66-7570	F4D	FP167
66-0425	RF4C	FP901		66-7575	F4D	FP220
66-0427	RF4C	FP750		66-7578	F4D	FP161
66-0430	RF4C	FP900		66-7579	F4D	FP263
66-0433	RF4C	FP789		66-7582	F4D	FP265
66-0444	RF4C	FP811		66-7583	F4D	FP290
66-0446	RF4C	FP843		66-7587	F4D	FP303
66-0452	RF4C	FP816		66-7588	F4D	FP430
66-0456	RF4C	FP740		66-7589	F4D	FP291
66-0461	RF4C	FP674		66-7591	F4D	FP156
66-0463	RF4C	FP737		66-7593	F4D	FP221
66-0470	RF4C	FP328		66-7595	F4D	FP472
66-0473	RF4C	FP749		66-7596	F4D	FP264
66-0474	RF4C	FP728		66-7604	F4D	FP353
66-0475	RF4C	FP734		66-7605	F4D	FP374
66-0476	RF4C	FP615		66-7607	F4D	FP398
66-0478	RF4C	FP774		66-7609	F4D	FP246
66-7456	F4D	FP552		66-7610	F4D	FP143
66-7457	F4D	FP479		66-7614	F4D	FP400
66-7458	F4D	FP096		66-7615	F4D	FP209
66-7459	F4D	FP456		66-7619	F4D	FP142
66-7460	F4D	FP412		66-7623	F4D	FP401
66-7461	F4D	FP082		66-7625	F4D	FP327
66-7464	F4D	FP457		66-7629	F4D	FP154
66-7466	F4D	FP391		66-7633	F4D	FP127
66-7467	F4D	FP299		66-7634	F4D	FP375
66-7469	F4D	FP555		66-7635	F4D	FP481
66-7470	F4D	FP289		66-7640	F4D	FP413
66-7471	F4D	FP478		66-7641	F4D	FP218
66-7472	F4D	FP537		66-7642	F4D	FP255
66-7475	F4D	FP339		66-7644	F4D	FP160
66-7477	F4D	FP121		66-7645	F4D	FP144
66-7478	F4D	FP459		66-7647	F4D	FP549
66-7484	F4D	FP462		66-7648	F4D	FP323
66-7485	F4D	FP569		66-7649	F4D	FP343
66-7486	F4D	FP440		66-7650	F4D	FP169
66-7487	F4D	FP516		66-7652	F4D	FP275
66-7488	F4D	FP262		66-7656	F4D	FP159
66-7489	F4D	FP535		66-7658	F4D	FP431
66-7490	F4D	FP534		66-7659	F4D	FP196
66-7491	F4D	FP503		66-7660	F4D	FP356
66-7496	F4D	FP546		66-7662	F4D	FP477
66-7497	F4D	FP562		66-7663	F4D	FP361
66-7498	F4D	FP406		66-7664	F4D	FP418
66-7502	F4D	FP350		66-7667	F4D	FP399
66-7506	F4D	FP267		66-7668	F4D	FP266
66-7509	F4D	FP422		66-7669	F4D	FP257

66-7674	F4D	FP145		66-8753	F4D	FP282
66-7676	F4D	FP149		66-8762	F4D	FP168
66-7677	F4D	FP238		66-8782	F4D	FP190
66-7679	F4D	FP281		66-8783	F4D	FP475
66-7681	F4D	FP162		66-8786	F4D	FP186
66-7683	F4D	FP372		66-8788	F4D	FP165
66-7684	F4D	FP254		66-8789	F4D	FP128
66-7685	F4D	FP336		66-8794	F4D	FP150
66-7687	F4D	FP225		66-8802	F4D	FP228
66-7688	F4D	FP432		66-8804	F4D	FP280
66-7689	F4D	FP091		66-8805	F4D	FP166
66-7692	F4D	FP279		66-8808	F4D	FP371
66-7694	F4D	FP474		66-8813	F4D	FP232
66-7696	F4D	FP352		66-8815	F4D	FP184
66-7698	F4D	FP271		66-8816	F4D	FP216
66-7699	F4D	FP347		66-8817	F4D	FP240
66-7701	F4D	FP146		66-8819	F4D	FP203
66-7702	F4D	FP405		66-8821	F4D	FP403
66-7704	F4D	FP153		66-8824	F4D	FP227
66-7705	F4D	FP345		66-8825	F4D	FP332
66-7706	F4D	FP326		67-0218	F4E	FP388
66-7708	F4D	FP424		67-0223	F4E	FP439
66-7710	F4D	FP237		67-0229	F4E	FP389
66-7711	F4D	FP420		67-0241	F4E	FP360
66-7712	F4D	FP402		67-0245	F4E	FP446
66-7714	F4D	FP105		67-0246	F4E	FP726
66-7716	NF4D	FP647		67-0256	F4E	FP304
66-7718	F4D	FP331		67-0263	F4E	FP376
66-7720	F4D	FP103		67-0275	F4E	FP310
66-7721	F4D	FP455		67-0288	F4E	FP369
66-7722	F4D	FP421		67-0299	F4E	FP727
66-7723	F4D	FP241		67-0311	F4E	FP296
66-7725	F4D	FP324		67-0320	F4E	FP435
66-7726	F4D	FP367		67-0324	F4E	FP316
66-7728	F4D	FP157		67-0328	F4E	FP319
66-7729	F4D	FP215		67-0333	F4E	FP333
66-7730	F4D	FP370		67-0337	F4E	FP380
66-7731	F4D	FP269		67-0343	F4E	FP769
66-7733	F4D	FP325		67-0349	F4E	FP639
66-7735	F4D	FP100		67-0356	F4E	FP654
66-7738	F4D	FP283		67-0364	F4E	FP622
66-7739	F4D	FP318		67-0370	F4E	FP743
66-7741	F4D	FP270		67-0390	F4E	FP642
66-7742	F4D	FP197		67-0428	RF4C	FP699
66-7749	F4D	FP242		67-0433	RF4C	FP813
66-7751	F4D	FP193		67-0438	RF4C	FP754
66-7754	F4D	FP098		67-0441	RF4C	FP766
66-7755	F4D	FP349		67-0442	RF4C	FP753
66-7759	F4D	FP473\N428FS		67-0443	RF4C	FP775
66-7760	F4D	FP199		67-0444	RF4C	FP830
66-7765	F4D	FP355		67-0448	RF4C	FP686
66-7766	F4D	FP147		67-0450	RF4C	FP724
66-7767	F4D	FP288		67-0453	RF4C	FP784
66-7768	F4D	FP476		67-0454	RF4C	FP676
66-7771	F4D	FP201		67-0455	RF4C	FP341
66-7772	F4D	FP268		67-0458	RF4C	FP672
66-7773	F4D	FP337		67-0459	RF4C	FP298
66-8693	F4D	FP158		67-0462	RF4C	FP698
66-8698	F4D	FP448		67-0464	RF4C	FP377
66-8699	F4D	FP608		67-0465	RF4C	FP383
66-8700	F4D	FP226		67-0467	RF4C	FP759
66-8705	F4D	FP276		67-0469	RF4C	FP782
66-8709	F4D	FP164		68-0303	F4E	FP660
66-8710	F4D	FP222		68-0308	F4E	FP621
66-8715	F4D	FP277		68-0317	F4E	FP501
66-8719	F4D	FP185		68-0320	F4E	FP539
66-8722	F4D	FP229		68-0324	F4E	FP540
66-8723	F4D	FP354		68-0340	F4E	FP578
66-8727	F4D	FP192		68-0343	F4E	FP511
66-8728	F4D	FP272		68-0357	F4E	FP768
66-8732	F4D	FP223		68-0371	F4E	FP505
66-8733	F4D	FP213		68-0375	F4E	FP556
66-8735	F4D	FP200		68-0379	F4E	FP559
66-8739	F4D	FP202		68-0385	F4E	FP697
66-8745	F4D	FP214		68-0388	F4E	FP557

68-0389	F4E	FP656	69-0357	RF4C	FP896
68-0391	F4E	FP643	69-0358	RF4C	FP364
68-0410	F4E	FP696	69-0359	RF4C	FP741
68-0423	F4E	FP510	69-0361	RF4C	FP329
68-0447	F4E	FP648	69-0362	RF4C	FP636
68-0450	F4E	FP653	69-0363	RF4C	FP752
68-0452	F4E	FP554	69-0368	RF4C	FP342
68-0460	F4E	FP499	69-0369	RF4C	FP776
68-0462	F4E	FP661	69-0370	RF4C	FP886
68-0463	F4E	FP466	69-0374	RF4C	FP330
68-0464	F4E	FP582	69-0376	RF4C	FP805
68-0507	F4E	FP500	69-0378	RF4C	FP878
68-0509	F4E	FP514	69-0379	RF4C	FP675
68-0511	F4E	FP657	69-0382	RF4C	FP889
68-0536	F4E	FP553	69-0384	RF4C	FP806
68-0548	RF4C	FP890	69-7201	F4G	FP828
68-0551	RF4C	FP884	69-7209	F4G	FP849
68-0552	RF4C	FP876	69-7214	F4G	FP862
68-0553	RF4C	FP885	69-7216	F4G	FP807
68-0555	RF4C	FP897	69-7218	F4G	FP598
68-0557	RF4C	FP777	69-7231	F4G	FP838
68-0561	RF4C	FP887	69-7233	F4G	FP833
68-0562	RF4C	FP655	69-7234	F4G	FP761
68-0564	RF4C	FP881	69-7252	F4E	FP601
68-0565	RF4C	FP888	69-7254	F4G	FP855
68-0567	RF4C	FP792	69-7257	F4G	FP844
68-0568	RF4C	FP735	69-7258	F4G	FP861
68-0569	RF4C	FP690	69-7260	F4G	FP866
68-0571	RF4C	FP755	69-7262	F4G	FP817
68-0572	RF4C	FP877	69-7268	F4G	FP788
68-0574	RF4C	FP677	69-7287	F4G	FP857
68-0576	RF4C	FP668	69-7288	F4G	FP859
68-0578	RF4C	FP594	69-7289	F4G	FP867
68-0580	RF4C	FP781	69-7293	F4G	FP797
68-0581	RF4C	FP756	69-7300	F4G	FP851
68-0582	RF4C	FP751	69-7303	F4G	FP839
68-0583	RF4C	FP793	69-7566	F4G	FP787
68-0584	RF4C	FP758	69-7581	F4G	FP827
68-0585	RF4C	FP624	69-7582	F4G	FP786
68-0589	RF4C	FP664	69-7583	F4G	FP599
68-0592	RF4C	FP895	71-0237	F4E	FP644
68-0593	RF4C	FP858	71-0238	F4E	FP635
68-0595	RF4C	FP170	71-0239	F4E	FP669
68-0596	RF4C	FP171	71-0240	F4E	FP645
68-0599	RF4C	FP205	71-0243	F4E	FP595
68-0600	RF4C	FP188	71-0247	F4E	FP469
68-0602	RF4C	FP172	71-0248	RF4C	FP891
68-0605	RF4C	FP194	71-0251	RF4C	FP880
68-0606	RF4C	FP189	71-0252	RF4C	FP799
68-0607	RF4C	FP173	71-0254	RF4C	FP800
68-0608	RF4C	FP206	71-0255	RF4C	FP808
68-0609	RF4C	FP195	71-0259	RF4C	FP893
69-0236	F4G	FP606	71-1073	F4E	FP649
69-0237	F4G	FP731	71-1075	F4E	FP670
69-0239	F4G	FP607	71-1076	F4E	FP617
69-0242	F4G	FP765	71-1079	F4E	FP591
69-0244	F4G	FP823	71-1081	F4E	FP628
69-0245	F4G	FP762	71-1083	F4E	FP691
69-0246	F4G	FP847	71-1084	F4E	FP681
69-0250	F4G	FP798	71-1085	F4E	FP687
69-0251	F4G	FP832	71-1086	F4E	FP630
69-0254	F4G	FP852	71-1087	F4E	FP684
69-0255	F4G	FP764	71-1088	F4E	FP671
69-0267	F4G	FP860	71-1089	F4E	FP616
69-0269	F4G	FP738	71-1092	F4E	FP588
69-0270	F4G	FP682	71-1397	F4E	FP605
69-0274	F4G	FP763	72-0122	F4E	FP517
69-0279	F4G	FP848	72-0124	F4E	FP507
69-0281	F4G	FP865	72-0128	F4E	FP596
69-0284	F4G	FP864	72-0135	F4E	FP720
69-0290	F4E	FP600	72-0136	F4E	FP620
69-0292	F4G	FP874	72-0139	F4E	FP558
69-0304	F4G	FP856	72-0140	F4E	FP564
69-0352	RF4C	FP757	72-0141	F4E	FP705
69-0356	RF4C	FP702	72-0142	F4E	FP685

72-0143	F4E	FP587		74-1040	F4E	FP721
72-0144	F4E	FP898		74-1041	F4E	FP437
72-0145	RF4C	FP879		74-1042	F4E	FP493
72-0147	RF4C	FP892		74-1043	F4E	FP641
72-0149	RF4C	FP883		74-1044	F4E	FP869
72-0150	RF4C	FP894		74-1045	F4E	FP836
72-0152	RF4C	FP809		74-1047	F4E	FP604
72-0153	RF4C	FP822		74-1048	F4E	FP868
72-0154	RF4C	FP802		74-1049	F4E	FP760
72-0155	RF4C	FP801		74-1050	F4E	FP845
72-0156	RF4C	FP882		74-1052	F4E	FP846
72-0159	F4E	FP666		74-1053	F4E	FP395
72-0161	F4E	FP590		74-1055	F4E	FP835
72-0162	F4E	FP584		74-1057	F4E	FP428
72-0165	F4E	FP631		74-1059	F4E	FP451
72-0166	F4E	FP592		74-1060	F4E	FP834
72-0167	F4E	FP579		74-1061	F4E	FP567
72-0168	F4E	FP566		74-1620	F4E	FP683
72-1477	F4E	FP718		74-1621	F4E	FP618
72-1478	F4E	FP580		74-1622	F4E	FP575
72-1479	F4E	FP695		74-1623	F4E	FP611
72-1482	F4E	FP703		74-1624	F4E	FP693
72-1483	F4E	FP638		74-1625	F4E	FP397
72-1484	F4E	FP576		74-1626	F4E	FP512
72-1485	F4E	FP708		74-1627	F4E	FP652
72-1489	F4E	FP450		74-1628	F4E	FP572
72-1490	F4E	FP710		74-1629	F4E	FP583
72-1493	F4E	FP589		74-1630	F4E	FP665
72-1494	F4E	FP561		74-1631	F4E	FP651
73-1160	F4E	FP667		74-1634	F4E	FP662
73-1164	F4E	FP722		74-1635	F4E	FP706
73-1165	F4E	FP602		74-1636	F4E	FP581
73-1166	F4E	FP719		74-1637	F4E	FP506
73-1167	F4E	FP519		74-1638	F4E	FP573
73-1168	F4E	FP899		74-1640	F4E	FP497
73-1171	F4E	FP715		74-1642	F4E	FP465
73-1172	F4E	FP603		74-1643	F4E	FP714
73-1173	F4E	FP840		74-1644	F4E	FP619
73-1176	F4E	FP870		74-1645	F4E	FP586
73-1181	F4E	FP560		74-1648	F4E	FP614
73-1182	F4E	FP632		74-1649	F4E	FP725
73-1183	F4E	FP873		74-1650	F4E	FP650
73-1184	F4E	FP713		74-1651	F4E	FP679
73-1185	F4E	FP597		74-1652	F4E	FP707
73-1186	F4E	FP393		74-1653	F4E	FP717
73-1187	F4E	FP434		150438	F4N	8F112
73-1188	F4E	FP692		150440	F4N	8F090
73-1189	F4E	FP378		150492	F4N	8F154
73-1193	F4E	FP565		150643	F4N	8F121
73-1195	F4E	FP453		150651	F4N	8F089
73-1196	F4E	FP441		151003	F4N	8F086
73-1197	F4E	FP629		151401	F4N	8F055
73-1198	F4E	FP633		151431	F4N	8F141
73-1199	F4E	FP634		151444	F4N	8F111
74-0643	F4E	FP429		151446	F4N	8F143
74-0644	F4E	FP701		151482	F4N	8F120
74-0645	F4E	FP447		151502	F4N	8F055
74-0646	F4E	FP520		151514	F4N	8F136
74-0647	F4E	FP445		151977	RF4B	8F325
74-0648	F4E	FP498		151978	RF4B	8F328
74-0650	F4E	FP704		151980	RF4B	8F333
74-0652	F4E	FP716		151983	RF4B	8F330
74-0653	F4E	FP613		152208	F4N	8F101
74-0654	F4E	FP612		152246	F4N	8F152
74-0655	F4E	FP483		152278	F4N	8F144
74-0656	F4E	FP875		152288	F4N	8F107
74-0657	F4E	FP694		152293	F4N	8F138
74-0659	F4E	FP663		152295	F4N	8F133
74-0662	F4E	FP680		152300	F4N	8F171
74-0663	F4E	FP841		152306	F4N	8F170
74-0664	F4E	FP837		152317	F4N	8F116
74-0665	F4E	FP709		152318	F4N	8F132
74-0666	F4E	FP723		152965	F4N	8F114
74-1038	F4E	FP678		152970	F4N	8F125
74-1039	F4E	FP593		152975	F4N	8F129

152981	F4N	8F131		155545	F4S	8F241
152991	F4N	8F087		155550	F4S	8F286
153007	F4B	8F175		155555	F4S	8F291
153012	F4N	8F150		155558	F4S	8F233
153027	F4N	8F142		155559	F4S	8F277
153039	F4N	8F145		155560	F4S	8F273
153050	F4N	8F127		155566	F4S	8F311
153084	DF4J	8F199		155567	F4S	8F247
153091	RF4B	8F337		155570	F4S	8F223
153092	RF4B	8F339		155575	F4S	8F309
153093	RF4B	8F338		155731	F4S	8F196
153094	RF4B	8F335		155732	F4S	8F257
153095	RF4B	8F327		155733	F4S	8F198
153096	RF4B	8F329		155735	F4S	8F173
153102	RF4B	8F336		155736	F4S	8F312
153103	RF4B	8F332		155739	F4S	8F323
153105	RF4B	8F334		155743	F4S	8F200
153106	RF4B	8F326		155746	F4S	8F220
153107	RF4B	8F331		155747	F4S	8F177
153108	RF4B	8F324		155748	F4J	8F163
153776	F4J	8F128		155753	F4S	8F167
153777	F4J	8F164		155754	F4S	8F283
153779	F4S	8F256		155761	F4S	8F261
153780	F4S	8F300		155765	F4S	8F272
153784	F4S	8F284		155766	F4S	8F202
153796	F4J	8F184		155767	F4S	8F229
153800	F4S	8F259		155773	F4S	8F168
153805	F4S	8F270		155779	F4S	8F210
153807	F4S	8F188		155781	F4S	8F252
153812	F4J	8F104		155783	F4S	8F248
153825	F4S	8F195		155784	F4S	8F260
153826	F4S	8F249		155787	F4S	8F285
153827	F4S	8F214		155792	F4S	8F208
153828	F4S	8F244		155794	F4S	8F305
153841	F4J	8F182		155799	F4S	8F186
153842	F4S	8F212		155801	F4S	8F213
153845	F4S	8F314		155805	F4S	8F302
153847	F4S	8F228		155812	F4S	8F204
153848	F4S	8F190		155820	F4S	8F197
153852	F4J	8F106		155821	F4S	8F271
153855	F4S	8F169		155822	F4S	8F306
153857	F4S	8F298		155823	F4J	8F279
153859	F4S	8F243		155825	F4S	8F278
153861	F4S	8F165		155827	F4S	8F237
153862	F4S	8F268		155830	F4J	8F262
153864	F4S	8F230		155835	F4J	8F189
153872	F4S	8F172		155838	F4S	8F266
153873	F4S	8F174		155839	F4S	8F250
153881	F4S	8F218		155844	F4S	8F317
153890	F4S	8F239		155849	F4S	8F290
153891	F4S	8F303		155851	F4S	8F296
153893	F4S	8F294		155852	F4S	8F215
153896	F4S	8F191		155854	F4S	8F206
153898	F4S	8F231		155858	F4S	8F253
153900	F4S	8F192		155859	F4S	8F287
153903	F4S	8F263		155862	F4S	8F235
153907	F4S	8F175		155863	F4S	8F292
153908	F4S	8F166		155871	F4S	8F254
153910	F4S	8F299		155874	F4S	8F281
153914	F4N	8F139		155876	F4S	8F280
154781	F4S	8F307		155879	F4S	8F288
154782	F4S	8F236		155880	F4S	8F224
154785	F4J	8F180		155881	F4S	8F319
155504	F4S	8F219		155883	F4S	8F293
155505	F4J	8F183		155888	F4S	8F301
155515	F4S	8F265		155891	F4S	8F289
155519	F4S	8F209		155893	F4S	8F246
155522	F4S	8F255		155896	F4S	8F304
155525	F4S	8F226		155897	F4S	8F221
155528	F4S	8F267		157245	F4S	8F205
155530	F4S	8F203		157248	F4S	8F201
155532	F4S	8F193		157249	F4S	8F242
155539	F4S	8F222		157254	F4S	8F282
155541	F4S	8F194		157257	F4S	8F178
155542	F4S	8F297		157258	F4J	8F295

157260	F4S	8F258		157308	F4S	8F232
157264	F4J	8F216		157345	RF4B	8F340
157268	F4J	8F264		157309	F4S	8F238
157269	F4S	8F211		158348	F4S	8F207
157272	F4S	8F313		158350	F4S	8F227
157276	F4S	8F275		158352	F4S	8F320
157278	F4S	8F318		158354	F4S	8F269
157281	F4S	8F316		158356	F4J	8F181
157283	F4S	8F274		158363	F4J	8F187
157286	F4J	8F308		158365	F4J	8F185
157287	F4S	8F276		158369	F4S	8F321
157291	F4S	8F225		158372	F4S	8F315
157292	F4J	8F245		158373	F4S	8F310
157296	F4J	8F251		158376	F4S	8F322
157300	F4S	8F217		158378	F4S	8F179
157301	F4S	8F234				

LTV F8 CRUSADER (29)

144614	RF8G	2F435		146882	RF8G	2F437
144618	RF8G	2F423		146889	RF8G	2F431
145608	RF8G	2F422		146985	F8K	2F281
145609	RF8G	2F425		149145	F8J	2F440
145622	RF8G	2F430		149149	F8J	2F445
145623	RF8G	2F432		149201	F8J	2F444
145627	RF8G	2F420		149204	F8J	2F441
145633	RF8G	2F436		149210	F8J	2F442
145645	RF8G	2F429		149215	F8J	2F439
145647	RF8G	2F419		150658	F8J	2F443
146835	RF8G	2F426		150680	F8J	2F446
146845	RF8G	2F433		150683	F8J	2F438
146847	RF8G	2F419		150845	F8J	2F448
146863	RF8G	2F416		150871	F8J	2F447
146865	RF8G	2F421				

GRUMMAN F9 COUGAR (1)

147283	TF9J	3F311

DOUGLAS F10 SKYNIGHT (4)

124610	F10B	VH001		124663	EF10B	VH003
124620	EF10B	VH002		127047	EF10B	VH005

GRUMMAN F14 TOMCAT (32)

158622	F14A	1K052		159006	F14A	1K...
158979	F14A	1K013		159009	F14A	1K036
158980	F14A	1K015		159010	F14A	1K032
158982	F14A	1K028		159014	F14A	1K034
158986	F14A	1K033		159015	F14A	1K041
158987	F14A	1K018		159017	F14A	1K...
158988	F14A	1K016		159020	F14A	1K037
158989	F14A	1K...		159021	F14A	1K046
158991	F14A	1K...		159422	F14A	1K035
158994	F14A	1K...		159425	F14A	1K038
158991	F14A	1K011		159427	F14A	1K043
158994	F14A	1K...		159429	F14A	1K055
159000	F14A	1K014		159434	F14A	1K...
159002	F14A	1K008		159435	F14A	1K...
159003	F14A	1K...		159436	F14A	1K...
159004	F14A	1K...		159442	F14A	1K...

McDONNELL DOUGLAS F15 EAGLE (46)

73-0090	F15A	FH...		75-0021	F15A	FH030
73-0092	F15A	FH...		75-0022	F15A	FH045
73-0096	F15A	FH006		75-0031	F15A	FH055
73-0103	F15A	FH007		75-0032	F15A	FH035
74-0086	F15A	FH016		75-0034	F15A	FH046
74-0092	F15A	FH017		75-0036	F15A	FH049
74-0096	F15A	FH027		75-0038	F15A	FH038
74-0099	F15A	FH019		75-0042	F15A	FH039
74-0100	F15A	FH028		75-0046	F15A	FH021
74-0103	F15A	FH047		75-0050	F15A	FH022
74-0104	F15A	FH033		75-0060	F15A	FH037
74-0105	F15A	FH020		75-0062	F15A	FH052
74-0110	F15A	FH029		75-0073	F15A	FH054
74-0111	F15A	FH005		75-0074	F15A	FH048
74-0115	F15A	FH023		75-0077	F15A	FH018
74-0118	F15A	FH034		75-0089	F15B	FH025
74-0123	F15A	FH053		76-0038	F15A	FH...
74-0127	F15A	FH032		76-0051	F15A	FH...
74-0128	F15A	FH012		76-0076	F15A	FH014
74-0134	F15A	FH031		76-0089	F15A	FH015
74-0135	F15A	FH024		76-0096	F15A	FH...
74-0138	F15B	FH036		76-0100	F15A	FH...
75-0020	F15A	FH044		76-0104	F15A	FH...

GENERAL DYNAMICS F16 FIGHTING FALCON (18)

79-0331	F16A	FG016		79-0405	F16A	FG...
79-0335	F16A	FG013		80-0496	F16A	FG...
79-0336	F16A	FG...		80-0510	F16A	FG015
79-0340	F16A	FG012		80-0512	F16A	FG...
79-0349	F16A	FG...		80-0529	F16A	FG...
79-0352	F16A	FG...		80-0530	F16A	FG...
79-0363	F16A	FG...		80-0625	F16B	FG...
79-0395	F16A	FG018		90-0951	F16B	FG...\92616 PAF
79-0396	F16A	FG...		92-0405	F16A	FG...\92736 PAF

REPUBLIC F84 THUNDERSTREAK (2)

51-1725	F84F	FA245		53-7600	F84F	FA144

NORTH AMERICAN F100 SUPER SABRE (95)

55-2794	F100D	FE280		56-3110	F100D	FE365
55-2800	F100D	FE283		56-3133	QF100D	FE466
55-2830	QF100D	FE329		56-3179	F100D	FE416
55-2881	F100D	FE510		56-3190	F100D	FE304
55-2919	F100D	FE...		56-3194	F100D	FE418
55-2942	QF100D	FE524		56-3203	F100D	FE278
55-3504	F100D	FE330		56-3204	F100D	FE282
55-3567	F100D	FE401		56-3211	F100D	FE274
55-3665	F100D	FE609		56-3221	F100D	FE317
55-3741	F100D	FE566		56-3254	F100D	FE279
55-3745	F100D	FE331		56-3279	F100D	FE447
55-3804	F100D	FE584		56-3331	F100D	FE390
55-3812	QF100D	FE549		56-3385	F100D	FE522
56-2917	F100D	FE444		56-3425	F100D	FE442
56-2932	F100D	FE596		56-3427	F100D	FE320
56-2980	QF100D	FE586		56-3463	F100D	FE348
56-3018	F100D	FE295		56-3726	F100F	FE359
56-3022	F100F	FE642		56-3727	F100F	FE360
56-3029	QF100D	FE297		56-3738	F100F	FE620
56-3038	F100D	FE290		56-3746	F100F	FE624
56-3046	F100D	FE480		56-3751	QF100F	FE408
56-3093	F100D	FE430		56-3754	F100F	FE616
56-3099	F100D	FE293		56-3760	F100F	FE575

56-3762	F100F	FE516		56-3868	QF100F	FE439
56-3763	F100F	FE511		56-3880	QF100F	FE639
56-3768	F100F	FE377		56-3882	F100F	FE591
56-3773	F100F	FE520		56-3883	QF100F	FE521
56-3787	QF100F	FE472		56-3891	F100F	FE519
56-3791	F100F	FE494		56-3893	F100F	FE553
56-3794	F100F	FE363		56-3898	F100F	FE515
56-3795	F100F	FE625		56-3905	QF100F	FE...
56-3805	F100F	FE619		56-3906	F100F	FE622
56-3812	F100F	FE311		56-3907	F100F	FE626
56-3813	F100F	FE598		56-3910	QF100F	FE517
56-3814	F100F	FE355		56-3912	F100F	FE323
56-3818	QF100F	FE495		56-3915	F100F	FE514
56-3819	F100F	FE552		56-3917	F100F	FE567
56-3822	QF100F	FE638		56-3922	F100F	FE604
56-3824	F100F	FE...		56-3928	F100F	FE603
56-3825	QF100F	FE640		56-3929	F100F	FE381
56-3830	F100F	FE606		56-3956	F100F	FE554
56-3832	QF100F	FE637		56-3962	F100F	FE322
56-3836	F100F	FE610		56-3982	F100F	FE374
56-3837	QF100F	FE631		56-3990	F100F	FE618
56-3840	F100F	FE623		56-3994	F100F	FE611
56-3859	F100F	FE496		56-4001	F100F	FE612
56-3860	QF100F	FE555		N404FS	F100F	FE630
56-3861	QF100F	FE641				

McDONNELL F101 VOODOO (5)

57-0342	F101F	FF418		58-0282	F101B	FF328
57-0436	F101B	FF402		58-0330	F101B	FF400
58-0281	F101B	FF398				

CONVAIR F102 DELTA DAGGER (6)

53-1811	F102A	FJ240		56-1515	F102A	FJ216
56-1047	F102A	FJ155		56-2350	TF102A	FJ383
56-1455	F102A	FJ194		57-0821	F102A	FJ211

LOCKHEED F104 STARFIGHTER (1)

57-1320	F104D	FB053

REPUBLIC F105 THUNDERCHIEF (17)

59-1731	F105D	FK040		60-5385	F105D	FK042
59-1759	F105D	FK054		61-0071	F105D	FK049
59-1771	F105D	FK048		61-0164	F105D	FK046
59-1774	F105D	FKO35		62-4344	F105G	FK051
59-1822	F105D	FKO30		62-4432	F105G	FK093
60-0452	F105D	FK061		63-8266	F105G	FK014
60-0471	F105D	FK070		63-8288	F105F	FK012
60-0492	F105D	FK045		63-8305	F105G	FK015
60-0504	F105D	FK044				

CONVAIR F106 DELTA DART (69)

56-0457	F106A	FN084		57-2455	F106A	FN168
56-0458	F106A	FN051		57-2466	F106A	FN167
56-0461	F106A	FN074		57-2467	F106A	FN175
56-0463	F106A	FN060		57-2476	F106A	FN148
56-0466	F106A	FN091		57-2480	F106A	FN178
57-0244	F106A	FN015		57-2505	F106A	FN170
57-0245	F106A	FN052		57-2506	F106A	FN169
57-2453	F106A	FN135		57-2508	F106B	FN042

57-2509	F106B	FN196		59-0044	F106A	FN184
57-2510	F106B	FN048		59-0046	F106A	FN193
57-2512	F106B	FN00.		59-0047	F106A	FN188
57-2517	F106B	FN177		59-0048	F106A	FN183
57-2522	F106B	FN147		59-0049	F106A	FN190
57-2524	F106B	FN176		59-0051	F106A	FN098
57-2530	F106B	FN160		59-0063	F106A	FN081
57-2532	F106B	FN04.		59-0065	F106A	FN003
57-2539	F106B	FN034		59-0079	F106A	FN013
57-2541	F106B	FN041		59-0080	F106A	FN027
57-2546	F106B	FN021		59-0090	F106A	FN067
57-2547	F106B	FN097		59-0093	F106A	FN073
58-0773	F106A	FN057		59-0094	F106A	FN038
58-0790	F106A	FN080		59-0095	F106A	FN089
58-0900	F106B	FN056		59-0105	F106A	FN059
58-0902	F106B	FN055		59-0106	F106A	FN058
58-0903	F106B	FN046		59-0109	F106A	FN104
58-0904	F106B	FN044		59-0133	F106A	FN064
59-0003	F106A	FN070		59-0137	F106A	FN062
59-0010	F106A	FN069		59-0147	F106A	FN017
59-0012	F106A	FN077		59-0149	F106B	FN197
59-0016	F106A	FN066		59-0150	F106B	FN030
59-0025	F106A	FN072		59-0153	F106B	FN130
59-0032	F106A	FN088		59-0158	F106B	FN019
59-0033	F106A	FN187		59-0161	F106B	FN038
59-0037	F106A	FN182		59-0164	F106B	FN050
59-0040	F106A	FN192				

GENERAL DYNAMICS F111 (116)

63-9777	F111A	FV008		67-0163	FB111A	BF009
63-9781	F111A	FV012		67-7192	FB111A	BF010
63-9783	FB111A	FV016		67-7195	FB111A	BF004
65-5702	F111A	FV...		67-7196	F111G	FV156
65-5704	F111A	FV006		68-0009	F111E	FV...
65-5705	F111A	FV005		68-0010	F111E	FV143
65-5707	F111A	FV009		68-0014	F111E	FV...
65-5708	F111A	FV007		68-0021	F111E	FV169
65-5710	F111A	FV028		68-0028	F111E	FV...
66-0053	F111A	FV037		68-0031	F111E	FV172
67-0045	F111A	FV...		68-0033	F111E	FV...
67-0053	F111A	FV070		68-0035	F111E	FV...
67-0061	F111A	FV062		68-0038	F111E	FV170
67-0062	F111A	FV033		68-0039	F111E	FV...
67-0065	F111A	FV067		68-0046	F111E	FV171
67-0071	F111A	FV026		68-0051	F111E	FV173
67-0074	F111A	FV068		68-0053	F111E	FV...
67-0076	F111A	FV077		68-0062	F111E	FV...
67-0077	F111A	FV...		68-0085	F111D	FV040
67-0079	F111A	FV029		68-0087	F111D	FV039
67-0081	F111A	FV...		68-0089	F111D	FV038
67-0084	F111A	FV065		68-0090	F111D	FV149
67-0085	F111A	FV027		68-0096	F111D	FV128
67-0086	F111A	FV...		68-0100	F111D	FV126
67-0087	F111A	FV071		68-0101	F111D	FV043
67-0088	F111A	FV058		68-0104	F111D	FV154
67-0090	F111A	FV069		68-0107	F111D	FV147
67-0091	F111A	FV075		68-0108	F111D	FV125
67-0094	F111A	FV...		68-0111	F111D	FV...
67-0095	F111A	FV072		68-0112	F111D	FV114
67-0096	F111A	FV...		68-0117	F111D	FV042
67-0101	F111A	FV032		68-0120	F111D	FV080
67-0104	F111A	FV...		68-0121	F111D	FV135
67-0106	F111A	FV076		68-0122	F111D	FV138
67-0107	F111A	FV066		68-0123	F111D	FV148
67-0108	F111A	FV073		68-0124	F111D	FV046
67-0110	F111A	FV031		68-0126	F111D	FV045
67-0115	F111E	FV166		68-0127	F111D	FV150
67-0119	F111E	FV...		68-0128	F111D	FV063
67-0121	F111E	FV...		68-0129	F111D	FV044
67-0122	F111E	FV...		68-0134	F111D	FV...
67-0161	FB111A	BF003		68-0137	F111D	FV142
67-0162	F111G	FV168		68-0143	F111D	FV131

68-0149	F111D	FV144		68-0246	FB111A	BF008
68-0150	F111D	FV145		68-0248	FB111A	BF...
68-0151	F111D	FV119		68-0249	FB111A	BF018
68-0162	F111D	FV074		68-0250	FB111A	BF013
68-0165	F111D	FV146		68-0252	F111G	FV079
68-0171	F111D	FV081		68-0254	FB111A	FV165
68-0172	F111D	FV129		68-0256	FB111A	BF017
68-0174	F111D	FV...		68-0257	F111G	FV164
68-0175	F111D	FV144		68-0258	FB111A	BF014
68-0176	F111D	FV141		68-0260	F111G	FV161
68-0177	F111D	FV134		68-0262	FB111A	BF016
68-0179	F111D	FV132		68-0269	FB111A	BF015
68-0240	FB111A	BF012		68-0288	FB111A	BF011
68-0241	F111G	FV167		69-6504	F111G	FV050
68-0244	F111G	FV078		69-6513	FB111A	BF007

BELL AH1 HUEY COBRA (31)

157757	AH1J	7H183		157789	AH1J	7H181
157758	AH1J	7H179		157790	AH1J	7H218
157759	AH1J	7H222		157795	AH1J	7H180
157760	AH1J	7H216		157796	AH1J	7H176
157761	AH1J	7H214		157798	AH1J	7H215
157768	AH1J	7H219		157799	AH1J	7H177
157769	AH1J	7H175		157800	AH1J	7H189
157770	AH1J	7H213		157801	AH1J	7H225
157772	AH1J	7H201		157803	AH1J	7H211
157773	AH1J	7H182		157804	AH1J	7H223
157774	AH1J	7H184		157805	AH1J	7H224
157775	AH1J	7H217		159214	AH1J	7H231
157778	AH1J	7H221		159221	AH1J	7H233
157783	AH1J	7H178		159225	AH1J	7H232
157785	AH1J	7H212		3-4412	AH1J	7H090
157787	AH1J	7H220				

BELL UH1 IROQUOIS (153)

58-2084	UH1A	XA005		63-12930	UH1B	HF087
60-3578	UH1B	HF127		63-12974	UH1H	XA365
60-3592	UH1B	HF092		63-12976	UH1H	XA366
60-3608	UH1B	XA183		63-12986	UH1H	XA340
60-3618	UH1B	HF136		63-12998	UH1H	XA319
61-0709	UH1B	HF135		63-13146	UH1P	HF032
61-0752	UH1B	HF121		63-13584	UH1H	XA341
61-0755	UH1B	XA104		64-13591	UH1H	
62-1885	UH1B	HF134		64-13592	UH1H	
62-1971	UH1B	XA267		64-13621	UH1H	XA335
62-1974	UH1B	HF133		64-13626	UH1H	XA326
62-2039	UH1B	XA116		64-13646	UH1H	XA371
62-2075	UH1B	HF105		64-13654	UH1H	XA357
62-2098	UH1B	XA098		64-13740	UH1H	XA318
62-2101	UH1B	XA267		64-13751	UH1H	XA367
62-2106	UH1B	XA315		64-13814	UH1H	XA325
62-4571	UH1B	HF132		64-13816	UH1H	XA348
62-4582	UH1B	HF131		64-13827	UH1H	XA336
62-4596	UH1B	HF137		64-13896	UH1H	XA359
62-12372	UH1H	XA357		64-13919	UH1B	HF093
62-12524	UH1B	XA141		64-13975	UH1B	
63-8512	UH1B	XA266		64-13989	UH1B	XA019
63-8551	UH1B	XA131		64-14006	UH1B	
63-8645	UH1B	XA128		64-14020	UH1B	XA072
63-8677	UH1B	HF124		64-14027	UH1B	HF140
63-8681	UH1B	HF097		64-14029	UH1B	HF120
63-8/53	UH1H	XA317		64-14074	UH1B	HF112
63-8778	UH1H	XA368		64-14091	UH1B	XA058
63-8782	UH1H	XA344		65-7918	UH1F	HF146
63-8805	UH1H			65-7925	UH1F	HF...
63-8819	UH1H	XA343		65-7962	UH1F	HF174
63-8836	UH1H	XA345		65-9533	QUH1M	XA307
63-8850	UH1H	XA369		65-9592	UH1H	XA370
63-12923	UH1B	HF086		65-9603	UH1H	XA346

65-9630	UH1H	XA332		66-1236	TH1F	HF...
65-9632	UH1H	XA312		66-1237	TH1F	HF173
65-9648	UH1H	XA327		66-1238	UH1F	HF...
65-9671	UH1H	XA331		66-1240	TH1F	HF152
65-9736	UH1H	XA320		66-1244	UH1F	HF...
65-9751	UH1H			66-1246	UH1F	HF...
65-9752	UH1H	XA313		66-1247	UH1F	HF153
65-9753	UH1H	XA322		66-1249	UH1F	HF156
65-9814	UH1H	XA374		66-15023	QUH1M	
65-9816	UH1H	XA352		66-15056	QUH1M	
65-9825	UH1H	XA353		66-15124	QUH1M	XA308
65-9832	UH1H	XA314		66-15218	QUH1M	XA306
65-9853	UH1H			66-16354	UH1H	XA328
65-9857	UH1H	XA354		66-16466	UH1H	XA333
65-9878	UH1H	XA358		66-16618	UH1H	XA355
65-9977	UH1H	XA349		67-17691	UH1H	
65-9980	UH1H	XA330		68-15440	UH1H	XA339
65-10007	UH1H	XA334		151848	UH1E	7H172
65-10012	UH1H	XA321		151853	UH1E	7H168
65-10106	UH1H	XA350		152426	UH1E	7H174
65-10108	UH1H	XA364		152431	UH1E	7H170
65-12739	QUH1M	XA304		154757	UH1E	7H163
65-12858	JUH1H	XA337		154767	UH1E	7H169
65-12886	UH1H	XA361		154774	UH1E	7H173
66-0645	QUH1M	XA305		157179	HH1K	7H185
66-0848	UH1H	XA356		157180	HH1K	7H186
66-0850	UH1H	XA329		157181	HH1K	7H187
66-0888	UH1H	XA360		157182	HH1K	7H191
66-0908	UH1D			157183	HH1K	7H193
66-1001	UH1H	XA338		157184	HH1K	7H194
66-1043	UH1H	XA324		157190	HH1K	7H196
66-1116	UH1H	XA347		157194	HH1K	7H192
66-1220	UH1P	HF...		157198	HH1K	7H195
66-1224	UH1F	HF150		157199	HH1K	7H197
66-1225	TH1F	HF154		157841	TH1L	7H159
66-1227	TH1F	HF165		157845	TH1L	7H140
66-1228	UH1F	HF151		157858	UH1L	7H158
66-1229	UH1F	HF167		158551	VH1N	7H198
66-1231	TH1F	HF169		158552	VH1N	7H199
66-1232	TH1F	HF172		158556	VH1N	7H200
66-1233	TH1F	HF155		158557	VH1N	7H190
66-1234	TH1F	HF...				
66-1235	UH1F	HF...				

KAMAN H2 SEASPRITE (8)

149753	SH2F	8H023		151304	SH2F	8H006
150143	SH2F	8H019		151308	SH2F	8H007
150173	SH2F	8H024		151311	SH2F	8H020
150179	SH2F	8H022		151314	SH2F	8H...

SIKORSKY H3 SEA KING (11)

63-9687	CH3E	HH027		149682	HH3A	9H016
65-5693	CH3E	HH022		151553	HH3A	9H017
65-5696	CH3E	HH021		152696	SH3G	9H...
65-5697	CH3E	44006		154107	SH3D	9H022
67-14705	HH3E	HH...		154118	SH3D	9H...
148051	SH3G	9H...		1496	HH3F	
149003	SH3G	9H039				

SIKORSKY H34 CHOCTAW (2)

56-4316	VH34C	3H264		154900	UH34D	3H231

SIKORSKY H52 SEAGUARD (6)

1373	HH52A	46006	1421	HH52A	46009
1379	HH52A	46002	1424	HH52A	46016
1419	HH52A	46003	1465	HH52A	46015

SIKORSKY H53 SEA STALLION (24)

152395	CH53A	HC...	153302	CH53A	HC004
152408	CH53A	2J031	153309	CH53A	2J045
152409	CH53A	2J042	153708	CH53A	2J043
152410	CH53A	2J044	153711	CH53A	2J037
152411	CH53A	2J040	153713	CH53A	HC015
152412	CH53A	HC011	153720	CH53A	2J032
153279	CH53A	2J036	153726	CH53A	2J033
153287	CH53A	2J030	153732	CH53A	2J029
153291	CH53A	2J038	154863	CH53A	2J034
153292	CH53A	2J041	154877	CH53A	2J039
153293	CH53A	HC009	154878	CH53A	2J035
153294	CH53A	HC010	158684	RH53D	2J060

BELL H57 SEA RANGER (26)

157361	TH57A	4H013	157388	TH57A	4H017
157364	TH57A	4H019	157392	TH57A	4H018
157365	TH57A	4H003	157394	TH57A	4H030
157367	TH57A	4H004	161696	TH57B	4H040
157370	TH57A	4H027	161697	TH57B	4H041
157373	TH57A	4H011	162038	TH57C	4H033
157375	TH57A	4H005	162039	TH57C	4H039
157376	TH57A	4H006	162046	TH57C	4H032
157378	TH57A	4H029	162055	TH57C	4H034
157379	TH57A	4H007	162667	TH57C	4H037
157381	TH57A	4H015	162673	TH57C	4H035
157382	TH57A	4H023	162674	TH57C	4H036
157385	TH57A	4H016	162820	TH57C	4H038
157387	TH57A	4H024			

CESSNA O2 SUPER SKYMASTER (136)

67-21299	O2A	HV...	67-21403	O2A	HV112
67-21309	O2A	HV156	67-21407	O2A	HV128
67-21310	O2A	HV116	67-21410	O2A	HV161
67-21313	O2A	HV120	67-21416	O2A	HV219
67-21321	O2A	HV179	67-21417	O2A	HV173
67-21324	O2A	HV158	67-21421	O2A	HV114
67-21330	O2A	HV143	67-21424	O2A	HV154
67-21342	O2A	HV159	67-21433	O2A	HV131
67-21345	O2A	HV216	67-21435	O2A	HV118
67-21346	O2A	HV127	67-21454	O2B	HV162
67-21352	O2A	HV126	67-21470	O2B	HV152
67-21353	O2A	HV138	68-6873	O2A	HV117
67-21354	O2A	HV111	68-6874	O2A	HV186
67-21355	O2A	HV164	68-6875	O2A	HV123
67-21356	O2A	HV133	68-6876	O2A	HV147
67-21360	O2A	HV115	68-6877	O2A	HV153
67-21363	O2A	HV224	68-6880	O2A	HV121
67-21366	O2A	HV176	68-6881	O2A	HV122
67-21371	O2A	HV177	68-6889	O2A	HV188
67-21372	O2A	HV225	68-6892	O2A	HV149
67-21373	O2A	HV119	68-6893	O2A	HV137
67-21380	O2A	HV...	68-6895	O2A	HV140
67-21383	O2A	HV160	68-6896	O2A	HV133
67-21387	O2A	HV130	68-6897	O2A	HV144
67-21392	O2A	HV180	68-6900	O2A	HV191
67-21393	O2A	HV217	68-6903	O2A	HV189
67-21397	O2A	HV183	68-10829	O2A	HV276
67-21398	O2A	HV144	68-10830	O2A	HV195

68-10831	O2A	HV155	69-7605	O2A	HV243
68-10832	O2A	HV163	69-7606	O2A	HV252
68-10833	O2A	HV124	69-7607	O2A	HV244
68-10838	O2A	HV...	69-7608	O2A	HV190
68-10840	O2A	HV207	69-7609	O2A	HV245
68-10849	O2A	HV277	69-7611	O2A	HV232
68-10857	O2A	HV211	69-7612	O2A	HV246
68-10865	O2A	HV212	69-7615	O2A	HV258
68-10868	O2A	HV279	69-7623	O2A	HV226
68-10869	O2A	HV255	69-7624	O2A	HV234
68-10870	O2A	HV256	69-7625	O2A	HV221
68-10872	O2A	HV248	69-7626	O2A	HV239
68-10967	O2A	HV165	69-7627	O2A	HV240
68-10971	O2A	HV187	69-7628	O2A	HV259
68-10976	O2A	HV145	69-7631	O2A	HV175
68-10977	O2A	HV171	69-7632	O2A	HV223
68-10983	O2A	HV220	69-7635	O2A	HV...
68-10989	O2A	HV181	69-7636	O2A	HV257
68-11003	O2A	HV230	69-7637	O2A	HV182
68-11028	O2A	HV184	69-7638	O2A	HV286
68-11046	O2A	HV172	69-7639	O2A	HV272
68-11050	O2A	HV170	69-7640	O2A	HV254
68-11057	O2A	HV166	69-7641	O2A	HV270
68-11068	O2A	HV268	69-7642	O2A	HV278
68-11123	O2A	HV263	69-7643	O2A	HV235
68-11125	O2A	HV262	69-7644	O2A	HV229
68-11126	O2A	HV267	69-7645	O2A	HV275
68-11135	O2A	HV265	69-7646	O2A	HV271
68-11152	O2A	HV168	69-7649	O2A	HV236
68-11155	O2A	HV196	69-7650	O2A	HV241
68-11157	O2A	HV203	69-7651	O2A	HV260
68-11158	O2A	HV266	69-7653	O2A	HV261
68-11163	O2A	HV269	69-7654	O2A	HV...
68-11167	O2A	HV264	69-7655	O2A	HV237
68-11168	O2A	HV208	69-7656	O2A	HV110
68-11169	O2A	HV174	69-7657	O2A	HV285
68-11171	O2A	HV198	69-7662	O2A	HV247
68-11172	O2A	HV199	69-7663	O2A	HV242
68-11173	O2A	HV193	69-7664	O2A	HV...
69-7601	O2A	HV287	69-7665	O2A	HV233
69-7604	O2A	HV251	69-7666	O2A	HV238

LOCKHEED P2 NEPTUNE (3)

147952	SP2H	1P355	148340	SP2H	1P403
147963	SP2H	PA002			

LOCKHEED P3 ORION (72)

149673	UP3A	2P078	152146	P3A	2P046
150527	UP3A	2P041	152148	P3A	2P029
150528	UP3A	2P043	152153	P3A	2P... EdA
150604	P3A	2P021	152154	P3A	2P031
150608	UP3A	2P006	152157	P3A	2P052
151353	UP3A	2P060	152162	P3A	2P053
151354	UP3A	2P074	152163	P3A	2P055
151356	UP3A	2P025	152167	P3A	2P039
151358	P3A	2P020	152168	P3A	2P034
151360	P3A	2P...	152173	P3A	2P038
151366	UP3A	2P015	152174	P3A	2P048
151368	UP3A	2P073	152175	P3A	2P033
151378	P3A	2P028	152176	P3A	2P044
151383	UP3A	2P023	152177	P3A	2P049
151386	UP3A	2P010	152179	UP3A	2P009
151388	UP3A	2P018	152180	P3A	2P037
151389	UP3A	2P013	152181	P3A	2P047
151393	P3A	2P026	152183	P3A	2P030
151396	UP3A	2P007	152184	P3A	2P050
152140	P3A	2P051	152185	P3A	2P057
152142	P3A	2P040	152186	P3A	2P032
152143	P3A	2P056	152187	P3A	2P035

152718	P3B	2P095	152756	P3B	2P...
152721	P3B	2P069	152758	P3B	2P094
152723	P3B	2P063	152760	P3B	2P081
152726	P3B	2P091	152762	P3B	2P...
152730	P3B	2P083	153417	P3B	2P...
152732	P3B	2P...	153418	P3B	2P036
152734	P3B	2P068	153423	P3B	2P065
152736	P3B	2P066	153430	P3B	2P087
152737	P3B	2P080	153432	P3B	2P06.
152743	P3B	2P079	153434	P3B	2P082
152744	P3B	2P...	153439	P3B	2P090
152746	P3B	2P093	164467	P3C	2P... Pakistan 25
152750	P3B	2P086	164468	P3C	2P... Pakistan 26
152753	P3B	2P084	164469	P3C	2P... Pakistan 27

GRUMMAN S2 TRACKER (33)

133351	US2C	1S724	152347	S2G	1S580
136433	US2B	1S752	152808	S2G	1S519
136643	US2B	1S731	152810	S2G	1S567
144725	US2A	1S746	152815	S2G	1S563
149259	S2E	1S239	152817	S2G	1S531
149265	S2E	1S244	152820	S2G	1S570
149268	S2E	1S242	152824	S2G	1S532
149848	S2E	1S241	152825	S2G	1S568
149854	S2E	1S566	152826	S2G	1S571
149855	S2E	1S353	152828	S2G	1S529
149862	S2E	1S359	152838	S2G	1S593
149873	S2E	1S441	152841	S2G	1S595
151640	S2E	1S401	153568	S2G	1S545
151654	S2E	1S338	153570	S2G	1S552
152337	S2G	1S526	153573	S2G	1S558
152341	S2G	1S500	153579	S2G	1S587
152345	S2G	1S569			

GRUMMAN S3 VIKING (17)

158869	S3A	2S...	160585	S3A	2S...
159395	S3A	2S027	160586	S3A	2S...
159396	S3A	2S030	160591	S3B	2S...
159410	S3A	2S...	160593	S3A	2S017
159417	S3A	2S...	160594	S3A	2S...
159749	S3A	2S...	160595	S3A	2S...
160568	S3A	2S014	160597	S3A	2S...
160570	S3A	2S016	160598	S3A	2S...
160574	S3A	2S018			

LOCKHEED T1 SEASTAR (1)

142263	T1A	1T044

ROCKWELL T2 BUCKEYE (41)

152472	T2B	2T170	156717	T2C	2T200
153538	T2B	2T177	156721	T2C	2T199
153553	T2B	2T178	156722	T2C	2T206
153555	T2B	2T173	157035	T2C	2T189
155213	T2B	2T...	157037	T2C	2T207
155221	T2B	2T174	157042	T2C	2T194
155224	T2B	2T...	157047	T2C	2T198
155227	T2B	2T179	157063	T2C	2T195
155229	T2B	2T172	158317	T2C	2T183
155230	T2B	2T175	158331	T2C	2T203
155231	T2B	2T171	158576	T2C	2T190
155232	T2B	2T...	158582	T2C	2T...
155233	T2B	2T176	158602	T2C	2T202
155234	T2B	2T169	158608	T2C	2T184
155240	T2C	2T201	158885	T2C	2T209
156714	T2C	2T205	158897	T2C	2T204

159152	T2C	2T181	159172	T2C	2T208
159154	T2C	2T191	159704	T2C	2T188
159158	T2C	2T182	159708	T2C	2T197
159161	T2C	2T193	159723	T2C	2T186
159164	T2C	2T192			

NORTH AMERICAN T28 TROJAN (8)

137726	T28B	5T218	138246	T28B	5T178
138155	T28B	5T176	138272	T28B	5T082
138231	T28B	5T189	146279	T28C	5T153
138245	T28B	5T074	146283	T28C	5T227

CONVAIR T29 (7)

51-5127	T29B	CS051	52-1162	T29C	TB314
51-5133	T29B	CS026	53-3461	T29D	TB315
51-5165	T29B	TB312	53-3477	T29D	TB316
52-1160	T29C	TB313			

LOCKHEED T33 SHOOTING STAR (58)

52-9757	T33A	TCB92	58-0536	T33A	TCD28
52-9803	T33A	TCD30	58-0540	T33A	
52-9848	T33A	TCD19	58-0555	T33A	TCD31
53-4900	T33A	TCD24	58-0610	T33A	TCD34
53-4915	T33A	TCB80	58-0612	T33A	
53-4918	T33A	TCB81	58-0618	T33A	
53-4988	T33A	TCB93	58-0628	T33A	TCD52
53-5012	T33A	TCB82	58-2098	T33A	TCB29
53-5180	T33A	TCB83	58-2099	T33A	TCD20
53-5201	T33A	TCB84	136828	T33B	3T087
53-5203	T33A	TCB85	137936	T33B	TCC92
53-5248	T33A	TCB87	137940	T33B	TCC93
53-5249	T33A	TCB88	137959	T33B	TCC94
53-5901	T33A	TCD13	137970	T33B	TCC95
53-5970	T33A	TCD29	137976	T33B	TCD05
55-4385	T33A	TCB00	137993	T33B	TCC10
55-4386	T33A	TCD27	138034	T33B	TCC96
56-1716	T33A	TCD23	138040	T33B	TCC97
56-3679	T33A	TCD26	138056	T33B	TCD06
57-0563	T33A	TCD18	138064	T33B	TCC98
57-0592	T33A		138074	T33B	CC99
57-0595	T33A	TCD22	138078	T33B	TCD00
57-0611	T33A	TCD32	138977	T33B	TCD07
57-0711	T33A		138979	T33B	TCD01
57-0719	T33A	TCD25	138983	T33B	3T028
57-0739	T33A		138998	T33B	TCD02
57-0759	T33A	TCD21	141509	T33B	TCD03
58-0494	T33A	TCD33	141516	T33B	3T305
58-0529	T33A	TCD35	141526	T33B	TCD04

BEECH T34 MENTOR (17)

140676	T34B	4T089	140946	T34B	4T167
140678	T34B	4T161	143986	T34B	4T110
140698	T34B	4T092	144003	T34B	4T083
140709	T34B	4T091	144051	T34B	4T074
140845	T34B	4T168	144095	T34B	4T087
140865	T34B	4T105	144097	T34B	4T114
140880	T34B	4T162	144099	T34B	4T078
140910	T34B	4T107	144115	T34B	4T081
140945	T34B	4T086			

CESSNA T37 TWEETY BIRD (61)

56-3493	T37B	TE011	57-2311	T37B	TE...
56-3510	T37B	TE...	57-2313	T37B	TE026
56-3517	T37B	TE014	57-2316	T37B	TE007
56-3537	T37B	TE031	57-2317	T37B	TE028
56-3545	T37B	TE053	57-2331	T37B	TE032
56-3557	T37B	TE...	57-2336	T37B	TE061
56-3566	T37B	TE...	57-2337	T37B	TE...
56-3579	T37B	TE017	57-2341	T37B	TE012
56-3581	T37B	TE005	57-2343	T37B	TE022
56-3583	T37B	TE015	57-2345	T37B	TE...
56-3587	T37B	TE...	57-2350	T37B	TE027
57-2231	T37B	TE058	58-1862	T37B	TE...
57-2233	T37B	TE...	58-1872	T37B	TE001
57-2236	T37B	TE013	58-1874	T37B	TE...
57-2239	T37B	TE018	58-1887	T37B	TE064
57-2243	T37B	TE...	58-1938	T37B	TE...
57-2247	T37B	TE019	58-1939	T37B	TE...
57-2253	T37B	TE024	58-1962	T37B	TE065
57-2264	T37B	TE...	58-1977	T37B	TE047
57-2267	T37B	TE009	59-0289	T37B	TE006
57-2269	T37B	TE023	59-0338	T37B	TE008
57-2273	T37B	TE...	59-0350	T37B	TE004
57-2278	T37B	TE025	59-0390	T37B	TE020
57-2281	T37B	TE062	60-0072	T37B	TE030
57-2291	T37B	TE029	60-0087	T37B	TE...
57-2292	T37B	TE...	60-0103	T37B	TE037
57-2295	T37B	TE016	60-0136	T37B	TE...
57-2296	T37B	TE021	60-0143	T37B	TE063
57-2305	T37B	TE...	60-0190	T37B	TE...
57-2307	T37B	TE...	64-13426	T37B	TE066
57-2309	T37B	TE002			

NORTHROP T38 TALON (66)

60-0548	T38A	TF001	61-0886	AT38B	TF...
60-0556	T38A	TF022	61-0891	AT38B	TF...
60-0562	T38A	TF020	61-0899	AT38B	TF...
60-0564	T38A	TF023	61-0904	AT38B	TF...
60-0565	T38A	TF015	61-0911	AT38B	TF...
60-0571	T38A	TF017	61-0917	AT38B	TF101
60-0575	T38A	TF018	61-0940	AT38B	TF060
60-0577	T38A	TF014	61-0947	AT38B	TF...
60-0579	T38A	TF026	62-3614	AT38B	TF...
60-0589	AT38B	TF...	62-3627	AT38B	TF...
60-0591	AT38B	TF...	62-3641	AT38B	TF...
60-0594	AT38B	TF...	62-3660	AT38B	TF...
60-0595	AT38B	TF...	62-3678	AT38B	TF...
61-0804	AT38B	TF...	62-3738	AT38B	TF112
61-0806	AT38B	TF...	62-3752	AT38B	TF...
61-0807	AT38B	TF...	63-8149	AT38B	TF...
61-0809	AT38B	TF...	63-8164	AT38B	TF106
61-0812	AT38B	TF...	63-8172	AT38B	TF...
61-0818	AT38B	TF...	63-8211	AT38B	TF...
61-0831	AT38B	TF...	63-8214	AT38B	TF...
61-0835	AT38B	TF...	63-8215	AT38B	TF...
61-0836	AT38B	TF...	64-13188	AT38B	TF...
61-0842	AT38B	TF...	64-13215	AT38B	TF...
61-0845	AT38B	TF105	64-13245	AT38B	TF...
61-0847	AT38B	TF...	64-13252	T38A	TF...
61-0852	AT38B	TF...	64-13276	AT38B	TF...
61-0857	AT38B	TF...	64-13280	AT38B	TF...
61-0860	AT38B	TF...	64-13292	AT38B	TF...
61-0864	AT38B	TF...	65-10330	T38A	TF052
61-0866	AT38B	TF...	65-10365	T38A	TF053
61-0875	AT38B	TF103	65-10439	AT38B	TF...
61-0876	AT38B	TF...	65-10456	AT38B	TF...
61-0880	AT38B	TF...	68-8140	AT38B	TF...

NORTH AMERICAN ROCKWELL T39A SABRELINER (76)

59-2869	CT39A	TG033	62-4462	CT39A	TG046
59-2872	CT39A	TG015	62-4468	CT39A	TG020
60-3479	CT39A	TG082	62-4469	CT39A	TG064
60-3481	CT39A	TG085	62-4474	CT39A	TG027
60-3482	CT39A	TG016	62-4475	CT39A	TG084
60-3484	CT39A	TG024	62-4483	CT39A	TG055
60-3487	CT39A	TG021	62-4486	CT39A	TG050
60-3490	CT39A	TG062	62-4490	CT39A	TG079
60-3491	CT39A	TG009	62-4491	CT39A	TG081
60-3492	CT39A	TG007	62-4493	CT39A	TG076
60-3493	CT39A	TG057	62-4497	CT39A	TG053
60-3494	CT39A	TG094	62-4501	CT39A	TG073
60-3497	CT39A	TG066	150543	T39D	7T027
60-3499	CT39A	TG037	150544	T39D	7T006
60-3500	CT39A	TG030	150546	T39D	7T014
60-3501	CT39A	TG093	150547	T39D	7T021
60-3508	CT39A	TG042	150548	T39D	7T008
61-0636	CT39A	TG089	150549	T39D	7T011
61-0637	CT39A	TG035	150969	T39D	7T026
61-0638	CT39A	TG096	150973	T39D	7T013
61-0641	CT39A	TG036	150974	T39D	7T015
61-0642	CT39A	TG045	150975	T39D	7T007
61-0643	CT39A	TG022	150976	T39D	7T016
61-0648	CT39A	TG017	150978	T39D	7T009
61-0652	CT39A	TG087	150979	T39D	7T017
61-0653	CT39A	TG071	150980	T39D	7T004
61-0656	CT39A	TG010	150981	T39D	7T012
61-0657	CT39A	TG023	150982	T39D	7T022
61-0665	CT39A	TG028	150983	T39D	7T023
61-0666	CT39A	TG014	150984	T39D	7T010
61-0668	CT39A	TG051	150988	T39D	7T005
61-0676	CT39A	TG049	150990	T39D	7T024
61-0682	CT39A	TG032	150991	T39D	7T003
61-0683	CT39A	TG025	151336	T39D	7T025
62-4451	CT39A	TG060	151340	T39D	7T018
62-4454	CT39A	TG018	151341	T39D	7T019
62-4457	CT39A	TG002	151342	T39D	7T020
62-4459	CT39A	TG041	157353	CT39E	7T...

FAIRCHILD T46 (2)

84-0493	YT46A	TM002	85-1596	T46A	TM001

BEECH U8 SEMINOLE (1)

56-3714	U8D	HU003

GRUMMAN U16 ALBATROSS (3)

141278	HU16D	HP084	2132	HU16E	HP079
1016	HU16E	HP090			

GRUMMAN OV1 MOHAWK (55)

59-2621	OV1B	YA040	61-2685	OV1C	YA016
59-2622	OV1B	YA104	61-2689	OV1C	YA098
59-2623	OV1B	YA054	61-2695	OV1C	YA101
59-2627	OV1B	YA087	61-2696	OV1D	YA...
59-2628	OV1B	YA027	61-2697	OV1C	YA090
59-2629	OV1B	YA...	61-2710	OV1C	YA099
59-2630	OV1B	YA078	61-2714	OV1C	YA081
59-2633	OV1B	YA...	61-2717	OV1C	YA021
59-2636	OV1B	YA055	61-2718	OV1C	YA059
60-3757	OV1C	YA069	61-2721	OV1C	YA102
60-3759	OV1C	YA079	61-2728	OV1C	YA100
60-3760	OV1C	YA089	62-5857	JOV1C	YA077
61-2677	OV1C	YA012	62-5861	OV1B	YA022

62-5862	OV1B	YA109	66-18883	OV1C	YA...
62-5869	OV1B	YA106	66-18884	OV1C	YA103
62-5870	OV1B	YA014	66-18887	OV1C	YA035
62-5871	OV1B	YA063	66-18889	OV1C	YA091
62-5872	OV1D	YA...	66-18890	OV1C	YA097
62-5882	OV1B	YA110	66-18891	OV1C	YA010
62-5883	OV1B	YA023	66-18892	OV1C	YA085
62-5884	OV1B	YA105	66-18893	OV1C	YA044
62-5892	OV1B	YA031	66-18894	OV1C	YA014
62-5893	OV1B	YA107	67-18903	OV1D	YA113
62-5895	OV1B	YA108	68-15940	OV1D	YA...
62-5901	OV1B	YA053	68-15942	OV1D	YA115
62-5905	OV1B	YA017	68-15960	OV1D	YA...
64-14249	RV1D	YA009	69-17004	OV1D	YA...
64-14251	OV1B	YA066			

ROCKWELL OV10 BRONCO (13)

67-14604	OV10A	HA005	155457	OV10A	1V032
67-14618	OV10A	HA007	155459	OV10A	1V024
67-14675	OV10A	HA006	155471	OV10A	1V...
68-3797	OV10A	HA008	155475	OV10A	1V027
155405	OV10A	1V033	155480	OV10A	1V025
155428	OV10A	1V034	155496	OV10A	1V023
155454	OV10A	1V026			

UNCLASSIFIED

Boeing 707\720 (127)

N1R	023B	CZ...	N795TW	131B	CZ...
N62TA	123B	CZ073	N796TW	131B	CZ...
N160GL	321B	CZ119	N798TW	131B	CZ...
N245AC	138B	CZ...	N799TW	131B	CZ...
N342A	338C	CZ...	N880PA	321B	CZ...
N402PA	321B	CZ...	N881PA	321B	CZ...
N495PA	321B	CZ097	N885PA	321B	CZ...
N496PA	321B	CZ100	N944JW	321B	CZ...
N497PA	321B	CZ101	N1181Z	321B	CZ...
N519GA	123B	CZ...	N2464C	051B	CZ...
N707GE	321B	CZ107	N2464K	051B	CZ076
N720AC	023B	CZ...	N3161	047B	CZ068
N720BC	030B	CZ...	N3162	047B	CZ062
N720BG	023B	CZ127	N3163	047B	CZ061
N746TW	131B	CZ024	N3164	047B	CZ066
N747TW	131B	CZ041	N3165	047B	CZ060
N748TW	131B	CZ035	N3833L	047B	CZ074
N749TW	131B	CZ...	N4450Z	059B	CZ118
N750TW	131B	CZ020	N5487N	707	CZ120
N752TW	131B	CZ025	N5038	123B	CZ...
N754TW	131B	CZ021	N5517Z	321B	CZ...
N755TW	131B	CZ...	N6721	131B	CZ...
N756TW	131B	CZ018	N6722	131B	CZ030
N758TW	131B	CZ019	N6723	131B	CZ039
N759TW	131B	CZ034	N6724	131B	CZ...
N770BE	024B	CZ...	N6727	131B	CZ045
N773TW	331B	CZ083	N6728	131B	CZ043
N774TW	331B	CZ...	N6729	131B	CZ026
N775TW	331B	CZ109	N6763T	131B	CZ023
N778PA	139B	CZ102	N6764T	131B	CZ037
N779TW	331B	CZ...	N6771T	131B	CZ040
N780TW	331B	CZ110	N6789T	131B	CZ027
N781TW	131B	CZ030	N6790T	131B	CZ028
N782TW	131B	CZ042	N7509A	123B	CZ072
N783TW	131B	CZ033	N7551A	123B	CZ065
N784TW	131B	CZ022	N7552A	123B	CZ...
N785TW	131B	CZ...	N7554A	123B	CZ...
N793NA	138B	CZ069	N7564A	323C	CZ114
N793TW	331B	CZ111	N7570A	123B	CZ124

N7571A	123B	CZ010	N8733	331B	CZ150
N7572A	123B	CZ...	N8735	331B	CZ...
N7573A	123B	CZ126	N8737	331B	CZ...
N7574A	123B	CZ059	N8738	331B	CZ...
N7575A	123B	CZ008	N15711	331C	CZ...
N7576A	123B	CZ004	N16738	131B	CZ055
N7577A	123B	CZ016	N16739	131B	CZ...
N7578A	123B	CZ005	N18702	331B	CZ096
N7579A	123B	CZ015	N18703	331B	CZ081
N7580A	123B	CZ002	N18704	331B	CZ...
N7581A	123B	CZ003	N18706	331B	CZ...
N7582A	123B	CZ006	N18707	331B	CZ094
N7583A	123B	CZ064	N18708	331B	CZ...
N7584A	123B	CZ012	N18710	331B	CZ104
N7588A	123B	CZ071	N18711	331B	CZ112
N7589A	123B	CZ...	N18712	331B	CZ108
N7590A	123B	CZ009	N18713	331B	CZ105
N7591A	123B	CZ017	N24666	051B	CZ075
N7592A	123B	CZ013	N28714	331C	CZ...
N8417	323C	CZ...	N86740	131B	CZ032
N8705T	331B	CZ...	N86741	131B	CZ029
N8725T	331B	CZ093	G-BFEO	323C	CZ...
N8729	331B	CZ...	G-SAIL	323C	CZ...
N8730	331B	CZ103	RP-C1886	351C	CZ...
N8732	331B	CZ...			

Piper PA-48 Enforcer (1)

N482PE	PA48	AG001

Mikoyan MiG (11)

822	MiG15		511	MiG17
...	MiG15		634	MiG17
038	MiG15UTI		905	MiG17
010	MiG17		917	MiG17
303	MiG17		1524	MiG17
406	MiG17			

GRAND TOTAL	4116

NASA

Best known for its launching of moon-rockets and space-shuttle vehicles, NASA also operate a large number of military and civil aircraft throughout the United States as support, communications and pure research vehicles. Nominally civil-registered, usually in the NxxxNA range, many of these aircraft merely display the NASA logo and the three digit number of its civil registration (e.g. Boeing 747-123, the shuttle transporter, is N905NA or NASA 905). In the following list we give ONLY the NASA-digits serial if the allotted civil registration IS in the NxxxNA series, but DO quote the civil registration if this does NOT follow the usual practice in any way. Some aircraft carry BOTH serial presentations simultaneously. A few aircraft retain their US military serials. Although these can be found listed in the main body of the book, they are included at the end of the NASA serials for completeness.

Each major NASA location is allocated a batch of one hundred numbers for its aircraft. The allocation is as follows:-

NASA Wallops I.	Wallops Island	400 series
NASA Langley	Langley AFB, Va.	500 series
NASA Lewis	Cleveland-Hopkins IAP, Oh	600 series
NASA Ames	Moffett Field NAS, San Jose, Ca.	700 series
NASA Dryden	Edwards AFB, Ca.	800 series
NASA Johnson	Ellington ANGB, Houston, Tx.	900 series

The individual serials allocated are:-

1		Grumman G-1159A Gulfstream III	c\n 309	ex N18LB
2	(1)	Grumman G-159 Gulfstream I	c\n 98	to N29AY
2	(2)	Grumman G-159 Gulfstream I	c\n 96	ex N1NA
3		Grumman G-159 Gulfstream I	c\n 92	
4	(1)	Lockheed 1329 Jetstar I	c\n 5015	to N172L
4	(2)	Grumman G-159 Gulfstream I	c\n 151	
5	(1)	VT29B	c\n 321	USAF 51-7909
5	(2)	Grumman G-159 Gulfstream I	c\n 125	ex N10NA
6		Beech C45H Expediter	c\n AF-817	SoC
7	(1)	Beech B65-80 Queenair	c\n LD-77	SoC
7	(2)	Beech 200 Super King Air	c\n BB-997	
8	(1)	Beech B65-80 Queenair	c\n LD-79	
8	(2)	Beech 200 Super King Air	c\n BB-950	
9	(1)	Beech B65-80 Queenair	c\n LD-49	
9	(2)	Beech 200 Super King Air	c\n BB-1091	
10	(1)	C47J Skytrain		USN Bu17268
10	(2)	Grumman G-159 Gulfstream I	c\n 125	to NASA5
12		Bell 206A JetRanger	c\n 508	to NASA950
14		Lockheed 1329 Jetstar I	c\n 5003	to NASA814
19		Convair 340	c\n 3	to NASA707
20		C121G Constellation	c\n 4143	USAF 54-4065 to NASA420
21		C121G Constellation	c\n 4159	USAF 54-4076 to NASA421
27		C54G Skymaster	c\n 36009	USAF 45-0556 to NASA427
28		C118A Liftmaster	c\n 43575	USAF 51-3828 to NASA428
29		UH1B Iroquois		SoC
218		JB57A Canberra		USAF 52-1418 scrapped
224		T33A Shooting Star	c\n 6941	USAF 51-9157 to MASDC
230		AJ2 Savage		USN Bu134069 SoC
231		C54 Skymaster		SoC
232		C54G Skymaster	c\n 36090	USAF 45-0637 to NASA432
234		Bell X14A		US Army 56-4022 to NASA704
237		B57B Canberra		USAF 52-1576 to NASA809
238		C54G Skymaster	c\n 36031	USAF 45-0578 to NASA438
414		UH1D Iroquois	c\n 428	US Army 62-1908 ex N732NA
415	(1)	UH1B Iroquois	c\n 914	US Army 63-8689
415	(2)	UH1H Iroquois	c\n 5129	US Army 65-10085
416		UH1B Iroquois	c\n 584	US Army 62-2064
417		UH1B Iroquois		
418		JUH1M Iroquois	c\n 1646	US Army 66-0664
420	(1)	C121G Constellation	c\n 4143	ex NASA20
420		UH1H Iroquois	c\n 5148	US Army 65-10104
421	(1)	C121G Constellation	c\n 4159	ex NASA21
421	(2)	F5A Freedom Fighter	c\n N.6004	USAF 63-8367 SoC
422		VC121A Constellation	c\n 2605	USAF 48-0613 preserved Ft.Rucker
424		UH1B Iroquois	c\n 1031	US Army 64-13907
425		CT39E Sabreliner	c\n 282-95	USN Bu158380
426		P3B Orion	c\n 5175	USN Bu152735
427	(1)	C54G Skymaster	c\n 36009	ex NASA27 to N4958M
427	(2)	EC130Q Hercules	c\n 4901	USN Bu161495

428	(1)	C118A Liftmaster	c\n 43575	ex NASA28 SoC
428	(2)	NP3A Orion	c\n 1003	USN Bu148276
				to NASA927
429		Lockheed L-188 Electra	c\n 1103	ex N97
430		Short SC-7 Skyvan	c\n SH.1844	ex N30DA
431	(1)	T39D Sabreliner	c\n 285-2	USN Bu150970
431	(2)	CT39A Sabreliner	c\n 265-16	USAF 60-3488
432		C54G Skymaster	c\n 36090	ex NASA232 SoC
438		C54G Skymaster	c\n 36031	ex NASA438 SoC 1981
501	(1)	C47D Skytrain		USAF 43-49526
				to NASA636
501	(2)	Grumman AA-1B Trainer	c\n AA-1B-0001	
502		T28A Trojan	c\n 174-263	USAF 51-3725 SoC
503	(1)	Aero Commander 680	c\n 577-222	SoC
503	(2)	Cessna 402B	c\n 402B-031	ex NASA719
504		Beech C23 Musketeer	c\n M-1608	
505		U3A Blue Canoe	c\n 38076	USAF 57-5921
506		Beech B80 Queenair	c\n LD-507	
507		Cessna 172K Skyhawk	c\n 59729	
508		DHC-6 Twin Otter 100	c\n 4	to NASA607
510		T34C Mentor	c\n GL-108	
511		T38A Talon	c\n N.5748	USAF 64-10329
512		OV-1B Mohawk		US Army 62-5880
514		T38A Talon	c\n N.5747	USAF 65-10328
				ex\to NASA908
515		Boeing 737-130	c\n 19437\1	ex N73700
516	(1)	B57 Canberra		SoC
516	(2)	F16A Fighting Falcon	c\n 61-...	USAF
519		Piper PA-28RT Cherokee	c\n 28R-7635243	
520		XV6A Kestrel		ex XS692\64-18266
				pres. Hampton, Va
521		XV6A Kestrel		ex XS689\64-18263
				preserved NASM
522		LTV-Hiller-Ryan XC142A		USAF 62-5924 preserved
				AFM
530		Bell 204B	c\n 2017	
531		Lockheed XH51N	c\n 1003	
532		Bell OH4A	c\n 004	US Army 62-4204 SoC
533		YCH46C Sea Knight		US Army 58-5514 SoC
535		UH19D Chicasaw		US Army 56-1552 SoC
537		H13H Sioux		US Army 57-6207 SoC
538		SH3A Sea King		ex NASA 933
540		OH58A Kiowa	c\n 41564	US Army 71-20703
541		AH1G Huey Cobra		
542		Bell 204		
544		CH47 Chinook		
545		Sikorsky S-72		
550		F5F Tiger II	c\n W.1001	USAF 73-0889
566		LearJet 25	c\n 25-064	ex N266GL
600		F106A Delta Dart		SoC
605		Grumman G-1159 Gulfstream II	c\n 318	ex N650PF
607	(1)	NF106B Delta Dart		USAF57-2507:N607NM
				wfu Langley
607	(2)	DHC6 Twin Otter	c\n 4	ex NASA508
614		T34 Mentor		
616	(1)	NF106B Delta Dart		USAF 57-2516
616	(2)	LearJet 25	c\n 25-035	ex N33TR
631		Aero Commander		SoC
635		O2A Super Skymaster		
636	(1)	C47D Skytrain		USAF 43-48086
				to NASA638
636	(2)	C47D Skytrain		ex NASA501
636	(3)	OV10A Bronco	c\n 305-1	USMC Bu155390
637	(1)	OV1B Mohawk		US Army 64-14244
637	(2)	OV10A Bronco	c\n 321-125	USAF 68-3799
638		C47D Skytrain		ex NASA636 SoC
650		Grumman G-1159 Gulfstream II	c\n 118	to NASA945
670		EC121K Warning Star		USN Bu145937
				to MASDC 5.71
671		EC121K Warning Star		USN Bu143201
				to MASDC 5.71
672		CH21C Shawnee		USAF 56-4342 SoC
673		CH21C Shawnee		USAF 56-4344 SoC
701	(1)	Gates Learjet 23	c\n 23-049	to NASA933
701	(2)	Beech B200 King Air	c\n BB-1164	

702		XV15A\Bell 301	c\n 0001	
703	(1)	F100C Super Sabre	o\n 214-1	USAF 53-1709
				preserved San Jose
703	(2)	XV15A\Bell 301	c\n 0002	w.o.20.8.92
704	(1)	Bell X14A		ex NASA234
704	(2)	YAV8B Harrier		USMC Bu158934
705	(1)	XV5B Vertifan		USAF 62-4506
				pr. Ft. Rucker, Al
705	(2)	LearJet 24A	c\n 24A-102	ex N365EJ
706	(1)	Hiller UH12E	c\n 2265	
706	(2)	Lockheed ER2		USAF 80-1063
707	(1)	Convair 340	c\n 3	ex NASA19 to N8048X
707	(2)	NC130B Hercules	c\n 3507	USAF 58-0712 ex N929NA
708		Lockheed U2C		USAF 56-6681
				preserved Ames
709	(1)	Lockheed U2C		USAF 56-6682
				preserved Robins
709	(2)	Lockheed ER2		USAF 80-1097
710		CV990A Coronado	c\n 30-10-29	to NASA713
711		CV990A Coronado	c\n 30-10-1	w.o.13.4.73
712		CV990A Coronado	c\n 30-10-37	destroyed 8.85
713		CV990A Coronado	c\n 30-10-29	
714		L.300\NC141A Starlifter	c\n 6110	
715	(1)	Hiller L4		SoC
715	(2)	DHC-5\C8A QSRA Buffalo	c\n 2	USAF 63-13687
716	(1)	C8A Buffalo	c\n 1	USAF 63-13686
716	(2)	AV8A HarrierUSMC Bu158385		
717	(A)	T38A Talon	c\n N.5776	ex NASA915
717	(B)	Douglas DC-8-72	c\n 46082\458	ex N801BN
718	(1)	YO3A Q-Star	c\n 011	USAF 69-18010
718	(2)	YOV10A Bronco	c\n 300-3	USN Bu152881
719	(1)	Cessna 402B	c\n 402B-0313	to NASA503
719	(2)	AV8A Harrier		
730		AH1E Huey Cobra	c\n 22106	US Army 77-22768
731		OH6A Cayuse	c\n 0604	US Army 67-16219
732	(1)	OH1D Iroquois	c\n 428	US Army 62-1908
				to NASA414
732	(2)	OH58A Kiowa	c\n 41564	US Army 71-20703
				returned to US Army
733		UH1H Iroquois	c\n 11519	US Army 69-15231
734		UH1H Iroquois	c\n 4335	US Army 64-13628
735		SH3A Sea King		USN Bu149723
736		AH1G Huey Cobra	c\n 20004	US Army 66-15248
737		CH47B Chinook	c\n B396	US Army 66-19138
				to CH47D as 89-0176
740		Sikorsky S-72	c\n 72-001	
741		Sikorsky S-72	c\n 72-002	
748		Sikorsky UH60A Blackhawk	c\n 70-572	US Army 82-23749
801		Aero Commander 680F	c\n 1288-131	ex N6297X
802		F6A Skyray		USN Bu139208
				pres. Sidney, Oh
802		F8C Crusader		USN Bu141350
803	(1)	Northrop M2-F3 lifting body	c\n NLB-101	preserved NASM
803	(2)	Pik-20		
804	(1)	Northrop HL10	c\n NLB-102	
804	(2)	Mooney M20		
805		Ames AD-1		
808		Piper PA30-160B Twin Comanche	c\n 30-1498	
809		B57B Canberra		ex NASA237
810	(1)	F8A Crusader		USN Bu141353
810	(2)	Schweizer SGS1-36		
811		F104N Starfighter	c\n 4045	
812		F104N Starfighter	c\n 4053	stored
813		F104N Starfighter	c\n 4058	w.o.8.6.66
814		Lockheed 1329 Jetstar I	c\n 5003	ex NASA14
815		Lockheed T33A Shooting Star	c\n 9795	USAF 55-4351 sold
816	(1)	NTF8A Crusader		USN Bu143710 dumped
816	(2)	F106B Delta Dart		USAF 57-2516
817		C47H Skytrain		USN Bu17136
818		YF104A Starfighter	c\n 183-1007	USAF 55-2961
				preserved NASM
819		F104B Starfighter	c\n 283-5015	USAF 57-1303
820		NF104A Starfighter	c\n 183-1078	USAF 56-0790
				preserved Dryden
821		T38A Talon	c\n N.5772	ex NASA911

822	(1)	Bell 47G-3B1	c\n 6670	
822	(2)	T38A Talon		
824		TF104G Starfighter		WGAF 2733 stored
825		TF104G Starfighter		WGAF 2808
826		F104G Starfighter		WGAF 2464
827		C47B Skytrain		preserved Fairchild AFB
831		SR71B Blackbird	c\n 2007	USAF 68-17956
833		Boeing 720-027	c\n 18066	destroyed
834		F14A Tomcat		USN 158613
				to NSWC Dahgren, Va
835		F15A Eagle		USAF 71-0287 HIDEC
840		YFA18A Hornet		USMC Bu160780
841		FA18A Hornet		USMC Bu161216
843		FA18A Hornet		USMC Bu161251
844		FA18A Hornet		USMC Bu161213
845		YFA18A Hornet		USMC Bu160781
846		FA18A Hornet		USMC Bu161355
848		F16XL Fighting Falcon	c\n 61-3	USAF 75-0747
849		F16XL Fighting Falcon	c\n 61-5	USAF 75-0749
901		T38A Talon	c\n N.5960	USAF 66-8381
				w.o.28.2.71
902		T38A Talon	c\n N.5540	USAF 63-8193
903		T38A Talon	c\n N.5547	USAF 63-8200
904		T38A Talon	c\n N.5551	USAF 63-8204
905	(1)	T38A Talon	c\n N.5278	USAF 61-0912 SoC
905	(2)	Boeing 747-123	c\n 20107\86	based at Dryden
906		T38A Talon	c\n N.5745	USAF 64-10326
907		T38A Talon	c\n N.5746	USAF 64-10327
908		T38A Talon	c\n N.5747	USAF 64-10328
				to\ex NASA514
909		T38A Talon	c\n N.5770	USAF 65-10351
910		T38A Talon	c\n N.5771	USAF 65-10352
911	(1)	T38A Talon	c\n N.5772	USAF 65-10353
				to NASA821
911	(2)	T38A Talon	c\n N.5748	USAF 65-10329
911	(3)	Boeing 747-SR46		
912		T38A Talon	c\n N.5773	USAF 65-10354
913		T38A Talon	c\n N.5774	USAF 65-10355
914		T38A Talon	c\n N.5775	USAF 65-10356
915		T38A Talon	c\n N.5776	USAF 65-10357
				to NASA717
916		T38A Talon	c\n N.5961	USAF 66-8382
917		T38A Talon	c\n N.5962	USAF 66-8383
918		T38A Talon	c\n N.5963	USAF 66-8384
919		T38A Talon	c\n N.5964	USAF 66-8385
920		T38A Talon	c\n N.5965	USAF 66-8386:N920NS
921		T38A Talon	c\n N.5966	USAF 66-8387:N921NS
923		T38A Talon	c\n N.5934	USAF 66-8355
924		T38A Talon	c\n T.6027	USAF 67-14825
925		WB57F Canberra		USAF 63-13501
				Pima Museum
926	(1)	Convair 240	c\n 149	SoC
926	(2)	WB57F Canberra		USAF 63-13503
927		NP3A Orion	c\n 1003	ex NASA428
928	(1)	C133A Cargomaster		USAF 54-0136 to USAF
928	(2)	WB57F Canberra		
929		NC130B Hercules	c\n 3507	USAF 58-0712 to NASA707
930		KC135A Stratotanker	c\n 17969	USAF 59-1481\N98
933	(1)	SH3A Sea King		USN Bu149723
				to NASA538
933	(2)	LearJet 23	c\n 23-049	ex NASA701
934		T33A Shooting Star	c\n 1450	USAF 57-0721 SoC
935		T33A Shooting Star	c\n 1451	USAF 57-0722 SoC
936		T33A Shooting Star	c\n 1720	USAF 58-0671 MASDC
937		T33A Shooting Star	c\n 1155	USAF 56-3671 MASDC
938		T33A Shooting Star	c\n 9830	USAF 55-4386 SoC
939		T33A Shooting Star	c\n 1173	USAF 56-3689 SoC
940	(1)	T33A Shooting Star	c\n 9732	USAF 53-6011 MASDC
940	(2)	Boeing 377 Super Guppy	c\n 15938	N940NS stored AMARC
941		T33A Shooting Star	c\n 1150	USAF 56-3666 MASDC
942		T33A Shooting Star	c\n 7303	USAF 52-9338 MASDC
943		T33A Shooting Star	c\n 1449	USAF 57-0720 SoC
944		Grumman G-1159 Gulfstream II	c\n 144	ex HB-ITR
945	(1)	T33A Shooting Star	c\n 8738	USAF 53-5400
945	(2)	Grumman G-1159 Gulfstream II	c\n 118	ex NASA650

946		Grumman G-1159 Gulfstream II	c\n 146	ex N897GA
947	(1)	Bell 47G-3B	c\n 6665	SoC
947	(2)	Grumman G-1159 Gulfstream II	c\n 147	ex N898GA
948		Grumman G-1159 Gulfstream II	c\n 222	ex N5253A
949		Bell 47G-3B2	c\n 6754	
950		Bell 206A Jet Ranger	c\n 508	ex NASA12:N950NS
952		Bell LLTV		preserved Ft. Rucker, Al
955		T38A Talon	c\n T.6232	USAF 69-7082
956		T38A Talon	c\n T.6234	USAF 69-7084
957		T38A Talon	c\n T.6236	USAF 69-7086
				w.o.11.5.72
958		T38A Talon	c\n T.6238	USAF 69-7088
959		T38A Talon	c\n T.6240	USAF 70-1550
960		T38A Talon	c\n T.6242	USAF 70-1552
961		T38A Talon	c\n T.6245	USAF 70-1555
962		T38A Talon	c\n T.6246	USAF 70-1556
963		T38A Talon	c\n T.6262	USAF 70-1572
52-0008		NB52B Stratofortress	c\n 16498	Dryden
57-2545		F106B Delta Dart		Ames
82-0976		F16A Fighting Falcon	c\n 61-569	
63-9766		F111A	c\n 1	Dryden
63-9778		NF111A	c\n 13	Dryden
63-9783		YFB111A	c\n 18	
82-0003		X29A		Dryden\AFFTC\DARPA
68-17230		OH6A Cayuse	c\n 1190	Dryden
83-24126		Aero Commander 680W	c\n 1772-10	
142693		A4B Skyhawk		Ames
144904		A4B Skyhawk		Ames
157991		F14A Tomcat		Dryden
160775		FA18A Hornet		Dryden\spares
160777		FA18A Hornet		Dryden
160785		FA18A Hornet		Dryden
161250		FA18A Hornet		Dryden
161494		EC130Q Hercules	c\n 4896	Wallops Island

GOVERNMENT AGENCIES

N1	Grumman G-1159C Gulfstream IV	c\n 1071	ex N410GA
N2	Learjet 31A	c\n 31A-063	ex N27
N3	Grumman G-159 Gulfstream I	c\n 160	
N4	Learjet 31A	c\n 31A-038	ex N5016V
N13	Beech F90 King Air	c\n LA-124	
N14	Beech F90 King Air	c\n LA-131	
N15	Beech F90 King Air	c\n LA-138	
N16	Beech C90 King Air	c\n LJ-893	
N17	Beech C90 King Air	c\n LJ-896	
N18	Beech F90 King Air	c\n LA-145	
N19	Beech C90 King Air	c\n LJ-909	
N20	Beech C90 King Air	c\n LJ-912	
N21	Beech C90 King Air	c\n LJ-902	ex N18\N5
N22	Grumman G-1159C Gulfstream IV	c\n 1042	ex N400GA
N26	Cessna 560 Citation V	c\n 560-0113	ex N6803L
N27	Cessna 560 Citation V	c\n 560-0109	ex N2
N29	McDonnell Douglas DC-9-15	c\n 45732\41	ex N119
N35	Beech 200 King Air	c\n BB-88	ex N79\N4
N37	Bell UH1H Iroquois	c\n 12649	ex 70-16344
N40	Boeing 727-25	c\n 19854\628	ex N8171G
N41	Westwind 1121B	c\n 143	ex N81
N42	Convair 880	c\n 55	ex N112
N42	Westwind 1121B	c\n 142	ex N82, cancx 10.91
N43	Westwind 1121B	c\n 131	ex N83, cancx 9.92
N46	Boeing 727-30	c\n 18360\24	ex N77\N97
N48	Grumman G-159 Gulfstream I	c\n 67	ex N376
N52	Sabre 75A	c\n 380-10	
N53	Sabre 75A	c\n 380-14	
N58	Sabre 75A	c\n 380-24	
N60	Sabre 75A	c\n 380-28	
N62	Sabre 75A	c\n 380-31	
N66	Beech 300 King Air	c\n FF-1	
N67	Beech 300 King Air	c\n FF-2	
N68	Beech 300 King Air	c\n FF-3	
N69	Beech 300 King Air	c\n FF-4	
N70	Beech 300 King Air	c\n FF-5	
N71	Beech 300 King Air	c\n FF-6	
N72	Beech 300 King Air	c\n FF-7	
N73	Beech 300 King Air	c\n FF-8	
N74	Beech 300 King Air	c\n FF-9	
N75	Beech 300 King Air	c\n FF-10	
N76	Beech 300 King Air	c\n FF-11	
N77	Beech 300 King Air	c\n FF-12	
N78	Beech 300 King Air	c\n FF-13	
N79	Beech 300 King Air	c\n FF-14	
N80	Beech 300 King Air	c\n FF-15	
N81	Beech 300 King Air	c\n FF-16	
N82	Beech 300 King Air	c\n FF-17	
N83	Beech 300 King Air	c\n FF-18	
N84	Beech 300 King Air	c\n FF-19	
N86	Sabre 40	c\n 282-86	ex N2255C, cx 9.92
N87	Sabre 40	c\n 282-87	ex N2256B, cx 9.92
N88	Sabre 40	c\n 282-88	ex N2237C
N89	Sabre 40	c\n 282-89	
N94	BAe125-800A\C29A	c\n 258129	ex N269X\88-0269
N95	BAe125-800A\C29A	c\n 258134	ex N271X\88-0270
N96	BAe125-800A\C29A	c\n 258131	ex N270X\88-0271
N97	BAe125-800A\C29A	c\n 258154	ex N272X\88-0272
N98	BAe125-800A\C29A	c\n 258156	ex N273X\88-0273
N99	BAe125-800A\C29A	c\n 258158	ex N274X\88-0274

FEDERAL BUREAU of INVESTIGATION - FBI

N8520L	Beech A90 King Air	c\n LJ-156	ex N22

UNITED STATES BORDER PATROL - USBP

N67BP	Hughes OH-6A Cayuse	ex 65-12986
N97BP	Hughes OH-6A Cayuse	ex 66-17787

N58HP	Bell UH1B Iroquois	c\n 401	ex 62-1881
N59HP	Bell UH1B Iroquois	c\n 423	ex 62-1933
N91LC	Lockheed P3(AEW) Orion	c\n 5409	ex A9-299 to N145CS
N145CS	Lockheed P3(AEW) Orion	c\n 5409	ex N91LC
N147CS	Lockheed P3B Orion	c\n 5162	ex Bu152722
N193BK	MBB Bk-117B-1	c\n 7158	
N283B	Beech C12A Huron	c\n BC-35	ex 76-22558\N7067B
N586RE	Cessna 550 Citation II	c\n 550-0199	ex N67983
N752CC	Cessna 550 Citation II	c\n 550-0018	ex (N3225M)
N753CC	Cessna 550 Citation II	c\n 550-0109	ex N2665N
N797CW	Cessna 550 Citation II	c\n 550-0232	ex N929DS
N1200N	Cessna 550 Citation II	c\n 550-0681	
N1254X	Cessna 550 Citation II	c\n 550-0494	
N1255K	Cessna 550 Citation II	c\n 550-0505	
N1257B	Cessna 550 Citation II	c\n 550-0497	ex (N12549)
N2351K	Cessna 550 Citation II	c\n 550-0594	ex N1302X
N2663Y	Cessna 550 Citation II	c\n 550-0602	ex (N1303T)
N2734K	Cessna 550 Citation II	c\n 550-0595	ex (N13024)
N3262M	Cessna 550 Citation II	c\n 550-0652	
N4614N	Cessna 550 Citation II	c\n 550-0659	
N5056D	GAF N22S Nomad	c\n 164	
N5314J	Cessna 550 Citation II	c\n 550-0663	
N5408G	Cessna 550 Citation II	c\n 550-0666	
N6302W	GAF N22S Nomad	c\n 159	
N6305U	GAF N22S Nomad	c\n 160	
N6328	GAF N22S Nomad	c\n 163	
N6338C	GAF N22S Nomad	c\n 165	
N6507B	Beech 200 King Air	c\n BB-498	ex N23707
N6637G	Cessna 550 Citation II	c\n 550-0670	
N6763L	Cessna 550 Citation II	c\n 550-0673	
N6775C	Cessna 550 Citation II	c\n 550-0677	
N6776T	Cessna 550 Citation II	c\n 550-0680	
N7064B	Beech C12A Huron	c\n BC-2	ex 73-22251
N7068B	Beech C12A Huron	c\n BC-16	ex 73-22265
N7066D	Beech C12A Huron	c\n BC-40	ex 76-22563
N7074G	Beech C12A Huron	c\n BC-17	ex 73-22266
N7166P	Beech 200 King Air	c\n BB-482	ex N6017
N9085U	Piper PA42 Cheyenne IIIA	c\n 42-5501034	
N12549	Cessna 550 Citation II	c\n 550-0501	
N26494	Cessna 550 Citation II	c\n 550-0605	
N26496	Cessna 550 Citation II	c\n 550-0607	
N26621	Cessna 550 Citation II	c\n 550-0593	ex (N1302V)
N37201	Cessna 550 Citation II	c\n 550-0655	
77-22943	Beech C12A Huron	c\n BC-54	
82-0667	Cessna U-26A		
77-22718	UH60A Blackhawk	c\n 70-012	
78-22982	UH60A Blackhawk	c\n 70-045	Davis Monthan AFB
78-23010	UH60A Blackhawk	c\n 70-073	March AFB
79-23297	UH60A Blackhawk	c\n 70-114	Miramar NAS
79-23299	UH60A Blackhawk	c\n 70-116	Davis Monthan AFB
79-23321	UH60A Nlackhawk	c\n 70-138	
79-23350	UH60A Blackhawk	c\n 70-167	
80-23423	UH60A Blackhawk	c\n 70-181	
80-23465	UH60A Blackhawk	c\n 70-223	
81-23576	UH60A Blackhawk	c\n 70-297	Homestead AFB
81-23577	UH60A Blackhawk	c\n 70-298	Jacksonville NAS
82-23670	UH60A Blackhawk	c\n 70-363	Homestead AFB
83-23927	UH60A Blackhawk	c\n 70-752	
86-24548	UH60A Blackhawk		
86-24557	UH60A Blackhawk		
86-24558	UH60A Blackhawk		
87-24641	UH60A Blackhawk		
150514	Lockheed P3A Orion	c\n 5040	Corpus Christi NAS
151390	Lockheed P3A Orion	c\n 5103	Corpus Christi NAS
151395	Lockheed P3A Orion	c\n 5108	Corpus Christi NAS
152170	Lockheed P3A Orion	c\n 5140	Corpus Christi NAS
154575	Lockheed P3(AEW) Orion	c\n	under conversion

UNITED STATES FORESTRY SERVICE - USFS

N104Z	Beech B90 King Air	c\n LJ-42	

N107Z	Beech B200C King Air	c\n BL-124	
N131FF	Lockheed C130A Hercules	c\n 3138	ex 56-0530
N132FF	Lockheed C130A Hercules	c\n 3142	ex 56-0534
N133FF	Lockheed C130A Hercules	c\n 3143	ex 56-0535
N134FF	Lockheed C130A Hercules	c\n 3146	ex 56-0538
N135FF	Lockheed C130A Hercules	c\n 3148	ex 56-0540
N136FF	Lockheed C130A Hercules	c\n 3149	ex 56-0541
N137FF	Lockheed C130A Hercules	c\n 3092	ex 56-0484
N138FF	Lockheed C130A Hercules	c\n 3227	ex 57-0520
N429DF	Bell UH1H Iroquois		ex 69-15145
N436DF	Grumman TS2A Tracker		
N463DF	Cessna O2A Super Skymaster		

UNITED STATES IMIGATION SERVICE - USIS

| N5805 | Fairchild SA226T Merlin IIIB | c\n T-324 | ex N77UU |

UNITED STATES MARSHALS SERVICE - USMS

N9AX	Cessna 500 Citation I	c\n 500-0016	ex N711CR\N7110K
N12AB	Beech 90 King Air	c\n LJ-45	ex HB-GBK
N57MS	Mitsubishi Mu2K	c\n 285	ex N111JE\N11SJ
N111JD	LearJet 23	c\n 23-006	ex N23CH
N113	Boeing 727-30	c\n 18935\234	ex VR-CBA\N18G
N127MS	Sabre 75A	c\n 380-18	ex N55
N131MS	Sabre 75A	c\n 380-22	ex N132MS
N225MS	Beech 200 King Air	c\n BB-496	ex N180S
N2200A	Sabre 75A	c\n 380-26	ex N128MS
N2777	Boeing 727-61	c\n 19176\290	ex N27
N7145V	Lockheed 1329 Jetstar 731	c\n 5001	ex N1\N11
N7148J	Sabre 75A	c\n 380-33	ex N63\N129MS
N8540D	Boeing C97L Stratofreighter	c\n 16984	
N12659	Sabre 75A	c\n 380-16	ex N126MS
N48318	Grumman HU16G Albatross		ex 51-7187
N71460	Sabre 75A	c\n 380-5	ex N233LP
N71543	Sabre 75A	c\n 380-29	ex N132MS

UNITED STATES DEPARTMENT of AGRICULTURE - USDA

N104Z	Beech B90 King Air	c\n LJ-472	
N107Z	Beech 200C King Air	c\n BL-124	
N187Z	Fairchild T26 Merlin IIA	c\n T26-29	ex N22CE
N318W	Beech 200 King Air	c\n BB-402	ex N400QK
N773W	Sabre 75A	c\n 380-20	ex N56
N774W	Sabre 75A	c\n 380-37	ex N65

UNITED STATES DEPARTMENT of COMMERCE\NOAA

N42RF	Lockheed P3C Orion		ex 159773
N43RF	Lockheed P3C Orion		ex 159875
N46RF	Beech C90 King Air	c\n LJ-572	ex N57074\N4PS
N52RF	Cessna 550 Citation II	c\n 550-0021	ex N900LJ
N53RF	Rockwell Commander 690A	c\n 11153	ex N57074
N56RF	Bell UH1H Iroquois	c\n 8935	ex 66-16741
N57RF	Bell UH1H Iroquois	c\n 9431	ex 67-17233
N58RF	Bell UH1H Iroquois	c\n 10992	ex 68-16339
N59RF	Bell UH1C Iroquois	c\n 1421	ex 65-9521

UNITED STATES DEPARTMENT of ENERGY

N10EG	Cessna 550 Citation II	c\n 550-0055	ex (N1466K)\N2JZ
N20EG	Beech A100	King Air	c\n B-179ex N25747
N29AF	McDonnell Douglas DC-9-15F	c\n 45826\79	ex CF-TON
N79SL	McDonnell Douglas DC-9-15F	c\n 47011\102	ex N60AF
N135DE	Learjet 35A	c\n 35A-667	ex N91566
N185XP	Beech B200 King Air	c\n BB-952	ex N1852B
N2748X	Beech B200 King Air	c\n BB-1258	
N6451D	Beech 200 King Air	c\n BB-1009	

N7233R	Beech B200C King Air	c\n BL-69	ex N2811B
N7806M	Fairchild FH227D	c\n 515	
N63791	Beech B200 King Air	c\n BB-1100	

UNITED STATES DEPARTMENT of the INTERIOR

N611	Grumman Commander 900	c\n 15018	
N615	Rockwell Commander 690A	c\n 11317	
N618B	Rockwell Commander 690A	c\n 11232	ex N11NP\N618
N1547A	Rockwell Commander 690A	c\n 11337	ex N81476
N81470	Rockwell Commander 690A	c\n 11335	

UNITED STATES DEPARTMENT of JUSTICE\DRUG ENFORCEMENT AGENCY

N40BP	Sabre 40	c\n 282-40	ex N715MR
N104CA	CASA212 Aviocar	c\n 356	
N119CA	CASA212 Aviocar	c\n 357	
N379CA	CASA212 Aviocar	c\n 379	
N824MB	MBB Bo105S	c\n S-768	
N2709Z	Fairchild SA227AT Merlin IVC	c\n AT-695B	
N45818	Fairchild SA226T Merlin III	c\n T-235	ex XC-UTE

UNITED STATES GOVERNMENT

N6040W	Beech 200 King Air	c\n BB-493

UNITS

Throughout this publication aircraft operating units have been quoted where known for all Air Force, Army, Navy and Marine aircraft. Below can be found a list of these units.

Where abbreviations are used in the main text (e.g.AW = Airlift Wing) the full title is given at the beginning of each group of units lists. An additional list of more general abbreviations and acronyms can be found elswhere.

UNITED STATES AIR FORCE

Some USAF aircraft carry a two-letter code. Originally these were allocated on a squadron basis, so a base wing comprising three squadrons would have three codes allocated. More recently allocation has been on a base WING basis, regardless of the number of component squadrons co-located. In the case of ANG units, codes are allocated on a STATE basis, regardless of the number of ANG squadrons in the state. Since the recent reorganisation of the USAF command structure, codes have been appearing on large bomber and transport aircraft as well as the fighter and trainer types to which they had (generally) previously been confined.

Where a unit operates at wing or group level, the component squadrons are show in parentheses.

Command Structure

AIR COMBAT COMMAND	HQ- Langley AFB, Va.
USAF Air Warfare Center (HQ - Eglin AFB)	85TES, 475WEG.
USAF Weapons & Tactics Center (HQ - Nellis AFB)	ADS, FWS, 57FW, 66RQS.
1st Air Force (HQ - Tyndall AFB)	responsible for four US Air Defense Sectors most aircraft drawn from ANG units plus: 56RQS, 57FS.
8th Air Force (HQ - Barksdale AFB)	2W, 5W, 27FW, 28BW, 96W, 319BW, 384BW, 410BW, 509BW.
9th Air Force (HQ - Shaw AFB)	1FW, 4W, 23W, 31FW, 33FW, 42BW, 347FW, 363FW, 416BW.
12th Air Force (HQ - Davis Monthan AFB)	9W, 55W, 310ALS, 49FW, 92BW, 93BW, 314AW, 355W, 366W, 388W, 552ACW, IAAFA.

AIR MATERIEL COMMAND	HQ- Wright-Patterson AFB
Aeronautical Systems Center (HQ - Wright-Patterson AFB)	4950TW
Air Force Development Test Center (HQ - Eglin AFB)	46TW
Air Force Flight Test Center (Edwards AFB)	412TW, 545TG
Air Logistic Center	Ogden, Oklahoma City, Sacramento, San Antonio, Warner-Robbins.

Aerospace Maintenance and Regeneration Center
(Davis-Monthan AFB)

USAF Museum
(Wright-Patterson AFB)

US Military Training Mission\Embassy Flights
(Dhahran\various)

AIR MOBILITY COMMAND	HQ- Scott AFB

| 15th Air Force
(HQ - Travis AFB) | 19ARW, 22ARW, 43ARW, 305ARW, 380ARW.
60AW, 62AW, 63AW, 375AW, 463AW. |
| 21st Air Force
(HQ - McGuire AFB) | 89AW, 317AW, 436AW, 437AW, 438AW. |

AIR EDUCATION & TRAINING COMMAND

HQ- Randolph AFB

| 2nd Air Force
(HQ - Keesler AFB) | TTC Lackland\37TrW, TTC Sheppard\82TrW. |
| 19th Air Force
Randolph AFB) | 12FTW, 14FTW, 47FTW, 58FTW, 64FTW, 71FTW, 80FTW, (HQ -
97AMW, 325FTW, 542CTW. |

PACIFIC AIR FORCES (PACAF)   HQ- Hickam AFB, Hi.

Direct Reporting   15ABW.

5th Air Force (HQ - Yokota AB, Japan)	18W, 374AW, 432FW.
7th Air Force (HQ - Osan AB, Korea)	8FW, 51W.
11th Air Force (HQ - Elmendorf AFB)	3W, 343W.
13th Air Force (HQ - Andersen AB, Guam)	497FTS, 633ABW.

SPACE COMMAND   HQ- Vandenberg AFB, Ca

| 14th Air Force
(HQ - Vandenberg AFB) | responsible for all ICBM activity
1SW, 37RQS, 47RQS, 54RQS, 76RQS, 744ALF. |
| 20th Air Force
(HQ - F.E. Warren AFB) | |

US AIR FORCES EUROPE (USAFE)   HQ- Ramstein AB, FRG

3rd Air Force (HQ - RAF Mildenhall, UK)	20FW, 48FW, 100ARW.
16th Air Force (HQ - Aviano AB, Italy)	401FW.
17th Air Force (HQ - Sembach AB, FRG)	32FS, 36FW, 52FW, 86W, 435AW.

AIR FORCE ACADEMY   HQ- Colorado Springs, Co

AIR FORCE RESERVE   HQ- Robins AFB

| 4th Air Force
(McClellan AFB) | 64ALS, 71SOS, 95ALS, 96ALS, 301RQS, 304RQS,
349AW, 433AW, 445AW, 446AW, 711SOS, 731ALS,
757ALS. |
| 10th Air Force
(HQ - Bergstrom AFB) | 45FS, 46FTS, 47FS, 63ARS, 72ARS, 74ARS, 77ARS,
78ARS, 79ARS, 302FS, 303FS, 314ARS, 336ARS, |

457FS, 465FS, 466FS, 704FS, 706FS.

22nd Air Force	300ALS, 317ALS, 326ALS, 327ALS, 328ALS, 335ALS,
(HQ - Dobbins AFB)	337ALS, 356ALS, 357ALS, 700ALS, 701ALS, 702ALS,
	707ALS, 709ALS, 732ALS, 756ALS, 758ALS, 815WS.

AIR NATIONAL GUARD   HQ - Washington, DC.

see below for individual ANG units. There is no intermediate command structure within the ANG, except where a unit reports to another regular AF unit.

unit locations, codes and equipment

ABW-Air Base Wing

| 15ABW | Hickam AFB, Hi | C135C |

ACCS-Air Command and Control Squadron

1ACCS	Offutt AFB, Nb		E4B	
7ACCS	Keesler AFB, Mo	KS	EC130E	:reports to 552ACW
9ACCS	Hickam AFB, Hi		EC135J	:reports to 15ABW

ACW-Airbourne Control Wing

| 552ACW | Tinker AFB, Ok | OK | C135E, E3B, E3C | (8ADCS\963\964\965\966ACS) |

ADS-Air Demonstration Squadron

| | Nellis AFB, Nv | F16A, F16B | THE THUNDERBIRDS |

ALF-Airlift Flight

| 744ALW | Peterson AFB, Co | C21A | :Space Command |

ALS-Airlift Squadron

13ALS	Kadena AB, Japan	C12F	:controlled by 18W
17ALS	Elmendorf AFB, Ak	C130H	:reports to 616AG
58ALS	Ramstein AB, FRG	C12F, C20A, C21A, UH1N, T43A	:controlled by 86W
63ALS	Selfridge ANGB, Mi	C130E	:AFRES - reports to 927AG
64ALS	Chicago O'Hare IAP, Il	C130H	:AFRES - reports to 440AW
68ALS	Kelly AFB, Tx	C5A	:AFRES - controlled by 433AW
95ALS	Mitchell Field, Milwaukee, Wi	C130H	:AFRES - reports to 440AW
96ALS	Minneapolis St. Paul IAP, Mn	C130E	:AFRES - reports to 302AW
105ALS	Nashville MAP, Tn	C130H	:ANG - reports to 118AW
109ALS	Minneapolis St. Paul IAP, Mn	C130E	:ANG - reports to 133AW
115ALS	NAS Point Mugu, Ca	C130E	:ANG - reports to 146AW
130ALS	Yaeger AP, Charleston, WV	C130H	:ANG - reports to 130AG
135ALS	Baltimore State AP, Md	C130E	:ANG - reports to 135AG
137ALS	Stewart-Newburgh MAP, NY	C5A	:ANG - reports to 105AG
139ALS	Schenectady County AP, NY	C130H, LC130H, C12J	:ANG - reports to 109AG
142ALS	Greater Wilmington AP, De	C130H	:ANG - reports to 166AG
143ALS	Quonset Point AP, RI	C130E	:ANG - reports to 143AG
144ALS	Kulis ANGB\Anchorage IAP, Ak	C130H	:ANG - reports to 176AG
154ALS	Little Rock ANGB, Ar	C130E	:ANG - reports to 189AG
155ALS	Memphis IAP, Tn	C141B	:ANG - reports to 164AG
156ALS	Douglas MAP,Charlotte, NC	C130B	:ANG - reports to 145AG
158ALS	Savannah MAP, Ga.	C130H	:ANG - reports to 165AG
164ALS	Mansfield Lahm MAP, Oh	C130H	:ANG - reports to 179AG
165ALS	Standiford Field, Ky	C12F,C130H	:ANG - reports to 123AW
167ALS	Martinsburg-WV RAP, WV	C130E	:ANG - reports to 167AG
180ALS	Rosecrans MAP, St. Joseph, Mo	C130H	:ANG - reports to 139AG
181ALS	Dallas NAS, Tx	C130H	:ANG - reports to 136AW
183ALS	Thompson Field, Jackson, Ms	C141B	:ANG - reports to 172AG
185ALS	Oklahoma City, Ok.	C130H	:ANG - reports to 137AW
187ALS	F.E. Warren AFB, Wy	C130B	:ANG - reports to 153AG
200ALS	Buckley ANGB, Co	CT43A	:ANG - reports to 140FW
201ALS	Andrews AFB, Md	C21A, C22B	:ANG - reports to 113FW

303ALS	March AFB, Ca	C130B, C130E	
	:AFRES - reports to 445AW		
310ALS	Howard AFB, Panama, CZ	C21A, C27A, C130H,	
		CT43A	:reports to 24W
327ALS	Willow Grove NAS, Pa	C130E	:AFRES - reports to 913AG
328ALS	Niagara Falls IAP, NY	C130E	:AFRES - reports to 94AW
345ALS	Yokota AB, Japan	C130E	:reports to 374AW
356ALS	Rickenbacker ANGB, Oh	C130E	:AFRES - reports to 907AG
357ALS	Maxwell AFB, Al	C130H	:AFRES - reports to 908AG
457ALS	Andrews AFB, Md	C12F, C21A	:controlled by 89AW
458ALS	Scott AFB, Il	C12F, C21A	:controlled by 375AW
700ALS	Dobbins AFB, Ga	C130H	:AFRES - reports to 94AW
731ALS	Peterson AFB, Co	C130E	:AFRES - reports to 302AW
757ALS	Youngstown MAP, Oh	C130B	:AFRES - reports to 440AW
758ALS	Gtr Pittsburgh IAP, Pa	C130H	:AFRES - reports to 94AW

AMW-Air Mobility Wing

| 97AMW | Altus AFB, Ok | C5B, C141B, T37B | (56\57ALS) |

ARS-Airbourne Refueling Squadron

11ARS	Altus AFB, Ok	KC135R	:reports to 4570G
63ARS	Selfridge ANGB	KC135E	:AFRES
71ARS	Barksdale AFB, La	KC10A	:reports to 4580G
72ARS	Grissom AFB, In	KC135R	:AFRES
74ARS	Grissom AFB, In	KC135R	:AFRES
93ARS	Castle AFB, Ca	KC135R	:reports to 3980G
108ARS	O'Hare IAP, Chicago, Il	KC135E	:ANG
116ARS	Fairchild AFB, Wa	KC135E, C12J	:ANG
117ARS	Forbes ANGB, Topeka, Ks	KC135E	:ANG
126ARS	Mitchell Field, Milwaukee, Wi	KC135R	:ANG
132ARS	Bangor IAP, Me	KC135E	:ANG
133ARS	Pease ANGB, NH	KC135E	:ANG
141ARS	McGuire AFB, NJ	KC135E, C26A	:ANG
145ARS	Rickenbacker ANGB, Oh	KC135R	:ANG
146ARS	Greater Pittsburgh IAP, Pa	KC135E	:ANG
147ARS	Greater Pittsburgh IAP, Pa	KC135E	:ANG
150ARS	McGuire AFB, NJ	KC135E	:ANG
151ARS	McGhee-Tyson AP, Knoxville, Tn	KC135E	:ANG
153ARS	Key Field, Meridian, Ks	KC135R, C26A	:ANG
166ARS	Rickenbacker ANGB, Oh	KC135R, C26A	:ANG
168ARS	Eielson AFB, Ak	KC135D, KC135E, C26B:ANG	
191ARS	Salt Lake City IAP, Ut	KC135E	:ANG
197ARS	Phoenix Sky Harbor IAP, Az	KC135E	:ANG
203ARS	Hickam AFB, Hi	KC135R	:ANG
306ARS	Altus AFB, Ok	KC135R	:reports to 4570G
314ARS	Beale AFB, Ca	KC135E	:AFRES
336ARS	March AFB, Ca	KC135E	:AFRES
351ARS	RAF Mildenhall, UK	KC135R	:reports to 100ARW

ARW-Airbourne Refueling Wing

19ARW	Warner Robins AFB, Ga	KC135R	(99\912ARS)
	McConnell AFB, Ks	KC135R	(384ARS)
22ARW	March AFB, Ca	KC10A, C12F, C21A,	
		KC135A, T38A	(6\9ARS\459ALS)
43ARW	Malstrom AFB, Mt	KC135R, T38A	(91\97ARS)
	Beale AFB, Ca	KC135R	(350ARS)
	Dyess AFB, Tx	KC135R	(917ARS)
	Ellsworth AFB, SD	KC135R	(28ARS)
	Minot AFB, ND	KC135R	(906ARS)
100ARW	RAF Mildenhall, UK	KC135R	(351ARS)
319ARW	Grand Forks AFB, ND	KC135R	(46\905ARS)
380ARW	Plattsburgh AFB, NY	KC135Q, KC135R, T37B	(310\380ARS)
	Loring AFB, Me	KC135R	(42ARS)
	Griffiss AFB, NY	KC135R	(509ARS)

AW-Airlift Wing

60AW	Travis AFB, Ca	C5A, C5B, C141B	(22\21\19\20ALS)
		aircraft also used by 349AW AFRES	
62AW	McChord AFB, Wa	C141B	(4\7\8ALS)
89AW	Andrews AFB, Md	C9C, C12C, C12F, C20B, C20C,	
		C21A, VC25A, C135B, C137C,	

			UH1N	(1HS\1\99\457ALS)
314AW	Little Rock AFB, Ar	LK	C130E	(50\53\61\62ALS)
374AW	Yokota AB, Japan	YJ	C9A, C21A, C130E,	(30\36\459ALS)
			C130H, UH1N	
375AW	Scott AFB, Il		C9A, C12F, C21A	(11ALS\375FTS\458ALS)
433AW	Kelly AFB, Tx		C5A	:AFRES (68ALS)
435AW	Rhein Main AB, FRG		C130E	(37\75\76ALS)
	Ramstein AB, FRG		C9A	(55AAS)
436AW	Dover AFB, De		C5A, C5B	(3\9\31ALS)
			aircraft also used by 512AW AFRES	
437AW	Charleston AFB, SC		C141B	(14\15\16\17ALS)
			aircraft also used by 514AW AFRES	
438AW	McGuire AFB, NJ		C141B	(6\13\18ALS)
439AW	Westover AFB, Ma		C5A	:AFRES (337ALS)
445AW	March AFB, Ca		C130H, C141B	:AFRES (729\730ALS)
				303ALS)
446AW	McChord AFB, Wa		C141B	:AFRES (97\313\728ALS)
	Scott AFB, Il		C9A	(73AAS)
459AW	Andrews AFB, Md		C141B	:AFRES-controls 756ALS

BG-Bomb Group

319BG	Grand Forks AFB, ND	GF	B1B, T38A	(46BS)

BW-Bomb Wing

2BW	Barksdale AFB, La	LA	B52H	(96BS)
5BW	Minot AFB, ND	MT	B52H, T38A	(23BS)
28BW	Ellsworth AFB, SD	EL	B1B, T38A	(37\77BS)
42BW	Loring AFB, Me	LZ	B52G, T37B	(69BS)
92BW	Fairchild AFB, Wa	FC	B52H, T37B, UH1N	(325BS\36RQS)
93BW	Castle AFB, Ca	CA	B52G, T37B	(328BS\329CCTS)
384BW	McConnell AFB, Ks	OZ	B1B, T38A	(28BS)
410BW	K.I. Sawyer AFB, Mi	KI	B52H, T37B	(644BS)
416BW	Griffiss AFB, NY	GR	B52H, T38A	(668BS)
509BW	Whiteman AFB, Mo	WT	B2A	(393BS) forming

CTW-Crew Training Wing

542CTW	Kirtland AFB, NM		MC130H, HC130N, HC130P,HH1N, HH3E, MH53J.	(550\551FTS)

FS-Fighter Squadron

18FS	Eielson AFB, Ak	AK	A10A	:reports to 343FW
19FS	Osan AB, S. Korea	OS	OV10A	
25FS	Eielson AFB, Al	AK	OV10A	:reports to 343FW
25FS	Suwon AB, S. Korea	SU	A10A	:reports to 51FW
27FS	George AFB, Ca	VV	OV10A	
32FS	Soesterberg AB, Neth.	CR	F15A, F15C, F15D	:reports to 32FG
45FS	Grissom AFB, In	IN	A10A	:AFRES
47FS	Barksdale AFB, La	BD	A10A	:AFRES
48FS	Langley AFB, Va	LY	F15A, F15B	
57FS	Keflavik AB, Iceland	IS	F15C, F15D	:reports to 35W
89FS	Wright-Patterson AFB, Oh	DO	F16A, F16B	:AFRES
93FS	MacDill AFB, Fl	FM	F16A, F16B	:AFRES
101FS	Otis ANGB, Ma		C12J, F15A, F15B	:ANG
det.1	Bangor AP, Me		F15A, F15B	:ANG
103FS	NAS Willow Grove, Pa	PA	A10A, OA10A, C26A	:ANG
104FS	Baltimore State AP, Md	MD	A10A, OA10A	:ANG
107FS	Selfridge ANGB, Mi	MI	C26B, F16A, F16B	:ANG
110FS	St. Louis-Lambert Fd, Mi SL		C12F, F15A, F15B	:ANG
111FS	Ellington ANGB, Tx		C26A, F16A, F16B	:ANG
det.1	Holloman AFB, NM		F16A, F16B	:ANG
112FS	Toledo Express IAP, Oh	OH	F16C, F16D	:ANG
113FS	Hulman Fd Terre Haute In	HF	F16C, F16D	:ANG
114FS	Kingsley Field AP, Or		F16A, F16B	:ANG
118FS	Bradley ANGB, Ct	CT	A10A, C12F	:ANG
119FS	Atlantic City IAP, NJ	NJ	F16A, F16B	:ANG
120FS	Buckley ANGB, Co	CO	C26B, F16A, F16B	:ANG
121FS	Andrews AFB, Md	DC	F16A, F16B	:ANG
122FS	NAS New Orleans, La		F15A, F15B, C130H	:ANG
123FS	Portland IAP, Or		C26A, F15A, F15B	:ANG
det.1	McChord AFB, Wa		F15A, F15B	:ANG
124FS	Des Moines, IAP, Ia	IA	C12J, F16C, F16D	:ANG
125FS	Tulsa, IAP, Ok	OK	A7D, A7K, C26B	:ANG

127FS	McConnell AFB, Ks		F16C, F16D	:ANG
128FS	Dobbins AFB, Ga	GA	F15A, F15B	:ANG
131FS	Barnes MAP,Westfield, Ma	MA	A10A	:ANG
134FS	Burlington IAP, Vt	VT	F16A, F16B, C26B	:ANG
det.1	Langley AFB, Va	VT	F16A, F16B	:ANG
136FS	Niagara Falls IAP, NY		F16A, F16B	:ANG
det.1	Charleston AFB, SC		F16A, F16B	:ANG
138FS	Hancock Field, Syracuse	NY	F16A, F16B	:ANG
148FS	Tucson IAP, Az	AZ	F16A, F16B (USAF)	:ANG
			F16A, F16B (KLu)	
149FS	Byrd IAP, Richmond, Va	VA	C26A, F16C, F16D	:ANG
152FS	Tucson IAP, Az	AZ	C26A, F16A, F16B	:ANG
157FS	McEntire ANGB, SC	SC	C130H, F16A, F16B	:ANG
159FS	Jacksonville IAP, Fl	FL	C130H, F16A, F16B	:ANG
160FS	Dannelly Fd Montgomery Al	AL	C130H, F16A, F16B	:ANG
161FS	McConnell AFB, Ks		C12J, F16A, F16B	:ANG
162FS	Springfield-Beckley MAP	OH	F16C, F16D	:ANG
163FS	Baer MAP, Fort Wayne, In	FW	C26A, F16C, F16D	:ANG
169FS	Greater Peoria AP, Il	IL	C26A, F16A, F16B	:ANG
170FS	Capitol AP Springfield IlSl		F16A, F16B	:ANG
171FS	Selfridge ANGB, Mi		C26A, F16A, F16B	:ANG
det.1	Seymour Johnson AFB, NC		F16A, F16B	:ANG
172FS	Battle Creek ANGB, Mi	BC	A10A, OA10A	:ANG
174FS	Sioux City MAP, Ia	HA	F16C, F16D	:ANG
175FS	Foss Fd, Sioux Falls, SD	SD	C12F, F16C, F16D	:ANG
176FS	Truax Field, Madison, Wi	WI	C26B, F16C, F16D	:ANG
177FS	McConnell AFB, Ks		F16A, F16B	:ANG
178FS	Hector Field, Fargo, ND		C26B, F16A, F16B	:ANG
179FS	Duluth IAP, Mn		C26B, F16A, F16B	:ANG
det.1	Tyndall AFB, Fl		F16A, F16B	:ANG
182FS	Kelly AFB, Tx	SA	F16A, F16B	:ANG
184FS	Fort Smith MAP, Ar	FS	C12F, F16A, F16B	:ANG
186FS	Great Falls IAP, Mt		C26B, F16A, F16B	:ANG
det.1	Davis Monthan AFB, Az		F16A, F16B	:ANG
188FS	Kirtland AFB, NM	NM	C26B, F16C, F16D	:ANG
190FS	Gowen Field, Boise AT Id		C26A, RF4C, F4G	:ANG
194FS	Fresno Air Terminal, Ca		C26A, F16A, F16B	:ANG
det.1	March AFB, Ca		F16A, F16B	:ANG
195FS	Tucson IAP, Az	AZ	F16A, F16B	:ANG
198FS	San Juan IAP, P.R	PR	C26A, F16A, F16B	:ANG
199FS	Hickam AFB, Hi		C130H, F15A, F15B	:ANG
302FS	Luke AFB, Az	LR	F16C, F16D	:AFRES
303FS	Richards Gebaur AFB, Mi	KC	A10A, OA10A	:AFRES
457FS	Carswell AFB, Tx	TH	F16A, F16B	:AFRES
465FS	Tinker AFB, Ok	SH	F16A, F16B	:AFRES
466FS	Hill AFB, Ut	HI	F16A, F16B	:AFRES
704FS	Bergstrom AFB, Tx	TX	F16A, F16B	:AFRES
706FS	NAS New Orleans, La	NO	F16A, F16B	:AFRES

FTS-Flying Training Squadron

375FTS	Scott AFB, Il		C12F, C21A	:controlled by 375AW
557FTS	Colorado Springs Academy, Co		T41A, T41B, T41C, UV18B and various gliders\motor gliders	:Air Force Academy

FTW-Flying Training Wing

12FTW	Randolph AFB, Tx	RA	C21A, T1A, T37B, T38A, AT38B, CT43A.	(332ALF\99\558\ 559\560FTS)
14FTW	Columbus AFB, Mi	CB	T37B, T38A, AT38B	(37\50FTS)
47FTW	Laughlin AFB, Tx		T1A, T37B, T38A	(85\86\87FTS)
58FTW	Luke AFB, Az	LF	F15E, F16C, F16D	(63\310\311\314\425 461\555FS)
64FTW	Reese AFB, Tx	LB	T1A, T37B, T38A	(35\52\54FTS)
71FTW	Vance AFB, Ok	VN	T37B, T38A	(8\25FTS)
80FTW	Sheppard AFB, Tx	EN	T37B, T38A, AT38B	(88\89\90FTS)
325FTW	Tyndall AFB, Fl	TY	F15A, F15B, F15C, F15D	(1\2\95FS)

FW-Fighter Wing

1FW	Langley AFB, Va	FF	F15C, F15D, UH1N, HH60G, HC130P, C21A	(27\71\94FS\72HS\ 41\71RQS)
8TFW	Kunsan AB, Korea	WP	F16C, F16D	(35\80FS)

20FW	RAF Upper Heyford, UK	UH	F111E	(55FS)
27FW	Cannon AFB, NM	CC	EF111A, F111E, F111F	(430ECS\428\522\ 523\524FS)
31FW	Aviano AB, Italy		F16C, F16D	forming
33FW	Eglin AFB, Fl	EG	F15C, F15D	(58\59\60FS)
48FW	RAF Lakenheath, UK	LN	F15E	(492FS)
49FW	Holloman AFB, NM	HO	F4E, F117A, HH60G, AT38B	(7\8\9\415\416\ 417\435FS\48RQS)
51FW	Osan AB, Korea	OS	OA10A, C12F, F16C, F16D, HH60G	(25\36FS\55ALF\ 38RQS)
52FW	Spangdahlem AB, FRG	SP	A10A, OA10A, F4G, F15C, F16C, F16D	(22\23\81\510FS)
57FW	Nellis AFB, Nv	WA	A10A, F15C, F15D, F16C, F16D, F111F	(422TES)
347FW	Moody AFB, Ga	MY	F16C, F16D	(68\69\307\308FS)
354FW	Eielson AFB, Ak	AK	OA10A, F16C, F16D	(18\353FS)
363FW	Shaw AFB, SC	SW	F16C, F16D	(17\19\21\309FS)
388FW	Hill AFB, Ut	HL	F16C, F16D	(4\34\421FS)
432FW	Misawa AB, Japan	MJ	F16C, F16D, HH60G	(13\14FS\39RQS)

OG-Operating Group

1530G	Fairchild AFB, Wa	KC135R	(43\92ARS)
4570G	Altus AFB, Ok	KC135R	(11\306ARS)
4580G	Barksdale AFB, La	KC10A	(2\32\71ARS)

OTS-Officer Training School

	Hondo MAP, Tx	T41A

RQS-Rescue Squadron

33RQS	Kadena AB, Japan		HH3E	:reports to 18W
36RQS	Fairchild AFB, Wa		UH1N	:reports to 92BW
37RQS	F.E. Warren AFB, Wy. det.10		UH1N	:reports to 90MW
38RQS	Osan AB, S. Korea		HH60G	:reports to 51W
39RQS	Misawa AB, Japan		HH60G	:reports to 432FW
41RQS	Patrick AFB, Fl		HH60G	:reports to 1FW
47RQS	Whiteman AFB, Mo		HH1H	:reports to 351MW
48RQS	Holloman AFB, NM		HH60G	:reports to 49FW
54RQS	Minot AFB, ND		HH1H	:reports to 91MW
56RQS	NAS Keflavik, Iceland		HH3E	:reports to 35W
66RQS	Nellis AFB,Nv		MH60G	:reports to 570G
71RQS	Patrick AFB, Fl		HC130P	:reports to 1FW
76RQS	Vandenberg AFB, Ca		UH1N	:reports to 30SW
79RQS	Grand Forks AFB, ND		HH1H	:reports to 321MW
102RQS	Suffolk County Airport, NY		HC130P, HH60G	:ANG
129RQS	NAS Moffet Field, Ca		HC130P, HH60G	:ANG
210RQS	Kulis ANGB\Anchorage IAP, Ak		HC130N, MH60G	:ANG
301RQS	Patrick AFB, Fl		HC130N, HC130P,HH60G	:AFRES
304RQS	Portland Airport, Or	PD	HC130P, HH60G	:AFRES

RS-Reconnaissance Squadron

106RS	Smith ANGB\B'ham MAP	BH	RF4C	:ANG
173RS	Lincoln MAP, Nb.		C12F, RF4C	:ANG
192RS	May ANGB\Reno IAP, Nv		C12J, RF4C	:ANG
196RS	March AFB, Ca.	CA	RF4C	:ANG

RW-Reconnaissance Wing

9RW	Beale AFB, Ca	BB	T38A, U2RT,U2R	(1\99RS)
55RW	Offutt AFB, Ne	OF	NKC135A, EC135C, C135E, KC135E, RC135M, RC135U, RC135V, RC135W, TC135W, WC135B, C21A	(2ACCS\24\38\343RS\ 55WRS\11ALF)

SOG-Special Operations Group

193SOG	Harrisburg IAP, Pa	nil	:controls 193SOS
352SOG	RAF Alconbury, UK	nil	:controls 7\21\67SOS

353SOG	Kadena AB, Japan		nil	:controls 1\17\31SOS

SOS-Special Operations Squadron

1SOS	Kadena AB, Japan		MC130E	:reports to 353SOG
7SOS	RAF Alconbury, UK		MC130H	:reports to 352SOG
8SOS	Hurlburt Field, Fl		MC130E, AC130H	:reports to 16SOW
9SOS	Eglin AFB, Fl		HC130N, HC130P	:reports to 16SOW
15SOS	Hurlburt Field, Fl		MC130H	:reports to 16SOW
16SOS	Hurlburt Field, Fl		AC130H	:reports to 16SOW
17SOS	Kadena AB, Japan		HC130N, HC130P	:reports to 353SOG
20SOS	Hurlburt Field, Fl		MH53J	:reports to 16SOW
21SOS	RAF Alconbury, UK		MH53J	:reports to 352SOG
31SOS	Osan AB, Korea		MH53J	:reports to 353SOG
55SOS	Eglin AFB, Fl		HH60G	:reports to 16SOW
67SOS	RAF Alconbury, UK		HC130N, HC130P	:reports to 352SOG
71SOS	Davis-Monthan AFB, Az		HH60G	:AFRES
				- reports to 919SOW
193SOS	Harrisburg IAP, Pa	PA	EC130E(RR)	:ANG
				-reports to 193SOG
711SOS	Duke Field, Fl		AC130A, C130A	:AFRES
				- reports to 919SOW

SOW-Special Operations Wing

16SOW	Hurlburt Field, Fl		HH53H	:controls 8\9\15\16\20\55SOS
919SOW	Duke Field, Fl		nil	:AFRES - controls 71\711SOS

SW-Space Wing

1SW	Tinker AFB, Ok		C135E	:operated by 552ACW

TES-Test and Evaluation Squadron

85TES	Eglin AFB, Fl	OT	RF4C, F15A, F15C, F16A, F16C, F16D.
det.6	Nellis AFB, Nv	OT	F4G.

TG-Test Group

545TG	Hill AFB, Ut		C130B, HC130H, NC130H, HH53A, NCH53A.

TW-Test Wing

46TW	Eglin AFB, Fl	ET	A10A, F4D, F4E, F15A, F15B, F15C, F15D, F16A, F16C, F111E, HH1N, T38A.
412TW	Edwards AFB, Ca	ED	OA37B, B1B, B2A, B52G, B52H, C17A, MC130H, AC130U, F15A, F15D, F15E, YF16A, F16A, YF16B, F16C, F16D, YF22A, HH1H, HH53C, NT33A, T38A, YT46A,U6A.
4950TW	Wright Patterson AFB, Oh.		EC18B, C21A, NC135A, NKC135A, C135C, C135E, EC135E, NKC135E, NC141A, NT39A.

W-Wing

3W	Elmendorf AFB, Ak	AK	C12F, C130H, F15C, F15D, F15E.	(517ALS\43\54\90FS)
4W	Seymour-Johnson AFB, NC	SJ	F15E, KC10A, T38A	(334\335\336FS\ 344\911ARS)
7W	Dyess AFB, Tx	DY	B1B, C130H, KC135A, KC135Q, T38A	(9BS\337BS\39\ 40ALS)
18W	Kadena AB, Japan	ZZ	C12F, F15C, F15D, KC135R, HH3E.	(13ALS\12\44\67FS\ 909ARS\33RQS)
23W	Pope AFB, NC	FT	A10A, OA10A, C130E, F16C, F16D	
86W	Ramstein AB, FRG	RS	C12F, C20A, C21A, CT43A, F16C, F16D	(2\41ALS\74\75FS) (74\75\76ALS\ 512\526FS)
355W	Davis-Monthan AFB, Az	DM	EC130H, OA10A, A10A	(41\43ECS\333\354\357\358FS)
366W	Mountain Home AFB, Id	MO	B52G, KC135A, F15C, F15D, F15E, F16C, F16D, EF111A	(34BS\22ARS\390\391 \389\429ECS)

WEG-Weapons Evaluation Group

475WEG	Tyndall AFB, Fl	E9A, F100D, QF100D, YQF100D, F100F,QF100F,	QF106A,
QF106B			
det.6	Holloman AFB, NM	QF106A, QF106B	

UNITED STATES ARMY

Airborne Division - AbDiv

101AbDiv	Ft. Campbell, Ky	various

Area Exploitation Batalion - AEB

A\151AEB	Dobbins AFB, Ga	OV1D GA NG

Armored Cavalry Regiment - ACR

prefix shows Batallion

4-2ACR	Feucht, FRG	(UE 26xAH1F,27xOH58C,17xUH60A,3xEH60C)	
	VII Corps		
3ACR	Ft. Bliss, Tx	AH1S	
4-11ACR	Fulda, FRG	(UE 26xAH64A,27xOH58C,3xEH60C,17xUH60A)	
	V Corps		
4-107ACR	Akron\Canton AP, Oh	UH1H,OH58A	OH NG
163ACR	Salt Lake City MAP2, Ut	C7A,C12D,AH1S,UH1H	UT NG
278ACR	Robinson AAF, Ar	various	

Armored Cavalry Squadron - ACS

4\7ACS	Co.A Camp Gary Owen, RoK	
	Co.B Camp Stanton, RoK	
	Co.C Camp Stanton, RoK	
	Co.D Camp Stanton, RoK	
1\17ACS	Ft. Bragg, NC	UH60A
117ACS	Picatinny, NJ	AH1S
124ACS	Austin, Tx	AH1G
126ACS	Quonsett Point, RI	OH6A

Armored Division - AD

1AD	4Bde Ansbach, FRG	(HQ Unit)	VII Corps
2AD	3Bde Lemwerder, FRG	(UE 2xUH1H,4xOH58A) III Corps	
3AD	4Bde Hanau, FRG	(HQ unit)	V Corps

Army Command - ARCOM

63ARCOM	Los Alamitos AFRC, Ca	UH1H,U21A	Reserve
77ARCOM	Newburgh\Stewart AP, NY	UH1H	
79ARCOM	NAS Willow Grove, Pa	UH1H,OH58A,B65	
81ARCOM	Dobbins AFB, Marietta, Ga	UH1H	
83ARCOM	Port Columbus, Oh	UH1H	
86ARCOM	NAS Glenview, Il	UH1H,OH58A,U8F,B65,B80 Reserve	
87ARCOM	Olathe, Ks	UH1H	
88ARCOM	St. Paul, Mn	UH1H	
89ARCOM	Olathe, Ks	UH1H	
90ARCOM	San Antonio, Tx	AH1G	
94ARCOM	Ft. Devens, Md	UH1H	
96ARCOM	Salt Lake City IAP, Ut	UH1H,U21D	
97ARCOM	Ft. Meade, Md	UH1H	
99ARCOM	Washington, Pa	UH1H	
102ARCOM	Scott AFB, Il	UH1H	
120ARCOM	Columbia, SC	UH1H	
121ARCOM	Fell City, Al	UH1H	
122ARCOM	Robinson AAF, Ar	UH1H	
123ARCOM	Indianapolis, In	UH1H	
124ARCOM	Paine Field, Everett, Wa	UH1H,T42A,U21A\D	
125ARCOM	Nashville, Tn	UH1H	

Army Security Agency Company - ASA Co

138ASA Co Orlando, Fl		RU21A
144ASA Co		RV1D
352ASA Co New Iberia		UH1H
524ASA Co Ft. Carson, Co		UH1H

Army Training Center - ATC

7ATC	Grafenwohr, FRG	(UE 3xUH1H)

Arctic Reconnaissance Group - ARG

207ARG	Bryant AAF, Ak	UV18A

Attack Helicopter Batallion - AHB

5AHB	Ft. Pplk\Polk AAF, Ky	UH1	
116AHB	Helena, Mt	UH1H	
1-126AHB	Quonset State AP, RI	AH1S	RI NG 26InfDiv
2-126AHB	Otis ANGB, Ma	UH1H	MA NG 26InfDiv
149AHB	Austin, Tx	AH1S	
3\159AHB	Barstow\Daggett AP, Ca	JUH1H,U21A	
184AHB	Wheeler AFB, Hi	UH60A	
390AHB	Camp Page, RoK	AH1S	
419AHB	Lakeland, Fl	UH1H	

Assault Helo Company - AHC

17AHC	Wheeler AFB, Hi	UH60A

Aviation Company - AvCo

18AvCo	Ft. Bragg, NC	OH58A	
51AvCo	McEntire ANGB, SC	UH1H	
53AvCo	Truax Field, Wi	UH1H	
55AvCo	Camp Page, RoK	UH60A	
56AvCo	Coleman Barracks, FRG	(UE 2xC12C,14xUH1H,?xOH58C,9xU21A)21TAACOM	
76AvCo	R.E. Byrd IAP, Va	OH6A	
104AvCo	Ft. Indiantowngap, Pa	UH1, OH6, CH54	PA NG
120AvCo	Ft. Wainwright, Ak	UH1H	
121AvCo	Ft. Benning, Ga	UH1H	
136AvCo	Dallas NAS, Tx	CH47D	
138AvCo	Orlando IAP, Fl	RU21A,RU21B,RU21C	
147AvCo	Barbers Point, Hi	CH47D	
167AvCo	C Tp\1	UH1M, OH58A	NE NG
179AvCo	Ft. Carson, Co	CH47D	
190AvCo	Olathe, Ks	CH47D Reserve	
198AvCo	Wilmington, De	UH1H, OH6A	
200AvCo	Ft. Bliss, Tx	UH1H	
201AvCo	Camp Page, RoK	UH60A	
207AvCo	Heidelberg, FRG	(UE 7Xc12C\D,10xUH1H,2UH60A)	
		21TAACOM\assigned to USAREUR	
219AvCo	Scott AFB, Il	UH60A	
228AvCo	Wilmington, De	CH47D	
242AvCo	Ft. Wainwright, Ak	CH47D	
275AvCo	Pyong Teak, RoK	UH1H	
281AvCo	Scott AFB, Il	UH1H	
323AvCo	Jackson, Ms	UH1H,OH6A	
327AvCo	Ft. Meade, Md	UH1H	
328AvCo	Trenton, NJ	OH6A	
1898AvCo	Bryant AAF, Ak	UH1H	

Aviation Detachment - AvDet

1AvDet	Stuttgart-Echterdingen, FRG	(UE 2xC12C,4xUH1H)	assigned to USEUCOM
2AvDet	Stewart IAP, Newburgh, NY	(UE 1xC12A,1xUH1H)	assigned to USMA
5AvDet	Maastricht, Netherlands	(UE 3xUH1H)	
6AvDet	Vicenza, Italy	(UE 4xC12C,7xUH1H)	assigned to SETAF
357AvDet	Chievres, Belgium	(UE 1xUH1H,2xUH60A) assigned to SHAPE	

Aviation Regiment - AVN

prefix shows (-) Batallion or (\) Company

2-1AVN	Ansbach, FRG	(UE 13xOH58C,18xAH64A) 1AD	Co.A\B\C

Unit	Location	Equipment	
3-1AVN	Ansbach, FRG	(UE 13xOH58C,18xAH64A) 1AD	Co.A\B\C
A\7-1AVN	Ansbach, FRG	(UE 12xUH60A)	1AD
B\7-1AVN	Ansbach, FRG	(UE 6xUH1H,6xOH58D,3xUH60A) 1AD	
9-1AVN	Ansbach, FRG		
2AVN	Camp Casey, RoK	UH60A	
1-3AVN	Ft. Hood\Hood AAF, Tx	OH58,UH60A,AH64A	2AD
2-3AVN	Giebelstadt, FRG	(UE 3xUH1H,13xOH58C) 3InfDiv	
3-3AVN	Giebelstadt, FRG	(UE 3xUH1H,13xOH58C) 3InfDiv	
4-3AVN	Ft. Hood, Tx	UH1, EH60	2AD
I\3AVN	Giebelstadt, FRG	(UE 2xUH1H)	3InfDiv
4AVN	Ft. Carson, Co	AH1S	
3-4AVN	Mainz-Finthen, FRG	(UE 21xAH1F,3xUH1H,13xOH58C) 8InfDiv	
2-15AVN	Simmons AAF\Ft. Bragg, NC	CH47D	
24AVN	Hunter AAF, Ga	UH60A	
25AVN	Wheeler AFB, Hi	AH1G	
1-25AVN	Wheeler AFB, Hi	AH1, OH58	
H\25AVN	Wheeler AFB, Hi	UH1, AH1, OH58	
A\26AVN	Bradley ANGB	AH1G	
B\26AVN	Otis ANGB	OH6A	
C\26AVN	Craig MAP, Fl	AH1G	
D\26AVN	Craig MAP, Fl	AH1G	
28AVN	Ft. Indiantowngap, Pa	UH1H	
	det. Raleigh-Durham, NC	UH1H	
44AVN	Decatur, Il	UH1H	
45AVN	Camp Humphreys, RoK	UH60A	
A\47AVN	det.1 St. Paul, Mn	UH1H	
A\47AVN	det.2 Boone MAP, Ia	UH1H	
B\47AVN	Davenport, Ia	UH1H	
C\47AVN	det.1 Midway AP, Il	UH1H	
C\47AVN	det.2 Decatur, Il	UH1H	
D\47AVN	Truax Field, Wi	UH1H, AH1S	
E\47AVN	Minneapolis\St. Paul, Mn	UH1H	
53AVN	Wheeler AFB, Hi	UH1, OH58, EH60, UH60	
81AVN	Spokane, Wa	UH1H	
1-82AVN	Simmons AAF\Ft. Bragg, NC	UH60A,AH64A	
2-82AVN	Simmons AAF\Ft. Bragg, NC	UH60A,AH64A	
92AVN	Paine Field AP, Everett, Wa	CH47C Reserve	
96AVN	Ft. Bragg, NC		
1-106AVN	Greater Peoria RAP, Il	CH47D IL NG	
107AVN	Ohio State University AP	AH1S,UH1H,OH58A	OH NG
B\108AVN	Salem\McNary Field, Or	UH1H,OH58A	OR NG
1-111AVN	Craig Municipal AP, Fl	UH1H,OH58A,UH60A,AH64A FL NG	
D\113AVN	Reno\Stead AP, Nv	CH54A,U21A	NV NG
F\126AVN	Otis ANGB	MA NG	
1-130AVN	Raleigh-Durham IAP, NC	OH58A,UH60A\L,AH64A NC NG	
1-131AVN	Birmingham IAP, Al	UH1H,OH6A	AL NG
A\1-131AVN	Montgomery-Dannelly Field	UH1H AL NG	
A\131AVN	Mobile Downtown AP, Al	UH1H AL NG	
B\1-132AVN	Davison AAF\Ft. Belvoir	UH1H DC NG	
1-135AVN	Whiteman AFB, Mo	AH1E,UH1H,OH58A	MO NG
2-135AVN	Buckley ANGB	AH1F,UH1H,OH6A,OH58ACO NG	
D\135AVN	Salina MAP, Ks	(UE 9xUH1H, 6xOH58A)KS NG	
138AVN	Orlando IAP, Fl	RU21A\B	Reserve
1-140AVN	Los Alamitos, Ca	UH1H,OH58A,UH60A,T42A CA NG	
	det. Mather AFB, Ca	UH1H CA NG	
D\140AVN	Buckley ANGB, Co	OH6A CO NG	
3-140AVN	Stockton MAP, Ca	AH1S,UH1H\M,CH47C,	CA NG
B\2-142AVN	Long Is. MacArthur AP, NY	(UE 23xUH1H)	NY NG
F\142AVN	Long Is. MacArthur AP, NY	(UE 2xUH1H)	NY NG
1-149AVN	Ellington ANGB, Tx	OH58A,UH60L,AH64A	TX NG
E\1-149AVN	NAS Dallas, Tx	(UE 5xUH60L)	TX NG
G\1-149AVN	NAS Dallas, Tx	(UE 1xUH1H,11xCH47D)TX NG	
1-150AVN	Trenton\Mercer County AP	UH1H,AH1S,OH6A	NJ NG
A\150AVN	Bradley IAP, Vt	UH1H,OH6A,	VT NG
1-151AVN	McEntire ANGB, SC	OH58A,UH60L,AH64A	SC NG
158AVN	Ft. Campbell, Ky	UH1H	
1-158AVN	NAS Dallas, Tx	UH1H,OH58A	Reserve
2-158AVN	Ft. Hood\Hood AAF, Tx	CH47D Reserve	
2-158AVN	Ft. Sill, Ok	CH47D	
A\3-158AVN	Selfridge ANGB, Mi	UH1H,OH6A	Reserve
5-158AVN	Giebelstadt, FRG	(UE 15xOH58A)	V Corps
5-158AVN	Wiesbaden, FRG	(UE 1xC12C,15xUH1H,15xUH60A,1xU21A) V Corps	
6-158AVN	Paine Field AP, Wa	CH47D Reserve	
C\6-158AVN	Coleman Barracks, FRG	(UE 1xUH1H,16xCH47D)V Corps	

418

```
8-158AVN       Weisbaden, FRG              UH1H,UH60A              V Corps
159AVN         Simmons AAF\Ft. Bragg, NC  CH47D
1-159AVN       Simmons AAF\Ft. Bragg, NC  UH1H
A\5-159AVN     Schwabish Hall, FRG        (UE 1xUH1H,16xCH47D)VII Corps
6-159AVN       Schwabish Hall, FRG        (UE 15xUH60A)           VII Corps
C\6-159AVN     Akrotiri, Cyprus           (UE 4xUH60A)           VII Corps
7-159AVN       Nellingen HP, FRG          (UE 4xUH1H)            VII Corps
HHC\159AVN     Simmons AAF\Ft. Bragg, NC  C12C,U21A
1-167AVN       Lincoln Municipal AP, Ne   AH1P,UH1H,OH58A         NE NG
169AVN         Albany, NY                 OH6A
1-183AVN       Boise Air Terminal, Id     UH1H\M,OH58A           ID NG
185AVN         Keesler AFB, Ms            CH54A  MS NG
1-185AVN       Lemmons MAP, Tupelo, Ms    UH1H,OH58A             MS NG
E\185AVN       Key Field, Meridian, Ms    CH54A  MS NG
192AVN         Birmingham IAP, Al         CH47D  AL NG
E\1-192AVN     Montgomery-Dannelly Field  OH58A  AL NG
193AVN         Wheeler AFB, Hi            UH1,AH1,OH58           HI NG
1-211AVN       Salt Lake City MAP2, Ut    OH58A,UH60A,AH64A      UT NG
B\214AVN       NAS Barbers Point, Hi      CH47
E\214AVN       Wheeler AFB, Hi            UH1
1-214AVN       Los Alamitos AFRC          UH1H,OH58A             Reserve
222AVN         Ft. Wainwright, Ak         UH1H
224AVN         Guthrie, Al                AH1G
1-224AVN       Weide AAF, Md              AH1F,UH1H,OH58A        MD NG
B\2-224AVN     Byrd Field, Richmond IAP,  (UE 15xUH60A)          VA NG
1-227AVN       Ft. Hood\Hood AAF, Tx      AH64A
2-227AVN       Hanau, FRG                 (UE 13xOH58C,3xUH60A,18xAH64A) 3AD
3-227AVN       Hanau, FRG                 (UE 13xOH58C,3xOH60A,18xAH64A) 3AD
I\227AVN       Hanau, FRG                 (UE 2xUH1H)            3AD
B\1-228AVN     Howard AFB, CZ             C12F
D\1-228AVN     Columbia MAP, SC           UH1H,OH58A,QA65        SC NG
B\2-228AVN                                OH58A  IL NG
A\4-228AVN     Camp Pickett\Soto Cano AB  UH60A  Honduras
B\4-228AVN     Camp Pickett\Soto Cano AB  UH1H    Honduras
C\4-228AVN     Camp Pickett\Soto Cano AB  CH47D  Honduras
+\4-228AVN     Camp Pickett\Soto Cano AB  UH60A  Honduras
229AVN         Ft. Campbell, Ky           AH1G
4-229AVN       Illesheim, FRG             (UE 13xOH58C,3xUH60A,18xAH64A)
                                                                 VII Corps
5-229AVN       Ft. Hood\Hood AAF, Tx      OH58A\C
1-238AVN                                  AH1S
A\1-244AVN     N.Orleans\Lakefront AP, La  UH1H   LA NG
G\244AVN       Hunter AAF, Ga             CH54A  GA NG
1-285AVN       Marana\Silver Bell AHP, Az  AH1F,AH1S,OH58A,AH64A AZ NG
355AVN         N. Orleans\Lakefront AP, La UH1H
C\385AVN       Papago AAF, Az             (UE 15xUH1H)           AZ NG
461AVN         Santa Fe MAP               (det.UE 6)            NM NG
477AVN         Ft. Bragg, NC              AH1S
E\502AVN       Aviano AB, Italy           (UE 1xUH1H,16xCH47D) 21TAACOM


Brigade

Berlin         Berlin-Templehof, FRG      (UE 1xU21A,6xUH1H)


Cavalry Air Brigade - CAB

A\3\158CAB     Selfridge ANGB, Mi         UH1    Reserve


Cavalry Regiment - Cav

               prefixed by -Batallion OR \Squadron


1\1Cav         Ansbach, FRG               (UE 8xAH1F,1xUH1H,12xOH58A) 1AD
2\1Cav         Lemwerder, FRG             (UE 4xAH1F,6xOH58C)
3\1Cav         Lafayette Regional AP      AH1S,UH1H\S,OH58A
3Sq\4Cav       Wheeler AFB, Hi
4\4Cav         Schweinfurt, FRG           (UE 8xAH1F,1xUH1H, 12xOH58A) 3InfDiv
2-6Cav         Illesheim, FRG             (UE 13xOH58C,3xUH60A,20xAH64A) VII Corps
3\6Cav         OH58A\C,CH47C
4\6Cav         Ft. Hood, Tx               UH60A,AH64A
5-6Cav                                    (UE 13xOH58C,3xUH60A,20xAH64A) V Corps
6-6Cav         Illesheim, FRG             (UE 13xOH58C,3xUH60A,20xAH64A) VII Corps
7\6Cav         Conroe\Mongomery Cty.AP    (UE 15xAH1S,3xUH1H,12xOH58A) Reserve
6Cav           Ft. Hood, Tx               AH64A
3\7Cav         Coleman Barracks, FRG      (UE 8xAH1F,1xUH1H,12xOH58A) 8InfDiv
```

Unit	Location	Equipment	Notes	Command
4\7Cav	Budingen, FRG	(UE 8xAH1F,1xUH1H,12xOH58C)3AD		
5\9Cav	Wheeler AFB, Hi	(UE 1xUH1,8xAH1,14xOH58)		
1\17Cav	Simmons AAF\Ft. Bragg, NC AH1F			
C\1-101Cav	Long Is. MacArthur AP,NY	(UE 6xOH6A, 4xAH1S) NY NG		
124Cav	Austin\Mueller MAP, Tx	UH1H,AH1S,OH58A\C,UH60L		TX NG
C\1-167Cav	Lincoln MAP, Ne	(UE 8xAH1P)		NE NG
D\1-167Cav	Lincoln MAP, Ne	(UE 12xOH58A)		NE NG

Corps - Cor

Unit	Location	Equipment	Notes
I Cor	Ft. Lewis, Wa	C12C	
V Cor	Weisbaden, FRG	C12, U21	controls 3AD, 8InfDiv
VII Cor	Grafenwohr, FRG	C12C controls 1AD, 3InfDiv	

Divisional Training Group - DivTrg

Unit	Location	Equipment
70DivTrg	Selfridge ANGB, Mi	UH1H
76DivTrg	Ft. Devens, Md	UH1H
78DivTrg	Willow Grove NAS, PA	UH1H
80DivTrg	Ft. Eustis, Va	UH1H
84DivTrg	Mitchell Field, Wi	UH1H
91DivTrg	Hamilton Field, Ca	UH1H
95DivTrg	Norman, Ok	UH1H
98DivTrg	Hancock Field, Syracuse	UH1H
100DivTrg	Bowman Field, Ky	UH1H
104DivTrg	Salem, Or	UH1H
108DivTrg	Columbia, SC	UH1H

Engineer Group - EngGrp

Unit	Location	Equipment	Command
240EngGrp	Bangor IAP, Me	UH1, OH58	ME NG
348EngGrp	Fell City, Al	UH1H	

Infantry Brigade - InfBde

Unit	Location	Equipment	Command
41InfBde	Salem, Or	UH1H	
48InfBde	Macon, Ga		GA NG
53InfBde	Tampa, Fl	UH1H	
197InfBde	Ft. Benning, Ga		
256InfBde	Lafayette, La		LA NG

Infantry Division - InfDiv

Unit	Location	Equipment	Command
1InfDiv	Ft. Riley, Marshall AAF, Ks	AH1F,UH1H,OH58A\C,UH60A,EH60C	
1InfDiv	3Bde Goppingen, FRG	(UE 2xUH1H,4xOH58A)	
3InfDiv	Giebelstadt, FRG	(HQ unit)	VII Corps
5InfDiv	Ft. Polk, La	various	
6InfDiv	Ft. Richardson, Ak	C12	
7InfDiv	Ft. Ord, Ca	U21	
8InfDiv	Bad Kreuznach, FRG	(HQ unit)	V Corps
9InfDiv	Ft. Lewis, Wa	OH58A	

Infantry Group - InfGrp

Unit	Location	Equipment
2067InfGrp	Bryant AAF, Ak	UH60A

Medical Batallion - MedBtn

Unit	Location	Equipment
34MedBtn	Ft. Benning, Ga	UH1H
57MedBtn	Ft. Bragg, NC	UH1H
85MedBtn	Ft. Meade, Md	UH1H
326MedBtn	Ft. Campbell, Ky	UH1H

Medical Command - MEDCOM

Unit	Location	Equipment
7MEDCOM	Darmstadt, FRG	(HQ unit)

Medical Company - MedCo

Unit	Location	Equipment	Command
7MedCo	Landstuhl, FRG	UH60A	
24MedCo	Lincoln MAP, Ne	(UE 12xUH1H)	NE NG
45MedCo	Nellingen, FRG	(UE 15xUH60A)	7MEDCOM
112MedCo	Bangor, Me	UH1 ME NG	
114MedCo	Howard AFB, Panama, CZ	UH1H	
123MedCo	Key Field, Ms	UH1H	

126MedCo	Mather AFB, Ca	UH1V CA NG	
148MedCo	Dobbins AFB, Ga	UH1H\V GA NG	
159MedCo	Wiesbaden, FRG	(UE 15xUH60A)	7MEDCOM
236MedCo	Landstuhl, FRG	(UE 15xUH60A)	7MEDCOM
307MedCo	Port Columbus, Oh	UH1H	
498MedCo	Columbia MAP, SC	UH1H\V Reserve	
507MedCo	Ft. Sam Houston\Kelly AHP	UH1V	
812MedCo	New Orleans Lakefront AP	UH1V LA NG	
det.	Santa Fe MAP, Ks	(UE 6xUH1H)	NM NG
1159MedCo	Concord Municipal AP, NH	UH1H\V NH NG	
1255MedCo	Reno\Stead AP, Nv	UH1H\V NV NG	
1259MedCo	McEntire ANGB	UH1V SC NG	
1267MedCo	Lincoln MAP, Ne (det.1)	(UE 6xUH1H\V)	NE NG

Medical Detachment - MedDet

36MedDet	Ft. Devens, Md	UH60A	
54MedDet	Ft. Lewis\Gray AAF, Wa	UH1H\V, U21A	
57MedDet	Simmons AAF\Ft. Bragg, NC	UH60A	
115MedDet	Reno, Nv	UH1H	
142MedDet	Bismarck, ND	UH1H	
145MedDet	Dobbins AFB, Marietta, Ga	UH1H	
247MedDet	Barstow\Dagget AP, Ca	UH60A	
273MedDet	Conroe\Montgomery Cty.AP	UH1V Reserve	
316MedDet	Elyria AP, Oh	UH1V Reserve	
321MedDet	Salt Lake City IAP, Ut	UH1V	
336MedDet	Newburgh\Stewart IAP, NY	UH1V	
343MedDet	Hamilton Field, Ca	UH1	
346MedDet	Orlando, Fl	UH1H	
347MedDet	N. Perry AP, Miami, Fl	(UE 6xUH1V)	
348MedDet	Orlando IAP, Fl	UH1V	
349MedDet	Orlando, Fl	UH1H	
354MedDet	Columbus, Oh	UH1H	
357MedDet	Ft. Carson, Co	UH1H	
374MedDet	Robinson AAF, Ar	UH1H	
412MedDet	Bowman Field, Ky	UH1H\V 81ARCOM	
427MedDet	Wheeler AFB, Hi	UH1H	
441MedDet	Frankfort, Ky	UH1H	
507MedDet	Ft. Hood, Tx	UH1V	
659MedDet	McEntire ANGB, SC	UH1H	
872MedDet	New Iberia	UH1H	
1133MedDet	Dannally Field, Al	UH1H	
1134MedDet	Dannally Field, Al	UH1H	
1136MedDet	Ellington ANGB, Tx	UH1H	

Military Intelligence Batallion - MIBtn

1MIBtn	Wiesbaden, FRG	(UE 8xRC12K,8xOV1D) V Corps	
3MIBtn	Desiderio AAF\Camp Humphries		
	Pyongtaek, Rok	RC12D,OV1D,RV1D	
15MIBtn	Ft. Hood\Rober Gray AAF	C12C,U21A,RU21H,OV1D	
224MIBtn	Hunter AAF, Ga	RC12D,U21A,RU21H,O\RV1D	
641MIBtn	Salem\McNary Field, Or	OV1D,U21A	OR NG
1042MIBtn	Salem, Or	OV1D	

Military Intelligence Company - MI Co

159MI Co	7Btn Dobbins AFB, Marietta	OV1D

Scouting Batallion - ScBtn

5ScBtn	Bryant AAF, Ak	UH60A

Special Operations Command - SOCOM

ASOCOM	McDill AFB, Fl	C12
1SOCOM	Fort Bragg\Simmons AAF	C12, PA31T

Special Forces Group - SFG   (UE 6 x UH1)

```
1SFG        JFK Center              UH1H
5SFG        JFK Center              UH1H
11SFG       Ft. Meade, Md           UH1H
12SFG       Alrington Heights, Il   UH1H
19SFG       Salt Lake City, Ut      UH1H
20SFG       Birmingham, Al          UH1H
```

Special Operations Aviation Group - SOAG

```
1-245SOAG Tulsa, Ok                 MH6J    OK NG
```

Special Operations Aviation Regiment - SOAR

```
160SOAR     Co.C\D Ft. Campbell, Ky   MH60K
            Co.A\B Ft. Campbell, Ky   MH6J
```

Support Company - SUPCOM

```
2SUPCOM     Nellingen AHP, FRG      (UE 2xUH1H)              VII Corps
3SUPCOM     Wiesbaden, FRG          (UE 2xUH1H)              V Corps
```

Transportation Batallion - TptBtn

```
70TptBtn    HHC Coleman Barracks,   (UE 2xUH1H)              21TAACOM
            Co.B Coleman Barracks,  (UE 3xUH1H)              21TAACOM
```

United States Army Consolidated Training Facility - USACTF

```
6USACTF     Hamilton Field, Ca      C12, U21, UH1
```

```
FORSCOM     Forces Command

HQ Flt Det  Atlanta\Fulton County AP   C12C\L
```

```
LANDSOUTHEAST  Land Forces South Eastern Command
HQ Flt.        (UE 2xC12,2xUH1H)
               Izmir-Cigli, Turkey
```

```
MEDDAC
HQ Flt.     Ft. Hood\Robert Gray AAF   UH1V
```

```
REDCOM      Readyness Command
HQ Flt.     MacDill AFB, Fl         C12
```

```
SETAF       Southern Europe Task Force
            6AvDet assigned         (UE 4xC12C,7xUH1H)
            Vicenza, Italy
```

```
SHAPE       Supreme Headquarters Allied Powers Europe
            357AvDet assigned       (UE 1xUH1H,4xUH60A)
            Chievres, Belgium
```

```
TAACOM      Theater Army Area Command
            HQ unit                 nil
            Coleman Barracks, FRG
```

```
TRADOC      Training and Doctrine Command
            HQ unit                 C12C\D,U21A
            Langley AFB, Va
```

```
USAADTA     United States Army Aviation Development Test Activity
            HQ unit                 Mi-8
```

Fort Worth\Meacham Field, Tx

USAAEFA	United States Army Aviation Engineering Flight Activity	
	HQ unit	AH1F\S,UH1H,JCH47C,CH47D,OH58D,MH60G,UH60A,AH64A
	T34C(USN),U21A	
	Edwards AFB, Ca	
USAICS	United States Army Intelligence Center and School	
	HQ unit	C12C,UH1H,EH60C,O2A(USAF),
	U21A,OV1D	
	Ft. Huachuca\Libby AAF, Az	
USAREUR	United States Army Europe	
	207AvCo assigned	(UE 7xC12C\D,10xUH1H,2xUH60A)
	Heidelberg, FRG	
USEUCOM	Unites States European Command	
	1AvDet assigned	(UE 2xC12C,4xUH1H)
	plus USAF Flight Section	(UE 3xC21A)
	Stuttgart-Etcherdingen, FRG	
USMA	United States Military Academy	
	2AvDet assigned	(UE 1xC12A,1xUH1H)
	(at Newburgh\StewartlAP, NY)	
	West Point, NY	
WSETCOM		
	HQ Flt	C12, G159
	Wheeler AFB,Hi	

ARMY RESERVE

USARASF	United States Army Reserve Aviation Support Facility

Georgia	Dobbins AFB, Ga	
Iowa	Des Moines IAP	
Pennslyvania	NAS Willow Grove	
Texas	NAS Dallas	90th
Washington	Paine Field, Everett, Wa	

NATIONAL GUARD

Aviation Classification Repair Activity Depot - AVCRAD

California	Fresno Air Terminal	C23B,OH58,U8
Connecticut	Groton\New London AP	C23B,AH1S,UH1H,U8
Mississippi	Gulfport-Biloxi Regional AP	C23B
Missouri	Springfield	C23B

Mobilisation AVCRAD Control Element - MACE

Maryland	Weide AAF	C23B

State Area Command - STARC

Alabama	Montgomery	C12F, C23B
Colorado	Buckley ANGB	C26B, T42A
Hawaii	Wheeler AFB	U21A
Louisiana	New Orleans Lakefront AP	C12D
Maryland	Weide AAF	U21D
New Mexico	Santa Fe MAP	C12D
Puerto Rico	San Juan	C23B
South Carolina	McEntire ANGB	C12A
Texas	NAS Dallas	U8F
Utah	West Jordan	
Virginia	Byrd Field, Richmond IAP	C12F

State NG units:-

Alabama	Birmingham Municipal AP	2AASF, 1-131AVN,1-192AVN

	Mobile Downtown AP	A\131AVN
	Dannelly Field AP	C12F,UH1H\V,OH6A
Arizona	Marana\Silver Bell AHP	1-285AVN, WAATS
	Papago AAF	C\385AVN
Arkansas		U21
California	Mather AFB	126MedCo
	Los Alamitos AFRC	1AASF, D\E\F\140AVN
	Stockton Metropolitan AP	3-140AVN
Colorado	Buckley ANGB	2-135AVN
Connecticut	Bradley IAP	UH1H,CH54A,UH60A,U8F
Delaware	Gtr. Wilmington	UH1H,T42A
D. of Columbia	Davison AAF, Va	B\1-132AVN
Florida	Craig Municipal AP	1-111AVN
	Lakeland Municipal AP	UH1, OH58
Georgia	Dobbins AFB	148MedCo, 151AEB
	Hunter AAF	G\244AVN
	Macon	48InfBde
Hawaii	Wheeler AFB	193AVN
	(UE 3xUH1,15xAH1,14xOH58)	
Idaho	Boise Air Terminal	1-183AVN
Illinois	Decatur AP	C12, UH1
	Greater Peoria RAP	1-106AVN, 2-228AVN
Kansas	Olathe\Johnson Cty. Ind. AP	190AvCo
	Ft. Riley \Marshall AAF	OH60A,AH64A
	Salina MAP	D\135AVN
Kentucky	Frankfort\Capital City AP	UH1, OH58, UH60, T42
Louisiana	New Orleans\Lakefront AP	A\1-224AVN, 812MedCo
	Lafayette	256InfBde
Maine	Bangor IAP	240EngGrp
	Bangor IAP	112MedCo
Maryland	Weide AAF	1-224AVN
Massachusetts	Otis ANGB	26InfDiv, 2-126AHB, F\126AVN
Mississippi	Haawkins Field, Jackson	1AASF
	Tupelo MAP	2AASF, 1-185AVN
	Meridian\Key Field AP	3AASF, E\185AVN
	Keesler AFB	185AVN
Missouri	Whiteman AFB	1AASF, 1-135AVN
Montana	Helena RAP	AH1S,UH1H\M,OH58A
Nebraska	Lincoln MAP	C\1-167AVN,D\1-167Cav,24MedCo
Nevada	Reno\Stead AP	D\113AVN, 1255MedCo
New Hampshire	Concord MAP	1159MedCo\U21A
New Jersey	Trenton\Mercer County AP	1-150AVN
New Mexico	Santa Fe MAP	det.461AVN, det.812MedCo
New York	Albany County AP	AH1, OH6
	Long Is. MacArthur AP	C\1-101Cav, B\2-142AVN, F\142AVN
	Rochester\Monroe AP	AH1, OH6
	Syracuse\Hancock Field	6AASF
North Carolina	Raleigh\Durham IAP	1-130AVN
North Dakota		UH1
Ohio	Ohio State University AP	107AVN
	Akron-Canton RAP	4-107ACR
Oklahoma	Tulsa IAP	1-245SOAG
	Norman\Westheimer Airpark Flt.Det. C12D, C310	
Oregon	Salem\McNary Field	B\108AVN, 641MIBtn
Pennsylvania	Ft. Indiantowngap	104AvCo
Puerto Rico	Isla Grande AP	UH1H\M
Rhode Island	Quonsett State AP	1-126AHB\U8F
South Carolina	McEntire ANGB	1-151AVN
	Columbia MAP	D\1-228AVN
South Dakota	Rapid City Regional AP	UH1, OH6, T42, U3, C310
Texas	Austin\Mueller Municipal AP	124Cav\C12
	Dallas NAS	3AASF, E\1-149AVN, G\1-149AVN
	Ellington ANGB	4AASF, 1-149AVN
Utah	Salt Lake City MAP2	163ACR, 1-211AVN
Vermont	Burlington IAP	A\150AVN
Virginia	Byrd Field, Richmond IAP	B\2-224AVN
Washington	Ft. Lewis\Gray AAF	66AvBde
West Virginia		UH1
Wisconsin	Madison\Truax Field	(UE 1xC12D,15xAH1S,3xUH1H, 14xOH58A)

Miscellaneous locations cross reference

Aberdeen ranges	Phillips AAF, Md
Camp Humphries	Desiderio AAF, RoK

Camp Robinson	Robinson AAF, Ar		
Edgewood Arsenal		Weide AAF, Md	MD NG
Fort Belvoir	Davison AAF, Va	DC NG, MDW	
Fort Benning	Ga	197InfBde	
Fort Bliss	Biggs AAF, Tx		
Fort Bragg	Simmons AAF, NC	82AbDiv, 1SOCOM, NC NG	
Fort Campbell	Ky	101AbDiv, 160AVN	
Fort Carson	Co		
Fort Chaffee	Ar		
Fort Devens	Md		
Fort Dix	McGuire AFB, NJ		
Fort Eustis	Felker AAF, Va	ADFTS, ATL, USAALS, USATM,	USATS
Fort Hood	Hood AAF, Tx	1-3AVN, 2-158AVN, 1-227AVN	5-229AVN
Fort Hood	Robert Gray AAF	MEDDAC, 15MIBtn	
Fort Huachuca	Libby AAF, Az	ATSC, USAICS	
Fort Indiantowngap	Muir AAF, Pa	EAATS, PA NG	
Fort Irwin	NTC		
Fort Jackson	SC		
Fort Knox	Godman AAF, Ky		
Fort Lewis	Gray AAF, Wa	66AvBde\WA NG	
Fort Meade	Tipton AAF, Md		
Fort Ord	Fritzche AAF, Ca	7InfDiv	
Fort Pickett	Blackstone AAF, Va		
Fort Polk	Polk AAF, La	5AHB, 507MedCo	
Fort Richardson	Ak	6InfDiv	
Fort Riley	Marshall AAF, Ks	1InfDiv	
Fort Rucker	Dothan, Al	AATC, USAAM	
Fort Rucker	Cairns AAF	AATC	
Fort Rucker	Guthrie AHP		
Fort Rucker	Hanchey AHP	AATC	
Fort Rucker	Knox AHP		
Fort Rucker	Lowe AHP	AATC	
Fort Rucker	Shell AHP	AATC	
Fort Sam Houston	C.L. Kelly AHP	507MedCo	
Fort Sheridan	Il		
Fort Sill	Henry Post AAF, Ok	FAC, 2-158AVN	
Fort Stewart	Wright AAF, Ga	24InfDiv	
Hunter AAF	Ga	224MIBtn, G\244AVN	

ABBREVIATION and ACRONYMS

AAD	Anniston Army Depot	UH1H
AASF	Army Aviation Support Facility	(NG)
AATC	Army Aviation Training Center	various
ADFTS	Aviation Division Flight Test School CH47C,AH58D,OH58D,AH64A	
ALS	Aviation Logistics School	CH47, OH58, UH60, AH64
AML	Aero Medical Laboratory	U21G
ATL	Applied Technology Laboratory	
ATSC	Aviation Training Support Co	OV1
ATTC	Aviation Technical Test Center	UH60A,AH64A
AVIM	Aviation Interim Maintenance	carried out by SUPCOM
CCAD	Corpus Christi Army Depot	UH1H,U21A
EAATS	Eastern ArmyNG Aviation Training Site	UH1,OH6,CH54,OH58,T42,U8,U21
FAC	Field Artillery Center	C12C,UH1H,OH58A
MDW	Military District of Washington	C12C,UH1H,UH60A,U21A
NTC	National Training Center	Barstow-Daggett AP, Ca
USAALS	US Army Aviation Logistics School various	
USAAM	US Army Aviation Museum	various
USAFSA	US Army Field Station Augsburg	(UE 3xUH1H) Gamblingen, FRG
USATM	US Army Transportation Museum	various
USATS	US Army Transportation School	various
WAATS	Western Army NG Aviation Training Site	UH60A

UNITED STATES NAVY

Nearly all USN aircraft, both rotary and fixed wing, carry a two letter code (single letter for training aircraft) followed by a three digit number. The primary purpose of the two letter code is to identify each Carrier Air Wing (CVW) and group together squadrons attached to it. The three digit number acts as an individual aircraft identification (both at Squadron and Wing level, by allocating specific batches of numbers for each carrier role). Permanently shore-based aircraft use a modified version of this, the code sometimes being allocated to a group of squadrons, sometimes to individual units. The single letter training codes are allocated to ALL squadrons in each training WING. It should be noted that if a squadron is reassigned from one CVW to another, although still at the SAME shore base, the two letter code WILL change. Shore bases operating communications aircraft are allocated a number/letter code.

HC-Helicopter Combat Support Squadron

HC-1	North Island NAS, Ca	UP	SH3G, SH3H, CH53E
HC-2	Jacksonville NAS, Fl	HU	SH3G, CH53E
HC-3	North Island NAS, Ca	SA	CH46D, CH46E, HH46A
HC-4	Norfolk NAS, Va	HC	CH53E
HC-6	Norfolk NAS, Va	HW	VH3A, CH46A, HH46A, UH46A, CH46D
HC-8	Norfolk NAS, Va	BR	CH46A, HH46A, UH46A, CH46D, UH46D
HC-11	North Island NAS, Ca	VR	CH46A, HH46A, UH46A, CH46D, CH46E
HC-16	Pensacola NAS, Fl	BF	SH3D, UH1N, HH46A

HCS-

HCS-4	Norfolk NAS, Va	NW	HH60H
HCS-5	Point Mugu NAS, Ca	NW	HH60H

HM-Helicopter Mine Countermeasure Squadron

HM-12	Norfolk NAS, Va	DH	RH53D, MH53E
HM-14	Norfolk NAS, Va	BJ	RH53D
HM-15	Alameda NAS, Ca	BT	MH53E
HM-18	Norfolk NAS, Va	NW	RH53D
HM-19	Alameda NAS, Ca	NW	RH53D

HS-Helicopter Anti-Submarine Warfare Squadron

HS-1	Jacksonville NAS, Fl	AR	SH3D, SH3G, SH3H	
HS-2	North Island NAS, Ca	NG	SH60F	
HS-3	Jacksonville NAS, Fl	AA	SH60F	
HS-4	North Island NAS, Ca	NL	SH60F	
HS-5	Jacksonville NAS, Fl	AG	SH3D, SH3H	
HS-6	North Island NAS, Ca	NH	SH3G, SH3H	
HS-7	Jacksonville NAS, Fl	AC	SH3H	
HS-8	North Island NAS, Ca	NK	SH3H	
HS-10	North Island NAS, Ca	RA	SH60F	
HS-11	Jacksonville NAS, Fl	AB	SH3D, SH3H	
HS-12	North Island NAS, Ca	NE	SH3G, SH3H	
HS-14	North Island NAS, Ca	NE	SH3H	
HS-15	Jacksonville NAS, Fl	AE	SH60F	
HS-75	Jacksonville NAS, Fl	NW	SH3D	:Reserve

HSL-Light Helicopter Anti-Submarine Warfare Squadron

HSL-30	Norfolk NAS, Va	HT	SH2F	
HSL-32	Norfolk NAS, Va	HV	SH2F	
HSL-33	North Island NAS, Ca	TF	SH2G	
HSL-34	Norfolk NAS, Va	HX	SH2F	
HSL-37	Barbers Point NAS, Hi	TH	SH2F, SH60B	
HSL-40	Jacksonville NAS, Fl	HK	SH60B	
HSL-41	North Island NAS, Ca	TS	SH60B	
HSL-42	Mayport NAS, Fl	HN	SH60B	
HSL-43	North Island NAS, Ca	TT	SH60B	
HSL-44	Mayport NAS, Fl	HP	SH60B	
HSL-45	North Island NAS, Ca	TZ	SH60B	
HSL-46	Mayport NAS, Fl	HQ	SH60B	
HSL-47	North Island NAS, Ca	TY	SH60B	
HSL-48	Mayport NAS, Fl	HR	SH60B	
HSL-49	North Island NAS, Ca	TX	SH60B	
HSL-51	Atsugi NAF, Japan	TA	SH60B	
HSL-84	North Island NAS, Ca	NW	SH2F	:Reserve
HSL-94	Willow Grove NAS, Pa	NW	SH2F	:Reserve

HT-Helicopter Training Squadron

HT-8	Whiting Field NAS, Fl	E	TH57B, TH57C	:reports to TW-5
HT-18	Whiting Field NAS, Fl	E	TH57B, TH57C	:reports to TW-5

TW-Training Wing

TW-5 HT-18	Whiting Field NAS, Fl	E	T34C, TH57B, TH57C	:controlsVT-2, VT-3, VT-6, HT-8,

VA-Attack Squadron

VA-34	Oceana NAS, Va	AG	KA6D, A6E
VA-35	Oceana NAS, Va	AA	KA6D, A6E
VA-36	Oceana NAS, Va	AJ	KA6D, A6E
VA-42	Oceana NAS, Va	AD	A6E, TC4C
VA-52	Whidbey Island NAS, Wa	NL	KA6D, A6E
VA-65	Oceana NAS, Va	AJ	KA6D, A6E
VA-75	Oceana NAS, Va	AC	KA6D, A6E
VA-85	Oceana NAS, Va	AB	KA6D, A6E
VA-95	Whidbey Island NAS, Wa	NH	KA6D, A6E
VA-115	Atsugi NAF, Japan	NF	KA6D, A6E
VA-128	Whidbey Island NAS, Wa	NJ	A6B, A6E, TC4C
VA-145	Whidbey Island NAS, Wa	NE	KA6D, A6E
VA-165	Whidbey Island NAS, Wa	NG	KA6D, A6E
VA-196	Whidbey Island NAS, Wa	NK	KA6D, A6E
VA-205	Atlanta NAS, Ga	AF	KA6D, A6E	:Reserve
VA-304	Alameda NAS, Ca	ND	KA6D, A6E	:Reserve

VAK-Tactical Aerial Refueling Squadron

VAK-208	Alameda NAS, Ca	AF	KA6D	:Reserve

VAQ-Tactical Electronic Warfare Squadron

VAQ-33	Key West NAS, Fl	GD	KA3B, TA3B, EKA3B, ERA3B, EA4F, TA4J, EA6A, TA7C, EP3A, EP3J	:controlled by FTRG
VAQ-34	Lemoore NAS, Ca	GD	FA18A, FA18B	:controlled by FTRG
VAQ-35	Whidbey Island NAS, Wa	GD	EA6B	:controlled by FTRG
VAQ-129	Whidbey Island NAS, Wa	NJ	EA6B	
VAQ-130	Whidbey Island NAS, Wa	AC	EA6B	
VAQ-131	Whidbey Island NAS, Wa	NE	EA6B	
VAQ-132	Whidbey Island NAS, Wa	AA	EA6B	
VAQ-134	Whidbey Island NAS, Wa	NL	EA6B	
VAQ-135	Whidbey Island NAS, Wa	NH	EA6B	
VAQ-136	Whidbey Island NAS, Wa	NF	EA6B	
VAQ-137	Whidbey Island NAS, Wa	AB	EA6B	
VAQ-138	Whidbey Island NAS, Wa	NG	EA6B	
VAQ-139	Whidbey Island NAS, Wa	NK	EA6B	
VAQ-140	Whidbey Island NAS, Wa	AG	EA6B	
VAQ-141	Whidbey Island NAS, Wa	AJ	EA6B	
VAQ-209	Norfolk NAS, Va	AF	EA6A	:Reserve
VAQ-309	Whidbey Island NAS, Wa	ND	EA6A	:Reserve

VAW-Airborne Early Warning Squadron

VAW-78	Norfolk NAS, Va	AF	E2C	:Reserve
VAW-88	Miramar NAS, Ca	ND	E2C	:Reserve
VAW-110	Miramar NAS, Ca	NJ	TE2C, E2C, C2A	
VAW-112	Miramar NAS, Ca	NG	E2C	
VAW-113	Miramar NAS, Ca	NK	E2C	
VAW-114	Miramar NAS, Ca	NL	E2C	
VAW-115	Atsugi AB, Japan	NF	E2C	
VAW-116	Miramar NAS, Ca	NE	E2C	
VAW-117	Miramar NAS, Ca	NH	E2C	
VAW-120	Norfolk NAS, Va	AD	E2C, C2A	
VAW-121	Norfolk NAS, Va	AG	E2C	
VAW-122	Norfolk NAS, Va	AE	E2C	
VAW-123	Norfolk NAS, Va	AB	E2C	
VAW-124	Norfolk NAS, Va	AJ	E2C	
VAW-125	Norfolk NAS, Va	AA	E2C	
VAW-126	Norfolk NAS, Va	AC	E2C	

VC-Composite Squadron

VC-8	Roosevelt Roads NAS, PR	GF	A4L, TA4F, TA4J, SH3H	
VC-10	Guantanamo Bay NAS, Cuba	JH		EA4F, TA4J

VF-Fighter Squadron

VF-1	Miramar NAS, Ca	NE	F14D
VF-2	Miramar NAS, Ca	NE	F14D
VF-11	Oceana NAS, Va	NK	F14D +

VF-14	Oceana NAS, Va	AC	F14A	
VF-21	Atsugi, Japan	NF	F14A	
VF-24	Miramar NAS, Ca	NG	F14A	
VF-32	Oceana NAS, Va	AC	F14D +	
VF-33	Oceana NAS, Va	AB	F14A	
VF-41	Oceana NAS, Va	AJ	F14A	
VF-43	Oceana NAS, Va	AD	A4E, A4F, TA4J, F5E, F5F, T2C	
VF-45	Key West NAS, Fl	AD	A4E, TA4F, TA4J, T39D	
VF-51	Miramar NAS, Ca	NL	F14A	
VF-74	Oceana NAS, Va	AA	F14A	
VF-84	Oceana NAS, Va	AJ	F14A	
VF-101	Oceana NAS, Va	AD	F14A	
VF-102	Oceana NAS, Va	AB	F14A	
VF-103	Oceana NAS, Va	AA	F14A	
VF-111	Miramar NAS, Ca	NL	F14A	
VF-124	Miramar NAS, Ca	NJ	F14A	
VF-126	Miramar NAS, Ca	NJ	A4E, A4F, TA4J, F5E, T2C, T38A,F16N	
VF-142	Oceana NAS, Va	AG	F14A	
VF-143	Oceana NAS, Va	AG	F14A	
VF-154	Atsugi, Japan	NF	F14A	
VF-201	Dallas NAS, Tx	AF	F14A	:Reserve
VF-202	Dallas NAS, Tx	AF	F14A	:Reserve
VF-211	Miramar NAS, Ca	NG	F14A	
VF-213	Miramar NAS, Ca	NH	F14A	
VF-301	Miramar NAS, Ca	ND	F14A	:Reserve
VF-302	Miramar NAS, Ca	ND	F14A	:Reserve

VFA-Fighter Attack Squadron

VFA-15	Cecil Field NAS, Fl	AJ	FA18C	
VFA-25	Lemoore NAS, Ca	NK	FA18C	
VFA-81	Cecil Field NAS, Fl	AA	FA18C	
VFA-82	Cecil Field NAS, Fl	AB	FA18C	
VFA-83	Cecil Field NAS, Fl	AA	FA18C	
VFA-86	Cecil Field NAS, Fl	AB	FA18C	
VFA-87	Cecil Field NAS, Fl	AC	FA18C	
VFA-94	Lemoore NAS, Ca	NH	FA18C	
VFA-97	Lemoore NAS, Ca	NL	FA18A, FA18B	
VFA-105	Cecil Field NAS, Fl	AE	FA18C	
VFA-106	Cecil Field NAS, Fl	AD	FA18C	
VFA-113	Lemoore NAS, Ca	NK	FA18A, FA18B	
VFA-125	Lemoore NAS, Ca	NJ	FA18A, FA18B, O2A	
VFA-127	Fallon NAS, Nv	NJ	F5E, F5F, FA18A, FA18B	
VFA-131	Cecil Field NAS, Fl	AG	FA18C	
VFA-136	Cecil Field NAS, Fl	AG	FA18C	
VFA-146	Lemoore NAS, Ca	NG	FA18A, FA18B	
VFA-147	Lemoore NAS, Ca	NG	FA18C	
VFA-151	Lemoore NAS, Ca	NE	FA18A, FA18B	
VFA-192	Atsugi AB, Japan	NF	FA18A, FA18B	
VFA-195	Atsugi AB, Japan	NF	FA18A, FA18B	
VFA-204	New Orleans NAS, La	AF	FA18A, FA18B	
VFA-303	Lemoore NAS, Ca	ND	FA18A, FA18B	:Reserve
VFA-305	Lemoore NAS, Ca	ND	FA18A, FA18B	:Reserve

VFC-

VFC-12	Oceana NAS, Va	AD	FA18A, FA18B	:Reserve
VFC-13	Miramar NAS, Ca	ND	FA18A, FA18B	:Reserve

VP-Patrol Squadron

VP-1	Barbers Point NAS, Hi	YB	P3C
VP-4	Barbers Point NAS, Hi	YD	P3C
VP-5	Jacksonville NAS, Fl	LA	P3C
VP-8	Brunswick NAS, Me	LC	P3C
VP-9	Barbers Point NAS, Hi	PD	P3C
VP-10	Brunswick NAS, Me	LD	P3C
VP-11	Brunswick NAS, Me	LE	P3C
VP-16	Jacksonville NAS, Fl	LF	P3C
VP-17	Barbers Point NAS, Hi	ZE	P3C
VP-22	Barbers Point NAS, Hi	QA	P3C
VP-23	Brunswick NAS, Me	LJ	P3C
VP-24	Jacksonville NAS, Fl	LR	P3C
VP-26	Brunswick NAS, Me	LK	P3C
VP-30	Jacksonville NAS, Fl	LL	P3A, TP3A, UP3A, P3C

VP-40	Brunswick NAS, Me	QE	P3C	
VP-45	Jacksonville NAS, Fl	LN	P3C	
VP-46	Jacksonville NAS, Fl	RC	P3C	
VP-47	Barbers Point NAS, Hi	RD	P3C	
VP-49	Jacksonville NAS, Fl	LP	P3C	
VP-60	Glenview NAS, Il	LS	P3B	:Reserve
VP-62	Jacksonville NAS, Fl	LT	P3C	:Reserve
VP-65	Point Mugu NAS, Ca	PG	P3C	:Reserve
VP-66	Willow Grove NAS, Fl	LV	P3B	:Reserve
VP-68	Patuxtent River NAS, Md	LW	P3C	:Reserve
VP-69	Whidbey Island NAS, Wa	PJ	P3C	:Reserve
VP-91	Moffett Field NAS, Ca	PM	P3C	:Reserve
VP-94	New Orleans NAS, La	LZ	P3B	:Reserve

VPU-Patrol Support Squadron

VPU-1	Brunswick NAS, Me		P3B	
VPU-2	Barbers Point NAS, Hi		P3B, P3C	

VQ-Air Reconnaissance Squadron

VQ-1	Agana NAS, Guam	PR	EP3E, UP3A	
	Misawa NAF, Japan	det	PR	EP3E
VQ-2	Rota AB, Spain	JQ	EP3E	
VQ-3	Barbers Point NAS, Hi	TC	E6A	
VQ-4	Patuxent River NAS, Md	HL	E6A	
VQ-5	Agana NAS, Guam	SS	ES3A	
VQ-6	Cecil Field NAS, Fl	ET	ES3A	

VR-Logistic Support Squadron

VR-46	Atlanta NAS, Ga	JS	C9B	:Reserve
VR-48	Andrews AFB, Md	JR	KC130T	:Reserve
VR-49	Glenview NAS, Il	RV	C9B	:Reserve
VR-52	Willow Grove NAS, Pa	JT	C9B	:Reserve
VR-53	Martinsburg NAS, WV	WV	KC130T	:Reserve
VR-54	New Orleans NAS, La	CW	KC130T	:Reserve
VR-55	Alameda NAS, Ca	RU	C9B	:Reserve
VR-56	Norfolk NAS, Va	JU	C9B	:Reserve
VR-57	North Island NAS, Ca	RX	C9B	:Reserve
VR-58	Jacksonville NAS, Fl	JV	C9B	:Reserve
VR-59	Dallas NAS, Tx	RY	C9B	:Reserve
VR-60	Memphis NAS, Tn	RT	C9B	:Reserve
VR-61	Whidbey Island NAS, Wa	RS	C9B	:Reserve
VR 62	Detroit NAF, Mi	JW	C9B	:Reserve

VRC-Carrier Transport Squadron

VRC-30	North Island NAS, Ca	RW	C2A, CT39E	
VRC-40	Norfolk NAS, Va	JK	CT39E, C2A	

VS-Anti Submarine Warfare Squadron

VS-21	Atsugi NAS, Japan	NF	S3A	
VS-22	Cecil Field NAS, Fl	AC	S3A	
VS-24	Cecil Field NAS, Fl	AJ	S3A	
VS-27	Cecil Field NAS, Fl	AD	S3A, S3B	
VS-29	North Island NAS, Ca	NH	S3A	
VS-30	Cecil Field NAS, Fl	AA	S3B	
VS-31	Cecil Field NAS, Fl	AG	S3A	
VS-32	Cecil Field NAS, Fl	AB	S3A	
VS-33	North Island NAS, Ca	NG	S3A, S3B	
VS-35	North Island NAS, Ca	NK	S3B	
VS-37	North Island NAS, Ca	NL	S3A	
VS-38	North Island NAS, Ca	NE	S3A	
VS-41	North Island NAS, Ca	NJ	S3A	

VT-Training Squadron

VT-2	Whiting Field NAS, Fl	E	T34C	:reports to TW-5
VT-3	Whiting Field NAS, Fl	E	T34C	:reports to TW-5
VT-4	Pensacola NAS, Fl	F	T2C, TA4J	:reports to TW-6
VT-6	Whiting Field NAS, Fl	E	T34C	:reports to TW-5
VT-7	Meridian NAS, Ms	A	TA4J	:reports to CTW-1

VT-10	Pensacola NAS, Fl	F	T2B, T34C, TA4J:reports to TW-6
VT-19	Meridian NAS, Ms	A	T2C	:reports to CTW-1
VT-21	Kingsville NAS, Tx	B	T45A	:reports to TW-2
VT-22	Kingsville NAS, Tx	B	TA4J	:reports to TW-2
VT-23	Kingsville NAS, Tx	B	T2C	:reports to TW-2
VT-27	Corpus Christi NAS, Tx	G	T34C	:reports to TW-4
VT-28	Corpus Christi NAS, Tx	G	T44A	:reports to TW-4
VT-31	Corpus Christi NAS, Tx	G	T44A	:reports to TW-4
VT-86	Pensacola NAS, Fl	F	TA4J, T39N,T47A:reports to TW-6

VX-Air Test and Evaluation Squadron

VX-1	Jacksonville NAS, Fl	JA	SH2F, SH3A, SH3D, SH3H, P3C,SH60B, S3A
VX-4	Point Mugu NAS, Ca	XF	F4S, F14A, F16N, FA18A
VX-5	China Lake NAS, Ca	XE	TA4J, A6E, A7E, A4M, AV8B, AH1J,AH1W, FA18A, UH1N,
OV10A			

VXE-Antarctic Development Squadron

VXE-6	Point Mugu NAS, Ca	XD	LC130F, LC130R, UH1N

VXN-Oceanographic Squadron

VXN-8	Patuxent River NAS, Md	JB	P3B, RP3D

OTHER UNITS

Station Flight aircraft codes:-

Atlanta NAS, Ga	7B	UC12B
Norfolk NAS, Va	7C	UC12M
Dallas NAS, Tx	7D	UC12B
Jacksonville NAS, Fl	7E	UC12B
Brunswick NAS, Me	7F	UC12B
Whidbey Island NAS	7G	UC12B, SH3G
Fallon NAS, Nv	7H	UH1H
Alemeda NAS, Ca	7J	UC12B
Memphis NAS, Tn	7K	UC12B
Point Mugu NAS, Ca	7L	UC12B
North Island NAS, Ca	7M	UC12B
Washington NAF, Md	7N	UC12B
Key West NAS, Fl	7Q	UC12B
Oceana NAS, Va	7R	UC12B
Lemoore NAS, Ca	7S	UC12B
Moffett Field NAS, Ca	7T	UC12B
Cecil Field NAS, Fl	7U	UC12B
Glenview NAS, Il	7V	UC12B
Willow Grove NAS, Pa	7W	UC12B
New Orleans NAS, La	7X	UC12B
Detroit NAS, Mi	7Y	UC12B
South Weymouth NAS, Ma	7Z	UC12B
Atsugi NAS, Japan	8A	UC12B
	8B	
Sigonella NAF, Italy	8C	UC12M
Rota AB, Spain	8D	UC12M
Roosevelt Roads, PR	8E	UC12M
Guantanamo Bay, Cuba	8F	UC12B
RAF Mildenhall, UK	8G	UC12M
Kadena AB, Japan	8H	UC12M
Agana NAS, Japan	8J	UC12M
Bahrain IAP, UAE	8K	UC12B
Misawa AB, Japan	8M	UC12B
El Centro NAS, Ca	8N	UC12B

FTRG   FLEET TACTICAL READINESS GROUP (NS Little Creek, Va)

NAWC   NAVAL AIR WARFARE CENTER (Arlington, Va)

NAWC-AD   Aircraft Division   NAS Patuxent River

Flight Test & Engineering Group-FTEG (controlling)

Strike Aircraft Test Directorate-SATD
Rotary Wing Test Directorate-RWTD

```
                    Test Pilots School-TPS
                    NAS Lakehurst
                    NAS Trenton
NAWC-WD             Weapons Divison          NAWS Point Magu
                    NAWS China Lake
                    White Sands
```

USS   INTREPID- Floating Navy\Marines Museum moored at Pier 62, New York, NY

CVW-Carrier Air Wing

Listed below are the current carrier air wings and their component USN/USMC squadron allocations

CVW-1\AB   USS America\CV-66
             VA-85, VAW-123, VF-33, VF-102, VFA-82, VFA-86, VS-32, HS-11.

CVW-2\NE   USS Ranger\CV-61
             VA-145, VAQ-131, VAW-116, VF-1, VF-2, VS-38, HS-14.

CVW-3\AC   USS John F. Kennedy\CV-67
             VA-75, VAQ-130, VAW-126, VF-14, VF-32, VFA-37, VFA-105, VS-22, HS-7.

CVW-5\NF   Dis-established.

CVW-6\AE   Dis-established.

CVW-7\AG   USS George Washington\CVN-73
             VA-34, VAQ-140, VAW-121, VF-142, VF-143, VFA-131, VFA-136, VS-31, HS-5.

CVW-8\AJ   USS Theodore Roosevelt\CVN-71
             VA-35, VAQ-141, VAW-126, VF-41, VF-84, VFA-15, VFA-87, VS-24, HS-9.

CVW-9\NG   USS Nimitz\CVN-68
              VA-165, VAQ-138, VAW-112, VF-24, VF-211, VFA-146, VFA-147, VS-33, HS-2.

CVW-10\NM   Dis-established.

CVW-11\NH   USS Abraham Lincoln\CVN-72
             VA-95, VAQ-135, VAW-117, VF-213, VFA-22, VFA-94, VMFA-314, VS-21, HS-6.

CVW-13\AK   Dis-established.

CVW-14\NK   USS Carl Vinson\CVN-70
             VA-196, VAQ-139, VAW-113, VF-11, VFA-25, VFA-113, VS-37, HS-8.

CVW-15\NL   USS Kitty Hawk\CV-63
             VA-165, VAQ-134, VAW-112, VF-51, VF-111, VFA-146, VFA-147, VS-33, HS-4.

CVW-17\AA   USS Saratoga\CV-60
             VAQ-137, VAW-125, VF-74, VF-103, VFA-81, VFA-83, VS-30, HS-3.

Unassigned Carriers.

USS Coral Sea	CV-43	
USS Forrestal	AVT-59	Aviation Training Ship
USS Independence	CV-62	SLEP maintenance
USS Constellation	CV-64	SLEP maintenance
USS Enterprise	CVN-65	
USS Dwight D. Eisenhower	CVN-69	
USS Lexington	AVT-16	preserved, Corpus Christi NAS, Tx
USS Hornet	CVS-12	laid up, Bremerton, Wa. since 1970
USS Bennington	CVS-20	laid up, Bremerton, Wa. since 1970
USS Bon Homme Richard	CVA-31	laid up, Bremerton, Wa. since 1971
USS Oriskany	CV-34	laid up, Bremerton, Wa. since 1976
USS Midway	CV-41	laid up, Bremerton. Wa. since 1992
USS Intrepid		Museum, pier 66-New York City, NY

UNITED STATES MARINE CORPS

Two letter codes, followed (usually) by two numbers are allocated to EVERY squadron in the USMC operating aircraft. Some headquarters units not currently operating aircraft are also allocated codes. As with the USN, shore bases operating communications aircraft are allocated a number/letter code. The digits following the two letter code are individual aircraft identification.

USMC aircraft occasionally embark with USN carriers at sea. When this happens, the aircraft are re-issued with codes AND numbers WITHIN the CVW range, and totally discard their USMC code identities.

HMA-Marine Attack Helicopter Squadron

HMA-773	Atlanta NAS, Ga	MP	AH1J

HMH-Marine Heavy Helicopter Squadron

HMH-361	Tustin MCAS, Ca	YN	CH53D, CH53E	
HMH-362	New River MCAS, NC	YL	CH53D	
HMH-363	Tustin MCAS, Ca	YZ	CH53D	
HMH-461	New River MCAS, NC	CJ	CH53D, CH53E	
HMH-462	Tustin MCAS, Ca	YF	CH53D	
HMH-463	Kaneohe Bay MCAS, Hi	YH	CH53D	
HMH-464	New River MCAS, NC	EN	CH53E	
HMH-465	Tustin MCAS, Ca	YJ	CH53E	
HMH-466	Tustin MCAS, Ca	YK	CH53E	
HMH-772	Willow Grove NAS, Pa	MT	CH53D	:Reserve
HMH-772	det.A Atlanta NAS, Ga	MS	CH53D	:Reserve
HMH-772	det.B Dallas NAS, Tx	QM	CH53D	:Reserve

HML-Marine Light Helicopter Squadron

HML-767	New Orleans NAS, La	MM	UH1N	:Reserve
HML-771	South Weymouth NAS, Ma	QK	UH1N	:Reserve
HML-776	South Weymouth NAS, Ma	QL	UH1N	:Reserve

HMLA-Marine Light Attack Helicopter Squadron

HMLA-167	New River MCAS, NC	TV	AH1J, AH1T, AH1W, UH1N
HMLA-169	Tustin MCAS, Ca	SN	AH1T, AH1W, UH1N
HMLA-267	Camp Pendleton MCAS, Ca	UV	AH1J, AH1W, UH1N
HMLA-269	New River MCAS, NC	HF	AH1J, AH1T, UH1N
HMLA-367	Tustin MCAS, Ca	VT	AH1J, AH1T, UH1N
HMLA-369	Camp Pendleton MCAS, Ca	SM	AH1J, AH1T, AH1W, UH1N

HMM-Marine Medium Helicopter Squadron

HMM-161	Tustin MCAS, Ca	YR	UH1N, CH46E, CH46F	
HMM-162	New River MCAS NC	YS	UH1N, CH46D, CH46E, CH46F	
HMM-163	Tustin MCAS, Ca	YP	UH1N, CH46E	
HMM-164	Tustin MCAS, Ca	YT	UH1N, CH46D, CH46E	
HMM-165	Kaneohe Bay MCAS, Hi	YW	UH1N, CH46D, CH46E, CH46F	
HMM-166	Tustin MCAS, Ca	YX	CH46E	
HMM-261	New River MCAS NC	EM	UH1N, CH46E, CH46F	
HMM-262	Futenma NAS, Japan	ET	UH1N, CH46E, CH46F	
HMM-263	New River MCAS NC	EG	CH46E, CH46F	
HMM-264	New River MCAS NC	EH	UH1N, CH46E, CH46F	
HMM-265	Kaneohe Bay MCAS, Hi	EP	CH46E, CH46F	
HMM-266	New River MCAS NC	ES	UH1N, CH46D, CH46E, CH46F	
HMM-268	Tustin MCAS, Ca	YQ	UH1N, CH46D, CH46E, CH46F	
HMM-364	Kaneohe Bay MCAS Hi	PF	CH46D, CH46E, CH46F	
HMM-365	New River MCAS NC	YM	CH46D, CH46E	
HMM-764	Tustin MCAS, Ca	ML	CH46D, CH46E	:Reserve
HMM-774	Norfolk NAS, Va	MQ	CH46D, CH46E	:Reserve

HMT-Marine Helicopter Training Squadron

HMT-204	New River MCAS, NC	GX	CH46D, CH46E, CH46F, CH53A, CH53D
HMT-301	Camp Pendleton MCAS, Ca	SU	CH46D, CH46E
HMT-302	Tustin MCAS, Ca	UT	CH53A, CH53E
HMT-303	Camp Pendleton MCAS, Ca	QT	AH1J, UH1N

HMX-Marine Helicopter Development Squadron

HMX-1	Quantico NAS, Va	MX	UH1N, VH1N, VH3D, CH46F, VH53D, VH60A

H&MS-Headquarters and Maintenance Squadron (now MALS)

H&MS-11	El Toro MCAS, Ca	TM	TA4F, OA4M

H&MS-12	Iwakuni AB, Japan	WA	TA4F, OA4M	
H&MS-13	Yuma MCAS, Az	YU	TA4F, OA4M	
H&MS-24	Kaneohe Bay MCAS, Hi	EW	TA4F, TA4J	
H&MS-31	Beaufort MCAS, SC	EX	TA4F, TA4J	
H&MS-32	Cherry Point MCAS, NC	DA	OA4M	
H&MS-36	Futemna NAF, Japan	WX	OV10A, OV10D	
H&MS-49	Willow Grove NAS, Pa	QZ	TA4J	:Reserve

MHWS-Marine Wing Headquarters Squadron

MWHS-4	New Orleans NAS, La	EZ	UC12B, CT39G	:Reserve

VMA-Marine Attack Squadron

VMA-124	Memphis NAS, Tn	QP	TA4J, A4M	:Reserve
VMA-131	Willow Grove NAS, Pa	QG	A4E, A4M	:Reserve
VMA-142	Cecil Field NAS, Fl.	MB	A4F, TA4J, A4M :Reserve	
VMA-211	Yuma MCAS, Az	CF	AV8B	
VMA-214	Yuma MCAS, Az	WE	AV8B	
VMA-223	Cherry Point MCAS, NC	WP	AV8B	
VMA-231	Cherry Point MCAS, NC	CG	AV8B	
VMA-311	El Toro MCAS, Ca	WL	AV8B	
VMA-513	Yuma MCAS, Az	WF	AV8B, TAV8B	
VMA-542	Cherry Point MCAS, NC	WH	AV8B	

VMAQ-Marine Tactical Electronic Warfare Squadron

VMAQ-1	Iwakuni AB, Japan	CB	EA6B	
VMAQ-2	Cherry Point MCAS, NC	CY	EA6B	
VMAQ-3	Cherry Point MCAS, NC	MD	EA6B	
VMAQ-4	Cherry Point MCAS, NC	RM	EA6B	

VMAT-Marine Attack Training Squadron

VMAT-102	El Toro MCAS, Ca	SC	A4M	
VMAT-202	Cherry Point MCAS, NC	KC	A6E, TC4C	
VMAT-203	Cherry Point MCAS, NC	KD	AV8B, TAV8B	

VMFA-Marine Fighter Attack Squadron

VMFA-112	Dallas NAS, Tx	MA	FA18A	:Reserve
VMFA-115	Beaufort MCAS, SC	VE	FA18A	
VMFA-121	El Toro MCAS, Ca	VK	FA18D	
VMFA-122	Beaufort MCAS, SC	DC	FA18A	
VMFA-134	El Toro MCAS, Ca	MF	FA18A	:Reserve
VMFA-212	Kaneohe Bay MCAS, Hi	WD	FA18A	
VMFA-224	Beaufort MCAS, SC	WK	FA18D	
VMFA-225	El Centro NAS, Ca	CE	FA18D	
VMFA-232	Kaneohe Bay MCAS, Hi	WT	FA18C	
VMFA-235	Kaneohe Bay MCAS, Hi	DB	FA18C	
VMFA-242	El Toro MCAS, Ca	DT	FA18D	
VMFA-251	Beaufort MCAS, SC	DW	FA18A	
VMFA-312	Beaufort MCAS, SC	DR	FA18C	
VMFA-314	El Toro MCAS, Ca	VW	FA18A	
VMFA-321	Andrews AFB, Md	MG	FA18A	:Reserve
VMFA-323	El Toro MCAS, Ca	WS	FA18A	
VMFA-332	Beaufort MCAS, SC	EA	FA18D	
VMFA-451	Beaufort MCAS, SC	VM	FA18A	
VMFA-533	Beaufort MCAS, SC	ED	FA18D	

VMFAT-Marine Fighter Attack Training Squadron

VMFAT-101	Yuma MCAS, Az	SH	FA18A, FA18D	

VMFT-Marine Fighter Training Squadron

VMFT-401	Yuma MCAS, Az	--	F5E, F5F	

VMGR-Marine Air Refueling\Transport Squadron

VMGR-152	Futemna NAF, Japan	QD	KC130F	
VMGR-234	Glenview NAS, Il	QH	KC130F, KC130T :Reserve	
VMGR-252	Cherry Point MCAS, NC	BH	KC130F, KC130R	
VMGR-352	El Toro MCAS, Ca	QB	KC130F, KC130R	

VMGR-452	Stewart Airport, Newburg	NY	KC130T

VMGRT-Marine Air Refueling\Transport Training Squadron

VMGRT-253	Cherry Point MCAS, NC	GR	KC130F

VMO-Marine Observation Squadron

VMO-2	Camp Pendleton MCAS, Ca	UU	OV10A, OV10D	
VMO-4	Atlanta NAS, Ga	MU	OV10A	:Reserve

Station Flight aircraft codes

Camp Smith AAF, Ar	BZ	UC12B	
Kaneohe Bay NAS, Hi	KB	HH46D	
Washington NAF, Md	5A	UC12M	HQ USMC
Beaufort MCAS,SC	5B	UC12M	
Cherry Point MCAS, NC	5C	UC12M	
New River MCAS, NC	5D	UC12M	
Futenma NAF, Japan	5F	UC12M	
Iwakuni AB, Japan	5G	UC12M	
El Toro MCAS, Ca	5T	UC12M	
Yumas MCAS, Az	5Y	UC12M	

UNITED STATES COAST GUARD

The USCG operates from the following locations:-

Arcata, Ca	nothing assigned
Astoria, Or	HH65A, HU25A
Barbers Point, Hi	HC130H
Borinquen, Puerto Rico	HH65A, HC130H
Brooklyn, Floyd Bennett Field, NY	HH65A
Cape Cod, Ma	HU25A
Cape May, NJ	HH65A
Chicago, Glenview NAS, Il	HH65A
Clearwater, Fl	HC130H
Corpus Christi, Tx	HH65A, HU25A
Detroit, Mi	HH65A
Elizabeth City, NC	VC4A, HC130H, HH60J, HH65A, HU25A
	Aircraft Repair and Supply Center
	Aviation Technical Training Center
Houston, Tx	HH65A
Humboldt Bay, Ca	HH65A
Kodiak, Ak	HC130H
Los Angeles, Los Angeles IAP, Ca	HH65A
Miami, Opa Locka AP, Fl	C212, RG8A, HH65A, HU25A
Mobile, Al	HH60J, HH65A, HU25A
	Air Training Center
New Orleans, La	HH65A
Norfolk NAS, Va	nothing assigned
North Bend, Or	HH65A
Port Angeles, Wa	HH65A
Sacramento, Ca	HH65A, HC130H
San Francisco, San Francisco IAP, Ca	HH60J
Savannah, Ga	HH65A
San Diego, Ca	HH65A
Sitka, Ak	nothing assigned
Traverse City, Mi	HH60J
Washington, National AP, DC	VC11A

UNITED STATES CUSTOMS SERVICE - USCS

Mojave, Ca	PA31 (1), UH60A (1)
North Island NAS, Ca	C550 (1), C12A (1), N22 (1)
Phoenix, Az	B200 (1), UH60A (1)
Davis Monthan AFB, Az	C550 (2), PA31 (1), C12A (1), UH60A (1)
El Paso, Tx	C550 (1), B200 (1)
Albequerque, NM	PA31 (1), UH60A (1)
San Angelo, Tx	PA31 (1), UH60A (2), C12A (1)
San Antonio, Tx	C550 (1), B200 (1)
Corpus Christi NAS, Tx	P3A (4), C12A (1), N22 (1)

Houston, Tx	C550 (1), PA31 (1), B200 (1), UH60A (1)
New Orleans, La	C550 (1), PA31 (1), C12A (1), N22 (1), UH60A (2)
Tampa, Fl	N22 (1)
Jacksonville NAS, Fl	P3S (1), PA31 (1), N22 (1), UH60A (1)
Homestead AFB, Fl	P3A (2), C550 (2), C12A (1), N22 (1), UH60A (2)
New York, NY	N22 (1)

BASES

BASE AIRFIELDS - UNITED STATES

This section has been expanded and rearranged from previous years to include ALL airfields in the United States with regular military connections. The 'user' index after the airport title indicates the military service, Government department or service, manufacturer, overhaul facility, product support company or whatever where aircraft listed in this publication are to be seen. For convenience in trip-planning by intending visitors, the locations are listed in State order. Where one airfield is known by more than one title, these are cross-referenced to the major user or owner (e.g. Lockheed Marietta\Atlanta NAS\Dobbins AFB).

ALABAMA - (Al)

Birmingham MAP\Smith ANGB	civil, ANG, Army NG, Hayes Industries
Cairns AAF\Fort Rucker	Army, Army Reserve, Aviation
Museum	
Craig Field AP	civil, Beech
Fell City-St. Clair County AP	civil, Army Reserve
Guthrie AAF\Fort Rucker	Army
Hanchey AHP\Fort Rucker	Army
Knox AHP\Fort Rucker	Army
Maxwell AFB	USAF, AFRES
Mobile-Bates Field CGAS	civil, USCG
Mobile-Downtown AP	civil, Army NG
Montgomery-Dannelly Field AP	civil, ANG, Army NG
Redstone Arsenal	Army
Smith ANGB (see Birmingham MAP)	ANG
Troy Airport	Sikorsky

ALASKA - (Ak)

Anchorage IAP\Kulis ANGB	civil, ANG
Casco Cove CGAS	USCG
Eielson AFB	USAF, ANG
Elmendorf AFB	USAF
Galena Airport	USAF det
King Salmon AP	USAF det
Kulis ANGB (see Anchorage IAP)	ANG
Shemya AFB	USAF\FOL

ARIZONA - (Az)

Davis-Monthan AFB	USAF, AMARC, USCS
Deer Valley Municipal AP	civil, Thunderbird Avn.
Gila Bend AAF	ranges
Laguna AAF\Yuma proving ground	USMC
Libby AAF\Fort Huachuca	Army, Army NG
Luke AFB	USAF, AFRES, Singapore DF
Marana\Silver Bell AHP (see Pinal)	Army NG
Mesa-Falcon Field	civil MDD Helicopters
Phoenix-Sky Harbor IAP	civil, ANG
Pinal Air Park\Evergreen\Silver Bell	civil, Army NG
Tucson IAP	civil, ANG
Williams AFB	USAF
Yuma MCAS	civil, USMC, MDD test

ARKANSAS - (Ar)

Blythsville AFB	USAF
Fort Smith MAP	civil, ANG
Little Rock AFB	USAF, ANG
Robinson AAF	

CALIFORNIA - (Ca)

Alameda NAS\Nimitz Field	Naval Dockyard, NAS, NARF, USMC
Amedee AAF\Sierra Ordnance Depot	
Barstow Daggett AP	civil, Army, Army\NTC
Beale AFB	USAF
Camp Pendleton MCAS	USMC
Camp San Louis Osbispo	Army
Castle AFB	USAF
China Lake NAS	NWC
Edwards AFB	AFFTC, NASA\Dryden, USAAEFA
El Centro NAF	FDS

438

El Toro MCAS	USMC
Fresno Air Terminal	civil, ANG, AVCRAD
Fritzsche AAF\Fort Ord	Army
George AFB	USAF
Hamilton Field	civil, Army
Humboldt Bay	USCG
Lemoore NAS	USN
Long Beach AP	civil, MDD
Los Alamitos AAF	Army NG, Army Reserve
Los Angeles IAP\CGAS	civil, USCG
March AFB	USAF, AFRES, ANG
Mather AFB	USAF, AFRES, Army
McClellan AFB\Sacramento CGAS	USAF, ALC\Sacramento, USCG
Miramar NAS	USN, NFWS
Moffett Field NAS	USN, NASA\Ames, ANG
Mojave AP	civil, Flite Systems, USCS
North Island NAS	USN, NARF, USCS
Norton AFB	USAF
Ontario Airport	Lockheed AS
Palmdale, AF plant 42	Lockheed, Northrop, Rockwell
Point Mugu NAS	ANG, USN, PMTC
Roberts AAF	
Sacramento CGAS (see McClellan AFB)	USCG
San Diego IAP\CGAS	civil, USCG
San Francisco IAP\CGAS	civil, USCG
Stockton Metropolitan AP	civil, Army NG
Travis AFB	USAF
Tustin MCAS\Santa Ana	USMC
Vandenburg AFB	USAF, missiles

COLORADO - (Co)

Buckley ANGB	ANG, Army
Butts AAF\Fort Carson	
Colorado Springs Academy Field	USAF\Academy
Colorado Springs MAP\Petersen AFB	civil, AFRES
Lowry AFB	TTC
Petersen AFB (see Colorado Springs MAP)	AFRES

CONNECTICUT - (Ct)

Bradley-Windsor Locks IAP\Bradley ANGB	civil, ANG, Army NG
Bradley ANGB (see Bradley IAP)	ANG, Army NG
Groton\New London Airport	civil, Army NG, AVCRAD

DELAWARE - (De)

Dover AFB	USAF
Greater Wilmington\Newcastle County AP	civil, ANG, Army NG, Boeing

DISTRICT OF COLUMBIA - (DC)

Bolling AFB	USN, USMC

FLORIDA - (Fl)

Cecil Field NAS	USN, USMC
Clearwater CGAS (see St. Petersburg)	USCG
Craig Municipal Airport	civil, Army NG
Duke Field (Eglin Aux. no.3)	AFRES
Eglin AFB	USAF
Homestead AFB	USAF, AFRES, USCS
Hurlburt Field	USAF
Jacksonville IAP	civil, ANG
Jacksonville NAS	USN, NARF
Key West NAS	USN
Lake City MAP	civil, Aero Corp,
Lakeland MAP	civil, Army NG
Macdill AFB	USAF, REDCOM, USASOCOM
Mayport NAS	USN
Melbourne Regional AP	civil, Grumman
Miami CGAS (see Opa Locka)	USCG
Miami-North Perry Airport	civil, Army Reserve
Opa Locka Airport\Miami CGAS	civil, USCG
Orlando International Airport	civil, Army, Army Reserve

Panama City NCSC Heliport	NCSC
Patrick AFB	USAF, FORSCOM det,
Pensacola NAS\Chevalier Field	USN, NARF, USN Museum,
St. Augustine Field	civil, Grumman
St. Petersburg-Clearwater IAP\CGAS	civil, USCG
Stuart Airport	civil, Grumman
Tyndall AFB	USAF
Whiting Field NAS	USN

GEORGIA - (Ga)

Anniston Army Depot (see Anniston\Calhoun	Army
Anniston\Calhoun County AP	civil, Army
Atlanta-Fulton County AP	civil, FORSCOM det
Atlanta NAS (see Marietta)	USN
Augusta\Bush Field	Army
Dobbins AFB (see Marietta)	AFRES, ANG, Army NG
Hunter AAF	Army, Army NG
Marietta Field	Lockheed, AFRES, ANG, Army NG,
·USN	
Lawson AAF\Fort Benning	Army
Moody AFB	USAF
Robins AFB	USAF, ALC\Warner-Robins
Savannah Municipal Airport\CGAS	civil, Grumman, ANG, USCG
Wright AAF\Fort Stewart	Army

HAWAII - (Hi)

Barbers Point NAS	USN, CINCPAC, Army
Barking Sands PMRF, Kauai	USAF, USN, Army
Hickam AFB (see Honolulu IAP)	ANG, AFSC
Honolulu IAP\Hickam AFB	civil, ANG, AFSC
Kaneohe Bay MCAS	USMC
Wheeler AFB	USAF, Army, Army NG

IDAHO - (Id)

Boise Air Terminal\Gowen Field	civil, ANG
Mountain Home AFB	USAF

ILLINOIS - (Il)

Capital Airport\Springfield	civil, ANG
Chanute AFB	TTC
Chicago CGAS (see Glenview NAS)	USCG
Chicago-Midway AP	civil, Army NG
Chicago-O'Hare IAP	civil, AFRES, ANG
Decatur Airport	civil, Army NG
Glenview NAS\Chicago CGAS	USN, USCG
Grayling AAF\Camp Grayling	
Greater Peoria Airport	civil, ANG
Haley AAF\Fort Sheridan	
Scott AFB	USAF

INDIANA - (In)

Fort Wayne MAP\Baer Field	civil, ANG
Grissom AFB	USAF, AFRES, ANG
Shelbyville MAP	civil, Army NG
Terre Haute-Hulman Regional AP	civil, ANG

IOWA - (Ia)

Des Moines MAP	civil, ANG
Sioux City MAP	civil, ANG

KANSAS - (Ks)

Marshall AAF\Fort Riley	
McConnell AFB	USAF, ANG
Sherman AAF\Fort Leavenworth	
Topeka-Forbes Field AP\ANGB	civil, ANG
Wichita-Mid Continent AP	Boeing, Cessna, LJ

KENTUCKY - (Ky)

Campbell AAF\Fort Campbell	Army
Frankfort-Capital City AP	Army NG
Godman AAF\Fort Knox	
Louisville-Bowman Field AP	civil, Army Reserve
Louisville-Standiford Field AP	civil, ANG

LOUISIANA - (La)

Barksdale AFB	USAF, AFRES
Layfayette Regional AP	civil, Army Reserve
New Orleans-Lakefront AP	civil, Army NG
New Orleans NAS\CGAS-Alvin Callender Field	USN, USMC, AFRES, ANG, USCG
Polk AAF\Fort Polk	

MAINE - (Me)

Bangor IAP	civil, ANG
Brunswick NAS	USN, General Offshore Corp.
Loring AFB	USAF

MARYLAND - (Md)

Aberdeen Proving Ground	ranges
Andrews AFB\Washington NAF	USAF, ANG, AFRES, USN, FLSW, USMC
Baltimore-Glenn L. Martin State AP	civil, ANG
Patuxtent River NAS	USN, NATC, NRL, NTPS
Tipton AAF\Fort Meade	Army
Washington NAF (see Andrews AFB)	USN, FLSW, USMC

MASSACHUSSETTS - (Ma)

Cape Cod CGAS (see Otis ANGB)	USCG
Hanscomb AFB	USAF
Moore AAF\Fort Devens	
Otis ANGB\Cape Cod CGAS	ANG, Army NG, USCG
Salem CGAS	USCG
South Weymouth NAS	USN, USMC
Westfield-Barnes MAP	civil, ANG
Westover AFB	AFRES

MICHIGAN - (Mi)

Alpena\Phelps-Collins AP	civil, ANG training
Battle Creek ANGB	ANG
Detroit CGAS (see Selfridge ANGB)	USCG
Detroit NAF (see Selfridge ANGB)	USN
Kellog Regional AP\Battle Creek ANGB	civil, ANG
K.I. Sawyer AFB	USAF
Lansing-Abrams MAP	civil, Army NG
Selfridge ANGB\Detroit NAF and CGAS	AFRES, ANG, USN, USCG
Traverse City CGAS	USCG
Wurtsmith AFB	USAF

MINNESOTA - (Mn)

Duluth IAP	civil, ANG
Minneapolis-St. Paul IAP	civil, AFRES, ANG
St. Paul Downtown\Holman Field	civil, Army, Army NG

MISSISSIPPI - (Ms)

Columbus AFB	USAF
Gulfport-Biloxi Regional AP	civil, Army NG, AVCRAD
Jackson-Allen C. Thompson Field	civil, ANG
Keesler AFB\CGAS	USAF, AFRES, USCG
Meridian\Key Field AP	civil, ANG, Army NG
Meridian NAS	USN

MISSOURI - (Mo)

Lambert-St. Louis IAP	civil, ANG, MDD
Richards-Gebaur AFB	USAF

St. Joseph-Rosecrans Memorial AP	civil, ANG
Whiteman AFB	USAF, missiles

MONTANA - (Mt)

Great Falls IAP	civil, ANG
Malstrom AFB	USAF

NEBRASKA - (Nb)

Lincoln MAP	civil, ANG
Offutt AFB	USAF, Army, Army NG

NEVADA - (Nv)

Fallon NAS	USN, NSWC, USMC
Groom Lake\Watertown strip	CIA, USAF (covert operations)
Indian Springs AFAF	USAF
Nellis AFB	USAF
Reno-Cannon IAP	civil, ANG
Reno-Stead AP	civil, Army NG
Tonopah AFS	CIA, USAF

NEW HAMPSHIRE - (NH)

Pease AFB	USAF, ANG

NEW JERSEY - (NJ)

Atlantic City AP\Bader Field	civil, ANG
Lakehurst NAS\NAEC	USN, NAEC, NATTC, Army
McGuire AFB	USAF, ANG
Morristown MAP	civil, Army

NEW MEXICO - (NM)

Cannon AFB	USAF
Holloman AFB	USAF, Raytheon
Kirtland AFB	USAF, ANG

NEW YORK - (NY)

Albany County AP	civil, Army NG
Bethpage AP	Grumman
Brooklyn CGAS\Floyd Bennett Field	USCG
CalvertonGrumman	
Griffiss AFB	USAF
Long Island-McArthur AP	civil, Army NG
Newburg-Stewart AP	civil, ANG, USMC
Niagara Falls IAP	civil, AFRES, ANG
Plattsburgh AFB	USAF
Schenectady County AP	civil, ANG
Seneca AAF	
Suffolk County AP\West Hampton	civil, ANG
Syracuse-Hancock Field	civil, ANG
Wheeler-Sac AAF	

NORTH CAROLINA - (NC)

Charlotte-Douglas IAP	civil, ANG
Cherry Point MCAS	USMC, NARF, SOES
Elizabeth City CGAS	USCG, CG\TTC, AFSC
Mackall AAF\Camp Mackall	
New River MCAS	USMC
Pope AFB	USAF, USAABNSOTBD
Seymour Johnson AFB	USAF
Simmons AAF\Fort Bragg	Army

NORTH DAKOTA - (ND)

Grand Forks AFB	USAF
Fargo\Hector Field	civil, ANG
Minot AFB	USAF

OHIO - (Oh)

Akron-Canton Regional AP	civil, Army NG
Cleveland-Lewis Research Center	NASA
Lorain County Regional Airport	civil, Army Reserve
Mansfield-Lahm MAP	civil, ANG
Port Columbus IAP	civil, Army Reserve
Rickenbacker ANGB	AFRES
Springfield MAP	civil, ANG
Toledo Express Airport	civil, ANG
Wright-Patterson AFB	USAF, AF Museum, AFOG, AFRES, ASD
Youngstown-Eiser MAP	civil, AFRES

OKLAHOMA - (Ok)

Altus AFB	USAF
Henry Post AAF\Fort Sill	
Oklahoma City-Will Rogers World AP	civil, ANG
Tinker AFB	USAF, AFRES, ALC\Oklahoma City
Tulsa IAP	civil, ANG
Vance AFB	USAF

OREGON - (Or)

Astoria CGAS	USCG
Kingsley Field AP	civil, ANG
Portland IAP	civil, AFRES, ANG

PENNSYLVANNIA - (Pa)

Greater Pittsburgh IAP	civil, AFRES, ANG
Harrisburg IAP\Olmsted Field	ANG
Johnsonville NAS	USN, NADC
Muir AAF\Fort Indian Town Gap	
Philadelphia Boeing-Vertol	
Warminster NASNADC	
Willow Grove NAS	USN, USMC, AFRES, ANG

PEURTO RICO - (PR)

Borinquen CGAS\Ramey AFB	USCG
Muniz ANGB (see San Juan IAP)	ANG
Roosevelt Roads NAS	USN
San Juan IAP\Muniz ANGB	civil, ANG

RHODE ISLAND - (RI)

Quonset State Airport	civil, ANG, Army NG

SOUTH CAROLINA - (SC)

Beaufort MCAS	USMC
Charleston AFB	USAF
Columbia Metropolitan AP	civil, Army, Army NG
McEntire ANGB	ANG
Myrtle Beach AFB	USAF
Shaw AFB	USAF, AFRES

SOUTH DAKOTA - (SD)

Ellsworth AFB	USAF
Sioux Falls-Joe Foss Field	civil, ANG

TENNESSEE - (Tn)

Arnolds AFB	
Knoxville-McGhee Tyson AP	civil, ANG
Memphis IAP	civil, Army NG
Memphis NAS	USN, NATTC, USMC
Nashville Metro AP	civil, ANG
Smyrna AP	civil, Army NG

TEXAS - (Tx)

Arlington	Bell Flight Research
Austin-Robert Meuller MAP	civil, Army NG

Bergstrom AFB	USAF, AFRES
Biggs AAF\Fort Bliss	Army
Brooks AFB (NOT an airfield!)	USAF
Carswell AFB	USAF, AFRES
Corpus Christi NAS	USN, Army, USCG, USCS
Dallas NAS\Hensley Field	USN, NARU, USMC, ANG, Army NG,
LTV	
Dyess AFB	USAF
Ellington ANGB\Houston CGAS	ANG, USCG, NASA\Johnson
Fort Worth (see Carswell AFB)	General Dynamics
Fort Worth-Meacham Airport	civil, Page Avjet
Greenville	civil, E Systems
Hondo Municipal Airport	civil, OTS
Houston CGAS (see Ellington ANGB)	USCG
Houston-D.W. Hooks Memorial AP	civil, Army Reserve
Kelly AFB	AFRES, ALC\San Antonio, ANG
Kingsville NAS	USN
Lackland AFB	TTC
Laughlin AFB	USAF
Martindale AAF\San Antonio	Army NG
Randolph AFB	USAF
Reese AFB	USAF
Robert Gray AAF\Fort Hood	Army
Sheppard AFB	TTC
Waco-TSTI Airport	civil, E-Systems

UTAH - (Ut)

Hill AFB	USAF, AFRES, AFSC, ALC\Ogden
Michael AAF\Dugway Proving Ground	
Salt Lake City IAP	civil, ANG
Salt Lake City No.2 MAP	civil, Army NG

VERMONT - (Vt)

Burlington IAP	civil, ANG, Army NG

VIRGINIA- (Va)

Blackstone AAF\Fort Pickett	
Davison AAF\Fort Belvoir	
Felker AAF\Fort Eustis	Army, MTPS
Langley AFB	USAF, TRADOC, NASA
Norfolk NAS	USN, NARF, USMC, FMFLANT
Oceana NAS	USN
Quantico MCAS	USMC
Richmond-Richard E. Byrd IAP	civil, ANG
Wallops Island	NASA

WASHINGTON - (Wa)

Boeing Field\King County AP, Seattle	civil, Boeing delivery
Everett (see Snohomish County)	Boeing
Fairchild AFB	USAF, ANG
Gray AAF\Fort Louis	
McChord AFB	USAF, ANG
Port Angeles CGAS	USCG
Renton Field, Seattle	civil, Boeing
Snohomish County\Paine Field, Everett	Army Reserve, Boeing
Whidbey Island NAS\Ault Field	USN

WEST VIRGINIA - (WV)

Charleston-Yeager AP	civil, ANG
Dawson AAF\Camp Dawson	
Martinsburg-Eastern WV Regional AP	civil, ANG

WISCONSIN - (Wi)

Madison-Dane County Reg.AP\Truax Field	civil, ANG
McCoy AAF\Fort McCoy	
Milwaukee\Gen. Billy Mitchell Field	civil, AFRES
Volk Field ANGB	ANG
Waukesha County AP	civil, Army Reserve
West Bend MAP	civil, Army NG

WYOMING - (Wy)

Cheyenne MAP   civil, ANG
F.E. Warren AFB   USAF, missiles

BASE AIRFIELDS - OVERSEAS

AZORES

Lajes Field   civil, USAF det

BELGIUM

Chievres   USAF\NATO, Army

CANAL ZONE

Albrook AFB, Panama   USAF
Howard AB, Panama   USAF, IAAFA

CUBA

Guantanamo Bay NAS   USN

GERMANY

Ansbach   Army
Bad Kreuznach   Army
Bad Tolz   Army
Berlin-Templehof   Army
Bitburg AB   USAF
Bremen-Lemwerder   Army
Coleman Barracks   Army
Fulda   Army
Garlstedt   Army
Giebelstadt   Army
Grafenwohr   Army
Hanau   Army
Heidelberg   Army
Illesheim   Army
Karlsruhe   Army
Kitzingen   Army
Landstuhl   Army
Nellingen   Army
Ramstein AB   USAF, Army
Rhein Main AB   USAF
Schwabisch Hall   Army
Schweinfurt   Army
Sembach AB   USAF
Spangdahlem AB   USAF
Stuttgart-Echterdingen   Army
Stuttgart IAP   USAF, Army
Wertheim   Army
Wiesbaden   Army

GREECE

Athens IAP   USAF

GREENLAND

Sondrestrom AB   USAF, USN
Thule AB   USAF, USN

GUAM

Agana NAS	USN
Guam CGAS	USCG

ICELAND

Keflavik AB	USAF, USN

ITALY

Aviano AB	USAF
Naples IAP	USN
NAS Sigonella	USN
Vicenza	Army

JAPAN

Atsugi NAS	USN
Camp Zama	Army
Futenma MCAS	USMC
Iwakuni MCAS	USMC
Kadena AB, Okinawa	USAF, USN
Misawa AB	USAF
Yokota AB	USAF

MARSHALL ISLANDS

Bucholz AAF\Kwajalein Atoll	Army
Dyess AAF\Roi-Namur	Army

NETHERLANDS

Soesterberg AB	USAF

SINGAPORE

Payar Lebar AB	USAF\PACAF

SOUTH KOREA

Camp Casey	Army
Camp Gary Owen	Army
Camp Laguardia	Army
Camp Page	Army
Camp Red Cloud	USAF det
Camp Stanley	Army
Camp Stanton	Army
Desiderio AAF\Camp Humphries	Army
Kunsan AB	USAF
Osan AB	USAF
Pyong Teak	Army
Suwon AB	USAF
Taegu AB	USAF

SPAIN

Rota AB,	USN

TURKEY

Ankara IAP	USAF
Incirlik AB, Adana	USAF
Izmir\Cigli AAF	Army
Sinop AAF	Army

UNITED KINGDOM

RAF Alconbury	USAF
RAF Fairford	USAF
RAF Greenham Common	USAF\NATO
RAF Lakenheath	USAF
RAF Mildenhall	USAF, USN

RAF Sculthorpe	USAF
RAF Upper Heyford	USAF

PRESERVED AIRCRAFT

The following sections list alphabetically by state/town, the location of US Military aircraft which have been preserved. The word 'preserved' should be interpreted in its widest possible sense, since individual aircraft conditions vary from derelict to pristine. Obsolete aircraft used for instructional or training purposes are also included where appropriate.
The formal aircraft serial presentation is given, followed by the official type designation (which are grouped in Bomber, Cargo, Fighter etc. order giving preference to the post 1962 system). In the third column any spurious serials displayed are given in quotation, as are currently carried civilian registrations if no other markings are readily visible. Also any generally correct, but unusually presented serials are noted here (without parentheses).
Following the list of US based aircraft, a list of US military aircraft at overseas locations is given.

ALABAMA			62-12508	OH23F Raven	
			51-16616	H25A Retriever	
			50-1840	XH26A	Light collapsable
Alabama Welcome Center - Highway 231\Florida border			2-5837	YH30 (McCullough MC4C)	
			55-4965	YH32 Hornet	
62-2018	UH1B Iroquois		53-4526	CH34A Choctaw	
			56-4320	VCH34A Choctaw	
Atmore - town			55-0644	CH37B Mojave	
			49-2890	XH39	(Sikorsky S.59)
51-8604	T33A Shooting Star		55-4459	XH40	(Bell 204)
			56-6723	YH40	(Bell 204)
Bayou la Batre - town			56-4244	YH41 Seneca	(Cessna CH.1B)
			60-3451	CH47A Chinook	
51-6794	T33A Shooting Star		65-7992	CH47A Chinook	
			Bu151262	XH51A Aerogyro	
Birmingham - Donnally Field Municipal Airport, Southern Museum of Flight			Bu151263	XH51A Aerogyro	
			64-18200	TH55A Osage	
			67-16879	TH55A Osage	
Bu152996	F4N Phantom II	'64049'	67-16944	TH55A Osage	
63-8502	UH1B Iroquois		67-16955	TH55A Osage	
Daleville - Cairns AAF, Fort Rucker US Army Aviation Center			67-16963	TH55A Osage	
			67-16966	TH55A Osage	
			66-8830	AH56A Cheyenne	c\n 1005
57-6532	YAO3	Inflatoplane	68-16687	OH58A Kiowa	
57-3080	YC7A Caribou		73-21656	YUH61A	UTTAS
51-11638	C45G Expeditor		73-22012	YCH62	
Bu39767	UC45J Expeditor		73-22247	YAH63A	(Bell 409)
Bu12436	C47H Skytrain	ex 42-23757	73-22249	YAH64A Apache	
48-0613	VC121A Constellation		40-3141	L1A Vigilant	
51-6998	LC126A	(Cessna 195)	42-15174	L4A Cub	
73-1653	YE5A Eagle	(Windecker)	43-0515	L4B Cub	
44-72990	F51D Mustang		42-99103	U19A\L5 Sentinel	
66-15246	AH1G HueyCobra		45-34985	L5G Sentinel	
N309J	AH1J HueyCobra	'23095'	46-0159	L13A	
78-23095	AH1S HueyCobra		47-0429	YL15 Scout	
59-1686	UH1A Iroquois		47-0924	L16A	(Aeronca 7BC)
59-1695	UH1A Iroquois	01695	47-1344	U18A\L17A Navion'A1744'	
60-3553	UH1B Iroquois		48-1046	U18B\L17B Navion	
62-1297	UH1B Iroquois		52-2536	L18C Super Cub	
62-1884	UH1B Iroquois		50-1327	O1A\L19A Bird Dog	
62-2099	UH1B Iroquois		51-4651	O1A\L19A Bird Bog	
63-8505	UH1B Iroquois		51-11972	O1A\L19A Bird Dog	
65-12740	UH1C Iroquois		51-4943	TO1A\TL19A Bird Dog	
60-6030	YUH1D Iroquois		55-4681	TO1D\TL19D Bird Dog	
73-22099	UH1H Iroquois		51-6263	YL20 Beaver	
58-1496	YOH3A		51-15782	TL21A Super Cub	
62-4201	YOH4A Jet RangerLOH comp.		'21700'	L23A Seminole	mock-up
62-4206	YOH5A	LOH comp.	52-2540	YL24 Courier	(Helio H.391)
62-4213	YOH6A Cayuse		69-18000	YO3A QStar	
68-17340	OH6A Cayuse		Bu131485	AP2E Neptune	
48-0827	H13B Sioux		44-72990	P51D Mustang	
48-0845	H13B Sioux		43-46521	R4B Hoverfly	
51-2468	OH13D Sioux		47-0482	R5F Dragonfly	
51-14193	OH13E Sioux		46-0001	XR9B	
67-17024	TH13T Sioux		51-3612	T28A Trojan	'03612'
49-2888	XH18A	(Sikorsky S.52)	Bu137747	T28B Trojan	
51-14272	UH19C Chickasaw		Bu144048	T34B Mentor	'50-735'
55-5239	UH19D Chickasaw		56-3466	T37A Tweety Bird	
56-2040	CH21C Shawnee\Workhorse		61-0685	CT39A Sabreliner	
51-3975	H23A Raven		67-15000	T41B Mescalero	
51-16142	OH23B Raven		67-15006	T41B Mescalero	
55-4064	OH23C Raven		67-15161	T41B Mescalero	
55-4109	OH23C Raven		57-6136	U1A Otter	

450

57-5863	U3A\L27A Blue Canoe	
55-4640	U4A\L26B Aero Commander (model	
560A)		
NASA705	XU5B Twin Courier	
58-1987	U6A Beaver	
56-3712	U8D SEminole	'263712'
58-1359	RU8D Seminole	
63-12902	NU8F Seminole	c\n LG-1
52-6219	YL26\YU9A Aero Commander (model	
520)		
63-13176	U10B Super Courier	
57-6539	YOV1A Mohawk	
62-5860	OV1C Mohawk	
53-4016	XV1	
54-0148	XV3A	(Bell 200)
64-18264	XV6A Kestrel	(Hawker P1127)
67-15345	X26B Schweizer\Lockheed missile	
56-6941	VZ-3RY Vertiplane	
NASA705	VZ5B Vertifan	
58-5508	VZ-7AP Aerial Jeep	

Daleville - Guthrie AAF

Bu154895	UH34D Seahorse
Bu124352	HO3S Dragonfly
65-12685	T42A Cochise
57-6135	U1A Otter
59-2229	U1A Otter
56-4026	U9C Aero Commanderc\n 344-36
57-6183	NRU9D Aero Commander
59-2631	OV1B Mohawk
61-2962	OV1C Mohawk

Florala - town

53-4938	T33A Shooting Star

Huntsville - Space Museum

64-13954	UH1B Iroquois
NASA952	Bell LLTV

Mobile - USS Alabama Memorial

44-31004	B25J Mitchell	'02344'
55-71	B52D Stratofortress	
44-76326	VC47D Skytrain	
51-2993	F86L Sabre	
54-102	F105B Thunderchief	
51-15859	CH21B Shawnee	
44-74216	P51D Mustang	
2129	HU16E Albatross USCG	
Bu79593	F6F Hellcat	

Montgomery - Dannelly Field Airport

52-7249	RF84F Thunderflash
66-7745	F4D Phantom II

Montgomery - Highway 80

53-847	F86K Sabre

Montgomery - Maxwell AFB

44-30649	B25J Mitchell	'253373'
55-57	B52D Stratofortress	
N214GB	C47	Skytrain
65-660	F4D Phantom II	
49-1301	F86A Sabre	'12760'
55-3678	F100D Super Sabre	'63436'
61-176	F105D Thunderchief	
59-1601	T38A Talon	

Ozark - Municipal Airport

51-1748	F84F Thunderstreak
56-3718	U8D Seminole
57-3097	U8D Seminole

Robertsdale - town

51-6707	T33A Shooting Star

Russellville - town

51-6554	T33A Shooting Star

Tuscaloosa - town

unknown	T33A Shooting Star '54109'

Tuskegee - Chappie James Museum

44-30649	B25J Mitchell

ALASKA

Anchorage - Anchorage IAP\Kulis ANGB

N41	C47A Skytrain	
56-4390	C123J Provider	
49-1849	F80C Shooting Star	
49-1195	F86A Sabre	'12807'
49-3015	T6G Texan	'34555'
57-606	T33A Shooting Star '35403'	

Anchorage - Wasilla Airport, Transportation Museum of Alaska

43-15200	C47A Skytrain
56-1282	F102A Delta Dagger
49-2001	H5H Hoverfly
53-4362	CH21B Workhorse

Anchorage - Elmendorf AFB

64-0890	F4C Phantom II
56-1274	F102A Delta Dagger
52-8696	H21B Workhorse

Fairbanks - Eielson AFB

64-0905	F4C Phantom II
68-11003	O2A Super Skymaster

Merrill Field

58-2048	U6A Beaver
Bu138202	T28B Trojan

ARIZONA

Apache Junction - town

unknown	T33A Shooting Star '66008'

Avra Valley - Avra Valley Airport

48-0609	C121A Constellation 'N494TW'

Casa Grande - town

51-5776	JF89C Scorpion	

Chandler - Chandler Bvd.

Bu134451	F9F8	Cougar
51-6261	F86D Sabre	'210115'

Chandler - Williams AFB

52-2844	F86E Sabre	
unknown	T33A Shooting Star	'70606'

Douglas - town

52-9608	T33A Shooting Star

Fort Huachuca - Museum

67-18930	OV1C Mohawk

Gila Bend - Gila Bend Aux. AF

52-6503	F84F Thunderstreak
56-2924	F100D Super Sabre
59-1739	F105F Thunderchief
57-0495	T33A Shooting Star

Glendale - Harry Bosnall Park, junction 59th and Bethany Home

54-2281	F100C Super Sabre

Glendale - Luke AFB

67-0327	F4E Phantom II	'31175'
Bu151510	F4N Phantom II	'37411'
73-0108	F15B Eagle	
52-6782	F84F Thunderjet	'6779'
52-5323	F86E Sabre	'24530'
53-1716	F100C Super Sabre	'42009'
56-0892	F104C Starfighter	'13243'
58-0745	T33A Shooting Star	

Globe - town

51-5915	F86D Sabre

Kingman - Airfield, Army Museum

51-11596	C45G Expediter	'N39M
62-1970	UH1B Iroquois	
unknown	H34A Chocktaw	

Marana - Pinal Air Park, Evergreen Museum

44-35439	A26C Invader	'N74833'
44-83785	VB17G Fortresss	'N207EV'
Bu92095	FG1D Corsair	'N67HP'
56-2992	F100D Super Sabre	
56-3141	F100D Super Sabre	
42-17137	PT13D Kaydet	'N450UR'
Bu131502	SP2F Neptune	'N202EV'
44-53186	P38L Lightning	'N38EV'
42-9749	P40K Warhawk	'N293FR'
44-63576	P51D Mustang	'N51DH'
Bu91726	TBM3E Avenger	'N5260V'
42-43973	AT6C Texan	'N33CC'
Bu138334	T28B Trojan	'NX394W'

Mesa - Pioneer Park, East Main St.

Bu125316	F9F5	Cougar

Mesa - Falcon Field Airport

'Champlin Fighter Museum' - WW1 hangar

contains twelve aircraft and replicas (all civil registered) of the period as well as a Spitfire T9, N8R\G-AVAV.

WW2	Hangar	
42-83731	A36A Apache	'N251A'
Bu88454	F2G Corsair	'NX4324'
Bu41930	F6F Hellcat	'NX103V'
Bu1262	F8F Bearcat	'NL700A'
Bu74560	FM2 Wildcat	'NL90523'
51-13371	F86F Sabre	
44-53097	P38M Lightning	'NL3JB'
unknown	P40 Warhawk	'NL10626'
42-8205	P47D Thunderbolt	'NX14519'
45-11628	P51D Mustang	'NL151X'

also contains two German and one Russian built aircraft. other outside exhibits

44-83514	B17G Fortress	'N9323Z'
43-35972	B25J Mitchell	'N9552Z'
43-17253	C54D Skymaster	'N99AS'
Bu153016	F4N Phantom II	
42-86092	AT6D Texan	'N3246G'
42-17163	PT13D Kaydet	'N4813V'

elsewhere

52-7519	H19B Chicasaw	
52-7548	H19B Chicasaw	
52-7579	H19B Chicasaw	
57-5948	H19D Chicasaw	
unknown	H19 Chicasaw	'XB067'

Peoria - 83rd and Peoria

51-9313	F84F Thunderjet	'26675'

Phoenix - Sky Harbor International Airport\ANGB

55-3818	F86F Sabre	
56-0891	F104C Starfighter (at ANGB gate)	
44-74494	P51D Mustang	

Phoenix - East McDowell Road, Arizona Military Museum

64-14156	UH1M Iroquois

Scottsdale - town

53-5915	T33A Shooting Star

Tucson - Davis Monthan AFB

A small display is situated on the main base just inside the main gate. In addition, the AMARC facility within the base has established a large collection of representative aircraft stored, past and present, within its own compound. See AMARC entry elsewhere.

'Warrior Air Park' main gate display:-

68-8229	GA7D Corsair	
79-0116	A10A Thunderbolt	'77-0117'
56-0659	B52D Stratofortress	
56-0493	C130A Hercules	
64-0699	F4C Phantom II	
Bu150639	F4N Phantom II	'40829'
51-6071	F86L Sabre	
56-3951	F100F Super Sabre	
61-0159	F105D Thunderchief	
63-8285	F105F Thunderchief	
65-5692	CH3E	

452

57-0741	T33A Shooting Star	
56-6716	U2C	
	OV10A	

Tucson - ANGB\International Airport

75-0394	A7D Corsair II	
53-80xx	F86D Sabre	
56-3055	F100D Super Sabre	
58-1207	F100F Super Sabre nose only	
56-1134	F102A Delta Dagger	
Bu145687	CH37C Mojave	

Tucson - Pima Air Museum

Bu135018	EA1F Skyraider	
Bu130361	YEA3A Skywarrior	
Bu142928	TA4B Skyhawk	
Bu148571	A4C Skyhawk	'N401FS'
70-0943	A7D Corsair II	
Bu149289	RA5C Vigilante	
44-35372	A26C Invader	
Bu158975	AV8A Harrier	
Bu159241	AV8A Harrier	
Bu39307	RB1 Conestoga	'XB-DUZ'
44-85828	B17G Fortress	'231892\N9323R'
38-0593	B18A Bolo	'N66267'
39-0051	B23 Dragon	'N534J'
44-44175	B24J Liberator	'HE477\N7866'
40-2347	B25B Mitchell	
43-27712	B25J Mitchell	
43-22494	DB26C Invader	
64-17653	RB26K Invader	'41-39378'
44-70016	TB29 Superfortress	
53-2135	EB47E Stratojet	
49-0372	KB50J Superfortress	
52-0003	NB52A Stratofortress	
55-0067	B52D Stratofortress	
58-0183	B52G Stratofortress	
53-3982	EB57D Canberra	
55-4274	B57E Canberra	
63-13501	RB57F Canberra	'N925NA'
61-2080	B58A Hustler	
55-0395	WB66D Destroyer	
72-1873	YC14A	
72-1875	YC15A	
42-56638	UC36A Electra	'N4963C'
Bu29585	C45G Expediter	'N75018'
Bu29646	C45G Expediter	43-50222
Bu39213	YC45J Expediter	
Bu51184	C45G Expediter	'N40090'
44-77635	C46D Commando	
44-78019	C46D Commando	'N32229'
41-7723	C47 Skytrain	
45-1074	C47B Skytrain	
Bu50837	C47J Skytrain	
42-72488	C54D Skymaster	
Bu12481	C60\R5O-5Lodestar'N15SA'	
42-94549	C69 Constellation	'N90831'
42-39162	UC78B\JRC1 Bobcat'N66794'	
44-23006	C82A Packet	'N6997C'
52-2626	C97G Stratofreighter'HB-ILY'	
53-0151	KC97G Stratofreighter	
Bu50826	C117D Skytrooper'43-49663'	
53-3240	VC118A Liftmaster	
49-0132	C119C Boxcar	
49-0157	C119C Boxcar	
48-0614	C121A Constellation	
53-0548	EC121T Warning Star	
53-0554	EC121T Warning Star	
55-4505	C123B Provider	
54-0580	C123K Provider	'N3142D'
52-1004	C124C Globemaster	
48-0636	YC125A Raider	'XB-GEY\N2573B'
57-0493	C130D Hercules	

52-5803	C131B Samaritan	
59-0527	C133B Cargomaster	
63-8057	EC135J Stratotanker	
61-2489	VC140B Jetstar	
Bu147227	E1B Tracer	
Bu139531	AF1E Fury	
Bu145221	F3B Demon	
64-0673	F4C Phantom II	
Bu151497	YF4J Phantom II	
Bu134748	F6A Skyray	
Bu141363	RF8A Crusader	
Bu144427	DF8F Crusader	
Bu141121	TAF9J Cougar	
Bu144426	RF9J Cougar	
Bu147397	TF9J Cougar	
Bu124629	TF10B Skyknight	
Bu141824	F11A Tiger	
45-8612	F80C Shooting Star	
45-59554	F84B Thunderjet	
47-1433	F84C Thunderstreak	
52-6563	F84F Thunderstreak	
51-1944	RF84F Thunderflash	
53-1525	F86H Sabre	
53-0965	F86L Sabre	
53-2674	F89J Scorpion	
51-5623	F94C Starfire	
54-1823	F100C Super Sabre	
57-0282	F101B Voodoo	
56-0214	RF101C Voodoo	
56-0011	RF101H Voodoo	
56-1393	F102A Delta Dagger	
54-1366	TF102A Delta Dagger	
57-1323	F104D Starfighter	
61-0086	F105D Thunderchief	
62-4427	F105G Thunderchief	
59-0003	F106A Delta Dart	
55-5118	YF107A	
Bu97142	F4U4 Corsair	'Bu97349'
Bu66237	F6F3 Hellcat	
Bu80375	F7F3 Tigercat	
Bu125183	F9F4 Panther	
Bu125177	F9F5 Panther	
63-13141	UH1F Iroquois	
64-13845	UH1D Iroquois	
65-9430	UH1M Iroquois	
66-1211	UH1P Iroquois	
48-0548	H5G Dragonfly	'N9845Z'
USCG232	HO3S Dragonfly	'N4925E'
Bu145842	TH13N Sioux	
52-5737	UH19B Chicasaw	'N2256G'
56-2159	CH21C Shawnee	
Bu134434	UH25B Retriever	
Bu147595	UH25C Retriever	
57-1684	VH34C Choctaw	
58-1005	CH37B Mojave	
59-1583	HH43H Huskie	
Bu139974	OH43D Huskie	
CG1390	HH52A Sea Guard 'N8224Q'	
CG1450	HH52A Sea Guard	
64-18017	TH55A Osage	
64-18133	TH55A Osage	
64-18203	TH55A Osage	
64-18273	TH55A Osage	
64-18350	TH55A Osage	
67-15418	TH55A Osage	
unknown	YAH56A Cheyenne'80548'	
Bu32976	J4F2 Widgeon	
USCG1059	JRS1	'NC16934'
43-26402	L2M Grasshopper	'N59068'
43-27206	L3B Grasshopper	'N46067'
44-16907	L5B Sentinel	'N4981V'
Bu133817	HUM1 (McCulloch MC4A)	
68-6901	O2A Super Skymaster 'N37581'	
69-18006	YO3A QStar	
Bu135620	AP2H Neptune	

44-53247	P38L Lightning	'N90813'
43-11727	P63E King Cobra	'N9003R'
45-8612	P80B Shooting Star	
Bu122071	PBM5A Mariner	'N3190G'
44-21819	PQ14B Cadet	
41-0969	PT17 Kaydet	
41-14675	PT19 Cornell	'N52819'
41-15736	PT22 Recruit	'N53963'
Bu10530	PT26 Cornell	'N1180C'
Bu37257	PV2 Harpoon	'N1270N'
Bu136468	S2F Tracker	'N7255C'
Bu147552	US2A Tracker	'N8225E'
64-17951	SR71A Blackbird	
Bu144200	T1A Seastar	
Bu90822	AT6D Texan	
41-17246	T6B Texan	'49-2908'
Bu92908	T6G Texan	
42-2438	AT7 Navigator	
42-56888	AT9A Jeep	
41-9577	AT11 Kansan	
42-42353	BT13A Valiant	
42-13788	AT17A Crane	
Bu140481	T28B Trojan	
Bu140570	T28C Trojan	
Bu140575	T28C Trojan	
52-1183	T29A	
51-7906	VT29B	
53-6145	T33A Shooting Star	
Bu136810	T33B Shooting Star	
55-2267	T37A Tweety Bird	
61-0854	T38A Talon	
62-4449	CT39A Sabreliner	
Bu69472	TBM3 Avenger	'N9593C'
58-2107	U3A Blue Canoe	
55-4595	U6A Beaver	'C-KMMS'
56-3701	U8D Seminole	
Bu149067	U11A Aztec	
51-0022	HU16A Albatross	
62-5903	OV1B Mohawk	
61-2724	OV1C Mohawk	

also contains a number of civil registered aircraft.

Yuma - MCAS Yuma

Bu158595	AV8A Harrier	
Bu148367	F4B Phantom II	'SH00'
57-0700	T33A Shooting Star	

ARKANSAS

Blythesville - Blythesville AFB

51-8965	T33A Shooting Star

Fort Smith - Municipal Airport\ANGB

51-11292	RF84F Thunderflash
52-3653	F86D Sabre
56-3434	F100D Super Sabre

Granette - town

53-6073	T33A Shooting Star

Jonesboro - town

53-5933	T33A Shooting Star

Little Rock - Camp Robinson

56-0057	RF101C Voodoo

Little Rock - Little Rock AFB

52-0595	B47E Stratojet
53-3841	B57C Canberra
53-8084	C119J Boxcar
53-7543	RF84F Thunderflash
56-0231	RF101C Voodoo
63-8261	F105F Thunderchief
51-9080	T33A Shooting Star

Little Rock - town

53-5342	T33A Shooting Star

Magnolia - town

51-6678	T33A Shooting Star

Wilson - town

53-1047	F86D Sabre

CALIFORNIA

Alameda - NAS Alameda

Bu148610	A4C Skyhawk	
	A4C Skyhawk	'ND300'
Bu152992	F4N Phantom II	

Alhambra - town

53-0563	F86L Sabre

Auburn - town

51-13012	F86L Sabre

Bakersfield - Kern County Airport

51-17059	XF84H Thunderstreak

Banning - town

51-13067	F86E Sabre

Barstow - Yesterday's Air Force

44-83663	B17G Fortress
44-61669	B29 Superfortress

Big Bear Lake - town

52-9172	T33A Shooting Star

Boron

53-1785	YF102	Delta Dagger

Burbank - George Izay Park

57-1333	F104D Starfighter

Campbell - town

51-6825	T33A Shooting Star

Caramillo - town

52-1230	F86H Sabre

China Lake - NAWC China Lake

Bu137814	A4A Skyhawk	
Bu138647	F11A Tiger	
Bu124587	XF4D1 Skyray	
56-1431	F102A Delta Dagger	
57-0777	F102A Delta Dagger	

Chino - Planes of Fame Museum

42-30604	B17F Fortress	
44-83683	B17G Fortress	
44-83684	B17G Fortress	
	B25 Mitchell	'N3675G'
	B25J Mitchell	'N1042B'
	B25J Mitchell	'N8163H'
43-22374	A26B Invader	
	A26C Invader	'N7705C'
44-35323	A26C Invader	'N8026E'
41-35075	B26C Marauder	
46-0010	B50A Superfortress	
Bu91264	TBM3 Avenger	
Bu51131	C45G Expeditor	
	F4B3	'NX3360G'
'88297'	F4U Corsair	
	F4U1 Corsair	'NX83782'
Bu93879	F6F5 Hellcat	'N4994U'
	FM2 Wildcat	'NX7835C'
	F8F Bearcat	
Bu145336	TF8A Crusader	
Bu149218	NF8E Crusader	
Bu126277	F9F5 Panther	
Bu141868	F11F1 Tiger	
'39529'	F80C Shooting Star	
45-59487	YF84A Thunderjet	
45-59566	F84B Thunderjet	
52-7265	RF84F Thunderflash	
'22074'	F86 Sabre	
	F86 Sabre	'N196B'
49-1318	F86A Sabre	
51-6012	F86D Sabre	
51-2849	NF86E Sabre	
Bu531351	QF86H Sabre	
Bu531383	QF86H Sabre	
56-1413	F102A Delta Dagger	
54-4001	H21B Shawnee	
33-123	P26A Peashooter	
44-23314	P38J Lightning	'N2'
42-18547	P39N Airacobra	
44-7369	P40N Warhawk	'NL85104'
'P18723'	P40N Warhawk	
42-28487	P47D Thunderbolt	
42-25254	P47G Thunderbolt	
43-6251	P51A Mustang	'NX4235Y'
44-13334	P51D Mustang	
44-14151	P51D Mustang	
44-14888	P51D Mustang	'NL5441V'
42-108777	YP59A Airocomet	
'4733866'	P80C Shooting Star	
Bu147436	T2A Buckeye	
	T33A Shooting Star	'NX12413'
53-5156	T33A Shooting Star	
53-5541	T33A Shooting Star	
54-0717	XT37 Tweety Bird	
61-0935	T38B Talon	

Chino - Highway 83, Alberts Dairy Equipment yard

Bu137852	UH34G Seabat	HD034
Bu143865	SH34J Seabat	HD075
Bu145752	UH34D Seabat	HD040
63-13196	CH34C Choctaw	HD017

Clovis

57-0775	F102A Delta Dagger	

Dixon - town

51-2950	F86D Sabre	

El Centro - NAF El Centro

Bu158698	AV8A Harrier	

El Toro - MCAS El Toro

Bu148492	A4F Skyhawk	
-	C117D Skytrooper	
Bu124988	F2D Banshee	
Bu148373	F4B Phantom II	
Bu128596	UH25B Retriever	

Fresno - Air Terminal

43-15579	VC47A Skytrain	
49-1272	F86A Sabre	
53-0642	F86L Sabre	
63-1804	F102A Delta Dagger	
59-0146	F106A Delta Dart	
44-73972	P51D Mustang	
52-9640	T33A Shooting Star	

Gaviota - town

51-13082	F86F Sabre	

Guadalupe - town

51-6923	T33A Shooting Star	

Hawthorne - Airfield

51-4424	T33A Shooting Star	

Hemet - Airfield

51-8827	T33A Shooting Star	

Hollywood - Universal Film Studios

51-16992	T33A Shooting Star	

Lancaster - Fox Field (Antelope Valley Aero Museum)

Bu137602	EA-1F Skyraider	'35300'
'150033'	A4E Skyhawk	
Bu156763	A7A Corsair	'160122'
43-21844	A20J Havoc	
41-13251	B25C Mitchell	
53-0272	C97G Stratofreighter	
54-1353	F102A Delta Dagger	

Lemoore - NAS Lemoore

Bu137062	EA1F Skyraider	
Bu142094	TA4B Skyhawk	
Bu156763	A7C Corsair	'156815'

Los Alamitos - ARFC airfield

'590097'	UH1A Iroquois	

Los Angeles - California Museum of Science and Industry

58-1196	T38A Talon	

Los Angeles - Traveltown Park

Bu129655	F7U3 Cutlass	
Bu127420	F9F6 Cougar	
Bu124359	P2V3 Neptune	

Los Angeles - Van Nuys MAP\ANGB

52-4513	F86F Sabre
56-0932	F104C Starfighter
57-0751	T33A Shooting Star

Madera

52-9769	T33A Shooting Star

Marxville - Beale AFB

44-35724	A26C Invader	'434517'
43-28222	B25J Mitchell	
52-1526	EB57B Canberra	
52-10604	C45H Expeditor	
42-23668	VC47A Skytrain	
53-0230	KC97L Stratofreighter	
56-6714	U2C	

Merced - Castle AFB

71-1367	YA9A	
43-38635	B17G Fortress	
37-0029	B18 Bolo	
39-0045	B23 Dragon	
44-41916	B24M Liberator	
44-86891	B25J Mitchell	'402344'
44-34766	A26A Invader	
41-39472	A26B Invader	'4435648'
44-70064	B29 Superfortress	'4461535'
47-0008	B45A Tornado	
52-0166	B47E Stratojet	
49-0351	WB50D Superfortress	
56-0612	B52D Stratofortress	
42-1230	BT13A Valiant	
'44588'	UC45J Expediter	'N87681'
44-77575	C46D Commando	
43-15977	C47N Skytrain	
'12473'	C60A Lodestar	'N1020V'
'435776'	UC78 Bobcat	
53-0354	KC97L Stratofreighter	
'2217'	UH13 Sioux	
62-4513	HH43F Huskie	
55-4512	C123K Provider	
52-1927	F89J Scorpion	
57-0412	F101B Voodoo	
57-1318	F104D Starfighter	
57-5837	F105B Thunderchief	'75887'
'4759'	L2 Grasshopper	
'22318'	L4 Cub	
'76511'	L5 Sentinel	'N57783'
51-15713	L21A Super Cub	
67-21413	O2A Super Skymaster	
68-10848	O2A Super Skymaster	
42-16691	PT17 Kaydet	
42-49384	PT23A	
'02684'	AT6 Harvard	
Bu146253	T28C Trojan	
50-0735	YT34 Mentor	
57-5849	U3A Blue Canoe	
54-1707	U6A Beaver	

Merced - Municipal Airport

53-3850	F80J Shooting Star

Mill Valley - town

53-3682	F86D Sabre

Miramar - NAS Miramar

Bu148149	E2B Hawkeye
Bu150536	E2B Hawkeye

Bu143755	TF8A Crusader
Bu138321	T28B Trojan

Mojave - Airport

56-0753	F104A Starfighter
56-0826	F104A Starfighter
57-1314	F104D Starfighter

Montclare - town

52-10170	F86D Sabre

Muroc - Edwards AFB

73-1664	YA10B Thunderbolt II	
56-0585	B52D Stratofortress	
44-91000	ZXF81	
44-91001	ZXF81	
52-5241	F86F Sabre	
58-0288	F101B Voodoo	
56-0760	NF104A Starfighter	
56-0756	NF104A Starfighter	'N820NA'
56-0801	F104A Starfighter	
61-0146	F105D Thunderchief	
63-9766	F111A	
Bu151874	UH1E Iroquois	'N39AM'
44-22633	XP59B	
52-9846	T33A Shooting Star	
61-0810	T38A Talon	
60-3505	CT39A Sabreliner	
64-1826	XV6A Kestrel	
46-0063	X1E	
'86652'	X24A	

Norton - Norton AFB

61-0069	F105D Thunderchief
61-0674	CT39A Sabreliner

Novato - town

52-2102	F86H Sabre

Oakland - Metropolitan Airport (Western Aerospace Museum)

Bu147666	KA3B Skywarrior	
39-0033	B23 Dragon	'N747M'
44-84029	B29 Superfortress	'N91329'
Bu80532	F7F3 Tigercat	
Bu111966	AT6F Texan	'N9828L'
Bu138349	T28B Trojan	'N4014'
Bu91453	TBM3E Avenger	

Palmdale - Air Force Plant no.42

52-2054	F86H Sabre

Paso Robles - town

51-8415	F86D Sabre

Port Hueneme - town

51-6781	T33A Shooting Star

Redding - town

52-5459	F86F Sabre

Rio Linda - town

53-1160	F86D Sabre

Riverside - March AFB

71-1368	YA9A	
44-6393	B17G Fortress	'230092'
42-50551	B24 Liberator	
44-34030	UB25J Mitchell	
44-35224	A26C Invader	
44-44272	A26 Invader	
44-61669	SB29 Superfortress	
55-0679	GB52D Stratofortress	
52-1519	EB57B Canberra	
53-0466	RB66B Destroyer	
'2365'	BT13 Valiant	
'833538'	C60 Lodestar	
53-0363	KC97L Stratofreighter	
54-0612	C123K Provider	
63-7611	F4C Phantom II	'82163'
47-1595	F84C Thunderjet	
51-9432	F84F Thunderstreak	
53-1304	F86H Sabre	
52-10949	F89J Scorpion	
54-1786	F100C Super Sabre	
59-0418	F101B Voodoo	
57-5803	F105B Thunderchief	
62-4383	F105D Thunderchief	
64-15480	UH1F Iroquois	
	L5 Sentinel	'N63085'
67-21465	O2B Super Skymaster	
	P40 Warhawk	'N9387A'
44-22614	P59A Airacomet	
'4216388'	PT13D Kaydet	
'49379'	PT13D Kaydet	
'8813627'	AT6 Havard	
	T28 Trojan	'N99394'
58-0513	T33A Shooting Star	
62-4465	CT39A Sabreliner	
56-6721	U2CT	
58-2163	U3A Blue Canoe	'38137'
51-7234	HU16E Albatross	'1293'

Rosamund - El Dorado Aircraft Supplies

53-0994	F86L Sabre
57-0393	F101B Voodoo
65-9638	UH1H Iroquois
66-1008	UH1H Iroquois
68-15365	UH1H Iroquois
66-15143	UH1M Iroquois
72-21379	OH58A Kiowa
68-8131	T38A Talon
Bu152879	YOV10A Bronco

Sacramento

57-0352	F101F Voodoo
57-0393	F100F Voodoo

Sacramento - Executive Airport

58-0491	T33A Shooting Star

Sacramento - Mather AFB

53-3489	T29C
68-10771	X25B

Sacramento - McClellan AFB

Bu132463	A1E Skyraider
70-0998	A7D Corsair II
44-35617	RB26C Invader
42-65281	B29 Superfortress
44-70049	B29 Superfortress
44-84084	B29 Superfortress
Bu141309	EC121K Warning Star

64-0706	F4C Phantom II	
45-8704	P80B Shooting Star	
51-1772	F84F Thunderstreak	
51-13002	F86F Sabre	
51-2968	F86L Sabre	
52-1959	F89D Scorpion	
55-3733	F100D Super Sabre cockpit only	
56-3288	F100D Super Sabre	'32288'
57-0427	F101B Voodoo	
56-1140	F102A Delta Dagger	'55431'
57-1303	F104B Starfighter	
54-0100	F105B Thunderchief	'40105'
62-4301	F105D Thunderchief	
61-0065	F105D Thunderchief	
63-8325	F105F Thunderchief	
63-9770	GF111A	
Bu152714	F111B	
65-5690	HH3E	
'5745'	L2B Grasshopper	'N53792'
Bu138327	T28B Trojan	
53-5205	T33A Shooting Star	
61-0660	T39A Sabreliner	

San Diego - Aerospace Museum, Balboa Park

	A4B Skyhawk	'NP302'
60-6933	A11A Blackbird	
51-2910	F86E Sabre	
Bu42874	F6F3 Hellcat	
Bu135763	F2Y1 SeaDart	
Bu33594	J2F6 Duck	
54-1619	X13A Vertijet	

San Francisco - International Airport, University Dept. of Aeronautics

53-4997	T33A Shooting Star
58-2158	U3A Blue Canoe

San Jose - Municipal Airport

53-1709	F100C Super Sabre 'NASA703'

San Jose - Moffet Field NAS

Bu128393	DP2E Neptune

San Lois Obispo - Camp San Lois Obispo

'612408'	C47A Skytrain
56-2258	OH23C Raven
51-7312	O1A Bird Dog
51-13793	OH13E Sioux
59-4973	UH19D Chicasaw
63-4534	CH34A Choctaw
53-32187	U6A Beaver

Santa Ana - MCAS Tustin

Bu150219	UH34D Choctaw
Bu138990	UH43D Huskiea

Santa Maria

56-0995	F102A Delta Dagger

Sierra Masre - town

51-13064	F86D Sabre

Travis - Travis AFB

'412446'	B17G Fortress
56-0696	B52D Stratofortress
42-65281	WB29 Superfortress

52-10865	C45H Expeditor	
41-19729	C56	Lodestar
Bu131602	C118B Liftmaster	
52-1000	C124C Globemaster	
5788	HC131A SamaritanUSCG	
56-1247	F102A Delta Dagger	
62-4299	F105D Thunderchief	
52-8688	HH21B Shawnee	
68-10848	O2A Super Skymaster	
68-18007	YO3A QStar	
60-3483	CT39A Sabreliner	
62-4452	CT39A Sabreliner	
'180'	PT19 Cornell	

Tulare

44-85738	B17G Fortress

Victorville - George AFB

53-1515	F86H Sabre
54-2299	F100D Super Sabre
56-0934	F104C Starfighter
61-0165	F105D Thunderchief
62-4416	F105G Thunderchief

West Covina - town

52-3784	F86D Sabre

Woodland - town

51-9025	T33A Shooting Star

Yucaipa - town

51-12985	F86E Sabre

COLORADO

Boulder - town

53-6028	T33A Shooting Star

Brighton - town

53-1067	F86D Sabre

Colorado Springs - Air Force Academy

55-0083	B52D Stratofortress	
64-0799	F4C Phantom II	
66-7463	F4D Phantom II	
75-0748	GF16A Fighting Falcon	
55-2967	F104A Starfighter	
60-0482	F105D Thunderchief	
59-1602	T38A Talon	
46-0676	X4	'6677'

Colorado Springs - Peterson AFB

55-4279	EB57E Canberra	
52-3425	EC121T Warning Star	
53-0782	F86L Sabre	
52-1941	F89J Scorpion	
50-1006	F94C Starfire	
58-0274	F101B Voodoo	
56-1109	F102A Delta Dagger	'01109'
56-0936	F104C Starfighter	'60808'
42-105927	P40N Warhawk	
44-89425	P47N Thunderbolt	
57-0575	T33A Shooting Star	

Denver - Buckley ANGB

'612884'	F86F Sabre	
53-1578	F100C Super Sabre	'41897'

Denver - Lowry AFB

Bu148613	NA4E Skyhawk
44-69729	B29 Superfortress
52-0005	RB52B Stratofortress
63-7551	F4C Phantom II
52-5308	F86F Sabre
56-3299	GF100D Super Sabre
56-3417	F100D Super Sabre
58-0271	F101B Voodoo
56-0984	GF102A Delta Dagger
56-0910	F104C Starfighter
60-0508	GF105D Thunderchief
59-0134	GF106A Delta Dart
53-4379	CH21B
Bu150213	UH34D Choctaw
56-1710	GT33A Shooting Star
57-5845	U3A Blue Canoe

Englewood - town

50-0731	F86D Sabre

Pueblo - Memorial Airport Aircraft Museum

Bu147702	A4C Skyhawk	
44-35892	A26C Invader	
44-62022	B29 Superfortress	
53-2104	RB47E Stratojet	
	C47J Skytrain	'N64605'
Bu131688	C119F Boxcar	
Bu134936	F6A Skyray	
Bu145349	F8A Crusader	
47-1562	F84C Thunderjet	
55-3503	F100D Super Sabre	
53-2418	JF101A Voodoo	
53-4347	CH21B Shawnee	
Bu148002	SH34J Choctaw	
Bu128402	SP2E Neptune	
'91872'	T33A Shooting Star	
Bu137939	T33B Shooting Star	

Yuma - town

51-8667	T33A Shooting Star

CONNECTICUT

Hartford - Airport

Bu138048	T33B Shooting Star

Meriden - town

53-1367	F86D Sabre

Windsor Locks - Bradley Airport\ANGB

Bu142246	A3B Skywarrior	
Bu142219	A4A Skyhawk	
Bu125739	AD4N Skyraider	
Bu22275	AM1	Mauler
44-85734	B17G Fortress	
43-4999	B25H Mitchell	
43-22499	A26C Invader	'N86481'
44-61975	TB29A Superfortress	

51-2360	WB47E Stratojet	
52-1488	RB57A Canberra	
59-0529	C133B Cargomaster	
Bu147217	E1B Tracer	
Bu145310	F4A Phantom II	
Bu146968	F8H Crusader	
Bu147030	F8K Crusader	
47-1513	F84C Thunderjet	
47-1530	F84C Thunderjet	
52-7374	RF84F Thunderflash	
58-0263	F86A Sabre	
52-1896	F89J Scorpion	
52-5761	F100A Super Sabre	
53-1580	F100A Super Sabre fuselage only	
55-3805	F100D Super Sabre	
55-3450	F102A Delta Dagger	
56-1221	F102A Delta Dagger	
56-0901	F104C Starfighter	
57-5778	F105B Thunderchief	
Bu134836	F4D1 Skyray	
Bu79192	F6F5 Hellcat	
Bu80759	XF4U4 Corsair	
Bu74120	FM2 Wildcat	
62-12550	UH1B Iroquois	
49-2007	H5H Dragonfly	
56-4257	H19D Chicasaw	
52-3812	OH23G Raven	
Bu145717	LH34D Chocktaw	
60-0289	HH43F Huskie	
Bu130063	HUP1 Retriever	'9602'
Bu131427	SP2E Neptune	
Bu33966	PBY5A Catalina	
51-0025	HU16B Albatross	
7228	HU16E Albatross USCG	
45-49458	P47D Thunderbolt '420344'	
43-11117	RP63C Kingcobra	
43-46503	R4B Hoverfly	
Bu140063	T28C Trojan	
57-2570	U6A Beaver	

DELAWARE

Delaware City - town

56-1650	T33A Shooting Star	

Dover - Dover AFB

42-32075	B17G Fortress	
51-7431	T33A Shooting Star	

Wilmington - New Hanover County Airport

53-1296	F86H Sabre	

DISTRICT OF COLUMBIA

Washington - Bolling AFB

61-0138	F105D Thunderchief	
Bu137756	T28B Trojan	

Washington   Smithsonian   Institute   (National
Aeronautics and Space Museum)

Bu148314	A4C Skyhawk	
41-31771	B26B Marauder nose only	
52-1551	EB57B Canberra	
57-0460	C130A Hercules	
55-2961	YF104A Starfighter 'NASA818'	

Bu9241	F4B4	
Bu9056	F9C2 Sparrowhawk	
'15392'	F4F Wildcat	
Bu111759	FH1 Phantom I	
Bu50375	F4U1D Corsair	
65-9446	UH1M Iroquois	
57-2729	UH13J Sioux	
Bu148768	UH34D Choctaw	
42-13610	XO60	
41-13574	P40E Warhawk	
42-108784	XP59A Airacomet	
44-83020	P80 Shooting Star	
41-18874	XR4 Hoverfly	
Bu54605	SBD6 Dauntless	
NASA521	XV6A Kestrel	
46-0062	X1	
56-6670	X15A	

This facility also contains a large number of civilian
aircraft exhibits and also foreign military aircraft.

FLORIDA

Arcadia - town

52-9696	T33A Shooting Star	

Bay Harbor

54-1881	F100C Super Sabre	

Callaway

57-0417	F101B Voodoo	'60417'

Cecil Field - NAS Cecil Field

Bu152670	A7B Corsair	'156790'

Cross City - town

51-4200	T33A Shooting Star	

Dade - town

53-0728	F86L Sabre	

Defuniak Springs - Airport

51-8959	T33A Shooting Star	

Eau Galle

55-2824	F100D Super Sabre	

Florida City - Highway 27

53-0998	F86L Sabre	

Homestead - Homestead AFB

54-2294	F100D Super Sabre	
66-0267	F4D Phantom II	

Homstead   - town\Highway US1

66-0273	F4D Phantom II	

Indian Rocks Beach - town

51-6059	F86D Sabre	

Jacksonville - International Airport

53-5325	T33A Shooting Star

Jacksonville - NAS Jacksonville

Bu149946	A6E Intruder
	F11A Tiger
	F11A Tiger
57-0817	F102A Delta Dagger

Key West - NAS Key West

Bu156612	RA5C Vigilante	'156610'
Bu153019	F4N Phantom II	
56-1374	F102A Delta Dagger	

Kissimmee - SST Aviation Exhibit Center

	B25 Mitchell
Bu135765	XF2Y1 SeaDart
56-0005	F101C Voodoo
52-9633	T33A Shooting Star
51-7144	HU16B Albatross

Macdill AFB

66-0302	F4E Phantom II

Maitland - town

54-0396	F89H Scorpion

Mary Esther - Hurlburt Field AFB

Bu132598	A1E Skyraider
64-17666	A26A Intruder
44-77424	C46D Commando
42-100510	C47A Skytrain
55-4533	C123K Provider
64-15493	UH1P Iroquois
56-4208	O1E Bird Dog
67-21368	O2A Super Skymaster
49-1663	T28A Trojan
63-3606	U10A Super Courier

Mayport - NAS Mayport

Bu142150	A4A Skyhawk
Bu161173	YSH60B Blackhawk

Milton - town\Highway 98

Bu138353	T28B Trojan	'136000'

Orange Park - town

53-4063	F86L Sabre

Orlando - Airport

56-0687	B52D Stratofortress
59-0400	F101B Voodoo

Panama City - Gulf Coast Community College

57-0438	F101B Voodoo

Panama City - Tyndall AFB

52-1516	EB57B Canberra	dumped
52-10137	F86D Sabre	'45244'
52-10133	F86L Sabre	
52-1862	F89J Scorpion	'48422'
56-3008	F100D Super Sabre	

57-0332	F101B Voodoo	
57-0858	F102A Delta Dagger	'57858'
56-0919	F104A Starfighter	
59-0145	F106A Delta Dart	
58-0619	T33A Shooting Star	'55225'

Pensacola - Museum of Naval Aviation, NAS Pensacola

Bu135300	A1H Skyraider	
Bu135418	A3A Skywarrior	
Bu137830	A4A Skyhawk	
Bu142929	A4B Skyhawk	
Bu143675	A4J Skyhawk	
Bu147788	A4C Skyhawk	
Bu154983	A4F Skyhawk	
Bu153525	TA4F Skyhawk	
Bu155000	A4F Skyhawk	
Bu156621	RA5C Vigilante	
Bu156624	RA5C Vigilante	
Bu137813	XA4D1 Skyhawk	
Bu130418	AJ2 Savage	
Bu136754	C1A Trader	
Bu49771	RC45J Expediter	
Bu12418	C47H Skytrain	
Bu50811	C47J Skytrain	
Bu56533	C54Q Skymaster	
Bu50821	C117D Skytrooper	
Bu128424	VC118B Liftmaster	
Bu128431	C118B Liftmaster	
Bu143221	EC121K Warning Star	
Bu141015	C131F Samaritan	
Bu148146	E1B Tracer	
Bu139486	F1E Fury	
Bu149457	F4B Phantom II	
Bu153915	F4N Phantom II	
Bu158355	F4S Phantom II	cockpit
Bu134806	F6A Skyray	
Bu145347	F8A Crusader	
Bu149134	F8H Crusader	
Bu141828	F11A Tiger	
Bu141884	F11A Tiger	Gate Guard
72-1570	YF17A	'201570'
Bu9351	FF1	
Bu94203	F6F5 Hellcat	
Bu92246	FG1D Corsair	
Bu86690	FM2 Wildcat	
Bu86749	FM2 Wildcat	
Bu128911	TH13L Sioux	
Du142377	TH13M Sioux	
Bu145864	CH37C Mojave	
Bu138101	OH43D Huskie	
1355	HH52A Seaguard	USCG
Bu157363	TH57A Sea Ranger	
1235	HO3S1G Dragonfly	USCG
Bu39047	HNS1 Hoverfly	
Bu75615	HOS1 Hoverfly II	
Bu125519	HO5S1	
Bu128647	HTE1	
Bu33581	J2F6 Duck	USCG
V212	J4F1 Widgeon	USCG
A5483	MF boat	
Bu43156	N2S5 Kaydet	
A8605	N2Y1 Fleet 1	
A2294	NC4	
Bu9681	NS1	
A8588	NT1	
2693	N3N3 Yellow Peril	
Bu5926	OS2U3 Kingfisher	
Bu33840	P80C Shooting Starex 47-1387	
Bu141234	SP2H Neptune	
Bu135523	SP5B Marlin	
Bu89082	XP2V1 Neptune	
Bu08317	PBY5 Catalina	
Bu37537	PV2 Ventura	

Bu39047	R4B Hoverfly	
Bu75615	R6A Hoverfly	
Bu9206	RR5	Trimotor
Bu83479	SB2C5	Helldiver
Bu06701	SBD3	Dauntless
'42-60817'		
Bu51968	SNJ5	Texan
Bu112121	SNJ6 Texan	
A5858	S4C1 Scout	
Bu153542	T2B Buckey	
Bu138223	T28B Trojan	
Bu138271	T28B Trojan	
Bu140041	T28B Trojan	
Bu140661	T28B Trojan	
Bu138073	T33B Shooting Star	
Bu140784	T34B Mentor	
Bu140816	T34B Mentor	
Bu144067	T34B Mentor	
Bu151339	T39D Sabreliner	
Bu53593	TBM3 Avenger	
Bu144672	U1B Otter	
57-6194	RU9D	
7236	HU16E Albatross	USCG

Pensola - Regional Airport

F11A Tiger

Pinellas Park

44-83663	B17G Fortress
53-0658	F86D Sabre
51-6838	T33A Shooting Star

Pompano Beach - town

54-0318	F89H Scorpion

St. Augustine

7214	HU16E Albatross	USCG

St. Petersburg\Clearwater - Yesterday's Air Force

Bu148592	A4C Skyhawk
44-83663	B17G Fortress
Bu145308	F4A Phantom II
56-0986	F102A Delta Dagger
	F3H Demon
1343	GUH34J Choctaw USCG '06494'
	P2V Neptune
58-0700	T33A Shooting Star
67-15006	T41A Mescalero
52-6135	U6A Beaver
60-3758	OV1C Mohawk

Tallahassee - Municipal Airport

Bu151858	UH1E Iroquois
58-0470	T33A Shooting Star

Tampa - MacDill AFB

66-0302	F4E Phantom II

Valparaiso - Eglin AFB, Armament Museum

Bu77231	PB-1W\B17G Fortress '46106'
	ex 44-83863
44-30854	JB25J Mitchell '02344'
53-4296	RB47H Stratojet
58-0185	B52G Stratofortress
	B57 Canberra
44-76486	C47K Skytrain '43-010'
53-0310	KC97L Stratofreighter

53-7821	C131B Samaritan	
64-0817	F4C Phantom II	'40813'
68-15796	F51D Mustang	ex 44-13571
49-0432	F80C Shooting Star '9713'	
51-9495	F84F Thunderstreak	
52-5513	F86F Sabre	'12831'
53-2610	F89J Scorpion	
54-1986	F100C Super Sabre '52951'	
56-0250	F101B Voodoo	
53-1799	F102A Delta Dagger	
57-1331	F104B Starfighter	
58-1155	F105D Thunderchief '58771'	
68-6864	O2A Super Skymaster 'N37555'	
44-89320	P47N Thunderbolt 'N345GP'	
49-0432	P80C Shooting Star '4713'	
64-17959	SR71A Blackbird	
53-5947	T33A Shooting Star	

Whiting Field - NAS Whiting Field

Bu154759	UH1E Iroquois	
Bu157807	TH1L Iroquois	
Bu157362	TH57A Ranger	
Bu138144	T28B Trojan	'138701'
Bu140861	YT34C Mentor	

Winter Haven - town

54-0337	F89H Scorpion

GEORGIA

Atlanta - NAS Atlanta

'135000'	A4 Skyhawk
68-15140	AH1G Huey Cobra

Atlanta - Dobbins AFB

63-7559	F4C Phantom II	'66-689'
56-2928	F100D Super Sabre	
62-4425	F105G Thunderchief	
63-8345	F105G Thunderchief	
41-14094	P40F Warhawk	

note: Atlanta NAS and Dobbins AFB are both located on the Lockheed airfield at Marietta.

Augusta - town

51-6746	T33A Shooting Star

Calhoun - World Aircraft Museum

52-6476	F84F Thunderstreak
51-5896	F86L Sabre
52-9574	T33A Shooting Star

Douglas - town

53-6132	T33A Shooting Star

Fort Benning - Lawson AAF

'093790'	C47A Skytrain
	UH1B Iroquois

Griffin - Spalding County Municipal Airport

53-6096	T33A Shooting Star

Hahira - town		
51-6727	T33A Shooting Star	
Lakeland - town		
51-8555	T33A Shooting Star	
Macon - Municipal Airport, ANG Base		
52-3651	F86L Sabre	
Marietta - Cobb County Youth Museum		
51-9382	F84F Thunderstreak	
Ozark - Blackwell Airport, Aviation & Technical College		
51-1748	F84F Thunderstreak	
67-17034	TH13T Sioux	
64-18138	TH55A Osage	
66-18314	TH55A Osage	
56-3718	U8D Seminole	
57-3097	U8D Seminole	
Richland - town		
49-2449	F89B Scorpion	
Savannah - International Airport		
48-0741	F84D Thunderjet	
51-5891	F86L Sabre	
Thomasville - town		
52-9191	T33A Shooting Star	
Valdosta - town center		
52-10057	F86L Sabre	
Warner Robins - Watson Bvd.		
52-9225	T33A Shooting Star	

Warner Robins AFB

44-84053	B29B Superfortress	
55-0085	B52D Stratofortress	
52-1457	RB57A Canberra	
55-0392	WB66D Destroyer	
63-9756	C7B Caribou	
Bu50811	C47J Skytrain	'4349442'
Bu150190	C47J Skytrain	'4348957'
45-0579	C54G Skymaster	
52-2604	KC97L Stratofreighter	
51-2566	C119C Boxcar	
54-0633	UC123K Provider	
51-0089	C124C Globemaster	
74-1686	YMC130H Hercules	
61-2488	VC140B Jetstar	
45-8357	F80C Shooting Star	
45-1604	F84E Thunderjet	
52-7244	RF84F Thunderflash	
53-1513	F86H Sabre	'31511'
53-2463	F89J Scorpion	
54-1851	F100C Super Sabre	
58-0276	F101F Voodoo	
57-0906	F102A Delta Dagger	
61-0099	F105D Thunderchief	
62-4438	F105F Thunderchief	
65-7959	UH1F Iroquois	
Bu142376	TH13M Sioux	
Bu143143	UH13P Sioux	

55-3228	UH19D Chicasaw	
52-8685	H21B Shawnee	
56-4021	OH23C Raven	
Bu148963	HH34J Choctaw	
58-1833	H43A Huskie	
52-3314	O1E Bird Dog	
Bu147954	SP2H Neptune	'44037'
Bu140553	T28B Trojan	
Bu140649	T28C Trojan	
49-1938	T29A	
52-9533	T33A Shooting Star	
Bu144033	T34B Mentor	
62-4461	CT39A Sabreliner	
Bu150986	T39D Sabreliner	
42-16365	PT17 Kaydet	
60-6052	U3B Blue Canoe	
52-6087	U6A Beaver	
51-7144	HU16B Albatross	
Waynesboro - town		
51-6635	T33A Shooting Star	
Willacoochie - town		
51-6612	T33A Shooting Star	
Wright AAF		
64-9497	UH1M Iroquois	
65-12741	UH1M Iroquois	

HAWAII

Barbers Point - NAS Barbers Point

Bu152291	F4N Phantom II	
Bu126275	F9F5 Panther	
Bu150279	SP2H Neptune	
Bu147870	S2D Tracker	

Honolulu - Bishop Museum

43-49852	C47B Skytrain	
44-9041	C54M Skymaster	

Honolulu - Hickam AFB

44-35596	RA26C Invader	
44-31504	B25J Mitchell	
63-7450	F4C Phantom II	
50-0653	F86E Sabre	
51-2841	F86E Sabre	
52-4191	F86L Sabre	
54-1373	F102A Delta Dagger	
55-3366	F102A Delta Dagger	
68-11166	O2A Super Skymaster	
51-6533	T33A Shooting Star	
67-14636	OV10A Bronco	

Kaneohe Bay - MCAS Kaneohe Bay

Bu146973	F8K Crusader	

Wheeler AFB

58-0324	F101F Voodoo	
68-11144	O2A Super Skymaster	
'39411'	P40 Warhawk	

IDAHO

Blackfeet - town

51-6074   F86D Sabre

Boise - Gowen Field

53-1812   F102A Delta Dagger
53-1816   F102A Delta Dagger

Burley - town

52-9594   T33A Shooting Star

Idaho Falls - Airport

53-1022   F86L Sabre

Mountain Home - Mountain Home AFB

52-6470   F84F Thunderstreak'26567'
54-1748   F100C Super Sabre
55-2784   F100D Super Sabre
63-9776   RF111A

Napa - City Park

49-2457   F89B Scorpion

Pocatello - town

52-10079   F86D Sabre

Twin Falls - Airport

56-1660   T33A Shooting Star

ILLINOIS

Atlamont - town

51-6211   F86D Sabre

Brookfield - town

53-0700   F86L Sabre

Centralia - town

52-9651   T33A Shooting Star

Chicago - NAS Glenview

           UH1B Iroquois
           UH1B Iroquois
68-16705   OH58A Kiowa
57-6142   U6A Beaver

Highland - town

53-4914   T33A Shooting Star

Hillsboro - town

51-6126   F86D Sabre

Mattoon - City Park

51-9339   F84F Thunderstreak

Odin - town

51-9343   F84F Thunderstreak

Peoria - Airport

51-9313   F84F Thunderstreak
68-11160   O2A Super Skymaster

Quincy - park

52-9770   T33A Shooting Star

Rantoul - Chanute AFB

44-35204   A26C Invader   '434414'
44-30635   B25N Mitchell   '02344'
45-21748   B29 Superfortress
51-13730   RB36H Peacemaker'492065'
56-0066   XB47 Stratojet   '2278'
52-8714   RB52B Stratofortress
56-0708   GB52E Stratofortress
52-1584   GNB57B Canberra
55-0666   YRB58A Hustler   '12059'
53-0412   GRB66B Destroyer
43-49336   VC47 Skytrain
52-0898   GC97G Stratofreighter
Bu141311   EC121K Warning Star
54-0692   C123K Provider
55-0037   C130A Hercules
56-2009   GC133A Cargomaster
45-8501   F80B Shooting Star'91830'
45-59494   YP84A Thunderjet '80656'
51-9531   F84F Thunderstreak
47-0614   F86A Sabre   '12910'
54-1784   F100C Super Sabre
56-0273   F101B Voodoo
54-1351   GTF102A Delta Dagger
56-1264   F102A Delta Dagger
56-0732   F104A Starfighter
63-8287   F105F Thunderchief
63-9767   GF111A
44-64265   P51H Mustang
52-9797   T33A Shooting Star
51-7200   HU16B Albatross
61-0686   UH1B Iroquois
Bu154778   UH1E Iroquois
67-21411   O2A Super Skymaster
54-2739   T37A Tweety Bird
62-4494   CT39A Sabreliner

Scott AFB

Bu131617   C118B Liftmaster
60-3495   CT39A Sabreliner

Southern - Airport

51-4374   T33A Shooting Star

Springfield

51-1822   F84F Thunderstreak

West Union - town

52-9604   T33A Shooting Star

INDIANA

Anderson   -   town

52-2004   F86H Sabre

Brownsburg   -   town

50-0975   F94C Starfire

Churubusco - City Park

53-1298   F86H Sabre

Covington - town

52-9326   T33A Shooting Star

Elkhart - town

51-9350   F84F Thunderstreak

Fort Wayne - Baer Field Municipal Airport

67-0389   F4E Phantom II

Gary - town

51-9079   T33A Shooting Star

Huntingdon - town

51-6745   T33A Shooting Star

Indianapolis - Stout Field Military Museum

44-76457   C47D Skytrain
54-1993   F100C Super Sabre
58-0321   F101B Voodoo
56-0898   F104C Starfighter

Mooresville - town

51-6782   T33A Shooting Star

Muncie - town

51-8609   T33A Shooting Star

Peru - Grissom AFB

44-83690   B17G Fortress   '4231255'
44-86843   B25J Mitchell
51-2315   B47B Stratojet   '20271'
43-49270   C47D Skytrain
52-2697   KC97L Stratofreighter
52-5850   C119G Boxcar
51-9381   F84F Thunderstreak
             F100D Super Sabre
57-1322   F104D Starfighter
61-0088   F105D Thunderchief
68-6871   O2A Super Skymaster
52-9563   T33A Shooting Star
57-5922   U3A Blue Canoe

Roanoake - town

53-5392   T33A Shooting Star

Terre Haute - Hulman Field

54-2181   F100D Super Sabre
56-3320   F100D Super Sabre '54181'

IOWA

Emmetsburg - town

52-9626   T33A Shooting Star

Iowa City - Airport

53-0750   F86L Sabre

Newton - town

52-3640   F86D Sabre

Oetwin - town

51-4406   T33A Shooting Star

Rock Valley - town

52-9612   T33A Shooting Star

Sergeant Bluffs - town

52-7267   RF84E Thunderflash

Sheldon - town

52-9614   T33A Shooting Star

Sioux City - Airport

52-6418   F84F Thunderstreak
54-2005   F100C Super Sabre

Urbandale - town

53-0924   F86L Sabre

Waverley - town

53-4951   T33A Shooting Star

KANSAS

Augusta - town

52-9488   T33A Shooting Star

Colby - town

51-6617   T33A Shooting Star

Hiawatha - town

53-0718   F86L Sabre

Independence - town

52-9258   T33A Shooting Star

Manhattan - Municipal Airport

51-5676   F94C Starfire

Olathe\Johnson County - England AFB

Bu129565	F7U Cutlass	

Topeka   -   Forbes Field and
Combat Air Museum

44-34104	A26B Invader	'N99420'
	B25J Mitchell	'N3161G'
52-1480	RB57A Canberra	
55-4260	EB57E Canberra	
44-76582	C47B Skytrain	'N7102'
52-3418	EC121H Warning Star	
51-1059	F84E Thunderjet	
52-6458	F84F Thunderstreak	
Bu141811	F11A Tiger	
	F86K Sabre	
57-0410	JF101B Voodoo	'N8234'
'481482'	L17A Navion	
Bu136486	US2A Tracker	
42-44709	T6D Texan	'N4292C'
Bu137759	T28B Trojan	'N49914'
51-8614	T33A Shooting Star	
52-9632	T33A Shooting Star	
58-2069	U6A Beaver	'N217GB'
58-1338	U8D Seminole	
58-1358	RU8D Seminole	

Wichita - Midcontinent Airport\Interstate 54

53-4218	B47 Stratojet	

Wichita - Mid Continent Airport\Cessna Wallace Plant

54-0717	XT37 Tweety Bird	

Wichita - McConnell AFB

55-0094	B52D Stratofortress	
'24256'	F86H Sabre	
52-4253	F105D Thunderchief	
53-8366	F105F Thunderchief	

KENTUCKY

Capital City

56-0125	RF101C Voodoo	

Fort Campbell

62-2010	UH1B Iroquois	

Fort Knox

60-3554	UH1B Iroquois	
64-14005	UH1B Iroquois	

Louisville - Municipal Airport

56-0001	RF101H Voodoo	

Sturgis - town

52-3694	F86D Sabre	

LOUISIANA

69-6234	A7D Corsair	
73-1667	GYA10A Thunderbolt	
63-8296	F105G Thunderchief	

Gretna - town

53-0822	F86L Sabre	

Kapian - town

52-9492	T33A Shooting Star	

Mansfield - town

51-6601	T33A Shooting Star	

Minden - town

60-0743	F89C Scorpion	

New Orleans - NAS New Orleans

	F11A Tiger	
56-3020	F100D Super Sabre	
51-16336	OH23 Raven	

Shevreport - Barksdale AFB

43-83884	B17G Fortress	'4338289'
44-48781	B24J Liberator	
53-2276	B47E Stratojet	
56-0629	B52D Stratofortress	
'4316130'C47	Skytrain	'N3753C'
43-5314	UC64A Norseman	
53-0240	KC97L Stratofreighter	
51-1396	F84F Thunderstreak	
	F111	
67-21440	O2B Super Skymaster	
44-14570	P51D Mustang	

White Castle - town

53-1032	F86L Sabre	

MAINE

Brunswick - NAS Brunswick

Bu128392	P2V5 Neptune	

Cambridge - Hanscomb AFB

'15085'	F86H Sabre	

Proctor

59-0407	F101F Voodoo	

Waterville - town

52-1856	F89J Scorpion	

MARYLAND

Andrews AFB

53-1574	F100A Super Sabre	
61-0041	F105D Thunderchief	
57-2523	F106B Delta Dart	
62-12554	UH1B Iroquois	
Bu77722	F6F5	Hellcat

Annapolis - US Naval Air Academy

Bu139968	A4A Skyhawk
Bu148275	F4A Phantom II
Bu3022	
N3N3	

Baltimore - Glenn L Martin State Airport

53-1411	F86H Sabre	
51-7183	HU16E Albatross	'10004'

Clinton - town

53-1348	F86H Sabre

Crisfield - town

52-2023	F86H Sabre

Cumberland - town

51-4157	T33A Shooting Star

Ellicott City - town

52-2048	F86H Sabre

Gaithersburg - park

51-6988	T33A Shooting Star

Lanhan - town

52-3864	F86H Sabre

Patuxent River - NAS Patuxent River

Bu146697	A5A Vigilante	'156697'
Bu156643	RA5C Vigilante	
Bu152658	A7A Corsair	
Bu153071	F4J Phantom II	
Bu134764	F6A Skyray	
Bu149240	S2D Tracker	

Pocomoke City - Route 31

52-9650	T33A Shooting star

Princess Anne - town

52-2066	F86H Sabre

Silver Hill - Smithsonian National Air & Space Museum storage facility

Bu132463	A1E Skyraider	
40-3097	B17D Fortress	nose only
44-86292	B29 Superfortress	
36-0353	XC35	
53-0243	KC97L Stratofreighter nose only	
Bu145307	F4A Phantom II	
Bu146860	RF8G Crusader	
48-0260	F86A Sabre	

56-3245	F100D Super Sabre	
56-3440	F100D Super Sabre	
56-0119	RF101C Voodoo	nose only
60-0445	F105D Thunderchief	
Bu136849	XFY1	Pogostick
Bu138899	XF8U1	Crusader
	F8F Bearcat	'N1111L'
Bu138652	H32	Hornet
Bu61064	N2S5	Kaydet
51-11963	O1A Bird Dog	
Bu5909	OS2U3	Kingfisher
42-67762	P38J Lightning	
44-32961	P47D Thunderbolt	
42-78846	XP55	
43-47954	XR5	Hoverfly
Bu51398	SNJ4	Texan
58-7055	VZ9 Avrocar	

MASSACHUSETTS

Cape Cod - Cape Cod CGAS

7250	HU16E Albatross	USCG

Falmouth - town

51-5692	F94C Starfire

Hanscom AFB

'15085'	F86H Sabre
Bu127074	TF10B Skyknight
58-1196	T38A Talon

Monument Beach - town

50-1054	F94C Starfire

Otis ANGB

52-6382	F84F Thunderstreak
53-1113	F86F Sabre
56-2995	F100D Super Sabre
51-4335	T33A Shooting Star
51-7250	HU16E Albatross

South Weymouth - NAS South Weymouth

Bu142940	A4B Skyhawk
Bu51483	SNJ4 Texan

MICHIGAN

Adrian - town

51-4322	T33A Shooting Star

Alpena

61-0103	F105D Thunderchief

Auburn - Oak Park

52-7262	RF84K Thunderflash

Battle Creek - Kellog Airport

52-1426	RB57A Canberra

51-1664	F84F Thunderstreak	
56-3894	F100F Super Sabre	
54-2730	T37B Tweety Bird	

Belleville - town

52-7259	RF84K Thunderflash	

Birmingham - town

53-4028	F86D Sabre	

Brekenridge - town

51-4067	T33A Shooting Star	

Calumet - Calumet AFS

51-4159	T33A Shooting Star	

Centre Line - town

51-4263	T33A Shooting Star	

Detroit - Selfridge ANGB

Bu142761	A4B Skyhawk
52-1485	RB57A Canberra
57-0514	C130A Hercules
51-1664	F84F Thunderstreak'1604'
51-1896	RF84F Thunderstreak
49-1095	F86A Sabre
56-3849	F100F Super Sabre
56-3025	F100D Super Sabre
56-0048	RF101C Voodoo
56-0817	F104A Starfighter
Bu98085	FG1D Corsair
67-21340	O2A Super Skymaster
Bu144721	US2A Tracker
53-6099	T33A Shooting Star
58-3111	U3B Blue Canoe

Detroit - Ypsilanti\Willow Run, Yankee AF Museum

55-0677	B52D Stratofortress
44-76716	C47B Skytrain
52-7259	RF84F Thunderflash
56-0235	NF101B Voodoo
Bu80382	F7F3N Tigercat
Bu141872	F11F Tiger
Bu155226	T2B Buckeye
49-1700	T28A Trojan
51-8786	T33A Shooting Star

Grayling - town

51-8673	T33A Shooting Star	

Grosse Point Woods - town

52-7275	RF84F Thunderflash	

Holland - town

51-5611	F94C Starfire	

Iron Mountain - town

53-5610	T33A Shooting Star	

Ishpemming - town

50-0453	T33A Shooting Star	

Kalamazoo

52-6486	F84K Thunderstreak	

Livonia - town

52-7260	RF84F Thunderflash	

Midland - town

53-6081	T33A Shooting Star	

Mount Clemens - town

53-0663	F86D Sabre	

Oak Park - town

51-9359	F84F Thunderstreak	

Otsego - town

51-9078	T33A Shooting Star	

Sebewaing - town

53-5073	T33A Shooting star	

Tri-Cities

52-1463	RB57A Canberra	

MINNESOTA

Albert Lea - town

53 5158	T33A Shooting Star	

Chisholm - town

51-13560	F94C Starfire	

Luverne - town

51-6038	F86D Sabre	

Minneapolis/St. Paul - Airport

43-49322	C47B Skytrain
53-2677	F89J Scorpion
51-13563	F94C Starfire
	SP2H Neptune
68-15795	P51D Mustang
55-3025	T33A Shooting Star

Minnesota Lake - town

51-8814	T33A Shooting Star	

West Duluth - town

51-13556	F94C Starfire	

MISSISSIPPI

Biloxi - Keesler AFB

52-5755	YF100	Super Sabre
56-0068	RF101C Voodoo	
56-0938	F104C Starfighter	
60-0535	F105D Thunderchief	
Bu138389 T28	Trojan	'13389'
58-0567	T33A Shooting Star	'3380'

Columbus AFB

54-2737	T37B Tweety Bird	'08502'
	T38A Talon	'8702'

Greenville - town

53-0554	T33A Shooting Star

Gulfport - town

51-5908	F86D Sabre

Hattiesburg - Airport

53-7636	RF84F Thunderflash

Hazelhurst - town

53-1061	F86D Sabre

Jackson - Allen C Thomas Field

44-34559	A26B Invader

Meridian - Key Field

53-9636	RF84F Thunderflash	'37832'
56-0217	RF101C Voodoo	

Meridian - McCain Field, NAS Meridan

Bu147522	T2A Buckeye

Meridian - town\Highway 39

Bu147516	T2B Buckeye

MISSOURI

Carrollyon - town

51-9021	T33A Shooting Star

Caruthersville - town

53-5044	T33A Shooting Star

Flat River - town

55-4349	T33A Shooting Star

Independence - Airfield

52-9469	T33A Shooting Star

Jackson - town

52-9249	T33A Shooting Star

Knob Noster - Whiteman AFB

51-2120	B47B Stratojet
65-7941	UH1F Iroquois

La Plata - town

52-1983	F86D Sabre

Richmond - town

53-5990	T33A Shooting Star

St. Joseph

53-0310	KC97L Stratofreighter

St. Louis - Lambert Field

55-3667	F100D Super Sabre

St. Louis - Goodfellow Bvd. Army Base

64-14185	UH1M Iroquois

St. Louis - Museum of Transportation

43-15635	VC47A Skytrain

Trenton - town

52-9709	T33A Shooting Star

MONTANA

Great Falls - International Airport

'472637'	F86A Sabre
53-2547	F89J Scorpion
56-1116	F102A Delta Dagger

Great Falls - Lion Park

56-1105	F102A Delta Dagger

Malmstrom AFB

52-1505	EB57B Canberra
53-0360	KC97L Stratofreighter
59-0419	F101F Voodoo
65-7956	UH1F Iroquois

St. Louis - Lindberg Field Int. Airport

68-0338	F4E Phantom II

NEBRASKA

Beatrice - Municipal Airport

51-8880	T33A Shooting Star

Bellevue - Offutt AFB, SAC Museum

44-83559	DB17P Fortress	'23474'
44-30363	TB25J Mitchell	

44-28738	TB25N Mitchell	
44-34665	VB26B Invader	
44-84076	TB29B Superfortress	
52-2217	B36J Peacemaker	
48-0017	NRB45C Tornado	
52-1412	B47E Stratojet	
52-8711	RB52B Stratofortress	
55-4244	B57E Canberra	
61-2059	B58A Hustler	
43-48098	C47A Skytrain	
42-72724	C54D Skymaster	
53-0198	KC97G Stratofreighter	
51-8024	C119G Boxcar	
49-0258	C124A Globemaster	
59-0536	C133B Cargomaster	
61-0287	EC135A Stratotanker	
51-1714	F84F Thunderstreak	
46-0524	XF85 Goblin	
53-1375	F86H Sabre	
59-0462	F101B Voodoo	
54-1405	F102A Delta Dagger	
68-0267	FB111A	
53-4426	SH19B Chicasaw	
52-8676	CH21B Shawnee	
64-17964	SR71A Blackbird	
50-0190	T29A	
58-0548	T33A Shooting Star	
62-4487	CT39A Sabreliner	
56-6701	U2C	
51-0006	HU16B Albatross	

Boys Town - town

53-5173	T33A Shooting Star	

Franklin - town

52-9205	T33A Shooting Star	

Humbolt - town

51-9111	T33A Shooting Star	

Lincoln - Municipal Airport

64-0998	RF4C Phantom II	
51-11259	RF84F Thunderflash	
53-0531	F86L Sabre	'523760'
52-9264	T33A Shooting Star	

McCook - town

53-1503	F86H Sabre	

NEVADA

Indian Springs - Indian Springs AFB

51-1776	F84F Thunderstreak'1776'	

Las Vegas North - Nellis AFB

64-0806	F4C Phantom II	

Reno - Cannon International Airport

59-0483	RF101B Voodoo	

NEW HAMPSHIRE

Antrim - town

53-9333	T33A Shooting Star	

Portsmouth - Pease AFB

44-61671	B29A Superfortress	
52-0410	EB47E Stratojet	'24100'
56-0683	B52D Stratofortress	
53-0327	KC97L Stratofreighter	

NEW JERSEY

Atce - town

52-9367	T33A Shooting Star	

Berlin - town

53-5307	T33A Shooting Star	

Brigantine - town

53-5907	T33A Shooting Star	

Camden - town

51-6949	T33A Shooting Star	

Englewood Cliffs - town

52-9284	T33A Shooting Star	

Flemington - town

51-5645	F94C Starfire	

Lakehurst - NAS Lakehurst

Bu142106	A4B Skyhawk	
Bu146698	RA5C Vigilante	
Bu148273	F4A Phantom II	
Bu145397	F8A Crusader	
	F11A Tiger	
Bu140740	T34B Mentor	

McGuire AFB

67-0270	F4E Phantom II	

Millville - town

49-0919	T33A Shooting Star	

Ocean City - town

51-6703	T33A Shooting Star	

Wrightstown - McGuire AFB

53-3300	C118A Liftmaster	
54-0629	C123K Provider	
57-5776	F105B Thunderchief	

44-27183	P38L Lightning	

NEW MEXICO

Alamogordo - City Park

49-0710	F80C Shooting Star	

Alamogordo - Holloman AFB

63-7535	F4C Phantom II	
51-9396	F84F Thunderstreak	
51-13028	F86E Sabre	'53154'
56-3220	F100D Super Sabre	
56-0886	F104C Starfighter	
61-0145	F105D Thunderchief	'10041'

Albuquerque - Kirtland AFB

52-0013	B52B Stratofortress	
53-1532	F100A Super Sabre	
61-0107	F105D Thunderchief	
65-5709	F111A	
64-15495	UH1F Iroquois	'57956'
Bu138499	UH19F Chicasaw	'13893'
52-8691	CH21B Shawnee	'34366'
52-8706	CH21B Shawnee	
59-1578	HH43F Huskie	
'8555'	H56 Dragonfly	'N92808'
Bu46457	PBY6A Catalina	'N4582U'
1280	HU16E Albatross	USCG'1008'

Clovis

53-1576	F100A Super Sabre	

Clovis - Cannon AFB

39-0025	B18A Bolo	'39-522'
'58490'	F80 Shooting Star	
51-1810	F84F Thunderstreak	
51-13010	F86E Sabre	'27523'
53-1251	F86H Sabre	'31404'
56-2940	F100D Super Sabre	'56141'
56-0187	RF101C Voodoo	three false serials
'11027'	F104G Starfighter	
63-9771	F111A	three false serials
58-0503	T33A Shooting Star	

Melrose

53-1533	F100A Super Sabre	

Rio Grande - City Park

51-9022	T33A Shooting Star	

Sante Fe

53-1600	F100A Super Sabre	

NEW YORK

Binghampton - town

47-1454	F84C Thunderjet	

Buffalo - Naval Park

58-0338	F101F Voodoo	
	FJ1 Fury	

Clay - town

58-0200	F86A Sabre	

Fabius - town

53-6149	T33A Shooting Star	

Garden City - Cradle of Aviation Museum

57-5783	F105B Thunderchief	
44-89444	P47N Thunderbolt	

Griffiss AFB

58-0225	B52G Stratofortress	

Hempstead - town

51-13562	F94C Starfire	

Islip - town

51-2986	F86D Sabre	

Monroe\Waterloo - Crane Park

52-10052	F86D Sabre	

Newburg - Stewart AFB

52-3702	F86L Sabre	

New York City - Fifth Avenue\118 St.

53-1272	F86A Sabre	

New York City - pier 86\W46 St., USS Intrepid

Flight Deck

Bu142833	A4B Skyhawk	'11984'
Bu156621	RA5C Vigilante	
Bu159232	AV8C Harrier	
Bu125413	XA3D1 Skywarrior	
Bu147212	E1B Tracer	
Bu150628	F4N Phantom II	
51-1658	RF84F Thunderflash	
52-7066	F84F Thunderstreak	
'24803'	F6F5 Hellcat	
Bu141117	F9F8 Cougar	
Bu127074	YF10B Skynight	
Bu141783	F11F1 Tiger	
Bu141884	F11A Tiger	
Bu133566	F3H2N Demon	
-	UH1B Iroquois	
69-15076	UH1H Iroquois	
62-3759	OH23D Raven	
57-1698	CH34A Choctaw	
1429	HH52A Seaguard USCG	
Bu135868	FJ3 Fury	
Bu131542	SP2E Neptune	
Bu133264	TS2A Tracker	
Bu151664	S2E Tracker	
Bu141538	T33B Shooting Star	
Bu24803	TBM1	
7216	HU16E Albatross USCG	

Hangar Deck

'11984'	A4B Skyhawk
Bu147867	NA6A Intruder

Entrance Pier

58-1192	YT38 Talon

Niagara Falls - International Airport

64-0660	F4C Phantom II
56-2993	F100D Super Sabre
56-0185	RF101C Voodoo
59-0413	F101F Voodoo
53-6064	T33A Shooting Star

Plattsburgh - Plattsburgh AFB

53-2385	B47E Stratojet

Shortsville - town

53-1337	F86L Sabre

Syracuse - Hancock Field

53-1519	F86K Sabre	
50-0877	YF94C Starfire	
56-1219	F102A Delta Dagger	'81801'
56-1365	F102A Delta Dagger	
62-4444	F105G Thunderchief	
56-1702	T33A Shooting Star	

Westhampton Beach - Suffolk County ANGB

53-1308	F86H Sabre
57-0788	F102A Delta Dagger

White Plains - Park

53-1348	F86H Sabre

NORTH CAROLINA

Charlotte - Douglas Airport

52-4142	F86L Sabre

Elizabeth City - Elizabeth City CGAS

7215	HU16E Albatross	USCG
7247	HU16E Albatross	USCG
-	PBY5 Catalina	

Ikin - town

51-8548	T33A Shooting Star

Fayetteville - picnic area

51-5576	F94C Starfire

Fort Bragg - Simmons AAF, 82 Airborne Div. Museum

44-78573	C46F Commando
43-48932	C47A Skytrain
53-8087	C119L Boxcar
54-0609	C123K Provider
59-1711	UH1A Iroquois

Maxton - town

53-0919	F86D Sabre

Mayodan - town

53-5169	T33A Shooting Star

New River - MCAS New River

Bu147191	UH34D Choctaw

Pope AFB

44-76462	TC47B Skytrain	'118427'
54-0669	C123K Provider	

Reedsville - town

53-6095	T33A Shooting Star

Rich Square - town

51-4505	T33A Shooting Star

Seymour-Johnson AFB

64-0770	F4C Phantom II	
53-1370	F86F Sabre	'112972'
61-0056	F105D Thunderchief	

Southern Pines - town

53-6024	T33A Shooting Star

NORTH DAKOTA

Dickinson - town

53-5078	T33A Shooting Star

Fargo - Hector Field

44-35523	A26C Invader	
53-2465	F89J Scorpion	'32469'
56-3208	F100D Super Sabre	
58-0311	F101F Voodoo	
56-1502	F102A Delta Dagger	
56-0926	F104C Starfighter	
44-74407	P51D Mustang	

Grand Forks - Grand Forks AFB

44-35493	A26C Invader	'N94445'
44-38834	B25J Mitchell	
58-0315	F101B Voodoo	
65-7946	UH1F Iroquois	
57-5959	H19D Chicasaw	

Hatton - park

51-6721	T33A Shooting Star

Jamestown - town

53-1253	F86H Sabre

Minot - Minot AFB

56-1505	F102A Delta Dagger
56-0460	F106A Delta Dart
66-1215	UH1F Iroquois

58-0466	T33A Shooting Star

Velva - town

51-9100	T33A Shooting Star

OHIO

Bradnel\Rising Sun - town

53-1023	F86L Sabre

Columbus - Defence Construction Supply Center

55-2884	F100D Super Sabre

Dayton - Wright Patterson AFB, Air Force Museum

Bu132469	A1E Skyraider	
69-6192	A7D Corsair II	'00970'
71-1370	YA10A Thunderbolt	
43-22200	A20G Havoc	
44-35733	A26C Invader	
42-83665	A36A Apache	
62-5951	YA37A Dragonfly	
41-12150	AT9 Fledgling	
41-27561	AT11 Kansan	'237493'
76-0174	B1A	
42-72843	B17G Fortress	
44-83624	B17G Fortress	
37-0469	B18A Bolo	
39-0037	B23 Dragon	
42-72843	B24D Liberator	
43-3374	B25B Mitchell	'02344'
43-34581	B26G Marauder	'2595857'
64-17676	B26K Invader	
44-27297	B29 Superfortress	
52-2220	B36J Peacemaker	
48-0010	B45C Tornado	
53-2280	B47E Stratojet	
49-0310	WB50D Superfortress	
49-0389	KB50J Superfortress	'80114'
53-0394	B52B Stratofortress	
56-0665	B52D Stratofortress	
52-1499	EB57B Canberra	
59-2458	B58A Hustler	
53-0475	RB66B Destroyer	
62-0001	XB70 Valkerie	
41-23075	BT13A Valiant	
42-90629	BT13B Valiant	
66-7943	VC6A King Air	
62-4193	C7B Caribou	
38-0515	C39A	
44-76068	UC43 Traveler	'39-139'
52-10893	C45H Expeditor	
44-78018	C46D Commando	
43-49507	C47D Skytrain	
41-20095	C53C Skytrooper	
43-16445	C60A Lodestar	
44-70296	UC64A Norseman	'470534'
42-71626	UC78B Bobcat	
45-57735	C82A Packet	
42-68852	XUC86A	
52-2630	KC97L Stratofreighter	
46-0595	VC118A Liftmaster	'6-505'
51-8037	C119J Boxcar	
53-0555	EC121D Warning Star	
53-7855	VC121E Constellation	
56-4362	C123K Provider	
51-0135	C124A Globemaster	
54-1626	C130A Hercules	
55-0301	C131D Samaritan	

56-2008	C133A Cargomaster	
55-3123	NKC135A Stratrotanker	
61-2492	C140B Jetstar	
62-5924	XC142A	'NASA522'
62-12200	YF4E Phantom II	
64-0829	F4C Phantom II	
64-0683	F4C Phantom II	
59-4989	YF5A Freedom Fighter	
60-6935	YF12A	
72-0119	F15A Eagle	
75-0745	F16A Fighting Falcon	
49-0986	F80C Shooting Star	
44-65168	F82B Twin Mustang	
50-1143	F84E Thunderjet	
52-6526	F84F Thunderstreak	
49-2430	YRF84F Thunderflash	
46-0523	XF85 Goblin	
49-1067	F86A Sabre	'49-1236'
50-0477	F86D Sabre	'23863'
53-1352	F86H Sabre	
52-1911	F89J Scorpion	'32509'
46-0680	XF91 Thunderceptor	
46-0682	XF92A	
49-2498	F94A Starfire	
50-0980	F94C Starfire	'01054'
53-1558	F100A Super Sabre	
54-1753	F100C Super Sabre	
55-3754	F100D Super Sabre	
58-0325	F101B Voodoo	
56-0166	RF101C Voodoo	
56-1416	F102A Delta Dagger	
56-0914	F104C Starfighter	
57-5793	F105B Thunderchief	
53-8320	F105G Thunderchief	
56-0451	F106A Delta Dart	'90082'
58-0787	F106A Delta Dart	
55-1119	XF107A	
45-27948	CG4A Hadrian	
64-15476	UH1P Iroquois	
43-46620	YH5A Dragonfly	
57-2728	UH13J Sioux	
52-7587	UH19B Chicasaw	'13893'
55-3228	UH19D Chicasaw	
51-15857	CH21B Shawnee	
46-0689	XH20	
52-8685	CH21B Shawnee	
Bu148963	UH34J Choctaw	
60-0263	HH43B Huskie	
Bu33857	J2F6 Duck	'8563'
41-19039	L1A Vigilant	
'326753'	L2M Grasshopper	
42-36200	L3B Grasshopper	
42-36446	L4 Grasshopper	'NC42050'
'298667'	L5 Sentinel	
43-2680	L6 Grasshopper	
47-1347	L17A Navion	'8928'
51-11917	O1G Bird Dog	
67-21331	O2A Super Skymaster	
33-0324	O38F	
35-0179	O46A	
39-0112	O47B	'328'
40-2763	O52	Owl
32-0261	P6E Hawk	
31-0559	P12E	
36-0404	P35A	
38-001	P36A Hawk	
44-53232	P38L Lightning	'42-67855'
44-3887	P39Q Airacobra	'17073'
'18731'	P40E Warhawk	
45-49167	P47D Thunderbolt	'44-32718'
42-23278	P47D Thunderbolt	
44-74936	P51D Mustang	'44-15174'
42-38353	XP56	
44-22650	P59B Airacomet	
43-8353	P61C Black Widow	'239468'

43-11728	P63E Kingcobra	
44-44553	P75A Eagle	
44-85200	P80R Shooting Star	
42-17800	PT13D Kaydet	
41-25284	PT17 Kaydet	
'41-14666'	PT19A Cornell	
41-1572	PT22 Recruit	
Bu46595	PBY6A Catalina	'433879'
43-46506	R4B Hoverfly	
43-45379	R6A Hoverfly II	
Bu38944	S4C Scout	
50-1279	T6G Texan	'41279'
49-1494	T28A Trojan	
Bu140048	T28B Trojan	
53-5974	T33A Shooting Star	
53-3310	T34A Mentor	
54-2732	T37B Tweety Bird	
62-4478	CT39A Sabreliner	
56-6722	U2A	
58-2124	U3A Blue Canoe	
51-16501	U6A Beaver	
66-14360	U10D Super Courier	'14374'
51-5282	HU16B Albatross	
64-18262	XV6A Kestrel	
48-1385	X1B	'NASA1385'
48-2892	X3 Stiletto	
46-677	X4	
50-1838	X5	
54-1620	XF13A Vertijet	
56-6671	X15A	
66-13551	X24B	
68-10770	X25A	

In addition to the aircraft listed, this collection contains a number of civil designed and registered aircraft, aircraft of World War One origin, and aircraft, both American and others, in the insignia of foreign air arms.

Donnelsville - town

51-8623	T33A Shooting Star

Franklin - town

53-1058	F86D Sabre

Grafton - town

51-9184	T33A Shooting Star

Haskins - town

52-9756	T33A Shooting Star

Kettering - town

52-9788	T33A Shooting Star

Lunkin - town

53-1528	F86H Sabre

Mansfield - Lahm Airport

52-7021	F84F Thunderstreak

Marengo - town

52-6444	F84F Thunderstreak

Marietta - park

52-9785	T33A Shooting Star

Massillen - town

51-5610	F94C Starfire

Newark - Newark AFB

53-5998	T33A Shooting Star

Newark - town

51-9173	T33A Shooting Star

Oberlin - FAA Cleveland ARTCC

58-0328	F101B Voodoo

Orrville - park

51-6546	T33A Shooting Star

Parma Heights - town

51-6816	T33A Shooting Star

Parma\Newbury - park

53-0959	F86L Sabre

Springfield - Airport

51-1797	F84F Thunderstreak
51-0791	F84G Thunderstreak '51791'
53-1559	F100A Super Sabre

Springfield - town

51-0791	F84F Thunderstreak

Struthend - town

52-3719	F86D Sabre

Tiffin - town

51-3048	F86D Sabre

Toledo - Express Airport

51-9525	F84F Thunderstreak '9525'
55-2855	F100D Super Sabre '58855'

Uhrichsville - park

52-3814	F86L Sabre

Vandalia - town

53-5895	T33A Shooting Star

Van Wert - town

53-0944	F86L Sabre

OKLAHOMA

Altus - Altus AFB

Bu131573	C118B Liftmaster

Altus - Park Ave\Highway 62N

51-7071	B47E Stratojet

Del City - park

52-9762	T33A Shooting Star

Elk City - town

51-8896	T33A Shooting Star

Fort Sill

63-8510	UH1B Iroquois	'12345'

Kingfisher - town

53-0773	F86D Sabre

McAlester - town

52-10142	F86L Sabre

Oklahoma City - state fairground

51-2387	B47 Stratojet
57-0038	B52F Stratofortress

Oklahoma City - Tinker AFB

53-4257	RB47E Stratojet
44-27343	B29 Superfortress
56-0695	B52D Stratofortress
'2150761'	C47 Skytrain
55-0552	C121H Constellation
62-4360	F105D Thunderchief
62-4088	UH1B Iroquois

Sperry - park

51-9183	T33A Shooting Star

Stroud - park

51-8277	F86D Sabre

Tulsa

54-1443	F100A Super Sabre

Tulsa - Airport

51-8409	F86D Sabre
55-3650	F100D Super Sabre

Vance AFB

62-4242	F105D Thunderchief
49-1689	T28A Trojan
51-4301	T33A Shooting Star
41-1436	BT13 Valiant

OREGON

Albany - Municipal Airport

51-6055	F86D Sabre

Astoria - town

51-3024	F86L Sabre

Clackamas   Camp Withycombe

55-3818	QF86F Sabre

Corvallis - town

51-8467	F86D Sabre

Eugene - Municipal Airport

53-2524	F89J Scorpion

Kingsley Field - Airport\ANGB

63-7479	F4C Phantom II

Portland

56-1368	F102A Delta Dagger

Portland - Airport

51-9480	F84F Thunderstreak
54-0322	F89H Scorpion
58-0301	F101B Voodoo
51-7244	HU16B Albatross

Vale - park

53-0781	F86L Sabre

Woodburn - Route 99E

51-6653	T33A Shooting Star

PENNSYLVANIA

Annville - Fort Indiantown Gap

56-2346	TF102A Delta Dagger

Avon - town

52-2065	F86H Sabre

Camp Hill - town

51-13577	F94C Starfire

Carlisle - town

51-5620	F94C Starfire

Corry - park

51-5671	F94C Starfire

Dallastown - town

52-9405	T33A Shooting Star

Delmont - Newhouse Park

51-6742	T33A Shooting Star

Erie - Airport

53-1338	F86H Sabre

Freeport - park

52-2043	F86H Sabre

Greensburg - Sports Field

53-5251	T33A Shooting Star

Highspire - town

51-5622	F94C Starfire

Imperial - route 22

53-0665	F86D Sabre

McAdoe - town

51-5998	F94C Starfire

Mechanicsburg - park

51-13555	F94C Starfire

Mount Joy - town

51-5593	F94C Starfire

New Kensington - Memorial Park

51-8513	T33A Shooting Star

Newville - town

51-13537	F94C Starfire

Philipsburgh - town

53-1316	F86H Sabre

Pittsburgh - Greater Pittsburgh IAP

51-1508	F84F Thunderstreak	
53-0894	F86D Sabre	
56-1415	F102A Delta Dagger	
44-84900	P51D Mustang	'48490'

Sadsburyville - town

52-10046	F86D Sabre

Warminster NAS

Bu145550	F8K Crusader

Willow Grove - NAS Willow Grove

Bu143568	AF1E Fury
Bu143806	F8A Crusader
Bu135764	XF2Y1 Sea Dart
Bu139642	F7U3 Cutlass
62-1920	UH1B Iroquois
Bu33824TV1	

PUERTO RICO

San Juan - University of Puerto Rico

57-0928	F104C Starfighter

SOUTH CAROLINA

Aiken - town

51-9664	T33A Shooting Star

Beaufort - MCAS Beaufort

Bu152270	F4N Phantom II	
Bu146963	F8K Crusader	
Bu149220	F8J Crusader	
Bu141376	FJ1 Fury	'5841'

Charleston - Charleston AFB

43-49355	VC47D Skytrain	'42-100972'
54-0180	EC121S Warning Star	
52-1072	C124C Globemaster	
57-0230	F106A Delta Dart	
58-0671	T33A Shooting Star	

Columbia - Airport

51-9169	T33A Shooting Star

Columbia - McEntire ANGB

53-1386	F86H Sabre
53-1064	F86L Sabre
56-0985	F102A Delta Dagger
56-0920	F104C Starfighter
44-11807	P51D Mustang
51-6915	T33A Shooting Star

Florence - Air & Missile Museum

64-17671	RB26K Invader
44-70113	B29 Superfortress
50-0062	B47B Stratojet
52-1458	RB57A Canberra
52-1459	RB57D Canberra
53-0431	WB66D Destroyer
Bu04959	BTD1 Destroyer
Bu141790	F11A Tiger
52-6553	F84F Thunderstreak
52-5737	F86H Sabre
53-2646	F89J Scorpion
52-2624	KC97G Stratofreighter
50-0128	C119C Boxcar
Bu141292	NC121K Constellation
56-0243	F101F Voodoo
53-1788	YF102A Delta Dagger
57-1301	F104B Starfighter
54-4003	CH21B Shawnee
55-4496	H34A Choctaw
40158	QH50C Gyrodyne-DASH drone
69-16365	OH58C Kiowa
Bu125506	HO4S1
53-6089	T33A Shooting Star
51-7212	HU16B Albatross

Greenville - park

52-1976	F86H Sabre

Patriot's Point - USS Yorktown

Bu142240	A3B Skywarrior	
'877949'	B25J Mitchell	'N2XD'
Bu147225	E1B Tracer	
Bu146939	F8K Crusader	
Bu147385	TF9J Cougar	
Bu141864	F11A Tiger	
Bu132057	FJ2 Fury	
Bu147171	UH34D Choctaw	
Bu151567	S2E Tracker	

TBM3   Avenger 'N60393'

Sumter - Shaw AFB

54-0465	EB66C Destroyer
63-7574	RF4C Phantom II '68708'
56-0099	RF101C Voodoo
'64-507'	CH3E
68-10962	O2A Super Skymaster '80962'

SOUTH DAKOTA

Bridgewater - town

51-6926	T33A Shooting Star

Canistota - park

53-6093	T33A Shooting Star

Huron - town

53-6100	T33A Shooting Star

Lake Norden - town

51-8665	T33A Shooting Star

Marlon - town

51-13563	F94C Starfire

Rapid City - Ellsworth AFB

44-33030	VB25J Mitchell
64-17640	B26K Invader
-	B29 Stratofortress
53-4296	RB47H Stratojet
56-0657	B52D Stratofortress
52-1548	EB57B Canberra
55-0292	C131D Samaritan
52-8886	F84F Thunderstreak
52-5737	F86H Sabre
53-2677	F89J Scorpion
59-0426	F101B Voodoo
57-5839	F105B Thunderchief
65-7951	UH1F Iroquois
67-21422	O2A Super Skymaster
51-8629	T33A Shooting Star

Sioux Falls - City Park

51-11443	F89D Scorpion

Sturgis - town

51-6071	F86D Sabre

TENNESSEE

Chattanooga

59-0412	F101B Voodoo

Crossville   -   town

51-6756	T33A Shooting Star

Dayton - City Park

51-6861	T33A Shooting Star

Jackson - town

51-6989	T33A Shooting Star

Johnson City - town

53-6009	T33A Shooting Star

Knoxville - City Park

52-3679	F86D Sabre

Knoxville - McGhee Tyson Airport

56-0890	F104C Starfighter
62-4375	F105D Thunderchief

Lookout Mountain

52-3840	F86D Sabre

Memphis - Memphis Belle Memorial

41-24485	B17F Fortress	
-	C47 Skytrain	
-	P40 Warhawk	'N9950'
-	PBY5 Catalina	'N45827'

Memphis - NAS Memphis

Bu156608	RA5C Vigilante

Nashville - Metro Airport

53-7529	RF84F Thunderflash

Pulaski - Airport

51-8775	T33A Shooting Star

TEXAS

Abilene - Dyess Elementary School

65-0796	F4D Phantom II

Abilene - Dyess AFB

44-35913	A26C Invader	'29243'
44-85599	DB17P Fortress	'05599'
52-0412	EB47E Stratojet	'24120'
56-0685	B52D Stratofortress	
52-1504	EB57D Canberra	
42-108808	C47A Skytrain	
42-71714	UC78B Bobcat	
53-0282	KC97L Stratofreighter	
54-0604	C123K Provider	
55-0023	C130A Hercules	
51-11293	RF84F Thunderflash	
51-9364	F84F Thunderstreak	
53-4035	F86L Sabre	
54-0298	F89H Scorpion	
54-1752	F100C Super Sabre	
57-0287	F101F Voodoo	
56-0748	F104A Starfighter	'56748'
59-1738	F105F Thunderchief '00517'	

67-21326	O2A Super Skymaster
52-1175	T29C
51-4300	T33A Shooting Star
Bu140810	T34B Mentor
61-0634	CT39A Sabreliner
51-7251	HU16E Albatross

Austin - Bergstrom AFB

53-0466	EB66B Destroyer	
63-7763	RF4C Phantom II	
64-1000	RF4C Phantom II	
52-7409	RF84F Thunderflash	
56-0135	RF101C Voodoo	'560119'

Austin - town

| 52-3770 | F86D Sabre |

Beeville - town

| '475' | A4 Skyhawk |

Bowie - park

| 51-6639 | T33A Shooting Star |

Breckenridge - town

| 53-3693 | F86D Sabre |

Chase Field - NAS Chase Field

| Bu142874 | A4B Skyhawk | '155103' |

Claude - town

| 51-1617 | F84F Thunderstreak |

Clear Lake City - Legion Post

| 52-6455 | F84F Thunderstreak |

Corpus Christi - NAS Corpus Christi

'84250'	UH1B Iroquois
'4934'	PBY5 Catalina
Bu140056	T28C Trojan
Bu140468	T28C Trojan

Crane - town

| 53-6086 | T33A Shooting Star |

Cresson - Pate Museum of Transportation

'28866'	C117D Skytrooper	
51-2675	C119G Boxcar	
Bu140448	YF8C Crusader	
Bu131063	F9C Cougar	
53-7595	RF84F Thunderflash	
53-1329	F86H Sabre	
59-0471	F101B Voodoo	
53-4324	CH21B Shawnee	
51-16386	OH23B Raven	'116260'
51-16387	OH23B Raven	
58-1841	HH43B Huskie	
Bu140659	T28C Trojan	
58-0621	T33A Shooting Star	

| Dallas | - | NAS Dallas |

Bu152267	F4N Phantom II
Bu148693	F8H Crusader
53-1030	F86F Sabre

| Bu148764 | UH34D Choctaw |
| 45-8607 | P80B Shooting Star |

Del Rio - Airport

| 52-3749 | F86D Sabre |
| 53-6124 | T33A Shooting Star |

Dennison - Legion Post

| 51-6144 | F86L Sabre |

Dennison\Paris - town

| 52-4239 | F86D Sabre |

Eagle Pass - town

| 53-6102 | T33A Shooting Star |

Farmers Branch - Valwood Parkway

| 53-5011 | T33A Shooting Star |

Fort Hood - Robert Gray AAF

| 59-1625 | UH1A Iroquois |
| 61-0693 | UH1B Iroquois |

Fort Sam Houston

| 60-6031 | UH1B Iroquois |

Fort Worth - Carswell AFB, Southwest Aerospace Museum

52-2827	B36J Peacemaker	
55-0063	B52D Stratofortress	
53-0283	KC97L Stratofreighter	
66-8714	F4D Phantom II	
51-6091	F86L Sabre	
52-1868	F89J Scorpion	
56-3154	F100D Super Sabre	
60-0500	F105D Thunderchief	
61-0100	F105D Thunderchief	
61-0152	F105D Thunderchief	
63-8343	F105F Thunderchief	
-	T6 Texan	
56-1767	T33A Shooting Star	
-	BT13 Valiant	
51-7176	HU16B Albatross	
'201287'	type unknown	

Fort Worth - Meacham Airport

| 55-0668 | TB58A Hustler |

Galveston - Sea Wolf Park

| Bu139516 | F1E Fury |

Grayson County

| 51-6091 | F86L Sabre |

Houston - Ellington ANGB

44-83872	B17 Fortress
52-6455	F84F Thunderstreak
58-0269	F101F Voodoo
56-1252	F102A Delta Dagger
52-9223	T33A Shooting Star

Iowa Park - Airport

51-8892	T33A Shooting Star	

Kingsville - NAS Kingsville

	A4 Skyhawk	
Bu147473	T2A Buckeye	

Lake Jackson - town

52-9463	T33A Shooting Star	

Laredo - Airport

51-6739	T33A Shooting Star	

Laughlin AFB

52-1509	EB57B Canberra	
52-9060	F84F Thunderstreak	
51-8595	T33A Shooting Star '18629'	
Bu144112	T34A Mentor	
56-6707	U2C	
Bu23774	UC45J Expediter '22274'	
49-1682	T28A Trojan	

Lubbock - Reese AFB

44-86880	TB25J Mitchell	
49-1679	T28A Trojan	
53-5346	T33A Shooting Star '29658'	

Midland - Odessa, Confederate Air Force

Bu132443	NA1E Skyraider	
53-4257	RB47E Stratojet	
52-1456	RB57A Canberra	
Bu23774	UC45J Expediter	
43-49355	VC47D Skytrain	
42-72675	C54D Skymaster	
53-0332	KC97G Stratofreighter	
51-2566	C119C Boxcar	
52-9060	F84F Thunderstreak	
53-2677	F89J Scorpion	
49-1938	VT29A	
51-7186	HU16B Albatross	
51-7193	HU16B Albatross	

Mount Pleasant - park

53-6103	T33A Shooting Star	

Nocoma - Airport

53-5535	T33A Shooting Star	

Plainview - Municipal Airport

51-6753	T33A Shooting Star	

Runge - park

53-6117	T33A Shooting Star	

San Angelo - Goodfellow AFB

53-1573	F100A Super Sabre	

San Antonio

53-8381	F5A Freedom Fighter	

San Antonio - Brooks AFB

58-1232	F100F Super Sabre	
57-5789	F105B Thunderchief '78003'	

57-5797	F105B Thunderchief '78005'	
57-5835	F105B Thunderchief '78001'	
61-0044	F105D Thunderchief '78002'	
61-0110	F105D Thunderchief '78004'	

San Antonio - Kelly AFB

44-62220	B29A Superfortress	
56-0692	B52D Stratofortress	
59-2437	B58A Hustler	
43-52436	XC99	
63-7515	F4C Phantom II	
56-3000	F100D Super Sabre	
-	F101 Voodoo	
56-1268	F102A Delta Dagger	
52-4387	F105D Thunderchief	
57-2533	F106B Delta Dart	
68-6865	O2A Super Skymaster '97602'	
41-32406	AT6 Texan	
57-0760	T33A Shooting Star	
Bu138019	T33B Shooting Star	

San Antonio - Lackland AFB

44-83512	B17G Fortress	
44-51228	B24M Liberator	
43-5103	B25H Mitchell	'N36766'
'435918'	B26C Invader	
55-0068	B52D Stratofortress	
52-1482	RB57A Canberra	
55-0390	WB66D Destroyer	
42-9637	UC45J Expediter	
44-76671	C47D Skytrain	
51-17640	VC118A Liftmaster	
Bu131600	GC118B Liftmaster '6313'	
51-2567	C119C Boxcar	
54-0155	EC121S Warning Star	
54-0593	C123K Provider	
54-0668	C123K Provider	
Bu149421	F4B Phantom II	
'08123'	F5B Freedom Fighter	
44-85123	EF80A Shooting Star	
44-85125	F80A Shooting Star	
46-0262	F82E Twin Mustang	
46-0600	F84B Thunderjet	
47-1486	EF84C Thunderjet	
52-6461	F84F Thunderstreak	
52-8889	F84F Thunderstreak	
52-8973	F84F Thunderstreak	
52-9089	F84F Thunderstreak	
47-0605	F86A Sabre	
49-2434	EF89A Scorpion	
48-0356	YF94A Starfire	
52-5759	F100A Super Sabre	
52-5770	F100A Super Sabre	
52-5773	F100A Super Sabre	
53-1629	F100A Super Sabre	
53-1684	F100A Super Sabre	
53-1712	F100C Super Sabre	
56-0241	F101B Voodoo	
58-0290	F101F Voodoo	
53-1797	F102A Delta Dagger	
53-1817	F102A Delta Dagger	
56-1151	F102A Delta Dagger	
56-2337	TF102A Delta Dagger	
56-0929	F104C Starfighter	
57-0915	F104C Starfighter	
54-0105	JF105B Thunderchief	
54-0107	F105B Thunderchief	
57-5813	F105B Thunderchief	
61-0106	F105D Thunderchief	
61-0108	F105D Thunderchief	
61-0115	F105D Thunderchief	
61-0199	F105D Thunderchief	
62-4228	F105D Thunderchief	

62-4259	F105D Thunderchief
62-4279	F105D Thunderchief
62-4318	F105D Thunderchief
62-4346	F105D Thunderchief
62-4353	F105D Thunderchief
60-3601	UH1B Iroquois
68-11164	O2A Super Skymaster
49-1679	T28A Trojan
49-1682	T28A Trojan
49-1689	T28A Trojan
49-1695	T28A Trojan
51-9282	T33A Shooting Star
54-0718	XT37 Tweety Bird
59-1605	T38A Talon

San Antonio - Randolph AFB

42-84560	AT6D Texan	
49-1695	T28A Trojan	'17782'
53-6147	T33A Shooting Star	
Bu144090	T34B Mentor	'576552'
61-0838	T38A Talon	

Sweetwater - town

51-4380	T33A Shooting star

Texarkana - park

51-4025	T33A Shooting Star

Tulia - airfield

60-0593	F86E Sabre

Webb - airfield

61-0436	F89D Scorpion

Whitesboro - park

61-8885	T33A Shooting Star

Wichita Falls - Sheppard AFB

additional aircraft serve with the Technical Training Center here.

52-5950	GYA37A Dragonfly	
54-0681	C123K Provider	
63-7751	RF4C Phantom II	
54-2151	F100D Super Sabre	
56-0009	RF101C Voodoo	
58-0329	GF101B Voodoo	
57-0826	F102A Delta Dagger	
56-0912	F104C Starfighter	'00912'
61-0151	F105D Thunderchief	
63-9773	F111A	'39772'
44-89348	P47N Thunderbolt	
44-64376	P51H Mustang	
45-57295	RP63G Kingcobra	
Bu51584	SNJ4 Texan	
49-1611	GT28A Trojan	
49-1934	GT29A	'01934'
51-5172	VT29B	
51-9157	T33A Shooting Star	
52-9497	T33A Shooting Star	
53-5255	T33A Shooting Star	
Bu156157	QT33A Shooting Star	
55-0206	T34A Mentor	
61-0858	T38A Talon	
63-8125	T38A Talon	
62-4482	CT39A Sabreliner	
51-5303	HU16B Albatross	

UTAH

Ogden - Hill AFB

44-86408	B29 Superfortress
52-1492	RB57A Canberra
52-10862	C45H Expediter
43-49281	C47B Skytrain
52-2109	C119G Boxcar
53-7814	C131B Samaritan
63-7424	F4C Phantom II
Bu150445	F4N Phantom II
51-1640	F84F Thunderstreak
52-3242	F84G Thunderstreak
52-2295	F86A Sabre
51-6055	F86L Sabre
54-0322	F89H Scorpion
52-5777	F100A Super Sabre
57-0252	F101B Voodoo
57-0833	F102A Delta Dagger
59-1743	F105D Thunderchief
62-4347	F105D Thunderchief
62-4440	F105G Thunderchief
68-10853	O2A Super Skymaster
43-46592	R4B Hoverfly
51-9271	T33A Shooting Star
41-25284	PT17 Kaydet

Provo - Airport

53-0809	F86L Sabre

Salt Lake City - International Airport

49-1273	F86A Sabre
57-5814	F105B Thunderchief

VERMONT

Burlington - Airport

52-1500	EB57E Canberra
52-1883	F89J Scorpion
55-3462	F102A Delta Dagger

VIRGINIA

Chatham - park

51-9119	T33A Shooting Star

Hampton

54-2145	F100D Super Sabre
56-0246	F101F Voodoo
51-9086	T33A Shooting Star

Hampton - Aerospace Park

72-1567	YF16 Fighting Falcon
51-1786	F84F Thunderstreak '61786'

51-3064	F86D Sabre	'53094'
52-2129	F89J Scorpion	'52129'

Hampton - Felker AAF, Fort Eustis Museum

61-0788	UH1B Iroquois	'61788'
66-0648	UH1M Iroquois	
46-0234	H13B Sioux	
51-14010	OH13E Sioux	
51-4275	UH19D Chicasaw	
56-2077	CH21C Shawnee	
56-16168	OH23C Raven	'0111'
Bu130043	UH25B Retriever	
57-1725	VH34C Choctaw	
57-1651	CH37B Mojave	
64-14203	YCH54A Tarhe	
66-8832	AH56A Cheyenne	
-	L4 Cub	
'18573'	L21 Super Cub	
'80377'	L23D Seminole	
63-12745	O1E Bird Dog	
55-3270	U1A Otter	
57-5869	U3A Blue Canoe	
58-1997	U6A Beaver	
58-3051	U8D Seminole	
59-4991	RU8D Seminole	
52-6218	YU9A Aero Commander	
56-6942	VZ4	

Langley AFB, TAC Memorial Park

71-0281	YF15A Eagle	
78-0001	F16A Fighting Falcon	
52-5747	F86H Sabre	'31483'
61-0073	F105D Thunderchief	
61-0188	F105D Thunderchief	'10217'

Norfolk - NAS Norfolk

Bu150541	E2B Hawkeye
Bu141351	F8A Crusader
Bu143701	TF8A Crusader

Oceana - NAS Oceana

'09102'	AD1 Skyraider	
Bu148261	F4B Phantom II	'152310'
Bu145322	F8A Crusader	
Bu149150	F8J Crusader	
Bu134950	F4D1 Skyray	
Bu123612	F9F2 Panther	
Bu127693	F2H3 Banshee	

Quantico - MCAS Quantico, Marine Corps Aviation Museum

Bu142226	A4A Skyhawk	
Bu132261	AD4B Skyraider	
Bu158695	AV8A Harrier	
Bu85890	TBM3 Avenger	
43-15952	C47A Skytrain	
Bu50834	C117D Skytrooper	
Bu143388	F4A Phantom II	
Bu139177	F6A Skyray	
Bu124618	EF10B Skyknight	
Bu12114	F4F4 Wildcat	
-	F6F Hellcat	'N41476'
Bu80375	F7F3E Tigercat	
Bu141872	F11F Tiger	
Bu13486	FG1	Corsair
Bu127693	F2H3	Banshee
'277546' F3H Demon		
Bu136119 FJ3	Fury	
Bu97369	F4U4	Corsair
Bu154760	UH1E Iroquois	

-	HH43B Huskie	
Bu147161	VH34D Choctaw	
Bu124344	HO3S1 Dragonfly	
-	PV1 Harpoon'N151V'	
Bu17421	SBD5 Dauntless	
Bu140557	T28C Trojan	
Bu140602	T28C Trojan	
Bu152880	YOV10A Bronco	

Richmond - R.E. Byrd Airport

52-8837	F84F Thunderstreak
61-0050	F105D Thunderchief

Surrey - town center

52-2005	F86H Sabre

Tappahannock - Airport

51-6690	T33A Shooting Star

WASHINGTON

Everett - Paine Field\Snomish County Airport

53-3798	UH13G Sioux
56-4023	U9C

McChord AFB

37-505	B18A Bolo
44-76502	VC47D Skytrain
52-0994	C124C Globemaster
56-0459	F106A Delta Dart
58-2106	T33A Shooting Star

Seattle

59-4987	YF5A Freedom Fighter
Bu152391	T2B Buckeye
Bu152473	T2B Buckeye
Bu155235	T2B Buckeye

Spokane - Fairchild AFB

56-0676	B52D Stratofortress
NASA827	C47B Skytrain
57-0439	F101B Voodoo
58-0335	F101B Voodoo
57-5823	F105B Thunderchief

Spokane - Felts Field

55-4335	T33A Shooting Star

Spokane - International Airport

51-14030	UH13E Sioux
51-4259	T33A Shooting Star'29667'

Spokane - town

55-2866	T33A Shooting Star

WEST VIRGINIA

Charleston - Kanawha Airport

44-72948	F51D Mustang

WISCONSIN

Appleton - Legion Post

51-5938	F86H Sabre

Camp Douglas - Volk Field ANGB

52-0905	KC97L Stratofreighter
51-9365	F84F Thunderstreak
53-1358	F86H Sabre
54-2106	F100C Super Sabre
56-1273	F102A Delta Dagger
56-2353	TF102A Delta Dagger
57-5838	F105B Thunderchief
67-21426	O2A Super Skymaster
44-72989	P51D Mustang

Franklin - EAA Museum

51-9456	F84F Thunderstreak

Milwaukee - Mitchell Field

44-30444	B25J Mitchell
53-5476	T33A Shooting Star

Oshkosh

53-1553	F100A Super Sabre

WYOMING

Casper - Airport

51-2826	F86E Sabre

Cheyenne - F.E. Warren AFB

65-7953	UH1F Iroquois

OVERSEAS

ALBANIA - Gjirokaster

51-4413	T33A Shooting Star

BELGIUM - Bruxelles

68-0590	RF4C Phantom II

CANADA - Calgary

Bu144671	U1B Otter

CANADA - Goose Bay

53-5413	T33A Shooting Star

FEDERAL REPUBLIC of GERMANY   -Berlin-Gatow

45-0557	C54G Skymaster

FEDERAL REPUBLIC of GERMANY - Bitburg AB

63-7421	F4C Phantom II
63-7576	F4C Phantom II

FEDERAL REPUBLIC of GERMANY - Frankfurt\Rhein Main AB

45-0951	C47D Skytrain	'N1350M'
61-2491	VC140B Jetstar	

FEDERAL REPUBLIC of GERMANY - Hahn AB

64-0879	F4C Phantom II
78-0016	F16A Fighting FalconBDRT
58-0318	F101F Voodoo
63-8357	F105F Thunderchief

FEDERAL REPUBLIC of GERMANY - Hermeskeil

63-7421	F4C Phantom II
63-7583	F4C Phantom II
68-0587	RF4C Phantom II
62-4417	F105G Thunderchief

FEDERAL REPUBLIC of GERMANY - Leipheim

63-7467	F4C Phantom II

FEDERAL REPUBLIC of GERMANY - Norvenich AB

63-7536	F4C Phantom II

FEDERAL REPUBLIC of GERMANY - Ramstein AB

64-0917	F4C Phantom II
74-1641	F4E Phantom II
57-0386	F101F Voodoo
63-8362	F105F Thunderchief

FEDERAL REPUBLIC of GERMANY - Sembach AB

52-5372	F86F Sabre
64-0922	F4C Phantom II
58-0267	F101B Voodoo
62-4471	CT39A Sabreliner

FEDERAL REPUBLIC of GERMANY - Spangdhalem AB

66-0308	F4E Phantom II
67-0260	F4E Phantom II
62-4446	F105G Thunderchief

FEDERAL REPUBLIC of GERMANY - Speyer

63-7423	F4C Phantom II

FEDERAL REPUBLIC of GERMANY - Zweibrucken AB

63-7583	F4C Phantom II
58-0322	F101B Voodoo

GUAM - Andersen AFB

55-0100	B52D Stratofortress

ICELAND - Keflavik AB

Bu17191	C117D Skytrooper

JAPAN - Kadena AB

52-4341	F86F Sabre	
52-5756	F100D Super Sabre	'52812'

| 62-4418 | F105F Thunderchief |
| 62-4484 | CT39A Sabreliner |

JAPAN - Yokota AB

| 61-0675 | CT39A Sabreliner '10475' |

NETHERLANDS - Soesterberg AB

-

PANAMA\CANAL ZONE - Albrook AFB

| 45-0967 | C47D Skytrain | GIA-IAAFA |
| '33100' | C117D Skytrooper | GIA-IAAFA |

PANAMA\CANAL ZONE - Howard AFB

| 54-0663 | GC123K Provider | GIA |
| 52-5385 | F86F Sabre | |

SPAIN - Zaragosa AB

| 55-3971 | F86F Sabre |

UNITED KINGDOM - RAF Alconbury

63-7419	F4C Phantom II	BDRT
Bu153008	F4N Phantom II	BDRT
56-0312	F101B Voodoo	BDRT
56-6692	U2CT	GIA

UNITED KINGDOM - Coventry, Baginton Airfield

54-2174	F100D Super Sabre
62-4535	HH43B Huskie
51-4419	T33A Shooting Star

UNITED KINGDOM - Duxford Airfield, Imperial War Museum

77-0259	A10A Thunderbolt II
56-0689	B52D Stratofortress
54-2165	F100D Super Sabre
51-7899	VT29B
51-4286	T33A Shooting Star

UNITED KINGDOM - Dumfries, Heathall

| 54-2163 | F100D Super Sabre |

UNITED KINGDOM - RAF Fairford

| 63-7699 | F4C Phantom II | BDRT |

UNITED KINGDOM - Flixton, Museum

| 54-2196 | F100D Super Sabre |
| 55-4433 | T33A Shooting Star |

UNITED KINGDOM - Hailsham

| 51-9252 | T33A Shooting Star |

UNITED KINGDOM - Hendon, RAF Museum

| 51-7473 | T33A Shooting Star |

UNITED KINGDOM - RAF Lakenheath

63-7471	F4C Phantom II	BDRT
63-7610	F4C Phantom II	BDRT
54-2269	F100D Super Sabre '63319'
62-4434F105G	Thunderchief	BDRT

UNITED KINGDOM - Lashenden, Headcorn Airfield

| 56-3938 | F100F Super Sabre |

UNITED KINGDOM - Leicester, L & L Aircraft Museum

| 54-2239 | F100D Super Sabre |

UNITED KINGDOM - RAF Mildenhall

| 57-0524 | C130A Hercules | BDRT |
| 64-0707 | F4C Phantom II | BDRT |

UNITED KINGDOM - RAF Odiham

| 61-2414 | CH47A Chinook | GIA |

UNITED KINGDOM - Rhoose, Cardiff Airport

| 54-2160 | F100D Super Sabre |
| 52-9963 | T33A Shooting Star |

UNITED KINGDOM - Usworth\Sunderland

| 54-2157 | F100D Super Sabre |
| 55-4439 | T33A Shooting Star |

UNITED KINGDOM - RAF Upper Heyford

63-7449	F4C Phantom II	BDRT
54-2212	F100D Super Sabre gate
62-4428	F105G Thunderchief